Advances in Citrus Nutrition

Anoop Kumar Srivastava
Editor

Advances in Citrus Nutrition

Springer

Editor
Anoop Kumar Srivastava
Soil Science
National Research Centre for Citrus
Nagpur, Maharashtra, India

ISBN 978-94-007-4170-6 ISBN 978-94-007-4171-3 (eBook)
DOI 10.1007/978-94-007-4171-3
Springer Dordrecht Heidelberg New York London

Library of Congress Control Number: 2012939386

Printed on acid-free paper

Springer is part of Springer Science+Business Media (www.springer.com)

Preface

A guesstimate proclaims over 900 million people in the world are undernourished, and malnutrition alone is responsible for 3.5 million deaths annually. Plant nutrition is a complex process that has developed over the course of plant evolution with the discovery of fundamental importance of plant nutrition, only second to the discovery of photosynthesis as an effective via-medium to bolden plant defence mechanism (Chap. 1). The accumulated biochemical and molecular evidences have incredibly confirmed that the nutrient stress is invariably associated with changes in antioxidant system (Chap. 2). Under such nutrient-induced stresses, phytophenolic nutrients are first to be affected (Chap. 3). Later, with the universal acceptability to the concept of essentiality of nutrients by Aron and Stout, the investigations on the anatomical, histological, and biochemical nutritional disorders became distinctly understandable through a variety of diagnostics. For such regulatory systems to function, nutrient conditions need to be sensed, signals need to be transduced, gene expression need to be transcriptionally and posttranscriptionally regulated, transporters be properly trafficked through endomembrane system, and cell cycles need to be coordinated. Such a wide range of responses may be a reflection of the very sophisticated systems that have evolved in plants over time.

Citrus fruits are produced in many countries around the world, although production shows geographical concentration in certain areas, but still citrus fruits rank first in international fruit trade in terms of value, evolving from a producer-driven to a more consumer-oriented market. In the backdrop of demography-driven diminishing per capita availability of arable land, plant nutrition has gained phenomenal significance in meeting the challenge of sustaining productivity over changing resource outputs. Indeed, from soil and plant diagnosis to suggestions for appropriate fertilizer applications (Chap. 4), current levels of citrus production would never have been possible without the knowledge of plant nutrition. A definite credit in this context could be accredited to developments in analytical techniques in both leaf (Chaps. 5 and 6) and juice (Chap. 7) analysis. Of late, trunk nutrition gained some momentum (Chap. 8) where conventional methods of nutrient supply have not been able to put forth the desired results spaced over time. In addition, proximal sensing of nutrient stress (Chaps. 9 and 10) and spectrum of soil enzymes as dictum of soil fertility changes (Chap. 11) have further provided some authoritative progress towards precise diagnosis of nutrient stresses. Such breakthroughs will go a long way in developing early warning system in the years to come to enable the redressal to genesis of any nutritional disorder within current growth cycle of crop.

Occurrence of nutrient constraints is as old as history of citrus cultivation. Any nutrient constraint at various crop phenophases on nutrient deprived soils has always baffled citrus nutritionists that could well jeopardize the incentives accruing through otherwise balanced fertilization in highly diversified nutrient demanding citrus cultivars. The current state of knowledge on the subject is very fragmentary. The subject becomes still very complex in the absence of knowledge on kinetics, and co-kinetics of different nutrients being partitioned across different growth stages so that growth stagewise nutrient demand is precisely modulated. Accordingly, type and source of nutrients are fed synchronizing with physiological nutrient demand (Chap. 12). Two major processes of nutrient cycling, viz, mineralization and immobilization of nutrients via litter fall, offer a lion's share in meeting out the crop nutrient demand in perennial canopy framework of citrus. Development of microbial consortium (microbial reactor) exploiting the native and natural microbial synergisms (with twin role as growth promoter and antagonistic against soil borne pathogens) is one of the popular methods of providing the desired dynamism to nutrient dynamics within the rhizosphere (Chap. 13). Such rhizosphere-specific consortia could further engineer rhizosphere's nutrient demand and supply through loading with organic manures in much value-added form, e.g., biodynamic soil fertility management (Chap. 14). The efforts such as these could only meet their objectivity unless duly supported by methods leading to improved nutrient use efficiency (Chaps. 15 and 16) including the intervention of genomics with metaloenzymes (Chap. 17) and variable rate fertilization (Chap. 18).

Development of nutrient norms using crop-specific plant parts in citrus cultivars needs a thorough revisit and to be field validated in order to provide their wider application down to orchard level. However, the major point of discontent still remains to be warded off with respect to whether nor not different nutrient norms are required as per cultivar within the same variety. The biggest constraint on the other hand in making soil test ratings more purposeful is the non-redressal of spatial variation in soil fertility. Conjoint use of geoinformatics (Geographical Information System, Global Positioning System, and Remote Sensing) with nutriomics, site-specific nutrient management strategy, fertigation (Chap. 19), comparatively new concept of open field hydroponics (Chap. 20), and exploiting nutrient–harmone synergy (Chap. 21) have collectively yielded definite edge over conventional methods of nutrient management. Ironically, one of the most profoundly researched nutritional disorder, popularly known as lime-induced iron deficiency, still needs multi-pronged strategy with regard to management of citrus on calcareous soils (Chap. 22). Well known mycorryizal dependency of citrus still remains an unexploited issue (Chap. 23). The concepts such as organic soil fertility management (Chap. 24) and integrated nutrient management utilizing collective efficacy of organic manures, inorganic fertilizers, and microbial diversity (Chap. 25) have taken this important issue a step forward towards sustainable nutrient management. Such approaches have given birth to a concept like best management practices (Chap. 26) duly validated through economic analysis (Chap. 27). The entire gamut of citrus nutrition remains an unfinished exercise unless dealt with the issues like assessing soil salinity (Chap. 28) and aluminium toxicity (Chap. 29) on soluble salt-rich (high pH) and divalent bases deprived (low pH) soils, respectively, considering the extreme sensitivity of citrus under both the soil conditions. Despite all these concerns, application of sensor-based technology has further added a new dimension in estimating the fruit yield (Chap. 30) in an authentic manner so that sustainable productivity vis-à-vis nutrient management strategies go hand in hand in offering an alternative source of nutritional security in an era of soils sick of multiple nutrient deficiencies taking their severe toll on human nutrition.

This is probably a maiden effort to consolidate the information related to different aspects of citrus nutrition in such a holistic manner. I am indebted to my colleagues numbering as many as 71 (all of them are highly respected for their excellent contribution), well represented by as many as 19 frontline citrus growing countries. This has been a truly educative experience while the course of this book. I am sure the book will serve as a novel source of information for the students, teachers, and researchers, as well.

Nagpur, India Anoop Kumar Srivastava

Editor Biography

Dr. A.K. Srivastava, having received his M.Sc. (Ag) and Ph.D in Soil Science from famous Banaras Hindu University, in 1984 and 1988, respectively, is working as Principal Scientist (Soil Science) at National Research Centre for Citrus, Nagpur. He has been extensively working on different aspects of citrus nutrition like nutrient constraints analysis of citrus orchards using DRIS-based soil-plant nutrient diagnostics, site specific nutrient management using spatial variability in soil fertility, development of citrus rhizosphere specific microbial consortium and INM module, fertigation scheduling, nutrient mapping using geospatial tools, nutrient dynamic studies, transformation of soil microbial biomass nutrients within citrus rhizosphere, fertility map as decision support tool for fertilizer recommendation through large number of externally funded projects.

Dr. Srivastava is credited with large of publications including 91 Research Papers (69 papers in Indian Journals and 21 papers in Foreign Journals) and 30 Policy Review Papers (23 in Indian Journals and 7 in Foreign Journals) and Awards like S.N.Ranade Award for Excellence in Micronutrient Research, FAI Silver Jubilee Award, International Plant Nutrition Institute-FAI Award, Netaji Subhash Chandra Bose Award for Excellence, National Magnum Foundation Award for Excellence, Bharat Jyoti Award etc.

Dr. Srivastava is Life Member of as many as 20 Academic Societies besides Honorary Member of World Association of Soil and Water Conservationists. Author of books like Citrus: Soil and Climate, Citrus Nutrition published by IBDC, Lucknow and Editor of book entitled Advances in Citrus Nutrition by Springer-Verlag, Netherlands. Elected Fellow of seven professionally Academic Societies (Maharashtra Academy of Sciences, National Environmental Science Academy, Environmental Research Academy, Indian Society of Citriculture, Indian Society of Soil Science, United Writers Association, and Indian Society of Agricultural Chemists).

Dr. Srivastava is Editor-in-Chief of Research Journal on Earth Sciences and International Journal of Horticultural and Crop Science Research; Honorary Editor, Agricultural Science Digest; Regional Editor of International Journal of Food, Agriculture and Environment and Member of Editorial Board of five other international journals including the Journal of Plant

Nutrition and Communications in Soil Science and Plant Analysis and seven national journals. He is regular Paper Setter on advance courses on Soil Fertility and Soil Chemistry for six agricultural universities.

He was invited as Citrus Expert by Govt. of Nepal in 1999 under Indo-Nepal MOU and Keynote Speaker in World Citrus Congress held at Wuhan, China in 2008.

Contents

Contributors

Maria Rosa Abenavoli Dipartmento de Biotechnologie per il Monitoraggio Agrovalimentare ed Ambientale (Bio. M.A.A.), Universitá degli Studi "Mediterranea", di Reggio Calabria, Italy

Francesca Antonucci CRA-ING (Agricultural Engineering Research Unit of the Agricultural Research Council), Monterotondo Scalo (RM), Italy

Vicent Arbona Ecofisiologia i Biotecnologia. Departament de Ciències Agràries i del Medi Natural, Universitat Jaume I, Castelló de la Plana, Spain

Okezie I. Aruoma School of Biomedical Sciences and School of Pharmacy, American University of Health Sciences, Signal Hill, CA, USA

Theeshan Bahorun National Research Chair, ANDI Centre of Excellence for Biomedical and Biomaterials Research, University of Mauritius, Réduit, Republic of Mauritius

Davide Barreca Dipartimento di Chimica Organica e Biologica, Università di Messina, Messina, Italy

Franc Bavec Faculty of Agriculture and Life Sciences, University of Maribor, Hoče pri Mariboru, Slovenia

Martina Bavec Faculty of Agriculture and Life Sciences, University of Maribor, Hoče pri Mariboru, Slovenia

José A.J. Berni CSIRO Marine and Atmospheric Research, Canberra, Australia

Brian Boman Indian River Research and Education Centre, University of Florida, Fort Pierce, FL, USA

Emmanuel Bourdon Laboratoire de Biochimie et Génétique Moléculaire (LBGM), Groupe d'Etude sur l' Inflammation Chronique et l'Obesité (GEICO), Université de La Reunion, Saint Denis Messag, Cedex 09, La Réunion, France

Stefano Canali CRA-RPS (Soil Plant System Research Centre of the Agricultural Research Council), Rome, Italy

C.W. Chan Malaysian Agricultural Research and Development Institute (MARDI) Electronics and Building Environmental System, Mechanisation and Automation Research Center, Kualampur, Malaysia

Li-Song Chen College of Resources and Environmental Sciences, Fujian Agriculture and Forestry University, Fuzhon, PR China

Corrado Ciaccia CRA-RPS (Soil Plant System Research Centre of the Agricultural Research Council), Rome, Italy

W. Conejero Centro de Edafología y Biología Aplicada del Segura (CSIC), Espinardo (Murcia), Spain

Corrado Costa CRA-ING (Agricultural Engineering Research Unit of the Agricultural Research Council), Monterotondo Scalo (RM), Italy

Christine Robert Da Silva Laboratoire de Biochimie et Génétique Moléculaire (LBGM), Groupe d'Etude sur l' Inflammation Chronique et l'Obesité (GEICO), Université de La Reunion, Saint Denis Messag, Cedex 09, La Réunion, France

F. Ferreres Centro de Edafología y Biología Aplicada del Segura (CSIC), Espinardo (Murcia), Spain

Stagno Fiorella CRA-ACM (Citrus and Mediterranean Crops Research Centre of the Agricultural Research Council), Acireale (CT), Italy

A. Galindo Centro de Edafología y Biología Aplicada del Segura (CSIC), Espinardo (Murcia), Spain

Giuliana Gambetta Ecofisiología de Citrus, Departemento de Produccíon Vegetal, Facultad de Agronomía, Universidad de la República, Montevideo, Uruguay

Jose M. Garcia-Mina CIPAV- R&D Department-Timac Agro Int (Roullier Group), Department of General Chemistry and Soil Chemistry, Research and Development Division, Faculty of Sciences, University of Navarra, Inabonos SA-Roullier Group, Orcoyen, Spain

Stephen E. Gareau Education and Technology Program, Department of Computer Information Systems, State University of New York College of Buffalo, Buffalo, NY, USA

Giuseppe Gattuso Dipartimento di Chimica Organica e Biologica, Università di Messina, Messina, Italy

A. Gil-Izquierdo Centro de Edafología y Biología Aplicada del Segura (CSIC), Espinardo (Murcia), Spain

Aurelio Gómez-Cadenas Ecofisiologia i Biotecnologia. Departament de Ciències Agràries i del Medi Natural, Universitat Jaume I, Castelló de la Plana, Spain

Alfredo Gravina Ecofisiología de Citrus, Departemento de Produccíon Vegetal, Facultad de Agronomía, Universidad de la República, Montevideo, Uruguay

E.A. Hanlon University of Florida, Immokalee, FL, USA

Francesco Intrigliolo CRA-ACM (Citrus and Mediterranean Crops Research, Centre of the Agricultural Research Council), Acireale (CT), Italy

A. Jiménez Centro de Edafología y Biología Aplicada del Segura (CSIC), Espinardo (Murcia), Spain

Davie Kadyampakemi Soil and Water Science Department, Southwest Florida Research and Education Centre, University of Florida, Immokalee, FL, USA

Ahmad Sattar Khan Institute of Horticultural Sciences, University of Agriculture, Faisalabad, Pakistan

Ashok Kumar Vice Chancellor, Chhatrapati Shahu Ji Maharaj University, Kanpur, Uttar Pradesh, India

Francisco Legaz Citriculture and Vegetable Production Department, Instituto Valenciano de Investigaciones Agrarias (IVIA), Apartado Oficial, Moncada, Valencia, Spain

Eliana G.M. Lemos Departmento de Tecnologia, Facultdade de Ciências Agráriase Veterinárias, Universidade Estadual Paulista, São Paulo, Brazil

Fusheng Li College of Agriculture, Guangxi University, Nanning,Guangxi, China

Jackson Marcondes Departmento de Tecnologia, Facultdade de Ciências Agráriase Veterinárias, Universidade Estadual Paulista, São Paulo, Brazil

Belén Martínez-Alcántara Citriculture and Vegetable Production Department, Instituto Valenciano de Investigaciones Agrarias (IVIA), Apartado Oficial, Moncada, Valencia, Spain

S. Medina Centro de Edafología y Biología Aplicada del Segura (CSIC), Espinardo (Murcia), Spain

Paolo Menesatti CRA-ING (Agricultural Engineering Research Unit of the Agricultural Research Council), Monterotondo Scalo (RM), Italy

Regina Menino Unidade de Investigação de Ambiente e Recursos Naturais, Instituto Nacional dos Recursos Biológios, Quinta do Marqués, Oeiras, Portugal

Kelly T. Morgan Soil and Water Science Department, Southwest Florida Research and Education Center, University of Florida, Immokalee, FL, USA

R. Mylavarapu University of Florida, Gainesville, FL, USA

Vidushi Neergheen-Bhujun Department of Health Sciences and ANDI Centre of Excellence for Biomedical and Biomaterials Research, University of Mauritius, Réduit, Republic of Mauritius

T.A. Obreza University of Florida, Gainesville, FL, USA

Ibrahim Ortas Department of Soil Science and Plant Nutrition, Faculty of Agriculture, University of Cukurova, Adana, Turkey

M.F. Ortuño Centro de Edafología y Biología Aplicada del Segura (CSIC), Espinardo (Murcia), Spain

Federico Pallottino CRA-ING (Agricultural Engineering Research Unit of the Agricultural Research Council), Monterotondo Scalo (RM), Italy

Eduardo Primo-Millo Citriculture and Vegetable Production Department, Instituto Valenciano de Investigaciones Agrarias (IVIA), Apartado Oficial, Moncada, Valencia, Spain

Ana Quiñones Citriculture and Vegetable Production Department, Instituto Valenciano de Investigaciones Agrarias (IVIA), Apartado Oficial, Moncada, Valencia, Spain

Deena Ramful-Baboolall Department of Agricultural and Food Science, Faculty of Agriculture, University of Mauritius, Réduit, Republic of Mauritius

Eran Raveh Department of Fruit Tree Sciences, Agricultural Research Organisation, Negev, Israel

Fernando Rivas Programa Nacional de Investigación en Producciôn Citricola, Instituto Nacional de Investigación Agropecuaria (INIA), Canelones, Uruguay

Giancarlo Roccuzzo CRA-ACM (Citrus and Mediterranean Crops Research Centre of the Agricultural Research Council), Acireale (CT), Italy

Philippe Rondeau Laboratoire de Biochimie et Génétique Moléculaire (LBGM), Groupe d'Etude sur l' Inflammation Chronique et l'Obesité (GEICO), Université de La Reunion, Saint Denis Messag, Cedex 09, La Réunion, France

Kenshi Sakai International Center, Environmental and Agricultural Engineering Department, Faculty of Agriculture, Tokyo University of Agriculture and Technology, 3-5-8 Fucha, Tokyo, Japan

Johannes Scholberg Biological Farming Systems, Wageningen University, Wageningen, The Netherlands

F. Sevilla Centro de Edafología y Biología Aplicada del Segura (CSIC), Espinardo (Murcia), Spain

Mahmoud M. Shaaban Fertilization Technology Department, National Research Centre, Dokki, Cairo, Egypt

Zora Singh Curtin Horticulture Research Laboratory, Department of Environment and Agriculture, School of Science, Faculty of Science and Engineering, Curtin University, Perth, WA, Australia

Sylwester Smoleń Department of Soil Cultivation and Fertilization of Horticultural Plants, Faculty of Horticulture, University of Agriculture in Krakow, Kraków, Poland

Agostino Sorgonà Dipartmento de Biotechnologie per il Monitoraggio Agrovalimentare ed Ambientale (Bio. M.A.A.), Universitá degli Studi "Mediterranea", di Reggio Calabria, Italy

Anoop Kumar Srivastava National Research Centre for Citrus, Nagpur, Maharashtra, India

Lola Suárez School of Mathematical and Geospatial Sciences, RMIT University, GPO Box 2476V Melbourne, VIC, 3001 Australia

Evelyne Tarnus Laboratoire de Biochimie et Génétique Moléculaire (LBGM), Groupe d'Etude sur l' Inflammation Chronique et l'Obesité (GEICO), Université de La Reunion, Saint Denis Messag, Cedex 09, La Réunion, France

Fabio Tittarelli CRA-RPS (Soil Plant System Research Centre of the Agricultural Research Council), Rome, Italy

A. Torrecillas Centro de Edafología y Biología Aplicada del Segura (CSIC), Espinardo (Murcia), Spain

Matjaž Turinek Faculty of Agriculture and Life Sciences, University of Maribor, Hoče pri Mariboru, Slovenia

Shalini Verma Department of Biotechnology and School of Biotechnology, Faculty of Science, Chhatrapati Shahu Ji Maharaj University Kanpur and Banaras Hindu University, Varanasi, Uttar Pradesh, India

Qiang-Sheng Wu College of Horticulture and Gardening, Yangtze University, Jingzhou, Hubei, China

Lijuan Yang Land and Environment College, Shenyang Agricultural University, Shenyang, Liaoning, China

Xujun Ye School of Biosystems Engineering and Food Science, College of Biosystems Engineering and Food Science, Zhejiang University, Hangzhou 310058, Zhejang Province, PR China

Yulong Zhang Land and Environment College, Shenyang Agricultural University, Shenyang, Liaoning, China

Plant Nutrition and Defense Mechanism: Frontier Knowledge

1

Jose M. Garcia-Mina

Abstract

Nutrition level of plants and their defense mechanism are highly interrelated. A large number of studies showed the beneficial effects of some mineral nutrients on both health and natural defense of different crops in response to the action of diverse types of pathogens. The main biochemical pathways and mechanisms involved in natural plant defense response against the attack of pathogens through schematic presentation is necessary in order to evaluate the potential role of certain mineral nutrients in the correct expression of this plant response. Defining the biological-chemical character of the action of different mineral nutrients on the activation of plant defense mechanisms has lend strong support in favor of existing synergies. As a function of the biological character of these effects, the action of each nutrient will be included in a specific level or class enclosed in the above-mentioned general classification.

Keywords

Mineral nutrition • Macronutrients • Secondary nutrients • Micronutrients • Plant protection • Natural plant defense mechanisms • Systemic acquired resistance • SAR • Phosphate • Phosphorous acid-based products

1.1 Introduction

Numerous studies have shown the influence of, practically, all mineral nutrients on the development of diverse plant diseases caused by different types of both soil and aerial pathogens (Datnoff et al. 2007a). These studies showed these effects of mineral nutrients in crops cultivated under both controlled greenhouse conditions and field (Reuveni and Reuveni 1998; Datnoff et al. 2007a). Likewise, these effects were observed when these nutrients were applied either alone or along with pesticides (Reuveni and Reuveni 1998). However, the mechanisms of action involved in the action of these nutrients, improving plant development in the presence of pathogens and reducing the intensity of the pathogen-mediated disease, are, normally, unclear (Datnoff et al. 2007a).

In this context, we could differentiate different levels of modes of action, which can be expressed due to the action of the mineral nutrient separately or in a synergetic way:

- In a first level (1L), we could explain the beneficial effects of diverse mineral nutrients on plant protection as a consequence of the positive effects of a well-equilibrated mineral nutrition on plant health and the potential plant strength or ability to control the attack of pathogens. This action is general and unspecific, and it could be defined as a nutritional-related effect, improving plant potential strength – health against pathogen attacks.
- In a second level (2L), we could explain the beneficial effects of specific mineral nutrients on plant protection because

J.M. Garcia-Mina (✉)
CIPAV- R&D Department-Timac Agro Int (Roullier Group), Department of General Chemistry and Soil Chemistry, Faculty of Sciences, University of Navarra
e-mail: jgmina@timacagro.es; jgmina@inabonos.com

A.K. Srivastava (ed.), *Advances in Citrus Nutrition*,
DOI 10.1007/978-94-007-4171-3_1,

these elements are directly related to the biosynthesis and/or functional activity of molecules directly involved in natural plant defense mechanisms. This case considers elements that intervene in either the synthesis or chemical composition of specific factors, cofactors or enzymes, phytoregulators, which intervene in diverse pathways included in the expression (activation) and/or regulation of natural plant defense mechanisms. Thus, these mineral nutrients might affect the expression of natural plant defense mechanisms by several general pathways: (1) the correction of a deficiency in this element; (2) the potentiation of plant-defense response due to the high bioavailability of this element within the plant. This type of action will only be expressed under a situation of pathogen attack that causes the activation by the plant of its natural defense system. In this case, the effect of the mineral nutrient is more specific, and it could be considered as a nutritional-physiological-related action.

- In a third level (3L), we could explain the beneficial effects of specific mineral nutrients on plant protection because these elements have the ability to directly activate one or several pathways involved in natural plant defense mechanisms. This action, which is very similar to that of natural elicitors, for instance, oligosaccharides (John et al. 1997), will be expressed even in the case of the absence of the attack of a specific pathogen. In this case, this ability of the mineral element should not be included in its nutritional or nutritional-physiological effect, as it involves a special biostimulation process that is different from that of a nutrient and is closer in nature to that of molecules with elicitor activity. We could name this level as an elicitor-related action.
- In a fourth level (4L), we could explain the beneficial effects of specific mineral nutrients on plant protection, because these elements are able to inhibit the growth and development of the pathogen due to a direct biocide (toxic) action. It becomes clear that this type of action should not be considered as related to the nutritional role of the element. In this case, we could consider this mineral nutrient as a type of pesticide. Therefore, this level could be named as a pesticide-related action.
- We consider the above-presented classification as rather important, because it permits us to differentiate a specific, new action inducing the activation of plant defense mechanisms as an elicitor (elicitor-related action, 3L) from those actions directly derived from the nutritional-physiological functions of the mineral nutrients within the plant (1L and 2L).
- Before describing the effect on plant protection of some, more relevant, mineral nutrients, we briefly describe the most important biochemical and molecular pathways involved in the expression of natural plant defense mechanisms.

1.2 Brief Description of Natural Plant Defense Mechanisms

In general, we could differentiate three different steps in the expression and activation of natural plant defense mechanisms (acquired systemic resistance; SAR) (Fig. 1.1) (Ryals et al. 1996; Jones and Dangl 2006; Gen-Ichiro et al. 2011).

A first step involved in the specific plant reaction is response to the penetration of the pathogen in plant tissues. This step could be defined as elicitation step (Boller and Felix 2009). A number of studies have shown that the perception of the presence of the pathogen by the plant is associated with the release of specific molecules, which are able to trigger the cascade of biochemical events included in the activation of plant defense mechanisms or systemic acquired resistance process (SAR). These molecules can be produced by the plant (endogen elicitors) or can derive from the pathogen as a consequence of the activation of certain enzymatic process in the plant–pathogen interaction site (Ryals et al. 1996; Boller and Felix 2009). All these biochemical processes involved in the elicitation/pathogen recognition step seem to involve specific elicitors-receptors within the plant (as a function of the pathogen type and plant species) and a quite sophisticated biochemical and molecular regulation both at transcriptional and posttranscriptional levels (Boller and He 2009; Deslandes and Rivas 2011).

We can also consider a second step, which is activated and regulated by specific second messenger molecules produced as a result of the elicitor molecular recognition (the elicitor–plant receptor interaction). These molecular messengers include chemical elements as Ca^{++} and inorganic (nitric oxide) or organic (cyclic nucleotides such as cGMP and cAMP) molecules (Ma et al. 2009). The action of these molecules can be regulated (potentiated or attenuated) by the action of specific plant regulators, such as cytokinins, auxins, polyamines, or ethylene (Gen-Ichiro et al. 2011). The complementary and interconnected action of these molecular messengers and plant regulators activates the main molecular pathways involved in plant defense mechanisms.

In general, we could discriminate between two main pathways: the salicylic acid–regulated pathway (SA) and jasmonic acid–regulated (JA) pathway (Gen-Ichiro et al. 2011). Both pathways seem to be coregulated. Thus, when a specific pathway is activated, the other one is inhibited (Reymond and Farmer 1998; Niki et al. 1998; Lorenzo and Solano 2005; Gen-Ichiro et al. 2011). However, this coregulation does not occur in all cases and circumstances. Thus, some studies have reported specific pathogen–plant interactions involving the activation of both pathways (Mur et al. 2006; Gen-Ichiro et al. 2011). Likewise, the SA pathway and the JA pathway seem to be coaffected or coregulated by different types of plant regulators. Thus, cytokinins and auxins are able to potentiate

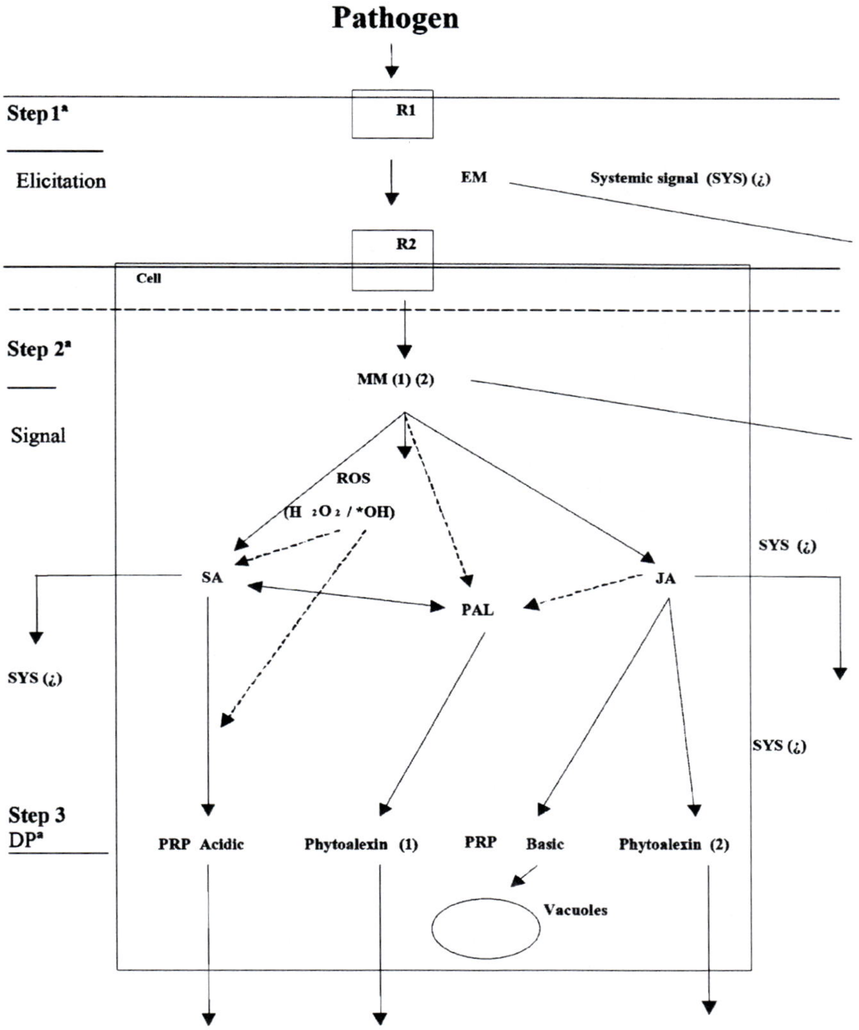

Fig. 1.1 Simplified representation of natural plant defense mechanisms (*SA* salicylic acid, *JA* jasmonic acid, *MM* messenger molecules (calcium, cyclic nucleotides, NO), *PRP* proteins related to pathogenesis, *EM* elicitor molecule, *DP* defense-related molecules production)

SA pathway and inhibit JA pathway, while, conversely, ABA and ethylene potentiate JA pathway and inhibit SA pathway (Ohashi and Ohshima 1992; Xu et al. 1994; Niki et al. 1998; Gen-Ichiro et al. 2011) (Fig. 1.2).

Genetic-related investigations reported that the SA pathway is principally involved in the regulation of both locally expressed basal resistance and SAR. Likewise, this pathway is activated against the attack of biotrophic pathogens, or insects that do not cause important damage in plant tissues (Gen-Ichiro et al. 2011). On the other hand, the JA pathway is activated against the attack of necrotrophic pathogens, wounding, and insects that can cause relevant damage in plant tissues (Gen-Ichiro et al. 2011).

Both pathways seem to be directly related to the activation of the biosynthesis of defensive molecules with different chemical nature and structure. Thus, the biosynthesis of proteins or chemical compounds with the ability to have a biocide (toxic) effect on pathogen development and survival would be the third step of natural plant defense mechanisms or SAR. Simplifying, we can differentiate fourth main types of defensive molecules (Blée 1998):

- Pathogenic-related proteins with acidic character (acidic PRP): A number of studies have characterized up to nine different classes of acidic PRP. These proteins normally involve enzymes with the ability to digest specific parts of the pathogen walls (glucanases, chitinases, etc.). Normally, these proteins are located in the extracellular medium (Blée 1998). The synthesis of these proteins is coregulated by both the SA pathway and reactive oxygen species (ROS). The production of ROS, in turn, seems to be

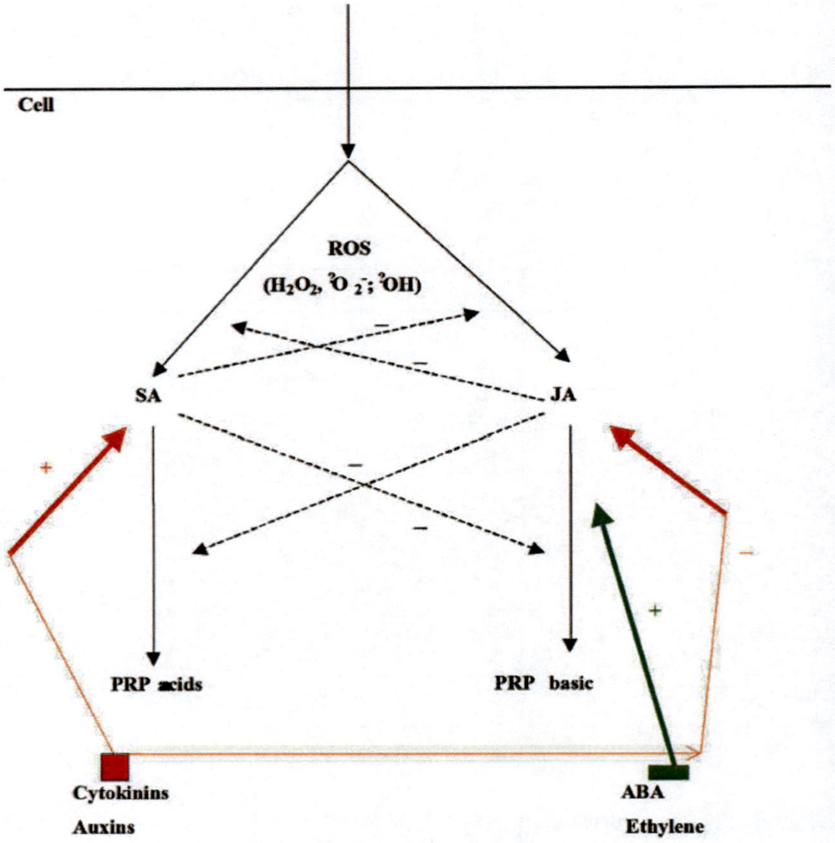

Fig. 1.2 Coregulation and interactions between salicylic acid (*SA*) and jasmonic acid (*JA*) pathways (*PRP* pathogenic-related proteins, *ABA* abscisic acid)

related to the oxidation of specific polyamines (Gen-Ichiro et al. 2011).

- Pathogenic-related proteins with basic character (basic PRP): These proteins are functionally similar to acidic PRP. They also include enzymes with the ability to hydrolyze specific parts of pathogen walls (glucanase and chitinase activities, etc.). However, these proteins are accumulated inside the cellule, in vacuoles (Blée 1998). In this case, the synthesis of these PRP class seems to be regulated by the JA pathway (Blée 1998).
- Another type of defense-related molecules is an ensemble of relatively simple chemical compounds, with low or medium molecular weight, with diverse structural configurations that are named phytoalexins (PTX). These molecules have the ability to directly inhibit the pathogen development and action due to their biocide effect.
- A large number of PTX families, with diverse chemical nature and metabolic origin, have been characterized. On many occasions, the chemical nature of these PTX families is directly related to plant species and plant–pathogen interaction types (Blée 1998). We can underline two main types: the PTX families with phenolic-flavonoid chemical character (PTX 1), which are linked, at least some of them, to the activity of L-phenylalanine-ammonia-lyase pathway (PAL) and chalcone-synthase pathway (CHS). These types of PTX seem to be mainly activated by the SA pathway, although the JA pathway may also activate this enzyme, as well as specific chemical compounds that can act independently on SA or JA regulation.

Other types of phytoalexins are several families with lipidic-terpenoid chemical character (PTX 2). The synthesis of these classes of PTX seems to be induced or regulated by the JA pathway (Blée 1998; Gen-Ichiro et al. 2011) (Figs. 1.1 and 1.2).

1.3 Potential Roles of the Different Mineral Nutrients in the Control of Plant Diseases

Classification of this plant protection-related activity as a function of its main chemical and biological features (levels of action: 1L, 2L, 3L, or 4L) is attempted in this section, and we will briefly present the possible roles and mechanisms of action of the most relevant mineral nutrients in the control of plant diseases (subject thoroughly reviewed by Datnoff et al. 2007a).

As a function of the biochemical-physiological character of the action mechanism, we will classify each nutrient within the different above-presented levels of action (1L–4L) (see Sect. 1.1).

1.3.1 Macronutrients

1.3.1.1 Nitrogen (N)

Numerous studies have shown that N nutrition strongly affects the intensity and expression of diverse plant diseases in crops cultivated under different experimental conditions: laboratory (pot soil or hydroponics), greenhouse, and field (Huber and Thompson 2007). However, these effects of N fertilizers on plant protection are rather unspecific, as we can find in the literature studies that describe a potentiation of disease intensity and studies that describe a reduction in the disease affection (Huber and Thompson 2007). On many occasions, these apparent contradictions might be explained as a function of the type of N nutrition (nitrate, ammonium, urea, or mixtures). However, many of these studies do not describe the N nutrition type (Huber and Thompson 2007).

In any case, the effect of N nutrition is mainly expressed through its effect on the synthesis of diverse metabolites, principally amino acids and proteins, which are directly linked to the synthesis of molecules involved in plant defense mechanisms: plant hormones, mainly cytokinins and auxins (Sakakibara et al. 2006; Garnica et al. 2010); and PRP and/or phytoalexins (Huber and Thompson 2007). This fact indicates that the effect of N nutrition on plant defense mechanisms has a predominant nutritional-physiological-related character that can be included in 1L and 2L classes.

In the case of ammonium, some studies reported that this molecule can have a direct biocide (toxic) effect on pathogen development (Huber and Thompson 2007). Consequently, ammonium can also be considered in 4L class (pesticide-related action) (Huber and Thompson 2007). Thus, the action of N forms on the activation of natural plant defense mechanisms (SAR, SIR, etc.) seems to be mainly indirect, through the action on the biosynthesis of metabolites that are precursors of molecules directly implied in plant defense mechanisms.

Another type of effect of N nutrition on plant disease is through an action on plant rhizosphere biological and physicochemical properties: pH value (ammonium produces an acidification while nitrate tends to alkalinize the medium), nutrient bioavailability, microbial populations, siderophore synthesis, and so on (Marschner 1995; Huber and Thompson 2007). Also, in this case, the effect of N forms on plant disease may be considered as indirect, and it can be included in the nutritional-related action levels (1L and 2L).

1.3.1.2 Potassium (K)

As in the case of N, K plays important roles in a large number of physiological processes, such as water homeostasis and the transport within the plant of different nutrients (Fe, Mg, NO_3, etc.) and metabolites (citrate, malate, etc.) as well (Marschner 1995). However, the effects of K on the expression of plant defense mechanisms seem to be rather unspecific and principally related to an indirect action on plant physiology and the biosynthesis of metabolites that could be associated with some aspects of plant defense mechanisms (Prabhu et al. 2007a). Thus, K deficiency has a deleterious effect on plant health and defense ability in response to pathogen attacks, increasing plant susceptibility to disease development (Prabhu et al. 2007a). This fact seems to be related to the correct function of other nutrients, such as phosphorus or silicon, which have more relevant direct roles in plant defense mechanisms (Prabhu et al. 2007a). All these results, taken together, indicate that we could include the action of K on plant defense mechanisms into 1L and 2L modes of action (nutritional-physiological-related action).

As for the possibility of a direct of role of K in the activation of plant defense mechanisms, several studies have shown that the foliar application of phosphate-potassium salts decreased the severity of several diseases (mainly powdery mildew) in some fruit trees (Manandhar et al. 1998; Reuveni and Reuveni 1998; Prabhu et al. 2007b). Other authors have reported that this action of phosphate-K foliar products was accompanied by an increase in the expression of the genes that codify the synthesis of specific PRPs, which are also activated by the application of the fungi (Tamás and Huttová 1996). However, it is not clear if this effect is due to either phosphate or K action, or both.

Certain studies have reported a potential indirect toxic effect of K by affecting arginine concentration in leaves, which has a biocide (toxic) action on the development of some aerial pathogens (4L class). However, this action is quite rare (Prabhu et al. 2007a).

1.3.1.3 Phosphorus (P)

As it has been indicated above, several studies have shown the efficiency of monopotassium phosphate foliar application in the control of foliar diseases caused by aerial pathogens, such as powdery mildew, in fruit trees and other crops (Reuveni and Reuveni 1998). Thus, Reuveni et al. (1998a, b) observed that the foliar application of monopotassium phosphate contributed to control powdery mildew development in both apple trees and pepper plants either applied alone or combined with systemic fungicides. Similar results had also been previously observed in cucumbers and maize (Reuveni and Reuveni 1998). Other studies have reported that the foliar application of monopotassium phosphate plus an adequate surfactant caused the activation in control noninfected plants of PRP genes that are also specifically activated by pathogen

Fig. 1.3 Effect of different formulations on the in vitro mycelium growth of *Penicillium expansum* (*MPPT* phosphonate-based product, *OF* *o*-phenylphenol, *TB* thio-bendazol, *BN* benomile, *NF* natural fungicide) (Adapted from Cenoz et al. 1998)

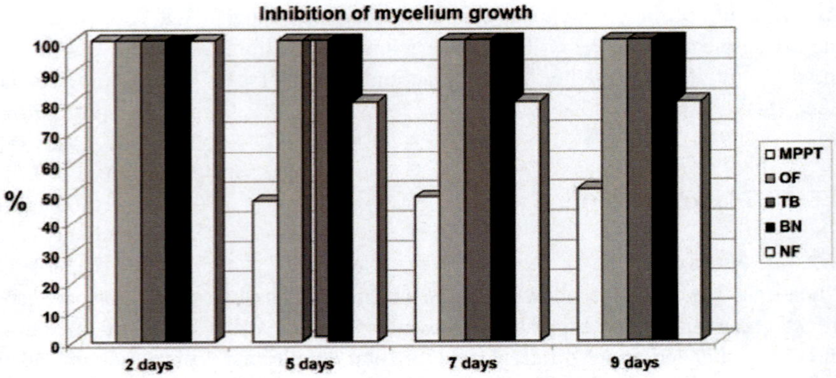

infection (Tamás and Huttová 1996). This fact indicates that the association of K and phosphate has the ability to positively affect plant defense mechanisms, besides their nutritional actions. In consequence, the action of these nutrients, considered together, could be classified within the 3L class (elicitor-related action). However, the mechanisms responsible for this effect of phosphate-K association are not clearly elucidated yet.

In addition to this elicitor-related action of phosphate-K association, several studies have reported that phosphate fertilizers have a nutritional-related action improving plant protection, mainly when they are applied in the soil. These effects are expressed in situation of P deficiency in the plant, as well as by the action of P on the rhizosphere properties (pH value, nutrient bioavailability, and microbiota) (Prabhu et al. 2007b). In these cases, P action should be considered within 1L and 2L classes. However, with the information now available, it is difficult to elucidate whether the above-explained 3L-class action of phosphate-K compounds is not derived from 1L and 2L classes–related actions.

There are other families of P-derived salts that have shown high efficiency in the control of *Phytophthora* sp. in different plant species, both fruit trees and horticultural (Guest and Grant 1991). These P-product types are derived from phosphorous acid (H_3PO_3) (Guest and Grant 1991). There exists in the literature an intense debate about the biological-nutritional character of the action of these types of compounds (Rickard 2000). Our studies in wheat seedlings cultivated in hydroponics have shown that the application of phosphite anion in the nutrient solution was not accompanied by a metabolic utilization of this P source as a nutrient, alternative to phosphate (Zamarreño et al. 2002). In fact, plant growth and some foliar and stem symptoms in phosphite-fed plants were similar to those developed in P-deficient plants (plants cultivated without any P source) (Zamarreño et al. 2002). These results confirmed previous results obtained by other authors in plants fed with phosphite as P source via roots (Rickard 2000; McDonald et al. 2001).

The foliar application of phosphite and phosphate–phosphite mixtures revealed that the presence of phosphite inhibited the metabolic use of phosphate (Zamarreño et al. 2002). This result was in line with those results that showed that the presence of phosphite anion blocks the mechanisms of the plant that are activated under P deficiency, although this P source is not used as a nutrient (Carswell et al. 1996). However, our study showed that a fraction of phosphite seems to be transformed into phosphate in the leaf (Zamarreño et al. 2002). All these results, taken together, indicate that the role of phosphite as a P-nutritional source, alternative to phosphate, is rather questionable.

However, a number of studies have shown that doses of phosphite that did not cause detrimental nutritional or physiological effects were able to help the plant to develop in the presence of diverse families of aerial and soil-borne pathogens, which mainly (but not only) belong to *Phytophthora* sp. (Guest and Grant 1991; Saindrenan and Guest 1995). Some studies have reported that phosphite has a direct action on plant defense mechanisms by the activation of the PAL activity and the biosynthesis of phytoalexins (Saindrenan and Guest 1995).

Studies carried out in our laboratory, in vitro and in Fuji apple fruit skin wall, with *Penicillium expansum*, in the presence of different compounds – synthetic fungicide (SF), natural fungicide (essential oils, NF), and a formulation containing monoethanolamine phosphonate (phosphite) (MPPT) (García-Mina et al. 1999; Cenoz et al. 1998) – showed that MPPT did not have relevant and sustainable effect on the growth of the pathogen mycelium in vitro (Fig. 1.3).

However, in Fuji apple skin walls infected with the pathogen, the pretreatment with the MPPT had a significant effect on pathogen development, which was reflected in a significant reduction of the disease intensity (Cenoz et al. 1998). In order to study more in depth the mechanism of action of MPPT, we investigated the stimulation of PAL activity in the exocarp and mesocarp of infected and noninfected Fuji apple fruit skin regions. The results showed that MPPT was able to

Table 1.1 Effect of different formulations on PAL activity (mg of *trans*-cinnamic acid per g fresh weight) in diverse regions of the skin of apple fruits (*Fuji* sp.) uninfected (control) and infected with *Penicillium expansum* (each treatment was carried out in triplicate)

Treatment	Uninfected	Infected
A. Exocarp		
Control (water)	0.079	0.087
SF	0.075	0.116
MPPT	0.090*	0.159*
NF	0.092*	0.127
B. Mesocarp		
Control (water)	0.081	0.113
SF	0.059	0.116
MPPT	0.057	0.174*
NF	0.083	0.117

Adapted from Cenoz et al. (1998)
*Significant difference with respect to the control. Fisher test for $p<0.05$

increase PAL activity in the exocarp in noninfected fruits but not in the mesocarp (Table 1.1). However, MPPT potentiated PAL activation in infected fruits in both regions of the fruit (skin) (Table 1.1) (Cenoz et al. 1998).

These results confirmed that phosphonate (phosphate)-based products did not present a relevant direct biocide (toxic) effect on pathogen development in vitro, while they are able to reduce the intensity of the disease caused by the pathogen in vivo through the activation of natural plant defense mechanisms (via the stimulation of PAL activity and phytoalexin synthesis).

In order to investigate whether the phosphonate-based product MPPT was able to activate the expression of those genes involved in the biosynthesis of PRP, we studied in tobacco plants the expression of diverse PRP-related genes in the presence and absence of MPPT treatment. The results showed that MPPT was able to increase the expression of the genes that codify both several PRP (PRP1 and 3) and PAL (Lemenager et al., unpublished results).

All these results indicate that phosphonate (phosphite)-based product has the ability to stimulate or activate plant defense mechanisms by affecting different (complementary) pathways. However, these products did not present relevant direct biocide (toxic) action on pathogen development and survival. Consequently, we can classify the action of phosphite or phosphonate anion within the 3L class (elicitor-related action).

1.3.1.4 Calcium (Ca)

Numerous studies have shown the important role of Ca^{++} and Ca-calmodulin complex as second messengers of the elicitor-induced response and the activation of plant defense mechanisms (Ma et al. 2009). This signal role of Ca^{++} is normally coupled to the action of complementary signal molecules as NO and cyclic nucleotides (cGMP and cAMP) (Ma et al. 2009). In this context, it becomes clear that an adequate availability of calcium within the plant is fundamental for the correct functioning of plant defense mechanisms. However, this effect is clearly related with the general nutritional-physiological role of Ca in the plant. Therefore, we can classify this Ca action within 1L and 2L classes. In this same context, we can consider all calcium effects related to the mechanical strength of cellular and organ membranes and walls, which directly affect skin (fruit or leaf) resistance, or root membrane resistance, against pathogen invasion (Marschner 1995; Rahman and Punja 2007) (1L and 2L classes). Likewise, the action of calcium fertilizer products (calcium carbonate or oxide, and calcium sulfate) on rhizosphere features (pH value-nutrient bioavailability, microbiota, etc.) (Marschner 1995) may be ascribed to the nutritional-related effects of calcium (1L and 2L classes).

1.3.1.5 Magnesium (Mg)

A number of studies have reported that Mg status in the plant affects the incidence of plant diseases (Jones and Huber 2007). However, this action of Mg is rather unspecific and seems to be mainly related to the general nutritional-physiological functions of Mg into the plant (Jones and Huber 2007). Therefore, we can classify the action of Mg on plant disease development within the classes involving a nutritional-physiological-related action (1L and 2L).

1.3.1.6 Sulfur (S)

The direct fungicide effect of elemental sulfur is very well known from many years ago (Haneklaus et al. 2007). This action seems to be associated with the ability of elemental S to disrupt the correct functioning of respiration and redox processes within the pathogen (Haneklaus et al. 2007). This ability of elemental S to inhibit pathogen development is clearly a pesticide-related action (4L). In addition to this S direct biocide (toxic) action, several studies have reported that a correct S nutrition is fundamental for the expression of some pathways included in plant defense mechanisms (Haneklaus et al. 2007). This fact is due to the involvement of sulfate in the biosynthesis of several amino acids (mainly methionine and cysteine) directly implied in the production of both certain phytoalexin classes, such as certain compounds with biocide activity in certain plant species (glucosinolates), and plant regulators (ethylene and polyamines) (Marschner 1995; Haneklaus et al. 2007). However, this action of S may be ascribed to the general nutritional-physiological roles of sulfate within the plant (1L and 2L). This is also the case of the effects of S-based fertilizers on rhizosphere properties, such as pH value-nutrient bioavailability and microbiota, which can also affect both plant resistance against pathogen attack and the survival of the pathogen in the soil (Marschner 1995; Haneklaus et al. 2007).

1.3.1.7 Main Micronutrients and Special Nutrients

In general, there are in the literature many studies that show the influence of micronutrients in plant disease intensity (Datnoff et al. 2007a). However, in all cases, except in the case of Cu, these effects are mainly expressed under conditions of micronutrient deficiency. These results indicate that the action of these micronutrients on plant defense mechanisms should mainly be classified within the classes ascribed to nutritional-physiological-related actions (1L, 2L). This fact was expected because all these micronutrients intervene in the activity of chemical processes (redox process, ROS production, etc.) or enzyme biosynthesis (Mn), which, in turn, severely affects the correct functioning of plant defense mechanisms in response to pathogen attack.

Thus, Fe is directly involved in the production of ROS through its role in the Fenton reaction (Bolwell and Wojtaszek 1997). Likewise, the availability in the soil of Fe is very important for pathogen survival, and the role of siderophores and phytosiderophores in this process seems to play an important role in the pathogen infection ability (Expert 2007). In the case of Mn, the deficiency of this micronutrient very significantly increases the susceptibility of the plant to pathogen attack. This fact may be directly related to Mn structural role in PAL structure (Reuveni and Reuveni 1998; Thompson and Huber 2007).

As for Zn, this micronutrient is also involved in the regulation of the production and control of the concentration of ROS within the plant (Duffy 2007). This fact influences the regulation of the SA pathway and the biosynthesis of both acidic PRP and some types of phytoalexins (Gen-Ichiro et al. 2011).

In the case of B, a number of studies have reported that its deficiency in plants increases the severity of plant diseases, and plant susceptibility to disease development as well (Stangoulis and Graham 2007). This fact could be associated with the relevant role of B in wall-membrane formation and mechanical strength (Stangoulis and Graham 2007).

Molybdenum may also influence the intensity of some plant diseases. However, the mechanism responsible for this Mo effect is unclear. It could be related to an effect of Mo availability within the plant on the K-N plant balance and nutritional efficiency (Graham and Stangoulis 2007).

Finally, some forms of Cu have toxic properties that explain the direct biocide (fungicide) effect of some of its inorganic compounds (mainly hydroxides and oxides) on pathogen survival (Marschner 1995; Evans et al. 2007). This direct effect of Cu-related compounds on pathogen development should be ascribed to the pesticide-related action class (4L). However, Cu also plays important roles in disease resistance because it is involved in enzyme activities (chalcone synthase), ROS production, and the regulation of genes; processes which are directly associated with the biosynthesis of diverse types of PRPs and phytoalexins (Evans et al. 2007). However, these actions on plant defense mechanisms are indirect in nature, and they are linked to Cu nutritional-physiological roles within the plant. Therefore, we should classify these actions within the classes linked to the nutritional-physiological-related actions (1L and 2L).

Silicon (Si) and Nickel (Ni). A number of studies have reported the beneficial effects of Si application in the production and quality of some crops, mainly cereals (rice) and horticulture (cucumber) (Datnoff et al. 2007b). Besides other actions of Si in plants improving their development under abiotic stress, these beneficial effects of Si have also been attributed to an increase in crop resistance to pathogen attacks. The most studied mechanism of Si for improving plant disease resistance is based on the relevant role of Si in the composition of root cell and leaf cell walls and membranes. This action increases the mechanical strength of the walls, which act as physical barriers against pathogen invasion (Datnoff et al. 2007b).

However, other studies have found some relationships between Si application and the accumulation of phytoalexins (phenolic and terpenoids) (Rodrigues et al. 2004; Datnoff et al. 2007b). This fact suggests a potential action of Si on PAL activity and the terpenoid pathway. In this sense, Maekawa et al. (2002) observed an important increase in ROS production, 15 min after inoculation with *M. grisea*. Dann and Muir (2002) reported an increase in chitinase and β-1,3-glucanase activities in healthy pea plants treated with Si in the absence of any pathogen. All these results strongly suggest that Si might play some important, direct, role in the activation of plant defense mechanisms. However, the nature and molecular basis of this mechanism remains to be elucidated yet. In any case, we may accept a potential effect of Si on plant protection that could be classified within the elicitor-related action class (3L).

As for Ni, many studies have reported that Ni deficiency increases both plant susceptibility to disease and disease severity (Wood and Reilly 2007). Likewise, Ni application improves plant resistance to several pathogens (Wood and Reilly 2007). These effects might be due to a direct action on plant defense mechanisms. However, there is no experimental evidence to support this hypothesis. Another possible mechanism, which may explain Ni action, might be related to its structural and functional role in the active center of urease and urease activity, and its effects on the whole N plant metabolism (Wood and Reilly 2007). All these results are compatible with an action of Ni derived from its nutritional-physiological functions in the plant (1L, 2L). Other studies have reported a direct, biocide, toxic effect on pathogen development and survival (Wood and Reilly 2007). It becomes clear that this specific action should be classified within the pesticide-related action class (4L).

In conclusion, experimental data support the following classification of the action of diverse mineral nutrients on the ability of the plant to face pathogen attacks and plant diseases:

1L. N, K, P-phosphate, Ca, Mg, S, Fe, Zn, Mn, Cu, Mo, B, Si, and Ni
2L. N, K, P-phosphate, Ca, Mg, S, Fe, Zn, Mn, Cu, Mo, B, Si, and Ni
3L. P-phosphate (along with K (?)) and P-phosphite (in this case, the nutritional role is unclear), and Si (?)
4L. S (mainly elemental sulfur), Cu (mainly basic compounds, hydroxides, oxides), ammonium, K-arginine accumulation, and Ni.

In summary, we could only classify as nutrients with an elicitor-related action, besides their nutritional-physiological-related action, P-phosphate (along with K ?), P-phosphite (also named phosphonate) (its nutritional role is unclear), and probably Si (?). However, from a practical viewpoint, the compound that is present in the market and is extensively used to manage several diseases in commercial crops or fruit trees orchards is P-phosphite or phosphorous acid-based products (Liñan 2011).

1.4 Application of Nutrient-Related Products to Disease Control in Citrus

As we have mentioned above, the main product family that is used to control specific diseases in commercial crops (horticulture) and fruit tree orchards is that including phosphorous acid-based formulations (Liñan 2011). These formulations may be liquid or solid, and they could include salts of K and/or micronutrients (mainly Fe, Mn, and Zn) (Liñan 2011). In the case of citrus, these products are used to control aerial and root diseases caused by pathogens that belong to *Phytophthora* sp. (mainly *Phytophthora citrophthora*).

Studies carried out in collaboration with Tuset's group (IVIA-Valencia) showed the efficiency of a specific phosphorous acid–based product (MPPT) to control the affection caused by *Phytophthora citrophthora* in citrus seedlings (Lapeña et al. 2003). In this experiment, young citrus plants (*Citrus sinensis*) cv. pineapple (4–5 months old) cultivated under greenhouse conditions were inoculated with zoospores of *P. citrophthora* isolated from citrus plants affected by this pathogen and cultivated under field conditions. The plants were treated before and after pathogen inoculation with different doses of several formulations, as described in Lapeña et al. (2003). A number of 24 plants were used for each treatment and dose. Four different formulations were used in the study: a metal phosphite (a liquid formulation containing an iron salt of phosphorous acid), a solid formulation of potassium phosphite, a solid formulation of fosetyl-Al 80%, and MPPT (García-Mina et al. 1999).

Two different doses were used for the metal phosphite, the K-phosphite and MPPT: 0.3% and 0.4%. fosetyl-Al was used at 0.3%. The different products were applied in irrigation.

Table 1.2 Effect of different products on the development of gummosis brown rot disease in young plants of *Citrus sinensis* cv. pineapple inoculated with zoospores of *Phytophthora citrophthora* after product treatment

Treatment (dose)	Affected plants (%)	Death plants (%)
Control	100	100
Fosetyl-Al (0.3%)	0	0
Eurofit (0.3%)	0	0
MPPT (0.4%)	0	0
Iron phosphite (0.3%)	66.67	0
Iron phosphite (0.4%)	57.14	0
K-phosphite (0.3%)	83.33	0
K-phosphite (0.4%)	85.71	0

Adapted from Lapeña et al. (2003)

Table 1.3 Effect of different products on the development of gummosis brown rot disease in young plants of *Citrus sinensis* cv. pineapple inoculated with zoospores of *Phytophthora citrophthora* before product treatment

Treatment (dose)	Affected plants (%)	Dead plants (%)
Control	100	100
Fosetyl-Al (0.3%)	100	88.90
Eurofit (0.3%)	100	87.50
MPPT (0.4%)	100	100
Iron phosphite (0.3%)	100	100
Iron phosphite (0.4%)	100	77.80
K-phosphite (0.3%)	100	100
K-phosphite (0.4%)	100	100

Adapted from Lapeña et al. (2003)

Twenty-one days after treatment, the degree of the disease in roots and shoots was evaluated. Some of the results obtained (Tables 1.2 and 1.3) in this experiment are presented. As can be observed in Table 1.2, only the treatments with MPPT and fosetyl-Al were associated with the absence of infection in plants inoculated after treatments. However, the Fe-phosphite and the K-phosphite presented a high degree of disease development, thus showing the relative inefficiency of some types of phosphites to control high intensity cases of gummosis brown rot disease. These differences in efficiency could be related to the pH and the ability of the anion to enter into the leaf.

When plants were inoculated before treatments, none of the products had the capacity of controlling the development of the disease (Table 1.3).

This fact demonstrates the importance of applying these types of products on plants before pathogen invasion and infection, as a preventive treatment. This fact is coherent with the type of activity of these types of products. Taking into account that these products activate plant defense mechanism by an elicitor-related action, it is very important to apply them before pathogen attack in order to induce the production of defense-related molecules by the plant, thus preparing the plant to pathogen arrival. However, when the pathogen has already infected the plant, the efficiency of these types of products (activators of plant defense mechanisms by an elicitor-related pathway) declines very significantly.

We obtained results in line with those described in Lapeña et al. (2003), in citrus plant orchards affected by *P. citrophthora* (Ferrari and Garcia-Mina 2004).

In this field study, we investigated the degree of infection and evolution of the disease caused by *P. citrophthora* in citrus trees (gummosis brown rot disease) cultivated under field conditions (Fig. 1.4). The study was carried out in Basilicata (Southern Italy), during two consecutive years (2000 and 2001) (Ferrari and Garcia-Mina 2004).

The main crop data are as follows:

- Crop: *Citrus sinensis*
- Variety: Navel
- Rootstock: Seville orange
- Planting date: 1990
- Planting distance: 5×6 m
- Irrigation: Sprinkling

Three different formulations were used: MPPT, fosetyl-Al and potassium phosphite. All products were used at 2.5 l per ha and applied as foliar spray. The experimental design was randomized complete blocks, fourth replications per treatment (control-untreated, MPPT, fosetyl-Al, and potassium phosphite), and five trees per replication.

The disease was evaluated using the following parameters: the number of trees presenting cankers (%), and the vigor of the trees expressed as a 0–10 scale according to the biomass development.

Some results obtained in this study are presented in Figs. 1.4, 1.5, and 1.6. As can be observed in these figures, the different treatments were efficient in controlling the development of cankers under field conditions in the 2-year experiment. In fact, there were not significant differences among treatments, although MPPT presented slightly better results. The significant efficiency of potassium phosphite under field conditions, much better than in the case of Tuset's study (Lapeña et al. 2003), can be explained as a consequence of the lower intensity of the disease in this experiment (natural infection). In fact, the degree of infection in untreated plants was around 50% and 55%.

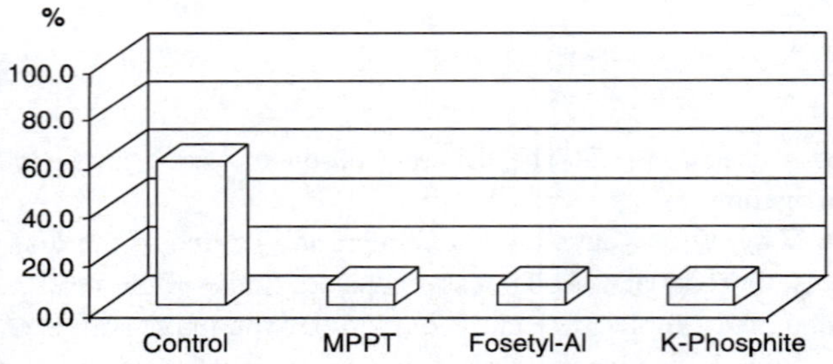

Fig. 1.4 Percentage of citrus trees affected by gummosis brown rot disease in 2000, experiment (Adapted from Ferrari and Garcia-Mina 2004)

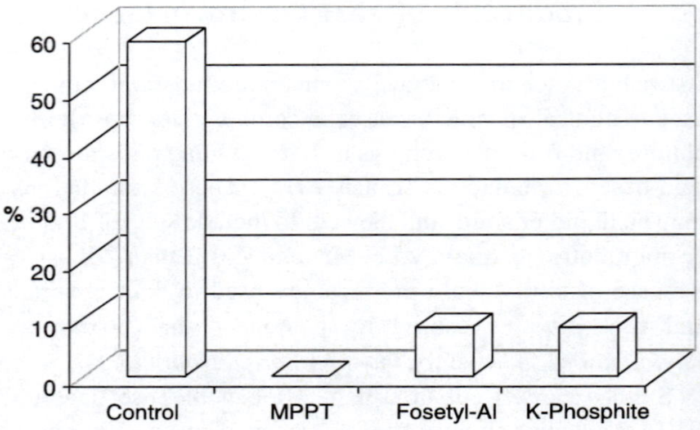

Fig. 1.5 Percentage of citrus trees affected by gummosis brown rot disease in 2001, experiment (Adapted from Ferrari and Garcia-Mina 2004)

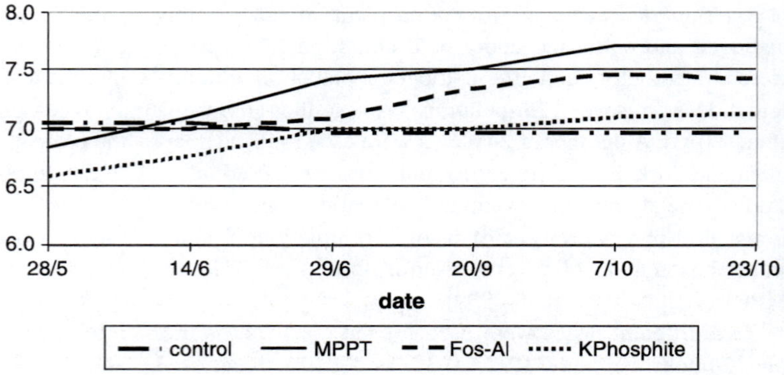

Fig. 1.6 Crop vigor. Results are expressed as a function of a 0-10 scale made according to biomass development (Adapted from Ferrari and Garcia-Mina 2004)

As for crop vigor, MPPT showed the best evolution of this parameter, probably due to the presence of some biostimulation associated with its higher efficiency in the control of the disease (Fig. 1.5).

In conclusion, the results obtained by Lapeña et al. (2003) and Ferrari and Garcia-Mina (2004) demonstrate the efficiency of the integrated work of MPPT and the other phosphonate-based products on the different levels of plant defense mechanisms, in the control of gummosis brown rot disease in citrus plants of different ages, cultivated under either greenhouse artificial inoculation or field-natural inoculation conditions.

Furthermore, the significant differences in the efficiency of the products depending on the moment of application, before or after inoculation, in the Tuset's study (Lapeña et al. 2003), also indicate that the mode of action of these products is clearly related to natural plant defense inducers with elicitor-mediated activity. These results also indicate that these products are more efficient when they are used preventively, before pathogen attack.

1.5 Future Research

All these above-discussed results, taken together, suggest that the specific study of the effects of those mineral nutrients that show some beneficial action in plant disease control on the transcriptional (the expression of salicylic acid– and jasmonic acid–regulated genes) and posttranscriptional (protein–enzyme synthesis and activity) mechanisms involved in plant natural defense response system would be of great interest. This research should be completed by complementary studies dealing with both the efficiency of treatments with these mineral nutrients on disease control in diverse plant species and the potential pesticide-type direct effect on in vitro pathogen development. All these studies will permit the development of new strategies to control plant diseases, involving special fertilizers with the ability to enhance crop resistance against pathogen attacks.

References

Blée E (1998) Phytooxylipins and plant defense reactions. Prog Lipid Res 37:33–72

Boller T, Felix GA (2009) Renaissance of elicitors: perception of microbe-associated molecular patterns and danger signals by pattern-recognition receptors. Annu Rev Plant Biol 60:379–406

Boller T, He SY (2009) Innate immunity in plants: an arms race between pattern recognition receptors in plants and effectors in microbial pathogens. Science 324:742–744

Bolwell GP, Wojtaszek P (1997) Mechanisms for the generation of reactive oxygen species in plant defence – a broad perspective. Physiol Mol Plant Pathol 51:347–366

Carswell Ch, Grant B, Theodorou M et al (1996) The fungicide phosphonate disrupts the phosphate-starvation response in *Brassica nigra* seedlings. Plant Physiol 110:105–110

Cenoz S, García Cantera R, García-Mina JM (1998) Effect of diverse formulations on PAL activity and the in vitro and in vivo development of *Penicillium expansum* in apple fruits (*Fuji* sp.) Spanish (Efecto de diferentes productos de origen natural sobre la actividad *PAL* y el control del crecimiento del hongo *in vitro* e *in vivo* en manzanas (*Fuji* sp.) infectadas por *Penicillium expansum*) Inab-Rapp. No 148. Roullier Group, Pamplona (Spain)

Dann EK, Muir S (2002) Peas grown in media with elevated plant-available silicon levels have higher activities of chitinases and β-1,3 Glucanase, are less susceptible to a fungal leaf spot pathogen and accumulate more foliar silicon. Australas Plant Pathol 31:9–13

Datnoff EL, Elmer WH, Huber DM (2007a) Mineral nutrition and plant disease. The American Phytopayhological Society, St. Paul

Datnoff LE, Rodrigues FA, Seebold KW (2007b) Silicon and plant disease. In: Datnoff LE et al (eds) Mineral nutrition and plant disease. The American Phytopayhological Society, St. Paul

Deslandes L, Rivas S (2011) The plant cell nucleus: a true arena for the fight between plants and pathogens. Plant Signal Behav 6:42–48

Duffy B (2007) Zinc and plant disease. In: Datnoff LE et al (eds) Mineral nutrition and plant disease. The American Phytopathological Society, St Paul, pp 155–176

Evans I, Solberg E, Huber DM (2007) Copper and plant disease. In: Datnoff LE et al (eds) Mineral nutrition and plant disease. The American Phytopathological Society, St. Paul

Expert D (2007) Iron and plant disease. In: Datnoff LE et al (eds) Mineral nutrition and plant disease. The American Phytopathological Society, St. Paul

Ferrari D, Garcia-Mina JM (2004) Efficacy evaluation of eurofit viti and max formulations against *Phytophthora citrophthora* on *Citrus* sp. in southern Italy. Spanish (Eficiencia del producto Eurofit Max en el control de la afección causada por *Phytophthora citrophthora* en naranjos cultivados en condiciones de campo). Lev Agr 371:217–220

García-Mina JM, Cenoz S, García Cantera R et al (1999) Composición capaz de estimular el mecanismo de defensa adquirida de las plantas. Spanish Patent 2 134 167

Garnica M, Houdusse F, Zamarreño AM et al (2010) Nitrate supply enhances active forms of cytokinins and indole acetic content and reduces abscisic acid in wheat plants grown with ammonium. J Plant Physiol 167:1264–1272

Gen-Ichiro A, Ozawa R, Massimo EM (2011) Recent advances in plant early signaling in response to herbivory. Int J Mol Sci 12:3723–3739

Graham RD, Stangoulis JCR (2007) Molybdenum and plant disease. In: Datnoff LE et al (eds) Mineral nutrition and plant disease. The American Phytopathological Society, St. Paul

Guest D, Grant B (1991) The complex action of phosphonates as antifungal agents. Biol Rev 66:159–187

Haneklaus S, Bloem E, Schnug E (2007) Sulfur and plant disease. In: Datnoff LE et al (eds) Mineral nutrition and plant disease. The American Phytopathological Society, St. Paul

Huber DM, Thompson IA (2007) Nitrogen and plant disease. In: Datnoff LE et al (eds) Mineral nutrition and plant disease. The American Phytopathological Society, St. Paul

John M, Röhrig H, Schmidt J et al (1997) Cell signaling by oligosacharides. Trends Plant Sci 2:111–115

Jones JD, Dangl JL (2006) The plant immune system. Nature 444: 323–329

Jones JF, Huber DM (2007) Magnesium and plant disease. In: Datnoff LE et al (eds) Mineral nutrition and plant disease. The American Phytopathological Society, St. Paul

Lapeña I, Tuset JJ, García-Mina JM (2003) Effect of diverse phosphorous acid-based formulations on the disease caused by *Phytophthora citrophthora* in orange plant seedlings Spanish (Efecto fungitóxico del acido fosforoso en naranjo dulce a la infección con zoosporas de Phytophthora citrophthora). Bol San Veg Plagas 29:4 13–420

Liñan C (2011) Vademecum de productos fitosanitarios y nutricionales. Spanish (Compendium of pesticides and nutritional products in Spanish market). Ediciones Agrotécnicas, Madrid

Lorenzo O, Solano R (2005) Molecular players regulating the jasmonate signalling network. Curr Opin Plant Biol 8:532–540

Ma W, Qi Z, Smigel A et al (2009) Ca2+, cAMP, and transduction of non-self perception during plant immune responses. Proc Natl Acad Sci USA 106:20995–21000

Maekawa K, Watanabe K, Kanto T et al (2002) Accumulation of silicon around penetration sites of *Magnaporthe grisea* and silicon-dependent promotion of superoxide generation after inoculation of rice leaf. In: Matoh T (ed) Second "silicon in agriculture" conference. Press-Net, Kioto

Manandhar HK, Jorgensen H, Mathur SB et al (1998) Resistance to rice blast induced by iron chloride, di-potassium hydrogen phosphate and salicylic acid. Crop Prot 17:323–329

Marschner H (1995) Mineral nutrition of higher plants. Academic, San Diego

McDonald AE, Grant BR, Plaxton WC (2001) Phosphite (phosphorous acid): its relevance in the environment and agriculture and influence on plant phosphate starvation response. J Plant Nutr 24:1505–1519

Mur LAJ, Kenton P, Atzorn R et al (2006) The outcomes of concentration-specific interactions between salicylate and jasmonate signaling include synergy, antagonism, and oxidative stress leading to cell death. Plant Physiol 140:249–262

Niki T, Mitsuhara I, Seo S et al (1998) Antagonistic effect of salicylic acid and jasmonic acid on the expression of PR protein genes in wounded mature tobacco leaves. Plant Cell Physiol 39:500–507

Ohashi Y, Ohshima M (1992) Stress-induced expression of genes for pathogenesis-relates proteins in plants. Plant Cell Physiol 33: 819–826

Prabhu AS, Fageria NK, Huber DM et al (2007a) Potassium and plant disease. In: Datnoff LE et al (eds) Mineral nutrition and plant disease. The American Phytopathological Society, St. Paul

Prabhu AS, Fageria NK, Berni RF et al (2007b) Phosphorus and plant disease. In: Datnoff LE et al (eds) Mineral nutrition and plant disease. The American Phytopathological Society, St. Paul

Rahman M, Punja ZK (2007) Calcium and plant disease. In: Datnoff LE et al (eds) Mineral nutrition and plant disease. The American Phytopathological Society, St. Paul

Reuveni R, Reuveni M (1998) Foliar fertilizer therapy. Crop Prot 17:111–118

Reuveni R, Dor G, Reuveni M (1998a) Local and systemic control of powery mildew (*Leveillula taurica*) on pepper plants by foliar spray of mono-potassium phosphate. Crop Prot 17:703–709

Reuveni R, Oppenheim D, Reuveni M (1998b) Integrated control of powery mildew on apple trees by foliar spray of mono-potassium phosphate and sterol inhibiting fungicides. Crop Prot 17:563–568

Reymond P, Farmer EE (1998) Jasmonate and salicylate as global signals for defense gene expression. Curr Opin Plant Biol 1:404–411

Rickard DA (2000) Review of phosphorous acid and its salts as fertilizer materials. J Plant Nutr 23:161–180

Rodrigues FA, McNally DJ, Datnoff LE et al (2004) Silicon enhances the accumulation of diterpenoid phytoalexins in rice: a potential mechanism for blast resistance. Phytopathology 94:177–183

Ryals JA, Neuenschwander UH, Willits M et al (1996) Systemic acquired resistance. Plant Cell 8:1809–1819

Saindrenan P, Guest D (1995) Involvement of phytoalexins in the response of phosphonate-treated plants to infection by phytophthora species. In: Daniel M, Purkasyastha RP (eds) Handbook of phytoalexin metabolism and action. Marcel Dekker, Inc, New York

Sakakibara H, Takei K, Hirose N (2006) Interactions between nitrogen and cytokinin in the regulation of metabolism and development. Trends Plant Sci 11:440–448

Stangoulis JCR, Graham RD (2007) Boron and plant disease. In: Datnoff LE et al (eds) Mineral nutrition and plant disease. The American Phytopathological Society, St. Paul

Tamás L, Huttová J (1996) Accumulation of pathogenesis-related proteins in barley induced by phosphate and salicylic acid. Biologia (Bratislaba) 51:479–484

Thompson IA, Huber DM (2007) Manganese and plant disease. In: Datnoff LE et al (eds) Mineral nutrition and plant disease. The American Phytopathological Society, St. Paul

Wood BW, Reilly CHC (2007) Nickel and plant disease. In: Datnoff LE et al (eds) Mineral nutrition and plant disease. The American Phytopathological Society, St. Paul

Xu Y, Chang P, Liu M et al (1994) Plant defense genes are synergistically induced by ethylene and methyl jasmonate. Plant Cell 6:1077–1085

Zamarreño AM, Urdaniz A, García-Mina JM (2002) Effects of formulations containing phosphite and phosphate on the mineral nutrition and growth of wheat plant seedlings. Spanish (Distribución de la fracción soluble de los aniones fosfato y fosfito en plantas de trigo tratadas con aplicaciones foliares de ambas fuentes de fósforo, y su relación con el desarrollo de las plantas). In: Abadia J et al (eds) Actas del IX Simposio Ibérico sobre nutrición mineral de las plantas. Institución Fernando el Católico (Diputación de Zaragoza), Zaragoza (Spain) No 2295

Antioxidant System in Citrus Under Nutrient Stress Conditions: Latest Developments

2

Vicent Arbona and Aurelio Gómez-Cadenas

Abstract

Nutritional imbalances are important factors that determine the distribution of natural populations of plants. However, in crop plants, alterations in nutrient levels limit vegetative growth, development and ultimately affect yield and production quality. In citrus, this is a relevant aspect due to the economic importance of this fruit tree. As other abiotic stress factors, nutrient imbalances affect the redox state of the cells by altering normal electron flow through transport chains (chloroplastic and/or mitochondrial) or impairing reactive oxygen species scavenging ability of plants. In this chapter, the effects of macro- and micro-nutrient imbalances on citrus cell redox state are reviewed, paying special attention to the defense and detoxification mechanisms involved. In addition, the role of reactive oxygen species as signaling compounds alone or associated with plant hormones is discussed.

Keywords

Reactive oxygen species • Oxidative stress • Oxidative damage • Hydrogen peroxide • Superoxide • Photosynthesis • Fertilization • Salt stress

2.1 Introduction

Natural distribution of mineral nutrients in the earth crust is a crucial factor that determines the distribution of plant species favoring the spreading of species with specific adaptations to nutrient deficiency or excess. Throughout years, humans have learned how to provide nutrients to poorly fertilized soils and, to some extent, how to regulate absorption of certain nutrients when these are present in excess in the substrate. However, this has led also to an overfertilization of certain areas therefore contaminating arable soils and aquifers. In plants, both excess and deficient fertilization induce alterations in metabolism growth and reproductive development. The extent to which the nutrient imbalance affects plant physiology depends on the specific nutrient and the relative tolerance of the plant species. In addition, nutrients might also interact with each other, depending on their relative concentration, leading to non-expected alterations in plants. For these reasons, nutrient imbalances are considered a source of abiotic stress in crop plants. As for other abiotic stress conditions such as salinity, soil waterlogging, heat, plants respond by modulating their photosynthetic metabolism and subsequently overproducing highly reactive oxidative subproducts as a secondary source of stress. In this sense, some of the macroscopic effects observed in plants subjected to nutritional imbalances (e.g., necrotic spots, chlorosis) could be a direct (or indirect) result of the production and accumulation of these substances. In this chapter, the effects of nutritional imbalances on oxidative stress are reviewed, focusing on citrus and providing examples on other crops or *Arabidopsis thaliana*.

V. Arbona (✉) • A. Gómez-Cadenas
Ecofisiologia i Biotecnologia, Departament de Ciències Agràries i del Medi Natural, Universitat Jaume I, Castelló de la Plana, Spain
e-mail: vicente.arbona@camn.uji.es; aurelio.gomez@uji.es; aurelio.gomez@camn.uji.es

A.K. Srivastava (ed.), *Advances in Citrus Nutrition*,
DOI 10.1007/978-94-007-4171-3_2, © Springer Science+Business Media B.V. 2012

2.2 Sources of Reactive Oxygen Species (ROS) in Plants Under Stress

Adverse environmental factors induce a number of physiological impairments that might end up with the accumulation of several toxic compounds that alters enzyme activity, protein functionality, as well as membrane and nucleic acid integrity. These compounds, derived from metabolic activity involving oxygen in cell organelles such as mitochondria, peroxisomes, and chloroplasts, are collectively known as reactive oxygen species (ROS). ROS are usually short-lived molecules, since they react readily with any surrounding molecule. Some of them can also initiate chain reactions that spread from the source and are only terminated when two initiating molecules react with each other, quenching the reaction. Depending on their chemical characteristics, ROS might be able to cross membranes or stay restricted in certain organelles. The main ROS, known to be extremely reactive, can be classified as free radicals: superoxide radical ($O_2^{\bullet -}$), hydroxyl radical ($HO^{\bullet}$), hydroperoxyl radical ($HO_2^{\bullet}$) and non-radical forms: hydrogen peroxide (H_2O_2) and singlet oxygen ($^1O_2$). The interaction of these primary ROS with biomolecules such as lipids (therefore generating lipid peroxide radicals) can initiate chain reactions (Halliwell 2006). Primary ROS originates from direct incomplete reduction of dioxygen (O_2), leading to highly reactive intermediates (Edreva 2005). In aerobic organisms such as higher plants, these reactions are associated to electron transport chains (ETC) and proton transport across membranes especially in chloroplasts and mitochondria (Blokhina et al. 2003; Edreva 2005). Indeed, photosynthesis is one of the major mechanisms giving rise to ROS in plants under stress, from photosynthetic pigments to light-harvesting complexes (Arbona et al. 2009; Blokhina et al. 2000; Blokhina and Fagerstedt 2010; Edreva 2005; López-Climent et al. 2008). Light absorption in higher plants takes place in the thylakoid membranes of chloroplasts. Under optimal conditions, light photons are absorbed by chlorophyll in antenna complexes which is transiently oxidized. Two electrons are then shuttled through a series of electron transporters (pheophytin, plastoquinone, cytochrome b6f, and plastocyanin) from chlorophyll in PSII to PSI. In turn, chlorophylls from PSI absorb light, and electrons are then transported through a second series of electron transporters including membrane-bound Fe-S proteins and ferredoxin (Fd). These two electrons are then used by NADP reductase to reduce two molecules of NADP. This mechanism is known as "Z" scheme; the energy of the electrons is lower at PSII and increases at PSI (Fig. 2.1). In addition, the chemiosmotic potential generated across membranes is also used to generate ATP in a process called photophosphorylation. During the whole process, PSII is able to oxidize water, yielding dioxygen and four hydrogen ions, therefore recovering the transferred electron through the ETC.

Abiotic stress conditions such as drought, salinity, nutrient imbalances, or soil waterlogging induce severe alterations of the photosynthetic metabolism that could end up with the leakage of electrons and the overproduction of ROS. There are two major sites for electron leakage in the ETC from PSI: membrane-bound Fe-S proteins and Fd. Traditionally, it was believed that in those situations where overloading

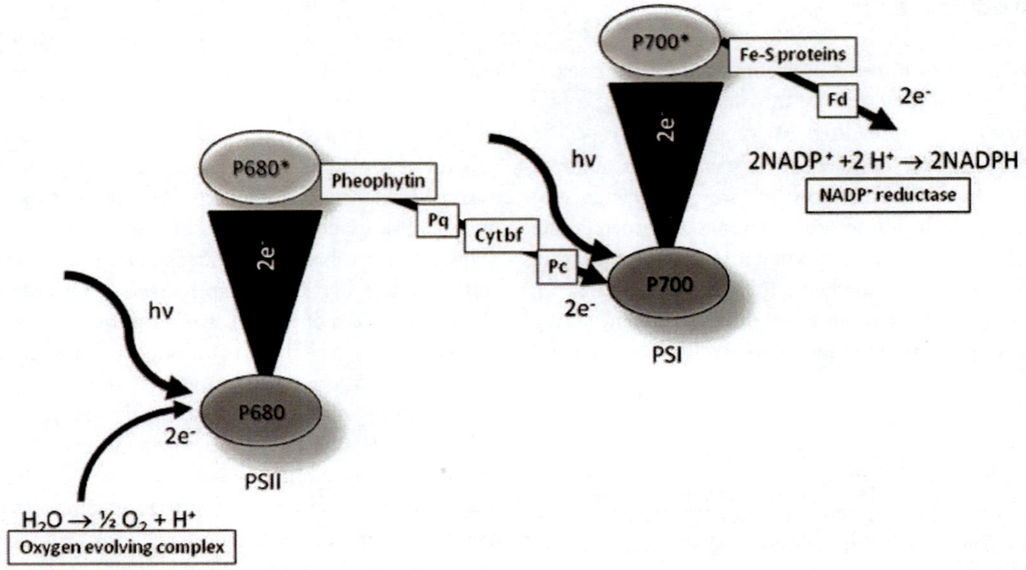

Fig. 2.1 Photosynthetic Z scheme. *PSII* and *PSI* photosystem II and I, *Pq* plastoquinone, *Cytbf* cytochrome bf complex, *Pc* plastocyanin, *Fe-S proteins* membrane-bound iron-sulfur proteins, *Fd* ferredoxin

(or malfunction) of ETC occurs, electrons are diverted to O_2 instead of producing reducing power, yielding $O_2^{\bullet-}$ in the so-called Mehler reaction. Increasing evidence points to quinones A and B at the ETC acceptor site of PSII as potential sites for electron leakage leading to $O_2^{\bullet-}$ overproduction and, indeed, PSII could be the main source for $O_2^{\bullet-}$ rather than PSI (Edreva 2005). Another important source of ROS in the chloroplast is the presence of chlorophyll photosensitizers (Edreva 2005; Garg and Manchanda 2009).

Molecular oxygen can be activated through light-dependent (or photodynamic) reactions that can be classified into two types: in type I photodynamic reactions, an excited pigment showing charge separation reacts with dioxygen, yielding $O_2^{\bullet-}$ and the oxidized pigment as products. In type II reactions, chlorophyll in the ground state (S_0) is activated, ejecting one electron from the electron pair, reaching a higher energetic state (doublet, S_2^*). Through energy dissipation as light or heat, the electron reaches lower energetic states S_1^* (singlet) and T^*(triplet); transfer of energy from S_1^* to T^* is accompanied by a reversal in electron spin. Therefore, transfer of energy from this triplet state to ground state dioxygen produces spin reversal of one electron and the formation of $^1O_2$ (Edreva 2005). Both superoxide and singlet oxygen are short-lived molecules (half-life ranging from 2 to 4 μs in a polar environment, Garg and Manchanda 2009). As mentioned above, ROS interact readily with almost any surrounding molecule, often yielding even more damaging subproducts. This is the case of $O_2^{\bullet-}$ that is (enzymatically or spontaneously) disproportionated into hydrogen peroxide (H_2O_2) which can diffuse some distance through membranes due to its non-charged chemical status and relative stability (half-life reaching 1 ms). Hydrogen peroxide itself can react with several molecules and, most importantly, with transition metals such as iron, copper present as prosthetic groups in several enzymes (e.g., superoxide dismutases), yielding hydroxyl radical ($HO^\bullet$) in a reaction known as "Fenton reaction" (H_2O_2 reacts with Fe^{2+}, yielding $HO^\bullet$, HO^-, and Fe^{3+}) or considering the global "Haber-Weiss reaction" in which H_2O_2 reacts with $O_2^{\bullet-}$, yielding O_2, $HO^\bullet$, and HO^-.

There are other cellular compartments in which production of ROS can occur as well. For instance, in mitochondria where biochemical reactions involve oxygen consumption and therefore electron transport takes place, several sites of electron leakage and $O_2^{\bullet-}$ and H_2O_2 release have been proposed, being the most important the cyanide-insensitive alternative oxidase (AOX).This enzyme catalyzes the four-electron reduction of O_2 by ubiquinone, competing for the electrons with the main respiratory chain and preventing the accumulation of over-reduced ubiquinone. This activity represents a branching of mitochondrial ETC that provides another possibility to divert the electron flow from the cytochrome oxidase pathway and to control coupling of respiration and ATP synthesis, subsequently controlling production of ROS by the ETC (Blokhina and Fagerstedt 2010). The endoplasmic reticulum as well as peroxisomes and glyoxisomes can be also important compartments for ROS production. Mixed-function oxygenases such as cytochrome P450 in endoplasmic reticulum, enzymes involved in the B-oxidation of fatty acids, and the C2 photorespiratory cycle in peroxisomes and glyoxisomes, xanthine and urate oxidases, and NADH oxidase can generate ROS when redox disbalance occurs (Perl-Treves and Perl 2002).

2.3 Molecular and Physiological Responses of Plants to Oxidative Stress: Defense Mechanisms

Plants posses an orchestrated system to cope with oxidative stress, namely antioxidant metabolites such as ascorbic acid and glutathione and also antioxidant enzymes localized in ROS production-prone cell compartments. These defenses can be either constitutively or environmentally induced (Arbona et al. 2008). The array of compounds of different chemical origins that directly react with ROS yielding less toxic products constitutes the nonenzymatic antioxidants. The amount of these metabolites in plant tissues is tightly controlled by its biosynthesis but, more importantly, by recycling through complex chain systems (Foyer and Noctor 2009). To this regard, ascorbate (Asa) and glutathione (GSH) represent the most important electron-donor molecules that regulate the cell redox status. These two metabolites exist as part of an oxidation-reduction cycle that keeps their levels almost constant and available to detoxify damaging ROS. Indeed, GSH is considered the "redox exchange coin" because it acts as a general cofactor of several recycling enzyme activities such as monodehydroascorbate reductase or phospholipid hydroperoxide glutathione peroxidase (PHGPX, Avsian-Kretchmer et al. 1999) therefore connecting Asa and lipid metabolism with GSH (Fig. 2.2).

Superoxide radical is produced in the PSI complex, and it is scavenged by SOD in the thylakoidal scavenging system (composed by thylakoid-bound Cu/Zn-SOD or Fe-SOD, APX, and Ferredoxin, Fd). The stromal scavenging system formed by stromal isoforms of SOD, APX, mono- and dehydroascorbate reductases (MDAR and DHAR), and glutathione reductase (GR) regenerates GSH and Asa from NADPH for stromal APX to detoxify H_2O_2 (adapted from Asada 1999).

Other antioxidant metabolites in plants include α-tocopherol or vitamin E, a lipophilic phenolic derivative that acts as chain terminator of free-radical reactions causing lipid peroxidation. Plants are able to synthesize vitamin E in inner chloroplast membranes, whereas animals need to acquire it through diet. Carotenoids are another important group of metabolites that protect plants from oxidative

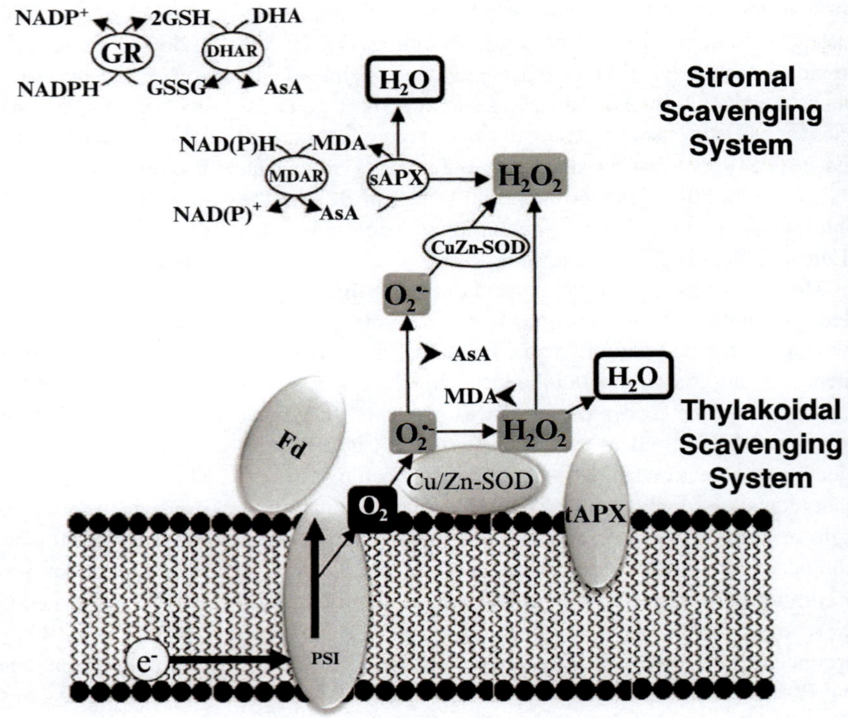

Fig. 2.2 Photoreduction of O_2 to H_2O in the water–water cycle. Superoxide radical is produced in the PSI complex, and it is scavenged by SOD in the thylakoidal scavenging system (composed by thylakoid-bound Cu/Zn-SOD or Fe-SOD, APX, and Ferredoxin *Fd*). The stromal scavenging system formed by stromal isoforms of SOD, APX, mono- and dehydroascorbate reductases (MDAR and DHAR), and glutathione reductase (*GR*) regenerates GSH and Asa from NADPH for stromal APX to detoxify H_2O_2 (Adapted from Asada 1999)

damage quenching triplet state chlorophyll and $^1O_2$, preventing initiation of lipid peroxidation. In addition, carotenoids are involved in energy transfer to other pigments in the antenna as secondary photosynthetic pigments and also in heat dissipation, as quencher of excitation energy. Particularly, xantophylls constitute a class of carotenoids that undergo prominent changes in response to intense light, allowing a better photoprotection (Perl-Treves and Perl 2002). Flavonoids are phenolic compounds that have been reported to exert antioxidant activity in vivo. They are synthesized from malonyl-CoA or 4-coumaroyl-CoA by chalcone synthase, yielding naringenin chalcone. This compound is further processed by chalcone isomerase that cyclicizes chalcones into flavanones (e.g., hesperetin, naringenin, or eriodictyol). The hydroxylation of flavanones by flavanone-3-hydroxylase yields dihydroflavonols (e.g., taxifolin and dihydrokaempferol), and the subsequent reaction catalyzed by flavonol synthase yields flavonols (e.g., isorhamnetin, quercetin, or kaempferol). Biosynthesis can be further derived into isoflavones (e.g., genistein or daidzein), flavanols (e.g., catechins and epicatechins), proanthocyanidins (x-mers of flavanols), and anthocyanidins formed by flavylium ion skeleton. These forms exist as sugar conjugates (e.g., glucosides, rutinosides, rhamnosides) or aglycones. It is not known how flavonoids exert their beneficial activity in plant tissues; however, indirect evidence supports their role in photoprotection from strong or UV light, biotic (Cevallos-Cevallos et al. 2009) and abiotic stresses (Djoukeng et al. 2008).

Several antioxidant enzyme activities are induced upon occurrence of oxidative stress. These enzyme activities are oriented to detoxify ROS, namely $O_2^{\bullet-}$ and H_2O_2. The main enzyme activities are:

1. Superoxide dismutases (SODs) are a group of metalloproteins that dismutate $O_2^{\bullet-}$ into H_2O_2. This enzyme activity is considered to be the "primary defense" against ROS. Dismutation is mediated by a metal ion (Cu, Mn, Zn, or Fe) located in the active site of the enzyme to which $O_2^{\bullet-}$ is properly attracted by positively charged amino acid residues. The different prosthetic groups found in SODs allow their classification in phylogenetically related Mn-SODs and Fe-SODs and unrelated Cu/Zn-SODs, all three being present in higher plants. Subcellular localization varies among species, but generally, Cu/Zn-SODs are cytosolic and chloroplastic, Mn-SODs are found in mitochondria, and Fe-SODs in chloroplasts, although several SOD isozymes have been identified in almost all cell compartments (Van Breusegem et al. 2001).
2. Catalases (CAT) act as H_2O_2:H_2O_2 oxidoreductases and are found mainly in peroxisomes and glyoxisomes. These enzymes catalyze the reduction of H_2O_2 or the oxidation of several substrates such as methanol, ethanol, formaldehyde, or formic acid. Catalases can be subdivided into class I, II, and III, being class I more common in photosynthetic tissues, class II are present in vascular bundle, and class III in seeds and young seedlings (Van Breusegem et al. 2001).

Table 2.1 Summary of antioxidant enzyme activities found in plant tissues, their substrates, cofactors, prosthetic groups, and subcellular localization

Enzyme	Substrate	Cofactors	Prosthetic groups	Subcellular localization
Superoxide dismutases	$O_2^{\bullet-}$	None	Mn, Zn, Cu, Fe	Cu/Zn-SOD chloroplasts/cytosol
				Mn-SOD mitochondria
				Fe-SOD chloroplasts
Catalases	H_2O_2	None	Heme	Peroxisomes and glyoxisomes
Ascorbate peroxidases	H_2O_2	Ascorbate	Heme	Cytosol, plasma membrane, thylakoid membranes, chloroplast stroma
Glutathione peroxidases	H_2O_2	Glutathione	None	All cell compartments

3. Ascorbate peroxidases (APX) are heme-containing enzymes that catalyze the detoxification of H_2O_2 using ascorbate as an electron donor yielding two molecules of water and monodehydroascorbate. APXs are ubiquitous enzymes and play a role in the defense against oxidative stress induced by abiotic stress. As extracted from DNA sequence data, at least seven different APX can be distinguished in plants: two soluble cytosolic forms, three types of cytosol membrane bound, one located in the stroma of the chloroplast, and one thylakoid membrane bound. These isoforms differ greatly in molecular and enzymatic properties probably in order to adapt to the different cell compartments and the distinct rates of ROS production.
4. Glutathione peroxidases (GPX) are a type of thioredoxin peroxidases or peroxiredoxins that exhibit the ability to reduce their N-terminus after reacting with peroxide which is regenerated through the formation of inter- or intramolecular disulfide bridges (Navrot et al. 2006), then the disulfide bridges are reduced via thiol-containing molecules: thioredoxin or glutaredoxin. These GPXs are present in almost all cell compartments and can efficiently reduce peroxides following the above mechanism (Avsian-Kretchmer et al. 1999; Navrot et al. 2006) (Table 2.1).

Other major recycling enzyme activities contribute to keep the cell redox status at optimal conditions: (1) monodehydroascorbate reductase (MDHAR) reduces monodehydroascorbate to ascorbate consuming NADPH as cofactor, linking photosynthetic performance and the maintenance of the redox status; (2) dehydroascorbate reductase (DHAR) reduces dehydroascorbate to ascorbate using GSH as cofactor; and (3) glutathione reductase (GR) that regenerates GSH from GSSG using NADPH as electron donor.

2.4 Oxidative Damage in Citrus Under Stress

Environmental cues cause imbalances in the production of ROS through an effect on different physiological mechanisms based on redox reactions. In higher plants, stress induces an uncoupling of some of the components of the ETC, resulting in an electron leakage ending up with the formation of lipoperoxides as well as other oxidative damage intermediates. Under these situations, a progressive over reduction of photosynthesis occurs and, therefore, photorespiration is activated in peroxisomes. The oxidative reactions that take place in there render large amounts of H_2O_2 that can cross membranes and affect any surrounding structure or molecule. Excess ROS produced under stress react with many different molecules, as pointed out above, yielding several oxidative by-products such as lipid hydroperoxides (noted as LOO•) that initiate chain reaction cascades with other lipid molecules (unsaturated fatty acids, noted as LOOH) that only end up when a terminator molecule reacts with LOO• (for instance, another LOO•, yielding LOOOOL). Lipid hydroperoxides also produce another subproduct named malondialdehyde (MDA, $CH_2(CHO)_2$) widely used as a marker for oxidative damage (Arbona et al. 2003, 2004, 2008; Hossain et al. 2009). Moreover, MDA can also react with deoxyadenosine and deoxyguanosine in DNA, forming several mutagenic DNA adducts. It has also been shown that the production of ROS under non-stressful conditions is involved in the control of cell differentiation and organogenesis in citrus (Faltin et al. 2010). Proteins also constitute a target for ROS effect, amino acids in active sites can be oxidized and therefore protein activity inhibited. Thiol groups are very susceptible sites of oxidation, and amino acids containing these groups are highly sensitive targets (e.g., cysteine or methionine). Lipid peroxidation products, such as 4-hydroxy-2-nonenal, can also damage proteins (Garg and Manchanda 2009).

The main effect of environmental stress in citrus is the reduction of gas exchange parameters (Arbona et al. 2005a, b) due to stomatal closure as well as a reduction in carboxylative efficiency (López-Climent et al. 2008; Arbona et al. 2009). As a general aspect, the shortage in CO_2 availability is likely to induce an uncoupling of the ETC that redirects electrons from the light-harvesting complexes to dioxygen rendering $O_2^{\bullet-}$. In addition, the higher O_2/CO_2 ratio existing under stress conditions favors photorespiration rather than photosynthesis resulting in an increased ROS production. In this sense, boron imbalances in citrus caused a reduction in net CO_2 assimilation accompanied by reductions in stomatal conductance (Sheng et al. 2010), as well as total chlorophyll content and soluble proteins (Han et al. 2009). The decrease in photosynthetic efficiency was also accompanied by an increase in the concentration of MDA and other thiobarbituric acid-reactive substances in plants subjected to B excess or deficit (Han et al. 2009). Other types of environmental cues such as salt stress affect mineral balances (Gimeno et al. 2010) as well as CO_2 assimilation and carbohydrate partition (Arbona et al. 2005b) with a concomitant effect on ROS production and oxidative stress in citrus plants, regardless the relative tolerance of different genotypes to stress (Arbona et al. 2003, 2004). In this sense, high concentrations of heavy metals such as Al^{3+} (Jiang et al. 2008) and Cd^{2+} (López-Climent et al. 2011) in the substrate reduce net photosynthetic rate as well as carboxylative efficiency in citrus, resulting in oxidative stress (unpublished results). In a recent publication, Forner-Giner et al. (2010) reported that iron starvation caused a decrease in catalase and peroxidase enzyme activities that could lead to an increase in oxidative damage in citrus plants.

In citrus, it has been reported that N supply regulates chlorophyll content and therefore photosynthesis and RubBPC activity, although no direct relationship to oxidative stress was assessed (Basiouny 2000). However, the reduction in chlorophyll concentration at the antennae is likely to reduce the intensity of electron flow from PSII and, therefore, the risk of electron leakage. On the contrary, in other crop plants such as *Coffea arabica*, reductions in N supply reduced PSII performance under high irradiance; however, there was no apparent effect on net carbon fixation. Under low N supply, coffee plants exhibited higher MDA and electrolyte leakage values than plants supplied with high concentrations of N, irrespective of the light fluency (Pompelli et al. 2010). Similar responses were recorded for mulberry shrubs (*Morus alba*; Tewari et al. 2007) and wheat (*Triticum aestivum*; Criado et al. 2007) under N deprivation, showing increases in MDA and decreases in chlorophyll and protein contents in comparison to well-fed controls.

Manganese is another important oligoelement involved in the regulation of many different physiological processes as well as part of the antioxidant enzyme Mn-SOD. In a recent publication, pummelo plants were grown under excessive Mn concentrations resulting in a severe impairment of the whole photosynthetic system from the donor side (PSII) to the reduction end of acceptor decrease, resulting in a reduction in the rate of CO_2 assimilation. On the contrary, the antioxidant system was enough to cope with overproduced ROS preventing further oxidative damage (Li et al. 2010). Changes in leaf and chloroplast ultrastructure linked to a decrease in electron transport rate and photochemical yield in response to different Mn levels have been reported in *Citrus volkameriana* (Papadakis et al. 2007a, b). In summary, although no evident changes in oxidative damage were recorded, imbalances in Mn supply could sensitize citrus plants to photoinhibition.

Sulfur is increasingly being considered the fourth major essential nutrient after N, P, and K. Adequate S nutrition improves photosynthetic performance, and its deficiency negatively regulates chlorophyll content of leaves, nitrogen accumulation, and photosynthetic enzymes. Sulfur has an especial relevance for the antioxidant system since it is incorporated into important molecules such as glutathione as thiol groups (–SH) or in amino acids like cysteine, which are important modulators of the stress response (Nazar et al. 2011). In rice, sulfur deprivation caused a reduction in chlorophyll concentration followed by decreases in carboxylative performance as well as an increase in the GSSG/GSH ratio, indicating an effect on the glutathione-dependent antioxidant system (Lunde et al. 2008). Nevertheless, the detrimental effect of S deprivation is aggravated when environmental conditions become adverse. For instance, Arabidopsis seedlings subjected to high irradiance in combination with S deprivation exhibited symptoms of senescence faster than well-fed controls (Wulff-Zottele et al. 2010). These phenotypes were correlated with a decrease in electron transport rate, an increase in non-photochemical quenching and a drop in sulfur-containing metabolites under high irradiance (Wulff-Zottele et al. 2010). It has been also proposed that the effect of oligonutrient deprivation occurs through an impairment of the acquisition of other major nutrients such as nitrogen, phosphorus, or potassium. For instance, sulfur deprivation caused severe growth arrest in plants of *Trifolium repens* correlated with a decrease in N fixation (Varin et al. 2010).

2.5 Involvement of Antioxidant Defenses in Stress Tolerance in Citrus

Plants subjected to environmental stress induce both antioxidant enzymes and metabolites to cope with ROS. It is widely accepted that a positive response of antioxidant defenses is associated to tolerance to abiotic stress (Hernández et al. 1995). However, as extracted from experiments carried out with citrus genotypes differing in abiotic stress tolerance, not always an increase in antioxidant defenses in plants subjected to stress is related to the actual tolerance. For instance, when Cleopatra mandarin and Carrizo citrange were subjected to identical salinity conditions, only the sensitive genotype, Carrizo, increased antioxidant defenses, whereas Cleopatra, the tolerant genotype, did not (Arbona et al. 2004). This apparent contradiction could be related to the lower exposure of the aerial part of this genotype to saline ions due to the exclusion process that takes place in roots (López-Climent et al. 2008). Moreover, no apparent increase in MDA, as oxidative damage marker, was observed. Under these conditions, Cleopatra follows a "conservative" strategy, reducing all photosynthetic and metabolic activity to prevent damage to important reaction centers, whereas Carrizo, as well as other citranges and also citrumelo, maintains metabolic and physiological processes unaltered until the intoxication with saline ions is too severe to keep an adequate physiological activity (López-Climent et al. 2008). Under soil flooding conditions, all tested citrus genotypes increased leaf MDA but showing clear differences in the onset of MDA accumulation associated to stress tolerance (Arbona et al. 2008). These differences were also associated to a more prominent activation of enzymatic antioxidant defenses and a less drastic effect of the stress on the photosynthetic apparatus (Arbona et al. 2009). These responses clearly link activation of antioxidant defenses with a previous effect on ROS overproduction and to impairment in photosynthetic activity. As explained above, alterations in mineral nutrient supply have an effect on photosynthetic activity and subsequently on oxidative metabolism.

2.5.1 Nitrogen

It has been reported that nitrogen supply has an effect on photosynthetic activities independent of light intensity (Basiouny 2000). Other crop plants, such as coffee (Pompelli et al. 2010), showed a general reduction in electrolyte leakage as well as in the production of H_2O_2 and MDA under high nitrogen supply. This reduction was not associated to an induction of antioxidant activities but rather to the activation of photoprotective mechanisms contributing to the dissipation of excess irradiative energy and resulting in no measurable effect on CO_2 fixation. In wheat plants, nitrogen deprivation caused a reduction in chlorophyll and rubisco contents as well as in CAT and APX activities, resulting in a moderate MDA increase (Criado et al. 2007). These results indicate that regulation of the photosynthetic system might be a common strategy to cope with oxidative stress associated to low N supply in coffee, wheat, and citrus.

2.5.2 Phosphorus

Phosphorus is another macronutrient whose deficiency has been reported to cause oxidative stress in several crop plants (Malusá et al. 2002; Li et al. 2007; Yao et al. 2010). In maize roots, a transcriptomics study conducted under P deprivation showed a major effect on C metabolism, mainly on carbon flow and lipid metabolism (Wasaki et al. 2003). Similar results were obtained by Li et al. (2007) after proteomic profiling by 2D-gel staining. About 36 protein spots associated to primary metabolism were differentially regulated by P starvation. Among which, a cytosolic pyruvate kinase, phosphoglucose isomerase, enolase, malate dehydrogenase, phosphogluconate dehydrogenase, and triosephosphate isomerase were upregulated, showing values raging between four- and sevenfold above well-feed controls, suggesting massive carbohydrate mobilization. Another strongly enriched category was secondary metabolism, with PAL showing the highest induction along with Caffeoyl O-methyltransferase and O-methyltransferase suggesting also the induction of lignin biosynthesis. Antioxidant defenses were only represented by glutathione metabolism-related proteins such as glutathione transferases and glutathione peroxidases which showed an induction above threefold in all cases. However, cytosolic APX expression was downregulated by P starvation (Li et al. 2007). In a recent study, Yao et al. (2010) showed that response of proteomic profiles to P starvation in tolerant and sensitive *Brassica napus* cultivars differs greatly. In general, the sensitive genotype exhibited the same response described above.

2.5.3 Potassium

In *Hordeum maritimum*, K starvation induces the antioxidant enzymes APX, GPX and the recycling activities GR, DHAR, and MDAR that contributed to the reduction in H_2O_2 concentration resulting in a moderate MDA increase (Hafsi et al. 2010). On the other side, K excess did not have any effect on MDA accumulation neither on antioxidant defenses in rice seedlings (Ding et al. 2008). In leaves of *Morus alba* shrubs, potassium deficiency had no effect on chlorophyll content, whereas carotenoids were reduced, resulting in a moderate increase in MDA content, due to the activation of SOD, APX, CAT, POD, and GR enzyme activities (Tewari et al. 2007). As stated by Cakmak (2005), K has a protective effect on plants through the maximization of photosynthesis and a correct carbohydrate partitioning in leaves, reducing therefore the susceptibility to ROS overproduction. In this sense, it has been shown that one of the early responses to oxidative stress in plants is the induction of massive K^+ efflux from root epidermal cells mediated by ROS (Cuin and Shabala 2007). Therefore, it is clear that correct potassium nutrition might contribute to palliate the strong K leakage induced by ROS. In general, K-starved plants are more susceptible to photooxidative stress; however, the antioxidant system is usually able to cope with ROS.

2.5.4 Boron

In grapefruit plants, B deficiency induces SODs as well as APX, MDAR, DHAR, and GR enzyme activities, whereas CAT activity decreases, in a way consistent with an increase in reduced ascorbate and glutathione pools, overall contributing to a moderate increase in MDA (Han et al. 2009). On the contrary, B excess did have a more dramatic impact on

Table 2.2 Main macro- and oligonutrients target physiological activities and specific effects on high and/or low nutrient supply

Nutrient	Nutrient type	Target physiological activity	Specific effect of nutrient imbalance
Nitrogen (N)	Macronutrient	Photosynthetic system	High N supply improves photoprotective mechanisms, no effect on CO_2 assimilation
			Low N supply induces reduction in rubisco and chlorophyll contents and a reduction in CAT and APX activities, resulting in a moderate MDA increase
Phosphorus (P)	Macronutrient	Carbon flow and lipid metabolism	Low P supply induces the upregulation of some glycolytic and TCA cycle enzymes, upregulation of secondary metabolism oriented to lignin biosynthesis, and downregulation of cytosolic APX
Potassium (K)	Macronutrient	Photosynthetic system and carbon flow	Low K supply induces the H_2O_2-scavenging enzymatic system, a moderate MDA increase, and a moderate reduction of carotenoids
Boron (B)	Micronutrient	Soluble antioxidants and their recycling activities	High B supply increases oxidative damage through the inhibition of ascorbate recycling activities
			Low B supply induces SODs, APX, and recycling activities, resulting in a moderate MDA increase
Manganese (Mn)	Micronutrient	Photosystem II activity and antioxidant enzyme activities	High Mn supply increases leaf SOD, CAT, and GPX enzyme activities, decreases DHAR activity, and subsequently reduces ascorbate pool in roots
Magnesium (Mg)	Micronutrient	Carbon fixation and partitioning	Low Mg supply induces ferritin1 and the massive accumulation of Fe and Cu. Symptoms are associated with Cu toxicity accompanied by decreased assimilation of other micronutrients
Sulfur (S)	Micronutrient	Photosynthesis and biosynthesis of glutathione	Low S supply results in the depletion of thiol-containing molecules, decrease in N assimilation, and an increase in glutathione and ascorbate recycling enzyme activities

oxidative stress, although its effect on photosynthetic machinery was not more severe than B deficiency. This more drastic effect could be associated to an apparent inhibition of ascorbate recycling activities (DHAR and MDAR) having therefore an impact on reduced ascorbate pool (Table 2.2).

2.5.5 Manganese

In a recent paper, Li et al. (2010) reported the effect of Mn excess on antioxidant mechanisms in citrus resulting in an increase in leaf total SOD activity as well as CAT and GPX respect to control conditions; however, DHAR activity was negatively affected by the treatment. In roots, most of the antioxidant enzymes but MDAR and DHAR remained unaffected resulting in a decrease in reduced ascorbate pools (Li et al. 2010). Interestingly, no effect on MDA accumulation neither in leaves nor in roots was observed, probably because the induction of antioxidant enzymes is enough to cope with increased ROS. Once again, alterations in Mn nutrition resulted in an effect on PSII activity (Papadakis et al. 2007a), probably mediated by changes in leaf and chloroplast ultrastructures (Papadakis et al. 2007b).

2.5.6 Magnesium

Although there is no information in the literature on the effect of Mg^{2+} deficiency on citrus physiology, in other plant systems, this deficiency has been associated to alterations in carbon partitioning between roots and shoots resulting in massive accumulation of carbohydrates in source leaves (Cakmak and Kirby 2008). Most of these symptoms are related to impairment of photosynthetic CO_2 fixation. Over reduction of electron transport favors the generation of ROS and the induction of antioxidant activities, especially under high light intensities inducing photooxidative processes (Cakmak and Kirby 2008). In a recent paper, Hermans et al. (2010) reported that in Mg-starved Arabidopsis, an increase in DHA/ASC ratio occurs in leaves and the opposite in roots, whereas glutathione homeostasis is maintained. In addition, results presented in this paper suggest that the effects of magnesium deficiency occur through the intracellular increase of iron and copper concentrations as suggested by the induction of Ferritin1, a chloroplastic iron-storage protein that accumulates upon iron excess and at the onset of senescence when ROS accumulate (Hermans et al. 2010). In addition, under these conditions, a vacuolar ABC transporter is also induced (MPR3). This protein can transport chlorophyll catabolites but also is involved in the detoxification of metals (Hermans et al. 2010). In turn, copper toxicity reduces assimilation of other nutrients such as Fe^{2+}, Zn^{2+}, K^+, and Ca^{2+} in plants (Bouazizi et al. 2010), resulting in a more severe phenotype.

2.5.7 Sulfur

It has been shown that the detrimental effects of sulfur starvation on plant physiology could be due to the limitation in the assimilation of other nutrients such as nitrogen (Varin et al. 2010). Under three different regimes of S supply to growth medium, N absorption followed an increasing trend with S in clover plants as well as the concentration of soluble proteins and nitrogen assimilation-related enzyme activities (Varin et al. 2010). Moreover, the depletion of thiol-containing molecules such as glutathione under sulfur deprivation occurs in short time, resulting in an increased incidence of oxidative stress. In rice, Lunde et al. (2008) found evidence of a strong upregulation of the GSH and AsA recycling activities under sulfur starvation oriented to ensure the availability of reduced forms in the absence of starting biosynthetic intermediates. However, this is limited in the long term since S starvation also affects photosynthetic activity and subsequently NADPH production.

2.6 Role of ROS as Signaling Molecules and Interaction with Plant Hormones

The production of ROS in plants under the effect of abiotic or biotic elicitors is a rapid and regulated process. These by-products of aerobic metabolism damage cell components, but they can also act as signal transduction molecules to counteract oxidative damage in stressed compartments (Møller and Sweetlove 2010). Most known ROS carry an electrostatic charge (as $O_2^{\bullet-}$) or are too reactive (as $^1O_2$ and $HO^\bullet$) to be considered as long distance signals, leaving H_2O_2 as the likely ROS messenger. This molecule has proved to be able to regulate genes in bacteria (Møller and Sweetlove 2010) and plants (Foyer et al. 2009). A clear evidence of H_2O_2 as signaling molecule in plants comes from experiences carried out in catalase-deficient tobacco plants (Vandenabeele et al. 2004). In these plants, it is possible to alter photorespiratory H_2O_2 levels by simply changing light fluency in growth chambers. Changes in gene expression were correlated with specific H_2O_2 levels also induced by abiotic stress (Vandenabeele et al. 2003, 2004; Vanderauwera et al. 2005). In citrus, Avsian-Kretchmer et al. (2004) reported that the promoter of the gpx1 gene (encoding for a glutathione peroxidase) responded specifically to hydrogen peroxide. This was evidenced by the fact that ectopic expression of uidA gene under the control of gpx1 promoter was induced by abiotic elicitors (NaCl, sorbitol, mannitol, KCl, and, to a lower extent, by Na_2SO_4) as well as by H_2O_2. The addition of catalase or diphenyleneiodonium chloride, an inhibitor of NADPH-oxidase, reduced H_2O_2-dependent expression of the gpx1 promoter but did not inhibit NaCl induction, suggesting two alternative pathways for hydrogen peroxide signaling: intra- and extracellular (Avsian-Kretchmer et al. 2004). At whole plant level, it is likely that ROS mediate major hormone signaling such as abscisic acid (ABA, Cho et al. 2009), nitric oxide (NO, Blokhina and Fagerstedt 2010), salicylic acid (SA, Ashraf et al. 2010; Gunes et al. 2007), and brassinosteroids (Ashraf et al. 2010). One of the main physiological responses when plants are subjected to environmental stress is stomatal closure. This is achieved by activation/inactivation of ion channels in the plasma membrane and endomembranes of guard cells to regulate flux of K^+, Cl^-, Ca^{2+}, and malate^{2-}. The activity of these channels is positively regulated by ABA through H_2O_2. In the transduction pathway, H_2O_2 acts upstream abi2-1, a protein phosphatase 2C, and downstream of abi1-1, encoding for a regulatory component of the ABA receptor (Cho et al. 2009).

Other plant hormones such as NO have also been studied for its involvement in signaling in plants under low oxygen conditions. The low O_2 availability causes several metabolic disturbances that lead to an increase in the production of NO from NOx either enzymatically or nonenzymatically (Blokhina and Fagerstedt 2010). In maize, salt stress caused imbalances in several nutrients and increased MDA accumulation that were counteracted by SA treatment. Interestingly, H_2O_2 levels increased in SA-treated maize seedlings, indicating that SA might inhibit the damaging effects of hydrogen peroxide on plant physiology. Treatment with SA also increased Cu^{2+}, Mn^{2+}, Fe^{2+}, and Mg^{2+} concentrations, in the same direction as salt stress, but decreased K^+, phosphorus, and Zn^{2+} concentrations contrastingly to NaCl treatment, suggesting that these changes could be at the basis for a better performance of maize under saline conditions (Gunes et al. 2007). Low nitrogen supply in wheat was associated to an early drop in isopentenyl adenosine (iPA) concentration and a later increase in ABA levels accompanied by a rise in H_2O_2 and MDA production. While iPA decrease was concomitant with reductions in protein concentration, ABA was correlated with senescence and the accumulation of ROS (Criado et al. 2007). Increasing evidence indicates that ROS mediate plant physiological changes to adapt to different environmental conditions (not necessarily stressful). Since photosynthesis and the energy metabolism are the principal factors being affected by changes in optimal growth conditions or even photoperiod, it is likely that these mechanisms act as environmental sensors and their by-products, i.e., ROS, as fine-tuning signals in interaction with plant hormones or members of the plant hormone signaling pathways. In this sense, it has been established that day length determines the response to H_2O_2 produced by photorespiration. In catalase-deficient Arabidopsis mutants, the accumulation of hydrogen peroxide under short-day conditions might cause upregulation of defense genes, whereas under long-day conditions H_2O_2 induces cell death, which is also mediated by SA (Foyer and Noctor 2009). In this context, nutritional or stress conditions that enhance photorespiratory H_2O_2 production could, in turn, be aggravated by environmental conditions such as light intensity altering the way the signals are transduced and, subsequently, the fate of the plant.

2.7 Concluding Remarks

Besides its obvious effects in growth and development, deficient mineral nutrition causes a severe impairment in the ability of plants to cope with adverse environmental conditions. It has been shown that plants subjected to nutrient starvation are more sensitive to high irradiance, salt stress, or other environmental factors. Moreover, several studies have shown that micronutrient shortage exerts its effects through the alteration of assimilation of other major nutrients, therefore suggesting the interconnection between micro- and macronutrients. In this sense, macronutrient starvation causes a less severe phenotype than micronutrient deficiency since apparently plants possess efficient mechanisms to cope with the alterations induced thereof. Research carried out in the processes that take place under macro- and micronutrient deficiency pointed out the involvement of the photosynthetic apparatus as the main source of ROS under these conditions; however, ROS scavenging machinery efficiently cleaves and eliminates all these toxic compounds, since no nutrient depletion seems to severely affect these mechanisms. Moreover, other nonenzymatic mechanisms such as flavonoid or anthocyanin biosynthesis or soluble antioxidants such as glutathione or ascorbate might take over the detoxification of ROS when the enzymatic system fails. Overall, nutrient depletion appears to be a situation quite common in natural environments, and for this reason, plants have development during their evolution several mechanisms to cope with it; for instance, under iron deprivation, plants recycle this metal from old leaves into new leaves, or when subjected to sulfur shortage, metabolites such as GSH are also recycled in order to override de novo biosynthesis. Another important aspect in ROS metabolism is their involvement as signaling molecules together with plant hormones. Rise in ROS production is a general feature under stress conditions, pointing at these molecules as direct mediators between perception of adverse environmental conditions and the physiological response. This could be, in turn, systemically modulated by plant hormones such as ABA, JA, or ethylene.

2.8 Future Research

A prospective research line is to assess whether plant hormones (or other molecules) act as mediators between the redox stimulus and whole plant responses such as growth arrest, stomatal closure, or proline accumulation or not. As short-lived molecules, ROS are thought to exert its action mediating stress perception at the cell level. An example is the release of fatty acids from membranes as a result of direct physical damage or ROS signaling. This is a common mechanism to animals and plants that leads to an inflammatory response mediated by eicosanoids or jasmonic acid-associated responses, respectively. In this sense, it is quite likely that other compounds derived from oxidative damage to cell walls, cell lysates, or other sources could act also as signaling molecules. The identification of these compounds using highly sensitive metabolite profiling techniques will provide links between stress perception at the organelle level and the physiological response.

Acknowledgments This work was supported by Ministerio de Ciencia e Innovación through grant AGL2010-22195-C03-01/AGR.

References

Arbona V, Flors V, Jacas J et al (2003) Enzymatic and non-enzymatic antioxidant responses of carrizo citrange, a salt-sensitive citrus rootstock, to different levels of salinity. Plant Cell Physiol 44:388–394

Arbona V, Jacas J, Gómez-Cadenas A (2004) Response of the antioxidant machinery of two citrus rootstocks to salt stress. Proc Int Soc Citric 2:644–648

Arbona V, Iglesias DJ, Jacas J et al (2005a) Hydrogel substrate amendment alleviates drought effects on young citrus plants. Plant Soil 270:73–82

Arbona V, Marco AJ, Iglesias DJ et al (2005b) Carbohydrate depletion in roots and leaves of salt-stressed potted *Citrus clementina* L. Plant Growth Regul 46:153–160

Arbona V, Hossain Z, López-Climent MF et al (2008) Antioxidant enzymatic activity is linked to waterlogging stress tolerance in citrus. Physiol Plant 132:452–466

Arbona V, López-Climent MF, Pérez-Clemente RM et al (2009) Maintenance of a high photosynthetic performance is linked to flooding tolerance in citrus. Environ Exp Bot 66:135–142

Asada K (1999) The water-water cycle in chloroplasts: scavenging of active oxygen species and dissipation of excess photons. Ann Rev Plant Physiol Plant Mol Biol 50:601–639

Ashraf M, Akram NA, Arteca RN et al (2010) The physiological, biochemical and molecular roles of brassinosteroids and salicylic acid in plant processes and salt tolerance. Crit Rev Plant Sci 29:162–190

Avsian-Kretchmer O, Eshdat Y, Gueta-Dahan Y et al (1999) Regulation of stress-induced phospholipid hydroperoxide glutathione peroxidase expression in citrus. Planta 209:469–477

Avsian-Kretchmer O, Gueta-Dahan Y, Lev-Yadun S et al (2004) The salt-stress signal transduction pathway that activates the gpx1 promoter is mediated by intracellular H_2O_2, different from the pathway induced by extracellular H_2O_2. Plant Physiol 135:1685–1696

Basiouny FM (2000) RuBP carboxylase, nitrate reductase and photosynthetic activities in citrus leaves as influenced by light intensities and nitrogen levels. Int Mult Con Int Fruit Veg Qual 1:98–105

Blokhina O, Fagerstedt K (2010) Reactive oxygen species and nitric oxide in plant mitochondria: origin and redundant regulatory systems. Physiol Plant 138:447–462

Blokhina O, Virolainen E, Fagerstedt K et al (2000) Antioxidant status of anoxia-tolerant and -intolerant plant species under anoxia and re aeration. Physiol Plant 109:396–403

Blokhina O, Virolainen E, Fagerstedt K (2003) Antioxidants, oxidative damage and oxygen deprivation stress: a review. Ann Bot 91: 179–194

Bouazizi H, Jouili H, Geitmann A et al (2010) Copper toxicity in expanding leaves of *Phaseolusvulgaris* L.: antioxidant enzyme response and nutrient element uptake. Ecotox Environ Saf 73: 1304–1308

Cakmak I (2005) The role of potassium in alleviating detrimental effects of abiotic stresses in plants. J Plant Nutr Soil Sci 168:521–530

Cakmak I, Kirby EA (2008) Role of magnesium in carbon partitioning and alleviating photooxidative damage. Physiol Plant 133:692–704

Cevallos-Cevallos JM, Rouseff R, Reyes-De-Corcuera JI (2009) Untargeted metabolite analysis of healthy and Huanglongbing-infected orange leaves by CE-DAD. Electrophoresis 30:1240–1247

Cho D, Shin D, Jeon BW et al (2009) ROS-mediated ABA signaling. J Plant Biol 52:102–113

Criado MV, Roberts IN, Echeverria M et al (2007) Plant growth regulators and induction of leaf senescence in nitrogen-deprived wheat plants. Plant Growth Regul 26:301–307

Cuin TA, Shabala S (2007) Compatible solutes reduce ROS-induced potassium efflux in Arabidopsis roots. Plant Cell Environ 30: 875–885

Ding YC, Chang CR, Luo W et al (2008) High potassium aggravates the oxidative stress inducedy by magnesium deficiency in rice leaves. Pedosphere 18:316–327

Djoukeng JD, Arbona V, Argamasilla R et al (2008) Flavonoid profiling in leaves of citrus genotypes under different environmental situations. J Agric Food Chem 56:11087–11097

Edreva A (2005) Generation and scavenging of reactive oxygen species in chloroplasts: a submolecular approach. Agric Ecosyst Environ 106:119–133

Faltin Z, Holland D, Velcheva M et al (2010) Glutathione peroxidase regulation of reactive oxygen species level is crucial for in vitro plant differentiation. Plant Cell Physiol 51:1151–1162

Forner-Giner MA, Llosá MJ, Carrasco JL et al (2010) Differential gene expression analysis provides new insights into the molecular basis of iron deficiency stress response in the citrus rootstock *Poncirus trifoliata* (L.) Raf. J Exp Bot 61:483–490

Foyer CH, Noctor G (2009) Redox regulation in photosynthetic organisms: signaling, acclimation, and practical implications. Antiox Redox Signal 11:862–878

Foyer CH, Bloom AJ, Queval G (2009) Photorespiratory metabolism: genes, mutants, energetics, and redox signaling. Ann Rev Plant Biol 60:455–484

Garg N, Manchanda G (2009) ROS generation in plants: boon or bane? Plant Biosyst 143:81–96

Gimeno V, Syvertsen JP, Rubio F et al (2010) Growth and mineral nutrition are affected by substrate type and salt stress in seedlings of two contrasting citrus rootstocks. J Plant Nutr 33:1435–1447

Gunes A, Inal A, Alpaslan M et al (2007) Salicylic acid induced changes on some physiological parameters symptomatic for oxidative stress and mineral nutrition in maize (*Zea mays* L.) grown under salinity. J Plant Physiol 164:728–736

Hafsi C, Romero-Puertas MC, del Río LA (2010) Antioxidative response of *Hordeum maritimum* L. to potassium deficiency. Acta Physiol Plant 33:193–202

Halliwell B (2006) Reactive species and antioxidants. Redox biology is a fundamental theme of aerobic life. Plant Physiol 141:312–322

Han S, Tang N, Jiang HX et al (2009) CO_2 assimilation, photosystem II photochemistry, carbohydrate metabolism and antioxidant system of citrus leaves in response to boron stress. Plant Sci 176: 143–153

Hermans C, Vuylsteke M, Coppens F et al (2010) Systems analysis of the responses to long-term magnesium deficiency and restoration in *Arabidopsis thaliana*. New Phytol 187:132–144

Hernández JA, Olmos E, Corpas FJ et al (1995) Salt-induced oxidative stress in chloroplast of pea plants. Plant Sci 105:151–167

Hossain Z, López-Climent MF, Arbona V et al (2009) Modulation of the antioxidant system in citrus under waterlogging and subsequent drainage. J Plant Physiol 166:1391–1404

Jiang HX, Chen LS, Zheng JG et al (2008) Aluminum-induced effects on Photosystem II photochemistry in citrus leaves assessed by the chlorophyll a fluorescence transient. Tree Physiol 28:1863–1871

Li K, Xu C, Zhang K et al (2007) Proteomic analysis of roots growth and metabolic changes under phosphorus deficit in maize (*Zea mays* L.). Plant Proteomics 7:1501–1512

Li Q, Cheng L-S, Jiang H-X et al (2010) Effects of manganese-excess on CO2 assimilation, ribulose-1,5-bisphosphate carboxylase/oxygenase, carbohydrates and photosynthetic electron transport of leaves, and antioxidant systems of leaves and roots in *Citrus grandis* seedlings. BMC Plant Biol 10:42–58

López-Climent MF, Arbona V, Pérez-Clemente RM et al (2008) Maintenance of a high photosynthetic performance is linked to flooding tolerance in citrus. Environ Exp Bot 62:176–184

López-Climent MF, Arbona V, Pérez-Clemente RM et al (2011) Effects of cadmium on gas exchange and phytohormone contents in citrus. Biol Plantarum 55:187–190

Lunde C, Zygadlo A, Simonsen HT et al (2008) Sulfur starvation in rice: the effect on photosynthesis, carbohydrate metabolism, and oxidative stress protective pathways. Physiol Plant 134:508–521

Malusá E, Laurenti E, Juszczuk I (2002) Free radical production in roots of *Phaseolus vulgaris* subjected to phosphate deficiency stress. Plant Physiol Biochem 40:963–967

Møller IM, Sweetlove LJ (2010) ROS signalling – specificity is required. Trends Plant Sci 15:370–374

Navrot N, Collin V, Gualberto J et al (2006) Plant glutathione peroxidases are functional peroxiredoxins distributed in several subcellular compartments and regulated during biotic and abiotic stresses. Plant Physiol 142:1364–1379

Nazar R, Iqbal N, Masood A et al (2011) Understanding the significance of sulfur in improving salinity tolerance in plants. Environ Exp Bot 70:80–87

Papadakis IE, Bosabalidis AM, Sotiropoulos TE et al (2007a) Leaf anatomy and chloroplast ultrastructure of Mn-deficient orange plants. Acta Physiol Plant 29:297–301

Papadakis IE, Giannakoula A, Therios IN et al (2007b) Mn-induced changes in leaf structure and chloroplast ultrastructure of *Citrus volkameriana* (L.) plants. J Plant Physiol 164:100–103

Perl-Treves R, Perl A (2002) Oxidative stress: an introduction. In: Van Montagu M, Inzé D (eds) Oxidative stress in plants. Taylor & Francis, London, pp 1–32

Pompelli MF, Martins SCV, Antunes WC et al (2010) Photosynthesis and photoprotection in coffee leaves is affected by nitrogen and light availabilities in winter conditions. J Plant Physiol 167: 1052–1060

Sheng O, Zhou G, Wei Q et al (2010) Effects of excess boron on growth, gas exchange, and boron status of four orange scion–rootstock combinations. J Plant Nutr Soil Sci 173:469–476

Tewari RK, Kumar P, Sharma PN (2007) Oxidative stress and antioxidant responses in young leaves of mulberry plants grown under nitrogen, phosphorus or potassium deficiency. J Int Plant Biol 49:313–322

Van Breusegem F, Vranová E, Dat JF et al (2001) The role of active oxygen species in plant signal transduction. Plant Sci 161: 405–414

Vandenabeele S, Van Der Kelen K, Dat J et al (2003) A comprehensive analysis of hydrogen peroxide-induced gene expression in tobacco. Proc Natl Acad Sci USA 100:16113–16118

Vandenabeele S, Vanderauwera S, Vuylsteke M et al (2004) Catalase deficiency drastically affects gene expression induced by high light in *Arabidopsis thaliana*. Plant J 39:45–58

Vanderauwera S, Zimmermann P, Rombauts S et al (2005) Genome-wide analysis of hydrogen peroxide-regulated gene expression in *Arabidopsis* reveals a high light-induced transcriptional cluster involved in anthocyanin biosynthesis. Plant Physiol 139:806–821

Varin S, Cliquet JB, Personeni E et al (2010) How does sulphur availability modify N acquisition of whiteclover (*Trifolium repens* L.)? J Exp Bot 61:225–234

Wasaki J, Yonetani R, Kuroda S et al (2003) Transcriptomic analysis of metabolic changes by phosphorus stress in rice plant roots. Plant Cell Environ 26:1515–1523

Wulff-Zottele C, Gatzke N, Kopka J et al (2010) Photosynthesis and metabolism interact during acclimation of *Arabidopsis thaliana* to high irradiance and sulphur depletion. Plant Cell Environ 33:1974–1988

Yao Y, Sun H, Xu F et al (2010) Comparative proteome analysis of metabolic changes by low phosphorus stress in two *Brassica napus* genotypes. Planta. doi:10.1007/s00425-010-1311-x

Phytophenolic Nutrients in Citrus: Biochemical and Molecular Evidence

3

Theeshan Bahorun, Deena Ramful-Baboolall, Vidushi Neergheen-Bhujun, Okezie I. Aruoma, Ashok Kumar, Shalini Verma, Evelyne Tarnus, Christine Robert Da Silva, Philippe Rondeau, and Emmanuel Bourdon

Abstract

Natural products as disease remedies have a history of near 5,000 years (India, China and Greece), and even today, in this advanced technological age, a revival of interest is being witnessed in the use of natural or plant-based therapeutic agents for the treatment of several pathological conditions. Citrus fruits have been utilised as a traditional medicine in India, China, Korea and Japan, and many studies have highlighted the various biological properties of their phytophenolics which are suggested to be responsible for the prevention of degenerative diseases such as diabetes and cancer. With the background of comprehensive studies conducted on Mauritian citrus fruits, this chapter reviews some of the literature data on the phytophenolic contents, vitamin C composition and antioxidant functions of citrus extracts and emphasises on their potential applications in nutrition management programmes for diabetes and cancer chemoprevention.

Keywords

Citrus • Phytophenolics • Vitamin C • Antioxidants • Diabetes • Cancer

T. Bahorun (✉)
National Research Chair, ANDI Centre of Excellence for Biomedical and Biomaterials Research, University of Mauritius, Réduit, Republic of Mauritius
e-mail: tbahorun@uom.ac.mu

D. Ramful-Baboolall
Department of Agricultural and Food Science, Faculty of Agriculture, University of Mauritius, Réduit, Republic of Mauritius
e-mail: d.ramful@uom.ac.mu

V. Neergheen-Bhujun
Department of Health Sciences and ANDI Centre of Excellence for Biomedical and Biomaterials Research, University of Mauritius, Réduit, Republic of Mauritius
e-mail: v.neergheen@uom.ac.mu

O.I. Aruoma
School of Biomedical Sciences and School of Pharmacy, American University of Health Sciences, 1600 East Hill Street, Signal Hill, CA 90755, USA
e-mail: oaruoma@auhs.edu

A. Kumar
Vice Chancellor, Chhatrapati Shahu Ji Maharaj University, Kanpur, Uttar Pradesh, India
e-mail: mamsjpr@gmail.com

S. Verma
Department of Biotechnology and School of Biotechnology, Faculty of Science, Chhatrapati Shahu Ji Maharaj University Kanpur and Banaras Hindu University, Varanasi 221 005, Uttar Pradesh, India
e-mail: ashalini27@gmail.com

E. Tarnus • C.R. Da Silva • P. Rondeau • E. Bourdon
Laboratoire de Biochimie et Génétique Moléculaire (LBGM), Groupe d'Etude sur l' Inflammation Chronique et l'Obesité (GEICO), Université de La Reunion, Plateforme CYROI, 15 Avenue Rene´ Cassin, BP 97715, Saint Denis Messag, Cedex 09, La Réunion, France
e-mail: Evelyne.TARNUS@univ-reunion.fr; robert@univ-reunion.fr; Philippe.Rondeau@univ-reunion.fr; ebourdon@univ-reunion.fr

A.K. Srivastava (ed.), *Advances in Citrus Nutrition*,
DOI 10.1007/978-94-007-4171-3_3, © Springer Science+Business Media B.V. 2012

Fig. 3.1 Equatorial cross section through a citrus fruit

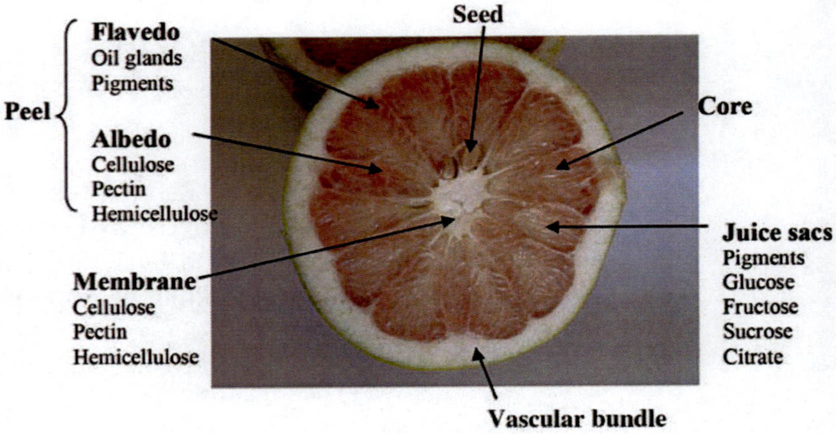

3.1 Introduction

The role played by dietary factors on health status has long been recognised, but it has been only recently that epidemiological and clinical data have provided a deeper insight on some of the intricate mechanisms of the effect of bioactive foods on human health. Citrus genus is one of the most important fruit tree crop in the world with an annual fruit production of approximately 102 million tons. Tropical countries like Mauritius enjoy the right balance of sunshine and rainfall for the growth of a wide range of exotic fruits including citrus which is the second most consumed fruit after bananas (Central Statistics Office 2008). More than 30 different varieties of citrus fruits are grown in Mauritius, either in backyard orchards or on a commercial basis, and consumed as fresh fruits or are processed into a variety of products such as juices, jams, marmalades and fruit pastes. Citrus fruits, in addition to providing an ample supply of vitamins, minerals, dietary fibres and pectins, contain a host of active phytochemicals including phytophenolics (e.g. flavanones, flavones, flavonols, phenolic acids) that can protect health. In fact, literature abounds of examples of citrus fruits, citrus fruit extracts and citrus flavonoids, exhibiting a wide range of promising biological properties including anti-atherogenic, anti-inflammatory and antitumor activity, inhibition of blood clots and strong antioxidant activity (Middleton and Kandaswami 1994; Montanari et al. 1998; Samman et al. 1998). Citrus is consumed mostly as fresh produce and juice, and most often, the peel is discarded. This represents a huge waste as citrus peels are reported to possess highest amounts of flavonoids compared to other parts of the fruit (Manthey and Grohman 2001). Citrus peels are subdivided into the epicarp or flavedo and mesocarp or albedo. The flavedo is the coloured peripheral surface of the peel, whilst the albedo is the white soft middle layer of the peel (Fig. 3.1). The beneficial health effects of citrus flavonoids would be particularly relevant in the Mauritian context considering the high incidence of cardiovascular diseases, diabetes and cancer on the island. This chapter emphasises on the phytophenolic, vitamin C and antioxidant screening of flavedo, albedo and pulp extracts of citrus fruits commonly consumed in Mauritius (Table 3.1). With this background, the antidiabetic and cancer chemopreventive potential of these extracts are discussed.

3.1.1 Phytophenolics and Vitamin C in Citrus Extracts

Most citrus species accumulate substantial quantities of flavonoids during the development of their different organs (Castillo et al. 1992). Four types of flavonoids occur in citrus species, namely the flavanones, flavones, flavonols and anthocyanins with the latter group occurring only in blood oranges. More than 60 individual flavonoids have been identified (Horowitz and Gentili 1977). Studies on the quantitative distribution of these flavonoids have shown that the flavanones predominate in all species of the genus, and they occur as glycosides, in which the aglycones are linked to a sugar moiety (Fig. 3.2) (Lewinsohn et al. 1989). Although flavones and flavonols have been found in low concentrations in citrus fruit tissues, they have been shown to be powerful antioxidants and free-radical scavengers with the highly methoxylated flavones exhibiting the highest biological activity (Benavente-Garcia et al. 1997).

Comprehensive studies conducted on 21 Mauritian citrus species demonstrated, with established correlations, that polyphenolic-rich extracts exhibited important antioxidant propensities in various test systems (Ramful et al. 2010a, b, 2011). Table 3.2 lists the citrus varieties having highest amounts of total phenolic, flavonoid and vitamin C in their flavedo, albedo and pulp extracts, respectively. The total phenolic content decreased in the following order: flavedo extracts > albedo extracts > pulp extracts. Gorinstein et al.

Table 3.1 Scientific and common names, variety and harvest dates of citrus fruits analysed

Scientific name	Common name	Variety	Harvest month	Variety and harvest code
Citrus sinensis	Orange	Valencia late	Aug	1
		Washington Navel	Mar	2A
			May	2B
Citrus unshiu	Satsumah	Owari	Mar	A
			May	B
Citrus clementina	Clementine	Commune	Mar	A
			May	B
Citrus reticulata	Mandarin	Fairchild	Apr	1A
			May	1B
		Dancy	May	2A
			Jun	2B
		Beauty	Jun	3A
			Aug	3B
		Suhugan	Aug	4
		Fizu	Aug	5
C. reticulata × *C. Sinensis*	Tangor	Elendale	Jun	A
			Aug	B
Citrus aurantium ssp. *bergamia*	Bergamot	–	Apr	–
Citrus meyeri	Lemon	Meyer	Apr	A
			May	B
C. reticulata × *C. paradisis*	Tangelo	Mineola	Jun	1A
			Aug	1B
		Orlando	Aug	2
		Ugli	Jun	3A
			Aug	3B
Fortunella margarita	Kumquat	Nagami	Apr	A
			Jun	B
Citrus mitis	Calamondin	–	Jun	A
			Aug	B
Citrus maxima	Pamplemousses (Pummelo)	Rainking	May	1A
			Aug	1B
		Kaopan	May	2A
			Aug	2B
		Pink	May	3A
			Aug	3B
		Chandler	Aug	4

(2001) also reported total polyphenols in the peels of lemons, oranges and grapefruits to be significantly higher than in the peeled fruits whilst quince peel extracts had threefold higher amounts than that of the pulp (Fattouch et al. 2007).

The levels of total phenolics in the Mauritian study are much higher than those measured in peels of similar varieties from Israel and New Zealand using the same methodology, indicating that the contents can be influenced by various factors such as genotypic differences, geographical and climatic conditions, cultural practices, harvest time, fruit maturity, environmental and growing conditions and extraction methods amongst others (Van der Sluis et al. 2001). Literature data on total phenolics of pulp extracts of citrus fruits are, however, comparable to the investigation on Mauritian citrus. Thus, Gorinstein et al. (2001) reported values for three varieties of citrus pulps in the range 1,350–1,640 μg/g FW, whilst the total phenolics content of pulp extracts in our study was between 406 and 1,694 μg/g FW.

The total flavonoid content of the extracts followed the same order as the total phenolics with highest levels in the flavedo extracts and lowest in the pulp extracts. Flavonoid derivatives, expressed in quercetin equivalents, in Mauritian citrus flavedos were generally high (>2,000 μg/g FW for the majority of samples analysed). Again, factors, including differences in variety and high sunlight conditions (a characteristic feature of tropical Mauritius), which can induce the accumulation of flavonoids (Li et al. 1993), are probably responsible for the relatively high yield. Using the same assay system but with catechin as standard, Gorinstein et al. (2004) reported that peeled Jaffa sweeties (a grapefruit

Fig. 3.2 Flavanone skeleton with substitution pattern (**a**) Flavanone aglycone (**b**) Rutinoside (Diglycoside) (**c**) Neohesperidoside (Diglycoside) (Adapted from Merken and Beecher 2000)

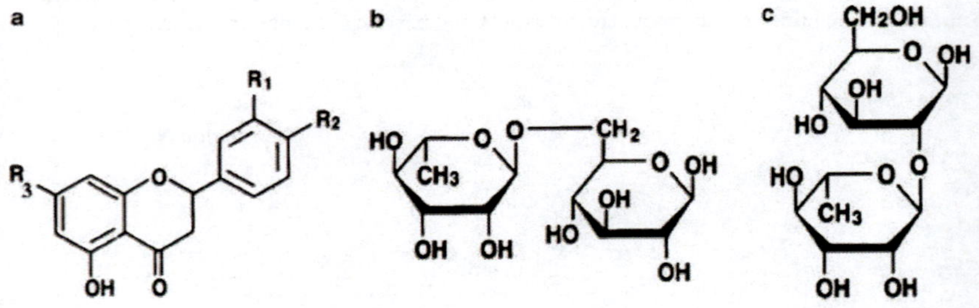

(a) Flavanone aglycone (b) Rutinoside (Diglycoside) (c) Neohesperidoside (Diglycoside)

Flavanone	R_1	R_2	R_3
Didymin	H	OMe	ORut [a]
Eriodictyol	OH	OH	OH
Hesperetin	OH	OMe	OH
Hesperidin	OH	OMe	ORut
Naringenin	H	OH	OH
Naringin	H	OH	ONeo [b]
Narirutin	H	OH	ORut
Neoeriocitrin	OH	OH	ONeo
Neohesperidin	OH	OMe	ONeo
Poncirin	H	OMe	ONeo

[a] rutinoside,
[b] neohesperidoside

Table 3.2 Mauritian citrus fruits with highest amounts of total phenolic, flavonoid and vitamin C in their flavedo, albedo and pulp extracts

	Flavedo	Albedo	Pulp
Total phenolic	*>5,500 µg/g FW*	*>5,500 µg/g FW*	*> 950 µg/g FW*
	~Clementine A	~Mandarin 2A and 2B	~Orange 1
	~Orange 2A and 2B	~Tangelo 1A	~Tangelo 1A, 3A and 3B
	~Mandarin 1A, 1B, 2A, 2B and 5	~Pamplemousses 2A, 3A and 3B	~Kumquat A and B
	~Tangelo 1A and 2	~Tangor A	~Calamondin A and B
	~Pamplemousses 2B		~Pamplemousse 2A
	~Tangor A and B		~Tangor A and B
Total flavonoids	*>3,600 µg/g FW*	*>3,600 µg/g FW*	*> 600 µg/g FW*
	~Clementine A	~Mandarin 2A	~Mandarin 1B
	~Mandarin 1A, 1B, 2A and 2B	~Tangelo 1A	~Tangelo 1A and 3A
	~Tangelo 2	~Pamplemousses 2A, 3A and 3B	~Pamplemousses 1A, 2A, 2B, 3A and 4
	~Pamplemousses 2B, 3A,and 4		
Total vitamin C	*>1,000 µg/g FW*	–	*>500 µg/ml*
	~Tangelo 3B		~Kumquat A
	~Pamplemousses 1A, 1B, 2B, 3A, 3B and 4		~Tangelo 3A
			~Pamplemousses 1B, 2A and 4

hybrid) and white grapefruits contained 471 and 377 µg/g FW, whilst 925 and 744 µg/g FW were measured in their respective peels.

Vitamin C content was higher in the peel extracts than in the pulp juice. Gorinstein et al. (2001) also reported the ascorbic acid content of three varieties of citrus peels to be significantly higher than that of peeled fruits. In our study, pamplemousses varieties showed relatively high vitamin C content in peel and pulp extracts. Cano et al. (2008), on the other hand, reported that orange varieties had higher vitamin C concentrations than mandarin, clementine, satsume and hybrid varieties, supporting the argument of a wide variation of vitamin C content in literature. Vitamin C levels in fruits and vegetables, in fact, can also be influenced by various factors such as genotypic differences, climatic conditions and cultural practices (Lee and Kader 2000).

Citrus fruits contain a wide range of flavonoid constituents which are encompassed in the flavanones, flavones and flavonols subclasses (Nogata et al. 2006; Mata Bilbao et al. 2007). HPLC analyses of nine flavedo extracts showed that, consistent with literature data (Londoño-Londoño et al. 2010), the flavanone glycoside, hesperidin, was present at highest concentrations (83–234 mg/g FW) in all the extracts except for a variety of pamplemousse. The flavanone glycosides poncirin, didymin, narirutin and flavone glycosides diosmin and isorhoifolin were present in all flavedo extracts, whereas the flavanone glycoside naringin was present only in one variety of Mandarin (1A). The presence of naringin was observed in Mandarin 1A despite its reported absence from mandarin varieties (Tomás-Barberán and Clifford 2000). Similar to flavedo extracts, hesperidin was the most abundant flavonoid glycoside in the albedo extracts where it was detected at concentrations ranging from 132 to 540 mg/g FW. Didymin, hesperidin and narirutin were ubiquitously present in all the nine albedo extracts analysed, whereas naringin and diosmin were not present in any of them. Rhoifolin was measured only in Mandarin 2B and Mandarin 5. Rutin was detected in albedo extracts of Clementine A, Mandarin 1A and Tangor A only. In line with literature data (Cano et al. 2008), hesperidin was the most abundant flavanone glycoside in the citrus pulp extracts. The latter was found at concentrations ranging from 7 to 27 mg/g FW, followed by narirutin (0.3–21 mg/g FW). Hesperidin was present in all the varieties except for Pamplemousse 2A, whilst narirutin was absent only in Tangor A.

The amount of flavonoid glycosides in the pulp extracts was much lower than in the flavedo and albedo extracts of Mauritian citrus fruits; a trend that was consistent with the phenolic and vitamin C contents, as well as the antioxidant activities (Ramful et al. 2011). Berhow et al. (1998) reported that the concentration of flavanones was greater in the citrus albedo, whilst the levels of flavones and flavonols decreased in the following order: flavedo > albedo > juice sacs. The data reported for flavonoid composition of citrus pulp juices are fairly heterogeneous as a result of the different techniques employed and the different units of measurement used by the various authors (mg 100/mg juice, mg 100/mg lyophilised juice, mg 100/mL juice, mg 100/mg fresh product) (Peterson et al. 2006a, b). In a survey of phenolic compounds in 35 citrus species, the same authors reported that the dominant neohesperidosyl flavanones were naringin (found at high concentrations in grapefruit, kumquat and pummelo), neoeriocitrin (found in bergamot and sour orange tissues) and neohesperidin (in tangelo). The dominant rutinosyl flavanones were hesperidin (in lemon, lime, mandarin and sweet orange), eriocitrin and narirutin. The flavanone profile of sweet orange is relatively simple and varies little among cultivars. It is generally agreed that orange fruit and juice contain hesperidin (Anis and Aminuddin 1981), narirutin (Rousseff et al. 1987), and didymin (Matsubara et al. 1985). Berhow et al. (1998) also found some orange cultivars to contain eriocitrin. Mandarin contains predominantly hesperidin, occasionally narirutin, and trace levels of didymin (Horowitz and Gentili 1977). Pummelo (pamplemousses) is reportedly one of the three species (in addition to sour orange and *Poncirus trifoliata*) that accumulate neohesperidosyl glycosides (Albach and Redman 1969) with naringin being the major flavanone in most pummelo cultivars (Park et al. 1983; Berhow et al. 1998).

3.1.2 Antioxidant Propensity of Citrus Organs

Given that the mechanisms of action of naturally occurring antioxidants can be diverse in vivo, a comprehensive prediction of the antioxidant efficacy initially in vitro requires a multiplicity of assessing methods with various implications for molecular targets (Aruoma 1994, 2003; Pérez-Jiménez et al. 2008). It is noteworthy that synergism and concentration may also bring effects that are not observed when individual constituents are tested (Kaur and Kapoor 2001). There is therefore no universal method that can measure the antioxidant capacity of all samples accurately and consistently. Clearly, matching radical source and system characteristics to antioxidant reaction mechanisms is critical in the selection of appropriate antioxidant capacity assay assessing methods (Prior et al. 2005).The antioxidant characterisation of the flavedo, albedo and pulp extracts of 21 varieties of Mauritian citrus fruits was evaluated using independent methods: the TEAC, FRAP and HOCl assays. From the initial results, nine different flavedo, albedo and pulp extracts, which showed highest antioxidant activities in these three assays, were further assessed for their ability to protect DNA from damage and their iron-chelating activity (Ramful et al. 2010a, b).

Citrus flavedo extracts had wide antioxidant potential ranges, thereby supporting their classification as low, moderate and high. Table 3.3 classifies the citrus fruits according to the antioxidant activities of their flavedo and pulp extracts, as measured by the TEAC, FRAP and HOCl scavenging assays. Some clementine, mandarin, tangor, tangelo and pamplemousse varieties were all classified in the high-level range with TEAC values greater than 40 μmol/g FW. The free-radical scavenging activities of the pulp extracts were much lower compared to the flavedo and albedo extracts with values in the range 2.6–9.9 μmol/g FW. These data are consistent with those of Wang et al. (2011) who reported the antioxidant activity of peel extracts of *Citrus sulcata* to be twice that of the pulp extracts using the TEAC assay. Flavedo extracts of Orange 2B, Clementine A, Mandarin 1, 2 and 5, Tangor and Tangelo 1A and 2 had FRAP values greater than 50 μmol/g FW. FRAP values for albedo extracts ranged from 5.8 to 55 μmol/g FW, whilst values for pulp extracts ranged from 3.3 to 10.4 μmol/g FW, clearly depicting that the ferric reducing efficacy of the extracts decreased in the order flavedo extracts > albedo extracts > pulp extracts. Flavedo extracts of Orange 2, Clementine A, Mandarin 1, 2A and 5, Tangor A and Pamplemousse 2B were characterised by low

Table 3.3 Classification of citrus fruits, according to the antioxidant activities of their flavedo, albedo and pulp extracts, as measured by the TEAC, FRAP and HOCl scavenging assays

	Flavedo			Pulp		
	Low	Medium	High	Low	Medium	High
TEAC	*<20 μmol/g FW*	*20–35 μmol/g FW*	*>35 μmol/g FW*	*<4 μmol/g FW*	*4–7 μmol/g FW*	*>7 μmol/g FW*
	Clementine B	Orange 1, 2A and 2B	Clementine A	Clementine	Orange 1, 2A and 2B	Tangelo 1A
	Mandarin 3A, 3B and 4	Mandarin 1B and 5	Mandarin 1A, 2A and 2B	Mandarin 1A, 2B, 3A, 3B and 4	Mandarin 1A, 1B, 2A and 5	Kumquat A and B
	Lemon A and B	Satsumah	Tangelo 1A and 2	Lemon	Satsumah	Calamondin
	Kumquat A and B	Tangelo 1B, 3A and 3B	Pamplemousse 2B	Tangelo 2	Tangor Bergamot	Pamplemousse 2A
	Calamondin B	Calamondin A	Tangor	Pamplemousse 1B, 3B and 4	Tangelo 1B, 2, 3A and 3B	
	Pamplemousse 1A, B and 2A	Pamplemousse 3A, 3B and 4		Bergamot	Pamp, 1, 2B, 3A, 3B and 4	
	Bergamot					
FRAP	*<30 μmol/g FW*	*30–50 μmol/g FW*	*>50 μmol/g FW*	*<4.5 μmol/g FW*	*4.5–7.5 μmol/g FW*	*>7.5 μmol/g FW*
	Mandarin 3A, 3B and 4	Clementine B	Clementine A	Clementine A	Clementine B	Mandarin 5
	Lemon	Orange 1 and 2A	Orange 2B	Satsumah A	Orange 1A, 2A and 2B	Tangelo 1A and 3
	Kumquat Calamondin	Satsumah Tangelo 1B and 3B	Mandarin 1A, 1B, 2A, 2B and 5	Mandarin 2B, 3A and 4	Satsumah B	Kumquat Tangor
	Tangelo 3A	Pamplemousse 2A and 2B	Tangelo 1A and 2	Lemon	Mandarin 1, 2A and 3B	
	Pamplemousse 1A, 1B, 3A, 3B and 4		Tangor	Pamplemousse 1A, 1B and 4	Tangelo 1B, 2 and 3B	
	Bergamot			Bergamot	Calamondin Pamplemousse 2 and 3	
HOCl	*>10 mg FW/ml*	*5–10 mg FW/ml*	*<5 mg FW/ml*	*>100 mg FW/ml*	*70–100 mg FW/ml*	*<70 mg FW/ml*
	Mandarin 3 and 4	Clementine B	Clementine A	Clementine	Mandarin 1A	Orange 1, 2
	Lemon	Orange 1	Orange 2	Mandarin 1B, 2A, 3, 4 and 5	Satsumah Tangelo 1B	Tangelo 1A, 3
	Kumquat Calamondin	Mandarin 2B	Mandarin 1, 2A and 5	Lemon	Calamondin	Kumquat Pamplemousse 1A, 2
	Pamplemousse 1	Satsumah Tangelo 1 and 3	Tangelo 2	Tangelo 2	Pamplemousse 1B, 3A and 4	
	Bergamot	Pamplemousse 2A, 3 and 4	Pamplemousse 2B	Pamplemousse 3B	Tangor B	
		Tangor B	Tangor A	Tangor A		
				Bergamot		

IC_{50} values (3.70–4.41 mg FW/mL), indicating the high effectiveness of the extracts to scavenge hypochlorite. Pulp extracts were relatively poor scavengers of hypochlorite with IC_{50} values in the range 52.5–175 mg FW/mL. No literature data is available on the assessment of the HOCl scavenging activity of citrus fruits. However, the anti-inflammatory effects of *Mangifera indica* L. (Martynez et al. 2000), the brown algae *Laminaria japonica* (Zhao et al. 2004) and the medicinal plant *Hypercum androsaerum* (Valentão et al. 2002) have also been assessed using the HOCl assay. Among these, interesting radical scavenging capacities were observed in *Mangifera indica* L. (Vimang) with an IC_{50} value of 0.04%, thereby supporting its use in traditional medicine as an anti-inflammatory and cancer preventive agent (Martynez et al. 2000).

The metal complex copper 1,10-phenanthroline is known to promote hydroxyl radical formation from molecular oxygen by redox cycling and is therefore a suitable agent for the stimulation of oxidative DNA damage (Aruoma 1993, Halliwell 1997). Indeed DNA fragmentation detected in different cells treated with copper phenanthroline is considered to result from direct attack upon DNA by the hydroxyl radical (Tsang et al. 1996). DNA damage, such as single- and double-strand breakage, base modification, cross-linking of DNA with other biomolecules particularly proteins, is reported to be early events of cancer, cardiovascular diseases, diabetes, cataract and neurological disorders (Cadet et al. 1997), and phytochemicals have profound chemopreventive effects through modulation of molecular events that damage DNA (Bisht et al. 2008). The level of protection against copper-phenanthroline-mediated oxidative DNA damage was in the following order for the flavedo extracts: Tangelo 1A > Clementine A > Tangor A > Pamplemousse 2B > Mandarin 5 > Mandarin 1A ≈ Orange 2B > Mandarin 2A > Tangelo 2.

Among the transition metals, iron is known as the most important lipid pro-oxidant due to its high reactivity. Benherlal and Arumughan (2008) reported that phytochemicals/extracts with high antioxidant activity but without iron chelation capacity failed to protect DNA in Fenton's system, suggesting that iron chelation was an essential requirement for extracts studied here to retard HO• generation by Fenton's reaction. In the study conducted on Mauritian citrus, Clementine A, Tangor A and Mandarin 1A and 5 were the most potent $Fe2^{+}$ ion chelator.

3.1.3 Antidiabetic Potential

Oxidative stress and alterations in glucose metabolism are important risk factors for diabetes and its related complications. Advanced glycated end products and their carbonyl derivatives are believed to contribute significantly to the pathogenesis of type 2 diabetes by their interaction with specific cell membrane receptors triggering, for instance, the nuclear factor-*Kappa* B (NF-κ B) signalling pathway to induce the expression of pro-inflammatory mediators and elicit oxidative stress which exacerbate diabetic complications (Stern et al. 2002). A great deal of effort has been focused on the identification of useful inhibitors of protein AGEs to delay or prevent glycation so as to alleviate the phenotype of these diseases (Pashikanti et al. 2010). Numerous AGEs inhibitors, including aminoguanidine, improved diabetic complications in animal models and clinical trials with, however, a number of adverse effects (Ho et al. 2010). It is suggested that AGEs inhibitors from natural foods/dietary biofactors may reasonably serve as valuable adjuvants. Using a diabetes-like oxidative stress model, the potential protective effect of antioxidant citrus fruit extracts on human adipocytes was evaluated (Ramful et al. 2010a, b). In spite of the determinant role of adipocytes in the aetiology of obesity-related disorders, there are very few reports on the effect of natural antioxidants on adipocyte-response to oxidative stress. The extracts were tested on SW872 liposarcoma cells subjected or not to H_2O_2 or AGEs. Cell viability, carbonyl accumulation, free-radical formation, tumour necrosis factor α (TNFα) and apolipoprotein E (apoE) secretions were assessed in treated cells (Ramful et al. 2010a, b). Data for two citrus species, namely Tangor and Tangelo showed pronounced abilities to delay free-radical-induced hemolysis in the red blood cell hemolysis test, thus providing complementary evidence of their antioxidative potency. This is the first report on the antioxidant propensity of nutritional compounds assessed by this red blood cell hemolysis test system patented in 1992 (Prost 1992). The low dose–response data therefore represent favourable applicable conditions to the in vivo environment, without affecting cellular viability and physiology.

Adipocyte cell viability was examined in the presence of different concentrations of tangelo and tangor flavedo, albedo and pulp extracts. Only the flavedo extract produced toxic effects at high concentrations (>0.75%). The phenolic richness of the extract could contribute to this observation. Analogous reports have previously been made, whereby phenolic-rich plant extracts exerted modulatory effects. These results indicated that phenolic constituents of complex plant mixtures possess the character of a 'Janus' (anti)genotoxicant, a term used to designate compounds behaving as genotoxin or antigenotoxin, depending upon the plant extract concentrations used (De Flora and Ramel 1988). The toxicity is suggested to be related to hydrogen peroxide formation arising from the auto-oxidation of phenolic molecules. In another work, Patil et al. (2009) showed that compounds purified from Mexican lime juice could induce apoptosis in human pancreatic cells (review in Roche et al. 2008) with the effects being shown to be proportionately linked to the flavonoid content. Our previous studies have shown that

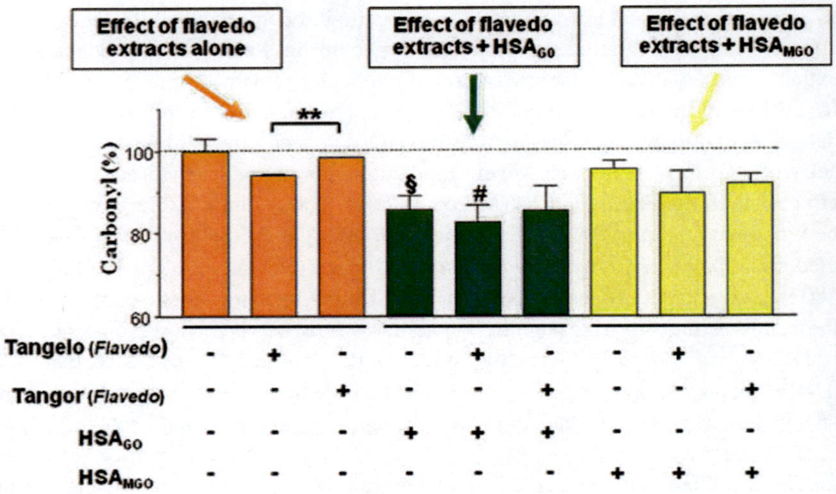

Fig. 3.3 Effect of citrus flavedo extracts on carbonyl accumulation in AGEs-treated adicpocytes. Results expressed as % of control cells treated with only 1% (v/v) DMSO. *Bars* represent mean ± SEM of two independent experiments performed in triplicate. Significance using student's *t* test for unpaired samples are: §$P=0.08$, #$P=0.07$ (vs. control); **$P<0.01$

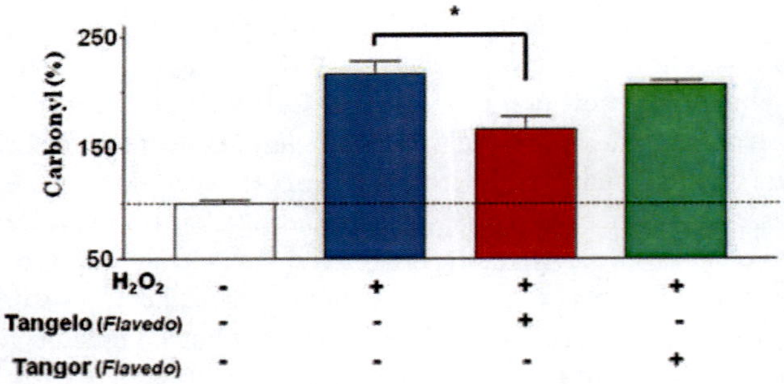

Fig. 3.4 Effect of citrus flavedo extracts on carbonyl accumulation in H_2O_2-treated adicpocytes. Results expressed as % of control cells treated with only 1% (v/v) DMSO. *Bars* represent mean ± SEM of two independent experiments performed in triplicate. Significance was assessed using one-way ANOVA followed by Dunnett's multiple comparison test; *$P<0.05$

native albumin had strong antioxidant activities (Bourdon et al. 1999). Our data show an increase in the half time of AAPH-induced hemolysis in the presence of native albumin, whilst a significant reduction is observed with MGO-mediated glycation. Similar results were obtained on MGO-modified BSA by Faure et al. (2005). In the same vein, another work performed by our group showed oxidative damages in adipocytes subjected to oxidative stress induced by glycated albumin (Roche et al. 2009). The reduction of carbonyl formation at the adipocyte level is clearly reflective of the antioxidant power of tangor and tangelo flavedos (Fig. 3.3). This antioxidant propensity is reinforced with co-treatment with native albumin, whilst glycated albumin is devoid of antioxidant power. Consistently, similar data are observed in adipocytes submitted to an oxidative stress generated by H_2O_2 (Fig. 3.4). A significant decrease in carbonyl formation was observed when cells pretreated with tangelo flavedo extracts were incubated in the presence of H_2O_2. Along this line, it has been reported that polyphenolics, more particularly anthocyanins, have the ability to protect 3T3-L1 adipocytes against H_2O_2-induced insulin resistance (Guo et al. 2008).

We further demonstrated that intracellular ROS formation is considerably lowered in cells pretreated with citrus flavedo extracts incubated in the presence or absence of H_2O_2 (unpublished data). Modulation of intracellular ROS production in the SW872 cells by citrus extracts was shown using the dichlorodihydrofluorescein diacetate (DCFH-DA) assay. This is a useful indicator of reactive oxygen species (ROS) and oxidative stress. The principle of the assay is summarised in Fig. 3.5. The nonpolar and nonionic DCFH-DA crosses cell membranes and is hydrolysed by intracellular esterases to non-fluorescent 2′,7′-dichlorofluorescin (DCFH). In the presence of ROS, such as hydrogen peroxide (H_2O_2), lipid hydroperoxides and peroxinitrite, DCFH is oxidised to fluorescent 2′,7′-dichlorofluorescein (DCF). In addition, DCFH can be oxidised by intracellular oxidases and oxidants formed during the reduction of H_2O_2. Altogether, these observations indicate that the oxidation of DCFH may be derived from several ROS intermediates (Wang and Joseph 1999). Therefore, DCFH is useful to indirectly measure the effect of intracellular antioxidant activities in scavenging ROS and in protecting DCFH from oxidation. Intracellular ROS formation was considerably lowered in cells pretreated with citrus flavedo extracts incubated in the presence or absence of H_2O_2 (Fig. 3.6). Hwang and Yen (2008) also reported that pretreatment of rat pheochromocytoma PC12 cells with citrus flavanones significantly eliminated the accumulation of intracellular ROS. Hesperetin and neohesperidin reduced the level of ROS by 16–24%, whilst hesperidin reduced the level of ROS by 32–48% in H_2O_2-indued PC12 cells (Hwang and Yen 2008). Other researchers have reported the use of the DCFH-DA assay in biological systems for the evaluation of natural antioxidants. Takamatsu et al. (2003) reported the screening of some flavonoids for their antioxidant properties on HL-60 cells using a DCFH-DA assay. Lu et al. (2004)

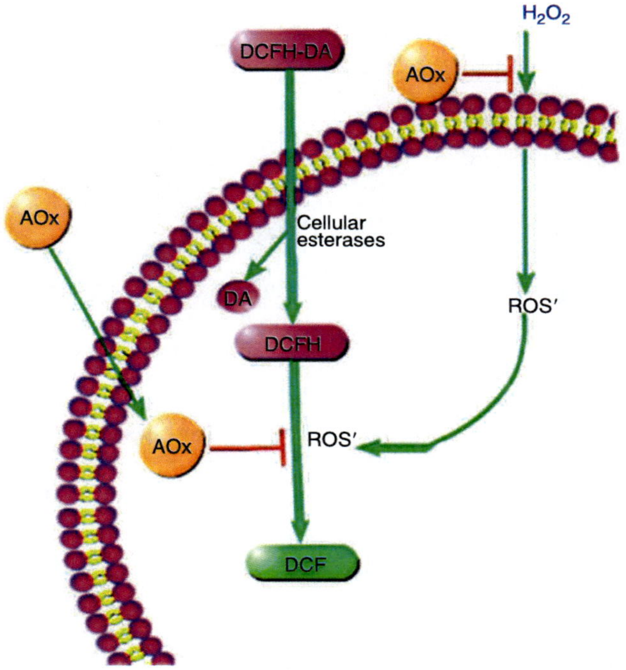

Fig. 3.5 Principle of the DCFH-DA assay. Cells were pretreated with citrus extracts followed by the addition of the DCFH-DA. The antioxidants in the citrus extracts bound to the cell membrane and/or passed through the membrane to enter the cell. DCFH-DA diffused into the cell where cellular esterases cleaved the diacetate moiety to form the more polar DCFH, which was trapped within the cell. Cells were treated with H_2O_2 which diffused into the cells acting as ROS. These ROS oxidised the intracellular DCFH to the fluorescent DCF. Antioxidants present in the citrus extracts prevented the oxidation of DCFH (Adapted from Wolfe and Liu 2007)

used a DCFH-DA assay to assess the antioxidant activities of procyanidins from grape seeds whilst Eberhardt et al. (2005) evaluated the antioxidant activities of broccoli extracts. Recently, Girard-Lalancette et al. (2009) developed a sensitive cell-based assay using DCFH oxidation for the determination of pro- and antioxidant properties of fruits and vegetable juices.

ApoE, which is a component of lipoproteins, e.g. chylomicrons, very low-density lipoprotein (VLDL), intermediate-density lipoproteins and high-density lipoprotein (HDL), is mainly produced and secreted by the liver (Beisiegel et al. 1988). ApoE is known to regulate both cellular and systemic cholesterol, as well as triglyceride metabolism (Mahley 1988; Tarnus et al. 2009), and has been extensively studied for its potential role in the aetiology of atherosclerosis, diabetes and obesity. ApoE, which was shown to exhibit anti-inflammatory, anti-atherogenic and antioxidant properties (Miyata and Smith 1996; Davignon 2005), has been found to be highly expressed by adipose tissue and adipocytes (Wassef et al. 2004; Zechner et al. 1991). However, if apoE expression by adipocytes has been known for many years, its importance in the adipose response to oxidative stress/antioxidants has never been thoroughly investigated. Significant reductions in apoE secretions were observed in albedo and pulp-extract-treated cells (Fig. 3.7). Recently, our group showed an increase in apoE secretion in SW872 cells subjected to oxidative stress induced by glucose or AAPH, a free-radical generator (Tarnus et al. 2009). We hypothesised that apoE may exert antioxidant effects at the adipocyte level, and its subsequent increase in expression may represent a defence response to oxidative stress (Tarnus et al. 2009). The decrease in apoE secretion in cells incubated with citrus extract seems to be an adaptive response to the presence of the exogenous citrus antioxidants.

Complementary to our data depicting the antidiabetic potential of citrus extracts, literature data on citrus phytophenolics suggest that naringenin is able to reduce glucose uptake and inhibit intestinal and renal Na+-glucose co-transporter (SGLT1) (Li et al. 2006) and that both naringin and hesperidin significantly increased the glucokinase mRNA level, whilst naringin reduced the mRNA expression of phosphoenolpyruvate carboxykinase and glucose-6-phosphatase in the liver (Jung et al. 2006). Recently, it was reported that a citrus extract of Dangyuja (citrus fruit from Korea), containing high levels of flavanone glycosides, could be used to control the blood glucose level of diabetic patients by inhibiting α amylase and α glucosidase in the intestinal tract (Gyo-Nam et al. 2009).

3.1.4 Cancer Chemoprevention Potential

The concept of chemoprevention by dietary means is gaining momentum in a number of chronic degenerative diseases primarily due to the dramatic rise of cancer and type 2 diabetes mellitus and the increasing incidence of cardiovascular diseases as major and interlinked healthcare problems. As it stands today, cancer is the second leading death cause in the world. In 2005, out of 58 million deaths worldwide, 7.6 million people died of cancer. Based on projections, cancer deaths will continue to rise with an estimated 9 million people dying from cancer in 2015, and 11.4 million dying in 2030. Cancer is a multifactorial and multistage process consisting of three distinct phases: initiation, promotion and progression phases. Whilst current clinical therapies including radiation, chemotherapy, immunosuppression and surgery are limited as indicated by the high morbidity and mortality rate from cancer, there is an imperative need for new treatment modalities. Chemoprevention which involves the use of pharmacological, dietary biofactors, phytochemicals and even whole plant extracts to prevent, arrest or reverse the cellular and molecular processes of carcinogenesis has been proposed due to its multiple intervention strategies.

The preventive mechanisms of tumour promotion by natural phytochemicals range from the inhibition of genotoxic effects, increased antioxidant and anti-inflammatory activity,

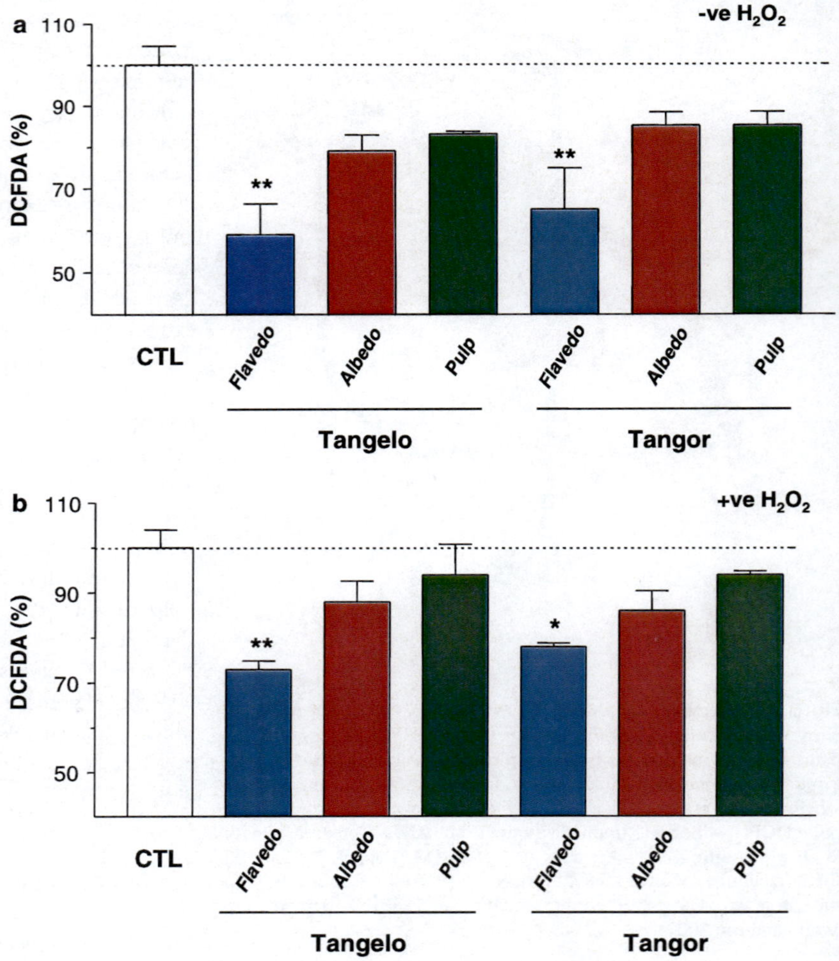

Fig. 3.6 Comparative antioxidant activities of citrus extracts in SW872 cells, as measured by the DCFH-DA assay, (**a**) in the absence of H_2O_2 and (**b**) in the presence of H_2O_2. Results are expressed as mean ± SEM of three independent *experiments* performed in triplicate using one-way ANOVA followed by Dunnett's multiple comparison test; $^{*}P<0.05$, $^{**}P<0.01$ vs. control

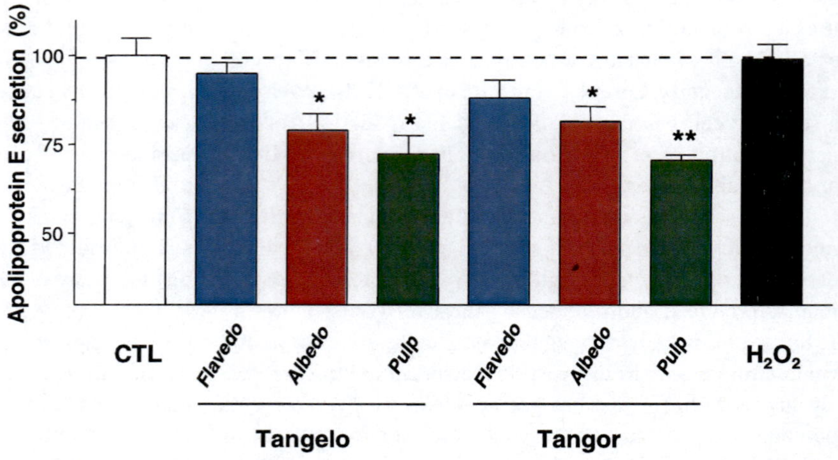

Fig. 3.7 Effect of citrus fruit extracts on ApoE secretion in SW872 cells. Results are expressed as percentage of the corresponding control cells treated with only 1% (v/v) DMSO and/or PBS

inhibition of proteases and cell proliferation, protection of intercellular communications to modulation of apoptosis and signal transduction pathways (Chen and Kong 2005; De Flora and Ferguson 2005; Holmes-McNary and Baldwin 2000; Aruoma et al. 2005; Soobrattee et al. 2008). Dietary polyphenols can induce the phase I and II detoxifying enzymes involved in the biotransformation and elimination of potential carcinogens. The natural chemopreventive

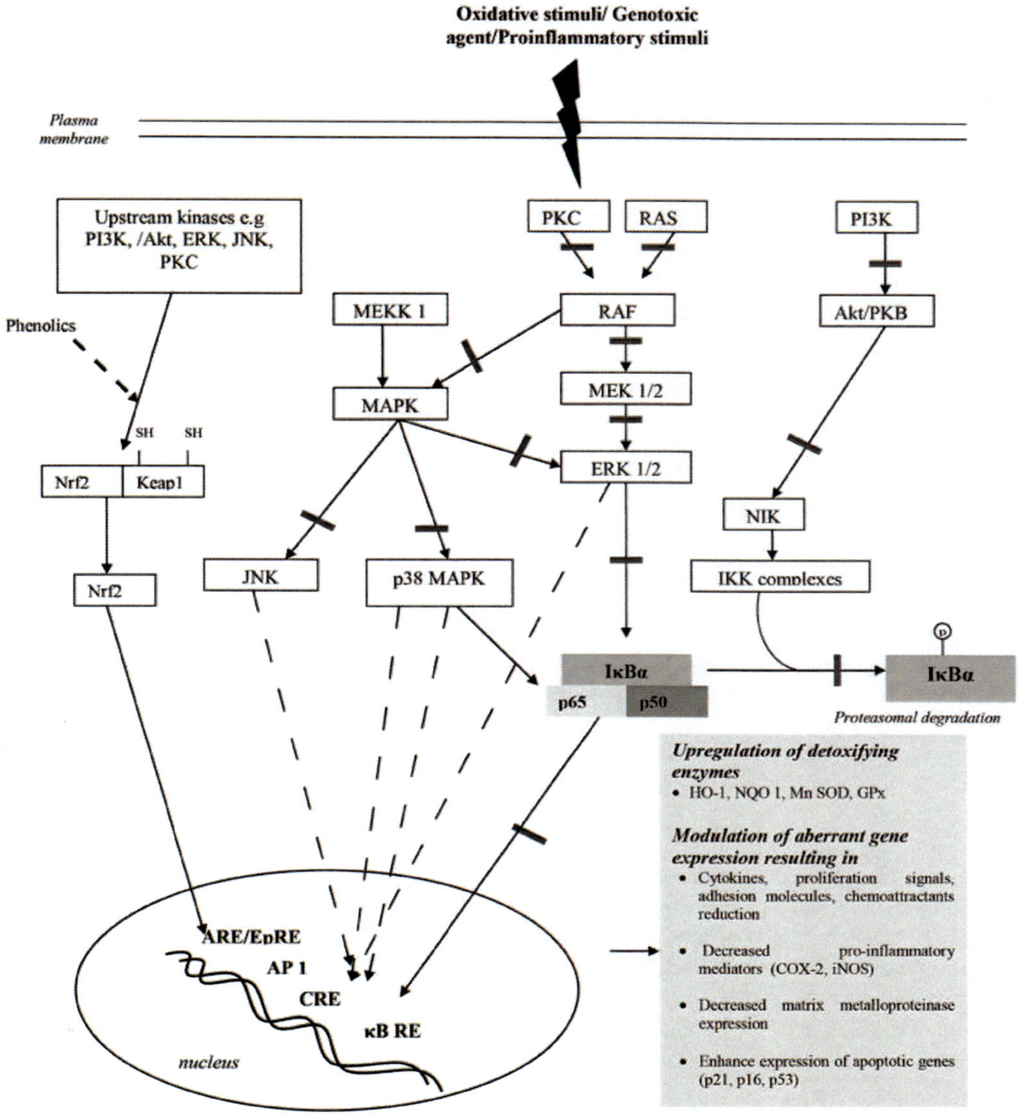

Fig. 3.8 Schematic representation of the intracellular signal transduction cascades activated by reactive oxygen species and converging on downstream transcription factors and the site where dietary phenolics possibly intervene. Activation of Nrf2 signalling and induction of phase II detoxifying and antioxidant genes by chemopreventive polyphenols represented by (——). (- - ►) indicates the sites where phenolic compounds have been reported to modulate and suppress the cell signal transduction cascades (Neergheen and Bahorun 2009)

compounds serve as transcriptional activators for the expression of glutathione S-transferase, NAD(P)H: quinine oxidoreductase (NQO), heme oxygenase 1 (HO 1), γ-glutamylcysteine synthetase (γGCS) and antioxidant enzymes via the antioxidant/electrophile response element (ARE/EpRE) (Neergheen and Bahorun 2009). The induction effects of phase II detoxifying agents by natural phytochemicals is mediated in part through the activation of Nrf 2 signalling pathways by upstream kinases (Fig. 3.8).

Citrus flavonoids have been reported to protect DNA by their ability to absorb ultraviolet light (Stapleton and Walbot 1994). The role of Naringin as an important modulator of superoxide dismutase and catalase activities and upregulator of gene expressions of superoxide dismutase, catalase and glutathione peroxidase in cholesterol-rich diet-fed rabbits has been highlighted (Jeon et al. 2001). It has been further suggested naringin might affect H_2O_2-induced expression of an apoptosis-associated gene or proteins as one of its pharmacological actions (Kanno et al. 2003). Orange juice containing hesperetin and naringenin delayed tumour development, suggesting their effectiveness to inhibit human breast cancer cell proliferation in vitro (So et al. 1996).

Naringenin was also found to inhibit proliferation of HT-29 colon cancer cell lines at concentrations of 0.71–2.85 mmol. These amounts are found to be effective in plasma and can be provided by drinking between 2 and 3 l of grapefruit juice daily. Taking into account its low bioavailability, higher volumes of grapefruit juice may however be required, thereby suggesting that naringenin in capsular form may be more practical and efficient. (Tripoli et al. 2007). Another study on naringenin showed that its administration to gastric carcinoma–induced rats largely up-regulated the redox status to decrease the risk of cancer. The authors concluded that up-regulation of antioxidants by naringenin treatment might be responsible for the anticancer effect in gastric carcinoma (Ekambaram et al. 2008).

Numerous reports indicate that citrus flavonoids affect cellular metabolism in various ways, thereby influencing cancer proliferation, e.g. inhibition of glycolysis (the most active metabolic pathway in tumoural cells (Manach et al. 1996)), depress production of lactate in leukaemia cell lines or in Ehrlich tumour cells (Suolinna et al. 1975), inhibition of the Na/K ATPase pump that could negatively influence the energetic metabolism, the synthesis of the proteins and DNA replication, by pH reduction of the cells (Hirano et al. 1989). Furthermore citrus flavonoids have been shown to potentiate the drug therapies effects against cancer. For example, quercetin prominently enhances the effect of Adriamycin in a multidrug-resistant MCF-7 human breast cancer cell line (Scambia et al. 1994) and cells of colon MCT-15 (Critchfield et al. 1994). The anti-metastatic and anti-invasive activities of citrus flavonoids, based on cell mobility inhibition (Bracke et al. 1991), have been observed in several human neoplastic cellular line proliferations: lymphoid and myeloid leukaemias (Larocca et al. 1990), gastric carcinoma (Yoshida et al. 1990), ovarian carcinoma (Scambia et al. 1990), prostate carcinoma (Peterson and Barnes 1993) and squamous cellular carcinoma (Kandaswami et al. 1991).

Nobiletin, which is a predominant methoxylated flavone in mandarins, has shown direct cytotoxicity on TMK-1, MKN-45, MKN-74 and KATO-III with a dose–response relationship. Loss of cell viability at low doses was found to be a consequence of apoptosis (Yoshimizu et al. 2004). A screening of 78 citrus species showed inhibitory effects of the Epstein-Barr virus antigen (EBV-EA) activation, indicating that citrus phytochemicals may inhibit susceptibility factors involved in the events leading to the development of cancer (Iwase et al. 1999). Hirata and his group, in a study conducted on citrus peels, isolated polymethoxyflavones and coumarin derivatives with anti-corpulence activities and the ability to inhibit the proliferation of human colon cancer HT-29 cells (Hirata et al. 2009).

More recently, the modulatory effects of hesperidin on attenuating the lipid peroxidation and down-regulation of key membrane bound marker enzyme activities and up-regulation of protein content affording an assurance for its potential use for the treatment of breast cancer have been reported (Nandakumar et al. 2011).

Extracts from lemon seed were also shown to have potent antioxidant activity and to induce apoptosis in MCF-7 cells, leading to the inhibition of proliferation, suggesting that aglycones and glucosides of the limonoids and flavonoid present may potentially serve as a chemopreventive agent for breast cancer (Kim et al. 2011). Recent advanced molecular studies focused to understand the structure-function relationship of citrus flavonoids in terms of their ability to alter the gene expression in the colon adenocarcinoma cells. Structurally related flavonoids (apigenin and quercetagetin) found in citrus were shown to have pronounced ability to inhibit colon cancer (SW480) cells as well as change the expression of apoptosis-related genes/proteins (Chidambara Murthy et al. 2011).

3.2 Conclusion

Never before has the focus on the health benefits of commonly available foods been so strong. The philosophy that food can be health promoting beyond its nutritional value is gaining acceptance within the public arena and among the scientific community as mounting research links diet/food components to disease prevention and treatment. The efficacy of citrus extracts is supported by conclusive evidence from animal models which have provided the concepts for underlying mechanisms of action (Fig. 3.9). However, research must go far beyond the simplistic claims of positive properties in vitro. It must be heavily supplemented by well-designed observational epidemiological studies, bioavailability investigations and intervention trials. The World Health Organisation (WHO) has declared that diabetes and cancer epidemics are underway. Although there are various antidiabetic and anticancer drug therapies available on the market, tolerance and side effects are still an issue with many of these. There is thus a clear opportunity to improve on the current standard of care. In that regard, citrus fruit extracts represent an excellent candidate to be developed into nutraceuticals and functional foods geared towards the management of diabetes and cancer.

3.3 Future Research

Further studies are needed to achieve a better understanding of the cellular and molecular mechanisms of nutritional antioxidants as well as their clinical effects. The outcome of the various above-mentioned ongoing studies in our group under the Centre for Biomedical and Biomaterials Research

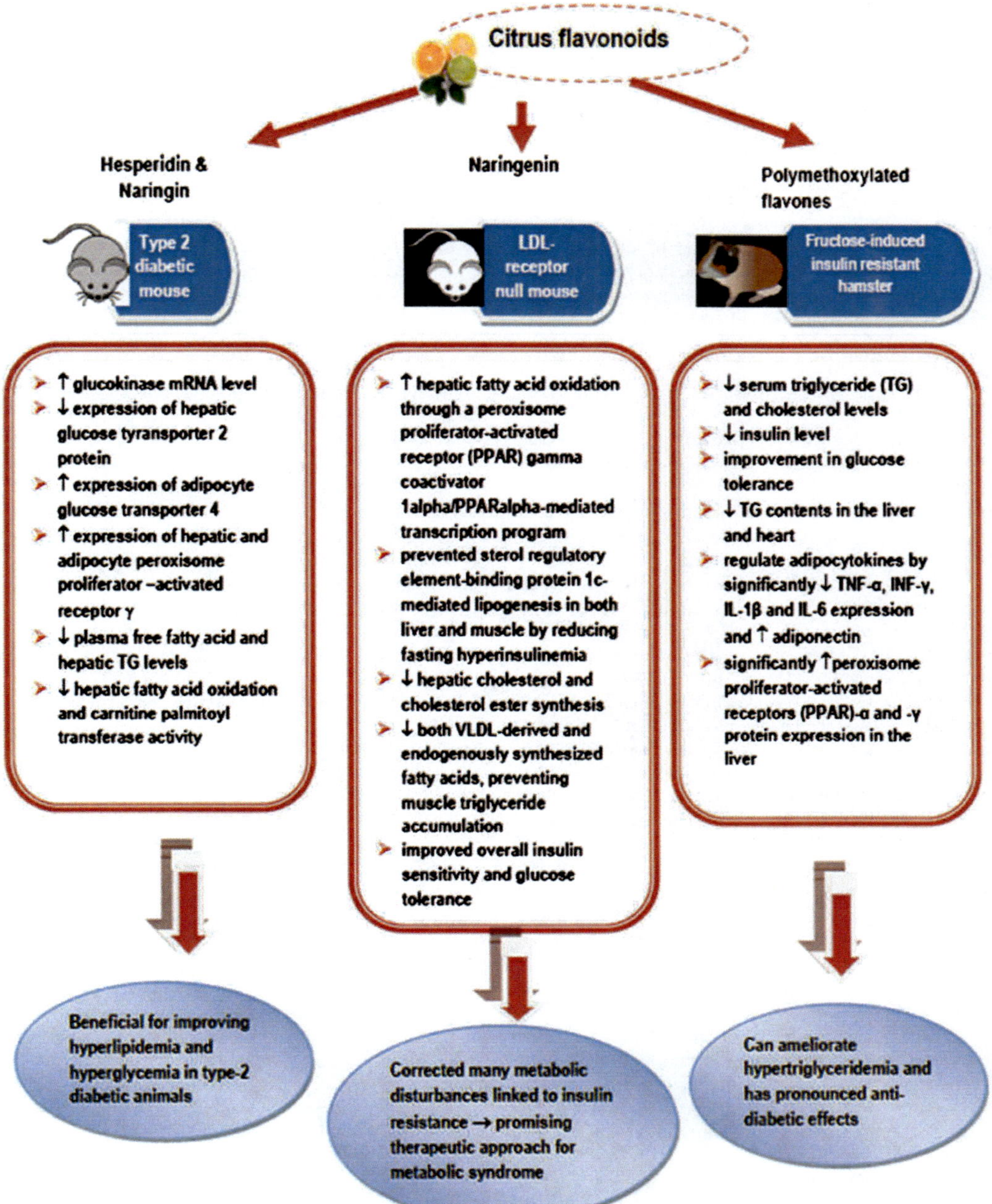

Fig. 3.9 Mechanisms of action of selected citrus flavonoids in animal models of diabetes

(CBBR) (an African Network for Diagnostics and Innovation (ANDI) centre of excellence) in partnership with established collaborations (University of Reunion, American University of Health Sciences (USA), Chhatrapati Shahu Ji Maharaj University, (Kanpur, India), Seoul National University, etc.) would thus constitute the basis for the selection of citrus fruit varieties with high polyphenolic content and antioxidant activities for the development of functional foods that would contain the right mix and amounts of antioxidant prophylactics to be used as supplements in a balanced diet within existing nutrition programmes.

Acknowledgments The authors are thankful to the University of Mauritius, University of Réunion (France) and Chhatrapati Shahu Ji Maharaj University Kanpur (India) for their full support. One of the authors (TB) is particularly thankful to Mr Gaurav Gaur, the Federation of Indian Chambers of Commerce & Industry (FICCI) and the

Government of India for the award of the CV Raman International Senior Fellowship for African Researchers. Part of this work has been carried out in the context of this fellowship.

References

Albach RF, Redman GH (1969) Composition and inheritance of flavanones in citrus fruit. Phytochemistry 8:127–143

Anis M, Aminuddin M (1981) Flavonoid patterns of leaves of some citrus species and their hybrids. Plant Biochem J 8(1):56–60

Aruoma OI (1993) Use of DNA damage as a measure of pro-oxidant actions of antioxidant food additives and nutritive components. In: Halliwel B, Arouma OL (eds) DNA and free radicals. London, Eliis Horwood, pp 315–317

Aruoma OI (1994) Nutrition and health aspects of free radicals and antioxidants. Food Chem Toxicol 32:671–683

Aruoma OI (2003) Methodological considerations for characterizing potential antioxidant actions of bioactive components in plant foods. Mutat Res 523–524:9–20

Aruoma OI, Bahorun T, Clement Y et al (2005) Inflammation, cellular and redox signalling mechanisms in cancer and degenerative diseases. Mutat Res 579:1–5

Beisiegel U, Weber W, Havinga JR et al (1988) Apolipoprotein E-binding proteins isolated from dog and human liver. Arteriosclerosis 8:288–297

Benavente-Garcia O, Castillo J, Marin FR et al (1997) Use and properties of citrus flavonoids. J Agric Food Chem 45:4506–4515

Benherlal PS, Arumughan C (2008) Studies on modulation of DNA integrity in Fenton's system by phytochemicals. Mutat Res 648:1–8

Berhow M, Tisserat B, Kanes K et al (1998) Survey of phenolic compounds produced in citrus. Technical Bulletin No 1856. United States Department of Agriculture, Agricultural Research Service, Washington, DC. http://www.ars.usda.gov/IS/np/phenolics.htm

Bisht K, Wagner KH, Bulmer AC (2008) Curcumin, resveratrol and flavanoids as anti-inflammatory, cyto-and DNA protective dietary compounds. Toxicology 278:88–100

Bourdon E, Loreau N, Blache D (1999) Glucose and free radicals impair the antioxidant properties of serum albumin. FASEB J 13:233–244

Bracke ME, Vyncke B, Opdenakker JM et al (1991) Effect of catechins and citrus flavonoids on invasion in vitro. Clin Exp Metastasis 9:13–25

Cadet J, Berger M, Douki T et al (1997) Oxidative damage to DNA: formation, measurement and biological significance. Rev Physiol Biochem Pharmacol 131:1–87

Cano A, Medina A, Bermejo A (2008) Bioactive compounds in different citrus varieties, discrimination among cultivars. J Food Compos Anal 21:377–381

Castillo J, Benavente-Garcia O, Delrio JA (1992) Naringin and neohesperidin levels during development of leaves, flower, buds and fruits of *Citrus aurantium*. Plant Physiol 99:67–73

Central Statistics Office (CSO) (2008) Digest of agricultural statistics. CSO, Port Louis. http://statsmauritius.gov.mu

Chen C, Kong ANT (2005) Dietary cancer-chemopreventive compounds: from signalling and gene expression to pharmacological effects. Trends Pharmacol Sci 26:318–326

Chidambara Murthy KN, Kim J et al (2011) Differential inhibition of human colon cancer cells by structurally similar flavonoids of citrus. Food Chem. doi:10.1016/j.foodchem.2011.10.014

Critchfield JW, Welsh CJ, Phang JM et al (1994) Modulation of adriamycin accumulation and efflux by flavonoids in HCT-15 colon cells: activation of P-glycoprotein as a putative mechanism. Biochem Pharmacol 48:1437–1445

Davignon J (2005) Apolipoprotein E and atherosclerosis: beyond lipid effect. Arterioscler Thromb Vasc Biol 25:267–269

De Flora S, Ferguson LR (2005) Overview of mechanisms of cancer chemo preventive agents. Mutat Res 591:8–15

De Flora S, Ramel C (1988) Mechanisms of inhibitors of mutagenesis and carcinogenesis: classification and overview. Mutat Res 202: 285–306

Eberhardt MV, Kobira Keck AS, Juvik JA et al (2005) Correlation analyses of phytochemical composition, chemical, and cellular measures of antioxidant activity of broccoli (*Brassica oleracea* L. Var. italica). J Agric Food Chem 53:7421–7431

Ekambaram G, Rajendran P, Magesh V et al (2008) Naringenin reduces tumor size and weight lost in N-methyl-N′-nitro-N-nitrosoguanidine-induced gastric carcinogenesis in rats. Nutr Res 28:106–112

Fattouch S, Caboni P, Coroneo V et al (2007) Antimicrobial activity of Tunisian quince (Cydonia oblonga Miller) pulp and peel polyphenolic extracts. J Agric Food Chem 55:963–969

Faure P, Troncy L, Lecomte M et al (2005) Albumin antioxidant capacity is modified by methylglyoxal. Diabetes Metab 31:169–177

Girard-Lalancette K, Pichette A, Legault J (2009) Sensitive cell-based assay using DCFH oxidation for the determination of pro- and anti-oxidant properties of compounds and mixtures: analysis of fruit and vegetable juices. Food Chem 115:720–726

Gorinstein S, Martin-Belloso O, Park YS et al (2001) Comparison of some biochemical characteristics of different citrus fruits. Food Chem 74:309–315

Gorinstein S, Cvikrova M, Machackova I et al (2004) Characterization of antioxidant compounds in Jaffa sweeties and white grapefruits. Food Chem 84:503–510

Guo H, Ling W, Wang Q et al (2008) Cyanidin 3-glucoside protects 3T3-L1 adipocytes against H2O2- or TNF-R induced insulin resistance by inhibiting c-Jun NH2-terminal kinase activation. Biochem Pharmacol 75:1393–1401

Gyo-Nam K, Jung-Geun S, Hae-Dong J (2009) Antioxidant and antidiabetic activity of Dangyuja (Citrus grandis Osbeck) extract treated with Aspergillus saitoi. Food Chem 117:35–41

Halliwell B (1997) Antioxidants: the basics – what they are and how to evaluate them. Adv Pharmacol 38:3–20

Hirano T, Oka K, Akiba M (1989) Effects of synthetic and naturally occurring flavonoids on Na+, K+–ATPase: aspects of structure–activity relationship and action mechanisms. Life Sci 45:1111–1117

Hirata T, Fujii M, Akita K et al (2009) Identification and physiological evaluation of the components from citrus fruits as potential drugs for anti-corpulence and anticancer. Bioorg Med Chem 17:25–28

Ho S, Wu S, Lin S, Tang Y (2010) Comparison of anti-glycation capacities of several herbal infusions with that of green tea. Food Chem 122:768–774

Holmes-McNary M, Baldwin ABS Jr (2000) Chemopreventive properties of transresveratrol are associated with inhibition of activation of the IκB kinase. Cancer Res 60:3477–3483

Horowitz R, Gentili B (1977) Flavonoids constituents of citrus. In: Nagy S, Shaw PE, Veldhuis MK (eds) Citrus science and technology, vol 1. AVI Publishing, Westport, pp 397–426

Hwang SL, Yen GC (2008) Neuroprotective effects of the citrus flavanones against H2O2-induced cytotoxicity in PC12 cells. J Agric Food Chem 56:859–864

Iwase Y, Takemura Y, Ju-ichi M et al (1999) Inhibitory effect of Epstein-Barr virus activation by citrus fruits, a cancer chemopreventor. Cancer Lett 139:227–236

Jeon SM, Bok SH, Jang MK et al (2001) Antioxidative activity of naringin and lovastatin in high cholesterol-fed rabbits. Life Sci 69:2855–2866

Jung UJ, Lee MK, Park YB et al (2006) Effect of citrus flavonoids on lipid metabolism and glucose-regulating enzyme mRNA levels in type-2 diabetic mice. Int J Biochem Cell Biol 38:1134–1145

Kandaswami C, Perkins E, Soloniuk DS et al (1991) Antiproliferative effects of citrus flavonoids on a human squamous cell carcinoma in vitro. Cancer Lett 56:147–152

Kanno SI, Shouji A, Asou K et al (2003) Effects of naringin on hydrogen peroxide-induced cytotoxicity and apoptosis in P388 cells. J Pharmacol Sci Jpn Pharmacol Soc 92:166–170

Kaur C, Kapoor HC (2001) Antioxidants in fruits and vegetables – the millennium's health. Int J Food Sci Technol 36(7):703–725

Kim J, Guddadarangavvanahally K, Uckoo RM, Patil BS (2011) Evaluation of chemopreventive and cytotoxic effect of lemon seed extracts on human breast cancer (MCF-7) cells. Food Chem Toxicol. doi:10.1016/j.fct.2011.10.057

Larocca LM, Piantelli M, Leone G et al (1990) Type II oestrogen binding sites in acute lymphoid and myeloid leukaemias: growth inhibitory effect of oestrogen and flavonoids. Br J Haematol 75:489–495

Lee SK, Kader AA (2000) Preharvest and postharvest factors influencing vitamin C content of horticultural crops. Postharvest Biol Technol 20:207–220

Lewinsohn E, Berman E, Mazur Y et al (1989) (7) Glucosilation and (1–6) rhamnosylation of exogeneous flavanones by undifferentiated citrus cell cultures. Plant Sci 61:23–28

Li Y, Ou-Lee TM, Raba R et al (1993) Arabidopsis flavonoid mutants are hypersensitive to UV-B irradiation. Plant Cell 5:171–175

Li JM, Che CT, Lau CB et al (2006) Inhibition of intestinal and renal Na^+-glucose cotransporter by naringenin. Int J Biochem Cell Biol 38:985–995

Londoño-Londoño J, Delima VR, Lara O et al (2010) Clean recovery of antioxidant flavonoids from citrus peel: optimizingan aqueous ultrasound-assisted extraction method. Food Chem 119:81–87

Lu Y, Zhao WZ, Chang Z et al (2004) Procyanidins from grape seeds protect against phorbol ester-induced oxidative cellular and genotoxic damage. Acta Pharmacol Sin 25:1083–1089

Mahley RW (1988) Apolipoprotein E: cholesterol transport protein with expanding role in cell biology. Science 240:622–630

Manach C, Regerat F, Texier O et al (1996) Bioavailability, metabolism and physiological impact of 4- oxo-flavonoids. Nutr Res 16:517–544

Manthey JA, Grohman K (2001) Phenols in citrus peel by products: concentrations of hydroxycinnamates and polymethoxylated flavones in citrus peel molasses. J Agric Food Chem 49:3268–3273

Martynez G, Delgado R, Perez G et al (2000) Evaluation of the in vitro antioxidant activity of *Mangifera indica* L. extract (Vimang). Phytother Res 14:424–427

Mata Bilbao ML, Andres-Lacueva C, Uregui O et al (2007) Determination of flavonoids in a citrus fruit extract by LC-DAD and LC-MS. Food Chem 101:1742–1747

Matsubara Y, Kumamoto H, Iizuka Y et al (1985) Structure and hypotensive effect of flavonoid glycosides in citrus unshiu peelings. Agric Biol Chem 49:909–914

Merken HM, Beecher GR (2000) Measurement of food flavonoids by high-performance liquid chromatography: a review. J Agric Food Chem 48:577–599

Middleton E, Kandaswami C (1994) Potential health-promoting properties of citrus flavonoids. Food Technol 48:115–119

Miyata M, Smith JD (1996) Apolipoprotein E allele-specific antioxidant activity and effects on cytotoxicity by oxidative insults and β-amyloid peptides. Nat Genet 14:55–61

Montanari A, Chen J, Widmer W (1998) Citrus flavonoids: a review of past biological activity against disease. In: Manthey JA, Buslig BS (eds) Flavonoids in the living system. Plenum Press, New York, pp 103–113

Nandakumar N, Jayaprakash R, Rengarajan T et al (2011) Hesperidin, a natural citrus flavonoglycoside, normalizes lipid peroxidation and membrane bound marker enzymes in 7, 12-Dimethylbenz (a) anthracene induced experimental breast cancer rats. Biomed Prev Nutr. doi:10.1016/j.bionut 2011.06.004

Neergheen VS, Bahorun T (2009) Therapeutic relevance of dietary polyphenols as natural antioxidants and modulators of cell signal transduction pathways in cancer and other degenerative diseases. In: Farombi EO (ed) Nutritional antioxidants in cancer and degenerative diseases. Transworld Research Network, Trivandrum, pp 131–152

Nogata Y, Sakamoto K, Shiratsuchi H et al (2006) Flavonoid composition of fruit tissues of citrus species. Biosci Biotechnol Biochem 70:178–192

Park GL, Avery SM, Byers JL et al (1983) Identification of bioflavonoids from citrus. Food Technol 37:98–105

Pashikanti S, de Alba DR, Boissonneault GA et al (2010) Rutin metabolites: novel inhibitors of nonoxidative advanced glycation end products. Free Radic Biol Med 48:656–663

Patil JR, Chidambara Murthy KN, Jayaprakasha GK et al (2009) Bioactive compounds from Mexican lime (*Citrus aurantifolia*) juice induce apoptosis in human pancreatic cells. J Agric Food Chem 57:10933–10942

Pérez-Jiménez J, Arranz S, Tabernero M et al (2008) Updated methodology to determine antioxidant capacity in plant foods, oils and beverages: extraction, measurement and expression of results. Food Res Int 41:274–285

Peterson G, Barnes S (1993) Genistein and biochain A inhibit the growth of human prostate cancer cells but not epidermal growth factor receptor tyrosine autophosphorylation. Prostate 22:335–345

Peterson JJ, Beecher GR, Bhagwat SA et al (2006a) Flavanones in grapefruit, lemons, and limes: a compilation and review of the data from the analytical literature. J Food Compos Anal 19:74–80

Peterson JJ, Dwyer JT, Beecher GR et al (2006b) Flavanones in oranges, tangerines (mandarins), tangors, and tangelos: a compilation and review of the data from the analytical literature. J Food Compos Anal 19:66–73

Prior RI, Wu X, Schaich K (2005) Standardized methods for the determination of antioxidant capacity and phenolics in foods and dietary supplements. J Agric Food Chem 53:4290–4302

Prost M (1992) Process for the determination by means of free radicals of the antioxidant properties of a living organism or a potentially agressive agents. US Patent 5,135,850, 4 Aug 1992

Ramful D, Tarnus E, Philippe R et al (2010a) Citrus fruit extracts reduces AGEs- and H_2O_2-induced oxidative stress in human adipocytes. J Agric Food Chem 58:11119–11129

Ramful D, Bahorun T, Bourdon E et al (2010b) Bioactive, phenolics and antioxidant propensity of flavedo extracts of Mauritian citrus fruits: potential prophylactic ingredients for functional foods application. Toxicology 278:75–87

Ramful D, Tarnus E, Aruoma OI et al (2011) Polyphenolics, Vitamin C composition and antioxidant propensities of Mauritian citrus fruit pulps. Food Res Int 44:2088–2099

Roche M, Rondeau P, Singh NR et al (2008) The antioxidant properties of serum albumin. FEBS Lett 582:1783–1787

Roche M, Tarnus E, Rondeau P et al (2009) Effects of nutritional antioxidants on AAPH- or AGEs-induced oxidative stress in human SW872 liposarcoma cells. Cell Biol Toxicol 25:635–664

Rousseff RL, Martin SF, Youtsef CO (1987) Quantitative survey of narirutin, naringin, hesperidin and neohesperidin in citrus. J Agric Food Chem 35:1027–1030

Samman S, Lyons Wall PM, Cook NC (1998) Flavonoids and coronary heart disease: dietary perspectives. In: Rice-Evans CA, Packer L (eds) Flavonoids in health and disease. Marcel Dekker, New York, pp 469–482

Scambia G, Ranelletti FO, Benedetti-Panici P et al (1990) Inhibitory effect of quercetin on OVCA 433 cells and presence of type II oestrogen binding sites in primary ovarian tumors and cultured cells. Br J Cancer 62:942–946

So FV, Guthrie N, Chambers AF et al (1996) Inhibition of human breast cancer cell proliferation and delay of mammary tumorigenesis by flavonoids and citrus juices. Nutr Cancer 26:167–181

Soobrattee MA, Bahorun T, Neergheen VS et al (2008) Phenolics content and antioxidant actions of the Rubiaceae, Ebenaceae, Celastraceae, Erythroxylaceae and Sterculaceae families of Mauritian endemic plants. Toxicol In Vitro 22:45–56

Stapleton AE, Walbot V (1994) Flavonoids can protect maize DNA from the induction of ultraviolet radiation damage. Plant Physiol 105:881–889

Stern DM, Yan SD, Yan SF et al (2002) Receptor for advanced glycation endproducts (RAGE) and the complications of diabetes. Ageing Res Rev 1:1–15

Suolinna EM, Buchsbaum RN, Racker E (1975) The effect of flavonoids on aerobic glycolysis and growth of tumor cells. Cancer Res 35:1865–1872

Takamatsu S, Galal AM, Ross SA et al (2003) Antioxidant effect of flavonoids on DCF production in HL-60 cells. Phytother Res 17:963–966

Tarnus E, Wassef H, Carmel JF et al (2009) Apolipoprotein E limits oxidative stress-induced cell dysfunctions in human adipocytes. FEBS Lett 583:2042–2048

Tomás-Barberán FA, Clifford MN (2000) Flavanones, chalcones and dihydrochalcones – nature, occurrence and dietary burden. J Sci Food Agric 80:1073–1080

Tripoli E, La Guardia M, Giammanco S et al (2007) Citrus flavonoids: molecular structure, biological activity and nutritional properties: a review. Food Chem 104:466–479

Tsang SY, Tam SC, Bremner L et al (1996) Copper-1,10-phenanthrolin induces internucleosomal DNA fragmentation in HepG2 cells, resulting from direct oxidation by the hydroxyl radical. Biochem J 317:13–16

Valentão P, Fernandes E, Carvalho F et al (2002) Antioxidant activity of Hypericum androsaemum infusion: scavenging activity against superoxide radical, hydroxyl radical and hypochlorous acid. Biol Pharm Bull 25(10):1320–1327

Van der Sluis AA, Dekker M, De Jager A et al (2001) Activity and concentration of polyphenolic antioxidants in apple: effect of cultivar, harvest year, and storage conditions. J Agric Food Chem 49:3606–3613

Wang H, Joseph JA (1999) Quantifying cellular oxidative stress by dichlorofluorescein assay using microplate reader. Free Radic Biol Med 27:612–616

Wang AY, Zhou MY, Lin WC (2011) Antioxidative and anti-inflammatory properties of citrus sulcata extracts. Food Chem 124:958–963

Wassef H, Bernier L, Davignon J et al (2004) Synthesis and secretion of apoC-I and apoE during maturation of human SW872 liposarcoma cells. J Nutr 134:2935–2941

Wolfe KL, Liu RH (2007) Cellular antioxidant activity (CAA) assay for assessing antioxidants, foods, and dietary supplements. J Agric Food Chem 55:8896–8907

Yoshida M, Sakai T, Hosokawa N et al (1990) The effect of quercetin on cell-cycle progression and growth of human gastric cancer cells. FEBS Lett 260:10–13

Yoshimizu N, Otani Y, Saikawa Y et al (2004) Anti-tumour effects of nobiletin, a citrus flavonoid, on gastric cancer include: antiproliferative effects, induction of apoptosis and cell cycle deregulation. Aliment Pharmacol Ther 20:95–101

Zechner R, Moser R, Newman TC et al (1991) Apolipoprotein E gene expression in mouse 3T3-L1 adipocytes and human adipose tissue and its regulation by differentiation and lipid content. J Biol Chem 266:10583–10588

Zhao X, Xue CH, Li Z-J et al (2004) Antioxidant and hepatoprotective activities of low molecular weight sulphated polysaccharide from *Laminaria japonica*. J Appl Phycol 16:111–115

Foliar Nutrition: Current State of Knowledge and Opportunities

4

Sylwester Smoleń

Abstract

The effectiveness of foliar nutrition is affected by numerous endogenous (related to leaf anatomic structure) as well as exogenous (nutrient concentration, soil type, pH) and environmental factors. Simultaneous application of foliar nutrition with plant growth and development biostimulators enables the increase of crop yield and the improvement of its quality. A significant trend in functional food production is plant biofortification with mineral nutrients – mainly Ca, Mg, microelements, and biogenic trace elements. Foliar nutrition can be used as a method of increasing crop level of these elements. Citrus plants, despite its thick cuticle layer of leaves, respond relatively well to foliar nutrition due to a high number of stomata on the lower leaf surface accompanied by a greater amount of cuticle pores (easing nutrient absorption) than any other epidermal cells. Foliar application treatments, when properly planned and conducted, may stimulate the flowering, increase yield, and improve nutritional and postharvest quality of citrus fruits.

Keywords

Foliar nutrition • Biostimulation • Biofortification • Point of deliquescence • Crop nutritional quality • Postharvest quality • Nutrient deficiency • Water quality

4.1 Introduction

"Feed the plants, not the soil." This statement represents one of the current views on plant mineral nutrition both by soil fertilization and foliar application. In the historical context of the development of plant nutrition science (since ancient times through the classic studies by Justus von Liebig in nineteenth century), the idea of plant fertilization through foliar application stood in strong opposition to commonly approved theories on plant mineral nutrition based on mineral uptake by roots. It is worth to mention that historical origins of foliar nutrition of citrus plants date back to 1844 (Srivastava and Singh 2003). Since the mid-twentieth century, a dynamic increase in the number of studies on foliar nutrition has been noted which contributed to the development and, consequently, implementation of agricultural recommendations for foliar fertilization in the cultivation of numerous crops.

In many works, plant treatment of supplying mineral nutrients by spraying is described as "foliar fertilization," "foliar nutrition," "foliar feeding," or "foliar application." A correct usage of any of these terms depends on the type and number of applied compounds. In common practice, "foliar nutrition" is mostly used although not always it is related strictly to mineral nutrient application. It should be emphasized that nutrients supplied by spraying are absorbed not only by leaves but also other green aerial parts of plants covered by epidermis (petioles, shoots, fruits) as well as scales which cover buds.

Foliar nutrition cannot replace or eliminate the natural way of nutrient uptake by plant roots. In general, foliar

S. Smoleń (✉)
Department of Soil Cultivation and Fertilization of Horticultural Plants, Faculty of Horticulture, University of Agriculture in Krakow, Al. 29 Listopada 54, 31-425 Kraków, Poland
e-mail: s.smolen@ogr.ur.krakow.pl; sylwester.smolen@interia.pl

A.K. Srivastava (ed.), *Advances in Citrus Nutrition*,
DOI 10.1007/978-94-007-4171-3_4, © Springer Science+Business Media B.V. 2012

application of mineral elements is not so effective as to cover total nutritional requirements of plants – even after foliar spraying with easily remobilized mineral elements (Marschner 1995; Michałojć and Szewczuk 2003; Starck 1997). It mainly results from restrictions in applying increased doses of mineral nutrients or its insufficient translocation from leaves to roots. Nevertheless, depending on plant species, a significant portion of plant requirements for mineral nutrients (mainly microelements) can be covered by foliar nutrition (Szewczuk and Michałojć 2003a).

Foliar fertilization is most commonly used for: supplying nutrient deficiency, improving nutritional status of plants and, thus, increasing crop yield and its quality. However, depending on plant species, environmental factors, and agrotechnique, this treatment can be applied for other purposes, such as mitigating the negative effects of stress conditions – drought, frost damages, etc.

Not only mineral elements can be supplied by plant spraying but also nutritional compounds (i.e., simple sugars, disaccharides), amino acids, peptide chains, organic acids, growth regulators and stimulators. The whole spectrum of organic compounds present in seed or marine algae extracts is applied with fertilizers. In such a case, "plant biostimulation" should be mentioned rather than "foliar nutrition."

Recently, a particular attention is put on plant biofortification with mineral nutrients, especially those characterized by insufficient distribution in food chain (soil–plant–consumer). This impaired transfer results in a deficient level of such macro- and microelements as Ca, Mg, Fe, Zn, Cu, I, and Se in the diet of many people in the world (Dayod et al. 2010; Hirschi 2009; White and Broadley 2009). In many cases, foliar application is considered the quickest and highly cost-effective method of its introduction into plants and, as a consequence, in human diet and animal fodder.

It needs to be mentioned that significant differences exist in the possibility of conducting foliar nutrition and its efficiency in the cultivation of annual (mono- and dicotyledonous), biennial, and perennial plants, as well as C3 and C4 plants. These dissimilarities result from differences in leaf anatomy, physiological and biochemical processes, and cultivation period – 1, 2, or many years. The efficiency of foliar treatment depends on numerous endo- and exogenous factors.

In spite of considerable progress in foliar nutrition research, still many problems are encountered related to the development of agricultural recommendations for conducting this treatment especially in the range of improving its efficiency, plant biostimulation, balancing reduced soil fertilization with foliar application, as well as adjusting current foliar nutrition programs for newly obtained cultivars.

To sum up, it can be stated that foliar nutrition is the fastest method of introducing mineral nutrients in the aerial parts of crop plants when compared to any other technique of soil fertilization, including fertigation.

4.2 Endogenous Factors Affecting Nutrient Absorption by Leaves and Other Aerial Parts of Plants

When endogenous factors are considered, the absorption efficiency for nutrients applied foliarly depends on the thickness of the cuticle covering epidermal cells (green shoots, lower and upper leaf surfaces) as well as the number of cuticular pores and ectodesmata (*syn.* ectoteichodes) located in this layer. Additionally, it is related to the distribution of trichomes and stomata in leaves accompanied by the highest occurrence of cuticle pores (Franke 1961; Kannan 2010; Marschner 1995; Michałojć and Szewczuk 2003).

4.2.1 The Cuticle

The cuticle protects above-ground parts of plants against excessive transpiration, pests, and infections as well as prevents leaching of particular compounds from leaves by precipitation. Cuticle thickness is species-dependent, but leaf position on the stem is also of significance. Citrus and coffee plants are characterized by a particularly thick leaf cuticle. What is more, leaves grown in the shade have thinner cuticle layer than those exposed to full sunlight (Marschner 1995).

The cuticle consists of three layers: (a) external – top coating of wax, (b) the cuticle proper – middle layer of cutin embedded with wax, and (c) inner layer composed of cutin and wax blended with carbohydrates (pectin, cellulose, etc.) – cuticular (cutinized) layer contacting cell walls. The cuticle proper is covered with a wax layer composed of highly polymerized saturated and unsaturated esters, ketones, fatty and hydroxy fatty acids, as well as alcohols (Fageria et al. 2009; Franke 1967, 1971; Holloway 1971; Michałojć and Szewczuk 2003; Weaver 1972). Waxy substances are of apolar–lipophilic character. Cutin, a main component of the middle part of the cuticle, contains numerous hydrophilic groups defining its polarity. The presence of carbohydrates in the most inner layer of cuticle furtherly increases its polar and hydrophilic character. As a consequence of described physicochemical properties of these three layers (differences in electrochemical potential), a distinct polarity gradient occurs which is mainly responsible for ion transport through cuticle (Komosa 1990; Michałojć and Szewczuk 2003; Schönherr 2006). Franke (1967) indicates that across the cuticle, polar substances soluble in water (cations, anions) are transported by "hydrophilic path," while apolar compounds soluble in lipids through "lipophilic path."

Open spaces can be formed in the cuticle layer favoring the process of intercuticular penetration. Its mechanism is based on the diffusion of organic and inorganic ions as well as undissociated molecules (Franke 1986; McFarlane and

Berry 1974; Norris and Bukovac 1972). Polar characteristics of cuticle and pectin layer, which is a part of an outer cell wall, determined mainly by its negative charge due to the presence of –OH and –COOH groups enable cation sorption (Franke 1986; Michałojć and Szewczuk 2003). Additionally, negative charge of these layers contributes to more efficient translocation of apolar molecules (*i.a.*, urea) and cations rather than anions. For that reason, a lower efficiency is observed for foliar nutrition with mineral nutrients in anion forms (NO_3^-, $H_2PO_4^-$, Cl^-, HPO_4^{2-}, SO_4^{2-}, $B_4O_7^{2-}$, MoO_4^{2-}, BO_3^{3-}) than with cations (NH_4^+, K^+, Ca^{2+}, Mg^{2+}, Fe^{2+}, Fe^{3+}, Mn^{2+}, Mn^{3+}, Cu^{2+}, Cu^+, Zn^{2+}). Rate of cation permeation through the cuticle declines with increasing valency. For both types of ions, cations and anions, nutrient translocation from the outer to the inner parts of leaves (influx) is greater than in the opposite direction – efflux. However, these differences are less distinct for organic molecules (Yamada et al. 1964).

The above-described mechanism of mineral nutrient absorption by epidermis of aerial parts of plants differs from processes occurring in roots. Typical antagonistic and synergistic interactions between cations and anions noted for the root uptake can also be assigned for foliar nutrition. Its strength and occurrence depend, however, on a dose and concentration of nutrients – particularly in the case of multielement fertilizers used for foliar nutrition. Due to a huge disproportion in macro- and micronutrient concentration in such fertilizers, it is recommended to use chelated forms of the latter ones which eases its absorption by leaves.

The efficiency of nutrient absorption is also dependent on the contact angle of liquid drop on leaf and stem epidermis surface. The lower the angle and more thorough the coating of epidermis surface by droplets of solution, the better translocation of dissolved nutrients into plant cells is observed. The value of the angle can be successfully decreased by the application of adjuvants and surfactants which reduce surface tension of liquid droplets (Komosa 1990; Michałojć and Szewczuk 2003).

Amide nitrogen in the form of urea [apolar organic compound –$CO(NH_2)_2$] undergoes the fastest transport across the cuticle. Foliar application of urea, even in low concentrations, hydrates the cuticle making its absorption (as well as other mineral nutrients) more effective (Marschner 1995; Michałojć and Szewczuk 2003; Smoleń and Sady 2008a). In the studies conducted by Bondada et al. (2006) on isolated grapefruit (*Citrus paradisi* Macf.) cuticle, maximum penetration rate of urea occurred approx. 40 min after droplet application, and the rate was about 2% of the amount applied per hour. Total urea penetration averaged approx. 35% and tended to decrease with increased cuticle weight until the cuticles were 6 months old. In comparison, in bean leaves (characterized by a relatively thin cuticle), 50% of foliarly applied urea was absorbed within 1–4 h (Wittwer and Teubner 1959).

4.2.2 Cuticular Pores and Ectodesmata – Stomata and Trichomes

Numerous models of nutrient transport across the cuticle have been described (Srivastava and Singh 2003). Basically, besides a direct penetration through this layer, mineral nutrients can be absorbed by cuticular pores and ectodesmata syn. ectoteichodes (Kannan 2010; Marschner 1995; Michałojć and Szewczuk 2003).

Relatively little is known on nutrient absorption through ectodesmata (Franke 1971). Kannan (2010) informs that ectodesmata were demonstrated in epidermal walls of mesophytic plant species. Michałojć and Szewczuk (2003) indicate that nutrient transport across the cuticle occurs in regions contacting with ectodesmata. There is no clear information in the literature whether these regions overlap with the distribution of cuticular pores and ectodesmata, in particular, plant species.

Cuticular pores, depending on its size, are permeable to compounds of various molecular weights, such as mineral nutrients, chelated microelements, and sugars (Marschner 1995). Studies conducted by Schönherr (2006) revealed that aqueous pores in plant cuticles arise by hydration of permanent dipoles and ionic functional groups. Aqueous pores preferentially occur in cuticular ledges, at the base of trichomes, and in cuticles over anticlinal walls and its radii range from 0.45 to 1.18 nm. Translocation of ionic compounds can be fairly rapid, and ions with molecular weights of up to 800 $g\,mol^{-1}$ can penetrate cuticles that possess aqueous pores with time.

Generally, the number of cuticle pores in stomata (guard and subsidiary cells) and trichomes exceeds its level in other epidermal cells. Thus, the density of stomata and trichomes on leaf surface (adaxial and abaxial) affects the efficiency of the uptake of mineral nutrients supplied by plant spraying (Komosa 1990; Marschner 1995). A significant role of stomata cells and trichomes in the absorption of nutrients applied through foliar nutrition results from a great number of ectodesmata in these cells. Trichome-related absorption decreases with the level of its cutinization and leaf aging (Komosa 1990). Leaves of citrus plants are characterized by several times higher stomatal density than observed in other plants (Table 4.1). In such a case, it may be stated that despite a relatively thick cuticle layer, foliar nutrition of these plants can be more successful than in other crop species.

Basic physiological role of stomata is to regulate gas exchange between plant and atmosphere. Nutrient uptake through stomata depends on its density (rather than the degree of stomatal opening), value of surface tension of applied liquids (solutions of lower tensions are more readily absorbed), and the morphology of stomatal cells. According to Marschner (1995), rate of mineral uptake through stomata is higher at night, when these structures are closed, than

Table 4.1 Number of stomata (stomata density) per unit area of adaxial and abaxial side leaves

Plant	Number of stomata (mm^{-2}) Adaxial side	Abaxial side	References
Citrus group			
Citrus plants – average	n.d.	400–700	Spiegel-Roy and Goldschmidt (1996)
Red Bush grapefruit	n.d.	477.9	Graham et al. (1992)
Marsh grapefruit	n.d.	438.9	Graham et al. (1992)
Swingle citrumelo	n.d.	498,3	Graham et al. (1992)
Valencia sweet orange	n.d.	654.2	Graham et al. (1992)
Orlando tangelo	n.d.	542.3	Graham et al. (1992)
Sour orange	n.d.	359.3	Graham et al. (1992)
Cleopatra mandarin	n.d.	530.5	Graham et al. (1992)
Others species			
Alfalfa	169	138	Piskornik (1994)
Apple tree	0	290–400	Piskornik (1994)
Arabidopsis	250	350	Coupe et al. (2006)
Bean	40	281	Piskornik (1994)
Cabbage	141	226	Piskornik (1994)
Cherry	0	250	Piskornik (1994)
Coffee	n.d.	138.8–184.4	Pompelli et al. (2010)
Corn	52	68	Piskornik (1994)
Geranium spp.	n.d.	7.75	Schletz (2008)
Lime	n.d.	370–420	Meinder and Mansfield (1968)
Oat	25	23	Piskornik (1994)
Pea	101	216	Piskornik (1994)
Peach	0	225	Piskornik (1994)
Soybeans	81–170	242–385	Ciha and Brun (1975)
Sunflower	58	156	Piskornik (1994)
Tomato	12	130	Piskornik (1994)
Wheat	50.7–57.5	34.9–39.7	Mohammady (2011)

n.d. not detected

during day when stomata are open. Studies conducted by Eichert et al. (2002) confirmed that stomata can be an important pathway for the uptake of foliar applied substances. In the case of *Sedum telephium* leaves, uptake rates increased by a factor of 1.5–36 when compared with leaves kept in darkness prior to the foliar application. Additionally, frequency of stomata per leaf surface reduces with increasing ploidy level (Beck et al. 2003).

Stomata number on leaf surface is not a constant value. According to Graham et al. (1992), stomatal density on leaves of various citrus species changes in relation to cultivation conditions – in field or greenhouse. Nevertheless, irrespective of cultivation site, this index was lower for leaves of full expansion than in leaves of two-third expansion.

Shaded leaves are usually greater in size and have thinner cuticle layer than leaves growing in full sunlight. Interesting is the total stomatal area calculated for both types of leaves as it affects the effectiveness of foliar nutrition – expressed as rate of nutrient absorption by leaves. Results of the studies conducted by Pompelli et al. (2010) indicated that coffee leaves exposed to full sunlight were characterized by higher stomatal density (184.4 mm^{-2}) than leaves grown at 50% of reduction in solar irradiation (138.8 mm^{-2}). However, with a significantly higher total leaf area, shaded leaves had a comparable stomatal area to full-sunlight leaves – 377.2 μm^2 and 374.7 μm^2, respectively. It therefore can be concluded that leaves developed in full sunlight as well as shaded ones take up mineral nutrients by leaves at a similar level while the efficiency of this process is the resultant of the following factors: cuticle thickness, stomatal density, and stomatal area.

Gas exchange occurring through stomata enables the uptake of mineral nutrients present in the atmosphere in gaseous forms. This applies mainly to SO_2, NH_3, and NO_2 which are rapidly included in metabolic pathways in mesophyll cells in leaves. The amount of nutrients absorbed in such way is difficult to estimate. The concentration of mentioned gases in the air depends on the degree of industrialization and transportation (emission of SO_2 and NO_2), agricultural technology (crop fertilization, livestock farming – ammonia emission), as well as microbiological processes occurring in soil. The excessive level of these gases in the air, particularly of SO_2, can induce plant damages (Malhotra and Hocking 1976).

To sum up, the efficiency of nutrient uptake by leaves is mainly dependent on leaf area (which relates to its developmental stage) as well as the number of stomata and trichomes along with cuticle pores accompanying these two types of cells.

4.2.3 Cell Wall and Cell Membrane

Apart from the cuticle, other barriers in nutrient uptake by aerial parts of plants are cell wall and cell membrane. Mineral nutrients and more complex compounds are transferred across cell wall through interfibrillar spaces and ectodesmata. The sole mechanism of nutrient transport in cell wall is based on diffusion. Donnan free space (DFS) (also called apparent free space AFS) plays a crucial role in cation sorption and translocation with diffusion (Komosa 1990). During nutrient absorption from leaf surface into deeper layers, cations are easily transported in this area, while anions are removed (Marschner 1995).

Nutrient translocation across cell membrane into cytoplasm is mainly regulated by the active mechanism with the use of protein carriers or ion pumps (Komosa 1990; Michałojć and Szewczuk 2003). Mineral nutrients transported into cytoplasm are utilized in various metabolic pathways. Further, they can be transported in ionic or organic form. Distribution of nutrients between mesophyll cells in leaf and into phloem and xylem occurs mainly by plasmodesmata, protein carriers, or ion pumps.

4.2.4 Leaf Growth and Development, Plant Age

Foliar fertilization requires higher leaf area index (LAI) for absorbing applied nutrient solution in sufficient amount (Fageria et al. 2009). However, young developing leaves (characterized by low LAI value) are a significant acceptor of mineral nutrients and photosynthetic products. Mineral nutrients applied at that stage of leaf development rapidly permeates through the cuticle. Mature leaves with the highest photosynthetic productivity remain the source of photosynthetic products for other parts of the plant. At that stage, compounds applied through foliar nutrition are translocated from leaves to shoots, flowers, fruits, and roots along with assimilates. In each development stage or phenologic phase, plant requirements toward particular mineral nutrients vary significantly. During early growth and leaf development, plants need higher nitrogen levels. In further stages, depending on plant species, increased requirements toward phosphorus are noted for flowering and seed setting while appropriate potassium and/or calcium nutrition is crucial for fruit growth and development. These variations in nutrient demands result from the changes in the dynamics of metabolic processes and plant productivity. The amount and proportion of applied mineral nutrients should be adjusted to particular development stages. What should also be taken into consideration is the efficiency of foliar nutrition of annual, biennial, and perennial plants (Magdziak and Kołodziej 2003), as well as trees and shrubs.

Efficiency of foliar nutrition decreases during leaf senescence. This process is characterized by remobilization of endogenous mineral nutrients from leaves to shoots. At the same time, higher rate of nutrient leaching from leaf surface by aqueous solutions (rain, fog, etc.) is observed as permeability of cell membranes for mineral nutrients significantly increases (Marschner 1995).

To sum up, if for certain nutrients foliar application at an early growth stage of citrus plants leaves is recommended, it is therefore necessary to reach a compromise between early application and allowing the crop to attain a leaf area large enough for the absorption of large amounts of nutrients (Srivastava and Singh 2003).

4.2.5 Leaf Turgor

Leaf turgor undergoes diurnal changes caused by direct influence of exogenous environmental factors – light, temperature, wind – affecting physiological and biochemical mechanisms regulating water relations in plants.

Leaves maintained at full turgor – during the morning and evening hours – present the highest ability to absorb and metabolize mineral nutrients applied foliarly. High temperature combined with light intensity and wind decreases leaf turgor, negatively affecting the rate of biochemical and physiological processes in plants. Leaves with lower turgor values are more prone to exhibit midday depression of net photosynthesis. This may result in weakened absorption and metabolism of mineral nutrients applied in the early morning hours particularly in hot and sunny days. Foliar nutrition is the more effective, the longer leaves maintain high turgor. If foliar application is conducted in the morning hours, it is achieved in cloudy days with moderate temperature and high air humidity.

4.2.6 Soil Fertilization, Nutritional Requirements of Plants, Nutritional Status of Plant vs. Efficiency of Foliar Nutrition

Soil applications of fertilizers are mainly done on the basis of soil tests, whereas foliar nutrient applications are mainly conducted on the basis of visual foliar symptoms or plant tissue tests. Hence, correct diagnosis of nutrient deficiency is fundamental for successful foliar fertilization (Fageria et al. 2009).

Szewczuk and Michałojć (2003a) inform that foliar nutrition gives better results if plant cultivation is conducted on soil with optimal pH value and level of mineral nutrients. Nevertheless, efficiency of foliar nutrition, particularly in the aspect of its influence on crop yield and its quality, is directly related to nutritional status of plant. Generally, the better is nutritional status of plant (mainly achieved by nutrient uptake from soil), the lower is the efficiency of mineral nutrient supplementation by foliar application (Wójcik 2004). It is a consequence of simple relations described by the classic Liebig's law of the minimum and Shelford's law of the tolerance. In plants with high nutritional status, additional foliar supplementation of mineral nutrients increases its leaf concentration to the level of "luxury consumption." In that case, a significant increase of crop yield or improvement of its quality should not be expected, and thus, this treatment is not economically justified. Adversely, additional foliar application of mineral nutrients is considered cost-effective if it is conducted to obtain crops biofortified with elements (particularly micro- and trace elements) that are deficient in the diet of human populations living in particular environments.

Foliar application of excessive doses of nutrients (in relation to plant requirements) may induce some negative effects on plant organisms. Toxicity of mineral nutrients applied foliarly can be observed irrespective of current nutritional status of plants as nutrients can be introduced in such forms (speciations) that cause plant damages even if applied in relatively low concentrations, *i.a.*, I^- and SeO_4^{2-} when compared to IO_3^- and SeO_3^{2-}.

A completely different aspect that should be considered together with foliar nutrition is the problem of mineral deficiency. Particular species of annual, biennial, perennial plants as well as trees and shrubs differ with respect to macro- and micronutrient requirements and therefore exhibit various tolerance toward nutrient deficiency (Szewczuk and Michałojć 2003b). It is important for microelements as its assimilation from soil is insufficient due to low level in soil, inappropriate pH, antagonistic interactions with other elements, soil humidity, or organic matter content.

If deficiency of a nutrient occurs in plants, its supplementation through foliar application will be more rapid than through soil fertilization (Marschner 1995). Foliar absorption of mineral nutrients is from 8 to 20 times more efficient than soil application. Nevertheless, such high efficiency is not commonly achieved in agricultural practice (Kuepper 2003).

It needs to be underlined that in many cases, nutrient deficiency in above-ground parts of plant (leaves, shoots, fruits) is brought about by impaired translocation, as for P and Ca (from roots or within leaves and fruits) rather than its low concentration in soil. A perfect example is symptoms of Ca deficiency noted in annual plants (tip burn) or citruses (blossom-end rot of citrus). Physiological causes of Ca deficiency in particular plant groups are complex and result not only from strongly acidic properties of soils but also disturbances in water relations – and therefore difficulties in Ca translocation to outer parts of leaves or fruit skin. In citrus plants, symptoms of calcium deficiency are usually associated with strongly acidic soils and climatic conditions characterized by numerous cloudy or rainy days. What is of particular importance, in many cases, these symptoms become visible only during postharvest storage. Foliar nutrition with Ca may prove to be efficient in mitigating and counteracting these physiological diseases. In this aspect, interesting are the results obtained by Xie and Zhang (2004) in citrus cultivation and foliar application of phosphorus which undergoes a relatively slow absorption by leaves. In these studies, an average of 24 h after treatment, the leaf retained 74.8% of original amount of ^{32}P applied, but fruit rind only retained 47.4% of the total ^{32}P applied. Within 72 h after application, the leaf exported 1.61 μmol g^{-1} dry weight tissue, whereas the rind exported 0.143 μmol g^{-1} dry weight tissue of the initial amount of ^{32}P applied. Some ^{32}P was transported into the adjacent fruit 5 days after the application of ^{32}P.

A significant issue in modern agriculture is the need of improving nitrogen utilization from mineral fertilizers mainly in order to control its loss from soil. Soil application of lower N doses may decrease crop yield, although this can be compensated by additional foliar nutrition. This relation was confirmed for annual, biennial, and herbaceous perennial plants (Rydz 2001; Jodełka et al. 2003; Wojciechowska et al. 2005; Smoleń and Sady 2008c). According to the reports of Rożek et al. (2000), Rydz (2001), and Wojciechowska et al. (2005), the efficiency of foliar nutrition is dependent on soil, climate, fertilizer type, and the amount of nitrogen used. In trees and shrubs, application of foliar nutrition for mentioned purpose seems less effective. It is the consequence of a limited amount of nitrogen that can be delivered foliarly when compared to nutritional requirements of this plant group.

4.3 Exogenous and Environmental Factors Affecting the Efficiency of Foliar Nutrition

Srivastava and Singh (2003) presented the list of the most important exogenous and environmental factors affecting foliar fertilization: light, temperature, wind, time of day, photoperiod, humidity, amount and intensity of precipitation, drought, osmotic potential of growing medium (or soil water), and nutrient stress. To factors associated with spray solution, quality of water used for foliar nutrition, nutrient concentration, and pH of working solution can be included. Each of these exogenous factors (particularly environmental ones) has a direct influence on physiological and biochemical processes in plants, at the same time impacting the efficiency of foliar nutrition. Effects induced by some of these factors have been discussed in the previous section.

If plant spraying is carried out in bright sunlight, droplets of working liquids on leaf surface act as miniature lenses and by focusing sun radiation cause leaf burn and necrosis. At high temperature, loss of plant turgor is observed which reduces the rate of biochemical reactions and impairs the absorption of applied compounds. What is more, with rising air temperature, relative humidity drops rapidly which increases the rate of water evaporation. As a consequence, nutrient penetration through the cuticle is significantly slowed down. Additionally, rapid drying of droplets increases the final concentration of nutrients on the leaf surface posing a risk particularly if highly concentrated fertilizers are applied. Therefore, in the conditions of high sunlight and temperature, plants may be unable to effectively take up and utilize compounds supplied through foliar spraying (Marschner 1995; Starck 1997; Kujawski 2005).

It should be emphasized that at elevated temperatures occurring together with increased intensity of solar radiation and length of daytime, significant changes are noted in the structure of cuticle waxes (transition to vertical configuration) and coverage of the leaf surface with these compounds. Such modifications improve nutrient absorption by leaves (Komosa 1990). Marschner (1995) informs that, in contrast to roots, the uptake of mineral nutrients by green parts of the plant is stimulated by light. Additionally, the rate of nutrient absorption by leaves is higher during daytime than at night which is a direct consequence of circadian rhythms of metabolic processes. Nevertheless, droplets of working liquid applied on leaf surface dry quicker during the day than in the evening hours.

With reference to the information presented above, it should be mentioned that there are also different opinions on the optimal timing of foliar application. Szewczuk and Michałojć (2003a, b) as well as Kujawski (2005) recommend to conduct this treatment during the late afternoon or evening hours, while Fageria et al. (2009) suggest the afternoon hours when air temperature is low (after 2–3 p.m.). Application of foliar nutrition in the evening hours contributes to a prolonged wetting of the plant surface favoring bacterial and fungal infections – if foliar nutrition is not combined with plant protection treatment. Additionally, active ingredients and formulas of fungicides might reduce the rate of nutrient absorption in leaves (Schönherr 2002).

In the context of discussed influence of light and temperature on the cuticle wax structure and stimulation of nutrient uptake by light (Komosa 1990; Marschner 1995), it seems unreasonable to conduct foliar nutrition in the afternoon and evening hours. After the sunset, the wax structure may return to its original shape, and most of metabolic processes slow down which, in consequence, may decrease the absorption rate and the efficiency of utilization of mineral nutrients supplied with foliar application. Nevertheless, for nutrients applied foliarly, ion uptake rate is higher at night (when stomata are closed) than during day (Marschner 1995).

A key factor affecting nutrient absorption by above-ground green parts of the plant is the drying time of working liquid droplets on leaf and shoots epidermis. A reduction in liquid evaporation from plant surface can be achieved by the application of humectants or antievaporants. Some of these chemicals are included in commercial foliar fertilizers or surfactants and adjuvants.

Rates of penetration are greatly affected by humidity over cuticles and hygroscopicity of salts (Schönherr 2002). Utilization of mineral nutrients supplied by foliar application is never 100% effective – even for urea, which is relatively easily translocated across the cuticle (Bondada et al. 2006). After droplet drying, a significant amount of applied compounds remains on epidermis surface as "dry deposit" – salt crystals or organic sediments (amino acids, saccharides, etc.). These solid-state forms can be redissolved in water vapor from humid air. The possibility of rehydration is determined by the point of deliquescence (POD) of the salt and humidity over the salt residue (Table 4.2). POD is defined as that humidity over a saturated solution containing solid salt. When humidity is above POD, the salt residue on the cuticle dissolves, while below, a solid residue is formed and penetration ceases (Schönherr 2002). In the case of foliar application of a mixture of various compounds (multielement fertilizers), ionic strength of the solution significantly changes (as well as cation–anion interactions) depending on the concentration of particular salts. As a consequence, values of POD are also different and difficult to determine experimentally. Generally, values of POD are not affected by temperature (Kolthoff et al. 1969).

Compounds with POD values above 90% are characterized by low ability of rehydration by air humidity. In such a case, efficiency of its absorption will be mainly influenced by the duration of leaf wetness after foliar application. Rehydration of dry deposit on leaf surface occurs also due to fog, dew, or small rainfall (short-time drizzle). Efficiency of the process depends on the solubility of dry deposit components and the amount of water supplied with precipitations. Intensive leaf wetting by rain, on the other hand, is responsible for the leaching of endogenous mineral nutrients from leaves as well as removal of dry deposit from endodermis. In the studies conducted by Bondada et al. (2006), simulation of rehydration of urea deposit on grapefruit leaves with water stimulated additional penetration of about 1%; however, this was not significant.

Potassium fertilizers (KNO_3 and KCl, in particular) have high values of POD, but foliar nutrition of citrus plants with K compounds gives satisfactory results in increasing yield and improving fruit quality (Srivastava and Singh 2003). It is more likely related to a large number of stomata and cuticular pores on citrus leaves facilitating nutrient absorption

Table 4.2 Characteristics of selected salts and organic compounds containing one or two mineral nutrients with respect to POD in the aspect of its applicability for foliar nutrition

Mineral nutrients	Salt/compound	POD (%)[a]	Assessment of salt/compound applicability for foliar nutrition
N	$CO(NH_2)_2$ – urea	n.d.	Very good
	NH_4NO_3	63	Average
N and P	$(NH_4)_2HPO_4$, $NH_4H_2PO_4$	n.d.	Average
P	H_3PO_4	n.d.	Average
K	$K_2CO_3 \cdot 2H_2O$	44	Good
K and Cl	KCl	86	Small
K and N	KNO_3	95	Small
K and P	K_2HPO_4	92	Small
	KH_2PO_4	95	Small
K and S	K_2SO_4	n.d.	Average
Ca	$CaCl_2 \cdot 6H_2O$	33	Very good
	Lactate of calcium	95	Small
	Acetate of calcium	100	Small
Ca and N	$Ca(NO_3)_2$	56	Very good
Mg and S	$MgSO_4 \cdot H_2O$	n.d.	Good
Mg and Cl	$MgCl \cdot 6H_2O$	33	Very good
Mg and N	$Mg(NO_3)_2 \cdot 6H_2O$	56	Good
Fe	$FeSO_4$	n.d.	Very good
	$FeCl_3 \cdot 6H_2O$	44	Good
	$Fe(NO_3)_3 \cdot 9H_2O$	54	Good
Mn	$MnSO_4$	n.d.	Very good
	$Mn(NO_3)_2 \cdot 4H_2O$	42	Very good
	$MnCl_2 \cdot 4H_2O$	60	Average
Zn	$ZnNO_3 \cdot 6H_2O$	42	Good
B	$Na_2B_4O_7 \cdot 10H_2O$ (borax)	n.d.	Good
Mo	Na_2MoO_4	n.d.	Very good
	$(NH_4)_6Mo_7O_{24} \cdot 4H_2O$	n.d.	Very good

n.d. not detected
[a]POD – Point of deliquescence by Schönherr (2002)

from these fertilizers. It should be mentioned, however, that the influence of foliar K application on citrus yield is dependent on the current state of plant nutrition with K (and N) as well as potassium level in soil (Srivastava and Singh 2003). Studies conducted by Schönherr and Luber (2001) revealed that penetration of potassium salts (5 $g\,l^{-1}$) across *Citrus* leaf cuticular membranes is higher than through pear leaves. It was demonstrated that K_2CO_3 was best suited for foliar applications as rates of penetration were large at 50% humidity and higher, while for other salts (KNO_3 and KCl), humidity was required to be 90–100% for maximum rates of penetration.

Air humidity not only affects the droplet drying or POD but also influences the rate of nutrient penetration across cuticular membrane, which generally increases with higher values of air humidity – as a direct consequence of a greater turgor (Schönherr 2001; Schönherr and Luber 2001).

A significant parameter determining the effectiveness of foliar nutrition is pH of applied solution. Marschner (1995) suggests that leaf damages are less severe if pH of sprayed solution is low – unfortunately, this author did not provide precise values. Komosa (1990) informs that nutrient absorption is higher with pH between 3 and 4. For particular compounds (mineral fertilizers), maximum rate of its absorption is noted at various pH values. What is more, with pH changing from acidic to alkaline, higher plant toxicity of increased concentration of foliarly applied NH_4^+ ions is observed. Results of the studies conducted by Schönherr (1976) revealed that average size of isolated pores of *Citrus aurantium* did not depend on pH, but the number of pores increased with pH from $5 \cdot 10^{10}$ (at pH 3) to $16 \cdot 10^{10}$ (at pH 9) per cm^2 of cuticular membrane. In further work, Schönherr (2002) points out that the main factor affecting the swelling of cuticular membrane, and thus the rate of salt penetration, is not solely pH but also air humidity. Zekri and England (2010) inform that the optimum pH of solutions for foliar nutrition of citrus plants should be between 5.0 and 7.5. If pH of spray solution is above or below critical values, leaf absorption of nutrient is poor, and the solution can cause leaf and fruit burn.

To sum up the discussion of the influence of exogenous and environmental factors on the efficiency of foliar nutrition treatments, it should be emphasized that low temperatures (particularly within the root zone) more strongly affect nutrient uptake by roots than leaves (Komosa 1990). In those conditions, application of foliar nutrition can improve nutrient status of plants.

4.4 Aims and General Rules for Foliar Nutrition

Foliar nutrition treatment is mostly considered an additional supplementation of mineral nutrients. In common practice, together with foliar supplementation of mineral nutrients, various compounds of biostimulation effect on plant growth and development are also applied. These compounds, apart from growth promotion, can improve nutrient absorption from leaves or its utilization in plants.

Currently, foliar nutrition with minerals can be combined with the application of marine algae (Durand et al. 2003; Mugnai et al. 2008) or seed extracts (Cwojdziński et al. 1996; Barczak and Cwojdziński 1998; Barczak et al. 2007) which are rich in natural plant metabolites such as sugars, amino acids, organic acids, phytohormones, peptide chains, etc. Additionally, for foliar biostimulation purposes, products of enzymatic hydrolysis of plant tissues or meat industry wastes (which contain large amounts of amino acids and peptide chains) may be applied. For that reason, a boundary between classic foliar nutrition and plant biostimulation – gaining more and more recognition – gets blurred. The issue whether foliar biostimulation can be considered as "foliar nutrition" should also be taken into discussion (see also Sect. 4.6).

In agricultural practice as well as in scientific research, foliar nutrition is applied for various purposes (Fageria et al. 2009; Smoleń and Sady 2008a, b; Smoleń and Szura 2008a, b; Smoleń and Ledwożyw-Smoleń 2011) including supplementation of mineral nutrient deficiency in plants, improvement of nutritional status of plants, counteracting or mitigating stress effects, growth stimulation (biostimulation), and crop quality control, as well as plant biofortification. In Table 4.3, general rules for foliar nutrition with respect to achieving desired goals are presented.

Table 4.3 Basics and general practical rules of foliar nutrition

Aim	Practical notes on foliar nutrition
Supplementation of nutrient deficiency	Nutrients should be applied in chemical forms easily taken up by plants. Apolar molecules (such as urea) are absorbed the quickest, and cation uptake is fastest than of anions
	Even if applied in low doses, urea hydrates the cuticle easing the absorption of urea itself as well as other mineral nutrients. For the improvement of the efficiency of foliar nutrition (with nutrients other than nitrogen), it is recommended to add 0.2–0.4% solution of urea into the working solution (apart from adjuvants and surfactants). It is particularly advised for plants with thick cuticular layer on leaves
	In the case of micronutrient deficiency, it is recommended to introduce chelated forms of these microelements – molybdenum and boron are the only micronutrients that are not chelated
	The number and frequency of foliar nutrition treatments depend on the intensity of deficiency symptoms – the maximum recommended fertilizer dose should never be exceeded
Improvement of nutritional status of plants	The optimal time for foliar nutrition shall be the stage of intensive plant growth and development which is related to the highest photosynthetic productivity and biomass increase
Counteracting and mitigating stress effects	Foliar nutrition with multielement fertilizers – also containing amino acids and sugars – is highly recommended. Fertilizers with biostimulators may be used if its application is advised by manufacturers to mitigate the effects caused by stress factors
	In the case of long-term low intensity of photosynthetically active radiation (PAR), it may be recommended to mix 0.5–2% solution of sugars (sucrose) or biostimulators containing sugars with the working solution. Sugars are the additional source of energy needed for the incorporation of mineral nutrients into organic compounds in metabolic processes responsible for plant growth and development
Biostimulation – growth stimulation and crop quality management	Prior to the application of fertilizers containing biostimulators, one should get acquainted with the rules of its usage. It is necessary to confirm whether a biostimulator can be applied together with plant protection products
	Recommended doses of biostimulators should never be exceeded as it may lead to various unfavorable effects. Application of these products should be conducted strictly according to manufacturer's recommendations
Biofortification	Plant biofortification concerns, among others, plant enrichment in microelements and/or trace elements. Practical notes of foliar biofortification of particular species are yet to be developed as they require conducting thorough vegetative experiments

4.4.1 Supplementation of Mineral Nutrient Deficiency in Plants

Insufficient nutrient status of plants is revealed by the development of physiological diseases on plant organs. The need of supplementing nutrient deficiency can result from incorrectly performed soil fertilization or impaired uptake of nutrients by roots. This is basically the main cause of physiological diseases observed on annual or biennial plants. In plantations of fruit trees and shrubs as well as perennials plants, soil depletion in mineral nutrient in subsequent years of cultivation most commonly contributes to the development of deficiency symptoms in plants. Impaired translocation of mineral nutrients from roots to above-ground parts of plants, especially fruits (as in the case of calcium), should not also be neglected.

4.4.2 Improvement of Nutritional Status of Plants

Foliar nutrition treatments can serve as an efficient method of improving the level of mineral nutrient utilization and, in consequence, plant fitness (Elmer et al. 2007) and yield (Biesiada and Kołota 1998). Higher nutritional status of plants can enhance its resistance to pathogen infection. What should also be emphasized is that poor plant nutrition is not always manifested by visible symptoms of deficiencies.

4.4.3 Counteracting or Mitigating Stress Effects

Environmental stress can seriously affect plant mineral nutrition on each step from nutrient uptake by roots through its transport to upper parts of plants to the mechanism of mineral metabolism processes. Stress factors can be related to the following:

- High air temperature often considered together with excessive light intensity
- Osmotic stress resulting from water deficit in soil – drought, salinity
- Excessive soil humidity due to long-term rainfalls and increased groundwater levels
- Long-term drop of air temperature below the range optimal for plant growth and development
- Long-term exposure to low levels of photosynthetically active radiation – noted mainly in the greenhouse production but also in the field cultivation of annual and biennial plants
- Frost damages of flowers, leaves, and young shoots of biennial, perennial plants, as well as trees and shrubs

A detailed description of the influence of particular stress factors on physiological and biochemical mechanisms of plants related to mineral nutrition is beyond the scope of foliar nutrition *sensu stricte*. Plants have developed natural mechanisms, protecting them from many of environmental stresses. Nevertheless, these defense processes require large amounts of metabolic energy for the synthesis of protective proteins during drought or heat stresses. It is worth to mention that foliar nutrition is considered the fasted method improving plant regeneration during these particular conditions.

4.4.4 Growth Stimulation (Biostimulation) and Crop Quality Control

In this place, it is necessary to make a distinction between the application of biostimulators *sensu stricte* and fertilizers containing compounds known for its biostimulative action (see also Sect. 4.6). In many countries, due to imperfect legal regulations, numerous "fertilizers" with a negligible amount of mineral nutrients and a relatively high concentration of stimulative compounds are introduced in the market.

The possibility of managing nutritional or postharvest quality of crop is a topic relatively rarely taken up in the research. It should be emphasized that foliar application of mineral nutrients, often simultaneously with plant biostimulation, allows to achieve the intended goal, for example, a reduction of nitrate level in crop (Smoleń and Sady 2008a, b, 2009; Smoleń and Ledwożyw-Smoleń 2011). Within the potential for crop quality management, it is crucial to obtain reproducible results which can be reached only through the development and experimental testing of foliar nutrition regimes. The first step, however, should be a thorough analysis of nutrient requirements and the specificity of biochemical and physiological processes occurring in particular plant species.

4.4.5 Plant Biofortification

Biofortification is defined as such a process that increases the content of biogenic elements such as Ca, Mg, Cu, Fe, Zn, I, or Se in edible parts of crop plants. As a result, improvement of consumer's health is expected to occur. Increased accumulation of biogenic elements in plants can be achieved through application of agronomic, genetic, or transgenic strategies (Dayod et al. 2010; White and Broadley 2009). Foliar application is one of the most efficient methods of plant enrichment in these elements and, thus, incorporation of its additional amounts into the food chain – human diet or animal fodder. What needs to be taken into consideration is the fact that either of mentioned elements, if applied excessively or in inappropriate chemical form, can become toxic for plants inducing growth inhibition, leaf chlorosis, and necrosis or, in extreme cases, plant death. Overall, such strategies are applicable to specific crops and mineral scenarios but cannot be universally applied as a strategy to boost the nutritional quality of foods (Hirschi 2009). Studies conducted by Machado et al. (2005) on citrus genomics allowed to state that the knowledge accumulated so far on metabolic pathways in citrus plants makes possible to begin biofortification programs focusing on the enhancement of the nutritional quality of citrus fruits and its derivatives.

In agricultural practice, depending on a current situation in crop cultivation and the course of climatic conditions, realization of the above-mentioned goals (listed from 1 to 5) will require the application of fertilizers differing with its composition, proportion of macro- and microelements, as well as containing stimulative compounds of various modes of action.

4.5 Influence of Foliar Nutrition on Physiological and Biochemical Processes in Plants

Foliar supplementation of mineral nutrients primarily improves nutritional status of above-ground parts of plants. Molecular, physiological, and biochemical influence of this treatment on plant organism, however, goes beyond the mineral nutrition of plants. It is particularly relevant for foliar application of biostimulators or fertilizers with stimulative compounds. A cascade of molecular and biochemical processes is activated, and the signals of these changes (of a chemical character – phytohormones, sugars, etc.) are transduced from leaves to other aerial as well as underground parts of the plant. The potential range of these interactions is broad, and even with the current progress in analytical techniques, it is not possible to conduct a global assessment of these relations. What can be performed is only evaluation of the influence exerted by foliar nutrition or biostimulation on selected molecular, biochemical, and physiological processes in controlled conditions – in phytotron experiments. Exogenous environmental factors additionally affect plant response to these treatments. It is repeatedly observed that if crop cultivation is conducted in field conditions, obtained results differ from those coming from experiments with controlled conditions. From a practical point of view, the latter ones are more informative as they provide a balanced description of foliar nutrition influence on plant metabolism, growth, and development, as well as crop yield and its quality.

Depending on the type, form, and concentration of applied nutrients (macro-, micro-, and trace elements), as well as plant development stage and environmental factors, foliar nutrition (or biostimulation) can contribute to the following:

1. Regulation of molecular and biochemical processes related to the metabolism of particular nutrients – its chemical reduction and incorporation into organic compounds
2. Induction or silencing of genes encoding transport proteins for mineral nutrients as well as those responsible for direct regulation of active transport of nutrients across cell membranes
3. Enhanced transport of exogenous mineral nutrients (applied foliarly) to roots, shoots, fruits, and seeds or its temporary deposition in sprayed leaves
4. Biosynthesis of phytohormones and transduction of chemical signals to other parts of the plant in order to regulate molecular and metabolic pathways related with mineral nutrition or other physiological processes in plants
5. Direct or indirect increase/reduction in the rate of active nutrient uptake by roots dependent on the current nutritional status of leaves – particularly in the case of nitrogen
6. Regulation or stimulation of enzymatic processes – mineral elements playing the role of enzyme activators/cofactors
7. Depending on nutrient (biostimulator) concentration – positive or negative influence on the intensity of physiological processes in plants – photosynthesis, cellular respiration, etc.

The complexity of above-mentioned interactions will be briefly presented in the example of selected steps of the metabolism of foliarly applied nitrogen along with foliar application of Ni, Mo, sucrose, cytokinins, salicylic acid, or exogenous amino acids.

The rate of uptake and penetration of various nitrogen forms into the cytosol of leaf cells is as follows: N–NH_2 (amide form) > N–NH_4 > N–NO_3.This order varies, however, from this described for uptake preference of roots.

The amide nitrogen from urea is not directly included into the metabolic path of this element in plant. The ammonium nitrogen, being the result of urea hydrolysis by the urease enzyme, is incorporated into simple organic compounds (Marschner 1995). Nickel is a cofactor for this enzyme, so urease activity, and therefore urea hydrolysis, strongly depends on Ni level in plants (Nicoulaud and Bloom 1998; Gheibi et al. 2009), simultaneous foliar nutrition of urea and nickel (Krogmeier et al. 1991), or foliar application of multi-element fertilizers containing this element. A positive interaction has been noted for citrus nutrition with nickel and urease activity which, in consequence, increases flowering and percentage of fruit set (Malavolta et al. 2006; Malavolta and Moraes 2007). Srivastava and Singh (2003) inform that the mechanism of flower development – flowering induction in citrus plants is related to NH_4^+ level in leaves as well as arginine biosynthesis – can be indirectly stimulated by foliar application of urea.

Ammonium ions occur in plant cells as a product of the reduction of NO_3^- (taken up both by roots and leaves through foliar nutrition) to NO_2^-, which process is catalyzed by nitrate reductase (NR). The next step is NO_2^- reduction to NH_4^+ by nitrite reductase (NiR) (Campbell 1999). Assimilation of ammonium nitrogen (taken up by roots or applied foliarly) takes place in GS-GOGAT cycle in chloroplasts. The crucial role in nitrogen assimilation process plays the alpha-oxoglutarate (created in the Krebs cycle) which is transformed by GDH to glutamine, which in turn enters GS-GOGAT cycle (Tischner 2000; Masclaux-Daubresse et al. 2002). Thus, the efficiency of the nitrogen assimilation process will be higher with greater photosynthetic productivity (Masclaux-Daubresse et al. 2002).

The process of NO_3^- reduction is under the control of numerous exo- and endogenous factors (Campbell 1999). Plants have the ability to accumulate relatively high levels of NO_3^- in leaves but not NH_4^+, which tends to be toxic in excessive concentrations. With a low photosynthetic productivity, the process of NO_3^- reduction and NH_4^+ assimilation is also inhibited. Another reason for a decreased rate of NO_3^- reduction is the accumulation of simple organic forms of nitrogen–glutamine and other amino acids. It basically hampers the active root uptake of nitrate and ammonium ions mediated by NO_3^- and NH_4^+ transport proteins – due to feedback regulation. It is worthy to mention that gene families encoding transport proteins for these two types of ions are divided into two systems: HATS – high-affinity transport system and LATS – low-affinity transport system (Forde 2000; Forde and Cole 2003; Gansel et al. 2001; Howitt and Udvardi 2000; Orsel et al. 2002). Some of genes belonging to HATS-NO_3^-, HATS-NH_4^+, LATS-NO_3^-, and LATS-NH_4^+ are expressed mainly in roots, others are ubiquitous in all (aerial and underground) parts of the plant (Forde and Cole 2003). Taking this into consideration, excessive foliar treatments with reduced forms of nitrogen (N–NH_2 and N–NH_4) but particularly amino acids (or mixtures containing peptide chains) may decrease the rate of nitrogen uptake by roots and, consequently, lower level of nitrogen nutrition expressed as N total content in plants (Wierzbińska 2009).

It needs to be underlined that a positive effect on NO_3^- reduction in leaves, and thus crop yield and its biological quality, was revealed for foliar application of molybdenum as NR cofactor (Williams et al. 2004), cytokinins (Campbell 1999), or sucrose (Kováčik 1999).

Interesting is the influence of exogenous sucrose applied foliarly on nitrogen metabolism. As reported by Marschner (1995), compounds with small molecule mass, for example, sugars (sucrose), do not penetrate through cuticular layer, and their assimilation takes place in the cuticular pores. Sucrose is transported between parenchymal cells of leaf with the participation of transporter proteins on the basis of sucrose/H^+ symport (Shakya and Sturm 1998; Starck 2003). The endogenous sucrose created in the process of photosynthesis in parenchymal cells along with exogenous sucrose (from foliar nutrition) is then transported by phloem to the acceptors, such as young developing leaves or storage roots (Starck 2003). Starck (2003) informs that sugar also plays the role of signal substance that indirectly informs cells or individual structures about the level of supply and demand for the photosynthetic products. Sugars together with hormones participate in the transduction of signals on the level of gene expression. Foliar application of sucrose was shown to decrease the level of NO_3^- (Kováčik 1999; Smoleń et al. 2010; Smoleń and Sady 2012). Reduction of nitrate content was obtained after spraying plants with an extract derived from *Lupinus angustifolius* seeds (Cwojdziński et al. 1996; Barczak et al. 2007; Barczak and Cwojdziński 1998). According to Cwojdziński et al. (1996), this extract, apart from nutritional components, proteins, and sugar, most probably contained dihydrozeatin (cytokinin). Another phytohormone belonging to the cytokinin group is benzyl adenine (BA). The physiological role of cytokinins is, apart from the others, connected with the regulation of biochemical processes by controlling enzyme activity (Borkowska 1997). Endo- and exogenous cytokinins are included to the group of factors stimulating NR activity (Yu et al. 1998).

In the study by Yaronskaya et al. (2006), exogenous cytokinins caused an increase in the content of alpha-linolenic acid (syn. 5-aminolevulinic acid – ALA; a direct precursor of chlorophyll) in barley seedlings, while in the research by Liu et al. (2006), they elevated chlorophyll level in *Spirodela polyrrhiza*. As a result, the application of exogenous cytokinins may result in a greater assimilative potential of plants, which in turn can indirectly contribute to better assimilation of nitrogen to organic compounds and N uptake by plants (Smoleń et al. 2010; Smoleń and Sady 2012). Still, it should be mentioned that cytokinins together with amino acids are involved in the feedback regulation of N uptake and metabolism in plants (Dluzniewska et al. 2006; Forde and Cole 2003).

Endogenous salicylic acid (SA) is considered a hormone-like substance, which plays an important role in the regulation of plant growth, development, and a variety of physiological processes in plants (Klessig and Malamy 1994). The study by Fariduddin et al. (2003) revealed a positive effect of exogenous salicylic acid on dry mass, net photosynthetic rate, and efficiency of carboxylation in *Brassica juncea*. A favorable effect of SA on the activity of NR was indicated by Jain and Srivastava (1981), Miguel et al. (2002), and Fariduddin et al. (2003).

In the summary, it is worth to add that simultaneous foliar application of nitrogen (in the form of urea), sucrose, Mo, BA, and SA can produce different effects that application of a single compound: urea, sucrose, BA, or SA (Smoleń et al. 2010; Smoleń and Sady 2012; Wierzbińska 2009).

4.6 Plant Biostimulation

An interesting trend in foliar nutrition of plants is the enrichment of fertilizers with substances of biostimulation activity (*syn.* stimulators, bioactivators, growth stimulants) for plant growth and development as well as selected metabolic processes. These compounds can be foliarly applied separately or together with mineral nutrients.

What are biostimulators? They mean inorganic and organic substances or its mixtures positively affecting plant development or other physiological processes in plants. One of the requirements for biostimulators is that they pose no risk for human, animal, or natural environment due to its application. Depending on legislation in a particular country, various classification of this group of compounds is provided. It is often that substances of stimulative character are included into the formulation of fertilizers for foliar nutrition, soil fertilization, or products designed for the nutrient solution preparation in the hydroponics.

The following substances can be included to this group: organic acids occurring in plants (salicylic acid, ALA syn. aminolevulinic acid, etc.), vitamins, amino acids, low-molecular-weight polypeptides, extracted phytohormones, phenolic compounds, sugars (mono-, di-, oligosaccharides), and many other organic compounds. They are obtained due to complex technological processes basing on extraction, enzymatic treatments, or microbiological processes. Taking into consideration the source, biostimulators can be divided into the following groups:

- Naturally occurring compounds – plant extracts
- Extracts from marine organisms – mainly phytoplankton
- By-products from the meat (food) industry wastes
- Synthetic organic compounds – products of chemical or pharmaceutical industry

In the case of biostimulators obtained from plant or animal tissues, they contain numerous organic and inorganic compounds naturally occurring in these organisms (Cwojdziński et al. 1996; Barczak and Cwojdziński 1998) and exhibit a broad spectrum of the physiological and biochemical influence on plants. The composition of the extracts is strictly dependent on raw material quality. Adversely, if biostimulators are derived from the extract by selective separation processes, as for several marine bioactive substances (MBS): N PRO (Durand et al. 2003), EXT1116, NA9158 and 251104 (Mugnai et al. 2008), or glycine betaine (Zamarreño et al. 1997), a stable composition and concentration of active substances can be achieved.

In many countries, plant biostimulation is gaining popularity. However, it should never be forgotten that plants are autotrophic and have the ability of synthesizing amino acids, vitamins, and organic acids. Among the skeptics on biostimulation, a question of the usefulness and the basic rule of this treatment arises. The answer is not simple as the group of biostimulators includes preparations containing compounds and substances (natural or synthetic) which, among others:

- Enhance plant growth and development through increasing the efficiency of soil fertilization or foliar nutrition with mineral elements
- Improve the functioning of physiological and biochemical processes in plant tissues
- Provide additional protection against unfavorable environmental conditions
- Increase plant resistance to diseases and pathogens

Plant biostimulation was shown to improve growth and development, quantity and quality of fruit yield (Basak and Mikos-Bielak 2008) also in citrus plants (Caronia et al. 2010; Fornes et al. 1995; Koo 1988; Santana et al. 2006).

Foliar application of biostimulators can be particularly effective during unfavorable environmental or stress conditions. Biostimulation efficiency (as well as for foliar nutrition *sensu stricte*) may be low if the treatment is conducted in near-optimum conditions (soil and environmental) additionally when manufacturer's recommendations are not followed. In such conditions, physiological changes in plants, regulated by environmental and endogenous factors (*i.a.*, phytohormones), proceed optimally.

It can be stated that, to a certain degree, plant biostimulation can be considered an alternative to genetically modified crops. A properly designed biostimulation strategy may allow to a controlled use of "native" genetic information in order to activate selected metabolic processes. The planning of foliar biostimulation program requires detailed information on chemical composition of applied substances as well as a thorough knowledge of the regulation and mechanism of individual biochemical and physiological processes in plants. The more complex is the composition of biostimulators (or fertilizers with these substances), the more difficult it is to predict the final result of its application. The reason is the complexity of plant reaction due to the simultaneous use of numerous stimulative compounds and mineral nutrients.

Another question of concern is the safety of plant biostimulation. A major disadvantage is that excessive frequency of treatments as well as application of too high doses may deregulate metabolic processes in plants. Many compounds used for biostimulation play an "indirect information role" in biochemical and physiological responses of plants. As a consequence of inappropriate application of biostimulators, a blockage, weakening, or change in the intensiveness of various processes may occur. That situation is particularly dangerous in the cultivation of fruit plants with excessive application of growth regulators (phytohormones *sensu stricte*). In such a case, a "complex dependence effect" may be noted which is characterized by, *i.a.*, impaired growth or flower and fruit setting due to discontinuation of several-year-long foliar spraying with exogenous growth regulators. A question arises whether the same reaction may be observed for foliar biostimulation. For that reason, plant biostimulation should be conducted strictly according to manufacturer's recommendations or be based on the results of independent scientific research – best if carried out in multiple growing seasons.

In the summary, it should be emphasized that foliar biostimulation of plants means the application of the compounds much larger than mineral nutrients. Thus, the efficiency of its absorption by leaves is lower, and the key factor affecting the rate of this process is the duration of leaf surface coverage by droplets of a working solution.

4.7 Foliar Nutrition vs. Crop Nutritional and Postharvest Quality Control

A separate issue is the influence of foliar application on nutritional, postharvest, and processing quality of crop yield. This problem is relatively complex, and the assessment of the effect of foliar nutrition on particular crop species requires consideration of various quality parameters (Biesiada and Kołota 1998; Berbeć et al. 2003; Jabłoński 2002). Depending on the crop, these characteristics include the level of nutritional and health-promoting compounds (*i.a.*, vitamins, antioxidants, essential oils), as well as those negatively affecting the consumer's health (*i.a.*, nitrates, heavy metals, mycotoxins). Significant is the fact that simultaneous foliar application of two or more mineral nutrients may lead to diverse results with respect to yield quality which can be modified by environmental conditions (Sugier 2003; Rożek et al. 2000; Rydz 2001; Wojciechowska et al. 2005). Thus it is difficult to propose the general relations between foliar nutrition and the quality of various crops.

Implementation of agricultural recommendations for foliar application of fertilizers in order to improve (in a repeatable manner under varying climatic conditions) not only the quantity but also the quality of crop yield needs conducting numerous vegetation experiments. An important aspect in ensuring the nutritional and processing value of yield is an appropriate selection of fertilizer composition for particular crops and cultivars. Development of the composition of specialty fertilizers designed for a single crop species should include both nutritional requirements as well as the level of those compounds (mainly microelements, *i.a.*, Cu, Mn, and Zn) that are introduced with plant protection products (Szewczuk et al. 2003).

Foliar nutrition – particularly when combined with the application of biostimulators or growth regulators – can be used for the management of plant growth and crop quality, such as the reduction of nitrate level in edible parts of plants (Smoleń and Sady 2009). Development of a program for crop quality management requires a thorough knowledge on mechanisms and regulation of key physiological and biochemical processes in plants. This is the only basis for a proper selection of biostimulators or its combination with growth regulators, the application of which should affect the above-mentioned processes in the most selective way. The efficiency of the strategy so developed should always be verified in vegetation experiments.

The issue of the improvement of postharvest quality of yield through foliar nutrition also falls within the scope of this field. By means of properly prepared and conducted programs of foliar nutrition and/or biostimulation, it is possible to positively affect yield suitability for storage. These treatments, if conducted during the preharvest period, may naturally protect fruits from skin damages, evaporation, fungal (Elmer et al. 2007), bacterial, or physiological diseases, as well as delay senescence during storage. For the improvement of the quality and postharvest stability of citrus fruits, foliar nutrition with K and Ca can be applied. Skin damages observed on citrus fruits during storage are mainly caused by potassium deficiency (Achilea et al. 2002). It is confirmed that foliar application of Ca reduces skin discoloration in citrus fruits (Crisosto et al. 1997, 2000).

4.8 Agrotechnique – Foliar Nutrition in Practice

Foliar application of nutrient solutions makes salt concentrations on a leaf surface higher than those of soil solutions (Wójcik 2004). It is possible due to the coverage of the aerial parts of plants with the cuticle not found on roots. This layer is, at the same time, a natural barrier to the transmission of mineral nutrients and other compounds applied foliarly. Despite this, nutrient absorption through leaves can be from 8 to 20 times more efficient than its uptake by roots (in optimal conditions). However, this rate of efficiency is rarely achieved in practice (Kuepper 2003). It is caused by quick evaporation of liquid droplets from epidermis surface, various thickness of the cuticle in particular species, or droplet drift from the cuticle.

Increase of the efficiency of nutrient absorption can be achieved by the use of adjuvants and surfactants which reduce the surface tension of working liquid as well as the application of humectants and antievaporants (Fernández et al. 2006; Komosa 1990; Michałojć and Szewczuk 2003; Srivastava and Singh 2003). Some practical advice on foliar nutrition treatment is included in Table 4.3.

A significant factor affecting the efficiency of foliar nutrition is the chemical and microbiological water quality. Water used for foliar nutrition purposes should contain the lowest level of mineral nutrients. It is confirmed that water from rivers, lakes, wells, etc., may have high amounts of mineral nutrients due to its leaching from soils – particularly in agricultural areas and around greenhouse farms (Breś 2009, 2010).

Foliar fertilization should be physiologically balanced containing all main and secondary plant nutrients and essential trace elements (Fritz 1978). With high concentration of mineral nutrients in water, it is possible to lower the dose of fertilizers applied foliarly – by adjusting the final level of nutrients in the working solution. An indirect indicator of high mineral content in water is the measurement of electrical conductivity (EC). EC values above 1.0–1.2 mS cm^{-1} inform

Table 4.4 Water EC values in relation to its salinity

Water quality[a]	EC (mS cm^{-1})	Salinity (mg dm^{-3})
Very high	0.3	<200
High	0.3–0.7	200–500
Sufficient	0.7–1.2	500–1,000

[a]The classification of water quality for hydroponics may serve as approximate information on its applicability for foliar nutrition

that water may contain relatively high amount of nutrients in a total number exceeding 1 g of salt dissolved in 1 dm^3 (Table 4.4). In the common practice, even when planning experiments with foliar nutrition of plants, an issue of nutrient content in water is usually neglected – except for dissolution of fertilizers or biostimulators in distilled water. Application of distilled water in vegetation experiments is, however, troublesome to perform, not to mention the practical foliar nutrition in large plantations, orchards, etc. In such cases, the effect of foliar application of fertilizers will be the result of the interaction between nutrients introduced with fertilizers and cations/anions previously present in water.

Another significant problem related to chemical quality of water is its hardness. High water hardness, and more precisely carbonate hardness (resulting from the presence of HCO_3^- anions and Ca^{2+}/Mg^{2+} cations), causes white spots on leaf surface due to drying droplets of working solution – this may negatively affect photosynthesis. Calcium content in hard waters exceeds the level of 120–140 mg dm^{-3} and that amount can be of significance when balancing the level of nutrients applied foliarly – for a comparison, 0.5% w/v water solution of $Ca(NO_3)_2$ contains approximately 950 mg $Ca \cdot dm^{-3}$. The process of Ca absorption by leaves may be impaired by too high pH of hard waters (mainly neutral or slightly alkaline). A decrease of water (working solution) pH by HNO_3 neutralizes HCO_3^- anions which can improve Ca absorption by leaves. Nevertheless, this treatment does not affect the total water hardness but only lowers the carbonate one. An issue of optimal pH level for the working solutions is discussed in Sect. 4.3. It is worth to mention that the solubility of phosphorus fertilizers in hard water may be significantly lowered by the precipitation of calcium phosphates.

Water used for foliar nutrition should be free from potential plant and human pathogens. In the case of its presence, water disinfection should be performed, but the application of products containing chlorine should be avoided. Spraying with chlorinated water may cause some developmental disorders or even kill beneficial microorganism colonizing leaf surfaces. High levels of chlorine are also unfavorable for plant growth. If chlorine is used for water disinfection, it is highly recommended to leave water in open containers for 12–24 h in order to remove chlorine by its natural evaporation (Kuepper 2003). Other methods of water disinfection (thermal, ozone, iodine, or selenium application) are either relatively expensive or potentially dangerous for plants due to excessive concentration of iodine or selenium in water. Sometimes it is recommended to use 35% solution of hydrogen peroxide in an approximate dose of 500 ml per 100 dm^3 of water. Foliar application of H_2O_2 (depending on a dose) may cause various effects on plants – from the enhancement of antioxidant status (when applied in low concentrations) to the increase of oxidative stress and occurrence of symptoms resembling a hypersensitive response when applied in high doses (Gechev et al. 2002).

When preparing working solution, powdery or granulated fertilizers should be totally dissolved so as not to clog sprayer nozzles. Solubility of solid fertilizers is affected by water temperature. Most of solid fertilizers, when dissolved, cause the drop of solution temperature – it is particularly observed for urea but also ammonium, potassium, and magnesium nitrates. Therefore it is advised to first dissolve liquid products (diluting of concentrated solutions in water is an exothermic process) and then powdery or granulated fertilizers. Calcium and sulfate fertilizers should never be mixed in high concentrations (risk of gypsum precipitation) as well as calcium and phosphate ones – as calcium phosphate may be the product. Fertilizer solutions must be prepared directly prior to its application to prevent the occurrence of unfavorable chemical reactions. In the case of simultaneous application of chemical plant protection products with foliar nutrition, they should be introduced into sprayers after dissolving fertilizers.

The efficiency of foliar application largely depends on the coverage of leaves with working solution – basically its lower surface where the highest number of stomata is located (Table 4.1). Leaf coverage with droplets is higher if the working solution is strongly atomized so important may be the overall functioning of the sprayer.

Additional significant factor affecting the efficiency of foliar nutrition is leaf turgor which depends on soil humidity, light intensity, air temperature, and humidity, as well as wind (Sect. 4.2.5). Therefore, it is recommended to conduct foliar nutrition in cloudy days with a moderate air temperature and high relative humidity.

The optimal solution would be to perform foliar application few days after rain or field irrigation – particularly during dry and hot season. A significant improvement of leaf turgor as well as metabolic activity of plant may be thus obtained. When it is not possible to irrigate the plantation, an alternative may be to spray plants with pure water 1 or 2 days before foliar nutrition treatment (depending on the current soil humidity and weather conditions). It is advised to use 50–100% higher amounts of water for spraying and further foliar nutrition than the standard applied. This strategy is recommended for foliar biostimulation of annual plants (Smoleń and Ledwożyw-Smoleń 2011). It can be assumed that in the case of citrus plants, such proceedings may be equally effective. Application of pure water temporarily increases leaf turgor and, consequently, activates metabolic processes. In a certain sense, this treatment is a kind of plant "preparation" for a proper foliar application of mineral nutrients and/or biostimulators in unfavorable environmental conditions caused by drought and high temperatures. Positive impacts of such a strategy observed in practice (Smoleń and Ledwożyw-Smoleń 2011) are most likely related to a "memory effect" in plants (Goh et al. 2003).

Acknowledgment I would like to sincerely thank my wife Iwona for her invaluable help with gathering the necessary literature and the preparation of English version of the manuscript.

References

Achilea O, Soffer Y, Raber D et al (2002) Bonus-NPK – highly concentrated, enriched potassium nitrate, an optimal booster for yield and quality of citrus fruits. Acta Hort 594:461–466

Barczak B, Cwojdziński W (1998) Content of macroelements in cucumber fruits sprayed with pesticides and the lupine extract. Sci Papers Agric Univ Poznań 27:11–18 (in Polish with English abstract)

Barczak B, Majcherczak E, Kozera W (2007) Effect of lupine extract and nitrogen fertilization on yield quality of celeriac. Veg Crop Res Bull 66:59–68. doi:10.2478/v10032-007-0008-6

Basak A, Mikos-Bielak M (2008) The use of some biostimulators on apple and pear trees. In: Sadowski A (ed) Biostimulators in modern agriculture - fruit crops. Editorial House Wieś Jutra, Warsaw

Beck SL, Dunlop RW, Fossey A (2003) Stomatal length and frequency as a measure of ploidy level in black wattle, *Acacia mearnsii* (de Wild). Bot J Linn Soc 141(2):177–181. doi:10.1046/j.1095-8339.2003.00132.x

Berbeć S, Andruszczak S, Łusiak J et al (2003) Effect of foliar application of Atonik and Ekolist on yield and quality of common thyme. Acta Agrophys 85:305–311 (in Polish with English abstract)

Biesiada A, Kołota E (1998) Application of multi-element fertilizers for foliar nutrition of celeriac. Fol Univ Agric Stein 72:29–33 (in Polish)

Bondada BR, Petracek JP et al (2006) Cuticular penetration characteristics of urea in citrus leaves. J Hort Sci Biotechnol 81(2):219–224

Borkowska B (1997) Cytokinins. In: Jankiewicz S (ed) The regulators of plant growth and development: property and function. PWN, Warsaw (in Polish)

Breś W (2009) Estimation of nutrient losses from open fertigation systems to soil during horticultural plant cultivation. Pol J Environ Stud 18(3):341–345

Breś W (2010) Influence of soilless cultures on soil environment. Ecol Chem Eng A 17(9):1069–1076

Campbell WH (1999) Nitrate reductase structure, function and regulation: bridging the gap between biochemistry and physiology. Annu Rev Plant Physiol Plant Mol Biol 50:277–303. doi:10.1146/annurev.arplant.50.1.277

Caronia A, Gugliuzza G, Inglese P (2010) Influence of L-proline on *Citrus sinensis* (L.) ['New Hall' and 'Tarocco Scirè'] fruit quality. Acta Hort 884:423–426

Ciha AJ, Brun W (1975) Stomatal size and frequency in soybeans. Crop Sci 15:309–313

Coupe SA, Palmer BG, Lake JA et al (2006) Systemic signaling of environmental cues in *Arabidopsis* leaves. J Exp Bot 57(2):329–341. doi:10.1093/jxb/erj033

Crisosto CH, Johnson RS, DeJong T (1997) Orchard factors affecting post-harvest stone fruit quality. Hort Sci 32(5):820–823

Crisosto CH, Day KR, Johnson RS et al (2000) Influence of in-season foliar calcium sprays on fruit quality and surface discoloration incidence of peach and nectarines. J Am Pomol Soc 54(3):118–122

Cwojdziński W, Barczak B, Polocyn J (1996) The effect of lupine extract on the decrease of nitrate nitrogen in selected vegetables. Zesz Prob Post Nauk Rol 440:45–50 (in Polish with English abstract)

Dayod M, Tyerman SD, Leigh RA et al (2010) Calcium storage in plants and the implications for calcium biofortification. Protoplasma 247(3–4):215–231. doi:10.1007/s00709-010-0182-0

Dluzniewska P, Gessler A, Kopriva S et al (2006) Exogenous supply of glutamine and active cytokinin to the roots reduces NO_3^- uptake rates in poplar. Plant Cell Environ 29:1284–1297. doi:10.1111/j.1365-3040.2006.01507.x

Durand N, Briand X, Meyer C (2003) The effect of marine bioactive substances (N PRO) and exogenous cytokinins on nitrate reductase activity in *Arabidopsis thaliana*. Physiol Plant 119(4):489–493. doi:10.1046/j.1399-3054.2003.00207.x

Eichert T, Burkhardt J, Goldbach HE (2002) Some factors controlling stomatal uptake. Acta Hort 594:85–90

Elmer PAG, Spiersa TM, Wood PN (2007) Effects of pre-harvest foliar calcium sprays on fruit calcium levels and brown rot of peaches. Crop Prot 26(1):11–18. doi:10.1016/j.cropro.2006.03.011

Fageria NK, Barbosa Filho MP, Moreira A et al (2009) Foliar fertilization of crop plants. J Plant Nutr 32:1044–1064. doi:10.1080/01904160902872826

Fariduddin Q, Hayat S, Ahmad A (2003) Salicylic acid influences net photosynthetic rate, carboxylation efficiency, nitrate reductase activity, and seed yield in *Brassica juncea*. Photosynthetica 41:281–284. doi:10.1023/B:PHOT.0000011962.05991.6c

Fernández V, Del Río V, Abadía J et al (2006) Foliar iron fertilization of peach (*Prunus persica* (L.) Batsch): effects of iron compounds, surfactants and other adjuvants. Plant Soil 289:239–252. doi:10.1007/s11104-006-9132-1

Forde BG (2000) Nitrate transporters in plants: structure, function and regulation. Biochim Biophys Acta 1465:219–235. doi:10.1016/S0005-2736(00), 00140-1

Forde BG, Cole JA (2003) Nitrate finds a place in the sun. Plant Physiol 131(3):395–400. doi:10.1104/pp. 016139

Fornes F, Sánchez-Perales M, Guardiola JL (1995) Effect of a seaweed extract on citrus fruit maturation. Acta Hort 379:75–82

Franke W (1961) Ectodesmata and foliar absorption. Am J Bot 48(8):683–691

Franke W (1967) Mechanisms of foliar penetration of solutions. Annu Rev Plant Physiol 18:281–300

Franke W (1971) The entry of residues into plants via ectodesmata. Resid Rev 38:81–115

Franke W (1986) The basis of foliar absorption of fertilizers with special regard to the mechanisms. In: Alexander A (ed) Foliar fertilization. Martinus Nijhoff Publishers, Dordrecht

Fritz A (1978) Foliar fertilization – a technique for improved crop production. Acta Hort 84:43–56

Gansel X, Muños S, Tillard P et al (2001) Differential regulation of the NO_3^- and NH_4^+ transporter genes *AtNrt2.1* and *AtAmt1.1* in *Arabidopsis*: relation with long-distance and local controls by N status of the plant. Plant J 26(2):143–155. doi:10.1046/j.1365-313x.2001.01016.x

Gechev T, Gadjev I, Van Breusegem F et al (2002) Hydrogen peroxide protects tobacco from oxidative stress by inducing a set of antioxidant enzymes. Cell Mol Life Sci 59(4):708–714. doi:10.1007/s00018-002-8459-x

Gheibi MN, Kholdebarin B, Ghanati F et al (2009) Urease activity in maize (*Zea maize* L. CV. 704) as affected by nickel and nitrogen sources. Iran J Sci Technol Trans A 33(A4):299–307

Goh CH, Nam HG, Park YS (2003) Stress memory in plants: a negative regulation of stomatal response and transient induction of *rd22* gene to light in abscisic acid-entrained *Arabidopsis* plants. Plant J 36(2):240–255. doi:10.1046/j.1365-313X.2003.01872.x

Graham JH, Gottwald TR, Riley TD et al (1992) Penetration through leaf stomata and growth of strains of *Xanthomonas campestris* in citrus cultivars varying in susceptibility to bacterial diseases. Phytopathology 82(11):1319–1325

Hirschi KD (2009) Nutrient biofortification of food crops. Annu Rev Nutr 29:401–421. doi:10.1146/annurev-nutr-080508-141143

Holloway PJ (1971) The chemical and physical characteristics of leaf surfaces. In: Prece TF, Dickinson CH (eds) Ecology of leaf surface micro-organisms. N.Y. Academic Press, London

Howitt SM, Udvardi MK (2000) Structure, function and regulation of ammonium transporters in plants. Biochim Biophys Acta 1465:152–170. doi:10.1016/S0005-2736(00)00136-X

Jabłoński K (2002) The advantages of potato foliar fertilization with microelements. Ziem Pol 2:22–29 (in Polish)

Jain A, Srivastava HS (1981) Effect of salicylic acid on nitrate reductase activity in maize seedlings. Physiol Plant 51:339–342. doi:10.1111/j.1399-3054.1981.tb05565.x

Jodełka J, Jankowski K, Ciepiela GA (2003) Estimation of nitrogen applied in foliar form used in system of meadow fertilization. Acta Agrophys 85:339–345

Kannan S (2010) Foliar fertilization for sustainable crop production: genetic engineering, biofertilisation, soil quality and organic farming. Sustain Agric Rev 4:371–402. doi:10.1007/978-90-481-8741-6_13

Klessig DF, Malamy J (1994) The salicylic acid signal in plants. Plant Mol Biol 26:1439–1458. doi:10.1007/BF00016484

Kolthoff IM, Sandell EB, Bruckenstein S (1969) Quantitative chemical analysis. Macmillan, London

Komosa A (1990) Influence of selected chemical properties of solutions and nutritional status of plants on the efficiency of foliar nutrition of greenhouse tomato. Dissertation, Scientific Papers of the Agricultural University of Poznan, p 210 (in Polish)

Koo RCJ (1988) Response of citrus to seaweed-based nutrient sprays. Proc Fla State Hort Sci 101:26–28

Kováčik P (1999) Effect of nitrogenous nutrition and sucrose foliar application on yield parameters of radish. Zahrad Hort Sci (Prague) 26:97–102 (in Czech, with English abstract)

Krogmeier MJ, McCarty GW, Shogren DR et al (1991) Effect of nickel deficiency in soybeans on the phytotoxicity of foliar applied urea. Plant Soil 135(2):283–286

Kuepper G (2003) Foliar fertilization. ATTRA. https://attra.ncat.org/attra-pub/summaries/summary.php?pub=286

Kujawski P (2005) Mineral nutrient uptake depends on climatic conditions. Hasło Ogrod 3:108–112 (in Polish)

Liu Q, Zhu Y, Tao H et al (2006) Damage of PS II during senescence of *Spirodela polyrrhiza* explants under long-day conditions and its prevention by 6-benzyladenine. J Plant Res 119(2):145–152. doi:10.1007/s10265-006-0259-1

Machado MA, Astua-Monge G, Targon MLPN et al (2005) Citrus genomics in Brazil (CitEST): experiences and future perspectives. In: Plant & animal genomes XIII conference, Town & Country Convention Center San Diego, January 15–19, 2005. http://www.intl-pag.org/13/abstracts/PAG13_W066.html

Magdziak R, Kołodziej B (2003) Effect of foliar fertilizers on the growth and development of American ginseng (*Panax quinquefolium* L.) in consecutive years of vegetation. Acta Agrophys 85:319–329 (in Polish with English abstract)

Malavolta E, Moraes MF (2007) Nickel – from toxic to essential nutrient. Better Crops 91(3):26–27

Malavolta E, Leão HC, de Oliveira SC et al (2006) Repartição de nutrientes nas flores, folhas e ramos da laranjeira cultivar Natal. Rev Bras Frutic 28(3):506–511 (in Portuguese with English abstract)

Malhotra SS, Hocking D (1976) Biochemical and cytological effects of sulphur dioxide on plant metabolism. New Phytol 76:227–237

Marschner H (1995) Mineral nutrition of higher plants, 2nd edn. Academic, London

Masclaux-Daubresse C, Valadier MH, Carrayol E et al (2002) Diurnal changes in the expression of glutamate dehydrogenase and nitrate reductase are involved in the C/N balance of tobacco leaves. Plant Cell Environ 25:1451–1462. doi:10.1046/j.1365-3040.2002.00925.x

McFarlane JC, Berry WI (1974) Cation penetration through isolated leaf cuticles. Plant Physiol 53:723–727

Meinder H, Mansfield TA (1968) Physiology of stomata. McGraw Hill, New York

Michałojć Z, Szewczuk C (2003) Theoretical aspects of foliar nutrition. Acta Agrophys 85:9–17 (in Polish with English abstract)

Miguel S, Gutiérrez RM, Larqué SA (2002) Low concentrations of salicylic acid increase nitrate accumulation in roots of *Pinus patula*. Phyton Int J Exp Bot 51:79–82

Mohammady DS (2011) Physiological characters associated with water-stress tolerance under pre anthesis water-stress conditions in wheat. Available online. http://www.shigen.nig.ac.jp/ewis/article/html/17/article.html

Mugnai S, Azzarello E, Pandolfi C et al (2008) Enhancement of ammonium and potassium root influxes by the application of marine bioactive substances positively affects *Vitis vinifera* plant growth. J Appl Physiol 20(2):177–182. doi:10.1007/s10811-007-9203-6

Nicoulaud BAL, Bloom AJ (1998) Nickel supplements improve growth when foliar urea is the sole nitrogen source for tomato. J Am Soc Hort Sci 123(4):556–559

Norris RF, Bukovac MJ (1972) Effect of pH on penetration of naphthalene acid and naphthaleneacetamide through pear leaf cuticle. Plant Physiol 49(4):615–618

Orsel M, Krapp A, Daniel-Vedele F (2002) Analysis of the *NRT2* nitrate transporter family in *Arabidopsis*: Structure and gene expression. Plant Physiol 129:886–896. doi:10.1104/pp. 005280

Piskornik Z (1994) Plant physiology for faculties of horticulture part I, 2nd edn. Wydawnictwo AR, Kraków (in Polish)

Pompelli MF, Martins SCV, Celin EF et al (2010) What is the influence of ordinary epidermal cells and stomata on the leaf plasticity of coffee plants grown under full-sun and shady conditions? Braz J Biol 70(4):1083–1088. doi:dx.doi.org/10.1590/S1519-69842010000500025

Rożek S, Sady W, Kasprzyk A (2000) The influence of foliar nutrition on quantity and quality of carrot field. Sci Pap Agric Univ Cracow 364:159–162 (in Polish)

Rydz A (2001) The effect of foliar nutrition urea on yield quality of broccoli cv. Lord F_1. Veg Crops Res Bull 54:61–64

Santana LM Gabriel R, Morales-Payan JP et al (2006) Effects of biostimulants on nursery growth of orange budded on volkamer lemon and swingle citrumelo. In: Proceedings of the 33rd Annual Meeting of the Plant Growth Regulation Society of America, Quebec City, July 2006. Plant Growth Regulation Society of America, LaGrange, pp 217–219

Schletz R (2008) Stomata densities of developing and mature leaves of geraniums. ESSAI 6(1):129–132. doi: http://dc.cod.edu/essai/vol6/iss1/42

Schönherr J (1976) Water permeability of isolated cuticular membranes: the effect of pH and cations on diffusion, hydrodynamic permeability and size of polar pores in the cutin matrix. Planta 128(2):113–126. doi:10.1007/BF00390312

Schönherr J (2001) Cuticular penetration of calcium salts: effect of humidity, anions and adjuvants. J Plant Nutr 164:225–231. doi:10.1002/1522-2624(200104)

Schönherr J (2002) Foliar nutrition using inorganic salts: laws of cuticular penetration. Acta Hort 594:77–84

Schönherr J (2006) Characterization of aqueous pores in plant cuticles and permeation of ionic solutes. J Exp Bot 57(11):2471–2491. doi:10.1093/jxb/erj217

Schönherr J, Luber M (2001) Cuticular penetration of potassium salts: effect of humidity, anions and temperature. Plant Soil 236(1): 117–122. doi:10.1023/A:1011976727078

Shakya R, Sturm A (1998) Characterization of source- and sink-specific sucrose/H+ symporters from carrot. Plant Physiol 118:1473–1480. doi:10.1104/pp. 118.4.1473

Smoleń S, Ledwożyw-Smoleń I (2011) Foliar nutrition of vegetables. Hasło Ogrod 3:32–37 (in Polish)

Smoleń S, Sady W (2008a) Foliar nutrition of carrot – the optimal time and products? Part I. Warzywa 5:22–28 (in Polish)

Smoleń S, Sady W (2008b) Foliar nutrition of carrot – the optimal time and products? Part II. Warzywa 6:32–35 (in Polish)

Smoleń S, Sady W (2008c) Effect of various nitrogen fertilization and foliar nutrition regimes on carrot (*Daucus carota* L.) yield. J Hort Sci Biotechnol 83(4):427–434

Smoleń and Sady (2012) The effect of foliar application of urea, molybdenum, benzyladenine, sucrose and salicylic acid on yield, nitrogen metabolism of radish plants and quality of edible roots. J Plant Nutr (in print). doi:10.1080/01904167.2012.676125

Smoleń S, Szura A (2008a) Compounds improving plant growth (part I) Growth regulators and resistance stimulators. Hasło Ogrod 5:12–16 (in Polish)

Smoleń S, Szura A (2008b) Compounds improving plant growth (part II). Growth stimulators and modern fertilizers. Hasło Ogrod 6:12–14 (in Polish)

Smoleń S, Sady W, Suppl (2009) The influence of nitrogen fertilization and Pentakeep V application on contents of nitrates in carrot. Acta Hort et Regiotec 12:221–223

Smoleń S, Sady W, Wojciechowska R (2010) The effect of foliar nutrition with nitrogen, molybdenum, sucrose and benzyladenine on the nitrogen metabolism in carrot plants. Veg Crop Res Bull 72:83–92. doi:10.2478/v10032-010-0008-9

Spiegel-Roy P, Goldschmidt EE (1996) Biology of citrus. Cambridge University Press, New York

Srivastava AK, Singh S (2003) Foliar fertilization in citrus – a review. Agric Rev 24(4):250–264

Starck JR (1997) Cultivation and fertilization of horticultural plants, 3rd edn. Państwowe Wydawnictwo Rolnicze i Leśne, Warszaw

Starck Z (2003) Transport and distribution of nutrient compounds in plants. SGGW Publisher, Warsaw (in Polish)

Sugier D (2003) Influence of the method plantation setting and foliage feeding on root yield and inulin content of common dandelion (*Taraxacum officinale* WEB). Acta Agrophys 85:331–337 (in Polish with English abstract)

Szewczuk C, Michałojć Z (2003a) Preface – foliar fertilization of plants. Acta Agrophys 85:7–8 (in Polish)

Szewczuk C, Michałojć Z (2003b) Practical aspects of foliar fertilization. Acta Agrophys 85:19–29 (in Polish with English abstract)

Szewczuk C, Sugier D, Studzińska-Jaksim P (2003) Influence of foliar application of fertilization on hop cones yield and quality. Acta Agrophys 85:297–304

Tischner R (2000) Nitrate uptake and reduction in higher and lower plants. Plant Cell Environ 23:1005–1024. doi:10.1046/j.1365-3040.2000.00595.x

Weaver RJ (1972) Plant growth substances in agriculture. WH Freeman and Company, San Francisco

White PJ, Broadley MR (2009) Biofortification of crops with seven mineral elements often lacking in human diets – iron, zinc, copper, calcium, magnesium, selenium and iodine. New Phytol 182(1):49–84

Wierzbińska J (2009) Effect of foliar application of nutrients combined with phytohormones and organic acids on yield quantity and biological quality of carrot (*Daucus carota* L.) yield. Dissertation. University of Agriculture in Krakow (in Polish with English abstract)

Williams CMJ, Maier NA, Bartlett L (2004) Effect of molybdenum foliar sprays on yield, berry size, seed formation, and petiolar nutrient composition of "Merlot" grapevines pages. J Plant Nutr 27:1891–1916. doi:10.1081/PLN-200030023

Wittwer SH, Teubner FG (1959) Foliar absorption of mineral nutrients. Annu Rev Plant Physiol 10:13–27

Wojciechowska R, Rożek S, Rydz A (2005) Broccoli yield and its quality in spring growing cycle as dependent on nitrogen fertilization. Folia Hort 17(2):141–152

Wójcik P (2004) Uptake of mineral nutrients from foliar fertilization. J Fruit Ornam Plant Res (special ed) 12:201–218

Xie S, Zhang Q (2004) Kinetics of uptake and export of foliar-applied radio-labeled phosphorus by leaf and fruit rind of Satsuma mandarin during fruit development. J Plant Nutr 27(2):223–237. doi:10.1081/PLN-120027651

Yamada Y, Wittwer SH, Bukovac MJ (1964) Penetration of ions through isolated cuticules. Plant Physiol 39(1):28–32

Yaronskaya E, Vershilovskaya I, Poers Y et al (2006) Cytokinin effects on tetrapyrrole biosynthesis and photosynthetic activity in barley seedlings. Planta 224:700–709. doi:10.1007/s00425-006-0249-5

Yu X, Sukumaran S, Marton L (1998) Differential expression of the *Arabidopsis Nia*1 and *Nia*2 genes. Plant Physiol 116:1091–1096. doi:10.1104/pp. 116.3.1091

Zamarreño A, Cantera RG, Garcia-Mina JM (1997) Extraction and determination of glycinebetaine in liquid fertilizers. J Agric Food Chem 45(3):774–776. doi:10.1021/jf960342h

Zekri M, England G (2010) Foliar feeding and SAR for citrus trees. University of Florida, http://citrusagents.ifas.ufl.edu/archived_presentations/fall_prg2010/Citrus%20Agents%20Web%20Site%20-%20Nutritional.pdf

5 Leaf Analysis in Citrus: Interpretation Tools

Regina Menino

Abstract

Occurrence of nutrient constraints is as old as history of citrus cultivation. Identifying multiple nutrient constraints has always perplexed the nutritionists due to lack of consistency in response to fertilization when executed in field. Such a discrepancy in diagnostic interpretation has emerged on account of a variety of interpretation tools made under usage. For example, when a leaf analysis dataset is subjected to different interpretation tools, highly overlapping diagnoses are obtained. Analysis of different tools used in leaf analysis data interpretation further warranted the necessity of working out the cultivar-specific nutrient diagnostic norms. Large number of interpretation tools have been put forward to interpret the leaf analysis data. Late DRIS (Diagnosis and Recommendation Integrated System), initially developed for rubber crop, found its great utility to many of the annual and perennial crops, citrus being one of them. Work done in India with reference to different citrus cultivars suggested a strong cultivar-specific dependence of the diagnoses that agreed with field diagnoses. Use of DRIS-based diagnoses provided replicating outcome under diverse soil types. The detailed account of different leaf analysis interpretation tools is further discussed.

Keywords

Leaf analysis • Interpretation tools • Critical nutrient concept • Nutrient concentration range • Nutrient balance • Crop logging • Boundary line concept • DRIS

5.1 Background Information

Perennial crops are quite different from annual crops in their nutritional requirement due to their plant size, density, rate of growth and rooting pattern, and phenomenon of bud differentiation and its relationship with the yield during the following season/year. Determination of the nutritional needs of fruit trees must be made prior to the renewed growth or the determination of potential yield. To ensure high economic productivity and to sustain the available soil nutrient status at a desirable level, correct doses of manures, biofertilizers, and chemical fertilizers must be applied, based on the use of reliable diagnostic tools. Considering energy, economy, and environment, it is imperative that manures, biofertilizers, and chemical fertilizers be used efficiently. The best diagnostic tool is one that recommends nutrient application in a direct economic response of the fruit crop. Diagnostic tools are designed to avoid nutrient shortage or excess, and if used properly, no decrease in fruit production or quality should occur. Leaf analysis seems to be the best method for identifying the need for application of nutrients.

The interpretation of leaf analysis is based on the premise that there is a significant biological relationship between the elemental content in leaf, plant growth, and fruit yield with a purpose to predict fertilizer requirement depending upon site characteristics. This is popularly known as 'critical value

R. Menino (✉)
Unidade de Investigação de Ambiente e Recursos Naturais
Instituto Nacional dos Recursos Biológicos, Quinta do Marquês, 2784-505, Oeiras, Portugal
e-mail: reginamenino@hotmail.com; reginamenino@gmail.com

A.K. Srivastava (ed.), *Advances in Citrus Nutrition*,
DOI 10.1007/978-94-007-4171-3_5, © Springer Science+Business Media B.V. 2012

approach' widely applied in citrus orchards with considerable success. These relationships normally reflect a sigmoidal response curve on which two critical values can be identified. These values for each nutrient are the value below and above which plant performance is reduced (Terblanche and Du Plessis 1992).

5.2 History of Leaf Analysis

Justus von Liebig, the German chemist (1803–1873), was the first to have an opinion that growth of plants was proportional to the amount of mineral substance available in the fertilizer and developed the "law of minimum" which states that the "growth of plants is limited by the nutrient present in the smallest quantity." As it can be seen in *The Book of the Rothamsted Experiments (1905)*, Hall (2009) envisaged plant analysis as a method for estimating the nutrient content of soils.

Leaf analysis is based on the premise that "the plant behaviour is related to concentrations of essential minerals in leaf tissue" (Smith 1966). Leaf analysis, as a method for assessing the nutrient requirements of crops, is based on the assumption that "within certain limits, there is a positive relation between doses of the nutrient supplied, leaf nutrient content and yield and/or quality" (Smith op. cit.). Nutrient supply in 1 year may have a major effect on both fruit tree nutrition and crop production in subsequent years as the plant responds to direct and residual soil fertility. The application of the same dose year after year is thus irrelevant (Bhargava and Chadha 1993). The crop, through leaf analysis, informs the growers that the power of his soil to supply the nutrient has not kept pace with the nutrient requirements of the crop.

The diagnosis of the nutritional status of citrus, based on the chemical analysis of some of its organs, has been generalized since the middle of the past century. Among the first and most complete works published on this subject, Chapman (1961) refers those from Thorpe, in 1868, from Wolf, in 1871–1880, from Ricciardi, in 1880, from Boschi, in 1895, and from Olivieri and Guerrieri, in 1895. Still, according to Chapman (op. cit.), in 1931, Barnette and his collaborators published the results from the chemical analysis of adult grapefruit trees, not only the chemical composition of the different parts of the trees but also its weight.

Based on the fact that the leaves have an active "assimilatory" function, being the "synthetic laboratory" that controls the plant nutrition, in the first half of the past century, Thomas (1945) referred that through the leaf analysis, it is possible to control the nutritional status of the plants. In face of that, the majority of the work done in this subject has incurred mainly in the leaves. However, more recently, several authors defend the analysis of other plant organs, such as flowers (Pestana et al. 2001) or fruits (Lacertosa et al. 2001), but there are not yet reference values or even consensus about the preference for one of those organs as the most adequate base for the management of the fertilization in the following cultural cycle or the correction of nutritional unbalances in the same year.

According to Bould (1984), leaf analysis is based on the following four assumptions: (1) leaf is the main site of plant metabolism; (2). changes in nutrient supply are reflected on the composition of the index tissue such as leaf or its petiole; (3) changes are more pronounced at certain stages of development than at others; and (4) concentration of the nutrients in the leaf at the specific growth stage is related to the performance of the crop.

For some authors, with the exception of N, that is slightly different for the various commercial species (Embleton et al. 1978), the chemical composition of the citrus is identical (Hanlon et al. 1995; Davies and Albrigo 1998). However, other authors, as Legaz et al. (1995) and Dias et al. (2002), indicate different reference values for N, P, and K for the various citrus species. Besides the species, the concentration of the nutrients in the leaves is dependent on, among other factors, the age, the type, and the localization of the leaves (Smith 1966; Embleton et al. 1978; Legaz et al. 1995; Davies and Albrigo 1998; Carranca 1999).

The concentration of the different nutrients is more stable in 4–7-month-old leaves and that is why this is the most adequate period referred for leaf sampling (Smith 1966; Embleton et al. 1978; Hanlon et al. 1995; Legaz et al. 1995; Davies and Albrigo 1998; Carranca 1999; Correia 2000). On the other hand, the presence of flowers or small growing fruits, in a branch, diminishes significantly the leaf concentration of the nutrients, as they are translocated to those organs, once the leaves, besides its photosynthetic function, also act as a storage organ (Smith 1966; Embleton et al. 1978; Legaz et al. 1995; Davies and Albrigo 1998; Carranca 1999). In fact, the nutrient concentration in nonfruiting shoot leaves seems to be the best indicator of the nutritional status of the tree, better than fruiting shoot leaves, since those are the branches that will carry the flowers and subsequently the fruits on the following year (Embleton et al. 1978; Legaz et al. 1995; Carranca 1999). According to Davies and Albrigo (1998), with the exception of South Africa (where the leaves for chemical analysis are sampled from fruiting shoots), in most part of the citrus production regions of the world, leaf samples are collected in nonfruiting shoots from spring flushes at 4–7 month old.

For foliar chemical analysis, Swietlik (1996), Alva and Tucker (1999), Carranca (1999), and Dias et al. (2002) recommend, per 2–4 ha, the sampling of 8–10 nonfruiting shoot leaves per tree, collected from the four quadrants of the canopy, in 15–20 trees randomly chosen and representative of the general status of the orchard. The diagnosis of the

Table 5.1 Reference values for leaf nutrients concentration, in nonfruiting shoot leaves, from the spring flushes (5–7 month old), for mature 'Valencia Late' orange trees

Nutrient	Deficient	Low	Optimum	High	Excess
N (g kg^{-1})	<22	22–23	24–26	27–28	>28
P (g kg^{-1})	<0.9	0.9–1.1	1.2–1.6	1.7–2.9	>3
K (g kg^{-1})	<4	4.0–6.9	7–10.9	11–20	>23?
Ca (g kg^{-1})	<16	16–29	30–55	56–69	>70?
Mg (g kg^{-1})	<1.6	1.6–2.5	2.6–6.0	7.0–11	>12?
B (mg kg^{-1})	<21	21–30	31–100	101–260	>260
Cu (mg kg^{-1})	<3.6	3.6–4.9	5–16	17–22?	>22?
Fe (mg kg^{-1})	<36	36–59	60–120	130–200?	>250?
Mn (mg kg^{-1})	<16	16–24	25–200	300–500?	>1,000?
Mo (mg kg^{-1})	<0.06	0.06–0.09	0.1–3.0	4–100	>100?
Zn (mg kg^{-1})	<16	16–24	25–100	110–200	>300

Adapted from Embleton et al. (1978)

Table 5.2 Reference values for leaf nutrients concentration, in nonfruiting shoot leaves, from the spring flushes (4–6 month old), for mature citrus trees

Nutrient	Deficient	Low	Optimum	High	Excess
N (g kg^{-1})	<22	22–24	25–27	28–30	>30
P (g kg^{-1})	<0.9	0.9–1.1	1.2–1.6	1.7–3.0	>3.0
K (g kg^{-1})	<7	7–11	12–17	18–24	>24
Ca (g kg^{-1})	<15	15–29	30–49	50–70	>70
Mg (g kg^{-1})	<2.0	2.0–2.9	3.0–4.9	5.0–7.0	>7.0
B (mg kg^{-1})	<20	20–35	36–100	101–200	>200
Cu (mg kg^{-1})	<3	3–4	5–16	17–20	>20
Fe (mg kg^{-1})	<35	35–59	60–120	121–200	>200
Mn (mg kg^{-1})	<17	17–24	25–100	101–300	>300
Mo (mg kg^{-1})	<0.05	0.06–0.09	0.1–1.0	2.0–5.0	>5.0
Zn (mg kg^{-1})	<17	17–24	25–100	101–300	>300

Adapted from Hanlon et al. (1995)

nutritional status of the trees must be made comparing the results from leaf chemical analysis with the reference values. However, these values must be used as a general indication, when compared with the results from the chemical analysis of citrus leaf samples, once, as already referred, may exist differences among rootstocks, cultivars, tree age, and phase of the growth cycle. The results of leaf analysis in young trees, for example, must be scrutinized very carefully, since the concentration of Fe and Zn is lower in these leaves (Swietlik 1996) and the content of N and K is higher, in relation to full production trees (Smith 1966; Swietlik 1996).

Most of the reference values of leaf analysis, existing on the bibliography, were established based on the relation between the concentration of the several leaf elements and the development of the plant, standardized through experimental studies on fertilization adequately delineated (Terblanche and Du Plessis 1992; Hanlon et al. 1995) for orchards at full production. Through the analysis of the results obtained in an experimental study with young Navel orange trees, Thompson et al. (2003) proposed that N concentration in young tree leaves, when adequately supplied with this element, should be around 28 g N kg^{-1} dry weight instead of 25 g N kg^{-1} dry weight, which is the value established by Embleton et al. (1978) as the optimum for adult trees.

In a long-term study (16 years) carried out in Portugal including 38 orange orchards of different varieties, aiming the establishment of critical levels for the several nutrients contained in the leaves, based on a correlation with some fruit quality indexes, Fragoso et al. (1990) observed that in a general way, the values were within the intervals adopted by Embleton et al. (1978) for mature 'Valencia Late' orange trees (Table 5.1), concluding that will be adequate to adopt them as reference values for orange trees in plain production.

More recently, Hanlon et al. (1995) published an actualization of the reference values for leaf content of nonfruiting shoot leaves for mature citrus trees (Table 5.2). These values were confirmed in the work of Alva and Tucker (1999) and are generally those adopted in Florida. In a recent publication (Kallsen 2002), where reference values of nutrients for nonfruiting shoot leaves in adult orange trees are available, it is possible to endorse the critical values (Table 5.1) that were considered uncertain by Embleton et al. (1978). The comparison between Tables 5.1 and 5.2 reveals some differences, being the greater for K, for which the values considered by Embleton et al. (1978), in any of the ranges, are lower than those considered by Hanlon et al. (1995). In relation to Mg, on the contrary, the values considered by Embleton et al. (1978) for the ranges optimum, high, and excessive are higher than those considered by Hanlon et al. (1995). Maximum limits for the elements that Embleton et al. (1978) considered uncertain (Table 5.1) were already established (Table 5.2), with a clear difference in case of the Mo.

In Portugal (Fragoso et al. 1990), as well as in Spain (Legaz et al. 1995; Agusti 2000), the tables from Embleton et al. (1978), based on values obtained in California, are the most adequate, probably due to the climatic identity. In any case, these values were established for full production trees, and there are no reference values for young nonbearing trees when the chemical composition of the plant is different from the mature tree (Smith 1966; Swietlik 1996).

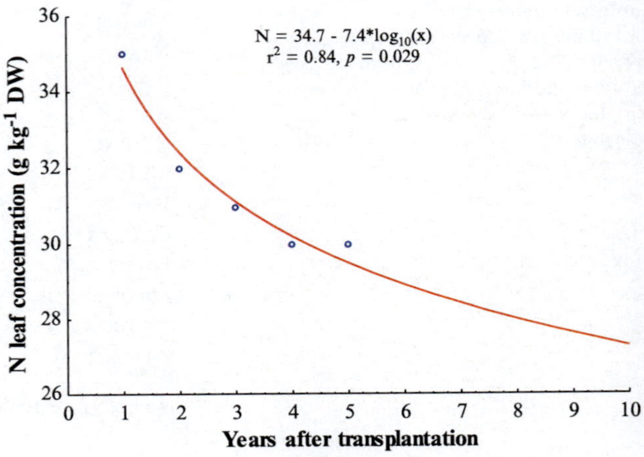

Fig. 5.1 Logarithmic adjustment for the mean N leaf concentration in 'Lane Late' orange trees, in the first 5 years after planting, with the estimation of values for the following 5 years

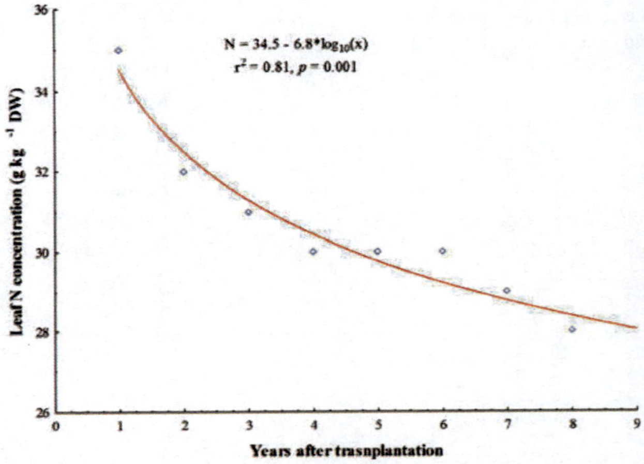

Fig. 5.2 Logarithmic adjustment for the mean N leaf concentration in 'Lane Late' orange trees, in the first 8 years after planting

5.2.1 Reference Values for N Confirmed by Foliar Analysis, from Planting till Full Production

As it was observed by Menino (2005) in a field experiment with 'Lane Late' oranges planted in a sandy soil located at Algarve (South of Portugal), during the first 4 years after transplanting the N leaf concentration, although increasing with increasing N rates, decreased from the first till the fourth year, suggesting that the optimum concentration values for leaf N content, for young nonbearing trees, with a vigorous growth, should not be defined, in a general way, by a unique value for all of the years (28 $gN\ kg^{-1}$ dry weight, as suggested by Weinert et al. (2002)) but by a series of decreasing values till the optimum recommended for mature trees (25 $gN\ kg^{-1}$ dry weight, according to Embleton et al. (1978)). Assuming that the mean values of leaf N concentration for all treatments correspond to the adequate nutritional status of the trees, the optimum leaf N concentration obtained in the experiment carried out by Menino (2005), expressed as $gN\ kg^{-1}$ dry weight, could be evaluated in accordance with the following logarithmic adjustment:

$$N = 34.7 - 7.4 \times \log_{10}(x), \text{ with an } r^2 = 0.84 \text{ for } p \le 0.029,$$

as it is shown in Fig. 5.1. According to this logarithmic adjustment, the reference values for leaf N concentration, during the first years in the field, would be the following:

34.7 $gN\ kg^{-1}$ dry weight, for the first year; 32.5 $gN\ kg^{-1}$ dry weight, for the second year; 31.2 $gN\ kg^{-1}$ dry weight, for the third year; 30.2 $gN\ kg^{-1}$ dry weight, for the fourth year; and 29.5 $gN\ kg^{-1}$ dry weight, for the fifth year.

Nevertheless, according to the abovementioned adjustment (Fig. 5.1), the critical value suggested by Weinert et al. (2002) for young trees (28 $gN\ kg^{-1}$ dry weight) would only be reached in the eighth year, which means that in the referred experimental conditions, the values were always much higher than the optimum. On the other hand, the optimum value referred by Embleton et al. (1978) for mature trees (between 24 and 26 $gN\ kg^{-1}$ dry weight) would only be reached in the 15th year.

In the prosecution of the mentioned study (Menino op. cit.) till the eighth year (Menino et al. 2008), the logarithmic adjustment (Fig. 5.2) maintains very close to the previous, with

$$N = 34.5 - 6.8 \times \log_{10}(x), \text{ with an } r^2 = 0.81 \text{ for } p \le 0.001,$$

as it is illustrated in Fig. 5.2. According to this logarithmic adjustment, the reference values for leaf N concentration, during the first years after transplanting in the field, would be the following:

34.5 gN kg^{-1} dry weight, for the first year; 32.5 gN kg^{-1} dry weight, for the second year; 31.3 gN kg^{-1} dry weight, for the third year; 30.4 gN kg^{-1} dry weight, for the fourth year; 29.7 gN kg^{-1} dry weight, for the fifth year; 29.2 gN kg^{-1} dry weight, for the sixth year; 28.8 gN kg^{-1} dry weight, for the seventh year; and 28.4 gN kg^{-1} dry weight, for the eighth year.

These values reveal a straight adjustment with those calculated earlier, for the first 5 years, and a perfect convergence for the reference values suggested by Weinert et al. (2002) for mature trees in full production.

5.3 Major Breakthroughs

A variety of interpretation tools (IT) have shown their application in leaf analysis of citrus. These are: critical nutrient concentration (Terblanche and Du Plessis 1992; Srivastava et al. 1999); nutrient concentration range (Parent and Dafir 1992); nutrient balance using factorial method (Cantarella et al. 1992), Kenworthy's balance index (Kenworthy 1973), Moller-Nielson balance concept (Moller Nielson and Friis-Nielson 1976b); crop logging (Abaev 1977); and boundary line concept (Walworth et al. 1986) – all suggesting only single value concentration and diagnosis and recommendation integrated system (DRIS) which considers the nutrient ratio (Walworth and Sumner 1987; Beverly 1987). The utility of these ITs in the past faced many limitations, especially in the context of alternative to identify the nutrient constraint at any growth stages during the season and, therefore, diagnoses found application only to a specified growth stage due to strong influence of leaf age.

Of different ITs, DRIS is claimed to have certain advantages over other conventionally used ITs (Malavolta et al. 1993; Li et al. 1999). The working premises of DRIS (Mourão Filho 2004) "are based on: (a) the ratios among nutrients are frequently better indicators of nutrient deficiencies than isolated concentration values; (b) some nutrient ratios are more important or significant than others; (c) maximum yields are only reached when important nutrient ratios are near the ideal or optimum values, which are obtained from high yielding-selected populations; (d) as a consequence of the stated in (c), the variance of an important nutrient ratio is smaller in a high yielding (reference population) than in a low yielding population, and" "the relations of significant nutrient ratios of high and low yielding populations can be used in the selection of significant nutrient ratios; (e) the DRIS indices can be calculated individually, for each nutrient, using the average nutrient ratio deviation obtained from the comparison with the optimum value of a given nutrient ratio, hence, as pointed by Jones (1981) and Walworth and Sumner (1987), the ideal value of the DRIS index for each nutrient should be zero." The efforts in the past have successfully established the DRIS norms for 'Valencia' orange in USA (Beverly et al. 1984; Wallace 1990), South Africa (Woods and Villiers 1992), Venezuela (Rodriguez et al. 1997), Brazil (Mourão Filho and Azevedo 2003); 'Verna' lemon in Spain (Cerda et al. 1995); 'Sicilian' lemon in Italy (Creste 1996) and 'Pera' sweet orange in Brazil (Creste and Grassi Filho 1998); acid lime (Varalakshmi and Bhargava 1998), 'Kinnow' mandarin (Hundal and Arora 2001), 'Nagpur' mandarin, 'Khasi' mandarin, and 'Mosambi' sweet orange in India (Srivastava et al. 2001; Srivastava and Singh 2003c, 2006).

Almost any conclusion can be drawn from the earlier attempts on the development of leaf nutrient diagnostics in countries like Argentina (Perez 1996), Australia (Jorgensen and Price 1978), Brazil (Quaggio et al. 1998), China (Koto et al. 1990), France (Marchal et al. 1978), India (Chahill et al. 1991; Srivastava et al. 1999; Srivastava and Singh 2002), Italy (Dettori et al. 1996), Japan (Terblanche and Du Plessis 1992), Turkey (Saatci and Mur 2000), Spain (Hellin et al. 1988), Costa Rica (Alvardo et al. 1994), USA (Chapman 1949; Koo et al. 1984; Swietlik 1996), employing a variety of diagnostic methods using different aged index leaves from fruiting as well as nonfruiting terminals. Such efforts have generated differential diagnostic capabilities in the absence of uniformity in guidelines used in diagnosing the nutrient constraints, e.g., a different set of optimum values are obtained when specific leaf analysis data are subject to contrasting ITs like multivariate quadratic regression analysis (MQRA) or diagnosis and recommendation integrated system (DRIS) using some commercial citrus cultivars of India (Table 5.3). DRIS-derived values were further very close to original values from high-performance elite orchards than values obtained from MQRA (Wallace 1990; Woods and Villiers 1992; Srivastava and Singh 2003c). Alves and Mourão Filho (2005) observed that conventional sufficient range approach (SRA) and DRIS were in agreement for nutritional diagnosis of K, while other nutrients like Cu, Mn, and Fe were diagnosed as deficient by DRIS and classified adequate to high by SRA in 'Valencia' sweet orange orchards on three different rootstocks in São Paulo, Brazil.

Arguments are often put forward to support the view that deficient, optimum, or excessive levels of nutrient concentration cannot be determined by means of absolute figures (critical levels) due to great deal of variation in vegetative activity. Hence, a new concept of the above evolutionary balance of bioelements and the critical area based on the links between nutrients and the balance of all the bioelements was proposed by Carpena-Artés (1978). According to this concept, the leaf level of any nutrient for a given moment is determined by the difference between the amount of nutrient which has reached the leaf and, from there, the amount transported to other plant organs. Hence, four diagnostic criteria of deficiency area,

Table 5.3 Optimum leaf nutrient levels for three commercial citrus cultivars of India using two most common ITs

	NM		MSO		KM	
Nutrients	MQRA	DRIS	MQRA	DRIS	MQRA	DRIS
N (%)	2.2–2.4	1.7–2.8	2.4–2.5	2.0–2.6	2.2–2.5	2.0–2.6
P (%)	0.07–0.10	0.09–0.15	0.13–0.15	0.09–0.17	0.10–0.11	0.09–0.10
K (%)	1.2–1.6	1.0–2.6	1.6–2.3	1.3–1.7	1.9–2.1	0.99–1.9
Ca (%)	1.3–1.5	1.8–3.3	2.6–3.2	1.7–3.0	2.1–2.3	2.0–2.5
Mg (%)	0.48–0.67	0.43–0.92	0.32–0.49	0.32–0.39	0.28–0.38	0.24–0.48
Fe (ppm)	110–132	75–113	132–148	70–137	148–180	85–249
Mn (ppm)	49–43	55–85	52–112	42–87	72–85	42–87.6
Cu (ppm)	8–14	10–18	7–10	7–16	10–19	2–14
Zn (ppm)	18–30	14–30	25–43	12–29	24–39	16–27
Yield (kg tree^{-1})	40–54	47–117	87–95	77–138	45–62	32–56

Adapted from Srivastava et al. (1999) and Srivastava and Singh (2003c)
NM 'Nagpur' mandarin, *MSO* 'Mosambi' sweet orange, *KM* 'Khasi' mandarin

critical area, normal area, and excess area were suggested for determining nutritional requirement.

Du Plessis and Koen (1992) advocated different leaf nutrient norms, based on climatic zones. The norms derived for the small fruit area, of cool climate, like Nelspruit in Mpumalanga province of South Africa (2.0–2.4% N, 0.95–1.50% K, and N/K ratio of 1.6–2.2) were different of those for the large fruit size area of Citrusdal in the Western Cape province (2.1–2.7% N, 0.70–0.90% K and N/K ratio of 3.0–4.5), having warm climate, with other nutrients like P (0.11–0.16%), Ca (3.5–5.5%), and Mg (0.30–0.55%) showing no significant difference. Most of the leaf nutrient diagnostics have, therefore, failed to find any universal applicability when tested over space and time under varying conditions. A cultivar-specific nutrient standard to suit regional growing conditions and comparing norms developed by different sampling methods appears to be the best approach. Conversion factors to relate norms that are developed for nonfruiting terminals with those for fruiting terminals and vice versa would be useful.

5.3.1 Critical Nutrient Concept

Over the past century, the pendulum of analytical diagnosis has swung from soil analysis to plant analysis. The outstanding contribution of Lundegardh (1951) has provided the usefulness of leaf analysis convincingly. Lagatu and Maume (1926) were the first to adopt the new approach to which they termed "Diagnostic Foliar." Macy (1936) introduced the concept of "critical nutrient percentage" in leaf dry matter and visualized three ranges or portions of a curve relating plant response to concentration percentages, namely the following: (1) a narrow minimal percentage range where both response and internal concentration remain fairly stable, (2) a poverty adjustment range where both response and internal concentration rise, and (3) a luxury consumption range where response is hardly noticed and concentration increases.

A critical nutrient level as "the range of concentrations at which growth of the plant are restricted in comparison with that of plants of higher nutrient level" was postulated by Ulrich (1948). This approach is based on the establishment of relationships between the concentration of different nutrients in the leaves and plant performance, with the sigmoidal response curve where two critical limits can be identified. These values for each element are respectively the values below and above which the plant growth is reduced (Terblanche and Du Plessis 1992). The interpretation of results of leaf analysis is based on the concept of critical values defined by Ulrich (1948) as the range of concentrations at which the growth of the plant is restricted in comparison to that of plants at a higher nutrient level. The results showed that plants with widely different nutrient consumption give similar yields as long as these nutrient concentrations are well above the critical level. Lundegardh (1951) described the physiological basis of leaf analysis and named assimilating plant leaf as "Central Laboratory of Nutrition." The leaf analysis became more widespread in 1955 when Kenworthy offered a service to fruit growers in Michigan for monitoring nutrient levels of their vineyards and orchards. In India, the first Leaf Analysis Laboratory was established in the Indian Institute of Horticultural Research at Hessaraghatta, Bangalore, in 1980 to conduct research and provide leaf analysis service to growers in India.

In order to make the best commercial use of leaf analysis, one must know the relative importance of the nutritional problems existing in a given orchard as well as the potential individual benefit or adverse effects of the nutrient's application. As early as Ulrich (1952) discussed the physiological basis for assessing the nutrient requirement for plants, the foundation on which the interpretation of leaf analysis was established.

The inability of the leaf analysis approach to deal adequately with the variation in nutrient concentration on a dry matter basis with age is probably its greatest disadvantage. To overcome this inconvenience, three approaches have been anticipated: in the first, sets of critical values for different stages of growth were proposed (Geraldson et al. 1973; Tserling 1974); in the second, the accumulation of dry matter with age is monitored in order to correct the nutrient concentration for increasing dry matter (Melsted et al. 1969); and in the third, sufficiency ranges were advanced such that the lower limit represents roughly the critical level while the upper is set at a value corresponding to an unusually high or toxic concentration (Small and Ohlrogge 1973). Although sufficiency ranges are purported to improve flexibility in diagnosis, they in fact decrease diagnostic precision because the limits are often too wide. Few of these attempts to improve the critical value approach have met with great success.

Critical value approach was established by Cate and Nelson (1965): the leaf analysis data are divided into two or more classes for the purpose of making the nutrient recommendation for a potential yield and fruit quality. However, the basis of defining different classes, i.e., deficient, very low, low, optimum, high, excessive, and/or toxic, is often subjective or arbitrary. A number of methods are available for setting the class limits. The procedure is to split the data into two groups, using the successive critical levels to ascertain that the particular critical level, which will maximize the overall predictive ability (R^2) with the means of two groups (classes) as the predictor value. An example for using the data, which is believed to be typical of this kind of problem, is presented. Several continuous correlation models can also be fitted to the same data. However, none gave as high as R^2 as a single low high split according to the procedure described by Bhargava and Singh (2001). The subsequent procedure is followed to work out the critical limit of a nutrient in an index tissue.

1. Correlation factor (CF) is calculated as $(\mathrm{GT})^2/n$.
2. Calculation of the total sum of squares (TSS) $(X_1 + X_2 + X_3 + X_4 + ...X_n) - \mathrm{CF}$.
3. Calculation of SSQ1 = $(X_1 + X_2) - \mathrm{CF}_1$ where $\mathrm{CF}_1 = (X_1 + X_2)_2/2$.
4. Calculation of the $\mathrm{SSQ}_2 = (X_3 + X_4 + X_5 + X_n)/2 - \mathrm{CF}_2$ where $\mathrm{CF}_2 = (X_3 + X_4 + ...X_n)_2/n$
5. $\mathrm{TSS} = (\mathrm{SSQ}_1 + \mathrm{SSQ}_2) = r$.
6. $R_2 = r \times 100/\mathrm{TSS}$.

According to the above technique, the orchards are divided in two groups: one in which is expected to have a relatively large response to a particular nutrient and another one in which is expected to have little or no response, assuming that the other nutrients were present in adequate amounts. A dividing line between the two categories might be determined approximately by a graphical technique in which the vertical and the horizontal lines are superimposed on the scatter diagrams so as to maximize the number of points in positive quadrants. The horizontal line is 90% probability line, and the vertical line is so drawn that the maximum points of the scatter diagram are on two positive quadrants. The vertical line, which is determined by the eye judgment, is known as critical levels.

Page and Martin (1964) suggested critical limit of leaf K as 0.28–0.44%. Bar-Akiva et al. (1967) recommended 0.07% as critical limit of leaf P content for Persian lime. Magnitskii and Takidze (1972) suggested critical level of 0.19% P and 1.6% K in mandarin trees in calcareous and podzolic soils of Georgia. Wang (1985) suggested critical limit of Ca, Mg, and Zn as mandarin grown on red earth in China. Aso (1967) suggested critical limit of less than 1.7% N as critical limit in Tucuman, Argentina. Bar-Akiva and Lavon (1968) recommended critical limit of leaf P as 0.075% for grapefruit. Primo et al. (1969) suggested critical limit of B, Mn, Fe, and Zn as less than 30, 18, 60, and 19 ppm, respectively, for sweet orange. While Ishihara et al. (1972) suggested critical limit based on various plant organs and accordingly, critical limit of Cu was found as 4.0, 3.0, 3.8, and 10.0 ppm for leaves, shoots, fruits, and fine roots, respectively. Rodriguez and Gallo (1961) observed the critical level of N, P, K, Ca, and Mg as 2.20%, 0.12%, 1.00%, 3.00%, and 0.30%, respectively. Coetzee (1980) suggested critical limit of K as 0.70–0.80% for Valencia orange.

In Brazil, critical level of N and K was observed as 2.66% and 1.87%, respectively, for foliar K diagnosis. In other studies, leaf K content of 1.0–1.7% has been suggested as optimum K concentration and 0.87% as critical level in 6–7-month-old leaves from fruit-bearing terminals of 'Valencia' orange (Rodriguez and Gallo 1961). In South Africa, critical limit of leaf K was observed as 0.9% in 7–9-month-old leaves from fruits bearing terminals and leaf K content of 0.55–0.80% associated with a N/K ratio of 3.3–4.1 for maximum production and fruit size of Valencia' orange in Citrusdal area (Du Plessis 1977). Studies conducted on 'Washington Navel' oranges in Australia indicated critical limit for leaf and juice K as 0.4% (0.80–1.10% optimum) and 11,314–1,373 ppm (1,424–1,575 ppm optimum), respectively, and suggested that fertilizer recommendations can be made on the basis of undigested juice K content (Gallasch et al. 1984). Shimizu and Morii (1985) suggested critical limit of Mg, Mn, Zn, and B as less than 0.13%, less than 16 ppm, less than 16 ppm, and 9–16 ppm, respectively, for 'Satsuma' mandarin.

Liu et al. (1984) suggested critical limit for Zn, Mg, and Ca as less than 25 ppm, 0.30%, and 3.0%, respectively, for major citrus species, viz., *Citrus reticulata, Citrus tankan, Citrus sinensis, Citrus tangerina,* and *Citrus grandis* of Fujian province (China). Singh and Tripathi (1985) suggested 20 ppm leaf Zn as a critical limit for distinguish chlorotic sweet orange from healthy trees, in Agra region of Uttar Pradesh.

5.3.2 Critical Nutrient Range

Evidence shows that plant with somewhat different nutrient concentration, well above the critical level, and changes in nutrient balance adversely affect the growth and productivity of the plant. The use of critical nutrient range (CNR) was, therefore, suggested by Dow and Robert (1981) rather than critical nutrient concentration (CNC).

There is generally a good relationship between concentration of the nutrients with the growth and yield. CNR is defined as the range of nutrient, at a specified growth stage, above the upper limit of which we are reasonably confident that the crop is amply supported and below the lower limit of which we are reasonably confident that the crop is deficient in the nutrient. It seems more practical to deal with critical concentration range rather than single concentration limit.

5.3.3 Nutrient Balance

5.3.3.1 Prevot's Factorial Method

Based on a considerable amount of research and experience with tropical crops, Prevot and Ollagnier (1961) developed a system based on Liebig's law of minimum, which takes nutrient balance, synergisms, and antagonisms into account using factorial experiments. The effect of increasing levels of one or several factors is calibrated, keeping all other conditions constant. From these calibrations, it is possible to determine the relative proportions of nutrient for balanced nutrition. In fact, the major problem with this approach is that there may be interactions between the factors being varied and those kept constant (Prevot and Ollagnier 1961). Although it is a definite improvement over the single factor approach used in the critical value system, it is unable to take into account simultaneously the many nutrient factors affecting growth, as is done in the DRIS approach (Bhargava and Chadha 1988).

5.3.3.2 Kenworthy's Balance Index

The Kenworthy's balance index (Kenworthy 1973) is calculated by the following procedure:

1. If a simple value (X) is smaller than standard values (S), then the balance index is calculated by:
   $B = (P + I)$
   $P = (X/S) \times 100$
   $I = (100 - P) \times (V/100)$
   where B = balance index; I = influence of variation; P = per cent of standard; V = coefficient of variation; S = standard value; X = value of sample under diagrams.
2. If sample value (X) is larger than the standards value (S), then calculation will be:
   $B = (P - I)$
   $P = (X/S) \times 100$
   $I = (100 - P) \times (V/100)$

These calculations tend to move the percent value toward a balance index of 100, based on the coefficient of variation for each nutrient.

Balance index has been used by Awasthi et al. (1979) to work out judicious nutrient doses for apple in Himachal Pradesh. The concept of nutrient balance is possible to understand when the composition values are converted into percent of the standard values. Therefore, a way of adjusting percent of the standard values is needed to account for normal variation. This may be done with the use of coefficients of variation for normal plants to develop a balance index. Awasthi et al. (1979) transformed the balance index into nutrient status of apples as indicated below:

Shortage – 17–50% index; below normal – 50–83% index; normal or optimum – 83–117% index; above normal – 117–150% index; excessive – 150–183% index.

Fertilizer treatments are not suggested until nutrient value approaches the level below which the tree performance may be reduced. In the case of N deficiency, applications may be increased approximately by the same percentage by which the balance index falls short of 100. A method was devised to convert the observed values into balance indexes that would eliminate these discrepancies of diagnosis. Mean of "normal range" is taken as 100. The coefficient of variation for each nutrient was selected as a means of adjusting the balance index according to variability in composition associated with normal plants. The method involved calculations that could adjust the deviations from 100 or standard values toward 100 in accordance to the coefficient of variation. If the sample value was below the standard value, the influence of variability was added to the percentage of standard to obtain the balance index. If the sample value was above the standard value, the influence of variability was subtracted from the percentage of standard.

5.3.3.3 Moller-Nielson Balance Concept

A diagnostic system, which attempted to address problems associated with physiological age and nutrient interactions, was proposed by Moller Nielson and Friis-Nielson (1976b) as follows:

- A series of curves are first generated from the nutrient response experiments, which relate nutrient concentration and accumulation of dry matter. The relationship of N concentration and dry plant mass at different times is determined from the trials and varied levels of added N. Similarly, those of P concentration and dry matter are worked out from P response experiments. With the aid of these curves, individual plant samples are collated with a standard plant mass. While this procedure theoretically eliminates variations in plant composition, due to stage of maturity, there is little evidence to suggest that the relationship of nutrient concentration and an increase of dry plant mass during aging is universal.

- At a second step of Moller Nielson's diagnostic method, standard nutrient values are derived through analysis of data from the factorial-designed fertilizer experiments. A boundary line approach is used to determine the optimal nutrient concentration. Only the uppermost points at each foliar nutrient concentration are used to draw the boundary line, which was called the "'pure'-effect nutrient" (Moller Nielson and Friis-Nielson 1976a).
- In the third step, boundary line curves for plots of interacting nutrients are used to determine the optimum levels of the other nutrients at the existing level of most limit nutrient, as done at the second step.
- The final step in this process consists of the calculations of the amount of the most limiting nutrient to be applied and its effect on the status of other nutrients. The method of calculating the required amount of nutrients is arbitrary, based on yield, level, plant uptake, soil fertility status, soil reactions, availability of irrigation water, and climate. After this, recommendations are made for the supplement of nutrients.

Moller Nielson's diagnostic system represents an innovative attempt to overcome two major problems in foliar diagnosis, i.e., effect of physiological maturity and nutrient interactions. On foliar composition and plant performance, however, several questions remain unanswered about the applicability of some of the relationships. Unfortunately, the amount of data required for accurately defining these relationships and the factors, which affect them, is extremely large which it is a serious obstacle to the widespread adoption of the system. Due to the requirement of extremely large data, it has not been used on a large scale. However, it has great potential for the proper diagnosis of problems associated with nutrition.

5.3.4 Crop Logging

Crop log may be defined as the manipulation of the biotic and abiotic factors relying on the information gleaned from the growing crop. In crop logging, the maximum frequency and thorough sampling are followed in 2–5 weeks. Leaf blade is normally used for the estimation of nitrogen, phosphorus, and potash. Root system is used for the information about the soil toxicities. Where salinity is a problem, water levels and electrical conductivity both are determined. All such data form the part of crop log. The sample collected may be used to monitor the nutritional status of an individual crop during its development, i.e., crop logging, therefore, ensures that its nutrient requirements are being met satisfactorily (Clements 1961; Gartell et al. 1979). Abaev (1977) suggested various level of N, P, K according to critical growth stages as: 2.1–2.3% N, 0.22% P_2O_5, and 1.8% K_2O at flowering; 2.4–2.7% N, 0.25–0.28% P_2O_5, and 1.85–2.0% K_2O at fruit formation; and 2.1–2.3% N, 0.25% P_2O_5, and 1.7–1.8% K_2O at fruit ripening stage, for lemon grown in Western Georgia.

5.3.5 Boundary Line Concept

A new approach to study the crop productivity has been developed by Webb (1972) in which the performance of the best, in the sample examined, is taken as a standard against which to judge the remainder, on the assumption that there are reasons other than chance which account for the inferior performance by a part of the population. The line defining the best performance in the population lies at the edge of any body of data, hence the name Boundary Line, and occurs wherever a cause-effect relationship between two variables exists. As stated by Walworth et al. (1986), "whether utilizing a critical value or a nutrient balance system such as the DRIS for interpreting plant tissue composition, determination of accurate optima is of paramount importance." In this same article, two procedures for determining such optima (within the Boundary Line Approach) are proposed, "one using the mean of a high yielding population, the second establishing yield maxima at all nutrient values." When the best performance can be quantified, the overall deficiency in yield, due to inferior performance, can be assessed. When it is allied to the knowledge of the components of yield, the position of the boundary line can be used to direct attention to the phase of growth most likely to respond to better management (Walworth et al. 1986; Bhargava and Sumner 1987).

Researchers have reasoned that if a unique relationship between a single growth factor and crop yield or quality can be defined, then optimizing that factor should permit the best crop performance. As a result, the literature is replete with regression relationships between such parameters as plant and soil analyses data and crop yield. These have often been used to establish critical values for diagnostic purposes. Unfortunately, most of relationships are developed under conditions where only one, two or, sometimes, three nutrients are tested at two or three levels. Consequently, the relationships so determined are specific to the condition unique to the experiment(s) involved and often do not hold valid under all conditions (Andrew 1968).

Nutritionists have attempted to identify and quantify the factors that are closely related to plant performance. If the expected relationship between growth factors and yield/quality can be defined (through techniques as, e.g., analysis of variance, correlation, and regression equations between plant analysis and crop yield), then optimizing those factors should permit the best crop performance. Using mean values of leaf analysis data of high-yielding population (Kenworthy 1973;

Cate and Nelson 1965; Schaffer et al. 1987), critical limits have been fixed for diagnostic purpose in horticultural crops, including citrus. Because the effect of a particular growth factor may change under varying conditions, due to interactions with other factors, critical values established in this way are not unique or universally applicable. This is quite obvious by wide variation in critical values published in literature (Chapman 1967; Kenworthy 1973; Cook and Wheeler 1978; Bhargava and Chadha 1988; Embleton et al. 1973). Use of sufficiency range rather than single value as critical limit (Dow and Robert 1981; Munsen and Nelson 1973; Bhargava and Chadha 1988) will alleviate this problem, although the dynamic nature of the relationship between mineral nutrition and dry matter over time is certainly responsible for some of the variations that exist.

Percentage yield (Nelson and Anderson 1977) has often been used in an attempt to overcome some of these difficulties. However, combining yields from different years or sites in this way largely ignores the complexity of the relationship between plant growth and environment. In the absence of identification and quantification of parameters affecting plant growth, regression approach is of limited value in interpolating to unknown situations and will simply remain on a posteriori approach to organizing data. The boundary line approach defines yields that may occur under a given set of conditions and can be used to determine plant tissue optima, offering an alternative to the conventional critical value system. In addition, the maximum yield that is possible at any given compositional value may be predicted from boundary lines. A comparison of the optima determined via the boundary line approach and those estimated by the mean of the high-yielding population revealed extremely small differences indicating that either method is acceptable for estimating these parameters (Walworth et al. 1986).

If one consciously sets about varying controllable growth factors as much as possible at many locations, a bank of observations that represent the variability encountered in the real world can be generated. This can also be achieved by sampling the variability that occurs naturally in a given crop industry under all conditions where that crop is produced. A scatter diagram of yield plotted against a plant growth factor for such data usually peaks at the optimum level of that particular growth factor. It became possible to develop a set of norms from such data bank, i.e., quantification of a growth factor to maximum yield level, that should be diagnostically precise and more universally applicable (Walworth et al. 1986). This concept is equally applicable to the critical value system, diagnosis, and recommendation integrated system and also to nutrient sufficiency level. The boundary line should be useful in diagnostic work in that the maximum possible yield consistent with any growth factor could be determined. The specificity of regression relationships of this type is due to the unique characteristics of the large number of plant growth factors existent in the individual plots for the particular growing season that produced the data. In different years, the regression equations may be different because other factors or interactions with other factors become more important.

Once a boundary line has been defined, it is a simple matter to locate the apex of that line, which corresponds to the optimal level of the growth factor in question. Alternatively, optima can be estimated by averaging the values of all observations if the population of observations is distributed normally. This method is essentially that used to calculate foliar norms for use in DRIS. Kenworthy (1967, 1973) used a similar technique to develop standard nutritional values for diagnosis of fruit tree foliage.

5.3.6 Diagnosis and Recommendation Integrated System

Development of soil-plant nutrient diagnostics has been the popular area of investigation, world over using a variety of diagnostic tools. A majority of the studies concentrated mainly on sweet orange cultivars. The scope of conventional diagnosis is limited due to strong influence of leaf age. The critical nutrient concentration and sufficiency range limit, developed by using index leaves as interpretation tools, provide little time in the growing season for fertilizer application to be really effective. Therefore, the currently available diagnostic methods are applicable only to narrowly specified developmental stage of crop. In this regard, it is difficult to draw any conclusion from the earlier attempts in several countries (Australia (Jorgensen and Price 1978), China (Wu et al. 1998), Turkey (Saatci and Mur 2000), Spain (Hellin et al. 1988), Costa Rica (Alvardo et al. 1994), USA (Swietlik 1996), Chile (Razeto et al. 1988), and various parts of India (Sharma and Mahajan 1990; Chundawat et al. 1990; Srivastava and Singh 2003b, 2004b)), employing a variety of diagnostic methods and using different aged index leaves from fruiting as well as nonfruiting terminals, amounting to many discrepancies in the diagnostic capabilities in the absence of any commonality in guidelines used in diagnosing the nutrient constraints. The orchards, hence, continue to produce suboptimally due to erroneous identification of the nutrient deficiency that made it further difficult to match the expanding gap emerged from the amount of nutrients added to that of annual demand with orchard age. The nutrient diagnostics available for commercial cultivars (e.g., 'Valencia' and 'Navel' (Perez 1996), 'Nagpur' mandarin (Srivastava et al. 2001; Srivastava and Singh 2003a), 'Satsuma' mandarin (Koto et al. 1990), 'Mosambi' sweet orange (Srivastava and Singh 2003b, 2004a)) have not found their universal applicability and often lacked severely in reproducibility when applied under different contrasting growing conditions because of distinct regional differences.

One of the major drawbacks of quantifying nutrient elements in terms of concentration on total leaf dry weight (DW) is that an increase in the leaf DW (as a result of sugar, starch, or other nutrient accumulation) will reduce the concentration of the nutrient for the same weight (dilution effect) with the opposite effect (concentration effect) when DW decreases. "A nutrient concentration which changes due to changing DW could then lead to inaccurate interpretation of nutrients results when comparing with standard published values or thresholds" (Schumann 2009). For this reason, in such efforts, the type of shoot, period of sampling, and number of leaves collectively affect the leaf sample and, therefore, the analytical result for nutrient concentration.

Besides physiological causes (e.g., Moreno et al. (1996) compared the DRIS indexes with standard methods to evaluate the effectiveness of DRIS in diagnosis Fe-chlorosis), also pathogenic effects have been reported as interfering in dilution/concentration effects. This is the case for the Huanglongbing (HLB) infection in citrus leaves where one of the recognized symptoms is a dramatic accumulation of starch. In diseased tissues, accumulations of starch capable of increasing the total leaf DW by nearly 50% have been recorded. Consequently, most leaf nutrient concentrations show apparent decline in HLB-infected blotchy-mottled leaves due to dilution by the added weight of accumulated starch. For example, in a replicated study of ten HLB-infected and ten healthy Hamlin orange trees, the sulfur concentration in symptomatic blotchy-mottled leaves was 13% lower than in asymptomatic leaves from the HLB trees or the healthy trees. Analysis of the data with DRIS revealed that the amounts of sulfur in the different leaf samples were not significantly different. The same conclusion was reached by converting the sulfur concentration data to a leaf area basis (milligrams per square meter). It was concluded that the sulfur "deficiency" in blotchy-mottled leaves was false and likely caused by the accumulation of starch (Schumann 2009).

Bias from undesirable nutrient dilution or concentration effects, due to uncontrollable changes in leaf tissue DW, is noticeably diminished when using nutrient interpretation with DRIS, since this method calculates ratios of nutrient concentrations, expressed as a fraction of DW, which being common to all of them is annulated. There are variants of the DRIS computations that can estimate an index of the leaf dry weight.

DRIS diagnoses generally agreed the diagnoses made by the sufficiency range method, with the advantage that DRIS reflected nutrient balance and identified the order in which nutrients are likely to become limiting. DRIS reflected changes in nutrient concentrations due to alternate-bearing or crop-load effects and agreed with the sufficiency range method when concentration changes were sufficient to affect this method (Beverly et al. 1984).

Diagnosis and recommendation integrated system (DRIS), although firstly developed for rubber trees (Beaufils 1973), is claimed to have certain advantages over other conventional interpretation tools (Beverly 1987; Malavolta et al. 1993; Li et al. 1999). DRIS method expresses results of plant nutritional diagnosis through indexes, which represent, in a continuous numeric scale, the effect of each nutrient in the nutritional balance of the plant. DRIS diagnoses generally agree with diagnoses made by the sufficiency range method, but with some additional advantages that DRIS reflects the nutrient balance (fluctuates narrowly across different crop developmental stage), identifies the order in which nutrients are responsible for limiting the fruit yield, and its ability to make diagnosis at any stage of crop development. These merits impart DRIS to be able to identify nutrient constraint early in crop growth and allow sufficient time for remediation of identified problem right in the same season of crop (Walworth and Sumner 1987). The efforts in the past have successfully established the DRIS norms for different citrus cultivars (Sumner 1977; Beverly et al. 1984; Varalakshmi and Bhargava 1998; Hundal and Arora 2001).

Several model modifications have been proposed to increase accuracy in the nutritional diagnosis for several crops. The calculation of the nutrient ratio functions is made according to one of three methods, namely the following: (1) the original method proposed by Beaufils (1973), (2) the Jones (1981) method, and (3) the Beaufils (1973) method, modified by Elwali and Gascho (1984). Although these nutrient function ratio calculation methods have been evaluated in some researches, there is not yet a clear definition for the best recommendation. The three methods applied to rubber trees revealed that Beaufils (1973) and Elwali and Gascho (1984) procedures presented similar results, and that Jones (1981) procedure showed dependence on the nutrient ratio (Bataglia and Santos 1990). In some citrus databases, the Beaufils (1973) method highlighted nutritional deficiencies, the Jones (1981) method had advantage for presenting more simple calculation and larger statistical formality, and the Elwali and Gascho (1984) method showed lesser interpretation errors (Santos 1997).

According to Beverly (1991), there are two ways for the second and last stage of DRIS indexes calculation (the function sum involving each nutrient), namely DRIS (Beaufils 1973) and M-DRIS (Hallmark et al. 1987; Walworth et al. 1986). The original DRIS method just uses the nutrient ratio functions. On the other hand, the M-DRIS method, a variation and expansion of original DRIS, foresees dry matter inclusion in the indexes calculation. The expressions are identical to the ordinarily used; however, in this case, the dry matter is treated as an additional constituent, and a new index is calculated, in the same way as for the other plant constituents. In fact, dry matter is, essentially, the sum of the concentration of three nutrients usually ignored in nutritional considerations:

C, H, and O. That additional index is the dry matter mass index, a good indicator of the sampled tissue maturity regarding the standard.

5.3.6.1 Brief Developments

DRIS has shown its application in both annual crops, viz., lettuce (Sanchez et al. 1991), tomato (Caron and Parent 1989), potato (Parent et al. 1994a), onion (Caldwell et al. 1994), cucumber (Mayfield et al. 2002), and carrot (Parent et al. 1994b) as well as perennial crops, viz., apple (Goh and Malakouti 1992), grapes (Bhargava and Raghupathi 1995), pecan (Beverly and Worley 1992), peach (Sanz 1999), mango (Schaffer et al. 1988), mango (Raghupathi and Bhargava 1999), pomegranate (Raghupathi and Bhargava 1998), banana (Angeles et al. 1993), sapota (Appa Rao et al. 2006), litchi (Hundal and Arora 1996), and papaya (Bowen 1992) with equally reproducing results. To overcome perceived weakness in the traditional DRIS approach to tissue nutrient analysis interpretation, three revisions and two new methods were applied to data for oranges cv Valencia. Use of logarithmic transformation, population parameters, and a single calculation method removed systematic errors, simplified the diagnostic method, and extended its applicability. The two new methods are individual nutrient concentrations, rather than the ratios. The changes produce diagnoses similar to those given by DRIS or the sufficiency range approach, but result in better recommendations, judging from a yield response test (Beverly 1987). The below described is the detail account of developments that have taken place with regard to application of DRIS in citrus.

5.3.6.2 Steps Involved for DRIS Norms

DRIS technique consists of describing the nutrient status of high-yielding populations, and to identifying variations from those conditions in unknown samples. The observations were divided into high- and low-yielding subpopulations, using 50 kg tree^{-1} (averaged yield level usually obtained at growers' field) as cut-off yield level to separate the subpopulations. For the two subpopulations, the mean ($\bar{x}$), standard deviation, and variance (S) were calculated for each nutrient concentration as well as all the ratios between nutrient concentrations (N/P, N/K, P/K, etc.). A variance ratio (S^2 for low-yielding population/S^2 for high-yielding population) was calculated for each nutrient concentration, and of two ratios involving each pair of nutrients, finally selecting the one with the larger variance ratio. The mean and coefficient of variation (CV) values in the high-yielding population for the selected ratios were used for calculating DRIS indexes. The nutrient with the most negative index is considered the most deficient and most limiting to fruit yield and *mutatis mutandis* in the opposite case.

The following procedure as initially developed by Beaufils (1973) and modified by Bhargava (2002) was used through a PC-based program for the development of DRIS norms, comprising: (1) definition of the parameters to be improved and the factors likely to affect them, (2) collection of all the reliable data available from the fields and experimental plots, (3) study of the relationship between yield and available nutrients in soil, (4) establishment of the relationship between yield and leaf nutrient composition, using the following steps: (a) each internal plant parameter is expressed in so many forms as possible (e.g., N/DM, N/P, P/N, N×P); (b) the whole population is divided into a number of subgroups based on the economic optimum; (c) the mean of each subpopulation is calculated for the various forms of expression; (d) if necessary, class interval limits between the average and the outstanding yields are readjusted so that the means of below average populations remain comparable; (e) chi-square test is performed to know if the populations confirm a normal distribution; (f) the variance ratios between the yield of subpopulations for all the forms of expressions are calculated together with the coefficient of variation; (g) the forms of expressions, for which significant variance ratios were obtained and essentially the same mean values for the population were selected in expression with common nutrient; and (h) the following equations were developed for the calculation of DRIS indexes based on leaf analysis:

1. N = 1/9 [f (N/P) + f (N/K) + f (N/Ca) + f (N/Mg) + f (N/Fe) + f (N/Mn) + f(N/Cu) + f (N/Zn)]

where $f(\mathrm{N/P}) = \left(\dfrac{\mathrm{N/P}}{n/p} - 1\right)\left(\dfrac{1{,}000}{\mathrm{CV}}\right)$ when N/P > *n*/*p*,

for example,

and $\left(1 - \dfrac{n/p}{\mathrm{N/P}}\right)\left(\dfrac{1{,}000}{\mathrm{CV}}\right)$ when N/P > *n*/*p*,

where N/P is the actual value of the ratio of N and P in the plant under diagnosis, *n*/*p* the value of the norm (the mean value of high-yielding orchards), and CV the coefficient of variation for population of high-yielding orchards.

1. P = 1/9 [-f (N/P) + f (P/K) + f (P/Ca) + f (P/Mg) + (P/Fe) + f (P/Mn) + f (P/Cu) + f(P/Zn)]
2. K = 1/9 [-f (N/K) + (K/P) + (K/Ca) + (K/Mg) + (K/Fe) + (K/Mn) + (K/Cu) + (K/Zn)]
3. Ca= 1/9 [-f (N/Ca) – f(P/Ca) – f(K/Ca) + f (Ca/Mg) + f (Ca/Fe) + f (Ca/Mn) + f (Ca/Cu) + f (Ca/Zn)]
4. Mg = 1/9 [-f (N/Mg) – f(P/Mg) – (K/Mg) – f(Ca/Mg) + f (Mg/Fe) + f (Mg/Mn) + f (Mg/Cu) + Mg/Zn)]
5. Fe = 1/9 [-f (N/Fe) – f (P/ Fe) – f (K/Fe) – f (Ca/ Fe) – f (Mg/Fe) + f (Fe/ Mn) + f (Fe/Cu) + f (Fe/ Zn)
6. Mn = 1/9 [-f (N/Mn) – f (P/Mn) – f (K/Mn) – f (Ca/Mn) – f (Mg/Mn) -f (Fe/Mn) + f(Mn/Cu) + f (Mn/Zn)]
7. Cu = 1/9 [-f (N/Cu) – f (P/Cu) – f (K/Cu) – f (Ca/Cu) – f (Mg/Cu) – f (Fe/Cu) – f(Mn/Cu) + f (Cu/Zn)]
8. Zn = 1/9 [-f (N/Zn) – f (P/Zn) – f (K/Zn) – f (Ca/Zn) – f (Mg/Zn) – f (Fe/Zn) – f (Mn/Zn) – f (Cu/Zn)]

The norms for classification of nutrients in leaves are derived using the mean of high-yielding orchards as the mean for optimum. The range for optimum is the value derived from –4/3 to +4/3 standard deviation from mean. The range for low was obtained by calculating –4/3 to –8/3 standard deviation from mean, and the value –8/3 standard deviation below mean was considered deficient. The value above +4/3 standard deviation from mean was considered as an excess (Bhargava 2002).

5.3.6.3 Leaf Analysis-Based Norms

Methods for nutritional diagnosis using leaf analysis consist of critical value, the SRA, and DRIS. The last method, since it uses the balancing concept (relationship among nutrients), might be more precise in detection of nutritional disorders.

Early studies on DRIS were carried out at California (USA) by Beverly et al. (1984) when preliminary reference values were derived for nutritional diagnosis of N, P, K, Ca, and Mg for 'Valencia' sweet orange. These values were also used for subsequent comparisons with the SRA and, overall, both methods presented similar results. However, the DRIS diagnosis was affected by the sample tissue type and maturation, and the indexes reflected the nutrient concentration change related to the yield alteration or to the presence of fruits in the shoots at sampling time. The DRIS indexes were in agreement with the SRA diagnosis, only when changes in nutrient concentration significantly affected the second method.

In a subsequent work, Beverly (1987) suggested three modifications on the DRIS method and proposed two new methods for nutritional diagnosis for 'Valencia' sweet orange. The logarithmic transformation, the use of standard populations, and the adoption of a unique calculus procedure are modifications introduced to avoid systematic errors and simplify the diagnosis method, broadening its application. The two new suggested methods were based on individual plant nutrient concentrations instead of nutrient ratios. The diagnosis resulted similar to the one obtained by DRIS or SRA, but provided more precise recommendations when evaluated by field tests. After this work, new researches involving data collecting during five more years revealed that SRA could be more advantageous than DRIS for 'Valencia' sweet orange (Beverly 1992). The author compared SRA, DRIS, and three modifications of DRIS. The SRA showed efficacy (not presenting false diagnosis) for N and P diagnosis status in 75% and 90% of the cases, respectively, compared to 50%, or less, obtained by the other methods.

The most advantageous of the three available procedures for the DRIS indexes calculations was that proposed by Jones (1981) which calculated the DRIS norms for N, P, K, Ca, Mg, and S, using a reference subpopulation with productivity equal or superior to 120 kg tree^{-1}. All tested methods showed efficacy for K diagnosis.

Wallace (1990) carried out studies on DRIS for 'Valencia' sweet orange, from the established by Beverly et al. (1984), investigating several N, P, and K ratios and interactions. This author observed a 23% yield increase in response to K supply which amount to a 69% increase when N and P were also added. DRIS reveals to be an effective method for nutritional diagnosis in this study. Woods and Villiers (1992), in a research work developed at South Africa, obtained well-succeeded DRIS results for 'Valencia' sweet orange, in disagreement with the results reported by Beverly (1992). Those authors observed good correlation between yield (kg tree^{-1}) and fruit quality (fruit mass; g), with DRIS indexes derived from 1,700 observations. The results were compared with the conventional diagnosis method. The DRIS norms were also evaluated in fertilization experiments, and the increase in yield and fruit quality (fruit mass) was consistent with DRIS diagnosis.

To develop DRIS norms for 'Verna' lemon nutritional diagnosis, research was carried out by Cerda et al. (1995) at Murcia and Alicante (Spain). The adopted reference population presented yield equal to above 125 kg tree^{-1}. The DRIS determinations were influenced by the rootstock/scion combination and leaf sampling period. The results of diagnosis agreed with those obtained by the SRA only when the analyzed leaves came from the same period of sampling than the ones for the DRIS norms. Under salinity conditions, DRIS was not effective in detecting, if the cause of nutrient deficiency, that is, whether the nutrient unbalance was due to high salinity or fertilization deficiency. Results obtained in hydroponics were used to establish a data bank for DRIS indexes calculation for several citrus rootstock/scion combinations in Spain (Moreno et al. 1996). Useful reference values were determined for Fe availability evaluation and its influence in the nutrition of studied citrus rootstock/scion combinations, under sufficient and deficient Fe supply. A lemon scion budded on *Citrus macrophylla* rootstock showed less Fe-chlorosis deficiency symptoms compared to the same lemon budded on sour orange. *Citrus volkameriana* induced higher Fe-deficiency tolerance than Cleopatra mandarin when used as rootstocks combined with sweet orange scions.

DRIS norms were developed for 'Valencia' sweet orange for a plant population with different plant ages, on various rootstocks, at several regions for the four most important citrus-producing states of Venezuela (Rodriguez et al. 1997). The reference population was obtained through the selection of the 20% most productive plants. The values obtained were comparable to the previously determined, as referred in the literature. The authors concluded that DRIS method might be a low cost, timesaving, and trustful alternative for the development of nutritional diagnosis norms.

In Brazil, there is a paucity of publications on DRIS method investigation, especially in fruit crops. Apart from the research carried out in other species, as, for instance, in banana (e.g., Teixeira et al. 2002), few studies are reported in citrus. Bataglia (1989) was probably the first author to report the application of this method for citrus nutritional diagnosis and indicated DRIS as an alternative diagnosis method, pointing out the need of using it together with other diagnoses criteria. Creste (1996) reported the first DRIS evaluation by comparison with the SRA in groves of Brazil, studying 'Siciliano' lemon. Data were obtained from the analysis of leaves of fruit branches of different plant ages and rootstocks, collected in several harvesting years. The reference population was derived from plants with productivity greater than 80 t ha^{-1}. After the DRIS norm calculations, the method was evaluated under field conditions. DRIS showed to be more advantageous over the SRA, mainly because it was able to discriminate the nutrient importance order of deficiency or excess. Santos (1997) evaluated the DRIS method using results of leaf analysis derived from a series of field experiments with N, P, K fertilization in commercial groves of the State of São Paulo. This author obtained superior results with the DRIS compared to SRA, for detecting yield limitation by nutrient deficiency. Mourão Filho and Azevedo (2003) established DRIS norms for the 'Valencia' sweet orange budded on Rangpur lime, 'Caipira' sweet orange, and *Poncirus trifoliata* rootstocks. The nutritional balance and indexes calculated by the derived norms were highly correlated with yield for the rootstock/scion combinations, from what it was inferred that DRIS norms might be applicable always that leaf sampling is collected from nonbearing fruit branches of irrigated-plant groves.

DRIS indexes developed for different citrus cultivars in India predicted optimum value of different nutrients as: 1.70–2.81% N, 0.09–0.17% P, 0.96–2.59% K, 1.73–3.43% Ca, 0.24–0.92% Mg, 69.5–249.0 ppm Fe, 21.0–87.6 ppm Mn, 2.13–17.6 ppm Cu, and 11.6–50.0 ppm Zn, in relation to fruit yield of 31.6–37.9 kg tree^{-1} and 15.7–19.4 kg tree^{-1} for mandarins and acid lime, respectively (Table 5.4).

Nutrient diagnostics popularly used in Australia differ widely as per the diversity in citrus-growing regions. Jorgensen and Price (1978) suggested leaf nutrient norms for central coastal areas of Queensland which recommended optimum limit of different nutrients reading 2.4–2.6% N, 0.14–0.16% P, 0.9–1.2% K, 3.0–6.0% Ca, 0.23–0.60% Mg, 60–120 ppm Fe, 25–100 ppm Mn, 5–10 ppm Cu, and 25–100 ppm Zn, while Gallasch and Pfeiler (1988) developed a comprehensive leaf nutrient standard for Riverland District of Victoria and Sunraysia District of New South Wales (Australia) which suggested optimum limit of 2.4–2.7% N, 0.14–0.17% P, 0.70–1.49% K, 50–129 ppm Fe, 6–15 ppm Cu, and 25–60% Zn ppm.

These limits turn out to be widely different in China using optimum values measuring 3.0–3.5% N, 0.15–0.18% P, 1.0–1.6% K, 2.5–5.0% Ca, 0.30–0.60% Mg, 50–120 ppm Fe, 25–100 ppm Mn, 4–100 ppm Cu, and 25–100 ppm Zn for Satsuma mandarin grown on quaternary red earth (Alfisols) using third leaf from vegetative terminals (Wang 1985). On the other hand, the leaf nutrient standards have been developed for citrus belts (concentrated in seven provinces) of contrasting climates (the cool and warm regions separately) and fruit sizes (small and large) in South Africa (Du Plessis and Koen 1992).

It remains to be seen that the diagnostic norms derived from specific index leaves and orchards, categorized into

Table 5.4 Leaf nutrient indexes (derived from DRIS-based analysis) for different commercial citrus cultivars of India

	Indexes		
Nutrients	Low	Optimum	High
Nagpur mandarin (*Citrus reticulata* Blanco)			
N (%)	1.12–1.69	1.70–2.81	2.82–3.38
P (%)	0.06–0.08	0.09–0.15	0.16–0.19
K (%)	0.22–1.01	1.02–2.59	2.60–3.38
Fe (ppm)	55.6–74.8	74.9–113.4	113.5–132.7
Mn (ppm)	40.2–54.7	54.8–84.6	84.2–98.7
Cu (ppm)	5.9–9.7	9.8–17.6	17.7–21.5
Zn (ppm)	5.5–13.5	13.6–29.6	29.7–37.7
Yield (kg tree^{-1})	12.9–47.6	47.7–117.2	117.3–152.1
Khasi mandarin (*Citrus reticulata* Blanco)			
N (%)	1.67–1.96	1.97–2.56	2.57–2.85
P (%)	0.06–0.08	0.09–0.10	0.11–0.13
K (%)	0.52–0.98	0.99–1.93	1.94–2.40
Fe (ppm)	22.6–84.5	84.6–249.0	249.1–331.3
Mn (ppm)	18.6–41.5	41.6–87.6	87.7–110.6
Cu (ppm)	1.83–2.12	2.13–14.4	14.5–20.6
Zn (ppm)	11.1–16.2	16.3–26.6	26.7–31.8
Yield (kg tree^{-1})	19.1–31.5	31.6–56.3	56.4–68.8
Mosambi sweet orange (*Citrus sinensis* Osbeck)			
N (%)	1.28–1.97	1.98–2.57	2.58–2.68
P (%)	0.050–0.090	0.091–0.17	0.18–0.21
K (%)	1.12–1.32	1.33–1.72	1.73–1.92
Ca (%)	1.09–1.72	1.73–2.98	2.99–3.62
Mg (%)	0.13–0.31	0.32–0.69	0.70–0.87
Fe (ppm)	25.9–69.4	69.5–137.1	137.2–200.1
Mn (ppm)	29.7–42.1	42.2–87.0	87.1–159.5
Cu (ppm)	2.0–6.5	6.6–15.8	15.9–20.5
Zn (ppm)	9.0–11.5	11.6–28.7	28.8–37.3
Yield (kg tree^{-1})	45.9–76.5	76.6–137.9	138.0–168.5
'Sathgudi' sweet orange (*Citrus sinensis* Osbeck)			
N (%)	1.32–2.00	2.01–2.42	2.43–2.60
P (%)	0.09–0.10	0.11–0.13	0.14–0.16
K (%)	0.72–1.11	1.12–1.82	1.83–2.01
Fe (ppm)	22.2–53.4	53.5–82.2	82.2–110.8
Mn (ppm)	18.2–48.6	48.7–79.3	79.4–116.2
Cu (ppm)	1.2–3.6	3.7–8.9	9.0–14.6
Zn (ppm)	10.2–16.4	16.5–23.2	23.3–35.8
Yield (kg tree^{-1})			

Adapted from Srivastava and Shyam Singh (2004a, 2008) and Varalakshmi and Bhargava (1998)

Table 5.5 Leaf analysis-based DRIS indexes for identifying nutrient constraints in different citrus cultivars

'Nagpur' mandarin ($n=57$)	Nutrients found deficient and low ($n=27$)				Nutrients found high and excess ($n=30$)						Yield (kg $tree^{-1}$)
	Zn	P	N	Fe	Cu	Mn	Mg	K	Ca		
Conc. (mg kg^{-1a})	9.2	0.06	1.56	68.3	19.2	91.6	0.92	2.62	3.34		32
DRIS indexes	−166	−60	−28	−20	16	42	55	63	98		
'Mosambi' sweet orange ($n=60$)	Nutrients found deficient and low ($n=32$)				Nutrients found high and excess ($n=28$)						Yield (kg $tree^{-1}$)
	N	Zn	K	P	Mg	B	Ca	Mo	Fe	Cu	
Conc. (mg kg^{-1a})	1.28	9.1	1.14	0.08	0.70	28.2	3.01	1.1	138.1	18.1	39
DRIS indexes	−185	−111	−82	−58	38	40	48	74	92	144	
'Khasi' mandarin ($n=108$)	Nutrients found deficient and low ($n=68$)						Nutrients found high and excess ($n=40$)				Yield (kg $tree^{-1}$)
	Zn	P	Ca	N	Mg	Cu	K	Mn	Fe		
Conc.(mg kg^{-1a})	10.5	0.06	1.66	1.60	0.18	1.9	1.98	94.2	268.1		22
DRIS indexes	−201	−101	−91	−86	−78	−42	104	219	276		

Adapted from Srivastava et al. (2007)
[a]Values of N, P, K, Ca, and Mg are given in %

deficient or optimum in different nutrients on the basis of nutrient concentration, have the same utility as that of norms developed through leaves sampled at other crop developmental stages (leaving the index sampling period) in order to make DRIS a more flexible monitoring tool without affecting the production at any crop stage. The overriding influence of physiography has a definite impact in dictating the relationship between the nutrient composition of index leaves, nutrient composition of soil, and the time of fruit maturity, when compared with the orchard conditions of valley versus hill slopes or floodplains versus hill slopes (Gualiya and Zonn 1990). Irrespective of such physiographical divergence, DRIS norms developed in one specific region may be applied to another region, if the elemental composition of high-yielding orchards is nearly identical, with normal skewness-free distribution of data.

5.3.6.4 Identification of Nutrient Constraints and Their Frequency Distribution

DRIS indexes are expressed by positive or negative values, which that the referred nutrient is in excess or deficiency range, respectively. The closer to zero are the indexes for all the nutrients. Occurrence of single or multiple nutrient deficiencies in citrus orchards is reported from all the six continents (Srivastava and Singh 2003a). These deficiencies are if not addressed in time through suitable diagnostic norms; the orchards coupled with reduced longevity continue to impart recurrent loss in production and imbalances the production economics. Works done for different commercial citrus cultivars in India showed nutrient deficiencies of Zn, P, N, and Fe due to their negative values in decreasing order (Table 5.5) using leaf analysis data. While, other nutrients, viz., Cu, Mn, Mg, K, and Ca with increasing positive indexes were observed in high to excess limit. A large positive nutrient index (more negative an index, the more lacking is the nutrient) indicates that the corresponding nutrient is present in relatively excessive quantity. Using the progressive nutrient diagnosis, if the first limiting factor Zn is corrected by its supply, the next nutrient that will limit the yield is P. Further, if Zn and P are satisfied, the next limiting nutrient is N followed by Fe (Table 5.5).

The frequency distribution of nutrient constraints as diagnosed through DRIS-based plant-soil nutrient diagnostics demonstrated a good complementarity between leaf and soil analysis. This was further evident from different correlation coefficient values between leaf and soil analysis values for N ($r=0.624$ $p=0.01$), P ($r=0.412$ $p=0.01$), K ($r=0.212$ $p=0.05$), Ca ($r=0.123$, nonsignificant), Mg ($r=0.181$, nonsignificant), Fe($r=0.416$ $p=0.01$), Mn ($r=0.512$ $p=0.01$), Cu ($r=0.458$ $p=0.01$), and Zn ($r=0.583$ $p=0.01$). The earlier studies have shown that nutritional problems of citrus orchards are better identified with the combined use of leaf and soil analysis than either of the two alone (Jorgensen and Price 1978; Srivastava et al. 2001). The nutrient Zn was estimated low to deficient (63.4–72.8%) followed by N (52.3–66.3%), K (28.3–35.3%), P (28.1–31.3%), and Fe (28.2–29.8%) irrespective of test methods used. Worldwide, Zn is claimed to be the single most frequently limiting nutrient impairing with the sustainable citrus production (Srivastava and Singh 2004a). These nutrient constraints laid the basis for fertilization to maximize the yield, and subsequently verify the ability of DRIS indexes in identifying the nutritional problems existing actually under the field conditions.

The utility of DRIS-based data is often questioned due to less dynamic in nature over a growth period limited by many interacting cofactors. This has resulted in limited work carried out in this direction. The studies undertaken with respect to commercial citrus cultivars in India demonstrated that mean DRIS indexes emerging deficient to low level of organic carbon, Zn, P, Fe, N, and K due to their negative values in decreasing order.

5.3.6.5 Validation of DRIS Indexes

Crop response studies are considered as the most reliable method of establishing the nutrient constraints occurring in the field. In an effort to validate DRIS norms in Nagpur mandarin through a fertilizer response experiment, various DRIS-based fertilizer treatments produced a significant response on both leaf nutrient composition and fruit yield (Table 5.6). The treatment (N, P, K, Zn, Fe) showed a significantly higher fruit yield (31.6 kg tree^{-1}) over other treatments such as $N_0P_1K_1Zn_1Fe_1$ (21.5 kg tree^{-1}), $N_1P_0K_1Zn_1Fe$ (23.2 kg tree^{-1}), and $N_1P_1K_0Zn_1Fe_1$ (22.2 kg tree^{-1}), supporting as an evidence to deficiency of N, P, and K, respectively. Such a fertilizer response became more evident from the changes in leaf nutrient concentration. The −N, −P, and −K respective treatments registered 1.89% N, 0.06% P, and 0.70% K, significantly lower over 2.04% N, 0.11% P, and 0.86% K.

The yield was further maximized to 43.3 kg tree^{-1} with an additional increment of N ($N_2P_3K_2Zn_1Fe_1$) registering over comparatively lower N rate ($N_1P_3K_2Zn_1Fe_1$) 36.4 kg tree^{-1}. Increasing K rate from K_2($N_2P_3K_2Zn_1Fe_1$) to K_4($N_2P_3K_4Zn_1Fe_1$) was associated with corresponding increase in leaf K from 1.19% to 1.54% imparting a corresponding increase in fruit yield from 43.3 to 58.8 kg tree^{-1}. The fruit yield was comparatively higher, 38.9 kg tree^{-1} with treatment containing Zn ($N_1P_2K_2Zn_1Fe_1$) against only 32.1 kg tree^{-1} with no Zn treatment ($N_1P_2K_2Zn_0Fe_1$). Likewise, absence of Fe treatment ($N_1P_1K_1Zn_1Fe_0$) showed a lower fruit yield of 25.8 kg tree^{-1} against 31.6 kg tree^{-1} with presence of Fe treatment ($N_1P_1K_1Zn_1Fe_1$), establishing the deficiency of both Zn as well as Fe.

DRIS indexes for various nutrients demonstrated remarkable changes under different treatments, e.g., $N_1P_1K_1Zn_1Fe_1$ treatment registered much lower index −20 versus −52 with treatment $N_0P_1K_1Zn_1Fe_1$ for N, −38 with $N_1P_0K_1Zn_1Fe_1$ versus −08 with $N_1P_1K_1Zn_1Fe_1$ for P, and −43 with $N_1P_1K_0Zn_1Fe_1$ versus −35 with $N_1P_1K_1Zn_1Fe_1$ for K. The K indexes reduced from −43 with $N_1P_1K_0Zn_1Fe_1$ to as much as −04 with $N_2P_3K_4Zn_1Fe_1$ indicating that each additional level of K brought down the negative index of K. Both the micronutrients, Zn and Fe, registered a negative DRIS index as −18 and −20 with $N_1P_2K_2Zn_0Fe_1$, and $N_1P_1K_1Zn_1Fe_0$, respectively, which significantly improved with those treatments supplying both the nutrients. It is important to recognize that an individual nutrient is not necessarily present in optimum concentration, if its indexes are equal to zero. A measure of total nutritional balance in a plant is indicated by the sum of nutrient indexes irrespective of sign. Increasing the nutrient doses reduced the sum of indexes from 164 ($N_0P_1K_1Zn_1Fe_1$) to as low as 38 ($N_2P_3K_4Z_1Fe_1$) which registered the highest yield (Table 5.6). The relationship between nutrient balance and yield thus became more visible. The fruit yield decreased substantially by increasing the sum of indexes ($r = -0.729$, $p = 0.01$) irrespective of their sign indicating that higher fruit yield is not necessarily obtained with large sum of indexes. Applying nutrient ratios instead of the isolated concentration values of each nutrient in the interpretation of leaf analysis, Mourão Filho and Azevedo (2003) reported a high correlation between the DRIS-based nutritional balance and fruit yield of 'Valencia' sweet orange orchards of São Paulo, Brazil.

5.4 Conclusion and Future Research

Often, the established standard sampling period many times occurs too late in the growing season so that fertilizer application will not be effective to correct the nutritional problem or may not match the sudden symptoms of a nutritional disorder when the producer mostly need the information (Walworth and Sumner 1987). To overcome this problem, there is a need for precise definition, of the sampling time, and important maturation stages of specific cultivar/variety. In addition to these limitations, little research has been developed to determine the influence of the cultivars in the nutrient concentration in a given maturation or developmental stage. Finally, factors that affect the tissue aging rate might also influence the relation between nutrient concentration and maturation. An option for these diagnostic methods is proposed through the DRIS (Beaufils 1973), which defined that, in general, nitrogen, phosphorus, and potassium concentrations decrease with tissue maturation. Therefore, the ratios N/P, N/K, and P/K (or reciprocal ratios) should remain constant. In the same way, because the concentrations of Ca and Mg generally increase with maturation, quotients between these nutrients (Ca/Mg or Mg/Ca) should result in constant values. Moreover, the product of two nutrients, with concentrations running in opposite directions with the time (N × Ca, for example), also should remain constant.

There are controversies regarding calculation procedures for the norms and DRIS indexes. One of the main questions is about the method application validation and the data universe that the norms are expected or supposed to represent. Most research results have indicated that the more specific is the universe for DRIS norms derivation, the more effective the method application is.

The criteria for the reference subpopulation definition also demand further studies and are, to a certain extent, specifically adjusted for each situation. In this way, DRIS norms should be developed for specific conditions, in which all factors to be correlated with yield or quality (or any other variable), to

Table 5.6 Progressive diagnosis with corresponding leaf nutrient concentration and fruit yield of Nagpur mandarin grown on Typic Haplustert soil type

Treatments					Leaf nutrient composition							DRIS indexes							Sum of indexes irrespective of sign	Mean fruit yield (kg tree^{-1})
N	P	K	Zn	Fe	N	P	K	Fe	Mn	Cu	Zn	N	P	K	Fe	Mn	Cu	Zn		
					(%)			(ppm)												
0	1	1	1	1	1.89	0.08	1.02	87.6	76.8	5.9	20.8	–52	–18	–12	42	18	20	02	164	21.5
1	0	1	1	1	1.97	0.06	0.86	89.6	49.1	6.5	17.9	–26	–38	–14	32	25	13	08	156	23.2
1	1	0	1	1	1.91	0.09	0.70	84.6	49.4	5.6	20.0	–28	–08	–43	42	20	11	06	158	22.2
1	1	1	1	1	2.04	0.11	0.86	88.5	48.6	5.9	19.7	–20	–07	–35	28	20	10	04	144	31.6
1	1	1	1	0	2.08	0.11	0.92	70.2	42.6	6.7	20.1	–06	–04	–28	–20	30	26	04	116	25.8
1	2	2	0	1	2.12	0.12	1.09	80.6	43.5	6.0	16.0	–14	–02	–20	18	20	16	–18	108	32.1
1	2	2	1	1	2.10	0.11	1.12	78.4	44.2	5.2	20.1	–20	–01	–14	10	17	10	–02	74	38.9
1	3	2	1	1	2.06	0.12	1.20	76.0	42.1	5.0	19.6	–18	05	–14	10	15	11	–04	72	36.4
2	3	2	1	1	2.26	0.13	1.19	82.4	50.6	6.7	19.8	–14	04	–13	08	10	09	–04	62	43.3
2	3	3	1	1	2.26	0.11	1.35	78.2	48.1	5.1	21.0	–12	03	–08	08	08	04	–03	46	49.3
2	3	4	1	1	2.32	0.11	1.54	86.2	48.7	5.5	27.0	–12	04	–04	06	06	03	–03	38	58.8
LSD (p=0.05)					0.16	0.01	0.14	6.8	NS	NS	2.5									5.2

Adapted from Srivastava and Singh (2008)

attain specific objectives, are known *per se*: cultivar, climate, soil and crop management, productivity, etc.

Finally, it is highlighted that researches, both in a world-wide basis, on DRIS method utilization are still in developing stage. Further investigations are necessary, on the identification and isolation of factors that significantly affect productivity under several citrus cultivar-specific management production systems, since DRIS reflects changes in nutrients concentration, due to alternate-bearing or crop-load effects in addition to age and type of tissue sampled.

DRIS-based nutrient diagnostic norms with further expansion under diverse applications to different critical growth stages of crop could prove an effective decision support system to address multiple nutrient constraints in order of decreasing or increasing influence on yield. A different nutrient diagnostic is required for prebearing to bearing orchards, and probably according to crop growth stages, since nutrients undergo a definite redistribution as per nutrient demand by developing fruits and supply form underground root system. Citrus decline is now a threatening problem in the world; application of DRIS holds a great promise in order to prioritize the impact of different nutritional disorders.

DRIS has further proven to be a good precision tool in extrapolating the nutrients level in relation to higher yield targets which have found a genuine validation through soil/leaf analysis values – crop response studies under long-term multilocation experiments. Studies of late have demonstrated good utility of DRIS for soil test studies, which will be very useful in predicting the soil test values-leaf nutrient interaction more gainfully to harness the consistency in production pattern of perennial crop like citrus. DRIS is warranted to be expanded to some nonessential nutrients like Na and Si.

Diagnosis of nutrient constraints based on DRIS analysis showed a good agreement between leaf and soil analysis data. All the nutrient constraints identified through original orchard data analysis further indicated a significant field response on fruit yield and improvement in respective nutrient concentration in leaves. These observations lend strong support for utility of DRIS in identification and management of nutrient constraints in citrus.

References

Abaev MN (1977) Some results of studies on foliar diagnosis of lemon trees. Subtropicheskie Kultury 5–6:150–153

Agusti M (2000) Citricultura. Ediciones Mundi-Presna SA, Madrid, p 416

Alva AK, Tucker DPH (1999) Soils and citrus nutrition. In: Timmer LW, Duncan LW (eds) Citrus health management, plant health management series. APS Press/University of Florida, St. Paul, pp 59–71

Alvardo A, Araya R, Bornemisza E, Hernandez RL (1994) Nutritional studies in eleven citrus cultivars and one local selection in the Atlantic zone of Costa Rica: plant analysis major and minor elements. Agronomia Costaricense 18:13–19

Alves FA, Mourão Filho A (2005) DRIS and sufficient range approaches in nutritional diagnosis of 'Valencia' sweet orange on three rootstocks. J Plant Nutri 28(4):691–705

Andrew CS (1968) Problems in the use of chemical analyses for diagnosis of plant nutrient deficiencies. J Aust Inst Agric Sci 34:154–162

Angeles DE, Sumner ME, Lahav E (1993) Preliminary DRIS norms for banana. J Plant Nutri 16:1059–1070

Appa Rao VV, Singh S, Sharma BD et al (2006) DRIS norms for sapota (*Manilkara achras* M.) in western plains of India. Indian J Hortic 63(2):145–147

Aso PJ (1967) Foliar nitrogen values and yields in Valenica oranges. Rev Indus Agric Tucuman 45(3):49–56

Awasthi RP, Dhuria HS, Karkara BK (1979) Nutritional status of apple orchards in Himachal Pradesh. I. Macronutrients. Indian J Hortic 36:57–63

Bar-Akiva A, Lavon R (1968) Peroxidase activity as an indicator of the iron requirement of citrus plants. Israel J Agric Res 18:145–153

Bar-Akiva A, Hiller V, Patt J (1967) Effect of phosphorus, chicken manure on yield, fruit quality and leaf composition of grapefruit trees. HortScience 2:51–53

Bataglia OC (1989) DRIS-Citros: uma alternativa pera avaliar a nutrição das plantas. Laranja 10:565–576

Bataglia OC, Santos WR (1990) Efeito do procedimento de cálculo e da população de referência nos índices do sistema integrado de diagnose e recomendação (DRIS). R Bras Ci Solo 14:339–344

Beaufils ER (1973) Diagnosis and recommendation integrated system (DRIS). Soil Sci Bull 1:1–132

Beverly RB (1987) Modified DRIS method for simplified nutrient diagnosis of 'Valenica' oranges. J Plant Nutri 10:1401–1408

Beverly RB (1991) A practical guide to the diagnosis and recommendation integrated system (DRIS). Micro-Macro, Athens, p 87

Beverly RB (1992) Prescient diagnostic-analysis shows sufficiency range approach superior to DRIS for citrus. Commun Soil Sci Plant Anal 23:2641–2649

Beverly RB, Worley RE (1992) Preliminary DRIS diagnostic norms for Pecan. HortScience 27:217–272

Beverly RB, Stark J, Ojala C, Embleton TW (1984) Nutrient diagnosis of Valencia oranges by DRIS. J Am Soc Hortic Sci 109(5):649–654

Bhargava BS (2002) Leaf analysis for nutrient diagnosis, recommendation and management in fruit crops. J Indian Soc Soil Sci 50(4):352–373

Bhargava BS, Chadha KL (1988) Leaf nutrient guide for fruits and plantation crops: a review. Fert News 33(7):21–29

Bhargava BS, Chadha KL (1993) Leaf nutrient guide for fruit crops. In: Chadha KL, Pareek OP (eds) Advances in horticulture, vol 2. Malhotra Publishing House, New Delhi

Bhargava BS, Raghupathi HB (1995) Current status and new norms for nitrogen nutrition of grapevine (*Vitis vinifera*). Indian J Agric Sci 65:165–169

Bhargava BS, Singh R (2001) Leaf nutrient norms in fruit crops. In: Singh R (ed) Proceedings of National Seminar horizons in production and post-harvest management of tropical and subtropical fruits, New Delhi, India, pp 113–114

Bhargava BS, Sumner ME (1987) Proposal for sampling grape (*Vitis vinifera* L) petiole for nutritional diagnosis. Commun Soil Sci Plant Anal 18(5):581–591

Bould C (1984) Diagnosis of mineral disorder in plants – principles. Chemical Publishing, New York, pp 1–96

Bowen JE (1992) Comparative DRIS and critical concentration interpretation of papaya tissue analysis data. Trop Agric 69:63–67

Cantarella H, Quaggio JA, Bataglia OC, van Raij B (1992) Response of citrus to NPK fertilization in a network of field trials in Sao Paulo state, Brazil. In: Tribulato E, Gentile A, Refergiato G (eds) Proceedings of International Society Citriculture, vol 2, Catania, Italy, 8–13 Mar 1992, pp 607–612

Caron J, Parent LE (1989) Derivation and assessment of DRIS norms for greenhouse tomatoes. Can J Plant Sci 69:1027–1035

Carpena-Artés O (1978) Diagnosis of the nutritive status of citrus. In: Cary PR (ed) Proceedings of International Society Citriculture, Sydney, Australia, 15–23 Aug 1978, pp 295–296
Carranca C (1999) Diagnóstico do estado nutritivo dos citrinos. Métodos de recomendação da fertilização. Agron Lusit 46:1–9
Cate RB, Nelson LA (1965) A rapid method for correlation of soil test analysis with crop response data. International soil testing series. Technical Bulletin 1, North Carolina State University, Agricultural Experiment Station, Raleigh, p 12
Cerda A, Nieves M, Martinez V (1995) An evaluation of mineral analysis of Verna lemons by DRIS. Commun Soil Sci Plant Anal 26:1697–1707
Chahill BS, Dhatt AS, Singh R et al (1991) Studies on the leaf nutrient standards in Kinnow. Indian J Hortic 48:315–320
Chapman HD (1949) Tentative leaf analysis standards. Calif Citrog 34:518
Chapman HD (1961) The status of present criteria for the diagnosis of nutrient conditions in citrus. In: Reuther W (ed) Plant analysis and fertilizer problems. American Institute of Biological Sciences, Washington, DC, pp 75–106
Chapman HD (1967) Plant analysis values suggestive of nutrient status of selective crops. In: Hardy GW et al (ed) Soil testing and plant analysis. Special publications 2. Soil Science Society of America, Madison, pp 77–92
Chundawat BS, Khiani RA, Kikani KP (1990) Nutritional survey of elite acid lime orchards in Gujarat. Indian J Hortic 48:183–186
Clements HF (1961) Crop logging of sugar cane in Hawaii. In: Reuther W (ed) Plant analysis and fertilizer problems. American Institute of Biological Sciences, Washington, DC, pp 131–147
Coetzee JGK (1980) The role of potassium in citrus production. Citrus Subtrop Fruit J 603:12–15
Cook JA, Wheeler WH (1978) Use of tissue analysis in viticulture. In: Reisenauer HM (ed) Soil and plant tissue testing in California. University of California Bulletin 1879, Berkeley, pp 14–16
Correia MP (2000) Caracterização fisiológica e nutritiva da clorose férrica em citrinos. Tese de Doutoramento em Ciências Agrárias, Faro, p 223
Creste JE (1996) Uso do DRIS na avaliação do estado nutricional do limoeiro Siciliano. UNESP/FCA, Botucatu, p 120
Creste JE, Grassi Filho H (1998) Estabelecimento de curvas de produtividade para três variedades e dois porta-enxertos cítricos na região sudoeste do Estado de São Paulo, com ênfase ao DRIS. In: Congresso Brasileiro de Fruticultura, 15., Poços de Caldas, Resumos. Poços de Caldas: Sociedade Brasileira de Fruticultura, 1:299
Davies FS, Albrigo LG (1998) Citrus. CAB International, Great Britain, p 244
Dettori A, D'hallewin G, Aggabio M (1996). S.E.M. studies on *Penicellium italicum* –'Star Ruby' grapefruit intercation as affected by hot water dripping. In: Manicom B, Robinson J, Du Plessis SF, et al (eds) Proceedings of International Society of Citriculture, vol 2, Sun City, 12–17 May 1996, pp 1288–1289
Dias JCS, Duarte L, Calouro F et al (2002) Produção integrada de citrinos. Fertilização e outras práticas culturais. Direcção-Geral de Protecção das Culturas (DGPC), Oeiras, p 48
Dow AI, Robert S (1981) Proposal: critical nutrient ranges for crop diagnosis. Agron J 74:401–403
Du Plessis SF (1977) Soil analysis as a necessary complement to leaf analysis for fertilizer advisory purposes. Proc Int Soc Citricult 1:15–19
Du Plessis SF, Koen TJ (1992) Leaf analysis norms for lemons [*Citrus limon* (L.) Burm]. In: Tribulato E, Gentile A, Refergiato G (eds) Proceedings of Seventh International Citrus Congress, vol 2. International Society of Citriculture, Acireale, pp 551–552
Elwali AMO, Gascho GJ (1984) Soil testing, foliar analysis, and DRIS as guide for sugarcane fertilization. Agron J 76:466–470
Embleton TW, Jones WW, Coggins CW Jr (1973) Aggregate effect of nutrients and gibberellic acid on Valencia orange crop value. J Am Soc Hortic Sci 98(3):281–285
Embleton TW, Jones WW, Platt RG (1978) Leaf analysis as a guide to citrus fertilization. In: Reisenauer HM (ed) Soil and plant-tissue testing in California. Division of Agricultural Sciences, University of California Bulletin 1879, Berkeley, pp 4–9
Fragoso MAC, Vicente MAF, Jordão PV et al (1990) Leaf analysis as fruit composition in orange trees: II – Foliar indices and fruit quality. Communicações do I Cong. Ibérico de Ciências Hortícolas 3:166–171
Gallasch PT, Pfeiler GR (1988) Developing a unique leaf analysis service, including a computer programme. In: Goren R, Mendel K (eds) Proceedings of sixth International Citrus Congress, International Society of Citriculture, vol 2. Balaban Publications, Rehovot, pp 689–697
Gallasch PT, Dalton GS, Ziersch J (1984) The use of juice analysis to define fertilizer requirements of citrus. Proc Int Soc Citricult 2:140–142
Gartell JW, Robson AD, Loneragan JF (1979) A new tissue test for accurate diagnosis of copper deficiency in cereals. J Agric West Aust (4th series) 20:86–88
Geraldson GM, Klacan GR, Lorenz OA (1973) Plant analysis as an aid in fertilizing vegetable crops. In: Walsh IM, Beaton JD (eds) Soil testing and plant analysis. American Society of Agronomy, Madison, pp 365–379
Goh KM, Malakouti MJ (1992) Preliminary nitrogen, phosphorus, potassium, calcium and magnesium DRIS norms and indexes for apple orchards in Canterbury, New Zealand. Commun Soil Sci Plant Anal 23:1371–1385
Gualiya MV, Zonn SV (1990) Genetic differences and productivity of red and yellow earths in relation to relief. Soviet Soil Sci J 22:16–27
Hall AD (2009) The book of the Rothamsted experiments (1905). Cornell University Library, Ithaca, New York, USA, 364 p
Hallmark WB, Mooy CJ, de Pesek J (1987) Comparison of two DRIS methods for diagnosing nutrient deficiencies. J Fert Issues 4:151–158
Hanlon EA, Obreza TA, Alva AK (1995) Tissue and plant analysis. In: Tucker DPH, Alva AK, Jackson LK et al (eds) Nutrition of Florida citrus trees. University of Florida, IFAS, Alfred, pp 13–16
Hellin E, Urena R, Sevilla F et al (1988) Effects of several Fe-chelates on the iron leaf content of Verna lemon trees. In: Goren R, Mendel K (eds) Proceedings of Sixth International Citrus Congress, International Society of Citriculture, vol 1. Balaban Publications, Rehovot, Israel, pp 555–560
Hundal HS, Arora CL (1996) Preliminary micronutrient foliar diagnostic norms for litchi (*Litchi chinesis* S.) using DRIS. J Indian Soc Soil Sci 44(2):294–298
Hundal HS, Arora CL (2001) Diagnosis and recommendation integrated system (DRIS) approach for diagnosing the nutrient status of Kinnow fruit trees. J Indian Soc Soil Sci 49:703–709
Ishihara M, Shimosako Y, Sakaguchi S et al (1972) Studies on copper deficiency of citrus. I. In: Methods of diagnosing copper deficiency from symptoms and from tree and soil analyses. Bulletin No 11. Horticulture Research Station (Hiratsuka), Kanagawa, pp 41–76
Jones WW (1981) Proposed modifications on the diagnosis and recommendation integrated system (DRIS) for interpreting plant analysis. Commun Soil Sci Plant Anal 12:785–794
Jorgensen KR, Price GH (1978) The citrus leaf and soil analysis system in Queensland. In: Cary PR (ed) Proceedings of International Citrus Congress, International Society Citriculture, vol 1, Griffith, New South Wales, Australia, pp 297–299
Kallsen C (2002) Fall leaf tissue samples important for maintaining citrus growth, fruit quality and yield. http://cekern.ucdavis.edu/Custom_Program

Kenworthy AL (1967) Plant analysis and interpretation of analysis for horticultural crops. In: Hardy GW et al (eds) Soil testing and plant analysis. Part II: Plant analysis, Special publication 2. Soil Science Society of America, Madison, pp 59–75
Kenworthy AL (1973) Leaf analysis as an aid in fertilizing orchards. In: Walsh LM, Beaton JD (eds) Soil testing and plant analysis. Soil Science Society of America, Madison, pp 381–392
Koo RCJ, Anderson CA, Stewart I et al (1984) Recommended fertilizers and nutritional spray for citrus. Florida Agricultural Experiment Station Bulletin 536D, Gainesville, Florida, USA
Koto M, Fujiwara S, Shengzhi S (1990) Leaf analysis standards for determining fertilizer needs and its computerization. In: Huang Bangyan, Yang Qien (eds) Proceedings of International Symposium Citriculture, Guangzhou, China, pp 383–389
Lacertosa G, Giuffrida A, Mennone C et al (2001) Stato nutrizionale e qualità dei fruti in agrumeti del metapontino. XVII Convegno SICA, pp 387–394
Lagatu H, Maume L (1926) Diagnostic de l'alimentation d'un végétal par l'evolution chimique d'une feuille convenablement choisie. C R Acad Sci Paris 182:633–655
Legaz F, Serna MD, Primo-Millo E (1995) Mobilisation of the reserve N in citrus. Plant Soil 173:205–210
Li J, Li XX, Peng JC et al (1999) Effect of five mineral elements on nutritional status of Navel orange tree. J Hunan Agric Univ 25(11):36–39
Liu XH, Zhao YQ, Zheng JJ et al (1984) A study of some mineral nutrients in the leaves of major cultivars of citrus in Fujian province. J Fujian Agric College 13(1):31–38
Lundegardh H (1951) Leaf analysis. Higher and Watts Ltd., London, pp 1–176
Macy P (1936) The quantitative mineral nutrient requirements of plants. Plant Physiol 11:749–764
Magnitskii KP, Takidze RM (1972) The interrelationship of ions in mandarin tree nutrition. Subtropicheskie Kul'tury 6:87–91
Malavolta E, Oliveira SA, Vitti GC (1993) The use of diagnosis recommendation integrated system (DRIS) to evaluate the nutritional status of healthy and blight affected citrus trees. In: Fragoso MAC, van Beusichem ML (eds) Proceedings of eighth international colloquium for the optimization of plant nutrition. Kluwer Academic, Dordrecht, 31 Aug–8 Sept 1992, pp 157–159
Marchal J, Cassin J, Favreau P et al (1978) Principal results of fertilizer studies on clementines in Corsica. Fruits (Paris) 33:822–827
Mayfield JL, Simonne EH, Mitchell CC et al (2002) Effect of current fertilization practices on nutritional status of double cropped tomato and cucumber produced with plasticulture. J Plant Nutri 25:1–15
Melsted SW, Moto HL, Peck TR (1969) Critical plant nutrient composition values useful in interpreting plant analysis data. Agron J 61:17–20
Menino MR (2005) Eficiência da utilização do azoto em jovens laranjeiras 'Lane Late' enxertadas em citranjeira 'Carrizo', na Campina de Faro. Tese de Doutoramento em Engenharia Agronómica. ISA, Lisboa, p 248
Menino MR, de Varennes A, Tomás JC (2008) Leaf N concentration on citrus trees during the first nine years in the field. In: Proceedings of 11th International Society of Citriculture 1:623–626
Moller Nielson J, Friis-Nielson B (1976a) Evaluation and control of the nutritional status of cereals II. Pure-effect of a nutrient. Plant Soil 45:339–351
Moller Nielson J, Friis-Nielson B (1976b) Evaluation and control of the nutritional status of cereals III. Methods of diagnosis and yield prognosis. Plant Soil 45:647–658
Moreno JJ, Lucena JJ, Carpena O (1996) Effect of the iron supply on the nutrition of different citrus variety/rootstock combinations using DRIS. J Plant Nutri 19:689–704
Mourão Filho FAA (2004) DRIS concept and applications on nutritional diagnosis in fruit crops. Sci Agric (Piracicaba, Brazil) 61(5):550–560
Mourão Filho FDeA, Azevedo JC (2003) DRIS norms for Valencia sweet orange on three rootstocks. Pesquisa Agropecuaria Brasileira 38:85–93
Munsen RD, Nelson WL (1973) Principles and practices in plant analysis. In: Walsh LM, Beaton JD (eds) Soil testing and plant analysis. Soil Science Society of America, Madison, pp 223–248
Nelson LA, Anderson RL (1977) Partitioning of soil test – crop response probability. Soil testing: correlating and interpreting the analytical results. ASA special publication 29. American Society of Agronomy, Crop Science Society of America and Soil Science Society of America 677, Madison, pp 19–38
Page AL, Martin JP (1964) Growth and chemical composition of citrus seedlings as influenced by Na additions to soils low in exchangeable K. Soil Sci 98:270–273
Parent LE, Dafir M (1992) A theoretical concept of compositional nutrient diagnosis. J Am Soc Hortic Sci 117:239–242
Parent LE, Cambouris AN, Muhawenimama A (1994a) Multivariate diagnosis of nutrient imbalance in potato crop. Soil Sci Soc Am J 58:1432–1438
Parent LE, Isfan D, Tremblay N et al (1994b) Multivariate nutrient diagnosis of carrot crop. J Am Soc Hortic Sci 119:420–426
Perez GC (1996) Argentina. In: Manicom B, Robinson J, Du Plessis SF et al (eds) Eighth International Citrus Congress, vol 2. International Society Citriculture, Sun City, p 1279
Pestana M, Correia PJ, de Varennes A et al (2001) The use of flora analysis to diagnose the nutritional status of orange trees. J Plant Nutr 24:1913–1923
Prevot P, Ollagnier M (1961) Law of the minimum and balanced mineral nutrition. In: Reuther W (ed) Plant analysis and fertilizer problem. American Institute of Biological Sciences Publication, Washington, DC 8:255–277
Primo E, Carrasco JM, Cunat P (1969) Trace element deficiencies in orange trees. I. Relationship between deficiency symptoms and trace metal content of orange tree leaves. Rev Agroquim Tecnol Aliment 9:240–253
Quaggio JA, Cantarella H, Van Raiz B (1998) Phosphorus and potassium soil test and nitrogen leaf analysis as a guide to citrus fertilization. Nutr Cycl Agroecosys 52(1):67–74
Raghupathi HB, Bhargava BS (1998) Diagnosis of nutrient imbalance in pomegranate by diagnosis and recommendation integrated system and compositional nutrient diagnosis. Commun Soil Sci Plant Anal 29:2881–2892
Raghupathi HB, Bhargava BS (1999) Preliminary nutrient norms for Alphonso mango using diagnosis and recommendation integrated system. Indian J Agric Sci 69:648–650
Razeto MB, Lengueire MJ, Rojas ZS, Reginato MG (1988) Correction of manganese and zinc deficiencies in orange. Agricultura Tecnica 48:347–352
Rodriguez O, Gallo JR (1961) Levantamento do estado nutricional de pomares cítricos de São Paulo pela análise foliar. Bragantia, Campinas (SP) 20:1183–1202
Rodriguez O, Rojas E, Sumner M (1997) Valencia orange DRIS norms for Venezuela. Commun Soil Sci Plant Anal 19:1461–1468
Saatci N, Mur Y (2000) Relationships between the concentration of iron, macro-and-micronutrients in satsuma mandarin leaves (*Citrus unshiu* Marc.). J Plant Nutri 23:1745–1750
Sanchez CA, Snyder GH, Burdine HW (1991) DRIS evaluation of the nutritional status of Crisphead. Lettuce. HortScience 26:274–276
Santos WR (1997) dos Avaliação do equilibrio nutricional dos macronutrientes em citros com diferentes adubações. Piracicaba: USP/ESALQ, p 112 (Dissertação-Mestrado)
Sanz M (1999) Evaluation of interpretation of DRIS system during growing season of the peach tree: comparison with DOP method. Commun Soil Sci Plant Anal 30:1025–1036
Schaffer AA, Sagee O, Goldschmidt EE et al (1987) Invertase and sucrose synthase activity, carbohydrate status and endogenous IAA levels during citrus leaf development. Physiol Plant 69(1):151–155

Schaffer B, Larson KD, Snyder GH, Sanchez CA (1988) Identification of mineral deficiencies associated with mango decline by DRIS. HortScience 23:617–619

Schumann A (2009) Potential use of DRIS for leaf nutrient diagnosis in Florida citrus. Citrus Industry. http://www.crec.ifas.ufl.edu/extension/trade_journals/2009/2009%20April%20use%20of%20DRIS%20for%20diagnosis.pdf

Singh V, Triphati BR (1985) Studies on chlorosis in sweet orange in Agra region of Uttar Pradesh. J Indian Soc Soil Sci 33:333–338

Sharma VK, Mahajan KK (1990) Studies on nutrient status of mandarin orchards in Himachal Pradesh. Indian J Hortic 47:180–185

Shimizu T, Morii M (1985) Influence of soil environment on nutrient deficiency of Satsuma mandarin (*Citrus unshiu*). Environ Con Biol 23(4):77–87

Small HG, Ohlrogge AJ (1973) Plant analysis as an aid in fertilizing soybeans and peanuts. In: Walsh IM, Beaton JD (eds) Soil testing and plant analysis. American Society of Agronomy, Madison, pp 315–328

Smith PF (1966) Leaf analysis of citrus. In: Childers NF (ed) Temperate to tropical fruit nutrition. Horticulture Publication, Rutgers University, New Brunswick, pp 208–228

Srivastava AK, Singh S (2002) Soil analysis based diagnostic norms for Indian citrus cultivar. Commun Soil Sci Plant Anal 33:1689–1706

Srivastava AK, Singh S (2003a) Diagnostics of citrus nutrition. In: Citrus nutrition. International Book Distribution Company, Lucknow, pp 8–236

Srivastava AK, Singh S (2003b) Plant and soil diagnosis norms for optimum productivity of Nagpur mandarin (*Citrus reticulata* Blanco) grown in central India. Fert News 48:47–54

Srivastava AK, Singh S (2003c). Diagnosis of nutrient constraints in citrus. In: Manual No. 2. National Research Centre for Citrus, Nagpur, pp 1–70

Srivastava AK, Singh S (2004a) Zinc nutrition, a global concern for sustainable citrus production. J Sustain Agric 25:1–20

Srivastava AK, Singh S (2004b) Nutrient diagnostics and management in citrus. In: Technical Bulletin no. 8. National Research Centre for Citrus, Nagpur, pp 106–111

Srivastava AK, Singh S (2006) Diagnosis of nutrient constraints in citrus orchards of humid tropical India. J Plant Nutri 29(6): 1061–1076

Srivastava AK, Singh S (2008) DRIS norms and their field validation in Nagpur mandarin (*Citrus reticulata* Blanco). J Plant Nutri 31(6):1091–1107

Srivastava AK, Kohli RR, Dass HC et al (1999) Evaluation of the nutritional status of Nagpur mandarin (*Citrus reticulata* Blanco) by foliar sampling. Trop Agric (Trinidad) 72(6):93–98

Srivastava AK, Singh S, Huchche AD et al (2001) Yield based leaf and soil interpretation for Nagpur mandarin in central India. Commun Soil Sci Plant Anal 32:585–599

Srivastava AK, Singh S, Tiwari KN (2007) Diagnostic tools for citrus: their use and implications in India. Better Crops India 1(1):26–29

Sumner ME (1977) Use of the DRIS system in foliar diagnosis of crops at high yield levels. Commun Soil Sci Plant Anal 6:269–280

Swietlik D (1996) Responses of citrus trees in Texas to foliar and soil Zn applications. In: Manicom B, Robinson J, Du Plessis SF et al (eds) Proceedings of eighth International Citrus Congress, vol 2. International Society of Citriculture, Sun City, pp 772–776

Teixeira LAJ, Santos WR, dos Bataglia OC (2002) The N and K diagnosis on banana plants using the diagnosis and recommendation integrated system (DRIS) a critical value approach. Rev Bras Frutic 24:530–535

Terblanche JH, Du Plessis SF (1992) Summary of workshop on leaf and soil analysis as a tool for determining fertilizer requirements of citrus. In: Tribulato E, Gentile A, Refergiato G (eds) Proceedings of International Society of Citriculture, vol 2, Acireale, 8–13 Mar 1992, pp 744–745

Thomas W (1945) Present status of diagnosis of mineral requirements of plants by means of leaf analysis. Soil Sci 59:353–374

Thompson TL, White SA, Walworth J et al (2003) Development of best management practices for fertigation of young citrus trees, 2003 Report. University of Arizona. Citrus and Deciduous Fruit and Nut Report.

Tserling VV (1974) In: Principal results of investigations on nutritive diagnosis of major crops in different natural zones of the USSR, vol 2. Seventh International Coll Plant and Fert Prob, pp 459–479

Ulrich A (1948) Plant analysis methods and interpretation of results. In: Diagnostic techniques for soil and crops. American Potash Institute, Washington, DC, pp 157–198

Ulrich A (1952) Physiological basis for assessing the nutritional requirements of plants. Ann Rev Plant Physiol 3:207–228

Varalakshmi LR, Bhargava BS (1998) Plant nutrient status and leaf diagnostic norms for acid lime (*Citrus aurantifolia*). Indian J Agric Sci 68:375–376

Wallace A (1990) Nitrogen, phosphorus, potassium interaction on Valencia orange yields. J Plant Nutri 13:357–365

Walworth JL, Sumner MW (1987) The diagnosis and recommendation integrated system (DRIS). Adv Soil Sci 6:150–188

Walworth JL, Letzsch WS, Sumner ME (1986) Use of boundary lines in establishing diagnostic norms. Soil Sci Soc Am J 50:123–127

Wang T (1985) Application of fertilizers to satsuma based on leaf analysis. J Soil Sci (Turang Tongbo) 16:275–277

Webb RA (1972) Use of the boundary line in the analysis of biological data. J Hortic Sci 47:309–319

Weinert TL, Thompson TL, White SA et al (2002) Nitrogen fertigation of young Navel oranges: growth, N status, and uptake of fertilizer N. HortScience 37:334–337

Woods DB, Villiers JM (1992) Diagnosing the nutrient status of 'Valencia' oranges in Southern Africa. In: Tribulato E, Gentile A, Refergiato G (eds) Proceedings Seventh International Citrus Congress. International Society of Citriculture, Acireale, Italy, pp 556–559

Wu Wer, LiuWengeng, LiVinguo et al (1998) Studies on relationship between fruit yield, quality and microelements in leaves of citrus. J Southwest Agric Univ 20:198–202

Leaf Analysis in Citrus: Developments in Analytical Techniques

6

E.A. Hanlon, K.T. Morgan, T.A. Obreza, and R. Mylavarapu

Abstract

The complexity and sophistication of instruments for plant tissue analysis continues to advance, both in research laboratory grade instruments as well as for field use. These advances are seen in direct measurement of nutrients and trace elements, as well as indirect measurement, especially as statistical approaches improve through modeling of relationships between direct and indirect methods. This chapter has given an overview of recent advances, often using review articles followed by selected applications to citrus leaf analysis, but also offering other venues for using these analytical tools to aid both research and precision citrus management.

Keywords

Leaf analysis • High temperature slow oxidation • Microwave digestion • Automated color analysis • Segment flow analysis • Flow injection technology • Discrete analysis • Inductively coupled plasma analysis • Isotope ratios • Sensors • Infrared spectroscopy • Laser-induced breakdown spectroscopy

6.1 Introduction

Plant tissue testing is a common tool of agronomists and horticulturalists for diagnosing nutrient requirements for optimum performance/production. Commonly, leaf tissue is the sampled plant part upon which selected analyses are conducted to determine nutrient or trace element concentrations. Subsequently, these results are interpreted, usually based upon research with the crop of interest allowing calibration of nutrient or trace element concentrations into deficiency, sufficiency, or toxicity. Much less common is research that generates crop response to added nutrients based upon plant tissue concentration interpretations during the growing season.

Citrus leaf tissue interpretations are discussed in other chapters of this book; however, publications by Smith (1966) and Embleton et al. (1973) are seminal works. Subsequent research into citrus leaf tissue interpretations has for the most part confirmed these early summaries.

Laboratory analyses leading to the interpretation of nutrient and/or trace elements are controlled by a number of factors, most of which will be covered in detail in other chapters. It is important, however, to review these pivotal results briefly to provide a framework for this analysis. For example, Ontermaa et al. (1996) discuss their observed reliability for citrus analyses applied to field production decisions and fertilization recommendations.

E.A. Hanlon (✉) • K.T. Morgan
Soil and Water Science Department, Southwest Florida Research and Education Center, University of Florida, 2685 State Rd 29 North, Immokalee, FL 34142, USA
e-mail: eahanlon@ufl.edu; Conserve@ufl.edu

T.A. Obreza
Soil and Water Science Department, University of Florida, 1038 McCarty Hall, Gainesville, FL 32611, USA
e-mail: obreza@ufl.edu

R. Mylavarapu
Soil and Water Science Department, University of Florida, 171 McCarty Hall, Gainesville, FL 32611-0290, USA
e-mail: raom@ufl.edu

A.K. Srivastava (ed.), *Advances in Citrus Nutrition*,
DOI 10.1007/978-94-007-4171-3_6, © Springer Science+Business Media B.V. 2012

Analysis of any tissue is essentially a measurement of what the plant has integrated through many uptake, distribution, metabolism, and stage of growth hormonal controls since planting. Said another way, tissue analysis will provide an insight into the physiological response of the plant to specific nutrient management throughout the previous 3–6 weeks before tissue sampling. This integration by the plant is influenced by the mobility of the nutrient(s) of interest within the plant. For example, consider quite mobile nitrogen (N) compared with immobile boron (B). One can readily see that the stage of leaf development for citrus is important since newly forming leaves are considered sinks while older leaves may act as sources if roots are unable to take up sufficient concentrations from the soil. For this reason, most diagnostic or monitoring protocols for orange tree leaf tissue samples are based upon newly mature leaves, for example, 4–6 months after a growth flush (Alva et al. 2006), while others have identified 6–7 months after a flush for mandarin (Manivannan and Chadha 2011). Alva and colleagues also reported citrus yield and quality response for a range of N and potassium (K) concentrations, but found no response to phosphorus (P) tissue concentrations from 0.8 to 2.4 $g\,kg^{-1}$ (Alva et al. 2006).

Similarly, human activities can affect the results of the laboratory analyses. For example, the preparation of leaves for analysis has an influence that appears to be inconsistent when leaves are washed by various methods to remove or minimize the effects of previously applied foliar sprays (Futch and Gallaher 1996; Alva and Tucker 1997). Futch and Gallaher (1996) focused on zinc (Zn) due to its affinity for sorption to the waxy cuticle on the upper surface of citrus leaves, finding advantages to specific washing methods, while Alva and Tucker (1997) note no benefits for washing with a detergent/acid/rinse procedure for all macro- and micronutrients. These researchers suggested that the time between micronutrient sprays and normal tissue sampling (approximately 6 months) was sufficient to avoid surface contamination in Florida, except for copper (Cu), a micronutrient known to bind tightly with organic compounds.

6.2 Intent of This Chapter

It is the intent of this chapter to present new analytical developments and deal specifically with analysis of selected nutrients or trace elements. Where possible, we have cited literature that summarizes earlier work and we then focused on new analytical equipment or innovative uses or combinations of instruments to enhance accuracy and precision. When a new or innovative approach could apply to citrus in our judgment, we have included information on that approach, even though citrus may have not been the research focus.

6.3 High-Temperature Slow Oxidation

The literature contains considerable information concerning leaf tissue preparation. Procedures vary based upon the type of sample processing in the analytical phase. Typically, ashing in a muffle furnace requires grinding to pass either a 1- or 2-mm screen depending on the quantity to be ashed (Campbell and Plank 1998). Dry ashing oxidation is affected by the internal surface of the muffle furnace since the surface either sorbs or releases specific elements, especially boron (B) (Miller 1998a).Other nutrients and trace elements, such as sulfur (S), arsenic (As), mercury (Hg), and selenium (Se), are subject to volatilization and must be processed by other means for quantification (Miller 1998a). Temperature selection is also well documented since other elements respond through potential volatilization, and typically, internal temperatures exceeding 500°C should be avoided. Controllers that allow so-called ramping up and down of internal temperatures are useful, permitting the oxidation to proceed sufficiently fast to ensure laboratory time efficiencies but slow enough to avoid rapid oxidation (burning) with resulting uncontrolled temperatures in the sample.

6.4 Microwave Digestion

As with muffle furnace digestion, abundant literature describes successful options for liquid digestion of plant tissue using microwave heating. Power requirements and the ability to control the input of energy for digestions have advanced through the years. Vessels have been explored that may be sealed or vented, and selected acids have also been tested using a number of plant tissues and compared with muffle furnace digestions (Soon and Kalra 1994; Kalra and Maynard 1998; Miller 1998b).

Different acids and acid/catalysts have been used, but nitric acid or sulfuric acid with hydrogen peroxide appear to be the most frequently tested (e.g., Esslemont et al. 2000).

Samples are heated by an oscillating electromagnetic field having electromagnetic radiation of frequencies 100–100,000 megacycles per second (Kalra and Maynard 1998). Because the radiation energy is applied directly to the digestion mixture, this radiation provides extremely rapid heating and good control of power and time in the process of plant tissue digestion (White and Douthit 1985).

Microwave digestion technique for sample dissolution/digestion uses microwave radiation in high-pressure Teflon PFA vessels, which are transparent to microwave radiation (Sah and Miller 1992). Teflon PFA is chemically inert and maintains physical integrity to a temperature as high as 260°C. The advantage of this technique includes rapid sample dissolution, minimizes contamination, and does not require $HClO_4$. Microwave digestion has been used successfully to extract organic samples for further analyses, such as atomic absorption, inductively coupled plasma mass spectroscopy (ICP-MS), and chromatography (Rea and Keeler 1998). Microwave dissolution gives excellent recovery of elements when analyzed by inductively coupled plasma optical emission spectrometry (ICP-OES) (Abu-Samra et al. 1975; Nadkarni 1984; White and Douthit 1985).

However, the speed at which the samples are heated in the microwave may result in exothermic reactions during the digestion process (Berghof Products and Instruments GmbH 2011). To avoid this undesirable result, selected sensors have been developed to monitor the pressure and temperature, which can be regulated by changes in the microwave power output or changes in digestion time.

6.5 Instrumentation

6.5.1 Automated Color Analysis

As analytical techniques were automated, those instruments based on absorption spectrophotometry were grouped under Automated Color Analysis technique. The basic technique called "continuous flow" invented by Skeggs, a clinical chemist, was commercialized in 1956 and sold using the brand name Technicon, still widely in use. The instrument was earlier also referred to as Sequential Multiple Analyzer or Sequential Multiple Analyzer with Computer and more currently is being sold as AutoAnalyzer. These instruments are capable of multichemistry analyses enabling determinations of nitrate, phosphate, etc., and their species, simultaneously. At least three major variations using this technique are currently available in the laboratories around the world. A survey done in Australia (Handson and Shelley 1993) revealed that at least one colorimetric procedure for plant analysis was employed in 70% of the labs.

6.5.1.1 Segmented Flow Analysis (SFA)

The underlying technique of AutoAnalyzers is called "segmented flow analysis," which has revolutionized laboratory analyses by significantly increasing the number of samples processed from a few dozen to a few hundred per day. A continuous stream of samples is separated by air bubbles into discrete segments, which are then transported into mixing coils through microtubings to various optional modules. The modules may include distillation, dialysis, extraction, heating, signal detector, etc., as appropriate. The introduced air or nitrogen gas (in some versions) bubbles act as barriers between the sample segments preventing cross-contamination or intersample dispersion during their transport and forming unique reaction segments. The bubbles also aid in chemical mixing of the reagents and sample in glass coils. Glass is generally preferred as it is inert, allows for visual checks, and is easy to clean. The diameter and the length of the coil control the rate and time of the reaction. Newer instrument models employ 1-mm coils, thus increasing the throughput of the samples. The unique feature of the SFA technique is that all the reactions are allowed to reach a completion or a steady state before the final measurement is recorded, thus increasing the sensitivity of the measurement as the chemical strength will be highest at the completion of the final reaction.

Ceballos et al. (2006) successfully used the SFA technique to determine total nitrogen in cassava root samples to estimate crude protein content. No significant difference ($p = 0.05$) was found when the cadmium column in the air-segmented continuous flow system was replaced by an enzymatic reduction method using soluble nitrate reductase purified from corn leaves for determination of nitrate+nitrite in natural water samples, providing a nontoxic alternative to cadmium (Patton et al. 2002).

6.5.1.2 Flow Injection Technology (FIA)

The second type of "continuous flow" instruments is built using a technique called the "flow injection analysis," invented in 1974 (Růžička and Hansen 1980). These systems use a design similar to that of SFA, but instead of air bubbles, the samples are separated by a carrier agent. The sample, carrier agent, and reagents are mixed to obtain a concentration gradient. By removal of air bubbles used in the SFA system, FIA instruments can be miniaturized considerably. In the FIA systems, the reactions are not required to reach completion or a steady state since both samples and reagents are allowed the same time to react. Dispersion of the sample during analysis renders the FIA systems unsuitable for kinetically slow reactions (Gardner and Malczyk 1983). In the FIA systems, precise timing of the reaction is crucial to the accuracy of the measurement.

The FIA technique has been further combined with other instruments to prepare the samples and to remove interferences with resulting improved sensitivities. Carrasco et al. (2007) compared colorimetric analysis of ammonium, phosphate, and nitrite in wastewater using multicommutated peristaltic and solenoid FIA systems. These researchers determined that the peristaltic system had more versatility, potential for miniaturization, and flexibility requiring no physical reconfiguration of the flow manifold (Carrasco et al. 2007).

6.5.1.3 Discrete Analysis

Use of discrete analyzers in soil and tissue testing started circa 2002, and the instruments have quickly become an attractive alternative in the automated color analyses arena due to their unique ability to offer hassle-free change among chemistries for multiple analytes, consistent and reproducible results, and reduced sample and reagent volume requirements. Analysis in these instruments is carried out in separate reaction chambers and does not include coil assemblages. This discrete setup allows for multiple tests on individual samples and/or different tests on different samples. Also, reagent addition and mixing for new samples are automated and can be performed simultaneously as analysis of the previous sample is being performed. This technique measures the concentrations of the analytes after the reaction reaches a steady state.

6.5.2 Atomic Absorption and Inductively Coupled Plasma Spectroscopy

Evans et al. (2003) and Husted et al. (2011) provide reviews for the use of atomic absorption spectrometry (AAS), which is a mature analytical method for plant tissue analyses. The functionality of AAS has been greatly increased by additions of gas selection, hydride generation methods, and automation of multiple light sources.

Advances using inductively coupled plasma (ICP) have allowed this technology to see widespread use for plant tissue analyses. While AAS is often limited to sequential analysis as a function of the light source(s), ICP can provide sequential or simultaneous analyses of multiple elements due to the higher energy associated with the plasma and the design of light bench/sensor array technology (Husted et al. 2011).

ICP spectrometry includes combinations of analytical techniques including ICP-optical emission spectrometry (ICP-OES) and ICP-mass spectrometry (ICP-MS), among which ICP-MS is the preferred analytical technique for routine analysis of plant tissues (Laursen et al. 2009), This technique has been successfully used in numbers of studies for analyzing citrus leaf tissues for different elements (Boaretto et al. 2008; Menesatti et al. 2010; Kraska and Breitenbeck 2010). However, full-quantitative ICP-MS analysis is time as well as labor consuming, and its capacity is not fully utilized. Therefore, semiquantitative ICP-MS has been recommended to be the strong alternative of the full-quantitative ICP-MS, which relies on a simple calibration procedure (Taylor 2001), for analyzing leaf tissues of various plants including citrus (García-Alonso et al. 1997).

Mass spectrometry is based on the physical properties of the atomic nucleus consisting of protons and neutrons (Becker 2008). By means of spectrometry, the mass of atoms and molecules can be determined by measuring the mass to charge ratio. The mass of an atom or molecule is measured in a multiple of the atomic mass constant, not in g or kg (Becker 2008). Samples are analyzed after microwave-induced digestion using closed vessels and acid/catalysts such as nitric acid, hydrogen peroxide, and fluoric acid. An ICP typically has five different components: sample introduction system (nebulizer), ICP torch, high-frequency generator, transfer optics and spectrometer, and computer interface. The touch assembly consists of three concentric tubes, usually made of silica, that are termed as outer loop, intermediate loop, and inner loop. The torch is situated within a water-cooled coil of a radio frequency (r.f.) generator. As flowing gases are introduced into the torch, the r.f. field is activated and the gas in the coil region becomes electrically conductive; this sequence of events results in formation of plasma (Bradford and Cook 1997).

6.5.3 Linking Together of Instrumentation for Greater Specificity

With linkages between or among selected instrumentation types, often referred to as multiple hyphenated instrumentation, for example, ICP-MS, the ability to analyze nutrients and/or trace elements has expanded. ICP-MS phosphorus (P) determination from a low-volume microwave digestion was compared with colorimetric flow injection analysis or aluminum block digestion and vanadomolybdate P colorimetric analysis (Esslemont et al. 2000). In addition to the fully quantitative methods that most laboratories use, Laursen et al. (2009) have employed a simultaneous ICP-MS (mass spectrometer) method for a reliable semiquantitative analysis of plant tissue digestions. The MS is used to estimate elements throughout the mass scan range, a small set of elements compose a suite of internal standards, and algorithms are used to estimate tissue concentrations. Determining concentrations for up to 70 elements at speeds fivefold that required for full quantification using traditional ICP-MS procedures, this process is useful for a number of field-scale management decisions and has been tested with a number of different plant species, including citrus (Laursen et al. 2009).

When high-pressure liquid chromatography (HPLC) is combined with hydride gas generation (HG) and atomic florescence spectroscopy (AFS), the resulting HPLC-HG-AFS system is useful for selected trace element analysis Nash et al. (2002) produced a review of this linkage focusing on antimony (Sb). Antimony (Sb) is of great health concern for human and animal food chains and is an excellent example of this linkage system. The use of HPLC-HG-AFS instrumentation to determine Sb has been refined for work with both solids (marine sediments, Potin-Gautier et al. 2005) and liquids (sea water, Gregori et al. 2005). The process uses the liquid column to create separation of the Sb from other constituents and among Sb species/compounds. The HG portion creates stibine gas, which is then introduced into the AFS detector. These two groups of researchers identified EDTA as an important chelate in the HPLC phase to stabilize trivalent Sb, the most reactive species. The chemistry and analytical technique should apply to citrus leaf tissue as well for other selected elements.

6.5.4 Isotope Ratios

Micronutrients are often used as a target for foliar applications. This application method affects, to some extent, the traditional analyses since the element may be sorbed in some fashion to leaf, but not functional with the plant system as a nutrient (see Sect. 6.1). The effectiveness of these foliar sprays is often recorded as the amount applied and subsequent yield improvement compared with a 0 application rate. Boaretto et al. (2011) compared boron (B) use in low-fertility tropical soils of Brazil. Fertigated 4-year-old "Valencia" sweet orange trees on "Swingle" citrumelo rootstock were treated with either foliar spray or soil-applied isotopically enriched ^{10}B. This technique allowed the researchers to compare the isotope ratios found in fruit that was solely from the foliar or soil applications, and reduce the interference of B from native soil sources. Soil applications allowed 21% recovery of the enriched B, while only 7% B originated from foliar applications (Boaretto et al. 2011). While enriched or depleted sources of nutrients and trace elements are useful for exploring research aspects that affect management decisions, use of this technique in citrus leaf diagnostic programs is unlikely.

6.5.5 Sensors

The development of sensors, often using a light source that impinges on individual plant parts or, more broadly, whole canopies, is an active field of research. Enhanced by state and federal programs in many countries, this research area continues to produce interesting results. Currently, this approach to make field-scale decisions based on representative sampling is not a rigorous analytical essay, but rather an indirect measurement. Since this topic is quite broad, we provide only a couple of examples.

Typical of sensors used for plant parts, Min and coworkers (2008) produced a sensor that indirectly measures nitrogen (N) in a single leaf. A light source impinges on the leaf, which is shielded from other ambient radiation. Filters pass selected wavelengths of reflect light from the leaf surface to the actual sensor. The wavelengths are often chosen for those that are sensitive to known leaf and nutrient properties, in this case, leaf chlorophyll and proteins, which relates well to N (Min et al. 2008). The sensor was measuring reflected light from 620 to 950 nm and 1,400–2,500 nm with a resolution of less than 30 nm. Comparing the calibrated sensor with more traditional chemical analysis results, the root mean square difference was 1.69 $g\,kg^{-1}$, which was sufficiently accurate to group citrus N concentrations into low, medium, or high categories with 70% accuracy (Min et al. 2008). This achievement is comparable with specific ion electrodes using sap expression for nitrate-N, but is considerably faster than the electrode method.

Whole grove indices can be developed using the so-called normalized difference vegetation index (NDVI), which compares near-infrared and visible reflectance difference to the sum of both. Fletcher et al. (2004) report on one such study for citrus using a commercially available sensor mounted in an aircraft. Tree health was easily differentiated between the stressed block and the unstressed block. The usefulness of this type of measurement is that entire blocks and groups of blocks can be sampled quickly and results using NDVI calculations are quickly analyzed and interpreted. Depending on the selectivity of the sensor, additional reasons for stress may be predicted.

Combining NDVI measurements with other information from the grove, Mann et al. (2011) produced productivity zones. The primary indicator was found to be canopy volume, which itself was ultrasonically recorded using another type of sensor. NDVI and soil characteristics were all strong indicators of yield. This integration for management of citrus is another indicator of the value of this type of sensor research and its applicability.

6.5.6 Near-Infrared Spectroscopy

Near-infrared (780–2,500 nm) instrumentation is being used throughout the citrus industry. Laboratory-grade instruments can be controlled via filters or equipped to scan through a range of wavelengths, often with a precision diffraction grating. In general, instruments are designed to focus on specific bonds, such as C–H, N–H, or O–H, that can be made to generate oscillations when an NIR source is used to stimulate overtones and combinations (Foley et al. 1998). For solid materials, like finely ground citrus leaves, NIR can penetrate the substance more deeply than visible or mid-infrared wavelengths, allowing a larger volume of material to respond to this energy input. The reflectance energy, either at specific filter settings or a spectrum scan, can be subjected to statistical analyses where the reflectance is related to standardized chemical analyses. For this reason, most NIR results are indirect measurements.

Portable equipment may also represent a combination of visual light and NIR. Citrus leaves (*Citrus sinensis* (L.) Osbeck cv. Tarocco) were analyzed individually ($n=50$) using a pen probe and portable spectrophotometer followed by chemical analyses (Menesatti et al. 2010). While a number of statistical comparisons were made, potassium (K, $r=0.991$) and magnesium (Mg, $r=0.883$) were correlated well while phosphorus (P, $r=0.481$) was poorly correlated (Menesatti et al. 2010). Variations observed in this study are typical of this type of indirect measurement that relies on statistical modeling of the resulting reflectance data to that of chemical analyses.

While the literature contains many references statistically calibrating NIR or NDVI to leaf tissue nutrient concentrations, this indirect technology can also be used in many other ways as both a research tool and as a quality control asset where qualitative information can be useful. For example, the quality aspects of citrus fruit can be approximated by NIR to assess quality-related attributes (Abbott 1999). Using portable NIR equipment, individual readings of citrus fruit were taken on trees and statistically regressed with refractometer readings (Zude et al. 2008). Findings indicated that fruit development and maturity were greatly affected by precipitation and that individual fruit maturity dates could be generated to predict grove harvesting windows for high fruit quality (Zude et al. 2008).

Blasco et al. (2007) expanded fruit-grading line cameras to include NIR and ultraviolet spectra in addition to the normal use of visual spectra. These innovations successfully identified imperfections that would have gone unnoticed if just visual spectra were used. These researchers reported increased profits from high-quality and acceptable-quality fruit being packed and shipped separately, receiving a premium for the high-quality fruit (Blasco et al. 2007)

6.5.7 Laser-Induced Breakdown Spectroscopy

Citrus leaves and many other plant parts can be analyzed using laser-induced breakdown spectroscopy (LIBS). After particulate size reduction (grinding), a pellet is lased and the resulting plasma emission is quantified using a spectrophotometer. Gomez et al. (2011) studied the effects of different types of tissue particle reduction. Ball milling and cryogenic grinding were more successful as measured by a 50% emission signal enhancement on LIBS measurements. These researchers found that pellets created with a particle size less than 7.5 μm, somewhat independent of grinding method but dependent on grinding time, proved to be superior. In this study, the laser source produced 25 $\mathrm{J\,cm^{-2}}$ at 1,064 nm for 25 pulses at 5 ns per pulse laser pulses per pellet site (Gomez et al. 2011). In contrast, other researchers tested a low-power laser that produced <1 mJ on raw citrus (*Citrus unshiu*) leaves (Ohta et al. 2009). The ablation process (dissipating leaf tissue moisture with the laser before taking a reading) was much more successful when metallic colloidal particles were applied to the raw leaf surface. The researchers identified the improvement as localized surface plasmon resonance that could be further enhanced by correctly selecting the size and material makeup of the colloidal particles.

6.6 Future Research

Instrumentation continues to both improve and become more automated increasing accuracy and precision while reducing analytical costs. With respect to citrus, analyses have aided in the documentation of plant responses, but the problems of scale are still an issue. For example, chemical and physical models of selected processes among cells and organs within cells do not translate well to whole trees, or from single trees to groves. Instrumentation, including 3-D printing, selected radiation tomography, and advances in computer algorithms, is likely to advance our knowledge regarding these issues of scale as these analyses and others are incorporated into current instrumentation in novel ways. In conclusion, the integration of scales must relate the citrus organism to its changing climate throughout the world. Analyses that aid in the understanding of the chemistry of plant responses to changing conditions are needed to maintain high-quality citrus production for generations to come.

Acknowledgments The authors extend their appreciation to S.S. Shukla and S. Barkataky for their assistance with manuscript preparation.

References

Abbott JA (1999) Quality measurement of fruits and vegetables. Postharvest Biol Technol 15:207–225

Abu-Samra A, Morris JS, Koirtyohann SR (1975) Wet ashing of some biological samples in a microwave oven. Anal Chem 47:1475–1477

Alva AK, Tucker DPH (1997) Surface decontamination of citrus leaves for macro and micro nutrient analysis. Proc Fla State Hortic Soc 110:86–88

Alva AK, Paramasivam S, Obreza TA et al (2006) Nitrogen best management practice for citrus trees I. Fruit yield, quality, and leaf nutritional status. Sci Hortic 107:233–244

Becker JS (2008) Trace, ultratrace and surface analysis. In: Becker JS (ed) Inorganic mass spectrometry: principles and applications. Willey, Chichester

Blasco J, Aleixos N, Gomez J et al (2007) Citrus sorting by identification of the most common defects using multispectral computer vision. J Food Eng 83:384–393

Boaretto RM, Quaggio JA, Filho FDAAM et al (2008) Absorption and mobility of boron in young citrus plants. Commun Soil Sci Plant Anal 39(17–18):2501–2514

Boaretto RM, Quaggio JA, Mattos D Jr et al (2011) Boron uptake and distribution in field grown citrus trees. J Plant Nutr 34(6): 839–849

Bradford T, Cook MN (1997) Inductively coupled plasma (ICP). http://www.cee.vt.edu/ewr/environmental/teach/smprimer/icp/icp.html. Consulted 20 Dec 2011

Campbell CR, Plank CO (1998) Preparation of plant tissue for laboratory analysis. In: Kalra YP (ed) Handbook of reference methods for plant analysis. CRC Press, Boca Raton, pp 37–50

Carrasco E, Bautista JAG, Mateo JVG (2007) Automated sequential monitoring of ammonium, phosphate, and nitrite in wastewater by multi-commutated peristaltic and solenoid pumped flow system – a comparative study. Chem Anal 52:757–770

Ceballos H, Sanchez T, Chavez AL et al (2006) Variation in crude protein content in cassava (Manihot esculenta Crantz) roots. J Food Compos Anal 19:589–593

Embleton TW, Jones WW, Labanouskas CK et al (1973) Leaf analysis as a diagnostic tool and guide to fertilization. In: Reuther W (ed) The citrus industry, vol 3. University of California, Division of Agricultural Sciences, Berkeley, pp 183–210

Esslemont G, Maher W, Ford P et al (2000) The determination of phosphorus and other elements in plant leaves by ICP–MS after low-volume microwave digestion with nitric acid. Atomic Spectrosc 21:42–46

Evans EH, Day JA, Price WJ et al (2003) Atomic spectrometry update: advances in atomic emission, absorption and fluorescence spectrometry, and related techniques. J Atomic Spectrom A8: 808–833

Fletcher RS, Escobar DE, Skaria M (2004) Evaluating airborne normalized difference vegetation index imagery for citrus orchard surveys. HortTechnol 14(1):91–94

Foley WJ, McIlwee A, Lawler I et al (1998) Ecological applications of near infrared reflectance spectroscopy – a tool for rapid, cost–effective prediction of the composition of plant and animal tissues and aspects of animal performance. Oecologia 116:293–305

Futch SH, Gallaher RN (1996) A comparison of citrus leaf wash methods for removal of zinc nutritional sprays. Proc Fla State Hortic Soc 109:43–46

García-Alonso JI, Camblor MG, Bayón MM et al (1997) Different quantification approaches for the analysis of biological and environmental samples using inductively coupled plasma mass spectrometry. J Mass Spectrom 32:556–564

Gardner WS, Malczyk JM (1983) Discrete injection segmented flow analysis of nutrients in small-volume water samples. Anal Chem 55:1645–1647

Gomez M, Junior DS, Nunes LC et al (2011) Evaluation of grinding methods for pellets preparation aiming at the analysis of plant materials by laser induced breakdown spectrometry. Talanta 85: 1744–1750

Gregori IE, Quiroz W, Pinochet H et al (2005) Simultaneous speciation analysis of Sb(III), Sb(V) and $(CH_3)_3SbCl_2$ by high performance liquid chromatography-hydride generation-atomic fluorescence spectrometry detection (HPLC-HG-AFS): application to antimony speciation in sea water. J Chromatogr A 1091:94–101

Handson PD, Shelley BC (1993) A review of plant analysis in Australia. Aust J Exp Agric 33:1029–1038

Husted S, Persson DP, Laursen KH et al (2011) Review: the role of atomic spectrometry in plant science. J Anal Atomic Spectrom 26:52–79

Berghof Products and Instruments GmbH (2011) Germany. http://www.berghof.com/multimedia /Downloads/BPI/Dokumente/Oeffentlicher_Bereich/Laborgeraete/Theorie/MW_Theorie_Probenvorbereitung_PT_en.pdf. Available on: 20 Dec 2011

Kalra YP, Maynard DG (1998) Microwave digestion of plant tissue in an open vessel. In: Kalra YP (ed) Handbook of reference methods for plant analysis. CRC Press, Boca Raton, pp 63–68

Kraska JE, Breitenbeck GA (2010) Simple, robust method for quantifying silicon in plant tissue. Commun Soil Sci Plant Anal 41(17): 2075–2085

Laursen KH, Hansen TH, Persson DP et al (2009) Recent developments in plant tissue analysis. In: Proceedings of the international plant nutrition colloquium XVI. University of California, Davis

Manivannan S, Chadha KL (2011) Standardization of time of sampling for leaf nutrient diagnosis on Kinnow Mandarin in northwest India. J Plant Nutr 34:1820–1827

Mann KK, Schumann AW, Obreza TA (2011) Delineating productivity zones in a citrus grove using citrus production, tree growth and temporally stable soil data. Precis Agric 12:457–472

Menesatti P, Antonucci F, Pallottino F et al (2010) Estimation of plant nutritional status by vis-NIR spectrophotometric analysis on orange leaves. Biosyst Eng 105:454

Miller RO (1998a) High-temperature oxidation: dry ashing. In: Kalra YP (ed) Handbook of reference methods for plant analysis. CRC Press, Boca Raton, pp 53–56

Miller RO (1998b) Microwave digestion of plant tissue in a closed vessel. In: Kalra YP (ed) Handbook of reference methods for plant analysis. CRC Press, Boca Raton, pp 69–74

Min M, Lee WS, Burks TE et al (2008) Design of a hyperspectral nitrogen sensing system for citrus. Comput Electron Agric 63:215–226

Nadkarni RA (1984) Applications of microwave oven dissolution in analysis. Anal Chem 56:2233–2237

Nash MJ, Maskall JE, Hill S (2002) Methodologies for determination of antimony in terrestrial environmental samples. J Environ Monit 2:97

Ohta T, Ito M, Kotani T et al (2009) Emission enhancement of laser-induced breakdown spectroscopy by localized surface plasmon resonance for analyzing plant nutrients. Appl Spectrosc 63(5): 555–558

Ontermaa E, Haines J, Reinholde V (1996) User perspective on laboratory analysis and its reliability. Proc Fla State Hortic Soc 109:81–85

Patton CJ, Fisher AE, Campbell WH et al (2002) Corn leaf nitrate reductase – a nontoxic alternative to cadmium for photometric nitrate determinations in water samples by air- segmented continuous-flow analysis. Environ Sci Technol 36:729–735

Potin-Gautier MF, Pannier W, Quiroz H et al (2005) Antimony speciation analysis in sediment reference materials using high-performance liquid chromatography coupled to hydride generation atomic fluorescence spectrometry. Anal Chim Acta 553:214–222

Rea AW, Keeler GI (1998) Microwave digestion and analysis of foliage for total mercury by cold vapor atomic fluorescence spectroscopy. Biogeochemistry 40:115–123

Růžička J, Hansen EH (1980) Flow injection analysis: Principles, applications and trends. Anal Chim Acta 114:19–44

Sah RN, Miller RO (1992) Spontaneous reaction for acid dissolution of biological tissues in closed vessels. Anal Chem 64(2):230–232

Smith PF (1966) Leaf analysis of citrus. In: Childers NF (ed) Nutrition of fruit crops. Horticultural Publications, Rutgers State University, New Brunswick, pp 208–228

Soon YK, Kalra YP (1994) A comparison of plant tissue digestion methods for nitrogen and phosphorus analyses. Can J Soil Sci 75:243–246

Taylor HE (2001) Inductively coupled plasma-mass spectrometry: practices and techniques. Academic Press, Boulder

White RT Jr, Douthit GE (1985) Use of microwave oven and nitric acid-hydrogen peroxide digestion to prepare botanical materials for elemental analysis by inductively coupled argon plasma emission spectroscopy. J Assoc Off Anal Chem 68:766–769

Zude M, Pflanz M, Kaprielian C et al (2008) NIRS as a tool for precision horticulture in the citrus industry. Biosyst Eng 99:455–459

7 Juice Analysis in Citrus: Latest Developments

Giuseppe Gattuso and Davide Barreca

Abstract

Citrus juices are very complex matrices, and a fast-growing number of compounds are currently being detected and identified, owing mostly to the steady advancements in analytical techniques and instrumental facilities. The aim of this chapter is to provide a survey of recent developments in the area of *Citrus* fruits analysis, with a special attention to those chromatographic and spectroscopic techniques employed in the identification, structural elucidation, and quantification of phenolics. In addition, an update is presented on latest developments in gas chromatography, liquid chromatography, capillary electrophoresis, FTIR, and RAMAN as tools to gain insight in the composition of these food matrices.

Keywords

Citrus juice • HPLC-MS-MS • NMR • FTIR • Capillary electrophoresis

7.1 Introduction

In the seventeenth century, a ship surgeon in the British Royal Navy, James Lind, attempting to find a remedy for the scurvy decimating the crew, succeeded in associating citrus juice consumption with scurvy prevention. This early evidence was later confirmed by the pioneering work of Albert Szent-Gyorgyi, who revealed that citrus juices are among the richest sources of vitamin C, one of the most important antioxidants in nature. Since then an uncountable number of analytical, biological, and epidemiological studies have more than confirmed that citrus juices play a definite role in human well-being.

Like no other beverages, citrus juices do not suffer from any human intolerance, and therefore, it is suggested in the nutrition of population of any age. Commercialization of *Citrus*-derived products is regulated by a strict set of laws to avoid adulteration of feed and beverages or commercial frauds, obviously laid out with the ultimate aim to guarantee consumers. Enforcement of these regulations, however, cannot be achieved without an in-depth knowledge of the composition, in both qualitative and quantitative terms, of the parent fresh *Citrus* fruits. In turn, such knowledge cannot be gained without mastering modern separation techniques that allow investigators to monitor quickly and precisely the composition of the raw starting materials. High-sensitivity analytical methodologies make possible to fine-tune fermentation processes, reveal microbial contamination, monitor storage effects, and evaluate the characteristic and highly diagnostic chromatographic fingerprint typical of these complex matrices. Furthermore, knowledge of the composition of a given food is essential for manufacturing processes, giving access to food manipulation aimed at the design of derived products, which may maintain organoleptic features but may, for instance, possess increased concentration of health promoting compounds.

Fresh hand-squeezed *Citrus* juices, beyond having always been ubiquitous in human diet, have attracted researchers' attention as they are a concentrated – and readily available – source of health promoting compounds, rich in secondary

G. Gattuso (✉) • D. Barreca
Dipartimento di Chimica Organica e Biologica, Università di Messina, Viale F. Stagno d'Alcontres 31, 98166 Messina, Italy
e-mail: ggattuso@unime.it; dbarreca@unime.it

A.K. Srivastava (ed.), *Advances in Citrus Nutrition*,
DOI 10.1007/978-94-007-4171-3_7,

metabolites, carbohydrates, minerals, proteins, vitamins, folate, and many other fundamental nutrient and nonnutrient substances essential for human diet (White 1990).

Citrus juices are composed of more than 95 wt% water, and more than 100 different compounds. Sucrose, glucose, and fructose are the main components of the carbohydrate fraction which, along with other species such as citric acid, accounts for about 75–90% of the total soluble solids in juice. Separation techniques coupled with mass spectrometry have played a key role in the elucidation of the remaining components of *Citrus* fruits, showing that aside from ascorbic acid, there are several other minor components with remarkable activity, such as polyphenols, phytosterols, carotenoids, furocoumarins, or limonoids.

Most of the recent investigations have focused on flavonoids as a consequence of their hypocholesterolemic effect, antioxidant, anti-inflammatory, analgesic, and anticancer properties, to name a few (Benavente-García and Castillo 2008). *Citrus* juices contain significant amount of these phenolics, with a single serving of juice providing up to 140 mg. To date, more than 40 different flavonoids derivatives have been identified in *Citrus* fruits, belonging to the flavanone, flavone, flavonol, and chalcone subclasses. HPLC-MS^2 and NMR, more than other techniques, allowed to demonstrate that they mostly occur as their glycosylated forms and, with very few exceptions, only D-glucose and L-rhamnose are found as saccharide substituents, either as monosaccharides or as the isomeric disaccharides, neohesperidose [α-(1→2)-L-rhamnopyranosyl-β-D-glucopyranose] and rutinose [α-(1→6)-L-rhamnopyranosyl-β-D-glucopyranose]. The disaccharides are generally found attached to the 7-position of the aglycone skeleton, whereas glucose has been found either at the 3- or 4′-position, as a classical *O*-linked substituent or – in flavones only – as a *C*-linked substituent at the 6- and/or 8-position. This subject has been recently reviewed (Gattuso et al. 2007b). Glycosylated flavonoids present in *Citrus* juices show very promising biological properties. These enigmatic but fascinating compounds have shown a broad range of activity from blocking cell division in several cancer cell lines to manifesting antibiotic effects against many pathological microorganisms. There is still a lot to be done in this area, especially in elucidating structure–activity relations with respect to their biological properties.

The future of research on *Citrus* fruits and the health benefits they provide is tightly linked to the analytical techniques employed, and it will proceed at the same pace as the development of novel methodologies that will allow new insight into their composition. The present chapter focuses on recent advances in separation, detection, and structural analysis of *Citrus* juice components, with particular emphasis on HPLC, capillary electrophoresis, LC-MS, and LC-NMR analytical technologies.

7.2 Sample Preparation

Many analytical procedures for the elucidation of *Citrus* juices composition require no preliminary sample manipulations (separation, extraction, centrifugation, etc.) of the crude products, and analyses can be performed directly on crude juices. This is especially true for some coupled LC-MS methods, which allow for the direct injection after minimal sample pretreatment (filtration). On the other hand, extraction procedures are mandatory when gross separations of plant materials are required prior to analysis to avoid interference with minor components determination. Over the years, several different sample pretreatment methods have been developed (Romanik et al. 2007). They generally rely on juice filtration and/or centrifugation or, more often, the analytes are isolated using liquid–liquid extraction (LLE) or solid-phase extraction (SPE). Other procedures currently employed for sample treatment are (1) pressurized liquid extraction (PLE), which takes advantage of nitrogen-carrier high pressure organic solvent flows; (2) microwave-assisted extraction (MAE), which allows for very fast high-pressure and high-temperature extraction of different plant components, with the additional advantage of very low solvent consumption; (3) supercritical-fluid extraction (SFE), which, using carbon dioxide as the extracting solvent, permits to entirely bypass the use of hazardous organic solvents; (4) matrix solid-phase dispersion (MSPD), a procedure particularly efficient when dealing with solid or highly viscous biological samples, also when very small amounts of samples are available; and (5) solid-phase microextraction (SPME), a smaller version of SPE that uses a short fused silica fiber coated with a polymeric stationary phase.

Extraction with solvents of graded polarity like methanol, ethanol, acetone, water, ethyl acetate and, to a lesser extent, propanol, dimethylformamide, and combinations of these are frequently used for extraction protocol, leading to the separation of compounds according to their different solubility.

7.3 High-Performance Liquid Chromatography

Chromatographic separation technologies have been extensively employed in the analysis of citrus juices and, in particular, high-performance liquid chromatography (HPLC) is, by far, the most powerful and utilized technique for citrus juices analysis. Owing to its versatility and high selectivity, HPLC techniques are capable of discriminating between very similar components, which differ only minimally in their substitution patterns or functional groups. Analysis of *Citrus* juices is usually carried out on reversed-phase (RP) C18- or

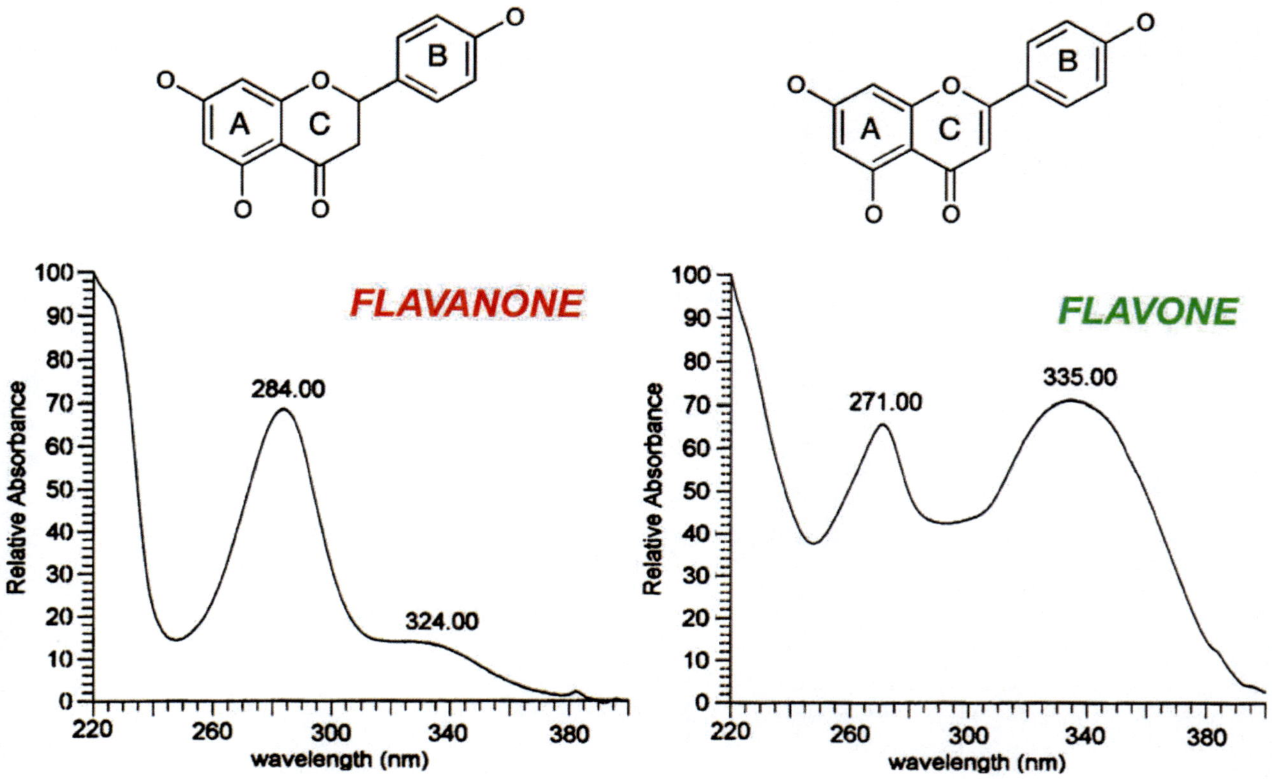

Fig. 7.1 Typical UV–VIS spectra of flavanones and flavones

C8-bonded silica columns, although other common phases employed include Sephadex, silica and polyamide, based on the chemical-physic properties of the analytes of interest. The internal diameter is about 3.9–4.6 mm with a length from 100 to 250 mm. Elution is based on both isocratic and gradient procedures utilizing generally, as a polar mobile phases, acetonitrile and/or methanol and/or tetrahydrofuran in combination with water-containing acetate buffer or formate buffer. The separation is usually carried out at 30°C, as thermostated columns yield more repeatable elution times, but high temperatures are sometimes recommended to reduce the time of analysis and to improve the column separation properties. HPLC separations are often enriched by using diode array detection (DAD) and electrospray ionization (ESI), fast atom bombardment mass spectrometry (FAB-MS), and atmospheric pressure chemical ionization (APCI) mass spectrometry methods to obtain more accurate information which may lead to identification of components of *Citrus* juices. In fact, MS detection may provide sufficient data for a complete structural elucidation of a given analyte, aside from providing molecular mass and assisting in the assignment of substitution pattern for each compound. The ESI or APCI sources are very soft ionization methods and are commonly utilized to obtain TIC chromatograms and mass spectra, generated either in positive or negative mode, showing pseudomolecular $[M+H]^+$ and $[M-H]^-$ ions together with other fragments depending on the voltage applied to the source. DAD data support the first information available to start the discrimination between the compounds present in *Citrus* juice, basing on the chromophore structure present in the analyte. For example, the UV spectra of flavanones and flavones, either as aglycone or as glycosyl derivatives, possess two strong absorptions commonly referred to as band I (300–380 nm) and band II (240–280 nm), with varying intensity resulting from substituents present on the A or B ring (Fig. 7.1).

Once the flavonoid aglycone is defined, being it flavanone, flavone, etc., MS and MS-MS spectra analysis yields fragmentation patterns which typically allow to determine whether glycosyl substituents are present, the nature of these saccharide units, and whether they are linked to the aglycone via *C*- or *O*-glycosyl bonds. In fact, MS/MS spectra of *C*-glycosylated flavonoids, focused on the $[M-H]^-$ ion, show a fragmentation pattern dominated by the presence of peaks assignable to $[M-H-18]^-$, $[M-H-90]^-$ and $[M-H-120]^-$ ions, as well as $[M-H-210]^-$ and $[M-H-240]^-$ fragment ions derived from the aglycone mass +113 and+83 amu (Fig. 7.2).

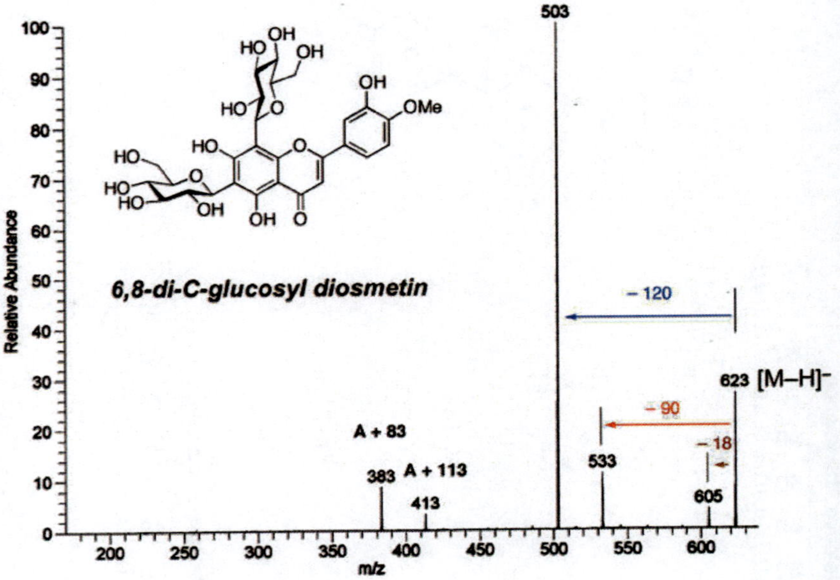

Fig. 7.2 Negative ion MS-MS spectra of 6,8-di-*C*-glucosyl diosmetin

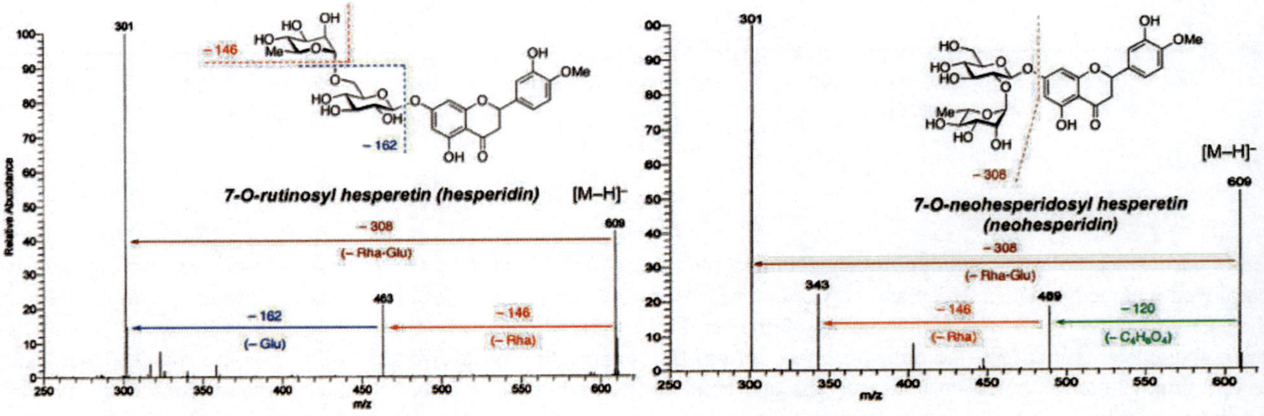

Fig. 7.3 Negative ion MS-MS spectra of hesperidin (*left*) and neohesperidin (*right*)

MS and MS-MS spectrometers are also essential tools to discriminate between *O*-rutinosides or *O*-neohesperidosides. The pseudomolecular ion at $[M–H–308]^-$ *m/z* is generally observed in the MS-MS spectrum for flavonoid *O*-rutinosides. The fragmentation patterns of flavonoid *O*-neohesperidosides are fairly different: they can be identified by the presence of the $[M–H–120]^-$ ion in the MS-MS spectrum, produced by a retro-Diels-Alder reaction on ring C (Fig. 7.3).

The flavonoid chromatographic profile is extremely important, not only for a clear taxonomical identification of species and cultivars, but also to easily identify adulteration of commercial juice. Orange juice, for instance, is characterized by the presence of hesperidin, narirutin, and didymin (Leuzzi et al. 2000), whereas bergamot juice by neoeriocitrin, naringin, and neohesperidin (Gattuso et al. 2006, 2007a), which are the same components of the closely related sour orange (Barreca et al. 2011a) and myrtle-leaved orange (Barreca et al. 2010, 2011b). Characteristic are, otherwise, the chromatographic fingerprint of lemon juice dominated by the presence of eriocitrin, hesperidin di-*C*-glucosyl-apigenin, and di-*C*-glucosyl-diosmetin (Caristi et al. 2003, 2006), while grapefruit juice is dominated by naringin, narirutin, and, to a lesser extent, hesperidin and neohesperidin.

In addition, the advent of the ultra performance LC system (UPLC), able to operate at high pressure (1,000 bar, compared with 400 bar in high-pressure LC), using 1–5 mm internal diameter (I.D.) columns with 1–2 μm range particles packing at high flow rates, has significantly reduced the times for separation and analysis of *Citrus* juice components with a concomitant increase in resolution and sensitivity (Widmer 2000; Plumb et al. 2004; Swartz 2005). Recently, Medina-Remón et al. described a rapid method for the separation and quantification of flavonoids in three *Citrus* fruit extracts by ultrahigh-performance liquid chromatography (UHPLC) using a photodiode array detector (Medina-Remón et al. 2001). This methodology is cheap and environmentally friendly, given that there is a *ca.* sixfold decrease in the solvent consumption with respect to classical HPLC. This new and very fast procedure allowed the simultaneous separation and quantification of 11 selected flavonoids (mainly naringin, rutin, quercetin, and isoquercitrin) in 5.5 min, 8.2 times faster than the corresponding HPLC analysis.

7.4 Multidimensional Chromatography

A development in LC is based on multicolumn or coupled column systems (CC-LC) (Fig. 7.4). This technique has proved to be effective in bioanalysis and, in particular, of *Citrus* juices, as it combines two or more different chromatographic systems (Russo et al. 2011; Dugo et al. 2009). This method requires the use of compatible mobile phase condition between the two or more chromatographic columns employed for a multidimensional separation. Recently, a protocol (Atuki et al. 2004) has been proposed for the complete analysis of flavanone glycosides in fresh hand-squeezed and commercial fruit juices by combining the quantitative estimation with the diastereomeric analysis. This study supports the use of flavanone glycosides as chemotaxonomic markers in quality control to identify adulterated processed juices. To this end, the major flavanone-7-*O*-glycoside constituents in *Citrus* fruit juices (naringin, hesperidin, neohesperidin, narirutin, and eriocitrin) were separated as diastereoisomers by multidimensional liquid chromatography, coupling two HPLC columns. The first one was a reversed-phase (RP18) column, used for the separation of flavanone glycosides which were, in turn, individually run through a carboxymethylated β-cyclodextrin (β-CD)-based column and resolved as their corresponding stereoisomers, which were finally identified by negative ion electrospray ionization (ESI-MS). The analysis revealed a significant difference in the diastereomeric ratios between freshly squeezed juices and juices from commercial sources.

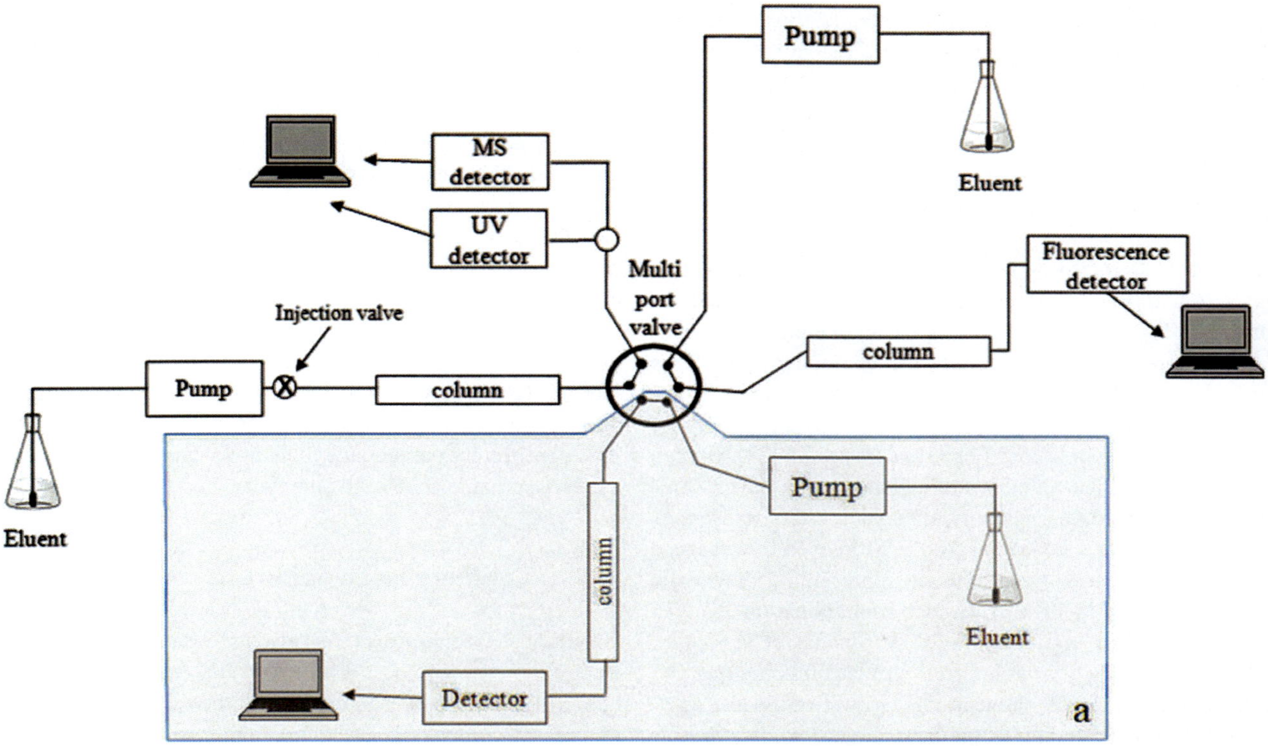

Fig. 7.4 Cartoon representation of a multidimensional system. The section (*a*) marked in the *gray* box represents a scheme that can be repeated *n* times

7.5 Nano-Liquid Chromatography (Nano-LC)

Nano-HPLC, sometimes also referred to as "nanobore HPLC" or "nano-scale HPLC," is a relatively new development in separation science driven by the attempt of producing column for liquid chromatography with ever smaller inner diameter. This allows for lower sample amount requirement and increased sensitivity. In fact, the use of chromatographic columns with smaller I.D., e.g., from 4.6 mm (classical column) to 100 μm (capillary column) may result in up to 2,000-fold increase in both sensitivity and resolution. This feature, together with the low costs owing to the use of considerably smaller amounts of chiral stationary phase, along with the limited eluent volumes, makes nano-LC a viable alternative to classical ones such as HPLC for the analysis of citrus juice (Fanali et al. 2011; D'Orazio et al. 2007). However, to date, very few papers have been published on the enantioselective separation, identification, and quantification carried out by nano-liquid chromatography. Recently, D'Orazio et al. reported on the utilization of chiral nano-LC-MS for amino acid determination as a useful tool for revealing *Citrus* juice adulteration, microbiological contamination, uncontrolled fermentation processes, etc. (D'Orazio et al. 2008). Si-Ahmed et al. reported also the use of a phenyl-carbamate-propyl-β-cyclodextrin stationary phase in nano-LC for the resolution of racemic hesperetin, not just in the juice, but also in human urine after ingestion of blood orange (Si-Ahmeda et al. 2010).

7.6 Gas Chromatography

Gas chromatography and coupled gas chromatography–mass spectrometry (GC-MS) and gas chromatography-olfactometry (GC-O) have been extensively used to determine the volatile composition and aroma-active compounds of *Citrus* juices and related commercial products. More than 300 components represent the main components characteristic of sweet orange flavor (Pérez-Cacho and Rouseff 2008). Orange juice, for instance, contain among the major flavor compounds hydrocarbons, alcohols, aldehydes, esters, and ketons (Selli et al. 2004; Jordán et al. 2005). Moreover, the presence of terpenes, also in trace amount, represents a diagnostic tool for detection of *Citrus* juice contamination by rind oil (Högnadóttir and Rouseff 2003; Smadja et al. 2005). HS-SPME-GC-MS volatile profiling platforms, among GC separation techniques, show promising application as a roadmap for future breeding or biotechnological applications as reported by González-Mas et al. (2011). This technique is capable to aid in selecting new varieties with better aroma, to monitor industrial processes, and to study the biosynthetic pathways leading to the production of volatiles in *Citrus*, basing on cluster and correlation analyses on more than 100 compounds.

GC techniques are also very interesting tools for the determination of norisoprenoids, which are useful markers of *Citrus* juice quality. They are volatile C9-C13 fragments derived from *in vivo* enzymatic degradation or postharvest thermal degradation of C40 carotenoids in foods, in particular α- and β-carotene, α- and β-cryptoxanthin, and neoxanthin. In a recent study, four norisoprenoids (α-ionone, β-ionone, β-cyclocitral, and β-damascenone) were identified in Valencia orange juice using time-intensity GC-O and GC-MS (Mahattanatawee et al. 2005).

7.7 Nuclear Magnetic Resonance

Nuclear magnetic resonance (NMR) spectroscopy is one of the most versatile and selective techniques to investigate *Citrus* juice components and monitor juice quality. Traditionally, it has been mostly employed in structural elucidation and to assess the purity of a given isolated analyte. However, driven by the emerging trend and economical benefit involved in foods quality control procedures and origin authenticity, it has rapidly expanded into the field of mixture analysis and multiple screening over the past few years (Spraul et al. 2008; Vogels et al. 1996). Moreover, the creation of a rich reference spectral database makes NMR spectroscopy combined with statistical multiparametric analysis an outstanding routine tool for *Citrus* juice quality certification, able to easily verify the presence of contaminations and frauds such as wrong labeling of the product or the type and origin of ingredients (Spraul et al. 2009) (Fig. 7.5). This fully automated multimarker/multi-aspect NMR screening approach allows, in a single analysis and in a few minutes, the evaluation of a multitude of parameters related to quality and authenticity such as, to name a few, the detection of frauds (addition of sugar or exhaustive enzymatic treatment), addition of citric acid or foreign juice, or extraction of orange peel. This analysis is based on the detection of more than 28 different compounds including sugars, amino acids, alcohols, organic acids, and aldehydes.

7.8 Capillary Electrophoresis

Capillary electrophoresis (CE), also known as capillary zone electrophoresis (CZE), is a relatively new technique, which has emerged as a powerful method for the separation of ionic and neutral compounds, providing high separation efficiency in short migration times. Separation of neutral compounds can be achieved by using micellar electrokinetic chromatography (MEKC) that relies on charged surfactants (such as

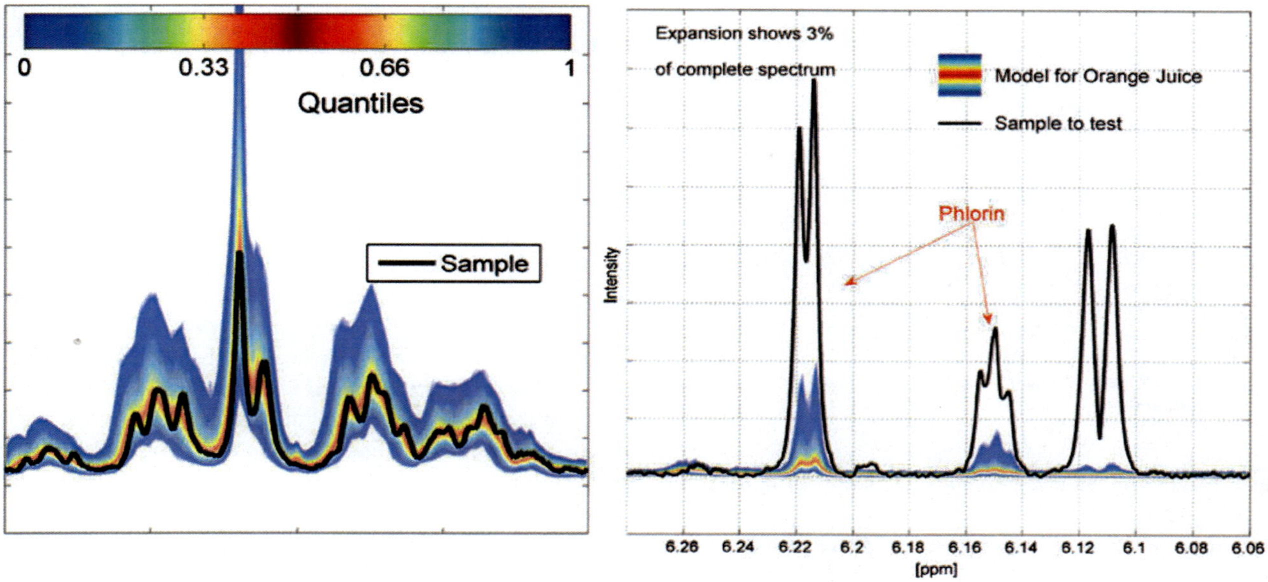

Fig. 7.5 *Left panel*: Nontargeted verification of the sample (*black line*) versus the quantile-plot of the respective reference database: apple juice (~1% of spectrum). *Right panel*: orange juice with high phlorin concentration (indicating over-extraction of the whole orange including its peel) (Reproduced from Spraul et al. 2009)

sodium dodecyl sulfate) to form charged micelles containing the target analyte. It has expanded considerably in the last decade, not only for *Citrus* juice analysis but also for other food matrices, due to the possibility of achieving rapid and accurate results with minimum sample preparation and solvent requirements. Moreover, no reconditioning period between analyses is required, and after a brief wash, a new sample can be analyzed. *Citrus* juices sample preparation requires only filtration and dilution, resulting then in the separation and identification of up to 30 compounds in a single 20 min of course. Separation in this case was achieved on uncoated 70 cm fused-silica capillary tubing, using a sodium borate buffer (35 mM pH 9.3) containing 5% (v/v) of acetonitrile at 21 kV and 23°C, with detection performed between 200 and 360 nm (Cancalon 1999; Cancalon 2001). Considering that the capillary diameter is 50 μm, samples can be run with very small amounts of buffer.

Several CE protocols have been developed to analyze a large variety of organic and inorganic compounds in *Citrus* juices. Good separation and identification are achieved for organic acids, flavonoids, and total vitamin C (ascorbic acid and dehydroascorbic acid) (Liao et al. 2001; Saavedra et al. 2000). Recently, five flavonoids (hesperidin, naringin, hesperetin, narigenin, and rutin) and ascorbic acid were separated and determined in grapefruit juice by capillary electrophoresis with electrochemistry detection (CE-ED) in a 75-cm length of 25-μm I.D. and 360-μm o.d. fused-silica capillary within 25 min in a 60-mM borate buffer (pH 9.0) (Wu et al. 2007). CE can be interfaced with various MS analyzers; however, TOF-MS is the most commonly used one due to its fast acquisition rates, which are necessary to statistically sample the narrow CE peaks. ESI is the ionization technique of choice for CE-MS, and it has been used in both targeted and nontargeted metabolite analysis.

7.9 Fourier Transform Infrared and Raman Spectroscopy

Near-infrared spectroscopy (NIRS) analysis has been often employed in the detection of sugar and organic acid in citrus juices. Already in 1996 this technique was harnessed to obtain an easy and accurate determination of glucose, fructose, sucrose, citric acid, and malic acid (Li et al. 1996). NIRS has been used also as a rapid and nondestructive technique for determining the total soluble solids content in fruits, being this one of the main features used for assessing citrus fruit and juice quality. Furthermore, a good correlation between the NIR spectra and the sugar content ($r = 0.97$, SEP = 0.05 °Brix) has been observed. The latest developments in Fourier transform infrared (FTIR) spectroscopy enabled also a rapid and noninvasive structural characterization of analytes in *Citrus* samples. In fact, FTIR may be used to provide reliable compositional and quantitative information, which has led to the possibility of determining the variety or origin of a given *Citrus* fruits, especially in the case of products of interest for the juice industry (Tewari et al. 2008). Moreover, the developments of modern

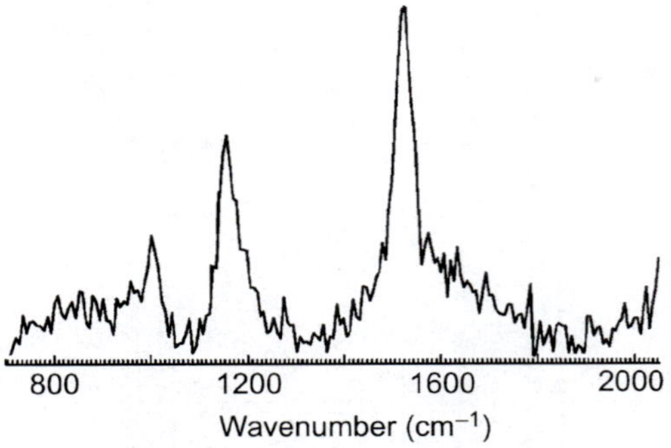

Fig. 7.6 Resonance Raman spectrum of β-carotene dissolved in tetrahydrofuran

accessories for sampling and advancements in multivariate (chemometric) statistical tools for analysis – including discriminant analysis (DA), partial least squares (PLS), and second derivative analysis – provided a decisive improvement in the sensitivity of FT-IR.

One of the most interesting applications of FT-IR is the identification, discrimination, and quantification of bacteria in citrus juice (Sivakesava et al. 2004; Stevens and Jaykus 2004; Oberreuter et al. 2003). Peaks may be assigned to specific chemical bonds appearing in bacterial strains, which may in turn be correlated to bacterial concentration (Jiang et al. 2004). Bacteria can be identified from spectral differences in any of the major spectral regions, such as that belonging to fatty acid (3,000–2,800 cm^{-1}), to proteins (amide I, amide II, 1,700–1,400 cm^{-1}), to carbohydrates (1,200–900 cm^{-1}), or from the fingerprint region (900–700 cm^{-1}) (Naumann 2000). The possibility to create a library for the spectral identification of bacteria makes this technique a very powerful tool for spotting juice contamination.

Even though they both are vibrational spectroscopies, Raman is a complementary technique to IR, as it provides different structural information. Raman spectroscopy is a very sensitive method for the nondestructive quantitative determination of compounds from many different natural sources. In *Citrus* juice – but not only in this product – Raman spectroscopy has been successfully employed for carotenoids detection. In fact, carotenoid molecules when irradiated at 488 nm provide three strong, high-frequency, Stokes-shifted signals (1,008, 1,159 and 1,525 cm^{-1}, Fig. 7.6) (Koyama 1995). Moreover, Raman measurements do not require any sample pretreatment (Ermakov et al. 2001), and the high water content of juice decreases the self-absorption of other chromophores. Comparison of Raman signals with the total carotenoid content (as estimated by HPLC) shows a direct correlation between the two methods, with the juice characterized by the presence of β-carotene, lycopene, and β-cryptoxanthin (Bhosale et al. 2004).

7.10 Quantitative Structure–Property Relationship Analyses

Quantitative structure–property relationship (QSPR) models are computational approaches based on the translation of the compound structure into molecular descriptors, which are then used as parameters to obtain correlation with experimentally determined properties. This type of data elaboration not only allows to predict specific properties but also, for instance, may allow to obtain predictions on amino acid composition, as it has been recently shown (Pomilio et al. 2010). Dragon theoretical descriptors were derived for a set of optimized amino acid structures, demonstrating the predictive power of the designed QSPR models to predict aminograms for 100% natural fresh juices and concentrates of Navel and Valencia oranges, and Eureka lemon, validated by HPLC procedure. An accurate determination of an aminogram concentration profile is of crucial importance in food industry, as it permits to find the characteristic fingerprint of the food under investigation, bringing to its unequivocal identification.

7.11 Ultrasound Technology

The velocity and attenuation of ultrasonic waves propagated through a sample yield information about physical properties as composition, structure, flow rate, physical state, and molecular properties. This provides a nondestructive and noninvasive technology for the evaluation of food sample quality (McClements 1997). The ultrasonic method lends itself to fast and economical measurement, and it has been employed to assess qualities such as fruit ripeness or solid fat content in oils, or for the evaluation frozen water percentage in frozen foods, an index of physic food quality/integrity (Lee et al. 2004). Ultrasonic velocity and attenuation varied as the temperature was decreased from 20°C to −50°C, indicating changes in the physical properties of the frozen samples, thus allowing to correlate damage to food products caused by freezing with the presence and size of ice crystals.

7.12 Denaturing Gradient Gel Electrophoresis

DNA markers have been largely employed for taxonomy study in many plants (Tingey and Tufo 1993; Clegg et al. 1994), providing extremely useful evidence both for species identification and for the evaluation plant-derived commercial products. Several DNA markers have been used in *Citrus*

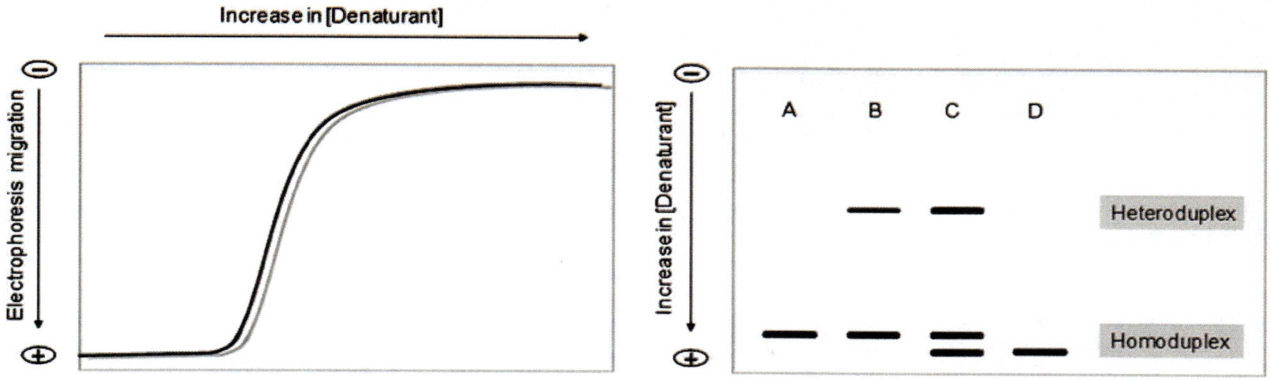

Fig. 7.7 Cartoon representation of perpendicular (*left*) and parallel (*right*) DGGE of pure orange juice and of an adulterated sample: *A* pure sample, *B* sample with small contamination, *C* sample with a remarkable amount of contaminant, and *D* contaminant

analysis, such as restriction fragment length polymorphism (RFLP), random amplified polymorphic DNA (RAPD), sequence-characterized amplified regions (SCARs), and chloroplast DNA (cpDNA). The last one, more than the others, is particularly efficient in phylogenetic analysis as a result of its evolutionary conservatism, of the relative abundance in plant tissue, small size, and predominant uniparental inheritance (Olmstead and Palmer 1994). The most common chloroplast gene employed to gather sequence data for plastidic plant analysis is the ribulose-1,5-bisphosphate carboxylase/oxygenase gene (*rbc*L) (Gielly and Taberlet 1994), although more rapidly evolving genes, introns, and spacers such as the noncoding region of the cpDNA (*trn*L intron) and the *trn*L-*trn*F intergenic spacer are better suited for taxonomic analysis (Brouat et al. 2001).

Recent studies described the use of *trn*L intron of cpDNA as a marker for differentiating *Citrus* species such as orange from mandarin, which revealed a significant variability between the two species (Liu and Shyu 2006). Moreover, the application of denaturing gradient gel electrophoresis (DGGE) to these samples allowed for an easy discrimination between pure and adulterated juices (Fig. 7.7). In fact, DGGE separates PCR amplicons of similar length with dissimilar nucleotide compositions, and DNA bands may reveal the addition of juice from different species than the one under investigation. Adulteration results in the appearance of at least two bands on the polyacrylamide gel pattern: the heteroduplex molecule type and the homoduplex one. The former can be observed when there is more than one DNA type in the same PCR reaction, whereas the latter has a mismatch in the DNA double-strand that induces a conformational distortion, which in turn causes DNA to denature at a low denaturant concentration. The increase in the mandarin/orange ratio results in an increase in heteroduplex molecules.

7.13 Conclusion

The enormous number of publications that appeared in the literature on the analysis of *Citrus* juices over the years underlines the relevance that these simple, yet ubiquitous dietary products possess in nutrition science. We have presented, in this chapter, a survey on recent advances made in separation science, with specific reference to applications in *Citrus* juice analysis. Sample preparation, separation, and detection of several classes of analytes such as flavonoids, carotenoids, amino acids, and nucleic acids, to name the most representative examples, have been discussed.

Innovative separation and identification technologies are changing the area of juice analysis. So far, HPLC coupled to (mostly) MS or MS-MS spectrometers has been the leading technique to undertake the elucidation of the composition of *Citrus* juices, both for research and product quality monitoring. Modern separation methodologies guarantee high-throughput, and a variety of coupled methods provided new and more sensitive means of detection, driving the advancement in knowledge seeking and quality control of these food matrices to the benefit of researchers and consumers alike.

7.14 Future Research

Recent advancements – such as UHPLC-ESI-TOF-MS coupled systems – have already determined significant improvements over classical instruments. In the next future, the increasing pace of technological improvement will surely lead to further developments in miniaturization, with the coupling of micro- and/or nano-LC to MS and NMR instruments. This will lead at once to several highly desirable results: the possibility of using very small amounts of samples, of carrying out analyses in the shortest possible time, and of dramatically reducing the amount of organic solvents employed, and hence the – nowadays unavoidable – environmental impact associated to the use of most analytical techniques.

References

Atuki Z, Brandi V, Sinibaldi M (2004) Separation of flavanone-7-o-glycoside diastereomers and analysis in citrus juices by multidimensional liquid chromatography coupled with mass spectrometry. J Agric Food Chem 52:5303–5308

Barreca D, Bellocco E, Caristi C et al (2010) Flavonoid composition and antioxidant activity of juices from chinotto (*Citrus × myrtifolia* Raf.) fruits at different ripening stages. J Agric Food Chem 58: 3031–3036

Barreca D, Bellocco E, Caristi C et al (2011a) Distribution of *C*- and *O*-glycosyl flavonoids, (3-hydroxy-3-methylglutaryl) glycosyl flavanones and furocoumarins in *Citrus aurantium* L. juice. Food Chem 124:576–582

Barreca D, Bellocco E, Caristi C et al (2011b) Elucidation of the flavonoid and furocoumarin composition and radical scavenging activity of green and ripe Chinotto (*Citrus myrtifolia* Raf.) fruit tissues, leaves and seeds. Food Chem 129:1504–1512

Benavente-García O, Castillo J (2008) Update on uses and properties of citrus flavonoids: new findings in anticancer, cardiovascular, and anti-inflammatory activity. J Agric Food Chem 56:6185–6205

Bhosale P, Ermakoviv IV, Ermakova MR et al (2004) Resonance Raman quantification of nutritionally important carotenoids in fruits, vegetables, and their juices in comparison to high-pressure liquid chromatography analysis. J Agric Food Chem 52:3281–3285

Brouat C, Gielly L, McKey D (2001) Phylogenetic relationships in the genus Leonardoxa (Leguminosae: Caesalpinioideae) inferred from chloroplast trnL intron and trnL-trnF intergenic spacer sequences. Am J Bot 88:143–149

Cancalon PF (1999) Analytical monitoring of citrus juices by using capillary electrophoresis. J AOAC Int 82:95–106

Cancalon PF (2001) Routine analysis of ascorbic acid in citrus juice using capillary electrophoresis. J AOAC Int 84:987–991

Caristi C, Bellocco E, Panzera V et al (2003) Flavonoids detection by HPLC-DAD-MS-MS in lemon juices from Sicilian cultivars. J Agric Food Chem 51:3528–3534

Caristi C, Bellocco E, Gargiulli C et al (2006) Flavone-di-*C*-glycosides in *Citrus* juices from southern Italy. Food Chem 95:431–437

Clegg MT, Gaut BS, Learn GH et al (1994) Rates and patterns of chloroplast DNA evolution. Proc Natl Acad Sci USA 91:6795–6801

D'Orazio G, Aturki Z, Cristalli M et al (2007) Separation of basic compounds of pharmaceutical interest by using nano-liquid chromatography coupled with mass spectrometry. J Chromatogr A 1150: 252–258

D'Orazio G, Cifuentes A, Fanali S (2008) Chiral nano-liquid chromatography–mass spectrometry applied to amino acids analysis for orange juice profiling. Food Chem 108:1114–1121

Dugo P, Giuffrida D, Herrero M et al (2009) Epoxycarotenoids esters analysis in intact orange juices using two-dimensional comprehensive liquid chromatography. J Sep Sci 32:973–980

Ermakov IV, Ermakova MR, McClane RW et al (2001) Resonant Raman detection of carotenoid antioxidants in human tissues. Opt Lett 26:1179–1181

Fanali C, Dugo L, D'Orazio G et al (2011) Analysis of anthocyanins in commercial fruit juices by using nano-liquid chromatography-electrospray-mass spectrometry and high-performance liquid chromatography with UV–vis detector. J Sep Sci 34:150–159

Gattuso G, Caristi C, Gargiulli C et al (2006) Flavonoid glycosides in bergamot juice (*Citrus bergamia* Risso). J Agric Food Chem 54:3929–3935

Gattuso G, Barreca D, Caristi C et al (2007a) Distribution of flavonoids and furocoumarins in juices from cultivars of *Citrus bergamia* Risso. J Agric Food Chem 55:9921–9927

Gattuso G, Barreca D, Gargiulli C et al (2007b) Flavonoid composition of citrus juices. Molecules 12:1641–1673

Gielly L, Taberlet P (1994) The use of chloroplast DNA to resolve plant phylogenies: non-coding versus *rbc*L sequence. Mol Biol Evol 11:769–777

González-Mas MC, Rambla JL, Alamar C et al (2011) Comparative analysis of the volatile fraction of fruit juice from different citrus species. PLoS One 6:1–11

Högnadóttir Á, Rouseff RL (2003) Identification of aroma active compounds in orange essence oil using gas chromatography–olfacotmetry and gas chromatography–mass spectrometry. J Chromatogr A 998:201–211

Jiang W, Saxena A, Song B et al (2004) Elucidation of functional groups on gram-positive and gram-negative bacterial surfaces using infrared spectroscopy. Langmuir 20:11433–11442

Jordán MJ, Goodner KL, Castillo M et al (2005) Comparison of two headspace solid phase microextraction fibers for the detection of volatile chemical concentration changes due to industrial processing of orange juice. J Sci Food Agric 85:1065–1071

Koyama Y (1995) Resonance Raman spectroscopy. In: Britton G, Liaaen-Jenson S, Pfander H (eds) Carotenoids, vol 1B. Birkhaeuser, Basel

Lee S, Pyrak-Nolte LJ, Cornillon P et al (2004) Characterisation of frozen orange juice by ultrasound and wavelet analysis. J Sci Food Agric 84:405–410

Leuzzi U, Caristi C, Panzera V et al (2000) Flavonoids in pigmented orange juice and second-pressure extracts. J Agric Food Chem 48:5501–5506

Li W, Goovaerts P, Meurens M (1996) Quantitative analysis of individual sugars and acids in orange juices by near-infrared spectroscopy of dry extract. J Agric Food Chem 44:2252–2259

Liao TC, Jiang CM, Wu MC et al (2001) Quantification of L-ascorbic acid and total ascorbic acid in fruits and spinach by capillary zone electrophoresis. Electrophoresis 22:1484–1488

Liu YS, Shyu YT (2006) Adulteration identification of citrus juice by denaturing gradient gel electrophoresis (DGGE). J Food Drug Anal 14:44–49

Mahattanatawee K, Rouseff R, Valim MF et al (2005) Identification and aroma impact of norisoprenoids in orange juice. J Agric Food Chem 53:393–397

McClements DJ (1997) Ultrasonic characterization of foods and drinks: principles, methods, and applications. Crit Rev Food Sci Nutr 37:1–46

Medina-Remón A, Tulipani S, Rotches-Ribalta M et al (2001) A fast method coupling ultrahigh performance liquid chromatography with diode array detection for flavonoid quantification in citrus fruit extracts. J Agric Food Chem 59:6353–6359

Naumann D (2000) Infrared spectroscopy in microbiology. In: Myers RA (ed) Encylopedia of analytical chemistry, vol 1. Wiley, Chichester

Oberreuter H, Brodbeck A, Von Stetten S et al (2003) Fourier-transform infrared (FT-IR) spectroscopy is a promising tool for monitoring the population dynamics of microorganisms in food stuff. Eur Food Res Technol 216:434–439

Olmstead RG, Palmer JD (1994) Chloroplast DNA systematics: a review of methods and data analysis. Am J Bot 81:1205–1224

Pérez-Cacho PR, Rouseff RL (2008) Fresh squeezed orange juice odor: a review. Crit Rev Food Sci Nutr 48:681–695

Plumb R, Castro-Perez J, Granger J et al (2004) Ultra-performance liquid chromatography coupled to quadrupole-orthogonal time-of-flight mass spectrometry. Rapid Commun Mass Spectrom 18:2331–2337

Pomilio AB, Giraudo MA, Duchowicz PR et al (2010) QSPR analyses for aminograms in food: citrus juices and concentrates. Food Chem 123:917–927

Romanik G, Gilgenast E, Przyjazny A et al (2007) Techniques of preparing plant material for chromatographic separation and analysis. J Biochem Biophys Methods 70:253–261

Russo M, Cacciola F, Bonaccorsi I et al (2011) Determination of flavanones in citrus juices by means of one- and two-dimensional liquid chromatography. J Sep Sci 34:681–687

Saavedra L, García A, Barbas C (2000) Development and validation of a capillary electrophoresis method for direct measurement of isocitric, citric, tartaric and malic acids as adulteration markers in orange juice. J Chromatogr A 881:395–401

Selli S, Cabaroglu T, Canbas A (2004) Volatile flavour components of orange juice obtained from the cv Kozan of Turkey. J Food Compos Anal 17:789–796

Si-Ahmeda K, Tazerouti F, Badjah-Hadj-Ahmed AY et al (2010) Analysis of hesperetin enantiomers in human urine after ingestion of blood orange juice by using nano-liquid chromatography. J Pharm Biomed Anal 51:225–229

Sivakesava S, Irudayaraj J, Debroy C (2004) Differentiation of microorganisms by FTIR-ATR and NIR spectroscopy. Trans ASAE 7: 951–957

Smadja J, Rondeau P, Sing ASC (2005) Volatile constituents of five citrus petitgrain essential oils from Reunion. Flavour Fragr J 20:399–402

Spraul M, Humpfer E, Schäfer H et al (2008) NMR-based mixture analysis on the example of fruit juice quality control using statistics and quantification. In: Holzgrabe U, Wawer I, Diehl B (eds) NMR spectroscopy in pharmaceutical analysis. Elsevier, Amsterdam

Spraul M, Schütz B, Rinke P et al (2009) NMR-based multi parametric quality control of fruit juices: SGF profiling. Nutrients 1:148–155. doi:10.3390/nu1020148

Stevens KA, Jaykus LA (2004) Bacterial separation and concentration from complex sample matrices: a review. Crit Rev Microbiol 30:7–24

Swartz ME (2005) UPLC: an introduction and review. J Liq Chromatogr Relat Technol 28:1253–1263

Tewari JC, Dixit V, Chob BK (2008) Determination of origin and sugars of citrus fruits using genetic algorithm, correspondence analysis and partial least square combined with fiber optic NIR spectroscopy. Spectrochim Acta A 71:1119–1127

Tingey SV, Tufo JP (1993) Genetic analysis with random amplified polymorphic DNA markers. Plant Physiol 101:349–352

Vogels JTWE, Terwel L, Tas AC et al (1996) Detection of adulteration in orange juices by a new screening method using proton NMR spectroscopy in combination with pattern recognition techniques. J Agric Food Chem 44:175–180

White DR (1990) Routine citrus analyses using HPLC with amperometric detection. Proc Fla State Hortic Soc 103:247–251

Widmer W (2000) Determination of naringin and neohesperidin in orange juice by liquid chromatography with UV detection to detect the presence of grapefruit juice: Collaborative study. J AOAC Int: 1155–1165

Wu T, Guan Y, Ye J (2007) Determination of flavonoids and ascorbic acid in grapefruit peel and juice by capillary electrophoresis with electrochemical detection. Food Chem 100:1573–1579

Trunk Nutrition

8

Mahmoud M. Shaaban

Abstract

Trunk nutrition is a method for fertilizing trees through xylem. This method was used along the time in small-scale studies to solve the problem of the uptake and/or translocation of a single element like iron or potassium. The previous studies on efficiency ratio of soil fertilization proved that a small portion of the added fertilizers is taken up by the plant roots while the great portion reached 62–85% of N, 80–95% of P and K is lost by leaching, volatilization, and fixation. Thus, it is recently borne in mind the idea of injecting fertilizers directly through trees trunk. Experiments with full trunk nutrition proved that only 5–10% of the fertilizer levels used as soil fertilization was sufficient for good growth and higher yields. Growth of the injection-fertilized mango (*Mangifera indica*) trees was 20–25% higher than soil-fertilized plants, while in grapevine (*Vitis vinifera*), fruit yield increases were 32–49% higher compared to soil fertilization. Fruit quality of grapevine clusters assessed of the plants fertilized through trunk was better than those fertilized through soil. Grapevine fresh juice content of the reduced sugars and ethanol increased by 7.5–11.9% and 41.4–50%, respectively while the total acidity decreased by 6.2–19.7%. Yield of 7-year-old guava and kaki trees fully fertilized through trunk increased by 84% and 89%, respectively, over soil-fertilized trees. Trunk nutrition of 3-year-old citrus trees showed that injected trees were more vigor and accumulated higher dry biomass compared to the soil-fertilized trees. It was also obvious from trunk nutrition studies that this method is simple and fairly economic, saves a huge amount of money, and keeps a clean environment. More studies should be done to establish this method for different species of fruit and ornamental trees in different environments.

Keywords

Dicots nutrition • Injection fertilization • Trees intravenous feeding

8.1 Introduction

Conventional methods of plant nutrition depend upon fertilization through soil (broadcasting, splitting, dressing, and fertigation). Foliar fertilization is considered as a complementary nutrient supplement through foliage in particular cases such as high pH values of the soil solution, high $CaCO_3$ content, high salinity, etc., that lead to less absorption and/or translocation of a single element or more. Sandy soils and poorly distributed high annual rainfall all over the world extremely contribute to significant leaching losses of nutrients from routine fertilization practices. Previous studies proved that only a small portion of the added fertilizers is taken up by the plant roots, especially those grown under sandy soil conditions, where the high permeability allowed fast leaching of fertilizers to underground water (Halliday and Trenkel 1992).

M.M. Shaaban (✉)
Fertilization Technology Department, National Research Centre, El-Tahrir Street, Dokki, Cairo, Egypt
e-mail: mshaaban1@hotmail.com

A.K. Srivastava (ed.), *Advances in Citrus Nutrition*,
DOI 10.1007/978-94-007-4171-3_8, © Springer Science+Business Media B.V. 2012

Generally, the lost portion of nitrogen was estimated as high as 62–85% of the added fertilizers (Dixon 2003). Morgan et al. (2006) found that N losses in citrus tree orchards grown on Florida sandy soil at high application rate of approximately 83 gN tree^{-1} were 60.7%, 68.6%, and 63.4% for March, May, and September, respectively, while mean percentage N losses for the lower rate were 83.6%, 82.6%, and 73.1%. In another study on citrus trees fertilization in California calcareous soils, Obreza et al. (2009) concluded that soil pH affects the rates of several reactions involving N and can influence the efficiency of N use by plants. They also found that nitrification, or the conversion of ammonium (NH_4^+) to nitrate (NO_3^-) by soil bacteria, is most rapid in soils with pH values between 7 and 8, then ammonium-N fertilizers applied to calcareous soils are converted within a few days to nitrate, which moves freely with soil water. The same authors added also that nitrogen loss through ammonia volatilization on calcareous soils is a concern when ammoniacal N is applied to the grove floor and remains there without moving into the soil. In the same study, surface application of urea was found to cause N loss if the urea is not incorporated or irrigated into the soil, regardless of initial soil pH.

Soil texture affects NO_3–N leaching. The coarser the soil (sandy in texture), the faster NO_3–N will leach from the soil. The finer the soil texture (clay and clay loam), the slower NO_3–N will leach from the soil. In a laboratory experiment, Gaines and Gaines (1994) found that most (90%) of the NO_3–N was leached from the sandy soil in the first 5 min or the first 100 ml of soil percolate after NO_3–N fertilizer had been applied, whereas most of the NO_3–N (90%) leached from the sandy clay loam soil in the first 10 h or the first 200 ml of soil percolate after NO_3–N fertilizer had been applied. Consumption of nitrogen fertilizers is growing rapidly all over the world (Chalk et al. 2001). Consequently, nitrate concentration of well water has shown rising trends in many countries to be a significant groundwater pollutant. In Europe, NO_3 concentrations exceeding the international norms for drinking water (10 mg l^{-1}) was found in groundwater under 22% of the cultivated land (Laegreid et al. 1999; WHO 1993). Reports by the Kansas Department of Agriculture (Emmons 2000) and the United States Geological Survey (Pope et al. 2001) have identified as many as 15% of ground water wells with NO_3–N concentration exceeding 10 mg l^{-1} (USEPA maximum containment level for drinking water quality). Drinking water with a large NO_3– may induce negative health effects, such as birth defects, cancer, and nervous system impairments (Keeney 1987; Jemison and Fox 1994). Methemoglobinemia known as "blue baby syndrome" as well as recurrent acute respiratory tract infections and diarrhea were found prevalent in all age groups with high nitrate concentrations in drinking water (Gupta et al. 1999, 2001).

Fixation of other nutrients like phosphorus and micronutrients in the form of low dissolved compounds in the soil is responsible for about another portion of the added fertilizers to be less available for the absorption by the plant roots (Horesh and Levy 1981). Phosphorous is one of the three primary macronutrients that shows one of the greatest problems in soil fertility related to its low availability in the soil and its high fixation. Only 5–20% of the applied phosphorous (P) is used by the plants, mainly due to fixation (Malavolta 1980; van Raij 1991). In a study on nutrient losses in some citrus orchards in China, Meng et al. (2001) found that percentage of NPK losses from the arable plots was 50–99%. In a lysimeters experiment in Southeast Nigeria, Wong et al. (1992) found that the leached exchangeable Ca and Mg from the soil reached 34% and 37%, respectively, while the leached exchangeable K reached about 10%. However, a great portion of potassium fertilizer can also be trapped within the structure of expanding clay minerals. This captured K is referred to as "fixed K," which is not presently available for plant uptake.

Micronutrients are required in very small quantities of only a few mg kg^{-1} in plant tissues. Micronutrients play many complex roles in plant nutrition and crop production. While most of the micronutrients participate in the functioning of a number of enzyme systems, there is considerable variation in the specific functions of the various micronutrients in plant growth processes. For example, copper, iron, and molybdenum are capable of acting as electron carriers in the enzyme systems that bring about oxidation–reduction reactions in plants. Such reactions are essential steps in photosynthesis and many other metabolic processes. Zinc and manganese function in many plant enzyme systems as bridges to connect the enzyme with the substrate upon which it is meant to act. During a period of 7 months, samples of various depths collected throughout from citrus, mango, guava, banana, and apple orchards in Pakistan and analyzed for available iron, copper, manganese, boron, and zinc status, Zia et al. (2006) found a widespread deficiency of zinc, boron, followed by iron throughout the country due to nutrient fixation in the soil and restricted mobility of iron, zinc, and boron in plant tissues.

The above-mentioned data show that a great portion of the added fertilizers reached 62–85% of N and 80–95% of P and K is lost by leaching, volatilization, and fixation. This means that a considerable portion of the world fortune is thrown into underground and air. To solve this problem, thinking is directed toward how to save these huge quantities of fertilizers to avoid the money loss and keep a clean environment. A partial solution came as the possibility of injecting special fertilizer formula (according to the fertilization requirements) directly into the trees veins in what is called trunk nutrition.

8.2 Facts About Trunk Nutrition

Mammalians including human beings can be nourished through veins. When a human being or an animal faced a problem preventing him to eat and drink from mouth, glucose and vitamins injected through veins can be enough to keep him/her healthy and active. Like human beings and animals have what is called blood vessels (veins and arteries), plants have the same (xylem and phloem). Congenitally, sometimes some of the farmers dig an iron nail on the tree trunk when they watch its leaf yellowing. The outer surface of the nail actually converted to iron hydroxide when immersed into the humid medium of the trunk tissues. Then the iron hydroxide can be solubilized in the wood sap and transport to the leaves, recovering their yellowing. This is a type of trunk nutrition came by natural thinking of the illiterate farmers who have no idea or information about the need of tees to iron or what is the role of iron in plant nutrition.

To understand the possibility of trunk nutrition, it is important to recognize the tree's anatomy and in what location on the trunk fertilizers can be injected? What happened to these fertilizers when injected? As previously mentioned, vascular plants contain two main types of conduction tissue, xylem and phloem, which extend from the leaves to the roots and are vital conduits for water and nutrient transport. The structure of xylem and phloem tissue depends on whether the plant is a flowering plant (including dicots and monocots) or a gymnosperm (polycots).

8.2.1 Trunk Anatomy of Dicot Plants

Xylem and phloem tissues are produced by meristematic cambium cells located in a layer just inside the bark of trees and shrubs. In dicot stems, the cambium layer produces phloem cells on the outside and xylem cells on the inside (Fig. 8.1). All the tissue from the cambium layer outward is considered bark, while all the tissue inside the cambium layer to the center of the tree is wood. Xylem tissue conducts water and mineral nutrients from the soil upward in plant roots and stems. It is composed of elongate cells with pointed ends called tracheids and shorter, wider cells called vessel elements (Fig. 8.2). The walls of these cells are heavily lignified, with openings in the walls called pits (Fig. 8.3). Tracheids and vessels become hollow, water-conducting pipelines after the cells are dead and their contents (protoplasm) have disintegrated. The xylem of flowering plants also contains numerous fibers, elongate cells with tapering ends and very thick walls.

8.2.2 Trunk Anatomy of Monocot Plants

Monocot stems, such as palms and bamboos, do not have a vascular cambium and do not exhibit secondary growth by the production of concentric annual rings. They cannot increase in girth by adding lateral layers of cells as in woody dicots. Instead, they have scattered vascular bundles composed of xylem and phloem tissue (Fig. 8.4). Each bundle is surrounded by a ring of cells called a bundle sheath (Fig. 8.5). The structural strength and hardness of woody monocots is due to clusters of heavily lignified tracheids and fibers associated with the vascular bundles.

8.2.3 In Which Plants Trunk Nutrition Can Be Successful?

According to the known anatomy of the flowering plants, trunk nutrition can be only successful in the stems of dicot trees like mango, citrus, guava, kaki, olive, grapes, peach,

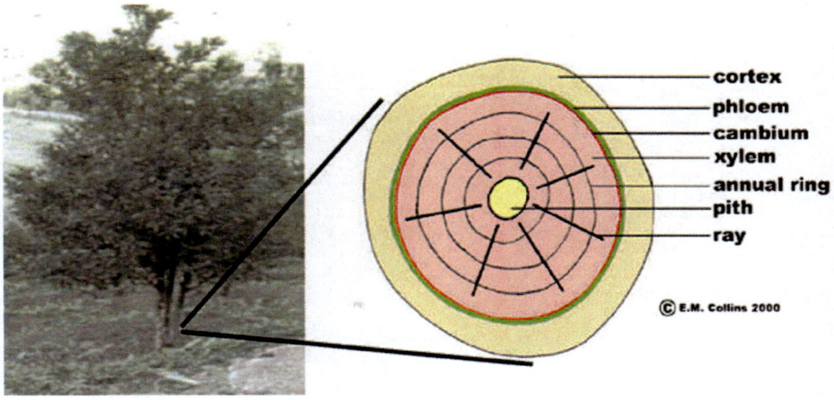

Fig. 8.1 Cross-section in dicot tree trunk

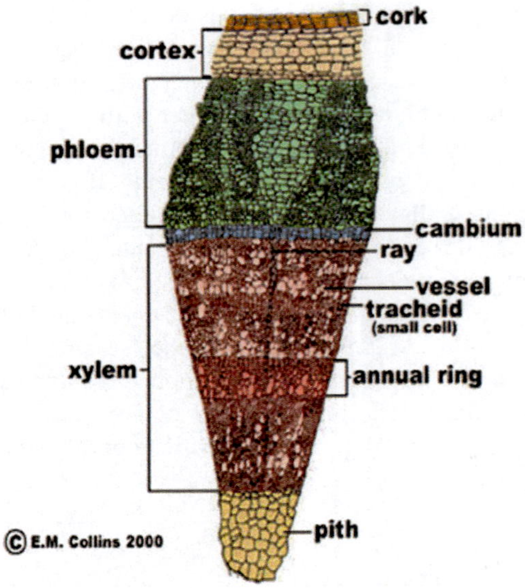

Fig. 8.2 A detailed cross-section in dicot trunk

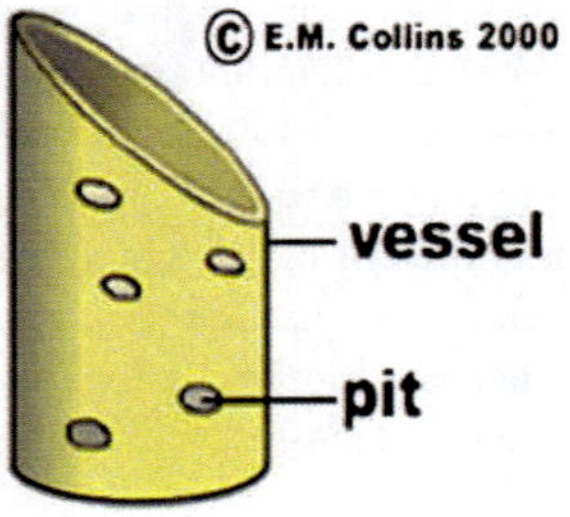

Fig. 8.3 Xylem vessel element

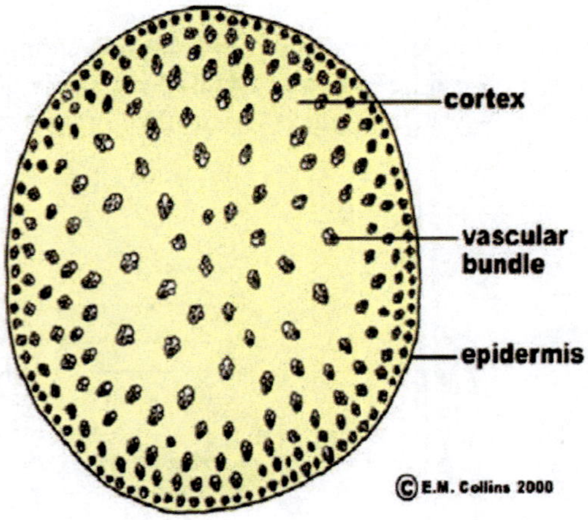

Fig. 8.4 Cross-section in monocot tree trunk

apricot, apple, pears, figs, avocado, etc., or in the roots of monocot trees like palms and banana. Nutrient solutions should be injected into xylem tissues which conduct water

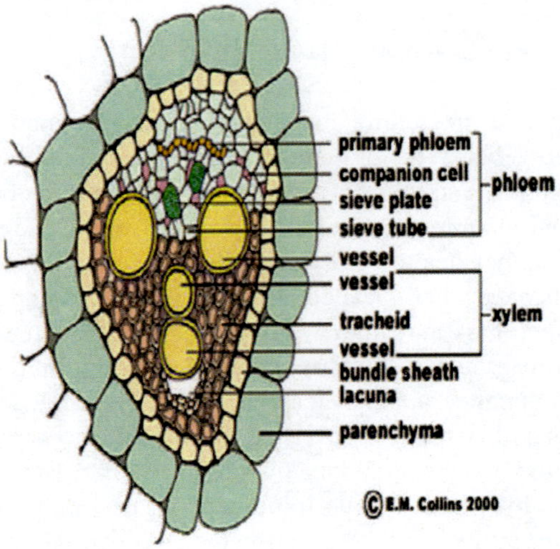

Fig. 8.5 Vascular bundle

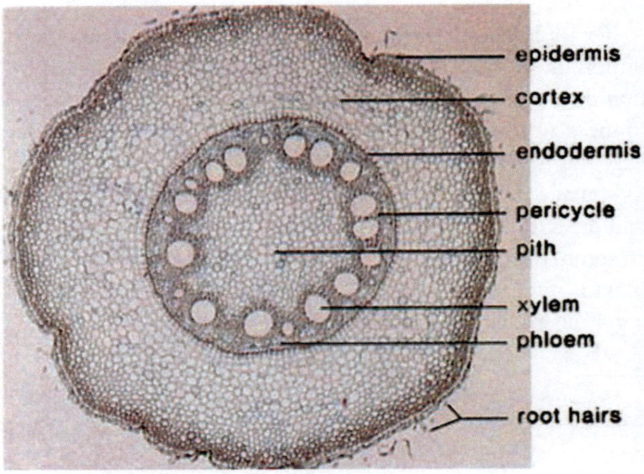

Fig. 8.6 Cross-section in monocot root

and mineral nutrients upward in plant stems. Because xylem in the trunk of the monocot plants is located inside the scattered bundles (Fig. 8.4) and too thin to be reached by an injection needle, it is not applicable to inject fertilizers in monocot stem xylem. However, the annual ring xylem in the stems of dicots is large enough to accept fertilizer injection and very easy to be reached by an injection needle. Root xylem of monocots, however, looks like stem xylem of dicots (Fig. 8.6) which render the possibility that fertilizers can be also injected into monocot roots. The only difference is that when we inject fertilizers into monocot roots, we need a longer injection needle compared to the needle needed for the injection into dicot stems because of the thick cortex of the root before the needle reaches xylem vessels.

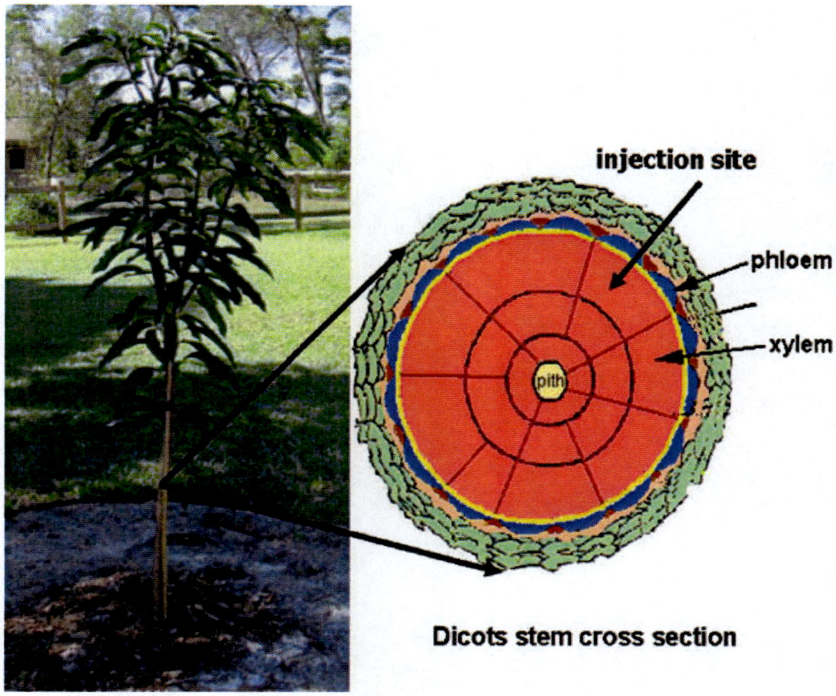

Fig. 8.7 Location of fertilizer injection in the trunk of dicot trees

8.2.4 How Can Fertilizers Be Injected into a Tree's Trunk?

Fertilizers can be injected into dicot trunk (Fig. 8.7) or monocot roots through a plastic injection needle (Fig. 8.8) attached to the tank containing the well-balanced and completely soluble nutrient solution (Shaaban 2009). The needle location on the trunk of dicots should be below the tree canopy and the tank in all cases should be approximately 1 m above the injection needle (Fig. 8.9). However, fertilizer injection cannot be successful unless there is a xylem layer thick enough to accept the injection needle to stick in and fixed inside. Normally, this desired xylem layer can be formed at the beginning of the third growing season of the trees. This is also the reason why injection fertilization through stem is very difficult and may be impossible for the dicot field crops, vegetables, and some of ornamental crops. These plants are thousands in the hectare and have very thin xylem layer unable to accept total fertilization through injection which makes this technique not applicable for these plants.

Fig. 8.8 Needle insertion as the primary step of trunk nutrition (Shaaban 2009)

8.2.5 Theoretical Aspects of Trunk Nutrition

Trunk nutrition or injection fertilization technique invented by Shaaban (2007) (Patent No. 23750/2007, Egyptian Academy of Scientific Research and Technology) depends upon: when a nutrient solution containing compounds of all the fertilizer requirements (as concentrated soluble solution) is injected into active xylem, the flow of the water stream in xylem vessels during passing the injection location should dilute this solution and transport the fertilizer elements to the leaves, exactly as they were absorbed by the roots to xylem vessels (Fig. 8.10). The difficulty here is how to insure that

Fig. 8.9 Location of fertilizer tank during trunk nutrition (Shaaban 2009)

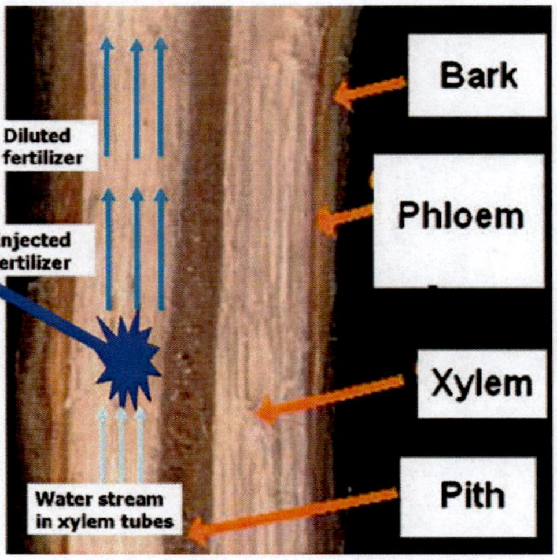

Fig. 8.10 Theory behind trunk nutrition (Shaaban 2007)

this technique provides an optimal dosing at different growth stages. Actually this is not a problem, because in active growth stages during maximum photosynthesis and translocation of solutes to storage organs, the flow rate of water stream from the roots is faster. Then this flow will carry the required amounts of fertilizers from the injection location in the trunk to leaves in the proper times. When the flow rate of water stream absorbed by the root slows down, it will carry fewer amounts of nutrients to the leaves exactly as done by the plant autoregulation in the periods of active growth and the periods when growth slows down.

The real challenge facing the full nutrition through trunk is how to combine a nutrient solution containing all the fertilizer elements (N, P, K, Ca, Mg, Fe, Mn, Zn, Cu, B, Mo, Co, and others may needed) from the commercial fertilizers in a balanced form without precipitation of any of these elements. On the other hand, this form of compounds should not be of acidic or alkaline reactions that harm the xylem tissues at the injection site or cause salinity effect on the plant tissues. This actually needs too much calculations and prolonged experiments with different trees varieties, ages, and locations, taking into consideration the different nutrient requirements of different trees varieties. However, partial trunk nutrition with one or two nutrient elements can be useful when the trees face a problem in absorbing these elements from the soil and foliar fertilization is not efficient, or not applicable.

8.3 Previous Studies on Trunk Nutrition

To study the response of two date palm trees cultivars (Khlas and Ruzaiz) to Fe trunk injection fertilization, Abo-Rady et al. (1987) found that injection with 100 g $FeSO_4.7H_2O$ tree^{-1} increased the Fe contents in the leaves of both cultivars and with 100 g Fe-EDDHA/tree in the leaves of Ruzaiz cultivar, while soil application of FeEDDHA was not effective. In a comparison study on the effect of fertilization method with potassium on the growth and fruit characteristics of date palm variety kabkab, Abdi and Hedayat (2010) used control treatment (no fertilization), soil surface application of 3 kg palm^{-1} as potassium sulfate (48% K_2O), foliar spray of potassium sulfate (2%), and potassium sulfate injection into the trunk of tree (2%). To inject fertilizer, they drilled the trees with a hand drill 1.5 m in height and 30 cm depth with 45 hermitages to down. They found that injection method was the best in achieving higher yield with better quality (Table 8.1). Mohebi et al. (2010) studied the effects of iron fertilization on yield and yield components of date palm "Sayer" in Al-Ahwaz research station of Khoozestan Province, Iran, on an 8-year-old date palm (*Phoenix dactylifera*) during 2002–2006. Their research treatments consisted of: (1) control treatment, (2) soil surface application of Fe in two levels, (3) application of Fe as localized placement method in two levels, and (4) Fe injection into the trunk of tree in four levels. They stated that injection of 25 g $FeSO_4$ tree^{-1} into the trunk showed the best results.

Table 8.1 Effect of different application methods of K on fruit setting, fruit drop percentage, and yield per palm of kabkab cultivar

	Fruit setting %									
	45 DAP		90 DAP		135 DAP		Fruit drop %		Yield/palm (kg)	
Treatments	2008	2009	2008	2009	2008	2009	2008	2009	2008	2009
Control	80 a[a]	75.2 a	45b	37 a	35 b	36b	19b	43.5a	91.37c	97.36c
K soil	79 a	75 a	46 b	41 a	40.2 b	41 b	19.3b	22.25b	103.31b	104.07b
K foliar	80.1 a	79.2 a	75.1 a	45a	61 a	59 a	26.24a	26.26b	103.5b	105.92b
K injection	77 a	76.1 a	55.1 b	45 a	33.1 b	61 a	21.57b	20.14b	112.4 a	115.37a

Adapted from Abdi and Hedayat (2010)
DAP Days after pollination
[a]Means within a column followed by the same letters are not significantly different by new Duncan's multiple range test ($P>0.05$)

These results of date palm trunk nutrition, however, are not absolutely acceptable because as we know, palms belong to the monocot plants which are too difficult to catch xylem vessels in their trunk to inject the fertilizers in. But the injected fertilizers in these experiments may be transported across the matrix to the conducting vessels upward to leaves. This may explain the relatively high quantity of the effective dose that led to increased Fe content of leaves or higher yields.

Peigang et al. (1993) stated that high-pressure trunk injection can supply mineral elements into the fruit trees. With apple trees, they found that Fe, Ca, B, Zn, and P injection improved these elements' content in leaves, and Fe injection into hawthorn trunk could eliminate iron deficiency. They concluded that high-pressure trunk injection is a new method for regulating fruit tree mineral nutrition, which is quite rapid, highly efficient, and fairly economic. In another trial with apple trees in China, Xiang et al. (2000) declared that iron fertilization by high-pressure trunk injection indicates that Fe^{2+} was transported along the central xylem, while most of Fe^{2+} was transported and stored in roots, and it was a little difficult for Fe^{2+} to be transported up to leaves. They found that the speed of transportation reached several hundred centimeters per hour when trunks were injected with iron fertilizer with high pressure. Marie (2009) mentioned that the Boulder County Business developed a system to treat chlorosis in dicot trees called "Tree IV" (Fig. 8.11). They described that they drill small holes in the tree's trunk near the base of the tree to reach xylem. Then, they attach injection plugs into the holes and place the injection needles into the tree's trunk. Using air pressure, they inject the tree with a liquid mix of chelated iron and magnesium. They noticed that recovery was proportional to the level of the severity of chlorosis at the time of treatment. In a preliminary study on trunk nutrient injection fertilization for old *Cinnamomum camphora* trees, with two different nutrient solutions, Ying et al. (2009) found that both of the two solutions could effectively promote the nutrient element contents in leaves of the old and weak *C. camphora* trees in a short time.

Fig. 8.11 All-natural intravenous solution "Tree IV" to treat iron chlorosis of Qoak trees (Adapted from Boulder County Business Marie 2009)

Shaaban (2009) studied injection fertilization through trunk for mango and grapevine trees. He stated that dicotyledonous vascular trees can be fully fertilized by injection through xylem. Only 5–10% of the fertilizer levels used as soil fertilization was sufficient for good growth and higher yield. Growth of the injection-fertilized mango (*Mangifera indica* var. "Sukkary white") trees was 20–25% higher than soil-fertilized plants (Fig. 8.12), while in grapevine (*Vitis vinifera* vars. "White Riesling" and "Spaet burgunder"), fruit yield increases were 32–49% higher compared to soil fertilization (Fig. 8.13). In the same study, fruit quality of grapevine clusters assessed (juice °Brix, pH, reduced sugars, total acidity, grape vinegar, apple vinegar, ethanol, and glycerin content) of the plants fertilized through injection was better than those fertilized through soil (Table 8.2). Grapevine fresh juice content of the reduced sugars and ethanol increased by 7.5–11.9% and 41.4–50%, respectively, while the total acidity decreased by 6.2–19.7%.

In a field experiment with 7-year-old guava and kaki trees grown on calcareous soils of Beheira Governorate, Egypt,

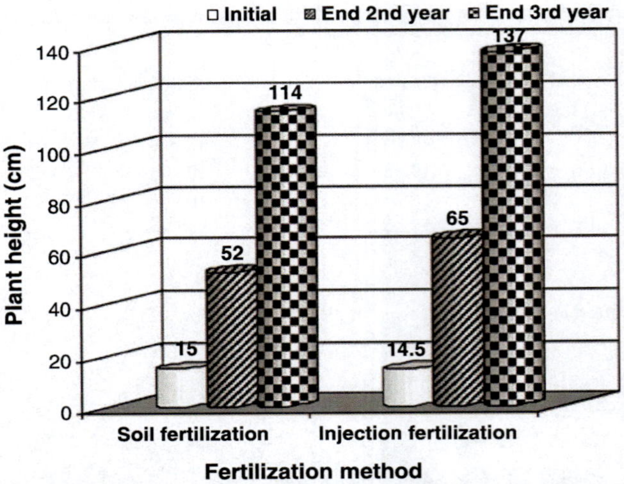

Fig. 8.12 Growth of mango trees fertilized through trunk injection throughout two growing sea sons compared to soil-fertilized plants (Adapted from Shaaban 2009)

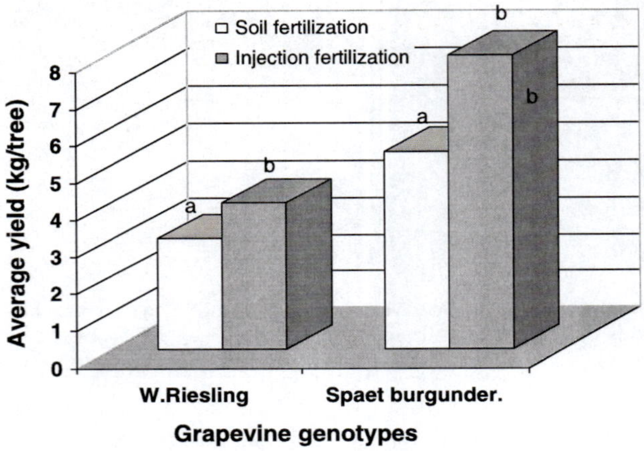

Fig. 8.13 Average yield of grapevine trees fertilized through trunk injection compared to soil-fertilized plants (Adapted from Shaaban 2009)

Shaaban (2011a) found that the yield of guava and kaki trees fully fertilized through trunk increased by 84% and 89%, respectively, over soil-fertilized trees (Fig. 8.14). He attributed the high ratio of yield increases achieved by trunk nutrition to the optimal supplement by phosphorus and micronutrients likely to be fixed and be of minor availability for absorption by the plant roots in the calcareous soil other than N and K supplements. To treat Huanglongbing (HLB) disease in citrus trees in China, Chen (1943) used nutrition elements such as zinc sulfate ($ZnSO_4$), copper sulfate ($CuSO_4$), boron (H_3BO_3), calcium sulfate ($CaSO_4$), ferric sulfate ($FeSO_4$), potassium hydrophosphate (KH_2PO_4), and magnesium sulfate ($MgSO_4$) to be filled (as powder) into trees trunk (Table 8.3). He found that trunk injection with these elements of considerable power to recover a reasonable ratio of the infected trees. In Japan, labeled ammonium phosphate was injected into one side only of 20-year-old "Satsuma" orange trees in the field and in pots by Tanaka (1960). He found that the absorbed P accumulated on the treated side, little or none being found on the other side, and was greatest in the upper part of the tree. He found also no difference due to age of leaves, and no radioactivity appeared in the fruit.

A method of introducing treatment agents comprising medicine or nutrients into a tree trunk (Fig. 8.15) was registered as US patent No. 6,216,388 B1 by Miller et al. (2001). This patent involved the boring of holes above ground into the trunk of navel orange trees. In this method, homogeneous plugs of solidified polyethylene glycol (PEG) with a treatment mixture of iron citrate or iron ammonium citrate (18 g), manganese sulfate (0.7 g), and zinc sulfate (0.6 g) were molded and inserted into the holes in the tree trunk and sealed with wax. The dissemination time of the treatment agent into the tree trunk was controlled by selection of the PEGs used, meaning that higher-molecular-weight PEGs take longer to dissolve and disseminate the treatment agent into the transpirational flow of the tree being treated than the lower-molecular-weight PEGs. According to the above-mentioned patent,

Table 8.2 Quality parameters of the grapevine fresh juice ($n=22$)

Cultivar	Fertilization method	pH °	Brix (g.l^{-1})	Reduced sugars Glucose (g.l^{-1})	Reduced sugars Fructose (g.l^{-1})	Total acidity (g.l^{-1})	Grape vinegar (g.l^{-1})	Apple vinegar (g.l^{-1})	Ethanol (g.l^{-1})	Glycerin (g.l^{-1})
White Riesling	Soil (control)	3.01	21.57	105.1	119.1	9.51	6.28	3.07	0.70	2.32
	Injection	3.12	23.19	111.6	129.4	8.96	5.59	2.75	0.99	3.59
	$LSD_{0.05}$	0.226	3.15	17.99	21.65	3.27	1.70	1.37	0.40	3.30
	% increase/decrease over/above control	+3.3	+7.4	+ 6.3	+ 8.6	−6.2	−13.3	−11.6	+41.4	+54.7
Spaet burgunder	Soil (control)	3.0	21.96	113.3	119.8	9.1	6.7	3.7	0.4	1.3
	Injection	3.2	24.14	127.9	133.0	7.6	5.5	3.8	0.6	0.4
	$LSD_{0.05}$	0.226	6.41	8.01	6.8	1.86	1.79	1.70	0.226	0.506
	% increase/decrease over/above control	+6.7	+9.0	+12.9	+11.0	−19.7	−17.9	−2.6	+50	−69.2

Adapted from Shaaban (2009)

18-year-old chlorotic navel orange trees having a trunk diameter of 9 in. were treated with four plugs of PEG-ferric ammonium citrate mixture having 0.1% iron. The treatment site was covered with grafting wax. The plugs found to totally dissolve in 5 days, and the symptoms of iron chlorosis gradually disappeared, and within 4 weeks, the leaves were noticeably greener. After 2 months from the injection, the trees were fully recovered and did not show any sign of phytotoxicity. Tian (2007) studied the effects of infusion fertilizer on the fruit quality of citrus. He applied the infusion fertilizer with different concentrations in the branches and near the root to investigate their effects on citrus quality. He found that applying infusion fertilizer had obvious promotion effects on the growth of plants, and the ratio of sugar to acid was enlarged and vitamin C content was increased. He also found that the effect of applying 6% infusion fertilizer in the branches was the best.

According to the patent of Shaaban (2007), orange trees were fully fertilized through trunk injection (Fig. 8.16). The trees were cultivated in sandy soil very poor in its nutrient content. The injected nutrient solution was combined from commercial fertilizers and contained all required fertilization nutrients. Injection fertilization started at the beginning of the third growing season upward. Other trees at the same age were soil fertilized for comparison. It was found that growth of the trunk-injected trees was more vigor, and the trees accumulated higher dry biomass compared to the soil-fertilized trees (Fig. 8.17). Vigor growth of the trunk-fertilized plants was attributed to the better nutrient balance realized by the injected formula. Concentrations of P, K, Mg, Mn, and Zn in the leaves of the trunk-injected plants were in the middle of the adequate range compared to soil-fertilized plants where the concentration of these nutrients were in the lower edge of the adequate range (Figs. 8.18 and 8.19). Even so, the soil-fertilized plants received the most optimized fertilizers as recommended by citrus nutritionists.

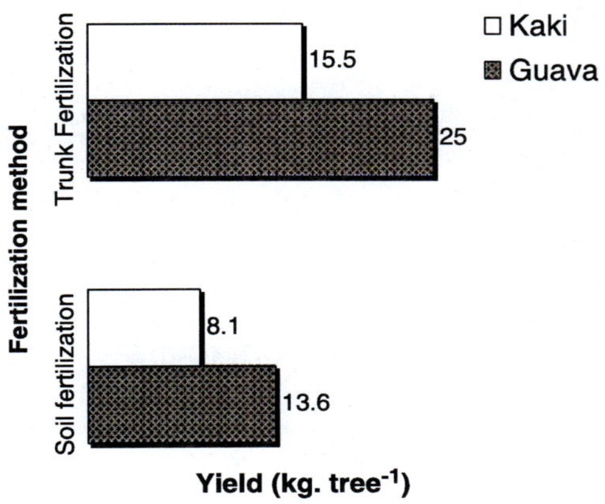

Fig. 8.14 Average yields of guava and kaki trees fertilized trough trunk injection compared to soil-fertilized trees (Adapted from Shaaban 2011a, unpublished data)

Table 8.3 Dose of nutrition element by trunk application of HLB citrus trees

Nutrition elements	$ZnSO_4$	$CuSO_4$	H_3BO_3	$CaSO_4$	KH_2PO_4	$MgSO_4$	Total dose
Dose (g) treatment	0.5	0.2	0.1	1.5	1.0	0.2	tree^{-1} (g)
I							3.5
II	+	+	+	+	+		3.3
III	+	+	+	+			2.3
IV	+	+	+				0.8

Adapted from Chen (1943)

8.4 Economics of Trunk Nutrition

- As previously mentioned, full fertilization through tree trunks saves about 85–90% of fertilizers ought to be added as soil fertilization. The money could be saved using trunk injection fertilization (Tables 8.4, 8.5, and 8.6) for citrus trees compared to fertigation technique (Shaaban 2007).
- On the other hand, trunk fertilization system is very simple, very cheap, and easy applicable.
- Trunk full fertilization saves also labor wedges paid for conventional fertilizer distribution along the trees' growth season.

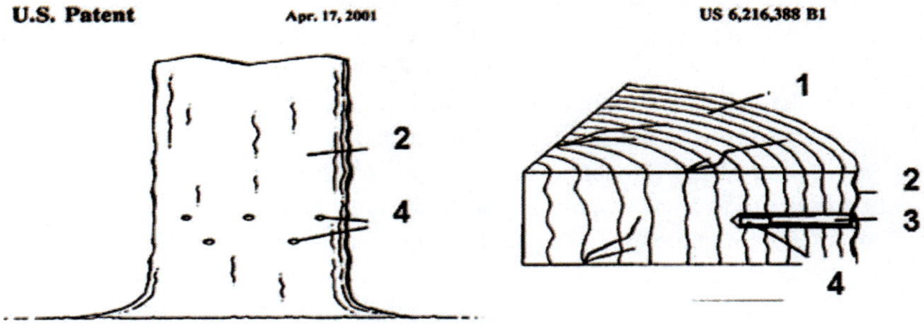

Fig. 8.15 A schematic diagram shows PEGs insertion into navel orange tree's trunk. *1* Insertion depth, *2* Insertion site, *3* Insertion hole, *4* Plugs location (US patent No. 6,216,388 B1)

Fig. 8.16 Orange tree fully fertilized through trunk injection (Adapted from Shaaban 2011b)

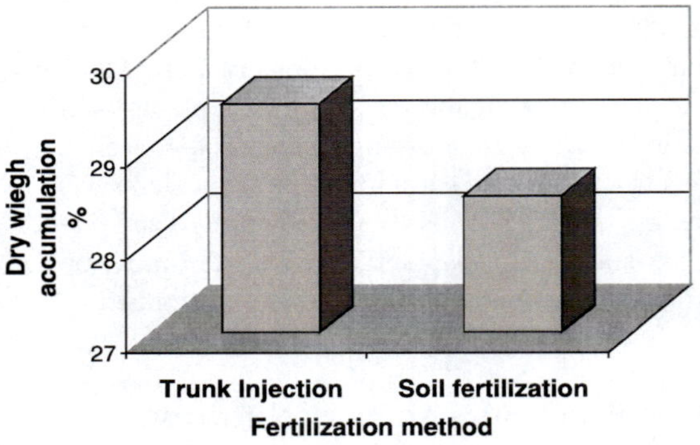

Fig. 8.17 Dry biomass accumulation (%) of 3-year-old orange trees as affected by fertilization method (Adapted from Shaaban 2011b, unpublished data)

- Following full fertilization through trunk injection, no fertilizers are added to soil. In such a case, the herbs will not compete with the trees' roots for the available soil nutrients, so there is no need to combat herbs manually or by herbicides. This also will save the money paid for labor, spraying machines, and purchasing herbicides.
- Using trunk injection fertilization allowed also adding the systemic pesticides at smaller quantities (if urgently needed) to the injected nutrient solution. This would saves also the money paid for the greater quantities that used as spray on the trees canopy and keep the outer environment around the trees free from pesticide hazards on human beings, animals, birds, and soil organisms. On the other hand, injecting pesticides through xylem would be of more and faster effect than spraying on the vegetative parts from outside.
- Application of the full fertilization through trunk injection in the rural areas would saves also invisible medical costs spent for treatment poisoning of people and animals by fertilizer, pesticides, and herbicides residuals in the drinking water. Moreover, the fruits produced by the trunk-fertilized trees found to contain no fertilizer or pesticides residues, which make them safer for the human consumption.

8.5 Trunk Nutrition Precautions

Some precautions were described by Shaaban (2007) to apply tree fertilization through trunk:

1. Citrus trees may not accept fertilization through trunk injection before the beginning of the third growing season
2. Shrubs can be injected through an injection needle from one side, but the big trees can be injected from two opposite sides
3. Injected fertilizers must be fully dissolved without any precipitations
4. Used water for fertilizer solubilization must be free from salts
5. Salinity and pH of the nutrient solution must be accurately adjusted to avoid the harmful effects of the highly alkaline or highly acidic solutions
6. In case abnormal symptoms appeared on the trees, nutrient solution taps must be switched off immediately and revise and/or recombine the nutrient solution
7. A routine passing through the trees' rows should be periodically done to ensure unleakage of the nutrient solution outside the injection needle and the flow of the nutrient solution is running normally
8. All parts of the used injection system are made from plastic materials. Because the system is continuously exposed to sunrays and other environmental factors, it can be damaged after some time. According to our experience, the system should be reexchanged once every 3 years. This is because the possible damages occurred and to change the injection site to the most recent xylem layer, as the old xylem layers became less conductible to water and dissolved nutrients.

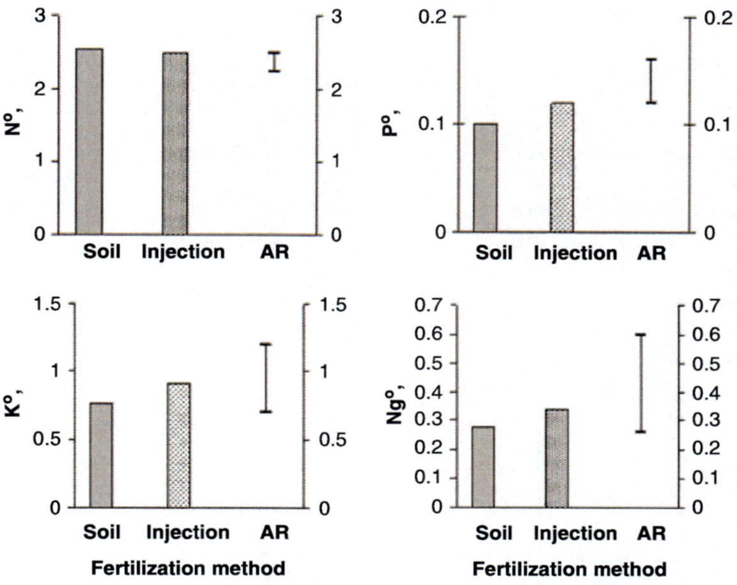

Fig. 8.18 Macronutrient concentrations in the leaves of orange as affected by fertilization method. *AR* Adequate range (Adapted from Shaaban 2011b, unpublished data)

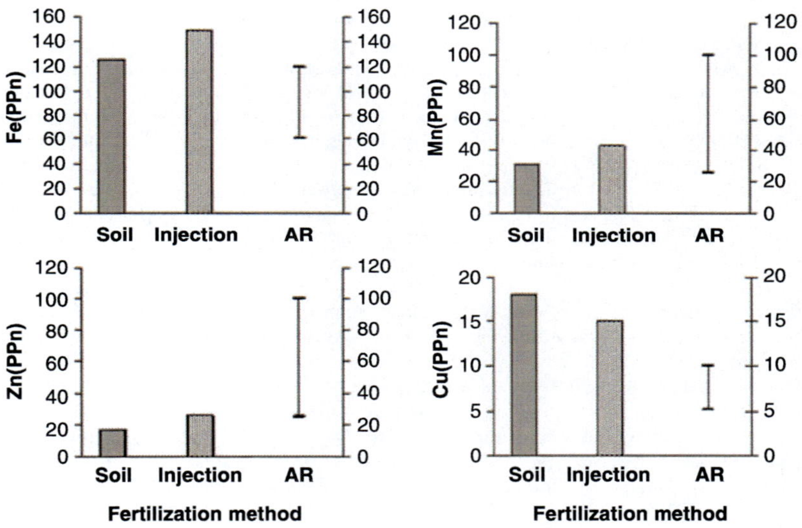

Fig. 8.19 Micronutrient balance in citrus leaves as affected by trunk injection compared to soil fertilization (Shaaban 2011 -unpublished data)-AR = Adequate range

Table 8.4 Injection fertilization costs for different ages of orange trees as compared to fertigation according to 2007 fertilizer prices US$

	Fertilizer cost for 1 ha (400 trees) US$		
Trees age (year)	Soil fertilization (fertigation)	Injection fertilization	Saving %
2	87.0	6.0	93
3	150.0	8.8	94
4	184.0	13.2	92.8
5	282.5	19.8	93
6	309.0	23.1	92.5
7	358.0	25.3	93.4
8 and more	389.0	28.6	92.6

Adapted from Shaaban (2007)

Table 8.5 Injection fertilization costs for different ages of mandarin and grape fruit trees as compared to fertigation according to 2007 fertilizer prices US$

	Fertilizer cost for 1 ha (400 trees) US$		
Trees age (year)	Soil fertilization (fertigation)	Injection fertilization	Saving %
2	92.5	9.5	89.7
3	125.0	11.5	90.8
4	165.5	16.8	89.8
5	205.5	24.4	88.1
6	231.5	26.0	88.7
7	262.5	30.6	88.3
8 and more	306.0	41.2	86.5

Adapted from Shaaban (2007)

Table 8.6 Injection fertilization costs for different ages of lemon (Benzher) trees as compared to fertigation according to 2007 fertilizer prices US$

	Fertilizer cost for 1 ha (400 trees) US$		
Trees age (year)	Soil fertilization (fertigation)	Injection fertilization	Saving %
2	66.5	9.0	86.4
3	165.0	17.3	89.5
4	218.0	24.7	88.6
5	255.0	28.8	88.7
6	267.5	32.0	88.0
7	310.5	36.2	88.3
8 and more	288.8	32.9	88.6

Adapted from Shaaban (2007)

References

Abdi GH, Hedayat M (2010) Yield and fruit physiochemical characteristics of 'Kabkab' date palm as affected by methods of potassium fertilization. Adv Environ Biol 4(3):437–442

Abo-Rady MD, Ahmed HS, Ghanem M (1987) Response of date palm to iron fertilization by trunk injection and soil application. Z Pflanzenernähr Bodenkd 150:197–200

Chalk PM, Heng LK, Moutonnet P (2001) Nitrogen fertilization and its environmental impact. In: Proceedings of the 12th international world fertilizer congress, Beijing, China, 3–9 Aug, pp 1–15

Chen QB (1943) A report of a study of Huanglongbing in Chaoshan. Q Bull New Agric 142–175

Dixon RC (2003) Foliar fertilization improves nutrient use efficiency. Fluid J Winter 2003:1–2

Emmons K (2000) Chemigation water sampling report—nitrates, years 1987–2000. KDA Division of Water Resources, Tech. Survey Section, Topeka, KS

Gaines TP, Gaines ST (1994) Soil texture effect on nitrate leaching in soil percolates. Commun Soil Sci Plant Anal 25(13&14): 2561–2570

Gupta SK, Gupta RC, Seth AK et al (1999) Epidemiological evaluation of recurrent stomatitis, nitrates in drinking water and cytochrome b5 reductase activity. Am J Gastroenterol 94:1808–1812

Gupta SK, Gupta RC, Seth AK et al (2001) Recurrent diarrhea in areas with high nitrate in drinking water. Arch Environ Health 56:369–374

Halliday DJ, Trenkel ME (eds) (1992) IFA world fertilizer use manual. International Fertilizer Industry Association, Paris, p 632

Horesh I, Levy Y (1981) Response of iron deficient citrus trees to foliar iron sprays with low surface tension surfactant. Sci Hort 15:227–233

Jemison JM, Fox RH (1994) Nitrate leaching from nitrogen fertilized and manured corn measured with zero-tension pan lysimeters. J Environ Qual 23:337–343

Keeney DR (1987) Nitrate in groundwater—agricultural contribution and control. In: Proceedings of the conference on agricultural impacts on ground water, Omaha, NE, 11–13 Aug 1986. National Water Well Association, Dublin, pp 329–351

Laegreid MO, Bockman C, Kaarstad O (1999) Agriculture, fertilizers and environment. CABI Publishing Associate, Wallingford. ISBN 0–85199–385–3

Malavolta E (1980) Elementos da nutrição mineral de plantas. Piracicaba Agronômica Ceres, São Paulo, p 254

Marie J (2009) An all-natural remedy for yellow-leaf disease. Boulder County Business Report, 4 Sept 2009

Meng HQ, Fu BJ, Yang LZ (2001) Effects of land use on soil erosion and nutrient loss in the Three Gorges Reservoir Area, China. Soil Use Manage 17:288–291

Miller GW, Awada S, Salam A (2001) Dissolving polymer plug for introducing nutrients and medicinal materials into tree trunks. US Patent No. 6216388B1, Apr 2001

Mohebi AH, Nabhani L, Dialami H (2010) Yield and yield components of "Sayer" date palm as affected by levels and methods of iron fertilization. Acta Hort (ISHS) 882:131–136

Morgan M, Scholberg JM, Obreza T (2006) Citrus nitrogen uptake and nitrification rates in sandy soils. In: Proceedings of the 18th world congress of soil science, Philadelphia, PA, 9–15 July 2006

Obreza TA, Alva AK, Calvert DV (2009) Citrus fertilizer management in calcareous soils. University of Florida, IFAS Extension, Gainesville, Publication #CIR1127

Peigang X, Xiang S, Dianbin G (1993) Studies on some mineral element injection through fruit tree trunk. Department of Horticulture, Shandong Agricultural University, Taiwan

Pope LM, Bruce BW, Hansen CV (2001) Ground-water quality in quaternary deposits of the central high plains aquifer, south central Kansas, 1999. Water-Resources Investigations Report 00–4259. USGS, Lawrence

Shaaban MM (2007) A method to fertilize trees and shrubs through trunk injection. Patent No. 23750. Acad Sci Res Technol, 31 July 2007

Shaaban MM (2009) Injection fertilization: a full nutritional technique for fruit trees saves 90–95% of fertilizers and maintains a clean environment. Fruit Veg Cereal Sci Biotechnol 3(1):22–27

Shaaban MM (2011a) Injection fertilization increases Kaka and Guava yields and saves about 90% of the fertilization costs (unpublished data)

Shaaban MM (2011b) Full fertilization of citrus tress through trunk injection compared to soil fert (unpublished data)

Tanaka Y (1960) Effects of phosphorus fertilization on *Citrus unshiu* Marc. I. A study on translocation and distribution of phosphorus applied to the soil or injected into the trunk of citrus trees using P^{32} as a tracer. J Hort Assoc Jpn 29:63–69

Tian J, Zhabg C, Xiong Y, Kong C (2007) Effects of infusion fertilizer on the fruit quality of citrus. J Anhui Agric Sci 34:98. doi: CNKI:SUN: AHNY .0.2007-34-098

van Raij B (1991) Fertilidade do solo e adubação. Agronômica Ceres, Piracicaba, pp 137–181

WHO (1993) Guidelines for drinking water quality recommendations, 2nd edn. World Health Organization, Geneva. ISBN 9241544600

Wong MT, van der Kruijs AC, Juo AS (1992) Leaching loss of calcium, magnesium, potassium and nitrate derived from soil, lime and fertilizers as influenced by urea applied to undisturbed lysimeters in south-east Nigeria. Fertil Res 31:281–289

Xiang CM, Jun XJ, Ping ZZ, Tai SZ, Zhao ZJ, Zhang FS, Li SH (2000) Application of iron fertilizer by high pressure trunk injection to remedy iron deficiency chlorosis and its mechanism in apple trees. Acta Agric Boreali-Sinica S436.6. doi:cnki:ISSN:1000-7091. 0.2000-S1-028

Ying L, Wei X, Hua XQ (2009) A preliminary study on trunk nutrient injection fertilization for old *Cinnamomum camphora* trees. J Southwest For Univ 29(4):29–32

Zia MH, Ahmad R, Khaliq I et al (2006) Micronutrients status and management in orchards soils: applied aspects. Soil Environ 25(1):6–16

Non-destructive Proximal Sensing for Early Detection of Citrus Nutrient and Water Stress

9

Paolo Menesatti, Federico Pallottino, Francesca Antonucci, Giancarlo Roccuzzo, Francesco Intrigliolo, and Corrado Costa

Abstract

This chapter reports the application of non-destructive optical-based technologies for the rapid and efficient assessment of the nutritional status and water stress detection improving their use efficiency. In the proximal sensing section, it was presented the use of spectral and hyperspectral imaging to evaluate the plant nutritional status. Proximal sensing offers the opportunity to rapidly collect a huge amount of crop canopy information. In the infrared thermography and thermometry section, results about their use to assess plant water stress analysing canopy and soil temperature variation were reported. Finally, the use of spectrophotometry and of the chlorophyll meter for the citrus nutrient detection is presented. The analyses of data were carried out by linear regressions and by multivariate statistics.

Keywords

Spectroscopy • Infrared thermography and thermometry • Nutritional plant • Water soil detection • Multivariate analysis

9.1 Introduction

Water and nutrient uptake are closely related and they are both essential for plant growth and productivity. The advent of precision production methods and the increase of both environmental and quality standard issues imply the necessity to improve nutrient and water use efficiency. These objectives and the related need for automation have led to the development of rapid methods for plant monitoring (Jones 2004). The current worldwide guidelines for citrus fertilisation for young and bearing trees are based on soil and leaf analyses and yield expectancy. In fact, soil testing and leaf analysis are the tools normally used to monitor the nutrient status of citrus orchards and modify the fertiliser management. Leaf analysis is the most widespread procedure, and it is performed since a long time by comparing results to well-established reference values of standard age spring-cycle leaves (Embleton et al. 1973a). Destructive sampling is necessary to examine dry mass and nutrient content of leaves, but this procedure is quite labour intensive and time consuming.

Hydric status in the soil-plant system can be monitored by means of soil water measurements (water content or potential), soil water balance calculations or plant stress sensing (tissue water status or physiological responses). Water potential, Ψ, is defined as the potential energy per unit mass of water with reference to pure water at zero potential, and it is expressed in units of pressure (MPa). In most biological systems, water has less potential energy than in the pure state, thus resulting in negative values for water potential.

The pressure chamber or Scholander bomb is the reference technique for Ψ determination. By means of this instrument, it is possible to measure the approximate water potential in plant tissues. A leaf attached to a stem

P. Menesatti (✉) • F. Pallottino • F. Antonucci • C. Costa
CRA-ING (Agricultural Engineering Research Unit of the Agricultural Research Council), Via della Pascolare 16, 00015 Monterotondo Scalo (RM), Italy
e-mail: paolo.menesatti@entecra.it; corrado.costa@entecra.it

G. Roccuzzo • F. Intrigliolo
CRA-ACM (Citrus and Mediterranean Crops Research Centre of the Agricultural Research Council), Corso Savoia 190, 95024 Acireale (CT), Italy
e-mail: giancarlo.roccuzzo@entecra.it; francesco.intrigliolo@entecra.it

A.K. Srivastava (ed.), *Advances in Citrus Nutrition*,
DOI 10.1007/978-94-007-4171-3_9, © Springer Science+Business Media B.V. 2012

(or the stem itself) is placed inside a sealed chamber, and pressurised gas is added to the chamber slowly. As the pressure increases, at some point, sap will be forced out of the xylem and will be visible at the cut end of the stem. The pressure that is required to do so is equal and opposite to the water potential of the leaf (Scholander et al. 1965). Stem water potential is highly correlated with stomatal conductance and consequently with transpiration rate and with assimilation rate that is known to limit yield. Leaf water potential and predawn leaf water potential are also highly correlated with stomatal conductance, but the correlation of stomatal conductance with stem water potential has been found to be higher than that with leaf water potential (Naor 2000). Surface temperature measured with infrared thermometers is an important tool to diagnose plant water stress and thus for irrigation scheduling which has been in practice for some decades. The introduction of thermal cameras made the wider use of such approaches feasible, especially when combined with automated image analysis (Jones 2004). This chapter reports the application of non-destructive optical-based (spectral and thermal) technologies, for ground-based proximal sensing that could be used in farm, for rapid, efficient and low-cost nutritional status and for the water stress detection in order to optimise orchards management.

9.2 Proximal Sensing

9.2.1 Ground-Based Hyperspectral Imaging as Proximal Sensing Technology

Spatial heterogeneity of physical and chemical properties represents a crucial factor with an impact on crop response and final production in terms of quantity and quality. Precision agriculture aims to a focused use of any input, in the right place, time and amount, in order to improve farmer's income and reduce the adverse environmental impact on crop production. There is a growing demand for rapid and non-invasive acquisition of fine-scale information on soil properties and plant physical-chemical status for site-specific management. Proximal sensing through different technologies can help offer the opportunity to rapidly collect a huge amount of information regarding the crop canopy and therefore to identify the plants needs. Many different types of proximal sensors can be used to measure plant, among which, multi- or hyperspectral imaging and thermographic imaging.

For nearly 20 years, the environmental monitoring techniques have made many important strides, thanks to innovations in electronics and information technology. In agriculture, e.g. collecting short optical parameters, i.e. spectral reflectance data, related to vegetation cover, allowed the determination of plant distress signals before their onset was visible to the naked eye (Massantini et al. 1992; Enterline 2001). The visible near-infrared (VIS-NIR) spectral analysis appears to be one of the most innovative and interesting being non-destructive. The information content of such a technique is very high being based on the principle that every molecule absorbs or reflects only specific wavelengths, hence it is possible to obtain correlations between the amount of a particular compound and the standard quantity of reflected light (reflectance), absorbed (absorbance) or released (transmittance) (Urbani et al. 2002). These techniques, coupled with precise optical systems calibrated to estimate the chlorophyll content of plants (SPAD), can provide important support to agriculture from both a technical and agronomic point of view and from an economic one; indeed, the real-time diagnosis of the nutritional status of plants leads to targeted interventions for the prevention and resolution of problems as well as to rationalised ones.

An example of how suitable these technologies can be is represented by the cases of study which follow.

9.2.2 Plant Nutritional Status Evaluation Through Canopy Spectral Image Analysis

Two distinct sets of tests are presented in the following case study, the first carried out in November 2001, the second in July 2002, respectively. The tests were conducted by the CRA-ING at the farm 'Palazzelli Experimental Institute' for citrus (Lentini, eastern Sicily) on nine adult plants of orange [*Citrus sinensis* (L.) Osbeck].

The system used in the research consists of an optical spectrometer that captures the picture cleanly and disperses, pixel by pixel, the spectrum on a camera matrix (V1-IMSPECTOR SPECIM Finland). For the acquisition of horizontal two-dimensional spectral images, the spectrometer is mounted on a horizontal controlled handling system (DV-Spectral Scanner Padova); a tripod (with electronic control) enables movement on a vertical axis of the spectrometer, to take images of very large objects varying the resolution and scan speed, and to synchronise the acquisition with the corresponding angle (Pan & Tilt-DV-Padova). Images have an optical resolution of 570 dots per 300 vertical lines. The reflectance values were measured only in regions of interest (ROI) corresponding to the densest part of the crown. The calibration of the instrument in this regard required special attention because it is a delicate operation which depends on the success and reliability of the measurements. In fact, the optical characteristics of the plants found in reflection strongly depend on the intensity and angle of incident sunlight. Since field surveys have no artificial lighting, it can happen to perform measurements with different lighting conditions due to the sun light variability and weather conditions. In order to reduce the variability of measurement due to different lighting conditions, the spectral optical system is calibrated to a black and white reference, which occupies the lower and upper limits of the dynamic response of the instrument. The blank sample is in turn standardised by accurate survey performed in the laboratory. This procedure led to good repeatability of measurement in the first test of an experimental survey. Moreover, to further evasluate the light variability, during the second test a white rectangular sheet was used as reference nearby each plant sample. Thus, each spectral image includes a constant white standard reference at about the same distance (and optical plane) of the object to be measured. The set of spectral information involved 115 different values of wavelengths from 400 to 970 nm with step 5 nm. The most interesting reports (minimum or maximum) between individual wavelengths were placed in conjunction with analytical measurements on leaves, plants and soil. Leaves were sampled from the nine plants, taken from non-interesting bearing terminal branches, and on these were carried out the analysis for the determination of N, P, K, Ca, Mg, Fe, Mn, Zn and intensity of staining Green (SPAD-502 Chlorophyll Meter, Minolta). Moreover, the plant hydric status (xylem index) and soil chemical-physical characteristics were determined. Afterwards, spectral images of the 20 leaves sampled from each of the plants subjected to in-field hyperspectral measurements were acquired. From the findings on the foliage of citrus for the first test, there were significant values of simple linear correlations between reflectance at individual wavelengths and N content of leaves (all lengths in the range 750–795 nm), in Mn (range 930–950 nm), and more importantly, P (range 400–700 nm, 880–970 nm).

The in-field spectral detection system showed promising performances, although the first test was not an easy to operate and calibrate. This also required a significant amount of light to work with a good signal-to-noise ratio. These difficulties have greatly limited the number of photographs. With the spectral range used, however, images of high spectral resolution were obtained. The spatial resolution was low, but enough to make the hyperspectral analysis afterwards. The data acquired during the second experimental trial are still being processed.

9.3 Infrared Thermography to Assess Plant Water Stress

Infrared thermography (IT) is a non-invasive and non-destructive method based on measuring specific electromagnetic radiation emitted by any object, according to the Stefan-Boltzmann's and Planck's laws (see Maldague 1994; Rahkonen and Jokela 2003 for basic theory). All objects emit heat (energy) waves. If an object is cold, its molecules vibrate slower and energy of longer wavelengths is emitted. When the temperature of the object rises, its molecules vibrate faster and the wavelength becomes shorter. Every particular energy wavelength has a temperature associated with it. Among the different image analysis techniques and technologies, thermography has the capability to associate to the image information, the thermal punctual information, which is the temperature of each single pixel, in order to operate comparisons between objects inside the same image. Moreover, a good thermographic system could record until 50 frames *per second*; in this way, it is possible to operate dynamic measurements observing temporal modification of the object. This information could also be used to study biological phenomena. Those systems could calculate the surface temperature of the object, operating corrections due to many factors: physical, environmental and instrumental. However, one of the most important factors for a correct infrared thermal measurement is the emissivity that measures the capability of an object to adsorb or emit the thermal radiation. This capability is equal to 1 only for ideal objects (black objects) but is, in general, very low (<0.4) for most metals, while it is high, to 0.85, for plastic or organic materials. The complement of emissivity is reflectivity, i.e. the capacity to reflect the incident radiation. The setting of correct value of emissivity depends also on the geometry of the material and the recording angle; for this reason, it is important in the quantitative thermal survey. Otherwise, the most important character that influences the emissivity value is the kind of material and its superficial structure. Thermal imagers are one of the predictive maintenance tools being widely used in commerce and industry. As spectral images, thermal images or thermal videos also can be processed by visual inspection or automated procedure.

9.3.1 Dynamic Thermography for Water Stress Detection

In this case study, 11 citrus tree canopies were monitored through a portable thermocamera in order to assess their hydric status and eventual stress. The instrument was used in acquiring dynamic videos of the canopies under study which were firstly cooled down using a white sheet, avoiding the sun exposure for a prefixed time lag. After the sheet removal, the warming up due to sun radiation of each tree was filmed for about 40 s. The canopy temperature was acquired by utilising a S40 FLIR thermocamera with the following characteristics: detector type, focal plane array (FPA) uncooled microbolometer; field of view (FOV), 24° (at distance of 1 m, the FOV is equal to 0.42×0.31 m); instantaneous field of view (IFOV), 1.3 mrad (the theoretical FOV of one pixel);

Table 9.1 Results of PLS prediction

Parameters	N	P	K	Ca	Mg	Fe	Mn	Zn
Model (75%)								
N° LV	15	9	15	7	13	10	5	8
Pre-processing X-Block	Baseline	Normalise	Baseline	Osc	Baseline	Gls weighting	Normalise	Baseline
Pre-processing Y-Block	Median centre	Mean centre	Mean centre	Mean centre	Median centre	Median centre	Median centre	Mean centre
r (observed vs predicted)	0.945	0.915	0.995	0.995	0.982	0.946	0.840	0.905
SEP	0.039	0.004	0.039	0.085	0.020	4.380	3.064	1.107
RMSE	0.039	0.003	0.039	0.085	0.020	4.348	3.042	1.099
Test (25%)								
r (observed vs predicted)	0.909	0.429	0.991	0.947	0.944	0.917	0.925	0.889
SEP	0.049	0.018	0.058	0.304	0.048	6.054	1.637	0.972
RMSE	0.051	0.019	0.057	0.614	0.048	6.413	1.683	0.986

In the table are reported the following: n° of latent vectors (LV), first and second pre-processing for the X-Block and one for the Y-Block, the correlation coefficient (r), the standard error of prevision (SEP) and the root mean square error (RMSE) for the model and test

image frequency, 60 Hz; spectral range, 7.5–13 μm; focus, automatic or manual; thermal sensitivity 50/60 Hz, 0.08°C at 30°C; temperature range –40°C to +120°C. Emissivity (ε) was set equal to 0.98 because the acquisition was done on biological materials.

The temperature estimation was carried out by a multivariate regression partial least squares (PLS) on the basis of thermographic data collected for 40 s step 1 s. The procedure (Wold et al. 2001; Menesatti et al. 2010) performed using PLS Toolbox in MATLAB V7.0 R14 (The Math Works, Natick, USA) included the following steps: (1) utilisation of thermographic data acquired second by second including 0 (41 total values) as X-Block variables; (2) extraction of mean PSI values as Y-Block variables; (3) random separation of data set into two subsets, one for the model (70% of the whole data set) and one for the external test (30% of the whole data set); (4) application of pre-processing algorithms to both X and Y; (5) application of the multivariate technique PLS: modelling and testing; and (6) calculation of efficiency parameter of prediction. To obtain the best prediction test, different X and Y pre-processing techniques were applied, from the simpler (none, mean centre, auto scale, median centre, baseline) to the more specific for thermal data (Savitsky Golay, Multiple Scatter Correction, Orthogonal Signal correction) (Table 9.1). The modelling included both calibration and validation. The predictive ability of the model is partially dependent on the number of the latent vectors (LV) used and was assessed by the statistical parameters root mean square error (RMSE), standard error of prevision (SEP) and correlation coefficient (r) between observed and predicted values. Finally, we recorded the ratio of percentage deviation (RPD), which is the ratio of the standard deviation of the laboratory measured (reference) data to the RMSE (Williams 1987). It is the factor by which the prediction accuracy has been increased compared with using the mean of the original data. The model chosen was for the number of LV that yielded the highest r, minimum SEP for predicted and known water content and maximum RPD. The multivariate analysis results (Table 9.2) showed a high r prediction value for the test equal to 0.99 with a RPD value major than 2.5 (3.5) indicating excellent modelling abilitys (Viscarra Rossel et al. 2007).

Table 9.2 Results of the partial least squares (PLS) for the prediction of the canopy temperature

Parameters	
Model (70%)	
N° LV	2
First pre-processing X-Block	Log1suR
Second pre-processing X-Block	Median centre
Pre-processing Y-Block	Autoscale
r (observed vs predicted)	0.6957
RPD	1.3703
SEP	1.9412
RMSE	1.8436
Test (30%)	
r (observed vs predicted)	0.9915
RPD	3.5006
SEP	1.4142
RMSE	1.2693

The prediction of water stress in citrus plants, carried out by canopy temperature estimation, produced good accurateness using dynamic thermography. Such a technique appears useful in substituting the standard PSI evaluation for hydric stress assessment, being fast and informative.

9.3.2 In-field Non-destructive Techniques

The increasing worldwide shortages of water are leading to an emphasis on developing methods of irrigation that maximise the water use efficiency. The maintenance of a slight plant water deficit can improve the partitioning of carbohydrate to reproductive structures and also control excessive vegetative growth, giving rise to what has been termed 'regulated deficit irrigation' (RDI) (Jones 2004). To achieve this objective, precise irrigation scheduling protocols are needed.

9.3.3 Thermometry to Assess Plant Water Stress

In several studies on citrus (Kaufmann 1968; Syvertsen and Levy 1982), measurements of predawn leaf water potential (Ψ_{Lpd}) with the pressure chamber provided a reliable index of plant water status. The predawn measurement with the pressure-chamber technique is relatively tedious, while midday measurement is a lengthy procedure (Chonè et al. 2001; Naor and Peres 2001). Classical methods for monitoring crop water status, including measurements of plant properties or meteorological variables to estimate the amount of water lost by the soil-plant system, give poor indications, while, with the advances in radiometry sap flow measurements, dendrometers and remote sensing, it could be possible to extend such plant-based methods to the field scale (Jones et al. 2002). Aim of a field study was to assess the relationship between the midday stem water potential measured with pressure chamber and the canopy temperature, an indirect physiological indicator measured with handheld infrared thermometer, in order to find an alternative to Scholander's chamber method. The experiment was carried out over 3 years (2003–2006), in a 35-year-old citrus orchard [*Citrus sinensis* (L.) Osbeck] cv. 'Valencia late', grafted on sour orange (*C. aurantium* L.). The irrigation water was supplied weekly through drip irrigation with six emitters per tree with a discharge rate of 4 $L h^{-1}$ each. Every year, the plot was fully irrigated, and tree water needs were calculated according to the soil water balance, using class A evaporation Pan and Kc values for citrus trees (Allen et al. 1998).

Ψ_{md} was determined weekly by means of pressure chamber. Stem water potential was determined in the northern sides of the canopy, near the trunk, on three non-fruiting shoots of the spring flush per index plant. The determinations were carried out on mature leaves enclosed in plastic bags covered with a silver foil for at least 90–120 min before measurements. Simultaneously, canopy temperature (IR_{md}) was measured, during midday, using a portable infrared thermometer (Infrared ag. Multimeter model 510, Everest Interscience Inc.; Fig. 9.1). Canopy temperature was measured by pointing the infrared thermometer to the canopy at a distance of 30 cm, in the same plants and canopy positions where Ψ_{md} determinations were carried out.

Ψ_{md} and IR_{md} showed similar trend with higher values during August and September (Fig. 9.2). The infrared permitted to observe at the same time large canopy areas and estimate hydric status of plants as the pressure chamber (Leinonen et al. 2006). Analysis of data allowed to calculate a linear function [$\Psi_{md} = f(IR_{md})$]. The correlation resulted to be highly significant, with a coefficient of correlation (r) of 0.865 and a coefficient of determination (R^2) of 0.748 (Fig. 9.3). The result shown confirmed that canopy temperature can be successfully used to predict plant water status (Leinonen and Jones 2004), even with a simplified approach.

Fig. 9.1 Portable infrared thermometer

This study demonstrated that it was possible to obtain a good estimate of tree midday water status utilising IR_{md} by means of an empirical model. Handheld infrared thermometer gave clear results, providing precise data and instantaneous information on crop water status useful for irrigation scheduling. The method seems particularly interesting for its practical and easy use at farm level. Infrared thermometry for diagnosis and quantification of plant stress will be greatly enhanced if used in combination with other sensing techniques such as spectral reflectance or fluorescence imaging.

9.4 Spectrophotometry Applied Within VIS and NIR Spectral Range for the Automated Detection of the Nutrient Status

There is an increasing need to review information on crop nutrition to adequately establish nutrient requirements and to fine-tune fertiliser rates. This is due to the need to optimise fertilisation programmes in order to maximise the yield of high-quality fruit (Embleton et al. 1973b, 1996; Koo 1989; Legaz-Paredes and Primo-Millo 1988), while minimising the amount of chemical fertilisers applied, to reduce the risks of environmental impact (Alva et al. 2003a; Davies 1997). Nutrients are essential for the proper metabolic functioning of trees and to ensure desirable commercial production (Davies and Albrigo 1994). They vary considerably with citrus-growing region, soil type, cultural techniques, leaf

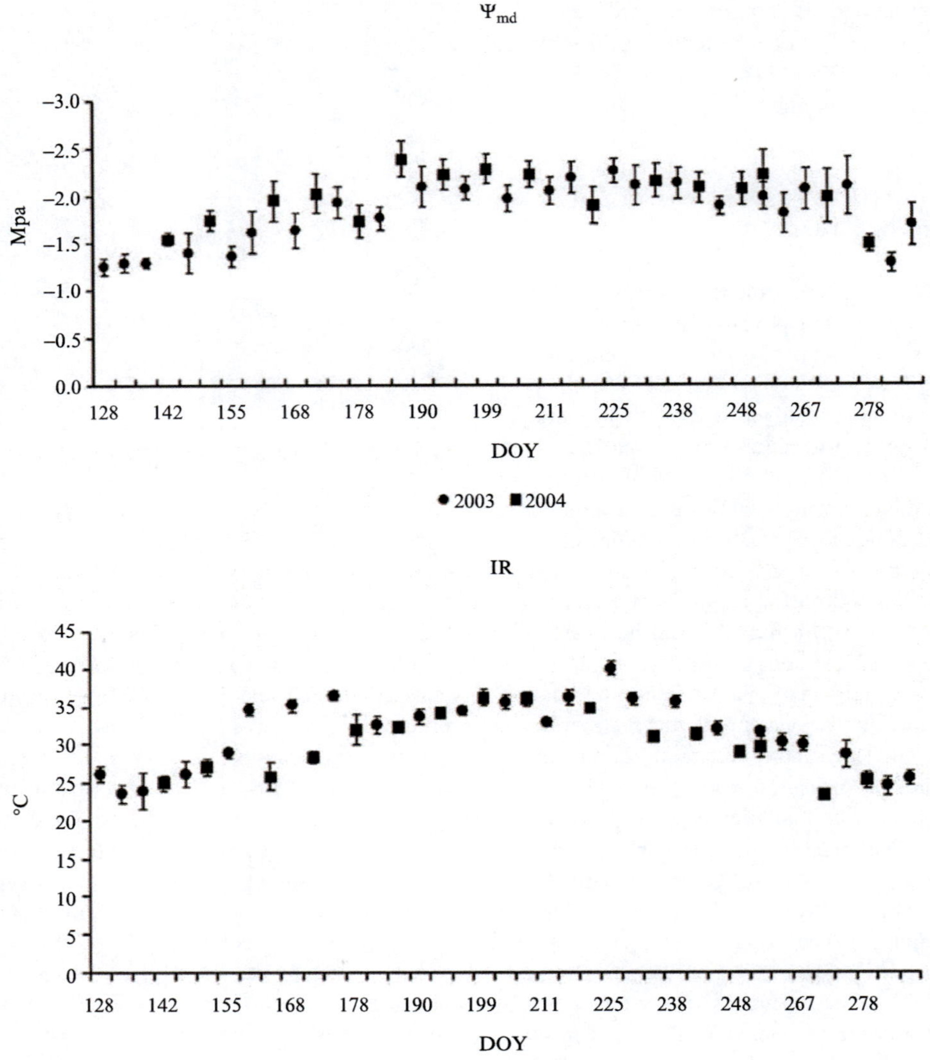

Fig. 9.2 Seasonal variation for the first 2 years of canopy temperature (IR_{md}) and midday stem water potential (Ψ_{md}) during the irrigation season

age and position on the tree, age of the tree and rootstock/scion combination. Nitrogen (N), phosphorus (P), potassium (K), calcium (Ca), magnesium (Mg), copper (Cu), iron (Fe), zinc (Zn) and manganese (Mn) are macro, meso and micro-nutrients of citrus leaves associated to growth, yield and quality factors, with relationships that vary with different elements (Embleton et al. 1973b; Hanlon et al. 1995). Leaf analysis is the most important tool for evaluating nutrient status of citrus and for guiding its fertilisation. Although other organs within the plant may act in a similar manner, the leaf is the most readily available source of tissue for analysis; it is metabolically very active, being the site of photosynthesis, which determines the primary processes occurring within the plant; and the leaf is a major site of carbohydrate and mineral storage (Embleton et al. 1973b). Analytical evaluations are performed more frequently, and with different aims, directly on the fruit as reported by many researches (Steuer et al. 2001; McGlone et al. 2003; Gomez et al. 2006; Cayuela 2008).

Results of the chemical analysis allow interpretation of plant nutritional status and identification of nutrient disorders caused by mineral excess or deficiency and serve as guide for balanced fertilisation programmes (Obreza et al. 1992; Ferguson et al. 1995; Intrigliolo et al. 1998). These analyses are normally compared to well-established standard values referred to as standard age spring-cycle leaves, taken from non-fruiting terminals of mature, fruit-bearing citrus trees. Embleton et al. (1973a) reported the leaf analysis standards for mature, bearing orange trees based on 5- to 7-month-old spring-cycle leaves from non-fruiting terminals. The values are within the range varying from deficient to

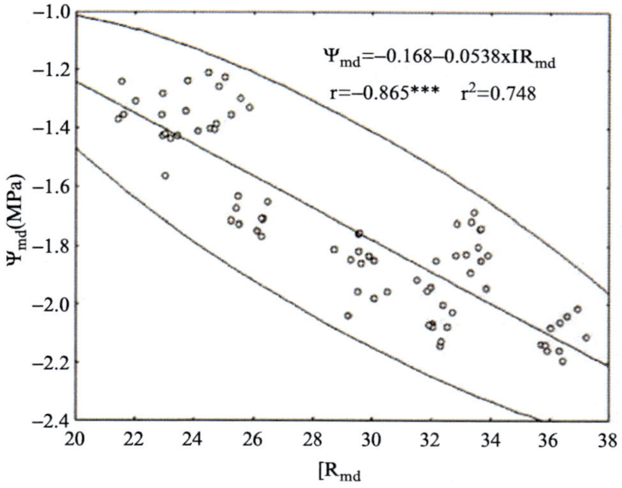

Fig. 9.3 Linear regression between midday canopy temperature and stem water potential

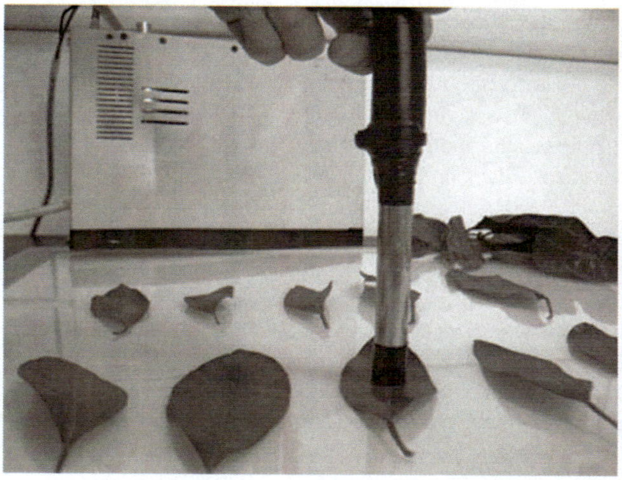

Fig. 9.4 Performing VIS-NIR spectral measurement on citrus leaves

excess categories as suggested by the guidelines for interpretation of leaf analysis (Obreza et al. 1992; Intrigliolo et al. 1999). The generalised lowering of the costs of the miniaturised spectrophotometers provides the possibility of using portable devices directly in the orchard for monitoring the maturity state of fruit. Using instruments directly in-field involves different interferences due to the environmental conditions, e.g. illuminations (kind and strength) and temperatures that should be taken into account during data processing as suggested by Ventura et al. (1998). Although such differences were shown, portable spectrometers are still currently used in many applications (Temma et al. 2002; Hernandez Sanchez et al. 2003; Saranwong et al. 2003a, b; Miller and Zude-Sasse 2004). Another well-known and established technique that must be considered while exploring the nutritional status of a plot of land is based on the SPAD Chlorophyll Meter (Piekielek et al. 1995; Lee et al. 1999; Read et al. 2002). Such a technique is based on the determination of leaf chlorophyll being this directly correlated with the leaf nitrogen content (Esposti et al. 2003).

Menesatti et al. (2010) proposed a study to determine orange leaves' nutritional status with a rapid, non-destructive, cost-effective technique utilising a VIS-NIR portable spectrophotometer and comparing its results with those from standard chemical analyses. The study was conducted on the experimental farm 'Palazzelli' of CRA-ACM (Eastern Sicily, 37°17′56″76 N, 14°50′29″76 E), in an irrigated 'Tarocco' blood orange orchard [*Citrus sinensis* (L.) Osbeck], planted in sandy loam soil. Two different clones of Tarocco were tested for leaves nutrient content: 'Arcimusa' and 'NL Meli' both grafted on sour orange [*C. aurantium* (L.)]. Tree nutritional status was evaluated by foliar analysis performed on 50 leaves of the index trees, placed in the middle of the plots. During the month of October, in the external side of the canopy, 5/7-month-old leaves of the year's spring flush were collected from non-fruiting twigs, according to the procedures of Embleton et al. (1973b) adapted to Italian conditions by Intrigliolo et al. (1999). Thirty leaves of each sample were analysed at the chemistry laboratory of the CRA-ACM. The remaining 20 leaves were analysed at the CRA-ING using spectrophotometric techniques. The acquisition of the raw spectral curve through the spectrophotometer took about 2 s for each leaf. Chemical analysis from leaves regarded the following elements: N, P, K, Ca, Mg, Fe, Zn, Mn. Nutrient concentrations were expressed as a percentage or parts per million (ppm) of the tissue dry matter. For the VIS-NIR measurements, a (portable) single channel spectrophotometer was used (Fig. 9.4). To acquire spectra, the 'pen' probe was used to measure the spectral reflectance response on each single leaf (spot area ≈ 10 mm^2). The reflectance measure is acquired by an optical quartz fibre (0.7 mm in diameter) fixed at 45° inside a circular aperture of 4 mm in diameter. Because the surface of the leaf was soft, it was possible to exclude all extraneous light from the probe. Spectral measurements were performed in laboratory following a white calibration (small variations in the level of external light), the instrumental integration time (light acquisition time) and subtraction of the background noise (variable with the instrument temperature). Mean reflectance values of all leaves for each treatment were compared by chemometric multivariate methods to each single reference chemical value. The procedure included the following steps: (1) extraction of raw spectra (X-Block variables); (2) extraction of measured values (Y-Block variables); (3) random separation of data set into two subsets, one for the model (75% of the whole data set) and one for the external validation test (25%); (4) application of pre-processing algorithms to both X and Y; (5) application of the chemometric technique PLS: modelling and testing; (6) calculation of efficiency parameter of prediction. The prediction of the nutrients content of leaves was performed using the PLS regression model.

The results found by Menesatti et al. (2010) of the PLS chemical values prediction are shown in Table 9.1. The best model ($r=0.995$) and the test ($r=0.991$) were obtained for K with a baseline for the X-Block pre-processing algorithm and a mean centre pre-processing for the Y-Block. The prediction ability of such a model was shown to be high with low values for the errors, having SEP=0.039 and RMSE=0.039. The results also showed a high efficiency in the estimation of N leaf content. Both the model and test PLS prediction showed a high value of r (0.945 and 0.909, respectively) (Table 9.2). The model of this parameter also had low values of SEP (0.039) and RMSE (0.039). The lowest values and results of PLS prediction were found for P ($r=0.429$).

9.5 Perspectives and Applied Implications

As reported by Menesatti et al. (2010), citrus trees require large quantities of mineral nutrients to attain adequate growth and yield, the needs of these varying with soil fertility and type (Koo et al. 1984). Although the mineral nutrition of citrus trees has been studied intensively, additional information has been frequently published, especially after the introduction of new fertigation technologies and innovative fertilisers (Alva et al. 2003b). The results of this study showed that a system based on a portable spectrophotometer can provide better knowledge of nutritional status of Tarocco orange bearing plants, achieving a more detailed and focused information in a shorter period and over wider areas. This makes the proposed methods suitable for use in precision farming (Alchanatis et al. 2005). Furthermore, the possibility of acquiring more detailed information, varying either in space and time, when compared with the standard chemical analysis, should prove to be a useful tool to increase fruit quality and to optimise the use of fertilisers, especially in organic farming systems. Esposti et al. (2003) reported that although the SPAD Chlorophyll Meter proved to be efficient in estimation of the N content in leaves, it could not reveal the content of other chemical compounds which the multi-parametric methods proposed here successfully estimated. Furthermore, such a technique could be able to provide a detailed analytical view of nutrient content, leading to more efficient fertigation planning in citrus orchards. Many researchers successfully used spectral systems to evaluate the N status of different crops (Sui et al. 1998; Tumbo et al. 2002a, b, c). However, even if N can be considered a key nutrient to monitor, the nutritional status of a crop is complex and is given by several parameters. At the beginning of the study, numerous standard chemical analyses were carried out, allowing a multiple correlation with the spectral data. This led to the possibility to develop a proper fertilisation strategy to improve the plant nutritional status and reduce the impact on the environment.

9.5.1 SPAD Chlorophyll Meter

In different plant species, some authors observed that direct and indirect methods used to assess chlorophyll contents in leaves were significantly correlated with leaf nitrogen levels (Duce et al. 1997; Intrigliolo et al. 2000; Jifon et al. 2005; Wood et al. 1992a). Total chlorophyll content in leaves (THCL) can be assessed directly by chemical analysis (Inskeep and Bloom 1985; Moran and Porath 1979) or by indirect methods using devices that are easy to operate in the field and non-destructive (Duce et al. 1997; Intrigliolo et al. 2000; Jifon et al. 2005). The portable Chlorophyll Meter SPAD-502 (Minolta), one of the devices used to determine indirectly chlorophyll levels, enables rapid, non-destructive in-field determinations of leaf greenness that is related to intervals in order to chlorophyll content (Monje and Bugbee 1992; Wood et al. 1992b). The values are calculated by determining the amount light transmitted by the leaf in two spectral bands (red and infrared). Further investigations revealed that SPAD readings can be performed on the entire leaf blade without intercepting the midrib, and that these readings showed similar results (Monje and Bugbee 1992; Roccuzzo et al. 2000). Monje and Bugbee (1992) and Wood et al. (1992a, b) revealed that correlation values between SPAD and N leaf content must be adjusted for each plant species, as the equation cannot be extended even to genetically similar species. In addition, the same authors reported that the correlation was lower in the presence of elevated leaf nitrogen content. Evident correlation between SPAD values, N levels and other nutrients was observed in apple and vine, even if differences were observed between species (Singha and Townsend 1989). Trials on citrus performed by Intrigliolo et al. (2000) using multiple regression showed that nitrogen is the only nutrient significantly correlated with THCL and SPAD readings. In agreement with diagnostic leaf sampling procedures (Embleton et al. 1996; Intrigliolo et al. 1999), leaves presenting malformations, parasite attacks and chlorotic symptoms were not used (Monje and Bugbee 1992). In a field study, the correlation between SPAD readings and leaf nitrogen content was assessed at monthly intervals in order to detect in which period of the year this correlation was highly significant.

Field determination of the N nutritional status of citrus [*Citrus sinensis* (L.) Osbeck] cv 'Valencia late' was realised in Italy (Sicily) and in Spain (Valencia). Leaf sampling was

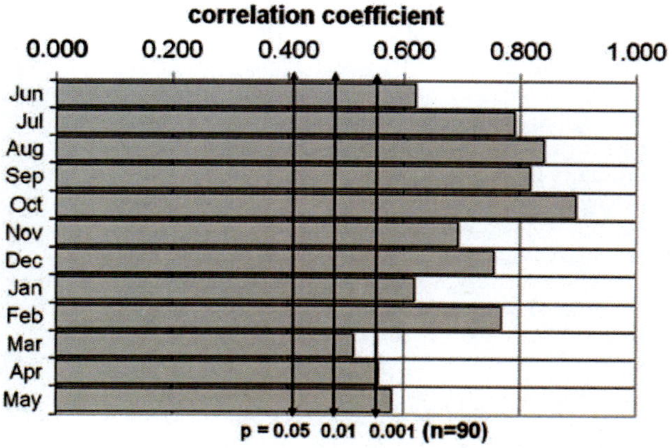

Fig. 9.5 Correlation coefficients between leaf N content and SPAD values over the year

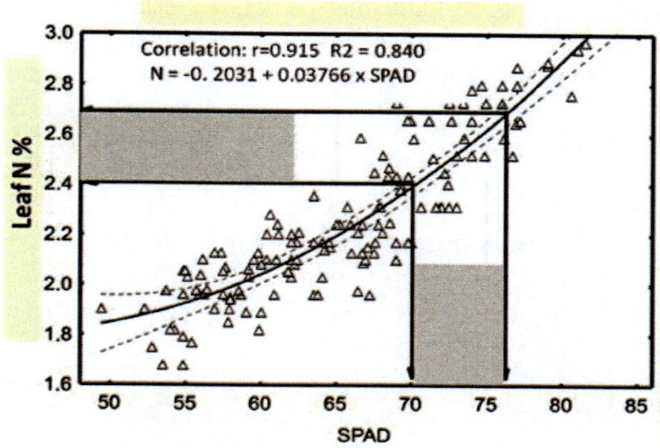

Fig. 9.6 Correlation between leaf N content and SPAD values in the period September–October

performed in both countries on a total of 30 orange trees. Trees were annually fertilised with increasing amounts of nitrogen (0–800 g tree^{-1}) in order to determine different leaf nitrogen levels and hence differences in leaf greenness. SPAD measurements were performed in field on leaf samples. The same leaves were immediately transferred to the laboratory where they were washed and dried in an oven (65°C) until constant weight, in order to determine N content using the Kjeldhal method. Correlation and regression analyses were performed on the obtained data using an SPSS package. Correlation values were analysed every month to detect when SPAD and leaf N levels presented major correlation.

Mean monthly SPAD values presented pattern similar to N concentrations especially in the June–February period. The close correlation between SPAD and leaf N values was observed (Fig. 9.5) throughout the year (significance constantly exceeded 0.001 and *r* values were between 0.514 and 0.897). The association between SPAD and N values may present limitations, depending on the sampling period. The absence of association could also be due to the fact that leaf N allocation is not exclusive of pigment-protein/reaction centre complex and that growth environment plays a central role (Jifon et al. 2005). The SPAD leaf N content regression curve (Fig. 9.6) in the September–October period shows that during this time interval, SPAD can be used to determine N status in 'Valencia late' oranges. Furthermore, the range of SPAD values (70–76) equivalent to optimal nitrogen content was defined (Embleton et al. 1973a; Intrigliolo et al. 1999). Nevertheless, more in-depth investigations are required to verify the equivalence between leaf N content and SPAD leaf N values in other nutritional classes both in deficiency and in excess. In fact, N values exceeding 2.7% or lower than 2.1% showed very high dispersion of points around the theoretical line and confidence intervals.

9.6 Predicting Modelling to Develop Models for Nutrient and Hydric Status Estimation

Multivariate analyses are generally divided into two main categories: unsupervised and supervised (Costa et al. 2011). For unsupervised techniques, grouping or clustering methods for multivariate elements (X-Block) are based on functional relationships among the same elements (distances, variances). They do not need for an a priori knowledge of the class categories. Differently, in supervised techniques, the class attribution is given by a single or multiple variables (Y-Block). In this way, multivariate methods are forced to cluster into a priori–established classes. Unsupervised methods are mainly applied in an exploratory sense, when the aim is to analyse or visualise non-forced aggregating relationships (unsupervised) among elements (Forina 2006). Concerning supervised techniques, it is possible to distinguish two main analytical approaches: modelling and classification. Supervised methods are derived from the observation and then the use of known classes, called the training set. The derived classification criteria can then be used to classify each new object within a test set. This can be applied for both classification and the computing of efficiency parameters (Costa et al. 2011). Classification analysis needs a decision rule, called the 'classification criterion', to distinguish objects into classes on the basis of selected quantitative features (Jayas et al. 2000). For modelling, it is instead possible to attribute objects into one or more classes, but also to none (i.e. in this case the object is an outlier). Modelling techniques calculate the 'prediction probability' with a classification threshold for each modelled class. The modelling efficiency is indicated by statistical parameters such as 'sensitivity' and 'specificity'. Sensitivity represents the percentage of the objects of a category accepted by the modelled class. Specificity is the percentage of objects different from the modelled classes, as rejected by this classification criterion. On the other hand, for the classification, a matrix of correct classification can be used (Forina 2006).

9.7 Conclusion and Future Research

The nutritional and water status of a crop is complex and given by several parameters. Generally, a fine analysis carried out with conventional methods (i.e standard chemical analyses) would require a lot of manual and laboratory work and high cost for the numerous samplings needed. Researchers have investigated several approaches in order to automate these procedures and to overcome the critical aspect of crop management in the collection of representative samples. This led to the possibility to develop a proper fertilisation and/or irrigation strategy to improve the plant nutritional status and to reduce the impact on the environment. Moreover, the monitoring of different nutrients is essential due to the relationships existing among them. For these reasons, methods increasing the acquisition of a high number of sample variables at a relatively low cost and time, such as vehicle-mounted optical sensing devices, represent promising applicative perspectives. These multi-device systems could include mobile instruments (i.e. visible-near and near-infrared spectrophotometers, infrared thermometers and thermocameras) and could be paired with multivariate statistics in order to improve the performance of the detection analysis.

References

Alchanatis V, Schmilovitch Z, Meron M (2005) In-field assessment of single leaf nitrogen status by spectral reflectance measurements. Precis Agric 6(1):25–39

Allen RG, Pereira LS, Raes D et al (1998) Crop evapotranspiration: guidelines for computing crop water requirements, vol 56, FAO irrigation and drainage paper. Food and Agriculture Organization of the United Nations, Rome, pp 15–27

Alva AK, Paramasivam S, Graham WD et al (2003a) Best nitrogen management practices for citrus production in sandy soils. Water Air Soil Pollut 14:139–154

Alva AK, Fares A, Dou H (2003b) Managing citrus trees to optimize dry mass and nutrient partitioning. J Plant Nutr 26:1541–1559

Cayuela JA (2008) Vis/NIR soluble solids prediction in intact orange (*Citrus sinensis* L.) cv. Valencia Late by reflectance. Postharvest Biol Technol 47:75–80

Chonè X, Van Leeuwen C, Dubourdieu D et al (2001) Stem water potential is a sensitive indicator of grapevines water status. Ann Bot 87:477–483

Costa C, Antonucci F, Pallottino F et al (2011) Shape analysis of agricultural products: a review of recent research advances and potential application to computer vision. Food Bioprocess Technol. doi:10.1007/s11947-011-0556-0

Davies FS (1997) Literature review of research related to nitrogen nutrition, fertilization and groundwater pollution of citrus. In: Florida Department of Agriculture and Consumer Services nitrogen BMP Technologies Group, Tallahassee, FL, USA, pp 70

Davies FS, Albrigo LG (1994) Citrus. In: Atherton J, Rees A (eds) Crop production sciences in horticulture, vol 2. CAB International, Wallingford

Duce P, Arca B, Spano D et al (1997) Uno strumento per la misura non distruttiva del contenuto in clorofilla: applicabilità dello strumento agli agrumi. Italus Hortus 4(4):26–31

Embleton TW, Jones WW, Labanauskas CH et al (1973a) Leaf analysis as a diagnostic tool and guide to fertilization. In: Reuther W (ed) The citrus industry, vol 3. University of California, Division of Agricultural Science, Berkeley, pp 183–210

Embleton TW, Reitz HJ, Jones WW (1973b) Citrus fertilization. In: Reuther W (ed) The citrus industry, vol 3. University of California, Division Agricultural of Science, Berkeley, pp 122–182

Embleton TW, Coggins CW, Witney GW (1996) What is most profitable use of citrus leaf analysis? Proc Int Soc Citric 2:1261–1264

Enterline J (2001) Agriculture remote sensing: getting a closer look. Earth Obs Mag 9(2):45–47. ISSN:1076–3430

Esposti MDD, Siqueira DLD, Pereira PRG et al (2003) Assessment of nitrogenized nutrition of citrus rootstocks using chlorophyll concentrations in the leaf. J Plant Nutr 26(6):1287–1299

Ferguson JJ, Davies FS, Tucker DPH et al (1995) Fertilizer guidelines. In: Tucker DPH, Alva AK, Jackson LK, Wheaton TA (eds) Nutrition of Florida citrus trees. University of Florida, Gainesville, pp 21–25

Forina M (2006) I metodi chemometrici di classificazione: a cosa servono e come sono utilizzati. In: Proceedings of the Nir Italia 2006, Ferrara, 25–26 May 2006

Gomez AH, Yong H, Pereira AG (2006) Non-destructive measurement of acidity, soluble solids and firmness of 'Satsuma' mandarin using VIS/NIR-spectroscopy techniques. J Food Eng 77:313–319

Hanlon EA, Obreza TA, Alva AK (1995) Tissue plant and analysis. In: Tucker DPH, Alva AK, Jackson LK et al (eds) Nutrition of Florida citrus trees. University of Florida, Gainesville, pp 13–20

Hernandez Sanchez N, Lurol S, Roger JM et al (2003) Robustness of models based on NIR spectra for sugar content prediction in apples. J Near Infrared Spectrosc 11:97–107

Inskeep WP, Bloom PR (1985) Extinction coefficients of chlorophyll a and b in N, N-dimethylformamide and 80% acetone. Plant Physiol 77:483–485

Intrigliolo F, Tittarelli F, Roccuzzo G et al (1998) La fertilizzazione degli agrumi. L'informatore agrario 9:79–84

Intrigliolo F, Roccuzzo G, Lacertosa G et al (1999) Agrumi: modalità di campionamento per terreno, foglie, acque d'irrigazione e frutti; valori analitici di riferimento. C.U.E.C.M, Catania

Intrigliolo F, Giuffrida A, Calabretta ML et al (2000) SPAD as an indicator of nitrogen status in Citrus. In: Proceeding of the IXth international citrus congress, Orlando, FL, pp 665–667

Jayas DS, Paliwal J, Visen NS (2000) Multi-layer neural networks for image analysis of agricultural products. J Agric Eng Res 77(2): 119–128

Jifon JL, Syvertsen JP, Whaley E (2005) Growth environment and leaf anatomy affect non destructive estimates of chlorophyll and nitrogen in Citrus sp. leaves. J Am Soc Hortic Sci 130(2):152–158

Jones HG (2004) Irrigation scheduling: advantages and pitfalls of plant-based methods. J Exp Bot 55:2427–2436

Jones HG, Stoll M, Santos T et al (2002) Use of infrared thermography for monitoring stomatal closure in the field: application to grapevine. J Exp Bot 53(378):2249–2260

Kaufmann MR (1968) Evaluation of the pressure chamber method for measurement of water stress in citrus. J Am Soc Hortic Sci 93: 186–190

Koo RCJ (1989) Citrus. In: Plucknett DL, Sprague HB (eds) Detecting mineral nutrient deficiencies in tropical and temperate crops, vol 3, Westview tropical agriculture series. Westview Press, Boulder

Koo RCJ, Anderson CA, Stewart I et al (1984) Recommended fertilizers and nutritional sprays for citrus. Fla Agric Exp Stn Bull 536D, pp 30

Lee W, Searcy SW, Kataoka T (1999) Assessing nitrogen stress in corn varieties of varying color. ASAE paper no. 993034. ASAE, St. Joseph

Legaz-Paredes F, Primo-Millo E (1988) Nòrmas para la fertilizaciòn de los agrios. Generalitat Valenciana, Cons. agricoltura i Pesca, Fullets Divulgaciò 5(88)

Leinonen I, Jones HG (2004) Combining thermal and visible imagery for estimating canopy temperature and identifying plant stress. J Exp Bot 55:1423–1431

Leinonen I, Grantm OM, Tagliavia CPP et al (2006) Estimating stomatal conductance with thermal imagery. Plant Cell Environ 29: 1508–1518

Maldague X (1994) Non-destructive testing monographs and tracts. In: Maldague X (ed) Infrared methodology and technology, vol 7. Gordon and Breach, New York

Massantini F, Masoni A, Mariotti M et al (1992) Influenza della concimazione azotata sulla riflettanza, assorbanza e trasmittanza della foglia a bandiera del frumento tenero (*Triticum aestivum* L.). Rivista di Agronomia 4:489–497

McGlone VA, Fraser DG, Jordan RB et al (2003) Internal quality assessment of mandarin fruit by Vis/NIR spectroscopy. J Near Infrared Spectrosc 11:323–332

Menesatti P, Antonucci F, Pallottino F et al (2010) Estimation of plant nutritional status by VIS-NIR spectrophotometric analysis on orange leaves [*Citrus sinensis* (L) Osbeck cv Tarocco]. Biosyst Eng 105:448–454

Miller WM, Zude-Sasse M (2004) NIR-based sensing to measure soluble solids content of Florida citrus. Appl Eng Agric 20: 321–327

Monje OA, Bugbee B (1992) Inherent limitations of non destructive chlorophyll meters: a comparison of two types of meters. HortScience 27:69–71

Moran R, Porath D (1979) Chlorophyll determination in intact tissues using N, N-dimethylformamide. Plant Physiol 65:478–479

Naor A (2000) Midday stem water potential as a plant water stress indicator for irrigation scheduling in fruit trees. Acta Hortic 537: 447–454

Naor A, Peres M (2001) Pressure-increase rate affects the accuracy of stem water potential measurements in deciduous fruit trees using the pressure-chamber technique. J Hortic Sci Biotech 76(6): 661–663

Obreza TA, Alva AK, Hanlon EA et al (1992) Citrus grove leaf tissue and soil testing: sampling, analysis, and interpretation. Cooperative extension service bulletin SL 115. University of Florida, Gainesville, pp 1–4

Piekielek WP, Fox RH, Toth JD et al (1995) Use of a chlorophyll meter at the early dent stage of corn to evaluate nitrogen sufficiency. Agron J 87:403–408

Rahkonen J, Jokela H (2003) Infrared radiometry for measuring plant leaf temperature during thermal weed control treatment. Biosyst Eng 86(3):257–266

Read JJ, Tarpley L, McKinio JM et al (2002) Narrow-waveband reflectance ratios for remote estimation of nitrogen status in cotton. J Environ Qual 31:1442–1452

Roccuzzo G, Rapisarda P, Giuffrida A et al (2000) I livelli di clorofilla nelle foglie di limone come indicatore dello stato nutrizionale azotato. V Giornate Scientifiche SOI 2:577–578

Saranwong S, Sornsrivichai J, Kawano S (2003a) On-tree evaluation of harvesting quality of mango fruit using a hand-held NIR instrument. J Near Infrared Spectrosc 11:283–293

Saranwong S, Sornsrivichai J, Kawano S (2003b) Performance of a portable near infrared instrument for Brix value determination of intact mango fruit. J Near Infrared Spectrosc 11:175–181

Scholander PF, Hammel HJ, Bradstreet A et al (1965) Sap pressure in vascular plants. Science 148:339–346

Singha S, Townsend C (1989) Relationship between chromaticity values and chlorophyll concentration in apple grape, and peach leaves. HortScience 24:1034–1038

Steuer B, Schulz H, Lager E (2001) Classification and analysis of citrus oils by NIR spectroscopy. Food Chem 72:113–117

Sui R, Wilkerson JB, Hart WE et al (1998) Integration of neural networks with a spectral reflectance sensor to detect nitrogen deficiency in cotton. ASAE paper no. 983104. ASAE, St. Joseph

Syvertsen JP, Levy J (1982) Diurnal changes in citrus leaf thickness, leaf water potential and leaf to air temperature difference. J Exp Bot 33:783–789

Temma T, Hanamatsu K, Shinoki F (2002) Development of a portable near infrared sugar-measuring instrument. J Near Infrared Spectrosc 10:77–83

Tumbo SD, Wagner DG, Heinenann PH (2002a) Hyperspectral-based neural network for predicting chlorophyll status in corn. Trans ASAE 45:825–832

Tumbo SD, Wagner DG, Heinenann PH (2002b) On-the-go sensing of chlorophyll status in corn. Trans ASAE 45:1207–1215

Tumbo SD, Wagner DG, Heinenann PH (2002c) Hyperspectral characteristics of corn plants under different chlorophyll levels. Trans ASAE 45:815–823

Urbani G, Menesatti P, Solaini S (2002) Tecnologie opto-elettroniche in agricoltura: nuove prospettive e indirizzi di ricerca. Mondo Macchina 6:35–38

Ventura M, De Jager A, De Putter FH et al (1998) Non-destructive determination of soluble solids in apple fruit by near infrared spectroscopy (NIRS). Postharvest Biol Technol 14:21–27

Viscarra Rossel RA, Taylor HJ, McBratney AB (2007) Multivariate calibration of hyper-spectral γ-ray energy spectra for proximal soil sensing. Eur J Soil Sci 58:343–353

Williams PC (1987) Variables affecting near-infrared reflectance spectroscopic analysis. In: Williams P, Norris K (eds) Near-infrared technology in the agricultural and food industries. American Association of Cereal Chemists, St Paul, pp 143–166

Wold S, Sjostrom M, Erikssonn L (2001) PLS-regression: a basic tool of chemometrics. Chemom Intell Lab Syst 58:109–130

Wood CW, Tracy PW, Reeves DW et al (1992a) Determination of cotton nitrogen status with a portable chlorophyll meter. J Plant Nutr 15(9):1435–1448

Wood CW, Reeves DW, Duffield RR et al (1992b) Field chlorophyll measurements for evaluation of corn nitrogen status. J Plant Nutr 15(4):87–500

Spectral Response of Citrus and Their Application to Nutrient and Water Constraints Diagnosis

10

Lola Suárez and José A.J. Berni

Abstract

The early diagnosis of nutrient and water stress is a key issue for the proper management of orchards. Unfortunately, the measurement of indicators in the field to assess those deficiencies is expensive and time consuming. Remote sensing provides the possibility of assessing water and nutrient stress over every single stand on the whole area of interest. Leaf and canopy spectra are affected by every vegetation process or change in leaf pigments. However, the effect of each nutrient deficiency on the canopy spectrum is very subtle and the wavelength ranges affected by the different nutrients are generally overlapping. Then, an analysis in detail is needed in order to properly discriminate between the different stress causes. This chapter presents a review of previous studies dealing with remote sensing of nutrient and water stress, defining the spectral regions affected and the methodologies that have been applied for vegetation stress assessment. Specific multispectral indices related to leaf pigment content and physiological vegetation processes, thermal imagery, and fluorescence studies are presented. Finally, some results obtained from radiative transfer modeling simulations highlight the importance of the spatial resolution for precise agriculture purposes.

There is extensive future research needed focusing on dissembling the overlapping effects of different nutrient deficiencies on the electromagnetic spectrum. Generally, vegetation indices used for nutrient assessment are affected by more than one nutrient content; a specific diagnosis is difficult to solve. As an example, pertinent indices can be developed in order to perform a proper stress diagnosis. Additionally, leaf-level radiative transfer models can be improved to account for differences in specific pigments. If leaf reflectance can be modeled for different nutrient levels, tree nutrient status can be assess through modeling inversion. Modeling inversion techniques are especially interesting because they do not rely on site-specific empirical relationships, being applicable to extensive areas.

Keywords

Remote sensing • Nutrient • Water • Stress • Vegetation indices • Thermal imagery • Fluorescence

L. Suárez (✉)
School of Mathematical and Geospatial Sciences, RMIT University, GPO Box 2476V Melbourne, VIC, 3001 Australia
e-mail: lola.suarezbarranco@rmit.edu.au

J.A.J. Berni
CSIRO Marine and Atmospheric Research, Canberra, Australia
e-mail: jose.jimenez-berni@csiro.au

A.K. Srivastava (ed.), *Advances in Citrus Nutrition*,
DOI 10.1007/978-94-007-4171-3_10, © Springer Science+Business Media B.V. 2012

10.1 Introduction

The condition of an orchard can be monitored by walking or driving around while doing a visual examination of every stand (Barros et al. 1983). Nevertheless, visual ratings of size, stress, or injury are frequently dependent on the intuitive knowledge or experience of the observer rather than on the accuracy of the systems. The objective measure of the plant status in the field consists in the use of instruments to measure the symptoms of the pathology, such as stem water potential in the case of water stress or tissue analysis in the laboratory in the case of nutrient stress. This field assessment is labor and time consuming making it unaffordable to be done at a stand level. The use of imagery to monitor vegetation status makes the individual stand assessment possible using a nondestructive technique. Moreover, it permits the future reinterpretation of the images (Blazquez 1988).

The use of remote sensing techniques to assess citrus status started in the late 1970s with aerial color infrared photographs (ACIR). In that moment, the scientists were conscious of the gap between science and users and designed the imagery acquisition system needed to assess general vegetation status with ACIR (Blazquez and Horn 1980). The difficulties rose when the specific diagnosis was needed as different pathologies can lead to the same effect on the reflectance signal (Blazquez 1988). Every purpose of study requires specific methodology and remote sensing data characteristics. Physiological stresses are generally impacting the visible part of the spectrum where the pigment absorption is governing the reflectivity. The reflectance at the NIR part of the spectrum is function of vegetation structure and volume (Nilsson 1995), and regions in the near and middle infrared are affected by dry matter compounds and water content in the leaf (Hunt and Rock 1989; Jacquemoud et al. 1996). Thermal imagery is used to estimate the evapotranspiration based on the temperature rise due to the decrease of the evaporative cooling with the stomata closure (Sepulcre-Cantó et al. 2006). Finally, vegetation fluorescence related to the dissipation of energy under stress conditions is normally measured on narrow atmospheric absorption bands (McFarlane et al. 1980) and has been used to assess heavy metal contamination, virus and fungi infections, and photosynthetic efficiency (Chaerle and Van der Straeten 2000). Sankaran et al. (2010) present several remote sensing techniques for plant disease detection.

Some authors have focused their studies on citrus species. Citrus canker caused by the bacteria *Xanthomonas axonopodis* pv. *Citri* has been assessed by using the reflectance at near-infrared and red regions (Marcassa et al. 2006; Qin et al. 2009), and Fletcher et al. (2001) used the visible and near infrared to assess *Phytophthora* foot rot infections in citrus trees. Furthermore, the management of pesticides doses has been improved by using variable rate technology based on image information reporting savings up to 80% (Du et al. 2008). Linear spectral unmixing was applied to multispectral and thermal imagery to map the doses needed in the whole field and apply specific amount of pesticides.

Min and Lee (2005) evaluated the impact of N concentration on the reflectance of orange leaves. Spectral bands located at the blue, red, near-infrared, and shortwave-infrared regions of the spectrum were selected as the most suitable ones for N detection. Some years later, Min et al. (2008) developed a hyperspectral nitrogen sensing system for N detection in orange leaves based on the results of the previous study. The detection of nutritional stress using remote sensing data has been widely accomplished by estimating the leaf chlorophyll content for many tree species. Sari et al. (2005) studied the effect of chlorophyll content on the spectra of orange leaves for single bands located at the blue and red regions and concluded that the different tree phenological stages affected the relationship between reflectance and chlorophyll content. Although the number of studies on the detection of nutritional stress in citrus species is low, there are many studies focused on other perennial species. The remote detection of nutrient stress has a dedicated section within this chapter. Dzikiti et al. (2009) presented a model to estimate the leaf water content using an inversion of the leaf scale radiative transfer model PROSPECT (Jacquemoud and Baret 1990) and water indices in mandarins. Suárez et al. (2010a) used airborne thermal and multispectral data to assess canopy water stress in an orange orchard. Thermal data, related with evaporative cooling and stomata conductance, was highly correlated with stem water potential measured in the field. The photochemical reflectance index (PRI, Gamon et al. 1992) was found to track physiological changes in the trees due to the water stress status and related to final quality of the fruits. As with nutritional stress, a dedicated section on remote sensing of water stress can be found later in this chapter. Finally, mandarin yield has been also estimated by Ye et al. (2006) who developed a model based on neural networks using visible and NIR airborne data.

10.2 Description of a Leaf Reflectance Spectrum

Little of the incident visible (0.4–0.7 mm) or near-infrared (0.7–1.3 mm) energy is reflected directly from the outer surface of a leaf because the cuticular wax layer is nearly transparent to radiation at these wavelengths (Knipling 1970). Leaf reflectance is low in the visible, starting with very low values in the blue (0.4–0.5 mm), slightly higher in the green (0.5–0.6 mm), and again reaching a minimum in the red (0.6–0.7 mm) (Jackson 1986). The main responsible of the leaf low reflectance in the visible part of the electromagnetic spectrum is the leaf pigment pool while the influence of pigment composition does not affect the near-infrared region significantly (Gates and Tantraporn 1952). Chlorophyll is mainly absorbing in the red part and partly contributing to the absorption in the blue and the green together with other pigments as carotenes and xanthophylls (Jackson 1986). In the near-infrared region, leaf absorption/reflection is mainly dependent on the structural discontinuities; meanwhile, in the mid-infrared region (1.3–3 mm), water and other compound concentrations play a major role (Peñuelas and Filella 1998).

Figure 10.1 depicts the reflectance, transmittance, and absorption proportional values of a leaf specifying the spectral regions affected by pigment absorption, cell structure, and water content. The chlorophyll absorption spectrum is also presented with two characteristic peaks, in the blue and red regions.

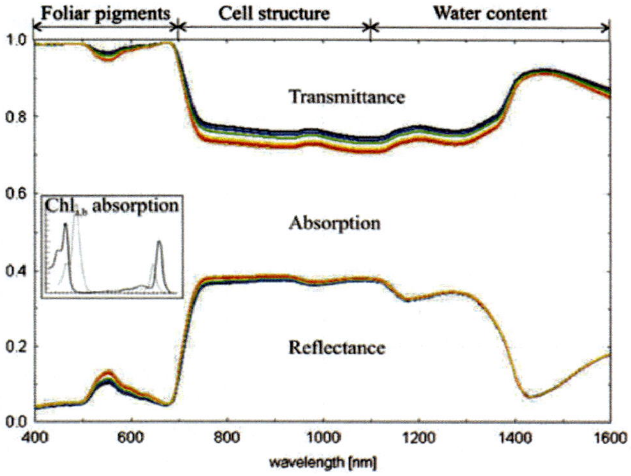

Fig. 10.1 Reflectance, transmittance, and absorption of a leaf, the chlorophyll *a* and *b* absorption in the visible, and the regions affected by foliar pigments, cell structure, and water content

10.3 Remote Sensing of Nutrient Stress

Leaf spectrum is affected by nutrient deficits. Figure 10.2 presents the reflectance and transmittance spectra of a healthy leaf and the spectra of leaves with nitrogen, phosphorous, and potassium deficiencies. As the impact of different nutrients in the electromagnetic spectrum generally overlaps, it is important to find out spectral regions where differences are driven by individual nutrients for a proper pathology assessment. Nitrogen, apart from being the main nutrient to monitor in crop management, has the strongest effect on leaf reflectance.

Most of the authors base the vegetation nutrient assessment on the estimation of leaf chlorophyll content, others study the effect of individual nutrient deficits on leaf spectra, and others develop spectral indices for specific nutrient detection. The following sections present the remote detection of nitrogen and phosphorous stress.

10.3.1 Remote Sensing of N Content in the Leaf

Nitrogen is the element that has the greatest effect on citrus production, and citrus needs more nitrogen than any other nutrient (Reuther et al. 1952; Blackmer et al. 1994; Bausch and Duke 1996). Nitrogen is a component of chlorophyll (the green pigment in leaves) and is associated with important tree functions such as growth, leaf production, flower initiation, fruit set, and fruit development and quality. In the past decade, some authors have studied the effect of nitrogen content on the leaf spectrum. Some of them started determining which wavelength were more significant to assess N content in the leaf, examining the spectra of leaves collected from

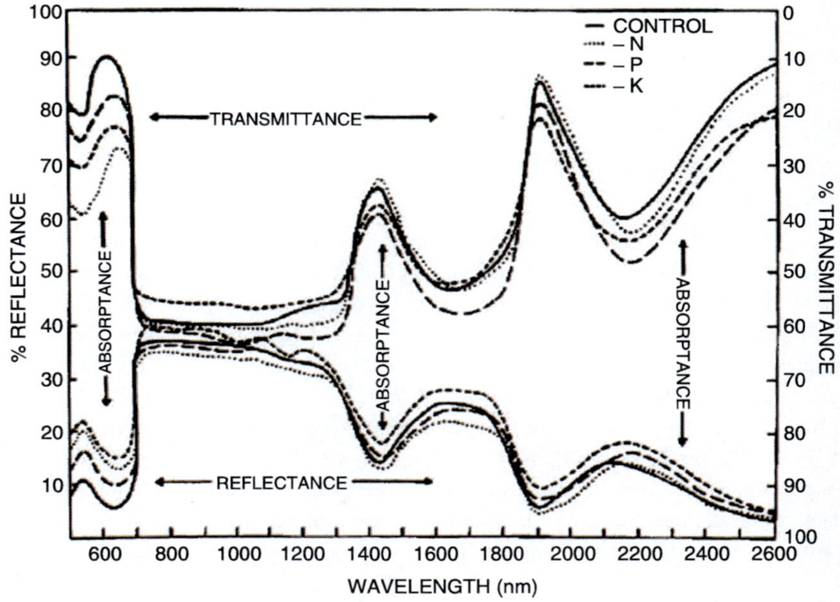

Fig. 10.2 Reflectance and transmittance leaf spectra corresponding to healthy, nitrogen-, phosphorous-, and potassium-deficient leaves (Adapted from Al-Abbas et al. 1974)

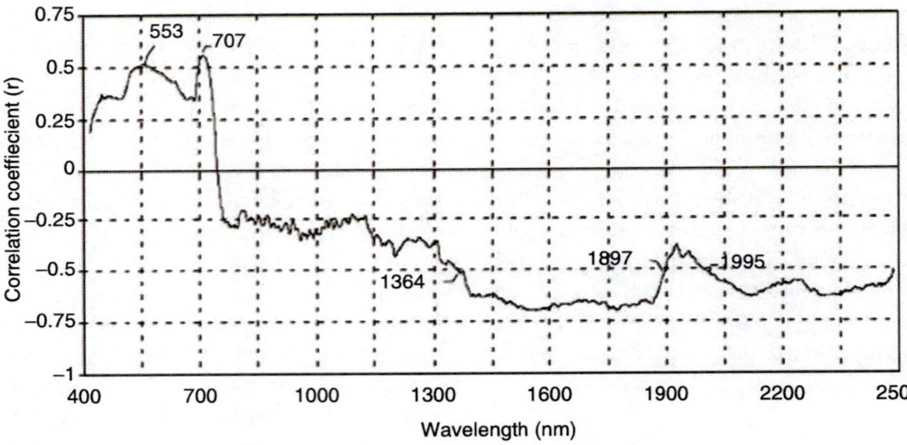

Fig. 10.3 Correlation coefficients found between N content and leaf spectra by Min and Lee (2005). The highest correlated bands were those located at 553, 707, 1,364–1,897, and 1,995–2,485 nm

Table 10.1 List of remote sensing indices developed for N or pigment content estimation

Index	Acronym	Formula	Biophysiological parameter	References
Normalized difference red edge	NDRE	(R790 − R720)/(R790 + R720)	Nitrogen	Barnes et al. (2000)
Normalized difference nitrogen index	NDNI	[log (1/R1510) − log (1/R1680)]/[log (1/R1510) + log (1/R1680)]	Nitrogen	Fourty et al. (1996)
Normalized difference lignin index	NDLI	[log (1/R1754) − log (1/R1680)]/[log (1/R1754) + log (1/R1680)]	Lignin content	Serrano et al. (2002)
Normalized pigment chlorophyll index	NPCI	(R680 − R430)/(R680 + R430)	Chlorophyll	Peñuelas et al. (1994)
Zarco-Tejada and Miller	ZTM	R750/R710	Chlorophyll	Zarco-Tejada et al. (2001)
Triangular vegetation index	TVI	0.5 × [120 × (R750 − R550) − 200 × (R670 − R550)]	Chlorophyll	Broge and Leblanc (2000)
Modified simple ratio	mSR	(R750 − R445)/(R705 − R445)	Chlorophyll	Sims and Gamon (2002)
Transformed chlorophyll reflectance index	TCARI	3 × [(R700 − R670) − 0.2 × (R700 − R550) × (R700/R670)]	Chlorophyll	Haboudane et al. (2002)
Modified chlorophyll reflectance index	MCARI	[(R700 − R670) − 0.2 × (R700 − R550)] × (R700/R670)	Chlorophyll	Daughtry et al. (2000)
Structural independent pigment index	SIPI	(RNIR − RR)/(RNIR − RB)	Chlorophyll and carotenes	Peñuelas et al. (1995)
Carotenoid reflectance index	CRI	1/R510 − 1/R550	Carotenoids	Gitelson et al. (2002)
Photochemical reflectance index	PRI	(R570 − R531)/(R570 + R531)	Xanthophyll DPS	Gamon et al. (1992)
Simple ratio pigment index	SRPI	R430/R630	Pigment content	Peñuelas et al. (1995)

citrus orchards where different N treatments were applied (Benedict and Swindler 1961; Osborne et al. 2002; Rodriguez et al. 2006). The leaf spectra were measured in the laboratory using an integrating sphere and compared to the N concentration. They reached the conclusion that the N-highest correlated spectral bands were located at 553, 707, 1,364–1,897, and 1,995–2,485 nm by examining the correlation coefficient (Fig. 10.3), stepwise multiple regression, and partial least squares regression procedures. Some other authors have demonstrated the sensitivity of reflectance at 550 nm to leaf nitrogen and chlorophyll content (Blackmer et al. 1994; Gitelson and Merzlyak 1994). Modeling studies have demonstrated that leaf reflectance at wavelengths where pigment absorption coefficients are high is more sensitive to low pigment concentration (Jacquemoud and Baret 1990; Yamada and Fujimura 1991). At canopy scale, similar results have been found by Filella et al. (1995) and O'Neill et al. (2002). As there is a close correlation between chlorophyll and N content, many authors assess N content by estimating the leaf chlorophyll concentration (Peñuelas and Filella 1998).

There are some indices specifically created for nitrogen detection (Table 10.1), which is the case of the nitrogen

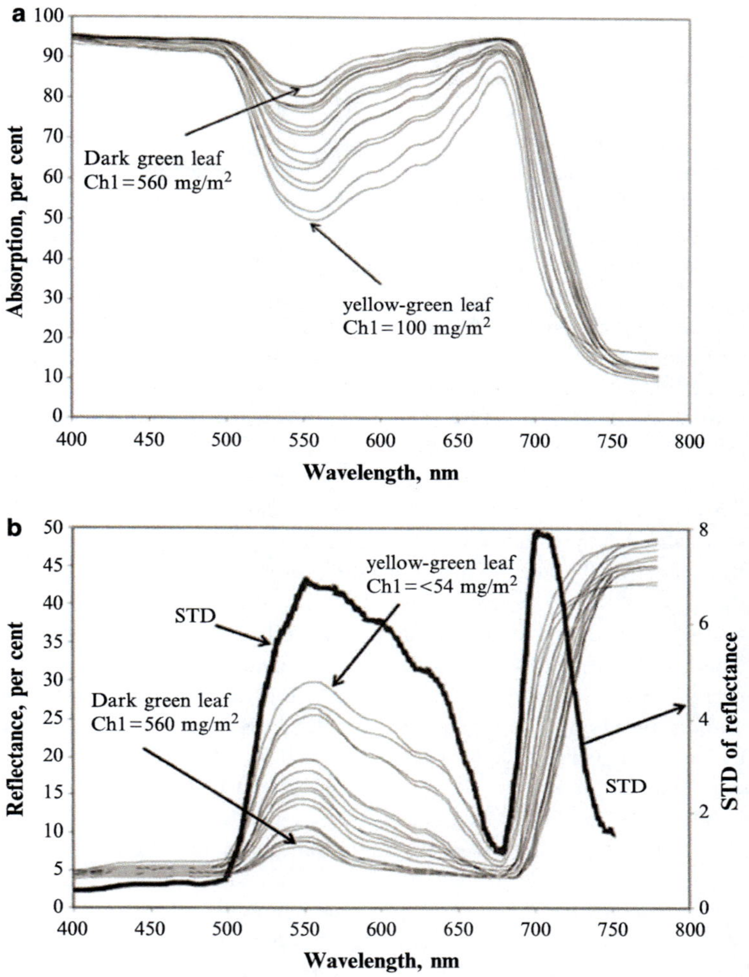

Fig. 10.4 (**a**) Absorption spectra measured in leaves with different chlorophyll concentrations. (**b**) Leaf reflectance for different chlorophyll concentration and the standard deviation of the spectra as function of the wavelength (Adapted from Hatfield et al. 2008)

reflectance index (NRI), the normalized difference red edge index (NDRE), and normalized difference nitrogen index (NDNI) (Fourty et al. 1996). Nevertheless, the nitrogen content in the leaf is mainly allocated in the chlorophyll molecule, and that is why many authors assess N content by estimating canopy chlorophyll content. Figure 10.4 shows the spectra measured on leaves with different chlorophyll content.

The maximum differences are found in the bands around the green peak and in the red edge region (Hatfield et al. 2008). A number of vegetation indices have been developed for nitrogen or pigment content estimation (Table 10.1).

Many authors use the far red region to assess chlorophyll content (examples can be found in Gates et al. 1965; Horler et al. 1983; Chappelle et al. 1992; Vogelmann 1993; Gitelson and Merzlyak 1994; Datt 1998; Curran et al. 1990), and others combine near-infrared and blue/green regions (as Peñuelas et al. 1994; Broge and Leblanc 2000; Daughtry et al. 2000; Sims and Gamon 2002; Haboudane et al. 2002), while the estimation of carotenoids is assessed using bands located in the visible (Gamon et al. 1992; Gitelson et al. 2002). It has been confirmed that the indices using multiple bands in their computations are more successfully applied to a wide range of species (Blackburn 2007).

10.4 Nitrogen Deficiencies at Canopy Scale

The indices developed at leaf scale are not always sensitive at canopy scale due to the confounding structural effects. The interaction of photons within the canopy layers depends on the leaf area index and the disposal of the leaves (leaf angle distribution) within the canopy (Norman et al. 1985). Those structural effects modify the photon trajectory and the overall signal. Moreover, the structure is also affecting the proportional soil signal influence within the surface reflectance of a pixel.

Additionally, spectral ratios help normalizing for differences in illumination intensity resulting from slope changes. The reflectance at a given wavelength is dependent on the incoming radiation, consequently on the sun angle. When using the ratio between two wavelengths, the effect of such sun angle is minimized (Lillesand et al. 2008). Blackmer et al. (1996) and later Osborne et al. (2004) concluded that the ratio between NIR and the green band was a good estimator of N content at the canopy scale. They computed the spectra of corn canopy undergoing N deficiencies relative to healthy corn canopy spectra and evaluated the influence of N deficiency along the electromagnetic spectrum. Gautam and Panigrahi (2007) used the same index combined with textural information extracted from infrared and red reflectance from imagery.

To overcome the vegetation structural effects when applying at canopy scale chlorophyll indices developed at leaf scale, some authors use ratios between an index sensitive to leaf chlorophyll content and a structural index. In this way, the effectiveness of the chlorophyll index is not masked out by the structure of the vegetation and the soil background, which is the case of the ratio between the transformed chlorophyll absorption reflectance index (TCARI) divided by the optimized soil-adjusted vegetation index (OSAVI) developed by Haboudane et al. (2002) or the normalized difference red edge (NDRE) divided by the normalized difference vegetation index (NDVI) by Tilling et al. (2006).

Alternatively to the use of vegetation indices to estimate chlorophyll content, the inversion of radiative transfer models can be used. At the top of the canopy, the interaction of the incoming radiation within the vegetation depends on the contribution of several components such as leaves, stems, soil, illumination, and view properties of each canopy element, as well as on their number, area, orientation, and position in space (Goel and Thompson 2000; Koetz et al. 2004). Radiative transfer models simulate the interaction of the photons through the leaf and within the canopy architecture and give as result the top of canopy reflectance for a given condition. At the leaf level, one of the inputs used to characterize the vegetation is the leaf chlorophyll content (Jacquemoud and Baret 1990). Canopy reflectance extracted from multispectral imagery can be inverted with coupled leaf-canopy radiative transfer models to obtain leaf chlorophyll content (Jacquemoud et al. 1995). This methodology has the advantage of being applicable to different species, sensors, and geometric conditions as it does not rely on empirical relationships obtained between vegetation indices and biophysical parameters. Leaf chlorophyll content has been derived from modeling inversion for grasslands (Darvishzadeh et al. 2008), olive (Zarco-Tejada et al. 2004a; Berni et al. 2009a), peach (Kempeneers et al. 2008), vineyards (Zarco-Tejada et al. 2005), and black spruce (Zarco-Tejada et al. 2004b).

10.5 Remote Sensing of Phosphorous Deficiency

There is not much documentation about remote sensing of nutrient deficiencies apart from nitrogen. Although N is considered the most important nutrient to monitor in crops, other element deficiencies as phosphorous can lead to lower shoot and root growth and eventually to a decrease in yield (Milton et al. 1991; Osborne et al. 2004). The biggest spectral differences due to leaf P concentration have been found in the 500–650 nm and near-infrared regions (Milton et al. 1991; Osborne et al. 2002; Yaryura et al. 2009). Phosphate-deficient leaves have less inorganic phosphorous in the tissue water, lower photosynthetic and stomatal conductance rates, and higher number of small cells per unit area (Jacob and Lawlor 1991). The impact on the near infrared can be attributed to the increase in the number of smaller cells in the palisade as leaf near-infrared reflectance is mainly responding to structure. The limited photosynthetic rate and the decrease of stomatal conductance are symptoms that are more affected by water or nitrogen stress. Changes in the green-red region can be due to a decrease of the pigment pool due to severe stress (Yaryura et al. 2009).

10.6 Remote Sensing of Water Stress

The early detection of water stress is a key issue to avoid yield loss, which can be affected even by short-term water deficits (Hsiao et al. 1976). It is well known that severe water deficits affect many physiological processes and have a strong impact on yield (Hsiao et al. 1976). However, even moderate water deficits, which are not easy to detect, can also have important negative effects on yield (Hsiao and Bradford 1983). It is important to be able to assess the level of stress through some pertinent indicators. The previsual detection of water stress has been successfully achieved with remote sensing data using thermal infrared radiation since long ago (Idso et al. 1978, 1981; Idso 1982a, b; Jackson et al. 1977, 1981, 1983; Jackson and Pinter 1981; Leinonen and Jones 2004; Wanjura et al. 2004; Cohen et al. 2005; Sepulcre-Cantó et al. 2006, 2007) and more recently being suggested the visible spectral region with the PRI index as an indicator of stress (Thenot et al. 2002; Serrano and Peñuelas 2005; Suárez et al. 2008, 2009, 2010a; Peguero-Pina et al. 2008).

It has long been known that tree water deficits affect fruit quality parameters (Veihmeyer 1927). However, when water deficits are imposed as in "regulated deficit irrigation" (RDI) treatments (Fereres and Soriano 2007), yield and fruit size are not affected (Girona et al. 2002), while some quality parameters such as total soluble sugars and total acidity

Table 10.2 Remote sensing indices used to estimate leaf water content

Index	Acronym	Formula	Biophysiological parameter	References
Normalized difference water index	NDWI	(R860 – R1240)/(R860 + R1240)	Water content	Gao (1996)
Reflectance water index	WI	R900/R970	Water content	Peñuelas et al. (1997)
Simple ratio water index	SRWI	R858/R1240	Water content	Zarco-Tejada et al. (2003)
Moisture stress index	MSI	R1599/R819	Water content	Hunt Jr and Rock (1989)
Normalized difference infrared index	NDII	(R819 – R1649)/(R819 + R1649)	Water content	Hardisky et al. (1983)

increase (Crisosto et al. 1994; Girona et al. 2003; Mills et al. 1994). The responses to RDI are variable depending on the timing and severity of water deficits (Marsal and Girona 1997; Girona et al. 2003) which vary within a given orchard, thus the need for remote sensing tools that could assist in monitoring stress over entire orchards. Additionally, the changes in irrigation depths with time and the lack of uniformity in water application during the irrigation period emphasize the need for a methodology that would cover the entire season, integrating the short-term variations in tree water status.

Water status of the trees has been traditionally measured using instruments in the field to measure stem water potential or water transport. Nevertheless, those measurements are time consuming and generally taken on a limited number of trees which represent the whole field, not having into account the existing spatial heterogeneity.

There are many factors affecting such spatial heterogeneity:

- Soil type/texture and the capacity to retain water
- Variable irrigation doses driven by maladjustments in the irrigation system
- Slope of the terrain
- Local problems of competence for water due to development of weeds
- Other factors (physical damage, heterogeneous treatment application, etc.)

In order to properly assess the water status of an orchard, spatially continuous data is needed; in this way, the spatial heterogeneity can be mapped. Moreover, field data collection is time and energy consuming, not being affordable for a proper monitoring of large areas.

There are a number of spectral indices that have been developed to estimate leaf water content (Table 10.2).

Previous studies have estimated leaf water content using remote sensing data through vegetation indices (Gao 1996; Ceccato et al. 2002) and through modeling inversion (Zarco-Tejada et al. 2003).However, leaf water content is only affected when the water stress is severe leading to leaf wilting and structural changes.

At that stage of water stress, yield may have been seriously affected. It is important then to be able to detect water stress in previous stages, when the stress is still "previsual." Previsual water stress can only be assessed through indicators that dynamically change with plant water status.

10.6.1 Thermal Remote Sensing of Water Stress

Water stress in vegetation results in stomatal closure which reduces the transpiration rate. As a result, the evaporative cooling decreases, resulting in a gradual increase of the canopy temperature which can be detected by means of infrared thermometers or thermal imaging sensors. This approach at detecting water stress became very popular in the 1970s and 1980s with the advent of handheld thermometers (Fuchs and Tanner 1966; Idso et al. 1978, 1981; Jackson et al. 1977, 1981; Jackson 1982). Unfortunately, there is not a direct relationship between canopy temperature and water stress given that there are a number of factors that have an effect on the canopy temperature. Some of these factors include incoming radiation, air temperature and humidity, or wind speed, among others.

The need to overcome those effects and finding some normalizing technique for canopy temperature led to the development of the crop water stress index (CWSI) (Idso et al. 1981; Jackson et al. 1981). The CWSI consisted in relating the actual difference between canopy and air temperatures (T_c and T_a, respectively) to the difference between the $T_c - T_a$ values of a nonwater-stressed (NWS) baseline and an upper $T_c - T_a$ limit, both being a function of the atmospheric vapor pressure deficit (VPD) (Idso et al. 1981). An index ranging from 0 to 1 is thus obtained which was found to be proportional to the stress level in many crops, provided that the NWS baseline is known for the crop and local conditions (Fig. 10.5).

However, this approach has the limitation that an empirical determination of the baseline and upper limit is required for each crop and even location. New theoretical and practical approaches have been proposed to overcome the need for the empirical retrieval of the NWS baseline required in the CWSI calculation. Jones (1999), for example, estimated the components of the CWSI equation (Eq. 10.1), solving the energy balance equations for two different conditions: (a) when the surface has no stomatal conductance and thus all the

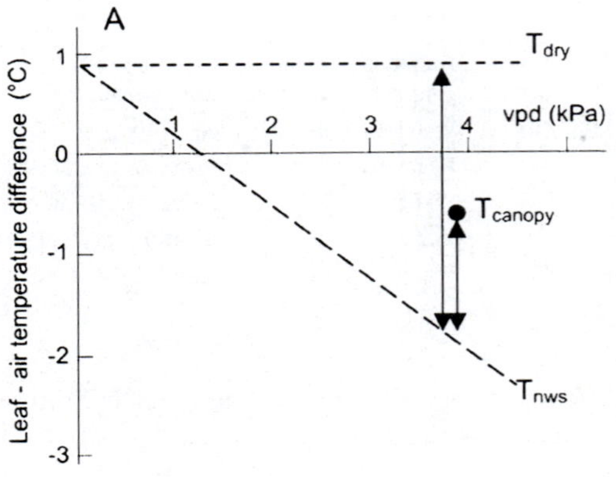

Fig. 10.5 Graphical representation on how the crop water stress index is calculated: $CWSI=(T_{canopy}-T_{nws})/(T_{dry}-T_{nws})$ (Adapted from Jones et al. 2002)

available energy is transferred into sensible heat, the temperature of equilibrium will be T_{dry}; (b) when the surface has the maximum stomatal conductance and thus most of the available energy is transferred into evaporation T_{wet}:

$$\mathrm{CWSI} = \frac{T_{canopy} - T_{wet}}{T_{dry} - T_{wet}} \tag{10.1}$$

The general equation for calculating the temperature of equilibrium of an evaporating surface can be defined as in Eq. 10.2:

$$T_c - T_a = \frac{r_a R_n}{\rho C_p} \cdot \frac{\gamma(1 + r_c / r_a)}{\Delta + \gamma(1 + r_c / r_a)} - \frac{\mathrm{VPD}}{\Delta + \gamma(1 + r_c / r_a)} \tag{10.2}$$

where T_c and T_a are the canopy and air temperatures, respectively, r_a is the aerodynamic resistance, R_n is the net radiation, ρ is the air density, C_p the specific heat of air, γ is psychrometric constant, r_c is the canopy resistance, Δ is slope of saturation vapor pressure curve at mean air temperature, and VPD is the water vapor pressure deficit.

Using Eq. 10.2, it is possible to calculate analytically the situation when the canopy is not transpiring (T_{dry}) and when it is transpiring at a maximum rate (T_{wet}). Substituting those values and the actual canopy temperature in Eq. 10.1, it is possible to obtain an analytical solution of CWSI.

The use of theoretical equations of CWSI based on the energy balance equation is limited by the need to estimate net radiation and aerodynamic resistance which is not trivial especially in heterogeneous canopies like the case of tree orchards. Some authors overcome this problem by using dry and wet references that account for the CWSI upper and lower limits, respectively, allowing the estimation of CWSI with a minimum of meteorological measurements (Jones et al. 2002; Cohen et al. 2005; Grant et al. 2007; Möller et al. 2007). For example, Jones et al. (2002) used Vaseline to coat grape leaves and prevent transpiration in order to account for the upper limit (T_{dry}) while the lower limit (T_{wet}) was estimated by spraying water on leaves. In other cases, T_{dry} is set to 5°C above air temperature (Irmak et al. 2000) while the lower limit is estimated by using artificial wet references made of polystyrene soaked in water (Cohen et al. 2005). However, the use of such reference surfaces is a clear limitation for the practical and extensive use of this methodology from a remote sensing point of view.

The major advantage of the analytical approach is that once aerodynamic resistance and net radiation can be measured or modeled, it allows the calculation of canopy conductance (Smith 1988; Lhomme and Monteny 2000; Leinonen et al. 2006). This approach was followed by Berni et al. (2009b) in order to calculate the CWSI from the canopy resistance estimated by means of minimum meteorological observation and physical models:

$$r_c = \frac{r_a(e_c^* - e_a)}{\gamma\left(\frac{r_{aR_n}}{\rho C_p} - (T_c - T_a)\right)} \tag{10.3}$$

where e_c^* is the water vapor pressure of saturation at the canopy temperature and e_a is the actual vapor pressure of air. See Eq. 10.2 for the rest of the terms.

Once the canopy conductance is estimated and provided that the potential conductance in absence of water stress is known (r_{cp}), the CWSI can be calculated in an analytical way according to Jackson et al. (1981):

$$\mathrm{CWSI} = 1 - \frac{E}{E_p} = \frac{\gamma(1 + r_c / r_a) - \gamma^*}{\Delta + \gamma(1 + r_c / r_a)} \tag{10.4}$$

where:

$$\gamma^* = \gamma(1 + r_{cp} / r_a)$$

Physiological studies have shown that the potential canopy conductance ($1/r_{cp}$) for a particular species depends on a number of environmental factors such as solar radiation, air temperature and humidity, or even the cultivar (Wright et al. 1995; Lu et al. 2003; Oguntunde et al 2007). There are two major ways to account for r_{cp} (Oguntunde et al. 2007), an empirical approach using a linear regression analysis of the major meteorological factors or a Javis-type nonlinear model that uses semiempirical functions for the estimation of the drivers of stomatal control.

Any of the abovementioned techniques for water stress monitoring based on canopy temperature could potentially be applied to tree orchards using both infrared thermometers and at image level using thermal cameras or airborne thermal imagery. However, the use of CWSI as a stress indicator has not been widely adopted for two main reasons (Cohen et al. 2005): (1) temperature remotely sensed from readily available satellite platforms or airborne sensors lacks the necessary spatial resolution for the accurate separation of canopy temperature from the sunlit and shaded soil background, which is especially critical in tree orchards like the case of citrus; and (2) the different equations of NWS baseline published are site dependent, since the VPD normalization procedure used for obtaining the CWSI does not account for differences in net radiation and aerodynamic resistance which are known to affect the index (Hipps et al. 1985; Jackson et al. 1988; Jones 1999). Avoiding the first issue is not easy, since currently available satellite thermal imagery is limited to Landsat TM and ASTER scanners, yielding 120 and 90 m, respectively, and MODIS or AVHRR, with 1 km pixel size. The medium-low spatial resolution of such satellite thermal scanners makes mapping water stress only potentially suitable for regional scales if successfully accounting for canopy heterogeneity (Moran et al. 1994; Norman et al. 1995). Sepulcre-Cantó et al. (2009) used satellite thermal information in combination with 3D radiative transfer models to understand scene components thermal effects on large ASTER pixels; however, this methodology relies on the availability of high-resolution visible imagery in order to obtain additional information for the modeling that is not necessarily available at regional scale.

Alternatively, airborne thermal imagery has been proved suitable for mapping water stress on discontinuous canopies (Sepulcre-Cantó et al. 2006, 2007), provided that the spatial resolution allows the detection of isolated tree crowns (<2 m pixel size). New thermal imaging sensors onboard unmanned aerial platforms provide submeter spatial resolution (Herwitz et al. 2004; Sugiura et al. 2005; Berni et al. 2009a) enabling the retrieval of pure canopy temperature, thus minimizing soil thermal effects. High-resolution thermal imagery would make possible the retrieval of energy fluxes from pure vegetation on open canopies, such as tree orchards, where most remote sensing–based methodologies do not perform well.

The use of high-resolution thermal imagery in combination of any of the abovementioned methodologies will allow farmers to obtain a snapshot of the spatial variability of water stress within the orchard and understand the underlying effects of soil heterogeneity and uniformities of the irrigation system on the crop performance. This will become a very powerful tool since the ultimate goal is not only to maximize water use efficiency but also to attain the maximum homogeneity of water application and thus homogeneous fruit size and quality.

However, despite thermal imagery delivers a good picture of the spatial component of water stress, the temporal evolution of water stress can be even more important for applications such as irrigation scheduling. It is hard to imagine that even with the use of unmanned aerial vehicles, it will be cost-effective to obtain routinely thermal images of the orchard with enough frequency to program the irrigation. The solution may come with the combination of spatial techniques with the use of infrared thermometers distributed along the orchard and operating within sensor networks together with a small weather station. This will allow estimating the irrigation needs in real time while at the same time the spatial heterogeneities could be taken into account and the estimates from the different sensors weighted properly.

For the particular case of citrus, there are very limited reports on the monitoring of water stress using thermal remote sensing. Sepaskhah and Kashefipour (1995) established a relationship between $T_c - T_a$ measured 1 h after midday, and daily evapotranspiration (*ET*) has been established for sweet lime. Recently, Stagno et al. (2011) reported good estimates of water status using an empirical relationship between canopy temperature at noon and stem water potential.

10.7 Use of Remote Sensing Data Related to Vegetation Physiology

Notwithstanding the advances in thermal detection, the visible part of the spectrum has also been useful for previsual water stress detection based on indices that use bands located at specific wavelengths where photosynthetic pigments are affected by stress condition. This is the case of the photochemical reflectance index (PRI, Gamon et al. 1992) that has been proposed to assess vegetation water stress based on xanthophyll composition changes (Thenot et al. 2002; Peguero-Pina et al. 2008; Suárez et al. 2008, 2009).

The photochemical reflectance index (PRI) was suggested by Gamon et al. (1992) as an indicator of the de-epoxidation state of the xanthophyll pigments related with photosynthetic processes (Fig. 10.6). Xanthophylls are pigments participating in an enzymatic cycle which plays a key role in stimulating energy dissipation within light-harvesting antenna proteins by nonphotochemical quenching, a mechanism to reduce the amount of energy that reaches the photosynthetic reaction centers. The xanthophylls contained in the vegetation and involved in the de-epoxidation processes are in the form of three different compounds: (1) violaxanthin (V), with two epoxides absorbing incident photons and transmitting them to the reactive centers to conduct the photosynthesis; (2) antheraxanthin (A), with one epoxide; and (3) zeaxanthin (Z) with no epoxides, by which the incoming energy is released and hence is not transmitted to the photosystems, and therefore not being used to conduct photosynthesis.

Nonphotochemical quenching is one of the main ways of protecting against photoinhibition which is light-induced reduction in the photosynthetic capacity of a plant. When light absorption exceeds light utilization in photosynthetic

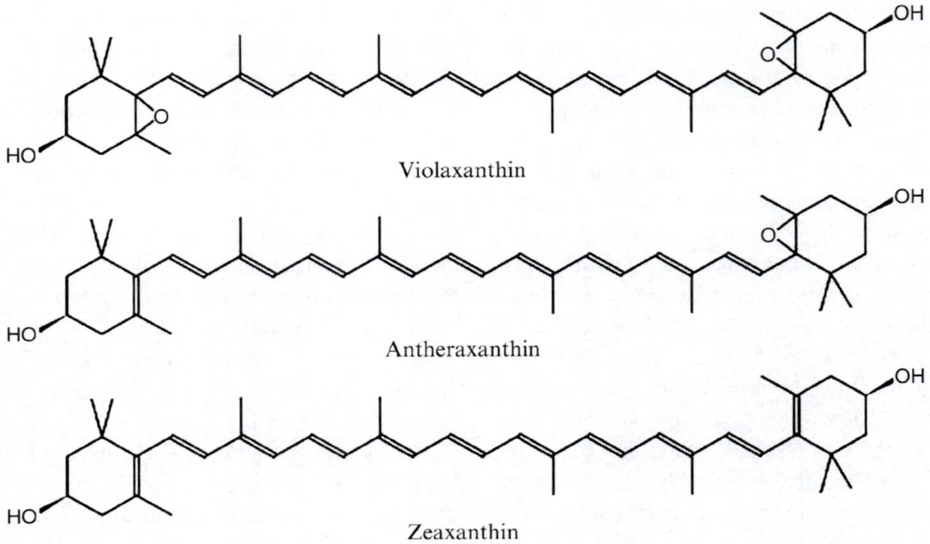

Fig. 10.6 Xanthophyll molecules participating in the xanthophyll cycle. Under stress conditions, violaxanthin de-epoxidize to antheraxanthin and the latter to zeaxanthin avoiding photosystems damage. The opposite happens under nonstress conditions

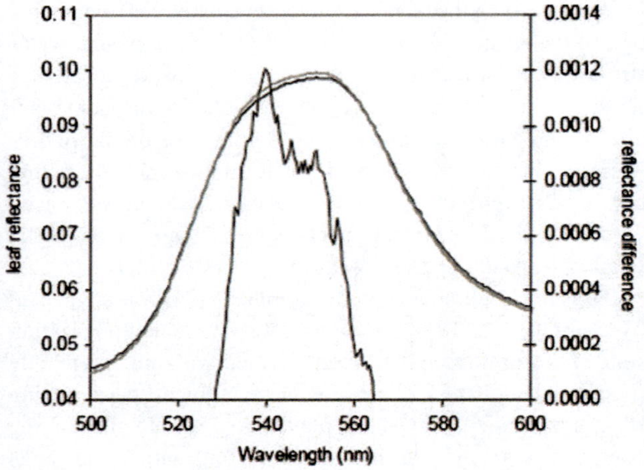

Fig. 10.7 Leaf reflectance of dark-adapted leaves (*dotted line*, first measurement) and under steady-state condition (*continuous line*, last measurement), showing the reflectance difference associated with xanthophyll pigment cycle

electron transport, excess absorbed light results. This can arise from exposure to a high photosynthetic photon flux density (PFD) or a low requirement for electron transport under environmental stresses.

The de-epoxidation of violaxanthin is related to an absorption feature in the visible part of the electromagnetic spectrum located around 530 nm. Figure 10.7 represents the reflectance difference between a dark-adapted leaf and a leaf excited by a light pulse at the xanthophyll-influence area.

The PRI is based on a normalized difference of the 530 nm band where xanthophyll pigment absorption occurs and a reference band located at 570 nm (Table 10.3).

As the xanthophyll pigments are related to light absorption mechanisms, the PRI index has been extensively linked to light-use efficiency (LUE) at leaf scale (Guo and Trotter 2004; Sims et al. 2006; Nakaji et al. 2006) and at canopy scale using field spectrometers (Trotter et al. 2002; Strachan et al. 2002; Nichol et al. 2000, 2002) and using satellite imagery such as EO-1 Hyperion (Asner et al. 2005), MODIS (Drolet et al. 2005) and AVIRIS (Fuentes et al. 2006).

The estimation of LUE through the remote sensing PRI index is demonstrated as a direct link to photosynthesis rate assessment (Nichol et al. 2000, 2006; Guo and Trotter 2004; Sims et al. 2006). In addition, photosynthesis has also been related to PRI through chlorophyll fluorescence and nonphotochemical quenching (Evain et al. 2004; Nichol et al. 2006).

The bands that are needed to calculate PRI are available in current sensors on board airborne and satellite platforms. The satellite sensors MODIS and Hyperion and the hyperspectral airborne sensors AHS, HyMap, AVIRIS, CASI, and ROSIS provide spectral information on the xanthophyll-influence area. An overview of the sensors that have been used to assess PRI can be found in Table 10.3.

Although PRI has been widely used at different ground spatial resolutions on homogeneous and heterogeneous study sites, it cannot be readily used to map vegetation stress without considering leaf and canopy structural effects on the

Table 10.3 Overview of the sensors on airborne and satellite platforms used to acquire PRI information in previous studies, the spectral bands used to calculate the index, and the ground spatial resolution of the imagery

Platform	Sensor	Spectral bands (nm)	Spatial resolution	References
Satellite-AQUA	MODIS	531,551	1 km	Drolet et al. (2005)
Satellite-EO 1	Hyperion	529,569	30 m	Asner et al. (2004, 2005)
Airborne	AHS	542,571	2 m	Suárez et al. (2008)
Airborne	HyMap	543,573	5 m	Kooistra et al. (2008)
Airborne	AVIRIS	517,567	2–16 m	Fuentes et al. (2006)
Airborne	CASI	528,567	2 m	Zarco-Tejada et al. (2000c)
Airborne	ROSIS	531,570	1 m	Zarco-Tejada et al. (2005)

index. PRI bands at 531 and 570 nm are affected by both leaf and canopy parameters such as chlorophyll content (Cab), dry matter (Cm), leaf thickness, leaf area index (LAI), and leaf angle distribution function (LADF), among others (Barton and North 2001; Suárez et al. 2008). Thus, PRI maps obtained over canopies with variable LAI mask the sensitivity of the index to stress, mostly tracking the spatial variation of the canopy leaf area density and structure (Barton and North 2001; Suárez et al. 2008). Consequently, modeling work at leaf and canopy scale is needed to enable an operational application of PRI to map water stress in nonhomogeneous canopies where structural changes play the main role in the reflectance signature (Suárez et al. 2009).

10.7.1 Chlorophyll Fluorescence

The light absorbed by the vegetation can be either used to perform photosynthesis or driven through a dissipation mechanism. Chlorophyll fluorescence is a way of dissipating excess energy accumulated in plant tissue. Solar-induced chlorophyll fluorescence emission is then used as a physiological indicator for remote assessment of prevision water stress (Flexas et al. 1999, 2000, 2002; Moya et al. 2004; Zarco-Tejada et al. 2009) due to the strong correlation demonstrated between steady-state chlorophyll fluorescence (Fs) and stomatal conductance. In addition, several studies have assessed the relationship of chlorophyll fluorescence with photosynthesis and plant physiological status (Papageorgiou 1975; Schreiber and Bilger 1987; Krause and Weis 1984; Lichtenthaler and Rinderle 1988; Lichtenthaler 1992; Larcher 1994; Schreiber et al. 1994). As the chlorophyll fluorescence signal is relative and needs normalization in order to be assessed (Maxwell and Johnson 2000), a set of parameters have been defined in order to measure chlorophyll fluorescence. The chlorophyll fluorescence parameters most frequently used to characterize the functioning of the photosynthetic apparatus under nonstress and stress conditions are Fv/Fm, ΦPSII (Yield), qP, Φexc., and NPQ (or qN) (full reviews can be found in Morales et al. 1991, 1998, 2000, 2006; Abadía et al. 1999). Although comparatively steady-state chlorophyll fluorescence Fs has received much less attention, its potential importance for chlorophyll fluorescence detection using remote sensing methods has been recently emphasized (Soukupová et al. 2008), along with increasing scientific interest during the past years. Nevertheless, the relationships found are not universally applied, and it has been suggested that C_3 and C_4 plants dissipation mechanisms are different due to the contribution of other processes to electron use in C_3 plants (Lichtenthaler et al. 1998; Flexas et al. 2000). Actually, not only the carbon fixation mechanism but also leaf roughness and thickness and the vegetation structure are influencing the fluorescence signal (Rosema et al. 1991; Olioso et al. 1992; Moya et al. 2006).

Several publications have described reflectance indices potentially related to incremental effects of fluorescence emission as an addition to leaf and canopy reflected signal. In particular, as an attempt to minimize structural effects, indices related to fluorescence emission bands located in the red and far red have been proposed to assess the relationship between apparent reflectance and chlorophyll fluorescence at both leaf and canopy levels (Olioso et al. 1992; Zarco-Tejada et al. 2000a, b). It has been suggested that while changes in chlorophyll fluorescence emission in the red are related to nutrient stress leading to lower chlorophyll content, blue/red and blue/near-infrared fluorescence ratios are sensitive to abiotic stress (Lichtenthaler et al. 1998). Solar-induced chlorophyll fluorescence can be also assessed by the in-filling method (McFarlane et al. 1980; Carter et al. 1990, 1996, 2004) which makes use of the atmospheric absorption bands located in the red and near-infrared regions. This method, based on the radiance difference of fluorescing and nonfluorescing targets in and out the atmospheric absorption band, has been applied using leaf (Meroni et al. 2008), near-canopy measured radiance (Evain et al. 2004; Liu et al. 2005a, b; Moya et al. 2004; Pérez-Priego et al. 2005), and airborne imagery (Zarco-Tejada et al. 2009).

Although it looks both PRI and Fs are similarly tracking changes in vegetation photosynthetic capacity affected by plant stress, they complement each other describing the vegetation internal functioning and are suggested to be used together (Daumard et al. 2010). Toward that direction, new systems are being developed and tested in the field which permits the simultaneous acquisition of sun-induced fluorescence data and PRI (Zarco-Tejada et al. 2009; Daumard et al. 2010).

10.8 Importance of Spatial Resolution

The spatial resolution of the imagery we use for remote sensing purposes must be related with the spatial sampling frequency needed to adequately represent the object of study (Duveiller and Defourny 2010). When the study area is heterogeneous and we need a pure signal, the level of detail of the required input information is higher than if we are retrieving a global scale parameter (Teillet et al. 1997). The heterogeneity of the landscape and the ground spatial resolution will determine the purity signal of a single pixel. When we are estimating a leaf physiological parameter as EPS, pure vegetation signal is needed as the sensitivity of the spectral signal to quick physiological changes is weak. Suárez et al. (2010a, b) simulated the effect of using medium spatial resolution versus very high spatial resolution imagery to estimate EPS.

While a good correlation is found when using very high spatial resolution imagery ($r^2=0.57$, Fig. 10.8a), the correlation is worse when aggregating soil and shadows in the overall pixel signal ($r^2=0.44$, Fig. 10.8b). Moreover, the PRI range of variation depending on the soil type can make the relationship completely inexistent. Analogously, the orientation of the rows in a row-structured crop determines the proportional shadow fraction, altering the signal of a mixed pixel. Analogously, the relationship between high-resolution and medium-resolution signal is affected when having heterogeneous soil type and row orientation (Fig. 10.9).

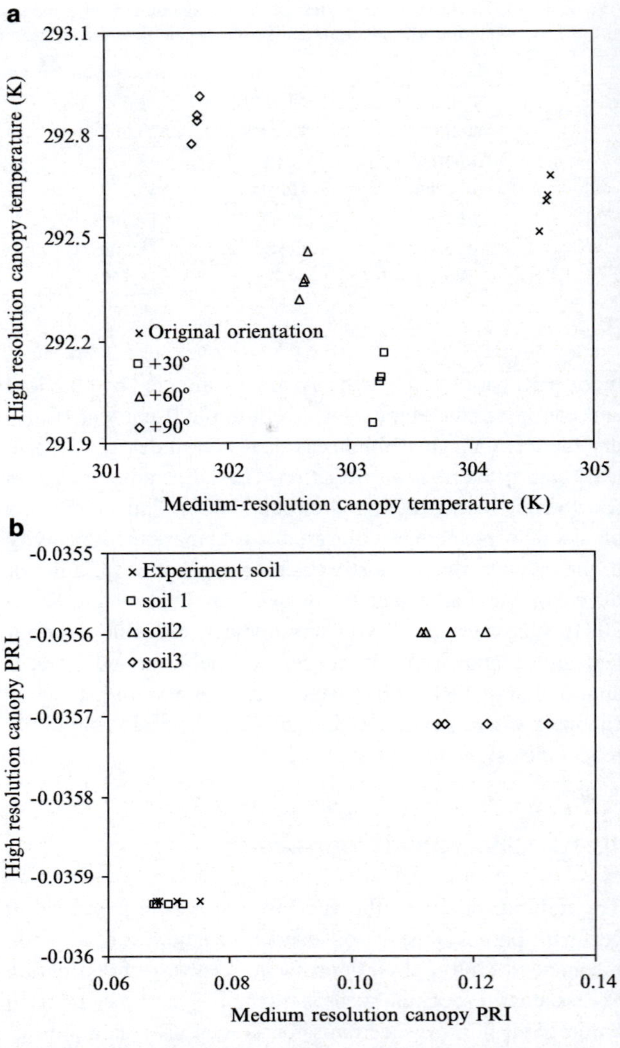

Fig. 10.9 Relationship between high-resolution and medium-resolution temperature and PRI for three types of soils and four different orientations in a row-structured crop (Adapted from Suárez et al. 2010b)

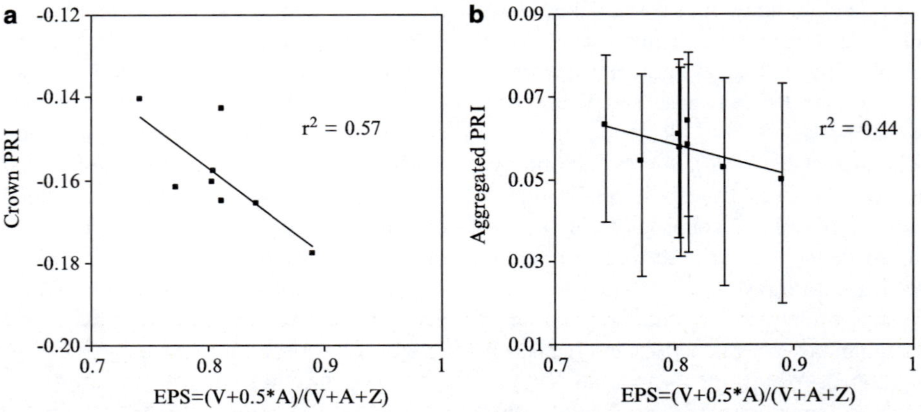

Fig. 10.8 Relationship between the PRI index extracted from high spatial resolution imagery (15 cm) for each crown and the EPS corresponding to leaf samples (Adapted from Suárez et al. 2010a)

10.9 Conclusions

Traditional orchard diagnosis has been carried out by observation of symptoms and posterior verification by analyzing tissue for nutrient stress or measuring water potential/stomatal conductance for water stress. Remote sensing gives the possibility of assessing citrus water and nutrient status at a large scale. That permits individual stand diagnosis over large areas without needing labor consuming visual assessment. The first step includes the detailed study of the different deficiencies' impact on leaf reflectance. Then, vegetation indices are developed based on specific spectral ranges affected. Methodologies developed to assess stress at leaf scale are later upscaled for canopy stress assessment. Some indices developed from studies at leaf scale cannot be used at canopy scale because they are highly sensitive to soil, shadows, or canopy architecture artifacts. To overcome those effects, some authors use combinations of pigment indices and structural indices; others use physical radiative models which are considering those elements within the simulated reflectance. Vegetation indices can be used to get a qualitative map of the status of vegetation based on reflectance response to stress. Nevertheless, in order to get quantitative information, area-specific algorithms that are built empirically are needed. The limitation of the methodologies based on area-specific algorithms is that the former cannot be universally applied or extrapolated to other areas. Methodologies based on algorithms developed using radiative transfer models are ready to be universally applied as they have into account vegetation structure and solar geometry effects on reflectance. However, model inversion requires higher computing power and time, not to mention the difficulties of performing proper algorithm training.

Although stress assessment has been proved to be achieved using remote sensing data, there is still extensive work and further research to do. There is a need for specific indices that assess individual nutrient stress and that are ready to be applied at canopy scale. Moreover, if leaf radiative transfer models were improved to account for differences in pigments/nutrients other than chlorophyll, there would be a notorious improvement in the remote sensing of vegetation status.

References

Abadía J, Morales F, Abadía A (1999) Photosystem II efficiency in low chlorophyll, iron-deficient leaves. Plant Soil 215:183–192

Al-Abbas AH, Barr R, Hall JD et al (1974) Spectra of normal and nutrient-deficient maize leaves. Agron J 66:16–20

Asner GP, Nepstad D, Cardinot G et al (2004) Drought stress and carbon uptake in an Amazon forest measured with spaceborne imaging spectroscopy. Proc Natl Acad Sci USA 101(16):6039–6044

Asner GP, Carlson KM, Martin RE (2005) Substrate age and precipitation effects on Hawaiian forest canopies from spaceborne imaging spectroscopy. Remote Sens Environ 98:457–467

Barnes EM, Clarke TR, Richards SE et al (2000) Coincident detection of crop water stress, nitrogen status and canopy density using ground-based multispectral data. In: Robert PC, Rust RH, Larson WE (eds) Proceedings of the 5th international conference on precision agriculture and other resource management. ASA/CSSA/SSSA, Madison

Barros SM, Davis JE, Blazquez CH et al (1983) Citrus grove mapping by microcomputer. Proc Fla State Hortic Soc 96:1–3

Barton CVM, North PRJ (2001) Remote sensing of canopy light use efficiency using the photochemical reflectance index. Model and analysis. Remote Sens Environ 78:264–273

Bausch WC, Duke HR (1996) Remote sensing of plant nitrogen status in corn. Trans ASAE 39(5):1869–1875

Benedict HM, Swindler R (1961) Non destructive method for estimating chlorophyll content of leaves. Science 113:2015–2016

Berni JAJ, Zarco-Tejada PJ, Suarez L et al (2009a) Thermal and narrow-band multispectral remote sensing for vegetation monitoring from an unmanned aerial vehicle. IEEE Trans Geosci Remote 47(3):722–738

Berni JAJ, Zarco-Tejada PJ, Sepulcre-Cantó G et al (2009b) Mapping stomatal conductance and CWSI in olive orchards using high resolution thermal remote sensing imagery. Remote Sens Environ 113:2380–2388

Blackburn GA (2007) Hyperspectral remote sensing of plant pigment. J Exp Bot 58(4):855–867

Blackmer TM, Schepers JS, Varvel GE (1994) Light reflectance compared to other nitrogen stress measurements in corn leaves. Agron J 86:934–938

Blackmer TM, Schepers JS, Varvel GE et al (1996) Nitrogen deficiency detection using reflected shortwave radiation from irrigated corn canopies. Agron J 88:1–5

Blazquez CH (1988) Comparisons between densitometric measurements, image analysis and photo interpretation readings of aerial color infrared photographs of citrus trees. Proc Fla State Hortic Soc 101:66–69

Blazquez, CH, Horn FW Jr (1980) Aerial color infrared photography: Applications in citriculture. National aeronautics and space administration, scientific and technical information branch, reference publication 1067. Washington DC

Broge NH, Leblanc E (2000) Comparing prediction power and stability of broadband and hyperspectral vegetation indices for estimation of green leaf area index and canopy chlorophyll density. Remote Sens Environ 76:156–172

Carter GA, Theisen AF, Mitchell RJ (1990) Chlorophyll fluorescence measured using the Fraunhofer line-depth principle and relationship to photosynthetic rate in the field. Plant Cell Environ 13:79–83

Carter GA, Jones JH, Mitchell RJ et al (1996) Detection of solar-excited chlorophyll a fluorescence and leaf photosynthetic capacity using a Fraunhofer line radiometer. Remote Sens Environ 55:89–92

Carter GA, Freedman A, Kebabian PL et al (2004) Use of a prototype instrument to detect short-term changes in solar-excited leaf fluorescence. Int J Remote Sens 25:1779–1784

Ceccato P, Gobron N, Flasse F et al (2002) Designing a spectral index to estimate vegetation water content from remote sensing data: Part 1: theoretical approach. Remote Sens Environ 82(2–3):188–197

Chaerle L, Van Der Straeten D (2000) Imaging techniques and the early detection of plant stress. Trends Plant Sci 5(11):495–501

Chappelle EW, Kim MS, McMurtrey JE III (1992) Ratio analysis of reflectance spectra (RARS): an algorithm for the remote estimation of the concentrations of chlorophyll a, chlorophyll b, and carotenoids in soybean leaves. Remote Sens Environ 39:239–247

Cohen Y, Alchanatis V, Meron M et al (2005) Estimation of leaf potential by thermal imagery and spatial analysis. J Exp Bot 56:1843–1852

Crisosto CH, Johnson RS, Luza JG et al (1994) Irrigation regimes affect fruit soluble solids concentration and rate of water loss of 'O'Henry' peaches. Hortic Sci 29(10):1169–1171
Curran PJ, Dungan JL, Gholz HL (1990) Exploring the relationship between reflectance red edge and chlorophyll content in slash pine. Tree Physiol 7:33–48
Darvishzadeh R, Skidmore A, Schlerf M et al (2008) Inversion of a radiative transfer model for estimating vegetation LAI and chlorophyll in a heterogeneous grassland. Remote Sens Environ 112(5):2592–2604
Datt B (1998) Remote sensing of chlorophyll a, chlorophyll b, chlorophyll a+b, and total carotenoid content in eucalyptus leaves. Remote Sens Environ 66(2):111–121
Daughtry CST, Walthall CL, Kim MS et al (2000) Estimating corn leaf chlorophyll concentration from leaf and canopy reflectance. Remote Sens Environ 74:229–239
Daumard F, Champagne S, Fournier A et al (2010) A field platform for continuous measurement of canopy fluorescence. IEEE Trans Geosci Remote 48(9)
Drolet GG, Huemmrich KF, Hall FG et al (2005) A MODIS-derived photochemical reflectance index to detect inter-annual variations in the photosynthetic light-use efficiency of a boreal deciduous forest. Remote Sens Environ 98:212–224
Du Q, Chang NB, Yang C et al (2008) Combination of multispectral remote sensing, variable rate technology and environmental modeling for citrus pest management. J Environ Manage 86:14–26
Duveiller G, Defourny P (2010) A conceptual framework to define the spatial resolution requirements for agricultural monitoring using remote sensing. Remote Sens Environ 114:2637–2650
Dzikiti S, Verreynne S, Strever A et al (2009) Detecting citrus tree water status by integrating hyperspectral remote sensing and physiological data in a water flow-storage model. Geosci Remote Sens Symp IEEE Int 5:V-348–V-350. IGARSS 2009. ISBN: 978-1-4244-3394-0
Evain S, Flexas J, Moya I (2004) A new instrument for passive remote sensing: 2. Measurement of leaf and canopy reflectance changes at 531 nm and their relationship with photosynthesis and chlorophyll fluorescence. Remote Sens Environ 91:175–185
Fereres E, Soriano M (2007) Deficit irrigation for reducing agricultural water use. J Exp Bot 58:147–159
Filella I, Serrano L, Serra J et al (1995) Evaluating wheat nitrogen status with canopy reflectance indices and discriminant analysis. Crop Sci 35:1400–1405
Fletcher RS, Skaria M, Escobar DE et al (2001) Field spectra and airborne digital imagery for detecting *phytophtora* foot rot infections in citrus trees. Hortic Sci 36:94–97
Flexas J, Escalona JM, Medrano H (1999) Water stress induces different levels of photosynthesis and electron transport rate regulation in grapevines. Plant Cell Environ 22:39–48
Flexas J, Briantais JM, Cerovic Z et al (2000) Steady-state and 719 maximum chlorophyll fluorescence responses to water stress in grapevine leaves: a new remote sensing system. Remote Sens Environ 73:282–297
Flexas J, Escalona JM, Evain S et al (2002) Steady-state chlorophyll fluorescence (Fs) measurements as a tool to follow variations of net CO_2 assimilation and stomatal conductance during water-stress in C-3 plants. Physiol Plant 114(2):231–240
Fourty T, Baret F, Jacquemoud S et al (1996) Leaf optical properties with explicit description of its biochemical composition: direct and inverse problems. Remote Sens Environ 56:104–117
Fuchs M, Tanner CB (1966) Infrared thermometry of vegetation. Argon J 58:597–601
Fuentes DA, Gamon JA, Cheng Y et al (2006) Mapping carbon and water vapour fluxes in a chaparral ecosystem using vegetation indices derived from AVIRIS. Remote Sens Environ 103:312–323
Gamon JA, Peñuelas J, Field CB (1992) A narrow-wave band spectral index that track diurnal changes in photosynthetic efficiency. Remote Sens Environ 41:35–44
Gao BC (1996) NDWI—a normalized difference water index for remote sensing of vegetation liquid water from space. Remote Sens Environ 58:257–266
Gates DM, Tantraporn W (1952) The reflectivity of deciduous trees and herbaceous plants in the infrared to 25 microns. Science 115:613–616
Gates DM, Keegan HJ, Schleter JC et al (1965) Spectral properties of plants. Appl Opt 4(1):11–20
Gautam RK, Panigrahi S (2007) Leaf nitrogen determination of corn plant using aerial images and artificial neural networks. Can Biosyst Eng 49:7.1–7.10
Girona J, Mata M, Fereres E et al (2002) Evapotranspiration and soil water dynamics of peach trees under water deficits. Agric Water Manage 54:107–122
Girona J, Mata M, Arbonès A et al (2003) Peach tree response to single and combined regulated deficit irrigation regimes under swallow soils. J Am Soc Hortic Sci 128(3):432–440
Gitelson A, Merzlyak MN (1994) Spectral reflectance changes associated with autumn senescence of *Aesculus hippocastanum* L. and *Acer platanoides* L. leaves. Spectral features and relation to chlorophyll estimation. J Plant Physiol 143:286–292
Gitelson AA, Zur Y, Chivkunova OB et al (2002) Assessing carotenoid content in plant leaves with reflectance spectroscopy. Photochem Photobiol 75:272–281
Goel NS, Thompson RL (2000) A snapshot of canopy reflectance models and a universal model for the radiation regime. Remote Sens Rev 18(2):197–225
Grant OM, Tronina L, Jones HG et al (2007) Exploring thermal imaging variables for the detection of stress responses in grapevine under different irrigation regimes. J Exp Bot 58:815–825
Guo J, Trotter CM (2004) Estimating photosynthetic light-use efficiency using the photochemical reflectance index: variations among species. Funct Plant Biol 31:255–565
Haboudane D, Miller JR, Tremblay N et al (2002) Integration of hyperspectral vegetation indices for prediction of crop chlorophyll content for application to precision agriculture. Remote Sens Environ 81(2–3):416–426
Hardisky MA, Klemas V, Smart RM (1983) The influences of soil salinity, growth form, and leaf moisture on the spectral reflectance of *Spartina alterniflora* canopies. Photogramm Eng Remote Sens 49:77–83
Hatfield JL, Gitelson AA, Schepers JS et al (2008) Application of spectral remote sensing for agronomic decisions. Agron J 100:117–131
Herwitz L, Johnson S, Dunagan R et al (2004) Imaging from an unmanned aerial vehicle: agricultural surveillance and decision support. Comput Electron Agric 44:49–61
Hipps L, Asrar G, Kanemasu E (1985) A theoretically-based normalization of environmental effects on foliage temperature. Agric For Meteorol 35:113–122
Horler DNH, Dockray M, Barber J (1983) Red edge measurements for remotely sensing plant chlorophyll content. Adv Space Res 3(2):273–277
Hsiao TC, Bradford KJ (1983) Physiological consequences of cellular water deficits. In: Taylor HM, Jordan WR, Sinclair TR (eds) Limitations to efficient water use in crop production. ASA/CSSA/SSSA, Madison, pp 227–265
Hsiao TC, Fereres E, Acevedo E et al (1976) Water stress and dynamics of growth and yield of crops. Water and plant life: problems and modern approaches. Springer, Berlin/Heidelberg/New York
Hunt ER Jr, Rock BN (1989) Detection of changes in leaf water content using near- and middle-infrared reflectances. Remote Sens Environ 30:43–54

Idso SB (1982a) Humidity measurement by infrared thermometry. Remote Sens Environ 12:87–91

Idso SB (1982b) Non-water-stressed baselines: a key to measuring and interpreting plant water stress. Agric Meteorol 27:59–70

Idso SB, Jackson RD, Reginato RJ (1978) Extending the "degree day" concept of phenomenological development to include water stress effects. Ecology 59:431–433

Idso SB, Jackson RD, Pinter PJ et al (1981) Normalizing the stress-degree-day parameter for environmental variability. Agric For Meteorol 24:45–55

Irmak A, Jones JW, Mavromatis T et al (2000) Evaluating methods for simulated soybean cultivar responses using cross validation. Agron J 92(6):1140–1149

Jackson RD (1982) Canopy temperature and crop water stress. Adv Irrig 1:43–85

Jackson RD (1986) Remote sensing of biotic and abiotic stress. Annu Rev Phytopathol 24:265–287

Jackson RD, Pinter PJ Jr (1981) Detection of water stress in wheat by measurement of reflected solar and emitted thermal IR radiation. In: Spectral signatures of objects in remote sensing. Institut National de la Reserche Agronomique, Versalles, pp 399–406

Jackson RD, Idso SB, Reginato RJ et al (1977) Crop temperature reveals stress. Crop Soils 29:10–13

Jackson RD, Idso SB, Reginato RJ et al (1981) Canopy temperature as a crop water-stress indicator. Water Resour Res 17:1133–1138

Jackson RD, Slater PN, Pinter PJ Jr (1983) Discrimination of growth and water stress in wheat by various vegetation indices through clear and turbid atmospheres. Remote Sens Environ 13:187–208

Jackson RD, Kustas WP, Choudhury BJ (1988) A reexamination of the crop water stress index. Irrig Sci 9:309–317

Jacob J, Lawlor DW (1991) Stomatal and mesophyll limitations of photosynthesis in phosphate deficient sunflower, maize and wheat plants. J Exp Bot 42:1003–1011

Jacquemoud S, Baret F (1990) PROSPECT: a model of leaf optical properties spectra. Remote Sens Environ 34:75–91

Jacquemoud S, Baret F, Andrieu B et al (1995) Extraction of vegetation biophysical parameters by inversion of the PROSPECT+SAIL model on sugar beet canopy reflectance data – application to TM and AVIRIS sensors. Remote Sens Environ 52:163–172

Jacquemoud S, Ustin SL, Verdebout J et al (1996) Estimating leaf biochemistry using the PROSPECT leaf optical properties model. Remote Sens Environ 56(3):194–202

Jones HG (1999) Use of infrared thermometry for estimation of stomatal conductance as a possible aid to irrigation scheduling. Agric For Meteorol 95:139–149

Jones HG, Stoll M, Santos T et al (2002) Use of infrared thermography for monitoring stomatal closure in the field: application to grapevine. J Exp Bot 53:2249–2260

Kempeneers P, Zarco-Tejada PJ, North PRJ et al (2008) Model inversion for chlorophyll estimation in open canopies from hyperspectral imagery. Int J Remote Sens 29(17–18):5093–5111

Knipling EB (1970) Physical and physiological basis for the reflectance of visible and near-infrared radiation from vegetation. Remote Sens Environ 1:155–159

Koetz B, Schaepman M, Morsdorf F et al (2004) Radiative transfer modeling within heterogeneous canopy for estimation of forest fire fuel properties. Remote Sens Environ 92:332–344

Kooistra L, Wamelink W, Schaepman-Strub G et al (2008) Assessing and predicting biodiversity in a floodplain ecosystem: assimilation of net primary production derived from imaging spectrometer data into a dynamic vegetation model. Remote Sens Environ 112:2118–2130

Krause GH, Weis E (1984) Chlorophyll fluorescence as a tool in plant physiology. II. Interpretation of fluorescence signals. Photosynth Res 5:139–157

Larcher W (1994) Photosynthesis as a tool for indicating temperature stress events. In: Schulze ED, Caldwell MM (eds) Ecophysiology of photosynthesis. Springer, Berlin, pp 261–277

Leinonen I, Jones HG (2004) Combining thermal and visible imagery for stimulating canopy temperature and identifying plant stress. J Exp Bot 55:1423–1431

Leinonen OM, Grant CPP, Tagliavia MM et al (2006) Estimating stomatal conductance with thermal imagery. Plant Cell Environ 29:1508–1518

Lhomme JP, Monteny B (2000) Theoretical relationship between stomatal resistance and surface temperatures in sparse vegetation. Agric For Meteorol 104:119–131

Lichtenthaler HK (1992) The Kautsky effect: 60 years of chlorophyll fluorescence induction kinetics. Photosynthetica 27:45–55

Lichtenthaler HK, Rinderle U (1988) The role of chlorophyll fluorescence in the detection of stress conditions in plants. CRC Crit Rev Anal Chem 19(suppl 1):529–585

Lichtenthaler HK, Wenzel O, Buschmann C et al (1998) Plant stress detection by reflectance and fluorescence. Ann N Y Acad Sci 851:271–285

Lillesand T, Kiefer R, Chipman J (2008) Remote sensing and image interpretation, 6th edn. Wiley, New York

Liu J, Judith AC, William BR et al (2005a) Comparison of surface radiative flux data sets over the Arctic ocean. J Geophys Res 110:C02015. doi:10.1029/2004JC002381

Liu L, Zhang Y, Wang J et al (2005b) Detecting solar-induced chlorophyll fluorescence from field radiance spectra based on the Fraunhofer line principle. IEEE Trans Geosci Remote 43:827–832

Lu P, Yunusa IAM, Walker RR et al (2003) Regulation of canopy conductance and transpiration and their modeling in irrigated grapevines. Funct Plant Biol 30:689–698

Marcassa LG, Gasparoto MCG, Belasque Junior J et al (2006) Fluorescence spectroscopy applied to orange trees. Laser Phys 16(5):884–888

Marsal J, Girona J (1997) Relationship between leaf water potential and gas exchange activity at different phonological stages and fruit loads in peach trees. J Am Soc Hortic Sci 122:415–421

Maxwell K, Johnson GN (2000) Chlorophyll fluorescence – a practical guide. J Exp Bot 51(345):659–668

McFarlane JC, Watson RD, Theisen AF et al (1980) Plant stress detection by remote measurement of fluorescence. Appl Opt 19:3287–3289

Meroni M, Rossini M, Picchi V et al (2008) Assessing steady-state fluorescence and PRI from hyperspectral proximal sensing as indicators of plant stress: the case of ozone exposure. Sensors 8:1740–1754

Mills TM, Behboudian MH, Tan PY (1994) Plant water status and fruit quality in "Braeburn" apples. Hortic Sci 29(11):1274–1278

Milton NM, Eiswerth BA, Ager CM (1991) Effect of phosphorous deficiency on spectral reflectance and morphology of soybean plants. Remote Sens Environ 36:121–127

Min M, Lee WS (2005) Determination of significant wavelengths and prediction of nitrogen content for citrus. Trans Am Soc Eng 48(2):455–461

Min M, Lee WS, Burks TF et al (2008) Design of a hyperspectral nitrogen sensing system for orange leaves. Comput Electron Agric 63:215–226

Möller M, Alchanatis V, Cohen Y et al (2007) Use of thermal and visible imagery for estimating crop water status of irrigated grapevine. J Exp Bot 58(4):827–838

Morales F, Abadía A, Abadía J (1991) Chlorophyll fluorescence and photon yield of oxygen evolution in iron-deficient sugar beet (*Beta vulgaris* L.). Plant Physiol 97:886–893

Morales F, Abadía A, Abadía J (1998) Photosynthesis quenching of chlorophyll fluorescence and thermal energy dissipation in iron-deficient sugar beet leaves. Aust J Plant Physiol 25:403–412

Morales F, Belkhodja R, Abadía A et al (2000) Photosystem II efficiency and mechanism of energy dissipation in iron-deficient, field grown pear trees (*Pyrus communis* L.). Photosynth Res 63:9–21

Morales F, Abadía A, Abadía J (2006) Photoinhibition and photoprotection under nutrient deficiencies, drought and salinity. In: Photoprotection, photoinhibition, gene regulation, and environment. Springer, Dordrecht, pp 65–85

Moran M, Clarke T, Inoue Y et al (1994) Estimating crop water deficit using the relation between surface–air temperature and spectral vegetation index. Remote Sens Environ 49:246–263

Moya I, Camenen L, Evain S et al (2004) A new instrument for passive remote sensing 1. Measurements of sunlight-induced chlorophyll fluorescence. Remote Sens Environ 91:186–197

Moya I, Daumard F, Moïse N et al (2006) First airborne multiwavelength passive chlorophyll fluorescence measurements over La Mancha (Spain) fields. In: Proceedings of the 2nd international symposium on recent advances in quantitative remote sensing, Torrent, Spain, pp 820–825

Nakaji T, Oguma H, Fujinuma Y (2006) Seasonal changes in the relationship between photochemical reflectance index and photosynthetic light use efficiency of Japanese larch needles. Int J Remote Sens 27(3):493–509

Nichol CJ, Huemmrich KF, Black TA et al (2000) Remote sensing of photosynthetic-light-use efficiency of boreal forest. Agric Meteorol 101:131–142

Nichol CJ, Lloyd J, Shibistova O et al (2002) Remote sensing of photosynthetic-light-use-efficiency of a Siberian boreal forest. Tellus B 54B:677–687

Nichol CJ, Rascher U, Matsubara S et al (2006) Assessing photosynthetic efficiency in an experimental mangrove canopy using remote sensing and chlorophyll fluorescence. Trees 20:9–15

Nilsson HE (1995) Remote sensing and image analysis in plant pathology. Can J Plant Pathol 17:154–166

Norman JM, Welles JM, Walter EA (1985) Contrasts among bidirectional reflectance of leaves, canopies, and soils. IEEE Trans Geosci Remote 23(5):659–667

Norman JM, Kustas WP, Humes KS (1995) Source approach for estimating soil and vegetation energy fluxes in observations of directional radiometric surface temperature. Agric For Meteorol 77:263–293

O'Neill AL, Kupiec JA, Curran PJ (2002) Biochemical and reflectance variation throughout a Sitka spruce canopy. Remote Sens Environ 80:134–142

Oguntunde PG, van de Giesen N, Savenije HHG (2007) Measurement and modelling of transpiration of a rain-fed citrus orchard under sub-humid tropical conditions. Agric Water Manage 87:200–208

Olioso A, Méthy M, Lacaze B (1992) Simulation of canopy fluorescence as a function of canopy structure and leaf fluorescence. Remote Sens Environ 41(2–3):239–247

Osborne SL, Schepers JS, Francis DD et al (2002) Detection of phosphorus and nitrogen deficiencies in corn using spectral radiance measurements. Agron J 94(6):1215–1221

Osborne SL, Schepers JS, Schlemmer MR (2004) Detecting nitrogen and phosphorus stress in corn using multi-spectral imagery. Commun Soil Sci Plant Anal 35(3–4):505–516

Papageorgiou G (1975) Chlorophyll fluorescence: an intrinsic probe of photosynthesis. In: Govindjee (ed) Bioenergetics of photosynthesis. Academic, New York, pp 319–371

Peguero-Pina JJ, Morales F, Flexas J et al (2008) Photochemistry, remotely sensed physiological reflectance index and de-epoxidation state of xanthophyll cycle in *Quercus coccifera* under intense drought. Oecologia 156(1):1–11

Peñuelas J, Filella I (1998) Visible and near-infrared reflectance techniques for diagnosing plant physiological status. Trends Plant Sci 3(4):151–156

Peñuelas J, Gamon JA, Fredeen AL et al (1994) Reflectance indices associated with physiological changes in nitrogen- and water-limited sunflower leaves. Remote Sens Environ 48:135–146

Peñuelas J, Baret F, Filella I (1995) Semi-empirical indices to assess carotenoids/chlorophyll a ratio from leaf spectral reflectance. Photosynthetica 31(2):221–230

Peñuelas J, Pinol J, Ogaya R et al (1997) Estimation of plant water content by the reflectance water index WI (R900/R970). Int J Remote Sens 18:2869–2875

Pérez-Priego O, Zarco-Tejada PJ, Sepulcre-Cantó G et al (2005) Detection of water stress in orchard trees with a high-resolution spectrometer through chlorophyll fluorescence *in-filling* of the O_2-A band. IEEE Trans Geosci Remote 43:2860–2869

Qin J, Burks TF, Ritenour MA et al (2009) Detection of citrus canker using hyperspectral reflectance imaging with spectral information divergence. J Food Eng 93:183–191

Reuther W, Embleton TW, Jones WW (1952) Mineral nutrition of tree crops. Ann Rev Plant Physiol 9:175–206

Rodriguez D, Fitzgerald GJ, Belford R et al (2006) Detection of nitrogen deficiency in wheat from spectral reflectance indices and basic crop eco-physiological concepts. Aust J Agric Res 57:781–789

Rosema A, Verhoef W, Schroote J et al (1991) Simulating fluorescence light-canopy interaction in support of laser-induced fluorescence measurements. Remote Sens Environ 37(2):117–130

Sankaran S, Mishra A, Ehsani R (2010) A review of advanced techniques for detecting plant diseases. Comput Electron Agric 72:1–13

Sari M, Sonmez NK, Karaca M (2005) Relationship between chlorophyll content and canopy reflectance in Washington navel orange trees (*Citrus sinensis* (L.) Osbeck). Pak J Bot 37(4):1093–1102

Schreiber U, Bilger W (1987) Rapid assessment of stress effects on plant leaves by chlorophyll fluorescence measurements. In: Plant response to stress. Springer, Berlin, pp 27–53

Schreiber U, Bilger W, Neubauer C (1994) Chlorophyll fluorescence as a non-intrusive indicator for rapid assessment of in vivo photosynthesis. In: Ecophysiology of photosynthesis ecological studies, vol 100. Springer, Berlin/Heidelberg/New York, pp 49–70

Sepaskhah AR, Kashefipour SM (1995) Evapotranspiration and crop coefficient of sweet lime under drip irrigation. Agric Water Manage 27(3–4):331–340

Sepulcre-Cantó G, Zarco-Tejada PJ, Jiménez-Muñoz JC et al (2006) Within-field thermal variability detection as function of water stress in *Olea europaea* L. orchards with high spatial remote sensing imagery. Agric Meteorol 136:31–44

Sepulcre-Cantó G, Zarco-Tejada PJ, Jiménez-Muñoz JC et al (2007) Monitoring yield and fruit quality parameters in open-canopy tree crops under water stress. Implications for ASTER. Remote Sens Environ 107:455–470

Sepulcre-Cantó G, Zarco-Tejada PJ, Sobrino JA et al (2009) Detecting water status in open canopies with thermal ASTER imagery and DART radiative transfer simulation. Agric Meteorol 149:962–975

Serrano L, Peñuelas J (2005) Assessing forest structure and function from spectral transmittance measurements: a case study in a Mediterranean holm oak forest. Tree Physiol 25:67–74

Serrano L, Peñuelas J, Ustin SL (2002) Remote sensing of nitrogen and lignin in Mediterranean vegetation from AVIRIS data: decomposing biochemical from structural signals. Remote Sens Environ 81:355–364

Sims DA, Gamon JA (2002) Relationships between leaf pigment content and spectral reflectance across a wide range of species, leaf structures and developmental stages. Remote Sens Environ 81:337–354

Sims DA, Luo H, Hastings S et al (2006) Parallel adjustment in vegetation greenness and ecosystem CO_2 exchange in response to drought in a Southern California chaparral ecosystem. Remote Sens Environ 103:289–303

Smith RCG (1988) Inferring stomatal resistance of sparse crops from infrared measurements of foliage temperature. Agric For Meteorol 42:183–198

Soukupová J, Cséfalvay L, Urban O et al (2008) Annual variation of the steady-state chlorophyll fluorescence emission of evergreen plants in temperate zone. Funct Plant Biol 35:63–76

Stagno F, Giuffrida A, Intrigliolo F (2011) Canopy temperature as an indicator of water status in citrus Trees. In: Proceedings of 6th IS on irrigation of horticultural crops. Acta Hort. 889, ISHS

Strachan IB, Pattey E, Boisvert JB (2002) Impact of nitrogen and environmental conditions on corn as detected by hyperspectral reflectance. Remote Sens Environ 80:213–224

Suárez L, Zarco-Tejada PJ, Sepulcre-Cantó G et al (2008) Assessing canopy PRI for water stress detection with diurnal airborne imagery. Remote Sens Environ 112:560–575

Suárez L, Zarco-Tejada PJ, Berni JAJ et al (2009) Modelling PRI for water stress detection using radiative transfer models. Remote Sens Environ 113:730–744

Suárez L, Zarco-Tejada PJ, González-Dugo V et al (2010a) Detecting water stress effects on fruit quality in orchards with time-series PRI airborne imagery. Remote Sens Environ 114:286–298

Suárez L, Zarco-Tejada PJ, Berni JAJ et al (2010b) The effect of spatial resolution in remote sensing of water stress using optical and thermal imagery. In: 3rd international symposium on recent advances in quantitative remote sensing, Valencia, Spain, pp 355–357. ISBN: 978–84–370–7952–3

Sugiura R, Noguchi N, Ishii K (2005) Remote-sensing technology for vegetation monitoring using an unmanned helicopter. Biosyst Eng 90:369–379

Teillet PM, Staenz K, Williams DJ (1997) Effects of spectral, spatial and radiometric characteristics on remote sensing vegetation indices of forested regions. Remote Sens Environ 61:139–149

Thenot F, Méthy M, Winkel T (2002) The photochemical reflectance index (PRI) as a water-stress index. Int J Remote Sens 23(23):5135–5139

Tilling AK, O'Leary G, Ferwerda JG et al (2006) Remote sensing to detect nitrogen and water stress in wheat. Ground breaking stuff. In: Proceedings of the 13th ASA conference, Perth, Australia

Trotter GM, Whitehead D, Pinkney EJ (2002) The photochemical reflectance index as a measure of photosynthetic light use efficiency for plants with varying foliar nitrogen contents. Int J Remote Sens 23(6):1207–1212

Veihmeyer FJ (1927) Some factors affecting the irrigation requirements of deciduous orchards. Hilgardia 2:125–284

Vogelmann TC (1993) Plant tissue optics. Ann Rev Plant Physiol Plant Mol Biol 44:231–251

Wanjura DF, Maas SC, Winslow DR et al (2004) Scanned and spot measured canopy temperatures of cotton and corn. Comput Electron Agric 44(1):33–48

Wright IR, Manzi AO, de Rocha HR (1995) Surface conductance of Amazonian pasture: model application and calibration for canopy climate. Agric For Meteorol 75:51–70

Yamada N, Fujimura S (1991) Nondestructive measurement of chlorophyll pigment content in plant leaves from three-color reflectance and transmittance. Appl Opt 30(27):3964–3973

Yaryura P, Cordon G, Leon M et al (2009) Effect of phosphorus deficiency on reflectance and chlorophyll fluorescence of cotyledons of oilseed rape (*Brassica napus* L.). J Agron Crop Sci 195(3): 186–196

Ye X, Sakai K, Ortega Garciano L et al (2006) Estimation of citrus yield from airborne hyperspectral images using a neural network model. Ecol Model 198:426–432

Zarco-Tejada PJ, Miller JR, Mohammed GH et al (2000a) Chlorophyll fluorescence effects on vegetation apparent reflectance: I. Leaf-level measurements and model simulation. Remote Sens Environ 74(3): 582–595

Zarco-Tejada PJ, Miller JR, Mohammed GH (2000b) Chlorophyll fluorescence effects on vegetation apparent reflectance: II. Laboratory and airborne canopy-level measurements with hyperspectral data. Remote Sens Environ 74(3):596–608

Zarco-Tejada PJ, Miller JR, Mohammed GH et al (2000c) Optical indices as bioindicators of forest condition from hyperspectral CASI data. In: Casanova (ed) Proceedings of the 19th EARSeL symposium in remote sensing in the 21st century: economic and environmental applications, Rotterdam, pp 517–522. ISBN 9058090965

Zarco-Tejada PJ, Miller JR, Mohammed GH et al (2001) Scaling-up and model inversion methods with narrow-band optical indices for chlorophyll content estimation in closed forest canopies with hyperspectral data. IEEE Trans Geosci Remote 39(7):1491–1507

Zarco-Tejada PJ, Rueda CA, Ustin SL (2003) Water content estimation in vegetation with MODIS reflectance data and model inversion methods. Remote Sens Environ 85(1):109–124

Zarco-Tejada PJ, Miller JR, Morales A et al (2004a) Hyperspectral indices and model simulation for chlorophyll estimation in open-canopy tree crops. Remote Sens Environ 90(4):463–476

Zarco-Tejada PJ, Miller JR, Harron J et al (2004b) Needle chlorophyll content estimation through model inversion using hyperspectral data from boreal conifer forest canopies. Remote Sens Environ 89(2):189–199

Zarco-Tejada PJ, Berjón A, López-Lozano R et al (2005) Assessing vineyard condition with hyperspectral indices: leaf and canopy reflectance simulation in a row-structured discontinuous canopy. Remote Sens Environ 99:271–287

Zarco-Tejada PJ, Berni JAJ, Suárez L et al (2009) Imaging chlorophyll fluorescence from an airborne narrow-band multispectral camera for vegetation stress detection. Remote Sens Environ 113:1262–1275

Soil Enzyme Activities and Soil Fertility Dynamics

11

Lijuan Yang, Yulong Zhang, and Fusheng Li

Abstract

Soil fertility dynamics have been an interesting area in unravelling the diverse complexities involved with dynamic soil-plant relationship. Soil enzymes have been proven as the sensor against minutest edaphic, microbiological and biochemical changes in addition to soil contamination, often called nutrient cycling either as a result of intrinsic transformation or changes triggered by soil amendments for pollutants. Activities pertaining to soil enzymes in response to nutrient source have ably discriminated the sustainable impact on soil fertility and added a new dimension in developing reproducible soil quality indicators. Studies on soil enzymes with reference to the changes in response to fertilizer application, ground-cover crops, tillage, sol microbiological biomass, soil physico-chemical properties, air drying/storage, soil moisture fluctuations, soil compaction and elevated CO_2 have been thoroughly reviewed with emphasis on soil fertility dynamics.

Keywords

Soil enzyme • Fertility dynamics • Soil microbial biomass • Soil quality indicator • Environmental stress • Management practice

11.1 Introduction

Soil enzymes are biological catalysts of specific reactions depending upon a variety of factors such as pH, temperature and the presence (or absence) of inhibitors (Burns 1978, 1982). Other factors inclusive of climate, type of amendment, cultivation techniques, crop type and edaphic properties also affect enzyme catalyzed reactions. Soil enzymes are predominantly of microbial origin and are closely related to microbial abundance and or activity. They play an essential role in catalyzing reactions necessary for organic matter decomposition and nutrient cycling (Table 11.1). Activities of soil enzymes are greatly affected by organic matter content of soil (Dalal 1975) and often are used as indices of microbial activity and soil fertility (Kumar Jha et al. 1992).

Enzymes in soil may be extra- or intracellular. Extracellular enzymes are necessary for the breakdown of organic macromolecules, like cellulose, hemicelluloses or lignin, whereas intracellular enzymes are responsible for the breakdown of smaller molecules like sugars or amino acids. Soil enzyme activities (1) are often closely related to soil organic matter, soil physical properties and microbial activities or microbial biomass; (2) change much sooner than other parameters, thus provide early indications of the changes in soil health; and (3) involve simple procedures (Dick et al. 1996). In addition, soil enzyme activities can be used as effective measures of microbial activity, soil productivity, salinity and inhibiting effects of pollutants (Thanana and Nawar 1994; Tate 1995). Nannipieri et al. (1990) pointed out that enzymatic activities are substrate specific and related to specific reactions. For this reason, it is difficult to obtain an overall picture of soil fertility status from

L. Yang (✉) • Y. Zhang
Land and Environment College, Shenyang Agricultural University, Shenyang, Liaoning, China
e-mail: yljsau@126.com; ylzsau@163.com

F. Li
College of Agriculture, Guangxi University, Nanning, Guangxi 530005, China
e-mail: lpfu6@163.com

A.K. Srivastava (ed.), *Advances in Citrus Nutrition*,
DOI 10.1007/978-94-007-4171-3_11,

Table 11.1 Important soil enzymes and their functions

Enzyme	Substrate	Enzyme reaction	Soil function	Significance
Amidase	Carbon and nitrogen compounds	N-mineralization	Nutrient cycling	Plant available NH_4^+
β-glucosidase	Carbon compounds	Cellulose hydrolysis	Organic matter decomposition	Energy for microorganisms
FDA hydrolysis	Organic matter	Carbon and various nutrients	Organic matter decomposition and nutrient cycling	Energy and nutrients for microorganisms as measure of microbial biomass
Phosphatase	Phosphorous	Release of PO_4^{3+}	Nutrient cycling	Plant available P
Sulphatase	Sulphur	Release of SO_4^{2-}	Nutrient cycling	Plant available S
Urease	Nitrogen	Release of NH_3 and CO_2	Nutrient cycling	Plant available NH_4^+

Parts adapted from Bandick and Dick (1999), Baldrian (2009), Dick et al. (1994),Dick and Tabatabai (1994), Dick et al. (2000), Dodgson et al. (1982), and Kertesz and Mirleau (2004)

one enzymatic activity value. The simultaneous measurement of various enzymes, on the other hand, seems to be useful to evaluate soil biochemical activity and the process related to soil fertility dynamics (Pascual et al. 1998). Biochemical tools that allowed rapid measurement of soil enzyme activities made soil enzymology fashionable in the late 1960s, and such tools remained widely used for 20 years (Insam 2001).

Some enzymes in the soil only facilitate the breakdown of organic matter, e.g. hydrolase and glucosidase, while others are involved in nutrient mineralization, e.g. amidase, urease, phosphatase and sulphatase. In soil exposed to degradative processes, its biological properties are affected firstly, and so productive capacity diminishes (Garćia et al. 2000). Early indicators of ecosystem stress may function as the sensors, whose perturbation may provide a sensitive warning of soil degradation (Dick 1994). Soil enzyme activities are useful candidate sensors since they integrate information about the microbial status and soil physico-chemical conditions (Aon et al. 2001). Soil enzyme production as result of microbial metabolism is a sensitive indicator of soil microbial activity (Aon and Colaneri 2001).

Soil enzymes can be broadly divided into two groups: (1) hydrolytic enzymes responsible for the acquisition of C, N and P to support primary metabolism and (2) complex compounds like lignin in co-metabolic acquisition of nutrients. According to the result of García-Ruiz et al. (2009), the seasonal variability of individual soil enzymes ranged from 29% to 71% without a consistent temporal trend. Management system explained, on average, a maximum of 26.3% of the variability found for soil enzymes. The results further revealed the need for extensive comparative assessments to draw clear conclusions on the improvement of soil under sustainable management practices (García-Ruiz et al. 2009) in addition to soil fertility changes (Chen et al. 2003).

11.2 Soil Enzymes and Soil Fertility Relation

Measurements of several enzymatic activities have been used to establish the indices of soil biological fertility (Dick and Tabatabai 1992). Soil enzymes are frequently linked with fertility dynamics because of their utmost sensitivity to management practices, although they undergo distinct changes long before any detectable changes in soil quality indicators. Studies by Srivastava et al. (2010) demonstrated a strong sensitivity of soil enzymes in dictating fruit yield variation through the multiple-tier system of soil fertility evaluation in relation to fruit yield of citrus orchards in central India (Table 11.2). The fruit yield displayed significant correlations with available N ($r=0.532, p=0.01$), P ($r=0.412$, $p=0.01$), K ($r=0.389$, $p=0.05$), Fe ($r=0.508$, $p=0.01$), Mn ($r=0.489$, $p=0.01$), and Zn ($r=0.532$, $p=0.01$). While bacterial count and fungal count as the parameters of soil microbial population correlated more significantly with fruit yield ($r=0.561$ and $r=0.612$, $p=0.01$). These correlation values, however, further improved with finer parameters like microbial biomass nutrients such as C_{mic} ($r=0.582, p=0.01$), N_{mic} ($r=0.692$, $p=0.01$) and P_{mic} ($r=0.698$, $p=0.01$), suggesting that microbial biomass nutrients are more sensitive indicators than soil microbial population. Likewise soil microbial population proved more sensitive indicator than available nutrients in evaluating the soil fertility. But soil enzymes reflected the highest degree of correlationship with fruit yield (Table 11.2), thereby warranting much better effectiveness of soil enzymes over rest of the other soil fertility indices.

11.3 Profile of Soil Enzyme Functions

Studies of enzyme activities provide information on the biochemical processes occurring in soil. There is growing evidence that soil biological parameters are undoubtedly potential and sensitive indicators of soil ecological stress or restoration. Soil enzymes regulate ecosystem functioning and in particular, play a key role in identifying nutrient mining (Joachim and Patrick 2008). All soil contains a group of enzymes that determine soil metabolic processes which in turn, depend on its physical, chemical, microbiological and biochemical properties. Enzyme tests are most frequently used today, and these are urease (Tabatabai and Bremner 1972), phosphatase (Hoffmann 1967) and dehydrogenase

Table 11.2 Relationship of fruit yield with different soil fertility indices in citrus orchards of central India

	Soil available nutrients						
	N	P	K	Fe	Mn	Cu	Zn
Yield	0.532 ($p=0.01$)	0.412 ($p=0.01$)	0.389 ($p=0.05$)	0.508 ($p=0.01$)	0.489 ($p=0.01$)	0.312 (NS)	0.532 ($p=0.01$)
	Soil microbial population						
	Bacterial count		Fungal count				
Yield	0.561 ($p=0.01$)		0.612 ($p=0.01$)				
	Microbial biomass nutrients						
	C_{mic}	N_{mic}	P_{mic}				
Yield	0.582 ($p=0.01$)	0.692 ($p=0.01$)	0.698 ($p=0.01$)				
	Soil enzymes						
	Urease	Alkaline phosphatase	Dehydrogenase				
Yield	0.712 ($p=0.01$)	0.782 ($p=0.01$)	0.799 ($p=0.01$)				

Adapted from Srivastava et al. (2010)
C_{mic}, N_{mic} and P_{mic} stand for microbial biomass C, microbial biomass N and microbial biomass P, respectively
NS non-significant
($p=0.01$) and ($p=0.05$) stand for significance at 1% and 5%, respectively

(Trevors 1984), indicating the wider acceptance of the methods in the scientific community. The enzyme level in soil system varies in amounts primarily due to the difference in nature and properties of soils. For example, soil peroxidase activity was higher in rhizosphere of any mycorrhized citrus and vice-versa with respect to catalase enzyme irrespective of whether or not, soil was under water stress (Wu et al. 2008). Major enzyme activities naturally in soil are described below:

- Amylase is a starch hydrolyzing enzyme. It is known to be constituted by α-amylase and β-amylase. Research evidences suggest that α-amylase converts starch-like substrates to glucose and/or oligosaccharides, and β-amylase converts starch into maltose (Pazur 1965).
- Urease is involved in the hydrolysis of urea to carbon dioxide and ammonia, which can be assimilated by microbes and plants. It acts on carbon-nitrogen (C–N) bonds other than peptide linkage (Bremmer and Mulvaney 1978; Karaca et al. 2002).
- Arylsulphatase is the enzyme involved in the hydrolysis of arylsulphate esters by fission of the oxygen-sulphur (O–S). This enzyme is believed to be involved in the mineralization of ester sulphate in soils (Tabatabai 1994). Also, it may be an indirect indicator of fungi as only fungi (not bacteria) contain ester sulphate, the substrate of arylsulphatase (Bandick and Dick 1999; Kertesz and Mirleau 2004).
- β-glucosidases are widely distributed in nature, and their hydrolysis products as low-molecular-weight sugars are important source of energy for soil microorganisms. ß-glucosidase catalyzes the hydrolysis of ß-D-glucopyranoside and is one of the three or more enzymes involved in the saccharification of cellulose (Bandick and Dick 1999; Turner et al. 2002). These enzymes play an important role in soils because it is involved in catalyzing in the hydrolysis and biodegradation of various β-glucosides present in plant debris decomposing in ecosystem (Ajwa and Tabatabai 1994) with glucose as final product. Acosta-Martinez and Tabatabai (2000) reported β-glucosidase as sensitive to pH changes, which could be effectively used as good biochemical indicator for measuring ecological changes resulting from soil acidification in situation involving activities of this enzyme.
- Cellulase is an important enzyme considering the fact that 50% of the biomass synthesized by photosynthetic fixation of CO_2 (Eriksson et al. 1990). This enzyme is responsible for degrading cellulose into glucose, cellubiose and high molecular weight oligosaccharides (White 1982). It is accepted system comprises of three major types of enzymes, viz. endo-1(attacks cellulose chain at random), 4-β-glucanase (removes glucose or cellobiose from non-reducing end of cellulose chains) and β-o-glucanase (hydrolyzes cellobiose and other water soluble cellodextrins to glucose), although none of the hypotheses describing the mechanisms involved in the degradation of cellulose by cellulase is universally accepted out of so many hypotheses proposed from time to time.
- Dehydrogenase enzyme is known to oxidize soil organic matter by transforming protons and electrons from substrates to acceptors. These processes are part of respiration pathways of microorganisms and closely related to soil-air-water relations (Kandeler 1996). Studies by Brzezinska et al. (1998) suggested that soil water content and temperature influence the dehydrogenase activity indirectly affecting soil redox status. Additionally, dehydrogenase enzyme is often used as a measure of any disruption caused by pesticides, trace elements or management practices (Reddy and Faza 1989; Garcia and Hernandez 1997).

- Phosphatase is an enzyme of great agronomic value because it hydroles compounds of organic phosphorus and transforms them into different forms of inorganic phosphorus, which are assimilable by plants (Amador et al. 1997). Variation in phosphatase activity apart from the changes in the quantity and quality of a soil's phosphorated substrates is also a good indicator of its biological state (Pascual et al. 1998, 2002). Phosphatases are broad group of enzymes that are capable of catalyzing hydrolysis of esters and anhydrides of phosphoric acid playing critical role in soil cycles. Plants have developed many morphological and metabolic adaptations to tolerate low phosphatase availability in soil. Under P-stress in soil, acid phosphatase secretion by plant roots is increased to enhance solubilization and remobilization of phosphate, influencing the ability of plant to cope up with soil P-stress (Versaw and Harrison 2002).

11.4 Soil Microorganisms and Enzymatic Changes

Soil microorganisms and soil enzymes not only play an active role in influencing soil fertility as a result of their involvement in the cycle of nutrients like carbon and nitrogen, which are required for plant growth, but also are sensitive biological indicators for soil quality evaluation besides sensitively reflecting minute changes in soil environment. Microbial biomass (MB) is labile portion of the organic fraction of soils and serves as both an important source and sink for available nutrients (Garćia and Rice 1994). The chemical quantity of soil C pool reflects specific changes the land use in microbial community composition having far-reaching consequences on soil carbon storage and cycling. Soil microbial biomass carbon (C_{mic}), soil microbial biomass nitrogen (N_{mic}) and soil microbial biomass phosphorous (P_{mic}) are recycling and storing pools of soil C, N and P, respectively (Insam et al. 1991). Soil C_{mic} comprises 1–9% of soil organic C (Anderson and Domsch 1989), but due to its fast turnover (it is estimated to be 1–3 years), the microbial biomass pool plays a key role in controlling the nutrient cycling and energy flow in soil ecosystems (He et al. 1997). However, it is often questioned on the basis of the fact that soil enzyme activity provides a functional measure, which can vary independently of microbial structure (Waldrop et al. 2004).

Microbial biomass and enzyme divertase affected by management practices can be used as sensitive indicators of ecological stability (Ajwa and Tabatabai 1994). Soil enzymes and microbiological biomass are closely related because of the transformations of the important organic elements occur through microorganism (Hu and Cao 2007). These studies ably demonstrated the existence of strong correlation amongst soil enzyme activities, microbial biomass and various soil properties.

Many studies have reported significant correlations among soil enzyme activities, microbial biomass and various soil properties (Frankenberger and Dick 1983; Perucci et al. 1984; Dick et al. 1988a, b). Dick et al. (1988a) found a strong correlation between dehydrogenase activity and microbial biomass C (MBC). Frankenberger and Dick (1983) found significant correlation between MBC and the activities of acid and alkaline phosphatases and urease. In a long-term study on the effects of residue management on enzyme activities, Dick et al. (1988b) found significant correlation between burning effect of the residue and acid phosphatase activity at the top 20 cm of soils but weak correlations with several other soil enzymes. Long-term N fertilization significantly increased activities of ß-glucosidase and acid phosphatase but decreased urease activity. They further found that several soil enzyme activities can be used as indicators of ecological changes caused by N fertilization and long-term burning management practices. The relevance of these changes in surface soil to the long-term sustainability of this ecosystem needs further evaluation (Ajwa et al. 1999).

A major weakness of enzyme tests is that the actual microbial activity of a soil is not well reflected. Moreover, the tests show historic features of enzymes bound to soil organic matter or clay minerals. Therefore, Visser and Parkinson (1992) disputed the suitability of enzyme assays for microbial activity and soil quality assessments, with the exception of dehydrogenase because its biological properties make it unlikely to be present in soil in an extracellular state (Skujins 1978). To overcome the interpretation problems of single enzyme test, Beck (1984) proposed a soil microbiological index calculated from microbial biomass, reductase and hydrolase activities. This index, however, never became popular. In another attempt to propose such an index, Trasar-Cepeda et al. (1998) found a very close relation of total N with a linear combination of soil microbial biomass C, mineralized N, phosphomonoesterase, β-glucosidase and urease activity. This test set was proposed as a soil quality index closely related to soil sustainability. The problem with both Beck (1984) and Trasar-Cepeda et al. (1998) approaches is that the usefulness of the enzyme tests used as components of their indices is often disputed, and universally accepted enzyme tests are still lacking.

11.5 Soil Enzymes and Soil Properties

All soils contain a group of enzymes that determine soil metabolic processes (McLaren 1975) which in turn, will depend on its physical, chemical, microbiological and biochemical properties. The enzyme levels in soil system vary in amounts

Table 11.3 Influence of different soil properties on soil enzymes activity

Soil properties	Function
Soil clays	Most activity remains associated with clays, increases resistance to proteolysis and increases the temperature of inactivation
Soil organic matter	Provides stability to soil nitrogen compounds, exhibits pH and temperature changes and has inability to purify enzyme free of organic matters
Clay-organic matter complex	Lignin and bentonite protect enzymes against proteolytic track, and enzymes are first bound to organic matter, then to clays

primarily due to the fact that each soil type have different amounts of organic matter, composition and activity of its living organisms and intensity of biological processes. Soil enzymes, which are the main ingredients of the soil ecosystem, reflect the ability of nutrient to transform in soil. Therefore, soil enzymes, as an index of soil quality, can reflect changes in soil quality caused by time or other conditions (Yang and Wang 2002; He et al. 2002; Cao et al. 2003). Enzyme activities in the soil are not only closely related to the factors such as soil type, soil structure, organic matter and pH but also to the kinds of crops which are grown (Dick 1994; Doran and Parkin 1994). Any change in these parameters may cause consequent changes in soil fertility activities.

There are two kinds of soil enzymes, viz. constitutive (the enzymes that always remain present in constant amount in a cell, not affected by addition of any particular substrate since their functions are expressed by genes, e.g. pyrophosphate) and inducible (the enzyme that remain present in trace amounts, but quickly increases in concentrations when substrate is present, e.g. amidase). Influence of different soil properties on enzymatic functions can be summarized (Table 11.3) as below:

Many investigations have suggested that it is possible to predict the soil fertility response to soil enzyme activity (Bandick and Dick 1999; Deng and Tabatabai 1997; Doran and Parkin 1994). Most recent research has shown that the soil enzymes ß-D-glucosidase, urease, acid and alkaline phosphatase and arylsulphatase were the key enzymes involved in the mineralization of organic C, N, P and S in the soil. Fluorescein diacetate (FDA) hydrolysis by proteases, lipases and esterases are used to determine amounts of active fungi and bacteria (Schnürer and Rosswall 1982; Dick et al. 1996). According to Cookson (2002), 80% of the variation in the activities of alkaline phosphomonoesterase and phosphodiesterase was accounted for the differences in soil physicochemical properties besides inherent P level of soil.

The activities of invertase and acid phosphatase, MB, C and N, and C_{mic}/N_{mic} (MBC)/(MBN) were significantly correlated with latitude ($P<0.05$, $r^2=0.198$, 0.635, 0.558, 0.211 and 0.317, respectively), that is, increasing with the latitude. Significant positive correlations ($P<0.05$) were observed between invertase activity and the total N and available P and between acid phosphatase activity and the total C, C/N, available N, total P and available P. The urease, acid phosphatase and dehydrogenase activities were significantly correlated ($P<0.05$) with the soil pH and electrical conductivity (EC). C_{mic} and N_{mic} were positively correlated ($P<0.05$) with the total C, C/N and available P. The C_{mic}/N_{mic} ratio was positively correlated with the total C, total N, C/N and available N ($P<0.05$). The spatial distribution of soil enzyme activities and microbial biomass resulted from the changes in soil properties such as soil organic matter, soil pH and EC, partially owing to variations in temperature and rainfall along the latitudinal gradient (Liu et al. 2008). Understanding of soil enzymes activities is a critical factor in assuring that soil remains healthy for an integrated biological assessment of soils due to their crucial role in several biological activities, their ease of measurement and their rapid response to changes in soil management (Srivastava and Ngullie 2009).

11.5.1 Soil Enzymes as Soil Quality Indicator

Physical and chemical properties have been extensively used to measure soil quality (Parrr and Papendick 1997). However, these properties usually change on timescale (decades), which is too long for management purposes. In contrast, soil properties based on biological and biochemical activities, such as soil enzymes and the nematode community, have been shown to respond to small changes in soil conditions, thus providing information sensitive to subtle alterations of soil quality (Pascual et al. 2000). Therefore, soil enzyme activities have been suggested as suitable indicators of soil quality because of their intimate relationship with soil biology, ease of measurement and rapid response to change in soil management (Srivastava 2009).

The nitrogen mineralization capacity refers to the capability of soils to transform organic nitrogen compounds into ammonium/nitrate under optimum moisture and temperature conditions over a given period of time. This determination has been used to assess the influence of soil management on soil quality (Gil-Sotres et al. 2005). Among the hydrolases, acid phosphomonoesterase activity has been the most frequently used for estimating changes in soil quality due to either management of the presence of contaminants. It is a good index of the quality and quantity of organic matter in the soil (Bergstrom et al. 2000) and can be very high in arable soils as long as the levels of organic matter in the soil are maintained (Dick et al. 1994). The dehydrogenase activity of the soil is considered to be an indicator of the microbial redox system and of the oxidative activities of the soil

(Trevors 1984). Dehydrogenase activity has also been used to evaluate the degree of recovery of degraded soils, being considered as a good indicator, even in the soils that have been contaminated by petroleum (Margesin et al. 2000).

11.5.2 Enzymatic Changes Due to Storage

Soil properties are susceptible to the change under different sample storage and pre-treatment regimes, and as a consequence, biochemical parameters are usually determined in fresh samples, kept cold or frozen, in order to reduce possible alternations (Trasar-Cepeda et al. 2000). Nevertheless, air-dried soil and short incubations facilitate routine soil testing procedures and adopting a technique that uses dried soils may significantly reduce variability within the same soil samples and reduce the amount of refrigeration space necessary for storage of moist soil (Haney et al. 2004). Hence, air-drying greatly facilitates soil sample processing and the use of this pre-treatment encourages the adoption of biochemical parameters as part of a soil quality index (Bandick and Dick 1999). Li and Sarah (2003) examined the effect of air-drying on some enzyme activities along a climatic transect from the Judean Mountains in the east of Israel and found no differences related to field-moist samples in extracellular arylsulphatase, alkali phosphatase and acid phosphatase.

Bandick and Dick (1999) observed that the enzyme assays for α- and β-glucosidase, amidase, arylsulphatase and urease consistently had the same ranking of treatments between field-moist and air-dried samples in soil plots comparing different crops, agricultural practices and nearby pasture grass in USA. Chen (2003) used air-dried soils in determinations of soil phosphatase activity in Chinese fir plantation ecosystems of rhizosphere and bulk soil and found differences depending on sampling distance from tree stem and soil profile. In contrast, Speir and Ross (1975) found that phosphatase activity measured in air-dried soil in New Zealand pasture declined considerably related to field activity. In addition, Sparling et al. (1986) estimated phosphatase and phosphodiesterase activities before and after air-drying in samples from grassland soils from New Zealand and observed a decline in their activity with air-drying. Trasar-Cepeda et al. (2000), who assessed the best storage procedure for soils from NW of Spain taken from the organic soil material and Ah horizon, concluded that air-drying was the procedure that caused the greatest alteration in BBA-protease, urease, invertase and β-glucosidase activities. Short incubations of rewetted soil samples can produce fluctuations in these enzyme activities, mainly of acid phosphatase, and estimations in these conditions are not so representative of field-moist soil values (Zornoza et al. 2006).

11.5.3 Soil Moisture Changes and Enzymatic Activities

A greater degree of nutrient limitation to plant growth can further slow the regeneration process of soil enzymes under soil moisture stress conditions. Sardans and Peñuelas (2005) conducted an experiment of water availability manipulation to study the effects of enhanced drought on the activity of five soil enzymes. The drought treatment consisted of runoff exclusion by a ditch along the entire top edge of the upper part of treatment plots and partial rain exclusion by suspending PVC strips and funnels. A reduction of 10% of soil moisture produced by runoff exclusion decreased urease activity by 10–67%, protease activity by 15–66% and ß-glucosidase activity by 10–80%, depending on annual period and soil depth. The reduction of 21% of soil moisture produced by runoff and rainfall exclusion together reduced urease activity by 42–60%, protease activity by 35–45%, β-glucosidase activity by 35–83% and acid phosphatase activity by 31–40%. No significant effects were observed on alkaline phosphatase activity. The activities of the enzymes involved in the nitrogen cycle, protease and urease were the most affected by drought. These results show the link between drought and a slower nutrient turnover, which decreases the nutrient supply to plants (Sardans and Peñuelas 2005).

Flooding is one of the most important parameters; little attention has been paid to its effect on soil enzyme activities, especially in paddy soils. Pulford and Tabatabai (1988) showed that short-term waterlogging (7 days) inhibited the activities of phosphatase, urease, β-D-glucosidase, arylsulphatase and amidase in ten Iowa soils in the USA and that the inhibition effect was dependent on the soil. According to Wang and Lu (2006), compared with air-dried soil, waterlogging resulted in a significant decrease ($P \leq 0.05$) of FDA and β-D-glucosidase activities, and this effect was enhanced with increasing waterlogging time. Waterlogging also significantly inhibited ($P \leq 0.05$) arylsulphatase as well as alkaline and acid phosphatase activities but did not decrease the activities with the increase in waterlogging time. Short-term waterlogging did not affect urease activity, but prolonged waterlogging decreased it markedly. In contrast, the aerobic incubation significantly increased ($P \leq 0.05$) FDA and alkaline phosphatase increased with the increase of incubation time, whereas β-D-glucosidase activities did not increase with the increase of incubation time. With aerobic treatments, the activities of FDA and alkaline phosphatase increased with incubation time, whereas β-D-glucosidase activity decreased. A significant difference ($P \leq 0.05$) was usually observed between the treatments involving different levels of waterlogging for the activities of FDA as well as alkaline and acid phosphatase; however, β-D-glucosidase and urease were usually not significant ($P \leq 0.05$). No activity differences were observed between waterlogging and aerobic incubation for arylsulphatase and urease (Wang and Lu 2006).

11.5.4 Soil Compaction and Enzymatic Activities

Soil compaction, where the soil pore volume is reduced and pore size distribution may be changed, is one of the key soil disturbances (Powers et al. 1990). Impacts of intensive cultivation operations include soil compaction, surface soil structure degradation, decrease in nutrient availability due to biomass removals or erosion and increased sediment yield as a result of erosion (Moffat 1991). Moehring and Rawls (1970) reported that compaction reduced the survival of trees in replanted areas by as much as 57%. When planted in compacted soils, young seedlings are more susceptible to water stress and N deficiencies (Reisinger et al. 1988).

Both van der Linden et al. (1989) and Breland and Hansen (1996) found that soil compaction reduced N mineralization and N availability. Because of the possible creation of anaerobiosis under compacted soil condition, denitrification is generally thought to be the likely process of N loss from compacted soil conditions. The effect of soil compaction on the microbial activity varies. Soil compaction is generally thought to decrease microbial-activity-like soil respiration or enzyme activity, due to the air-filled pore spaces, but the results of the few studies conducted varies (Wronski and Murphy 1994; Startsev et al. 1998). Dick et al. (1988b) showed that soil phosphatase activity was reduced in compacted forest sites in Oregon. Other microbial properties like soil microbial biomass carbon were unaffected or decreased depending on environmental conditions. Soil compaction significantly affected microbial activity by reducing acid phosphatase. Severe soil compaction decreased young tree growth and reduced N fertilizer uptake (Jordan et al. 2003).

Soil enzymes, hence, possess undoubtedly clear-cut edge over other indices used in soil fertility evaluation. The role of soil enzymes in regulating soil fertility dynamics could be promisingly foreseen in developing soil quality indices for sustainable maintenance of soil health through optimized nutrient supply within crop rhizosphere.

11.6 Environmental Stress and Soil Enzymes

It has been predicted that elevated atmospheric CO_2 will increase enzyme activity as a result of CO_2-induced carbon entering the soil. Elevated atmospheric CO_2 directly affects plant photosynthesis (Vu et al. 1997). There are three plant-mediated mechanism by which elevated atmospheric CO_2 might influence soil microbial communities: (1) elevated atmospheric CO_2 stimulates plant photosynthesis, and consequently, net primary productivity increases, thus photosynthate below ground increases; (2) chemistry of green leaves is altered; generally, the carbon to nitrogen ratio increases in green leaf tissue; and (3) elevated atmospheric CO_2 reduces the stomatal conductance of plants which results in higher water use efficiency. At a plant community level, this often results in decreased stand transpiration and higher soil water content (Ebersberger et al. 2003). As a result, effects of elevated atmospheric CO_2 on plant metabolism and soil microorganisms would cause the alteration of soil enzyme, which plays a key role in nutrient cycling and turnover in the ecosystem.

Rosenzweig et al. (1995) investigated the potential impacts of global climate change on fruit yield in the USA were investigated through simulations of citrus. Simulated treatments included combinations of three increased temperature regimes (+1.5°C, +2.5°C and +5.0°C), and three levels of atmospheric carbon dioxide (440, 530 and 600 ppm) in addition to control runs representing current climatic conditions. Downton et al. (1987) recorded the response to elevated CO_2 of 3-year-old fruiting Valencia orange scions (*Citrus sinensis* (L.) Osbeck) on citrange rootstock (*C. sinensis* × *Poncirus trifoliata* (L.) Raf.). Fruit yield from the CO_2-enriched trees did not differ from the controls besides soluble solids content, dry weight, seed number or rind thickness. The progression of fruit coloration was more rapid for the CO_2-enriched trees. These results indicated that fruit yield will increase as global levels of CO_2-continue to rise, at least in those species that experience source limitation during fruit development.

11.6.1 Elevated CO_2 and Soil Enzymes

Atmospheric carbon dioxide concentrations increased dramatically during the last century and are expected to double by the end of this century mainly due to fossil fuel burning, deforestation and agricultural activity (IPCC 1996). Many studies have addressed the possible effects of elevated CO_2 on terrestrial ecosystems (Körner and Arnone 1992). A fertilizing effect of elevated CO_2 causing simulated primary productivity has been widely noted, at least in the short term (Hunt et al. 1991). However, recent studies have indicated that other factors (e.g. nutrient or water availability) may constrain the CO_2 fertilizing effect and hence lead to a negative feedback to the stimulated primary (Baxter et al. 1994). The main logic of these arguments lies in the below-ground processes, and whether enhanced primary productivity would increase or decrease the available nutrient pool in the soil.

Zak et al. (1993) have proposed that increased exudates from vegetation will enhance microbial mineralization activities and hence increase the nutrient pool (a positive feedback). In contrast, Diaz et al. (1993) have suggested that the increased microbial biomass would cause higher microbial nutrient uptake and decrease available nutrients to plants (a negative feedback). Thus, enzyme activities have been reported to represent general microbial activities as well as mineralization rates of specific compounds (Sinsabaugh et al. 1991). However, no significant differences were found in CH_4 emission or soil enzyme activities (β-glucosidase, phosphatase, and *N*-acetylglucosaminidase) in the bulk soil. Overall, the results suggested that the elevated CO_2 would increase the primary productivity of the fen vegetation and stimulate N_2O and CO_2 emissions as consequence of an enhanced DOC supply from the vegetation to the soil microbes (Kang et al. 2001). An elevated atmospheric CO_2 increased soil enzyme activity, which may be attributed to the following two factors: (1) elevated atmospheric CO_2 led to more plant biomass in the soil, which in turn stimulated soil microbial biomass and activity, and (2) elevated atmospheric CO_2 increased plant photosynthesis, thereby increasing plant-derived soil enzymes (Yuan et al. 2006).

11.6.2 Elevated CO_2 and Soil Fertility Changes

Soil fertility depletion is the fundamental biophysical cause of declining productivity (Pawlson 2005). Citrus-based farming system has high potential for sequestering carbon for mitigation of climate change (Srivastava and Kohli 1999). The perennial nature of citrus trees act as carbon sinks by sequestering the atmospheric carbon. Modifying fertilizer application to enhance nutrient availability, use of soil amendments to improve soil fertility providing irrigation at critical growth stages and conservation of soil moisture are important interventions (Srivastava and Singh 2009). Considerable studies have gone into questions of just how climate change will affect crop and soil productivity. However, the problem of predicting the future course of citrus cultivation in a changing world is confounded by the fundamental complexity arising on account of interplay of set of dynamic factors specific to each citrus belt (Srivastava and Singh 2008). Experimental studies of the long-term effects of CO_2 in more realistic field settings have not yet been done on a comprehensive scale (Rosenzweig and Hillel 1993).

The sensitivity of soil carbon to warming is a major uncertainty to the projections of carbon dioxide and climate. Experimental studies indicate increased soil organic carbon (SOC) decomposition at higher temperatures, resulting in increased CO_2 emissions from soils. But recent evidences favour against this theory. The initially increased CO_2 efflux returns to the pre-warming rates within 1–3 years, and turnover times of apparent carbon pool are insensitive to temperature with non-labile SOC showing more sensitivity to temperature than labile SOC (Knorr et al. 2005). Significantly more carbon is stored in the world's soils than in present in atmosphere. Disagreements exist regarding the effects of climate change on global soil carbon stocks. Despite much research, a consensus has not yet emerged on the temperature sensitivity of soil carbon decomposition due to differential kinetic properties of diverse soil organic compounds (Davidson and Janssens 2006)

It is contemplated that global warming may significantly alter soil composition at the molecular level and that such changes could have a major impact on atmospheric levels of CO_2. Global warming may change present-day decomposition patterns by altering the soil microbial communities and activities, thus changing the overall flow of carbon into and out of the soil, and affecting the soil fertility as well. The implication of the increased degradation of lignins is that less carbon remains in soil solid phase, and more CO_2 is released from soil into the atmosphere (Lance Frazer 2009). To understand the soil-climate interactions better, we need more soil research to focus the molecular level within an eye towards predicting both short- and long-range changes in system. In this context, the studies pertaining to changes in nutrient pool of citrus rhizosphere, microbial communities and soil carbon partitioning are highly imperative in response to elevated CO_2 and temperature.

11.7 Management Practices and Changes in Soil Enzymes

Management practices (e.g. crop rotation, mulching, tillage and application of fertilizers and pesticides) may have diverse effects on various enzymes and microbial activities of soil (Dick et al. 1987; Tabatabai 1994). Indeed, enzyme activities have been shown to be more sensitive total carbon concentration to soil management practices. Application of organic wastes, such as sewage sludge, as a source of organic matter is a common practice, especially to soils containing little organic matter, to maintain or improve soil quality and soil enzymatic activities (Giusquiani et al. 1995). Some studies evaluated the effects of various management practices on soil enzyme activities in soils (Saa et al. 1993; Eivazi and Bayan 1996). However, little is known about the long-term effects of burning and N fertilization on the relationship between soil enzyme activities in tallgrass prairie. Soil enzyme activities are especially significant because they control nutrient release for plant and microbial growth. Likewise, the labile pool of soil organic matter, which is closely related to soil microbial biomass, could be used as a sensitive indicator of short-term changes in soil management (Chantigny 2003).

Many long-term studies have shown that soil enzyme activities are sensitive in discriminating among soil management practices, such as fertilization by means of animal manure or green manures/crop residues (Dick et al. 1988b; Martens et al. 1992) and municipal refuse amendment (Perucci 1992), as well as among tillage treatments (Gupta and Germida 1988).

The enzyme such as β-D-glucosidase, urease, acid and alkaline phosphatases and arylsulphatase are also believed to be sensitive to field management practices inducing significant changes in soil quality (Dick et al. 1996, 2000; Yakovchenko et al. 1996; Bandick and Dick 1999; Adam and Duncan 2001). The response of soil enzyme activities to specific soil practices has been used to compare agricultural systems (combinations of soil practices) such as organic versus conventional farming (Benitez et al. 2006; Van Diepeningen et al. 2006; Melero et al. 2006). However, in these studies, the emphasis has been set on the comparative assessments of many dependent variables measured at a single-time in the selected systems rather than on a time series dynamic assessment. Seasonal fluctuations in, for instance, soil water potential, gas exchange and temperature, account for most of the microbiological variability, including production, survival and stabilization of soil enzymes within the soil matrix (Aon et al. 2001).

11.7.1 Fertilizer Application

Nitrogen fertilization may affect soil enzyme activities and microbial biomass of tallgrass prairie. Because organic matter modifies the development of enzyme producing microbial community (Browman and Tabatabai 1978), management practices that minimize the addition of organic matter to soils may reduce enzyme activities and thus affect the ability of soils to supply nutrients needed for plant growth. Therefore, changes in enzyme and microbial activities could alter the availability of nutrients for uptake (Dick et al. 1988a, b), and these changes are potentially sensitive indicators of soil quality (Dick 1994).

Yuan et al. (2006) observed that with elevated atmospheric CO_2, ß-glucosidase activity significantly decreased ($P \leq 0.05$) with low N application rates, had no significant effect with a normal N nitrogen application rate and significantly increased ($P \leq 0.05$) with a high N application rate. For urease activity, at low and normal N application rates (but not high N application rate), elevated atmospheric CO_2 significantly increased ($P \leq 0.05$) it. For acid phosphatase, elevated atmospheric CO_2 only had significant higher effects ($P \leq 0.05$) at high N application rates. Under different CO_2 concentrations, effects of N fertilization are also different. Invertase and acid phosphatase activities at elevated CO_2 concentration significantly increased ($P \leq 0.05$) with N treatments, but there has been no effect with the ambient CO_2 concentration. For urease activity, at ambient CO_2 concentration, N fertilization increased it significantly ($P \leq 0.05$), whereas at elevated CO_2 concentration, it was not significant. Additionally, with ß-glucosidase activity, there were no significant effects from N application.

Table 11.4 Changes in soil enzyme activities in response to fertilization treatments

	Dehydrogenase		β-glucosidase	
Treatments	March	September	March	July
Control	3.2	4.3	21.2	21.4
100% compost	5.1	7.2	34.4	30.0
50% NPK+50% compost	5.5	10.3	37.8	33.2
100% NPK	4.5	5.6	36.5	37.2

Adapted from Joa et al. (2010)
Data based on the time of peak activity
Units of dehydrogenase and β-glucosidase expressed in μg TPF g^{-1} 24 h^{-1} and μg PNP g^{-1} h^{-1}, respectively

Urease, the enzyme that catalyzes the hydrolysis of urea, has also been widely used in the evaluation of changes in soil quality due to soil management. Its activity increased due to organic fertilization (Pascual et al. 1999) and addition of cattle slurry (Kandeler and Eder 1993) and decreased as a consequence of ploughing (Saviozzi et al. 2001). Soil urease showed close relation with urea hydrolyzation and increased the utilization rate of nitrogen fertilizer (Cookson and Lipiece 1996; Klose and Tabatabai 1999). The activity of acid phosphatase speeded upon soil organic phosphorous decomposition and improved soil phosphorus validity, which is an important index to assess soil phosphorous bioavailability (Pascual et al. 2002). With regard to the enzymes involved in the carbon cycle, β-glucosidase is most widely used in the evaluation of soil quality in soils subjected to different management treatments.

Joa et al. (2010) demonstrated higher urease and dehydrogenase activity in 50% NPK plus 50% compost than 100% NPK in citrus grown on Andisol volcanic ash soil. Principal component analysis of the microbial community by phospholipid fatty acid (PLFA) profiling showed that PLFA profiles from 100% NPK and 50% NPK+50% compost were different in addition to compositional change in microbial community (Table 11.4).

11.7.2 Tillage Response

Soil tillage causes a rapid loss of organic matter content leading to low soil biological activity, a decline in aggregate stability and a reduction in crop productivity (Bayer et al. 2001). In addition, carbon sequestration in the soil may be reduced with some tillage practices (Aslam et al. 2000). Soil organic matter distribution, nutrient cycling and microbial activity are influenced by the type and the degree of soil tillage (Salinas-García et al. 2002). Changes in soil moisture,

Table 11.5 Changes in under-tree canopy soil enzymatic activities (μg g^{-1} soil) in response to different permanent cover crops in citrus orchard (Ultisol)

Cover crops	Amylase	Cellulase	Arylsulphatase	Acid phosphatase	Alkaline phosphatase
C_m–C_c	263	51	3.4	156	25.8
A_p–C_t	171	46	1.4	92	13.9
P_n–C_t	271	46	3.2	96	23.2
B_h–C_t	320	53	2.5	115	37.2
B_h–S_t	254	53	3.6	132	26.9
Mean	256	50	2.8	118	25.4

Adapted from Balota et al. (2011)
C_m *Calopogonium mucunoides*, A_p *Archis pintoi*, P_n *Paspalum notatum*, B_h *Brachiaria humidiciola*
C_t and S_t stand for conventional tillage and strip tillage, respectively

temperature and C input can also have a large effect on the soil microbial biomass and its activity, which, in turn, affect nutrient availability due to soil organic matter turnover (Ross 1987).

Organic matter is involved in the enhancement of soil quality since it acts on soil structure, nutrient storage and biological activity. The quality and quantity of soil organic matter normally changes very slowly, and many years (5–10 years) are required to detect the changes resulting from disturbances. The extent of soil organic matter turnover is mainly controlled by the size and activity of the microbial biomass, which may respond to disturbances over shorter time scales (months). Therefore, soil biological and biochemical parameters have a role as early and sensitive indicators of soil ecological stress and restoration (Roldán et al. 2003; Izquierdo et al. 2003). Bolinder et al. (1999) indicated that the soil microbial biomass was more sensitive to changes in soil quality than total organic carbon or nitrogen. Roldán et al. (2005) reported that no-tilled soil had increased the values of water-soluble C, dehydrogenase, urease, acid phosphatase activities, aggregate stability and glomalin compared to tilled soils, especially in the shallowest (0–5 cm) layer. The water regime had no effect on soil structural stability or total microbial activity (Roldán et al. 2005).

Loss of soil organic matter following tillage practices has been attributed to the destruction of macroaggregates and subsequent mineralization of labile soil organic matter. Thus, the structure of soil protects soil organic matter and influences organic matter turnover and soil fertility. Plants roots increase the stability of surrounding aggregates (Lynch and Bragg 1985) through several interacting mechanisms. Roots and associated mycorrhizal hyphae may form a three-dimensional network that enmeshes fine particles of soil into aggregates. Recent studies have also indicated that arbuscular mycorrhizal (AM) fungi produce a glycoprotein, glomalin, that acts as an insoluble glue to stabilize aggregates (Wright and Anderson 2000). These authors have found that glomalin changes quickly in response to crop rotations tillage management practices (Wu et al. 2004).

Use of plants to cover the inter-row space between the perennial crops is one of the most efficient single practices for reducing soil erosion due to reduction in erosive energy of raindrops impacting against the soil surface, thus avoiding the disaggregation of soil particles and facilitating greater water infiltration, thereby improving soil physico-chemical properties (Rutigliano et al. 2004) and microbial activity (Acosta-Martínez et al. 2010). Balota et al. (2004) reported much higher enzymatic activity (amylase 350–830 μg g^{-1} soil, arylsulphatase 4–113 μg g^{-1} soil and cellulose 67–220 μg g^{-1} soil) in soils having higher available nutrient supply level in soil. Studies by Balota et al. (2011) showed that soil tillage system and groundcover crops influenced the soil enzyme activities (amylase 171–320 μg g^{-1} soil, cellulose 46–53 μg g^{-1} soil, arylsulphatase 1.4–3.6 μg g^{-1} soil, acid phosphatase 92–156 μg g^{-1} soil and alkaline phosphatase 13.9–37.2 μg g^{-1} soil). Both under the tree canopy (Table 11.5) and in the inter-row space, the cultivation of *Brachiaria humidiciola* provided higher activity of amylase, arylsulphatase, acid phosphatase and alkaline phosphatase than other groundcover species. Strip tillage increased the activities of soil enzymes (arylsulphatase and acid phosphatase) than conventional tillage system using *Brachiaria humidiciola* groundcover crop, thereby suggesting better effectiveness of grasses over leguminous species.

11.8 Future Research

Summarizing, soil enzyme assays, including community-level physiological profiling, are never stand-alone parameters. They are quick monitoring tools but cannot give answers on functionality. To characterize the soil, or to understand differences among soils, additional methods addressing pools and fluxes are necessary (Insam 2001). A better understanding of changes in soil microbial biomass and enzyme activity in an ecosystem would allow greater understanding of the effect of a disturbance on soil community functions. Because some soil enzymes respond to sudden disturbances of the soil system, they can effectively aid developing sustainable land management practices.

When evaluating the efficiency of an in situ immobilization strategy, the following important questions needed to be answered: (1) are microorganisms with the desired characteristics and activities present at the remediated site and at what densities, (2) what is their activity level, and (3) how is the composition and function of the microbial community influenced by changes in environmental parameters (Geets et al. 2003). Although, there is a general consensus that efficiency of the soil remediation also depends on presence and activity of microorganisms, the ecological consequences of inorganic amendments for these features of microflora of metal contaminated soils have received little attention.

Is there any relationship between crop phenophases (physiological indices) and soil biochemical indices (soil enzymatic activities) indicating that soil biochemical characteristics are affected significantly by crop growth in the interaction system of citrus, soil properties and microorganisms? Clearly, there is a need for integrated studies of the interaction between the biology, chemistry and physics of soil nutrient dynamics if we are to manage soils for improved nitrogen use efficiency with emphasis on microbially mediated processes.

Biochemical properties can still not be considered as efficient diagnostic tools since neither the individual properties nor the indices obtained with them can be considered to be of general use as those properties or indices that appear to be valid in certain situations are totally ineffective in others. Standardization of the methods and the construction of databases of soil biological and biochemical properties will have to be made as core agenda involving enormous analytical efforts.

There are number of reasons that can explain the inability of biochemical approaches to quantify soil quality, which comprise of (1) lack of standard analysis methods accepted by all laboratories as a fundamental problem when interpreting the values of biochemical properties, (2) differences in sample collection, storage, pre-treatment and protocols for determining enzymatic activities and (3) high degree of variability between biochemical properties, both seasonal and edaphic factors, in addition to lack of reference value or broad databases for high-quality soils that could be used to make comparisons (Gil-Sotres et al. 2005).

References

Acosta-Martinez V, Tabatabai MA (2000) Enzyme activities in a limed agricultural soil. Biol Fertil Soils 31:85–91

Acosta-Martínez V, Dowd SE, Bell CV et al (2010) Microbial community composition as affected by dryland cropping systems and tillage in a semiarid sandy soil. Diversity 2:910–931

Adam G, Duncan H (2001) Development of a sensitive and rapid method for the measurement of total microbial activity using fluoresein diacetate (FDA) in a range of soils. Soil Biol Biochem 33:943–951

Ajwa HA, Tabatabai MA (1994) Decomposition of different organic materials in soils. Biol Fertil Soils 18:175–182

Ajwa HA, Dell CJ, Rice CW (1999) Changes in enzyme activities and microbial biomass of tallgrass prairie soil as related to burning and nitrogen fertilization. Soil Biol Biochem 31(51):769–777

Amador JA, Glucksman AM, Lyons JB et al (1997) Spatial distribution of soil phosphatase activity within a riparian forest. Soil Sci 162:808–825

Anderson TH, Domsch KH (1989) Ratios of microbial biomass carbon to total organic carbon in arable soil. Soil Biol Biochem 21: 471–479

Aon MA, Colaneri AC (2001) II. Temporal and spatial evolution of enzymatic activities and physico-chemical properties in an agricultural soil. Appl Soil Ecol 18:255–270

Aon MA, Cabello MN, Sarena DE et al (2001) I. Spatio-temporal patterns of soil microbial and enzymatic activities in an agricultural soil. Appl Soil Ecol 18:239–254

Aslam T, Choudhary MA, Saggar S (2000) Influence of land-use management on CO_2 emissions from a slit loam soil in New Zealand. Agric Ecosyst Environ 77:257–262

Baldrian P (2009) Microbial enzyme-catalyzed processes in sols and their analysis. Plant Soil Environ 55(9):370–378

Balota EL, Kanashiro M, Colozzi-Filho A et al (2004) Soil enzyme activities under long-term tillage and crop rotations system in subtropical agro-ecosystems. Braz J Microbiol 35:300–306

Balota EL, Machineski O, Truber PV et al (2011) Effect of tillage systems and permanent ground cover intercropped with orange trees on soil enzyme. Braz Arch Biol Technol 25:221–228

Bandick A, Dick RP (1999) Field management effects of soil enzyme activities. Soil Biol Biochem 31:1471–1479

Baxter R, Gantley M, Ashenden TW et al (1994) Effects of elevated carbon-dioxide on 3 grass species from montane pasture.2. Nutrient-uptake, allocation and efficiency of use. J Exp Bot 45:1267–1278

Bayer C, Martin-Neto L, Mielniczuk J et al (2001) Changes in soil organic matter fractions under subtropical no-till cropping systems. Soil Sci Soc Am J 65:1473–1478

Beck T (1984) Mikrobiologische und biochemische Charakterisierung landwirtschaftlich genutzter Bo¨den: I.Mit. Die Ermittlung der Bodenmikrobiologischen Kennzahl. Z Pflanzenernaehr Bodenkd 147:456–466

Benitez E, Nogales R, Campos M et al (2006) Biochemical variability of olive orchard soils under different management systems. Appl Soil Ecol 32:221–231

Bergstrom DW, Monreal CM, Tomlin AD et al (2000) Interpretation on soil enzyme activities in a comparison of tillage practices along a topographic and textural gradient. Can J Soil Sci 80:71–79

Bolinder MA, Angers DA, Gregorich EG et al (1999) The response of soil quality indicators to conservation management. Can J Soil Sci 79:37–45

Breland TA, Hansen S (1996) Nitrogen mineralization and microbial biomass as affected by soil compaction. Soil Biol Biochem 28:655–663

Bremmer JM, Mulvaney RL (1978) Urease activity in soils. In: Burns RG (ed) Soil enzymes. Academic, New York, pp 149–196

Browman MG, Tabatabai MA (1978) Phosphodiesterase activity in soil. Soil Sci Soc Am J 42:284–290

Brzezinska M, Stepninewska Z, Stepninewski W (1998) Soil oxygen status and dehydrogenase activity. Soil Biol Biochem 30:1783–1790

Burns RG (1978) Enzyme activity in soil. Some theoretical and practical considerations. In: Burns RG (ed) Soil enzymes. Academic, New York, pp 295–339

Burns RG (1982) Enzyme activity in soil: location and a possible role in microbial ecology. Soil Biol Biochem 14:423–427

Cao H, Sun H, Yang H et al (2003) A review: soil enzyme activity and its indication for soil quality. Chin J Appl Enviorn Biol 9(1):105–109

Chantigny MH (2003) Dissolved and water-extractable organic matter in soils: a review on the influence of land use and management practices. Geoderma 113:357–380

Chen HJ (2003) Phosphatase activity and P fraction in soils of an 18-year-old Chinese fir (Cunninghamia lanceolata) plantation. Forest Ecol Manage 178:301–310

Chen SK, Edwads CA, Subler S (2003) The influence of two agricultural biostimulants on nitrogen transformations, microbial activity, and plant growth in soil microcosms. Soil Biol Bichem 35:9–19

Cookson P (2002) Variation in phosphatase activity in soil: case study. Agric Sci 7:65–72

Cookson P, Lipiece AG (1996) Urease enzyme activities in soils of the Batinan region of the Sultanate of Oman. J Arid Enviorn 32:225–238

Dalal RC (1975) Urease activity in some Trinidad soils. Soil Biol Biochem 7:5–8

Davidson EA, Janssens EA (2006) Temperature sensitivity of soil carbon decomposition and feedbacks to climate change. Nature 7081:165–173

Deng SP, Tabatabai MA (1997) Effect of tillage and residue management on enzyme activities in soil: III phosphatase and arylsulphatase. Biol Fertil Soils 24:141–146

Diaz S, Grime JP, Harris J, McPherson E (1993) Evidence of a feedback mechanism limiting plant response to elevated carbon dioxide. Nature 364:616–617

Dick RP (1994) Soil enzyme activities as indicators of soil quality. In: Doran JW, Coleman DC, Bezdicek DF et al (eds) Defining soil quality for sustainable environment, SSSA Special publication 35. Soil Science Society of Amer, Madison, WI, pp 107–124

Dick WA, Tabatabai MA (1992) Potential uses of soil enzymes. In: Metting FB Jr (ed) Soil microbiology ecology: applications in agricultural and environmental management. Marcel Dekker, New York, pp 95–127

Dick SP, Tabatabai MA (1994) Cellulase activity in soils. Soil Biol Biochem 26:1347–1354

Dick RP, Rasmussen PE, Kerle EA (1987) Kinetic parameters of enzyme activities as influenced by organic residues and N fertilizer management. In: Agronomy abstracts, 79th annual meeting, Atlanta, GA. American Society of Agronomy, Madison

Dick RP, Myrold DD, Kerle EA (1988a) Microbial biomass and soil enzyme activities in compacted and rehabilitated skid trail soils. Soil Sci Soc Am J 52:512–516

Dick RP, Rasmussen PE, Kerle EA (1988b) Influence of long-term residue management on soil enzyme activity in relation to soil chemical properties of a wheat-fallow system. Biol Fertil Soils 6:156–164

Dick RP, Sandor JA, Eash NA (1994) Soil enzyme activities after 1500 years of terrace agriculture in the Coka Valley, Perú. Agric Ecosyst Environ 50(2):123–131

Dick RP, Brakewill D, Turco R (1996) Soil enzyme activities and biodiversity measurements as integrating biological indicators. In: Doran JW, Jones AJ (eds) Handbook of methods for assessment of soil quality. Soil Science Society of America, Madison, pp 242–272

Dick WA, Cheng L, Wang P (2000) Soil acid and alkaline phosphatase activity as pH adjustment indicators. Soil Biol Biochem 32:1915–1919

Dodgson KS, White G, Fitzgerald JW (1982) Sulphatase enzyme of microbial origin. Afr J Biotechnol 1:56–159

Doran JA, Parkin TB (1994) Defining and assessing soil quality. In: Doran JW, Coleman DC, Bezdicek DF (eds) Defining soil quality for a sustainable environment. Soil Science Society of America, Madison, pp 3–21

Downton WJS, Grant WJR, Loveys BR (1987) Carbon dioxide enrichment increases yield of Valencia orange. Aust J Plant Physiol 14(5):493–501

Ebersberger D, Niklaus PA, Kandeler E (2003) Long term CO_2 enrichment stimulates N mineralization and enzyme activities in calcareous grassland. Soil Biol Biochem 35:262–269

Eivazi F, Bayan MR (1996) Effects of long-term prescribed burning on the activity of selected soil enzymes in an oak-hickory forest. Can J Forest Res 26:1799–1804

Eriksson KEL, Blancbette RA, Ander P (1990) Biodegradation of cellulose. In: Eriksson KEL, Blancbette RA, Ander P (eds) Microbial and enzymatic degradation of wood and wood components. Springer, New York, pp 89–180

Frankenberger WT Jr, Dick WA (1983) Relationships between enzyme activities and microbial growth and activity indices in soil. Soil Sci Soc Am J 47:945–951

Garcia C, Hernandez T (1997) Biological and biochemical indicators in derelict soil subject to erosion. Soil Biol Biochem 29:171–177

Garćia FO, Rice CW (1994) Microbial biomass dynamics in tall-grass prairie. Soil Sci Soc Am J 58:816–823

Garćia C, Hernández T, Pascual J, Moreno JL, Ros M (2000) Microbial activity in soils of SE Spain exposed to degradation and desertification processes. Strategies and their rehabilitation. In: Garćia C, Hernández T (eds) Research and perspectives of soil enzymology in Spain. Consejo Superior de Investigaciones Científicas, Madrid, pp 93–143

García-Ruiz R, Ochoa V, Vinegla B et al (2009) Soil enzymes, nematode community and selected physico-chemical properties as soil quality indicators in organic and conventional olive farming: influence of seasonality and site features. Appl Soil Ecol 41:305–314

Geets J, vangronsveld J, Diels L, Van der Lelie D (2003) The application of molecular tools to followup bioremediation. J Soil Sediment 3:251

Gil-Sotres F, Trasar-Cepeda C, Leirós MC, Seoane S (2005) Different approaches to evaluating soil quality using biochemical properties. Soil Biol Bichem 37:877–887

Giusquiani PL, Pagliai M, Gigliotti G, Businelli D, Benetti A (1995) Urban waste compost: effects of physical, chemical and biochemical soil properties. J Environ Qual 24:175–182

Gupta VVSR, Germida JJ (1988) Distribution of microbial biomass and its activity in different aggregate size classes as affected by cultivation. Soil Biol Biochem 20:777–786

Haney RL, Franzluebbers AJ, Porter EB et al (2004) Soil carbon and nitrogen mineralization: influence of drying temperature. Soil Sci Soc Am J 68:489–492

He ZL, Yao HY, Chen GC (1997) Relationship of crop yield to microbial biomass in highly weathered soils of China. In: Ando T (ed) Plant nutrition for sustainable food production and environment. Kluwer Academic, Tokyo, pp 751–752

He WX, Jiang X, Bian YR et al (2002) Study on soil enzyme activity effected by dimehypo. J Northwest Agric For Tech Univ 30(1):13–17

Hoffmann G (1967) Eine photometrische Methode zur Bestimmung der Phosphatase-Aktivität in Böen. Z Pflanzenernaehr Bodenkd 118:193–198

Hu C, Cao Z (2007) Size and activity of the soil microbial biomass and soil enzyme activity in long term field experiments. World J Agric Sci 3:63–70

Hunt R, Hand DW, Hannah MA, Neal AM (1991) Response of CO_2 enrichment in 27 herbaceous species. Funct Ecol 5:410–421

Insam H (2001) Developments in soil microbiology since mid 1960s. Geoderma 100:389–402

Insam H, Mitchell CC, Dormaar JF (1991) Relationship of soil microbial biomass and activity with fertilization practices and crop yield of three ultisols. Soil Biol Biochem 24:459–464

IPCC (Intergovernmental Panel on Climate Change) (1996) Climate change 1995: the science of climate change. Cambridge University Press, Cambridge

Izquierdo I, Caravaca F, Alguacil MM et al (2003) Changes in physical and biological soil quality indicators in a tropical crop system

(Havana, Cuba) in response to different agroecological management practices. Environ Manage 32:639–645

Joa JH, Moon DG, Won HY et al (2010) Effect of different fertilization practices on soil microbial activities and community structure in volcanic ash citrus orchard soil. In: 19th world congress soil science, soil solutions for a changing world, Brisbane, Australia, 1–6 Aug 2010. http://www.inss.org/19th% 20SCSS/Symposium/pdf

Joachim HJRM, Patrick AN (2008) Selected soil enzymes: examples of their potential roles in the ecosystem. Afr J Biotechnol 7:181–191

Jordan D, Ponder F, Hubbard VC (2003) Effects of soil compaction forest leaf litter and nitrogen fertilizer on two oak species and microbial activity. Appl Soil Ecol 23:33–41

Kandeler E (1996) Nitrate. In: Schinner F, Ohllinger R, Kandeler E et al (eds) Methods in soil biology. Springer, Berlin/Heidelberg/New York

Kandeler E, Eder G (1993) Effect of cattle slurry in grassland on microbial biomass and on activities of various enzymes. Biol Fertil Soils 16:249–254

Kang H, Freeman C, Trevor WA (2001) Effects of elevated CO_2 on fen peat biogeochemistry. Sci Total Environ 279:45–50

Karaca A, Naseby DC, Lynch JM (2002) Effects of cadmium contamination with sewage sludge and phosphate fertilizer amendments on soil enzyme activities, microbial structure and available cadmium. Biol Fertil Soils 35:428–434

Kertesz MA, Mirleau P (2004) The role of soil microbes in plant sulphur nutrition. J Exp Bot 55:1939–1945

Klose S, Tabatabai MA (1999) Urease activity of microbial biomass in soils. Soil Biol Biochem 31:205–211

Knorr W, Prentice IC, House JI et al (2005) Long term sensitivity of soil carbon turnover to warming. Nature 7023:298–301

Körner C, Arnone JA III (1992) Response to elevated carbon dioxide in artificial tropical ecosystem. Science 257:1672–1675

Kumar Jha D, Sharma GD, Mishra RR (1992) Soil microbial population numbers and enzymes activities in relation to altitude and forest degradation. Soil Biol Biochem 24:761–767

Lance F (2009) Climate change: will warmer soil be fertile. Enviorn Health Prospect 117(2):59–61

Li X, Sarah P (2003) Enzyme activities along a climatic transect in the Judean Desert. Catena 53:349–363

Liu XM, Liang WJ, Jaing Y (2008) Distribution of soil enzyme activities and microbial biomass along a latitudinal gradient in farmlands of Songliano plain, Northeast China. Pedosphere 10(4):431–440

Lynch JM, Bragg E (1985) Microorganisms and soil aggregate stability. Adv Soil Sci 2:133–171

Margesin R, Zimmerbauer A, Schínner F (2000) Monitoring of bioremediation by soil biological activities. Chemosphere 40:339–346

Martens DA, Johanson JB, Frankenberger WT (1992) Production and persistence of soil enzymes with repeated additions of organic residues. Soil Sci 153:53–61

McLaren AD (1975) Soil as system of humus and clay immobilized enzymes. Chem Scripta 8:97–99

Melero S, Porras JC, Herencia JF et al (2006) Chemical and biochemical properties in a silty loam soil under conventional and organic management. Soil Tillage Res 90:162–170

Moehring DM, Rawls IW (1970) Detrimental effects of wet weather logging. J For 68:166–167

Moffat AJ (1991) Forestry and soil protection in the UK. Soil Use Manage 7:145–151

Nannipieri P, Greco S, Ceccanti B (1990) Ecological significance of the biological activity in soil. In: Bollag JM, Stotzky G (eds) Soil biochemistry, vol 6. Marcel Dekker, New York, pp 293–355

Parrr JF, Papendick RL (1997) Soil quality: relationship and strategies for suitable dryland farming system. Ann Arid Zone 36:181–191

Pascual JA, Hernández T, Ayuso M et al (1998) Enzymatic activities in an arid soil amended with urban wastes. Lab Exp Bioresour Technol 64:131–138

Pascual JA, García C, Hernández T (1999) Lasting microbial and biochemical effects of the addition of municipal solid waste to an arid soil. Biol Fertil Soils 30:1–6

Pascual JA, García C, Hernández T et al (2000) Soil microbial activity as a biomarker of degradation and remediation processes. Soil Biol Biochem 32:1877–1883

Pascual JA, Moreno JL, Hernandez T (2002) Persistence of immobilised and total urea and phosphatase activities in a soil amended with organic wastes. Bioresour Technol 82:73–78

Pawlson D (2005) Climatology: will soil amplify climate change. Nature 7023:204–205

Pazur JH (1965) Enzymes in synthesis and hydrolysis of starch. In: Paschal EF, Whistler R (eds) Starch: chemistry and technology. Academic, New York, pp 133–175

Perucci P (1992) Enzyme activity and microbial biomass in a field soil amended with municipal refuse. Biol Fertil Soils 14:54–60

Perucci P, Scarponi L, Businelli M (1984) Enzyme activities in clay-loam soil amended with various crop residues. Plant Soil 81:345–351

Powers RF, Alban DH, Millar RE et al (1990) Sustaining site productivity in North American Forests: problems and prospects. In: Gessel SP, Lacate DS, Weetman DG et al (eds) Proceedings of the 7th North American forest soils conference on sustained productivity of rest soils. Faculty of Forestry Publication, Vancouver, pp 47–79

Pulford ID, Tabatabai MA (1988) Effect of waterlogging on enzyme activities in soils. Soil Biol Biochem 20:215–219

Reddy GB, Faza A (1989) Dehydrogenase activity in sludge amended soil. Soil Biol Biochem 21:327–333

Reisinger TW, Simmons GL, Pope PE (1988) The impact of timber harvesting on soil properties and seedling growth in the south. South J Appl For 12:58–67

Roldán A, Caravaca F, Herńandez MT et al (2003) No tillage, crop residue additions and legume cover cropping effects on soil quality characteristics under maize Patzcuaro watershed (Mexico). Soil Tillage Res 72:65–73

Roldán A, Salinas-García JR, Alguacil MM et al (2005) Changes in soil enzyme activity, fertility, aggregation and C sequestration mediated by conservation tillage practices and water regime in a maize field. Appl Soil Ecol 30:11–20

Rosenzweig C, Hillel D (1993) Agriculture in a greenhouse world: potential consequences of climate change. Nat Geogr Res Explor 9:208–221

Rosenzweig C, Phillips J, Goldberg R et al (1995) Potential impacts of climate change on citrus and potato production in the US. Agric Syst 52(4):455–479

Ross DJ (1987) Soil microbial biomass estimated by the fumigation-incubation procedure: seasonal fluctuation and influence of soil moisture content. Soil Biol Biochem 19:397–404

Rutigliano FA, D'Ascoli R, De Santo AV (2004) Soil microbial metabolism and nutrient status in a Mediterranean area as affected by plant cover. Soil Biol Biochem 36:1719–1729

Saa A, Trasar-Cepeda MC, Gil-Sotres F, Carballas F (1993) Changes in soil phosphorus and acid phosphate activity immediately following forest fires. Soil Biol Biochem 25:1223–1230

Salinas-García JR, Velásquez-García JJ, Gallardo-Valdez M et al (2002) Tillage effects on microbial biomass and nutrient distribution in soils under rain-fed corn production in central-western Mexico. Soil Tillage Res 66:143–152

Sardans J, Peñuelas J (2005) Drought decreases soil enzyme activity in Mediterranean *Quercus ilex* L. forest. Soil Biol Biochem 37:455–461

Saviozzi A, Levi-Minzi R, Cardelli R et al (2001) A comparison of soil quality in adjacent cultivated, forest and native grassland soils. Plant Soil 233:251–259

Schnürer J, Rosswall T (1982) Fluorescein diacetate hydrolysis as a measure of total microbial activity in sol and litter. Appl Environ Microbiol 43:1256–1261

Sinsabaugh RL, Antibus RK, Linkins AE (1991) An enzymic approach to the analysis of microbial activity during plant litter decomposition. Agric Ecosyst Environ 34:43–54

Skujins J (1978) History of abiotic soil enzyme research. In: Burns RG (ed) Soil enzymes. Academic, New York, pp 1–49

Sparling GP, Speir TW, Whale KN (1986) Changes in microbial biomass C, ATP content, soil phosphomonoesterase and phosphodiesterase activity following air-drying of soil. Soil Biol Biochem 18:363–370

Speir TW, Ross DJ (1975) Effects of the storage on activities of protease, urease, phosphatase, and sulphatase in three soils under grazed pasture. N Z J Sci 18:231–237

Srivastava AK (2009) Integrated nutrient management: concept and application in citrus. In: Tennant P, Beakebhia N (eds) Citrus II. Tree For Sci Biotechnol 3 (Special Issue 1):32–58

Srivastava AK, Kohli RR (1999) Agro-ecological approach for land use planning of citrus. Agric Rev 20(1):41–47

Srivastava AK, Ngullie E (2009) Integrated nutrient management: theory and practice. Dyn Soil Dyn Plant 3(1):1–30

Srivastava AK, Singh S (2008) Citrus nutrition research in India: problems and prospects. Indian J Agric Sci 78:3–16

Srivastava AK, Singh S (2009) Citrus decline: soil fertility and plant nutrition. J Plant Nutr 32(2):197–245

Srivastava AK, Huchche AD, Dinesh Kumar (2010) Development of INM module for sustained productivity of Citrus. Annual report 2009–2010. National Research Centre for Citrus, Nagpur, Maharashtra, India, pp 40–41

Startsev NA, McNabb DH, Startsev AD (1998) Soil biological activity in recent clear cuts in west-central Alberta. Can J Soil Sci 78:69–76

Tabatabai MA (1994) Soil enzymes. In: Page AL, Miller RH, Keeney DR (eds) Methods of soil analysis. American Society of Agronomy, Madison, pp 775–833

Tabatabai MA, Bremner JM (1972) Assay of urease activity in soil. Soil Biol Biochem 4:479–487

Tate RL (1995) Soil microbiology. Wiley, New York, pp 182–189

Thanana E, Nawar A (1994) Salinity and mycorrhizal association in relation to carbohydrate status, leaf chlorophyll and activity of peroxidase and polyphenol oxidase enzymes in sour orange seedlings. Alex J Agric Res 39:263–280

Trasar-Cepeda C, Leiros C, Gilsotres F et al (1998) Towards a biochemical quality index for soils – an expression relating several biological and biochemical properties. Biol Fertil Soils 26:100–106

Trasar-Cepeda C, Leirós MC, Garćia-Hernández F et al (2000) Biochemical properties of the soils of Galicia (NW Spain): their use as indicators of soil quality. In: Garćia C, Hernández T (eds) Research and perspectives of soil enzymology in Spain. Consejo Superior de Investigaciones Científicas CSIC, Madrid, pp 175–206

Trevors J (1984) Dehydrogenase activity in soil: a comparison between the INT and TTC assay. Soil Biol Biochem 16:673–674

Turner BL, Hopkins DW, Haygarth PM et al (2002) ß-glucosidase activity in pasture soils. Appl Soil Ecol 20:157–162

Van der Linden AMA, Jeurissen LJJ, vanVeen JA et al (1989) Turnover of the soil microbial biomass as influenced by soil compaction. In: Hansen JA, Henriksen K (eds) Nitrogen in organic wastes applied to soils. Academic, London, pp 25–36

van Diepeningen AD, De Vos OJ, Korthals GW et al (2006) Effects of organic versus conventional management on chemical and biological parameters in agricultural soils. Appl Soil Ecol 31:120–135

Versaw WK, Harrison MJ (2002) A chloroplast phosphate transporter PHT2, Influences allocation of phosphate within the plant and phosphate starvation response. Plant Cell 14:1751–1766

Visser S, Parkinson D (1992) Soil biological criteria as indicators of soil quality: soil microorganisms. Am J Alt Agric 7:33–37

Vu JCV, Allen LH Jr, Boote KJ et al (1997) Effects of elevated CO_2 and temperature on photosynthesis and rubisco in rice and soybean. Plant Cell Environ 20:68–76

Waldrop MA, Zak DR, Sinsabaugh RL (2004) Microbial community response to nitrogen deposition in northern forest ecosystems. Soil Biol Biochem 36:1443–1451

Wang XC, Lu Q (2006) Effect of waterlogged and aerobic incubation on enzyme activities in paddy soil. Pedosphere 16(4):532–539

White AR (1982) Visualisation of cellulases and cellulose degradation. In: Brown RM Jr (ed) Cellulose and other natural polymer systems: biogenesis structure and degradation. Plenum Press, New York

Wright SF, Anderson RL (2000) Aggregate stability and glomalin in alternative crop rotation for the central Great Plans. Biol Fertil Soils 31:249–253

Wronski EB, Murphy G (1994) Response of forest crop to soil compaction. In: Sloane BD, Ouwerkerk CV (eds) Soil compaction in crop production. Elsevier, Amsterdam, pp 317–342

Wu XS, Zhou KB, Xia RX et al (2004) Investigation on natural development of arbuscular mycorrhiza in citrus trees. Subtrop Plant Sci 33:39–40 (in Chinese with English abstract)

Wu QS, Xia RX, Zou YN (2008) Improved soil structure and citrus growth after inoculation with three arbuscular mycorrhizal fungi under drought stress. Eur J Soil Biol 44:122–128

Yakovchenko V, Sikora L, Kaufmann D (1996) A biological based indicator of soil quality. Biol Fertil Soils 21:245–251

Yang WQ, Wang KY (2002) Advances on soil enzymology. Chin J Appl Enviorn Biol 8(5):564–570

Yuan X, Lin X, Chu H et al (2006) Effects of elevated atmospheric CO_2 on soil enzyme activities at different nitrogen application treatments. Acta Ecol Sin 26(1):48–53

Zak DR, Pregitzer KS, Curtis PS et al (1993) Elevated atmospheric CO_2 and feedback between carbon and nitrogen cycles. Plant Soil 51:105–117

Zornoza R, Guerrero C, Mataix-Solera J et al (2006) Assessing air drying and rewetting pre-treatment effect on some soil enzyme activities under Mediterranean conditions. Soil Biol Biochem 38:2125–2134

Surfactant and Nutrient Uptake in Citrus

12

Zora Singh and Ahmad Sattar Khan

Abstract

The availability of essential nutrients in appropriate amount ensures the growth, productivity and quality for sustainable production of tree fruits including citrus. The essential mineral element below the critical level adversely affects growth and yield even other minerals are within optimum range. A balanced and continuous supply of the essential nutrients is a key for citrus growers to ensure profitability. Soil application is recommended for most of the macronutrients such as N, P and K, as these are required in large quantities. However, the soil application leads to nutrient losses through runoff, leaching and fixation. Therefore, the foliar application is considered as an attractive alternative to the soil application of essential nutrients. Particularly, the trace elements which are required in small quantity are preferably applied through the foliar sprays. Nutrients applied through the foliar spray are directly absorbed though leaves and have several positive attributes. Foliar application can reduce overall fertiliser application rate and energy use, improve nutrient uptake and reduce underground water pollution. The efficiency of foliar application of nutrients can further be enhanced with the addition of surfactants or adjuvant into the spray solution. Surfactants can be classified into various groups on the basis of their composition such as cationic, anionic, non-ionic, zwitterionic or ampholytic. The surfactants are widely used in foliar application of both macro- and micronutrients. This chapter reviews the available information about the use of surfactants in the foliar application of mineral elements in citrus with special reference to nitrogen (N), calcium (Ca), iron (Fe), zinc (Zn), boron (B) or copper (Cu).

Keywords

Adjuvant • Absorption • Citrus • Cuticle • Essential nutrients • Foliar application • Macronutrients • Micronutrients • Nutritional disorders • Surfactants • Trace elements

12.1 Foliar Nutrition in Citrus

Generally roots not only provide anchorage and support to plants but also help to absorb water and nutrients from the soil. The roots and shoots of plants including citrus are functionally separate but mutually dependent. The roots absorb water and mineral nutrients from the soil and transport to the areal parts including shoots and leaves. In return the shoots supply photosynthates to the roots. Presently the need of absorption of inorganic nutrients through plant leaves has

Z. Singh (✉)
Curtin Horticulture Research Laboratory, Department of Environment and Agriculture, School of Science, Faculty of Science and Engineering, Curtin University, GPO Box U 1987, Perth 6845, WA, Australia
e-mail: z.singh@curtin.edu.au; zora@muresk.curtin.edu.au

A.S. Khan
Institute of Horticultural Sciences, University of Agriculture, Faisalabad 38040, Pakistan
e-mail: ahmad_khan157@yahoo.com

A.K. Srivastava (ed.), *Advances in Citrus Nutrition*,
DOI 10.1007/978-94-007-4171-3_12,

increased due to fixation of some essential nutrients in the soil which become unavailable to plant roots (Fernández et al. 2009; Srivastava and Singh 2003a). Absorption and translocation of some nutrient elements are slow by roots, and in certain cases, relatively large amount of inorganic nutrients is required for root supply which leads to the soil and water pollution. Hence, the foliar application of nutrients is more efficient, quick and economical method of nutrient application. It offers faster rate of absorption and photosynthetic activity, ultimately modifying the physiology of the leaves with delayed onset of leaf abscission (Kannan 1986).

12.2 Foliar Application of Mineral Elements

Foliar application of nutrients is an environmentally friendly and target-oriented nutrient application technique. Many factors such as physical and chemical properties of the nutrient, time of application, molecular weight, concentration of active ingredient and environmental conditions regulate the penetration of spray solution through leaf surface particularly the penetration through cuticle and stomata (Riederer and Schneider 1990; Fernandez and Ebert 2005; Srivastava et al. 2008; Fernandez and Eichert 2009). Earlier, the foliar application of both macro- and micronutrients has been extensively used in citrus. The effects of foliar applications of various nutrients such as N (Khan et al. 2009; Lovatt 1999), Mg (Embleton and Jones 1958), Ca (Dong et al. 2007), Fe (Papadakis et al. 2007; Pestana et al. 2001), Zn (Embleton et al. 1964), Mn (Papadakis et al. 2007), B (Dong et al. 2007; Papadakis et al. 2003) and Cu (Orbovc et al. 2007) have been investigated in different citrus species. Effectiveness of these foliar applications of nutrients is generally accessed in relation to its penetration, reduction or deficiency correction, improvement in yield and management of fruit quality (Khan et al. 2009; Leonard 1967; Lovatt 1999; Srivastava et al. 2008).

Foliar application of nutrients is an additional tool to supplement the normal soil feeding under special situation such as at the critical times of nutrient demand or in conditions of lower availability of these nutrients in the soil (Khan et al. 2009). The morphological and anatomical feature of citrus leaves affects the absorption and penetration of the spray material (Bondada et al. 2006). The role of number and morphology of stomata, structure and composition of cuticle affecting absorption and penetration of nutrients applied as a foliar spray has been investigated in citrus (Bondada et al. 2006; Orbović et al. 2001). Recently, growth and self-assembly of cuticular and epicuticular waxes have been examined and studied in detail with atomic force microscopy (Koch et al. 2004). Hence, characteristics of leaf surface and presence of epicuticular waxes determine the rate of retention, wettability and absorption of spray material (Koch and Ensikat 2008). In addition, the use of surfactant or adjuvant further facilitates penetration and absorption of the nutrients through leaves. Environmental factors such as temperature, light, relative humidity and wind speed also affect the amount and composition of epicuticular waxes, structure of leaf surface and stomata functioning (Ramsey et al. 2005). Addition of the adjuvant proves beneficial to modify the physico-chemical properties of spray solution to effectively wet the leaf surface (Hazen 2000).

12.3 Anatomy of Citrus Leaf

Like other angiosperms, a complete citrus leaf consists of petiole and leaf blade. Petiole may be broadly winged, winged, or wingless, depending upon the species and cultivars. The upper and lower surfaces of leaves are covered with epidermis. The upper epidermis of mature leaf is composed of tubular parenchyma cells covered with a thick layer of cuticle, whereas large calcium-oxalate-containing palisade cells protrude into the upper epidermis. Oil glands are also present in the thin-walled upper epidermal cells arranged in concentric order (Schneider 1968). The lower epidermis consisting of parenchyma cells interspersed with stomata. Stomata composed of a pore and two guard cells. A resinous plug is formed in the outer chamber of stomata where ridges of cuticle overhang the stomata opening, which prevent the penetration of water and other solutes through stomata (Turrell 1947). The cuticle covering the upper and lower epidermis is the first barrier for absorption of liquids through leaf surface which makes the leaf surface waterproof (Orbović et al. 2001).

The cuticle is composed of soluble cuticular lipids which are embedded into the matrix of lipid polyester cutins in citrus (Jeffree 1996). Fatty acids, primary alcohols, esters and hydrocarbons have been identified as major constituents of cuticle in citrus (Haas and Schönherr 1979). The changes in the quantity and chemistry of cuticular waxes throughout the development of citrus leaves influence the wettability and permeability of citrus leaf cuticles (Haas and Schönherr 1979; Jeffree 1996).

12.4 Mechanism of Nutrient Uptake

Absorption of nutrients applied through foliar sprays starts with surface absorption followed by passive penetration through cuticle and finally active absorption by leaf cell (Durstberger et al. 2008; Wittwer and Teubner 1959). The cuticular membrane present in the outer surface of leaf is the first barrier for the entry of solutes. It consists of a thin layer of cutin, wax and cellulose. Numerous channels and pores have been identified in the leaf cuticle. The inner surface of the leaf cuticle is attached with epidermal cells. Ions have to enter the cell wall before absorption by epidermal cells (Fernandez and Eichert 2009). Presence of apoplastic regions for transport in the cell wall facilitates absorption of the nutrients. The solutes after penetrating the cuticle and cell wall reach the intercellular spaces, and from here these are absorbed by leaf cells (Wojcik 2004). This entry of solutes across the cell membrane, plasmalemma, is an active phenomenon and is regulated by metabolic energy derived from the photosynthesis and respiration (Durstberger et al. 2008).

The surface of leaf cuticle is not smooth and covered by natural waxy layer, protuberances and structures like trichomes which increase surface tension. Hence, application of any additive into the solution reduces the surface tension and increases the area for the absorption of solute. Various surfactants or adjuvants are commercially used to improve the absorption of foliar applied nutrients in fruit crops including citrus (Neumann and Prinz 1974; Swietlik and Faust 1984).

12.5 Surfactants

The term surfactant or adjuvant is generally used for any surface-active agent added into the solution to improve its performance and effectiveness. In other words, the surfactants are those compounds that cause reduction in surface tension and alter the energy relationships at interfaces. Due to presence of both hydrophobic and hydrophilic group within the same molecule, these compounds adjust themselves as interfaces (Colwell and Rixon 1961). On the basis of presence and nature of electrical charge or absence of ionisation on the hydrophilic portion of the molecule, the surfactants are classified as anionic, cationic, non-ionic or ampholytic part of the molecules (Parr and Norman 1965).

12.6 Classification of Surfactants

Surfactants can be classified separately according to the composition of their tail and head structure (Table 12.1). Non-ionic surfactants have no charge groups in their heads, whereas the heads of ionic surfactants carry net charge. In ionic surfactants, if the charge is negative, the surfactant is more specifically called anionic; however, if the charge is positive, it is called cationic. In addition, if surfactant contains a head with two oppositely charged groups, it is called zwitterionic. These are compounds having an alkyl-substituted benzene, naphthalene or paraffinic chain ring that constitute hydrophobic part linked with a negatively charged carboxyl, sulphate, sulphonate or phosphate hydrophilic group. These mainly include alcohols and/or fatty acids, and due to lower surface tension, they improve spreading, sticking and uptake of the sprayed materials.

Cationic quaternary ammonium, arsonium, iodonium, phosphonium or sulphonium compounds have similar hydrophobic groups as in anionic surfactants. However, in addition they also linked with positively charged hydrophilic group. In non-ionic compounds, the hydrophobic group is associated with non-ionised hydrophilic groups as polymerised esters of polyhydric alcohols, or polyether alcohols ethylene oxide (Nelson and Garlich 1969; Parr and Norman 1965). The non-ionic surfactants are active surface-tension depressants, and due to lack of ionisation, chemically they are inert. When the concentration of surfactants either ionic or other types exceeds a critical level in an aqueous system, they form molecular aggregates or micelles. This critical level has been found to be associated with abrupt changes in many characteristics of the surfactant.

The cuticle is a heterogeneous, lipophilic contiguous extracellular polymeric membrane covering the upper and lower leaf epidermis (Bondada et al. 2006). Before absorption, the foliar applied nutrients must penetrate through this primary barrier (Bukovac et al. 1981). In certain citrus species, the epicuticular waxes occurred as platelets that increased with leaf age, and dewaxing the cuticles has been reported to significantly enhance the penetration rate of nutrients up to 60% (Bondada et al. 2006).

Zwitterionic (amphoteric) surfactants consist of primary, secondary or tertiary amines or quaternary ammonium cation with one of the mentioned groups, whereas ampholytic surfactants are compounds that consist of similar molecular arrangement as hydrophilic groups with the capacity to become cationic in an acidic medium and anionic in a basic medium. Due to lack of ionisation, these surfactants are inert and have greater application in biological systems since they work as surface-tension depressants (Corkill et al. 1961). Organo-silicones are usually blends of silicone to non-ionic or other surfactants: a few within this classification are composed entirely of silicone. These surfactants provide a tremendous reduction in surface tension and spread more than conventional surfactants. Esterified seed oil surfactants are produced by reacting fatty acids from seed oils with an alcohol to form esters. The methyl or ethyl esters produced by this reaction are combined with surfactants/emulsifiers to form esterified seed oil. These surfactants reduce surface tension and improve herbicide uptake by improving herbicide distribution on the leaf surface.

The concentration at which the surfactants reach to a critical level and form micelle is known as the critical micelle concentration (CMC). The maximum ability of a surfactant to lower the surface and interfacial tensions in an aqueous system is attained at CMC range. Hence, the micelles form in water, their tails form a core that can encapsulate an oil droplet and their ionic/polar heads maintain favourable contact with water. The aggregate formed by surfactants in oil is referred to as a reverse micelle. In this situation, the tails maintain favourable contact with oil, and heads are always in the core (Parr and Norman 1965).

Usually increase in the concentration of surfactant, the osmotic pressure also increased until a plateau is reached in

Table 12.1 Classification of surfactants

Based on the composition of surfactant tail		
(a) A hydrocarbon chain		Alkanes, alkenes, aromatic hydrocarbons and cycloalkanes
(b) An alkyl ether chain		Ethoxylated surfactants, propoxylated surfactants
(c) A fluorocarbon chain		Fluorosurfactants
(d) A siloxane chain		Siloxane surfactants
Based on the composition of surfactant head		
(a) Anionic surfactants	Carboxylates	Alkyl carboxylates
		i. Sodium stearate
		ii. Fatty acid salts
		Sodium lauroyl sarcosinate
		Carboxylate fluorosurfactants
		i. Perfluorooctanoate (PFOA)
		ii. Perfluorononanoate (PFNA)
	Phosphates	Alkyl ether phosphate (AEP)
		Alkyl aryl ether phosphate (AAEP)
	Sulphonates	Docusates
		i. Dioctyl sodium sulfosuccinate
		Sulphonate fluorosurfactant
		i. Perfluorooctanesulphonate (PFOS)
		ii. Perfluorobutanesulphonate
		Alkyl benzene sulphonates
	Sulphates	Alkyl sulphates
		i. Ammonium lauryl sulphate
		ii. Sodium lauryl sulphate
		iii. Sodium dodecyl sulphate (SDS)
		Alkyl ether sulphates
		i. Sodium laureth sulphate (SLES)
		ii. Sodium myreth sulphate (SMS)
(b) Cationic surfactants	pH-dependent primary, secondary or tertiary amines	Primary amines become positively charged at pH <10
		Secondary amines become charged at pH <4
		i. Octenidine dihydrochloride
	Permanently charged quaternary ammonium cation	Cetyl trimethylammonium bromide (CTAB)
	Alkyltrimethylammonium salts	Cetyl trimethylammonium chloride (CTAC)
		Cetylpyridinium chloride (CPC)
		Polyethoxylated tallow amine (POEA)
		Benzalkonium chloride (BAC)
		Benzethonium chloride (BZT)
		5-Bromo-5-nitro-1,3-dioxane (BND)
		Dimethyldioctadecylammonium chloride
		Dioctadecyldimethylammonium bromide (DODAB)
(c) Non-ionic	Polyoxyethylene glycol	Octaethylene glycol monododecyl ether
		Pentaethylene glycol monododecyl ether
	Fatty alcohols	Cetyl alcohol
		i. Stearyl alcohol
		ii. Cetostearyl alcohol
		iii. Oleyl alcohol
	Polyoxypropylene glycol alkyl ethers	
	Block copolymers of polyethylene glycol and polypropylene glycol	Poloxamers
	Polyoxyethylene glycol octylphenol ethers	i. Triton X-100
		ii. Triton B-1956
		iii. Tween 20
		iv. Tween 80

(continued)

Table 12.1 (continued)

Based on the composition of surfactant tail		
	Glucoside alkyl ethers	Decyl glucoside
		i. Lauryl glucoside
		ii. Octyl glucoside
	Polyoxyethylene glycol sorbitan alkyl esters	Polysorbates
	Sorbitan alkyl esters	Spans
	Cocamide MEA, cocamide DEA	
	Polyoxyethylene glycol alkylphenol ethers	Nonoxynol-9
	Glycerol alkyl esters	Glyceryl laurate
	Dodecyldimethylamine oxide	
(d) Zwitterionic (amphoteric)	Carboxylates	Amino acids
		i. Betaines
		ii. Imino acid
	Sulphonates	CHAPS (3-[(3-Cholamidopropyl)dimethylammonio]-1-propanesulphonate)
		Sultaines: cocamidopropyl hydroxysultaine
	Phosphates	Lecithin
(e) Ampholytic surfactants		
(f) Organo-silicones		
Esterified seed oils		

the CMC range. Surfactant molecules lose their independent existence in the solution and become part of the micellar aggregate when surfactant molecules added in excess as indicated by the osmotic pressure. Generally, these aggregates are arranged in a way that the hydrophobic part is always towards the outside and the hydrophilic towards the inner side. In certain cases, some surfactants may reach CMC at a concentration lower than 0.01% (Baur 1999).

12.7 Foliar Application of Nitrogen

Nitrogen (N) plays an important role in the tree nutrition for consistent bearing and quality citrus production. Lower N level in citrus trees is one of the limiting factor affecting fruit yield and as well as quality. Usually, like other fruit crops, in citrus, N is applied through soil application. However, efficacy of soil-applied N is affected by many unrelated factors, and also there is always risk of groundwater pollution (Albrigo 1999; Huett 1996). Hence, urea applied as a foliar spray is rapidly absorbed by citrus leaves (Ali and Lovatt 1992). Foliar application of low-biuret urea (LBU) below 0.5% has been reported to use as a foliar spray throughout the world for citrus without causing any phytotoxic effects (Albrigo 2002; Jones and Embleton 1954). Application of N fertilisation has been reported to greatly influence the quality of fruit, and particularly the foliar spray applications of urea are an efficient and cost-effective way to supply N to citrus, which enhanced the fruit size, peel thickness, juice content and yield in some citrus species (El-Otmani et al. 2002; Khan et al. 2009; Rabe 1994). When urea is sprayed onto the citrus leaves, it first must diffuse through the cuticles through micropores traversing the cuticular matrix (Schonherr and Schmidt 1979). However, in certain cases, urea itself can also act as a surfactant because it can increase the uptake of foliar applied nutrients (Swietlik and Faust 1984).

12.8 Foliar Application of Calcium

Calcium being an essential macronutrient plays an important role in plant growth and development. It stimulates cell division and elongation and builds the structure and permeability of cell membrane. Calcium is a major part of cell wall, especially in the middle lamella which stores 60% of calcium, and the rest of calcium is in the cell membrane (Huang et al. 2008; Poovaiah et al. 1988). It cross-links the pectin polysaccharide matrix and strengthens the structural rigidity of the cell wall. Therefore, calcium deficiency in the cell walls results in solubilisation of pectin, acceleration of senescence (Zaragoza et al. 1996) and development of some physiological disorders such as albedo breakdown in sweet oranges (Storey and Treeby 2002; Treeby and Storey 2002). Foliar applications of calcium solution have not always significantly increased the calcium concentration in citrus fruit as calcium is mobile in xylem and cuticles are the first barriers to prevent the penetration of calcium into leaves and fruit (Schonherr 2001; Treeby and Storey 2002).

Table 12.2 Effects of different surfactants added into an aqueous spray solution containing 2% $Ca(NO_3)_2$ on the concentrations of Ca in leaf, fruit rind and pulp tissues at 101 and 195 days after sprays (182 and 276 days after full bloom, respectively) in 'Washington Navel' orange

	Days after sprays (days after full bloom)					
	Ca (%)					
	Leaf		Rind		Pulp	
Treatments	101 (182)	195 (276)	101 (182)	195 (276)	101 (182)	195 (276)
Control	2.09b	2.29	0.38d	0.33c	0.13b	0.11b
$Ca(NO_3)_2$ 2%	2.17b	2.56	0.44d	0.36bc	0.22a	0.12b
$Ca(NO_3)_2$ 2% and Tween 20	2.74a	2.36	0.76a	0.53a	0.25a	0.14a
$Ca(NO_3)_2$ 2% and Tween 80	2.27b	2.62	0.60bc	0.44abc	0.23a	0.12b
$Ca(NO_3)_2$ 2% and Triton X100	2.32b	2.35	0.55cd	0.40bc	0.24a	0.12b
$Ca(NO_3)_2$ 2% and Tergitol	2.27b	2.51	0.66ab	0.47ab	0.25a	0.13ab
LSD ($P \leq 0.05$)	0.25	ns (0.13)	0.15	0.12	0.04	0.02

Adapted from Pham et al. (2010)
Within each column, means followed by different letters are significantly different at $P \leq 0.05$
$n=4$ replications, ns = not significant at $P \leq 0.05$. Values within the bracket in the last row represent standard errors of means (SE)

Calcium uptake can be improved by its foliar application in soluble form such as calcium nitrate or calcium chloride. In addition, absorption of Ca can further be enhanced by addition of surface-active agents which increased the efficiency of foliage-applied Ca (Wojcik 2004). It has been reported that surfactant can enhance the uptake of calcium into plant tissues through different pathways. First, surfactants enhanced the uptake of Ca ions due to improving a distribution of Ca ions as surfactants resulted in the lower contact angles of spray solution on the leaf surface (Schmitz-Eiberger et al. 2002). Second, surfactants also induce the penetration of solutes through the stomata, cuticular membranes and the cell wall, and eliminate or decrease the air layer between the liquid and leaf surfaces (Wojcik 2004). Third, surfactants improve the calcium uptake due to reduction in the drying of droplets, an increase in binding capacity of the cuticle to Ca^{2+} (Wojcik 2004). Finally, surfactants also remove sites of adsorption by damaging and extracting cuticle wax. Addition of surfactants in the spray solutions to improve the absorption and penetration of foliar applied Ca has been reported in fruit crops such as apple (Harker and Ferguson 1991; Saure 2005; Schönherr 2000) and mango (Singh et al. 2000) and citrus (Pham et al. 2010). Recently, foliar applications of aqueous solutions containing 2% $Ca(NO_3)_2$ and 'Tween 20' (0.05%), 'Tween 80' (0.05%), 'Triton X-100' (0.05%) or 'Tergitol' (0.05%) as a surfactant commencing at 81 days after full bloom with 10-day intervals significantly increased the Ca penetration through leaf, rind and pulp of 22-year-old 'Washington Navel' orange trees as compared to control (Table 12.2). However, among different surfactants, the 'Tween 20' (0.05%) was most effective in improving Ca uptake and reducing albedo breakdown in the 'Washington Navel' orange compared to all other surfactants (Pham et al. 2010). The surfactants added to an aqueous solution containing $Ca(NO_3)_2$ can be used to enhance the Ca uptake into sweet orange leaf and fruit due to the lower surface tension between droplet and surface of leaf and fruit leading to better distribution of Ca ion on the surface of leaf and fruit. The efficiency of surfactants on enhancing the Ca uptake into leaf and fruit may partially attribute to increased binding capacity of the cuticle for Ca ion due to the improved wetting on the leaf and fruit of surfactants. The efficiency of surfactants to enhance the Ca uptake was associated with the value of hydrophilic-lipophilic balance (HLB) of surfactants. Surfactants with higher HLB value were more effective in improving the Ca uptake into leaf and fruit (Wojcik 2004). Among different surfactants, the 'Tween 20' was the most effective in enhancing the uptake of Ca into leaf and fruit without affecting other fruit quality parameters such as percentage of juice, juice pH, soluble solids concentration, ascorbic acid and individual organic acids in 'Washington Navel' sweet oranges (Pham 2009).

12.9 Foliar Application of Iron

The deficiency of iron (Fe) is widespread nutritional disorder and limiting factor for successful commercial cultivation of various fruit trees including citrus. This problem is generally associated with accumulation of excessive amount of lime in the soil (Horesh and Levy 1981). Such types of conditions are characterised as 'lime-induced chlorosis'. Citrus trees affected with such chlorosis sometimes also develop unidentified gummosis on the limbs of the tree.

Iron deficiency symptoms are more prevalent in calcareous soil with higher soil pH (Fernández et al. 2009). The Fe deficiency in trees may cause interveinal leaf chlorosis, reduced shoot growth and excessive leaf fall, and under severe cases trees may die (Rombola and Tagliavini 2006). The deficiency of Fe also affects reproductive growth which can be judged with reduction in fruit set, fruit size, yield and poor fruit colour, reduced fruit firmness and acidity (Alvarez-Fernandez et al. 2004, 2006, 2007). Among various strategies used to correct Fe chlorosis in fruit crop including use of resistant rootstock, soil amendments and soil application of Fe chelates, the foliar Fe application has been found as more targeted, quick and economical strategy to correct Fe chlorosis. However, many plant, environmental and physico-chemcial factors may influence the plant response to Fe sprays (Fernandez and Ebert 2005).

The corrections of such deficiency are possible with application of Fe chelates (Embleton et al. 1973). However, the higher dose of chelates required to correct this deficiency makes it uneconomical. The soil application of iron in calcareous soils and particularly in the orchards where trifoliate orange (*Citrus trifoliata* L. Ref.) or its hybrids are used as rootstock may aggregate the situation as these rootstocks are sensitive to Fe deficiency. However, the foliar application of iron compounds such as ferrous sulphate with low surface tension surfactant (0.1% 1.77 silicon block copolymer) has been reported to effectively restore the photosynthesis and peroxidase activities and chlorophyll contents of chlorotic citrus trees as compared to the soil application of Fe 138 chelate (Horesh and Levy 1981; Koch et al. 2004).

Abiotic factors such as temperature, light and relative humidity have been found to play a vital role in the absorption of mineral nutrients through leaf surface. Physiological state of the plant, droplet drying and cuticular hydration also contribute towards the net absorption of nutrient elements through foliar sprays (Fernández et al. 2009). Increased relative humidity up to certain limit increases the cuticular permeability and delays the drying of salt deposits (Schonherr 2001). Foliar application of Fe fertilisation at 30–60% RH showed the beneficial effects on Fe-deficient citrus, pear, peach, apple, mango, plum and almond crops in terms of leaf re-greening and improved fruit yield and quality (Fernandez and Ebert 2005; Fernández et al. 2009).

Penetration of Fe-containing solutions through the cuticle at 100% RH did not seem to be influenced by change in temperature from 15°C to 35°C (Schönherr et al. 2005). Similarly, light also stimulates stomatal opening and various biochemical processes such as photosynthesis or xylem flux which may influence foliar Fe uptake at some stage (Fernández et al. 2009). Absorption of light by leaves promotes Fe penetration into the symplast *via* a light-dependent, plasma membrane bound Fe (III) reductase (Fernandez and Ebert 2005; Koch et al. 2004).

Upper and lower leaf epidermises are involved in the process of foliar nutrient uptake. Penetrability of the lower epidermis *vs.* the upper epidermis is mainly regulated by stomatal and cuticular variations between the two leaf surfaces (Wojcik 2004). The variation in the structure and composition of cuticle, size and distribution of stomata and differences among various plant species has been found to affect the uptake of foliar nutrients through leaves (Schonherr 2006). Young and partially open leaves have more penetration than older and expanded ones. The morphological and physiological states of the plant such as root osmotic potential and root temperature also influence the absorption of foliar applied iron. Addition of hygroscopic humectants in the spray solution has been reported to improve the performance of Fe sprays in arid and semi-arid areas (Schönherr et al. 2005). Adjuvant or surfactants increase wetting and spray droplet retention. However, the mechanism through which they influence the uptake of foliar sprays is complex and unclear (Liu 2004). Lime-induced Fe chlorosis can only be corrected with the foliar application of $FeSO_4$ with low-surface-tension surfactants at the adaxial (upper) surface than abaxial (under) surfaces of citrus leaves (Levy and Horesh 1984).

Application of electrospray-ionisation time-of-flight mass spectrometry (ESI-TOF MS) has enabled accurate functional end-group identification and quantification of ionisable compounds by using exact mass/charge (*m/z*) ratio determinations of surfactant and Fe-containing mixtures (Alvarez-Fernández et al. 2007).

Under severe deficiency of Fe, the leaves become light green to yellow or sometimes nearly white between the veins due to reduced level of chlorophyll. These symptoms are more prominent in young leaves. In the acidic soils, the correction of iron chlorosis in citrus orchards can be achieved with soil application of 20–30 g Fe per tree as iron ethylenediamine tetraacetic acid (FeEDTA). However, in calcareous soils, application of ferric ethylenediamine di(o-hydroxy phenyl) acetate (FeEDDHA) at 10–30 g per tree is quite expensive treatment and is not economical. As an alternative, the application of different water-soluble sources of Fe with dimethyl sulfoxide (DMSO) has been found to markedly improve the Fe content with rapid leaf greening and higher leaf chlorophyll contents of Fe chlorotic orange and grapefruit trees than without DMSO (Wallihan et al. 1964).

The choice of a surfactant to be used with foliar nutrient sprays depends upon its ability to be free from any type of phytotoxic effects. In addition to the phytotoxicity, the surface tension and ability to damage the cell membrane and spreading ability also influence its effectiveness. Previously, attempts to overcome the Fe deficiency in citrus trees with the foliar application of Fe have remained unsuccessful, despite the addition of surfactants into the spray solution (Wallihan et al. 1964). Penetration of spray solution through leaves is more effective and efficient when surface tension of a spray solution is sufficiently low. Hence, surfactants with lower surface tension such as fluorocarbon- and silicon-based chemical have been found to have less phytotoxic effect than by carbohydrate-based surfactants (Neumann and Prinz 1974). Foliar application and absorption of nutrients needs a non-destructive delivery system. It is mainly achieved by spraying these elements through adjuvant. Acute shortage of nutrients, such as Fe deficiency, can efficiently be recovered by using silicon-based adjuvant. Although most of the adjuvants used in foliar application of nutrients are not harmful to plant foliage, their effect on adsorption of nutrients by leaf under orchard conditions is very small. Radioisotope assays by using scanning electron microscope and mineral analyses clearly indicate that Ferti-Vant (a non-destructive adjuvant) is quite effective in enhancing the penetration of nutritional elements into citrus leaf under field conditions as compared to other common spraying adjuvants (Wiesman et al. 2002). The foliar application of Fe, P and K with Ferti-Vant has been found to increase their absorption by improving the surface areas covered by leaves and adherence with leaf surface in comparison to other common surfactants. No irreversible epicuticular wax-dissolving effect in citrus leaves treated by $Fe_2(SO_4)$ coated with Ferti-Vant was detected with scanning electron microscope or in vivo (Wiesman et al. 2002).

12.10 Foliar Application of Zinc

Deficiency of zinc (Zn) in citrus orchards has been observed globally and is one of the key issues for sustainable citrus production throughout the world (Srivastava and Singh 2003b). Various physiological, morphological and biochemical abnormalities are induced by Zn deficiency in citrus plants, which ultimately reduce flowering, fruit set and yield both quantitatively as well as qualitatively (Srivastava and Singh 2005). Generally, the deficiency of Zn in citrus is characterised by occurrence of interveinal chlorosis, which in certain cases is coupled with rosetting. The terms rosette, little leaf, frenching and mottle leaf, etc., are also used for deficiency of Zn (Srivastava and Srivastava 1992). Competition of Zn with some other mineral elements such as P, Fe, Mn and Ca may lead to its deficiency. Various sources such as inorganic compounds, synthetic and natural chelates and natural organic complexes are used for Zn application to citrus trees (Mortvedt 1991). Application of diethylene triamine pentaacetic acid (DTPA) has been reported as most efficient source for Zn application in alkaline soils, whereas in acid to neutral soils, hydroxyethylethylenediaminetriacetic acid (HEDTA) has been found as best chelate for soil application (Swietlik 1989; Swietlik and Faust 1984).

Poor mobility of Zn in soil has discouraged its soil application. The soil application of Zn is not very effective as roots of the fruit crop occupy deep soil layers and Zn does not easily move within the soil (Swietlik 2002). Majority of research work carried out on Zn nutrition in citrus has revealed better effectiveness of foliar application of Zn in combination with other trace elements such as Mn, Cu and Fe as compared to it alone or soil application (Srivastava and Singh 2005). Chelates of Zn have been found to have better penetration through citrus leaves than inorganic Zn salts. Citrus leaves absorb EDTA better than roots. Presence of chelates like EDTA not only improves absorption but also helps to improve translocation within plant (Arora et al. 1970).

12.11 Foliar Application of Boron

Boron (B) is another essential micronutrient which plays significant physiological functions during growth and development of citrus (Zekri and Obreza 2003). It is different from other micronutrients because it has very narrow threshold between deficiency and toxicity (Yua and Ryan 2008). It has been reported to improve pollen grain germination, pollen tube elongation consequently fruit set and finally yield in citrus (Abd-Allah 2006). B toxicity is a nutritional disorder, which decreases plant growth and productivity particularly in the arid and semi-arid citrus-growing areas of the world (Papadakis et al. 2003). Deficiency of B is more extensive than that of any other micronutrient which adversely affects yield and fruit quality in citrus (Zekri and Obreza 2003). Symptoms of B deficiency appear both on vegetative and reproductive parts of the citrus trees such as development of translucent or water-soaked flecks on leaves, defoliation of leaves, yellowing and enlargement of midrib, die back of twigs, gum formation in bark, poor fruit set, excessive fruit abortion, rind thickness and granulation (Chapman 1968; Karim et al. 1996; Maurer and Davies 1993).

Calcareous and alkaline soils limit the availability of B and induce deficiencies in citrus grown under these conditions. The B is also readily leached from soils and should be regularly replaced. Low remobilisation of B in plants is a problem compounding the management of this nutrient. Strategies should be aimed at applying a number of small amounts of B throughout the growing season. Soil and foliar application are equally used to supply essential nutrients to trees. There is evidently a connection between photosynthetic production of specific sugars and mobility of B from the xylem to the phloem and subsequent transfer to various plant parts and organs (Brown and Shelp 1997).

Foliar application of B has been found to increase fruit set in 'Redblush' grapefruit and 'Hamlin' oranges (Karim et al. 1996; Maurer and Davies 1993). In contrary, pre- and post-bloom foliar application of Solubor from 250 to 100 mg L^{-1} with Activator-90 (0.1% v/v) as a surfactant did not significantly affect the physical and biochemical characteristics of 'Navel' oranges than untreated trees (Maurer and Truman 2000). The application of B at higher concentrations may induce adverse impact on the growth and productivity of trees. Pre-bloom foliar application of Solubor at 2,000–2,500 mg L^{-1} had been reported to decrease fruit set in 'Kinnow' mandarins (Maurer and Taylor 1999).

12.12 Foliar Application of Copper

Copper (Cu) is an essential mineral element needed for plants in very small quantity. It is also an important component of fungicides used to control many fungal diseases in citrus such as Phytophthora rot, Alternaria rot, scab and blue and green moulds (Rogers and Timmer 2006). Previously, inappropriate use of adjuvant into foliar Cu sprays has been found to induce injuries in citrus. The use of oil as an adjuvant in the spray solution further enhances the phytotoxic effects of Cu in some citrus species (Albrigo et al. 1997). The penetration of foliar Cu through leaf surface has been reported to be improved by using different surfactants such as organosilicone L-77 (Field et al. 1992). Addition of silicon-based L-77 surfactant with Kocide fungicide suspension markedly increased the penetration of Cu through abaxial leaf surface and stomata pores in 'Marsh' grapefruit as compared to urea and spray oil (Orbović et al. 2007). L-77 surfactant has been found to increase the spread of droplets of the spray solution by lowering surface tension (Schonherr and Bukovac 1972). L-77 is considered as a relatively safe chemical for foliar sprays because it has minor negative effects on citrus photosynthesis, when applied alone or in combination with urea (Orbović et al. 2001). Mixing of organosilicone surfactant in Cu-based fungicides should be avoided because these may enhance phytotoxicity through increased penetration of Cu into citrus leaves by reduction in pH.

12.13 Conclusions and Future Research

Both soil and foliar applications of mineral elements such as N, K, Ca, Fe, Cu, Zn, Mn and B are used commercially to fulfil the nutritional requirements of citrus throughout the world. Under certain situations, the foliar fertilisation has been proved more effective than the soil application, particularly when nutrients are completely absent from the soil, severer deficiency of nutrients, condition of nutrient unavailability, nutrient imbalance and/or nutrient immobilisation. The foliar application of nutrients is also effective whether applied as a single element or in combination with other mineral elements. The foliar application of nutrient(s) is more effective than the soil fertilisation due to its higher efficiency, quick plant response and convenience in application, and it minimises underground water pollution. However, the absorption and penetration of foliar applied nutrients is influenced by the phonological stage, nutritional status of the tree, environmental factors (light, temperature, wind speed, humidity and spray time) and formulation of spray materials (concentration, application techniques, water quality, amount and nature of additives). The efficacy and utilisation of foliar application of nutrients have been improved with use to additives and surfactants in the spray solution to alleviate their nutritional deficiencies. However, a wide range of surfactants are available for foliar fertilisation; therefore, the selection of a surfactant for regular commercial foliar application of mineral nutrients should be done in relation to epicuticular and cuticular wax composition depending upon citrus species. Advanced fertilisation techniques such as electrostatic sprayers and sonic bloom, which have recently further expanded the scope of foliar nutrient application, need to be studied for its potential commercial application in citrus. Previously most of the research work conducted is mainly focusing on the corrective measure to overcome deficiencies of both macro- and micronutrients through foliar nutrition. In the current climate change scenario, the understanding of the role of additives in the foliar nutrition during various phenological stages of citrus, especially at the critical times, will prove beneficial to develop models under different citrus-based cropping systems. Along with the traditional foliar nutrition of citrus tree, there is need to develop some molecular and biochemical markers to facilitate round-the-year foliar nutritional management through a better use of precision-oriented informatics.

References

Abd-Allah AS (2006) Effect of spraying some macro and micro nutrients on fruit set, yield and fruit quality of Washington Navel orange trees. J Appl Sci Res 2:1059–1063

Albrigo LG (1999) Effects of foliar applications of urea or nutriphite on flowering and yield of Valencia orange trees. Proc Fla State Hortic Soc 112:1–4

Albrigo LG (2002) Foliar uptake of N-P-K sources and urea biuret tolerance in citrus. Acta Hortic 594:627–633

Albrigo LG, Timmer LW, Townsend KT et al (1997) Copper fungicides–residues for disease control and potential for spray burn. Proc Fla State Hortic Soc 110:67–70

Ali AG, Lovatt CJ (1992) Winter applications of foliar urea. Citrograph 78:7–9

Alvarez-Fernandez A, García-Lavina P, Fidalgo J et al (2004) Foliar fertilization to control iron chlorosis in pear (*Pyrus communis* L.) trees. Plant Soil 263:5–15

Alvarez-Fernandez A, Abadía J, Abadía A (2006) Iron deficiency, fruit yield and quality. Springer, Dordrecht

Alvarez-Fernández A, Orera I, Abadía J et al (2007) Determination of synthetic ferric chelates used as fertilizers by liquid chromatography-electrospray mass spectrometry in agricultural matrices. J Am Soc Mass Spectrom 18:37–47

Arora JS, Yoshitake T, Matsumoto K (1970) Studies on the absorption and translocation of 65$ZnCl_2$ by the Satsuma orange leaves. J Jpn Soc Hortic Sci 39:298–302

Baur P (1999) Surfactant effects on cuticular penetration of neutral polar compounds: dependence on humidity and temperature. J Agric Food Chem 47:753–761

Bondada BR, Petracek PD, Syvertsen JP et al (2006) Cuticular penetration characteristics of urea in citrus leaves. J Hortic Sci Biotechnol 81:219–224

Brown PH, Shelp BJ (1997) Boron mobility in plants. In: Bell RW, Rerkasem B (eds) Boron in soils and plants. Kluwer Academic Publishers, Dordrecht, pp 85–101

Bukovac MJ, Rasmussen HP, Shull VE (1981) The cuticle: surface structure and function. In: Johari O (ed) Scanning electron microscopy. SEM Inc AMF O'Hare, Chicago

Chapman HD (1968) The mineral nutrition of citrus. In: Reuther W, Batchelor LD, Webber HJ (eds) Citrus industry, vol 2. Division of Agricultural Science, University of California, Berkeley

Colwell CE, Rixon WE (1961) Consideration in the use of nonionic surface-active agents. Am Dyestuff Rep 50:39–42

Corkill JM, Goodman JF, Ottewill RH (1961) Micellization of homogenous non-ionic detergents. Trans Faraday Soc 57:1627–1636

Dong T, Xia R, Xiao Z et al (2007) Effect of pre-harvest application of calcium and boron on dietary fibre, hydrolases and ultrastructure in 'Cara Cara' navel orange (*Citrus sinensis* L. Osbeck) fruit. Sci Hortic 121:272–277

Durstberger T, Bolhàr-Nordenkampf HR, Meister MH et al (2008) The foliar uptake of micronutrients. Acta Hortic 804:631–637

El-Otmani M, Ait-Qubahou A, Taibi FZ et al (2002) Efficacy of foliar urea as N source in sustainable citrus production systems. Acta Hortic 594:611–617

Embleton TW, Jones WW (1958) Correction of magnesium deficiency of orange trees in California. Proc Am Soc Hortic Sci 71:280–288

Embleton TW, Wallihan EF, Goodall GE (1964) Effect of soil vs. foliar applied zinc, and of foliar applied manganese on California lemons. Proc Am Soc Hortic Sci 86:253–259

Embleton TW, Reitz NJ, Jones WW (1973) Citrus fertilization. In: The citrus industry, vol 3. University of California, Berkeley, p 145

Fernandez V, Ebert G (2005) Foliar iron fertilization – a critical review. J Plant Nutr 28:2113–2124

Fernandez V, Eichert T (2009) Uptake of hydrophilic solutes through plant leaves: current state of knowledge and perspectives of foliar fertilization. Crit Rev Plant Sci 28:36–68

Fernández V, Orera I, Abadía J et al (2009) Foliar iron-fertilisation of fruit trees: present knowledge and future perspectives – a review. J Hortic Sci Biotechnol 84:1–6

Field RJ, Dobson NC, Tisdall LJ (1992) Species-specific sensitivity to organosilicone surfactant-enhancement of glyphosate uptake. In: Foy CL (ed) Adjuvants for agrichemicals. CRC Press, Boca Raton, pp 424–431

Haas K, Schönherr J (1979) Composition of soluble cuticular lipids and water permeability of cuticular membranes from citrus leaves. Planta 146:339–403

Harker FR, Ferguson IB (1991) Effects of surfactants on calcium penetration of cuticles isolated from apple fruit. Sci Hortic 46:225–233

Hazen JL (2000) Adjuvants – terminology, classification, and chemistry. Weed Technol 14:773–784

Horesh I, Levy Y (1981) Response of iron-deficient citrus trees to foliar iron sprays with a low-surface-tension surfactant. Sci Hortic 15:227–233

Huang XM, Wang HC, Zhong WL et al (2008) Spraying calcium is not an effective way to increase structural calcium in litchi pericarp. Sci Hortic 117:39–44

Huett DO (1996) Prospects for manipulating the vegetative-reproductive balance in horticultural crops though nitrogen nutrition: a review. Aust J Agric Res 47:47–66

Jeffree CE (1996) Structure and ontogeny of plant cuticles. In: Kerstiens G (ed) Plant cuticles-an integrated functional approach. BIOS Scientific Publishers, Oxford, pp 33–82

Jones WW, Embleton TW (1954) Nitrogen fertilizer and orange production. Calif Citrogr 39:120–122

Kannan S (1986) Physiology of foliar uptake of inorganic nutrients. Proc Indian Acad Sci 96:457–470

Karim MR, Wright GC, Taylor KC (1996) Effect of foliar boron sprays on yield and fruit quality of citrus. Citrus research report. University of Arizona, College of Agriculture, Tucson, Series

Khan AS, Malik AU, Pervez MA et al (2009) Foliar application of low-biuret urea and fruit canopy position in the tree influence the leaf nitrogen status and physico-chemical characteristics of kinnow mandarin (*Citrus reticulata* blanco). Pak J Bot 41:73–85

Koch K, Ensikat HJ (2008) The hydrophobic coatings of plant surfaces: epicuticular wax crystals and their morphologies, crystallinity and molecular self assembly. Micron 39:759–772

Koch K, Neinhuis C, Ensikat HJ et al (2004) Self-assembly of epicuticular waxes on living plant surfaces imaged by atomic force microscopy (AFM). J Exp Bot 55:711–718

Leonard CD (1967) Use of dimethyl sulfoxide as a carrier for iron in nutritional foliar sprays applied to citrus. Ann NY Acad Sci 141:148–158

Levy Y, Horesh I (1984) Importance of penetration through stomata in the correction of chlorosis with iron salts and low-surface-tension surfactants. J Plant Nutr 7:279–281

Liu Z (2004) Effects of surfactants on foliar uptake of herbicides – a complex scenario. Colloids Surf B Biointerfaces 35:149–153

Lovatt CV (1999) Timing citrus and avocado foliar nutrient applications to increase fruit set and size. HortTechnol 9:607–612

Maurer MA, Davies FS (1993) Use of reclaimed water for irrigation and fertilization of young "Redblush" grapefruit trees. Proc Fla State Hortic Soc 106:22–30

Maurer MA, Taylor KC (1999) Citrus and deciduous fruit and nut research report. College of Agriculture and Life Sciences, the University of Arizona, Tucson

Maurer MA, Truman J (2000) Effect of foliar boron sprays on yield and fruit quality of Navel oranges in 1998 and 1999. Citrus and deciduous fruit and nut research report. College of Agriculture and Life Sciences, the University of Arizona, Tucson

Mortvedt JJ (1991) Micronutrient fertilizer technology. In: Mortvedt JJ, Cox FR, Shuman LM et al (eds) Micronutrients in agriculture. Soil Science Society of America, Madison

Nelson PV, Garlich HH (1969) Relationship of chemical classification and hydrophile-lipophile balance of surfactants to enhancement of foliar uptake of iron. J Agric Food Chem 17:148–152

Neumann PM, Prinz R (1974) Evaluation of surfactants for use in the spray treatment of iron chlorosis in citrus trees. J Sci Food Agric 25:221–226

Orbović V, Achor D, Petracek PD et al (2001) Air temperature, humidity and leaf age affect penetration of urea through citrus leaf cuticles. J Am Soc Hortic Sci 126:44–50

Orbović V, Achor D, Syvertsen JP (2007) Adjuvants affect penetration of copper through isolated cuticles of citrus leaves and fruit. HortSci 42:1405–1408

Papadakis IE, Dimassi KN, Therios IN (2003) Response of two citrus genotypes to six boron concentrations: concentration and distribution of nutrients, total absorption, and nutrient use efficiency. Aust J Agric Res 54:571–580

Papadakis IE, Sotiropoulos TE, Therios IN (2007) Mobility of iron and manganese within two citrus genotypes after foliar applications of iron sulfate and manganese sulfate. J Plant Nutr 30:1385–1396

Parr JF, Norman AG (1965) Considerations in the use of surfactants in plant systems: a review. Bot Gaz 126:86–96

Pestana M, Correia PJ, Varannes A et al (2001) Effectiveness of different foliar iron applications to control iron chlorosis in orange trees grown on a calcareous soil. J Plant Nutr 24:613–622

Pham TTM (2009) Pre-harvest factors affecting fruit quality in sweet oranges with an emphasis on albedo breakdown. PhD thesis, Curtin University of Technology, Perth

Pham TTM, Singh Z, Behboudian MH (2010) Different surfactants improve the calcium uptake into leaf and fruit of 'Washington Navel' sweet orange [*Citrus sinensis* (L.) Osbeck] and reduce albedo break down. In: Proceedings of the XIth international citrus congress, Wuhan, China, vol 1, pp 595–602

Poovaiah BW, Glenn GM, Reddy ASN (1988) Calcium and fruit softening: physiology and biochemistry. Hortic Rev 10:107–152

Rabe E (1994) Yield benefits associated with pre-bloom low biuret urea sprays on Citrus spp. J Hortic Sci Biotechnol 69:495–500

Ramsey RJL, Stephenson GR, Hall JC (2005) A review of the effects of humidity, humectants, and surfactant composition on the absorption and efficacy of highly water-soluble herbicides. Pestic Biochem Physiol 82:162–175

Riederer M, Schneider G (1990) The effect of the environment on the permeability and composition of Citrus leaf cuticles.II. Composition of soluble cuticular lipids and correlation with transport properties. Planta 180:154–165

Rogers M, Timmer LW (2006) Florida citrus pest management guide. Univ Fla Coop Ext Serv IFAS SP-43

Rombola AD, Tagliavini M (2006) Iron nutrition of fruit tree crops. In: Abadía J, Barton LL (eds) Iron nutrition and interactions in plants. Springer, Dordrecht, pp 61–83

Saure MC (2005) Calcium translocation to fleshy fruit: its mechanism and endogenous control. Sci Hortic 105:65–89

Schmitz-Eiberger M, Haefs R, Noga G (2002) Enhancing biological efficacy and rainfastness of foliar applied calcium chloride solutions by addition of rapeseed oil surfactants. J Plant Nutr Soil Sci 165:634–639

Schneider H (1968) The anatomy of citrus. In: Reuther W, Batchelor LD, Webber HJ (eds) Citrus industry, vol 2. Divison of Agricultural Science, University of California, Berkeley

Schönherr J (2000) Calcium chloride penetrates plant cuticles via aqueous pores. Planta 212:112–118

Schonherr J (2001) Cuticular penetration of calcium salts: effects of humidity, anions, and adjuvants. J Plant Nutr Soil Sci 164:225–231

Schonherr J (2006) Characterization of aqueous pores in plant cuticles and permeation of ionic solutes. J Exp Bot 57:2471–2491

Schonherr J, Bukovac MJ (1972) Penetration of stomata by liquids, dependence of surface tension, wettability, and stomatal morphology. Plant Physiol 49:813–819

Schonherr J, Schmidt HW (1979) Water permeability of plant cuticles. Planta 144:391–400

Schönherr J, Fernández V, Schreiber L (2005) Rates of cuticular penetration of chelated Fe(III): role of humidity, concentration, adjuvants, temperature and type of chelate. J Agric Food Chem 53:4484–4492

Singh Z, Janes J, Tan SC (2000) Effect of different surfactants on calcium uptake and its effect on fruit ripening, quality and postharvest storage of mango under modified atmosphere packaging. Acta Hortic 509:413–417

Srivastava AK, Singh S (2003a) Foliar fertilization in citrus. Agric Rev 24:250–264

Srivastava AK, Singh S (2003b) Zinc nutrition. In: Citrus nutrition. International Book Distributing Company, Lucknow, pp 752–813

Srivastava AK, Singh S (2005) Zinc nutrition, a global concern for sustainable citrus production. J Sustain Agric 25:5–42

Srivastava AK, Srivastava OP (1992) Zinc fractions and availability of applied zinc in salt affected soil amended with gypsum and pyrites. J Indian Soc Soil Sci 40:890–892

Srivastava AK, Singh S, Albrigo LG (2008) Diagnosis and remediation of nutrient constraints in citrus. In: Janick J (ed) Hort Rev. John Wiley Sons Inc, Hoboken, NJ, USA, pp 277–364

Storey R, Treeby MT (2002) Nutrient uptake into Navel oranges during fruit development. J Hortic Sci Biotechnol 77:91–99

Swietlik D (1989) Zinc stress on citrus. J Rio Gd Val Hortic 42:87–95

Swietlik D (2002) Zinc nutrition of fruit trees by foliar sprays. Acta Hortic 93:123–129

Swietlik D, Faust M (1984) Foliar nutrition of fruit crops. Hortic Rev 6:287–355

Treeby MT, Storey R (2002) Calcium-spray treatments for ameliorating albedo breakdown in 'Navel' oranges. Aust J Exp Agric 42:496–502

Turrell FM (1947) Citrus leaf stomata: structure, composition and pore size in relation to penetration of liquids. Bot Gaz 108:476–483

Wallihan EF, Embleton TW, Sharpless RG (1964) Response of chlorotic citrus leaves to iron sprays in relation to surfactants and stomata apertures. Proc Am Soc Hortic Sci 85:210–217

Wiesman Z, Luber M, Ronen A et al (2002) Ferti-Vant – a new non-destructive and long-lasting in vivo delivery system for foliar nutrients. Acta Hortic 594:585–590

Wittwer SH, Teubner FG (1959) Foliar absorption of mineral nutrients. Ann Rev Plant Physiol 10:13–30

Wojcik P (2004) Uptake of mineral nutrients from foliar fertilization (Review). J Fruit Ornam Plant Res 12:201–218

Yua SK, Ryan J (2008) Boron toxicity tolerance in crops: a viable alternative to soil amelioration. Crop Sci 48:854–865

Zaragoza S, Almela V, Tadeo ER et al (1996) Effectiveness of calcium nitrate and GA3 on the control of peel-pitting of 'Fortune' mandarin. J Hortic Sci 71:321–326

Zekri M, Obreza TA (2003) Plant nutrients for citrus trees. UF-IFAS, Soil Water Sci Dept Ext Pub SL 200, pp 1–5

Rhizosphere Microbial Communities: Isolation, Characterization, and Value Addition for Substrate Development

13

Qiang-Sheng Wu and Anoop Kumar Srivastava

Abstract

Rhizosphere, the soil region nearest to plant root system, inhabits various microorganisms varying in their community structure and diversity, still remains largely unexploited despite so much of breakthroughs in isolation and characterization, but lacks in addressing citrus specific rhizosphere properties. Exploiting microbial synergisms is one of the popular methods of substrate dynamics and associated changes in nutrient environment of rhizosphere. Soil microbial biomass dominates fungal mycelium accumulates and retains mineral nutrients, immobilizing as much as ~20% of the total soil N and P. The extraradical mycorrhizal mycelial (attains as much as 3% of root weight) networks are recognized as the hidden nutrient/water-absorbing interfaces (glomalin, a soil-based glycoprotein secreted by AMF as a major seat of activity) in addition to safeguarding citrus against a number of other biotic and abiotic stresses. The microbial diversity existing within top 0–20-cm citrus rhizosphere soil was characterized and isolated by the promising microbes, namely, *Bacillus polymyxa*, *Azotobacter chroococcum*, *Bacillus mycoides*, *Pseudomonas fluorescens*, and *Trichoderma harzianum*. Pure culture of these microbes in value-added form (*Bacillus polymyxa* $33–44 \times 10^6–10^7$ cfu mL^{-1}, *Azotobacter chroococcum* as $10–38 \times 10^6–10^7$ cfu mL^{-1}, *Bacillus mycoides* $7–22 \times 10^5–10^6$ cfu mL^{-1}), *Pseudomonas fluorescens* $14–28 \times 10^6–10^7$ cfu mL^{-1}, and *Trichoderma harzianum* $30–32 \times 10^6–10^7$ cfu mL^{-1} was developed and prepared a mixture called microbial consortium (MC). MC so developed displayed an excellent response under both nursery as well as grown-up plants in integrated nutrient management (INM) module. These results suggested two cardinal points, namely, (1) the developed microbial consortium holds a good promise as an independent use as well as use in format of integrated nutrient management and (2) deserves an application in providing resilience to rhizosphere against any possible depletion in either available supply of nutrients or loss of native soil microbial biomass as an indicator of rhizosphere health.

Keywords

Rhizosphere • Microbial communities • Mycorrhiza diversity • Microbial consortium • Citrus

Q.-S. Wu (✉)
College of Horticulture and Gardening, Yangtze University,
Jingzhou, Hubei 434025, China
e-mail: wuqiangsh@163.com

A.K. Srivastava
National Research Center for Citrus, Shankarnagar Post Office,
Post Box. No. 464, Nagpur, Maharashtra 440 010, India
e-mail: aksrivas2007@gmail.com

A.K. Srivastava (ed.), *Advances in Citrus Nutrition*,
DOI 10.1007/978-94-007-4171-3_13, © Springer Science+Business Media B.V. 2012

13.1 Introduction

Researchers have invested renewed efforts to assess the abundance and distribution of microbial species and relate this information on the microbial community structure to ecosystem function since the first estimate of prokaryotic abundance in soil (Kent and Triplett 2002). The rhizosphere refers to the soil region nearest to the plant root system and is characterized by a high microbial activity (de Ridder-Duine et al. 2005). Roots release large quantity of metabolites from living root hairs or fibrous root systems into rhizosphere (Nihorimbere et al. 2011). The rhizosphere inhabits lots of soil microorganisms, such as bacteria and fungi, which compete for water, nutrients, and space and sometimes improve their competitiveness by developing an intimate association with plant (Hartmann et al. 2009). Generally, about 3,000–5,000 kg of soil living organisms occur in one hectare of soil (Baar 2010).

Broadly, there are three distinct components recognized in the rhizosphere: the rhizosphere per se (soil), the rhizoplane, and the root itself. The rhizosphere is thus the zone of soil influenced by roots through the release of substrates that affect microbial activity. The rhizoplane is the root surface, including the strongly adhering soil particles. The root itself is a part of the system because certain endophytic microorganisms are able to colonize inner root tissues. The rhizosphere effect can thus be viewed as the creation of dynamic environment where microbes develop and interact (Bartelt-Ryser et al. 2005; Nihorimbere et al. 2011). These rhizosphere microorganisms can control ecosystem functioning through decomposition and nutrient cycling and may serve as indicators of land-use change and ecosystem health (Ros et al. 2006; Balser et al. 2010). Owing to the above function, rhizosphere microorganisms play the important roles in growth of their hosts. Understanding rhizosphere microbial communities is necessary to regulate plant growth.

According to Heerden et al. (2002), the occurrence of microbial diversity in citrus waste during thermic phase of composting mainly comprised the bacterial species *Bacillus licheniformis*, *B. macerans*, and *B. stearothermophilus* and, to a lesser extent, fungi such as *Absidia corymbifera*, *Aspergillus fumigatus*, *Emericella nidulans*, *Penicillum diversum*, *Paecilomyces variotii*, *Rhizomucor pusillus*, *Talaromyces thermophilus*, and *Thermomyces lanuginosus*. The bacteria prevalent in the final product (65–70 days of composting) included *B. licheniformis*, *B. macerans*, *Proteus vulgaris*, *Pseudomonas aeruginosa*, *P. fluorescens*, *P. luteola*, and *Serratia marcescens*, whereas fungi isolated most frequently comprised *Aspergilus puniceus*, *A. ustus*, *E. nidulans*, *Paecilomyces lilacinus*, *T. lanuginosus* yeasts, and a *basidiomycetous* sp. probably *Coprinus lagopus*. Compared to the fresh product, the mineral content of citrus waste compost was low but nevertheless of the same order as that of similar chemically unenriched composed materials. Preplant incorporation of the compost (2–8 kg m^{-2}) improved the growth of citrus trees much as 25% compared to control trees in non-amended soil.

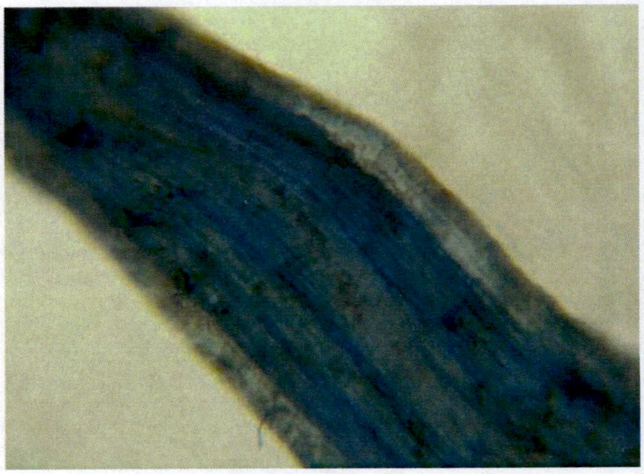

Fig. 13.1 The mycorrhizal structure in citrus trees in field

In rhizosphere, arbuscular mycorrhizal fungi (AMF), belonging to *Phylum Glomeromycota*, are naturally occurring soil beneficial microorganisms that form symbiotic association with roots of ~80% of land's plants. The arbuscular mycorrhizal (AM) symbiosis can provide water and nutrient to the host plant. In return, ~20% of plant-fixed carbon is transferred to the AM symbiosis (Parniske 2008). AMF are relevant members of the rhizosphere populations that are known to carry out many critical ecosystem functions such as improvement of plant establishment, enhancement of plant nutrient uptake, plant protection against environmental stresses, and improvement of soil structure (Rokni and Goltapeh 2011). Owing to the associated mycorrhizas in roots, the zone surrounding individual fungal hyphae of arbuscular mycorrhizas is defined as "mycorrhizosphere" (Andrade et al. 1998). Mycorrhizosphere communities have exhibited the important roles in both plant growth and soil fertility (Duponnois et al. 2008). AMF are the "normal" condition in citrus rhizosphere for forming AM structures in roots (Fig. 13.1). Sometimes, some horticultural practices, such as overuse of fertilizers and pesticides, soil sterilization, and plant production in sterile soilless media result in the non-AMF conditions of rhizosphere, which strongly inhibit citrus growth (Graham 1986). Therefore, AMF are essential for the growth of citrus in the field soils and possible role in developing integrated nutrient management (INM) module in grown-up orchards (Srivastava 2009, 2010).

The purpose of this chapter is to review some results about isolation, characterization, and value addition for substrate development of rhizosphere microbial communities in citrus.

This chapter reviews the discussion on mycorrhizosphere microbial communities in addition to other rhizosphere microorganisms with the ultimate purpose of rhizosphere-specific microbial consortium development and subsequent evaluation as substrate under nursery.

13.2 Rhizosphere Microbial Isolation

In general, a plate dilution method (Johnson and Curl 1972) can determine the population of microbes. Approximately 1 g rhizospheric soil is homogenized with 9 mL sterilized distilled water and shaken for 30 min. Fungi are grown on PDA medium, bacteria on beef extract peptone agar (pH 7.2), and actinomycetes on sodium caseinate agar (pH 6.7). Plates were incubated at 25°C, and cfu were counted after 5–7 days, except for the bacteria plates at 30°C for 30 h. Each rhizosphere bacterial isolate is submitted to examine the biochemical characterizations using methyl red, Voges-Proskauer, citrate utilization, and indole production test (Stainer et al. 1987), as well as physically through Gram's staining.

A technique for extraction of spores of AMF was successfully developed by Gerdemann and Nicolson (1963) using wet sieving and decanting and slightly modificated by Pacioni (1992) and by Smith and Skipper (1979) using sucrose centrifugation. Simply, a 1/10 (v/v) ratio of soil/water was mixed and suspended. The suspension is stirred with a magnetic stirrer for 10 min. A suspension of soil in water is passed through a series of metallic sieves (e.g., 450-, 100-, and 63-μm sieve) arranged in decreasing order of mesh width. The debris are poured into a 50-mL centrifuge tube containing 40% sucrose and centrifuged at 1,800 rpm. The debris at the bottom of the centrifuge tube are centrifuged again by adding 40% sucrose. The twice upper solutions are incorporated for examining spores under a stereoscopic microscope, and the spores are counted per microscopic field. The spores are collected, stained in a drop of Melzer's reagent by mixing with polyvinyl alcohol-lactic acid-glycerol in a ratio of 1:1, and then identified at the species level using spore color, size wall structure, and other morphological structures (Schenck and Perez 1990) and by the INVAM (http://invam.caf.wvu.edu/fungi/taxonomy/classification.htm). Other research methods in AMF are shown in Gaur and Varma (2007).

13.3 Rhizosphere Microbial Diversity

The rhizosphere region of soil is relatively rich in nutrients due to loss of as much as 40% of plant photosynthates from the roots (Lynch and Whipps 1991). The organic substances released from roots to the rhizosphere soil support higher microbial biomass, activity, and diversity in the rhizosphere than in the bulk soil (Nannipieri et al. 2007). Consequently, the rhizosphere supports large and active microbial populations capable of exerting a variety of influences on plant growth. The extent of the diversity of microorganisms in soil is seen to be critical to the maintenance of soil health and quality, as a wide range of microorganisms is involved in important soil functions. Two main drivers of soil microbial community structure, i.e., plant type and soil type, exert their function in a complex manner (Garbeva et al. 2004). Microbial communities influence plant productivity by modifying pathogenesis regulating growth and controlling the balance of toxic substances. In evolutionary terms, it can be expected that microorganisms will have influenced greatly the diversity of plants in natural vegetation and vice versa (Lynch 2002). In this context, two terms, namely, microbial diversity (defined as number of different species and their relative abundance in a given community in a given habitat) and community structure (implies that information is included on the number of individuals of different recognizable taxa) are often interchangeably used.

The importance of rhizosphere microbial populations for maintenance of the root health, nutrient uptake, and tolerance of environmental stress is now well recognized (Bowen and Rovira 1999; Cook 2002). These beneficial microorganisms can be the significant component of management practices to achieve attainable crop yield. The prospect of manipulating crop rhizosphere microbial populations by inoculation of beneficial microorganisms to increase plant growth has shown considerable promise (Bowen and Rovira 1999). Recent progress in our understanding of rhizosphere microbial diversity, colonization ability, mechanisms of action, formulation, and application should facilitate their development as reliable components in the sustainable management (Nelson 2004).

Perspectives for soil microbiologists are frequently envisioned through the molecular techniques offering new insight into the soil black box so that microbial community composition and microbial activities can be investigated and even localized on a microscale (Insam 2001). In situ hybridization is able to show where bacteria and fungi are existing (Lübeck et al. 2000) and even where they are active. Of particular interest for the future is the study of microbial hot spots, as they do occur in the guts of soil microfauna or around root surface. It is assumed that in hot spots, the majority of nutrient turnover processes take place and microbial loops are formed (Clarholm 1994). The microbial loop in soil has been described as the rhizosphere flora, where the bacteria that utilize the root-derived carbon bind inorganic carbon and inorganic nutrients and are transported to the roots by the mass flow and diffusion. Enzyme activity is generally higher in rhizosphere that in bulk soil as a result of a greater microbial activity sustained by root exudates or due to release of enzymes from roots.

Considering the wide number of biological and biochemical properties involved in the functioning of the soil,

Visser and Parkinson (1992) pointed out the need to consider different levels of study which entails the use of specific groups of properties. One level is that of the biotic community which implies the use of properties that are related to the structure of the microbial population and classically used to verify the composition and distribution of different functional groups of microorganisms of soil. A second level involves population studies, which considers the dynamics of specific organisms or communities of organisms (biological indicators). A third level, ecosystem level, considers the use of properties that are related to the cycles of bioelements (C, N, P, and S) especially when related to the transformation of the organic matter in the soil. In other words, properties of rhizosphere microbial diversity are related to the size, diversity, and activity of microbial biomass as well as to the activity of the soil hydrolytic enzymes (Srivastava and Ngullie 2009).

Owing to the highly complex microhabitats in rhizosphere, there are lots of bacteria, fungi, nematodes, protozoa, algae, and microarthrops inhabited in rhizosphere, resulting in microbial diversity. The rhizosphere microbial diversity is affected by the specific content of root exudates, plant species, plant developmental stage, and soil type (Nihorimbere et al. 2011). In addition, low N treatment seemed to increase the proportion of α-proteobacteria relative to the entire bacterial population near the root tip (Schallmach et al. 2000). So, rhizosphere microbial diversity is dependent on a number of abiotic and biotic factors. In addition, differences in amount and composition of root exudate will affect the survival of microbial species (Marschner and Timonen 2006).

The two bacterial strains, *Pseudomonas putida* 06909 and *Pseudomonas fluoriscens* 09906, were detected in a culture of *Phytophthora parasitica* isolated from citrus rhizosphere soil (Turney et al. 1992). They reduced fungal propagule densities of *Phytophthora parasitica* in the citrus rhizosphere through regulating hyphal colonization and pyoverdine production, implying that the two bacterial strains may be useful in biological control of citrus root rot (Yang et al. 1994).

The endophytic communities were assessed in surface-disinfected citrus branches by planting and denaturing gradient gel electrophoresis. The dominant isolates were characterized by fatty-acid methyl ester analysis as *Bacillus pumilus*, *Curtobacterium flaccumfaciens*, *Enterobacter cloacae*, and *Methylobacterium* spp. (including *Methylobacterium extorquens*, *M. fujisawaense*, *M. mesophilicum*, *M. radiotolerans*, *M. zatmanni*, *Nicardia* sp., *Pantoea agglomerans*, and *Xanthomonas compestris*) (Welington et al. 2002).

AM colonization has been shown to both decrease root exudation and affect root exudate composition (Marschner and Timonen 2006). And AMF themselves may release some special exudates that selectively influence rhizosphere microbial diversity, but the related studies are not conducted in citrus plants until now. It guesses that hyphae of AMF might create an additional habitat for soil microorganisms, i.e., root morphological and nutrient modification, regulation of rhizosphere pH, and improvement of soil structure. There has been increasing interest in the survey of AMF species in citrus rhizosphere. Nemec et al. (1981) examined 79 citrus orchards and nurseries in California and 66 citrus orchards and nurseries in Florida and found that *Glomus constrictus*, *G. etunicatus*, *G. fasciculatus*, *G. macrocarpus*, and *Sclerocystis sinuosa* were present in the two states. In addition, *Gigaspora margarita* and *G. mosseae* were present only in Florida and *G. microcarpus* and *G. monosporus* in California. An interesting phenomenon was that *G. fasciculatus* was associated with the 0–30-year-young citrus trees, whereas *G. constrictus* with the 30–70-year-old citrus trees. Schubert et al. (1993) also reported three spore types, belonging to the genus *Glomus*, isolated from 26 citrus orchards in Italy. These results suggest that diversity of AMF in citrus rhizosphere is dependent on environments and host plants.

AM were found in half of the 149 soil and citrus root samples collected in Apulia, Basilicata, Calabria, Messina, and Catania areas of Italy. *Gigaspora* and sour orange nurseries and 2 unidentified *Glomus*-like species were found in the other samples (Inserra et al. 1980). The studies for presence of AM on coffee, soya bean, lemon, and molasses grass (*Melinus minutiflora*) growing at 3 localities showed the occurrence of many species, namely, *Acaulospora trappet*, *A. scrobiculata*, *Glomus macrocarpus*, *G. occultum*, and *Gigaspora margarita* (Caldeira et al. 1983). Mycorrhiza species, *Glomus mosseae*, *Glomus clarum*, and *Glomus caledonium*, were commonly observed in citrus orchards of Italy (Palazzo et al. 1992). Camprubi and Calvet (1996a) suggested *G. mosseae* and *Glomus intraradices* to be the most common AM fungi in citrus soils of eastern Spain which can be introduced into soils by adopting crop rotation with mycorrhizal aromatic plants, namely, *Lavandula vera*, *L. angustifolia*, *Thymus vulgaris*, and *Rosmarinus officinalis* (Camprubi and Calvet 1996b). Other genera of *Gigaspora*, *Scutellospora*, and *Glomus* have also been commonly observed in citrus orchard soils of Japan (Ishii and Kadoya 1996).

In 25 citrus (*C. reticulata*) orchards of northern Thailand, a total of 22 species of AMF was found in the rhizosphere and was part of four genera: *Acaulospora*, *Gigaspora*, *Glomus*, and *Scutellospora* (Youpensuk et al. 2008). *G. etunicatum* and *Acaulospora scrobiculata* were the most frequently found in most samples. Additionally, soil P levels could influence diversity of AMF in each genus. For example, *Acaulospora morrowiae*, *G. invermium*, *Scutellospora coralloidea*, and *Scutellospora nigra* were only found in the soil with 250 mg P/kg or less. Another experiment conducted by Watanarojanaporn et al. (2011) showed that the rhizosphere soils of 22 citrus orchards in Thailand exhibited a total of 13 species of AMF in 6 genera: 7 species in *Glomus*, 2 species in *Acaulospora*, 1 species in *Entrophospora*, 1 species in *Scutellospora*, 1 species in *Gigaspora*, and 1 species in *Sclerocystis*. Of the individual species, *G. etunicatum* and *Acaulospora tuberculata* were commonly observed in all

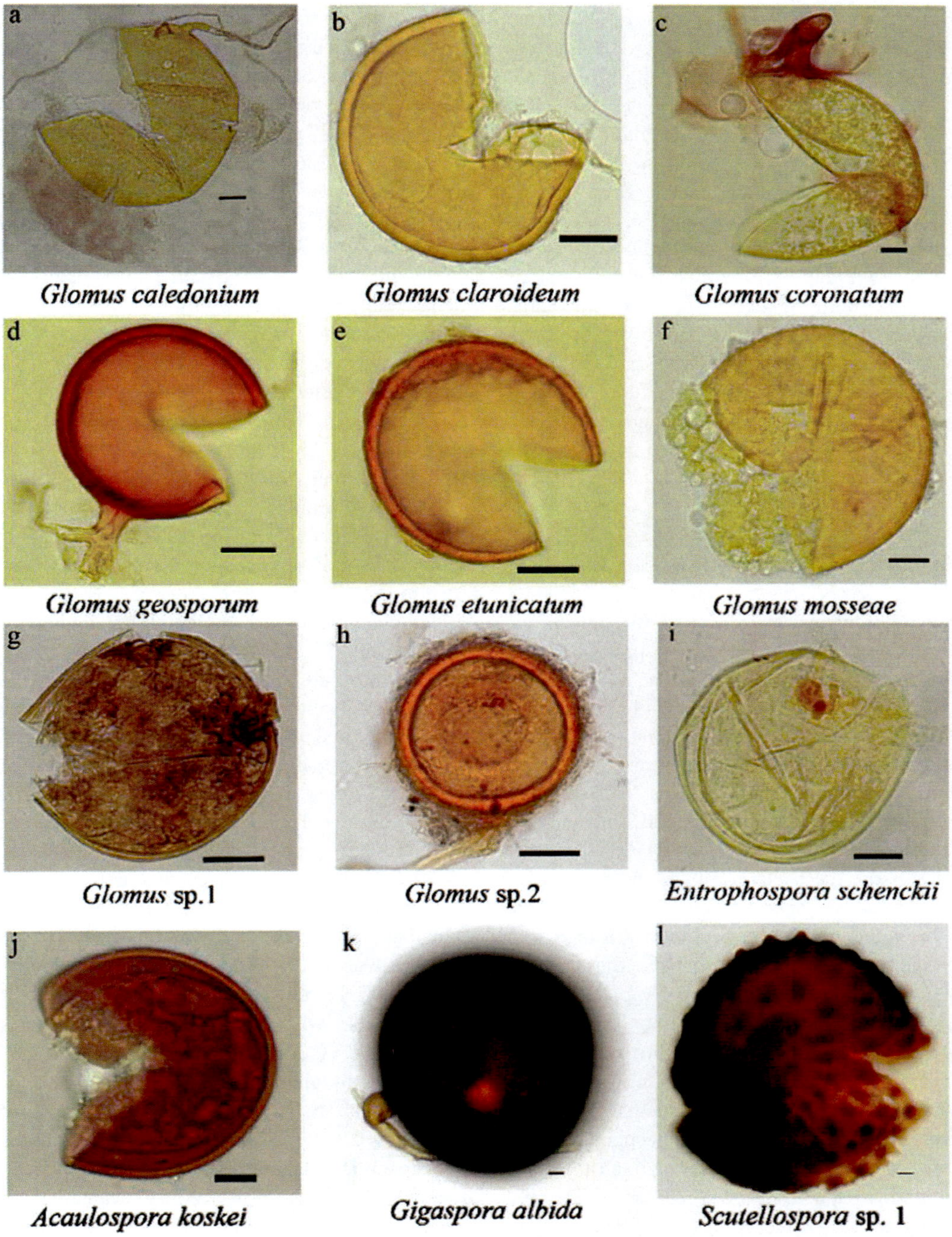

Fig. 13.2 AMF morphotypes in rhizosphere of navel orange grafted on trifoliate orange, 13–15 years old, located in Zigui county, Three Gorges Region, China. Bar = 30 μm (Adapted from Wang et al. (2011))

sampling sites, whereas *G. calospora* and *Sclerocystis* sp. were restricted in northeastern region of Thailand. The most common AM fungi found in citrus nurseries and orchards in the major citrus-growing areas of eastern Spain belonged to *Glomus* species, and *G. mosseae* and *G. intraradices* were the AMF most frequently associated with citrus roots (Camprubi and Calvet 1996a). In 20 citrus orchards (Newhall navel orange grafted on trifoliate orange, 13–15 years old) located in Zigui county, Three Gorges Region (China), *Glomus* species were the most common AMF in all soil samples, and other four genera, namely, *Acaulospora*, *Entrophospora*, *Gigaspora*, and *Scutellospora* were less frequent in occurrence (Wang et al. 2011). The isolated and identified AM fungi included *Glomus caledonium*, *G. claroideum*, *G. coronatum*, *G. geosporum*, *G. etunicatum*, *Entrophospora schenckii*, *Acaulospora koskei*, and *Gigaspora albida* (Fig. 13.2).

It is interesting that combined inoculation of AMF and PGPR seems to stimulate the microbial activity in the rhizosphere, compared to the sole AMF. For example, compared with the alone *Scutellospora* spp. CAM 3, the treatment with both *Scutellospora* spp. CAM 3 and *Pseudomonas fluoriscens* CPF 1 exhibited 65% higher spore count in rhizosphere of *Coleus forskohlii* (Priya and Kumutha 2009). In case of total fungal population, about 1.5–2 times increase was found in the combined inoculation over the inoculation of PGPR. It seems that the AMF and PGPR might present a positive interaction, which could be explained that extraradical mycorrhizal mycelium releases the glomalin for improving soil aggregates (Bedini et al. 2009), thus making for the activity of microorganisms such as bacteria, fungi, and actinomycetes. Until now, we do not know any related information about the interaction of AMF and other microorganisms in citrus.

13.4 Rhizosphere Microbial Population

A variety of microorganisms exists in citrus rhizosphere, such as bacteria, fungi, nematodes, actinomyces, protozoa, algae, etc. Generally, rhizosphere bacteria tend to be more abundant in citrus rhizosphere than other microbes. In the rhizosphere of the five citrus rootstocks (*C. unshiu*, *C. sinensis*, *C. reticulate*, *C. grandis*, and *Poncirus trifoliata*) grown in Sichuan province, China, the number of bacteria was 768.71×10^9 per gram dry soil, fungi 17,549 per gram dry soil, and actinomyces 12,462 per gram dry soil (Zou and Wang 1995). Compared with the non-root soil, the bacteria, fungi, and actinomyces quantities of the citrus rhizospheres were 67.6, 12.4, and 26.9 times. However, another study conducted by Rangaswami and Vasantharajan (1962a) showed that the fungi number was 3–6 times in the rhizospheres of the three 6-year-old citrus species in the Annamalai University Experimental Orchard than in soils, actinomycetes 2–6 times, and bacteria 40–90 times. However, sulfur-oxidizing bacteria were absent in citrus rhizosphere (Rangaswami and Vasantharajan 1962b). There was no significant difference of rhizosphere microorganisms among the three *Citrus* species, though the fungal population in the rhizosphere of the citrus variety Pummelo was slightly higher than that in the other two varieties (Rangaswami and Vasantharajan 1962a). Additionally, aspergilli, penicillia, and gram-negative non-sporing rods were higher in rhizospheres than in soil, and Mucor, Rhizopus, gram-positive rods, cocci, and sporeformers less (Rangaswami and Vasantharajan 1962b, c).

For AMF, the spore density varied between tillage and no-tillage 13- to 15-year-old citrus trees (Newhall navel orange grafted on trifoliate orange) located in Zigui county, Three Gorges Region (China), ranging from 1,267 spores per kg dry soil in the clean-tillage soils to 4,693 spores per kg dry soil in the no-tillage and sod culture soils (Wang et al. 2011). However, in Citrus Demonstration Farm, Sanxihe, and Dengjiapo of Zhigui county, the citrus trees, namely, Newhall navel orange grafted on trifoliate orange showed 470–628 spores per kg dry soil (Wu et al. 2004). In the *C. unshiu* cv Guoqing No. 1/No. 4 trees grafted on trifoliate orange grown in Wuhan, Hubei (China), the spore density varied from 862 to 2,448 spores per kg dry soil and was highest in June–August and lowest in December–February (Wu et al. 2006a). For the potted citrus plants, spore density of intergeneric somatic hybrids between *C. reticulata* and *Poncirus trifoliata* was 708 spores per kg dry soil, the somatic hybrids between *C. reticulata* and *C. jambhiri* 383, Troyer Citrange 378, *Swingle citrumelo* 374, and *Poncirus trifoliata* 347 (Wu et al. 2005). These above results suggest that spore population of AMF in citrus rhizosphere is dependent on host species and season.

Microbial biomass studies in citrus orchards of Zhejiang province of China showed a large variation in microbial biomass-C (1.62–3.16 mg kg^{-1}), microbial biomass-N (19.0–35.2 mg kg^{-1}), and microbial biomass-P (20.2–42.3 mg kg^{-1}) which constituted 1.61–2.60%, 1.2–2.5%, and 2.4–8.4% of total organic-C, respectively (He et al. 2002). Badiyala et al. (1990) observed an improvement in microbiological composition of Hapludalf under 6-year-old kinnow mandarin following grass mulch application. A classical review by Bhattacharya et al. (1999) showed the occurrence of *Bacillus polymyxa*, *Bacillus subtilis*, *Aspergillus terreus*, and *Trichoderma viridi* as phosphate solubilizers in citrus-growing soils of India, having phosphate-solubilizing capacity of 13.30% to 81.68% P_2O_5 through insoluble tricalcium phosphate. Population density of phosphate-solubilizing microorganisms has been observed to vary from 7×10^4 to 16×10^5 g^{-1} soil (Paliwal et al. 1999).

Studies by Kohli et al. (1997) showed a higher correlation of fruit yield with the population density of *Azotobacter* ($r = 0.692$, $p = 0.01$), ammonifiers ($r = 0.512$, $p = 0.01$), and phosphate-solubilizing bacteria ($r = 0.618$, $p = 0.01$) than with available N ($r = 0.489$, $p = 0.01$), P ($r = 0.316$, $p = 0.05$), and K ($r = 0.321$, $p = 0.05$) in soil, suggesting the possibility of using microbial biomass and its turnover as a potential diagnostic tool for soil fertility evaluation. Significantly higher population density of *Azotobacter*, ammonifiers, and phosphate-solubilizing bacteria was observed in rhizosphere ($0.4–16.5 \times 10^4$, $0.5–95.0 \times 10^5$, and $0.9–78.0 \times 10^4$ cfu g^{-1} soil) than non-rhizosphere zone ($0.3–10.5 \times 10^4$, $0.5–35.0 \times 10^5$, and $0.9–38.2 \times 10^4$ cfu g^{-1} soil, respectively) of Nagpur mandarin. These observations provide an evidence about the presence of favorable soil microbial buildup (Table 13.1), which could be effectively exploited in improving the availability of nutrients from soil native pool of immobilized nutrients in soil, despite suboptimum application of farmyard manure at an average rate of 40–50 kg $tree^{-1}$ $year^{-1}$, supplying

Table 13.1 Microbial composition of rhizosphere versus non-rhizosphere zones of mandarin orchards of central India

	Rhizosphere soil (cfu g^{-1} soil)			Non-rhizosphere soil (cfu g^{-1} soil)		
Locations	Azotobacter (× 10^4)	Ammonifiers (× 10^5)	P-solubilizers (× 10^5)	Azotobacter (× 10^4)	Ammonifiers (× 10^5)	P-solubilizers (× 10^5)
Kalmeshwer	8.0–16.5	0.5–70.0	0.9–56.0	4.0–10.5	0.5–32.0	0.9–32.0
Katol	1.5–16.0	5.0–55.0	11.0–53.0	1.0–10.2	2.0–31.1	4.0–38.2
Narkhed	1.0–15.0	3.0–95.0	8.0–78.0	1.0–8.2	1.0–31.0	3.0–31.0
Ramtek	0.4–8.0	4.0–52.0	7.0–38.0	0.5–7.0	8.0–32.0	6.0–20.0
Saoner	0.9–10.0	7.0–41.0	7.0–29.0	2.0–8.0	14.0–35.0	7.0–14.0
Hingna	3.0–10.0	10.0–30.0	7.0–20.0	0.3–2.6	9.0–24.0	1.1–34.0

only 200–300 gN tree^{-1} compared to recommended dose of 600–800 g tree^{-1}. Chen et al. (2004) suggested that soil microorganisms that grow in soil environment with more neutral soil pH (5.5–7.5) have a greater tolerance to soil pH changes than those growing in more acidic or alkaline soil pH conditions. Talukdar (1999) observed the population of *Glomus fasciulatum*, *Gigaspora margarita*, and *Acaulospora* sp. as 488–620 spores 100 g^{-1} and 134–480 spores 100 g^{-1} in citrus soil. Another study by Bhattacharya (1999) reported the distribution of AM population as 37–76 spores 100 g^{-1} soil with *Glomus* as most predominant genus in Nagpur mandarin-growing soils of central India. The AM population was higher during early stage (pre-bearing 1–5-year-old orchards) and declined during the next phase (5–10-year-old orchards).

Sanikidze (1970) observed an increase in population of microorganisms, namely, *Bacillus* spp., *Penicillium*, *Mucor*, *Trichorderma*, *Aspergillus* sp., *Actinomycetes*, *Clostridium pasteurianum*, *Azotobacter* sp., *Nitrobacter* sp., and cellulose-decomposing bacteria in the top soil of mandarin plantation following the application of lime and pure peat. The highest population of bacteria was observed in sweet orange followed by lemon tree rhizosphere on calcareous soils and mandarin trees growing on red soils of Georgia (Gochelashvili 1973). Another study by Saakashvili et al. (1971) reported most numerous saprophytic bacteria (65–70 million g^{-1} dry roots) on the roots and the least number in the outer rhizosphere (2 million g^{-1} soil). The saprophytic bacterial count was correlated directly with fruit yield. The role of blue green algae in citrus orchards has not been studied much. In subtropical Russian type orchards, Gochelashvili (1978) observed fixation of N up to 48–51 kg ha^{-1}. In Spain, Pomares et al. (1981) recorded improvement in the yield and quality of Salustiana orange using blue green algae as an inoculant over chemical fertilizers in sandy loam soil having pH 8.4–8.5 and $CaCO_3$ 6.7–7.2%.

Measurements of microbial activities by classical assays combined with measurements based on molecular techniques can improve our knowledge on rhizosphere processes and conduct to a better understanding of the meaning of measurements of microbial activity in soil. Studies on gene expression in the rhizosphere soil can permit a better understanding of the processes such as biological control stimulation of microbial activity by root exudates, competition between microorganisms and roots for nutrients, and molecular colloquia between microorganisms and between roots and microorganisms. Techniques for extracting and characterizing mRNA from soil are now available (Nannipieri et al. 2003) whereas soil proteomics is still in infancy even if the relative methodological problems and the potential applications have been highlighted (Nannipieri 2006). An advancement is linking between functional activity to community structure has been obtained by applying stable isotope probe (SIP) to soil (Manefield et al. 2006). It can be more rewarding to use labeled root exudates compounds and monitoring microorganisms of rhizosphere soil involved in the assimilation of the target compound and monitoring microorganisms of rhizosphere soil involved in the assimilation of target compound by the use of any SIP technique than to pulse the whole plant and then monitoring the labeled nucleic acids or other technique (phospholipids fatty-acid profiling) involving microorganisms of rhizosphere metabolizing the labeled root exudates. In addition, methodological improvement of the technique will allow designing new reporter bacteria to respond to specific root exudates, as it already occurs for specific signal involved in molecular colloquia (Sórensen and Nybroe 2006).

13.5 Mycorrhizosphere Extraradical Mycelial Networks: Hidden Nutrient/Water-Absorbing Interfaces

Microbial biomass of soil can accumulate and retain mineral nutrients, immobilizing as much as 20% and 18%, respectively, of the total soil N and P. Fungal mycelium dominates the microbial biomass of soil, commonly contributing 70–100 g dry wt m^{-2} (Watkinson et al. 2005). When the fungal mycelium reaches out of the roots, extraradical mycelium attains as much as 3% of root weight and partly or absolutely replaces root hairs to carry out the functions of root hairs (Fig. 13.3). Generally, the roots contain ~10–100 cm mycorrhizal mycelium per cm root (Cardoso and Kuyper 2006).

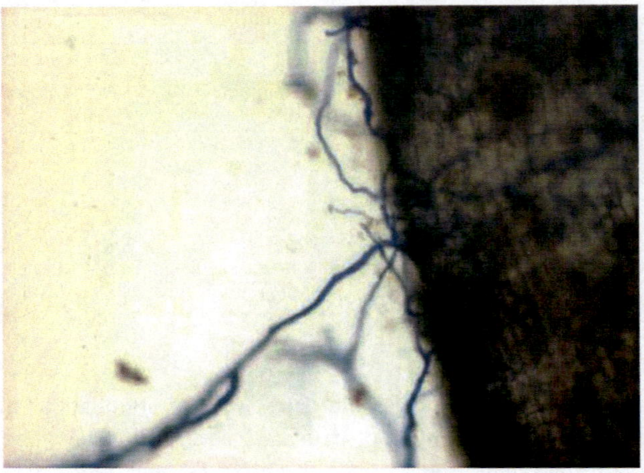

Fig. 13.3 The presence of extraradical hyphae of arbuscular mycorrhizae in citrus trees in field. The mycorrhizosphere is the zone of soil influenced by roots and their associated AMF

The extraradical mycorrhizal mycelium connects to the rhizosphere, which refers to mycorrhizosphere (Fig. 13.3). The rhizosphere extraradical mycelial networks are increasingly recognized as the main nutrient/water-absorbing interface of the plant-soil-mycorrhiza system and also provide especial pathways for uptaking soil water and nutrients (Leake et al. 2004).

In a root box, the densities of hyphae of *Gigaspora margarita* in the plots of trifoliate orange × bahiagrass and trifoliate orange × millet were higher than those of the sole trifoliate orange plat, suggesting that the other plants promote an interconnection with trifoliate orange constructing extraradical mycorrhizal mycelial networks (Cruz et al. 2002). On the other hand, it seemed that better extraradical mycorrhizal mycelial networks exist between citrus trees and other plants under an intercropping culture because extraradical mycorrhizal mycelial networks are able to colonize and interconnect contiguous plants by means of hyphae extending from one root system to another (Giovannetti et al. 2006). In contrast, a poor extraradical mycorrhizal mycelial network was established in the intercropping culture between citrus trees and plants (e.g., radish and cockscomb) less infected by AMF. For example, the extraradical mycorrhizal mycelial networks by *Gigaspora margarita* in the sod culture system were greater in the rhizosphere between rough lemon and *Vulpia myuros* than between rough lemon and bahiagrass (Ishii et al. 2007). In addition, sour orange was beneficial for establishing the mycelial networks with bahiagrass in cropping culture than trifoliate orange. The plants easily infected by AMF will more benefit the formation of extraradical mycorrhizal mycelial networks in the citrus orchards than the plants rarely infected by AMF. Probably, it is related to signal molecule substances released from the roots (Lambais 2006) or increase of AMF populations induced by the cropping system (Plenchette et al. 2005). Since extraradical mycorrhizal mycelial networks are involved in uptaking soil nutrients and water, it is reasonable to believe that in the citrus orchards, intercropped plants, which are highly mycorrhizal dependent, would have precedence over other intercropped plants in the intercropping culture.

13.6 Mycorrhiza and Soil Hydraulic, Glomalin, or Osmotic Stress

13.6.1 Mycorrhiza and Soil Hydraulics

Hydraulic lift is defined as the movement of water via roots from moist or deeper soil layer to relatively dry shallow soil layers (Scholz et al. 2010). Hydraulic lift is also regarded as an available reabsorption by the same plant (Meinzer et al. 2004) and a second ecological function in facilitating plant nutrient acquisition (Egerton-Warburton et al. 2008). Since extraradical mycorrhizal mycelial networks widely occur in rhizosphere of citrus in the field (Fig. 13.2), the mycorrhizal mycelium takes part in soil water uptake by soil hydraulic.

A previous experiment conducted by Levy et al. (1983) observed that inoculation with *G. intraradices* slightly reduced root hydraulic conductivity in rough lemon seedlings exposed to three drying cycles each of 5–7 days. It seems that AMF seedlings suffer from more severe drought condition and exhibit greater water use due to depleting soil water more quickly than non-AMF seedlings. A higher water-use efficiency (WUE) was observed in *G. mosseae*-colonized trifoliate orange seedlings exposed to 20%, 16%, and 12% water content of soil, and the increase of WUE by mycorrhization was gradually decreased with the decreasing soil water content, which is probably related to the extent of AMF colonization (Wu and Xia 2006b). Graham and Syversten (1984) reported >2 times root hydraulic conductivity per unit root length of Carrizo citrange and sour orange seedlings colonized by *G. intraradices* under well-watered conditions, compared with non-AMF seedlings. However, no significant differences of root hydraulic conductivity between mycorrhizal and non-mycorrhizal Carrizo citrange of equal phosphorus status under well-watered conditions were found. These results suggest that phosphorus enhancement due to mycorrhization is the primary approach for greater root hydraulic conductivity, and the contribution of extraradical mycorrhizal hyphae to water uptake is negligible under well-watered conditions.

A significant correlation was found between total/live hyphal length and efflux of hydraulical lift from extraradical mycelia or soil water potential in *Quercus agrifolia* grown in mesocosms (Egerton-Warburton et al. 2008). Even after prolonged drought (70–80 days), mycorrhizal hyphae were found in the soils with water potential values as low as −20 MPa

Table 13.2 Active hyphae, functional hyphae, and total hyphae in AMF trifoliate orange seedlings under ample water or drought stress conditions

Water status	AMF	Active hyphae (%)	Functional hyphae (%)	Total hyphae (%)
Ample water				
	G. diaphanum	15.4cd	19.7c	23.1c
	G. mosseae	31.44a	43.2a	67.2a
	G. versiforme	17.0bc	34.4b	46.9b
	Non-AMF	0	0	0
Drought stress				
	G. diaphanum	8.5e	10.3d	13.6c
	G. mosseae	20.6b	21.3c	24.8c
	G. versiforme	11.4de	20.4c	22.2c
	Non-AMF	0	0	0

Adapted from Wu et al. (2011) with slight modification
Data followed by the same letter within a column show no significant difference among treatments at $p<0.05$

(Querejeta et al. 2003). It is interesting that during the day, water flows from soil to AMF to plants, and during night, from plant to fungus to soil (Allen 2009). Allen (1982) determined water fluxes of extraradical mycelium and found that the mycelium (a 10-μm diameter) would transport 90 nL/h of water between the mycelium network and the root. Therefore, water from extraradical mycorrhizal mycelial networks may be a complement for water uptake during drought.

Mycorrhizal fungal hyphae include live and dead hyphae, and the former participates in water and nutrient transport. Herein, live hyphae consist of functional and active hyphae. Functional hyphae are regarded as an index of hyphal vitality, and active hyphae play an important role in biomass accumulation of host plants under drought stress. In general, mycorrhizal functional hyphae are higher than active hyphae in trifoliate orange inoculated with *G. diaphanum*, *G. mosseae*, and *G. versiforme* under ample water and drought stress conditions (Table 13.2, Wu et al. 2011). Drought stress generally significantly inhibited the active, functional, and total hyphae (Table 13.2). Therefore, the maintenance of AM hyphal activities under the conditions of drought stress will help citrus plants to sustain greater nutritional (especially P) uptake and water transport.

13.6.2 Mycorrhiza and Glomalin

Glomalin, a soil special glycoprotein released by living hyphae of AMF, is positively correlated with soil aggregate water stability (Rillig 2004). Since soil aggregation is important for maintaining soil surface integrity and allowing water to infiltrate (Franzluebbers et al. 2000), it is reasonable that glomalin might indirectly regulate plant water relations, thus enhancing adverse tolerance of the host.

In the rhizosphere of Newhall navel orange grafted on trifoliate orange, a fraction of glomalin-related soil protein (GRSP), total Bradford-reactive soil protein (T-BRSP), ranged from 8.4 to 12.1 mg g^{-1} soil (Wang et al. 2011). However, in the rhizosphere of *C. unshiu* grafted on trifoliate orange, T-BRSP is 0.47–0.81 mg g^{-1} soil, and easily extractable Bradford-reactive soil protein (EE-BRSP) is 0.34–0.58 mg g^{-1} soil (Wu et al. 2012). These BRSPs are obviously lower than the observations in other soils (Rillig et al. 2006). This may be due to a higher proportion of *Glomus* species in citrus rhizosphere, which tends to produce less glomalin per unit biomass (Treseder and Turner 2007).

In only an experiment conducted by Wu et al. (2008), drought stress slightly increased the concentrations of T-BRSP in rhizosphere of *G. versiforme*-, *G. mosseae*-, and *G. diaphanum*-colonized trifoliate orange, compared with ample water (Fig. 13.4). T-BRSP was significantly ($p<0.01$) positively correlated with >2 mm water-stable aggregates (WSA), 1–2 mm WSA, and >0.25 mm WSA. Owing to better soil aggregates in citrus soils by GRSP, mycorrhizal soils present better soil structure and soil aggregate water stability that, in turn, lead to better plant water relations. So, glomalin indirectly affects water relations of citrus by way of controlling soil aggregate water stability.

13.6.3 Mycorrhiza and Osmotic Stress

Osmotic stress caused by drought, salinity, or cold results in dehydration that limits plant productivity and geographical distribution (Zhang et al. 2010). At the same time, plants also involve in a variety of defense mechanisms such as osmotic adjustment and antioxidative defense systems. It is well documented that exogenous AMF applied into citrus soils are able to improve osmotic stress by enhancing antioxidative defense system and improving osmotic adjustment (Wu and Zou 2009b, 2011; Wu et al. 2009).

In general, AMF trifoliate orange seedlings exhibited higher net accumulations of K^+ in root, Ca^{2+} and Mg^{2+} in leaf, glucose in leaf and root, fructose in leaf, and sucrose

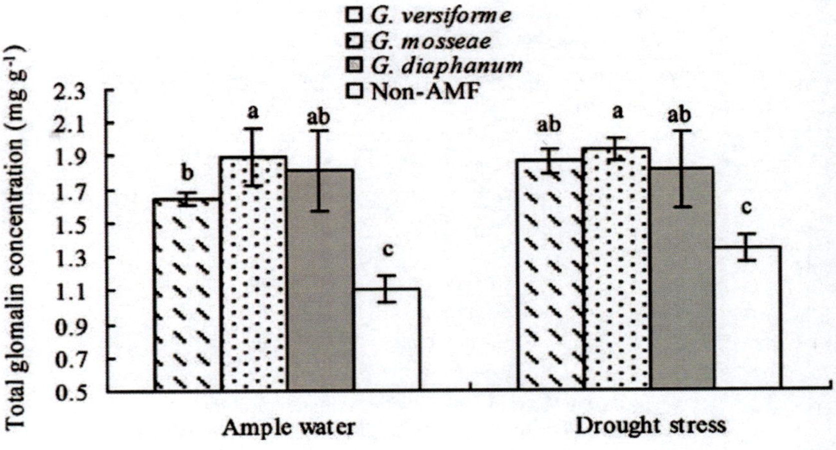

Fig. 13.4 Total glomalin concentration of AMF and non-AMF trifoliate orange seedlings exposed to ample water and drought stress (Adapted from Wu et al. (2008) with slight modification)

in leaf and root under drought stress conditions, compared with non-AMF seedlings (Table 13.2, Wu et al. 2007). Proline accumulation was lower in AMF than in AMF trifoliate orange and red tangerine seedlings under drought stress conditions (Table 13.2, Wu and Xia 2006a; Wu et al. 2007). It seems that AMF can improve osmotic adjustment of citrus plants, which is correlated with K^+, Ca^{2+}, Mg^{2+}, glucose, fructose, and sucrose net accumulation but not with proline.

Under the conditions of soil salinity, *G. mosseae* and *Paraglomus occultum* significantly decreased Na^+ concentrations of leaf and root and increased K^+ and Mg^{2+} concentrations of leaf and root (Wu and Zou 2009a; Wu et al. 2010a). For ionic balance, a higher ratio of K^+/Na^+, Ca^{2+}/Na^+, or Mg^{2+}/Na^+ was found in leaf and root of *G. mooseae*-colonized red tangerine seedlings under non-salinity conditions and a higher ratio of K^+/Na^+ or Mg^{2+}/Na^+ in leaf and root of *Paraglomus occultum*-colonized seedlings under salinity conditions (Wu and Zou 2009a; 2010a).

For reactive oxygen metabolism (ROS), AM symbiosis notably increased the guaiacol peroxidase (G-POD), glutathione reductase (GR), catalase (CAT), and ascorbate peroxidase (APX) activities of leaf; APX activity of root; and ascorbate (ASC) and glutathione (GSH) concentrations of leaf and root in trifoliate orange seedlings under drought stress conditions (Table 13.3, Wu et al. 2006b). Under the condition of salinity, red tangerine colonized by *G. mosseae* showed significantly higher superoxide dismutase (SOD) and CAT activities of leaf and ASC and GSH contents of leaf (Wu et al. 2010c). Since the AMF citrus plants recorded higher enzymatic and nonenzymatic antioxidative defense system, the lower malondialdehyde, hydrogen peroxide, and superoxide anion radical concentrations were found in AMF citrus plants than in non-AMF plants under osmotic stress conditions, resulting in a lower oxidative damage.

Table 13.3 Effects of the AM fungus, *Glomus versiforme*, on osmotic adjustment and reactive oxygen species metabolism of *Poncirus trifoliata* seedlings under ample water and drought stress conditions

		Ample water		Drought stress	
Osmotic stress	Parameter	Leaf	Root	Leaf	Root
Osmotic adjustment					
	K^+	↑	↔	↑	↑
	Ca^{2+}	↔	↑	↑	↑
	Mg^{2+}	↔	↔	↔	↔
	Proline	↓	↔	↓	↔
	Soluble sugar	↑	↑	↑	↔
	Soluble starch	↑	↔	↑	↑
	Nonstructural carbohydrate	↑	↑	↑	↑
	Glucose	↓	↑	↔	↑
	Fructose	↓	↑	↔	↔
	Sucrose	↔	↑	↑	↑
Reactive oxyge species metabolism					
	Malondialdehyde	↓	↓	↓	↓
	Hydrogen peroxide	↔	↓	↓	↓
	Superoxide anion radical	↔	↓	↓	↓
	Superoxide dismutase	↔	↔	↔	↓
	Guaiacol peroxidase	↑	↔	↑	↔
	Catalase	↔	↔	↑	↔
	Ascorbate peroxidase	↔	↑	↑	↑
	Glutathione reductase	↑	↔	↑	↔
	Soluble protein	↑	↑	↔	↑
	Ascorbate	↑	↔	↑	↑
	Glutathione	↑	↑	↑	↑

Reconsolidation from Wu et al. (2006b, 2007)
"↑," "↓," and "↔," respectively, mean that AMF significantly increased, significantly decreased, and did not significantly change the parameter

In mycorrhizal development, mycorrhizal colonization and arbuscles but not vesicles and entry points had a substantive direct effect on ROS (Wu and Zou 2009c).

For temperature stress, only a small quantity of studies showed that at high temperature stress (35°C), *G. mosseae* significantly enhanced SOD and CAT activities in leaf of trifoliate orange seedlings (Wu 2011); at low temperature stress (15°C), *G. mosseae* notably increased Ca concentration but did not affect antioxidative enzymatic activity of trifoliate orange seedlings (Wu and Zou 2010; Wu 2011). It seems that mycorrhizal alleviation of temperature stress in trifoliate orange seedlings is at high temperature, and the alleviation is obviously weakened at low temperature.

13.7 Rhizosphere Microbial Utilization: Biofertilizer

Biofertilizer is potentially identified as a substance which contains various living microorganisms (FNCA Biofertilizer Project Group 2006). When biofertilizer is applied to the soil, these microorganisms are able to spread the rhizosphere, enhance the microbial communities of the rhizosphere, colonize the interior of the plant, and thus promote plant growth by increasing the supply nutrient uptake. Generally, the microorganisms used for biofertilizers contain bacteria of *Bacillus*, *Pseudomonas*, *Lactobacillus*, photosynthetic bacteria, nitrogen-fixing bacteria, yeast, and fungi of *Trichoderma* and phylum Glomeromycota. Herein, fungi of phylum Glomeromycota (namely AMF), plant growth–promoting rhizobacteria (PGPR), and *Rhizobium* are widely studied and utilized as the biofertilizers. The beneficial soil microbes adding into the soils have shown to increase soil fertility and suppress a range of soil-borne plant pests and pathogens (Stewart et al. 2010). These biofertilizers have less environmental harm and act as both nutrient suppliers and soil structural regulators. The mycorrhizal biofertilizers in citrus are discussed in detail in the following sections.

Menge (1985), a famous mycorrhizal researcher of citrus, first advised the AMF as biofertilizers in citrus culture, which can be substituted for substantial amounts of some fertilizers. Subsequently, Graham (1986) also regarded AMF as biofertilizers, which are beneficial in citrus production. If citrus trees are planted in fumigated soils (namely, remove native AMF), there are massive phosphorus applications to citrus rhizosphere for normal growth. Menge (1985) estimated that when exogenous AMF were inoculated into the fumigated soil in citrus orchard, the AMF were the equivalent of 100–500 lb phosphorus per acre, suggesting that the mycorrhizal biofertilizers can partly substitute phosphorus fertilizers in citrus culture. Using mycorrhizal biofertilizers in citrus culture is rather important for normal growth of citrus trees and conserving energy and nonrenewable resources. Menge (1985) proposed a commercial scheme of AMF inoculum for field use in citrus production. The proposed scheme was as follows:

- Spores of native AMF are isolated from the rhizospheres of the free-pathogenic citrus trees in the field.
- These spores are inoculated into roots of sudangrass growing in a low nutrient sand in the greenhouse, which is fertilized once per week with one-half the standard Hoagland's solution minus phosphorus.
- The spores in the "pot cultures" are isolated by wet sieving, elutriation, or centrifugation and then surface disinfested with chloroamine T or sodium hypochlorite.
- The disinfested spores are again used to inoculate the roots of sudangrass grown under aseptic conditions in growth chambers for 1–4 weeks.
- After examining the root mycorrhizal infection and observing pathogenic organisms, these root segments and soils are used as a "mother culture" to produce inocula on a large scale for the field use.

In China, Shen and Wang (1994) also carried out a method for large-scale commercial use in citrus nursery. Simply, *G. citricolum*–infected citrus roots are inoculated around the roots of white clover in the sterilized sand bed. After 7 months, the shoots of the white clover are removed, and seeds of trifoliate orange are sown into the sand bed. The trifoliate orange seedlings are transplanted from the bed after 1–2 months as mycorrhizal seedlings for nursery production, and the seeding can be carried again.

As a rule, the mycorrhizal biofertilizers are given under the poor but not plenteous nutrition conditions. The inoculated dosage should be 8,900,000 propagules per hectare (Urban Creeks Council 2006). At any stage of plant growth, mycorrhizal biofertilizers alone or in combination with other biofertilizers such as PGPR can be used, and the best efficient stage is at seedlings stage before transplantation (FNCA Biofertilizer Project Group 2006). After 2–4 weeks, other non-biofertilizers can be applied, but phosphorus fertilizer must be limited. One-layer or two-layer inoculated method can be advised for applying mycorrhizal biofertilizers (Wu, data not published). Soil-vermiculite mix (1:1 by volume) is an effective growth substrate of mycorrhizal trifoliate orange seedlings (Wu et al. 2010b). Even now, these methods are not utilized on a large scale in field in a variety of agricultural soils and locations because all the approaches are inevitable infected by other microorganisms. Further studies will need to carry out large-scale field tests under normal agricultural conditions.

Inoculation of bacterium *Azospirillum brasilense* (3 g plant^{-1} as a root dip at transplanting) to sweet orange cv. 'Mosambi' plants substituted for at least one-fourth (25%) of total N requirement (Singh and Sharma 1993). A review on the role of biofertilizers in citrus by Bhattacharya et al. (1999) revealed the positive relation between fruit yield of Nagpur mandarin and population density of *Azotobacter*,

Azospirillum, and phosphate solubilizers (*Pseudomonas striata*). The review of the extensive data collected revealed 60–70% success with significant increase in yield in 10–30% of the trials.

13.8 Substrate Dynamics

Consistent efforts are being made to find alternatives to conventional fertilizers, media, and practices, although chemical properties of formulated substrates may affect plant growth and nutritional response in varied ways, namely, (1) improvement in soil hydraulic properties, (2) maintenance of better available pool of nutrients, and (3) establishment of dynamic soil microbial environment more suited to crop requirement (Dutt et al. 2002; Altland and Buamscha 2008). The origin of a substrate and its pH are considered two most important guiding principles in developing a substrate dynamic to plant's rhizosphere in addition to physical stability, ease in rewetting ability to withstand compression, and low shrinkage rate over time (Roose and Haase 2000; Altland 2006). Dutt and Sonawane (2006) observed excellent performance of chrysanthemum (*Chrysanthemum indicum* L.) on a substrate containing cocoa-peat-compost-rice husk. Recently, studies (Buamscha et al. 2007; Altland et al. 2008) documented that DFB (Douglas Fir Bark) alone provided sufficient micronutrients for annual vinca (*Catharanthus roseus* L.) grown at low pH (4.6–5.5) while Hernandez-Apaolaza et al. (2005) suggested that the use of pink bark in coconut (*Cocos nucifera* L.) coir-based media formulations served as one alternative of recycling waste materials. Fisher et al. (2006) suggested peat-based substrate treated with lime to adjust pH within an optimum range.

Coir dusts with a particle size distribution similar to peat showed comparatively higher aeration and lower capacity to hold total and easily available water. An air-water balance similar to that in peat became apparent in coir dust at a comparatively lower coarseness index (29% vs. 63% by weight in peat). Stepwise multiple regression analysis showed that particles with diameters in the range of 0.125–1 mm had a remarkable and highly significant impact on the physical properties studies, while particles <0.125 mm and >1 mm had only a slight or nonsignificant effect (Abad et al. 2005).

Four types of media [coir, 1 coir: 2 peat (by volume), peat, and sandy loam soil] were evaluated by Merhaut and Newman (2005) for their effects on plant growth and nitrate (NO_3^-) leaching in the production of oriental lilies (*Lillium* L.) 'Starfighter' and 'Casa Blanca.' Results indicated that the use of coir and peat did not significantly influence plant growth (shoot dry weight) relative to the use of sandy loam soil. However, substrate type influenced the amount of NO_3^- leached through the media and N accumulation in the shoots for 'Starfighter' but not for 'Casa Blanca.'

Various recipes for potting mixes exist that do not contain synthetic components (Kuepper and Adam 2003; Salifu et al. 2006). Koller et al. (2004) used several plant- and animal-based substrates in the production of vegetable transplants. They stipulated that plant-based substrates should be mixed into the potting medium 2 weeks before sowing seed to prevent damage. Worm castings of EF have been tested as a component of media for organic production to tomato, and it was found that seedling development improved as percent of worm castings in the medium increased (Ozores-Hampton and Vavrina 2002). Regardless of their origin, these materials and practices are generally referred to as being alternatives to conventional fertilizers, media, and practices. To be accepted as commonplace in the industry, alternative materials and practices must be compared to existing conventional materials and practices (Russo 2005).

For example, a typical substrate tested in azalea (*Rhododendron atlanticum*) and camellia (*Camellia japonica*) (Merhaut et al. 2006; Blythe et al. 2006) consisted of 5 sphagnum peat moss, 4 pine bark (6.7–9.5 mm), and 1 washed builders sand (by volume) amended with dolomite 65 at a rate of 0.59 kg m^{-3} and ultrafine calcium sulfate at a rate of 0.59 kg m^{-3}, mixing the substrate and amendment. The nutrient composition (mg L^{-1}) was observed as 1,306 Ca, 019 Mg, 2.62 Fe, 0.59 Mn, 0.75 Zn, 0.11 Cu, and 0.01 Mo. The substrate was later mixed with different controlled release fertilizers, CRF, namely, Osmocote (24-4-9), Nutricote (18-6-8), Multicote (17-5-11 + minor nutrients), and Polyon (17-5-11 + micronutrients), all having 365 days release formulations in terms of highly acidic pH and particle size distribution.

Crop residue is another option to be used as a strong support to substrate in any INM program. For example, in India, crop residues available are estimated to be 600 million Mg. Rice and wheat are two major crops, generating around 250 million Mg of residues (Selvakumar et al. 2008). Decomposing paddy straw is a problem because it contains approximately 40% cellulose, 20% hemicellulose, and 12% lignin and has high C:N. Several strains of mesophilic and thermophilic microorganisms were screened for utilization of paddy straw. Four fungi, *Phanerochaete sporium*, *T. viride*, *Aspergillus nidulus*, and *A. awamori*, were identified by Selvakumar et al. (2008) to carry out solid-state fermentation of paddy straw; all combinations were good. The process involved construction of perforated brick tanks for proper aeration for composting of paddy straw. The straw was supplemented with poultry droppings (8:1) or urea at 0.5% to bring down the C:N ratio of the straw. A tank of 1 m^3 can accommodate 80 kg of straw. Rock phosphate (1%) along with inoculum containing 4 fungi was applied at 0.5 kg ton^{-1} straw and mixed in the tank. Moistened paddy with sufficient water, and within 2–3 months, compost with a C:N ratio of 15:1 can be successfully obtained. Such an attempt needs to be replicated

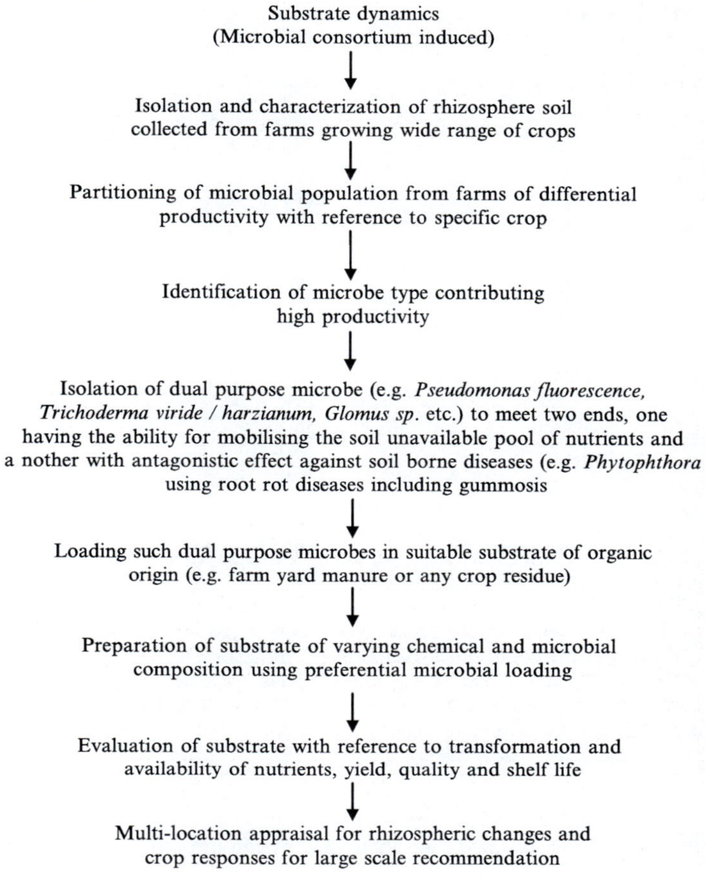

Fig. 13.5 Schematic representation of steps involved in development of microbially loaded substrate

using other crop residues as substrate. Siddiqui et al. (2008) observed that the application of *T. harzianum*–inoculated rice straw compost not only improved the morpho-physiological characters of okra but reduced the wet rot incidence compared to control and offered an environmentally friendly alternative to inorganic fertilizers/fungicides, resulting in higher yield.

Coinoculation or combined inoculation of different microbe types is another area which can be gainfully exploited in formulating the microbially rich substrate, provided that information on the synergism between different microbes is known (Marschner et al. 2004). In the past, a number of studies have suggested the coinoculation of different microbes, which can be summarized as *A. brasilense* – *P. striata*/*B. polymyxa* in sorghum (Alagawadi and Gaur 1992), *A. lipoferem* – *Agrobacterium radiobacter*/*A. lipoferem* – *Arthrobacter mysorens* in barley (Belimov et al. 1995), *A. brasilense* – *Rhizobium* in lentil (Yadav et al. 1992) and chickpea (Fabbrie and Gallo Del 1995), *A. brasilense* – *A. chroococcum* – *Klebsiella pneumoniae* – *R. meliloti* in alfalfa (Hassouna et al. 1994), *A. brasilense* – *R. leguminosarum* in soybean (Neyra et al. 1995), and *A. brasilense*/*Streptomyces mutabilis* – *A. chroococcum* in wheat (Elshanshoury 1995). Many studies on coinoculation of microbes involving AM fungi and bacteria have also been suggested for improvement in both yield and quality. These include *A. brasilense* – *G. fasciculatum* in wheat (Gori and Favilli 1995) and strawberry (Bellone and de Bellone 1995), *A. brasilense* – *Pantoea dispersa* in sweet pepper (del Amor et al. 2008), and *A. chroococcum* – *G. mosseae* in pomegranate (Aseri et al. 2008).

Various steps involved in preparation of dynamic substrate have been further depicted through a flow diagram (Fig. 13.5) in fulfilling rhizosphere's diverse requirements.

13.8.1 Rhizosphere Microbial Variability: A Hot Spot

The potential effects of perennial trees on soil properties have been the focus of large number of studies across the globe, and the associated plant-soil interactions provide important feedbacks that regulate a crop-specific rhizosphere transformation (Porazinska et al. 2003; Binkley and Menyailo 2005; Kulmatiski et al. 2008). Earlier studies on soil-plant

Table 13.4 Biometric performance, within the rhizosphere zone (0–15 cm depth) of different citrus species/varieties (pooled data of 2 seasons)

				Soil microbial communities (cfu g^{-1})	
Sr. no.	Species/varieties	Canopy volume[a] (m^3)	Root density (g cc^{-1} soil)	Bacterial count	Fungal count
1.	LL	0.559	0.498	94×10^6	82×10^7
2.	EL	0.399	0.178	79×10^6	64×10^7
3.	ML	0.446	0.291	84×10^6	80×10^7
4.	BL	1.221	0.523	102×10^6	86×10^7
5.	Mu	0.309	0.142	24×10^4	41×10^5
6.	DT	0.353	0.168	21×10^4	46×10^5
7.	Cl	0.119	0.078	20×10^4	48×10^5
8.	It	0.252	0.098	16×10^4	40×10^5
9.	OV	0.267	0.104	26×10^4	44×10^5
10.	MB	0.238	0.098	22×10^4	40×10^5
11.	RN	0.314	0.104	28×10^4	41×10^5
12.	CCN	0.220	0.092	12×10^4	48×10^5
13.	Tr	0.398	0.156	28×10^4	52×10^5
14.	RR	0.211	0.271	22×10^4	42×10^5
15.	EG	0.502	0.487	92×10^6	84×10^7
	CD ($p=0.05$)	0.046	0.018	4	14

Adapted from Ngullie et al. (2008, 2009, 2010)
LL (Lisbon lemon – *Citrus limon* Lisbon); EL (Eureka lemon – *Citrus limon* 'Eureka'); ML (*Citrus aurantifolia* Swingle cv Mexican lime); BL (Bears lime – *Citrus latifolia* Tanaka) (Persian lime); Mu (Murcott – *Citrus reticulata* 'Murcott'); DT (Daisy tangerine – *Citrus tangerine*); Cl (Clemenule – *Citrus reticulata* 'Clemenules'); It (Itaborai – *Citrus sinensis* L. Osbeck); OV (Olinda Valencia – *Citrus sinensis* 'Olinda' nucellar Valencia); MB (Moro blood – *Citrus sinensis* 'Moro'); RN (Ruby nucellar – *Citrus sinensis* 'Ruby'); CCN (Cara Cara Navel – *Citrus sinensis* 'Cara Cara'); Tr (Trovita – *Citrus sinensis* 'Trovita'), RR (Rhode Red – *Citrus sinensis* 'Rhod-E-Red'); EG (Early Gold – *Citrus sinensis* 'Early Gold')
cfu stands for colony forming units
[a]Total increase over initial canopy

interaction has revealed that plant species can not only give significant impacts on soil physicochemical and biological properties within the rhizosphere through litter nutrient dynamics (Ngullie et al. 2008) but also on the composition of soil microbial community (Ushio et al. 2008) who later found that the compositions of soil microbial communities can converge under canopy of the same species. Studies of macroscale ecology have often treated a perennial crop floor as relatively homogenous in terms of substrate quality, microbial community, and mineralization process or if not studies that simultaneously consider the spatial patchiness of soil physicochemical and microbial properties as rare occurrence (Saetre and Bääth 2000). In essence, use of compost in combinations with mineral fertilizers brings benefits in terms of yield, crop protection, microbial diversity, water management, and soil carbon sequestration (Binkley and Giardina 1988; Binkley and Valentine 1991; Ross et al 2009).

In a study carried out on Typic Aqualf in northeast India, all the citrus varieties showed a differential expression of canopy volume, varying from 0.119 to 1.22 m^3 within two seasons (Table 13.4). The maximum canopy growth was observed in BL followed by LL, EG, and ML, with poorest canopy expressed by Cl. These differential biometric responses were well supported by variation in root density (0.078–0.323 gcc^{-1}) as evident from high coefficient of correlation ($r=0.769$, $p=0.01$), with BL displaying excellent performance, while CCN/RR being the poor performer. Much higher fungal and bacterial count within the rhizosphere of varieties, namely, BL, LL, EG, ML, etc., compared to rhizosphere of rest of the varieties warrant strong support in favor of characterizing rhizospheric features that develop while interaction with specific agroclimate (Ngullie et al. 2010).

The strategy of introduction of microorganisms in citrus soil is a recent adoption and requires in-depth comprehensive studies to improve the process of application for harnessing higher benefits. In a separate study, Kalita et al. (1996) observed the occurrence of *Bacillus subtilis*, *Bacillus polymyxa*, *Aspergillus terreus*, and *Trichoderma viridi* in citrus orchard soils of central India. These microbes could play a significant role in controlling *Xanthomonas campestris* pv. *citri*, the incitant of citrus canker.

13.8.2 Isolation and Efficacy of Soil Microbes

In nature, microorganisms do not live in isolation in a certain space and time. They coexist with many different microorganisms, establishing relationship that have an effect in the biological adequacy of all interacting species. Microorganism selection remains largely based on pure culture evaluation and relies in the use of single strains.

Table 13.5 Effect of different soil properties on predictability of fruit yield through step-down multivariate regression analysis

	Equations	Contribution ($R^2 \times 100$)
1.	$Y=31.73+0.30X_1+2.56X_2+0.12X_3+6.12X_4+4.32X_5-3.89X_6-16.82X_7+1.21X_8+2.04X_9+26.27X_{10}+2.89X_{11}-3.11X_{12}$	89.89
2.	$Y=25.82+0.23+4.11X_2+0.32X_3-5.12X_4+3.89X_5-14.39X_7+1.31X_8+3.11X_9+32.17X_{10}+2.12X_{11}-9.23X_{12}$	88.39
3.	$Y=22.12+0.49X_1+6.13X_2+5.18X_4+4.11X_5-24.38X_7+3.34X_8-8.14X_9-32.17X_{10}-3.94X_{11}-8.14X_{12}$	87.82
4.	$Y=21.39+0.51X_1+4.39X_2-6.39X_4-24.30X_7+4.18X_8-10.12X_9-44.21X_{10}+3.12X_{11}+6.12X_{12}$	87.10
5.	$Y=20.12+1.11X_1-5.30X_4+31.11X_7+5.04X_8-11.12X_9+1.39X_{10}+3.10X_{11}-5.10X_{12}$	86.34
6.	$Y=24.09+1.16X_1+30.17X_7+6.04X_8-12.12X_9-2.12X_{10}+4.12X_{11}-2.39X_{12}$	86.19
7.	$Y=26.19-2.12X_1+22.12X_7-3.18X_8-10.19X_9-3.18X_{10}-3.89X_{11}$	85.13
8.	$Y=21.89-1.32X_1-1.69X_7+4.12X_8-12.89X_9+2.89X_{10}$	84.10
9.	$Y=28.14-2.11X_7+3.89X_8-12.15X_9+1.94X_{10}$	84.00
10	$Y=23.43-3.82X_8-4.15X_9+2.04X_{10}$	83.12

Adapted from Srivastava et al. (2010)
X_1 = available N; X_2 = available P; X_3 = available K; X_4 = available Fe; X_5 = available Mn; X_6 available Cu; X_7 = available Zn; X_8 = bacterial count; X_9 = fungal count; X_{10} = organic carbon; X_{11} = inorganic carbon; X_{12} = total carbon; Y = fruit yield

Table 13.6 Evaluation of asymbiotic nitrogen-fixing bacteria on changes in soil available nitrogen in different soils (incubation study period: 45 days)

	Available N (mg kg^{-1})								
	Acid soil			Neutral soil			Alkaline soil		
N-fixing microbes	Con.	Tr.	Dif.	Con	Tr.	Dif.	Con.	Tr.	Dif.
Azotobacter chroococcum	128.4	156.4	28.0	107.9	132.3	24.4	84.3	94.3	10.0
Azospirillum brasilens	127.3	140.8	13.5	108.2	122.1	13.9	85.2	112.4	27.2
CD ($p=0.05$)	–	–	3.2	–	–	2.2	–	–	4.7

Adapted from Srivastava et al. (2010)
Efficiency computed on the basis of changes over control
Acid soil (pH 4.8 and texture sandy loam), neutral soil (pH 7.4 and texture clay), and alkaline soil (pH 8.3 and texture loam)
Con. control, *Tr.* treated, *Dif.* difference

The step-down multiple regression analysis data (Table 13.5) indicated that 83.12% variation in fruit yield was observed due to bacterial count, fungal count, and organic carbon content. On including different available nutrients and inorganic/total carbon, no appreciable increase in predictability could be achieved as R^2 increased up to 89.83% only. It is thus evident that soil microbial (bacterial and fungal count) population and organic carbon content are the main indices of soil fertility affecting the fruit yield predictability.

Soil microbial communities are often difficult to fully characterize mainly because of their immense phenotypic ad genotypic diversity, heterogenicity, and crypticity. For example, bacterial populations in the soil top layers can go up to more than 10^9 cells g^{-1} soil (Torsvik and Ovreas 2002), and most of these cells are generally unculturable. The fraction of cells making up the soil microbial biomass that have been cultured and studied in any detail are negligible, often less than 5% (Torsvik et al. 1990).

A large number of microbes were isolated from rhizosphere of citrus orchards established on acid soils of northeast India and neutral alkaline soils of central India and northwest India. These microbes comprised of N-fixers (*Azotobacter chroococcum* and *Azospirillum brasilens*), P-solubilizing fungi (*Trichoderma harzianum* and *Trichoderma virdie*), and P-solubilizing bacteria, namely, *Pseudomonas fluorescens*, *Pseudomonas striata*, *Bacillus subtilis*, *Bacillus mycoides*, *Bacillus polymyxa*, *Bacillus stearothermophilus*, *Bacillus cereus*, *Bacillus coagulans*, *Bacillus licheniformis*, *Bacillus circulans*, *Bacillus pumilus*, and *Bacillus sphaericus* (Fig. 13.4). These soil microbes were tested for their efficacy under different soil environment through lab-oriented incubation study.

13.8.2.1 Evaluation of N-Fixing Bacteria

An incubation study was carried out to evaluate the nitrogen-fixing capacity (as evident from changes in soil available N) at field capacity soil moisture level under three different soil types (acid soil with pH 4.8, neutral soil with pH 7.4, and alkaline soil with pH 8.3) under three different sets of experiments (Table 13.6). The data on changes in soil available hence generated were summarized and discussed. Although, there is significant increase in available N with both the N-fixing bacteria when compared with control, and is invariably so on all the three soil types of contrasting properties. On acid soil and alkaline soil, *Azotobacter chroococcum* was

Table 13.7 Evaluation of phosphate-solubilizing microbes on changes in soil available phosphorus in different soils (incubation study period: 45 days)

	Available P (mg kg^{-1})								
	Acid soil			Neutral soil			Alkaline soil		
PSM	Con.	Tr.	Dif.	Con.	Tr.	Dif.	Con.	Tr.	Dif.
Pseudomonas fluorescens	4.2	5.1	0.9	14.8	19.9	5.1	12.2	17.4	5.2
Pseudomonas striata	4.8	8.9	4.1	14.2	18.1	3.9	11.3	13.2	1.9
Trichoderma harzianum	5.1	6.1	1.0	15.1	17.1	4.0	11.8	18.1	6.3
Trichoderma viride	4.9	10.2	5.3	14.9	15.2	0.3	11.9	12.1	0.2
CD ($p = 0.05$)	–	–	1.4	–	–	1.1	–	–	1.8

Adapted from Srivastava et al. (2010)
Efficiency computed on the basis of changes over control
Acid soil (pH 4.8 and texture sandy loam), neutral soil (pH 7.4 and texture clay), and alkaline soil (pH 8.3 and texture loam)
Con. control, *Tr.* treated, *Dif.* difference
PSM stands for phosphate-solubilizing microbes

most efficient (evident from increase in available N by 28.0 and 24.4 mg kg^{-1}, respectively, over control compared to increase by 13.5 and 13.9 mg kg^{-1} over control with *Azospirillum brasilense*). However, on alkaline soil, *Azospirillum brasilense* proved more effective (increase in available N over control by 27.2 mg kg^{-1}) than *Azotobacter chroococcum* (increase in available N over control by only 10.0 mg kg^{-1}). These observations lend strong support that soil-based microbial efficacy needs to be worked out if microbial biofertilizers have to find a due place in effective INM program.

13.8.2.2 Evaluation of Phosphate-Solubilizing Microbes

A total of four phosphate-solubilizing microbes, two as bacterial species (*Pseudomonas fluorescens* and *Pseudomonas striata*) and two as fungal species (*Trichoderma harzianum* and *Trichoderma viride*), were tested on three different soil types (acid soil, neutral soil, and alkaline soil) as three separate experiments with respect to phosphate-solubilizing capacity under complete randomized block design (CRD). The results of these three incubation studies have been summarized (Table 13.7), which are subsequently discussed. Changes in available P (Olsen-P) induced by these phosphate-solubilizing microbes showed highest efficacy of *Pseudomonas striata* (increase in available P by 4.1 mg kg^{-1} over control) and *Trichoderma viride* (increase in available P by 5.3 mg kg^{-1} over control) on acid soil. While on one hand on neutral soil, *Pseudomonas fluorescens* (increase in available P by 5.1 mg kg^{-1} over control), *Pseudomonas striata* (increase in available P by 3.9 mg kg^{-1} over control) and *Trichoderma harzianum* (increase in available P by 4.0 mg kg^{-1} over control) displayed their best performance on phosphate-solubilizing capacity, on the other hand on alkaline soil, *Pseudomonas fluorescens* (increase in available P by 5.2 mg kg^{-1} over control) and *Trichoderma harzianum* (increase in available P by 6.3 mg kg^{-1} over control) proved as most effective microbes. These observations are of immense utility with the objective of developing soil type–based microbial consortium to ensure their best performance.

13.8.2.3 *Bacillus* Diversity and Phosphate-Solubilizing Capacity

As many as ten *Bacillus* sp. (Fig. 13.6) were isolated from soils of citrus orchards representing large variation in physicochemical properties. These *Bacillus* species were then evaluated for phosphate-solubilizing capacity (interpreted on the basis of increase in available P over control) on three different soil types (acid soil, neutral soil, and alkaline soil) under three different incubation experiments using CRD design. Results are discussed below in summarized form (Table 13.8).

On acid soil, *Bacillus polymyxa* (increase in Bray's-P by 4.1 mg kg^{-1} over control), Bacillus cereus (increase in Brays-P by 2.5 mg kg^{-1} over control), and *Bacillus pumilus* (increase in Brays-P by 1.6 mg kg^{-1} over control) were observed as most effective phosphate-solubilizing *Bacillus* species.

On neutral soil, *Bacillus polymyxa* (increase in Olson-P by 4.4 mg kg^{-1} over control), *Bacillus mycoides* (increase in Olson-P by 3.8 mg kg^{-1} over control), *Bacillus subtilis/Bacillus cereus Bacillus sphaericus/Bacillus licheniformis* (increase in Olson-P by 3.2 mg kg^{-1} over control) were observed as highly efficient *Bacillus species* for phosphate-solubilizing ability.

On alkaline soil, results were quite different. The species such as *Bacillus coagulans/Bacillus stearothermophilus/Bacillus sphaericus* (increase in Olson-P by 3.4–3.8 mg kg^{-1} over control) were observed as highly efficient phosphate-solubilizing Bacillus species. Such information paved the way for development of an effective microbial consortium based on the nature and properties of soil.

A potassium-solubilizing microbe type *Bacillus mycoides* was isolated from high yielding orchard No. 2. This microbe was not observed in any of the soil samples subjected to microbial analysis. The potassium-solubilizing capacity of the isolated microbe was studied as per following procedure:

- The microorganism culture was prepared by taking 10 mL nutrient broth and inoculated microbe at 28°C in incubator for 2 days.
- Centrifuged the culture at 4,000 rpm and rinsed with distilled water and retrieved the microbial cells settled at the bottom.
- Took 10 g three different soil types in quadruplicate in 250-mL conical flash, added 200 mL Bromfield liquid

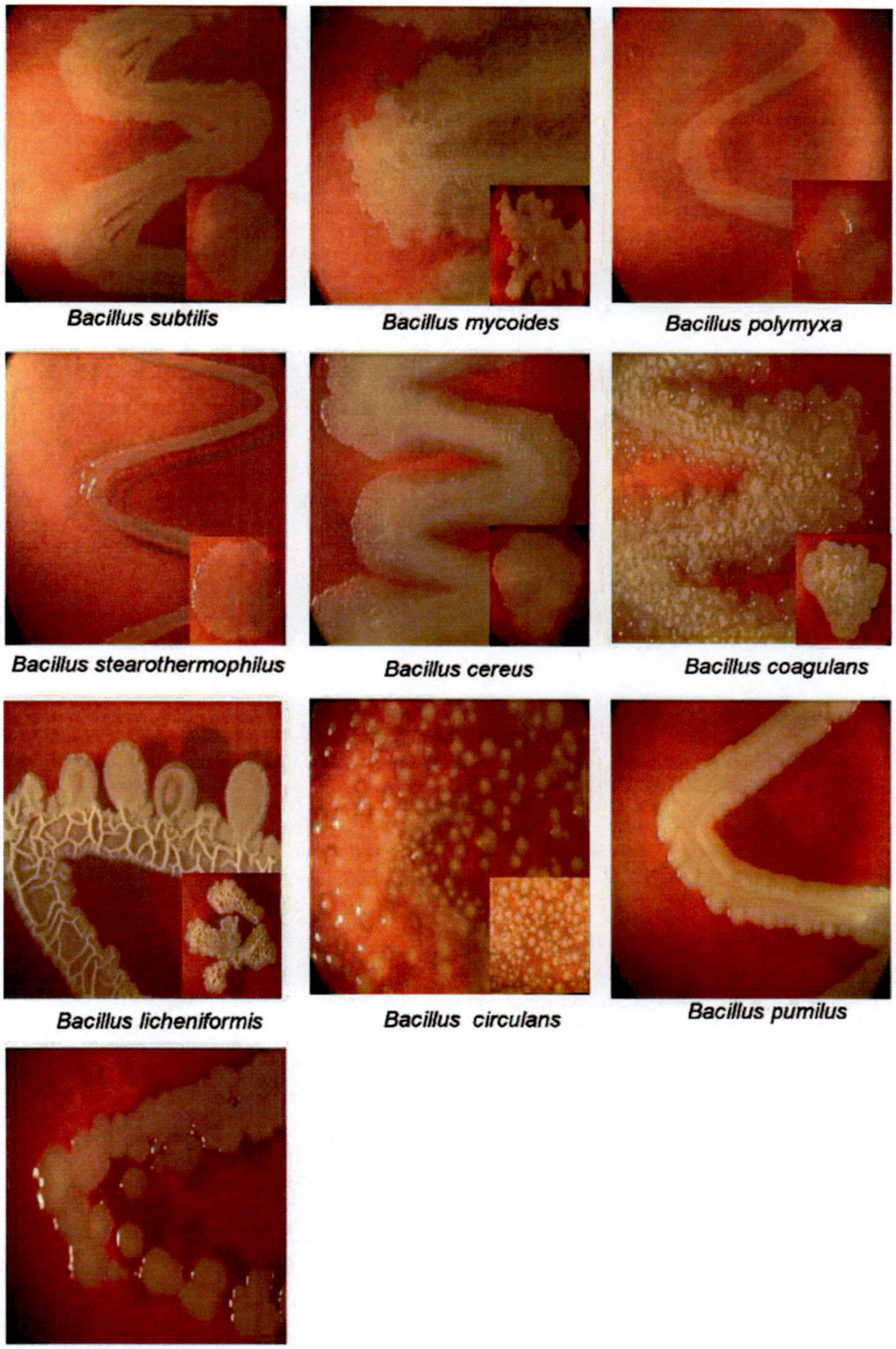

Fig. 13.6 Diversity in *Bacillus* species isolated from rhizosphere soil of different citrus cultivars

media, followed by addition of microbial cells, and kept for 30 days at ambient conditions. Then, the mixture was incubated at 80°C for 2 h.

- Added 10 mL of 6% H_2O_2 and kept for 4 h, filtered it through Whatman filter No. 40, then added 10 mL of 30% H_2O_2 in the filtrate. The mixture was kept on hot plate for an hour to let entire organic matter oxidize, and at last clear solution was obtained.
- Potassium content was directly measured to assess the K solubilization.

Table 13.8 Evaluation of *Bacillus* sp. for phosphate-mobilizing capacity under different soil types (incubation study period: 45 days)

	Soil available phosphorus (mg kg^{-1})								
	Acid soil			Neutral soil			Alkaline soil		
Bacillus sp.	Con.	Tr.	Dif.	Con.	Tr.	Dif.	Con.	Tr.	Dif.
1. *Bacillus subtilis*	6.0	6.8	0.8	13.6	16.8	3.2	11.6	14.4	2.8
2. *Bacillus mycoides*	6.2	6.4	0.2	13.4	17.2	3.8	11.4	12.4	1.0
3. *Bacillus polymyxa*	6.1	10.2	4.1	13.8	18.2	4.4	11.2	13.3	2.1
4. *Bacillus stearothermophilus*	6.4	7.0	0.6	13.6	14.1	0.5	11.4	14.8	3.4
5. *Bacillus cereus*	6.3	8.8	2.5	13.7	16.8	3.1	11.3	12.2	3.9
6. *Bacillus coagulans*	6.4	6.6	0.2	13.4	14.8	1.4	11.4	15.2	3.8
7. *Bacillus licheniformis*	6.4	6.9	0.5	13.6	16.4	2.8	11.4	14.5	3.1
8. *Bacillus circulans*	6.2	6.9	0.7	13.2	15.8	2.6	11.5	14.2	2.7
9. *Bacillus pumilus*	6.2	7.8	1.6	13.4	14.0	0.6	11.7	12.0	0.3
10. *Bacillus sphaericus*	6.3	7.1	0.8	13.4	16.3	2.9	11.8	15.3	3.5
CD ($p=0.05$)	–	–	0.30	–	–	0.40	–	–	0.50

Adapted from Srivastava et al. (2010)
Efficiency computed on the basis of changes over control
Acid soil (pH 4.8 and texture sandy loam), neutral soil (pH 7.4 and texture clay), and alkaline soil (pH 8.3 and texture loam)
Con. control, *Tr.* treated, *Dif.* difference

Table 13.9 Potassium-solubilizing capacity of *Bacillus mycoides*

	Available K (mg kg^{-1})		
Treatment	Soil 1	Soil 2	Soil 3
Control	182.4	214.6	114.8
Treated	310.8	248.2	210.9
T test ($p=0.05$)	14.3	16.4	18.8

Adapted from Srivastava et al. (2010)
Soil 1 (texture: 42.4% clay, 31.3% silt, 26.3% sand)
Soil 2 (texture: 59.4% clay, 22% silt, 18.5% sand)
Soil 3 (texture: 46.3% clay, 34.9% silt, 18.8% sand)

Data obtained on potassium-solubilizing capacity (Table 13.9) showed that in all three soil types, a significant improvement in available K (from 182.4 to 310.8 mg kg^{-1} in soil 1, from 214.6 to 248.2 mg kg^{-1} in soil 2, and from 114.8 to 210.9 mg kg^{-1} in soil 3) was observed as a result of incubation of soil with this microbes, confirming its K solubilizing behavior.

13.8.3 Development of Microbial Consortium and Evaluation

Formation of associations with other organisms can promote protection from potentially inhibitory environmental factors where such associations reflect synergistic lifestyles facilitating more effective and efficient growth and biogeochemical cycles than individual populations as a community. Such associations are often called microbial consortium in which members of the consortium maintain metabolic and ecological compatibility for individual niches to exist in the close proximity in soil (Radianingtyas et al. 2003; Lazdunski et al. 2004). Such microbial consortium is more resistant to environmental changes and can compete much better than single microorganism. As a result, different species of microbial consortium inside an ecosystem propagate with different dynamics depending upon their genetic potentiality as well as capacity of adjustment to the microenvironmental conditions giving better yield and quality (Bashan 1998). If the microorganism interactions are evaluated and included in such selection process, a microbial consortium may outperform the results achieved by pure cultures. When different microbial strains are made into an inoculum consortium, each of the constituent stains of the consortium not only outcompete with others for rhizosphere establishments but complement functionally for plant growth promotion.

Different microbial components in a microbial consortium should possess (1) high rhizosphere competence, (2) high competitive saprophytic ability, (3) ease for mass multiplication, (4) safe to environment, (5) broad spectrum of action, (6) excellent and reliable efficacy, and (7) compatible with other rhizosphere microbes and in able to tolerate other abiotic stresses (Rainey 1999; Date 2001).

The efficient microbes identified (*Azotobacter chroococcum*, *Bacillus polymyxa*, *Pseudomonas fluorescens*, *Trichoderma harzianum*, and *Bacillus mycoides*) through previous incubation studies were brought forward in consortium form (Fig. 13.7) and evaluated for their colony multiplication behavior with an objectivity. Are they compatible to each other?

13.8.3.1 Evaluation in Consortium Mode

The efficient microbes isolated through soil (*Bacillus polymyxa* at population of 12×10^3 cfu g^{-1}, *Pseudomonas fluorescens* at population of 5×10^3 cfu g^{-1}, *Trichoderma harzianum* at population of 12×10^3 cfu g^{-1}, *Azotobacter*

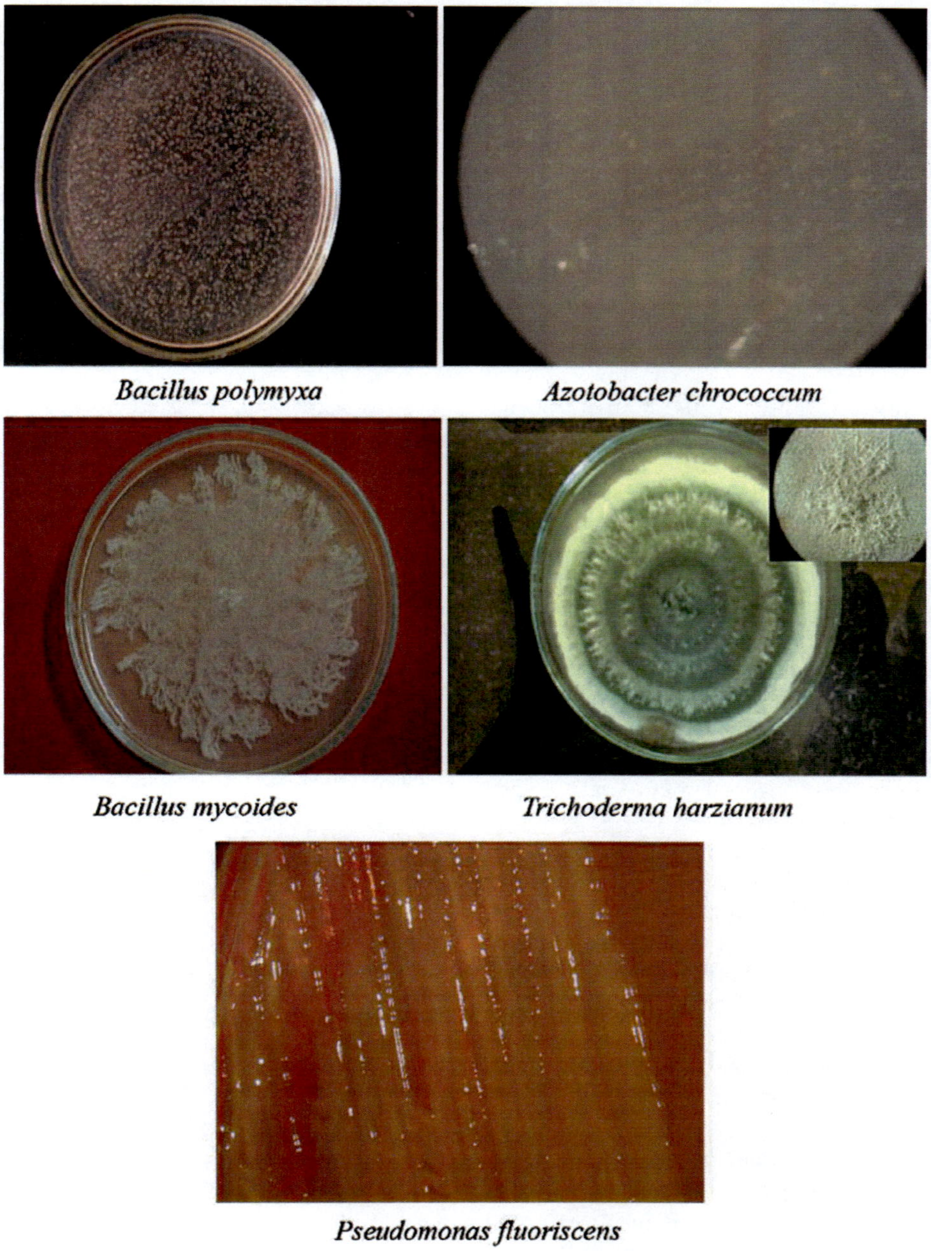

Fig. 13.7 Microbial consortium developed for Nagpur mandarin (*Citrus reticulata* Blanco) grown on montmorillonite-rich black clay soil in central India

chroococcum at population of 16×10^3 cfu g^{-1}, and *Bacillus mycoides* at population of 3×10^3 cfu g^{-1}) were brought in broth in order to achieve much high population as a substantial value addition (*Bacillus polymyxa* at population of 33×10^7 cfu g^{-1}, *Pseudomonas fluorescens* at population of 14×10^7 cfu $g^{-1,}$ *Trichoderma harzianum* at population of 32×10^7, *Azotobacter chroococcum* at population of 10×10^6, and *Bacillus mycoides* at population of 7×10^5 cfu g^{-1}) and then evaluated for 3 weeks for population changes at weekly interval in a complete consortium mode (Table 13.10). The population growth of all 5 microbes in consortium mode at first week, during 2009–2010, displayed slight reduction (*Bacillus polymyxa* 42×10^7 cfu mL^{-1}, *Pseudomonas fluorescens* 58×10^6 cfu mL^{-1}, *Trichoderma harzianum* 50×10^6 cfu mL^{-1}, *Azotobacter chroococcum* 7×10^6 cfu mL^{-1}, and *Bacillus mycoides* 12×10^4 cfu $mL-^1$), but after second week, the population growth started building up (*Bacillus polymyxa* 61×10^6 cfu mL^{-1}, *Pseudomonas fluorescens*

Table 13.10 Evaluation of different microbes in consortium mode through incubation studies

Nature of microbes	Microbial population in soil (cfu g^{-1})	Microbial population in broth (cfu mL–^{1})	Microbial population in consortium mode after (cfu mL^{-1}) 7 days	14 days	21 days
2009–2010					
1. *Bacillus polymyxa*	12×10^3	33×10^7	42×10^5	61×10^6	66×10^7
2. *Pseudomonas fluorescens*	5×10^3	14×10^7	58×10^6	88×10^6	18×10^7
3. *Trichoderma harzianum*	12×10^3	32×10^7	50×10^6	81×10^6	38×10^7
4. *Azotobacter chroococcum*	16×10^3	10×10^6	7×10^6	16×10^6	22×10^7
5. *Bacillus mycoides*	3×10^3	7×10^5	12×10^4	10×10^4	14×10^4

Nature of microbes	Microbial count in soil (cfu g^{-1})	Microbial count in broth (cfu mL^{-1})	Microbial count in consortium mode after 15 days (cfu mL^{-1})	30 days (cfu mL^{-1})	45 days (cfu mL^{-1})	60 days (cfu mL^{-1})
2010–2011						
B. polymyxa	16×10^3	44×10^6	48×10^6	52×10^6	68×10^6	62×10^6
P. fluorescens	11×10^3	22×10^6	28×10^6	48×10^6	72×10^6	74×10^6
T. harzianum	14×10^3	28×10^6	30×10^6	41×10^6	48×10^6	44×10^6
A. chroococcum	22×10^3	38×10^7	42×10^7	52×10^7	69×10^7	82×10^7
B. mycoides	10×10^3	22×10^6	28×10^6	30×10^6	40×10^6	32×10^6

Adapted from Srivastava et al. (2010, 2011)
cfu stands for colony forming units

88×10^6 cfu mL^{-1}, *Trichoderma harzianum* 81×10^6 cfu mL^{-1}, *Azotobacter chroococcum* 16×10^6 cfu mL^{-1}, and *Bacillus mycoides* 10×10^4 cfu mL^{-1}), and at third week, population was nearly the same to initial growth for microbes, namely, *Pseudomonas fluorescens* (18×10^7 cfu mL^{-1}) and *Trichoderma harzianum* (38×10^6 cfu mL^{-1}). Other three microbes displayed a higher population growth (*Bacillus polymyxa* 66×10^7 cfu mL^{-1}, *Azotobacter chroococcum* 22×10^6 cfu mL^{-1}, and *Bacillus mycoides* 14×10^4 cfu mL^{-1}) compared to initial population. These observations provided a strong value-added database support that such microbial consortium holds very good promise in nursery as well as grown-up orchards.

In subsequent study during 2010–2011, the microbial activity of all the five component microbes was followed as 15 days interval up to 60 days of storage to find out the shelf life of such microbial consortium. It was observed that *Bacillus polymyxa* and *Bacillus mycoides* population which initially recorded the population of 44×10^6 and 22×10^6 cfu mL^{-1}, respectively, increased to as high as 62×10^6 and 32×10^6 cfu mL^{-1}, respectively, up to 60 days of storage. Similarly, the population of *Pseudomonas fluorescens* increased from initial population of 22×10^6 to 74×10^6 cfu mL^{-1} after 60 days of storage. *Azotobacter chroococcum* observed an increase from initial concentration of 38×10^7 cfu mL^{-1} to as high as 82×10^7 cfu mL^{-1}. The concentrations of *Trichoderma harzianum* registered an increase from initial population of 28×10^6 to 44×10^6 cfu mL^{-1} up to 60 days of incubation. These observations provided strong evidence that microbial activity of consortium remains active up to 60 days offering better avenues of its effectiveness under diverse utility conditions.

Table 13.11 Evaluation of microbial consortium in nursery plants (pooled data of 2 seasons)

Treatments	Root weight (g)	Shoot weight (g)	Root:shoot	Stem diameter (mm)
Seedlings (period: 45 days)				
Control	2.99	9.08	1.1.037	8.61
Treated	9.59	24.86	1:2.85	11.9
$t_{P=0.05}$	3.65	5.63	–	1.43
Buddlings (period: 124 days)				
Control	4.10	10.72	1:2.63	20.20
Treated	11.76	26.41	1:2.20	28.51
$t_{P=0.05}$	2.03	5.635		2.02

Adapted from Srivastava et al. (2010, 2011)

13.8.3.2 Evaluation of Microbial Consortium in Nursery

The developed microbial consortium was evaluated in nursery plants, both on seedlings for 45 days and in buddlings for 124 days, using a total of 354 plants. Out of these 354 plants, 172 plants in 13 replications (each with 4 units) were treated, and other 172 plants in 13 replications were kept as untreated control (Table 13.11). The response of microbial consortium on rough lemon seedlings showed a significant increase in various growth parameters (9.59 g root weight, 24.86 g shoot weight, and 11.9 mm stem diameter on per plant basis) over control (2.99 g root weight, 9.08 g shoot weight, and 8.6 mm stem diameter on per plant basis). Similar observations were made on buddlings also. There was a significantly higher growth with microbial consortium treated buddlings (11.76 g root weight, 26.41 g shoot weight, and 28.51 mm stem diameter on per plant basis) compared to untreated

Table 13.12 Changes in soil fertility indices in response to inoculation with microbial consortium (pooled data of 2 seasons)

	Soil fertility (mg kg^{-1})						
	N	P	K	Fe	Mn	Cu	Zn
Control	116.2	13.2	166.7	8.8	6.7	1.12	0.62
Treated	123.4	16.2	169.7	13.7	10.2	1.16	0.88
$t_{P=0.05}$	3.95	2.0	NS	1.75	1.35	NS	0.12
	Microbial biomass nutrients (mg kg^{-1})						
	C_{mic}			N_{mic}			P_{mic}
Control	119.8			21.8			13.5
Treated	147.7			34.1			17.8
$t_{P=0.05}$	9.85			2.5			1.25

Adapted from Srivastava et al. (2010, 2011)
Computed on the basis of analysis after days of inoculation
C_{mic}, N_{mic}, and P_{mic} stand for microbial biomass-C, microbial biomass-N, and microbial biomass-P, respectively

control (4.10 g root weight, 10.72 g shoot weight, and 20.20 mm stem diameter). These observations confirmed the effectiveness of developed microbial consortium in growth-promoting abilities.

13.8.3.3 Changes in Soil Nutrient and Microbial Pool

The inoculation with microbial consortium brought a significant change in available supply of different nutrients in soil and microbial biomass nutrients (Table 13.12). A significantly higher soil fertility status with microbial consortium treated plants (123.4 N – 16.2 P – 13.7 Fe – 10.2 Mn – 0.88 Zn mg kg^{-1}) was observed compared to untreated control (116.2 N – 13.2 P – 8.8 Fe – 6.7 Mn – 0.62 Zn mg kg^{-1}). Similarly, microbial biomass nutrients were higher in the rhizosphere treated with microbial consortium (147.7 mg kg^{-1} C_{mic}, 34.1 mg kg^{-1} N_{mic}, and 17.8 mg kg^{-1} P_{mic}) than untreated control (119.8 mg kg^{-1}, C_{mic}, 21.8 mg kg^{-1} N_{mic}, and 13.5 mg kg^{-1} P_{mic}). The above observations strongly supported the effectiveness of microbial consortium in improving chemical and biological indices of citrus rhizosphere.

The treatment combination of ¾P + AM + N was observed the best treatment with reference to better growth and yield of high-quality fruits of 'Mosambi' sweet orange, suggesting the compatibility of biofertilizers (AZO) and AM inoculation in combination with chemical fertilizers for better growth, yield, and fruit quality. Such observations in the long term are expected to cut down the cost of chemical fertilizers, particularly N and P, and building up fertility by maintaining better soil physical conditions (Singh et al. 2000). High efficiency of Azospirillum for fixing nitrogen and better mobilization of fixed phosphorus by AM even at high temperatures can make these highly suited for Mosambi sweet orange (Manjunath et al. 1983).

13.9 Conclusion

Citrus rhizosphere possesses a relative amount of microorganisms, which are affected by various abiotic and biotic factors, such as root exudates, plant species, plant developmental stages, soil environments, etc. The rhizosphere microbial biomass contains mineral nutrients, as much as ~20% of the total soil N and P. Extraradical fungal mycelium attains as much as 3% of root weight. The variation in microbial communities in terms of structure as well as diversity offers a strong perspective to develop a rhizosphere-specific microbial consortium that can be later uploaded with any organic manure, called substrate to engineer rhizosphere (often referred as rhizosphere engineering), for mitigating dynamic nutrient demand across crop phenophases. Rhizosphere extraradical mycorrhizal mycelial networks are regarded as the main nutrient/water-absorbing interfaces. Contributions of mycorrhizal hyphae to water uptake are dependent on soil water status. AMF can enhance the tolerance of osmotic stress by osmotic adjustment, antioxidative defense systems, and glomalin. The rhizosphere AMF and other native microbe of diverse nature could lead a definite way forward to synthesize an integrated biofertilizer for using citrus culture. But before such concepts come into practice, large-scale field responses are must to verify their promise.

13.10 Future Research

Plants have a distinct impact on characteristic activity of resident soil microbial communities and therefore play an important role in determining the development of the disease-suppressive state. Likewise, plant genotype will modulate these same biological communities and should be considered when developing strategies to exploit the potential of such a natural disease control system. Implementation of consistently effective practices to manage these resources in an economically and environmentally feasible manner will require more detailed investigation of these biologically complex systems and refinement of currently available methodologies (Mazzola 2004).

The very small soil sample sizes in the range of a few milligrams allow a very small-scale spatial resolution. Soil will no longer be a black box, but we will be able to see where the microbes live, what their role in soil processes is, and how their abundance and activity is influenced by soil physical and chemical properties. Thus, today, in soil microbiology, questions like who is active and where are the activities located are answered that have been asked many years ago. Soil management may aim to successfully establish desired microbial populations. Such microorganisms may be degraders of xenobiotics, nitrogen fixers, or pathogen antagonists.

In the not-so-far future, a single key biological player in soil may be altered in a desired way, thus altering soil functions in a beneficial way to man. This will hopefully increase the sustainability of agricultural systems on the long run and also enable us to successfully remediate polluted soils and protect natural ecosystems (Insam 2001).

Still not clear-cut understanding is worked out about the presence of functionality mode between crop species, changes in rhizosphere microbial communities, and mineralization processes. A pertinent question in this regard emerges as do soil enzyme activities and substrate quality differ within rhizosphere of different crop species. The answer is if yes, then what are the relationships among soil enzyme activities, soil physicochemical properties, substrate quality, and soil microbial community. Nutrient transformations often rely on appropriate cellular and microenvironmental or microzonal redox conditions. The spatial and temporal requirements for microenvironmental overlap among microbial groups involved in nutrient transformations necessitates close proximity and diffusional exchange with other biogeochemically distinct complimentary, microbial groups (Paert and Pinckney 1996; Radianingtyas et al. 2003).

The potential of rhizosphere microbial isolates is formulated using different organic and inorganic carriers either through solid or liquid fermentation technologies. Some of the rhizosphere microbes possess threat to infect human beings as opportunistic pathogens have to be clarified before large-scale commercial acceptance, registration, and adoption of rhizosphere microbes for pest and disease management (Garbeva et al. 2004). Though AMF are not host specific, further study will be needed to select an efficient mycorrhizal fungus from these surveyed AMF species for use with citrus rootstock so that AMF could be efficiently infused in such attempt of providing rhizosphere resilience.

Considering the complexity involved with the analysis of microbial diversity in soil, there needs to first understand the freshwater microbial diversity, and then scale up to soil, culture-independent methods of assessing soil microbial diversity such as PCR-based methods and alternative methods to PCR approaches (examining physiological or metabolic characteristics of microbial communities) are yet to be clinically worked out. Studies are yet to be initiated focusing on the rhizosphere microbial diversity versus plant nutrition or fruit quality issues. Microarray technology will soon enable us to assess the community diversity in soils by directly exposing and hybridizing oligonucleotides fixed on membranes (Ogram 2000) in addition to relating community structure with community function using messenger RNA combined PCR amplification (Gottschal et al. 1997).

Studies on gene expression in the rhizosphere soil can permit a better understanding of processes such as biological control, stimulation of microbial activity by root exudates, competition between microorganisms and roots for nutrients, and molecular colloquia between microorganisms, between roots, and between roots and microorganisms. Techniques for extracting and characterizing mRNA from soil are now available (Nannipieri et al. 2003) whereas soil proteomics is still in its infancy (Nannipieri 2006). An advancement in linking between functional activity to community structure has been obtained by applying stable isotope probe to soil (Manefield et al. 2006). Reporter technology has been used to assess several functions in the rhizosphere soil including gene expression even at the single cell level (Sórensen and Nybroe 2006). The ever-increasing knowledge of the promoter and regulator gene along with the refinement of reporter gene insertion techniques will allow using the reporter gene technique for monitoring induction, expression, and regulation of virtually any gene in the rhizosphere.

These studies will lead to understand (1) induction of shifts in microbial communities by altering the microenvironment, (2) isolation of specific microorganisms or group of microorganisms for signature markers and then detecting the signatures after induction, (3) detection of shifts in microbial community nutritional status with alteration in the environment, (4) detection of specific microorganisms and their activity, and (5) consequences of specific predation on the composition and diversity (community structure) in soil microbial communities (White 1988).

Acknowledgments These studies pertaining to mycorrhiza were supported by the National Natural Science Foundation of China (30800747; 31101513), the Key Project of Chinese Ministry of Education (211107), and the Science-Technology Research Project of Hubei Provincial Department of Education, China (Q20111301). Part of information generated by Dr. A.K. Srivastava on isolation and efficacy of soil microbes was worked out under project entitled "Development of INM Module for Citrus" under Indian Council of Agricultural Research, New Delhi (India).

References

Abad M, Fornes F, Carolina C et al (2005) Physical properties on various coconut coir dusts compared to peat. HortScience 40:2138–2144

Alagawadi AR, Gaur AC (1992) Inoculation of *Azospirillium brasilense* and phosphate-solubilizing bacteria on yield of sorghum [*Sorghum biocolor* (L.) Moench] in dry land. Trop Agric 69:347–350

Allen MF (1982) Influence of vesicular-arbuscular mycorrhizae on water movement through *Bouteloua gracilis*. New Phytol 91:191–196

Allen MF (2009) Water relations in the mycorrhizosphere. In: Luttge U, Beyschlag W, Budel B, Francis D (eds) Progress in botany 70. Springer, Berlin/Heidelberg, pp 257–276

Altland JE (2006) Substrate pH, a tricky topic. Digger 50:42–47

Altland JE, Buamscha MG (2008) Nutrient availability form Douglas Fir Bark in response to substrate pH. HortScience 43:478–483

Altland JE, Buamscha MG, Horneck DA (2008) Substrate pH affects nutrient availability in fertilized Douglas Fir Bark substrates. HortScience 43:2171–2178 del Amor FM, Serrano-Martinez A, Fortea MI et al (2008) The effect of plant-associative bacteria (*Azospirillum* and *Pantoea*) on the fruit quality of sweet pepper under limited nitrogen supply. Sci Hortic 117:191–196

Andrade G, Linderman RG, Bethlenfalvay GJ (1998) Bacterial associations with the mycorrhizosphere and hyphosphere of the arbuscular mycorrhizal fungus *Glomus mosseae*. Plant Soil 202:79–87

Aseri GK, Jain N, Panwar J et al (2008) Biofertilizers improve plant growth, fruit yield, nutrition, metabolism and rhizosphere enzyme activities of pomegranate (*Punica granatum* L.) in Indian Thar desert. Sci Hortic 117:130–135

Baar J (2010) Development of soil quality metrics using mycorrhizal fungi. Span J Agric Res 8:137–143

Badiyala SD, Awasthi RP, Gupta RD (1990) Effect of fertilisers and management practices on microflora in Alfisols growing citrus. J Indian Soc Soil Sci 38:537–540

Balser TC, Wixon D, Moritz LK et al (2010) The microbiology of natural soils. In: Dixon GR, Tilston EL (eds) Soil microbiology and sustainable crop production. Springer, Heidelberg, pp 27–58

Bashan Y (1998) Inoculants of plant growth promoting bacteria for use in agriculture. Biotechnol Adv 16:729–770

Bartelt-Ryser J, Joshi J, Schmid B et al (2005) Soil feedbacks of plant diversity on soil microbial communities and subsequent plant growth. Perspect Plant Ecol Evol Syst 7:27–49

Bedini S, Pellegrino E, Avio L et al (2009) Changes in soil aggregation and glomalin-related soil protein content as affect by the arbuscular mycorrhizal fungi species *Glomus mosseae* and *Glomus intraradices*. Soil Biol Biochem 41:1491–1496

Belimov AA, Kojemiakov AP, Chuvarliyeva CV (1995) Interaction between barley and mixed cultures of nitrogen fixing and phosphate solubilizing bacteria. Plant Soil 173:29–37

Bellone CH, de Bellone SC (1995) Morphogenesis of strawberry roots infected by *Azospirillium brasilense* and VA mycorrhiza. NATO, Adv Study Inst, Series G (Ecol Sci) 37:251–255

Bhattacharya P (1999) Use of biofertilizers in citrus. In: Singh S, Naqvi SAMH (eds) Citriculture. National Research Centre for Citrus, Nagpur, pp 194–204

Bhattacharya P, Kumar R, Jain RK (1999) Biofertilizers use in citrus – prospects and strategies. In: Abstract, International symposium on citriculture, Nagpur, 17–19 Nov 1999, p 83

Binkley D, Giardina C (1988) Why do tree species affect soils? The warp and woof of tree-soil interactions. Biogeochemistry 42:89–106

Binkley O, Menyailo O (2005) Tree species effects on soil: implications for global change. Springer, NATO Sci Series IV, Earth & Environ Sci 55:182–195

Binkley D, Valentine D (1991) Fifty-year biogeochemical effects of green ash, white pine and Norway spruce in a replicated experiment. For Ecol Manag 40:13–25

Blythe EK, Merhaut DJ, Newman JP et al (2006) Nutrient release from controlled release fertilizers in acid substrate in a greenhouse environment: II. Leachate, calcium, magnesium, iron, manganese, zinc copper, and molybdenum concentrations. HortScience 41:788–793

Bowen GD, Rovira AD (1999) The rhizosphere and its management to improve plant growth. Adv Agron 66:1–102

Buamscha MG, Altland JE, Sullivan DM et al (2007) Micronutrient availability in fresh and aged Douglas fir bark. HortScience 42:152–156

Caldeira SF, Chaves GM, Zambolim I (1983) Observation on endotrophic vesicular-arbuscular mycorrhiza in different plant species. Revista Ceres 30(167):19–24

Camprubi A, Calvet C (1996a) Isolation and screening of mycorrhizal fungi from citrus nurseries and orchards and inoculation studies. HortScience 31(3):366–369

Camprubi A, Calvet C (1996b) A field inoculation system, for citrus nurseries using pre-cropping with mycorrhizal aromatic plants. Fruits (Paris) 51:133–137

Cardoso IM, Kuyper TW (2006) Mycorrhizas and tropical soil fertility. Agr Ecosyst Environ 116:72–84

Chen GC, He ZL, Wang YJ (2004) Impact of pH on microbial biomass carbon and microbial biomass phosphorus in red soils. Pedosphere 14(1):9–15

Clarholm M (1994) The microbial loop in soil. In: Ritz K, Dighton J, Giller KE (eds) Beyond the biomass. Wiley, New York, pp 221–230

Cook RJ (2002) Advances in plant health management in the twentieth century. Ann Rev Phytopathol 38:95–116

Cruz AF, Ishii T, Matsumoto I et al (2002) Network establishment of vesicular-arbuscular mycorrhizal hyphae in the rhizospheres between trifoliate orange and some plants. J Jpn Soc Hortic Sci 71:19–25

Date RA (2001) Advances in inoculation technology: a brief review. Austral J Exp Agric 41:321–325

De Ridder-Duine AS, Kowalchuk GA, Klein Gunnewiek PJA et al (2005) Rhizosphere bacterial community composition in natural stands of *Carex arenaria* (sand sedge) is determined by bulk soil community composition. Soil Biol Biochem 37:349–357

Duponnois R, Galiana A, Prin Y (2008) The mycorrhizosphere effect: a multitrophic interaction complex improves mycorrhizal symbiosis and plant growth. In: Siddiqui ZA, Akhtar MS, Futai K (eds) *Mycorrhizae*: sustainable agriculture and forestry. Springer, Heidelberg, pp 227–240

Dutt M, Patil PT, Sonawane PC (2002) Effect of various substrates on growth and flowering of chrysanthemum. Indian J Hortic 59:191–195

Dutt M, Sonawane PC (2006) Nutrient uptake in chrysanthemum grown on various substrates. Indian J Hortic 63:66–69

Egerton-Warburton LM, Querejeta JI, Allen MF (2008) Efflux of hydraulically lifted water from mycorrhizal fungal hyphae during imposed drought. Plant Signal Behav 3:68–71

Elshanshoury AR (1995) Interactions of *Azotobacter chroococcum*, *Azospirillum brasilense* and *Streptomyces mutabilis* in relation to their effect on wheat development. J Agron Crop Sci 175:119–127

Fabbrie P, Gallo Del M (1995) Specific interaction between chickpea (*Cicer arietinum*) and three chick-pea-*Rhizobium* strains inoculated singularly and in combination with *Azospirillum brasilense*. NATO, Adv Study Inst, Series G (Ecol Sci) 37:267–267

Fisher P, Huang J, Argo W (2006) Modeling lime reaction in peat-based substrates. Acta Hortic 718:461–468

FNCA Biofertilizer Project Group (2006) Biofertilizer manual. Japan Atomic Industrial Forum, Tokyo

Franzluebbers AJ, Wright SF, Stuedemann JA (2000) Soil aggregation and glomalin under pastures in the southern piedmont USA. Soil Sci Soc Am J 64:1018–1026

Garbeva P, van Veen JA, van Elsas JD (2004) Microbial diversity in soil: selection of microbial populations by plant and soil type and implications for disease suppressiveness. Ann Rev Phytopathol 42:243–270

Gaur A, Varma A (2007) Research methods in arbuscular mycorrhizal fungi. In: Varma A, Oelmuller R (eds) Soil biology: advanced techniques in soil microbiology, vol 11. Springer, Berlin, pp 377–396

Gerdemann JW, Nicolson TH (1963) Spores of mycorrhizal endogone extracted from soil by wet sieving and decanting. Trans Br Mycol Soc 46:235–244

Giovannetti M, Avio L, Fortuna P et al (2006) Self recognition and non-self incompatibility in mycorrhizal networks. Plant Signal Behav 1:1–5

Gochelashvili ZA (1973) The total number of bacteria in mandarin orange and lemon trees rhizosphere. Subtropicheskie Kultury 4:54–57

Gochelashvili ZA (1978) The fixation of free nitrogen in soil and on roots of citrus trees in subtropical of the Georgian SSr. Mikro-biologiya 47:860–865

Gori A, Favilli F (1995) First results on individual and dual inoculation with *Azospirillum* – *Glomus* on wheat. NATO, Adv Study Inst, Series G (Ecol Sci) 37:245–249

Gottschal JC, Meijer WG, Oda Y (1997) Use of molecular probing to assess microbial ecology. In: Megusar F, Gantar G (eds) Perspectives in microbial ecology. Proceeding of the 4th international symposium on microbial ecology, Ljubljana, Slovenia, 1997, pp 455–459

Graham JH (1986) Citrus mycorrhizae: potential benefits and interactions with pathogens. HortScience 21:1302–1306

Graham JH, Syvertsen JP (1984) Influence of vesicular-arbuscular mycorrhiza on the hydraulic conductivity of roots of two citrus rootstocks. New Phytol 97:277–284

Hartmann A, Schmid M, van Tuinen D et al (2009) Plant-driven selection of microbes. Plant Soil 321:235–257

Hassouna MG, Hassan MT, Madkour MA (1994) Increased yield of alfalfa (*Medicago sativa*) inoculated with N_2-fixing bacteria and cultivated in a calcareous soil of Northwestern Egypt. Arid Soil Res Rehabil 8:389–393

He ZL, Chen G, Yao H et al (2002) Microbial biomass and its turnover: a potential diagnosing tool for soil fertility quality. In: Transactions of the 17th world congress soil science, Thailand, 2002, pp 14–21

Heerden IV, Cronje C, Swart SH et al (2002) Microbial, chemical and physical aspects of citrus waste composting. Bioresour Technol 81:71–76

Hernandez-Apaolaza L, Gasco AM, Gasco JM et al (2005) Reuse of waste materials a growing media. Comm Soil Sci Plant Anal 24:349–363

Insam H (2001) Developments in soil microbiology since mid 1960s. Geoderma 100:389–402

Inserra RN, Nemec S, Giudice VL (1980) A survey of endomycorrhizal fungi in Italian citrus nurseries. Rivista dell Ortoflorofrutticoltura Italiana 64(1):83–88

Ishii T, Kadoya K (1996) Utilisation of vesicular-arbuscular mycorrhizal fungi in citrus orchards. In: Manicom B, Robinson J, Du Plessis SF (eds) Proceedings of the international society citriculture, Sun City, South Africa, 12–17 May 1996, vol 2, pp 777–780

Ishii T, Matsumura A, Horii S et al (2007) Network establishment of arbuscular mycorrhizal hyphae in the rhizospheres between citrus rootstocks and *Paspalum notatum* or *Vulpia myuros* grown in sand substrate. Biol Fertil Soils 44:217–222

Johnson LF, Curl EA (1972) Methods for research on ecology of soil-borne pathogens. Burgess Publishing Company, Minneapolis

Kalita P, Bora LC, Bhagbali KN (1996) Phytoplane microflora of citrus and their role in management of citrus canker. Indian Phytopathol 49:234–237

Kent AD, Triplett EW (2002) Microbial communities and their interactions in soil and rhizosphere ecosystem. Ann Rev Microbiol 56:211–236

Kohli RR, Srivastava AK, Huchche AD et al (1997) Relationship of leaf nutrient status, soil available nutrients and microbial composition with fruit yield of Nagpur mandarin. In: Abstract, National symp osium on citriculture, Nagpur, 17–19 Nov 1997, pp 53

Koller M, Alfoldt T, Siegrist M et al (2004) A comparison of plant and animal based fertilizer for the production of organic vegetable transplants. Acta Hortic 631:209–215

Kuepper G, Adam K (2003) Organic potting mixes for certified production. Horticulture. Horticulture Technical Note No. 112, Appropriate Technology Transfer for Rural Areas, National Sustainable Agriculture Information Service, Fayetteville, Arkansas, pp 48–52

Kulmatiski A, Beard KH, Stevens JR et al (2008) Plant-soil feedbacks: a meta-analytical review. Ecol Lett 11:980–992

Lambais MR (2006) Unraveling the signaling and signal transduction mechanisms controlling arbuscular mycorrhizal development. Sci Agric 63:405–413

Lazdunski AM, Ventre I, Sturgis JN (2004) Regulatory circuits and communications in gram-negative bacteria. Nat Rev Microbiol 2:581–592

Leake J, Johnson D, Donnelly D et al (2004) Networks of power and influence: the role of mycorrhizal mycelium in controlling plant communities and agroecosystem function. Can J Bot 82:1016–1045

Levy Y, Syvertsen JP, Nemec S (1983) Effect of draught stress and vesicular-arbuscular mycorrhiza on citrus transpiration and hydraulic conductivity of roots. New Phytol 93:61–66

Lübeck PS, Hansen M, Sorensen J (2000) Simultaneous detection of the establishment of seed-inoculated *Pseudomonas fluoriscens* strain DR54 and native soil bacteria on sugar beet root surfaces using fluorescence antibody and in situ hybridization techniques. FEMS Microbiol Ecol 33:11–19

Lynch JM (2002) Plant diversity and microbial diversity – the role of the rhizosphere. In: Lynch JM, Schepers JS, Unver I (eds) Innovative soil-plant systems for sustainable agricultural practices. OECD, Paris, France, pp 147–154

Lynch JM, Whipps JM (1991) Substrate flow in the rhizosphere. In: Keister DL, Cregan B (eds) The rhizosphere and plant growth (Beltsville symposia in agricultural research). Kluwer, Dordrecht, pp 15–24

Manefield M, Griffiths RI, Bailey MJ et al (2006) Stable isotope probing: a critique of its role in linking phylogency and function. In: Nanniperi P, Smalla K (eds) Nucleic acids and proteins in soil. Springer, Heidelberg, pp 205–255

Manjunath A, Mohan R, Bagyaraj DJ (1983) Responses of citrus to VAM inoculation in unsterile soils. Can J Bot 61:2729–2732

Marschner P, Crowley D, Yang CH (2004) Development of specific rhizosphere bacterial communities in relation to plants species, nutrition and soil type. Plant Soil 261:199–208

Marschner P, Timonen S (2006) Bacterial community composition and activity in rhizospheres of roots colonised by arbuscular mycorrhizal fungi. In: Mukerji KG, Manoharachary C, Singh J (eds) Microbial activity in the rhizosphere. Springer, Berlin, pp 139–154

Mazzola M (2004) Assessment and management of soil microbial community structure for disease suppression. Ann Rev Phytophathol 42:35–59

Meinzer FC, Brooks JR, Bucci S et al (2004) Converging patterns of uptake and hydraulic redistribution of soil water in contrasting woody vegetation types. Tree Physiol 24:919–928

Menge JA (1985) Mycorrhiza agriculture technologies. In: Innovative biological technologies for lesser developed countries – workshop proceedings, Office of Technology Assessment, Washington, pp 185–203

Merhaut DJ, Blythe EK, Newman JP et al (2006) Release from controlled fertilizers in acid substrate greenhouse environment: I. Leachate electrical conductivity, pH and nitrogen, phosphorus, and potassium concentration. HortScience 41:780–787

Merhaut D, Newman J (2005) Effects of substrate type on plant growth and nitrate leaching in cut flower production on Oriental Lily. HortScience 40:2135–2137

Nannipieri P (2006) Role of stabilized enzymes in microbial ecology and enzyme extraction from soil with potential applications in soil proteomics. In: Nannipieri P, Smalla K (eds) Nucelic acids proteins in soil. Springer, Heidelberg, pp 75–94

Nannipieri P, Ascher J, Cecchenrine MT et al (2003) Microbial diversity and soil functions. Eur J Soil Sci 54:655–670

Nannipieri P, Ascher J, Ceccherini MT et al (2007) Microbial diversity and microbial activity in the rhizosphere. Ci Suelo (Argentina) 25:89–97

Nelson LM (2004) Plant growth promoting rhizobacteria: prospects for new inoculants. Online Crop Manag. doi:10.1034/CM-2004-301-05-RV

Nemec S, Menge JA, Platt RG et al (1981) Arbuscular mycorrhizal fungi associated with citrus in Florida and California notes on their distribution and ecology. Mycologia 73:112–127

Ngullie E, Reddy LS, Singh AK et al (2008) Rhizosphere transformations and response of citrus. In: Shyam S, Shivankar VJ, Srivastava AK et al. (eds) Proceedings of the national symposium on citriculture: emerging trends, National Research Centre for Citrus, Nagpur, 24–26 July 2008, pp 193–196

Ngullie E, Singh AK, Reddy LS et al (2009) Rhizospheric changes favouring growth of citrus. In: 19th world congress of soil science, Brisbane (under Division Soil Use and Management Division and sub-section Integrated Nutrient Management), 1–6 Aug 2009

Ngullie E, Reddy S, Singh AK et al (2010) Rhizosphere properties and pre-bearing response of citrus. In: Abstract, 28th international horticultural congress semester (category – modelling perennials), Lisboaal, 22–27 Aug 2010

Nihorimbere V, Ongena M, Thonart MSP (2011) Beneficial effect of the rhizosphere microbial community for plant growth and health. Biotechnol Agronomie Socié Environ 15:327–337

Neyra CA, Atkinson A, Olubayi O (1995) Coaggregation of *Azospirillum* with other bacteria: basis for functional diversity. NATO, Adv Study Inst, Series G (Ecol Sci) 37: 429–439

Ogram A (2000) Soil molecular microbial ecology at age 20: methodological challenges for the future. Soil Biol Biochem 32:1499–1504

Ozores-Hampton MP, Vavrina CS (2002) Worm castings: an alternative to sphagnum peat moss in organic tomato (*Lycopersicon esculentus* Mill.) transplant production. In: Proceedings of the international composting and compost utilization, Columbus, 6–8 June 2002, pp 218–230

Paert HW, Pinckney JL (1996) A mini review of microbial consortia: their roles in aquatic production of biogeochemical cycling. Microbiol Ecol 31:225–247

Pacioni G (1992) Wet-sieving and decanting techniques for the extraction of spores of vesicular-arbuscular fungi. In: Norris JR, Read DJ, Varma AK (eds) Methods in microbiology: techniques for the study of mycorrhiza, vol 24. Academic Press Limited, Tokyo, pp 317–322

Palazzo D, Pommerening B, Vanadia S (1992) Effect of soil sterilization and vesicular arbuscular mycorrhiza on growth of sour orange (*Citrus aurantium* L.) seedlings. In: Tributato E, Gentile A, Refergiato G (eds) Proceedings of the international society of citriculture, Acireale, Italy, 8–13 Mar 1992, vol 2, pp 621–623

Paliwal MK, Mowade Sarita, Bhattacharya P (1999) Studies on the distribution pattern and performance of P-solubilising microorganisms and VA mycorrhiza in citrus soil of Nagpur. In: Abstract, International symposium on citriculture, Nagpur, 23–27 Nov 1999, pp 85–86

Parniske M (2008) Arbuscular mycorrhiza: the mother of plant root endosymbioses. Nat Rev Microbiol 6:763–775

Plenchette C, Clermont-Dauphin C, Meynard JM et al (2005) Managing arbuscular mycorrhizal fungi in cropping systems. Can J Plant Sci 85:31–40

Pomares F, Tarazona F, Estela M et al (1981) Evaluation of a commercial blue green algae inoculants as fertilizer on citrus. In: Matsumoto K, Oogaki C, Kozaki I et al (eds) Proceedings of the international society of citriculture, Tokyo, 9–12 Nov 1981, vol 2, pp 583–585

Priya LS, Kumutha K (2009) Effect of arbuscular mycorrhizal inoculation on enzyme activities and microbial population in the rhizosphere of *Coleus forskohlii* Briq. Mycorrhiza News 20:14–21

Porazinska DL, Bardgett RD, Blaauw MB et al (2003) Relationships at the aboveground-belowground interface: plants, soil biota and soil processes. Ecol Monogr 73:377–395

Querejeta JI, Egerton-Warburton LM, Allen MF (2003) Direct nocturnal water transfer from oaks to their mycorrhizal symbionts during severe soil drying. Oecologia 134:55–64

Radianingtyas H, Robinson GK, Bull AT (2003) Characterization of a soil derived bacterium consortium degrading chloroaniline. Microbiology 149:3279–3287

Rainey PB (1999) Adaptation of *Pseudomonas fluorescens* to the plant rhizosphere. Environ Microbiol 1:243–257

Rangaswami G, Vasantharajan VN (1962a) Studies on the rhizosphere microflora of citrus trees: I. Quantitative incidence of microorganisms in relation to root and shoot growth. Can J Microbiol 8:473–477

Rangaswami G, Vasantharajan VN (1962b) Studies on the rhizosphere microflora of citrus trees: II. Qualitative distribution of the bacterial flora. Can J Microbiol 8:479–484

Rangaswami G, Vasantharajan VN (1962c) Studies on the rhizosphere microflora of citrus trees: III. Fungal and actinomycete flora of the rhizosphere. Can J Microbiol 8:485–489

Rillig MC (2004) Arbuscular mycorrhizae, glomalin, and soil aggregation. Can J Soil Sci 84:355–363

Rillig MC, Hoye AT, Carran A (2006) Minimal direct contribution of arbuscular mycorrhizal fungi to DOC leaching in grassland through losses of glomalin-related soil protein. Soil Biol Biochem 38:2967–2970

Rokni N, Goltapeh EM (2011) Diversity of arbuscular mycorrhizal fungi associated with common sugarcane varieties in Iran. J Agric Technol 7:1017–1022

Ros M, Klammer S, Knapp B et al (2006) Long term effects of compost amendment of soil on functional and structural diversity and microbial activity. Soil Use Manag 22:209–218

Ross MS, Ruiz PL, Sah JP et al (2009) Chilling damage in a changing climate in coastal landscapes of the subtropical zone: a case study from south Florida, global change. Biology 15:1817–1832

Roose R, Haase DL (2000) The use of coir as a containerized growing medium for Douglas fir seedlings. Nat Plant J 1:107–111

Russo VM (2005) Organic vegetable transplant production. HortScience 40:623–628

Saakashvili MA, Gochelashvili ZA, Mgaloblishvili TS (1971) The effect of different rates of nitrogenous fertilizers on the soil microflora in an orange plantation. Subtropicheskie Kul'tury 1:86–92

Saetre P, Bääth E (2000) Spatial variation and pattern of soil microbial community structure in a mixed spruce-birch stand. Soil Biol Biochem 32:909–917

Salifu FK, Nicodemus MA, Jacobs DF et al (2006) Evaluating chemical indices of growing media for nursery production of *Quercus rubra* seedlings. HortScience 41:1342–1346

Sanikidze GS (1970) The quantity of microorganism in the soil of a mandarin plantation in relation to the application of peat-mineral-ammonium fertilizers. Subtropicheskie Kul'tury 3:65–71

Schallmach E, Minz D, Jurkevitch E (2000) Culture-independent detection of changes in root-associated bacterial populations of common bean (*Phaseolus vulgaris* L.) following nitrogen depletion. Microb Ecol 40:309–316

Schenck NC, Perez Y (1990) Manual for the identification of VA mycorrhizal fungi. Synergistic Publications, Gainesville, 283

Scholz FG, Bucci SJ, Hoffmann WA et al (2010) Hydraulic lift in a *Neotropical savanna*: experimental manipulation and model simulations. Agri Forest Meteorol 150:629–639

Schubert A, Aiassa A, Palazzo D et al (1993) Occurrence of mycorrhiza in citrus orchards in the metaponto area of Basilicata (Italy). Acta Hortic 324:61–66

Selvakumar G, Mishra PK, Bisht JK (2008) Composting agroresidues with fungal inoculant. ICAR Newslett 14:5–6

Siddiqui Y, Meon S, Ismail R et al (2008) Bio-efficiency of compost extracts on the wet rot incidence, morphological and physiological growth of okra (*Abelmoschus esculentus* [L.] Moench). Sci Hortic 117:9–14

Singh C, Sharma BB (1993) Leaf nutrient composition of sweet orange as affected by combined use of bio- and chemical fertilizer. Indian Hortic 41:131–134

Singh C, Saxena SK, Goswami AM et al (2000) Effect of fertilizers on growth, yield and quality of sweet orange (*Citrus sinensis* Osbeck) cv Mosambi. Indian J Hortic 57:114–117

Sórensen S, Nybroe O (2006) Reporter gene in bacterial inoculants can monitor life conditions and functions in soil. In: Nannipieri P, Smalla K (eds) Nucleic acids proteins in soil. Springer, Heidelberg, pp 375–395

Srivastava AK (2009) Integrated nutrient management: concept and application in citrus. In: Tennant P, Beakebhia N (eds) Citrus II. Tree Forest Sci Biotechnol 3 (Special Issue 1), pp 32–58

Srivastava AK (2010) Integrated nutrient management concept and applications in citrus in south east Asia. In: Deng X, Xu J, Lin S

et al (eds) Proc eedings of the 11th international citrus congress, international society citriculture, Wuhan, 26–30 Oct 2008, vol 1, pp 726–733
Srivastava AK, Huchche AD, Kumar D (2010) Development of INM module for sustained productivity of citrus. Ann Rep 2009–2010, National Research Centre for Citrus, Nagpur, pp 40–41
Srivastava AK, Huchche AD, Kumar D (2011) Development of INM module for sustained productivity of citrus. Ann Rep 2010–2011, National Research Centre for Citrus, Nagpur, pp 40–41
Srivastava AK, Ngullie E (2009) Integrated nutrient management: theory and practice. Dyn Soil Dyn Plant 3(1):1–30
Shen TH, Wang SL (1994) A practical study on the propagation and inoculation technique of VA mycorrhizal fungi in citrus. Acta Agric Jiangxi 6:25–30
Smith GW, Skipper HD (1979) Comparison of method to extract spores of vesicular arbuscular mycorrhizal fungi. Soil Sci Soc Am J 43:722–725
Stainer RY, Adelberg EA, Ingraham JL (1987) General microbiology, 4th edn. Macmillan, London
Stewart A, Brownbridge M, Hill RA et al (2010) Utilizing soil microbes for biocontrol. In: Dixon GR, Tilston EL (eds) Soil microbiology and sustainable crop production. Springer, Heidelberg, pp 315–371
Talukdar NC (1999) Vesicular-arbuscular mycorrhizal fungi (VAMF) biofertilizer prospects and potential in N.E. Region of India. In: Singh T, Yadav AK, Roychoudhury S (eds) Use and development of microbial inoculants. NERCB, Guwahati, pp 85–89
Torsvik V, Goksoyr J, Daae FD (1990) High diversity in DNA in soil bacteria. Appl Environ Microbiol 56:782–787
Torsvik V, Ovreas L (2002) Microbial diversity and function in soil: from genes to ecosystem. Curr Opin Microbiol 5:241–245
Treseder KK, Turner KM (2007) Glomalin in ecosystems. Soil Sci Soc Am J 71:1257–1266
Turney JK, Menge JA, Yang CH et al (1992) Biocontrol of citrus root rot using soilborne bacteria. Phytopathology 82:1080
Ushio M, Wagai R, Balser TC et al (2008) Variations in the soil microbial community composition of a tropical montane forest ecosystem: does tree species matter. Soil Biol Biochem 40:2699–2702
Urban Creeks Council (2006) Biofertilizers and mycorrhizae. http://www.urbancreeks.org/Biofertilizers_Mycorrhizae.pdf
Visser S, Parkinson D (1992) Soil biological criteria as indicators of soil quality: soil microorganisms. Am J Altern Agric 7:33–37
White DC (1988) Validation of quantitative analysis for microbial biomass, community structure and metabolic activity. Arch Hydrobiol 31:1–18
Wang P, Liu JH, Xia RX et al (2011) Arbuscular mycorrhizal development, glomalin-related soil protein (GRSP) content, and rhizospheric phosphatase activity in citrus orchards under different types of soil management. J Plant Nutr Soil Sci 174:65–72
Watanarojanaporn N, Boonkerd N, Wongkaew S et al (2011) Selection of arbuscular mycorrhizal fungi for citrus growth promotion and *Phytophthora* suppression. Sci Hortic 128:423–433
Watkinson SC, Boddy L, Burton K et al (2005) New approaches to investigating the function of mycelial networks. Mycologist 19:11–17
Welington LA, Marcon J, Maccheroni W Jr et al (2002) Diversity of endophytic bacterial populations and their interaction with *Xyella fastidiosa* in citrus plants. Appl Environ Microbiol 68:4906–4914
Wu QS (2011) Mycorrhizal efficacy of trifoliate orange seedlings on alleviating temperature stress. Plant Soil Environ 10:459–464
Wu QS, He XH, Zou YN et al (2012) Spatial distribution of glomalin-related soil protein and its relationships with root mycorrhization, soil aggregates, carbohydrates, activity of protease and β-glucosidase in the rhizosphere of *Citrus unshiu*. Soil Biol Biochem 45:181–183
Wu QS, Levy Y, Zou YN (2009) Arbuscular mycorrhizae and water relations in citrus. In: Tennant P, Beakebhia N (eds) Citrus II. Tree Forest Sci Biotech 3(Special Issue 1), pp 105–112
Wu QS, Xia RX (2006a) Arbuscular mycorrhizal fungi influence growth, osmotic adjustment and photosynthesis of citrus under well-watered and water stress conditions. J Plant Physiol 163:417–425
Wu QS, Xia RX (2006b) Effects of arbuscular mycorrhizal fungi on leaf solutes and root absorption areas of trifoliate orange seedlings under water stress conditions. Front For China 3:312–317
Wu QS, Xia RX, Guo WW et al (2005) Effect of citrus rootstocks and rootstocks/scions on development of arbuscular mycorrhizas. Soil 37:95–99 (in Chinese with English abstract)
Wu QS, Xia RX, Zou YN (2006a) Arbuscular mycorrhizal fungal growth on citrus roots and its correlations with soil available phosphorus content and phosphatase activity. Chin J Appl Ecol 17:685–689 (in Chinese with English abstract)
Wu QS, Xia RX, Zou YN (2006b) Reactive oxygen metabolism in mycorrhizal and non-mycorrhizal citrus seedlings subjected to water stress. J Plant Physiol 11:1101–1110
Wu QS, Xia RX, Zou YN (2008) Improved soil structure and citrus growth after inoculation with three arbuscular mycorrhizal fungi under drought stress. Eur J Soil Biol 44:122–128
Wu QS, Xia RX, Zou YN et al (2007) Osmotic solute responses of mycorrhizal citrus (*Poncirus trifoliata*) seedlings to drought stress. Acta Physiol Plant 29:543–549
Wu QS, Zhou KB, Xia RX et al (2004) Investigation on natural development of arbuscular mycorrhiza in citrus trees. Subtrop Plant Sci 33:39–40 (in Chinese with English abstract)
Wu QS, Zou YN (2009a) Arbuscular mycorrhizal symbiosis improves growth and root nutrient status of citrus subjected to salt stress. ScienceAsia 35:388–391
Wu QS, Zou YN (2009b) Arbuscular mycorrhizas improve water relations of plants exposed to drought. In: Hemantaranjan A (ed) Advances of plant physiology, vol 11. Scientific Publishers, Jodhpur, pp 23–52
Wu QS, Zou YN (2009c) Mycorrhiza has a direct effect on reactive oxygen metabolism of drought-stressed citrus. Plant Soil Environ 55:436–442
Wu QS, Zou YN (2010) Beneficial roles of arbuscular mycorrhizas in citrus seedlings at temp stress. Sci Hortic 125:289–293
Wu QS, Zou YN (2011) Citrus mycorrhizal responses to abiotic stresses and polyamines. In: Hemantaranjan E (ed) Advances in plant physiology, vol 12. Scientific Publishers, Jodhpur, pp 31–56
Wu QS, Zou YN, He XH (2010a) Contribution of arbuscular mycorrhizal fungi to growth, photosynthesis, root morphology and ionic balance of citrus seedlings under salt stress. Acta Physiol Plant 32: 297–304
Wu QS, Zou YN, He XH (2011) Differences of hyphal and soil phosphatase activities in drought-stressed mycorrhizal trifoliate orange (*Poncirus trifoliata*) seedlings. Sci Hortic 129:294–298
Wu QS, Zou YN, Liang Q (2010b) Efficient growth substrate selection of Glomus mosseae-colonized trifoliate orange (*Poncirus trifoliata*) seedlings. In: Zhang Y (ed) The 2nd conference on key technology of horticulture. London Science Publishing, London, pp 65–68
Wu QS, Zou YN, Liu W et al (2010c) Alleviation of salt-stress in citrus seedlings inoculated with mycorrhiza. Changes in leaf anitoxidant defense systems. Plant Soil Environ 56:470–475
Yadav K, Prasad V, Mandal K et al (1992) Effect of coinoculation (*Azospirillum* and *Rhizobium* strains) on nodulation, yield, nutrient uptake and quality of lentil [*Lens culinaris*]. Lens Newslett 19:29–31
Yang CH, Menge JA, Cooksey DA (1994) Mutations affecting hyphal colonization and pyoverdine production in pseudomonads antagonistic toward *Phytophthora parasitica*. Appl Environ Microl 60:473–481
Youpensuk S, Lordkaew S, Rerkasem B (2008) Arbuscular mycorrhizal fungi associated with tangerine (*Citrus reticulata*) in Chiang Mai province, northern Thailand, and their effects on the host plant. ScienceAsia 34:259–264
Zhang H, Murzello C, Sun Y et al (2010) Choline and osmotic-stress tolerance induced in *Arabidopsis* by the soil microbe *Bacillus subtilis* (GB03). Am Phytopathol Soc 23:1097–1104
Zou HQ, Wang ZS (1995) Population dynamics of citrus rhizosphere microorganisms and rhizosphere effect. Eco-Agric Res 3:39–47 (in Chinese with English abstract)

Biodynamic Soil Fertility Management

14

Matjaž Turinek, Martina Bavec, and Franc Bavec

Abstract

Biodynamic (BD) agriculture became the subject of interest in the last years as an increasing number of researchers, professionals, and farmers are starting to explore and practice this way of farming. Basic BD practices and their background are presented in this chapter, backed with research results and findings from scientific literature. One of the more known and at the same time controversial topic of BD farming, the BD preparations, are explored more into detail, with a focus on processes in the compost, the soil, and the buildup of soil fertility with their use. Moreover, effects of BD practices on perennial crops are exemplified on the case of the vine since there is yet no research published on other perennial crops. The importance of the complex soil-plant system and the potential influence of BD practices are put into perspective.

Keywords

Soil fertility • Biodynamic tools • Biodynamic preparations • Soil-plant interaction • Recommendations

14.1 Introduction

Biodynamic (BD) agriculture is one of the sustainable agricultural systems, and its foundations were laid in the so-called "Agricultural Course" held by the Austrian philosopher Rudolf Steiner in 1924 (Turinek et al. 2009). It presents a holistic approach toward farming, where the main focus stands on perceiving the world around us as a complex system, where there is not only the visible, material world but also the invisible world of energies and the spiritual dimension. Next to that, the development of the farm and the farmer through space and time and the idea of a farm organism or farm individuality are some of the core principles of BD agriculture. This usually indicates that farm management should minimize nutrient and energy inputs in order to make the farm self-supporting and autonomous, which is true also for organic farms. But it also encompasses a broader idea of the farm placement in its surroundings, the involvement of the people working on the farm, a balance between the subsystems or "organs" of the farm (arable crops, pastures, livestock, horticulture, etc.) and the elements of nature, such as forests, heaths, moors, and watercourses (Vereijken et al. 1997). The main principles of modern organic farming, such as composting, green manures, closed nutrient, and life cycles, among others, were taken after BD farming principles (Conford 2001) and are still at the core of BD farm management. The methods and proposals developed from the Agricultural Course present the basis for most of BD farmers, whereas there have been some regional modifications on the application of the proposed methods, especially regarding the use of BD

M. Turinek (✉) • M. Bavec • F. Bavec
Faculty of Agriculture and Life Sciences, University of Maribor, Pivola 10, Hoče pri Mariboru 2311, Slovenia
e-mail: matjaz.turinek@uni-mb.si; martina.bavec@uni-mb.si; franci.bavec@uni-mb.si

A.K. Srivastava (ed.), *Advances in Citrus Nutrition*,
DOI 10.1007/978-94-007-4171-3_14, © Springer Science+Business Media B.V. 2012

preparations (i.e., different ways of storing and using the preparations, "new" preparations being added to the original eight, selling vs. on-farm making of the preparations, etc.). Most of research done and literature published (scientific and professional) is available in German since the German-speaking countries present the origin of BD farming and also have the longest history of practicing it.

14.2 Geographical Developments

BD agriculture is practiced on every continent of the Earth and is thus practically applicable in every climatic, cultural, economic, and social environment. However, not all BD farms apply for certification of their products according to BD standards, also known under the private trademark "Demeter," which is managed by the nonprofit Demeter International Association, seated in Germany. Thus, there are around 4,500 BD farms in 43 countries, whose area of approximately 142,000 ha is certified according to Demeter standards (Demeter 2011). Most of that area lies in Germany (66,000 ha), as the development of BD agriculture began there and has the longest tradition, followed by France (7,500 ha) and India (7,273 ha). Next to that, there are many farmers who apply BD methods; however, they do not apply for Demeter certification. In Slovenia, for example, there are only 18 certified Demeter farms; however, over 1,500 small gardeners, farmers, and beekeepers have joined the BD "Ajda" association, which promotes BD farming practices and organizes lectures, workshops, and training for those interested in learning this way of farming. Moreover, Australian BD farmers established their own BD association (which also certifies their products but is not acknowledged by Demeter International), with over 1,400 members who farm on around 800,000 ha of land using BD methods.

14.3 Biodynamic Tools

One of the main "tools" to improve agriculture, given in the Agricultural Course, are the BD preparations (Table 14.1 and Fig. 14.1), which were further developed and tested in the last 85 years. They are made from medicinal plants, cow dung, and minerals using the individual preparation procedures (Koepf et al. 1996; von Wistinghausen et al. 2005, 2007). The interested reader is kindly referred to professional literature for a detailed description of preparing and using the preparations. However, the thoughts behind the preparations are unconventional and sometimes difficult to understand as they are to be used as homeopathic preparations for the soil and plants. BD 500, made out of cow dung and buried over winter in the soil (Fig. 14.2), is regarded as a soil fertility preparation, which enhances soil biological functions through the stimulation of soil life. BD 501, made out of finely ground quartz or pure silica and buried in the soil over summer, is meant to enhance plant resilience and resistance by stimulating the establishment of balance in the plant, making it more active and at the same time "aware" of its surroundings. BD 502-507 have a similar function of helping to establish well-balanced and mature compost fertilizers for the soils, helping to build long-term humus complexes and adding to the long-term soil fertility. The underlying natural science mechanistic principle of BD preparations is still under investigation, whereas some attempts have been made to explain the mode of action in the past. Effects were firstly explained as a normalization (normalizing yields under low yielding conditions) or compensation (BD preparations compensating for lower N fertilization) effect, where both explanations leave many open questions (Raupp and Konig 1996).

Table 14.1 BD preparations, main ingredients, type of use, and mentioned areas of influence

Number of preparation	Main ingredient[a]	Use	Mentioned in the agricultural course in connection with
BD 500	Cow manure	Field spray	Soil biological activity
BD 501	Silica	Field spray	Plant resilience
BD 502	Yarrow flowers (*Achillea millefolium* L.)	Compost preparation	K and S processes
BD 503	Chamomile flowers (*Matricaria recutita* L.)	Compost preparation	Ca and K processes
BD 504	Stinging nettle shoots (*Urtica dioica* L.)	Compost preparation	N management
BD 505	Oak bark (*Quercus robur* L.)	Compost preparation	Ca processes
BD 506	Dandelion flowers (*Taraxacum officinale* Web.)	Compost preparation	Si management
BD 507	Valerian extract (*Valeriana officinalis* L.)	Field spray, compost preparation	P and warmth processes

[a]The procedure of preparation and fermentation is in detail described in various sources (Steiner 1924; Sattler and Wistinghausen 1989; von Wistinghausen et al. 2005, 2007). BD preparations are designed to be used together on a farm/farming system

Fig. 14.1 BD preparations 500, 502–507 are usually stored in ceramic or glass containers, surrounded and covered by peat, together in a larger box in a cool and dark place. BD 501 is stored in a glass container and kept on a sunny place

Fig. 14.2 Cow manure in cow horns for BD 500 before being buried in the soil over winter

A systems response and adaptation model was suggested as a possible explanation, where the effects of BD preparations do not depend only on their properties and mode of application. Foremost, properties of soils and plants, environmental conditions, and how they interact are suggested as factors which determine the effects of BD preparations to the greatest extent (Raupp and Konig 1996). Moreover, BD preparations are applied in small quantities of 4–160 g/ha, where physical or biological effects seem unlikely (Reganold 1995). In addition, BD preparations were also shown to have hormone-like effects (Goldstein et al. 2004). The latest indirect, and most prominent, explanation on their mode of action was given by Montagnier et al. (2010), who have found some bacterial and viral DNA sequences to induce low-frequency electromagnetic waves in high aqueous dilutions, which then invoke the "creation" of the same DNA sequences in the "clean" medium, although none of the original DNA sequence exists in the dilution. Put simply, this means that water acts as a transmitter of information being dissolved in it through electromagnetic waves and this information (whatever it may be) is then transmitted to the recipient of this aqueous dilution – in our case to the soil or plants – and is then responsible for the creation of substances in it. This, however, presupposes that the information to be transferred from the BD preparations is already in balance and adapted to soil/plant conditions – thus the active preparation procedures with burying them in biologically active soils and/or fermentation.

The aforementioned regional differences between applications of BD preparations can be seen on Australian BD farms reported in studies and other BD field trial comparison studies and farm comparisons. Namely, in Australia only preparation BD 500 was applied 1–2 times each year (Ryan and Ash 1999). Also, Nguyen and Haynes (1995) report only preparation of BD 500 being used on a BD farm in New Zealand. As discussed in professional literature, however, the preparations were designed to be used together and only as such they can have the desired, wholesome, effectiveness, i.e., help to create balance in the soil and plants.

Moreover, Steiner (1924) mentioned positive effects of a full moon in an agricultural context. On this basis, Spiess (1990a, b) scientifically researched the effects of lunar rhythms and proved the influence of several rhythms on growth, yield, and quality of little radish and rye. His results, however, were contrasting to findings of Thun (1994), who found one single rhythm to be the most important one. On the basis of a reanalysis of Spiess' data, Kollerstrom and Staudenmaier (2001) argue that Spiess' experimental results comply with the findings and recommendations of Thun, therefore confirming a lunar influence on crop growth and development. However, the influence of other astrological bodies (i.e., planets) on crop growth and development is difficult to research and therefore also difficult to scientifically reject or prove an influence. One cannot say to live without a certain planet for a year and then see the influence this planet had on crop growth and development. And with our current knowledge, it is also impossible to shield areas of planet Earth from just certain, not all, influences planets have. Therefore, one would have to come up with an innovative and at the same time trustworthy idea to be able to conduct scientific research in this highly interesting area. However, Thun (1994) conducted field research on her own experimental fields for the last 50 years on this matter, and she yearly publishes the "Astronomical planting calendar," which is nowadays translated into over 40 languages worldwide.

There are also other approaches and/or tools that BD farmers use on their farms, but most of them are already known from good organic farming husbandry. Nevertheless, some of the most important are:

- Use of mature composts for fertilization
- Mixed farms with rearing of animals
- Diverse crop rotations with the inclusion of leguminous plants
- Green manures and cover crops
- Use of stable, locally adapted seeds and/or varieties

Then there are some practices, which are strongly endorsed, however, not mandatory for Demeter-certified farms, such as:

- Building on the local economy
- Use of minimum tillage or preservation tillage for arable land
- Working in the social area of a farm (handicapped people, kindergartens, schools, education, and seminars)

14.4 Effects of BD Preparations on Soil Fertility

As for the BD preparations, which are the greatest peculiarity of BD farming, experimental results show their effects not only on yields (Fig. 14.3) but also on some ongoing processes in compost piles and in the long term in the soil. Carpenter-Boggs et al. (2000) report higher average temperatures (3.4°C higher compared to the control pile) throughout the active composting period, whereas Zaller (2007) measured no significant differences in the average temperature of BD and conventional (CON) compost piles. BD-treated compost also contained 65% more nitrate in the final samples, respired carbon dioxide (CO_2) at a 10% lower rate, and had a larger dehydrogenase enzyme activity to CO_2 production ratio (Carpenter-Boggs et al. 2000). Carpenter-Boggs et al. (2000) suggest that BD preparations caused these effects through their bioactive ingredients or by serving as microbial inoculants. In addition, the microbial population in BD preparations was found to be substantial (Rupela et al. 2003), where bacteria population ranged from 3.45 to 8.59 $\log_{10}$ g^{-1}. Also, a population of fungi was found in the preparations 502 and 506 (5.30 and 4.26 $\log_{10}$ g^{-1}, respectively). Several bacteria and fungi strains found in BD preparations showed a potential for suppressing fungal plant pathogens (Rupela et al. 2003). This population could also be the reason for the significant and clear-cut difference in dehydrogenase, protease, and phosphatase activity with respect to the farming systems in the DOK (Biodynamic, Organic and Conventional (CON) agriculture long-term comparison) trial, where highest values were measured for the BD system (Maeder et al. 2002). Similar, but not as accentuated, differences were found also in another long-term trial (Raupp 2001), whereas Zaller and Köpke (2004) found no differences between the BD and organic plots (Fig. 14.4).

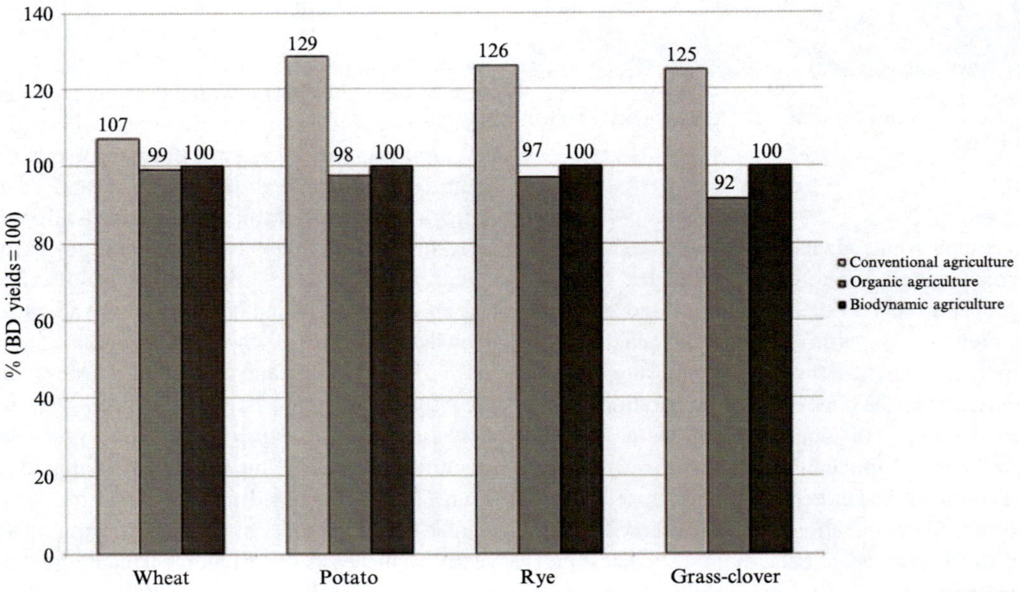

Fig. 14.3 Relative yields of wheat, potato, rye, and clover grass depending on farming system, based on results from published scientific research trials

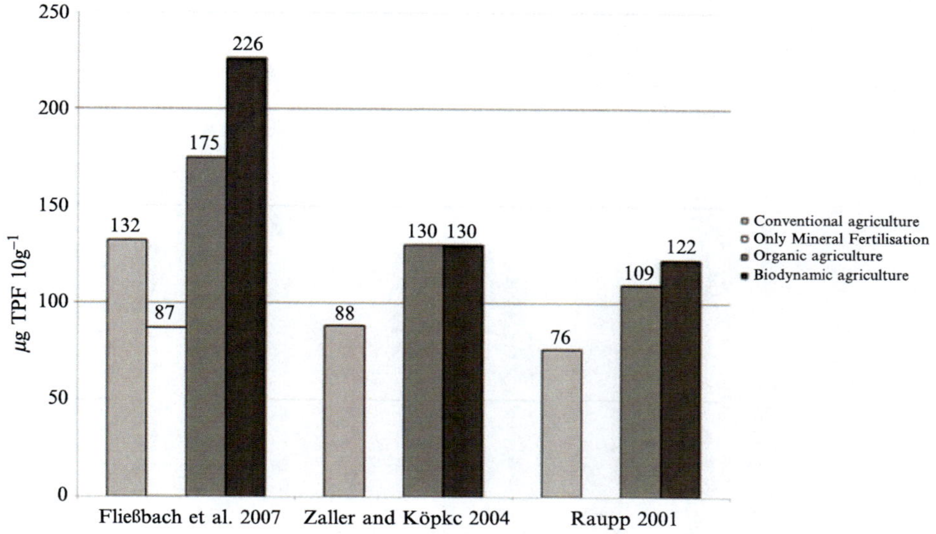

Fig. 14.4 Dehydrogenase activity in three long-term field trials in Europe depending on the farming system

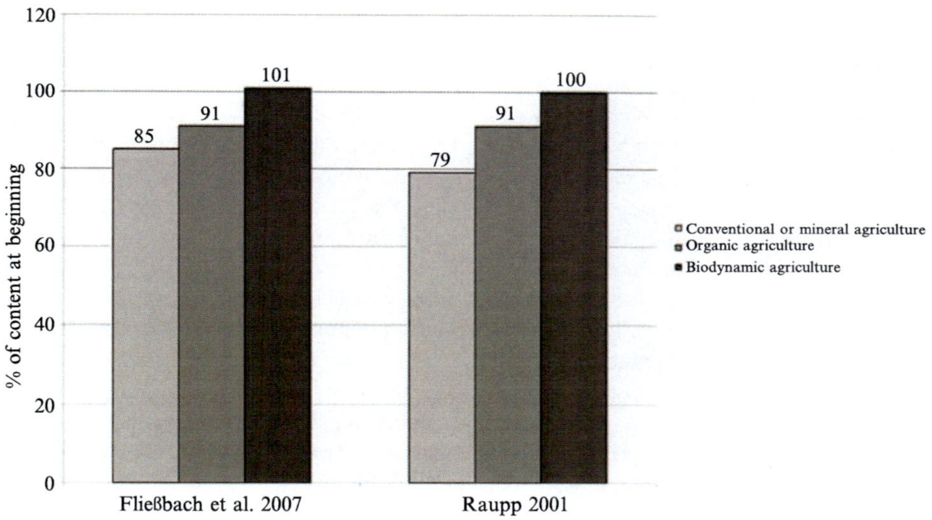

Fig. 14.5 Change in soil organic matter carbon in two long-term trials after 20 years depending on the farming system

Moreover, higher soil organic matter contents on BD plots were found in two separate field trials (Fig. 14.5) (MIN-ORG trial in Germany and the DOK trial in Switzerland), as well as in two studies on BD farms in New Zealand and in the Netherlands, where BD farming practices have been applied for several years (Reganold et al. 1993; Droogers and Bouma 1996; Raupp 2001; Maeder et al. 2002). Microbial biomass nitrogen also differed significantly in the DOK trial and accounted highest in the BD system with 59% more than in the CON farmyard manure (FYM) system (Fließbach et al. 2007). Furthermore, the microbial biomass carbon was 35% higher in the BD system, compared to the CON-FYM system (Maeder et al. 2002; Oehl et al. 2004). In contrast, Zaller and Köpke (2004) report no differences between treatments in regard to microbial biomass carbon, where untreated FYM and FYM treated with BD preparations were applied (Fig. 14.6). In both cases, microbial biomass carbon was significantly higher than on control plots (Zaller and Köpke 2004), which leads to the conclusion that FYM had an important effect on the soil microbial biomass buildup. Wada and Toyota (2007) went a step further and discovered that FYM applications add to the stability of soil biological functions,

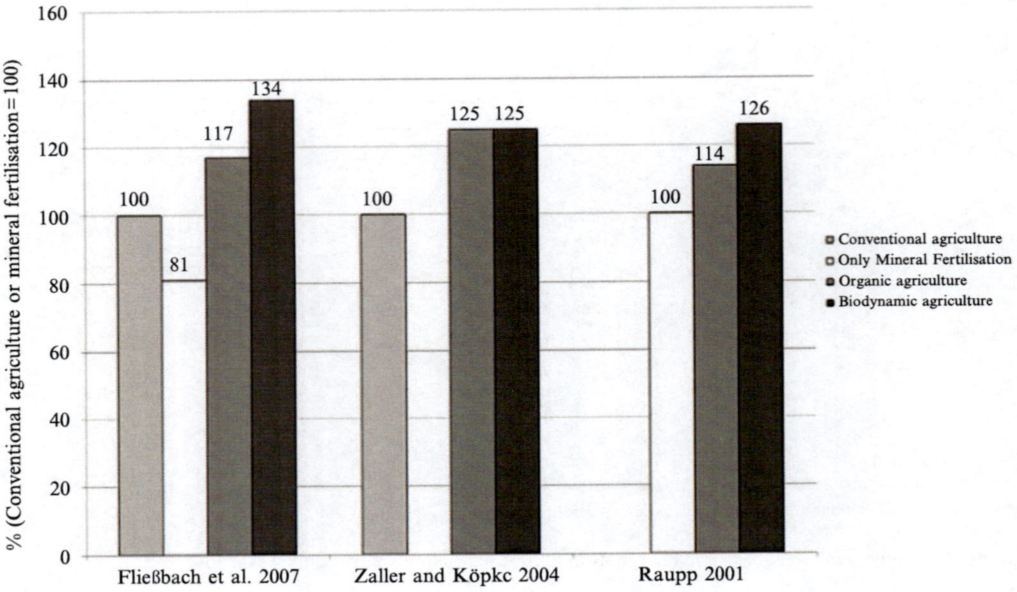

Fig. 14.6 Microbial biomass carbon content in three long-term trials, where also BD agriculture was included

where microbial and fungal populations show resilience and resistance against disinfection. In addition, FYM contributes toward a changed soil nitrogen composition and higher rates of protein amino acids, which bind nitrogen in the soil (Scheller and Raupp 2005). However, differences between treatments do not seem to depend solely on amino acid supply from manure. An altered amino acid metabolism in the soil also influences soil amino acid composition and contents. Soils receiving FYM with BD preparations have a lower catabolism:anabolism ratio than soils receiving non-prepared FYM, which results also in a more intensive humidification process. The explanation for the influence of BD preparations on anabolism is yet to be found (Scheller and Raupp 2005).

14.5 Importance of the Plant-Soil Interaction

Additionally, when science makes progress in discovering the interconnectedness of the many Earth's systems and the complexity of nature, many of the statements given in the "Agricultural Course" can be understood or at least give food for thought. For example, it was suggested that "…if you work the soil as just explained, then the plant will be ready to attract 'things' in its wider surroundings. The plant can take benefit not just from the contents of the field, where it grows, but also of the contents of the soil in the neighboring pasture, if the plant needs it. The plant can also benefit from the soil in the neighboring forest, if it is made sensible in the described way…" (Steiner 1924, p. 160). And indeed, with today's knowledge on the existence of extensive mycelial networks in soils, which have been proved to connect individual species, genera, and even families of plants (He et al. 2003), in connection with results indicating the improvement of arbuscular mycorrhizal fungi by BD preparations (Ryan and Ash 1999; Maeder et al. 2002), we can understand and confirm this assumption given over 85 years ago. Moreover, nutrient mobilization from soil minerals is not the only benefit of arbuscular mycorrhizal fungi. Frey-Klett et al. (2007) report that fixation of atmospheric nitrogen and protection of plants against root pathogens are also among the myriad benefits of arbuscular mycorrhizal fungi and mycorrhiza helper bacteria. Raupp (2001) also reports of a higher density of roots on plots treated with BD preparations. Mycorrhiza helper bacteria could be the possible reason for this effect, as they have been proved to stimulate lateral root formation and thus increase potential root-mycorrhiza interaction points (Frey-Klett et al. 2007).

An active interaction between the soil and the plant is also mentioned in the "Agricultural Course." Today, as we try to understand the complexity of plant nutrition, organic and BD practices advocate "feeding" (fertilizing) the soil, so the plant can feed itself from it indirectly. In this sense, we attained some interesting results in trials on red beet (Bavec et al. 2010), where higher levels of malic acid were measured in samples from BD and control plots. A study of Rudrappa et al. (2008) hints toward one of the possible reasons for this phenomenon, observed also in other studies. It was demonstrated that malic acid, selectively excreted through roots, signals beneficial rhizobacteria and encourages their interaction with plants. Beneficial soil bacteria have been found to confer immunity against a wide range of foliar diseases by activating plant defenses. Organic acids (as well as phenolic compounds) have been also found to participate in leveling out P deficiency by being excreted through plant roots (Badri and Vivanco 2009). The aforementioned potential role of organic acids and phenolic compounds in leveling out P deficiency is partly reflected in the malic acid concentrations and the total phenolic content (TPC) in our trial. However, the BD system deviates from this assumption in both cases – despite relatively high levels of P added also high values for malic acid and TPC were measured. Reasons for this deviation could be sought in a changed microbial structure, enzyme activity, or amino acid metabolism found in BD systems (Turinek et al. 2009). Plant-microbial interactions and plant-soil interactions are increasingly being researched and seem to play an important role in providing plants with nutrients and activating resilience against pests and diseases, whereas a consequence food products can also gain some beneficial constituents/compounds (Badri and Vivanco 2009). Reganold et al. (2010), for example, found more than 200 different unique strains of microorganisms in organic soils, as compared to only two in conventional soils for strawberry production.

14.6 Effects on Perennial Crops

Up-to-date scientific research on perennial crops has been only done and/or published on vines since BD wine grape production is increasingly attracting attention as some of the world's prestigious wine producers have started to use BD practices in the last decade (Reeve et al. 2005). Experimental results suggest BD practices have an effect on wine grape canopy and chemistry, where a more balanced canopy and wine composition were measured for BD wine production. However, no significant effects on soil fertility parameters were shown in the same 6-year on-farm comparison trial between ORG and BD practices in an organic vineyard in California (Reeve et al. 2005). Probst et al. (2008), however, measured significant differences in soil fertility between CON and BD soils in vineyards with a long history of BD (since 1981) and CON cultivation in Germany, which correlate with the aforementioned results on field trials on annual crops (Fig. 14.5). There is still an ongoing research comparison trial including integrated, organic, and BD (with different BD preparation use frequency) wine production at the research station Geisenheim in Germany. First results indicate that not only the use of but also the timing in plant growth stage and the kind of preparation (BD 500 or BD 501) have an influence on grape and consequently wine quality (Meissner 2011), for example, overuse of the BD 500 preparation resulted in unbalanced and unripe "green" wines. There are also some professional/research associations dealing with BD fruit production, where apple production is the main focus. The most organized and visible one is the International Working Group on Biodynamic Fruit Production (www.biodynamicfruit.org), where useful resources, links, and contacts to professional literature and practitioners can be found.

14.7 Summarized Biodynamic Recommendations

A biodynamic farmer strives for balance on his farm, soils, fields, meadows, orchards, animals, and himself. As manifold demonstrated in nature and also in research presented in this chapter, a balanced organism is the basis for long-term stability and successful organic and biodynamic production. What is then the added value of the BD method? Certainly, results show better yields and healthier plants with the use of BD preparations – be it in compost, soils, or plants. Another important aspect is the personal development of the farmer through time and space through detailed observation and reflection, also called the "Goethean phenomenological approach" since it was successfully practiced by Goethe. Through dedicated observation of a phenomenon in the time span of several months or years, one can, of course, only with the subsequent reflection of the observed, eventually understand the phenomenon itself. This approach may not be something new, but it is certainly something that is not done consciously and with discipline anymore, be it by farmers, researchers, teachers, etc. So raising awareness and understanding for the processes in the world around us is another important aspect and recommendation for practical and research work, which is central in BD farming. Since only with the understanding of the specific conditions each farm is positioned in, one can choose and use the right measures at the right time in the right way at that specific farm. This brings us to another important aspect already mentioned at the beginning of this chapter. The BD "method" is not a "one-size-fits-all" recipe. The beauty and advantage of it is its adaptability to local conditions all over the world. So in reflection, the BD farming practice can also be regarded as a path of personal development for those engaged with it. However, one is not obliged or forced to take this path. The positive effects of the BD approach mentioned in this chapter are not conditioned with any personal development.

The interested reader is kindly referred to the numerous professional literature on the topic of BD agriculture, where more details on the use and practice of BD practices can be found. Even if most of it is in German, there are also some quality English books.

14.8 Future Research with Concluding Remarks

But what are some of the future research challenges we are faced with? What about the energy efficiency or ecological impact of BD production on a wider scale (production to consumption)? Do we need to include economic feasibility into our studies? Then there are some more detailed questions regarding BD preparations. Does it make a difference if they are made on farm or bought from a distant location? Does this affect the effectiveness of the preparations? Must the making of the preparations with the use of animal organs stay as given by Steiner? Or do we need to move forward, explore new possibilities, and develop an understanding for the reasons behind given procedures? What about research on farm animals? Moreover, is there a difference between BD prepared compost of animal and plant origin? How does this affect soil fertility and health? How do we need to change and adapt our soil fertilization and tillage systems in order to get balanced soils? Clearly, there is a need for more research on the effects and use of BD practices in perennial crops. However, how to approach this matter? Do we need to make more production systems comparison trials for that matter? If yes, how well defined are the systems to be compared? And what are the areas of interest to compare? Soil quality and long-term fertility are certainly of high interest as they present the bases for healthy plants and high-quality produce. As a continuation, food quality is still a highly discussed and debated area, which would also deserve more attention on this account.

A working group of researchers and professionals, who gathered in an active process to exchange thoughts, experiences, and research results (Hurter 2007), is one of the way signs into the future. More such informal groups and networks are being created all over the world. Also a web portal on biodynamic research (http://biodynamic-research.net), which was not long ago put into practical existence, could facilitate the exchange of ideas, thoughts, and results. A worldwide network of farmers, researchers, advisors, teachers, and others interested in BD farming could contribute toward naming and addressing questions from everyday practice. For it to work efficiently, however, one would need dedicated and motivated persons who would actively participate in the creative process.

References

Badri DV, Vivanco JM (2009) Regulation and function of root exudates. Plant Cell Environ 32:666–681. doi:10.1111/j.1365-3040.2009.01926.x

Bavec M, Turinek M, Grobelnik-Mlakar S et al (2010) Influence of industrial and alternative farming systems on contents of sugars, organic acids, total phenolic content, and the antioxidant activity of red beet (*Beta vulgaris* L. ssp. vulgaris Rote Kugel). J Agric Food Chem 58:11825–11831. doi:10.1021/jf103085p

Carpenter-Boggs L, Reganold JP, Kennedy AC (2000) Effects of biodynamic preparations on compost development. Biol Agric Hortic 17:313–328

Conford P (2001) The origins of the organic movement. Floris Books, Edinburgh

Demeter IEV (2011) Demeter-International E.V. – A World-wide Network. http://www.demeter.net/. Accessed 1 July 2011

Droogers P, Bouma J (1996) Biodynamic vs. conventional farming effects on soil structure expressed by simulated potential productivity. Soil Sci Soc Am J 60:1552–1558

Fließbach A, Oberholzer HR, Gunst L et al (2007) Soil organic matter and biological soil quality indicators after 21 years of organic and conventional farming. Agr Ecosyst Environ 118:273–284. doi:10.1016/j.agee.2006.05.022

Frey-Klett P, Garbaye J, Tarkka M (2007) The mycorrhiza helper bacteria revisited. New Phytol 176:22–36. doi:10.1111/j.1469-8137.2007.02191.x

Goldstein W, Barber W, Carpenter-Boggs L et al (2004) Comparisons of conventional, organic, and biodynamic methods. Michael Fields Agricultural Institute, East Troy

He X-H, Critchley C, Bledsoe C (2003) Nitrogen transfer within and between plants through common mycorrhizal networks (CMNs). Crit Rev Plant Sci 22:531–567. doi:10.1080/0735268031878237

Hurter M (2007) Zur vertiefung der biologisch-dynamischen landwirtschaft. Verlag am Goetheanum, Dornach

Koepf HH, Schaumann W, Haccius M (1996) Biologisch-dynamische landwirtschaft eine einführung. Ulmer (Eugen), Stuttgart

Kollerstrom N, Staudenmaier G (2001) Evidence for lunar-sidereal rhythms in crop yield: a review. Biol Agric Hortic 19:247–259

Maeder P, Fliessbach A, Dubois D et al (2002) Soil fertility and biodiversity in organic farming. Science 296:1694–1697. doi:10.1126/science.1071148

Meissner G (2011) Grundbetrachtung der Rebe. Bio-dynamische landwirtschaft III. Lehr- und Forschungsgemeinschaft für bio-dynamische Lebensfelder, Edelschrott, pp 61–72

Montagnier L, Aissa J, Del Giudice E et al (2010) DNA waves and water. arXiv:1012 5166. doi: http://arxiv.org/abs/1012.5166

Nguyen ML, Haynes RJ (1995) Energy and labour efficiency for three pairs of conventional and alternative mixed cropping (pasture-arable) farms in Canterbury, New Zealand. Agr Ecosyst Environ 52:163–172. doi:10.1016/0167-8809(94)00538-P

Oehl F, Frossard E, Fliessbach A et al (2004) Basal organic phosphorus mineralization in soils under different farming systems. Soil Biol Biochem 36:667–675. doi:10.1016/j.soilbio.2003.12.010

Probst B, Schüler C, Joergensen R (2008) Vineyard soils under organic and conventional management – microbial biomass and activity indices and their relation to soil chemical properties. Biol Fertil Soils 44:443–450. doi:10.1007/s00374-007-0225-7

Raupp J (2001) Research issues and outcomes of a long-term fertilization trial over two decades: a contribution to assessing long-term agronomic trials. Ber Landwirtsch 79:71–93

Raupp J, Konig UJ (1996) Biodynamic preparations cause opposite yield effects depending upon yield levels. Biol Agric Hortic 13:175–188

Reeve JR, Carpenter-Boggs L, Reganold JP et al (2005) Soil and winegrape quality in biodynamically and organically managed vineyards. Am J Enol Vitic 56:367–376

Reganold JP (1995) Soil quality and profitability of biodynamic and conventional farming systems: a review. Am J Altern Agric 10:36–45

Reganold JP, Palmer AS, Lockhart JC et al (1993) Soil quality and financial performance of biodynamic and conventional farms in New Zealand. Science 260:344

Reganold JP, Andrews PK, Reeve JR et al (2010) Fruit and soil quality of organic and conventional strawberry agroecosystems. PLoS One 5:e12346. doi:10.1371/journal.pone.0012346

Rudrappa T, Czymmek KJ, Paré PW et al (2008) Root-secreted malic acid recruits beneficial soil bacteria. Plant Physiol 148:1547–1556. doi:10.1104/pp. 108.127613

Rupela OP, Gopalakrishnan S, Krajewski M et al (2003) A novel method for the identification and enumeration of microorganisms with potential for suppressing fungal plant pathogens. Biol Fertil Soils 39:131–134. doi:10.1007/s00374-003-0680-8

Ryan M, Ash J (1999) Effects of phosphorus and nitrogen on growth of pasture plants and VAM fungi in SE Australian soils with contrasting fertiliser histories (conventional and biodynamic). Agric Ecosyst Environ 73:51–62. doi:10.1016/S0167-8809(99), 00014-6

Sattler F, von Wistinghausen E (1989) Der landwirtschaftliche Betrieb – Biologisch-Dynamisch, 2nd edn. Ulmer (Eugen), Stuttgart

Scheller E, Raupp J (2005) Amino acid and soil organic matter content of topsoil in a long term trial with farmyard manure and mineral fertilizers. Biol Agric Hortic 22:379–397

Spiess H (1990a) Chronobiological investigations of crops grown under biodynamic management.1. Experiments with seeding dates to ascertain the effects of lunar rhythms on the growth of winter rye (*Secale cerale*, cv Nomaro). Biol Agric Hortic 7:165–178

Spiess H (1990b) Chronobiological investigations of crops grown under biodynamic management. 2. Experiments with seeding dates to ascertain the effects of lunar rhythms on the growth of little radish (*Raphanus sativus*, cv Parat). Biol Agric Hortic 7:179–189

Steiner R (1924) Geisteswissenschaftliche grundlagen zum gedeihen der landwirtschaft: landwirtschaftlicher kurs. Rudolf Steiner Verlag, Dornach, p 160

Thun M (1994) Hinweise aus der konstellationsforschung für Bauern, Weinbauern, Gärtner und Kleingärtner, 8th edn. Aussaattage Thun-Verlag, Biedenkopf

Turinek M, Grobelnik-Mlakar S, Bavec M et al (2009) Biodynamic agriculture research progress and priorities. Renew Agric Food Syst 24:146–154. doi:10.1017/S174217050900252X

Vereijken JFHM, van Gelder T, Baars T (1997) Nature and landscape development on organic farms. Agri Ecosyst Environ 63:201–220. doi:10.1016/S0167-8809(97), 00013-3

von Wistinghausen C, Scheibe W, Heilmann H et al (2005) Anleitung zur anwendung der biologisch-dynamischen feldspritzpräparate und düngerpräparate, 3rd edn. Lebendige Erde, Darmstadt

von Wistinghausen C, Scheibe W, Wistinghausen E et al (2007) Anleitung zur Herstellung der biologisch-dynamischen Präparate, 4th edn. Lebendige Erde, Darmstadt

Wada S, Toyota K (2007) Repeated applications of farmyard manure enhance resistance and resilience of soil biological functions against soil disinfection. Biol Fertil Soils 43:349–356. doi:10.1007/s00374-006-0116-3

Zaller JG (2007) Seed germination of the weed rumex obtusifolius after on-farm conventional, biodynamic and vermicomposting of cattlemanure. Ann Appl Biol 151:245–249. doi:10.1111/j.1744-7348.2007.00172.x

Zaller JG, Köpke U (2004) Effects of traditional and biodynamic farmyard manure amendment on yields, soil chemical, biochemical and biological properties in a long-term field experiment. Biol Fertil Soils 40:222–229. doi:10.1007/s00374-004-0772-0

15 Nutrient Use Efficiency in Citrus

Johannes Scholberg and Kelly T. Morgan

Abstract

Citrus production systems are inherently complex, which may hamper efficient resource management. This chapter aims to use a systems approach for structuring more efficient management options for citrus production. It starts by providing a brief overview of production trends in major production regions and outlining the use of a systems-based approach for structuring more sustainable management practices. Starting with the physiological processes and mechanisms controlling nutrient uptake of young seedlings, resource capture and utilization are presented including root growth dynamics and nutrient interception capacity as affected by diurnal crop water use, temperature, and nutrient supply. Uptake processes are scaled-up to a tree level and linked to specific environmental, crop development, and management aspects and integrated into generic conceptual models with special reference to the interactive effects of irrigation and fertility management as related to crop nutrient interception capacity. Then the scope is expanded to look at nutrient management at the field level on an annual basis including tree nutrient allocation, environmental emissions, and development of annual nutrient budgets. In the final section, future perspectives are provided for more effective use of modeling approaches for system design and more cost-effective and sustainable resource use.

Keywords

Nutrient use efficiency • Systems analysis • System design • Resource management • Modeling approaches

15.1 Introduction

During the past decades, commercial citrus producers throughout the world have been facing serious challenges including restrictions on water, fertilizer and agrochemical use, potential treats of devastating diseases, and fierce competitions for global markets. In an era of increased focus on sustainable development, smart technologies, and resource conservation, there has been gradual shift from increasing yield and maximizing profits to making better use of technical innovations to ensure more efficient use of finite resources including land, energy, water, and nutrients. Within this context, it is relevant for scientists to critically reflect on how to effectively integrate information related to physiological processes, pedoclimatic conditions, and management practices to increase nutrient use efficiency (NUE) in citrus production systems. Understanding the processes controlling nutrient uptake is critical for development of management practices that enhance NUE while reducing nutrient losses and potential environmental impacts. However, one of the

J. Scholberg (✉)
Biological Farming Systems, Wageningen University,
Post Box 563, 6700 AN, Wageningen, The Netherlands
e-mail: jscholberg@yahoo.com

K.T. Morgan
Soil and Water Science Department, Southwest Florida Research and Education Center, University of Florida,
2685 State Road, 29 North, Immokalee, FL 34142, USA
e-mail: conserv@ufl.edu

A.K. Srivastava (ed.), *Advances in Citrus Nutrition*,
DOI 10.1007/978-94-007-4171-3_15,

key challenges in enhancing NUE is related to lack of effective integration of the underlying processes. Soil biochemical processes feature numerous transformations among different nutrient fractions, and corresponding pool sizes are typically difficult to quantify. Moreover, compared to agronomic systems, it is rather challenging to effectively map the response of long-lived perennial crops (e.g., citrus trees) to changing environmental and/or management conditions.

Citrus-based systems typically feature fruit-bearing scions that are grafted upon rootstock seedlings and interactions among rootstock and scion add to the overall complexity of the system. Trees may require 10–12 years to reach maturity, and in mature orchards, each year a number of trees need to be replaced. Tree characteristics and spatial variability within and among different orchard units thus may vary greatly. Moreover, citrus trees have complex above- and below-ground growth patterns and internal nutrient reserves are rather large and thus form well-buffered systems, which may (initially) mask more subtle responses to changes in nutrient management. Responses to specific treatments is rather slow while seasonal perturbations during sequential years further complicate matters, thus hampering detection of consistent trends.

Due to the large size of mature trees and the inherent variability in citrus orchards outlined above, it is rather challenging, time-consuming, and costly to effectively assess NUE in mature citrus trees. To address some of these constraints, we propose the use of a systems-based approach for integrating fragmented knowledge at a higher aggregation level. The objective of this chapter therefore is to provide a systems approach based on the DEED methodology outlined by Giller et al. (2008) as follows: (a) describe how physiological, environmental, and management conditions affect nutrient use efficiency (NUE) at the tree and field scale; (b) explain underlying processes that control NUE; (c) explore options for enhancing NUE as related to both tactical and strategic production practices; and (d) design measures that may enhance NUE in citrus orchards. This review thus aims to provide a synthesis of existing literature and selected scientific studies and to integrate both scientific knowledge and management-based perspectives into an overall conceptual framework for more efficient resource use in citrus production systems with special reference to nitrogen.

15.2 Citrus Production

15.2.1 Citrus Production Systems

The various citrus species are believed to be native to the subtropical and tropical regions of Asia and the Malay Archipelago (Webber and Batchelor 1943). Sweet orange was first introduced to the North American continent in 1493, and since this time it has been introduced to all continents within the (sub)tropics. Globally, Brazil is the largest citrus producer accounting for 19% of world production, followed by China, USA, Mexico, and Spain. In terms of citrus crops, oranges and tangerines are the most commonly grown, and they account for 65% and 21% of the total citrus production, respectively (Pumphrey 2006). Most of the citrus production in Brazil is located in the state of São Paulo of which 25% were just planted this century (Mattos et al. 2006). Currently, Italy is the largest organic citrus producer followed by Cuba (as related to the reduced availability of imported inorganic fertilizer), while in the USA, less than 1% of the citrus production is produced organically (Willer and Lukas 2009). Citrus production systems may be divided into fresh-market and processing operations. In the continental USA, California is the major producer of fresh-market oranges and accounts for 26% of the US production and 42% of the product value. Florida dominates the US processing market and accounts for 71% of the total US production and 55% of the product value. In Florida, 230,290 ha of citrus is grown translating to a production of 12 million metric tons thereby (Florida Agricultural Statistics Service 2010). During the past years, the citrus production area and total production has slowly declined. This is related to outbreaks of invasive diseases, increasing fertilizer costs, restrictions on water and nutrient use, and increased urbanization. Moreover, to remain viable, Florida citrus producers must sustain yields while adhering to maximum daily loads for N and P, which can only be realized by increasing overall NUE (Obreza and Sartain 2010; Obreza and Schumann 2010).

15.2.2 General Tree Growth Characteristics

Citrus is an evergreen fruit-bearing tree. Under Florida conditions, citrus trees do not exhibit dormancy nor do trees shed their leaves during the winter. Turrell et al. (1969) developed generic logistic tree growth equations for citrus based on growing conditions and cultural practices in California. Similar relationships were developed for Florida conditions as well (Fig. 15.1). Based on this graph, it appears that citrus trees may require 12 years to reach maturity. Maximum fruit yield, under well-managed conditions, may be on the order of 80–90 MT ha^{-1}. Mature groves are pruned frequently to ensure compact and continuous orchard tree canopies, which facilitates high production and easy harvesting. In the northern subtropics, new shoot growth follows distinct cycles (flushes) with two to four cycles occurring yearly. The first and usually largest flush starts in the early spring (late February to early March), the second from early June to early July, and the third in late summer (August to September). The principal blooming period for all commercial species is early spring and usually lasts approximately 6 weeks (mid-February to late March). Fruit ripening for

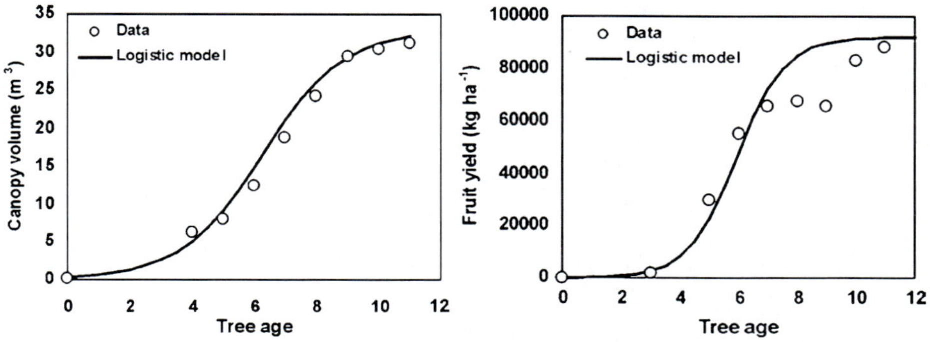

Fig. 15.1 Relationship between tree age and canopy size (*left figure*) and tree age and fruit production for citrus trees in Florida (*right figure*)

most citrus varieties occurs in late fall and winter, proceeding to the next spring bloom. However, late maturing varieties such as "Valencia" may require 12–15 months to mature (Morgan 2004).

15.2.3 Tree Growth, Dry Weight Partitioning, and Nutrient Demand

Of all nutrients, nitrogen has the most pronounced effect on overall citrus tree growth, yield, and fruit quality (Obreza and Schumann 2010). Low leaf N values will reduce assimilate production, which in turn may reduce tree growth, whereas excessively high leaf N concentrations do not result in enhanced growth, and NUE tends to decrease as N supply increases (Lea-Cox and Syvertsen 1996). Under Florida conditions, growth response to P applications are rare, and inherent tree and soil P reserves typically suffice to ensure optimal production (Obreza and Schumann 2010). However, in Brazil soil P supply may be limiting and use of soil P test is desirable (Quaggio et al. 1998). In terms of nutrient demand, although soil nutrient supply may govern tree growth and production to a large extent, active growth in turn also defines sink strength and regulates nutrient uptake capacity through feedback mechanisms and thereby indirectly also affects NUE. Leaves tend to form strong sinks for nitrogen since they contain high concentrations of metabolic active and protein-rich compounds (Lea-Cox et al. 2001; Lea-Cox and Syvertsen 1996; Kato 1986; Weinbaum et al. 1978). Recently formed leaves may contain up to 3–3.5% N, but values tend to drop from 2.5–2.8% (at full leaf expansion) to 1.5–2.0% prior to leaf senescence (Obreza and Morgan 2008). Roots and branches can be differentiated into more metabolic active outer tissues (e.g., bark) and inner (e.g., xylem) tissue that provides structural strength and transport functions. Outer tissues contain 1.2–1.5% N, whereas woody tissue only contains 0.35–0.5% N (Cameron and Appleman 1933). As the diameter of roots and branches increase, the ratio of outer tissue to inner tissue and thereby overall N content declines.

Quiñones et al. (2003) reported that in 8-year-old navel orange trees, 21–26% of the total tree DM was accumulated in the fruit, compared to 13–16% in the leaves, and 31% in the roots. In a study of 6-year-old trees by Mattos (2000) in Florida, DM partitioning to leaves, branches, trunk, lateral roots, and fibrous roots constituted 14%, 37%, 9%, 10%, and 14% of the total dry biomass, respectively. Cameron and Compton (1945) analyzed 8-year-old "Valencia" trees grown in California and found that dry matter allocation to leaves, twigs, branches, trunk, main root, lateral roots, and fibrous roots were 17%, 10%, 43%, 2%, 2%, 14%, and 2%, respectively. Feigenbaum et al. (1987) analyzed 22-year-old Shamouti orange trees fertilized at two annual N rates, and biomass percentages for leaves were 6.4% and 8.4% for the low and high N treatments, respectively. At low N fertilizer application rates, the DM allocation to roots was 34% vs 28% for the high N rate. Kato et al. (1984) reported that for 21-year-old Satsuma mandarin trees in Japan, DM allocation to leaves, twigs, branches, trunk, lateral roots, and fibrous roots were 16%, 5%, 31%, 23%, 20%, and 3%, respectively. Morgan et al. (2006c) developed generic relationships that define DM allocation as a function of either trunk cross-sectional area (TCSA) or tree canopy volume (TCV). Starting from 3 to 4-year-old trees, as trees attain maximum size, DM allocation to leaves decreases from 20% to 12%, the corresponding values for root DM allocation went from 39% to 29%, while the DM allocation to branches increased from 15% to 24% (Morgan et al. 2006b).

15.2.4 Fertilization in Citrus Production

Nitrogen is one of the most important plant nutrients (Kato 1986). The annual worldwide inorganic N fertilizer use toward the end of the past century was on the order of

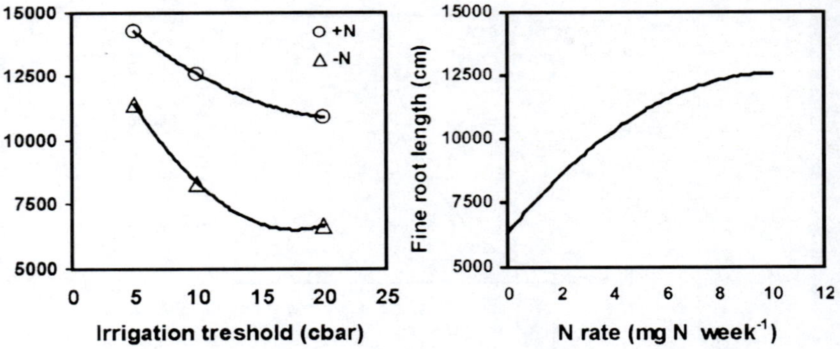

Fig. 15.2 Effect of N fertigation application rate (*left*) and soil water status (*right*) on fine root length; it is of interest that N application will enhance root growth even under relatively dry soil conditions

73×10^6 t N (Mengel 1992). Since that time, there has been a gradual global shift with increased use in specific regions in Asia and Latin America, stagnating use in Africa, and reduction in fertilizer use in Europe and the USA due to more stringent environmental regulations. Adequate nitrogen supply is required for optimal tree growth and high citrus yields (Obreza and Schumann 2010). Understanding factors that control citrus nutrient uptake can help growers to tailor their fertilizer management to maximize profits while reducing nutrient losses and environmental emissions, and thereby provides a sound basis for sustainable citrus production (Scholberg et al. 2002).

15.2.4.1 Nutrient Supply

Depending on the soil type and soil organic matter content, the soil may provide between 65 and 130 kg of N ha^{-1} for plant uptake, and the remainder of tree N demand will have to come from supplementary fertilizer application (Scholberg et al. 2000). Nutrients can be applied in either organic (e.g., manure or compost) or inorganic forms (e.g., ammonium nitrate). The injection of fertilizer in the irrigation system (fertigation) may provide growers with an option to tailor nutrient supply to actual tree demand and should facilitate higher NUE. With the use of reclaimed water containing 7 mg NL^{-1}, about 1.8 kg Nha^{-1} is supplied to the trees per 25 mm irrigation water. However, the fraction of this nitrogen that is taken up by the plant is typically low, especially at excessively high irrigation rates (Scholberg et al. 2002). Dasberg et al. (1984) showed that overall N losses in citrus systems increased from 8% to 48% as N application rates increased from 87 to 393 kg Nha^{-1}. Lea-Cox et al. (2001) showed that NUE decreased linearly with increasing N rate. More frequent application of small amounts of fertilizer reduces excessively high peak values of residual labile soil nutrients, can decrease the risk of potential leaching losses, and thus may enhance NUE (Alva et al. 1998; Paramasivam et al. 2001; Scholberg et al. 2002; Quiñones et al. 2007). An overview of the overall N cycle for a citrus production system is provided (Fig. 15.2).

15.2.4.2 Inorganic Sources

Inorganic fertilizer nitrogen can be broadcasted as granular water-soluble fertilizer, injected into the irrigation water (fertigation), or be applied by foliar application of spray grade (low biuret) urea (Obreza and Schumann 2010). Granular forms of inorganic fertilizer are usually complete fertilizers which, in addition to nitrogen, provide most of the essential nutrients in the right proportions. Most commonly used N source includes ammonium nitrate, ammonium sulfate, and urea (Obreza and Schumann 2010). A field study with 6-year-old trees showed that NUE of ammonium nitrate was 40% compared to 24% for urea (Mattos et al. 2003). Lower NUE for urea may have been related to N volatilization losses (Obreza and Schumann 2010). However, results from studies with lysimeters (e.g., Syvertsen and Smith 1996; Quiñones et al. 2007) show that potential NUE can be much higher (e.g., 61–75%). This discrepancy may be related that weighing lysimeters allow researchers to keep a close check on overall water balances and thereby N leaching losses can be greatly reduced. Based on results from a lysimeter study, it was concluded that eight fertigation applications via drip irrigation increased NUE by 20% compared to the use of two split application combined with flood irrigation (Quiñones et al. 2007). Similar findings were also reported in other studies (Alva and Paramasivam 1998; Dasberg et al. 1988; Boman 1996). However, in previous field studies, mixed results have been obtained with no proven yield benefits with increased number of dry soluble fertilizer split applications on mature trees (Koo 1984). In a more recent study, use of weekly fertigation increased yield, canopy volume, leaf N concentration, and total soluble solid content of juice compared to three split applications for tress less then 5 years of age, while for trees 8–10 years of age it only increased tree canopy volume and soluble solids content of the fruit (Morgan et al. 2009).

Under Florida conditions, the use of enhanced efficiency fertilizers (EEFs) including slow-release (e.g., IBDU) and controlled-release (e.g., sulfur coated urea) fertilizer may be beneficial since nutrients are released gradually over time, which can reduce nitrogen losses due to leaching (Paramasivam and Alva 1997; Obreza and Schumann 2010). Typically, these materials are more expensive but can be beneficial especially for newly planted trees and/or to reduce nitrogen leaching (Scholberg et al. 2002). Nutrient release characteristics, as affected by soil temperature and soil water status, must be considered to make sure that the plant gets its nitrogen when it needs it. Currently, foliar nitrogen application is not widely used for citrus production in Florida and only limited amounts of nutrients may be applied in this manner (5–10 kg N/application). However, foliar application of nitrogen allows fast and efficient uptake of nitrogen and may provide a viable alternative for soil applications since it can reduce groundwater contamination. Uptake is optimal when temperatures are 25–31°C, RH values are high, and solution pH = 7–8 (Orbovic et al. 2001). Applied as a 3% foliar urea spray (higher concentrations may cause leaf burn especially on new growth), approximately 10–15 kg of N ha^{-1} can be supplied to the crop per application. Applying urea at higher concentrations in January (50 kg ha^{-1}) causes non-toxic leaf stress, which has been shown to enhance flowering (Scholberg et al. 2000).

15.2.4.3 Organic Amendments

Organic nitrogen sources typically have low nitrogen content (they are "bulky"), and most of the nitrogen only becomes gradually available over time depending on mineralization rates as related to their composition (C:N ratio) of the material, application method, and pedoclimatic conditions (Obreza and Schumann 2010; Vitousek 1982). Organic amendments can be either broadcast or applied to the row middles (e.g., sandwich system) which was described by Scholberg et al. (2010). Under Florida conditions, mineralization rates are fast, and approximately 50–75% of organic nutrients may become available via mineralization (Obreza and Schumann 2010). However, in many developing regions use of inorganic nutrient sources may be hampered due to their high cost and/or limited availability, and many citrus growers in such regions may be "organic by default." At higher application rates, such nutrient sources may improve inherent soil fertility, soil texture, and soil water holding capacity and thus can provide a valuable asset especially on soils with low inherent water and nutrient retention capacities. The use of cover crops such as perennial peanuts may be beneficial since they can reduce weed growth and soil degradation while retaining nutrients and enhancing soil quality (Scholberg et al. 2010). In this manner, use of organic amendments may enhance overall NUE.

15.3 Nutrient Use Efficiency in Citrus Systems

15.3.1 Defining Nutrient Use Efficiency

Citrus nutrient use efficiency (NUE) pertains to the ability of a tree crop to effectively utilize available nutrients to sustain optimal tree growth and fruit yields. In the literature, a number of different terminologies and definitions may be used. The term nitrogen use efficiency may be expressed as (a) total N accumulation; (b) total DM accumulation per unit of nutrient taken up, which is also called N efficiency ratio and is the inverse of N tissue N concentration; (c) total plant dry weight expressed on a nitrogen concentration basis, which is sometimes also referred to as N utilization efficiency; or (d) total nutrient accumulation per unit of root dry weight (Sorgona et al. 2006). Although this may be interesting from a physiological perspective, it bears little relevance from a practical perspective because overall tree growth is greatly reduced. This is exemplified for both water use efficiency (WUE) and nitrogen use efficiency (NUE) for seedling trees (Fig. 15.3). What is pertinent here is that under severely N-limiting conditions, overall NUE values appear to be the greatest. Scholberg et al. (2002) preferred the term N fertilizer uptake efficiency which was defined as the fraction of applied fertilizer that is effectively taken up by a tree. In this case, the focus is on the fraction that is recovered by the tree and is similar to the apparent nitrogen recovery (ANR) rate sometimes used in fertilizer rate studies. From a production perspective, agronomic efficiencies may thus be more relevant (e.g., extra growth or yield per additional unit of fertilizer being added) (Alva et al. 2006).

For the purpose of this review, we will use the conceptual approach presented by Giller et al. (2006) since it facilitates a more process-based approach and more clearly captures the underlying processes affecting NUE and can be defined as:

$$\text{NUE} = \text{Interception efficiency} \times \text{Uptake efficiency} \times \text{Conversion efficiency} \tag{15.1}$$

The terms interception and uptake efficiency may be combined and referred to as "capture efficiency" because interception and uptake efficiency is more difficult to document on a tree or field scale (Zotarelli et al. 2009). The *interception efficiency* refers to the more static component (e.g., root distribution), which is determining what fraction of nutrients is being intercepted by the root systems prior to being leached (nitrate and to a lesser extend ammonium and potassium) and/or are transformed to less available forms (phosphorus). In a way, the below-ground interception capacity mirrors the above-ground canopy development (Eissenstat and Duncan

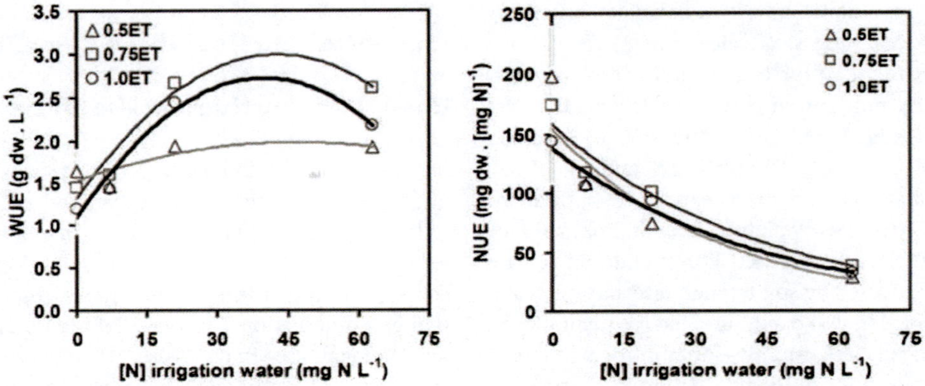

Fig. 15.3 Interactive effect of N fertigation concentration and water application rate on water use efficiency (WUE, *left graph*) and N use efficiency (NUE, *right graph*). Maximum WUE efficiencies occurred at intermediate N concentrations whereas maximum observed NUE occurred at low N concentrations (due to a N tissue dilution effect) and at mild water deficits

1992) and is also related to the overall sink capacity (e.g., tree size), and this component only changes gradually over time (Morgan et al. 2006c).

The *uptake efficiency* refers to a more dynamic component (e.g., root nutrient uptake per unit root length). This aspect of NUE may show appreciable variations over short time intervals since source supply (e.g., labile soil nutrient pool size), environmental conditions (temperature, radiation and rainfall), and sink demand capacity (as related to metabolic processes) are interactive and rather dynamic processes. Overall uptake efficiency may change rapidly and appreciably, and this is especially the case for poorly buffered sandy soils where labile nitrogen pools are being most prone to large fluctuations. In this case soil testing for mineral N content is not considered to be meaningful.

The *conversion efficiency* is related to the "building" cost of developing metabolic active vs more structural and/or storage components that typically require lower amounts of nutrients. The conversion efficiency may be assessed by taking the inverse of the tissue nutrient content.

15.3.2 Assessing NUE

Use of ^{15}N and other labeled compounds revolutionized tree physiology research and facilitated improved understanding of internal nitrogen cycling in tree systems (Lea-Cox and Syvertsen 1996; Quiñones et al. 2007). However, use of this method on a field scale is very expensive both in terms of cost of the material and subsequent tissue analysis (Scholberg et al. 2001). Another approach is use of small seedling trees to measure and monitor physiological processes in a more cost-effective manner, especially since it is easier to ensure adequate tree uniformity at a younger development stage (Scholberg et al. 2001, 2002). Yet another approach is complete excavation of entire trees, which is rather invasive and very labor intensive (Morgan et al. 2006c). Modeling approaches, including N balances, may also be used to integrate processes at different scales and/or levels of complexity and may elucidate more complex dynamics over time. We aim to effectively integrate these different approaches, so processes may be more effectively studied at different tree development stages and spatial scales.

Lea-Cox and Syvertsen (1996) reported lower NUE with higher N application rate for greenhouse grown seedlings. The NUE reported ranged from 47% to 60% after an uptake period of 31 days. Syvertsen and Smith (1996) estimated NUE to be 61–83% for 4-year-old grapefruit trees grown in lysimeters while NUE decreased with increased N application rates. Using lysimeters, Boaretto et al. (2010) found that NUE of 3-year-old lemon trees was 52% compared to 36% for orange trees, although both were grafted on the same rootstock and had similar initial size; final tree weight after 1 year was 43% greater for lemon trees. Feigenbaum et al. (1987) reported that the NUE for a ^{15}N-labeled KNO_3 applied to 22-year-old "Shamouti" orange was 40%. For the purpose of this review, the main emphasis will be on commercial citrus production system with special reference to Florida conditions. The rationale for this is that Florida soils have low inherent water and nutrient retention capacities and therefore are ideal to study nitrogen dynamics.

15.3.3 Effect of Soil Physical Aspects and Nutrient Sources on NUE

15.3.3.1 Water and Nutrient Interactions

Water and nutrient efficiencies are intrinsically linked (Morgan and Hanlon 2007). Excess application of N fertilizer invariably results in reduced NUE and in combination with excess irrigation and/or rainfall in displacement of labile nutrients.

Water Supply and Tree Growth

Water is one of the most important factors that limits plant growth (Syvertsen 1996), and assimilate production is closely linked to plant transpiration. Water and nutrient supply have pronounced effects on crop yield and fruit quality of citrus (Alva and Paramasivam 1998). Soil moisture affects nutrient uptake in various ways. The first factor being adequate moisture levels are required to facilitate the diffusion of nutrients to the root surface which affects active uptake. Under soil conditions, the diffusion coefficients are lower due to the tortuosity effect of soil pores which hamper N movement to the roots, with this effect being most pronounced under dry soil conditions. The second factor being that passive uptake of nutrients is proportional to overall ET rates. Prolonged periods of drought affect nutrient uptake by reducing plant growth (nutrient demand) and assimilate availability (nutrient uptake capacity). A reduction in soil water potential thus can both reduce water uptake (Syvertsen 1985) and root growth (Bevington and Castle 1985). Inadequate soil moisture supply may result in a more pronounced reduction in shoot growth compared to root growth. Although total assimilate production is reduced during drought stress, root-shoot ratios typically increase (Castle 1978; Syvertsen 1985). Reduced water availability to part of the root zone may result in increased water uptake by the remainder of the root system (Ferrier and Alexander 1991; Simonneau and Habib 1994). Increasing the threshold value for irrigation on a loamy soil from 20 to 100 cbar resulted in increased root proliferation at greater soil depth (Cahoon et al. 1964). Excessively high irrigation rates, on the other hand, can reduce root longevity and root density (Menocal-Barberena 2000). Citrus root systems typically acclimate readily to changes in wetted areas (Dasberg 1992). However, it may take up to 6 years for the deepest roots to fully adjust to changes in irrigation practices (Cahoon et al. 1964).

Soil Water Status and Nutrient Uptake

Water is of central importance in the transport of solutes in soils or plants, whether by diffusion or mass flow (Tinker and Nye 2000). Crop production is thus closely related to leaf photosynthesis and canopy size. Water stress will reduce leaf area expansion and leaf photosynthesis (Kramer and Boyer 1995).Under water-limiting conditions, crop production increases proportionally to irrigation supply (Bar-Yosef and Sagiv 1982; Kramer and Boyer 1995). Citrus are evergreens and therefore require water for transpiration throughout the year. Citrus leaves are thick and waxy, resulting in high cuticular resistance to transpiration (Mills et al. 1999). Koo (1963) and Koo and Sites (1955) stated that water requirements of grapefruit are generally higher than orange or mandarin varieties for trees of equal size.

Table 15.1 Effects of prior irrigation rate on N uptake of two seedling types (expressed as uptake per citrus seedling per 12 h) as affected by irrigation management 1 week prior to N application during the end of the drying cycle and after rewetting the soil

Irrigation rate	N uptake (mg N pit^{-1}) Swingle Dry	Swingle Rewet	Volk Dry	Volk Rewet
1.00 ET	14.1 (100%)	12.2 (100%)	13.7 (100%)	13.2 (100%)
0.33 ET	11.0 (84%)	9.8 (80%)	10.7 (78%)	13.0 (98%)
0.00	9.5 (67%)	8.4 (68%)	7.7 (56%)	10.1 (71%)

Table 15.2 Nitrogen uptake per root compartment for different citrus seedling types (expressed as uptake per citrus seedling per 144 h) as affected by soil water status of corresponding root compartment for split-rooted seedlings

N uptake mg N plt^{-1}			
Swingle		Volk	
Dry (>30 cb)	Wet (<10 cb)	Dry (>30 cb)	Wet (<10 cb)
43.9 (65%)	67.7 (100%)	48.7 (57%)	74.8 (100%)

Nutrient and water use efficiency are typically intrinsically linked to the plant's ability to respond to changes in rhizosphere environments. Overall nutrient uptake is a combination of short-term dynamics (e.g., uptake rate) and long-term adaptation (e.g., differential root proliferation). Studying changes in root growth induced by N placement and localized drought stress are helpful to improve our understanding of mechanisms that underlie water and nutrient uptake efficiencies. Studying short-term nutrient uptake dynamics of citrus seedlings via Soil-N Uptake Monitoring Systems (Scholberg et al. 2001) showed that moderate water stress 1 week prior to a fertigation event reduced N uptake by 12–16%, whereas more severe water stress resulted in uptake reductions of 33–44% (Table 15.1). After a full wetting cycle of 24 h, this reduction was completely reversible for Volkameriana rootstocks provided that the drought stress was not too severe, whereas for Swingle rootstocks subsequent N uptake after rewetting remained lower. It may be concluded that more pronounced long-term reduction may result in reduced growth and assimilate supply associated with prior water stress. This experiment was repeated using split root systems with dry and wet compartments, which showed that applying fertigation to a dry (>30 cb) soil compartment had 35–43% lower N uptake compared to application to a moist (<10 cb) soil compartment (Table 15.2). Since citrus roots are lignified and can withstand prolonged periods of drought and wet and dry compartments were reversed during the course of the experiment, it is concluded that soil conditions do have a very large impact on overall uptake, which may be related to reduced soil infiltration and diffusion of applied fertilizer materials.

In terms of irrigation management, more water (over-irrigation) does not translate into higher yields but rather less efficient use of water and nutrients (Scholberg et al. 2002). If water is applied according to crop water requirements, leaching losses can be minimized, and trees can make optimal use of fertilizer as well. Soil moisture monitoring devices placed at different soil depths can be a useful tool to monitor soil water content. Limiting soil water depletion to 25–50% will prevent excessive crop water stress or fertilizer volatilization (excessively dry conditions) and leaching, shallow root systems, root rot, and denitrification (excessively wet conditions). Citrus trees can modify root growth in response to changing/unfavorable soil conditions. However, this takes time and may result in reduced growth and/or fruit yield. An even distribution of fertilizer combined with moist, but not excessively irrigated, root zone promotes both optimal root growth and nutrient uptake, thus high NUE (Paolillo et al. 1999; Scholberg et al. 2002).

Under soil conditions, soil solution concentration need to be twice as high to sustain similar uptake rates compared to solution culture conditions (Cerezo et al. 1997). Other studies also showed that optimal growth under solution culture occurred at N concentrations between 0.52 and 0.7 mmol compared to 1–1.5 mmol for sand cultures (Chapman and Liebig 1941; Maust and Williamson 1994). It could be argued that the intense mixing of an aerated solution culture system facilitates more efficient nitrate uptake especially at low N concentrations. Part of this may be related to pronounced changes in root morphology for roots formed in solution cultures. Although root growth continues throughout most of the year, root growth occurs in flushes that alternate with shoot growth (Castle 1978). The median life span of fibrous roots for citrus trees is on the order of 100 days (Eissenstat and Anchor 1999). Even under dry soil conditions, citrus trees will sustain previously formed roots for more than 2 months (Kosola and Eissenstat 1994). On deep sandy soils, the surface 0–30 cm of the root zone can contain up to 40–50% of the fibrous root weight, and root densities tend to be highest beneath the drip line of the tree (Menocal-Barberena 2000).

15.3.3.2 Interactive Effects of Water and Nutrient Supply on Root Distribution

A reduction in soil water potential can reduce water uptake (Syvertsen 1985) and root growth (Bevington and Castle 1985; Kosola and Eissenstat 1994). Inadequate soil moisture supply may result in a more pronounced reduction in shoot growth compared to root growth. Although total assimilate production is reduced during drought stress, root-shoot ratios typically increase (Castle 1978; Syvertsen 1985). Reduced water availability to part of the root zone may result in increased water uptake by the remainder of the root system (Ferrier and Alexander 1991; Simonneau and Habib 1994). Increasing the threshold value for irrigation on a loamy soil from 20 to 100 cbar resulted in increased root proliferation at greater soil depth (Cahoon et al. 1964). Excessively high irrigation rates, on the other hand, can reduce root density (Menocal-Barberena 2000). Citrus root systems typically acclimate readily to changes in wetted areas (Dasberg 1992). Increases in fertilizer N can increase root growth to a considerable depth, but the largest effects generally occurred near the surface (Ford 1952, 1953; Smith 1956, 1965; Ford et al. 1957).

In order to assess interactions between soil water and nitrogen status on changes in root growth dynamics, Paolillo et al. (1999) used split-rooted citrus seedlings. This study showed that under drier soil conditions, roots became thicker resulting in reduced specific root length (SRL) which may be related to increased thickening of the exodermis and development of suberin and lignin layers in the wall of the exodermis (Eissenstat and Anchor 1999). Under field conditions, however, SRL values also decreased at excessively high irrigation rates (Menocal-Barberena 2000). Overall root length of citrus seedlings was highest when soil water tension was kept in the 4–8 cbar range and root length increased quadratically with N application (Fig. 15.3). Control treatments with no N being applied had lower root lengths compared to treatments receiving N, but overall assimilate partitioning to roots was greater under N and water limiting conditions. In the absence of N application, root length was evenly distributed between soil compartments as long as soil water tension remained between 4 and 8 cbar. Under dry soil conditions, root partitioning to the moist (8 cbar) compartment increased by about 11%. Proliferation of fine roots due to N application was most pronounced in the upper 7.5 cm of the soil profile. Application of N to a specific soil compartment resulted in an increase in root partitioning to that compartment of 9%, 22%, and 24% for the 4, 8, and 16 cbar irrigation treatments, respectively. This is an important result since it shows an interaction between N supply and drought stress and demonstrates the ability of citrus root systems to concentrate roots where there is adequate supply of nutrients. Although citrus roots can potentially make use of fertilizer applied to a relatively dry part of the root zone uptake, efficiency may be still lower since dry conditions hamper uptake as was shown in the first section. Moreover, root proliferation may require a couple of weeks, and fertilizer may be prone to leaching and/or volatilization prior to increased uptake via increased rooting density. Overall, plant growth and total root length of seedling trees was highest if the soil was kept moist and N application was spread evenly over the entire wetted zone (Paolillo et al. 1999). On a field scale, it was shown that excessive N rates, on the other hand, may reduce root growth (Ford et al. 1957; Castle 1978).

15.3.3.3 Water and Nutrient Retention Versus NUE

Adequate water supply is critical in maximizing nutrient use efficiency, crop production, and quality of most horticultural crops. However, on soils with poor water retention, application of excess water may promote displacement of mobile nutrients such as nitrate before complete uptake has occurred (Morgan et al. 2006b; Scholberg et al. 2002). The ability of crop plants to take up and utilize N efficiently is key to providing adequate N for crop growth while reducing N leaching. On sandy soils, a rainfall event of 25 mm may displace nitrate by up to 30 cm deep. Moreover, in such soils after one pore volume (~120 mm) water passes through the root zone, approximately 92–96% of the residual nitrate N not taken up the root system may be lost due to leaching (Scholberg et al. 2001). Root distribution studies by Morgan et al. (2007) showed that for young trees (canopy volume <3 m^3) grown under microsprinkler irrigation, 85% of the overall roots occurred in the upper 30 cm of the soil profile, and that this percentage decreases linearly to 57% when trees reach their maximum size. Appropriate irrigation scheduling and matching irrigation amounts with the water holding capacity of the effective root zone thus is essential to minimize the risk of potential N leaching associated with over-irrigation and/or potential leaching rainfall events. Scholberg et al. (2002) found N uptake of greenhouse-grown seedlings to be proportional to soil temperature, ET_o, and canopy biomass. As soil N concentrations increase, the time duration for near-complete uptake increases almost proportionally. As a result, the risk of N displacement below the effective root zone and the risk of potential N leaching and/or groundwater contamination may also increase greatly when soil N levels are excessively high. Increasing the residence time from 2 to 8 h increases NUE from 36% to 82% and 17% to 34% for low vs high N application rates, respectively (Scholberg et al. 2002).

15.4 Physiological Aspects and Nutrient Demand

From a physiological perspective, NUE is often linked to photosynthetic traits (Lea-Cox and Syvertsen 1996). In this context, the relative fraction of leaf nitrogen allocated to photosynthetic capacity is a key element of overall economy of assimilate production (Hikosaka 2004), which ultimately drives fruit yield. In NUE evaluations, adequate N acquisition and formation of photosynthetic capacity may be regarded as key investment costs for ensuring adequate production capacity. It was shown that NUE from a photosynthetic perspective is closely related to specific leaf N levels and leaf structure. Investing in relatively thick, long-lived leaves, as in the case with citrus, comes at an assimilate cost since a relative large fraction of both N and assimilates are required for formation of structural tissues (Hikosaka 2004). Similar evaluations may be made for roots to assess which root structure and longevity traits may be desirable from the perspective of overall efficiency of nutrient acquisition. Thinner roots have more surface area and higher uptake capacity per unit carbon invested, and such roots are quite effective in exploiting fertile soil regions. However, under adverse conditions, effectiveness may be more important than efficiency (Eissenstat and Volder 2005). In the next section, we will exemplify selected aspects of crop physiological processes and link these to overall NUE.

15.4.1 Nutrient Interception Capacity

Excess application of nutrients may result in high residual soil nutrient concentrations and displacement of labile nutrients prior to complete uptake (Quiñones et al. 2007; Scholberg et al. 2002). Nutrient interception capacity is a combination of spatial and temporal root growth and uptake dynamics. In the next section, we will describe these components in more detail.

15.4.1.1 Root Distribution and Functionality

Understanding growth patterns of root systems of citrus trees provides unique opportunities for designing more efficient water and nutrient management guidelines. Citrus root distribution is governed by genetics, crop development stage, management, and pedoclimatic conditions (Morgan et al. 2006b). Although thicker roots (>4 mm) may constitute a large fraction of the overall root mass, the distribution of fibrous roots (<1–2 mm) plays a more critical role in overall water and nutrient uptake. Fibrous roots are also the most vulnerable part of the root system, and their development, function, and longevity are strongly influenced by soil characteristics, environmental changes, crop species, crop growth stage, and cultural practices. In terms of inherent rooting traits, Castle and Krezdorn (1975) described two general types of root systems, the first characterized as "extensive," featuring extensive lateral and vertical development, and the second as "intensive," with less extensive root expansion and higher fibrous root concentrations mainly confined to the upper soil layers. Trees on rough lemon, "Volkamer" lemon, and "Palestine" sweet lime (*C. limettioides* Tan.) rootstocks typified the extensive type of root system where 50% of the fibrous roots occurred below 70 cm in the soil with wider spreading lateral development. Examples of the intensive type were "Rusk" citrange and trifoliate orange, where few fibrous roots were found below 70 cm, and the root system was less developed laterally. Some rootstocks like sour orange and "Cleopatra" mandarin were classified as intermediate.

Root distribution is also affected by physical and chemical soil properties (Hillel 1971; Eissenstat 1991).

Anaerobic conditions associated with the presence of shallow (perched) water tables will constraint the vertical expansion of the root system. Most roots thus may be confined to the upper 45 cm of the soil profile and overall tree size may be constrained by actual rooting depth and fibrous root density (Ford 1954, 1972; Reitz and Long 1955; Calvert et al. 1967, 1977; Calvert and Phung 1972). More extensive lateral root development may occur on finer-textured soils (Kaufmann et al. 1972; Boswell et al. 1975). In this case, overall root densities may be lower, root distributions are more uniform, and root systems may be shallower with few roots found below a soil depth of 50–70 cm (Adriance and Hampton 1949; Kimball et al. 1950; Cahoon et al. 1956, 1959, 1961; Boswell et al. 1975; Mikhail and El-Zeflawi 1978). Menocal-Barberena (2000) observed that in well-drained sandy soil in Florida, about 40% of the fibrous roots occurred in the top 30 cm with 9–14% at each of the remaining 30 cm depth increments to 180 cm while few roots were found below 180 cm. Eissenstat and Duncan (1992) found that canopy and root growth dynamics are closely related. Within 30 days of canopy pruning, 20% of the fibrous roots in the upper 30 cm of the soil senesced, but overall root length density was restored within 63 days. However, this relatively short-term reduction in fibrous root density did adversely affect yield because of fruit abortion. Under water and N-limiting conditions, assimilate allocation to roots were reduced (Paolillo et al. 1999; Lea-Cox and Syvertsen 1996). In terms of phosphorus uptake efficiency, roots association of micorrhiza may be relevant because their very fine hyphae allow more effective surface area and thus increased interception and uptake of phosphorus. Under P-limiting conditions, presence of mycorrhiza can greatly enhance P-uptake and plant growth (Graham and Eissenstat 1998). However, when soil P availability is adequate, which is commonly the case in Florida, the carbon cost associated with their infection may not provide any benefits, and a positive growth or enhancement of NUE may be lacking (Graham and Eissenstat 1998).

15.4.1.2 Uptake Mechanisms

Soil N uptake is affected by a number of factors, with soil N concentration, root length density, and soil temperature being most important (Marschner et al. 1991; Scholberg et al. 2002). Although uptake kinetics of crops like barley, cotton, corn, and tomato have been studied extensively (Claasen and Barber 1974; Rao and Rains 1976; Bloom 1996), little information is available on the N uptake kinetics of citrus. Optimal soil solution N concentrations are greatly affected by the growing medium. Although concentrations of 200–400 mg N mg N L^{-1} are commonly used for fertigation of citrus grown in a potting medium, optimal growth could be attained with soil N concentrations of 15–20 mg N L^{-1} for an inert sand (Maust and Williamson 1994) and 10 mg N L^{-1} with

Table 15.3 Effects of soil solution concentration on N uptake (expressed as uptake per mg fresh fibrous root) and corresponding contributions of both active and passive uptake

		Uptake mechanism	
$N_{concn.}$ mg N L^{-1}	V_{max} (mg N g^{-1} h^{-1})	Active (%)	Passive (%)
3.5	1.5	91	9
22.5%	4.2%	83	17
64.5	6	63	37

solution culture (Chapman and Liebig 1941). To date, most uptake studies for citrus focused on growth response to N supply (Chapman and Liebig 1941; Lea-Cox and Syvertsen 1996; Dou and Alva 1998), long-term N uptake, and/or N partitioning within the plant (Legaz and Primo-Millo 1988; Lea-Cox and Syvertsen 1993, 1996). Although plant growth response to N supply has been studied extensively, the physiological basis for improving nutrient uptake efficiency (NUE) in crops remains poorly understood (Rufty et al. 1990).

15.4.1.3 Uptake Dynamics

Nitrogen can be taken up by either passive or active uptake mechanisms. Plants can take N up as either NO_3^- or NH_4^+. However, under (sub)tropical conditions, most soil N is readily converted to NO_3^-, and most of the N is taken up as NO_3^- (Scholberg et al. 2000). Nitrate movement into the root requires adequate metabolic energy supply, and conditions that hamper metabolic root activity will reduce N uptake (Clarkson 1986). Although N uptake was observed to be affected by transpiration rate (Lea-Cox and Syvertsen 1993), transpiration may not be required for N uptake (Tanner and Beevers 1990). Transpiration-induced mass flow contributed up to 50% of total N uptake by spruce (Marschner et al. 1991), but for citrus it was typically well below 30%, with its contribution to overall N uptake being greatest at higher soil N concentrations (Table 15.3) and increasing linearly with ET rate.

Michaelis-Menten enzyme kinetics are routinely used to capture nitrogen uptake dynamics. Overall uptake kinetics can be calculated using regression analysis of double reciprocal (Lineweaver-Burke) plots of uptake rates vs solution concentration (Claasen and Barber 1974). Thus, active N uptake can be defined using the following equation:

$$V = [\mathrm{N}] * V_{\mathrm{max}} / (K_{\mathrm{m}} + [\mathrm{N}]) \quad (15.2)$$

where V is the actual nitrogen uptake, [N] is the nitrate concentration in the soil solution, V_{max} is the maximum uptake, and K_m is the N concentration at half of the maximum uptake attained. Nitrate uptake is governed by independent transport systems, with a high affinity saturatable system (HATS) prevailing at concentrations <0.5 mm and a low affinity linear system (LATS) controlling uptake at higher concentrations (Aslam et al. 1997).

To attain maximum uptake capacity, roots require an induction period of the transport system. This induction time increased with concentration and the resulting induction stage was fairly labile and uptake potential dropped to constitutive levels within a few days (Clarkson 1986; Maeck and Tischner 1986). Efflux rates indirectly control net uptake with efflux values ranging from 10% to 70% of net influx rates (Clarkson 1986). Excessively high N supply may inhibit root growth and increase efflux rates (Bloom 1996). Under N-limiting conditions, uptake efficiency (K_m) decreased, whereas uptake capacity (V_{max}) initially increased but eventually dropped due to reduced growth rates (Clarkson 1986; Heins and Schenk 1986; Rufty et al. 1990). Maximum growth rates could be sustained at N concentrations below K_m values, if relative nutrient addition rates are adequate (Ingestad 1982; Clarkson 1985). Alternatively, constant high N supply increased amino acid concentrations in the phloem stream and resulted in downregulation of subsequent root N uptake (Youssefi et al. 1999, 2000). This downregulation of N uptake is also indirectly controlled by sink strength, with high relative growth (a strong sink demand for N) delaying feedback inhibition (Larsons et al. 1992). Seasonal variations of N uptake also occur, with uptake being highest during periods of active shoot growth (Weinbaum et al. 1978; Maust and Williamsons 1994).

Scholberg et al. (2001, 2002) developed Soil-N Uptake Monitoring (SUM) systems to assess short-term soil nitrogen uptake dynamics of seedling trees. This approach was also used to monitor biweekly uptake dynamics of grafted citrus seedlings and uptake results compare favorably with those of ^{15}N-based measures (Linares et al. 2010). The advantage of this approach is that it provides a cost-effective alternative to ^{15}N for near-continuous monitoring of N-uptake dynamics for soil-based systems. By exposing seedling trees to initial soil solution concentrations of 0.2, 1.5, 4.7 mmol NO_3–N, which translates to 3, 22, and 66 mg NL^{-1} and by using different retention times (e.g., 3–144 h), soil-based Michaelis-Menten coefficients could be calculated (Scholberg et al. 2001). Uptake for annual crops is often expressed per unit total root weight for small seedlings (Aslam et al. 1997; Cerezo et al. 1997). However, in the case of citrus, this seems inappropriate since structural roots account for 30–50% and 50–70% of the total root fresh and dry weight, respectively. From a physiological point of view, expressing uptake as a function of root length may be more meaningful. From a practical point, using root dry weight may be more desirable, although for mechanistic models the use of root length density may be more appropriate.

Complete N uptake occurred within 4–24 h when initial soil solution concentrations were 3 vs 22 mg NL^{-1}. More detailed measures with consecutive short uptake cycles showed that maximum uptake (actual uptake approaching potential uptake based on Michaelis-Menten uptake parameters) only occurred after 6–12, 3–6, and 0–3 h after initial exposure to a fertigation solution for nitrogen concentrations of 3, 22, and 66 mg NL^{-1}, respectively (Fig. 15.4). This apparent lag in maximum uptake is referred to as induction and is related to the induction of the carrier of the active uptake system (Clarkson 1986). Declines in subsequent nitrogen uptake appear to be most pronounced at high N application rates (Fig. 15.3). This may be related to a feedback mechanism by which high internal N levels resulting in downregulation of subsequent nitrate uptake. This decline in N uptake over time due to downregulation was also reported for wheat (Goyal and Huffaker 1986). It is of interest to observe that downregulation did not occur at lower N application rates. It may be argued that sink strength may be one of the controlling factors governing downregulation of nitrogen uptake and also NUE (Youssefi et al. 1999, 2000). Values for V_{max} (on a root dry weight basis) for citrus were twice as high compared to estimated values for peach trees (Ran et al. 1994), but similar to those for ammonium uptake for citrus (Hassan and Van Hai 1976). Nitrate uptake of citrus per unit root length was similar to those reported for maize (Heins and Schenk 1986). Uptake capacity (V_{max}) of citrus roots under soil conditions was similar to those for barley (Rao and Rains 1976), cotton (Aslam et al. 1997), and sugar-beet (Maeck and Tischner 1986) under solution culture conditions. Under solution culture conditions, citrus showed a biphasic uptake pattern with a V_{max} value of 0.25 μmol g^{-1} FW h^{-1} below

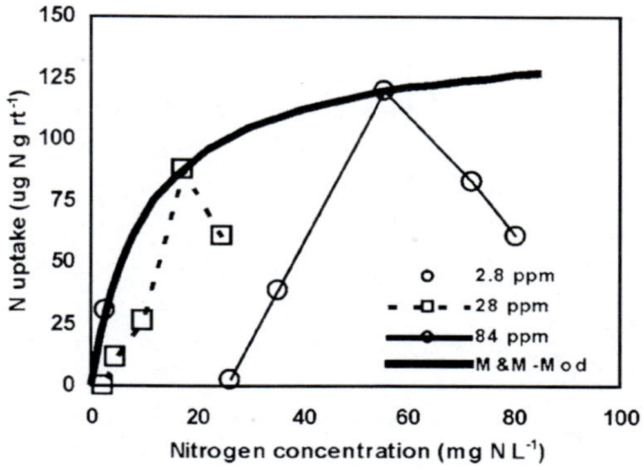

Fig. 15.4 Uptake dynamics of young citrus seedlings showing consecutive uptake cycles (right-most point within a data sequence is the first uptake cycle) with maximum uptake (actual uptake approaching the theoretical maximum based on Michaelis-Menten uptake parameters) only occurring after 6–12, 3–6, and 0–3 h after initial exposure to a fertigation solution for nitrogen concentrations of 2.8, 28, and 84 mg N L^{-1}, respectively as shown by uptake points increasing before merging towards the theoretical maximum uptake is related to induction. Subsequent nitrogen uptake declines due to downregulation with this effect being most pronounced (largest deviations from theoretical uptake *curve*) at high initial concentrations, as shown by uptake point diverting away from the theoretical maxima and uptake declining to values close to zero

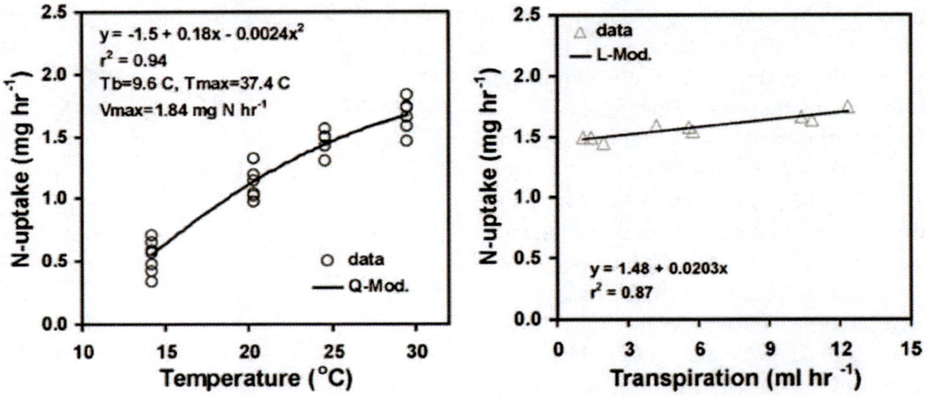

Fig. 15.5 Nitrogen uptake of citrus seedlings as affected by soil temperature (*left graph*) which is one of the key environmental factors affecting active uptake and the effect of N transpiration rate (modified by partial and full shading of the leaves) on N uptake

1 mmol followed by a low affinity liner uptake pattern between 1 and 10 mmol (Cerezo et al. 1997).

It is relevant to notice that the overall shape of the Michaelis-Menten as shown in Fig. 15.4 as related to saturation of active career system provides a physiological basis for explaining the law of the diminishing returns or reduced NUE at higher N application rates. This was exemplified by Sorgona et al. (2006) who demonstrated that both leaf area expansion and shoot dry weight accumulation as related to fertigation solution N concentration followed similar shaped saturation response curves.

Diurnal Uptake Patterns

Pronounced diurnal fluctuations in N uptake have been observed for a number of crops including barley (Rao and Rains 1976), maize (Keltjens and Nijenstein 1987), soybean (Delhon et al. 1996), and tomato (Le Bot and Kirkby 1992). Decreased nitrate uptake during the night coincided with decreased N reductase levels and were probably related to reduced assimilate supply (Delhon et al. 1996; Keltjens and Nijenstein 1987). Shading increased diurnal fluctuations whereas supplying roots with additional glucose offset uptake reductions during the dark period (Delhon et al. 1996). Differences between crops may be related to the amount of substrate in storage pools and retranslocation rates to the roots during the dark period (Kato 1986). Depending on solution N concentration, overall reductions in nitrate uptake during the night were on the order of 0%, 29%, and 35% at soil solution concentrations of 3, 22, and 66 mg N L^{-1}, respectively (Scholberg, unpublished). This reduction appears to be relatively small compared to maize and cotton where night uptake was reduced by 50–80% (Delhon et al. 1996; Keltjens and Nijenstein 1987). Since actual difference between day and night soil temperatures was relatively small (<2°C), diurnal difference in uptake rates seems to be most likely related to a reduction in soil nutrient mass flow to roots and/or assimilate supply from the shoots. It has been proposed that "apparent mass flow contribution" would be proportional to the product of soil solution concentration times the water influx into the root (Tinker 1968). Higher soil solution concentrations would therefore result in a greater contribution of mass flow on total nutrient (N) uptake, and also a more pronounced effect on diurnal uptake patterns. Although nitrate uptake during the light period was significantly greater at higher soil solution concentrations, no such differences were observed during the dark period. It appeared that overall nitrate uptake during the dark period appeared to be sink-limited and "saturated" at higher soil solution concentrations. Estimated contributions of both active and mass-flow-based uptake for citrus seedlings are further highlighted (Table 15.1).

Effects of Shading on Nitrate Uptake

Use of differential shading treatments (0%, 50%, and 100% of the leaves being completely shaded) for 24 and 96 h allowed assessment of the effects of light on tree N uptake of young seedlings. During the first day of shading, the reduction in N uptake during the daytime was 14% vs 25% during the daytime for 50% and 100% shading, respectively (Scholberg, unpublished). Corresponding reductions during the nighttime were 0% vs 10%. Shading for 4 days resulted in a reduction of N uptake for Swingle rootstocks of 67–77% compared to reductions of 22–44% for Volkamer rootstocks. The difference in uptake between shaded and nonshaded treatments during the day could be caused by a reduction in active uptake (due to reduced root metabolism and/or assimilate supply) and/or passive uptake (transpiration induced mass flow as shown in Fig. 15.5). Alternatively, differences between shading treatments during the dark period should be mainly related to differences in root metabolism. It was proposed that the presence of leaves, regardless of light levels, play an important role in nitrate uptake due to their roles as potential N sinks and/or storage reservoirs for substrates (Kato 1986; Weinbaum et al. 1978). The reductions in nitrate uptake due to shading were most pronounced for trees with lower shoot to root ratios resulting in a more rapid depletion of assimilate reserves. Cotton plants showed a much greater reduction in nitrate uptake due to partial shading (Delhon et al. 1996) compared to citrus, which may be related to citrus being efficient in adapting to shaded conditions (Syvertsen 1984). Based on relatively high N uptake rates with intermediate shading, it may be possible that assimilate supply for nitrate uptake should have been adequate (nonlimiting) during the prior (light) cycle, and the reduction in nitrate uptake for the intermediate shading treatment during the light cycle was mainly related to a downregulation of transpiration, and it appears that the contribution of passive influx to total N uptake would be on the order of 30% or less, which is in agreement with findings reported in the literature for short-term uptake (Rao and Rains 1976; Tanner and Beevers 1990). Another study with citrus seedlings showed a much greater increase in N uptake with increased transpiration rates (Lea-Cox and Syvertsen 1993). However, closer inspection of their data showed that a large portion of the variation in both uptake and transpiration rate may be related to differences in plant size. It thus appears that young citrus trees will maintain relatively high (70–100% of potential uptake rates) nitrate uptake rates for several days even if low light levels and/or high humidity reduce transpiration.

Temperature Effects

Temperature has a direct effect on metabolic processes including tree growth and active nutrient uptake. The Q_{10} values differ between species and temperature sensitivity of NO_3-N transport was greatest at lower temperatures (Clarkson and Warner 1979). The base temperatures for nitrate uptake ranged from near or below zero for cold-adapted crops like barley (MacDuff and Jackson 1991) to around 10°C for subtropical crops like citrus (Fig. 15.5). Temperature effects on overall N uptake of young citrus trees appeared to be much more pronounced compared to the effects of transpiration (Fig. 15.3). Maximum uptake appears to occur around 30–37°C, whereas uptake ceased below 10°C. Kato et al. (1982) found a tenfold increase in ^{15}N uptake of "Satsuma" mandarin during summer (mean temperature 23°C) compared with the winter season (mean minimum temperature 3°C). Quiñones et al. (2003) thus argued that low NUE in citrus systems may be partly related to excessive N application rates during the winter.

15.4.1.4 Uptake Dynamics and NUE

In the context of NUE, K_m is indicative of uptake efficiency at lower N concentrations and V_{max} relates to uptake capacity. So it may be argued that citrus rootstocks that have high NUE should have low K_m values (effective in scavenging remaining soil N) and high V_{max} values (capable of quickly depleting soil nitrogen prior to displacement associated with soil water transfer after a leaching rainfall event and/or over-irrigation). Overall NUE values for small citrus seedlings for a range of soil solution concentrations was highest with more frequent application and/or at lower N concentrations, where as overall NUE increased linearly with residence time. Sorgona et al. (2006) noticed distinct difference in NUE efficiency among citrus rootstocks. In response to continuous fertigation concentration ranging from 0.1 to 10 mmol $N L^{-1}$, rough lemon showed the highest leaf area expansion and DM accumulation; sour and sweet orange had intermediate values, while Cleopatra mandarin had the (s)lowest growth. Total N accumulation followed similar trends. Use of N efficiency measure could not explain difference in performance in growth among rootstocks, whereas the N uptake efficiency (N uptake expressed per root weight) showed good correlation with actual growth performance (Sorgona et al. 2006).

15.4.2 Nitrogen Demand, Accumulation, and Partitioning

Knowledge of the nutritional need of different tree parts as well as the seasonal demand for nutrients can provide a physiological basis for crop fertilization (Legaz and Primo-Millo 1988). Lea-Cox et al. (2001) stated that newly developed leaves and fruits are major sinks for N and may accumulate 40–70% of the N taken up. Similar findings were reported by Legaz et al. (1982) with about 30% of the N take-up being accumulated in newly formed leaves. Quiñones et al. (2007) reported that 28–29% from N accumulated in the fruit came from N-fertilizer. Lea-Cox et al. (2001) demonstrated that under N-limiting conditions, heavy fruit loads will reduce tree N levels, internal tree N reserves may not be adequate for maximum spring flush development. In terms of overall N accumulation in different tree parts, in a study with 15-year-old "Valencia" trees in California, it was observed that the total vegetative biomass leaves, branches, and roots accounted for 45%, 35%, and 20% of total N (Cameron and Compton 1945). Corresponding values for 3.5-year-old trees were 62%, 21%, and 17% (Cameron and Appleman 1935). Morgan et al. (2006c) developed generic equation expressing tree N allocation within total vegetative biomass as a function of canopy size. It was shown that N accumulation to leaves decreased from 47% for young trees to 37% at maturity, whereas corresponding values for N accumulation in branches increased from 11% to 27%. Overall N allocation to fruit for mature trees is on the order of 14–20%, and with each ton of fruit harvested, approximately 1.2 kg N, 1.7 kg P, and 1.5 kg K are being removed from the orchard (Obreza

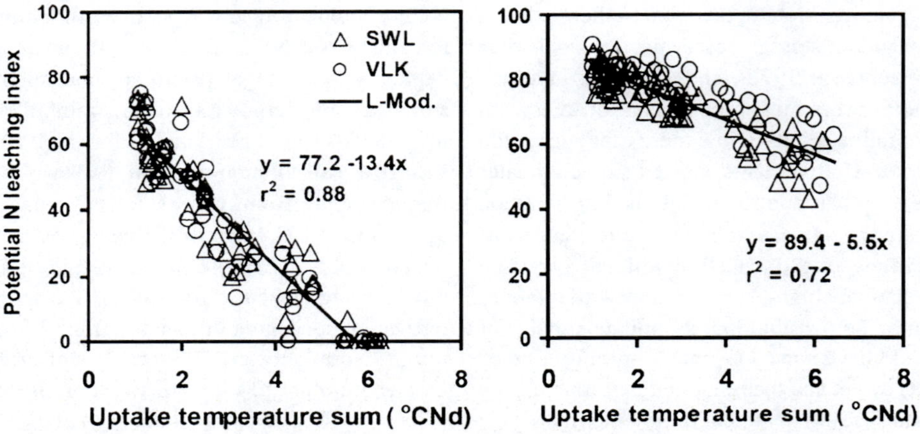

Fig. 15.6 Effect of the temperature uptake sum ($\Sigma[T_{soil}-T_{base}]/24$) on the risk of potential nitrogen leaching when initial soil solution concentrations are 3 mg NL^{-1} (*left graph*) vs 15 mg NL^{-1} (*right graph*)

and Schumann 2010). Seasonal changes in leaves, bark, twig, and root N concentration were greater than N changes in woody branch tissue. Cameron and Compton (1945) reported that trees contained more N just before initiation of growth flush during spring than at any other time of year. It was suggested that the reduction between December and February was the result of deposition of starch and possible other carbohydrates. A decrease of all tree tissue N concentration occurred during bloom, fruit set, and periods of active growth in the spring and early summer since fruits then form relatively strong N sinks. During the summer and autumn, N concentrations gradually increased to the midwinter maximum (Fig. 15.6).

15.4.3 Tree Nutrient Demand and Internal Cycling

As trees become larger and their capacity for storage increases, internal N cycling provides an essential component of overall tree N supply (Millard 1995). Moreover, this internal buffering capacity combined with efficient internal translocation may provide trees with a competitive edge in terms of NUE compared to annual crops. Dasberg (1987) found that 80% of the N in new growth came from stored rather than applied N, suggesting previous nutrition has significant influence on current season growth and fruit yield. Feigenbaum et al. (1987), working with 22-year-old "Shamouti" orange trees, showed that only about 20% of the leaf and fruit N originated from fertilizer N, suggesting considerable redistribution from stored reserves. Less than 14% of the labeled N was found in roots or large limbs. Numerous reports suggest that actively growing tissues act as a sink for N uptake, and that the young developing leaves and fruits constitute the strongest sink. Fruit load is inversely related to N allocation to root and leaf, which in turn affects overall growth of these tree components (Lenz 2000). Mooney and Richardson (1992) observed an N concentration gradient between the roots, trunk, and branches of citrus trees in New Zealand. High concentrations were found in the branches with lower concentrations in the roots. Nitrogen concentrations in the trunk were highest at bud break and declined steadily through fruit set and developed to a minimum at fruit harvest. Nitrogen concentration for all categories peaked at flowering and then decreased steadily until harvest.

Legaz et al. (1982) reported that at post-blossom, the N concentration in the spring leaves decreased due to this tissue becoming an N source for the developing fruit. Using 4-year-old "Valencia" orange trees, daily root N uptake was lower during dormancy, increased during flowering, was highest during fruit set, and later decreased toward the end of the summer and autumn flushes. The greatest accumulation of N absorbed from fertilizer (with respect to the total N absorbed from fertilizer in the whole tree) was found in the young leaves and roots, followed primarily by twigs and stems, then flowers and fruits. Kato et al. (1982) found that total N contents decreased in both bark and wood during the sprouting period of 21-year-old "Satsuma" mandarins. Greatest decreases in N were found in parts with higher concentrations of N (i.e., leaves, shoots, and fine roots). It was also concluded that the trunk and large roots are main N reservoirs for new shoot development. The N was stored mainly as protein, free proline, arginine, and asparagines. Protein decreased in all plant parts in proportion to total N in the plant part. Proline decreased mainly in the leaves and bark, arginine in wood of shoots and asparagines in bark of fine roots. Lea-Cox and Syvertsen (1996) reported that at high N supply, retranslocation from older to newer tissue ceased, whereas as soon as N becomes limiting to rapid growth, translocation did occur.

As leaves age, leaf N concentration may drop from 3.0% to 1.5% as they senesce (Obreza and Morgan 2008). So, approximately 50% of the N in leaves is retained within the tree, whereas the rest may be partly recovered upon mineralization. Within 1 year after initial application, 7–15% of the applied N may be accumulated in the litter layer due to senescence of leaves and small twigs (Quiñones et al. 2007; Morgan 2004). This amount may be even higher after a freeze or canopy pruning. Based on data provided by Quiñones et al. (2007), it was calculated that 6–9 months after application of inorganic N, 15–20% of the N occurs within the organic soil N pool. This N may be directly absorbed by the soil microbial community and/or organic soil compounds or returned to the soil after initial uptake via root exudates and decomposition of senescing tissues. It was proposed that more frequent N application may result in more efficient nutrient cycling (Quiñones et al. 2007).

Under Florida conditions, decomposition rates are fast and most nutrients from decomposing tree litter are recycled within the first year (Obreza and Schumann 2010; Dou and Alva 1998). Vitousek (1982) argued that application of N-rich litter material will result in rapid breakdown and that this may negatively affect overall NUE. Material with low C:N ratios would be more effective in trapping labile soil N and thereby transform this in a more stable form that is gradually released over time, and in this manner, nutrients may be used more effectively.

15.4.4 Synchronization of N Demand and Supply

15.4.4.1 Temporal Aspects

In terms of nutrient uptake and demand, two distinct scales should be considered. The first being overall tree age and the second being season uptake patterns. Although overall tree N accumulation reaches a maximum around 12–14 years (tree maturity), actual tree N demand may be highest prior to maturity due to high N demands associated with rapid canopy expansion rates (Fig. 15.6). Citrus trees have pronounced growth cycles. A period of extensive root growth is typically followed by a leaf flush, and application of nitrogen prior and during active leaf flushes can enhance tree growth. Hilgeman (1941) estimated N uptake by grapefruit in Arizona by determining changes in leaf N concentration seasonally. Maximum N uptake by the trees occurred in March and September relative to January due to higher mean soil temperature. In a 3-year study, Chapman and Parker (1942) determined N removed from solution culture and reported that the months of least N absorption were January and February.

Uptake rates increased during the period of late spring through early fall (May to October) with a maximum in July. Roy and Gardner (1946) in Florida reported similar results. In Israel, Dasberg (1987) also demonstrated that the highest rate of ^{15}N uptake by citrus trees occurred during fruit set and the lowest occurred during winter. Under Florida conditions, N requirements appear to be highest between January and June and at least 50–60% of the N is typically applied during this period (Morgan et al. 2011). Nitrogen uptake during periods when soil temperature values are 13°C vs 18°C were four and two times slower than under optimal conditions (Fig. 15.3).

Morgan et al. (2011) concluded that 75% of the vegetative biomass increase and 30% of the fruit biomass occurred during the spring. In this study, total fertilizer N application during the spring was equal to half of the total annual N application. Based on N balance calculations, it appears that insufficient N was applied in the spring and/or insufficient quantities were taken up leading to reduced leaf N concentrations. The net result would be translocation of N from storage organs (i.e., twigs, branches, and trunk), resulting in the reduced bark and wood tissue N concentrations. This underlines the critical importance of internal tree N redistribution from storage organs to ensure optimal tree performance during periods of high tree N demand. The presence of large internal nitrogen reserves and continuous canopy (conversion and storage capacity) combined with the presence of a well-developed root system (interception capacity) during the entire year for mature citrus trees may be the reason why this crop does not readily respond to split application of fertilizer. Fertilizer N applied after the spring period allowed internal N reserves to be restored as shown by a continuous increase in overall tissue N concentration in both developing tree tissues, which are presumable physiologically active, and storage tissues throughout the fall. These tissue N concentration data demonstrate the critical role of internal N storage capacity of citrus trees on soils with inherently poor soil fertility and very limited N retention capacities.

15.4.4.2 Spatial Aspects

The quantity of fibrous roots decreases with depth and lateral distance from the trunk. Over the past several decades, the use of microsprinklers has rapidly increased which in turn may confine the overall root system. Studies in Israel have shown that, with the use of fertigation, high citrus yields can be obtained even with area coverage as low as 35% (Dasberg et al. 1981). However, irrigating only 40% of the soil area resulted in a reduction of transpiration by 28% since the increased uptake from the irrigated zone could only partly compensate for reduced water uptake in the nonirrigated zone (Moreshet et al. 1983). A reduction in area coverage and total irrigation supply by 20% in citrus resulted in a yield reduction of 11–13% (Bielorai 1982). With the use of fertigation,

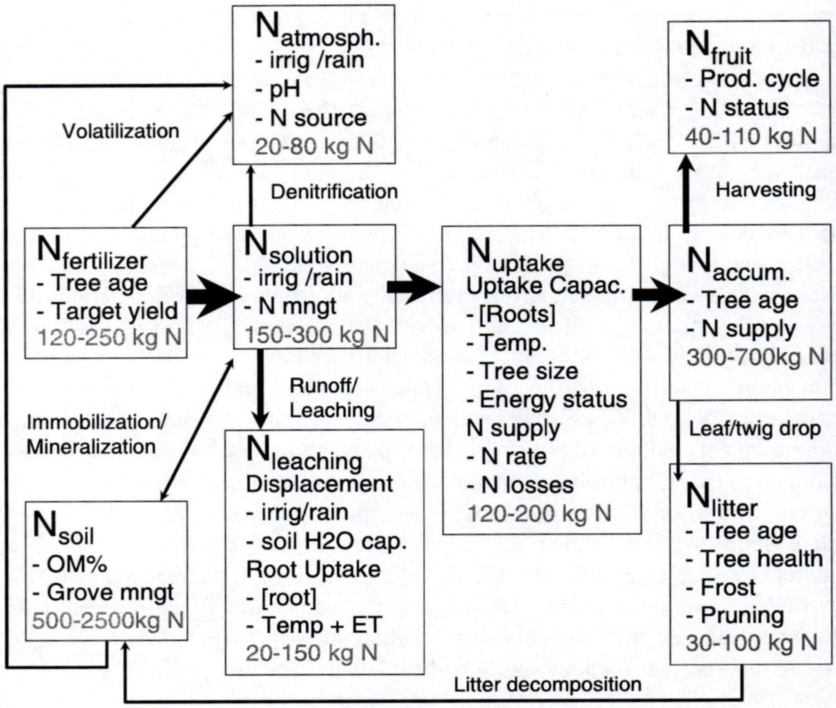

Fig. 15.7 Schematic overview of a nitrogen balance and internal N transfer processes, controlling factors, and corresponding N amounts (shown in gray and expressed as kg N/ha) for a mature citrus orchard in Florida

nitrogen uptake and fruit yield increased with increasing ground coverage (Koo 1984). These results indicate a relationship between area water and nutrient applications and tree growth and productivity.

Elezaby (1989) reported lateral fibrous root distribution to a depth of 180 cm of a 10-year-old "Valencia" tree grafted on "Volkamer" lemon grown on a soil with a deep sand profile and spaced at 4.5 × 6.0 m as follows: 9% of the fibrous roots between 0 and 60 cm from the trunk, 31% between 120 and 180 cm, and 21% between 240 and 300 cm. The vertical distribution was 42% of the fibrous roots between 0 and 30 cm from the soil surface and 14% or less at each 30 cm depth increment to 180 cm. Morgan et al. (2007) assessed spatial root distribution, and it was shown that for young trees (canopy volume <3 m^3), 68% of the roots were confined to the zone within 0.75 m from the tree trunk, and this value decreased linearly with three sizes to 21% for mature trees. Root concentration in the zone 75–125 cm from the tree trunk showed a quadratic response and increased from 32% for young trees to 40% for medium-sized trees, but declined to 28% for mature trees. The 125–175 cm zone showed an almost exponential increase over time with initial values remaining around 2–3% until trees reach medium size when values increased rapidly to 18% and reached 43% at tree maturity. In the most distant root zone, little or no root growth occurred until trees reached maximum tree size, and at this point this section accounted for 6–11% of the total root length.

15.4.5 Nutrient Budgets and Recommendations

Nitrogen balance studies in citrus can provide information on physiological tree N requirements and can be used to develop methods that minimize potential losses of N to groundwater and the atmosphere (Feigenbaum et al. 1987). Most nutrient budgets are based on N cycling diagrams and an example for citrus production in Florida is provided in Fig. 15.7. Nutrient recommendations (N_{rec}) integrate two intrinsically linked aspects, the first being the physiological crop nutrient requirement, which is the amount of nutrient that is needed to sustain optimal growth and tree performance (e.g., the crop N content at which N is no longer limiting growth or yield). This value is defined by inherent genetic traits, pedoclimatic conditions, and crop development stage but may also be prone to seasonal variation. The second component (NUE) is the overall fertilizer uptake efficiency for a specific production system which may be defined as the product of capture (including interception and uptake efficiency) and conversion efficiencies (Giller et al. 2006). The fertilizer uptake efficiency for a specific production system depends on environmental and management conditions including the application method, rate, frequency, timing, and nutrient source (Quiñones et al. 2007; Zotarelli et al. 2009).

Combining these components results in a general nitrogen recommendation as follows:

$$N_{rec} = \text{Crop N requirement} / \text{NUE} \quad (15.3)$$

Crop N requirement can be determined from overall tree N accumulation, and fruit N removal under optimal growing conditions and values may range depending on tree size, current tree N status, and target yields. As previously stated, NUE is rather complex in nature since it includes a number of crop, management, and pedoclimatic factors. Due to the dynamic and interactive nature of processes that control NUE, it is hard to ascertain, and use of decision support systems may be warranted (Morgan et al. 2006a). Although approximately 1.2 kg N is removed per ton of harvested fruit, yield response curves have shown that overall NUE under optimal management conditions may be on the order 2.2–2.6 kg N per ton of fruit (Alva et al. 2006). Differences between these values are probably related to NUE (approximately 0.6–0.75 for mature trees) and additional nutrient requirements for tree maintenance and canopy growth. Based on yield response curves for well-managed and highly productive groves, optimum rates appear to occur at 260 kg N ha^{-1}. Under Florida conditions, maximum N recommendations for mature very high yielding (>80–90 t ha^{-1}) trees thus are 280 kg N ha^{-1}, but overall recommendations are much lower and are on the order of 200 kg N ha^{-1}. At times, there are distinct differences between rootstocks. Mattos et al. (2006) reported that young orange trees grafted on rough lemon had higher NUE compared to those grafted on "Swingle" citrumelo or "Cleopatra" mandarin rootstocks. In terms of P applications, applications may only be needed when Mehlich-I soil test values drop below 30 ppm or tissue P levels drop below 0.12% (Obreza and Schumann 2010).

15.4.6 Nutrient Losses, Inefficiencies, and Environmental Emissions

It should be emphasized that N losses and potential N emission rates are inversely related to overall NUE and that efficient use of both water and nutrients is essential to minimize environmental impacts. The potential contribution of fertilizer N to the deterioration of ground water quality may be appreciable (Embleton et al. 1978). This impact is especially pronounced in Florida where the combination of high seasonal rainfall, sandy soils, and shallow water tables create conditions that favor potential groundwater contamination (Alva et al. 2006; Alva and Paramasivam 1998; Calvert and Phung 1972; Mansell et al. 1980). But nitrate pollution associated with excess fertilization of orange groves is also well-documented for other regions including Spain (Quiñones et al. 2007). For citrus, approximately 70–90% of the roots occur in the upper part of the soil profile (Morgan et al. 2007); thus, once nitrogen moves below this relatively shallow root zone, it may become a liability rather than an essential nutrient used to increase crop yield. On vulnerable soil types, optimizing NUE via improved irrigation and fertilizer management practices is thus essential to minimize environmental impacts of commercial citrus production operations. Most N balance studies have been unable to completely account for total N applied to the soil. Some authors attributed this fraction (usually 30–50%) to atmospheric loss.

Khalaf and Koo (1983) concluded that unaccounted for N was either incorporated into soil organic matter or stored in the tree and was supported by Dasberg (1987). Others made no attempt to fully account for the applied N (Mansell et al. 1980). In terms of estimated leaching losses under Florida field conditions, based on simulation with LEACHM model, unaccounted for N was estimated to be on the order of 10–35 kg N ha^{-1} which translates to 5–16% of the applied N (Paramasivam et al. 2001). Depending on weather conditions and irrigation rates, net annual recharge under Florida conditions are on the order of 386–710 mm (Alva et al. 2006). For each 10 kg N/ha, leached concentrations in recently added groundwater may be increased by 1.4–2.6 mg NO_3–N L^{-1}.

15.5 Designing High NUE Systems for Citrus

To improve yield and profits while minimizing the risk of potential environmental impacts, growers need to develop rational fertilization management practices that are based on improved understanding of underlying processes and close monitoring of system performance (Quiñones et al. 2003; Morgan et al. 2006a; Obreza and Schumann 2010). As discussed in previous sections, there is a multitude of interacting factors governing overall NUE and use of integrated measure is thus needed. In this context, we can differentiate between development of appropriate strategic and tactical management. The former are related to long-term design and overall orchard management. The second pertain to in-season resource management practices including close monitoring of system performance as related to overall production goals.

15.5.1 Strategic Measures

15.5.1.1 Orchard Infrastructure and Design

Use of Fertigation

Injection of the fertilizer in the irrigation system (fertigation) can be an efficient way of supplying some or all the nitrogen to the crop (Gardenas et al. 2005). Quiñones et al. (2007)

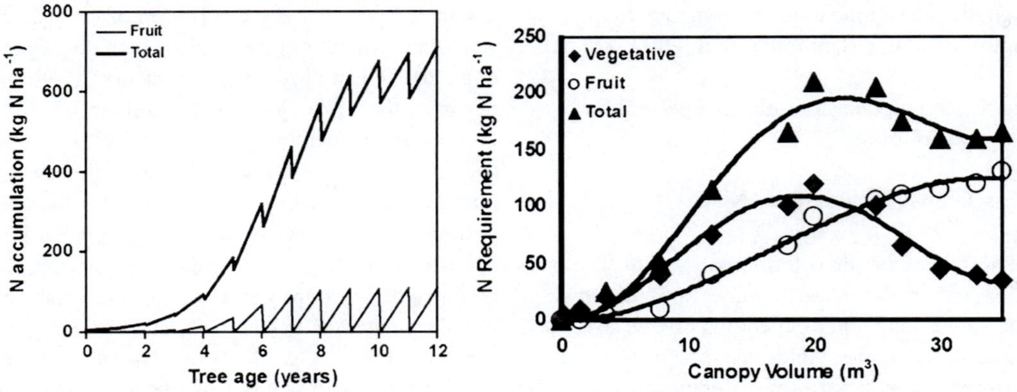

Fig. 15.8 Conceptual model for nitrogen accumulation in fruits and the entire tree as a function of tree age (*left figure*) and calculated tree N requirements and N allocation as related to tree size based on actual field measurements (*right figure*, Adapted from Morgan et al. 2006)

demonstrated that NUE with high-frequency N fertigation applied via drip irrigation was 75% vs 63% for low-frequency application combined with flood irrigation. However, fertigation works best with well-maintained and well-designed irrigation systems that will allow a relatively even wetting of at least 30% of the land area. In order to reduce potential N-leaching losses, fertilizer should be injected toward the end of the irrigation cycle (Gardenas et al. 2005) while allowing enough time for flushing of the emitters at the end of the line (Scholberg et al. 2002). The fraction of the fertilizer N that is taken up by the crop typically increases with more frequent fertigation. Benefits of more frequent application are greatest for young trees and with lower N rates. Citrus less than 5 years of age had significantly greater canopy volume and fruit yields with as little as 50% of recommended soluble N rates with controlled release fertilizer and higher frequency fertigation (Obreza and Rouse 1993). At high N rates, effects of frequency on tree growth and fruit yields may be small and/or inconsistent.

15.5.1.2 Target-Oriented Fertilization

Historic yield and tissue test results may be used to develop realistic target yields. Based on overall nutrient balances including nutrient export associated with target fruit yield and nutrient required for sustaining inherent tree nutrient reserves at optimal physiological levels, groove-specific nutrient recommendations may be developed. Linking tree growth analysis (Morgan et al. 2006c), fruit yield curves as a function of tree size (Fig. 15.1), with fruit N removal allows development of curves that show physiological tree N requirement as a function of tree canopy volume (Fig. 15.8). Based on this figure, it may be argued that N fertilizer supply should be greatest prior to maturity since at that time still substantial amount of N is required to build up the tree canopy and internal reserves.

15.5.1.3 Site-Specific Management

Groves that have a high percentage resets may benefit from site-specific N rates, and application of controlled-release fertilizers to the younger trees can prevent excessive N leaching. Solid fertilizer spreaders may be adjusted in such a manner that dry fertilizer materials are only applied to the rooted zone and not to the row middles. Remote sensing and variable rate technology may be used to apply fertilizer rate based on tree size to avoid overapplication of recently planted young trees that replaced an old tree (resets), and in this manner fertilizer use may be reduced by up to 40% (Mann et al. 2010). Satellite images may be used to assess variability in tree canopy use distributions, and combined with site-specific yield maps, this may be used to identify "high" vs "low" yield field sections that may be fertilized based on their actual yield potential (Mann et al. 2010). Other issues of concern are frequent monitoring of the uniformity of irrigation sprinkler output combined with frequent servicing of irrigation sprinklers, use of appropriate irrigation management zone based on uniform soil properties and tree crop water demands, and use of reduced volume emitters for recently replaced trees and to develop irrigation.

15.5.1.4 Frequent Use of Soil Organic Amendments

Florida sandy soils, which prevail in most Florida orange groves, have inherently low water and nutrient retention capacities (Morgan et al. 2007) and are prone to excess N leaching (Alva et al. 2006). There is a close relationship between soil organic matter (SOM) content and soil water retention. Hudson (1994) showed that as SOM for a sandy soil increased from 1% to 2% soil water storage increases by 37%. However, under Florida conditions, the breakdown of SOM is relatively rapid and annual application of large amounts (10–20 t ha^{-1}) may be required to increase SOM levels to 2–3%. In case of limited supply of organic materials, use of a sandwich system approach by which organic amendments are only applied to the irrigated section of the tree row may be an effective measure to make optimal use of available resources, especially if such system is combined with a perennial leguminous row middle cover (Scholberg et al. 2010).

15.5.2 Tactical Measure

15.5.2.1 Nutrient Sources

Especially with young trees, use of enhanced efficiency fertilizers (EEFs) may enhance NUE, and relatively low tree N demands combined with labor savings may render such measure cost effective. Although for mature trees, use of EEFs may be hampered due to their higher costs, potential leaching losses may be greatly reduced, while in five out of eight studies similar or better tree growth and yield could be attained with the use of EEFs (Obreza and Sartain 2010).

15.5.2.2 Soil and Tissue Testing

Under Florida conditions, soil analysis may include annual determination of soil organic matter content, pH, extractable P, calcium, and magnesium (Obreza and Schumann 2010). In other regions where historic fertilizer application are low and soil pH values may be less favorable, soil testing may also include other macro- and micronutrients. Under (sub)tropical conditions, most of the nitrogen is converted to the nitrate form (Obreza and Sartain 2010). Sato et al. (2009) suggested rapid ammonification of urea, and subsequent N transformation in sandy soils lead to increase NUE. Nitrification in sandy soils can occur is less than 2 weeks (Sato et al. 2009). Therefore, especially on soils with low inherent soil fertility, soil nitrogen concentrations typically drop to values below 5–10 ppm N within 1–2 months after fertilizer application (McNeal et al. 1994).

Soil testing for nitrogen is therefore not done on a routine basis in Florida, although it may be used in more arid regions including Spain (Legaz and Primo Millo 1988). Although soil testing provides a baseline, discrepancy between total soil nutrient content, nutrient availability, and actual uptake may limit the interpretation of results. For soil test results to be meaningful, use of appropriate soil extraction methods based on calibrated soil test are needed. However, for many crops and regions, such calibrated test and suitable extraction approaches and lab facilities may be lacking and the sampling is cumbersome. Therefore, use of leaf tissue testing may be preferable, and soil testing may be confined to assess lime requirements and/or to monitor excessive build up of soil nutrients (Quaggio et al. 1998). Moreover, specific soil conditions (excessively wet, dry or cold soils or poor root health due to poor drainage and/or infestation by pests and diseases) may still hamper effective uptake regardless of soil nutrient levels. Moreover, in the case of labile nutrients (e.g., nitrogen and potassium), current soil status may be poorly correlated with future tree uptake (Koo 1962). Therefore, use of soil testing should be complemented with frequent monitoring of diagnostic tissue sampling of leaves and/or fruits which can be used to "fine-tune" fertilizer recommendations (Obreza and Schumann 2010). Based on 6 years of field data, Alva et al. (2006) showed that for optimal fruit yield, diagnostic leaf N concentrations should be 2.3–2.8%, while corresponding values for K were 1.4–1.9% K. Quaggio et al. (1998) showed that relative yield in Brazil were highest when leaf tissue values were around 2.8% N, although overall agronomic NUE decreased as leaf N tissue values increased. It was also noted that resin extractable P and exchangeable K provided effective basis for fertilizer recommendations (Quaggio et al. 1998).

15.5.2.3 Improved Synchronization

The lack of synchrony between root distribution and fertilizer application will also negatively impact overall NUE values. Due to distinct rainy seasons in Florida, roots of trees at commercial spacing rapidly occupy the volume of soil outside the irrigated zone. However, since microsprinklers typically irrigate only 25–60% of the total ground area, broadcasting dry fertilizer to the entire production area will hamper uptake in nonirrigated soil areas during periods of low rainfall. Moreover, based on overall root distribution data of citrus trees over time, it is apparent that broadcasting fertilizer to young trees will result in a relatively small fraction of the fertilizer being intercepted be tree roots. Recent technical advances with custom-designed EEFs combined with improved assessment of release patterns allow better synchronization between nutrient supply and tree N demand. In the case of young trees, EEFs can be applied at planting, whereas for older trees nutrients EEFs should be applied during early spring before the initial spring flush and bloom (Obreza and Sartain 2010). Excessively high nitrogen applications prior to or during cold spells therefore are not advisable, especially since it may coincide with frost protection measures. In this case, the application of large volumes of water is required which will be leaching most of the nitrogen before the tree can take it up. Due to leaching rainfall and reduced crop nitrogen demand, nitrogen applications during the summer should also be low (Obreza and Schumann 2010; Scholberg et al. 2002).

15.5.2.4 Improved Irrigation Scheduling

Proper irrigation scheduling based on soil water storage capacity (as defined by soil characteristics and the effective root zone) and tree crop water demand (as related to tree size and environmental conditions) can greatly reduce the risk of N displacement below the effective root zone (Obreza and Schumann 2010). For practical purposes, combination of water budget/models (e.g., ET-based systems) can be used to assess total amount of irrigation that needs to be applied, whereas frequent monitoring of soil moisture (e.g., via commercially available time-domain reflectometry sensors) allows improvement of timing and/or frequency of irrigation application (Morgan and Hanlon 2007). Effective use of such measures should reduce excessive fluctuations in soil moisture levels, thereby reducing tree water stress and N leaching losses while facilitating more efficient use of both water and nutrients (Obreza and Schumann 2010). In terms of irrigation scheduling, the integration of current real-time soil moisture monitoring techniques with both short (e.g., based on radar images) and long-term (based on El Nino oscillations) expected precipitation amounts may be used to modify timing of N application and to develop deficit regimes that can reduce the risk of rainfall-induced N displacement.

15.5.2.5 Row Middle Management and Weed Control

Root distribution studies with mature navel and "Valencia" orange groves in California showed that yield was not related to the under-canopy root quantities but was correlated with the root quantities measured between adjacent rows where soil water contents were typically lower most of the year (Cahoon et al. 1956). Removal of grasses and/or weeds will result in lower annual ET_c values by 31–51% (Smajstrla et al. 1986), whereas Stewart et al. (1969) estimated annual bare soil evaporation and 2/3 sod cover ET_c to be 68% and 92% of full sod cover, respectively. Similarly, row middles may also compete with citrus trees for nutrient uptake.

A study with monocropped vs intercropped young citrus trees, Bermuda grass and perennial peanut showed that N uptake of grasses, especially during early spring could be double or more than that of young citrus trees, whereas perennial peanut had relatively low N uptake and only competed for water. Overall NUE was 47% for citrus systems compared to 84% for citrus system with perennial peanut as groundcover and 97% for citrus with Bermuda grass system. Although from a NUE perspective the last system seems to be ideal, young citrus trees have rather shallow and still poorly developed root systems (Morgan et al. 2007) and grasses due to their much higher growth rates can easily outcompete citrus trees. In organic managed orchards, applying organic soil amendments only to the tree row combined with integrated use of prostrate leguminous cover crops and/or applying mulches near young the trees may thus provide options for enhancing overall NUE (Scholberg et al. 2010).

15.5.3 Modeling Approaches

As outlined in the previous sections, processes that control NUE are complex, but development of water and nutrient budgets can provide a key component in enhancing overall NUE. Mathematical models may be used to assess the interactive effects of management practices and pedoclimatic conditions on crop water and N fertilizer requirements, soil moisture dynamics, crop water and N uptake, and potential nitrate leaching. Traditionally, there have been two classes of models, the first focuses to a larger extent on soil-based processes while the second on crop growth and yield. Increased environmental concerns pertaining to emissions associated with agricultural production has resulted in development and use of a large number of models to predict contaminant behavior in agricultural systems. Examples include the Nitrate Leaching and Economic Analysis Package (N-LEAP) (Follett et al. 1994), Groundwater Loading Effects of Agricultural Management Systems (GLEAMS) (Reck 1994; Reyes et al. 1994), and Leaching Estimation and Chemical Model (LEACHM) (Jemison et al. 1994). In terms of crop growth models, these include the Crop-Environment Resource Synthesis (CERES) developed to model growth and yield of grain crops (Jones and Kiniry 1987; Kiniry et al. 1997; Kiniry and Bockholt 1998; Saseendran et al. 1998; Lizaso et al. 2001), which formed the basis for the later developed APSIM model. The CROPGRO model was initially developed as a family of crop-specific models for the prediction of legume and vegetable crops but has been integrated into the Decision Support System for Agrotechnology Transfer (DSSAT) (Jones et al. 1991; Hoogenboom et al. 1994; Wagner-Riddle et al. 1997). These models are process-oriented and used to predict dry matter growth (Shen et al. 1998), crop development (Batchelor et al. 1994, 1997; Piper et al. 1996), and final yield (Batchelor et al. 1996; Heinemann et al. 2000) for a range of agronomic crops.

However, few models have been developed specifically for use in citrus production. Most models have been designed for specific applications including the prediction of potential crop damage caused by citrus pathogens (Timmer and Zitko 1996) or scale insects (Arias-Reveron and Browning 1995). Other citrus models have been used for irrigation scheduling (Xin et al. 1997), and predicting time of flowering (Bellows and Morse 1986). During the past years, there is increased interest in the use of computer models as a scientific basis for more efficient resource use (Morgan et al. 2003, 2006a). One of the reasons for developing a new DSS for citrus was that most of the existing models including DSSAT and APSIM were exclusively developed for annual crops and/or perennial grasses, whereas most tree models are mainly designed for assessing ecological processes and are ill-suited for resource management. Another issue was that models such as DSSAT originally were designed for simple agronomic systems and cannot capture more complex, spatially differentiated irrigation and nutrient application and uptake patterns. Adaptations of such models for citrus in their original form thus posed serious constraints (Morgan et al. 2003).

Scholberg et al. (2002) provided an initial conceptual framework for upscaling nutrient uptake kinetics from seedling to a field scale (Scholberg et al. 2002). Although this conceptual approach was rather novel and provided a scientific basis for development of decision tools for more efficient use of both water and nutrients for citrus, validation of such models under field conditions was essential. This work included development of functional root growth equations for mature trees based on detailed root growth assessment (Morgan et al. 2007) and improved assessment of biomass and nutrient accumulation as affected by tree size (Morgan et al. 2006c). These studies formed building blocks for the actual development and employment of a decision support system (DSS) for more efficient water and nutrient use in commercial citrus systems, and approaches used were outlined by Morgan et al. (2006a).

15.6 Future Research Recommendations

Based on the information provided in the previous sections, it is obvious that the processes underlying overall NUE are complex, especially for citrus systems. Improved integration of crop and soil processes at the field, farm, and watershed scale is essential to enhance our understanding of processes that control NUE. Since farmers will be required to provide a much broader array of ecological services, while traditional agricultural research approaches fail to address complex societal issues, a more interdisciplinary and systems-based approach is needed. Development of appropriate models will allow effective integration of interactive processes at different levels of aggregation and complexity. Such models may be used by scientists for *Ex ante* exploration of different management practices and scenarios and, in this manner, assess the impact of technical interventions at the farm level on targeted environmental quality standards at the watershed level (e.g., maximum daily loads and groundwater quality standards) in a transparent manner. These models may also be used to screen potential innovations and to develop "basket of technologies" that are most likely to generate required future agroecological services associated with different pedoclimatic settings and/or development scenarios. In this manner, viable technology niches and future research priorities may be explored more efficiently.

Currently, the lack of integration of private enterprises and farmers during the testing and development of new technologies by researchers hampers efficient transfer and adaptation of technical innovations. Most producers fail to accept scientific results and demand large-scale field demonstrations, while scientists are disconnected from farmers' realities and are forced to focus on imposed publication targets instead. These diverging mandates create inefficiencies in terms of use and sharing of resources, experience, and information. This issue needs to be addressed to ensure more effective involvement of end users during technological development, which may be facilitated by using a co-innovation approach. This requires a sharing of ownership and responsibilities during the innovation and technology transfer processes.

Growers and commercial companies thus should be actively involved during the design, development, and testing of decision support systems for enhancing NUE to ensure that these tools reflect management priorities and information management skills of end users.

References

Adriance CW, Hampton JE (1949) Root distribution in citrus as influenced by environment. Proc Am Soc Hortic Sci 53:103–108

Alva AK, Paramasivam S (1998) Nitrogen management for high yield and quality of citrus in sandy soils. Soil Sci Soc Am J 62: 1335–1342

Alva AK, Paramasivam S, Graham WD (1998) Nitrogen management for high yield and quality of citrus in sandy soils. Soil Sci Soc Am J 62:1335–1342

Alva AK, Paramasivam S, Obreza TA et al (2006) Nitrogen best management practice for citrus trees I. Fruit yield, quality, and leaf nutritional status. Sci Hortic 107:233–244

Arias-Reveron JM, Browning HW (1995) Development and mortality of citrus snow scale under constant temperature and relative humidity. Environ Entomol 24:1189–1195

Aslam M, Nielson K, Travis RL et al (1997) Nitrate uptake, efflux, and in vivo reduction by Pima and Acala cotton cultivars. Crop Sci 37:1795–1801

Bar-Yosef B, Sagiv B (1982) Response of tomato to N and water applied via a trickle irrigation system. I nitrogen. Agron J 74:633–637

Batchelor WD, Jones JW, Boote KJ et al (1994) Carbon-based model to predict peanut pod detachment. Trans ASAE 37:1639–1646

Batchelor WD, Jones JW, Boote KJ (1996) Comparisions of methods to compute peanut seed size distribution by crop growth models. Trans ASAE 39:737–744

Batchelor WD, Zeiss MR, Pedigo LP et al (1997) Development of a model to predict soybean pod color distribution. Trans ASAE 40:221–227

Bellows TS, Morse JG (1986) Modeling flower development in navel oranges. Sci Hortic 30:117–126

Bevington KB, Castle WS (1985) Annual root growth pattern of young citrus trees in relation to shoot growth, soil temperature, and soil water content. J Am Soc Hortic Sci 110:840–845

Bielorai H (1982) The effects of partial wetting of the root zone on yield and water use efficiency in a drip- and sprinkler-irrigated mature grapefruit grove. Irrig Sci 3:89–100

Bloom AH (1996) Nitrogen dynamics in plant systems. Life Support Biosph Sci 3:35–41

Boaretto RM, Mattos D, Quaggio JA et al (2010) Nitrogen-15 uptake and distribution in two citrus species. In: World congress of soil science, soil solutions for a changing world, Brisbane, Australia, pp 156–159

Boman BJ (1996) Fertigation versus conventional fertilization of flatwoods grapefruits. Fertil Res 44:123–128

Boswell SB, McCarty CD, Lewis LN (1975) Tree density affects large-root distribution of 'Washington' Navel orange trees. Hortic Sci 10:593–595

Cahoon GA, Harding RB, Miller DB (1956) Declining citrus root systems. Calif Agric 10:3–12

Cahoon GA, Morton ES, Jones WW et al (1959) Effects of various types of nitrogen fertilizers on root density and distribution as related to water infiltration and fruit yields of Washington Navel oranges in a long-term fertilizer experiment. Proc Am Soc Hortic Sci 74:289–299

Cahoon GA, Stolzy LH, Mortan ES (1961) Effect of mulching on soil moisture depletion in citrus orchards. Soil Sci 92:202–206

Cahoon GA, Stolzy LH, Garber MJ et al (1964) Influence of nitrogen and water on the root density of mature Washington Navel orange trees. Proc Am Soc Hortic Sci 85:224–231

Calvert DV, Phung HT (1972) Nitrate-nitrogen movement into drainage lines under different soil management systems. Soil Crop Sci Soc Fla Proc 31:229–232

Calvert DV, Koo RCJ, Ford HW (1967) Flood irrigation studies with citrus. Proc Fla State Hortic Soc 80:79–85

Calvert DV, Ford HW, Stewart EH et al (1977) Growth response of twelve citrus rootstock-scion combinations on a Spodosol modified by deep tillage and profile drainage. Proc Int Soc Citricult 1:79–84

Cameron SH, Appleman D (1933) The distribution of total nitrogen in the orange tree. Proc Am Soc Hortic Sci 30:341–348

Cameron SH, Appleman D (1935) The distribution of total nitrogen in the orange tree. Am Soc Hortic Sci 30:341–348

Cameron SH, Compton OC (1945) Nitrogen in bearing orange trees. J Am Soc Hortic Sci 46:60–68

Castle WS (1978) Citrus root systems: their structure, function, growth, and relationship to tree performance. Proc Int Soc Citricult 1:62–69

Castle WS, Krezdorn AH (1975) Effect of citrus rootstocks on root distribution and leaf mineral content of 'Orlando' tangelo tree. J Am Soc Hortic Sci 100:1–4

Cerezo M, Gracia-Agustin P, Serna D et al (1997) Kinetics of nitrate uptake by citrus seedlings and inhibitory effects of salinity. Plant Sci 126:105–112

Chapman HD, Liebig GF (1941) Nitrate concentration and ion balance in relation to citrus nutrition. Hilgardia 13:141–173

Chapman HD, Parker ER (1942) Weekly absorption of nitrate by young, bearing orange trees growing out-of –doors in solution culture. Plant Phys 17:366–376

Claasen N, Barber SA (1974) A method for characterizing the relation between nutrient concentration and flux into roots of intact plants. Plant Phys 54:564–568

Clarkson DT (1985) Factors affecting mineral nutrient acquisition by plants. Ann Rev Plant Physiol 36:77–115

Clarkson DT (1986) Regulation of the absorption and release of nitrate by plant cells: a review of current ideas and methodology. In: Lambers H, Neeteson JJ, Stulen I (eds) Fundamental, ecological, and agricultural aspects of nitrogen metabolism in higher plants. Martin Nijhof Publishers, Dordrecht, pp 3–27

Clarkson DT, Warner AJ (1979) Relationships between root temperature and transport of ammonium and nitrate ions by Italian and perennial ryegrass (*Lolium multiflorum* and *Lolium perenne*). Plant Phys 64:557–561

Dasberg S (1987) Nitrogen fertilization in citrus orchards. Plant Soil 100:1–9

Dasberg S (1992) Irrigation management and citrus production. In: Tribulato E, Gentile A, Refrigerato G (eds) Proceedings of the international society of citriculture, Catnia, Italy, 8–13 Mar, vol 3, pp 1307–1310

Dasberg S, Bielorai H, Erner Y (1981) Partial wetting of the root zone and nitrogen effects on growth, yield and quality of Shamouti oranges. Acta Hortic 119:103–108

Dasberg S, Erner Y, Bielorai H (1984) Nitrogen balance in a citrus orchard. J Environ Qual 13:353–356

Dasberg S, Bar-Akiva SA et al (1988) Fertigation versus broadcasting in an orange grove. Fertil Res 15:147–154

Delhon P, Gojon A, Tillard P et al (1996) Diurnal regulation of NO_3^- uptake in soybean plants. IV. Dependence on current photosynthesis and sugar availability to the roots. J Exp Bot 47:893–900

Dou H, Alva AK (1998) Nitrogen uptake and growth of two citrus rootstock seedlings in a sandy soil receiving different controlled-release fertilizer sources. Biol Fertil Soils 26:169–172

Eissenstat DM (1991) On the relationship between specific root length and the rate of root proliferation: a field study using citrus rootstocks. New Phytol 118:63–68

Eissenstat EM, Anchor DS (1999) Anatomical characteristics of roots of citrus rootstocks that vary in specific root length. New Phytol 141:309–321

Eissenstat DM, Duncan LW (1992) Root growth and carbohydrate responses in bearing citrus trees following partial canopy removal. Tree Physiol 10:245–257

Eissenstat DM, Volder A (2005) The efficiency of nutrient acquisition over the life of a root. In: Bassirrad H (ed) Nutrient acquisition by plants an ecological perspective. Springer, Berlin, pp 185–220

Elezaby AA (1989) Physiological and biological studies on root systems of some citrus stocks. PhD dissertation, Cairo University, Giza

Embleton JW, Mastumura M, Stolzy LH et al (1978) Citrus nitrogen fertilizer management, groundwater pollution, soil salinity and nitrogen balance. Appl Agric Res 1(1):57–64

Feigenbaum S, Bielorai H, Erner Y et al (1987) The fate of ^{15}N labeled nitrogen applied to mature citrus trees. Plant Soil 97:179–187

Ferrier RC, Alexander IJ (1991) Internal redistribution of N in Sitka spruce seedlings with partly droughted root systems. For Sci 3: 860–870

Florida Agricultural Statistics Service (2010) Florida citrus statistics 2008–2009. Florida Department of Agriculture and Consumer Services, Tallahassee, 108 pp

Follett RF, Shaffer MJ, Brodahl MK et al (1994) NLEAP simulation of residual soil nitrate for irrigated and non-irrigated corn. J Soil Water Conserv 49(4):375–382

Ford HW (1952) The distribution of feeder roots of orange and grapefruit trees on rough lemon rootstock. Citrus Indus Mag 14(11): 22–23

Ford HW (1953) Root distribution of chlorotic and iron-chelate-treated citrus trees. Proc Fla State Hortic Soc 66:22–26

Ford HW (1954) Root distribution in relation to the water table. Proc Fla State Hortic Soc 67:30–33

Ford HW (1972) Eight years of root injury from water table fluctuations. Proc Fla State Hortic Sci 85:65–68

Ford HW, Reuther W, Smith PF (1957) Effect of nitrogen on root development of 'Valencia' orange trees. Proc Am Soc Hortic Sci 70:237–244

Gardenas AI, Hopmans JW, Hanson BR et al (2005) Two-dimensional modelling of nitrate leaching for various fertigation scenarios under micro-irrigation. Agric Water Manage 74:219–242

Giller KE, Rowe EC, de Ridder N et al (2006) Resource use dynamics and interactions in the tropics: scaling up in space and time. Agric Syst 88:8–27

Giller KE, Leeuwis C, Andersson JA et al (2008) Competing claims on natural resources: what role for science? Ecol Soc 13:34–51

Goyal SS, Huffaker RC (1986) A novel approach and a fully automated microcomputer-based system to study kinetics of NO_3^-, NO_2^-, and NH_4^+ transport simultaneously by intact wheat seedlings. Plant Cell Environ 9:209–215

Graham JH, Eissenstat DM (1998) Field evidence for the carbon cost of citrus mycorrhizas. New Phytol 140:103–110

Hassan MM, Van Hai T (1976) Ammonium and potassium uptake by citrus roots. Physiol Plant 36:20–22

Heinemann AB, Hoogenboom G, Georgiev GA et al (2000) Center pivot irrigation management optimization of dry beans in humid areas. Trans ASAE 43(6):1507–1516

Heins B, Schenk M (1986) Nitrate-uptake characteristics of roots as affected by nitrate supply. In: Lambers H, Neeteson JJ, Stulen I (eds) Fundamental, ecological, and agricultural aspects of nitrogen metabolism in higher plants. Martin Nijhof Publishers, Dordrecht, pp 41–45

Hikosaka K (2004) Interspecific differences in the photosynthesis-nitrogen relationships: patterns, physiological causes, and ecological importance. J Plant Res 117:481–494

Hilgeman RH (1941) Nitrogen uptake by grapefruit trees in the Salt River Valley. Proc Am Soc Hortic Sci 39:119–124

Hillel D (1971) Soil and water physical principles and processes. Academic, New York

Hoogenboom G, White JW, Jones JW et al (1994) BEANGRO: a process oriented dry bean model with a versatile user interface. Agron J 86:182–190

Hudson BD (1994) Soil organic matter and available water capacity. J Soil Water Conserv 49:189–194

Ingestad T (1982) Relative addition rate and external concentration; driving variables used in plant nutrition research. Plant Cell Environ 5:443–453

Jemison JM, Jabro JD, Fox RH (1994) Evaluation of LEACHM: II. Simulation of nitrate leaching from nitrogen-fertilized and manured corn. Agron J 86:852–859

Jones CA, Kiniry JR (1987) CERES-MAIZE: a simulation model of maize growth and development. Texas A&M University Press, College Station

Jones JW, Dayan E, Allen LH et al (1991) A dynamic tomato growth and yield model (TOMGRO). Trans ASAE 34(2):663–672

Kato T (1986) Nitrogen metabolism and utilization in citrus. Hortic Rev 8:181–216

Kato TS, Kubota S, Bambang S (1982) Uptake of ^{15}N-itrogen by citrus trees in winter and repartitioning in spring. J Jpn Soc Hortic Sci 50:421–426

Kato T, Makoto Y, Tsukahara S (1984) Storage forms and reservoirs of nitrogen used for new shoot development in Satsuma mandarin trees. J Jpn Soc Hortic Sci 52:393–398

Kaufmann MR, Boswell SB, Lewis LN (1972) Effect of tree spacing on root distribution of 9-year-old 'Washington' Navel oranges. J Am Soc Hortic Sci 97:204–206

Keltjens WG, Nijenstein JH (1987) Diurnal variations in uptake, transport and assimilation of NO_3^- and efflux of OH^- in maize plants. J Plant Nutr 10:887–900

Khalaf HA, Koo RCJ (1983) The use of controlled-release nitrogen on container grown citrus seedlings. Citrus Veg Mag 46:10–32

Kimball MH, Wallace A, Mueller RT (1950) Changes in soil and citrus root characteristics with non-tillage. Calif Citrog 35:432–433

Kiniry JR, Bockholt AJ (1998) Maize and sorghum simulation in diverse Texas environments. Agron J 90:682–687

Kiniry JR, Williams JR, Vanderlip RL et al (1997) Evaluation of two maize models for nine U.S. locations. Agron J 89:421–426

Koo RCJ (1962) The use of leaf, fruit, and soil analysis in estimating potassium status of orange trees. Proc Fla State Hortic Soc 75:67–72

Koo RCJ (1963) Effects of frequency of irrigation on yield of orange and grapefruit. Proc Fla State Hortic Soc 74:1–5

Koo RCJ (1984) The importance of ground coverages by fertigation for citrus on sandy soils. Fertil Issue 1:75–78

Koo RCJ, Sites JW (1955) Results of research and response of citrus to supplemental irrigation. Proc Soil Sci Soc Fla 15:180–190

Kosola KR, Eissenstat DM (1994) The fate of surface roots of citrus in dry soil. J Exp Bot 45:1639–1645

Kramer PJ, Boyer JS (1995) Water relations of plant and soils. Academic, San Diego

Larsons CM, Mattsson M, Duarte P et al (1992) In: Mengel K, Pilbeam DJ (eds) Nitrogen metabolism of plants. Claradon Press, Oxford, pp 71–90

Le Bot J, Kirkby EA (1992) Diurnal effects of nitrate and potassium during the vegetative growth of tomato plants. J Plant Nutr 15: 247–264

Lea-Cox JD, Syvertsen JP (1993) Salinity reduces water use and nitrate-N-use efficiency of citrus. Ann Bot 72:47–54

Lea-Cox JD, Syvertsen JP (1996) How nitrogen supply affects growth and nitrogen uptake, use efficiency, and loss from citrus seedlings. J Am Soc Hortic Sci 121:105–114

Lea-Cox JD, Syvertsen JP, Graetz DA (2001) Springtime ^{15}nitrogen uptake, portioning and leaching losses from young bearing citrus trees of differing nitrogen status. J Am Soc Hortic Sci 126:242–251

Legaz F, Primo-Millo E (1988) Absorption and distribution of nitrogen-15 applied to young orange trees. In: Goren R, Mendel K (eds) Proceedings of the international society of citriculture, Rehovot, Israel, pp 643–661

Legaz F, Primo-Millo E, Primo-Yufera E et al (1982) Nitrogen fertilization in citrus, I: absorption and distribution of nitrogen in Calamondin trees (*C mitis* Bl.) during flowering, fruit set, and initial fruit development periods. Plant Soil 66:339–351

Lenz F (2000) Effects of fruit load on the nutrient uptake and distribution in citrus trees. Acta Hortic 531:115–120

Linares JC, Scholberg JMS, Graetz D et al (2010) Effects of perennial peanut and common bermudagrass on nitrogen and water uptake of young citrus trees. J Plant Nutr 33:200–218

Lizaso JI, Batchelor WD, Adams SS (2001) Alternate approach to improved kernel number calculation in CEREX-MAIZE. Trans ASAE 44:1001–1018

MacDuff JH, Jackson SB (1991) Growth and preference for ammonium or nitrate uptake by barley in relation to root temperature. J Exp Bot 42(237):521–530

Maeck G, Tischner R (1986) Nitrate uptake and reduction in sugar-beet seedlings. In: Lambers H, Neeteson JJ, Stulen I (eds) Fundamental, ecological, and agricultural aspects of nitrogen metabolism in higher plants. Martin Nijhof Publishers, Dordrecht, pp 33–36

Mann KK , Schumann AW, Obreza TA (2010) Delineating productivity zones in a citrus grove using citrus production, tree growth and temporally stable soil data precision. Agric doi: 10.1007/s11119-010-9189-y, Springer Science+Business Media, LLC 2010

Mansell RS, Selim HM, Calvert DV et al (1980) Nitrogen and water distribution in fertilized sandy soil during irrigation and drainage. Soil Sci Soc Am J 44:95–102

Marschner H, Hausling M, George E (1991) Ammonium and nitrate uptake rates and rhizosphere pH in non-mycorrhizal roots of Norway spruce [*Picea abies* (L.) Karst.]. Trees 5:14–21

Mattos D (2000) Citrus response functions to N, P, and K fertilization. PhD dissertation, University of Florida, Gainesville

Mattos D, Gaetz DA, Alva AK (2003) Biomass distribution and nitrogen-15 partitioning in citrus trees on a sandy Entisol. Soil Sci Soc Am J 67:555–563

Mattos D, Quagio JA, Cantarella H et al (2006) Response of young citrus trees on selected rootstocks to nitrogen, phosphorus, and potassium fertilization. J Plant Nutr 29:1371–1385

Maust BE, Williamson JG (1994) Nitrogen nutrition of containerized citrus nursery plants. J Am Soc Hortic Sci 119:195–201

McNeal BL, Stanley CD, Espinoza LE et al (1994) Nitrogen management for vegetables and citrus: some environmental considerations. Soil Crop Sci Soc Fla Proc 53:44–51

Mengel K (1992) Nitrogen: agricultural productivity and environmental problems. In: Mengel K, Pilbeam DJ (eds) Nitrogen metabolism of plants. Claradon Press, Oxford, pp 1–15

Menocal-Barberena OA (2000) Effect of rootstock on root distribution of citrus. MS thesis, University of Florida, Gainesville

Mikhail EH, El-Zeflawi BM (1978) Effect of soil types and rootstocks on root distribution and leaf composition of citrus trees. In: Cary PR (ed) Proceedings of the international society of citricult, New South Wales, Australia, vol 1, pp 214–216

Millard P (1995) Internal cycling of nitrogen in trees. Acta Hortic 383:3–14

Mills TM, Morgan KT, Parsons LR (1999) Canopy position and leaf age affect stomatal response and water use in citrus. J Crop Prod 2(2):163–179

Mooney PA, Richardson AC (1992) Seasonal trends in the uptake and distribution of nitrogen in Satsuma mandarins. In: Tribulato E, Gentile A, Refrigerato G (eds) Proceedings of the international society of citriculture, Catnia, Italy, 8–13 Mar, vol 1, pp 593–597

Moreshet S, Cohen Y, Fuchs M (1983) Response of mature "Shamouti" orange trees to irrigation of different soil volumes at similar levels of available water. Irrig Sci 3:223–236

Morgan KT (2004) Nitrogen and biomass distribution, and nitrogen and water uptake parameters for citrus. PhD dissertation, Soil and Water Science Department, University of Florida, Gainesville, pp 174

Morgan KT, Hanlon EA (2007) Improving citrus nitrogen uptake efficiency: effective irrigation scheduling. Fact Sheet SL-247. Soil Water Science Department, Florida Cooperative Extension Service, Institute of Food and Agricultural Sciences, University of Florida, Gainesville

Morgan KT, Jones JW, Scholberg JMS et al (2003) Development of TREEGRO: a fruit tree model for DSSAT. Paper #033071. ASAE Technical Library

Morgan KT, Beck HW, Scholberg JMS et al (2006a) In-season irrigation and nutrient decision support system for citrus production. In: Proceedings of the 4th world congress computers in agriculture, Orlando, July 2006, pp 640–654

Morgan KT, Obreza TA, Scholberg JMS et al (2006b) Citrus water uptake dynamics on a sandy Florida Entisol. Soil Sci Soc Am J 70:90–97

Morgan KT, Scholberg JMS, Obreza TA et al (2006c) Biomass and nitrogen relationships with orange tree growth. J Am Soc Hortic Sci 131:49–156

Morgan KT, Scholberg JMS, Obreza TA et al (2007) Sweet orange root distribution in time and space. J Am Soc Hortic Sci 132:262–269

Morgan KT, Wheaton TA, Castle WS et al (2009) Response of citrus to irrigation, nitrogen rate and application method. Hortic Sci 44: 145–150

Morgan KT, Scholberg JMS, Obreza TA et al (2011) Seasonal nitrogen storage dynamics and budgets of mature citrus trees on a sandy Entisol. J Plant Nutr (in press)

Obreza TA, Morgan KT (eds) (2008) Nutrition of Florida citrus trees, 2nd edn. Publication SL 253. Soil Water Science Department, Florida Cooperative Extension Service, Institute of Food and Agricultural Sciences, University of Florida, Gainesville, USA

Obreza TA, Rouse RE (1993) Fertilizer effects on early growth and yield of 'Hamlin' orange trees. Hortic Sci 28:111–114

Obreza TA, Sartain JB (2010) Improving nitrogen and phosphorus fertilizer use efficiency for Florida's horticultural crops. Keeping water and nutrients in the Florida citrus tree root zone. Hortic Technol 20:23–33

Obreza TA, Schumann A (2010) Keeping water and nutrients in the Florida citrus tree root zone. Hortic Technol 20:67–73

Orbovic V, Achor D, Petracek P et al (2001) Air temperature, humidity and leaf age affect penetration of urea through grapefruit leaf cuticles. J Am Soc Hortic Sci 126:44–50

Paolillo AM, Scholberg JMS, Parsons LR et al (1999) Water and nitrogen status modify root growth of two citrus rootstocks. Proc Fla State Hortic Soc 112:18–22

Paramasivam S, Alva AK (1997) Nitrogen recovery from controlled-release fertilizers under intermittent leaching and dry cycles. Soil Sci 162(6):447–453

Paramasivam S, Alva AK, Fares A et al (2001) Estimation of nitrate leaching in an Entisol under optimum citrus production. Soil Sci Soc Am J 65:914–921

Piper EL, Boote KJ, Jones JW et al (1996) Comparison of two penology models for prediction flowering and maturity date of soybean. Crop Sci 36:1606–1614

Pumphrey DA (2006) Situation and outlook for Citrus. Citrus-special feature article. USDA, Washington, DC, p 8

Quaggio JA, Cantarella H, van Raij B (1998) phosphorus and potassium soil test and nitrogen leaf analysis as a basis for citrus fertilization. Nutr Cycle Agroecosyst 52:67–74

Quiñones A, Banuls J, Primo Millo E et al (2003) Effects of 15N application frequency on nitrogen uptake efficiency in Citrus trees. J Plant Physiol 160:1429–1434

Quiñones A, Martinze-Alcantra B, Legaz F (2007) Influence of irrigation system and fertilization management on seasonal distribution of N in the soil profile and on N-uptake by citrus trees. Agric Ecosyst Environ 122:399–409

Ran Y, Habib H, Bar-Yosef B et al (1994) Root volume effects on nitrogen uptake and partitioning in peach trees. Agron J 86:530–534

Rao KP, Rains DW (1976) Nitrate absorption by barley. I. Influence of nitrate reductase activity. Plant Physiol 57:59–62

Reck WR (1994) GLEAMS modeling of BMP's to reduce nitrate leaching in middle Suwannee River area. In: Campbell KL, Graham WD (eds) Environmentally sound agriculture. Proceedings of the second conference', Orlando, 20–22 Apr 1994. ASAE Publication 04-94. American Society of Agricultural Engineers, St Joseph, pp 361–367

Reitz HJ, Long WT (1955) Water table fluctuation and depth of rooting of citrus trees in the Indian River area. Proc Fla State Hortic Soc 68:24–29

Reyes MR, Bengtson RL, Fouss JL (1994) GLEAMS-WT hydrology submodel modified to include subsurface drainage. Trans ASAE 37(4):1115–1120

Roy WR, Gardner FE (1946) Seasonal absorption of nutrient ions by orange trees in sand culture. Proc Am Soc Hortic Sci 47:107–118

Rufty TW, MacKown CT, Volk RJ (1990) Alterations in nitrogen assimilation and partitioning in nitrogen stressed plants. Physiol Plant 79:85–95

Saseendran SA, Hubbard KG, Singh KK et al (1998) Optimum transplanting dates for rice in Kerala, India, determined using both CERES v3.0 and ClimProb. Agron J 90:185–190

Sato S, Morgan KT, Ozores-Hampton M et al (2009) Spatial distributions in sandy soils with seepage irrigation: ammonium and nitrate. Soil Sci Soc Am J 73(3):1044–1052

Scholberg JMS, Parsons LR, Wheaton TA (2000) Citrus nitrogen nutrition: some production considerations for more efficient nitrogen use. Citrus Indus 81:18–19

Scholberg JMS, Parsons LR, Wheaton TA et al (2001) Procedures for determining the effects of environmental conditions on plant nitrogen uptake: an alternative approach. Soil Crop Sci Soc Fla Proc 60:40–49

Scholberg JMS, Parsons LR, Wheaton TA et al (2002) Soil temperature, N concentration, and residence time affect nitrogen uptake efficiency of citrus. J Environ Qual 31:759–768

Scholberg JMS, Dogliotti S, Zotarelli L et al (2010) Cover crops use in agrosystems: innovations and applications, chapter 3. In: Lichtfouse E (ed) Genetic engineering, biofertilisation, soil quality and organic farming. Sustainable Agriculture Reviews No 4. Springer, New York, pp 390

Shen J, Batchelor WD, Jones JW et al (1998) Incorporation of a subsurface tile drainage component into a soybean growth model. Trans ASAE 41:1305–1313

Simonneau T, Habib R (1994) Water uptake regulation in peach trees with split-root systems. Plant Cell Environ 17:379–388

Smajstrla AG, Parsons LR, Zazueta FS et al (1986) Water use and growth of young citrus trees. Paper No 86–2069. ASAE summer meeting, San Luis Obispo

Smith PF (1956) Effect of phosphate fertilization on root growth, soil pH, and chemical constituents at different depths in an acid sandy Florida soil. Proc Fla State Hortic Soc 69:25–29

Smith PF (1965) Effect of nitrogen source and placement on the root development of 'Valencia' orange trees. Proc Fla State Hortic Soc 78:55–59

Sorgona A, Abenavoli MR, Gringeri PG et al (2006) A comparison of nitrogen use efficiency definitions in Citrus rootstocks. Sci Hortic 109:389–393

Stewart EH, Browning JE, Burt EO (1969) Evapotranspiration as affected by plant density and water-table depth. Trans ASAE 12(5): 646–647

Syvertsen JP (1984) Light acclimation in citrus leaves. II. CO_2 assimilation and light, water, and nitrogen use efficiency. J Am Soc Hortic Sci 109:812–817

Syvertsen JP (1985) Integration of water stress in fruit trees. Hortic Sci 20:1039–1043

Syvertsen JP (1996) Water stress and carbon budgets. In: Manicom B, Robinson J, Du Plessis SF et al (eds) Proceedings of the international society of citriculture, Suncity South Africa, 12–17 May, vol 1, pp 46–50

Syvertsen JP, Smith MZ (1996) Nitrogen uptake efficiency and leaching losses from lysimeter-grown citrus trees fertilized at three nitrogen rates. J Am Soc Hortic Sci 121:57–62

Tanner W, Beevers H (1990) Does transpiration have an essential function in long-distance ion transport in plants. Plant Cell Environ 13:745–750

Timmer LW, Zitko SE (1996) Evaluation of a model for prediction of post bloom fruit drop of citrus. Plant Dis 80:380–383

Tinker PB (1968) The transport of ions in the soil around plant roots. In: Rorison IH (ed) Ecological aspects of the mineral nutrition of plants. Blackwell Scientific Publications, Oxford, pp 135–147

Tinker PB, Nye PH (2000) Solution movement in the rhizosphere. Oxford University Press, New York

Turrell FM, Garber MJ, Jones WW et al (1969) Growth equations and curves for citrus trees. Hilgardia 39(16):429–445

Vitousek P (1982) Nutrient cycling and nutrient use efficiency. Am Nat 119:552–572

Wagner-Riddle C, Gillespie TJ, Hunt LA et al (1997) Modeling a rye cover crop and subsequent soybean yield. Agron J 89:208–218

Webber HJ, Batchelor LD (1943) The citrus industry. In: Histroy, botany and breeding, vol I. University of California Press, Berkely

Weinbaum SA, Merwin ML, Muraoka TT (1978) Seasonal variation in nitrate uptake efficiency and distribution of absorbed nitrogen in non-bearing prune trees. J Am Soc Hortic Sci 103:516–519

Willer H, Lukas K (Eds.) (2009) The world of organic agriculture - statistics and emerging trends 2009. IFOAM, Bonn; FiBL, Frick; ITC, Geneva

Xin J, Zazueta FS, Smajstrla AG et al (1997) CIMS – an integrated real-time computer system for citrus microirrigation management. Appl Eng Agric 13:785–790

Youssefi F, Brown PH, Weinbaum SA (1999) Regulation of nitrogen uptake at the whole-plant level: a study in almond trees (*Prunus dulcis*). Hortic Technol 9(4):598–600

Youssefi F, Weinbaum SA, Brown PH (2000) Regulation of nitrogen partitioning in field-grown almond trees: Effects of fruit load and foliar nitrogen applications. Plant Soil 227:273–281

Zotarelli L, Avila L, Scholberg JMS et al (2009) Benefits of vetch and rye cover crops to sweet corn under no-tillage. Agron J 10:252–260

Nitrogen in Citrus: Signal, Nutrient, and Use Efficiency

16

Agostino Sorgonà and Maria Rosa Abenavoli

Abstract

Nitrogen (N) is the most important nutrient for growth, fruit yield, and quality of citrus plants. In order to reduce both the requirements for costly nitrogen fertilizers and environmental pollution of soil and water, the improvement of the nitrogen use efficiency (NUE) on citrus plants is fundamental in sustainable agriculture. In this chapter, a critical overview on the definitions of NUE and its components, nitrogen uptake (NUpE), and nitrogen utilization efficiency (NUtE) was provided, together with current knowledge and future challenges to understand and manipulate NUE in citrus plants. Further, the different N fertilizer use strategy in combination with irrigation to increase the NUE in citrus species was explained. The nitrogen content, the removal and the partitioning among the citrus organs, and the N availability in citrus soils provided a comprehensive picture of the N economy in citrus trees and soil orchards, and the basis of the NUE. However, an important approach for improving the NUE in citrus plants was to understand the regulation of the morpho-physiological and molecular mechanisms controlling plant nitrogen economy such as nitrogen uptake, translocation, assimilation, and remobilization. This approach accompanied by new techniques in molecular biology, root biology, plant-soil interactions, and modeling will provide an accurate criteria to discriminate between the nitrogen-efficient and inefficient citrus plants. Finally, the future challenges for improving NUE in citrus species considering both the "agronomic" and "physiological" approaches were discussed.

Keywords

Nitrogen uptake efficiency • Nitrogen utilization efficiency • Rootstock • Root architecture • Root morphology • Nitrate uptake transport system • Ammonium uptake transport system

16.1 Introduction

Nitrogen (N) is one of the most important nutrients for plant growth and development, and it is considered to be a major yield-limiting factor for many crops. As a constituent of proteins, nucleic acids, and secondary products, N consists of 1.50–6.00% of the dry weight of many crops (Benton 1998). Higher plants acquire N from the soil mainly in the mineral forms such as ammonium (NH_4^+) and nitrate (NO_3^-) and also in the organic form such as urea and amino acids. Because of the essential role of nitrogen in crop production, over the past six decades, high N fertilizer dose was applied allowing to double the agricultural food production worldwide to meet a growing population but, on the other hand, causing many environmental problems. Indeed, the intensive use of N fertilizers in agriculture (ninefold increase) due mainly to a low price of N fertilizers, progressively caused a major detrimental

A. Sorgonà (✉) • M.R. Abenavoli
Dipartimento di Biotecnologie per il Monitoraggio Agroalimentare ed Ambientale (Bio. M.A.A.), Università degli Studi "Mediterranea", di Reggio Calabria, Italy
e-mail: asorgona@unirc.it; mrabenavoli@unirc.it

A.K. Srivastava (ed.), *Advances in Citrus Nutrition*,
DOI 10.1007/978-94-007-4171-3_16, © Springer Science+Business Media B.V. 2012

impact on the biosphere such as eutrophication of freshwater (London 2005) and marine ecosystems (Beman et al. 2005) and an increase of the N oxides and toxic ammonia into the atmosphere (Ramos 1996; Stulen et al. 1998). Therefore, the challenge for the next years will be to develop a highly productive agriculture which preserves the quality of the environment and reduce the human risk development, thus an "eco-efficient agriculture." As reported by Keating et al. (2010), eco-efficient agriculture consists "more agricultural output, in terms of quantity and quality, for less input of land, water, nutrients, energy, labor or capital." This concept translates to the nitrogen means "a better use of the nitrogen for the crop growth and yield" or the "nitrogen use efficiency" (NUE).

In this chapter, we reported some aspects controlling the NUE in the citrus plants, the most economically important evergreen fruit crop in the world. Since nitrogen is the most important nutrient for citrus cultivation, fruit yield, and quality (Alva and Tucker 1999; Dasberg et al. 1984; Embleton and Jones 1978; Tucker et al. 1995), generally, the citrus farmer applied a heavy N fertilization which, combined with a not suitable N management practices, caused a severe groundwater contamination mainly attributed to NO_3^- leaching (Alva and Tucker 1999; Davies 1996; Embleton and Jones 1978; Embleton et al. 1986; Ramos et al. 2002). This has become a major environmental problem in Florida and Spain citrus production regions (Lamb et al. 1999; Fernàndez et al. 1998). Further, an application of N fertilization in excess in citrus production caused an increase of soil acidification (He et al. 1999; Cantarella et al. 2003) and ammonia volatilization (Cantarella et al. 2003) and a reduction of fruit quality (Legaz and Primo-Millo 1988).

In particular, in the first paragraph, the N partitioning among the plant organs of the seedlings and mature trees, bearing and no-bearing trees, and the N distribution between the plant and nursery and orchards soils were reported. Successively, starting from a general definition of NUE, a specific NUE definition together with data from experimental studies for citrus plants was provided. Further, other definitions such as "nitrogen uptake efficiency" (NUpE), "nitrogen utilization efficiency" (NUtE), "fertilizer use efficiency" (FUE), and "fertilizer N recovery" (FNR) were also reported. The fourth paragraph detailed on the NUE improvement in the citrus plants with particular focus on the growth and production responses to the rate, type, time, and frequency of fertilizer N application, soil type, and their interactions with irrigation management. Then, the nitrogen-efficient and -inefficient citrus rootstocks and the criteria adopted to facilitate the screening of citrus genotypes for improved efficiency were detailed. Further, the morphophysiological basis for improving the NUE such as the nitrate and ammonium transport systems, kinetic and energetics, and root morphological and architectural traits were analyzed.

16.2 Nitrogen Status of the Citrus Plant and Soil Nursery and Orchards

Nitrogen is an essential nutrient to sustain high growth in young citrus tree and plant vigor, fruit yield, and quality, in mature ones.

Nitrate and ammonium are the main source of nitrogen for citrus species, showing a seasonal uptake peaking during the periods of active shoot growth (Maust and Williamson 1994; Weinbaum et al. 1984). Generally, after absorption, the nitrate is translocated to the shoot in inorganic form, while ammonium was firstly reduced to amino acids, mainly glutamate, in roots and then translocated to the aboveground (Kato 1981, 1986). The N acquisition is needed to recover the N lost in harvested fruits, abscised fruitlets and flowers, senescent leaves, pruning wood, and root turnover. Further, it is necessary to replenish the N reserves used for ensuring the formation of new developing organs during the early stage of vegetative growth. For example, Chapman (1968) reported that 40 tons of orange discharged 47.2 kg of N, and Alva et al. (1998) pointed out a removal of 52.8, 56.2, 66.8, and 67.4 lb of N in 500 boxes of fruits (fresh wt basis) of Hamlin, Parson Brown, Valencia, and Sunburst orange varieties, respectively. Therefore, the citrus plants exhibited a different N status and partitioning among the different organs during the annual cycle. Table 16.1 (modified from Legaz et al. 1995) reported the N content (% with respect to the total tree) of the different organs in 3-year-old Valencia Late orange tree grafted on Troyer citrange during an annual cycle. In particular, Legaz et al. (1995) underlined the greatest N content in leaves (33.2–41.4%) and, at lesser extent, in roots (30.5–37.2%), considering the main organs of N reserves in citrus plants.

For a long time, the leaf tissue analysis was a useful tool to evaluate citrus N status by comparing the actual leaf N concentration with the critical one established from previous studies (Jones and Embleton 1969; Malavolta 1992; Terblanche and Du Plessis 1992; Hanlon et al. 1995; Quaggio et al. 1998; Kohli et al. 1998). The optimum level of N leaf content was between 25 and 27 $g\,kg^{-1}$ for orange (Koo et al. 1984; Alva et al. 2006), 22–23 $g\,kg^{-1}$ for grapefruit (He et al. 2003), and 27–29 $g\,kg^{-1}$ for Clementine mandarin trees (Hammami et al. 2010). The N leaf removal from the critical leaf concentration standards caused a nitrogen deficiency or excess status in the citrus plants.

The N-deficient citrus trees exhibited the following characteristics: (1) slow, weak and stunted growth, and irregular flushes; (2) decrease and sparse in flowering and fruiting; and (3) yellowing of the foliage, the most important visual symptom of the N starvation (Spiegel-Roy and Goldschmidt 1996; Davies and Albrigo 1994; Zekri and Obreza 2003). The chlorosis firstly appeared on the older leaves which became completely pale, while the new leaves are small,

Table 16.1 N content (% with respect to the total tree) of diverse organs of Valencia Late orange tree (3 years old) grafted on Troyer citrange at different growth stages

	Dormancy (%)	Flowering (%)	Fruit set (%)	Midsummer flush (%)	Early autumn flush (%)
Reproductive organs	0.0	4.2	3.0	2.6	6.1
Leaves	39.0	33.2	36.9	39.3	41.4
Twigs	0.0	2.1	2.4	2.8	3.4
Old branches + trunk	24.0	23.3	21.9	21.6	18.6
Roots	37.0	37.2	35.8	33.7	30.5

Modified from Legaz et al. (1995)

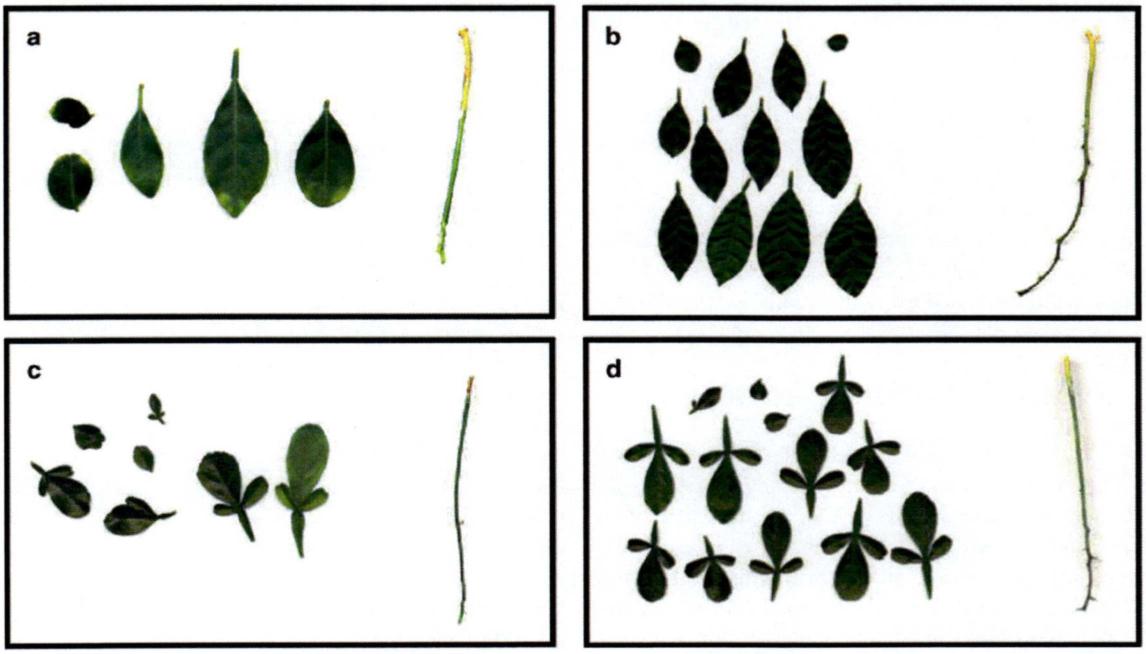

Fig. 16.1 Leaves and stems of Volkamer lemon (**a** and **b**) and Carrizo citrange (**c** and **d**), 120 days old, grown with two different nitrate concentrations: 5 μM (**a** and **c**) and 1,000 μM (**b** and **d**). To note (1) the *yellowing* of the old leaves in (**a**) and (**c**) and (2) the higher number of leaves in (**b**) and (**d**)

thin, and fragile. The leaves and stems of Volkamer lemon and Carrizo citrange seedlings, 120 days old, grown at two different nitrate concentrations (5 and 1,000 μM) are further shown (Fig. 16.1) (Sorgonà, unpublished data). Both citrus rootstocks, at low nitrate levels, exhibited a reduced number of leaves whose older leaves showed yellowing. Further, the stems of the N-starved citrus rootstocks appeared also chlorotic and lesser taller than those grown with 1,000 μM nitrate (Sorgonà, unpublished data).

The N excess also hampered the growth and production of citrus plants. Nitrogen fertilization over the recommendation rates such as 225 kg ha^{-1} $year^{-1}$ for young orange trees and 280 kg ha^{-1} $year^{-1}$ for mature trees (Obreza and Morgan 2008) caused a reduction of fruit yield and quality. Schumann et al. (2003) observed that the high rate of N fertilizer (240 kg ha^{-1} $year^{-1}$) produced a reduction of fruit number and yield, juice and soluble yield, and fruit mass of Hamlin orange trees. Further, Koo (1988) showed that the increase of N fertilization rates caused a reduction of fruit size, weight, and peel thickness of orange fruits. Recently, Hammami et al. (2010) demonstrated that N rates in excess of 192 kg ha^{-1} $year^{-1}$ caused fruit yield reduction in Clementine mandarin.

The main cause of N deficiency or excess in citrus trees depended on the N availability in the soil. Low N availability in both citrus orchards and nursery resulted from N losses mainly due to nitrate leaching and ammonium volatilization. Syvertsen and Smith (1996) reported that N losses from 4-year-old grapefruit trees grown in lysimeters averaged

11% and 20% for Volkameriana lemon and Sour orange rootstocks, respectively. Lea-Cox and Syvertsen (1996) observed nitrate leaching between 10% and 40% in Sour orange and Volkamer lemon seedlings (22 weeks old), grown in pot systems. Further, Lea-Cox et al. (2001) observed that the <30% and >50% of the total N applied were leached as ^{15}N nitrate at soil layers below 30 cm in Redblush grapefruit (4 years old) grafted on Volkamer lemon and Sour orange, respectively, grown for 29 days in tanks. Finally, in Valencia sweet orange (6 years old) orchard, the N inorganic content in the 20–60-cm soil depth layer accounted for up to 56 kg ha^{-1} (average of 4 years of fertilizer application) with annual N application rate of 260 kg ha^{-1} year^{-1} (Cantarella et al. 2003). Beside the leaching of nitrate, the volatilization of ammonium is a component of N losses from the soil. Indeed, NH_4^+ volatilization, between 26% and 44% was observed after the urea fertilizer application in field cultivated with Valencia sweet orange (10 years old) (Cantarella et al. 2003).

How the N status of the citrus plants and citrus soil in orchards and nursery is modified by different biotic and abiotic factors such as rootstocks, irrigation, doses and timing of N fertilizer application, and irrigation will be discussed later.

16.3 Nitrogen Use Efficiency Definitions

The word "efficiency" generally indicates "the level of output per unit of input." Plant system "efficiency" defines the "growth, physiological activity, yield, or harvested yield (output) per unit of land, water, nutrient, or energy (input)." While focusing the attention on the specific nutrient such as nitrogen, the term "nitrogen use efficiency" defines as "the plant growth, physiological activity, yield or harvested yield per unit of nitrogen." Even within this simple ratio, "nitrogen use efficiency" has been defined in many ways in diverse context (Clark 1990; Blair 1993) which are grouped in "agronomic" and "physiological" terms. With regard to agronomic terms, the NUE definition emphasized the productivity including "the plant biomass or yield or harvested yield per unit of available N in the soil" (together with the N residual present in the soil other than that applied by fertilizer) (Caradus 1990; Moll et al. 1987; Saric 1982; Saurbeck and Helal 1990) or "the plant biomass or yield or harvested yield per unit of nitrogen applied" (only fertilizer applied) (Balingar et al. 1990; Blair 1993; Thung 1988) (g of plant dry weight per mg of nitrogen or kg yield per kg fertilizer). With regard to the internal nutrient plant requirement, the NUE definition felt into the physiological group being defined as "plant biomass produced per unit nitrogen absorbed" (Balingar et al. 1990; Gerloff and Gabelman 1983; Glass 1989) or "amount of harvestable product per unit of N absorbed" (Moll et al. 1987) (g plant dry weight per mg of nitrogen or kg of yield per g of nitrogen).

Summarized experimental data of different citrus NUE pointing out their wide variability are shown (Table 16.2) because of genotypes, age, season, scion/rootstock, and experimental setup. In spite of this wide NUE variability, a common consideration should be done: the NUE increased with the plant aging. Indeed, the NUE (g plant DW g^{-1} N applied) of orange trees budded on Carrizo citrange were 5.25, 24.1, and 54.9 after the first, second, and third years, respectively (values extrapolated from Menino et al. 2007). In terms of fruit yield per fertilizer applied, the NUE showed a temporal pattern: the young (3–5 years old) trees of Ambersweet orange on Swingle citrumelo exhibited an average of 101.3 kg fruit yield per kg^{-1} fertilizer, while mature trees (8–10 years old) reached 382.5 kg fruit yield per kg^{-1} (extrapolated data from Morgan et al. 2009). Finally, similar results were obtained from Davies and Zalman (2002) in the Rohde Red Valencia orange grafted on different citrus rootstocks.

Nitrogen use efficiency is however a complex trait that according to Moll et al. (1982) can be dissected into "nitrogen uptake efficiency" (NUpE) and "nitrogen utilization efficiency" (NUtE) (Table 16.2). The NUpE referred to the ability of the plant to remove N from soil, and it was defined as "the nitrogen absorbed in the plant or in the yield per unit of N supplied or applied" (Maust and Williamson 1994). Generally, the NUpE was expressed as mg N per g N applied but, for a better comparison, was also reported as % of nitrogen respect to the N applied by fertilizer which was also termed as "fertilizer N recovery" (FNR) or "fertilizer use efficiency" (FUE). Syvertsen and Smith (1996) estimated the FNR of Redblush grapefruit grafted on Volkamer lemon and Sour orange to be 61% and 52%, respectively, averaged over the 2-year period and N rates. Furthermore, Scholberg et al. (2002) pointed out that the NUpE values in citrus rootstocks (Swingle citrumelo and Volkamer lemon, 10 weeks old), calculated by the difference of the N leaching losses between tree tank and no tree tank, ranged between 16.6% and 83.1% in relation to N rates and N residence times. However, this technique did not consider the N losses due to volatilization or immobilization from the soil (no tree tank). Conversely, the advantage of ^{15}N-labeled fertilizer technique was the ready N identification and estimation which improved the estimation of the nitrogen uptake efficiency in citrus plants. Wallace (1953) firstly used this technique in citrus plants, and afterward several authors benefited from the ^{15}N tracer for estimating the N uptake and remobilization. By this technique, Quiñones et al. (2007) reported that the NUpE values for Navelina orange (8 years old) on Carrizo citrange, in lysimeters, ranged between 62.7% and 75.1%. More recently, Boaretto et al. (2010) estimated the NUpE averaged in 36% and 52% for orange and lemon trees (3 years old), respectively, both grafted on Swingle citrumelo. Similar values ranging between 25% and 80% were obtained not lately in mature trees grown in the field (Dasberg et al. 1984;

Table 16.2 An update of nitrogen use efficiency (NUE), nitrogen uptake efficiency (NUpE), and nitrogen utilization efficiency (NUtE) values reported or extrapolated from literature for the citrus plants

Citrus plants (rootstocks or scion/rootstocks)	Age (year old)	Experimental treatments	References	NUE (g plant dry weight g^{-1} N applied)	NUpE (% of N absorbed with respect to the N applied)	NUtE (g plant dry weight mg^{-1} N absorbed)
Hamlin orange/ Swingle citrumelo	6	Fertilizer type	Mattos et al. (2003)	81.6–108.8[a]	25.5–39.4	0.276–0.320[a]
Valencia orange/ Swingle citrumelo;	3	Species	Boaretto et al. (2010)	28.4–40.6[a]	29.2–53.8	0.079–0.078[a]
Lisbon lemon/ Swingle citrumelo		N application				
Newhall navel orange/Carrizo citrange	1–2	N rates	Weinert et al. (2002)	–	2.2–5.3	–
		Fertigation frequency			10.9–23.5	
Rohde Red Valencia orange/ Swingle citrumelo	3	Rootstocks	Davies and Zalman (2002)	5.8–327.7[b]	–	–
Rohde Red Valencia orange/ Carrizo citrange	4	Age				
Rohde Red Valencia orange/ *Citrus volkameriana*	5	N rates				
Hamlin orange/ Cleopatra mandarin	25	N application	Alva et al. (2006)	283.9–681.2[b, c]	–	–
Clementine mandarin/ Sour orange	25	Different NK ratio	Hammami et al. (2010)	147.0–235.6[b, c]	–	–
Swingle citrumelo	30 weeks old	Type of fertilizers	Dou and Alva (1998)	6.9–16.7[a]	–	–
Cleopatra mandarin						
Swingle citrumelo	5 months old	Fertigation frequency	Melgar et al. (2010)	26.1–90.0	–	–
Pera sweet orange/ Rangpur lime	4	–	Boaretto et al. (2006)	76.2[a]	20–27	–
Swingle citrumelo	8 months old	Type of amendments	Syvertsen and Dunlop (2004)	51.9[a]–65.1[a]	55–81	–
Valencia orange/ Rough lemon	36	Type of fertilizers	Alva et al. (1998)	252.8–267.8[a, b, c]	–	–
Cleopatra mandarin	16 weeks old	Salinity stress	Lea-Cox and Syvertsen (1993)	–	13.8–14.4	27.2–32.4[a]
Volkamer lemon						
Hamlin orange/ Swingle citrumelo	4–5	Fertigation frequency and wetting pattern	Syvertsen and Sax (1999)	53.0–114.5[a, b]	12–44[a, d]	–
Hamlin orange/ Swingle citrumelo	6–7	Fertigation frequency	Syvertsen and Jifon (2001)	32.1–79.6[a, b]	24.1–41.5[d]	–
Redblush grapefruit/ Volkamer lemon	4	Rootstocks	Syvertsen and Smith (1996)	–	18–83[d]	–
Redblush grapefruit/ Sour orange		N rates				
Navelina/Carrizo citrange	8	N application	Quiñones et al. (2007)	–	62.7–75.1[e]	–

(continued)

Table 16.2 (continued)

Citrus plants (rootstocks or scion/rootstocks)	Age (year old)	Experimental treatments	References	NUE (g plant dry weight g^{-1} N applied)	NUpE (% of N absorbed with respect to the N applied)	NUtE (g plant dry weight mg^{-1} N absorbed)
Navelina/Carrizo citrange	8	N application	Quiñones et al. (2003)	274.9–328.2[a]	64–75[e]	6.13–6.25[a e]
Cleopatra mandarin	14–16 weeks old	Rootstocks	Lea-Cox and Syvertsen (1996)	40–100	26.8–60.0[e]	–
Swingle citrumelo		N rates				
Sour orange						
Volkamer lemon						
Orange/Carrizo citrange	2	Age	Menino et al. (2007)	5.2–54.9[a]	6–30[e]	0.087–0.183[a, e]
Redblush grapefruit/ Sour orange	4	Rootstocks	Lea-Cox et al. (2001)	24–155[a]	14.9–42.2[e]	–
		N rates				
Redblush grapefruit/ Volkamer lemon						
Swingle citrumelo	10 weeks old	Rootstocks	Scholberg et al. (2002)	–	16.6–83.1[d]	–
Volkamer lemon		N rates				
		Time of residence of N				
Rough lemon	132 days old	Rootstocks	Sorgonà et al. (2006)	–	24–80[f]	0.04–0.09
Sweet orange						
Cleopatra mandarin						
Sour orange						

[a]Extrapolated from experimental data
[b]NUE in terms of fruit yield: kg fruit yield kg^{-1} fertilizer
[c]Referred to ha of soil
[d]The NUpE was calculated as the difference between the N leached from tank without tree with that in presence of tree
[e]NUpE measured by ^{15}N technique
[f]NUpE calculated as total nitrogen accumulation divided by root dry weight

Feigenbaum et al. 1987; Weinbaum and Van Kessel 1998) and citrus rootstock seedlings (Lea-Cox and Syvertsen 1996). The young nonbearing citrus plant, conversely, exhibited lower values of FUE by <5% (Weinert et al. 2002) and <6% (Menino et al. 2007) in the first year after transplantation, while Lea-Cox et al. (2001) reported FNR values ranging from 14.9% to 39.3% in bearing Redblush grapefruit, 4 years old. Probably, these contrasting results could be due to the stronger influence of the fruit loading as sink on the nitrogen uptake. The NUpE was also defined as total nitrogen accumulation per unit of root dry weight (Elliot and Laüchli 1985). In this respect, Sorgonà et al. (2006) showed the NUpE values ranging from 24 to 80 mg N accumulated per g^{-1} root dry weight in different citrus rootstocks grown in pot at different nitrate levels.

The term "nitrogen utilization efficiency" or "NUtE" indicates "the ability of the plant to use N to produce biomass or yield or harvested yield." Often called the "nutrient efficiency ratio," it was evaluated by the total plant dry weight divided by nitrogen absorbed (g plant dry weight or kg yield per mg N absorbed) (Balingar et al. 1990; Gerloff and Gabelman 1983; Glass 1989; Moll et al. 1987). Lea-Cox and Syvertsen (1993) observed that the young Cleopatra mandarin and Volkamer lemon (16 weeks old) exhibited greater NUtE values by 32.4 and 27.2 g dry weight per mg N absorbed, respectively, than more mature (9 years old) and grafted citrus plants (Navelina orange on Carrizo citrange), whose NUtE ranged between 6.13 and 6.25 g dry weight per mg N absorbed (Quiñones et al. 2003).

The NUE, NUpE, and NUtE definitions have been also referred to the different plant organs providing a "partitioning of nitrogen efficiency" within the citrus plants. The N recovery efficiency measurement in the different plant organs aimed at estimating and monitoring the fate and transformations of N applied through the ^{15}N-enriched fertilizer in the soil-plant citrus system. In particular, the studies were focused on (1) the identification of plant organ where the ^{15}N fertilizer was mostly allocated, (2) the recognition of plant organ with the stronger N demand (sink organs), and (3) the influence of N rate and seasonal application, fertilizer types, rootstocks, and plant age on ^{15}N distribution patterns among the plant organs. For example, the N provision from March to June generally caused a preferential N allocation in young organs (Kubota et al. 1976; Akao et al. 1978; Lea-Cox et al. 2001; Martìnez et al. 2002; Quiñones et al. 2005, 2007), while a delay of the N application time, during late fall or winter, determined a higher N recovery in the old tissues (Legaz et al. 1983; Quiñones et al. 2005, 2007). The higher N recovery in young tissue than older ones in the orange trees on Carrizo citrange did not vary with increase of the plant age (Menino et al. 2007).

16.4 Improving the NUE in the Citrus

In the past 25 years, the experimental studies for improving NUE and/or NUtE, maintaining optimal citrus fruit yield and quality and minimizing nitrate leaching below the root zone, have been focused on two main aspects: (1) to develop the optimal "N and irrigation best management practices," i.e., the influence of the rate, type, time, and frequency of application of N fertilizer, soil type and their interactions with irrigation; (2) to provide information on the morphological, physiological, and molecular mechanisms that defined NUE-related traits which are associated with N-efficient citrus rootstocks.

16.4.1 Best Management Practices

A substantial work was carried out to define several strategies of a rational N fertilization in citrus trees aimed at maximizing the NUE or NUpE and, in parallel, reducing the N leaching losses in the soil.

An important first result was that the N status of the citrus trees affected both the NUE and the NUpE, i.e., the efficiency with which nitrogen was absorbed by its root system. Indeed, a negative correlation between total N plant content and the NUE (g DW g^{-1} N applied) in different citrus rootstocks (Sour orange and Volkamer lemon), grown at different N rates, was observed by Lea-Cox and Syvertsen (1996). Similar behavior was subsequently confirmed by Lea-Cox et al. (2001) with the same citrus rootstocks grafted with Redblush grapefruit. Furthermore, the N-starved conditions were positively correlated with NUpE which was higher than that of the N-sufficient trees. In fact, Dasberg (1987) showed that the N-deficient citrus trees exhibited a 57% NUpE which instead reached only a 40% value in trees grown at high N levels. This NUpE response to the N status of mature citrus trees was also evident in citrus rootstock seedlings. Indeed, at N rates of 18, 53, and 105 mg $week^{-1}$, Sour orange and Volkamer lemon (14 weeks old) exhibited 51%, 47% and 27%, and 50%, 49% and 32% of NUpE, respectively (Lea-Cox and Syvertsen 1996). These results suggested that the N uptake interacted with the N reserves to meet the N requirements for the growth and yield of the citrus plants.

A second result was that in relation to the different forms of N fertilizers applied to the soils, a variability in the NUE- or NUpE-related citrus responses was observed. Mattos et al. (2003) showed a higher NUpE in Hamlin orange (6 years old) supplied with the ammonium nitrate (39.4%) than the urea fertilizer (25.5%). An improvement of citrus NUE by fertigation management with respect to dry granular fertilizer application was also reported by Dasberg et al. (1988), Alva and Paramasivam (1998), Alva et al. (1998) and Quiñones et al. (2005). Further, Alva et al. (2006) demonstrated that NUE, expressed as an increment in fruit yield (kg fruit kg^{-1} N applied) of the Hamlin orange tree grafted on Cleopatra mandarin, was greater with the N applied as fertigation or water-soluble granules than with a mix of these fertilizers. Although, several authors indicated that the controlled-release N fertilizers (CRF) (both resin- and sulfur-coated N organic and inorganic forms) enhanced the growth and yield of citrus trees compared to that dry and soluble N fertilizers (Koo 1986; Dou and Alva 1998; Schumann et al. 2003; Morgan et al. 2009). However, the effects of CRF on the NUE are still lacking. Some studies reported a reduction of N losses in the soils with the CRFs fertilizers application, suggesting an indirect effect to improve the N uptake efficiency (Koo 1986; Alva and Tucker 1993; Dou and Alva 1998).

Finally, a third important result was to develop an optimal combination of irrigation and N management to improve N uptake efficiency of citrus trees. Drip irrigation determined a higher fertilizer N recovery (75.1%) of Navelina orange grafted onto Carrizo citrange with respect to flooding irrigation (62.7%) (Quiñones et al. 2007). A moderate irrigation rate increased the yield of young and mature Ambersweet orange trees with respect to lower one (Morgan et al. 2009). By fertigation, it was also possible to manage the frequency of N application which in turn positively influenced the NUE and NUpE. Quiñones et al. (2003, 2005, 2007) observed a higher NUE with 66 split application by drip irrigation with respect to five applications by flood irrigation in Navelina orange trees. Similar results were pointed out by Scholberg et al. (2002), Alva et al. (2006), Boman (1996), and Morgan et al. (2009), although several authors pointed that citrus rootstock seedlings (Melgar et al. 2010) and mature trees (Syvertsen and Jifon 2001; Weinert et al. 2002) showed no significant relationship between N application frequency by fertigation and NUE and NUpE.

16.4.2 Nitrogen-Efficient and -Inefficient Citrus Rootstocks: Root Morphology and Nitrogen Uptake Mechanisms

In its last review on citrus rootstocks, Castle (2010) argued that "….citrus rootstocks bring many advantages and profitability to commercial enterprise.. …citrus rootstocks are the sole determining element that allows citrus to be grown in particular circumstances," and he concluded "…as the knowledge base increases, perhaps new rootstocks designed in response to particular concerns could more readily be produced…." Synthesizing Castle's opinion and correlating them with the NUE in citrus species, we may assert that (1) the rootstocks are the main subject for enhancing the fitness of citrus plants to different N soil availabilities and (2) the morpho-physiological and molecular mechanisms of rootstocks are essential for the improvement of NUE in citrus plants.

The genotypic variability of citrus rootstocks, collected by Wutscher (1989), induced a different leaf nitrogen content on scion component. Wutscher also grouped the rootstocks in high- and low-induced N levels: Rough lemon, Sweet orange, Rusk citrange, Alemow and Rangpur lime, the high N level inducer rootstocks, and Sour orange, Trifoliate orange, Cleopatra mandarin, and grapefruit, the lower ones, whereby these results underlined that there is a different N acquisition capacity among the citrus rootstocks responsible consequently of a diverse citrus NUE. Syvertsen and Smith (1996) observed that N uptake efficiency of Redblush grapefruit budded on Volkamer lemon, high vigorous rootstock, was 61% averaged over the 2-year period and N rates, while that on Sour orange, low vigorous rootstock, was 52% only. Lea-Cox et al. (2001) confirmed that the stronger-induced vegetative growth rootstock absorbed the ^{15}N more than lower vigorous ones, showing the higher N uptake efficiency. However, this NUpE pattern was not observed on citrus rootstock seedlings (Scholberg et al. 2002).

Commonly, the single value of NUpEs or NUEs was used for comparing the behavior of different citrus rootstocks at diverse treatments (rate, type, time, and frequency of application of N fertilizer) with the aim to individuate the nitrogen-efficient and -inefficient citrus rootstocks. Gourley et al. (1994), comparing various criteria for defining crop NUE, demonstrated that single-value terms of NUE were not suitable, especially under low nutrient input, to discriminate between nitrogen-efficient and -inefficient germplasms. They suggested that the nutrient efficiency classification should take into account the plant performance either in presence and absence of the nutrient considered, and they proposed that "a well defined response curves are required for nutrient efficient differences to be determined." This approach enabled to estimate the maximum yield at non-limiting nutrient availability (α) and the nutrient concentration at which half-maximum yield is achieved (β) (Fig. 16.2), essential indices for determining nutrient efficiency in crop germplasms. Indeed, equivalent α and different β defined efficient/inefficient genotypes, while different α indicated genotypes with higher/lower genetic potential (Fig. 16.2) which as sustained by Gourley et al. (1994) exhibited "a greater overall genetic potential…due to factors other than those mechanisms specifically associated with nutrient acquisition…." However, these theoretical criteria were applied on different herbaceous germplasms (*Trifolium repens* L.) in response to different phosphorus levels (Gourley et al. 1994). More recently, Sorgonà et al. (2006), adopting the Gourley's criteria, characterized the nitrate efficiency in citrus rootstocks and compared the results with other nitrogen efficiency

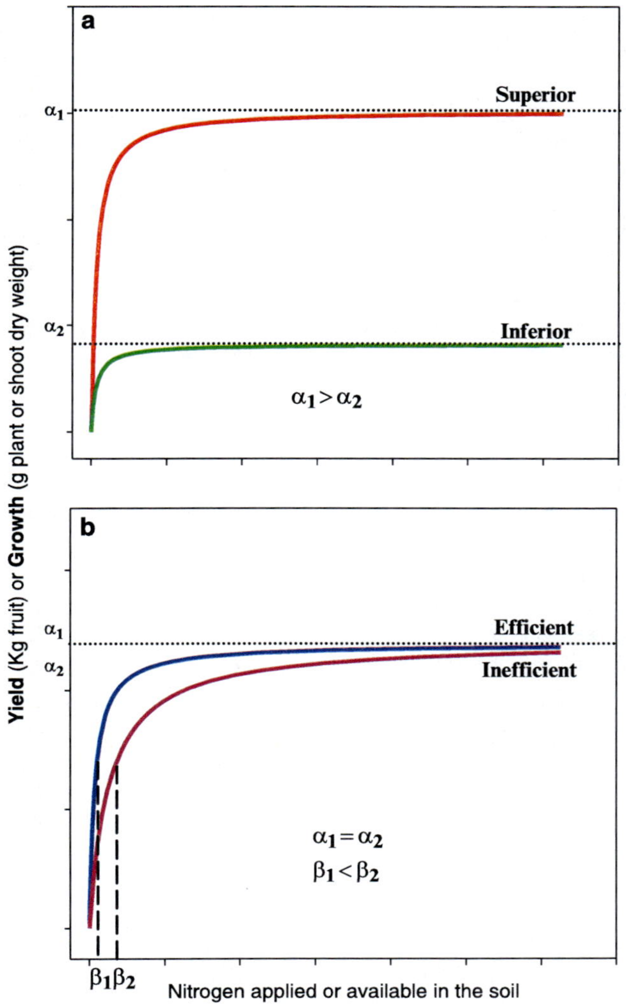

Fig. 16.2 Gourley's criteria adopted by Sorgonà et al. (2006) to identify the nitrogen efficiency in citrus rootstocks. (**a**) The maximum yield or growth obtained at non-limiting N availability (α) is different between the superior and inferior potential genetic of citrus genotypes. (**b**) The α's values are equal, while the β's values are different: the value of β of nitrogen-efficient citrus genotypes is lower than that of nitrogen-inefficient (Adapted from Gourley et al. 1994)

definitions. In agreement with Gourley et al. (1994), they demonstrated that (1) the estimation of the α and β indices by shoot dry weight response curves in response to increasing nitrate availability permitted, in a less ambiguous way, to discriminate the nitrate-efficient and inefficient citrus rootstocks, and (2) in addition to Gourley's criteria, they proposed the use of total leaf area response curves, parameter estimated by nondestructive techniques instead of shoot dry weight curves, for the early characterization of citrus rootstocks with high NUE. Assuming these criteria, it was possible to define the Rough lemon and Cleopatra mandarin as the rootstocks with superior and inferior genetic potential, respectively, while Sour orange and Sweet orange as the nitrate-efficient and -inefficient citrus rootstocks, respectively (Sorgonà et al. 2006). However, it will be needed to verify these results on the grafted citrus rootstocks for a wide and practical application.

Identification of citrus genotypes with different nitrogen efficiencies generally includes investigation of potential morphological, physiological, and molecular mechanisms involved. Nitrogen-efficient genotypes usually exhibited above- and below-ground traits which conferred them an improved and "aggressive" nitrogen acquisition from N-deficient soils. The understanding of their architectural, morphological, physiological, and molecular mechanisms involved in response to low nitrogen availability will make it possible to genetically manipulate the plant to improve its nitrogen use efficiency. The root nitrogen acquisition capacity depends on root biomass, morphology, age, and proliferation and on nitrogen transport mechanisms, but it was correlated to shoot and leaf structural and biochemical features also. For a better understanding of these processes, we will split the NUE-related above- and below-ground traits of citrus rootstocks into three parts: root architecture and morphology, root nitrogen transport, and above-ground structures.

In general, root morphology and architecture are important plant traits for nutrient uptake efficiency (Sattelmacher et al. 1993; Lynch 1995), and several studies, mainly on cereal species, demonstrated the close relationships between the root performances and plant growth and yield (for review and references therein, see Hirel et al. 2007; Garnett et al. 2009). Furthermore, some positive correlations among QTLs for N-uptake and root morphology and architecture clearly underlined the importance of an efficient root system in N acquiring to increase the NUE (Coque et al. 2008). The citrus rootstocks exhibited a genetic variability in root morphology and architecture: Rough lemon, Sour orange, and Cleopatra mandarin showed a vigorous and spreading root system; Sweet orange and Orlando tangelo was compacted; and Carrizo citrange was poorly developed (Castle and Youtsey 1977).

In order for this root variability to be exploited to improve the nitrogen acquisition of the citrus species, it was necessary to understand how the root morphology and architecture of citrus rootstocks responded to change of soil nitrogen availability. Sorgonà et al. (2005) showed that Volkamer lemon and Carrizo citrange seedlings modified their root morphology and architecture in relation to the nitrate availability. In particular, at low nitrate supply (5 μM), Volkamer lemon allocated more biomass toward the root, increasing the length of the tap and first-order lateral roots and was lesser branched than Carrizo which, on the other hand, exhibited a higher length of second-order lateral roots and a pronounced root proliferation. However, at the high nitrate level (1,000 μM), this effect disappeared. These first results indicated that the citrus rootstocks showed a root morphological

Fig. 16.3 Volkamer lemon (120 days old) grown at two nitrate supplies, 5 μM (*left seedling*) and 1,000 μM (*right seedling*). Note the different root architectures: herringbone-like (*left*) and dichotomous-like (*right*)

plasticity in response to nitrate supply which was obtained by a within-root modification of the morphology. How much was the root plasticity and which root order was more plastic in N-efficient citrus rootstocks in response to the change of nitrogen supply were discussed later by Sorgonà et al. (2007). In particular, they evidenced that (1) the second-order was more responsive than tap and first-order lateral roots, (2) the biomass allocation more than structural parameters (root fineness and tissue density) was the "morphological components" that drives the length variation of different root orders, and (3) the slow- and fast-growing citrus rootstocks adopted a different root morphological strategy to the soil nitrate changes. Indeed, Cleopatra mandarin, slow-growing rootstock, exhibited a root system highly plastic, characterized by long tap root and poor branching; conversely, Rough lemon, fast-growing rootstock, invested on the length of second-order lateral roots and on the root proliferation, especially at low nitrate availability (Sorgonà et al. 2007). All these morphological traits defined a different root architecture more responsive to the modification of N availability, highly able to explore the soil environment and, consequently, to acquire nitrogen from the soil.

Generally, the root architecture of citrus rootstocks can vary within two extreme types: the herringbone system, with branching confined to the main axis, and the dichotomous type with more random branching at low and high N availability, respectively. Figure 16.3 shows the shifting of root architecture of Volkamer lemon seedlings grown at two nitrate levels (5 and 1,000 μM) from herringbone (seedling on the left) to dichotomous (seedling on the right). Further, root architecture is correlated with soil resource exploitation efficiency: in low-fertility soils, the herringbone-like structure is more efficient in nutrient acquisition, but more expensive to construct than dichotomous root architecture (Fitter and Stickland 1991; Fitter et al. 1991). The different root architecture of Volkamer and Carrizo, herringbone- and dichotomous-like structure, respectively, grown at low N availability, revealed their different root strategy efficiency for taking up the nitrate. In particular, Carrizo citrange exhibited an optimal root architecture to acquire the nitrate in N-deficient soil (Sorgonà et al. 2005). In-depth study on the root architecture responses to N availability on Rough lemon and Cleopatra mandarin, superior and inferior genetic potential for nitrate acquisition, and Sour orange and Sweet orange, N-efficient and -inefficient rootstocks, was conducted (Sorgonà et al. 2007). Rough lemon and Sweet orange exhibited a higher degree of root architecture plasticity in response to different soil N levels, shifting from a herringbone-like to dichotomous-like root architecture at low nitrate and high nitrate levels, respectively. Conversely, Cleopatra mandarin and Sweet orange showed a lesser plastic root architecture in response to the nitrate availability (Sorgonà et al. 2007).

The root capacity for N transport was widely studied at physiological and molecular level, mostly on herbaceous species (see reviews and reference therein Forde and Clarkson 1999; Tischner 2000), but first results can be drawn for citrus species. Nitrate and ammonium are the nitrogen forms mainly absorbed by citrus rootstocks; therefore, we focused on the transport systems of both ions. Like the herbaceous species, the citrus rootstocks take up the nitrate by at least two different transport systems, a low affinity (LATS) and high affinity transport system (HATS) (Cerezo et al. 1997, 2000; Sorgonà and Cacco 2002; Sorgonà et al. 2005, 2006). The LATS has a low affinity for nitrate and is activated by external nitrate concentrations higher than 1,000 μM in Troyer citrange and Cleopatra mandarin (Cerezo et al. 1997, 2000) or 200 μM in Sour orange (Sorgonà et al. 2010). The HATS showed a high affinity for nitrate, operating at external concentrations up to 1,000 μM (Cerezo et al. 1997) or 200 μM (Sorgonà et al. 2010). This nitrate transport system, more interesting in N-deficient soils, is highly regulated and made up by the constitutive (cHATS) and the inducible (iHATS) high affinity transport systems. The cHATS is constitutively expressed in NO_3^--starved roots, mediating a constant nitrate uptake rate, while the iHATS is induced by NO_3^- and feedback regulated by downstream N metabolites (Sorgonà and Cacco 2002; Sorgonà et al. 2005, 2010). Sorgonà and Cacco (2002) showed that N-starved *Citrus volkameriana* seedlings exhibited a net nitrate uptake rate (NNUR) of cHATS by 0.085 μmol NO_3^- g^{-1} FW h^{-1} which increased (induction phase of iHATS) reaching, after 24 h of nitrate contact, the complete induction with 0.29 μmol $NO_3^-g^{-1}$ FW h^{-1} NNUR. A subsequent feedback inhibition caused a decline of the NNUR (decay phase of iHATS).

Further, an estimate of the half-time ($t_{1/2}$) of NNUR, during the induction and inhibition phases, indicated that the nitrate transport system of *Citrus volkameriana* was induced in 10.3 h and inhibited after 46.8 h of contact with anion (Sorgonà and Cacco 2002). The authors demonstrated that the half-time of the induction phase was negatively correlated ($r^2=0.855$) with the number of root tip (Sorgonà and Cacco 2002), suggesting that the root systems of citrus rootstocks characterized by elevated numbers of root tips showed a rapid induction in NNUR. The role of root tip as NO_3^--sensing region for the early soil exploration and belowground competition was recently confirmed by Sorgonà et al. (2010) which observed an earlier maximum induction of the iHATS of N-nitrate of the apical root segments of tap root compared to basal ones in Sour orange. The comparison of cHATS and iHATS pattern for nitrate among citrus rootstocks characterized by different root morphology and architecture could provide useful information on the diverse physiological basis of NUE. For example, Carrizo citrange, characterized by longer second-order lateral roots, pronounced root proliferation, and dichotomous-type architecture, showed a higher efficiency in nitrate uptake by higher full, and faster induction of the nitrate transport system than Volkamer lemon having a longer tap and first-order lateral roots and herringbone-type architecture (Sorgonà et al. 2005). These results further confirmed that the root systems of citrus rootstocks characterized by higher branching and proliferation were more "aggressive" for catching and taking up the nitrate in N-deficient soils and, hence, more efficient in nitrogen uptake.

Regarding the NH_4^+, the regulation of the transport systems of this ion in citrus rootstocks was investigated by Cerezo et al. (2001). As well as the nitrate, the ammonium uptake in Troyer citrange showed a biphasic pattern characterizing by two different transport systems: a low affinity non-saturable (LATS) and a high affinity saturable (HATS). The V_{max} and K_m, kinetic parameters of N-deficient Citrange troyer, were 12.5 μmol g^{-1} root DW h^{-1} and 170 μM, respectively. Over 1 mM external NH_4^+ concentration, the influx of this ion increased linearly, indicating that it was operating the LATS. The HATS and LATS for ammonium in Troyer citrange were regulated in an opposite manner: the N-deficient condition increased the activity of the HATS and decreased that of the LATS, while under NH_4^+ supply, the activity of the LATS was stimulated and repressed that of HATS (Cerezo et al. 2001). The molecular mechanism of NH_4^+ transport system in citrus rootstock was studied by Camañes et al. (2007), which identifies and isolates the CitAMT1 highly homologous to ammonium transporter AMT1 of other plant species. They further studied the regulation of the NH_4^+ uptake by light conditions and C status (Camañes et al. 2007).

The capacity to acquire nitrogen by the root systems was sustained by the above-ground plant traits, such as leaf area and/or stem height which was correlated with nitrate use efficiency. In herbaceous species, an alteration of leaf expansion in response to the N availability was observed (Ryser and Lambers 1995; Walch-Liu et al. 2005; Tian et al. 2007). In citrus species, Sour orange, nitrate-efficient rootstocks (Sorgonà et al. 2006), exhibited a greater leaf area and higher stem length than Sweet orange, nitrate-inefficient rootstock, especially at low nitrate availability (Sorgonà et al. 2011). This result suggested that the leaf area and stem height could be considered NUE-related traits in citrus rootstocks. However, the construction cost in terms of biomass of leaf area and the stem height could reduce the nitrogen efficiency of the citrus rootstock. In this respect, Sorgonà et al. (2011) found that Sour orange used more efficiently the biomass for constructing a unit of leaf area than Sweet orange, confirming that the former rootstock pointed out more adapted above-ground morphology for sustained an efficient nitrogen uptake from the soil.

16.5 Future Research

As detailed above, the improvement of NUE and its components, NUtE, and NUpE may be the primary goal over the next years to minimize the loss of N, reduce environmental pollution, and decrease the input cost in the citrus cultivation. In this respect, an exciting challenge will be to understand the following key aspects regarding either the impact of the agronomic management practices and citrus rootstock morpho-physiological and molecular mechanisms involved in NUE, NUtE, and NUpE:

1. The genotypic variability of the citrus rootstocks responses to different N regimes, especially to N limitations
2. The physiological basis of the citrus (species, rootstocks) responses to split application of N fertilizer during the growing season, and the interactive effect of the nitrogen with soil water status
3. The morpho-physiological and molecular traits (at development, growth, metabolic levels) controlling N use in citrus rootstocks, particularly in N limited availability, to develop, through molecular breeding and genetic engineering, citrus species with improved NUE

An increase of knowledge of these aspects together with the genomics, proteomics, and transcriptomic approaches will likely pave the way for engineering citrus rootstocks/scion combination able to give satisfactory economic yield under N-deficient soils or reduced N fertilizer inputs.

References

Akao S, Kubota S, Hayashida M (1978) Utilization of reserve nitrogen, especially autumn nitrogen, by Satsuma mandarin trees during the development of spring shoots (I). J Jpn Soc Hortic Sci 47:31–38

Alva AK, Paramasivam S (1998) An evaluation of nutrient removal by citrus fruits. Proc Fla State Hortic Soc 111:126–128

Alva AK, Tucker DPH (1993) Evaluation of a resin coated nitrogen fertilizer for young citrus trees on a deep sand. Proc Fla State Hortic Soc 106:4–8

Alva AK, Tucker DPH (1999) Soils and citrus nutrition. In: Timmer LW, Duncan LW (eds) Citrus health management. APS Press, St. Paul, pp 59–71

Alva AK, Paramasivam S, Graham WD (1998) Impact of nitrogen management practices on nutritional status and yield of Valencia orange trees and groundwater nitrate. J Environ Qual 27:904–910

Alva AK, Paramasivam S, Obreza TA et al (2006) Nitrogen best management practice for citrus trees. I. Fruit yield, quality, and leaf nutritional status. Sci Hortic 107:233–244

Balingar VC, Ducan RR, e Fageria NK (1990) Soil-plant interaction on nutrient use efficiency in plants. In: Balingar VC, e Ducan RR (eds) Crops as enhancers of nutrient use. Academic, San Diego, pp 351–373

Beman JM, Arrigo K, Matson PM (2005) Agricultural runoff fuels large phytoplankton blooms in vulnerable areas of the ocean. Nature 434:211–214

Benton JJ Jr (1998) Plant nutrition. Manual. CRC Press LLC, Boca Raton

Blair G (1993) Nutrient-efficiency – what do we really mean? Dev Plant Soil Sci 50:205–213

Boaretto AE, Ueta FZ, Trivelin PCO et al (2006) Efficiency of nitrogen fertilization on citrus orchards. In: Proceedings of the Vth international symposium in mineral nutrition of fruit plants, Talca, Chile. Acta Hortic 721:331–336

Boaretto RM, Mattos DJ, Quaggio JA et al (2010) Nitrogen-15 uptake and distribution in two citrus species. In: 19th world congress of soil science, soil solutions for a changing world, Brisbane, Australia, pp 156–159

Boman BJ (1996) Fertigation versus conventional fertilisation of Flatwoods grapefruit. Fertil Res 44:123–128

Camañes G, Cerezo M, Primo-Millo E et al (2007) Ammonium transport and CitAMT1 expression are regulated by light and sucrose in Citrus plants. J Exp Bot 58(11):2811–2825

Cantarella H, Mattos D Jr, Quaggio JA et al (2003) Fruit yield of Valencia sweet orange fertilized with different N sources and the loss of applied N. Nutr Cycl Agroecosyst 67:215–223

Caradus JR (1990) Mechanisms improving nutrient use by crop and herbage legumes. In: Balingar VC, e Ducan RR (eds) Crops as enhancers of nutrient use. Academic, San Diego, pp 253–311

Castle WS (2010) A career perspectives on citrus rootstock, their development, and commercialization. Hortic Sci 45(1):11–15

Castle WS, Youtsey CO (1977) Root system characteristics of citrus nursery trees. Proc Fla State Hortic Soc 90:39–44

Cerezo M, García-Augustín P, Serna MD et al (1997) Kinetics of nitrate uptake by *Citrus* seedlings and inhibitory effects of salinity. Plant Sci 126:105–112

Cerezo M, Flors V, Legaz F et al (2000) Characterization of the low affinity transport system for NO_3– uptake by *Citrus* roots. Plant Sci 160:95–104

Cerezo M, Tillard P, Gojon A et al (2001) Characterization and regulation of ammonium transport systems in Citrus plants. Planta 214:97–105

Chapman HD (1968) Mineral nutrition of citrus. In: Reuther W, Batchelor LD, Webber HJ (eds) The citrus industry, vol 2, Revisedth edn. University of California, Division of Agricultural Science, Berkeley, pp 127–189

Clark RB (1990) Physiology of cereals for mineral nutrient uptake, use and efficiency. In: Balingar VC, Duncan RR (eds) Crops as enhancers of nutrient use. Academic, San Diego, pp 131–209

Coque M, Martin A, Veyrieras J et al (2008) Genetic variation for N-remobilization and postsilking N-uptake in a set of maize recombinant inbred lines. 3. QTL detection and coincidences. Theor Appl Genet 117:729–747

Dasberg S (1987) Nitrogen fertilization in citrus orchards. Plant Soil 100:1–9

Dasberg S, Erner Y, Bielorai H (1984) Nitrogen balance in a citrus orchard. J Environ Qual 13:353–356

Dasberg S, Bar-Akiva A, Spazisky S et al (1988) Fertigation versus broadcasting in an orange grove. Fertil Res 15:147–154

Davies FS (1996) Literature review of research related to nitrogen nutrition, fertilization, and groundwater pollution of citrus. Technical report. Florida Department of Agriculture and Consumer Services, Tallahassee, FL

Davies FS, Albrigo LG (1994) Citrus. CABI Publishing, Wallingford, Oxfordshire, UK

Davies FS, Zalman GR (2002) Fertilization, rootstocks, growth and yields of young 'Rohde Red' Valencia orange trees. Proc Fla State Hortic Soc 115:14–17

Dou H, Alva AK (1998) Nitrogen uptake and growth of two citrus rootstock seedlings in a sandy soil receiving different controlled-release fertilizer sources. Biol Fertil Soil 26:169–172

Elliot GC, Laüchli A (1985) Phosphorus efficiency and phosphate–iron interactions in maize. Agron J 77:399–403

Embleton TW, Jones WW (1978) Nitrogen fertilizer management programs, nitrate pollution potential and orange productivity. In: Nielsen DR, MacDonald TG (eds) Nitrogen in the environment, vol 11. Academic, New York, pp 275–297

Embleton TW, Matsumura M, Stolzy LH et al (1986) Citrus nitrogen fertilizer management, groundwater pollution, soil salinity, and nitrogen balance. Appl Agric Res 1:57–64

Feigenbaum S, Bielorai H, Erner Y et al (1987) The fate of ^{15}N labelled nitrogen applied to mature citrus trees. Plant Soil 97:179–187

Fernàndez R, Lòpez J, Navarrete P (1998) Map of nitrate concentration in the underground waters in Spain. Geominer Technologic Institute of Spain (ITGE), Ministry of Environment, Madrid

Fitter AH, Stickland TR (1991) Architectural analysis of plant root systems 2. Influence of nutrient supply on architecture in contrasting plant species. New Phytol 118:383–389

Fitter AH, Stickland TR, Harvey ML et al (1991) Architectural analysis of plant root systems. I. Architectural correlates of exploitation efficiency. New Phytol 118:375–382

Forde BG, Clarkson DT (1999) Nitrate and ammonium nutrition of plants: physiological and molecular perspectives. Adv Bot Res 30:1–90

Garnett T, Conn V, Kaiser BN (2009) Root based approaches to improving nitrogen use efficiency in plants. Plant Cell Environ 32:1272–1283

Gerloff GC, Gabelman WH (1983) Genetic basis of inorganic plant nutrition. In: Laüchli A, Bilieski RL (eds) Encyclopaedia of plant physiology, vol 15B, New series. Springer, New York, pp 453–480

Glass ADM (1989) Physiological mechanisms involved with genotypic differences in ion adsorption and utilization. Hortic Sci 24:559–564

Gourley CJP, Allan DL, Russelle MP (1994) Plant nutrient efficiency: a comparison of definitions and suggested improvement. Plant Soil 158:29–37

Hammami A, Rezgui S, Hellali R (2010) Leaf nitrogen and potassium concentrations for optimum fruit production, qualità and biomass tree growth in Clementine mandarin under Mediterranean climate. J Hortic Forest 2(7):161–170

Hanlon EA, Obreza TA, Alva AK (1995) Tissue and plant analysis. In: Tucker DPH, Alva AK, Jackson LK (eds) Nutrition of Florida citrus trees. University of Florida, IFAS, Lake Alfred, pp 13–16

He ZL, Alva AK, Calvert DV et al (1999) Effects of nitrogen fertilization of grapefruit trees on soil acidification and nutrient availability in a Riviera fine sand. Plant Soil 206:11–19

He ZL, Calvert DV, Alva AK et al (2003) Thresholds of leaf nitrogen for optimum fruit production and quality in grapefruit. Soil Sci Soc Am J 67:583–588

Hirel B, Le Gouis J, Ney B et al (2007) The challenge of improving nitrogen use efficiency in crop plants, towards a more central role for genetic variability and quantitative genetics within integrated approaches. J Exp Bot 58(9):2369–2387

Jones WW, Embleton TW (1969) Development and current status of citrus leaf analysis as a guide to fertilization in California. In: Proceedings of the first international citrus symposium, Riverside, vol 3, pp 1669–1671

Kato T (1981) Major nitrogen compounds transported in xylem vessels from roots to top in citrus tree. Physiol Plant 52:275–279

Kato T (1986) Nitrogen metabolism and utilization in citrus. Hortic Rev 8:181–216

Keating BA, Carberry PS, Bindraban PS et al (2010) Eco-efficient agriculture: concepts, challenges, and opportunities. Crop Sci 50:S109–S119

Kohli RR, Srivastava AK, Huchche AD et al (1998) Diagnosis of leaf nutrient levels for optimum productivity of Citrus reticulate Blanco grow in black clay soils under a sub-humid tropical climate. Trop Agric Res Ext 1(2):81–86

Koo RCJ (1986) Controlled-release sources of nitrogen for bearing citrus. Proc Fla State Hortic Soc 99:46–48

Koo RCJ (1988) Fertilization and irrigation effects of fruit quality. In: Ferguson JJ, Wardowski WF (eds) Factors affecting fruit quality – citrus short course, proceedings 97. University of Florida, Cooperative Extension Service, Gainesville, pp 35–42

Koo RCJ, Anderson CA, Stewart I et al (1984) Recommended fertilizers and nutritional sprays for citrus. Fla Agric Exp Stn Bull 536D

Kubota S, Kato T, Akao S et al (1976) ^{15}N absorption and translocation by Satsumas mandarin trees. IV. Behaviour of nitrogen supplied in early summer. Bull Shikoku Agric Exp Stn 29:55–66

Lamb ST, Graham WD, Harrison CB et al (1999) Impact of alternative citrus management practices on groundwater nitrate in the Central Florida Ridge. Part I. Field investigation. Trans ASAE 42(6): 1653–1668

Lea-Cox JD, Syvertsen JP (1993) Salinity reduces water use and nitrate-N-use efficiency of citrus. Ann Bot 72:47–54

Lea-Cox JD, Syvertsen JP (1996) How nitrogen supply affects growth and nitrogen uptake, use efficiency, and loss from citrus seedlings. J Am Soc Hortic Sci 119:195–201

Lea-Cox JD, Syvertsen JP, Graetz DA (2001) Springtime ^{15}nitrogen uptake, partitioning, and leaching losses from young bearing citrus trees of differing nitrogen status. J Am Soc Hortic Sci 126:242–251

Legaz F, Primo-Millo E (1988) Guidelines for citrus fertilization. Technical report, No 5–88. Department of the Agriculture, Fish and Food of the Valencian Government, Valencia, Spain (in Spanish)

Legaz F, Primo-Millo E, Primo-Yu´fera E et al (1983) Dynamics of ^{15}N-labeled nitrogen nutrients in Valencia orange trees. In: Matsumoto K, Oogaki C, Kuzaki I et al (eds) Proceedings of the international society of citriculture, V international citrus congress, Tokyo, Japan, 9–12 Nov 1981, vol 2, pp 575–582

Legaz F, Serna MD, Primo-Millo E (1995) Mobilization of the reserve N in citrus. Plant Soil 173:205–210

London JG (2005) Nitrogen study fertilizes fears of pollution. Nature 433:791

Lynch JP (1995) Root architecture and plant productivity. Plant Physiol 109:7–13

Malavolta E (1992) Leaf analysis in Brazil – present and perspectives. In: Tribulato E, Gentile A, Refrigerato G (eds) Proceedings of the international society of citriculture, Catania, Italy, 8–13 Mar 1992, vol 2, pp 570–574

Martìnez JM, Bañuls J, Quiñones A et al (2002) Fate and transformations of ^{15}N labeled nitrogen applied in spring to Citrus trees. J Hortic Sci Biotechnol 77:361–367

Mattos D Jr, Graetz DA, Alva AK (2003) Biomass distribution and nitrogen-15 partitioning in citrus trees on a sandy entisol. Soil Sci Soc Am J 67:555–563

Maust BE, Williamson JG (1994) Nitrogen nutrition of containerized citrus nursery plants. J Am Soc Hortic Sci 119:195–201

Melgar JC, Schumann AW, Syvertsen JP (2010) Fertigation frequency affects growth and water and nitrogen use efficiencies of Swingle Citrumelo citrus rootstock seedlings. HortSci 45(8):1255–1259

Menino MR, Carranca C, de Varennes A (2007) Distribution and remobilization of nitrogen in young non-bearing orange trees grown under Mediterranean conditions. J Plant Nutr 30:1083–1096

Moll RH, Kamprath EJ, Jackson WA (1982) Analysis and interpretation of factors which contribute to efficiency of nitrogen utilization. Agron J 74:562–564

Moll RH, Kamprath EJ, Jackson WA (1987) Development of nitrogen-efficient prolific hybrids of maize. Crop Sci 27:181–186

Morgan KT, Wheaton TA, Castle WS et al (2009) Response of young and maturing citrus trees grown on a sandy soil to irrigation scheduling, nitrogen fertilizer rate, and nitrogen application method. HortSci 44(1):145–150

Obreza TA, Morgan KT (2008) Nutrition of Florida citrus trees. SL253. University of Florida, Lake Alfred

Quaggio JA, Cantarella H, van Raij B (1998) Phosphorus and potassium soil test and nitrogen leaf analysis as a base for citrus fertilization. Nutr Cycl Agroecosyst 52:67–74

Quiñones A, Bañuls J, Primo-Millo E (2003) Effects of ^{15}N application frequency on nitrogen uptake efficiency in Citrus trees. J Plant Physiol 160:1429–1434

Quiñones A, Bañuls J, Primo-Millo E et al (2005) Recovery of the ^{15}N-labeled fertiliser in Citrus trees in relation with timing of application and irrigation system. Plant Soil 268:367–376

Quiñones A, Màrtinez-Alcàntara B, Legaz F (2007) Influence of irrigation system and fertilization management on seasonal distribution of N in the soil profile and on N-uptake by citrus trees. Agric Ecosyst Environ 122:399–409

Ramos C (1996) Effect of agricultural practices on the nitrogen losses to the environment. Fertil Res 43:183–189

Ramos C, Agut A, Lidòn AL (2002) Nitrate leaching in important crops of the Valencian community region (Spain). Environ Pollut 118:215–223

Ryser P, Lambers H (1995) Root and leaf attributes accounting for the performance of fast- and slow-growing grasses at different nutrient supply. Plant Soil 170:251–265

Saric MR (1982) Theoretical and practical approaches to the genetic specificity of mineral nutrition of plants. In: Saric MR (ed) Genetic specificity of mineral nutrition of plants. Proceedings of the international symposium, Belgrade, pp 9–20

Sattelmacher B, Gerendas J, Thoms K et al (1993) Interaction between root growth and mineral nutrition. Environ Exp Bot 33:63–73

Saurbeck DR, Helal HM (1990) Factors affecting the nutrient efficiency of plants. In: El Bassam N, Dambroth M, Loghman BC (eds) Genetic aspects of plant mineral nutrition. Kluwer Academic Publishers, Dordrecht, pp 11–16

Scholberg JMS, Parsons LR, Wheaton TA et al (2002) Soil temperature, nitrogen concentration, and residence time affect nitrogen uptake efficiency in citrus. J Environ Qual 31:759–768

Schumann AW, Fares A, Alva AK et al (2003) Response of Hamlin orange to fertilizer source, annual rate and irrigated area. Proc Fla State Hortic Soc 116:256–260

Sorgonà A, Cacco G (2002) Linking the physiological parameters of nitrate uptake with root morphology and topology in wheat (*Triticum durum* Desf.) and in citrus rootstock (*Citrus volkameriana* Ten & Pasq). Can J Bot 80:494–503

Sorgonà A, Abenavoli MR, Cacco G (2005) A comparative study between two citrus rootstocks: effect of nitrate on the root morpho-topology and net nitrate uptake. Plant Soil 270:257–267

Sorgonà A, Abenavoli MR, Gringeri PG et al (2006) A comparison of nitrogen use efficiency definitions in Citrus rootstocks. Sci Hortic 109:389–393

Sorgonà A, Abenavoli MR, Gringeri PG (2007) Comparing morphological plasticity of root orders in slow- and fast-growing citrus rootstocks supplied with different nitrate levels. Ann Bot 100:1287–1296

Sorgonà A, Cacco G, Di Dio L et al (2010) Spatial and temporal patterns of net nitrate uptake regulation and kinetics along the tap root of *Citrus aurantium*. Acta Physiol Plant 32:683–693

Sorgonà A, Lupini A, Abenavoli MR (2011) Nitrate use-efficiency: a morphological analysis of the above- and below-ground functional traits in two citrus rootstocks. Glob J Plant Ecophysiol 1(1):26–37

Spiegel-Roy P, Goldschmidt EE (1996) Biology of Citrus. Cambridge University Press, New York

Stulen I, Prez-Soba M, De Kok LJ et al (1998) Impact of gaseous nitrogen deposition on plant functioning. New Phytol 139:61–70

Syvertsen JP, Dunlop JM (2004) Hydrophilic gel amendments to sand soil can increase growth and nitrogen uptake efficiency of citrus seedlings. HortSci 39(2):267–271

Syvertsen JP, Jifon JL (2001) Frequent fertigation does not affect citrus tree growth, fruit yield, nitrogen uptake, and leaching losses. Proc Fla State Hortic Soc 114:88–93

Syvertsen JP, Sax SM (1999) Fertigation frequency, wetting patterns and nitrate leaching from lysimeter-grown citrus trees. Proc Fla State Hortic Soc 112:9–14

Syvertsen JP, Smith ML (1996) Nitrogen uptake efficiency and leaching losses from lysimetric-grown citrus trees fertilized at three nitrogen rates. J Am Soc Hortic Sci 121:57–62

Terblanche JH, Du Plessis SF (1992) Summary of workshop on leaf and soil analysis as a tool for determining fertilizer requirements of citrus. In: Tribulato E, Gentile A, Refrigerato G (eds) Proceedings of the international society of citriculture, Catnia, Italy, 8–13 Mar 1992, vol 2, pp 744–745

Thung M (1988) Phosphorus: a limiting nutrient in bean (*Phaseolus vulgaris* L.) production in Latin America and field screening for efficiency and response. In: El Bassam N, Dambroth M, Loghman BC (eds) Genetic aspects of plant mineral nutrition. Kluwer Academic, Dordrecht, pp 501–521

Tian Q, Chen F, Zhang F et al (2007) Genotypic difference in nitrogen acquisition ability in maize plants is related to the coordination of leaf and root growth. J Plant Nutr 29(2):317–330

Tischner R (2000) Nitrate uptake and reduction in higher and lower plants. Plant Cell Environ 23:1005–1024

Tucker DPH, Alva AK, Jackson LK et al (1995) Nutrition of Florida citrus trees. Publication # SP 169. University of Florida-IFAS, Gainesville

Walch-Liu P, Filleur S, Gan Y et al (2005) Signaling mechanisms integrating root and shoot responses to changes in the nitrogen supply. Photosynth Res 83:239–250

Wallace A (1953) Nitrogen absorption and translocation by citrus cuttings at different root temperatures. Proc Am Soc Hortic Sci 61:89–94

Weinbaum S, Van Kessel C (1998) Quantitative estimates of uptake and internal cycling of ^{14}N-labeled fertilizer in mature walnut trees. Tree Physiol 18:795–801

Weinbaum SA, Klein I, Broadbent FE et al (1984) Effects of time of nitrogen application and soil texture on the availability of isotopically labeled fertilizer nitrogen to reproductive and vegetative tissue of mature almond tree. J Am Soc Hortic Sci 109:339–343

Weinert TL, Thompson TL, White SA (2002) Nitrogen fertigation of young navel oranges: growth, N status, and uptake of fertilizer N. HortSci 37(2):334–337

Wutscher HK (1989) Alteration of fruit tree nutrition through rootstocks. HortSci 24(4):578–583

Zekri M, Obreza T (2003) Macronutrient deficiencies in citrus: nitrogen, phosphorus, and potassium, SL210. Florida Cooperative Extension Service, Institute of Food and Agricultural Science, University of Florida, Gainesville. Downloaded at http://edis.ifas,ufl.edu

Nitrogen Metabolism in Citrus Based on Expressed Tag Analysis

17

Jackson Marcondes and Eliana G.M. Lemos

Abstract

Nitrogen metabolism in citrus arouses special interest among other plant nutrients due to its important role in plant growth and fruit production. Nitrogen is first assimilated in organic forms as glutamine, glutamate, aspartate, and asparagine from ammonium or nitrate. This vital process integrates pathways from energetic, central intermediary, and biosynthetic metabolism routes, culminating in the translocation of compounds by all parts of the plant. From a number of essential enzymes involved in this process, we depicted seven enzymes named nitrate reductase (NR), nitrite reductase (NiR), glutamine synthetase (GS), glutamate synthase (GOGAT), glutamate dehydrogenase (GDH), aspartate aminotransferase (AspAT), and asparagine synthetase (AS) for gene prospection based on EST sequencing of *Citrus sinensis*, *C. reticulata*, *C. aurantifolia*, *C. latifolia*, *C. aurantium*, and *Poncirus trifoliata*. These species are part of the CitEST Brazilian program which is focusing on efforts to obtain a great number of EST (expressed sequence tags) related to different citrus species and genera at different developmental stages or under biotic or abiotic stresses. Irrespective of origin, conditions, and nucleotide similarities, citrus species in general conserve a precise set of enzymes for nitrogen metabolism.

Keywords

Nitrogen assimilation • Ammonium assimilation • Amino acids biosynthesis • CitEST

17.1 Introduction

The development of genetics and genomics applied to citrus plants has been very important in elucidation of nitrogen metabolism in a global view. This knowledge is essential, considering the variety of crop conditions and the dissemination of devastating citrus diseases worldwide that arrest the production of a desirable husbandry. Citrus genome (*n*) is approximately 3.67 Mbp and is virtually inappropriate to be completely sequenced and analyzed. In this way, expressed sequence tags (ESTs) are particularly interesting, considering the potential of comparative functional genomics analyses for discovery of RNAs; those catalyses produce viable enzymes in different conditions (Talon and Gmitter 2008).

EST sequencing is a viable, fast, and inexpensive way to identify new genes in association with their pattern of expression and regulation, whose data can be used for construction of metabolic maps. Brazilian CitEST (Expressed Citrus genome) program had concentrated efforts in order to obtain a great number of EST related to different citrus species and genera at different developmental stages or under biotic or abiotic stresses (Table 17.1). More than 280,000 reads were generated from a total of 35 cDNA libraries corresponding to the largest citrus sequence database project in the world (Targon et al. 2007; Reis et al. 2007).

J. Marcondes (✉) • E.G.M. Lemos
Departmento de Tecnologia, Faculdade de Ciências Agrárias e Veterinárias, Universidade Estadual Paulista, São Paulo, Brazil
e-mail: jackson@fcav.unesp.br; egerle@fcav.unesp.br

Table 17.1 CitEST cDNA libraries

Species	Common name	Tissue	Conditions
Citrus sinensis	Sweet orange	Bark, flower, fruit, leaf	Healthy greenhouse and young plants, fruit development stages, infected or not with *Xylella fastidiosa*, infected or not with CiLV
Citrus reticulata	Mandarin	Fruit, leaf	Fruit development stages, infected or not with *Xylella fastidiosa*
Citrus limonia	Rangpur lime	Root	With or without hydric stress
Citrus latifolia	Tahiti lime	Leaf	Healthy greenhouse plant
Citrus aurantifolia	Key lime	Leaf	Healthy greenhouse plant
Citrus limettioides	Sweet lime	Leaf	Infected with CiLV
Citrus aurantium	Sour orange	Leaf	Healthy field plant
Poncirus trifoliata	Trifoliate orange	Bark, leaf, seed	Healthy greenhouse plant, fruit development stages, infected or not with *Phytophthora parasitica*, infected or not with CTV
Citrus sunki	Sunkat	Bark	Healthy greenhouse plant

Adapted from Targon et al. (2007)

17.2 Nitrogen for Citrus Production

World citrus production has been stimulated by progressive consumption of fruits and processed citrus products since the 1980s. Important areas for citrus production are distributed in Americas, Mediterranean countries, and Asian regions, even if São Paulo/Brazil and Florida/USA maintain the major positions as largest producers (FAO 2011).

The varieties of sweet orange (*Citrus sinensis* L. Osbeck) correspond to the most appreciated and produced citrus type in the world, followed by grapefruit (*C. paradisi* Macf.), pomelos (*C. maxima* Merr.), lemons (*C. limon* L. Burm.), limes (*C. aurantifolia* [Christm.] Swing), and mandarins (*C. reticulata* Blanco). There are other phylogenetic-related genera compatible with *Citrus,* and their production is encouraged by the diversity of edible fruit types – e.g., kumquats (*Fortunella japonica* (Thunb.) Swingle) – or by their importance for the improvement of rootstocks and scions – e.g., trifoliate orange (*Poncirus trifoliata* L. Raf.) (Talon and Gmitter 2008).

Nutrients are essential to proper metabolic functioning of plants and to assure reliable production in agroeconomic systems. Citrus trees consist mainly of carbon compounds and water in addition to nutrients that comprise a small percentage of the total fresh weigh. Nitrogen (N) is the element required in large quantities and can be considered very important together other macronutrients as P, K, Mg, Ca, and S, besides Mn, Cu, Zn, B, Fe, and Mo that act as micronutrients (Smith 1966a; Chapman 1968).

Nitrogen is a component particularly important for citrus growth and development, attaining acceptable commercial yields when present in adequate amounts.

Nitrogen fertilizer is commonly required in greater amounts by citrus, and its amount removed from an acre of soil corresponds to ~1.2 kg of N by 1.0 ton of fresh fruits (Erickson 1968). Nitrate is the major form for nitrogen absorption by citrus under normal arable soils conditions. In acidified or non-aerated soils, ammonia could be the predominant form to be assimilated by plants. There are relates that ammonia form can be used by citrus, but a general improvement of plant status was preferentially observed with nitrate (Kato 1986).

Nitrate is commonly deficient due to rapid leaching from the soil and uptake by the tree. In this case, application of N as urea, nitrate, ammonia, or combinations is indicated in all fertilizer programs worldwide (Sharples and Hilgeman 1969; Jackson and Davies 1984; DuPlessis and Koen 1988). Both in greenhouse and in the field, cultural practices in the nursery and during plant husbandry over-irrigate and over-fertilize the plants attending that trees will become profitable sooner (Castle and Rouse 1990; Maust 1992). This practice is not recommended because N excess is as dangerous as N deficiency for fruit yields and quality.

Fruit produced on low or nitrogen-deficient trees is generally smooth, but yields are low. On the other hand, nitrogen excess has direct effects on delaying fruit maturity prior to promoting vegetative growth and development (Bouma and Mcintyre 1963). An appreciable amount of nitrogen can be accumulated very rapidly by citrus roots, and the rate of absorption is largely influenced by temperature; absorption is less efficient mainly when soil temperature is lower (Kato et al. 1982; Kato 1986).

Classification of optimum nutrient status for sweet orange trees based on concentration of mineral elements in dry matter relies on 2.5–2.7% N in leaves from non-fruiting terminals (Smith 1966b). Ranges bellow or above this general guideline indicate deficiency or excess of N, and both cases present deleterious effects for orchards depending on every situation and cultivar.

17.3 Nitrogen Metabolism in Citrus Plants

Nitrogen (N_2) fixation by any way was not reported in citrus plants. Thus, assimilation of N is due to nitrate or ammonium uptake. Nitrogen seems to be translocated easily within the tree (Wallace et al. 1954), but ion uptake requires energy. At cellular level, carbon and nitrogen pathways become interconnected once carbohydrates metabolism provides energy for nitrogen assimilation. In this case, light affects these processes supplying reducing agents to drive both metabolisms. During darkness, substrate reserve storage in leaves acts as powerful reservoirs for sustainable metabolism (Neuhaus and Emes 2000).

Nitrate uptake can be reduced in presence of ammonium probably by proper inhibition of nitrate uptake or by reduction of nitrate reductase activity (Frith and Nichols 1975; Haynes and Goh 1978). Nitrate may be processed by booth root and leaf cells, and its accumulation in vacuoles reflects an important role as osmotic agent. In contrast, ammonium ions are toxic and have to be incorporated in organic compounds quickly (Chaillou et al. 1994). Variation in metabolic processing of nitrogen sources is defined by their chemical nature and diversity of plant and environmental conditions.

Independent of nitrogen sources, assimilation in organic compounds occurs by the same pathway and derives from ammonium assimilation into plant cells. So, additional energy is imputed in order to reduce nitrate to ammonium prior to its incorporation in organic compounds (Miflin and Lea 1977). Inorganic nitrogen is assimilated and primarily incorporated in organic forms into glutamine, glutamate, asparagines, and aspartate, which in turn constitute the main nitrogen carriers within the plants. These primary amino acids are the major compounds that are used as precursors for the synthesis of other amino acids, amides, and amines, culminating in the production of purine, pyrimidine, and other related reduced nitrogen-carbon compounds (Lam et al. 1996; Wickert et al. 2007).

The conversion of inorganic nitrogen to essential organic nitrogen involves the co-integration of carbon skeletons from tricarboxylic acid cycle and the participation of enzymes in various steps (Table 17.2). Inorganic nitrogen must be first reduced to ammonium (NH_4^+) so that it is incorporated into organic forms. Nitrate (NO_3^-) reduction to NH_4^+ is mediated by two key enzymes, nitrate reductase (NR) and nitrite reductase (NiR). NR catalyzes the conversion of NO_3^- into nitrite (NO_2) by reduction of two electrons, following the reduction of NO_2^- into NH_4^+ by NiR through transfer of six electrons. NH_4^+ is then converted by glutamine synthetase (GS) to glutamine, the first organic molecule containing nitrogen by this process (Fig. 17.1).

Glutamine is essential for organic nitrogen compounds synthesis through its first association with α-ketoglutarate by the action of glutamate synthase (GOGAT) yielding glutamate. Levels of α-ketoglutarate in turn are equilibrated by action of glutamate dehydrogenase (GDH) that still controls the rate of reducing agents (NAD^+/NADH) and NH_4^+. Finally, glutamate together with oxaloacetate can be used for generation of aspartate and asparagine by sequential action of aspartate aminotransferase (AspAT) and asparagine synthetase (AS) (Fig. 17.1). Glutamate and aspartate are ready to evolve all metabolic routes that will give rise to nitrogen organic molecules required for a sustainable plant growth and development.

Table 17.2 Biochemical reactions for nitrogen assimilatory enzymes in plants

Enzyme	Reaction
NR	$NO_3^- + NAD(P)H + H^+ + 2\ e^- = NO_2^- + NAD(P) + H_2O$
NiR	$NO_2^- + Fd\ (red) + 8\ H^+ + 6\ e^- = NH_4^+ + Fd\ (ox) + 2\ H_2O$
GS1/GS2	Glutamate + NH_4^+ + ATP = glutamine + ADP + Pi
Fd-GOGAT	Glutamine + 2-oxoglutarate + 2 Fd (red) = 2 glutamate + 2 Fd (ox)
NADH-GOGAT	Glutamine + 2-oxoglutarate + 2 NADH = 2 glutamate + NAD
GDH	Glutamate + H_2O + NAD/NADP = NH_4^+ + 2-oxoglutarate + NADH/NADPH
AspAT	Glutamate + oxaloacetate = aspartate + 2-oxoglutarate
AS	Glutamine + aspartate + ATP = asparagine + glutamate + AMP + PPi

Adapted from Lam et al. (1996)

Abbreviations: *NR* nitrate reductase, *NiR* nitrite reductase, *GS1* cytoplasmic glutamine synthetase, *GS2* chloroplastic glutamine synthetase, *Fd-GOGAT* ferredoxin-dependent glutamate synthase, *NADH-GOGAT* NADH-dependent glutamate synthase, *GDH* glutamate dehydrogenase, *AspAT* aspartate aminotransferase, *AS* asparagine synthetase

The data mining for genes acting in nitrogen metabolism was performed by Wickert and colleagues (2007) based on CitEST database for different species and conditions of citrus plants. The genes for the main enzymes, including nitrate reductase (NR), nitrite reductase (NiR), glutamine synthetase (GS), glutamate synthase (GOGAT), glutamate dehydrogenase (GDH), aspartate aminotransferase (AspAT), and asparagine synthetase (AS), were found in citrus plants, revealing a complete set of enzymes for the vital processes of nitrogen assimilatory pathways.

Apart from the nine species investigated by CitEST program (Table 17.1), enzymes related to nitrogen metabolism were found only in six citrus species: *Citrus sinensis* (CS), *C. reticulata* (CR), *C. aurantifolia* (CG), *C. latifolia* (LT), *C. aurantium* (CA), and *Poncirus trifoliata* (PT) (Table 17.3). Until now, investigation on CitEST database has revealed 399 reads related to the seven studied enzymes. *C. sinensis* was the citrus species with the highest number of sequences related to those enzymes (198), followed by *C. reticulata* with 110 reads, and *P. trifoliata* with 61 reads. *C. aurantium* showed the lowest number of sequences (6) related to nitrogen metabolism enzymes.

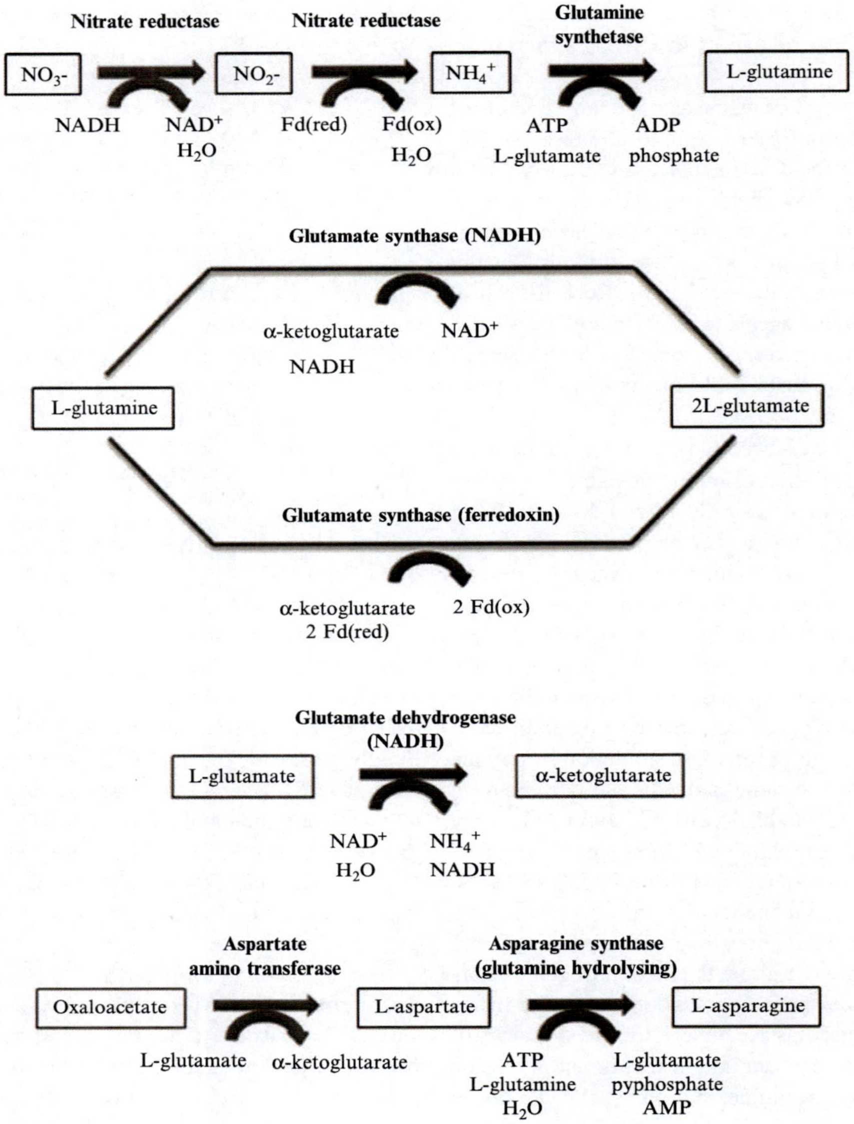

Fig. 17.1 Conversion of inorganic nitrogen to essential organic nitrogen compounds based on AraCyc, showing nitrate assimilation pathway and biosynthesis of glutamine, ammonium assimilation cycle and biosynthesis of glutamate, glutamate degradation, and biosynthesis of aspartate and asparagine (Wickert et al. 2007)

Entries for all enzymes investigated were signed in *P. trifoliata*, including different forms of GOGAT. By the time, NADH-GOGAT was found just in *P. trifoliata* and *C. aurantium*. On the other hand, Fd-GOGAT was present in all six species in which nitrogen metabolism genes were unveiled by EST sequencing, as well as cytoplasmic and chloroplastic forms of GS. Although *C. sinensis* and *C. reticulate* have all enzymes genes too, just Fd-GOGAT form was found for these species. GDH was not encountered in *C. latifolia* and *C. aurantium*; AspAT was absent in *C. aurantifolia* and *C. aurantium*; while AS was present only in *C. sinensis* and *C. aurantifolia*.

With respect to nitrate assimilation in our investigation, NR was not found in *C. latifolia* and *C. aurantium,* although present in *C. sinensis*, *C. reticulata*, *C. aurantifolia*, and *P. trifoliata*. Neither NiR was found in *C. latifolia*, *C. aurantium*, including *C. aurantifolia* too, but present in others.

Root libraries would have been the primary focus on the nitrogen assimilatory analysis, but they were so scarce that render little contribution (Wickert et al. 2007) even until now.

Table 17.3 EST distribution by CitEST libraries

Enzyme	Reads by plant species (%)						Reads by tissue or condition (%)					
	CS	CR	PT	CG	LT	CA	Leaf	Bark	Fruit	Root	Healthy	Sick
NR	14.3	38.1	33.3	14.3	–	–	81.0	–	4.8	–	57.1	23.8
NiR	57.1	14.3	28.6	–	–	–	57.1	–	42.9	–	28.6	28.6
GS1/GS2	46.1	26.3	18.0	4.8	3.0	1.8	70.1	3.6	26.3	–	34.1	34.7
Fd-GOGAT	47.7	37.5	9.1	2.3	1.1	2.3	59.1	1.1	42.0	–	23.9	27.3
NADH-GOGAT	–	–	50.0	–	–	50.0	100.0	–	–	–	50.0	50.0
GDH	57.1	9.5	28.6	4.8	–	–	38.1	19.0	42.9	–	52.4	4.8
AspAT	62.4	25.9	8.2	–	3.5	–	48.0	6.0	40.0	6.0	38.0	14.0
AS	87.5	–	–	12.5	–	–	37.5	–	62.5	–	25.0	12.5

Abbreviations: CS *Citrus sinensis*, CR *Citrus reticulata*, PT *Poncirus trifoliata*, CG *Citrus aurantifolia*, LT *Citrus latifolia*, CA *Citrus aurantium*

Bark libraries were poor too, but they have shown entries for more different enzymes. By the time, the leaf and fruit libraries were the most abundant in reads both in healthy and sick conditions (Table 17.3).

Furthermore, it is important to remember that EST sequencing corresponds to gene expression analysis based on availability and concentration of ribonucleic acids (RNAs) at the moment and conditions of tissues processing. In spite of synthesis and degradation, enzymes catalyzing reactions also are driven by activation and inactivation. These prerogatives must be considered during our further discussion.

17.4 Prospection of Enzymes Related to Nitrogen Metabolism Cycle

17.4.1 Assimilation of Nitrogen into Ammonium: Nitrate Reductase (NR) and Nitrite Reductase (NiR)

The extent to which nitrate assimilation occurs in root is a function of the species and growth conditions. The assimilation of nitrate begins with its uptake into the cell after it is taken from the soil by epidermal and cortical cells of the root. The initial step of inorganic nitrogen assimilation occurs in cytosol and is catalyzed by NR, through nitrate reduction to nitrite (Fig. 17.1). Many findings suggest that NR is present in various parts of citrus plants, even in a variety of fruit trees (Kato 1986). NR activities were reported in various organs of citrus plants, including germinating seeds, cotyledons, rootlets, shoots, leaves, and the flavedo and albedo of fruit (Bar-Akiva and Sagiv 1967).

The highest activity was reported in leaf fragments which is concordant with the abundance of reads in leaf libraries assessed by our investigation. Eighty-one percent of reads for NR was present in leaf, predominantly in healthy tissues (Table 17.3). NR activity is affected by a list of conditions as substrate supply, ammonium and amino acids, inorganic salts and ions, sugars and organic acids, growth regulators, seedling age and diurnal rhythms, temperature, light, hydric status, gaseous environment, and herbicides and fungicides, as well as antibiotics and other metabolic inhibitors (Srivastava 1980). Higher temperatures are so drastic to NR, like in most enzymes, that their activity could be inhibited and even not detected.

The regulation of both NR transcription and activity allows plants to adjust the amount of nitrate reduction. NR (EC 1.9.6.1) is an oxidoreductase described as an iron-sulfur molybdenum flavoprotein, depending on NAD^+ or $NADP^+$ as acceptors, while two protons (two electrons are required) are consumed in the reaction of nitrate assimilation. This complex metalloenzyme consists in homodimers and homotetramers and many different genes and forms for NR were already found both in mono- and dicotyledonous plants (Nason 1963; Srivastava 1992). NR transcription in photosynthetic tissues demonstrates a cyclic pattern, so that maximal transcript values are achieved just before the appearance of light (Crawford and Arst 1993). In citrus, NR was detected in leaf and fruit tissues and is probably mainly expressed in leaves. It was not detected in bark and root tissues, and there was reduced expression in sick plants (Table 17.3), in this case citrus variegated chlorosis (CVC) caused by infection with *Xylella fastidiosa*.

After nitrate reduction, the next step in the nitrate assimilation pathway consists in the reduction of nitrite to ammonium, which is catalyzed by NiR (Fig. 17.1). In plants, the source of electrons is reduced ferredoxin, produced in chloroplasts by photosynthetic noncyclic electron transfer. According to Kato (1986), who performed an excellent revision of nitrogen metabolism and utilization in citrus, NiR had not been reported in citrus yet. However, in CitEST database, NiR was observed predominantly in leaf and fruit tissues (Table 17.3) and probably has the same level of expression in both (Wickert et al. 2007). It was also observed both in healthy and sick plants manifesting CVC (*X. fastidiosa* infection), suggesting that its expression apparently was not affected by changes in plant health in this case. There was no entries for NiR in root or bark tissues. However, it is important to note the lowest number of reads identified to nitrogen metabolism in CitEST was for NiR.

NiR (EC 1.7.7.1) in plants is described as an iron protein that contains siroheme and [4Fe-4S] clusters, classified as an oxidoreductase. In contrast to the two electrons required for nitrate reduction by NR, six electrons are transferred by NiR to reduce nitrite into ammonium (Joy and Hageman 1966). There is substantial evidence that the reductant required for the conversion of nitrite to glutamate is generated by the oxidative pentose phosphate pathway (OPPP). Studies with isolated root amyloplasts have shown that the oxidation of glucose 6-phosphate by the OPPP is tightly coupled to the reactions catalyzed by NiR (Bowsher et al. 1989; Neuhaus and Emes 2000). NiR is regulated transcriptionally usually in coordination with NR.

Considering that nitrite is toxic, cells must contain enough NiR to reduce all the nitrite produced by NR. Thus, plants maintain an excess of NiR gene expression in response to light and nitrate. If NiR concentrations are diminished, either by mutation or antisense expression, plants accumulate nitrite and display chlorosis. In wild-type plants, the regulatory mechanisms that control NR activity are thought to assist in preventing nitrite accumulation (Wray 1993; Crawford et al. 2000).

17.4.2 Assimilation of Inorganic Nitrogen into Amino Acids

In plants, all inorganic nitrogen is first reduced to ammonium before it is incorporated into organic form beginning with amino acids. Apart from the primary role of amino acids in protein synthesis, they carry out significant role in global plant metabolism serving as precursors to primary and secondary products. Four amino acids are considered N-transport amino acids used to assimilate nitrogen and transport it from source organs to sink tissues. Inorganic nitrogen that was assimilated into glutamate and glutamine is readily disseminated by plant metabolism, once those organic forms provide nitrogen to the biosynthesis of amino acids, nucleic acids, and other N-containing compounds. Instead, glutamine and glutamate serve as nitrogen donor for the synthesis of aspartate and asparagines. Aspartate is a metabolically reactive amino acid, and nitrogen accumulated in this way can be used in various aminotransferase reactions. On the other hand, asparagine is relatively inert and is primarily used as a nitrogen transport and storage compound (Corruzi and Last 2000).

Many authors devoted importance to arginine, asparagine, and proline as the major free amino acids in citrus, probably due to their abundance in fresh tissues and processed citrus juices (Erickson 1968; Vandercook 1977, Kato 1982, 1986). In spite of that, glutamine, glutamate, aspartate, and asparagine are the major amino acids translocated in the phloem of most species and require special attention because of their metabolic importance. The major enzymes that synthesize these N-transport amino acids are glutamine synthetase, glutamate synthase, glutamate dehydrogenase, aspartate aminotransferase, and asparagine synthetase (Lam et al. 1996; Corruzi and Last 2000).

17.4.3 Glutamine Synthetase (GS), Glutamate Synthase (GOGAT), and Glutamate Dehydrogenase (GDH)

The primary route for the initial assimilation of ammonium into organic forms resides in the main enzymes GS, GOGAT, and GDH. Each of these enzymes occurs in multiple isoenzymatic forms encoded by distinct genes. The individual isoenzymes of GS, GOGAT, and GDH have been proposed to play roles in three major ammonium assimilation processes (Lam et al. 1996): primary nitrogen assimilation, reassimilation of photorespiratory ammonium, and reassimilation of recycled nitrogen (Fig. 17.1). GS catalyze the ATP-dependent assimilation of ammonium into glutamine, using glutamate as a substrate. GS function in a cycle with GOGAT, which in turn catalyzes the reductive transfer of the amide group from glutamine to α-ketoglutarate, forming two molecules of glutamate. GDH can catalyze both the synthesis of glutamate and its catabolism. In a sense, GDH catalyzes the amination of α-ketoglutarate, while in the reverse sense it catalyzes the deamination of glutamate to yield α-ketoglutarate and ammonium. Acting together, GS, GOGAT, and GDH close the main cycle of organic N-assimilation.

Biochemical and molecular studies have revealed that each enzyme is encoded by a gene family, wherein individual members encode distinct isoenzymes that are differentially regulated by environmental stimuli, metabolic control, developmental control, and tissue/cell-type specificity. There are molecular evidences that genes involved in nitrogen assimilation and metabolism are not constitutively expressed, but in turn they are carefully regulated by diverse factors (Lam et al. 1996).

Enzymatic nomenclature is based on the classes of isoenzymes characterized by their cellular localization and reaction specificity. Two classes of glutamine GS (EC 6.3.1.2) were postulated in function of subcellular localization. The distinct physiological roles of GS1 and GS2 have been implicated by their organ-specific distributions: GS1 is located in the cytosol while GS2 is present inside the chloroplast. For instance, because GS2 is the predominant isoenzyme in leaves, it has been proposed their major function in primary assimilation process of ammonium from nitrate and/or in the reassimilation of photorespiratory ammonium.

Because cytosolic GS1 is predominant in roots, it has been proposed to function in nitrogen assimilation, although root GS2 has also been implicated in this process. The finding that cytosolic GS1 is the predominant GS isoenzyme expressed during senescence in different plants species suggest that it plays a role in the mobilization of nitrogen for translocation and storage (Lea et al. 1990; Lam et al. 1996).

GS corresponds to the highest number of reads offered for nitrogen metabolism search in CitEST libraries. In spite of that, there is no clear evidence for distinction of plastidic and cytoplasmatic isoforms of GS on database. Most of the GS sequences searched on CitEST database were found in leaf tissues, meaning that probably these sequences belong to GS2 form. Interestingly, GS reads were equally distributed both in healthy and in sick tissues, mainly infected with *X. fastidiosa* but also with Citrus tristeza virus (CTV). The representation of cytosolic GS1 was possibly impaired in CitEST due to the few root libraries available (Table 17.3). GS1 and GS2 play separate and nonoverlapping metabolic roles and reveal entirely independent expression patterns in plants (Edwards et al. 1990).

In higher plants, there are two distinct forms of GOGAT (glutamate synthase named in function of glutamine-2-oxo-glutarate aminotransferase action) that use NADH (NADH-GOGAT: EC 1.4.1.14) or ferredoxin (Fd-GOGAT: EC 1.4.7.1) as the electron carrier (Lam et al. 1996). Both NADH and NADPH could be used by the first form, but due to the higher activity in the presence of NADH, enzyme is therefore termed NADH-GOGAT. It is located primarily in plastids of non-photosynthetic tissues such as roots (Suzuki and Gadal 1984), and again oxidation of glucose 6-phosphate by the OPPP is correlated to serve as electron donor to GOGAT (Bowsher et al. 1992). In nonleguminous plants, NADH-GOGAT may function in primary assimilation or reassimilation of ammonium released during amino acid catabolism (Miflin and Lea 1980). Contrary to NADH-GOGAT, Fd-GOGAT is located primarily in the leaf chloroplast where light leads to an increase in Fd-GOGAT protein and activity (Lea et al. 1990). These findings suggested that the physiological roles of Fd-GOGAT are related to light-inducible processes in leaves such as photosynthesis and photorespiration. Fd-GOGAT may also play a smaller role in non-photosynthetic tissues, because some Fd-GOGAT activity is associated with roots (Suzuki et al. 1982).

The searched libraries of CitEST presented the both forms of GOGAT. The NADH-GOGAT was represented only by two sequences, but they were located in leaves. However, Fd-GOGAT was represented by 88 sequences, all located in leaves and developing fruits, agreeing with the findings that this isoenzyme is uniquely found in photosynthetic organisms. Sickness conditions again have shown to not affect distribution of those enzymes either by CVC or CTV (Table 17.3).

Two major forms of GDH have been reported: an NADH-dependent form (NADH-GDH: EC 1.4.1.2) found in the mitochondria (Day et al. 1988) and an NADPH-dependent form (NADPH-GDH: EC 1.4.1.4) located at the chloroplast (Lea and Thurman 1972). The GDH enzyme is abundant in several plant organs (Cammaerts and Jacobs 1985). Moreover, the GDH isoenzymatic profile can be influenced by dark stress, natural senescence, or fruit ripening (Srivastava and Singh-Rana 1987). These studies suggest that GDH may play a specific or unique role in assimilating ammonium or catabolizing glutamate during these processes. GDH presents a K_m for ammonium much higher than that presented by GS, which placed GDH as the major enzyme for primary nitrogen assimilation (Kato 1986; Lam et al. 1996). However, studies with ^{15}N showed that the primary route of ammonium assimilation in citrus roots is glutamine synthesis by GS, suggesting that GS/GOGAT display the major assimilatory pathway even when ammonium is plentiful (Kato et al. 1982). An anaplerotic function for GDH in mitochondria could also be active in reassimilation of ammonium, particularly those released from photorespiration (Yamada and Oaks 1987). Instead, GDH develops a catabolic role supported by the fact that GDH activity is induced during germination and senescence, when amino acid catabolism occurs (Lea et al. 1990).

In CitEST, GDH enzyme was found in leaves, developing fruits, and healthy barks. By the way, it was the most representative enzyme in bark libraries, considering those related to nitrogen metabolism. However, it was not possible to differentiate the two forms in citrus database (Table 17.3). In addition, GDH was predominant in healthy tissues showing sole one read for *C. reticulata* infected with CVC.

17.4.4 Aspartate Aminotransferase (AspAT) and Asparagine Synthetase (AS)

Following the assimilation of ammonia into glutamine and glutamate, nitrogen can be distributed to many other compounds by the action of transaminases. Transamination reactions based on glutamate regenerate α-ketoglutarate, an important precursor for ammonium assimilation. In addition, synthesis of aspartate regenerates the carbon skeletons required for further nitrogen assimilation by transferring an amino group from glutamate to oxaloacetate (Corruzi and Last 2000) (Fig. 17.1).

So, AspAT (EC2.6.1.1) plays a key role in nitrogen assimilation in plants and its integration with carbon metabolism, besides the central role in both aspartate synthesis and catabolism. AspAT is also known as glutamate:oxaloacetate aminotransferase (GOT), a pyridoxal phosphate-dependent enzyme (Forest and Wightman 1973). The activities of various AspAT isoenzymes have been found in different tissues and different subcellular locations such as the cytosol, mitochondria, chloroplasts, glyoxysomes, or peroxisomes (Scultz and Coruzzi 1995; Wilkie et al. 1995). The subcellular compartmentation of AspAT isoenzymes suggests that the different forms of AspAT might serve distinct roles in plant metabolism. It is also important to note that individual AspAT isoenzymes respond differently to environmental conditions and metabolic status such as light treatment or nitrogen starvation, which suggests that they serve distinct roles (Lam et al. 1996).

In the C3 plant Arabidopsis, the entire gene family of AspAT isoenzymes has been characterized (Scultz and Coruzzi 1995). Five different AspAT cDNA clones were obtained, including those encoding the mitochondrial, plastidic, peroxisomal, and cytosolic forms of AspAT. Even though two of the five *ASP* genes (formally *AAT* genes) encode cytosolic forms of AspAT (*ASP2* and *ASP4*), only *ASP2* is expressed at high levels, especially in roots. The *ASP1* and *ASP3* genes, which encode a mitochondrial and a peroxisomal form of AspAT, respectively, are each expressed at relatively high levels in all organs examined (Scultz and Coruzzi 1995).

In the searched libraries of citrus, AspAT reads were abundant in leaf and fruit tissues, mainly in healthy condition, but also was found in bark tree. In addition, AspAT was the unique enzyme found in root libraries from CitEST database. Besides the sickness like CVC and CTV, AspAT was also found in libraries from hydric-stressed plants showing this condition was insufficient to affect downstream metabolism of glutamate (Table 17.3). The form *ASP3* was founded in healthy leaves and developing fruits.

Amidation of aspartate by glutamine or ammonium yields asparagine. Despite the fact that asparagine was the first amino acid discovered and isolated in asparagus, the mechanism of asparagine biosynthesis in plants has been elucidated only recently. The glutamine-dependent asparagine synthetase enzyme (AS:EC6.3.5.4) is now generally accepted as the major route for asparagine biosynthesis in plants (Richards and Schuster 1992). In an ATP-dependent reaction, AS catalyzes the transfer of the amido group from glutamine to aspartate, generating glutamate and asparagine. However, ammonium is also a possible AS substrate, particularly in the case of maize roots (Oaks and Ross 1984). In some cases, asparagine is believed to act as an ammonium detoxification product produced when plants encounter high concentrations of ammonium.

The hypothesis that asparagine serves to transport nitrogen in plants is supported by high levels of AS activity detected in nitrogen-fixing root nodules and in cotyledons of germinating seedlings (Lam et al. 1996). The first two cDNA clones encoding plant AS were obtained from a pea library, and both *AS1* and *AS2* genes are expressed in leaves as well as in roots. Subsequently, studies of AS cDNA clones isolated from *Arabidopsis* and asparagus have shown that AS genes in these plants are expressed primarily in the leaves or the harvested spears, respectively (Lam et al. 1994). Moreover, studies of these AS cDNA clones, together with the previous biochemical data, have suggested that asparagine metabolism is regulated by the carbon/nitrogen status of a plant (Lam et al. 1994). The levels of asparagine and AS activities are also controlled by environmental and metabolic signals. The first striking observation of AS gene expression in pea and *Arabidopsis* was the high level of AS mRNA in dark-grown or dark-adapted plants (Urquhart and Joy 1982; Tsai and Coruzzi 1990; Lam et al. 1996).

In CitEST database, AS reads were present in low number and represented in leaf and predominantly in fruit tissues (Table 17.3). Due to low yield, it is difficult to assume if a healthy condition is important in AS expression.

17.5 Similarity Analysis

Nucleotide similarity analysis for all enzymes discussed for nitrogen metabolism has revealed no differences based on origin and conditions, meaning that there is similarity among the enzymes of different tissues, healthy or sick, and especially in different stages of development and maturation. On the other hand, there are dissimilarities among enzymes from different species, particularly for those from *P. trifoliate* and species of *Citrus*. Therefore, these were not so different as to prevent the generation of consensus sequence for citrus (Wickert et al. 2007).

Dissimilarities were also found among the aligned sequences from CitEST and those deposited in the NCBI databanks. A comparative analysis of the consensus sequence of the selected enzymes in *Citrus* and sequences for other plants in public database was performed. It was found that all searched enzymes show alignments with dicotyledonous plants than those for monocotyledonous like rice (*Oryza sativa*), corn (*Zea mays*), and wheat (*Triticum aestivum*), supporting the idea that the nitrogen metabolism is conserved among the plant species.

Citrus consensus for NR showed more similarity to *Malus domestica* (apple) and segregated from *P. trifoliate* which aligned in another branch with *Glycine max* (soybean), *Prunus persica* (peach), and *Hordeum vulgare* (barley). For NiR, the greatest similarity was noticed between citrus and *Nicotiana benthamiana*, a close relative of tobacco (Fig. 17.2). For enzymes related to ammonium assimilation, GS showed highest similarity to *Gossypium hirsutum* (cotton) and *G. max*, GOGAT with *Lupinus angustifolius* (a blue lupine legume), while for GDH there is high similarity with *P. persica* and *M. domestica*. Finally, it noticed more similarities for AspAT with *Prosopis juliflora* (the shrub algarroba or mesquite) and for AS with *G. hirsutum* (Fig. 17.2).

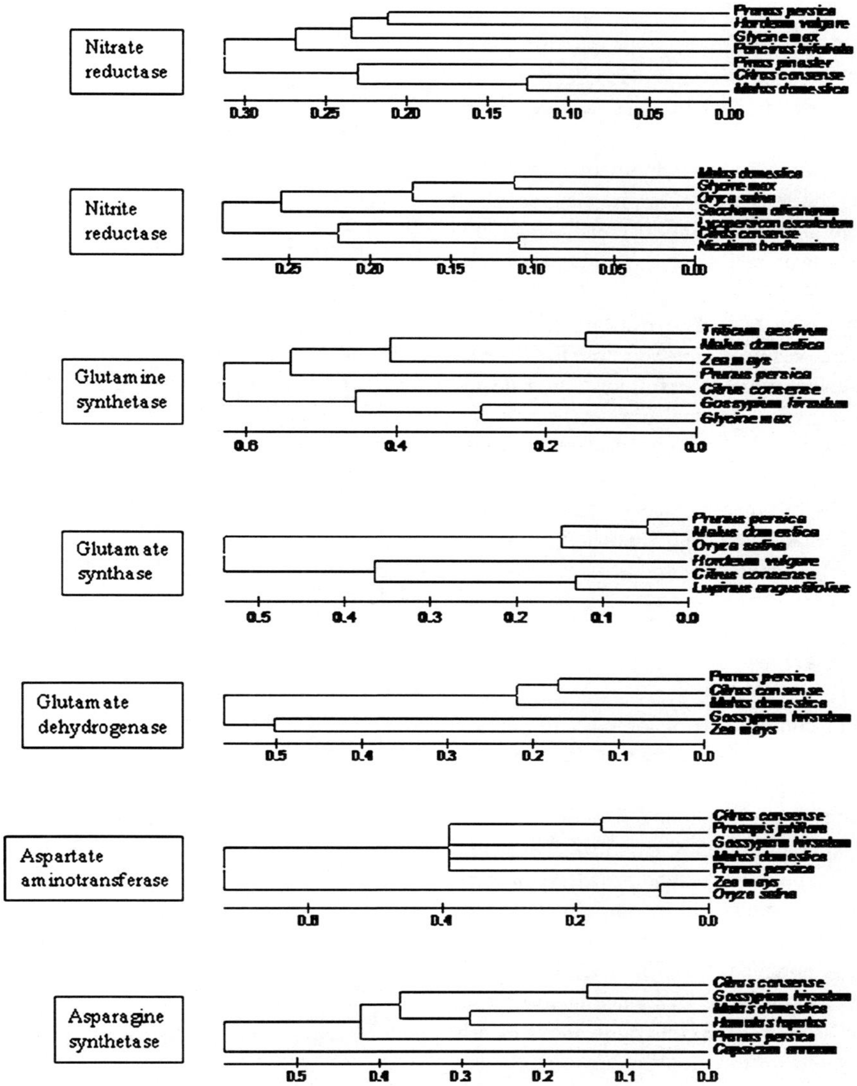

Fig. 17.2 General homology of the citrus consensus nucleotide sequences and other plants for enzymes related to nitrogen assimilation and metabolism

Mechanism of nitrogen assimilation seems to be highly conserved among plants, and that the differences found in DNA sequences are related to natural diversity not influenced by different stages of development, tissues, and conditions, besides the difference in levels of expression.

17.6 Conclusion

Biochemical, molecular, and genetic analysis of nitrogen metabolism have revealed important tools for citrus nutrition and fruit production. However, a study of nitrogen metabolism in different parts of the plant, including the fruit, based on gene expression pattern was necessary for a better comprehension of a comparative global view. This study was performed for us and included aspects relevant of phytosanitary status, which responds to more than 60% of production costs in our country. Brazilian researchers have concentrated efforts in CitEST program for citrus EST sequencing that allowed the development of a huge database for a collection of genes, physically located at Sylvio Moreira APTA Citrus Center (Agronomic Institute of Campinas), São Paulo/Brazil. The ESTs were generated from several libraries under biotic (*X. fastidiosa*, CTV, Citrus Leprosis Virus, *Phytophtora*, mite) and abiotic (drought) stresses for different species of citrus.

The similarity analysis of the sequences deposited in the CitEST database did not reveal significant differences between the different species of enzymes observed. Even though many mechanisms of regulation are involved in the transcriptional and translational control of these enzymes, different conditions of cultivation, development, and healthy did not present remarkable variations. Despite the great effort in the ESTs sequencing, no significant difference was found in the nitrogen metabolism among the studied sequences, which means that we have to keep up the efforts in more DNA sequencing, particularly for those from root libraries.

In spite of that, there was no doubt, and now there is further evidence for considerations that follow: citrus do not perform atmospheric nitrogen fixation; nitrate is not necessary but is sufficient for citrus nutrition; aerial parts display significant role in assimilation of nitrogen in organic compounds; sickness seems to not affect gene expression patterns; and finally, citrus species in general conserve a precise set of enzymes for nitrogen metabolism.

17.7 Future Research

Here we focus on the comprehension of part of nitrogen metabolism. Citrus plants are very different from leguminous plants and more complex than *Arabidopsis* model, so the prospection of genes must be expanded to those related to the metabolism of other amino acids and nitrogenous compounds, besides the mineral nutrition and growth regulators. The coverage of citrus genome is still incomplete. However, the advances in genomics based on the next generation DNA sequencing and the development of bioinformatics tools allow us to predict that citrus genome will be revealed in near future.

Genomics is so far used to solve constraints in citrus production, but this valuable information could be applied in taxonomic resolutions and breeding programs through comparative genomics, and in genetic manipulation through the identification of new genes and functions, beyond other perspectives. More strategies for consistent gene annotation and the conception of depositing information in the public domain are relevant topics to be addressed in this moment.

Apart from genomics, the citrus functional genomics is now an interesting reality which will be improved and moved forward to a global and systemic perception of citrus cell operation in all parts of the plant and in a diversity of growth and developmental conditions, as well as environmental interactions. In fact, this will be the improved knowledge which can be useful to citrus husbandry and industry.

Acknowledgments We are grateful to the cited Dr. E. Wickert and other authors, and also to Dr. M. A. Machado from Sylvio Moreira APTA Citrus Center (Agronomic Institute of Campinas), São Paulo/Brazil, which coordinated CitEST program and turned open the access to CitEST database.

References

Bar-akiva A, Sagiv J (1967) Nitrate reductase in the citrus plants: properties, assay conditions and distribution within the plant. Physiol Plant 20:500–506

Bouma D, Mcintyre GA (1963) A factorial field experiment with citrus. J Hortic Sci 38:175–198

Bowsher CG, Hucklesby DP, Emes MJ (1989) Nitrite reduction and carbohydrate metabolism in plastids purified from roots of *Pisum sativum* L. Planta 177:359–366

Bowsher CG, Boulton EL, Rose J et al (1992) Reductant for glutamate synthase is generated by the oxidative pentose phosphate pathway in non-photosynthetic root plastids. Plant J 2:893–898

Cammaerts D, Jacobs M (1985) A study of the role of glutamate dehydrogenase in the nitrogen metabolism of *Arabidopsis thaliana*. Planta 163:517–526

Castle WS, Rouse RE (1990) Total nutrient content of Florida citrus nursery plants. Proc Fla State Hortic Soc 103:42–44

Chapman HD (1968) The mineral nutrition of citrus. In: Reuther W, Batchelor LD, Webber HJ (eds) The citrus industry, vol 2, Anatomy, physiology, genetics, and reproduction. University of California, Division of Agricultural Sciences, Berkeley, pp 127–289

Chaillou S, Rideout JW, Raper CD, Morot-Gaudry JF (1994) Responses of soybean to ammonium and nitrate supplied in combination to the whole root system or separately in a split-root system. Physiol Plant 90:259–268

Corruzi G, Last R (2000) Amino acids. In: Buchanan BB, Gruissem W, Jones RL (eds) Biochemistry & molecular biology of plants. American Society of Plant Physiologists, Rockville, pp 358–410

Crawford NM, Arst HNJ (1993) The molecular genetics of nitrate assimilation in fungi and plants. Annu Rev Genet 27:115–146

Crawford NM, Kahn ML, Leustek T (2000) Nitrogen and sulfur. In: Buchanan BB, Gruissem W, Jones RL (eds) Biochemistry & molecular biology of plants. American Society of Plant Physiologists, Rockville, pp 786–849

Day DA, Salom CL, Azcon-bieto J et al (1988) Glutamate oxidation by soybean cotyledon and leaf mitochondria. Plant Cell Physiol 29:1193–1200

Duplessis SF, Koen TJ (1988) The effect of N and K fertilization on yield and fruit size of Valencia. In: Goren R, Mendal K (ed) Proceeding of sixth international citrus congress, Rehovot, pp 663–672

Edwards JW, Walker EL, Coruzzi GM (1990) Cell-specific expression in transgenic plants reveals nonoverlapping roles for chloroplast and cytosolic glutamine synthetase. Proc Natl Acad Sci U S A 87: 3459–3463

Erickson LC (1968) The general physiology of citrus. In: Reuther W, Batchelor LD, Webber HJ (eds) The citrus industry, vol 2, Anatomy, physiology, genetics, and reproduction. University of California, Division of Agricultural Sciences, Berkeley, pp 86–126

FAO (2011) Sugar, tropical beverage crops and fruits – citrus fruit. In: Medium-term prospects for agricultural commodities – Projections to the year 2010. Available at: http://www.fao.org/docrep/006/y5143e/y5143e12.htm#bm38

Forest JC, Wightman F (1973) Amino acid metabolism in plants. III. Purification and some properties of a multispecific aminotransferase isolated from bushbean seedlings (*Phaseo los* vulgaris L.). Can J Biochem 50:813–829

Frith GJT, Nichols DG (1975) Preferential assimilation of ammonium ions from ammonium nitrate solutions by apple seedlings. Physiol Plant 33:247–250

Haynes RJ, Goh KM (1978) Ammonium and nitrate nutrition of plants. Biol Rev 53:465–510

Jackson LK, Davies FS (1984) Mulches and slow-release fertilizers in a citrus young tree care program. Proc Fla State Hortic Soc 97:37–39

Joy KW, Hageman RH (1966) The purification and properties of nitrite reductase from higher plants, and its dependence on ferredoxin. Biochem J 100:263–273

Kato T (1982) Major nitrogen compounds transported in xylem vessels from roots to top in citrus trees. Physiol Plant 52:275–279

Kato T (1986) Nitrogen metabolism and utilization in citrus. Hortic Rev 8(5):181–216

Kato T, Kubota S, Rambang S (1982) Uptake and utilization of nitrogen by Satsuma mandarin trees in low temperature season. Bull Shikoka Agric Exp Station 40:1–5

Lam HM, Peng SSY, Coruzzi GM (1994) Metabolic regulation of the gene encoding glutamine-dependent asparagines synthetase in *Arabidopsis thaliana*. Plant Physiol 106:1347–1357

Lam HM, Coschigano KT, Oliveira IC et al (1996) The molecular-genetics of nitrogen assimilation into amino acids in higher plants. Annu Rev Plant Physiol Plant Mol Biol 47:569–593

Lea PJ, Thurman DA (1972) Intracellular location and properties of plant L-glutamate dehydrogenases. J Exp Bot 23:440–449

Lea PJ, Robinson SA, Stewart GR (1990) The enzymology and metabolism of glutamine, glutamate, and asparagine. In: Miflin BJ, Lea PJ (eds) The biochemistry of plants: intermediary nitrogen metabolism. Academic, New York, pp 121–159

Maust B (1992) Nutrition of citrus trees in the nursery. MSc thesis, University of Florida, Gainesville. Apud: Davies FS, Albrigo LG (1994) Citrus. Crop production science in horticulture, vol 2. CAB International, Wallingford

Miflin BJ, Lea PJ (1977) Amino acid metabolism. Annu Rev Plant Physiol 28:299–329

Miflin BJ, Lea PJ (1980) Ammonia assimilation. In: Miflin BJ (ed) The biochemistry of plants: amino acids and derivatives. Academic, New York, pp 169–202

Nason A (1963) Nitrate reductases. In: Boyer PD, Lardy H, Myrback K (eds) The enzymes, vol 7, 2nd edn. Academic, New York, pp 587–607

Neuhaus HE, Emes MJ (2000) Nonphotosynthetic metabolism in plastids. Annu Rev Plant Physiol Plant Mol Biol 51:111–140

Oaks A, Ross DW (1984) Asparagine synthetase in *Zea mays*. Can J Bot 62:68–73

Reis MS, Takita MA, Palmieri DA et al (2007) Bioinformatics for the Citrus EST Project (CitEST). Genet Mol Biol 30(3):1024–1029

Richards NGJ, Schuster SM (1992) An alternative mechanism for the nitrogen transfer reaction in asparagine synthetase. FEBS Lett 313:98–102

Scultz CJ, Coruzzi GM (1995) The aspartate aminotransferase gene family of *Arabidopsis* encodes isoenzymes localized to three distinct subcellular compartments. Plant J 7:61–75

Sharples GC, Hilgeman RH (1969) Influence of differential nitrogen fertilization on production, tree growth, fruit size and quality, and foliage composition of Valencia orange trees in central Arizona. Proc First Int Citrus Symp 3:1569–1578

Smith PF (1966a) Citrus nutrition. In: Childers NF (ed) Fruit nutrition. Horticultural Publications, New Brunswick, pp 174–207

Smith PF (1966b) Leaf analysis of citrus. In: Childers NF (ed) Fruit nutrition. Horticultural Publications, New Brunswick, pp 208–228

Srivastava HS (1980) Regulation of nitrate reductase activity in higher plants. Phytochemistry 19:725–733

Srivastava HS (1992) Multiple functions and forms of higher plant nitrate reductase. Phytochemistry 31:2941–2947

Srivastava HS, Singh-Rana P (1987) Role and regulation of L-glutamate dehydrogenase activity in higher plants. Phytochemistry 26: 597–610

Suzuki A, Gadal P (1984) Glutamate synthase: physicochemical and functional properties of different forms in higher plants and other organisms. Physiol Veg 22:471–486

Suzuki A, Vidal J, Gadal P (1982) Glutamate synthase isoforms in rice: immunological studies of enzymes in green leaf, etiolated leaf, and root tissues. Plant Physiol 70:827–832

Talon M, Gmitter FG (2008) Citrus genomics. Int J Plant Genomics ID 528361:17p

Targon MLPN, Takita MA, Amaral AM et al (2007) CitEST libraries. Genet Mol Biol 30(3):1019–1023

Tsai FY, Coruzzi GM (1990) Dark-induced and organ-specific expression of two asparagine synthetase genes in *Pisum sativum*. EMBO J 9:323–332

Urquhart AA, Joy KW (1982) Transport, metabolism, and redistribution of Xylem-borne amino acids in developing pea shoots. Plant Physiol 69:1226–1232

Vandercook CE (1977) Nitrogenous compounds. In: Nagy S, Shaw PE, Veldhuis MK (eds) Citrus science and technology. AVI, Westport, pp 239–265

Wallace A, Zidan ZI, Mueller RT et al (1954) Translocation of nitrogen in citrus trees. Proc Am Soc Hortic Sci 64:87–104

Wickert E, Marcondes J, Lemos MV et al (2007) Nitrogen assimilation in citrus based on CitEST data mining. Genet Mol Bio 30(3):810–818

Wilkie SE, Roper J, Smith A et al (1995) Isolation, characterization and expression of a cDNA clone encoding aspartate aminotransferase from Arabidopsis thaliana. Plant Mol Biol 27:1227–1233

Wray JL (1993) Molecular biology, genetics and regulation of nitrite reduction in higher plants. Physiol Plant 89:607–612

Yamada T, Oaks A (1987) Synthesis of glutamate by mitochondria – an anaplerotic function for glutamate dehydrogenase. Plant Physiol 70:749–756

Variable Rate Application of Fertilizer in Citrus

18

C.W. Chan

Abstract

Spatially variable application of fertilizer to citrus fields is dependent upon accuracy of the input variables. The accuracy in determining a variable rate application map and the application accuracy are important. An accuracy model is described using a 3.6-ha Florida citrus block. The effects of boundary determination, interpolation method, and global positioning system (GPS) location errors were studied for determining a variable rate nitrogen application map based upon citrus yield maps. Accuracy models can be developed considering interaction among GPS horizontal accuracy, differential GPS (DGPS) sampling frequencies, and machine delay times of a hypothetical variable rate applicator for nitrogen (N) fertilizer application based on an application map. Parameters studied included five GPS horizontal accuracy levels, two levels of DGPS horizontal accuracy, two DGPS sampling frequencies, and two machine delay times. Two integrated models were developed which documented the effects of the parameters. Machine delay time was the most important factor, and GPS horizontal accuracy was the second most important.

Keywords

Spatially variable • Fertilizer application • GPS • Machine dynamics • Accuracy requirements

18.1 Introduction

18.1.1 Variable Rate Application

Spatially variable crop production (SVCP) can improve economic returns and reduce the introduction of undesirable residues into the environment. Yield mapping is a logical starting point in SVCP providing valuable insights into potential farm problems and the starting point for spatially variable fertilizer or pesticide application. Both yield mapping and variable rate application have been widely studied and commercially implemented, particularly in agronomic crops.

Spatially variable rate application (VRA) has been extensively used in precision agriculture for many crops. There are two broad types of VRA, map-based and sensor-based, which may be used individually or in a complementary manner. Map-based application is the most common. A map-based VRA system usually consists of an applicator fitted with a differential global positioning system (DGPS) receiver, where the machine's field location determines application rates from an application map. An application map or treatment map shows the precise locations and quantities of the treatments or applications within the field. The application map is usually derived from geographical information system (GIS) analysis, using measurement of various relevant field parameters and established relationships among yield limiting and economic constraints. The application map is then transferred to a VRA controller located in the

C.W. Chan (✉)
Mechanisation and Automation Research Center, Malaysian Agricultural Research and Development Institute (MARDI) Electronics and Building Environmental System,
P.O. Box 12301, 50774 Kuala Lumpur, Malaysia
e-mail: cwchan@mardi.my

A.K. Srivastava (ed.), *Advances in Citrus Nutrition*,
DOI 10.1007/978-94-007-4171-3_18, © Springer Science+Business Media B.V. 2012

operator console. The applicator's field location will then determine the desired application rate from the digital map and trigger the application controller of the VRA.

18.1.2 Accuracy Issues

Most SVCP research and development has been directed toward hardware and software development and system implementation. However, there have also been some concerns expressed about accuracy, particularly for variable rate application of pesticides and fertilizers based on an application map (Goense 1997).

Accuracy in generating the application map used in a VRT is important as it affects the input to the machine controller during field application. Application maps derived mainly from yield maps are affected by errors introduced during the yield mapping process. The inaccuracies of individual components of the application map contribute to the overall map inaccuracy. Chan (2000) described a conceptually integrated model for the establishment of application maps in his analysis of the use of GPS and GIS in precision liquid fertilizer application for Florida citrus. The purpose of this model was to establish a theoretical relationship between all relevant parameters affecting the application map of a VRA system. In his case study, the model developed was used to assist users of VRA system in thetir assessment of the accuracy requirements for the application map.

18.1.3 An Example Case of Florida Citrus Fertilizer Application

Nitrogen (N) application has the greatest effect on Florida citrus production (Jackson et al. 1995). It is applied on a regular basis as it is readily leached from poor soils on the Florida central ridge (Entisols) growing areas. The excessive leaching of N can cause surface water contamination in the Flatwoods soils (Spodosols) near the Florida coastal areas. Spatially variable rate application (VRA) may improve economic returns in addition to reducing the introduction of undesirable chemical residues into the environment.

There is limited use of VRA for citrus fertilizer application in the scientific literature, but some Florida citrus growers are practicing sensor-based VRA by applying liquid fertilizer based on the tree canopy location. For example, one grower used a liquid fertilizer applicator mounted on a tractor (Fig. 18.1a, b) equipped with a sprayer controller (Fig. 18.2a) (MidTech TASC 6300, Midwest Technologies, Inc., Springfield, IL), a boom control switch box and three 2.23-m spray boom sections with 11 nozzles spaced 200 mm apart on each boom section (Fig. 18.2b). Fertilizer was applied with the left and right booms. Two photoelectric sensors located on the left and right sides of the tractor triggered spraying operation in the field when tree canopy was detected. Besides sensor-based VRA, some citrus growers/manufacturers have shown interest in using a map-based VRA for liquid fertilizer application. In 2008–2009, a citrus VRT fertilization and pesticide spraying systems were tested and licensed by the University of Florida to a company to be used by the growers (Schumann et al. 2010). The systems include feed-forward command controller, GPS guidance for navigation, optical sensors, and selective chemical applications. It is important to consider the integrated accuracy requirements of a map-based variable rate liquid N application. There are many studies highlighting the importance of accuracy in map-based VRA (Anderson and Humburg 1997; Chan et al. 2002; Goense 1997; Schueller and Wang 1994). Anderson and Humburg (1997) reported that the basis for spatially variable field treatments in soil and crop management maps are limited if VRA cannot accurately apply inputs according to the desired prescription. Integrated accuracy requirements of a map-based VRA require identification of the key factors. These can be classified, from an operational perspective, into the three major groups: application map, DGPS navigation, and variable rate application.

18.1.4 Application Map

A nitrogen application map can be derived based on established N recommendation and citrus yield map. Jackson et al. (1995) reported that N is required for tree growth and fruit production. Replacement of the N removed by fruit harvest is the main N requirement in a mature orchard. They suggested an N application rate of 0.18 kg N per Florida field box (40.8 kg) for oranges and 0.14 kg N per Florida field box (40.8 kg) of grapefruit harvested from the field. Currently, there is no widely accepted relationship defining various farm production inputs in maximizing yield potential of a citrus crop other than this. A complete citrus nutrition program requires considerations in many factors such as those described in Jackson et al. (1995) and Scholberg et al. (2000).

Several studies on development of citrus yield maps have been reported (Whitney et al. 1999; Chan et al. 1999a; Chan 2000). Manual harvesting is a common practice in Florida citrus. Workers use ladders and picking bags to harvest fruits and place the harvested fruit in field containers (tub or pallet bins) that normally have a capacity of ten Florida field boxes or 0.71 m^3 (Whitney et al. 1999). Each worker is paid by number of Florida field boxes and places the harvested fruit in one container for tallying purposes. A crew leader then tallies each tub filled by each worker and empties the contents of the fully filled tub into a specialized vehicle, termed a "goat," fitted with a hydraulically actuated loader boom (Fig. 18.3). The empty tub is placed back on the ground.

Fig. 18.1 (**a**) Side view of a chemical liquid sprayer used in a citrus orchard and (**b**) the same sprayer applying chemical in between two rows of citrus trees

Fig. 18.2 (**a**) Variable rate controller located in the tractor console and (**b**) a section of the spray boom

Fig. 18.3 An instrumented truck equipped with a GeoFocus Goat GPS (yield monitor) used in collecting manually harvested citrus field tubs

The crew leader then records the field position of the tub by actuating a switch that records the GPS position of the tub. Whitney et al. (1998) developed a citrus yield mapping system using a modified Crop Harvesting Tracking System (GeoFocus Inc., Gainesville, FL) (Schueller et al. 1999). The GPS data for tub locations were differentially corrected by post-processing using GeoFocus software. These points were then converted to shapefile format using the GIS computer program ArcView Version 3.1 (Environmental Systems Research Institute (ESRI), Redlands, CA).

Application map errors include mapping errors and map resolution during development of the map. Chan et al. (2002) reported that mapping errors of a yield-based variable rate nitrogen fertilizer application map for Florida citrus were caused by the field boundary offset, GPS horizontal accuracy, and surface interpolation method. The application map was derived from citrus yield data, collected using a yield monitor fitted with a GPS receiver, using GIS analysis and an established N recommendation equation. They reported that GPS horizontal accuracy was the most important variable among the three factors affecting the accuracy of an application map.

Application map resolution is another important factor affecting the accuracy of field application in a VRA system (Monson 1997). Usually, quicker data access from the application map to the machine controller can be achieved in a raster map format than a vector format (Anderson and Humburg 1997; Tyler et al. 1997).

18.1.5 Machine Navigation Using DGPS

A map-based VRA using a real-time DGPS receiver will introduce positional errors. DGPS accuracy is affected by such factors as signal latency, atmospheric conditions, GPS signal, quality of GPS receiver, local environmental conditions, and the type of differential correction signals (Barnes and Cross 1998).

To date, there have been no reported studies on the effect of machine navigation error and DGPS sampling frequency on the accuracy of VRA for tree-based crops.

18.1.6 Variable Rate Application Error

Machinery components in a VRA do not respond instantly to changes in desired application rate. Such factors as working speed, controller algorithm, design and configuration of the system controller and actuators, spraying width, and physical characteristics of the chemical used contribute to the overall dynamic behavior of a variable rate applicator. Paice et al. (1996) described both lateral and longitudinal spatial resolution as being important for spraying application. The lateral resolution is determined by the spraying width while the longitudinal resolution is a function of both the spraying system response time and machine forward speed. The importance of delay time in the accuracy of machine application has been highlighted in many studies (Anderson and Humburg 1997; Cahn and Hummel 1995; Qiu et al. 1998; Rockwell and Ayers 1996; Schueller and Wang 1994; Sudduth et al. 1995; Tyler et al. 1997; Zhu et al. 1998).

Feed-forward command has been used to minimize the effects of machine delay time (Schueller and Wang 1994). They used a transportation delay of 5 s with a 1 s time constant to model a liquid fertilizer VRA system operating at 18 km/h with a map resolution of 40 m · 40 m and an application width of 20 m. They introduced a 5.7 s feed-forward command time to reduce the effect of overall machine delay on application error but did not incorporate the machine navigation error and application map error. Qiu et al. (1998) studied the simulated performance of a direct injection sprayer for herbicide application by considering three types of nozzle, hose diameters, and flow rates. They reported that transportation time delay ranged from 10.1 to 22.2 s and 1.5 to 2.9 s without and with feed-forward command, respectively. Similar delay time without feed-forward command for a 10.7-mm hose diameter of an injection sprayer was reported by Zhu et al. (1998).

18.2 Integrated Accuracy Analysis Model

Goense (1997) derived a calculation method for a spatially variable rate fertilizer spreader integrating the various factors affecting the accuracy of a VRA system. His work involved 5 levels of application map resolution, 4 range values of semi-variogram of field variability, 11 different spreader effective working widths, 4 GPS navigation positional errors ranging from 0.1 to 2.0 m, and 5 different types of fertilizer spread pattern. The method was based on variance calculation between desired rate and applied rate using geostatistics. Spatial variability of the application rate was described by a semivariogram, using an exponential model. He found that the shape of the spreading pattern had little influence on the accuracy of application. Also, the effect of GPS navigation accuracy depended on the application map resolution and working width of independently controlled sections of the spreader. Goense (1997) did not study the effects of equipment dynamic response such as delay time, the interpolation method used in developing the application map, modeling of GPS error in deriving the desired application, or DGPS navigation error.

Modeling of GPS navigation error is a complex issue. Anonymous (1995) reported that the expected GPS position error distribution using 3 months' data collected from the GPS control segment monitor stations were similar to a Gaussian distribution. Chan (2000) evaluated two DGPS systems at the University of Florida, Gainesville, USA. Results from the static accuracy tests suggested a Gaussian distribution for both systems. Further navigation tests, using the two systems in a nearby citrus orchard, were affected by poor DGPS data recorded due to the inadequacy of the two DGPS systems as a stand-alone unit besides facing high electronic background noises for the DGPS receivers at the test site. Paz et al. (1997) simulated recording of GPS positions by a moving GPS receiver using a computer program.

They adopted a simpler method using a random GPS error where the simulated GPS receiver error direction has a uniform distribution between 0° and 360°. Two types of error simulations were used: the maximum GPS error for a uniform error distribution and the standard deviation in a normal error distribution. Goense (1997) used similar standard deviation in his error calculations.

18.2.1 Development of an Accuracy Model

Chan et al. (1999a, b) described a conceptual approach for studying the integrated effect of GIS/mapping, GPS/navigation, and machine dynamics for a case study in a Florida citrus orchard. A methodology was developed to analyze the error sources of a map-based variable rate liquid N application for Florida citrus (Chan 2000). A case study on yield-based variable rate N fertilizer application map was reported using that methodology (Chan 2000; Chan et al. 2002). They analyzed yield-based N fertilizer application maps derived using two rectangular field boundaries (B = original and L = offset) identified using two base maps, five GPS horizontal accuracy levels ranging from 0 to 5 m, and two surface interpolation methods (inverse distance weighted and ordinary kriging). The best N fertilizer application map was developed using an established yield-based N fertilizer model and a best yield map that correlated closely to the ground truth yield (Chan 2000). The best yield map was defined as one using a rectangular boundary B with zero GPS horizontal error ($G = 0$ m) and interpolation by kriging-Gaussian method. Finally, an error model was developed based on the analysis of each corresponding map square cell between the best and simulated application maps using a GIS software. The boundary offset was reported as having the greatest influence on the accuracy of application map followed by GPS horizontal error and the interpolation method.

A similar approach using the error analysis methodology developed by Chan (2000) can be used to analyze the integrated effect of spatial parameters for a map-based variable rate liquid N fertilizer application for Florida citrus considering application map, navigation error, and machine delay time of the VRA. The steps involved the field and simulated data collected from GIS/mapping, DGPS/navigation, and VRA machine dynamics/delay time. An integrated model using the following four error sources can be developed:

- Five GPS horizontal accuracy levels (G)
- Two DGPS navigation horizontal accuracy levels (T)
- Two DGPS sampling frequencies (F)
- Two VRA machine delay times (D)

A hypothetical map-based VRA similar to the above commercial fertilizer applicator fitted with a real-time DGPS receiver navigation system in a Florida citrus orchard was assumed in this study.

The 3.6-ha citrus orchard selected for this study was located in a Flatwoods area and included two soil series: Pomello and Waveland. These are moderately well and poorly drained soils, respectively. The orchard was centered at approximately 82.2775 W and 27.6423 N. The trees were "Hamlin" orange on "Carrizo" rootstock that were planted in 1986 with a spacing between rows of 9.14 m and an in-row spacing of 4.3 m. The trees had formed a continuous hedgerow at the time of this study and were 4–5 m high.

18.2.2 Application Maps

Five N application maps from the results in Chan (2000) and Chan et al. (2002) were used. They were derived from analysis using the rectangular boundary B obtained from a 1:30,000 scale, 0.32 m pixel resolution, and geo-referenced black and white aerial photograph (Fig. 18.4a, b); yield maps for developing application maps can be derived using ordinary kriging with a Gaussian model interpolation (kriging-Gaussian) method and five levels of G (0, 0.5, 1, 3, and 5 m) (Fig. 18.5a–e). Only G was selected to vary. The rationale for this selection was that the best base map required for rectangular boundary B, such as a farm survey map or accurate aerial photograph, could be identified when they are available. Besides boundary, comparing the interpolated yield map with the ground truth yield, especially when individual tree yield data are available, can identify the best interpolation method. The best N fertilizer application map described earlier in Chan (2000) and Chan et al. (2002) was used.

Application map resolution was assumed to be non-limiting, and a raster format with a square cell size of 0.457 m was selected. The choice of this cell size accounted for both the 0.5 m GPS horizontal error used in yield map interpolation and a uniform row spacing covered by the spray pattern. It was also assumed that the VRA was capable of applying all the desired rates in the application map.

18.2.3 Machine Delay Time

The range of VRA applicator machine delay times needed for the integrated model can be identified by conducting field tests. Two sets of representative machine delay times or sprayer response time were obtained using the same commercial fertilizer applicator described earlier. Field tests conducted using water and at five discharge rates ranging from 935.3 to 1590.1 l/ha. Spraying was triggered in two ways: by manually turning on the spraying switch in the operator's console while keeping the photoelectric sensor switch off and by triggering the activated photoelectric sensor by blocking its view with an object. The spraying test was videotaped. Sprayer response time was defined as from the switching on of the

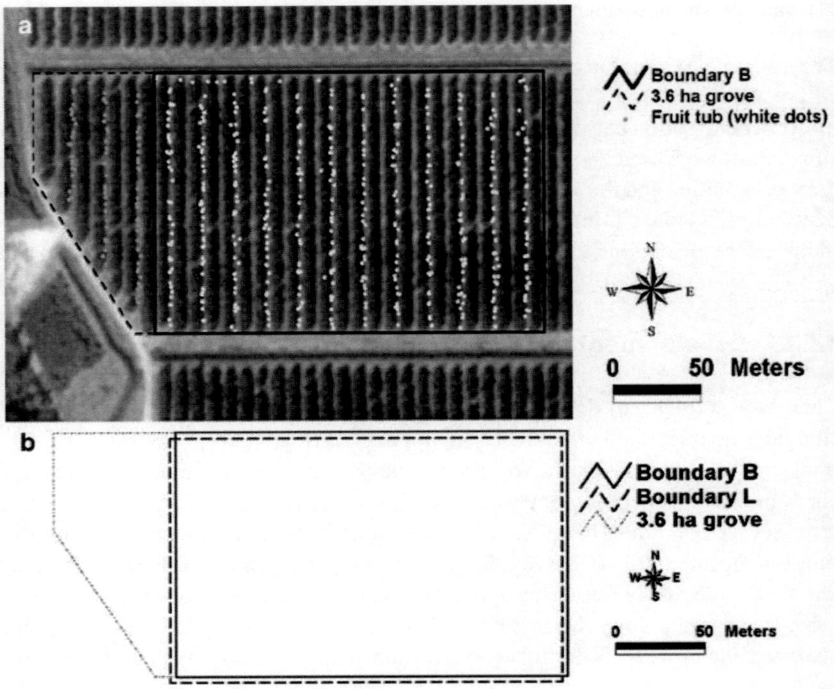

Fig. 18.4 (**a**) Original location of tubs of harvested fruit (*dots*) overlaid on a geo-referenced aerial photograph and (**b**) two selected rectangular boundaries B and L of the 3.6-ha orchard

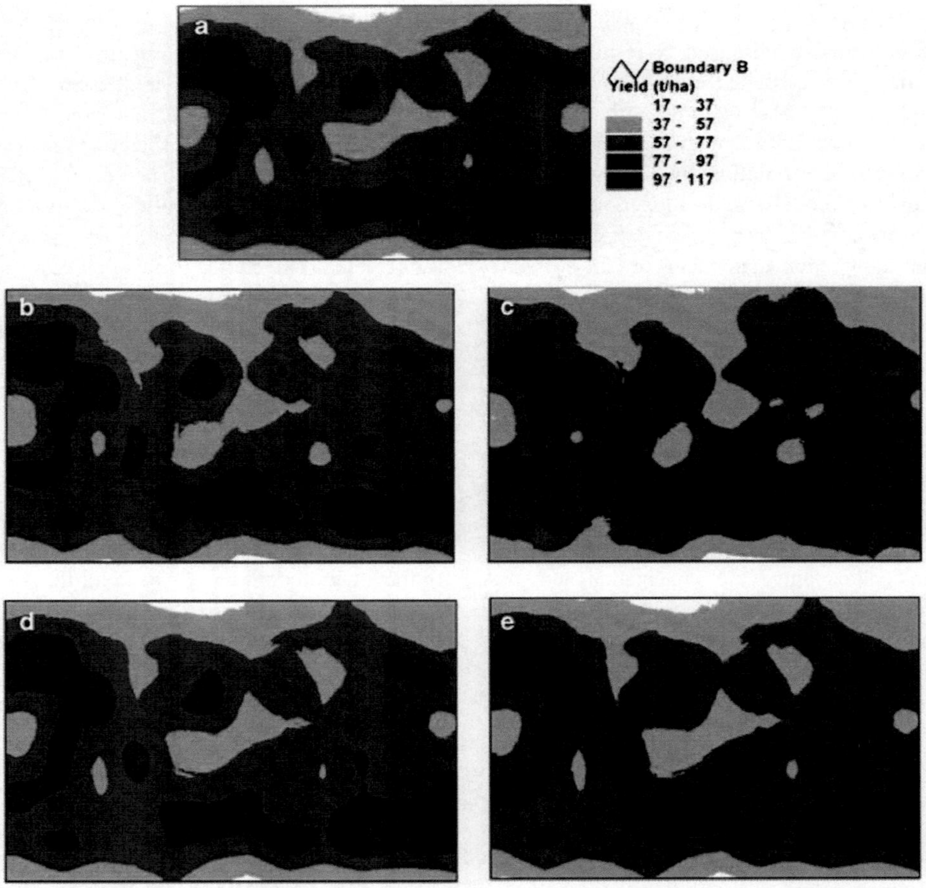

Fig. 18.5 Interpolated yield maps using kriging-Gaussian method in boundary B for GPS static horizontal error of (**a**) 0 m, (**b**) 0.5 m, (**c**) 1 m, (**d**) 3 m, and (**e**) 5 m

Table 18.1 Summary of field tests on machine response time for a liquid fertilizer sprayer

	Sprayer response time(s)	
	Optical sensor off	Optical sensor on
Spray rate (l/ha)	Mean/standard deviation(s)	Mean/standard deviation(s)
935.3	1.20/0.17	0.79/0.06
1,122.3	0.96/0.12	0.70/0.04
1,309.2	0.70/0.61	0.61/0.02
1,403.1	0.81/0.22	0.60/0.03
1,590.1	0.84/0.09	0.58/0.18

spraying to the time all the nozzles were observed spraying liquid. The response time was determined by playing back the recorded spraying test operation on a player capable of viewing video at 30 frames per second. The sprayer response time ranged from 0.70 to 1.20 s with the optical sensor switch off. It was slightly faster with the optical sensor on and ranged from 0.58 to 0.79 s (Table 18.1). The slightly slower response time, with the optical sensor switch off, could be due to errors in observing the operator's manual signaling of each switching on operation in the operator's console. This was reflected in the high standard deviations in response time for spray rates equal to 1309.2 l/ha and higher. Considering the ranges of response time results in Table 18.1, delay times of 0 and 2 s were selected with 0 s representing the system with a 2 s feed-forward command.

18.2.4 Modeling of Navigation Error (*T*), DGPS Sampling Frequency (*F*), and Spray Pattern

The same approach used by Paz et al. (1997), Chan (2000), and Chan et al. (2002) was used to model DGPS positions in this study. Two levels of *T* (0 and 2 m) were simulated. The maximum DGPS error of 2 m was used for $T=2$ m assuming a uniform error distribution, i.e., *G* was modeled in a randomly distributed manner around the original position (*G* ¼ 0 m) in a circle by randomizing the angular distribution of the horizontal error. The original position lay along the midpoint of the between tree rows spacing (9.14 m).

Two DGPS sampling frequencies (*F*) were selected at 0.5 and 10 Hz. This selection reflects the common range of sampling frequency for currently available commercial DGPS receivers. Most such receivers operate at 1 Hz. The choice of 0.5 Hz was assumed to be the lower limit of DGPS receiver update rate, when operating in a fringe area with poor real-time DGPS signal.

A total of 23 spray paths were needed to complete the fertilizer spraying in the test field covering an area defined by a rectangular boundary A. GPS tracking software, ArcView Tracking Analyst extension available from ArcView (Environmental Systems Research Institute (ESRI), Redlands, CA), was used to model the DGPS positions along each spray path. The spraying pattern was assumed to be uniform along a 9.14-m-width spray boom and modeled to form a rectangular spray pattern between two rows of trees. A constant tractor forward operating speed at 7.98 km/h traveling along the center between two rows of trees was used. The spraying operation was assumed to start exactly at the field boundary and stop exactly at the opposite field boundary, defined by the outer edges of the tree canopy. Different test runs using $T=0$ m at $F=0.5$ and 10 Hz were simulated. The applicator traverses from south to north direction for odd number paths and north–south direction for even number paths. Simulated field positions of the applicator obtained from using even number paths. Simulated field positions of the applicator obtained from using $T=0$ m for all the 23 machine paths, at $F=0.5$ and 10 Hz, were further selected for modeling, using the method described earlier, to represent each *T* at 2 m away from the respective 0 m point. Figure 18.6 details some examples of the modeled DGPS points along machine path 8 at $T=0$ and 2 m sampled at $F=0.5$ and 10 Hz. Each dot represents the field position of the tractor as detected by the DGPS receiver.

18.2.5 Simulated Treated Map

For each application map developed at $G=0$, 0.5, 1, 3, and 5 m, corresponding simulated treated maps were derived using combinations of $T=0$ and 2 m, $F=0.5$ and 10 Hz, and $D=0$ and 2 s. The N application map rate that fell within each modeled DGPS position was assigned to its corresponding sprayed rectangular area using the ArcView GIS program. This represents the VRA machine controller reading the input rates detected from the application map and the field location indicated by the DGPS receiver. In cases where $D=2$ s, the applied rate of N was shifted, using a spreadsheet computer program, by a time scale of 2 s along the sprayed path.

18.2.6 Integrated Model

Subsequent analyses of the integrated effects of *G*, *T*, *F*, and *D* on the simulated treated maps were undertaken to compare the applied N rate (kg/ha) of each spatial location of the simulated treated N map with the best application map. The application mapping error equation was used:

$$E = [\mathrm{AM}_{\mathrm{BEST}} - \mathrm{N}_{\mathrm{TREATED}}]$$

where E=error in simulated treated map, $\mathrm{AM}_{\mathrm{BEST}}$=best application map derived using an established yield-based N fertilizer model and a best yield map, $\mathrm{N}_{\mathrm{TREATED}}$=all the simulated treated maps using *G*, *T*, *F*, and *D*, and *E*, $\mathrm{AM}_{\mathrm{BEST}}$, and $\mathrm{N}_{\mathrm{TREATED}}$ are expressed as rate of N application.

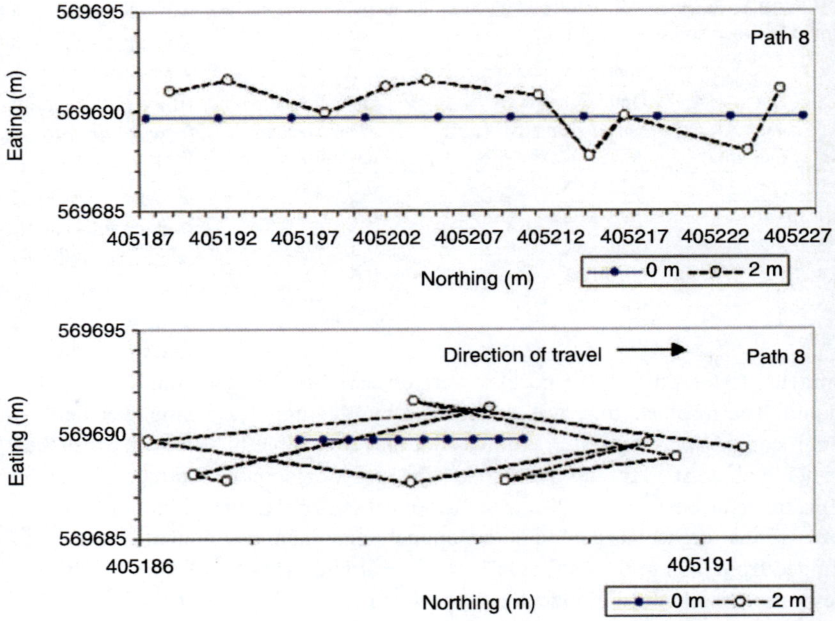

Fig. 18.6 Modeled machine path 8 with 0 and 2 m navigation error at (*top*) 0.5 Hz and (*bottom*) 10 Hz DGPS-sampling frequency traveling at 7.98 km/h

Spatial differences in N rate between maps were computed using a raster-based grid-cell geoprocessing system, Grid, available in the Spatial Analyst 1.1 program (Environmental Systems Research Institute (ESRI), Redlands, CA). A grid was first divided into uniform square cells, each representing an actual portion of geographic space. The VRA error in each cell of the resultant error map (*E*) expressed as rate of N incorrectly applied was converted to an absolute value:

$$AE = [AM_{BEST} \text{-} N_{TREATED}]$$

where AE = absolute error for a simulated treated map.

Each absolute error map was then exported to a file to be further analyzed, using a computer spreadsheet program, deriving a weighted mean absolute error in N rate for the whole map. A weighted mean absolute error value for each absolute error in a simulated treated map was computed from a total of 143,060 square cells, each 0.457 m · 0.457 m. The weighting factor was based on the number of square cells associated with each absolute error value. Multiple regression analysis was used to analyze the weighted mean absolute error in N rate for each of the 40 N simulated treated maps to develop a general model.

18.3 VRA Field-Treated Maps

18.3.1 Simulated Map

Figure 18.7 shows some simulated N-treated maps without machine delay time, obtained using application maps at *G* horizontal error of 0 and 5 m, respectively, using boundary B and kriging-Gaussian interpolation method using two levels of *T* and *F*. All the maps showed a somewhat blocky pattern as they reflect the rectangular spray pattern covering a 9.14 m width between two rows of trees. As expected, the spray pattern with *F* = 10 Hz is more highly resolved than with *F* = 0.5 Hz. This is due to the more frequent changes in the sprayer application rate because the machine position is updated more frequently at 10 Hz than at 0.5 Hz.

Due to the effect of a 2 s machine time delay (*D*), there are horizontal strips equivalent to a distance of 4.44 m in the applied map representing a 2-s interval where no fertilizer was applied (Fig. 18.8). This happened for both the 0.5 and 10 Hz DGPS sampling frequency along the south-north and north–south directions for all the 23 paths resulting in the pattern of unsprayed rectangles of 9.14 m · 4.44 m at the beginning of every sprayed path in the simulated treated map.

18.3.2 Absolute Error Maps

Figures 18.9 and 18.10 are some absolute error maps computed between the best application map and simulated treated map using two levels each of *T*, *F*, and *D*. Similar blocky patterns were observed due to the rectangular spray pattern. Generally, regions of high absolute error were observed in those absolute error maps computed with a combination of 2 s machine delay time coupled with a higher DGPS navigation error. The absolute error of N maps was classified into six levels in 30 kg/ha steps up to 150 kg/ha and then one step to 320 kg/ha. The weighted mean absolute error of N rate ranged from 5.88 to 42.29 kg/ha (Table 18.2). Using *F* = 10 Hz resulted in a

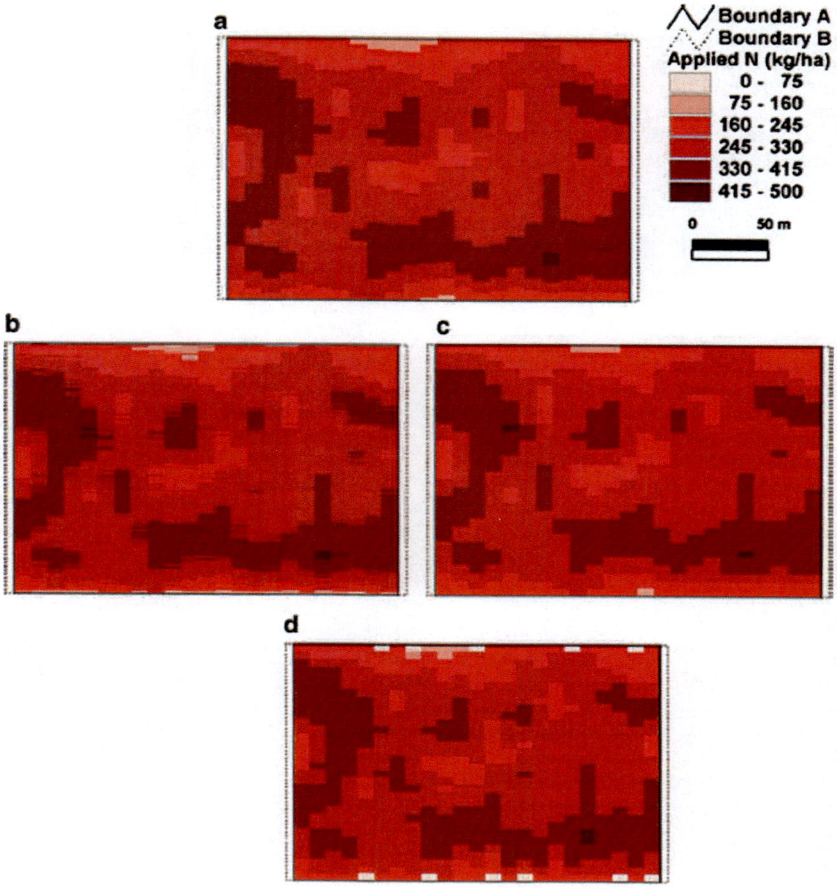

Fig. 18.7 Simulated treated N maps obtained using application map with $G=0$ m, sprayer operating at 7.98 km/h, without machine delay time, at (**a**) $F=10$ Hz, $T=0$ m, (**b**) $F=10$ Hz, $T=2$ m, (**c**) $F=0.5$ Hz, $T=0$ m, and (**d**) $F=0.5$ Hz, $T=2$ m (F DGPS sampling frequency, T DGPS navigation error)

higher variation in the weighted mean absolute error than using $F=0.5$ Hz over a range of G from 0 to 5 m.

With $T=0$ m, the variation of weighted mean absolute error was 2.19 kg/ha (18.80–16.61) at $F=10$ Hz whereas at $F=0.5$ Hz the variation was 1.1 kg/ha (14.84–13.74). Similarly with $T=2$ m, the variation of weighted mean absolute error was 0.31 kg/ha (10.27–9.96) at $F=10$ Hz whereas at $F=0.5$ Hz the range was 0.17 kg/ha (7.11–6.94). The more frequent sampling rate of 10 Hz than 0.5 Hz resulted in a higher variation in the resultant sprayed areas.

18.3.3 Integrated Effect of GPS, Navigation Error, DGPS Sampling Rate, and Machine Delay Time

The 40 weighted mean absolute error values of the 40 N simulated treated maps can be expressed as a function of the horizontal GPS error (G) of the desired application map in combination with the two DGPS navigation horizontal accuracy levels, two DGPS sampling frequencies, and two machine delay times. Assuming the effects of navigation error, DGPS sampling rate and machine delay time are additive, and when combined with the effect of G, the weighted mean absolute error for the data was approximated as

$$M_{\text{ID}} = 6.65 + 1.47\,T + 1.40\,F + 3.76\,D + 2.56\,G, \quad R^2 = 0.99$$

where M_{ID} is the weighted mean absolute error (kg/ha), T, F, D, and G are expressed in meters with $T=0$ and 2 m, $F=0.22$ m (10 Hz–7.98 km/h) and 4.44 m (0.5 Hz–7.98 km/h), $D=0$ m (0 s machine delay time) and 4.44 m (2 s machine delay time), and $G=0$, 0.5, 1, 3, and 5 m. The quantity 6.65 is the estimated average weighted mean absolute error at $T=F=D=G=0$ m. Since F cannot be zero, 6.65 cannot be interpreted by itself but only as part of the prediction equation. The coefficients for T, F, D, and G represent the estimated change in MID for a 1 unit change in each independent variable. In other words, we estimate MID increases by 2.94 (2×1.47) upon changing T from 0 to 2 m (holding F, D, and G fixed). All the estimated coefficients were significantly ($P<0.025$) greater than zero. Furthermore, the t values from

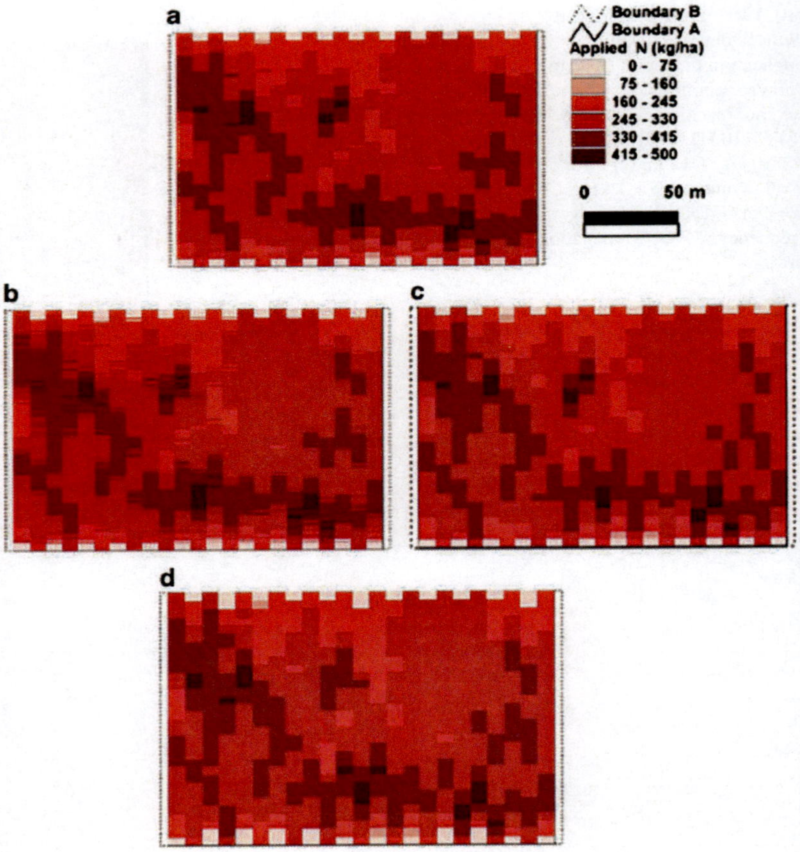

Fig. 18.8 Simulated treated N maps obtained using application map with $G=5$ m, sprayer operating at 7.98 km/h, having a 2 s machine delay time, at (**a**) $F=10$ Hz, $T=0$ m, (**b**) $F=10$ Hz, $T=2$ m, (**c**) $F=0.5$ Hz, $T=0$ m, and (**d**) $F=0.5$ Hz, $T=2$ m (*F* DGPS sampling frequency, *T* DGPS navigation error)

statistical analysis for *T*, *F*, *D*, and *G* were 4.4, 8.8, 24.9, and 14.2, respectively. Therefore, a 2 s machine delay time appeared to have the greatest influence on MID where *D* was approximately 5.7 times *T* (24.9/4.4), 2.8 times *F* (24.9/8.8), and 1.8 times *G* (24.9/14.2). Machine delay time, *D*, appeared to be the most important factor in the overall integrated linear model when the horizontal GPS error of the desired application map (*G*) ranged from 0 to 5 m. The importance of GPS accuracy in yield-based N fertilizer application map for citrus has been highlighted earlier (Chan et al. 2002). The influence of coefficient *G* (3.94) in their error model derived from absolute error of application maps on weighted mean absolute error (kg ha^{-1}) was significantly reduced to 2.56 when *T*, *F*, *D*, and *G* were considered.

18.3.4 Model with No Machine Delay Time

The effect of machine delay time can be reduced with the introduction of a feed-forward command in the VRA sprayer controller, assuming a negligible time delay in the triggering of spraying at the beginning of each trip when the optical sensors detected the tree canopies. Twenty weighted mean absolute error values of the 40 N simulated treated maps, without machine delay times, were analyzed. The weighted mean absolute error for the data was approximated as

$$M_I = 5.52 + 1.96\,T + 1.06\,F + 3.31\,G, \quad R^2 = 0.98$$

where M_I is the weighted mean absolute error (kg/ha). All the estimated coefficients were significantly ($P<0.0001$) greater than zero. The estimated average weighted mean absolute error at $T=F=G=0$ m was slightly reduced to 5.52. M_I increased faster by 3.92 (2×1.96) upon changing *T* from 0 to 2 m (holding *F* and *G* fixed) than before.

The *t* values from statistical analysis for *T*, *F*, and *G* were 7.5, 8.5, and 23.4, respectively. Therefore, *G* appeared to have the greatest influence on M_I where *G* is approximately 3.1 times *T* (23.4/7.5) and 2.8 times *F* (23.4/8.5). Horizontal GPS error (*G*) used to prepare the N fertilizer application map appears to be the most important factor in the overall integrated linear model. The influence of coefficient *G* (3.94) in the Chan et al. (2002) error model derived from absolute error of application maps on weighted

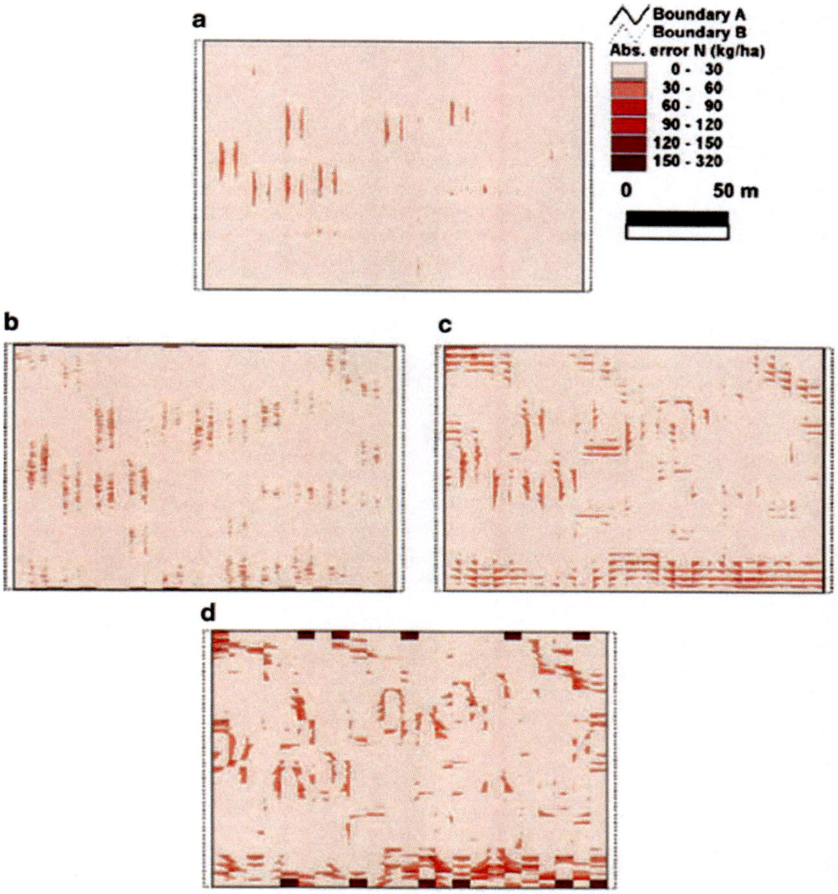

Fig. 18.9 Absolute error maps derived from differences between the best application map versus simulated treated map obtained with input from the same application map with no machine delay time, at (**a**) $F=10$ Hz, $T=0$ m, (**b**) $F=10$ Hz, $T=2$ m, (**c**) $F=0.5$ Hz, $T=0$ m, and (**d**) $F=0.5$ Hz, $T=2$ m (*F* DGPS sampling frequency, *T* DGPS navigation error)

mean absolute error (kg/ha) is still the most important factor but was slightly reduced to 3.31 when *T*, *F*, and *G* were considered.

18.4 Conclusions

An integrated model for a VRA system was developed based on the test conditions used in the present set of data. A 2 s machine delay time appeared to have the greatest influence on M_I where *D* was approximately 5.7 times *T*, 2.8 times *F*, and 1.8 times *G*. However, with electronics, the machine delay time, *D*, can be compensated with feed-forward command. An error model without machine time delay suggested the importance of *G* in the map-based VRA for Florida citrus liquid N fertilizer application in relation to the DGPS navigation error and its sampling frequency requirements. *G* was approximately 3.1 times *T* and 2.8 times *F*. This study also revealed the importance of machine delay time in a VRA system which needs to be identified and compensated. Finally, the significance of *G* in a VRA from the present study as well as those reported earlier suggested the need for a good and accurate GPS receiver.

18.5 Future Research

The integrated model can be used to assist users of a map-based VRT fertilizer application in a tree-based crop in selecting appropriate accuracy level of GIS, GPS, and machine dynamic parameters. Besides, it can be useful to other researchers working on improving these GIS, GPS, and machine dynamic.

The derivation of nitrogen application map based on yield data alone may not be sufficient. Other data layers such as soil, leaf analysis, and water regime collected using the GPS receiver should be similarly analyzed like the yield map.

The selection of a good GPS receiver in the application of a map-based VRT influence three of the four spatial parameters: static horizontal accuracy, machine navigation, and

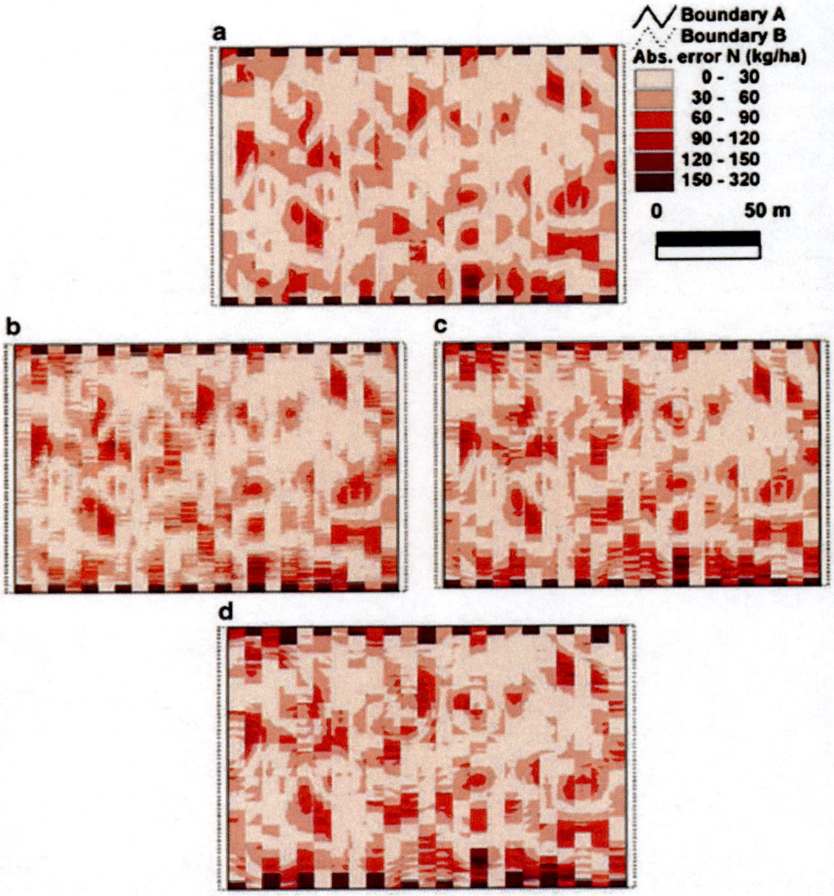

Fig. 18.10 Absolute error maps derived from difference between the best application map versus simulated treated map obtained with input from application map with $G=5$ m, having a 2 s machine delay time, at (**a**) $F=10$ Hz, $T=0$ m, (**b**) $F=10$ Hz, $T=2$ m, (**c**) $F=0.5$ Hz, $T=0$ m, and (**d**) $F=0.5$ Hz, $T=2$ m (F DGPS sampling frequency, T DGPS navigation error)

Table 18.2 Weighted mean absolute errors between the best N fertilizer application map versus simulated treated map using desired N application map at five levels of GPS horizontal error and two levels each for navigation error (T), DGPS sampling frequency (F), and machine delay time (D)

Parameters			Weighted mean absolute error of N (kg/ha)				
			GPS horizontal error, G (m)				
D (s)	F (Hz)	T (m)	0	0.5	1	3	5
0	10	0	5.88	6.79	7.96	15.54	24.68
		2	10.56	10.98	11.88	18.58	27.17
	0.5	0	11.80	11.87	12.87	18.72	26.64
		2	15.89	15.96	16.76	22.15	29.63
2	10	0	25.62	25.26	25.86	29.44	35.58
		2	26.66	26.36	26.97	30.73	36.93
	0.5	0	32.97	32.30	32.89	35.33	39.91
		2	35.18	34.85	35.13	38.13	42.29

sampling frequency. It contributes to significant sources of error to the overall model besides machine delay time.

It may be required to investigate the effects of using base map of different scale in the development of an application map for situation where the choice of aerial photograph or boundary selection involved additional cost.

Current yield map data is represented by boxes of harvested fruit per hectare. The effect of yield map based on actual weight of harvested fruit per individual tree basis should be investigated. Furthermore, yield data per individual tree basis should be obtained so that a comparison of the best interpolation method and its best choice of interpolation parameters can be accurately identified.

Nitrogen is usually applied in split application to reduce potential volatilization loss and leaching. Thus, the recommended rate in the desired map is divided into two or three applications. It is expected that the contribution of navigation error, DGPS sampling rate, and machine delay time will be compounded. The effects of split nitrogen application method should be investigated in the development of an integrated model.

Machine delay time of 2 s highly influence the mean absolute error in a map-based VRT fertilizer application system. This study suggested further improvements needed to the present map-based VRT sprayer design as studies by Qiu et al. (1998) indicated a delay time of 2.9 s despite using a feed-forward command.

Different spray patterns can be incorporated in future study to investigate the behavior of different types of the spray nozzles, spacings, and heights of the VRT system. Furthermore, a refined spray pattern should consider lateral resolution along the spray boom using estimated spatial position of spray nozzle which can vary its application rate at each nozzle along the spray boom. Finally, different field application speeds can also be investigated in the development of an integrated model.

Acknowledgments and Disclaimer We would like to thank the various supporters of this work, including the Malaysian Agricultural Research and Development Institute (MARDI) and the Florida Citrus Production Research Advisory Council. The assistance of Professor J. K. Schueller, Professor W. M. Miller, Professor J. D. Whitney, Professor J. David Martsolf, Professor J. Wayne Mishoe, Professor T. Adair Wheaton, Professor J. A. Cornell, Mary Beth Freeman, Greg Drouillard, and Douglass Thompson is gratefully acknowledged. Comments on the integrated model from Mr. Ahmad Shokri Othman are also acknowledged. The mention of commercial products is for information purposes only and does not constitute an endorsement.

References

Anderson NW, Humburg DS (1997) Application equipment for site specific management. In: Pierce FJ, Sadler EJ (eds) The state of site-specific management for agriculture. ASA CSSA/SSSA, Madison, pp 245–281

Anonymous (1995) Global positioning system standard positioning service signal specification. Annex B. Standard positioning service performance characteristics. US Coast Guard Navigation Center Technical Report, pp 21

Barnes JB, Cross PA (1998) Processing models for very high accuracy GPS positioning. Navigation 51(2):180–185

Cahn MD, Hummel JW (1995) A variable rate applicator for side dressing liquid N fertilizer. In: Rust RH, Larson WE, Robert PC (eds) Proceedings of the second international conference site – specific management for agricultural systems. ASA CSSA/SSSA, Madison, pp 683–689

Chan CW (2000) Integrated analysis of GPS, GIS and machine dynamic response in precision liquid fertilizer application. Unpublished PhD dissertation, University of Florida, Gainesville

Chan CW, Schueller JK, Miller WM et al (1999a) Interpolation errors in citrus yield mapping. ASAE paper no. 99-FL101, ASAE, St. Joseph

Chan CW, Schueller JK, Miller WM et al (1999b) Accuracy in spatially-variable crop production and an illustration in citrus yield mapping. In: Stafford JV (ed) Proceedings of the second European conference on precision agriculture. Sheffield Academic Press, Sheffield, pp 805–813

Chan CW, Schueller JK, Miller WM et al (2002) Error sources on yield-based fertilizer variable rate application maps. Precis Agric 3:81–94

Goense D (1997) The accuracy of farm machinery for precision agriculture: a case for fertilizer application. Neth J Agric Sci 45: 201–217

Jackson LK, Alva AK, Tucker DPH et al (1995) Factors to consider in developing a nutrition program. In: Tucker DPH, Alva AK, Jackson LK et al (eds) Nutrition of Florida citrus trees. IFAS University of Florida, Gainesville, pp 3–11

Monson RJ (1997) A methodology for quantifying the capabilities of a GPS navigation system. ASAE paper no. 97-3060. American Society Agricultural Engineering, St. Joseph, p 21

Paice MER, Miller PCH, Day W (1996) Control requirements for spatially selective herbicide sprayers. Comput Electron Agric 14: 163–177

Paz SM, Cugnasca CE, Saraiva AM et al (1997) Development of a simulation tool for a GPS receiver. In: Stafford JV (ed) Proceeding of the first European conference on precision agriculture. Sheffield Academic Press, Sheffield, pp 603–610

Qiu W, Watkins GA, Sobolik CJ et al (1998) A feasibility study of direct injection for variable-rate herbicide application. Trans ASAE 41(2):291–299

Rockwell AD, Ayers PD (1996) A variable rate, direct nozzle injection field sprayer. Appl Eng Agric 12(5):531–538

Scholberg JMS, Parsons LR, Wheaton TA (2000) Citrus nitrogen nutrition: some production considerations for more efficient nitrogen use. Citrus Ind 81(2):18–19

Schueller JK, Wang MW (1994) Spatially-variable fertilizer and pesticide application with GPS and DGPS. Comput Electron Agric 11:69–83

Schueller JK, Whitney JD, Miller WM, Wheaton TA and Turner AE (1999) Low-cost automatic yield mapping in hand-harvested citrus. Comput Electron Agric 23:145–153

Schumann AW, Hostler K, Tunbo S (2010) What's new in variable rate technology? Citrus Industry. Ocala, FL, Southeast Agnet 91:4

Sudduth KA, Borgelt SC, Hou J (1995) Performance of a chemical injection spraying system. Appl Eng Agric 11(3):343–348

Tyler DA, Roberts DW, Nielsen GA (1997) Location and guidance for site-specific management. In: Pierce FJ, Sadler EJ (eds) The state of site-specific management for agriculture. ASA CSSA/SSSA, Madison, pp 161–182

Whitney JD, Miller WM, Wheaton TA et al (1998) Precision farming applications in Florida citrus. ASAE paper no. 98-1097. ASAE, St. Joseph

Whitney JD, Miller WM, Wheaton TA et al (1999) Precision farming applications in Florida citrus. Appl Eng Agric 15(5):399–403

Zhu H, Fox RD, Ozkan HE et al (1998) Time delay for injection sprayers. Trans ASAE 41(3):525–530

Open Field Hydroponics: Concept and Application

19

Kelly T. Morgan and Davie Kadyampakemi

Abstract

Advanced citrus production systems that combine grove design, size limiting rootstocks, irrigation and nutrient management, and mechanical harvesting have the potential to make citrus more efficient and economically competitive. Recently, the open hydroponic system (OHS) of citrus production has been combined with orchard design to achieve these efficiencies. Open hydroponics are defined as a system of management practices aimed at increased productivity of citrus orchards by continuously applying a balanced nutrient mixture through the irrigation system, limiting the root zone by restricting the number of drippers per tree and maintaining the soil moisture in the rooted zone near field capacity. Concepts of OHS are to maximize water and nutrient use efficiency through improved nutrient availability to concentrate roots in the irrigated zone. These concepts are accomplished through intensive water and nutrient management and results in increased early growth, sustained yields, and reduced nutrient leaching. Additional horticultural principles employed include higher tree density with size-controlling rootstocks grown on soil ridges, if needed, for improved drainage. Limited published information is available for citrus grown under OHS conditions. However, one study reported that orchards on OHS have outgrown trees on conventional production systems and appear to be more productive. Canopy volumes increased by approximately three times and fruit volume by more than five times per unit of N after 4 years of intensive management compared with conventionally grown trees in replicated trials. Yield increases approaching 30% have been reported from several studies in many citrus-producing regions of the world. Use of this advanced production system may maintain higher levels of productivity through improved water and nutrient use efficiencies resulting in improved short- and long-term economic returns, particularly in citrus industries infected with diseases such as Citrus Greening.

Keywords

Rootstock • Size limiting • Intensive irrigation and nutrient management • Root zone • Nutrient use efficiency

19.1 Introduction

Production systems that combine grove design and irrigation management to increase yield and grove operational efficiency have been studied in many citrus-producing regions of the world and have many economic advantages (Roka et al. 2009). These production systems together with

K.T. Morgan (✉) • D. Kadyampakemi
Soil and Water Science Department, Southwest Florida Research and Education Center, University of Florida, 2685 State Road 29 North, Immokalee, FL 34142, USA
e-mail: conserv@ufl.edu; dkadyampakemi@gmail.com

A.K. Srivastava (ed.), *Advances in Citrus Nutrition*,
DOI 10.1007/978-94-007-4171-3_19, © Springer Science+Business Media B.V. 2012

mechanical harvesting have the potential to make citrus production more efficient and economically competitive. The two basic grove production components are the Advanced Production System (APS) and the Open Hydroponic System (OHS) (Stover et al. 2008). The fundamental concepts of APS/OHS for citrus production combine high tree density with intensive management to optimize tree performance. These concepts are designed to more fully and efficiently exploit a citrus tree's potential by providing optimal water and nutrient conditions. Those improvements are expressed in maximizing water and nutrient use efficiency and concentrating root within irrigation zones which should lead to less nutrient leaching. The OHS is an integrated system of practices, including irrigation, and nutrition practices that was developed in Spain in the early 1990s to contend with gravel based soils and the problem of low fertility (Martinez-Fuentes et al. 2004; Falivene et al. 2005). Tight control over water and nutrient-mediated plant growth and development by the OHS using irrigation to train the root system into a limited area and fertigates with daily requirements of nutrients (Stover et al. 2008).

Falivene (2005) defined the OHS as a system of management practices aimed at increased productivity of citrus orchards by continuously applying a balanced nutrient mixture through the irrigation system, limiting the root zone by restricting the number of drippers per tree, and maintaining the soil moisture near field capacity in a limited wetted zones. The combination of these practices is claimed to provide a greater control and manipulation of nutrient uptake at specific physiological stages and improved water uptake (Yandilla 2004). Many concepts associated with OHS has been successfully used in the production of peaches, almonds, grapes, citrus, avocados, and several vegetable crops in Spain, South Africa, Chile, Argentina, Morocco, and California (USA). In South Africa, commercial growers have adapted the OHS through use of drip fertigation on a daily basis during daylight hours (Pijl 2001; Schoeman 2002) resulting in increased citrus yield and fruit size (Kruger et al. 2000a, b; Kuperus et al. 2002). OHS principles were introduced in Australia as an intensive fertigation practice in citrus orchards (Falivene et al. 2005) but are somewhat less intensive than the original OHS developed in Spain to better meet the conditions and needs of the Australian industry. Thus, OHS has been modified to meet local cultural, weather, and soil conditions. In Florida, OHS-related management practices are being applied to higher density citrus plantings. The combination of OHS with high tree density is being promoted under the name APSs (Stover et al. 2008; Morgan et al. 2009b; Roka et al. 2009). The use of mechanical harvesting in these high-density low canopy volume orchards will not be a subject of this review, but contributes to the overall efficiency of the production system (Roka et al. 2009).

19.2 Open Hydroponic System Concepts

The goals of OHS are to (1) increase initial tree growth rate, (2) establish early sustained fruit production, (3) maximize efficiency of production inputs, and (4) improve return on investment to achieve profits in as short a period of time as possible. To accomplish these goals within the framework of environmental needs, four fertigation concepts have been developed: (1) maximization of water and nutrient use efficiency, (2) improvement in nutrient availability, (3) concentration of roots in the irrigated zone, and (4) reduction in nutrient leaching. These goals and concepts have been incorporated in the APS that utilizes OHS along with tree planting density, tree size control, and horticultural manipulation (pruning and girdling).

19.2.1 Maximize Water and Nutrient Use Efficiency

Several studies conducted over many years have revealed that it is possible to increase yields and efficiency of water use and nutrient use through water-saving irrigation methods. In a study on water use efficiency and nutrient uptake on low volume irrigated citrus in New South Wales in Australia on Tiltao sand, Falivene (2005) found that water uptake was limited by water availability rather than root density. Soils maintained at drained upper limit (field capacity) in the root zone resulted significantly in greater tree water use. Also, fertilizer injection with the microsprinkler system significantly increased the efficiency of N and P uptake compared with surface application, whereas leaf K levels were lower under low-volume irrigation. Multiple applications of N in relatively small amounts with drip irrigation resulted in lower soil residual mineral N concentrations and enhanced N uptake efficiency by the citrus roots (Alva and Paramasivam 1998; Paramasivan et al. 2001).

Alva and Paramasimam (1998) found groundwater N concentrations below citrus trees fertilized with soluble granular fertilizers ranged from 6.5 to 16.5 mg L^{-1} compared with groundwater N concentrations ranging from 1.4 to 6.9 mg L^{-1} below fertigated trees. In an investigation of soil N solutions below fertigated citrus trees, Paramasivan et al. (2001) found NO_3-N concentrations occasionally peaked at 12–100 mg L^{-1} at depths above 120 cm below the soil surface, but NO_3-N concentrations mostly remained below 10 mg L^{-1} at 240 cm. These studies indicate that, with proper irrigation scheduling, soil N can be maintained in citrus tree root zones and not leached to groundwater. In a 3-year study, Bryla et al. (2005) found that peach trees irrigated by surface and subsurface drip produced 19% higher prune weights and up to 24% greater marketable yield than those irrigated by microjets and furrow irrigation on a Hanford fine sandy loam in California. Drip irrigation systems, in particular, are known to improve irrigation and fertilizer use efficiency because water and nutrients are applied directly to the root zone (Camp 1998). The benefits of frequent fertigation and/or irrigations in achieving high water and nutrient use efficiency offered by drip irrigation can be negated by improper water placement as shown with total leaf area, and fibrous root length were reduced by 44–55% in grapefruits (Zekri and Parsons 1988). Therefore, careful placement of water in the root zone is important in fruit production to ensure that water and nutrient uptake are optimized.

19.2.2 Improved Nutrient Availability

Several fundamental aspects of citrus physiology and cultural management must be taken into consideration when implementing OHS. Elements to be considered include fertilizer requirements and irrigation scheduling, and have been researched using many tree sizes and soil conditions. The following section reviews reports to help guide us in interpreting likely responses of citrus to OHS. Highly intensive nutrition program of OHS with the goal of rapid tree growth and productivity typically results in a higher level of vigor. Considering current best management practices (BMPs) and potential improvements in nutrient use efficiency, a goal for successful adoption of OHS will have to be more efficient use of applied nutrients. To determine whether this is feasible, we must review studies on fertilization practices on citrus, which sometimes appear confusing and contradictory, due to different tree ages, soil types and application practices compared. Obreza and Rouse (1993) showed that an increase in fertilizer rate from 0.32 to 0.64 kg N $tree^{-1}$ in the third year after planting resulted in a decrease in total soluble solids concentration and soluble solids to acid ratio. Koo and Smajstrla (1984) made similar observations with annual N rates greater than 224 kg ha^{-1} using trickle irrigation and fertigation on 26-year-old "Valencia" orange on an Astatula fine sand in Florida. Furthermore, Koo (1980), in trials on sandy soil, found no significant differences due to fertigation frequencies (3 or 10 times a year) on 13-year-old "Valencia" orange. Similarly, Syversten and Jifon (2001) studied fertigation in 6-year-old "Hamlin" oranges in Florida at 12, 37, and 80 times per year and found that fertigation frequency did not affect leaf nutrient concentration, canopy size, fruit yield, or juice quality.

Morgan et al. (2009a) examined the effect of N fertilizer rates and methods of applying N on growth and productivity of young (3–5 years old) and maturing (8–10 years old) citrus trees on well-drained sandy Entisols of central Florida. They observed that in young trees controlled release fertilizer applied once a year and fertigation 30 times annually produced higher average yields (20.67 Mg ha^{-1}) and larger trees (7.9 m^3 $tree^{-1}$) compared with fertigation or dry granular fertilizer applied four times annually (18.77 Mg ha^{-1} and 7.5 m^3 $tree^{-1}$). In maturing trees, however, the dry granular fertilizer applied 4 times a year and fertigation 30 times annually produced similar yields (52.0 and 53.7 Mg ha^{-1}, respectively) and total soluble solids (9.14 and 9.15°Brix, respectively). However, canopy volumes for the same trees were significantly greater with fertigation treatment (25.0 m^3 $tree^{-1}$) compared with the dry granular fertilizer (23.2 m^3 $tree^{-1}$). They observed that increased number of split applications would likely promote tree growth, albeit, little increase in fruit yield may be obtained in mature citrus.

Alva et al. (2003) found that total annual N applications could be reduced by >10% by applying approximately half the annual N (224 kg ha^{-1}) in three foliar applications and the remaining half as fertigations with no significant reduction in yield. Thus, they proposed a combined use of foliar fertilizer application and fertigation as a BMP for N because these were effective in reducing nitrate leaching to surficial groundwater. Nevertheless, even the most intensive practices in the studies reviewed provide many fewer seasonal applications of fertilizer than a typical OHS in which three or more fertigation events per day are standard (Falivene et al. 2005). Proper irrigation system design is important in OHS to ensure that the system does not leak and/or fail at some point or over-irrigation causing nutrient leaching. There are two main types of irrigation scheduling programs in OHS: pulsing irrigation and continuous (Falivene et al. 2005). A pulsing irrigation management program involves short pulses (e.g., 20–30 min each) of irrigation provided to the trees at a number of times throughout the day while a continuous irrigation management program uses low output rates to match water use conditions in summer. The number and timings of pulses are based on a calculation of readily available water and average tree water use along with monitoring of irrigation scheduling devices like tensiometers, capacitance probes, and trunk diameter measuring devices. In a restricted root zone situation, up to nine or more pulses of irrigation needed to be scheduled throughout the day in summer (Falivene et al. 2005).

19.2.3 Concentrate Roots in Irrigated Zone

Michelakis et al. (1993), studying avocado water use in a Mediterranean climate in Greece under drip irrigation, found that root density was generally higher in the upper 50-cm soil layers and within 2 m from the drip line, with about 70% of the roots located in this region. They attributed the higher root percentage in the upper soil layers to biological factors and to the higher oxygen diffusion rate. In the study, Michelakis et al. (1993) applied irrigation water to each treatment using one drip lateral per row of trees with drippers of 4 $L h^{-1}$ discharge rate placed 70 cm apart. Coleman (2007) also observed that root length density in cottonwood, American sycamore, sweet gum, and loblolly pine was dependent upon depth and position relative to drip emitter when fertilizers were applied and is greatest at the surface and in proximity to the drip line. The factors controlling root length density in the woody species studied included age, depth, and proximity to the drip emitter. The principles underlying the restriction of the roots to the wetted zone using drip irrigation are also applicable to OHS.

19.2.4 Intensive Water and Nutrient Management

Kalmar and Lahar (1977) irrigated avocados with sprinklers at 7, 14, 21, and 28-day intervals on a grumusol with more than 60% clay from the soil surface to a depth of 150 cm and found that most water was absorbed from upper 60-cm soil layer, suggesting that this was where greater than 70% of roots were concentrated. Bryla et al. (2005) compared the effect of furrow and microsprinkler irrigation scheduled weekly or biweekly and surface and subsurface drip irrigation scheduled daily on production and fruit quality of peach on a Hanford fine sandy loam in California. They found that daily drip irrigations maintained significantly higher soil water content through the season, with significant reduction in soil water stress (−1.4 MPa average midday stem potential between drip irrigation events compared with −1.8 MPa between furrow irrigations). As a result of lower plant water status, higher marketable yields with larger fruits were produced. Schoeman (2002) also made similar observations in citrus using daily drip fertigation in South Africa. Schumann et al. (2003) compared fertilizers sources and rates in "Hamlin" orange on a Candler fine sand in Florida. The results showed that optimal soluble solids production for fertigation was obtained at a N level of 145 kg ha^{-1} while the N optima for dry granular and controlled release fertilizers were 180 and 190 kg ha^{-1}, respectively. The greater efficiency of fertigation amounted to a N saving of 35–45 kg ha^{-1} $year^{-1}$ and approximately 20% more soluble solids yield than the other fertilizer sources. Also, leaf N concentrations were significantly higher per unit of N applied for fertigation > dry granular > controlled release. Thus, Schumann et al. (2003) concluded that fertigation was the most efficient fertilizer source because of optimal placement in the root zone and optimal temporal distribution over the season.

Morgan et al. (2009a) made similar observations in young citrus trees on the sandy soils of central Florida. In theory, the growth and yield of citrus trees should be maximized if the demand for nutrients and water by the roots is always matched with an adequate supply from drip fertigation, thus avoiding even transient deficiencies. The daily timing for both water and nutrient delivery should coincide with the time of maximum transpiration flow, which is during the daylight hours. Thus, the common strategy in OHS systems worldwide is to pulse-fertigate during daytime hours and hopefully attain a high percentage of immediate uptake by the roots instead of temporarily storing the water and nutrients in the soil as with conventional fertilization and irrigation systems. Storage of water and nutrients in the soil before uptake by roots will increase losses and inefficiencies due to evaporation, leaching, adsorption, precipitation, and volatilization mechanisms as well as immobilization by microbes. All those processes are kinetically regulated, thus minimizing the duration of soil contact is one of the underlying principles of hydroponics.

Studies on tree root density distribution have been in done in Florida and other parts of the world. Castle and Krezdorn (1975) described two general types of citrus root systems, one characterized by extensive lateral and vertical development and the other by intensive higher fibrous root density near the soil surface. Trees on rough lemon (*C. jambhiri* Lush), Volkamer lemon (*C. volkameriana* Pasquale), and Palestine sweet lime (*C. limettioides* Tan.) rootstocks are typical of citrus trees with extensive root structure where 50% of the fibrous roots were at soil depths > 0.7 m. Large, high-yielding trees with extensive root systems dominated the citrus industry in Florida when trees were irrigated less intensively and planted at much lower densities. Unfortunately, rough lemon has been virtually eliminated as a commercial rootstock due to citrus blight disease (unknown etiology) in the 1970s and 1980s (Castle 1980). Carrizo citrange and Swingle citrumelo are examples of the intensive-type root systems with few fibrous roots below 0.7 m and less lateral development (Castle and Krezdorn 1975). These rootstocks now dominate the Florida citrus industry and are well suited for high-density, intensively managed plantings (Castle 1978). Morgan et al. (2007) found that average fibrous root length density (FRLD) decreased from 1.11 cm cm^{-3} at the soil surface to 0.27 cm cm^{-3} at 90 cm depth and decreased from 0.52 cm cm^{-3} near the tree trunk to 0.16 cm cm^{-3} at greater than 175 cm, resulting in mature trees with bimodal root systems. Also, the FRLD varied as a function of rootstock in which trees on Swingle citrumelo developed higher FRLD near the soil surface (1.39 cm cm^{-3}) than trees on Carrizo citrange (0.84 cm cm^{-3}) with similar FRLD below 0.3 m. Abrisqueta et al. (2008) studied root dynamics of young peach found that partial root zone drying and continuous deficit irrigation in Spain reduced root density in the nonirrigated zones by 42% and 73%, respectively. In the study, higher root length densities were recorded in non-limiting irrigation conditions than under deficit irrigation where root growth was reduced.

The use of OHS can limit root growth to within the irrigated zone. Research studies into restricted root zones using physical constraints have shown a reduction in yield in fruit and vegetables (Ismail and Noor 1996; Bar-Yosef et al. 1988; Boland et al. 2000). These studies attributed the yield reduction to reduced canopy growth. Reduced canopy growth or a reduction in yield per tree has not been observed to date in OHS (Boland et al. 2000; Falivene 2005). The wetted soil volume in OHS is considerably greater than the restricted root zone studies mentioned above where significant reductions in vegetative growth and yield have been reported (Falivene 2005). The study by Boland et al. (2000) on peach in Australia showed a significant reduction in growth and yield when the root zone was restricted to 3% of its potential. In contrast, the wetted soil volume in OHS is approximately 8–15% of the potential root volume (Falivene 2005). These studies envisage that in an OHS situation, the roots are redirected to grow more densely in a smaller volume of soil, but the soil volume is sufficiently large enough to support active root growth and a productive tree.

19.2.5 Reduced Nutrient Leaching

Many researchers have attempted to study nutrient leaching to sustain environmental quality (Warrick 1986). Paramasivan et al. (2001) found that nitrate-nitrogen leaching losses below the rooting depth accounted for 1–16% of applied fertilizer N and increased with increasing rate of N application (112–280 N ha^{-1} $year^{-1}$) and the amount of water drained. Paramasivan et al. (2001) also noted that the leached nitrate-nitrogen at 240 cm remained well below the maximum contaminant limit of 10 mg L^{-1}. They ascribed their observations to careful irrigation management, split fertilizer applications, and proper timing of the application. Thus, it should be possible to reduce nutrient leaching with an OHS, because in both scenarios, water and nutrients are applied in quantities approximating plant needs and close to the plant with less waste and at specific physiological stages of the plants (Mason 1990).

19.3 Principles Used in Advanced Production Systems

Certain principles of irrigation, nutrient, and horticultural management must be followed in a systematic approach to achieve the goals of OHS. The principles of production used in APS are currently being followed by some citrus growers, but require some modifications and more intensive management. The principles added to OHS to develop the APS are (1) higher tree densities, (2) size-controlling rootstock selection, (3) restricting root zones with drip irrigation, (4) intensive irrigation and nutrition management, and (5) horticultural manipulation.

Fig. 19.1 Densely spaced citrus trees grown on ridges. Trees grow into hedgerows within 3 years to maximize early yield (Photo by K. T. Morgan)

Fig. 19.2 Trees planted on ridges allow for good water drainage and air infiltration (Photo by K. T. Morgan)

19.3.1 Higher Tree Density (>725 trees ha^{-1})

The ideal grove is one in which there is dense planting of rapidly developing trees to bearing volume with sufficient bearing volume to support high levels of cropping (Fig. 19.1). The rapid growth of these trees and final size are critical to improvement in water and nutrient use efficiency. The use of planting ridges is likewise important, particularly in loam or clay loam soils (Fig. 19.2). The ridges allow for proper water drainage and air infiltration to maintain aerobic conditions in the drip irrigation zones. Such groves provide certain known advantages related to production, harvesting, and returns, but to be successful, smaller-sized, closely planted trees are essential (Roka et al. 2009). Changes in orchard design have occurred primarily in the deciduous fruit industries. Robinson et al. (2007) published results of planting densities ranging from 850 to 5,445 trees ha^{-1} for apple orchards in New York. They found that the optimum economic density was between 2,500 and 3,000 trees ha^{-1}. The optimum density achieved improved yield by >20% and quality coupled with lower costs of production. The practices and concepts that constitute the OHS are an excellent match with higher planting densities.

With the advent of the OHS for citrus, some data have demonstrated the performance of groves of closely spaced citrus trees have been managed with the OHS. Yields of Nova, Marisol, and Delite mandarins in Spain, planted at higher density (1,012 trees ha^{-1}) and grown using the OHS, were about 65–75 tons ha^{-1} in the sixth year which is higher than for a conventional orchard using low- to medium-density plantings (375–575 trees ha^{-1}) (Martinez-Fuentes et al. 2004; Falivene et al. 2005). In Florida, higher planting density (889 trees ha^{-1}) produced higher >33% early (4–8 years after planting) production compared with lower tree densities (370 trees ha^{-1}) (Wheaton et al. 1995; Parsons and Wheaton 2009). However, average annual fruit yields for the same high-density plantings were similar in at 9–13 years after planting when trees were maintained at a height of 5.5 m. When the trees were maintained at a 3.5 m height, the average annual yields were reduced by 50–60%. Those studies demonstrated the feasibility of higher density plantings for citrus, but the trials were conducted under lower tree planting densities than proposed for the future with OHS and under less-intensive management. Thus, there is a possibility of further increasing yield per unit area using OHS with densely planted citrus trees.

Fig. 19.3 The size of vigorous, rapidly growing trees must be controlled by hedge if nutrition is not altered to reduce vegetative growth (Photo by K. T. Morgan)

Fig. 19.4 The roots can clearly be seen beneath these citrus trees. The irrigation zone of each dripper is *outlined in white* delineating the highest root density with lower root densities outside these zones and nearer the soil surface (Photo by K. T. Morgan)

19.3.2 Size-Controlling Rootstock Selection

Rootstock selection along with tree planting density is a key element in the APS/OHS approach to the future. Citrus trees, like humans, need a certain amount of space to develop and flourish. When the allocated space is fixed, e.g., 1 ha of land, tree size becomes critical because the productive unit is the canopy and only a certain volume of canopy can be grown per unit land area. Vigorous, large trees are neither compatible with close spacing nor productive in their younger years (Fig. 19.3). Thus, in a world of economic necessity dictated by early and robust returns, small, closely spaced trees become a required component of the new production concepts (Stover et al. 2008; Morgan et al. 2009b). When that combination is achieved, the higher density grove will outperform the more conventional one especially in the early years.

19.3.3 Restricted Rootzones and Intensive Irrigation and Nutrient Management

Management of water and nutrients in the root zone of citrus is critical to establishing rapid growth and early sustained fruiting (Morgan et al. 2009b). Root density can be restricted to the wetted zone of each drip emitter (Fig. 19.4). Restriction of the roots in this manner allows for the soil surrounding the roots to be maintained at nearly field capacity. If the intensive management is provided, by pulsing drip applications at regular intervals on a daily basis, nutrients within the root zone will not be leached by excessive irrigation. The fertigation system must contain four basic

Fig. 19.5 Adequate filtration must be provided for the drip systems. This photo illustrates a manifold of sand filters from a surface water source (Photo by K. T. Morgan)

Fig. 19.6 Fertilizers can be custom blended using dry materials and multiple storage tanks connected to the fertigation system (Photo by K. T. Morgan)

parts of (1) filtration, (2) fertilizer mixing and/or storage, (3) fertilizer injection, and (4) fertigation control. Water filtration depends on the water quality and water source. For surface water, the system must contain a sand media filter adequately sized to remove biological materials as well as particulates depending on water quality. If the water source is a well, a screen or disk filter (Fig. 19.5) with an opening size of 0.05 mm or less is required. Fertilizers can be stored dry and mixed as needed in multiple tanks depending on requirements of the system (Fig. 19.6). A minimum of two tanks are recommended for best results. One tank would contain the nitrogen, potassium, and phosphorus components with the additional tanks containing desired

Fig. 19.7 Intensive management of fertigation demands the use of many injection valves with filters from the fertilizer supply tanks (Photo by K. T. Morgan)

Mg, Ca, and micronutrients. A manifold of valves from each fertilizer storage tank to the injection device must be provided (Fig. 19.7). The final element of the fertigation system is a controller or custom computer operating system depending on the size of the system to be used. The controller must be adequate to operate multiple injections with multiple drip pulses per day.

19.4 Horticultural Manipulation

High early production is essential for higher-density, shorter-cycle citrus production to be economically sound (Roka et al. 2009). Early cropping not only front-loads economic returns but also importantly competes with vegetative growth and helps keep trees smaller (Erner 1988; Takahara et al. 1980). Several horticultural practices to enhance early cropping have been explored and documented in citrus and other fruit crops, and many have been widely used in recent high-density citrus plantings in South Africa and likely other regions where intensive plantings have been utilized (Perez-Madrid et al. 2005). Not all scion/rootstock combinations will require horticultural intervention to accelerate cropping. Increasing citrus production usually means that the trees are forced to break juvenility and begin reproductive growth earlier than would naturally occur (He 1997). Trees can be manually manipulated using various horticultural techniques such as pruning and girdling to improve fruit set, yield, and quality (Fig. 19.7). These techniques have perhaps even greater potential to enhance fruit quality and also increase yield when applied to more easily managed small trees planted close together (APS) and intensively

Fig. 19.8 This tree has been girdled each year for 3 years to force the tree into early reproductive growth (Photo by K. T. Morgan)

managed using the OHS approach. Stover et al. (2008) suggested that use of APS/OHS systems should help control vegetative growth, keeping trees in check and reducing cost of pruning while also providing earlier cash flow (Roka et al. 2009).

The effects of girdling (Fig. 19.8) on crop performance depend on when it is done. Girdling in autumn enhances flowering in citrus (Goldschimidt and Colomb 1982), at full bloom improves fruit set in oranges (Monselise et al. 1972), and in summer, girdling increases grapefruit size by >10% (Fishler et al. 1983). In other cases, girdling was reported to limit nitrogen, phosphate, and calcium uptake in avocado in South Africa (Davie et al. 1995) and increased average fruit weight by 0.8 g but reduce the average soluble solids concentration at harvest by 0.8°Bx in grapes in California (Harrell and Williams 1987). However, Andrews et al. (1978) reported that girdling of peach trees in Florida increased first harvest fruit yield by an average of 45%, enhanced ripening by 3–5 days, but resulted in severe necrosis of leaves and gumming on the area of the cut. The pruning and girdling practices need to be carefully considered for use in high-density citrus plantings because benefits will need to be substantial to justify the high labor costs associated with these practices.

19.5 Results of OHS Applications

Limited information is available for citrus grown under OHS conditions. Kruger et al. (2000a, b) found 19% yield increases in Clementine mandarin for drip-irrigated citrus using OHS (47.86 Mg ha^{-1}) and microsprinkler-irrigated (40.22 Mg ha^{-1}) blocks with soluble fertilizer in a non-replicated OHS demonstration in South Africa. Similar results (25% increase) were found for drip-irrigated OHS-produced Midknight Valencia (55.3 Mg ha^{-1}) and microsprinkler-irrigated control (44.0 Mg ha^{-1}). Yield increases were contributed to a 27% increase in fruit number, but fruit size was not reduced. Likewise, Martinez-Valero (2004) reported third year and cumulative yields of 8.5 and 206.8, and 17.6 and 240.6 Mg ha^{-1} for OHS 6-year-old Nova hybrid mandarin (*C. reticulate* Blanco × Tangelo Orlando) and Marisol clementine (*C. reticulata* Blanco), respectively. No comparisons of treatments with common growing conditions were available. A study currently being conducted in Florida is comparing the effects of drip irrigation using OHS nutrient management with small-area and large-area microsprinkler irrigation systems on Hamlin and Valencia orange growth and production at three tree densities. Preliminary data after 4 years indicate that the OHS system improves growth (measured by increase in canopy volume) when drip-grown trees are compared with small-area and large-area microsprinkler-irrigated trees, respectively (Table 19.1; Morgan unpublished data). Yields in the same study increased by nearly 1.9 and 5.0 times when drip OHS is compared with small-area and large-area microsprinkler. During this study, water and fertilizer amounts were decreased by 20–50% using drip fertigation compared with microsprinkler.

19.6 Future Research

Key aspects of OHS include appropriate size-limiting rootstocks for selected scions grown at higher tree densities under soil and environmental conditions prevalent in each production area. This will require intensive long-term research in each production area of the world to determine the appropriate rootstock/scion combination for a range of in-row and between-row spacing. Determining this combination on a region-by-region basis is critical. Likewise, the range of tree densities depends on the field equipment used. Using large equipment currently available for wider-spaced trees will not be appropriate for closer-spaced trees. Thus, smaller equipment must be developed, and equipment that can spray, hedge, and harvest multiple rows simultaneously must also be considered. Such equipment will operate over the row and are available for many crops but not currently for citrus. While drip irrigation equipment is commonly used for citrus in many areas, microsprinkler irrigation is used in many areas, including Florida and Texas, for frost protection. Adaptation of OHS to use these microsprinkler systems will be necessary, as a dual drip fertigation and microsprinkler frost protection system is considered too expensive.

Table 19.1 Citrus tree growth (canopy volume) and yield for 5-year-old Hamlin and Valencia trees grown for 4 years under three irrigation and nutrient treatments

	Canopy volume (m^3)	Yield (Mg/ha)	Soluble solids (°Brix)	Canopy volume (m^3)	Yield (Mg/ha)	Soluble solids (°Brix)
Drip OHS	9.9 A (29.8 m^3/gN)	12.6 B (0.22 Mg/kgN)	3.8	14.0 A (42.5 m^3/gN)	14.5 A (0.26 Mg/kgN)	4.9
Small-area microsprinkler	7.7 B (14.0 m^3/gN)	18.7 A (0.16 Mg/kgN)	4.1	11.4 B (20.7 m^3/gN)	12.3 B (0.11 Mg/kgN)	5.0
Large-area microsprinkler	8.9 B (10.3 m^3/gN)	10.1 B (0.04 Mg/kgN)	4.1	12.0 B (13.9 m^3/gN)	11.6 C (0.05 Mg/kgN)	5.0

The drip and small-area microsprinkler treatments were irrigated daily and fertigated daily and weekly, respectively. Large-area microsprinkler-treated trees were irrigated at target soil water depletion and fertigated monthly

19.7 Conclusion

Intensive production systems to improve citrus production efficiencies will involve planting densities of 750 or more trees per ha on rootstocks that match final tree size to soil characteristics and planting density. Tree irrigation and nutrition using this new system will be linked through the use of management systems that will apply the appropriate ratio of nutrients to roots concentrated within the irrigated zone. The production system adopted by citrus producers should be operated in a manner that maintains current nutrient and water quality standards. Finally, the system will rely on selective horticultural manipulation of tree growth and fruiting through mechanical pruning and girdling as needed. This combined system of production will result in higher young tree growth rates, earlier fruit production, and may maintain high levels of productivity compared with current cultural practices especially in the presence of tree losses due to Citrus Greening.

References

Abrisqueta JM, Mounzer O, Álvareza S et al (2008) Root dynamics of peach trees submitted to partial rootzone drying and continuous deficit irrigation. Agric Water Manage 95:959–967

Alva AK, Paramasivam S (1998) An evaluation of nutrient removal by citrus fruits. Proc Fla State Hortic Soc 111:126–128

Alva AK, Paramasivam S, Graham WD et al (2003) Best nitrogen and irrigation management practices for citrus production in sandy soils. Water Air Soil Pollut 143:139–154

Andrews CP, Sherman WB, Sharpe RH (1978) Response of peach and nectarine cultivars to girdling. Proc Fla State Hortic Soc 91: 175–177

Bar-Yosef B, Schwartz S, Markovich T et al (1988) Effect of root volume and nitrate solution concentration on growth, fruit yield and temporal N and water uptake rates by apple trees. Plant Soil 107:49–56

Boland AM, Jerie PH, Mitchell PD et al (2000) Long-term effects of restricted root volume and regulated deficit irrigation on Peach: I. Growth and mineral nutrition. J Am Soc Hortic Sci 125(1): 135–142

Bryla DR, Dickson E, Shenk R et al (2005) Influence of irrigation method and scheduling on patterns of soil and tree water status and its relation to yield and fruit quality in peach. HortScience 40(7):2118–2124

Camp CR (1998) Subsurface drip irrigation: a review. Trans ASAE 41:1353–1367

Castle WS (1978) Citrus root systems: their structure, function, growth, and relationship to tree performance. In: Cary PR (ed) Proceedings of the international society for citriculture, Griffith, NSW, Australia, vol 1, pp 62–69

Castle WS (1980) Fibrous root distribution of 'Pineapple' orange trees on rough lemon rootstock at three tree spacings. J Am Soc Hortic Sci 105(3):478–480

Castle WS, Krezdorn AH (1975) Effect of citrus rootstocks on root distribution and leaf mineral content of 'Orlando' tangelo tree. J Am Soc Hortic Sci 100(1):1–4

Coleman M (2007) Spatial and temporal patterns of root distribution in developing stands of four woody crop species grown with drip irrigation and fertilization. Plant Soil 299:195–213

Davie SJ, Stassen PJC, van der Walt M (1995) Girdling for increased "hass" fruit size and its effect on carbohydrate production and storage. Proc World Avocado Congr 3:25–28

Erner Y (1988) Effects of girdling on the differentiation of inflorescence types and fruit set in 'Shamouti' orange trees. Israel J Bot 37:173–180

Falivene S (2005) Open hydroponics: risks and opportunities. Land and Water Australia through the National Program of Sustainable Irrigation. http://www.arapahocitrus.com/files/OHS_Stage_Publications.pdf

Falivene S, Goodwin I, Williams D et al (2005) Open hydroponics: risks and opportunities. stage 1: general principles and literature review. NSW Department of Primary Industries and Land and Water Australia, Brisbane NSW, Australia

Fishler M, Goldschimidt EE, Monselise SP (1983) Leaf area and fruit size in girdled grapefruit branches. J Am Soc Hortic Sci 108:218–221

Goldschimidt EE, Colomb A (1982) The carbohydrate balance of alternate bearing citrus trees and the significance of reserves for flowering and fruiting. J Am Soc Hortic Sci 107:206–208

Harrell DC, Williams LE (1987) The influence of girdling and gibberellic acid application at fruitset on Ruby seedless and Thompson seedless grapes. Am J Enol Vitic 38(2):83–88

He ShiJie (1997) Girdling technique for Hongjiangcheng orange cultivar. South China Fruits 26:25

Ismail MR, Noor KM (1996) Growth, water relations and physiological processes of starfruit (*Averrhoa carambola* L.) plants under root growth restriction. Sci Hortic 66:51–58

Kalmar D, Lahar E (1977) Water requirements of Avocado in Israel. I. Tree and soil parameters. Aust J Agric Res 28:859–868

Koo RCJ (1980) Results of citrus fertigation studies. Proc Fla State Hortic Soc 93:33–36

Koo RCJ, Smajstrla AG (1984) Effects of trickle irrigation and fertigation on Fred end-use quality of 'Valencia Orange'. Proc Fla State Hortic Soc 97:8–10

Kruger JA, Tolmay CD, Britz K (2000a) Effects of fertigation frequencies and irrigation systems on performance of Valencia oranges in two subtropical areas of South Africa. Proc Int Soc Citric Congr 9:232–235

Kruger JA, Britz K, Tolmay CD et al (2000b) Evaluation of an open hydroponics system (OHS) for citrus in South Africa: preliminary results. Proc Int Soc Citric Congr 9:239–242

Kuperus KH, Combrink N, Britz K et al (2002) Evaluation of an open hydroponics system (OHS) for citrus in South Africa. Annual research report. Citrus Research International (Pty) Ltd., Nelspruit, South Africa

Martinez-Fuentes AC, Mesejo C, Juan M et al (2004) Restrictions on the exogenous control of flowering in citrus. Acta Hortic 632: 91–98

Martinez-Valero R (2004) Preliminary results in citrus grove grown under the MOHT system. In: Proceedings of the international society for citriculture, Cape Town, South Africa, pp 316

Mason J (1990) Commercial hydroponics: how to grow 86 different plants in hydroponics. Kangaroo Press, Roseville

Michelakis N, Vougioucalou E, Clapaki G (1993) Water use, wetted soil volume, root distribution and yield of avocado under drip irrigation. Agric Water Manage 24:119–131

Monselise SP, Goren R, Wallerstein I (1972) Girdling effect on orange fruit set and young fruit abscission. HortScience 7:514–515

Morgan KT, Obreza TA, Scholberg JMS (2007) Characterizing orange tree root distribution in space and time. J Am Soc Hort Sci 132(2): 262–269

Morgan KT, Wheaton TA, Castle WS et al (2009a) Response of young and maturing citrus trees grown on a sandy soil to irrigation scheduling, nitrogen fertilizer rate, and nitrogen application method. HortScience 44(1):145–150

Morgan KT, Schumann AW, Castle WS et al (2009b) Citrus production systems to survive greening: horticultural practices. Proc Fla State Hortic Soc 122:114–121

Obreza TA, Rouse RE (1993) Fertilizer effects on early growth and yield of 'Hamlin' orange trees. HortScience 28(2):111–114

Paramasivan S, Alva AK, Fares A et al (2001) Estimation of nitrate leaching in an Entisol under optimum citrus production. Soil Sci Soc Am J 65:914–921

Parsons LR, Wheaton TA (2009) Tree density, hedging and topping, HS1026. Cooperative Extension Service. University of Florida, Gainesville

Perez-Madrid G, Almaguer-Vargas G, Maldonado-Torres R et al (2005) Girdling and gibberellic acid sprays in production, fruit quality and nutrient level in 'Monica' mandarin. Terra 23:225–232

Pijl I (2001) Drip fertigation: effects on water movement, soil characteristics and root distribution. M.S. thesis, University of Stellenbosch, Stellenbosch

Robinson TL, Hoying SA, DeMarree A et al (2007) The evolution towards more competitive apple orchard systems in New York. N Y Fruit Q 15(1):3–9

Roka MF, Muraro R, Allen RA et al (2009) Citrus production systems to survive greening: economic thresholds. Proc Fla State Hortic Soc 122:233–239

Schoeman SP (2002) Physiological measurements of daily daylight fertigated citrus trees. M.S. thesis, University of Stellenbosch, Stellenbosch

Schumann AW, Fares A, Alva AK et al (2003) Response of 'Hamlin' orange to fertilizer source, annual rate, and irrigated area. Proc Fla State Hortic Soc 116:256–260

Stover E, Castle WS, Spyke P (2008) The citrus grove of the future and its implications for huanglongbing management. Proc Fla State Hortic Soc 121:155–159

Syversten JP, Jifon JL (2001) Frequent fertigation does not affect citrus tree growth, fruit yield, nitrogen uptake, and leaching losses. Proc Fla State Hortic Soc 114:88–93

Takahara T, Okudai N et al (1980) Studies on the promotion of flowering and fruiting of juvenile citrus seedlings. I. Effect of girdling. Bull Fruit Tree Res Stn D Kuchinotsu 2:1–14

Warrick AW (1986) Soil water distribution. In: Nakayama B (ed) Trickle irrigation for crop production. Elsevier, Amsterdam

Wheaton TA, Whitney JD, Castle WS et al (1995) Citrus scion and rootstock, topping height, and tree spacing affect tree size, yield, fruit quality, and economic return. J Am Soc Hortic Sci 120(5):861–870

Yandilla (2004) Martinez Open Hydroponic Technology, Yandilla Park Limited, Renmark. http://www.yandillapark.com.au/Growers/ohs_main.htm

Zekri M, Parsons LR (1988) Water relations of grapefruit trees in response to drip, microsprinkler and overhead sprinkler irrigation. J Am Soc Hortic Sci 113:819–823

Fertigation: Concept and Application in Citrus

20

Ana Quiñones, Belén Martínez-Alcántara, Eduardo Primo-Millo, and Francisco Legaz

Abstract

Citrus is one of the most relevant crops worldwide with a yearly average production of 90×10^6 Mg. As in other crops, the use of fertilizers is crucial to maintain a high-quality yield. However, its abuse has led to environmental problems such as eutrophication, pollution, deforestation and loss of productive soil through erosion, in the last decades. Therefore, fertilizer together with other agricultural inputs (seeds, pesticides, water, machinery, etc.) should be managed more efficiently, enhancing, at the same time, the development of new technologies and government policy that ensures an agricultural systems that will produce safe and quality fruits. A rational fertigation programme should consider not only plant requirements but also a rational seasonal distribution of nutrients. This chapter describes different considerations to keep into account an optimal fertigation schedule in order to improve nutrient use efficiency, reduce nutrients losses and protect water quality.

Keywords

Nutritional requirements • Nutrient dose • Fertigation programme • Citrus

20.1 General Overview

Citrus is one of the most relevant crops worldwide with a yearly average production of 90×10^6 Mg in the last decade. In the Mediterranean countries, citrus is the second largest fruit crop after apples, in the European Union (EU). Spain is the leading producing country in the area with nearly 60% of tonnes produced in the whole EU (Ollier et al. 2009). In this country, citrus orchards cover around $300 \cdot 10^3$ ha ($6 \cdot 10^6$ Mg) of which up to 60% is developed in Comunidad Valenciana (CV). This area has remained more or less constant since 1990 (MARM 2010). The Comunidad Valenciana is the most important region, not only in acreage but also with respect to its long tradition of citrus farming.

The use of fertilizers is crucial to maintain a high-quality yield in this crop. However, its abuse has led to environmental problems such as eutrophication, pollution, deforestation and loss of productive soil through erosion, in the last decades. Therefore, fertilizer with other agricultural inputs (seeds, pesticides, water, machinery, etc.) should be used more efficiently, enhancing, at the same time, the development of new technologies and government policies aimed at ensuring agricultural systems that produce safe and high-quality fruits.

The main purpose of this chapter is to describe a methodology of a rational fertigation on citrus trees.

20.1.1 World Fertilizer Situation

According to FAO (2008) statistics, in a report that provides an overview of information on the world fertilizer situation in 2007/2008 and a forecast till 2011/2012, medium-term perspectives point to a slowing down of the world economy with a smoothing of the upward trend in emerging economies.

A. Quiñones (✉) • B. Martínez-Alcántara • E. Primo-Millo • F. Legaz
Citriculture and Vegetable Production Department, Instituto Valenciano de Investigaciones Agrarias (IVIA), Apartado Oficial, Moncada, 46113 Valencia, Spain
e-mail: quinones_ana@gva.es; martinez_belalc@gva.es; caballer_dol@gva.es; legaz_fra@gva.es

A.K. Srivastava (ed.), *Advances in Citrus Nutrition*,
DOI 10.1007/978-94-007-4171-3_20, © Springer Science+Business Media B.V. 2012

Table 20.1 Regional and subregional fertilizer consumption 2007/2008–2011/2012

	N		P		K	
Fertilizer	Share world consumptions (%)	Annual growth (%)	Share world consumptions (%)	Annual growth (%)	Share world consumptions (%)	Annual growth (%)
World		1.4		2.0		2.4
Africa	3.4	2.9	2.5	1.0	1.6	2.0
North America	13.5	0.3	12.0	0.5	17.1	0.7
Latin America	6.3	2.4	13.0	2.8	17.5	2.9
West Asia	3.5	1.7	3.3	1.0	1.4	2.4
South Asia	19.6	2.2	20.5	3.5	10.9	4.2
East Asia	38.3	1.3	36.1	1.9	35.2	3.3
Central Europe	2.7	1.8	1.5	1.2	2.4	1.0
West Europe	8.0	–0.3	5.6	–0.7	9.5	0.0
E Eur.+C Asia	3.0	2.4	2.0	4.5	3.1	1.6
Oceania	1.4	4.9	3.5	1.7	1.3	2.1

Source: Adapted from FAO (2008)

Global growth is seen as remaining sufficiently robust to sustain demand for food (especially high-value foods such as meat, fruits and vegetables) in emerging economies, thereby strengthening demand for fertilizers. In this sense, world fertilizer consumption is to grow annually at about 1.7% from 2007/2008 to 2011/2012, equivalent to an increment of about 15 million tonnes. About 69% of this growth will take place in Asia and 19% in America for the three major plant nutrients: nitrogen (N), potassium (K) and phosphorous (P).

'The expected share of world N, P and K consumption and the annual growth of this consumption are shown in Table 20.1. A revealing dimension of anticipated demand is the expected relative contribution to change in world consumption which is a function of the preceding two parameters. Low share of consumption combined with high consumption growth leads to higher contribution to changed world consumption than the share of consumption may suggest, e.g. Africa and Oceania in the case of N. Likewise, a high share of world consumption associated with low consumption growth means lower than may be expected contribution to world consumption, e.g. North America and West Europe for N. Evidently, if both share of consumption and consumption growth are high, the relative contribution to increased global demand will be high, e.g. South Asia for N' (FAO 2008).

20.1.2 Agriculture Fertilizers and the Environment

Fertilization in *citrus* orchards (as in other crops) has been based in the past on excessive rates and incorrect timing of element supply. These cultural practices have resulted in a low fertilizer use efficiency, an increase in the nutrient leaching to the environment, soil pollution and volatile emissions and, also, in a reduction in the fruit quality.

In the case of nitrogen fertilization, citrus farmers have applied high dosages of N because of poor fertilizing criteria and slight enhances in fruit yield when increasing N dosage. This has resulted in a deterioration in the commercial quality of the fruit (Chapman 1968), a reduction in the profitability of the citrus crops (Wild 1992) and an alarming increase in the concentration of the nitrate ion in underground waters (Burkart and Stoner 2002).

In this sense, farmers should use the information about the annual requirements established for citrus trees to adopt the optimum nutrient dose. These N requirements may be adjusted according to the nutrient uptake efficiency of fertilizers. So, it is of remarkable interest to set up a rational fertilization in a sustainable system that can ensure a competitive production with minor impact on the environment and conservation of the natural resource base. At the same time, the adoption of improved technology, new substrates, slow release fertilizers and new irrigation systems allows producers to increase fertilizer use efficiency and to use water more efficiently. In this context, fertigation has been raised as a valuable tool in recent years.

20.1.3 Fertigation: A Tool for Improving Nutrient Use Efficiency

The agricultural practice called fertigation appears in order to achieve a rational fertilizer application that can ensure a competitive production with minor impact. Fertigation (from the words fertilizer and irrigation) is an effective method of supplying crops in the field with fertilizers via the irrigation water (Bar-Yosef 1991). In the process of fertigation, the nutrient is applied to crop according to its daily demand of water and nutrients, which depends on its specific growth stage, to achieve maximum efficiency of the fertilizer applied (Kafkafi and Tarchitzky 2011). Therefore, nutrient dose supplied and its seasonal distribution is easily controlled.

Depending on the different localized irrigation systems, fertigation can be performed on surface or subsurface drip, spray, micro-jet and micro-sprinkler. These different techniques can be used, both in annual crops or fruit trees, according to soil type and the different characteristics of agricultural area. This versatility has led to a rapid expansion of fertigation in the world cultivated areas. Regarding the irrigation systems, in the citrus area of Spain, about 69% of citrus orchards are irrigated under fertigation, mainly with drip irrigation, and the remaining by flood irrigation (MARM 2010). Similar percentage are found in other citrus areas where localized (drip or mini-sprinklers) irrigation systems are mainly used for citrus and other tree crops (olives and deciduous trees), while sprinkler irrigation is dominant for fodder crops and some vegetables.

The advantages of fertigation are listed below (Burt et al. 1998):

- High water and nutrient use efficiency as a consequence of coupling fertilizer timing to the plant requirements and, therefore, minimized fertilizer/nutrient loss due to localized application and reduced leaching
- Reduced energy cost by the saving of labour and machinery and the efficient use of the costly chemicals to be applied
- Minimized soil erosion by avoiding heavy equipment traffic through the field to apply fertilizers

Moreover, fertigation allows safe use of recycled or saline water. Boman et al. (2005) affirmed that irrigation scheduling is a key factor in managing salinity. Increasing irrigation frequency and applying water exceeding the crop requirement are recommended to leach the salts and minimize its concentration in the root zone. Fertigation also reduces the risk of diseases since foliage remains dry. Schumann et al. (2009) observed a significative reduction in the number of trees infected by citrus greening disease or citrus canker by optimizing daily the nutrient levels for trees. Moreover, frequent and small water split with fertigation technique leads to a shallow and compact root system in comparison with a wide and deeper root system in flood-irrigated trees (Sne 2006), enhances N uptake efficiency by the fibrous roots and contributes to lower leaching below the root zone (Quiñones et al. 2007a).

20.1.4 Fertilization General Principles

With the aim of ensuring stable yields and optimum fruit quality in consonance with environment conservation, it is necessary to take into account the main laws governing soil fertility.

20.1.4.1 Law of Restitution

At the end of the crop cycle, soil characteristics and conditions should be similar to those seen at the beginning of the growth cycle. Therefore, it is essential that all the fertilizer elements removed from soil in harvested crops, leached by rain or excessive irrigation water and/or unavailable in soil solution, should be returned to prevent nutritional depletion (Voisin 1965).

20.1.4.2 Law of the Maximum

Law of the Maximum states that excess of an available element in the soil reduces the effectiveness of other elements and consequently lowers the harvested yield. Thus, in a proper nutrition balance, the effects of nutrient combination must be taken into account.

20.1.4.3 Law of the Minimum

Von Liebig's Law of the Minimum (1840) states that yield is proportional to the amount of the most limiting nutrient, whichever nutrient it may be. From this, it may be inferred that if the deficient nutrient is supplied, yields may be improved to the point that some other nutrient is needed in greater quantity than the soil can provide. However, von Liebig did not affirm his hypothesis in terms of proportionality relations.

Friis-Nielsen (1966) in an approach towards interpreting and controlling the nutrient status of growing plants by means of chemical plant analyses affirmed that either the Law of Maximum or the Law of Minimum may be decisive for plant growth, and their principles should be taken in consideration during the fertilization planning process.

20.1.4.4 Law of Diminishing Returns

Mitscherlich (1909) attempted to make clear the Law of Minimum and in the Law of Diminishing Returns asserts that the increase in any crop produced by a unit increment of a deficient nutrient is proportional to the decrement of that nutrient from the maximum (Ware et al. 1982).

20.1.5 Optimized Fertigation Programme

A rational fertigation programme should include not only plant requirements but also a rational seasonal distribution of nutrients. This chapter describes different considerations to keep into account an optimal fertigation schedule in order to improve nutrient use efficiency, reduce nutrients losses and protect water quality.

20.2 Citrus Nutritional Requirements

The purpose of fertilization is to increase the natural fertility of the soil in order to improve the nutritional status of a crop and ensure high fruit quality. Therefore, knowledge of citrus nutritional requirements (elements and dose) and seasonal uptake is key to establish optimum nutrient dose and application timing in a rational fertilization programme.

Table 20.2 Plant main macronutrients, form of uptake and their function in plant

Nutrient	Taken up as	Some important functions
Carbon (C)	CO_2	Forms the skeletons of all organic molecules, including both the cellulose and the lignin
Hydrogen (H)	H_2O	Component of carbohydrates
Oxygen (O)	CO_2, H_2O	Needed for plant respiration
Nitrogen (N)	NO_3^-, NH_4^+	Essential part of all amino acids, proteins, enzymes, coenzymes, chlorophyll, nucleotides, nucleic acids and others
Phosphorous (P)	$H_2PO_4^-$, HPO_4^{2-}	Found in nucleic acids, membrane phospholipids, attached to many sugars and has a central role in the ATP-ADP energy-providing system
Potassium (K)	K^+	Involved in osmotic and ionic regulation. Activates many enzymes in starch and protein metabolism
Calcium (Ca)	Ca^{2+}	Required for normal cell division and plays a major role in the maintenance of membrane integrity
Magnesium (Mg)	Mg^{2+}	Only mineral element in the chlorophyll molecule and is essential for photosynthesis to occur. Required for many enzymatic reactions
Sulphur (S)	SO_4^{2-}	Components of vitamins, coenzyme A and amino acids. Important in protein synthesis

Source: Adapted from Marschner (1995)

Table 20.3 Plant main micronutrients, form of uptake and their function in plant

Nutrient	Taken up as	Some important functions
Iron (Fe)	Fe^{2+}	Constituent of many enzymes, including cytochromes (respiratory enzymes) and ferredoxins involved in photosynthesis and N fixation
Zinc (Zn)	Zn^{2+}	Necessary for the optimum functioning of a range of enzyme systems, for the metabolism of auxin hormone and for the synthesis of nucleic acids
Manganese (Mn)	Mn^{2+}	Component of numerous enzymes
Copper (Cu)	Cu^{2+}	Essential in photosynthesis and respiration. Constituent of several enzymes
Boron (B)	BO_3^{3-}	Needed for protein synthesis, pollination and carbohydrate movement and metabolism
Molybdenum (Mo)	MoO_4^{2-}	Required for normal assimilation of N for nitrate reduction and N fixation
Chlorine (Cl)	Cl^-	During photosynthesis, enhances electron transfer from water to chlorophyll
Nickel (Ni)	Ni^{2+}	Constituent of the enzyme urease

Source: Adapted from Marschner (1995)

20.2.1 Essential Nutrient Elements

The fact that an element is present in plant tissues does not always mean that this element plays an essential role in plant development. According to Bennett (1993), an element becomes essential if it fulfils the following criteria:

- A lack of it makes it impossible for the plant to complete its vegetative or reproductive cycle.
- The function of the element is not replaceable by another element, and, therefore, symptoms of failure can only be prevented or eliminated by providing that nutrient.
- The item is directly involved in plant metabolism or must be a component of an essential plant constituent (e.g. magnesium is a component of a chlorophyll molecule).

Fourteen mineral elements are considered essential. By adding carbon (C), hydrogen (H) or oxygen (O), which are obtained from air and water, to the 14 minerals, a total of 17 are considered essential. With these elements and sunlight, a plant is able to synthesize all the needed compounds in plant metabolism. In addition to C, H, O, of which are easily available to plant, essential nutrients can be divided into two groups based on their relative needs in the tree, macronutrients or micronutrients. Macronutrients are generally required in the largest amounts, although with great variations between crops, in concentrations of 1,000 parts per million (ppm) or more. However, micronutrients are needed in small or trace quantities, at a concentration of 100 ppm or less. Different essential plant elements, form of uptake and their principal function in plants are depicted in Tables 20.2 and 20.3 (Marschner 1995).

Usually, only a small amount of an element is available in soil solution while a large amount is adsorbed on soil particles. Therefore, these natural sources should be supplemented with fertilizers, manures or sludges by fertilization programmes.

20.2.2 Symptoms of Nutrient Deficiencies

If the concentration of an essential nutrient in plant tissue drops below the level in which this nutrient is needed for optimal growth (Epstein and Bloom 2005). Nutrient deficiencies, when severe, manifest in more or less distinct symptoms, which are visually recognized (Futch and Tucker 2000; Quiñones et al. 2010).

20.2.2.1 Nitrogen

Deficit of this element is expressed by general chlorosis, light green in leaves, with mild deficiency, to yellow foliage as conditions intensify. Moreover, the midribs and lateral veins turn yellow, progressively, while the rest of the leaf remains at normal green colour, and tree growth and fruit production are reduced. The mature fruits are usually smaller with very thin peel and tend to premature colour break. The oldest parts of the plant are the first ones to show symptoms since nitrogen is translocated from older to younger plant compartments.

20.2.2.2 Phosphorus

This deficiency is unlikely to occur in citrus areas that have optimal levels of phosphorous in soils. Although rarely seen, trees present poor flowering, smaller fruits and with thicker peel, lower juice content and less consistent, and leaves may also exhibit a bronze appearance. However, it is remarkable that new plantings on previously uncropped land usually require substantial initial phosphorus to prevent phosphorous deficit.

20.2.2.3 Potassium

As a consequence of high mobility of this element in plants, potassium deficit affects, mainly, old leaves that wrinkle and curl. Fruits are smaller with smoother and thinner peel and tend to premature colour break. Sometimes, a fruit may be subject to splitting and/or dropping.

20.2.2.4 Magnesium

The first symptom is a yellowish green blotch next to the base of the leaf between the midrib and the outer edge. The yellow area enlarges until the only green remaining is at the tip and base of the leaf as an inverted V-shaped area on the midrib. With acute deficiency, leaves may become entirely yellow bronze and eventually drop. Fruits are minor size with thinner peel and lower acidity and soluble solids content.

20.2.2.5 Calcium

This deficiency rarely takes place in the Mediterranean area conditions. In trees presenting deficiency, plant growth is reduced; drying tips and defoliations are also observed.

20.2.2.6 Sulphur

Symptoms of sulphur deficiency are similar to those mentioned above with N deficit. Foliage colour turns light green; a curvature of leaf tip can also occur.

20.2.2.7 Iron

Because of the lack of mobility of iron in leaves, symptoms firstly appear on new foliage. Leaf veins are slightly darker green than interveinal areas that become increasingly yellow in severe cases. Trees may become partially defoliated with smaller foliage, number of fruits and fruit size. Iron deficiency is usually an indication of calcareous soil conditions and is more likely to be expressed on highly pH-sensitive rootstocks.

20.2.2.8 Zinc

At early stages, zinc deficit appears as small yellowish spots between green veins on the leaf that may become increasingly yellow except for the green veinal areas and smaller young leaves. In addition, yield is reduced and fruit size decreases, and low juice and solid soluble content are also observed.

20.2.2.9 Manganese

Both manganese and zinc deficiencies may simultaneously occur on calcareous soils and may become more severe on trees grafted on highly pH-sensitive rootstocks. Deficiency emerges as a dark green strip along the midrib, and main veins are surrounded by light green interveinal areas that turn to yellow as severity increases.

20.2.2.10 Copper

Copper deficiency is rarely found in citrus areas as a consequence of fungicide treatments. It is usually associated with large, dark green leaves on long soft angular shoots. Young shoots may develop into branches, which appear curved or 'S-shaped'. Twigs can develop blister-like pockets of clear gum at nodes, and reddish brown eruptions may occur in the outer portion of the wood as twigs mature.

20.2.2.11 Boron

Most characteristic symptoms of boron deficiency include translucent spots in leaves, darkish-coloured blotches and rubber bags in the fruit albedo and sometimes in the central core. Fruit may be somewhat misshapen with a lumpy surface.

20.2.2.12 Molybdenum

Molybdenum is required in the smallest quantities of any nutrient by trees. The most characteristic field symptoms are interveinal yellow spots, similar than those observed in N deficiency.

20.2.2.13 Chlorine

Clearly essential for tree growth, however, no confirmed deficiencies in normal soils have ever been reported.

20.2.2.14 Nickel

Nickel deficiency symptoms are not well established, although Ni-deficient seeds exhibit poor germination and vigour. Low nickel in leaves may result in necrosis initiating from the tips of the leaf.

Table 20.4 Annual nutrient needs for citrus trees

		Nutrient consumption (g)					Nutrients covered by old leaves reserves (%)					Net annual requirements (g)				
Tree age (year)	Dry weight (kg)	N	P	K	Mg	Fe	N	P	K	Mg	Fe	N	P	K	Mg	Fe
Young tree (2)	1.2	6.8	0.8	3.6	1.4	0.04	25	12	22	24	–	5.1	0.7	2.8	1	0.04
In development (6)	32	210	18	121	46	1.1	32	16	28	30	–	142	15	87	32	1.1
Adult (>12)	102	667	53	347	135	3.4	32	17	29	30	–	453	44	246	95	3.4

Source: Legaz and Primo-Millo (1988) and Martínez-Alcántara et al. (2011)

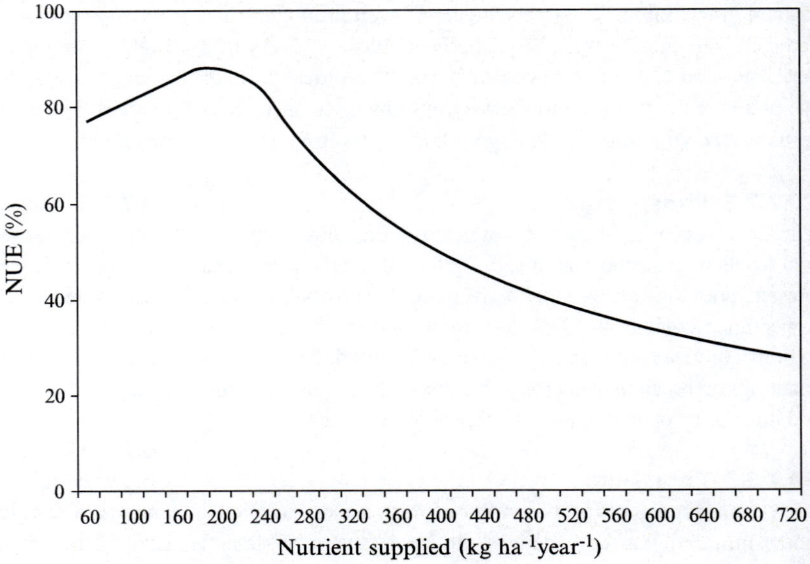

Fig. 20.1 Nutrient use efficiency curve

20.3 Citrus Nutrient Requirements

Annual nutrient needs of a crop are defined as the amount of nutrients consumed along the year by the plant, which are supposed to be enough for a correct plant development and fruit production. Its determination includes the needs of both new developing organs (reproductive and vegetative) and old permanent organs growth consumption. This demand does not include annual old leaves requirements because these leaves, at the beginning of a new fertilization programme, translocate mobile nutrients to different new organs, before its abscission.

In citrus, many years ago, quantitative determinations of nutrient consumptions were determined through chemical analysis of young or aerial plant tissues (Barnette et al. 1931; Cameron and Appelman 1933; Smith 1966). However, these data did not properly reflect the annual nutritional needs of the plant since neither elements accumulated in perennial tissues (roots, trunk and old branches) nor the nutrients supplied by the storage tissues (internal remobilization) can be determined without extracting plants from soil. Later on, Legaz and Primo-Millo (1988) and Martínez-Alcántara et al. (2011) determined the total amount taken up by a citrus tree along 1-year vegetative cycle by means of sequential destructive harvests of trees of different ages (2, 6 and 12 years old) along the cycle. In the case of N, these data were obtained by supplying nitrogen heavy isotope (^{15}N) in an inert soil-free medium (sand) or in soil. Annual nutrient requirement (Table 20.4) has shown that some nutrients are provided by the reserves of old leaves, except for Fe, which is scarcely mobile in the plant, and its translocation from old leaves to new developing organs can be considered negligible. The difference between new and old organs' nutrient demand and that covered by old leaves reserves represents net annual needs for citrus trees.

20.3.1 Nutrient Use Efficiency

To provide a rational nutrient dose, not only the amount of nutrients annually consumed by the trees but also nutrient use efficiency, the percentage of supplied nutrient that is actually removed by the crop, must be considered. This relationship is not linear and depends on the amount of nutrient applied. So, the efficiency increases with the nutrient dose supplied, up to a maximum, and decreases thereafter (Fig. 20.1) since plants down-regulate their transport mechanisms, absorbing only at rates sufficient to meet plant growth demand (Epstein and Bloom 2005).

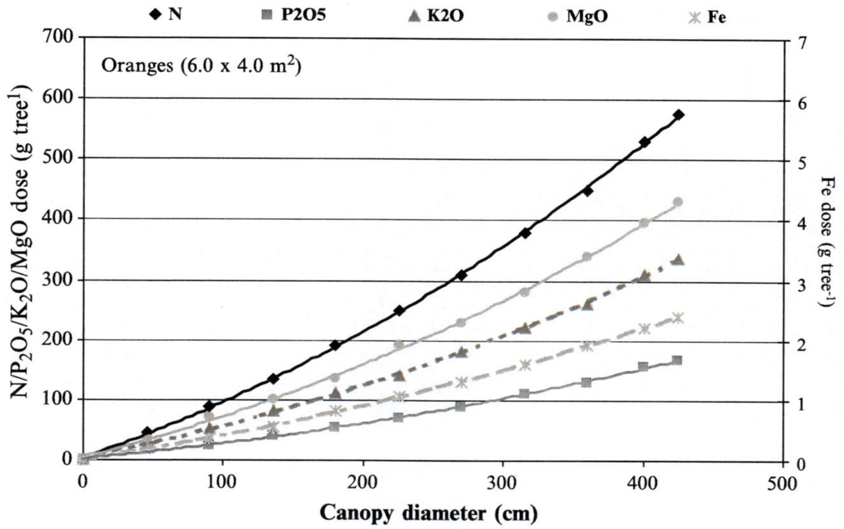

Fig. 20.2 Annual fertilizer dose for sweet oranges through fertigation

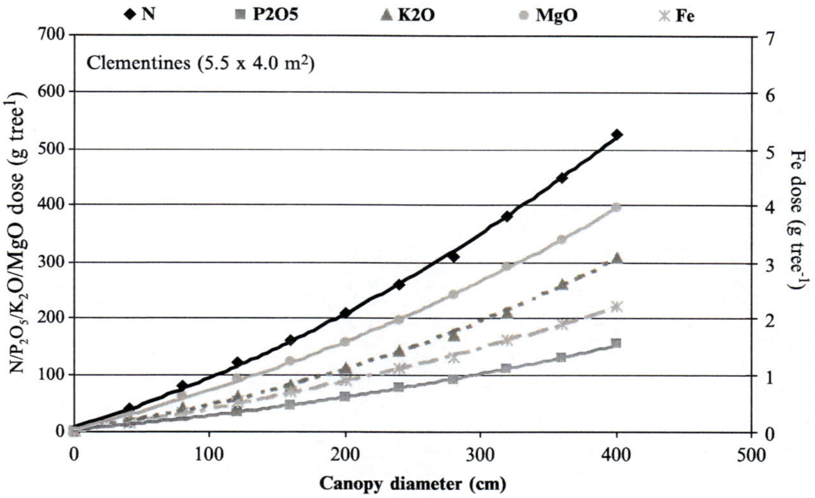

Fig. 20.3 Annual fertilizer dose for clementines through fertigation

20.3.2 Annual Fertilizer Dose

According to this concept, fertilizer recommendations should be obtained by multiplying annual net needs (Table 20.4) increased in nutrient use efficiency according to the formula (20.1):

$$\text{Annual fertilizer dose} = \text{Net annual needs} \times F_1 \times F_2 \quad (20.1)$$

where F_1 equals 100/percentage of fertilizer use efficiency in drip irrigation (from 40 to 60)F_2 is a conversion factor from elemental form (N, P, K, Mg) to fertilizer unit (N, P_2O_5, K_2O, MgO and Fe), N×1=N, P×2.3=P_2O_5, K×1.2=K_2O, Mg×1.7=MgO and Fe×1=Fe.

Fertilizer doses for typical plant densities for citrus varieties have been calculated (Figs. 20.2, 20.3, 20.4 and 20.5). These recommendations are mainly established according to canopy diameter since tree vigour is a result of rootstock-scion combination and agricultural practices rather than tree age (Quiñones et al. 2010). Nutrient dose increases with age until the maximum canopy diameter compatible with the planting frame; these are maximum nutrient doses (Table 20.5) to be applied regardless of tree age.

In addition, soil cation balance in irrigation bulb should be taken into account when setting fertilization criteria. In this sense, for MgO dose recommendations, the K/Mg (meq 100 g of soil^{-1}) ratio must be kept between 0.16 and 0.35 (Legaz et al. 1997). A value 0.35 in K/Mg ratio in fertilizer solution has been considered for data presented (Table 20.5).

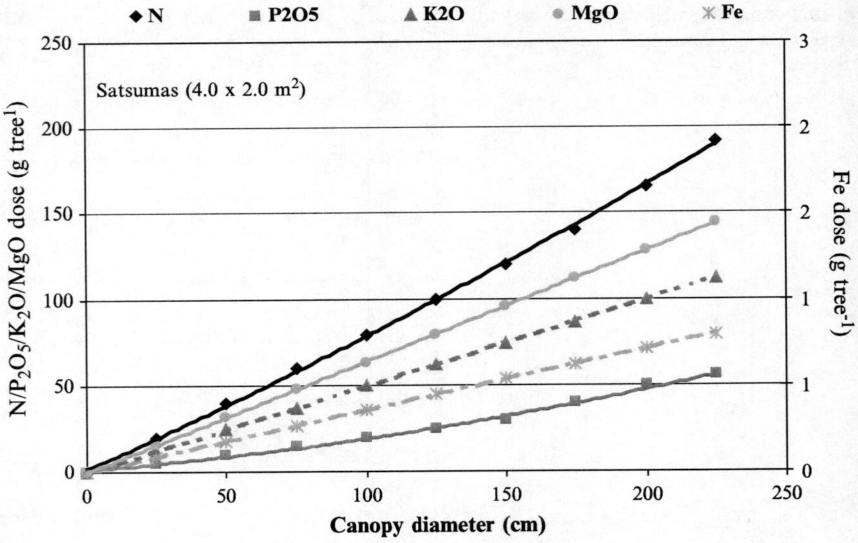

Fig. 20.4 Annual fertilizer dose for Satsumas through fertigation

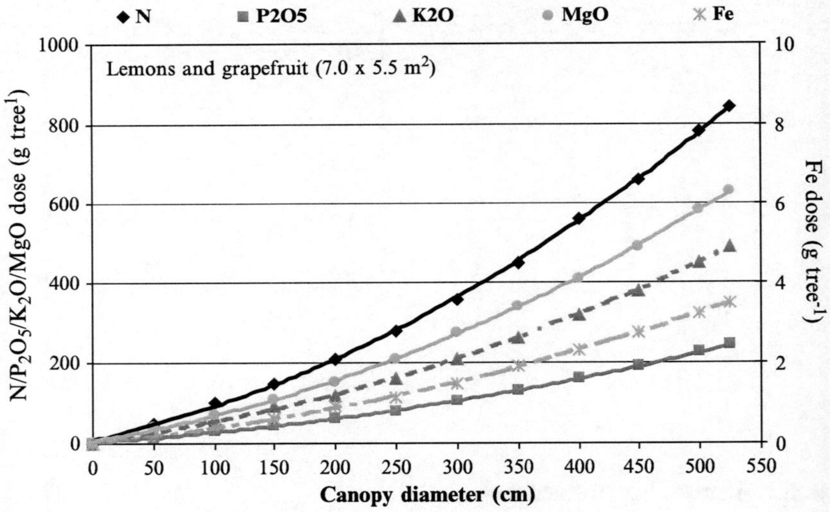

Fig. 20.5 Annual fertilizer dose for Lemons and Grapefruit through fertigation

Table 20.5 Maximum nutrient dose for citrus varieties according to most common plant densities

Citrus varieties	Oranges	Clementines	Satsumas	Lemon and grapefruit
Planting frame ($m \times m$)	6×4	5.5×4	4×2	7×5
Tree ha^{-1}	416	454	1,250	285
Nutrient dose	(g $tree^{-1}$)			
N	577	528	192	842
P_2O_5	168	154	56	245
K_2O	336	308	112	491
MgO	432	396	144	631
Fe	2.4	2.2	0.8	3.5

20.4 Optimization of Nutrient Dose

Annual fertilizer dose is, according to the criteria explained above, initially established for an optimum nutrient level, regardless of the elements that can be provided by soil or irrigation water. Therefore, soil and irrigation water analysis (once every 2 or 3 years) are crucial keys to adjust the recommended nutrient doses on the basis of local conditions. In addition, it is highly recommended to check tree nutritional status and hence tree response to fertilizer programme by means of foliar analysis (once a year) to prevent possible nutrient excess/deficiency that could result in irregular plant growth.

20.4.1 Nutrient Dose Correction According to Foliar Analysis

Foliar analysis is the most convenient method for evaluating the nutrient status of citrus trees and thus monitoring the availability of a nourishing element in soil solution. It provides accurate information on the amount of nutrients actually taken up and assimilated by the plant and reflects nutrient deficiency or excess states and an antagonism between elements.

20.4.1.1 Selection of Leaf Type and Definition of Appropriate Time for Leaf Sampling

Leaves are the most representative compartment of plant nutrient status (Embleton et al. 1973a) since it represents the seat of the active growth processes (Wallace 1943). Under natural conditions, citrus have three flushes of growth a year. In order to set which is the most suitable flush to sample, a huge number of analyses were carried out in commercial orchards located in Comunidad Valenciana, in the East of Spain, and comprising different combinations of cultural practices. Leaves on new flushes of growth (spring, summer and autumn flush) were sampled in adult trees from different cultivars grafted on several citrus rootstocks.

Spring flushes without terminal fruit were selected among all, as the most appropriate for foliar sampling because leaves present a higher reserve in mobile nutrients (due to greater biomass), compared to other flush tissues, to be remobilized during new organ development.

20.4.1.2 Establishing Optimal Leaf Sampling Time

The sampling time should coincide with the period when leaf nutrient levels are relatively stable (Bould et al. 1960; Cline and Mc Neill 1991), thus avoiding interpretation problems resulting from variations in nutrient and dry matter content throughout the vegetative cycle (Stephenson and Cull 1986). This is one of the most important factors when selecting the sampling time (Stephenson and Cull 1986); therefore, from the beginning of September to the end of November was considered the most suitable sampling period.

20.4.1.3 Criteria of the Standard Leaf Nutrient Thresholds

Accordingly with the results described above, in order to develop the standard foliar values, 7- to 9-month-old leaves from new spring flushes without terminal fruit were considered the most appropriate for sampling. Among these, the third leaf from the base of terminal shoots was collected to analyse plant nutrient status. In these leaves, foliar nutrient diagnostic criteria were established by Legaz and Primo-Millo (1988) and Legaz et al. (1995) in citrus of different ages. A great assay was carried out in almost 400 commercial citrus orchards located in the same area described in Sect. 20.4.1.1. During one growing season, at least 15 trees per plot were randomly selected and around 2 spring-flush leaves per tree were collected within the sampling period established. The nutritional status of the crop was determined by leaf tissue analyses and characterized by nutrient contents applying Kenworthy's balance index procedure. Kenworthy (1961) developed a set of standard values, calculated as mean of the concentrations of nutrient elements found in leaves from high-performing trees showing good growth, production and fruit quality. Kenworthy adjusted the mean values calculated to the balance index (*B*) by the following method:

If the concentration of the nutrient in the sample (*X*) was lower than the concentration in the concentration average (*S*):

$$B = [(X/S)\times 100] - [(100-(X/S))\times 100 \times (\mathrm{CV}/100)] \quad (20.2)$$

If the concentration of the nutrient in the sample (*X*) was greater than average (*S*):

$$B = [(X/S)\times 100] - [((X/S)\times 100) - (100\times(\mathrm{CV}/100))] \quad (20.3)$$

where CV is a coefficient of variation of the values averaged.

Values of the balance index (*B*) of each nutrient lower than 17 were defined as 'deficient', from 17 to 50 as 'low' and from 50 to 83 as 'normal'. Balance indexes from 83 to 117 were established as 'high', while values of 150–183 as 'excess'. With these values, standard foliar nutrient thresholds were established (Tables 20.6 and 20.7) and values from which yield and fruit quality may be compromised or cause toxicity were also established. In these situations, dose must be corrected.

Table 20.6 Standard leaf macronutrient concentrations on spring-flush leaves applying Kenworthy's balance index procedure[a]

Citrus varieties	Nutrient	Deficient	Low	Optimum	High	Excess
Oranges	N	<2.30	2.30–2.50	2.51–2.80	2.81–3.00	>3.00
	P	<0.10	0.10–0.12	0.13–0.16	0.17–0.20	>0.20
	K	<0.50	0.50–0.70	0.71–1.00	1.01–1.30	>1.30
Clementines	N	<2.20	2.20–2.40	2.41–2.70	2.71–2.90	>2.90
	P	<0.09	0.09–0.11	0.12–0.15	0.16–0.19	>0.19
	K	<0.50	0.50–0.70	0.71–1.00	1.01–1.30	>1.30
Satsumas	N	<2.40	2.40–2.60	2.61–2.90	2.91–3.10	>3.10
	P	<0.10	0.10–0.12	0.13–0.16	0.17–0.20	>0.20
	K	<0.10	0.10–0.60	0.61–0.90	0.91–1.15	>1.15
Oranges	Mg	<0.15	0.15–0.24	0.25–0.45	0.46–0.90	>0.90
Clementines	Ca	<1.60	1.60–2.90	2.91–5.00	5.01–6.50	>6.50
Satsumas	S	<0.14	0.14–0.19	0.20–0.30	0.31–0.50	>0.50

Source: Kenworthy (1961)

Table 20.7 Standard leaf micronutrient concentrations on spring-flush leaves applying Kenworthy's balance index procedure[a]

Citrus varieties	Nutrient	Deficient	Low	Optimum	High	Excess
Oranges	Fe	<35	35–60	61–100	101–200	>200
	Zn	<14	14–25	26–70	71–300	>300
Clementines	Mn	<12	12–25	26–60	61–250	>250
	B	<21	21–30	31–100	101–260	>260
Satsumas	Cu	<3	3–5	6–14	15–25	>25
	Mo	<0.06	0.06–0.09	0.10–3.00	3.01–100	>1,000

Source: Kenworthy (1961)

Table 20.8 Correction factors according to nutritional status for citrus in fertigation

Nutritional status	Deficient	Low	Optimum	High	Excess
N factor	1.5	1.4–1.1	1.0–0.9	0.8–0.6	0.5
P_2O_5 factor	2	1.9–1.1	1.0–0.6	0.5–0.0	0.0
K_2O factor	2	1.9–1.1	1.0–0.7	0.6–0.0	0.0
MgO factor	2	1.9–0.6	0.5–0.0	0.0–0.0	0.0
Fe factor	2	1.9–1.1	1.0–0.0	0.0–0.0	0.0

Each value corresponds to the extreme values for foliar concentration. Proportional coefficients must be used for intermediate values

20.4.1.4 Correction Factors

According to the tree nutritional status, annual fertilizer dose must be corrected by means of different factors (Table 20.8). In the case of iron or magnesium foliar concentrations above the optimum, these elements should be removed from nutrient solution.

20.4.2 Nutrient Dose Correction According to Soil Analysis

Soil analysis provides essential information about nutrients availability in addition to physical and chemical properties that affect soil's suitability for growing plants. Standard soil tests mainly comprise soil texture; electrical conductivity (an indicator of soil salinity); pH; available N, P, K, Ca, Mg and Na; cation exchange capacity and organic matter content (Walworth 2006). All these information must be taken into account to optimize the fertigation programme.

Soil texture reflects the proportion in which mineral particles of different sizes (sand, silt, clay) are found in a soil. Sandier soils allow water to move quickly through the soil but hold little water and may require frequent irrigation and fertilizer application, while clayey soils retain more nutrients and may absorb water very slowly.

Soil pH is a measure of the acidity or alkalinity of a soil. Citrus is mainly grown on alkaline soils. Fertigation programmes must be adjusted to this condition to prevent the non-availability of some nutrients.

Electrical conductivity of a soil extract is used to estimate the level of soluble salts, and this should be a routine test in arid zones and in citrus trees irrigated with desalinated or reclaimed water.

Nitrate analyses provide information on soil N availability since nitrate is the main form absorbed by plants. However, these analyses provide timely information since nitrate is easily washed out of the root zone, beyond the reach of plants.

Most soil P is strongly retained to soil particles. The amount of P removed during soil extraction is essentially dependent on the specific extractant used. The Olsen or bicarbonate extractant is appropriate for the most part of the citrus soils.

The four most abundant exchangeable cations in soil are K, Ca, Mg and Na. All these cations except Na are essential plant nutrients; however, Na is included because it plays an important role in soil physical properties and can limit water and nutrient uptake. Sodium levels can be expressed as sodium adsorption ratio (SAR), which represents the amount of Na relative to Ca and Mg in the water extracted from a saturated soil paste. On the other hand, cation exchange capacity (CEC) is an indicator of the ability of a soil to retain cations, some of which are plant nutrients. Soils with low CEC retain fewer cations and may require more frequent fertilizer applications than that showing high CEC values.

20.4.3 Correction According to Water Irrigation Analysis

Many studies have shown direct relationships between the addition of N in areas of intensive agriculture and the increase of nitrate (NO_3^-) concentration in groundwater (Babiker et al. 2004; de Paz and Ramos 2004). Drinking water and superficial groundwater monitoring also revealed indications of increasing trends of excess nitrate N in areas of the citrus-producing region in central Florida (Alva et al. 1998; Lamb et al. 1999). On the Mediterranean coast of Spain, where the cultivation of citrus fruits predominates, a severe increase in contamination by lixiviation of the nitrate ion has been observed in subterranean waters, with values clearly above 50 mg L^{-1} as nitrate (WQIGME 2012). Moreover, Mg is also usually found in measurable amounts. In these conditions, monoammonium phosphate and calcium and magnesium carbonates may cause precipitates (Boman et al. 2001), which can plug micro-irrigation system emitters.

Irrigation water thus provides an additional source of N and Mg for trees that should be assessed, in order to proportionally reduce supplied doses. The amount of N and Mg supplied by irrigation water is calculated as follows:

$$\text{Kg N ha}^{-1} = \frac{NO_3 \times V_r \times 22.6}{10^5} \times F \qquad (20.4)$$

where NO_3^- = nitrate concentration in irrigation water (ppm = mg L^{-1}), V_r = total irrigation volume (m^3 ha^{-1}), 22.6 = proportion of N in the nitrate ion, F = irrigation efficiency factor depending on run-off and percolation water and ranges from 0.6 to 0.9

$$\text{KgN ha}^{-1} = \frac{Mg^{++} \times V_r \times 1.66}{10^3} \times F_1 \times F_2 \qquad (20.5)$$

where Mg^{++} = magnesium concentration in irrigation water (ppm = mg L^{-1}), V_r = total irrigation volume (m^3 ha^{-1}), 1.66 = conversion factor from Mg to MgO, F_1 = irrigation efficiency factor that depends on run-off and percolation water and ranges from 0.6 to 0.9, F_2 = magnesium insolubility factor, which depends on intrinsic soil characteristics and ranges between 0.4 and 0.6

20.5 Research on the Improvement of Nutrient Use Efficiency in Fertigation

The purpose of fertilization is to increase the natural fertility of the soil in order to improve the nutritional status of crop plants. Citrus trees demand high amounts of fertilizers; unfortunately, farmers have applied excessive dosages of nutrients because of poor fertilizing criteria and slight enhances found in fruit yield when increasing the dosages. This has resulted in a deterioration in the commercial quality of the fruit (Chapman 1968), a reduction in the profitability of the citrus crops (Wild 1992) and a NO_3^- displacement, mainly, to deeper soil layers. In this case, many studies have shown direct relationships between this addition of N in areas of intensive agriculture and the alarming increase of NO_3^- concentration in groundwater (Singh and Kanehiro 1969; Bingham et al. 1971; Burkart and Stoner 2002; Babiker et al. 2004; de Paz and Ramos 2004). Monitoring of drinking water and superficial groundwater revealed increases in nitrate concentrations above the drinking water quality standards of 10 mg $NO_{3-}N$ L^{-1} (US Environmental Protection Agency 1994) in citrus-producing regions in central Florida (Alva et al. 1998; Lamb et al. 1999). Along the Mediterranean area, where the cultivation of citrus fruits predominates, a severe increase in contamination by leaching of the nitrate ion has also been observed in subterranean waters, above the limit of the World Health Organization (WHO) guideline (WHO 2004), of 50 mg L^{-1} as nitrate (Sanchís 1991).

Nowadays, efforts are being directed to understand the large number of processes in which nutrients are involved in the plant-soil system, like irrigation management, application frequency, timing of application as well as soil processes, in order to reduce rates and thus losses, which may result in surface and ground water pollution, maintaining crop productivity. This section compiles the results of several studies carried out by different authors with the aim of reevaluating current fertilization programmes. This information is necessary to deeply understand nutrient use efficiency and thus advance towards best management practices for citrus crops.

20.5.1 Nitrogen in Fertigation

Citrus trees demand high amounts of nitrogenous compounds as nitrogen (N) has a greater influence on growth and production than other nutrients (Smith 1966). So several research have been carried out about the amount of N uptake by citrus trees with different management practices.

The studies of N dynamics in laboratory soil columns (Vanden Heuvel et al. 1991; Esala and Leppänen 1998) and assays under greenhouse or field conditions (Westerman and Tucker 1979; Mansell et al. 1986; Recous et al. 1988; Bengtsson and Bergwall 2000) show important discrepancies which can be explained by the difference in experimental practicalities. However, the use of ^{15}N tracer opens the possibility to follow and accurately quantify this plant nutrient in different compartments of the system under study. Some authors have used this technique to determine N requirements of citrus trees and hence develop N fertilizer recommendations (Legaz et al. 1982; Kato et al. 1981; Mooney and Richardson 1994).

In citrus orchards, irrigation systems directly affect N absorbed from fertilizer (Naff) by the entire tree and the amount retained in soil or leached in drainage. Quiñones et al. (2005) obtained higher N recovery percentages in Navelina using drip irrigation (73%) than under flood irrigation (63%). These data are similar to those of Syvertsen and Smith (1996) who found a nitrogen use efficiency (NUE) value for lysimeter-grown citrus trees on the order of 61–68%. Further improvement of NUE by citrus with fertigation compared with dry granular fertilizer was reported by Dasberg et al. (1988), Alva and Paramasivam (1998) and Alva et al. (1998, 2003). Li et al. (2004) studied the influence of fertigation strategies on N distribution in soil profile with drip irrigation. For a given volume of water applied, increasing the application rate allows more water to distribute in the horizontal direction, as in drip irrigation, while decreasing the rate leads to more water in vertical direction and, therefore, nitrate leaching could be higher. In this line, Quiñones et al. (2007a) showed that the percentages retained in soil profile as NO_3^--N were significantly higher for the flood-irrigated (around 38% of the N retain) than for the drip-irrigated trees (8%). Nevertheless, no significant differences appeared in the amount of organic ^{15}N for both irrigation systems.

Frequency of N application also affects N distribution in plant-soil-leaching system. More frequent applications of dilute N solutions double NUE compared with less frequent application of more concentrated N solutions (Scholberg et al. 2002; Quiñones et al. 2005). In another study, Alva et al. (2006) demonstrated a slight increase in N uptake efficiency as a result of better management practices associated with N placement, timing of application and optimal irrigation scheduling when comparing fertigation (FRT-15N applications) versus water-soluble granular fertilizers (WSG-4N applications). Also increases in NUE were obtained by other authors expressed as increment in fruit yield. Boman (1996) reported a greater NUE (9% greater fruit yield) in grapefruit trees receiving a combination of one dry granular broadcast application (33% of the annual rate) and 18 fertigations at 2-week intervals compared to trees that received three applications of dry fertilizer. Alva et al. (2003) evaluated different combinations of irrigation and nitrogen management. Fruit yield of 36-year-old Valencia orange trees was greater with the application of N as fertigation compared to that of the trees which received a similar rate of N as three broadcast applications of granular product. In young trees, Morgan et al. (2009) found higher yields when compared controlled-released fertilizer and fertigation applied 30 times annually with dry granular fertilizer and fertigation applied four times.

Soil texture also constitutes a major factor determining N uptake efficiency and leaching losses (Alva et al. 2003; Gehl et al. 2005; Quiñones et al. 2007a). Quiñones et al. (2012) obtained significantly higher N uptake in trees grown in a sandy soil than in loamy soil. Concerning soil, nitrate concentration was used to evaluate the potential leaching of NO_3-N below the root zone of the trees. The citrus tree root system is comprised of a relative shallow, well-branched framework of woody laterals and fine fibrous roots (Castle 1980a). The fibrous roots are usually most densely concentrated near the soil surface (0–45 cm), while few roots are found below 90 cm (Castle 1980b; Zhang et al. 1996; Mattos et al. 2003). Nitrate concentration in soil in the upper 45 cm of the soil profile represents most of the N available for root uptake, while NO_3-N concentration detected below 90 cm depth could be an indication of potential NO_3^- leaching into groundwater. In this sense, the greater root development in sandy soil with respect to loamy soil (Martínez et al. 2002) accounts for higher rates of N uptake in the former since N uptake is proportional to root density (Scholberg et al. 2002).

With regard to N form, most plants can use either ammonium or nitrate (NO_3^- and NH_4^+) as nitrogen source (Hageman et al. 1992). In citrus, both could be absorbed; however, their tree uptake efficiency depends on numerous factors such as the ionic composition of the medium, pH, temperature, light and availability of carbohydrates (Kato 1986). Wallace (1953) studied the absorption of ^{15}N-labelled ammonium and nitrate ions in Valencia orange and Eureka lemon trees. In hydroponic medium, ammonium uptake was slightly higher than nitrate in an experimental period of 48 h. However, when plants were grown in soil, nitrate uptake ranged from 2 to 5 times higher than with ammonium fertilizer. In hydroponic culture, it is clear that citrus preferentially absorb ammonium fertilizers; however, the relative proportion absorbed in both ions is determined by several factors. First, the concentration ratio between two forms determines the pattern of uptake. Serna et al. (1992) studied the influence of different combinations of NO_3^- and NH_4^+ ions (100 /0, 75/25, 50/50, 25/75, 0/100) in nutrient solutions on the N uptake in citrus grown in hydroponic medium. These authors observed that nitrate uptake was saturated at 120 ppm NO_3^-, but the ammonium was not until 240 ppm NH_4^+. The form of N uptake is also related to pH of the solution. According to Bowling (1976), the anions are absorbed more rapidly than the cations from acid solutions and the opposite occurs when the pH increases. However, Wallace and Mueller (1957) found that the differential effect of pH on the relative uptake of NO_3^- and NH_4^+ ions depends on the N concentration in the nutrient solution, so with high concentrations of N in the medium (>112 ppm), there is a discriminatory effect of pH on the N uptake; this effect is negligible when these ions are at concentrations below 70 ppm. Temperature of the solution is another factor that seems to be involved in the relative uptake of these anions in citrus grown without soil. Sala et al. (1992, 1982) applied N solutions with different ratios of NO_3^-/NH_4^+ (11:0, 9:2 and 6:5 meq L^{-1}) to Washington Navel orange trees grown in sand. At temperatures below 15°C, N uptake was higher in trees irrigated with 6:5 solutions, while at higher temperatures, the greater N uptake corresponded to trees grown in 9:2 solution. In field conditions, preferential uptake of ammonium is not so evident and depends on type of soil, organic matter content and clay composition (Kato 1986). In well-aerated soils with a neutral pH, nitrification rapidly occurs in spring and summer, coinciding with greatest uptake period, and NO_3^- is thus the main uptake form in citrus. This is due to the fact that nitrate form is soluble in the soil solution and can be transported to the root zone, so it can be rapidly absorbed (Embleton et al. 1973b). The ammonium form can be adsorbed to colloids or fixed in clay soil, so it would be less available for plant uptake (Feigenbaum et al. 1994; Longeri et al. 2001). Quiñones et al. (2012) also found greater N recovery by whole tree in trees fertilized with potassium nitrate (40.1% and 37.0% in sand and loam soil, respectively) than those under ammonium fertilization (37.9% and 33.9% in sandy and loamy soil, respectively). Use of nitrification inhibitors (NI) could also affect NUE. Nitrate N fertilizers are absorbed more efficiently than ammonium N by citrus plants; however, ammonium fertilizers are recommended during the rainfall period. The addition of Ni to ammonium N fertilizers increases NUE (16%), resulting in lower N-NO_3^- content in the soil (10%) and in water drainage (36%).

20.5.2 Phosphorus in Fertigation

Unlike nitrogen, phosphorus has been less studied because, in general, the soils have enough phosphorus. In practice, the main important question the citrus growers could ask is whether there is enough available P in the soil solution to ensure a proper plant development (Kafkafi and Tarchitzky 2011).

In non-irrigated conditions, phosphorus shows very low mobility into the soil profile (Malavolta and Violante Netto 1989), and therefore, losses by leaching of this element are negligible (Coelho 1973). High fertigation frequency ameliorates this situation since there is a continuous forced mass flow, which goes from the surface into the soil. Increased saturation of P fixation sites in the soil due to high frequency and application rate results in higher amounts of P released to solution, which combined with the forced flow of water into the soil, facilitates the distribution and the consequent increased levels of P (Duenhas et al. 2002). In this sense, P application through drip irrigation can increase the movement of this nutrient in the soil profile, compared to the conventional application; moreover, the use of phosphoric acid provides increased mobility of soil P when compared to superphosphate (Vivancos 1996; Zanini et al. 2002). Anyway, phosphate rapidly reacts with Ca in basic soils and with Fe and Al in acid soils, being the distance travelled by applied P quite limited, even in sandy soils, as compared with the water (Ben-Gal and Dudley 2003). The low availability of P in the bulk soil limits hence plant uptake. In this sense, the efficiency of absorption of P can vary the order of 10% for furrow irrigation system, and up to 35% for irrigation (Papadopoulos 2001), because about 80% of the P becomes immobile and unavailable for plant uptake due to adsorption, precipitation or conversion to the organic form (Holford 1997).

In Florida, citrus orchards traditionally receive about 40 kg phosphorus per ha at planting, followed by applications of up to 100 kg ha^{-1} $year^{-1}$ until they enter the fruit-bearing years after the age of 4. From then onwards, citrus receive 20–50 kg ha^{-1} $year^{-1}$ (Tucker et al. 1990). However, according to Obreza (1990), there is a lack of fertilizer response in citrus trees newly planted in sandy soils. Similarly, adult citrus trees rarely respond to P fertilizer (Smith 1966), except when planted on soils with extreme P fixation capacity. In this sense, Cantarella et al. (1992) and Quaggio et al. (1998) observed positive yield responses of Valencia oranges and lemons to annual P fertilizer rates up to 62 kg ha^{-1} on a highly P-fixing Brazilian soil. On the contrary, Alva et al. (2003) found negligible effects of fertilization source (granular, controlled release formulation or liquid) and rates on citrus trees P content grown in a sandy soil.

20.5.3 Potassium in Fertigation

Citrus trees remove large amounts of potassium (K) compared with other nutrients; moreover, K enhances fruit set and thus yields, as well as affects, fresh fruit qualities. Potassium deficiency reduces fruit number and size; increases fruit creasing, plugging and drop; and decreases juice-soluble solids, acid and vitamin C content.

Potassium is present as component of rocks and soil (fixed position) or an exchangeable cation on all clay particles. Since the rate of K release from fixed position is slower than the rate of K demand by plants, additions of K in fertilizers are needed to normal plant development. This is especially important when drip irrigation is used since the volume of soil occupied by the active root is small and not all the soil volume contributes K to the growing plant (Kafkafi and Tarchitzky 2011). In soils containing appreciable amounts of organic matter or clay, mobility of K can be limited because the positive charge of K ion enables it to be held by the soils' negatively charged cation exchange complex. However, in sandy soils, with very low concentrations of clay or organic matter, the ability to hold K against leaching can be almost non-existent (Obreza 2003). According to this situation, and considering that citrus trees use large quantities of K, in a typical citrus fertilization programme, K is applied at relatively high rates. Potassium is applied at a K_2O rate equal to the N rate; however, this rate is increased by 25% when leaf K is consistently below optimum and especially in calcareous soils (Obreza 2003; Obreza and Morgan 2011). The efficiency of absorption of K can vary the order of 60% for furrow irrigation system and up to 90% for fertigation (Papadopoulos 2001).

20.6 Fertigation Scheduling Strategies

The aim of the fertigation programme is to cover the difference between crop demand and nutrient availability (dose) when this is really required (timing). As indicated above, by means of fertigation, fertilizers can be supplied to the crops in amount, form and at time when they are mostly needed. So scheduling of fertilizer application should match crop demand along different phenological stages. However, it is important to emphasize the difficulty of accurately quantifying the seasonal demand of citrus and especially that of adult trees. This is due to the fact that in order to quantify the nutrients in different plant compartments, destructive harvests of whole trees must be carried out in different phenological stages, which is a very heavy task. Furthermore, the evaluation of tree response to changes in nutrient management involves long-term field researches since this response could be also affected by fruit load, leaf nutritional status, varieties, growth parameters, frequency of application, etc.

Legaz and Primo-Millo (1992) and Martínez-Alcántara et al. (2011) by means of ^{15}N-isotope dilution technique suggested the seasonal distribution of nutrient in citrus in different growing conditions (Figs. 20.6, 20.7, 20.8, 20.9 and 20.10).

Martínez-Alcántara et al. (2011) studied the relative contribution of stored N from the previous year in supporting new vegetative growth and early fruiting of young citrus trees under different nitrogen supply conditions. These authors affirmed that low N application rates in early stages (flowering and fruit set) lead to higher translocation of N stored during the previous cycle to developing new organs.

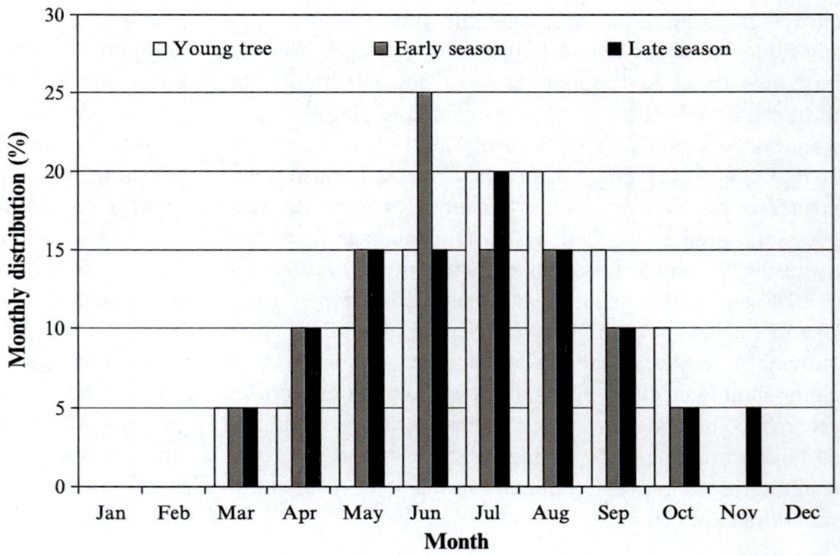

Fig. 20.6 Monthly distribution of nitrogen fertilizers in trees of different ages

Fig. 20.7 Monthly distribution of phosphorous fertilizers in trees of different ages

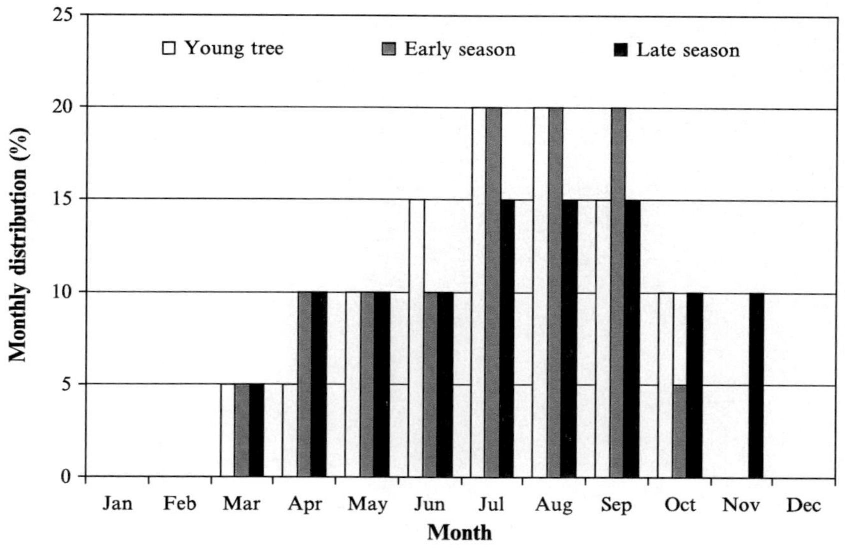

Fig. 20.8 Monthly distribution of potassium fertilizers in trees of different ages

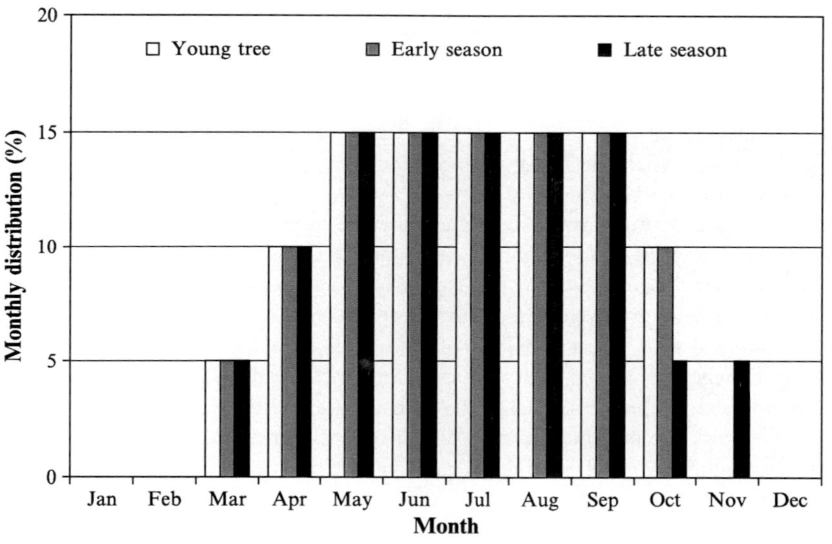

Fig. 20.9 Monthly distribution of magnesium fertilizers in trees of different ages

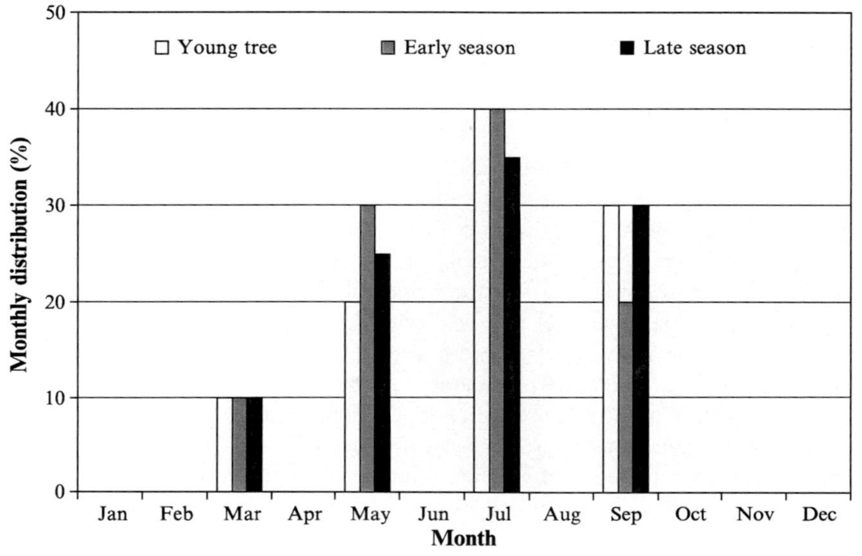

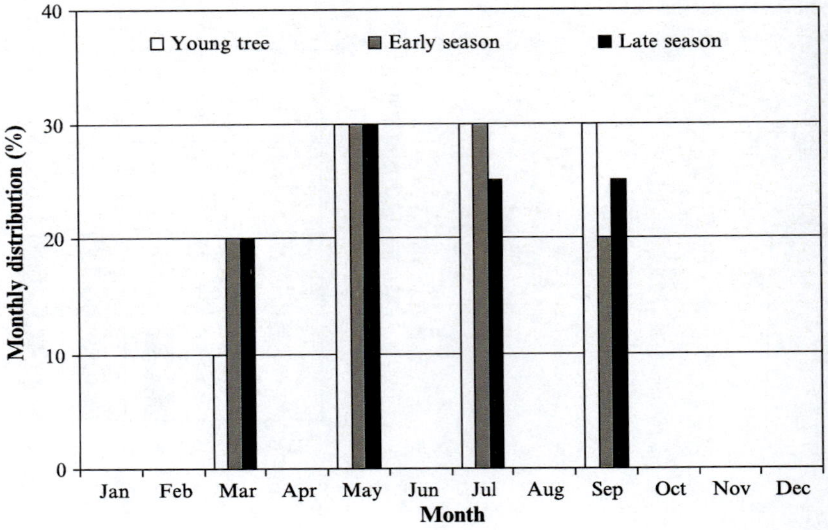

Fig. 20.10 Monthly distribution of iron fertilizers in trees of different ages

Hence, additional N dose in early spring supply is strongly recommended when nutritional status of young citrus trees is below the optimal thresholds.

Alva (2008) summarized results of recent evaluation of fertigation in citrus. On young citrus trees, no significant differences were found between fertigation and dry fertilizer broadcast treatments (Willis and Davies 1990). On the contrary, in adult trees, greater yields from fertigated trees were obtained when compared to dry granular fertilizer broadcast management (Schumann et al. 2003). In a parallel study, these authors compared different sprinkler coverage areas in citrus production. At the higher N rate, yield expressed in soluble solids and juice increased with increasing area of sprinkler.

Quiñones et al. (2007b) summarized the results of several studies based on ^{15}N techniques, in order to reevaluate current fertilization programmes organized according to different management practices. According to irrigation system, the dose of N and water commonly applied to commercial citrus orchards (up to 240 kgN ha^{-1} $year^{-1}$ and 5,000 m^{-3} ha^{-1} $year^{-1}$) can be markedly reduced (roughly 15%) with the use of drip-irrigated systems. In relation to time of N application, the N recovered by plants is 20% higher for summer N applications than for spring N when applied in flood irrigation trees. In drip irrigation, the highest NUE is obtained when N rate is applied following a monthly distribution in accordance with a seasonal absorption curve of N in which the maximum rates are supplied during summer. Concerning to split N application, several split N applications result in greater fertilizer use efficiency and smaller accumulations of residual nitrates in the soil. Lastly, N uptake efficiency is slightly lower in loamy than in sandy soils when N is applied both as nitrate and ammonium form; on the contrary, N retained in the organic and mineral fractions is higher in loamy soils that could be used in the next growing cycle.

20.7 Fertigation Fertilizer Source

In the process of fertigation, fertilizers are applied with the irrigation water; in this sense, there are some factors that have to be attended related to fertilizer characteristics:

- Solid fertilizers must show high water solubility and completely dissolve. Undissolved fertilizer precipitates in the tank and may plug irrigation tubing and emitters, reducing the available fertilizer for plants.
- Fertilizer salt index (i.e., a measure of the salt concentration that fertilizer induces in the soil solution) should remain low, to prevent crop injury on a particular soil. This index does not predict the exact amount of a nutrient, but it does permit comparisons of fluid formulations concerning their potential effects.
- Low content of insolubles, less than 0.02%.
- Compatibility with other fertilizers.
- Low corrosive effect to prevent damages in the irrigation system components.

Different show fertilizers commonly used in fertigation and the compatibility between them are further shown (Tables 20.9, 20.10 and 20.11). In the case of micronutrients, several metal micronutrient forms are relatively insoluble and, therefore, they are applied in sulphate form. These metal sulphates are water soluble; however, the metal ion has a strong electrical charge and becomes attracted to the cation

Table 20.9 Solid fertilizer commonly used in fertigation

Fertilizer	Main nutrients	Other nutrients	Solubility (g L^{-1})	CE (0.5 g L^{-1}) (dS m^{-1})
Ammonium nitrate	34.5% N		2,190	850
Calcium nitrate	15N	27% CaO	1,220	605
Urea	46% N		1,100	Irrigation water
Ammonium sulphate	21%N	28% SO_3	724	1,033
Magnesium nitrate	11% N	15% MgO	500	448
Monoammonium phosphate	12% N, 60% P_2O_5		400	455
Magnesium sulphate	16% Mg O		380	410
Monoammonium phosphate	12% N, 61% P_2O_5		365	880
Potassium nitrate	13% N, 46% K_2O		335	693
Potassium chloride	60% K_2O		340	948
Potassium sulphate	50% N	47% SO_3	110	
NPK compounds	High concentration	Sometimes	150/250	According formulae

Table 20.10 Liquid fertilizer commonly used in fertigation

Fertilizer	Main nutrients	Other nutrients	Density (g L^{-1})	pH
N solution 32	32% N		1,325	6/7
N solution 20	20% N		1,260	6/7
Acid nitric	13% N		1,360	Acid
Calcium nitrate solution	8% N	16% CaO	1,400	<4
Magnesium nitrate solution	7% N	9.5% MgO	1,300	<4
Phosphoric acid	54% P_2O_5		1,600	Acid
Potassium solution	10% K_2O		1,150	5

exchange sites of organic matter or clay particles, where it tends to sit near the soil surface (Boman and Obreza 2002). Therefore, they are normally provided by special fertilizers like quelate forms. An extensive work was carried out by Kafkafi and Tarchitzky (2011) about different fertilizer forms used in fertigation.

20.8 Conclusion

The agricultural practice called fertigation appears in order to achieve a rational fertilizer application that ensures a competitive production with minor impact. Since irrigation is also an important factor in the management of citrus trees, the application of nutrients in the irrigation water is shown as a sustainable method of supplying nutrients to crops. Fertigation should be scheduled not only to meet plant nutrient requirements but also to deliver them according to seasonal plant demand.

For this purpose, this chapter described these two pillars (dosage and timing) of an optimal fertilization programme towards a sustainable water and nutrient use. Then it should be taken account by citrus growers in order to improve nutrient use efficiency, reduce nutrients losses and protect water quality.

20.9 Future Trends in Fertigation

Nowadays, techniques and managements of agricultural production are directed towards the need to conserve resources, energy and a commitment to the environment. In this sense, fertigation has been raised as a valuable tool in recent years and has spread around the world in all agricultural areas, field and horticultural crops. This has led to an increase in fertilizer and water use efficiency. In the future, fertigation should continue to replace traditional flood irrigation.

Citrus are mainly grown in arid and semiarid region; in these areas, like in many regions of the world, the lack of water or lack of good water is a growing concern for the development of relevant agriculture since water is the most limiting factor for crop production. Furthermore, climatic conditions are characterized by low rainfall (400–600 mm $year^{-1}$) and irregular spatial and temporal distribution. On the other hand, the world's population has undergone an exponential growth, which has led to soaring food demand and, therefore, high natural-resource exploitation. Therefore, future trends in fertigation should be addressed to use another source of irrigation water like recycled sewage and/or desalination water.

In this context, improved water use efficiency (WUE), using different strategies, is also a key concept to solve this

Table 20.11 Fertigation fertilizer compatibility

	Ammonium nitrate	Calcium nitrate	Potassium nitrate	Magnesium nitrate	Monoammonium phosphate	Phosphoric acid	Acid nitric	Potassium sulphate	Magnesium sulphate	Micro sulphates	Micro-quelates	Urea	Ammonium sulphate	Potassium chloride
Ammonium nitrate	–	√	√	√	*	X	X	√	√	√	√	√	√	√
Calcium nitrate	√	–	X	√	X	X	X	*	X	X	X	√	X	√
Potassium nitrate	√	X	–	√	√	X	X	√	√	√	√	√	√	√
Magnesium nitrate	√	X	√	–	√	X	X	X	√	√	√	√	√	√
Monoammonium phosphate	X	X		√	–	X	X	R	√	X	X	√	√	√
Phosphoric acid	X	X	X	X	X	–	X	X	X	X	X	X	X	X
Acid nitric	X	X	X	X	X	X	–	X	X	X	X	X	X	X
Potassium sulphate	√	X	√	√	√	X	X	–	√	X	X	√	*	X
Magnesium sulphate	√	X	√	√	√	X	X	*	–	X	√	√	√	√
Micro-sulphates	√	*	√	√	X	X	X	R	√	–	*	√	√	√
Micro-quelates	√	X	√	√	X	X	X	X	X	X	–	√	√	√
Urea	√	√	√	√	√	X	X	√	√	√	√	–	√	√
Ammonium sulphate	√	X	√	√	√	X	X	*	√	√	√	√	–	√
Potassium chloride	√	√	√	√	√	X	X	X	√	√	√	√	√	–

√ compatible, *X* incompatible, *R* reduced compatibility, * at the moment water scarcity. So nowadays, efforts are being focussed on developing not only alternative irrigation methods but also new water management methods in order to reduce water dosages while maintaining maximum tree growth, without significantly affecting yield. In drip irrigation systems, subsurface drip irrigation (SDrI), where it is applied below the soil surface, using buried drip tapes, is being part of modern agriculture. Current commercial and grower interest levels indicate that future use of SDrI systems will continue to increase.

Improvement of WUE can be also achieved by means of DI. In this sense, it is possible to increase efficiency under different irrigation management methods based on deficit-irrigation (DI) programmes. These DI strategies are defined as a practise where the total water provided for the plant (irrigation plus effective rainfall) is below the crop's water needs in order to reduce gricultural water use, while simultaneously minimizing or eliminating negative impacts of stress on fruit yield or quality.

Lastly, nutrient use efficiency can be meliorated by using nitrification inhibitors or plant growth–promoting bio-effectors. Nitrification inhibitors restrict the microbial conversion of ammonium to nitrate that is mobile in soils and therefore leached. Thus, nitrification inhibitors have the potential to reduce nitrate leaching. Bio-effectors or biostimulants is a term that is used to describe microorganisms and active natural compounds involved in plant growth which are not a plant nutrient or pesticide but in some manner have a positive impact on plant health. The bio-stimulant may increase chlorophyll efficiency and production, enhance metabolism, increase antioxidants, enhance nutrient availability and increase the water holding capacity of the soil.

References

Alva AK (2008) Role of fertigation in horticultural crops: citrus. In: International fertigation symposium, Beijing, 20–24 Sept 2005

Alva AK, Paramasivam S (1998) Nitrogen management for high yield and quality of citrus in sandy soils. Soil Sci Soc Am J 62(5):1335–1342

Alva AK, Paramasivam S, Graham WD (1998) Impact of nitrogen management practices on nutritional status and yield of Valencia orange trees and groundwater nitrate. J Environ Qual 27:904–910

Alva AK, Paramasivam S, Graham WD (2003) Best nitrogen and irrigation management practices for citrus production in sandy soils. Water Air Soil Pollut 143:139–154

Alva AK, Paramasivam S, Obreza TA (2006) Nitrogen best management practice for citrus trees I. Fruit yield, quality, and leaf nutritional status. Sci Hort 107:233–244

Babiker IS, Mohamed AA, Terao H (2004) Assessment of groundwater contamination by nitrate leaching from intensive vegetable cultivation using geographical information system. Environ Int 29(8):1009–1017

Barnette RM, Debusk EF, Hester JB (1931) The mineral analysis of a nineteen-year-old Marsh Seedless grapefruit tree. Citrus Indus 12(3):5–6, 34

Bar-Yosef B (1991) Fertilization under drip irrigation. In: Palgrave DA (ed) Fluid fertilizers. Science and technology. Marcel Dekker, Inc, New York

Ben-Gal A, Dudley LM (2003) Phosphorous availability under continuous point source irrigation. Soil Sci Soc Am J 67:1449–1456

Bengtsson G, Bergwall C (2000) Fate of 15N labeled nitrate and ammonium in a fertiliser forest soil. Soil Biol Biochem 32:545–557

Bennett WF (1993) Plant nutrient utilization and diagnostic plant symptoms. In: Bennett WF (ed) Nutrient deficiencies and toxicities in crop plants. APS Press, St. Paul

Bingham FT, Davis S, Shade E (1971) Water relations, salt balance and nitrate leaching losses of a 960 acre citrus watershed. Soil Sci 112:410–418

Boman BJ (1996) Fertigation versus conventional fertilisation of Flatwoods grapefruit. Fert Res 44:123–128

Boman B, Obreza T (2002) Fertigation nutrient sources and application. Considerations for citrus. Circular 1410, Agricultural and Biological Engineering Department, Florida Cooperative Extension Service, Institute of Food and Agricultural Sciences, University of Florida. Original publication date April 2002. http://edis.ifas.ufl.edu. Accessed 15 July 2011

Boman BJ, Wilson PC, Ontermaa EA (2001) Understanding water quality parameters for citrus irrigation and drainage systems. Circular 1406, Agricultural and Biological Engineering Department, Florida Cooperative Extension Service, Institute of Food and Agricultural Sciences, University of Florida. Original publication date November 2001. http://edis.ifas.ufl.edu. Accessed 15 July 2011

Boman BJ, Zekri M, Stover E (2005) Managing salinity in citrus. Hort Technol 15(1):108–113

Bould C, Bradfield EG, Clarke GM (1960) Leaf analysis as a guide to the fruit crops. I-general principles, sampling techniques and analytical methods. J Sci Food Agr 11:229–242

Bowling DJF (1976) Uptake of ions by plant roots. Chapman and Hall/Halstead Press/Wiley, London

Burkart MR, Stoner JD (2002) Nitrate in aquifers beneath agricultural systems. Water Sci Technol 45(9):19–29

Burt C, O'Connor K, Ruehr T (1998) Fertigation. Irrigation training and research center. California Polytechnic State University, San Luis Obispo

Cameron SH, Appelman D (1933) The distribution of total nitrogen in the orange trees. Proc Am Soc Hortic Sci 30:341–348

Cantarella H, Quaggio JA, Bataglia OC et al (1992) Response of citrus to NPK fertilization in a network of field trials in São Paulo State, Brazil. In: Tribulato E, Gentile A, Refrigerato G (eds) International Citrus Congress, vol 2, Acireale, 8–13Mar 1992, pp 607–612

Castle WS (1980a) Citrus root systems: their structure, function, growth, and relationship of tree performance. In: Proceedings International Society Citriculture, Sydney, 15–23 Aug 1980, pp 62–69

Castle WS (1980b) Fibrous root distribution of "pineapple" orange trees on rough lemon root stock all three tree spacings. J Am Soc Hort Sci 105:478–480

Chapman HD (1968) The mineral nutrition of citrus. In: Reuther W, Batchelor LD, Webber MJ (eds) The citrus industry, 2nd edn. University of California, Division of Agricultural Sciences, Berkeley

Cline RA, Mc Neill B (1991) Leaf analyses for fruit crop nutrition. FactSheet 206/532. Queen's Printer for Ontario. ISSN 1198-712X

Coelho FS (1973) Fertilidade do solo, 1st edn. Instituto Campineiro de Ensino Agrícola, Campinas, 384 p

Dasberg S, Bar-Akiva A, Spazisky S (1988) Fertigation versus broadcasting in an orange grove. Fert Res 15:147–154

De Paz JM, Ramos C (2004) Simulation of nitrate leaching for different nitrogen fertilization rates in a region of Valencia (Spain) using a GIS-GLEAMS system. Agr Ecosyst Environ 103(1):59–73

Duenhas LH, Villas Bôas RL, Souza CMP et al (2002) Fertirrigação com diferentes doses de NPK e seus efeitos sobre a produção e qualidade de frutos de laranja (*Citrus sinensis* O.) "Valencia". Revista Brasileira de Fruticultura Jaboticabal 24(2° 1):214–218

Embleton TW, Jones WW, Labanauskas CK et al (1973a) In: Reuther WJ (ed) The citrus industry, vol 3. University of California, Division of Agricultural Sciences/EEUU, Berkeley, pp 183–211

Embleton TW, Reitz H, Jones WW (1973b). Citrus fertilization. In: Reuther W (ed) The citrus industry, revised edition, vol 3. University of California, Division of Agricultural Sciences, Berkeley, pp 122–182

Epstein E, Bloom AJ (2005) Mineral nutrition of plants: principles and perspectives, 2nd edn. Sinauer Associates, Sunderland

Esala M, Leppänen A (1998) Leaching of 15N-labeled fertiliser nitrate in undisturbed soil columns after simulated heavy rainfall. Comm Soil Sci Plant Anal 29:1221–1238

FAO (2008) Current world fertilizer trends and outlook to 2011/12. Food and Agriculture Organization of the United Nations, Rome

Feigenbaum S, Hadas A, Sofer M (1994) Clay-fixed labeled ammonium as a source of available nitrogen. Soil Sci Soc Am J 58:980–985

Friis-Nielsen B (1966) An approach towards interpreting and controlling the nutrient status of growing plants by means of chemical plant analyses. Plant Soil 24:63–80

Futch SH, Tucker DPH (2000) A guide to citrus nutritional deficiency and toxicity identification HS-797, one of a series of the Horticulture Sciences Department, Florida Cooperative Extension Services, Institute Food and Agricultural Sciences, University of Florida. Original publication date October 2000. http://edis.ifas.ufl.edu. Accessed 15 July 2011

Gehl RJ, Schmidt JP, Stone LR (2005) In situ measurements of nitrate leaching implicate poor nitrogen and irrigation management on sandy soils. J Environ Qual 34:2243–2254

Hageman RH et al (1992) Ammonium versus nitrate nutrition of higher plants. In: Hauck RD, Beaton JD, Goring CAI (eds) Nitrogen in crop production. American Society of Agronomy, Crop Science Society of America and Soil Society of America, Madison, pp 67–88

Holford ICR (1997) Soil phosphorus: its measurement, and its uptake by plants. Aust J Soil Res 35:227–239

Kafkafi U, Tarchitzky J (2011) Fertigation: a tool for efficient fertilizer and water management, 1st edn. International Fertilizer Association/International Potash Institute, Paris/Horgen, May 2011

Kato T (1986) Nitrogen metabolism and utilization in citrus. In: Janick J (ed) Horticultural reviews, vol 8. AVI Publishing Co, Westport, pp 81–216

Kato T, Kubota S, Tsukahra S (1981) ^{15}N absorption and translocation in Satsuma mandarin trees. VI. Uptake and distribution of nitrogen supplied in summer. Bull Shikoku Agri Exp Stn 36:1–6

Kenworthy AL (1961) Interpreting the balance of nutrient element in leaves of fruit trees. In: Reuther W (ed) Plant analysis and fertilizer problems, 8th edn. American Institute of BioSciences/EEUU, Washington, DC, pp 28–43

Lamb ST, Graham WD, Harrison CB (1999) Impact of alternative citrus management practices on groundwater nitrate in the Central Florida Ridge. Part I. Filed investigation. Trans ASAE 42(6):1653–1668

Legaz F, Primo-Millo E (1988). Normas para la fertlización de los agrios. Serie Fullets Divulgació n° 5–88. Consellería d´Agricultura i Pesca. Generalitat Valenciana, 29 pp

Legaz F, Primo-Millo E (1992) Influencia de la ferttilización nitrogenada en la contaminación por nitratos de las aguas subterráneas. Lev Agr 317–318:168–180

Legaz F, Primo-Millo E, Primo-Yúfera E (1982) Nitrogen fertilization in citrus. I. Absorption and distribution of nitrogen calamondin trees (*Citrus mitis* Bl.) during flowering fruit set and initial fruit development periods. Plant Soil 66:339–351

Legaz F, Serna MD, Ferrer P et al (1995) Análisis de hojas, suelos y aguas para el diagnóstico nutricional de plantaciones de cítricos. Procedimientos de toma de muestras. Servicio de Transferencia de Tecnología Agraria. Consellería d'Agricultura, Pesca i Alimentació. Generalitat Valenciana, 26 pp

Legaz F, Serna MD, Muñoz N et al (1997) Alteración de la relación K/Mg en la planta y en el medio de cultivo mediante el aporte de magnesio en riego localizado a goteo. Actas de Horticultura (SECH), 20, pp 536–543. I Congreso Ibérico y III Nacional de Fertirrigación, Murcia

Li J, Zhang J, Rao M (2004) Wetting patterns and nitrogen distributions as affected by fertigation strategies from a surface point source. Agr Water Manage 67:89–104

Liebig J (1840) Chemistry and its application to agriculture and physiology. Taylor and Walton, London

Longeri L, Vidal P, Fernández M (200) Fijacion de amonio en seis suelos de la VIII region de Chile. Agr Téc 61(2):180–191. Disponible en Internet: http://www.scielo.cl/scielo.php?script=sci_arttext&pid=S0365-28072001000200008&lng=es&nrm=iso. ISSN 0365-2807

Malavolta E, Violante Netto A (1989) Nutrição mineral, calagem, gessagem e adubação dos citros. Potafós, Piracicaba, 153 p

Mansell RS, Fiskell JGA, Calvert DV et al (1986) Distributions of labeled nitrogen in the profile of fertilised sandy soil. Soil Sci 141:120–126

MARM (2010) Ministry of environment, rural and marine. Agricultural report. http://www.magrama.es/es/. Accessed 26 Mar 2012

Marschner H (1995) Mineral nutrition of higher plants. Academic, London, p 889

Martínez JM, Bañuls J, Quiñones A et al (2002) Fate and transformations of ^{15}N labeled nitrogen applied in spring to citrus trees. J Hortic Sci Biotechnol 77:361–367

Martínez-Alcántara B, Quiñones A, Primo-Millo E et al (2011) Nitrogen remobilization response to current supply in young Citrus trees. Plant Soil 342:433–443

Mattos D, Graetz DA, Alva AK (2003) Biomass distribution and nitrogen-15 partitioning in citrus trees on a sandy entisol. Soil Sci Soc Am J 67:555–563

Mitscherlich EA (1909) Des Gesetz des Minimums und das Gesetz des abnehmended Bodenertrages. Landwirsch Jahrb 3:537–552

Mooney PA, Richardson AC (1994) Seasonal trends in the uptake and distribution of nitrogen in satsuma mandarins. In: Tribulato E, Gentile, Refrigerato G (eds) Proceedings of the 7th International Citrus Congress, vol 2, 1992, Acireale, pp 593–597

Morgan KT, Wheaton TA, Castle WS et al (2009) Response of young and maturing citrus trees grown on a sandy soil to irrigation scheduling, nitrogen fertilizer rate, and nitrogen application method. Hortic Sci 44:145–150

Obreza TA (1990) Young "Hamlin" orange tree fertilizer response in southwest Florida. Proc Fl St Hortic Soc 103:12–16

Obreza TA (2003) Importance of potassium in Florida citrus nutrition program. Better Crops 87:19–22

Obreza TA, Morgan KT (2011). Nutrition of Florida citrus trees. Document SL 253. Department of Soil and Water Sciences, 2nd edn. University of Florida, 96 p

Ollier C, Cardoso F, Dinu M (2009) Summary results of the EU-27 orchard survey. In: Statistics in focus, agriculture and fisheries. Eurostat. European Commission. epp.eurostat.ec.europa.eu/cache/ITY_OFFPUB/KS-SF-09-041/EN/KS-SF-09-041-EN.PDF.

Papadopoulos I (2001) Processo de transição da fertilização convencional para fertirrigação. In: Follegatti MV, Casarini E, Blanco FF, Brasil CRP, Resende RS (eds) Fertirrigação: flores, frutas e hortaliças, cap. 1. Guaíba: Agropecuaria, pp 9–69

Quaggio JA, Cantarella H, van Rajj B (1998) Phosphorous and potassium soil test and nitrogen leaf analysis as a base for citrus fertilization. Nutr Cycl Agroexcosyst 52:67–74

Quiñones A, Bañuls J, Primo-Millo E et al (2005) Recovery of the ^{15}N-labelled fertilizer in citrus trees in relation with timing of application and irrigation system. Plant Soil 268:367–376

Quiñones A, Martínez-Alcántara B, Legaz F (2007a) Influence of irrigation system and fertilization management on seasonal distribution of N in the soil profile and on N-uptake by citrus trees. Agric Ecosyst Environ 122:399–409

Quiñones A, Martínez-Alcántara B, Legaz F (2007b) Optimization of N fertilization management in citrus trees: ^{15}N as a tool in NUE improvement studies. In: Frank C (ed) Fertilizers: properties, applications and effects. Nova Science Publishing Inc, Hauppauge, pp 181–206

Quiñones A, Martínez-Alcántara B, Martínez JM et al (2012) Fate of ^{15}N labeled potassium nitrate in different citrus cultivated soils: influence of spring and summer application. Water Air Soil Pollut. doi:10.1007/s11270-011-1017-0 (in press)

Quiñones A, Martínez-Alcántara B, Primo-Millo E et al (2010) Abonado de los cítricos. In: Guía Práctica de la fertilización racional de los cultivos en España. Ministerio de Medio Ambiente, Medio Rural y Marino, Madrid (España), pp 193–204

Recous S, Machet JM, Mary B (1988) The fate of labeled ^{15}N urea and ammonium nitrate applied to a winter wheat crop. I. Nitrogen transformations in the soil. Plant Soil 112:205–214

Sala JM, Cuñat PM (1982) Influencia de la forma nítrica y amoniacal, en que se administra el nitrógeno, sobre la productividad y absorción de macronutrientes del naranjo W. Navel Rev Agroquim Tecnol Alimentos 22(4):589–597

Sala JM, Cuñat P, Collado M et al (1992) Effect on nitrogenous fertilization (quantity and nitrogen form). In: Tribulato E, Gentile, Refrigerato G (eds) Precocity of colour change of "Navelina" oranges. Proceedings of International Society of Citriculture, 8–13 Mar, Acireale, pp 598–602

Sanchís EJ (1991) Estudio de la contaminación por nitratos de las aguas de la provincia de Valencia. Origen, balance y evolución espacial y temporal. Barcelona´s University, pp 111–129 (in Spanish)

Scholberg JMS, Parsons LR, Wheaton TA et al (2002) Soil temperature, nitrogen concentration and residence time affect nitrogen uptake efficiency in citrus. J Environ Qual 31:759–768

Schumann AW, Fares A, Alva AK et al (2003) Response of "Hamlin" orange to fertilizer source, annual rate and irrigated area. Proc Fl St Hortic Soc 116:256–260

Schumann AW, Syvertsen JP, Morgan KT (2009) Implementing advanced citrus production systems in Florida-early results. Proc Fl St Hortic Soc 122:108–113

Serna MD, Borrás R, Legaz F, Primo-Millo E (1992) The influence of nitrogen concentration and ammonium/nitrate ratio on N-uptake, mineral composition and yield citrus. Plant and Soil 147:13–23

Singh BR, Kanehiro Y (1969) Adsorption of nitrate in amorphous and kaolinitic Hawaiian soils. (Ion exchange). Soil Sci Soc Am Proc 33(5):681–683

Smith PF (1966) Citrus nutrition, chap. VII. In: NF Childers (ed) Temperate and tropical fruit nutrition. Horticulture Publication, State University, New Brunswick, pp 174–207

Sne M (2006) Micro irrigation in arid and semi-arid regions. In: Kulkarni SA (ed) Guidelines for planning and design. International Commission on Irrigation and Drainage, New Delhi

Stephenson RA, Cull BW (1986) Standard leaf levels for bearing macadamia trees in Southeast Queensland. Scientia Hortic 30:73–82

Syvertsen JP, Smith ML (1996) Nitrogen uptake efficiency and leaching losses from lysimeter-grown citrus trees fertilized at three nitrogen rates. J Am Soc Hortic Sci 121(1):57–62

Tucker DPH, Davis RM, Wheaton TA et al (1990) A nutritional survey of south central, south west and east coast flat-woods citrus groves. Proc Fl St Hortic Soc 103:324–327

US Environmental Protection Agency (1994) National primary drinking water standards. EPA 810-F-94-001, p 2

Vanden Heuvel RM, Hoeft RG, Mulvaney RL et al (1991) Movement of nitrogen-15 labeled nitrate in large undisturbed columns of poorly drained soil. Comm Soil Sci Plant Anal 22:809–826

Vivancos AD (1996) Fertigación, 2nd edn. Ediciones Mundi-Prensa, Madrid, p 233

Voisin A (1965) Fertilizer application-soil, plant, animal. Crosby Lockwood and Son Limited, London, p 108

Wallace T (ed) (1943) The diagnosis of mineral deficiencies in plants by visual symptoms. University of Bristol. Agricultural and Horticulture Research Station, Long Ashton, Bristol/London

Wallace A (1953) Nitrogen absorption and translocation by citrus cuttings at different root temperatures. Proc Am Soc Hortic Sci 61:89–94

Wallace A, Mueller RT (1957) Ammonium and nitrate absorption from sand culture by rough lemon cuttings. Proc Am Soc Hortic Sci 69:183–188

Walworth JL (2006) Soil sampling and analysis. The University of Arizona Cooperative Extension publication AZ1412. University of Arizona, Tucson. http://cals.arizona.edu/pubs/crops/az1412.pdf. Accessed 9 July 2011

Ware GO, Ohki K, Moon LC (1982) The Mitscherlich, plant growth model for determining critical nutrient deficiency levels. Agron J 74:88–91

Water quality of groundwater from the Geo-Miner Istitute of Spain (WQIGME) (2012) www.igme.es/internet/default.asp. Accessed 26 Mar 2012

Westerman RL, Tucker TC (1979) In situ transformations of nitrogen-15 labeled materials in Sonoran desert soils. Soil Sci Am J 43:95–100

WHO (2004) Guidelines for drinking water quality. Recommendations, 3rd edn, vol 1, Geneva. http:// www.who.int/water_sanitation_health/dwq/GDWQ2004web.pdf.

Wild A (1992) Rusell's soil conditions and plant growth. 11th edn. Longman Group UK Limited (Ed)

Willis LE, Davies FS (1990) Fertilization, nitrogen leaching and growth of young « Hamlin » orange trees on two rootstocks. Proc Fl St Hortic Soc 103:30–37

Zanini JR, Villas Bôas RL, Feitosa Filho JC (2002) Uso e manejo da fertirrigação e hidroponía. Funep, Jaboticabal, 65 p

Zhang M, Alva AK, Li YC et al (1996) Root distribution of grapefruit trees under dry granular broadcast vs. fertigation method. Plant Soil 183:79–84

Nutrient-Hormone Interactions in Citrus: Physiological Implications

21

Alfredo Gravina, Giuliana Gambetta, and Fernando Rivas

Abstract

Citrus growth and development is regulated by complex but subtly tuned nutritional and hormonal interaction in response to environmental signalling. Flower induction is stimulated by low temperatures and water stress and mediated by previous fruit load, being endogenous gibberellins (GAs) content pointed as the main inhibiting hormones. The promoting role of carbohydrates or nitrogen (N) on flowering has not been demonstrated, but a minimum content of these compounds seems to be required for flower formation. Fruit set is a critical step for fruit production, being initially regulated by GAs content. Thereafter, during the onset of the source-sink competition, fruit set depends upon carbohydrates and N availability and fruitlets sink strength. During stage II of fruit development, fruit growth is promoted by auxins, carbohydrates and water accumulation. Afterwards, peel colour development, which is stimulated by low temperatures, is promoted by the decline of flavedo's GAs content and the steady-state level of ethylene, as carbohydrates and abscisic acid (ABA) increase and N decrease during this final stage.

Keywords

Carbohydrates • Competition • Flower induction • Fruit development • Fruit set • Gibberellins • Maturation • Nitrogen

21.1 Introduction

Growth and development in woody perennial plants has many differences regarding to annual crops. After overcoming the juvenile stage, in which species are not able to produce flowers, each annual cycle in perennials give rise to one or more flushes of vegetative and reproductive growth.

A. Gravina (✉) • G. Gambetta
Ecofisiología de Citrus, Departamento de Producción Vegetal, Facultad de Agronomía, Universidad de la República, Av. E. Garzón 780, C.P. 12900 Montevideo, Uruguay
e-mail: agravina@fagro.edu.uy; gambetta@fagro.edu.uy

F. Rivas
Programa Nacional de Investigación en Producción Citrícola, Instituto Nacional de Investigación Agropecuaria (INIA), Camino al Terrible s/n, Salto, Uruguay
e-mail: cfrivas@inia.org.uy

Under tropical conditions, *Citrus* annually brings about several sprouting and blooms, whereas growing in temperate climates, only the spring flush is reproductive. In temperate zones, although several flushes of growth may occur, these are mostly vegetative. The exception is provided by lemon trees (*C. limon* [L.] Burm.), which give rise to reproductive and vegetative shoots throughout the year.

Vast information is available relating nutrition and water supply with flowering and fruiting of *Citrus*. However, less information matching the controlling role of nutrients, carbohydrates, water and hormones during the developmental process is available. Therefore, hormonal and nutritional control of some *Citrus* developmental processes, with special emphasis on floral induction, fruit set, fruit development and maturation of *Citrus* has been focused in this chapter.

A.K. Srivastava (ed.), *Advances in Citrus Nutrition*,
DOI 10.1007/978-94-007-4171-3_21, © Springer Science+Business Media B.V. 2012

21.2 Floral Induction

Floral bud induction in *Citrus* is a complex process affected by both endogenous and exogenous factors, yet not fully elucidated. Extensive reviews on this subject have been published in the last decade of the twentieth century (Davenport 1990; Krajewsky and Rabe 1995; Srivastava et al. 2000). At present, the existence of substances directly involved in flowering induction of *Citrus* has not been demonstrated, whereas evidences of inhibiting molecules of this process are available. It has been proposed that *Citrus* buds are biologically reproductive and need to overcome some restrictions in order to express this potential (Goldschmidt and Monselise 1972; Agustí 1980). Microscopic studies have identified different stages of floral differentiation, in which no signs of histological changes could be observed before ecodormancy. At the microscopic level, the first signs of activity in the apical meristem can be observed from late December to mid-January in the Northern Hemisphere (NH). At the macroscopic level, the first signs of differentiation, recognizable by the appearance of sepal primordia, starts in January–February, coinciding with the onset of meristematic activity (Abbott 1935).

21.2.1 Exogenous Promoter Factors

Low temperature and water stress have been ascribed as key promoters of flower induction, which occurs 3–5 months before flowering under natural conditions.

Water stress is the most important exogenous factor inducing flowering in tropical zones. Under controlled conditions, *Citrus* plants growing in containers subjected to severe water stress of −3.5 MPa for periods of 4–5 weeks (Southwick and Davenport 1987), or −4.0 MPa for 10 weeks (Manzi 2011), produced significantly more flowers than irrigated plants after resuming the normal growing conditions. The longer the water stress period, the greater the return flowering after irrigation (Borroto et al. 1981).

Time exposure to low temperatures has been positively correlated with flower intensity (Moss 1976; Lovatt et al. 1988; García-Luis et al. 1992; Valiente and Albrigo 2004). Low temperature has been linked to endogenous changes such as the inhibition of the gibberellin-like substances translocation from root to the canopy in oranges (Eilati et al. 1969a). Furthermore, Satsuma (*C. unshiu* Marc.) plants growing at 4°C showed less stem gibberellin-like content than those growing at 25°C (Tamim et al. 1996). Low temperature during autumn and winter also reduces nitrogen (N) absorption (Chapman and Parker 1942), translocation (Wallace 1953) and leaf concentration (Poerwanto and Inoue 1990); however, no cause-effect relationship has been achieved with floral induction. A change on N metabolism by low temperatures has also been achieved by extending the exposure to cold stress, hence rising leaf NH_3-NH_4^+ concentration and bloom intensity. Low temperature also induces molecular changes regarding to flowering. Indeed, chilling temperatures during the flower induction period have also been shown to induce the expression of *Citrus FLOWERING LOCUS T* homologues, *CiFT* (Nishikawa et al. 2007).

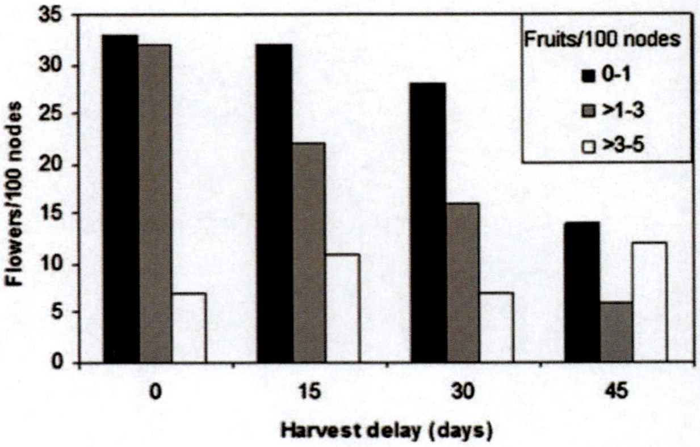

Fig. 21.1 Flowering intensity according to fruit number per 100 nodes and harvest delay in 'Ortanique' tangor (Gravina et al. 2004)

Although low temperature and water stress stimulate floral induction, these stimuli seem to be not enough to ensure the fate of floral meristem since fruit presence strongly inhibits the inductive process. Significant decrease in bloom has been reported after high crop load, especially in alternate bearing varieties (Moss 1977; Becerra and Guardiola 1984; García-Luis et al. 1986b, 1995; Gambetta et al. 2008). Fruit-mediated inhibitory effect is exerted in two ways: (1) indirectly, at early stages of fruit development by reducing summer and autumn sprouting, since their buds are more readily inducible than spring ones (Verreynne and Lovatt 2009) and (2) directly, at later stages of fruit development by reducing generative buds (Martínez-Fuentes et al. 2010). Therefore, the inhibition of flower bud induction caused by harvest delay has been reported (Borroto et al. 1981; García-Luis et al. 1986b). A short distance fruit effect on the return bloom has been evidenced, and its interaction with harvest delay has also been demonstrated (Fig. 21.1). In 'Ortanique' tangor (*C. sinensis* [L.] Osb. × *C. reticulata* Bl.), one fruit per 100 nodes inhibits flowering by delaying harvest 45 days. Two or three fruits per 100 nodes provoke a progressive inhibition of return bloom. More than three fruits per 100 nodes inhibit flowering independently of the harvest date (Gravina et al. 2004).

21.2.2 Hormonal Regulation

Fruit presence increases GA_1-GA_3 and GA_4-GA_{19} like substances on 'Valencia' sweet orange (*C. sinensis* [L.] Osb.)

pedicels during the inductive period (Plummer et al. 1989). Concentration of GA_{1+3} on Satsuma mandarin leaves was negatively correlated with the return bloom (Koshita et al. 1999). During colour break, GA_1 and GA_4 concentration diminishes in fruit flavedo, whereas it increases in shoot stem cortex of 'Washington' navel sweet orange (*C. sinensis* [L.] Osb.) (Gambetta et al. 2011). This could suggest a translocation of GAs from fruit tissues to the nearby branches during the inductive period. Considering the aforementioned evidences, the inhibitory role of GAs on floral induction has been widely accepted.

The role of abscisic acid (ABA) on flower bud induction remains controversial. In Satsuma mandarin, no significant differences in ABA leaf content from branches with or without fruit during the inductive period were found (Koshita et al. 1999). However, Okuda (2000) reported higher ABA concentrations in leaves from trees without fruit respect to fruit-bearing trees in alternating 'Aoshima' Satsuma mandarin. An increase of indole-3-acetic acid (IAA) concentration in branches without fruits at the end of winter has been reported. Girdling fruitful and unfruitful branches increases both ABA and IAA concentration in leaves, which was associated with the higher flowering intensity the following spring, thus suggesting a possible floral induction implication.

There is little experimental evidence on the possible role of other hormones in the process of floral induction, so this aspect needs further research until solid conclusions can be drawn.

21.2.3 Control of Floral Induction

The inhibitory effect of gibberellic acid (GA_3) on *Citrus* flowering is widely known (Goldschmidt and Monselise 1972; Iwahori and Oohata 1981; Guardiola et al. 1982; García-Luis et al. 1986b; Lord and Eckard 1987; Gravina et al. 1996). GA_3 sprays at 20 and 50 mg L^{-1} during the inductive period significantly reduce flowering intensity at the following spring. Moreover, a decrease in the proportion of generative inflorescences and an increase of leafy flower shoots is frequently observed. This change brings about a better sprouting vegetative-reproductive balance, which in unproductive parthenocarpic varieties can increase fruit set. Differences in bud sensitivity to GA_3 during the autumn–winter period have been observed. A greater sensitivity has been reported in November (NH) for mandarins and oranges. Other GA_3-sensitive period has been identified at the beginning of sprouting (Guardiola et al. 1982). However, the reduction of flowering by applying GA_3 has been shown to be negatively correlated with the flowering intensity; thus, the greater the expected flowering, the lower the effect of GA_3 on flowering inhibition. In 'Nova' mandarin (*C. reticulata* Bl. × [*C. paradisi* Macf. × *C. tangerina* Ort. ex Tan.]), the application of 40 mg L^{-1} of GA_3 in June (SH) is not efficient when flower intensity exceeds 200 flowers/100 nodes but significantly reduces flowering when control trees have 100–150 flowers/100 nodes (Gravina 2007). Despite this limiting response, the use of GA_3 to manage flower intensity has become a common practice in the citrus industry worldwide, being approved under sustainable and environmental friendly production systems.

Considering the anti-inductive role of GA_3, hormonal substances that limit or prevent their synthesis have been evaluated as promoters of flowering. Paclobutrazol (PBZ) is a growth retardant that inhibits the activity of kaurene oxidase and consequently the synthesis of GAs. Its application through irrigation increases blossom in lemon (Harty and Van Staden 1988), 'Salustiana' sweet orange (*C. sinensis* [L.] Osb.) (Greenberg et al. 1993; Martínez-Fuentes et al. 2002), Satsuma mandarin (Okuda et al. 1996), 'Hernandina', 'Marisol', 'Orogrande' Clementines (*C. clementina* Ort. ex Tan.) and 'Afourer' mandarin (*C. reticulata* Bl. × *C. sinensis* [L.] Osb.). However, there are some reports showing no effect of PBZ when applied to high bearing trees, resembling to some extent the limited effect of GA_3 under high flowering situations.

21.2.4 Carbohydrates and Nitrogen

The role of carbohydrates in the floral induction process of *Citrus* has also been controversial. Early studies report a putative correlation between carbohydrate content in leaves, stems and roots, and the return bloom. During years of high crop load ('on' year), carbohydrate accumulation and flowering are very low. After that, no direct effect on flower induction and differentiation processes could be found in mostly experimental evidences, although a quantitative but not determinant effect has been cited (Goldschmidt et al. 1985; García-Luis et al. 1995; Martínez-Fuentes et al. 2010). Most information has been generated in alternate bearing varieties, characterized by the lack of flowering after an 'on' year. By comparing carbohydrates accumulation in leaves of 'Salustiana' orange trees during 'off' and 'on' years, Monerri et al. (2011) were not able to find differences in carbohydrate concentrations in both situations, hence reinforcing the hypothesis that carbohydrates do not play a significant role in *Citrus* floral induction. In addition, these authors proved that in plants without fruits, root reserves accumulation takes place during the whole flowering period, whilst in fruited trees, it only takes place after harvest.

Girdling increased flowering by raising the availability of soluble carbohydrates and starch. However, since other substances (i.e. hormones) also accumulate as a cause of girdling, it has not been possible to clarify the direct effect of carbohydrates accumulation on flower induction

(Goldschmidt et al. 1985; Iwahori et al. 1990; Goldschmidt 1999; Koshita et al. 1999). Indeed, when girdling is performed on trees with high fruit load, flowering promoting effect is minimal or nil. However, although carbohydrates are not considered to trigger the floral inductive process, there is consensus that *Citrus* trees would need minimum levels of sugars so that floral differentiation could take place.

Starch accumulation in roots is drastically reduced during 'on' years. This decline diminishes root system activity, and consequently, leaf contents of N, phosphorous (P) and potassium (K) decrease. However, no effect on alternate bearing by maintaining leaf K with KNO_3 foliar sprays was found. Therefore, it appears unlikely that mineral nutrient could exert some influence on alternate bearing (Jones et al. 1975).

In addition, in fruit-bearing branches, no foliar reduction of NO_3-N was detected during peel colour break, whilst in non-bearing branches reduced forms of N (NH_3-NH_4^+) increased. However, considering that NO_3-N constitutes a small proportion of total nitrogen, these variations could be considered as a metabolic messenger rather than a direct nutritional regulation (Martínez-Fuentes et al. 2010).

Foliar sprays of urea before sprouting were shown to promote flowering in low N-content trees (Lovatt et al. 1988; Rabe 1994; El-Otmani et al. 2004). Similarly, a decrease in alternate bearing of 'Nour' Clementine mandarin was verified by foliar urea sprays during flower induction/differentiation stage in the 'off' years (Benhamou et al. 2004). By contrast, no relationship between foliar application of urea (1%) during the inductive period and return bloom were found (Ayalon and Monselise 1960). Similarly, when applied to well N-content trees, no flower improvement was achieved through urea sprays (Rabe and van Rensburg 1996). Therefore, N could play a role on the flower bud induction but before the onset of the ecodormancy stage.

Accordingly, the existing evidences cannot allow attributing any specific regulatory function to N on the floral induction and subsequent differentiation of *Citrus*.

In summary, GAs are the main endogenous factor related to floral induction. Low temperature and water stress reduce their endogenous content, promoting floral induction, whilst fruit presence increases their content, inhibiting the floral induction process. No conclusions can be drawn to the leading role of N and carbohydrates on the floral induction of *Citrus*.

21.3 Sprouting and Flowering

Sprouting, from bud break until the end of shoot extension, may last 3–4 weeks, varying according to weather conditions and horticultural practices. Shoot growth and elongation ends when the abscission of the apical meristem occurs, resuming growth from an axillary bud situated near the apex at the following sprouting period (Spiegel-Roy and Goldschmidt 1996). Soil temperature seems to be essential for vegetative growth; thus, in controlled conditions, the number of sprouted buds at 25°C doubles those obtained at 15°C (Khairi and Hall 1976).

N, P and K content in adult leaves diminish from the beginning of sprouting until full bloom, suggesting translocation of these elements in order to support the new growth; the lowest level coincides with the first flower abscission peak. By contrast, new leaves accumulate these nutrients from bloom until June drop (Sanz et al. 1987). Considering a longer period, Krueger and Arpaia (2008) report the same trend for N concentration, which decreases in mature leaves of 'Valencia' orange from winter to full bloom, implying a mobilization to the growing organs. In roots, trunk and branches, N concentration increases during the same period, suggesting that these organs would not export this nutrient to new growth. However, different results have been obtained for K and zinc (Zn), showing constant leaf concentration from winter to bloom, putting forward the key role of root absorption for the supply of these nutrients to the new developing organs (Krueger and Arpaia 2008). Carbohydrates increase from bud break to flowering in adult leaves and then decline in a constant way until June drop (Sanz et al. 1987).

Spring flush, especially at the early stages, is supported by carbohydrates, N, P and K reserves accumulated in leaves during the previous cycle, as well as by the photosynthetic activity of adult leaves. Low starch and high soluble sugar concentration in leaves from plants growing under K deficiencies have been found; α-amylase activity is induced by K shortage. This behaviour suggests that K is involved in carbohydrate status and metabolism of mature *Citrus* leaves (Lavon et al. 1995).

In terms of energetic balance, flowering is an important sink of carbohydrate in *Citrus*. In grapefruit, flowering demands about 14% of the total carbohydrates consumed during the reproductive cycle and 11% of the total annual consumption (Bustan and Goldschmidt 1998). In high flowering trees of 'Salustiana' orange ('on' year), the percentage of total dry matter used in the flowering process can exceed 16% of total consumption including harvested fruit (Monerri et al. 2011).

21.4 Fruit Set and Development

Fruit set is a key step of the reproductive process of *Citrus*. It is a complex process which comprises from biological to economical implications. The botanical definition of fruit set encloses the nutritional, physiological and biochemical processes which determine the transition from flower to developing fruitlet. The agricultural concept of fruit set refers to the final number of fruit obtained at harvest. Any factor which may disrupt the normal fruitlet growth during this stage can redirect flower or fruit to abscission, which may last even several weeks after petal fall. Although many factors can influence directly or indirectly fruit set or abscission, the following section will be focused in those regarding to hormonal and nutritional issues.

21.4.1 Intersink Competition

Fruit set and initial growth is a high-demanding energetic process. During this period, many thousands of fruitlets face a strong competition for carbohydrates against other tree growing organs (flowers, shoots and roots). It has been well established that fruit drop is inversely related to fruit growth rate (Zucconi et al. 1978; Ruiz and Guardiola 1994) which strongly depends on cell division and thus carbohydrates availability. Therefore, fruit survival depends upon regular carbohydrate supply, which depends on carbohydrates reserve mobilization (Goldschmidt and Koch 1996; Bustan and Goldschmidt 1998), current photosynthesis (Goldschmidt and Koch 1996; Kozlowski and Pallardy 1997; Iglesias et al. 2002), sink capacity and nutrients utilization. During this period, *Citrus* trees adjust sink demands to source potential by reducing individual growth or even triggering the abscission process. Accordingly, intersink competition during flowering and fruit set has been stated as one of the main factors affecting fruit survival and growth in *Citrus*. Regarding to this, the reduction of flowering intensity bring about greater ovary size at anthesis, revealing that competition can be stated even at early stages of flower development (Guardiola et al. 1984). Agustí et al. (1982) showed that the greater the flowering intensity, the earlier the starting of the abscission process in sweet orange.

During the spring flush period, *Citrus* give rise to different kinds of shoot being classified morphologically as (a) multiflowered leafless shoot, (b) single-flowered leafless shoots, (c) multiflowered leafy shoots, (d) single-flowered leafy shoots and (e) vegetative shoots, each one differing in their capacity to set fruits (Fig. 21.2).

Fig. 21.2 Reproductive shoots in *Citrus*. (**a**) Multiflowered leafless shoot, (**b**) single-flowered leafless shoot, (**c**) multiflowered leafy shoot and (**d**) single-flowered leafy shoot

Data of Fig. 21.3 reveal the importance of leaves for fruit set and evidence intersink competitions for survival, even when fruits are borne in the same shoot. Fruit arising from leafy shoots postponed abscission longer than those arising from leafless shoots. Besides, when comparing single-flowered shoots vs. multiflowered shoots, regardless leafy or leafless condition, data suggest that when born in multiflowered shoots, fruits present fewer chances for survival.

Additionally, fruit set increases dramatically when shoots are accompanied by leaves as shown in Fig. 21.3 (Rivas et al. 2006a).

Indeed, reports evidence a delay of abscission up to 1 month when shoot leaf number increases (Ruiz and Guardiola 1994; da Cunha Barros and Gravina 2006). This observation brings about major implications since, from an agronomic point of view, it has been reported that almost 75% and 25% of the fruits that reach harvest arise from leafy-flowered and leafless-flowered shoots, respectively (Agustí and Almela 1991). Furthermore, the greater fruit set in leafy inflorescences could rely upon their greater vascular area (Erner and Shomer 1996) and consequently better water supply (Erner et al. 2000) comparing with leafless inflorescence.

21.4.2 Source-Sink Communication: Physiological and Biochemical Holistic Approach

The role of the new leaves upholding fruit set has been ascribed to its capacity to be a source for organic nutrition. New leaves are able to export soluble carbohydrates about 3 weeks after flowering (Kriedemann 1969; Schaffer et al. 1985; Ruan 1993). Thereafter, leaves are able to counteract and respond to the nearby fruit demands, revealing a source-sink communication mediated by carbohydrate consumption/

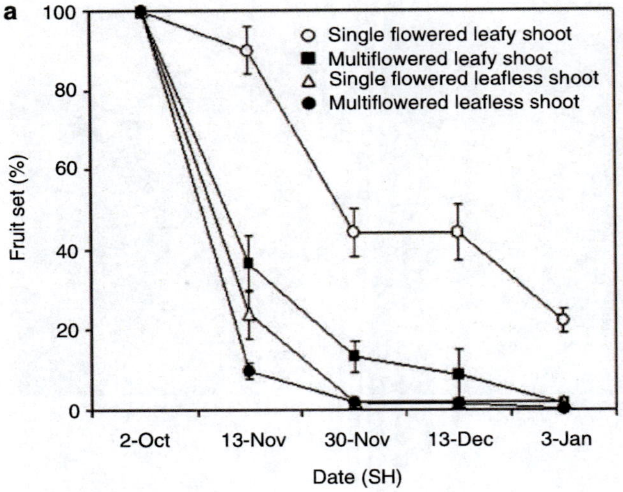

b

Shoot Type	Fruit Set (%)
Single flowered leafy shoot	22.0
Single flowered leafless shoot	2.0
Multiflowered leafy shoot	1.6
Multiflowered leafless shoot	0.4

Fig. 21.3 (**a**) Time course of abscission and (**b**) final fruit set regarding to shoot type in 'Nova' mandarin

accumulation and hormonal signalling. Down-regulation and up-regulation of photosynthetic capacity has been suggested in *Citrus* during fruit set (Rivas et al. 2007). Up-regulation mechanism can be triggered by an enhancement of sink utilization rate of sugars, releasing cytosolic inorganic phosphate (Pi) in source leaves, promoting photophosphorylation and increasing carbon dioxide (CO_2) fixation to triose-P by Rubisco (Leegood 1996; Paul and Pellny 2003). It has been suggested that maximal photosynthetic rates cannot be maintained by Pi-cycling via stromal starch synthesis only, and the chloroplast is often considered to be dependent on cytosolic sucrose synthesis for its Pi supply (Foyer and Galtier 1996). Consequently, the extent of feedback modulation of photosynthesis depends on the nature of the final synthesized carbohydrate. Furthermore, when sucrose is the main storage carbohydrate, leaves appear to show smaller decreases in CO_2 fixation than leaves accumulating starch (Goldschmidt and Huber 1992). As the major transported carbohydrate, sucrose can be quickly loaded to the phloem tissue and driven to the growing fruitlets in order to be consumed through the Krebs metabolic pathway, supporting constant growth energetic demands. As the number and size of the growing fruitlets increase, young leaves overcome carbohydrate requirement by raising quantum yield efficiency of PSII (Φ_{PSII}). This effect has been observed in many fruit trees species including *Citrus* (Layne and Flore 1995; Wünsche et al. 2000; Syvertsen et al. 2003; Urban et al. 2004; Rivas et al. 2007). Therefore, Rivas et al. (2007) pointed out that as a consequence of a higher fruit set, glucose and fructose in leafy flowered shoots are mainly diverted towards sucrose synthesis than to starch. On the other hand, when an upgrade of carbohydrate content at the canopy is induced by girdling, vegetative shoot diverts leave's glucose and fructose content towards starch synthesis. In this case, other than providing carbohydrate to support fruit set, Φ_{PSII} of vegetative shoots decreases as a clear signal of down-regulation of the photochemical process.

Recent work has stated that during bud sprouting and flowering, carbohydrates are used mainly from shoots; thus, at the time of full bloom, carbohydrate concentration falls to a minimum value. Afterwards, during fruit growth and abscission, carbohydrates supply is supported by current photosynthesis without evidencing root reserves contribution during fruit set (Monerri et al. 2011).

As stated previously, fruit set becomes a complex process in which sink strength can modulate to some extent, photosynthetic capacity, interacting to cope nutritional requirements for maintaining continuous fruit growth and promoting fruit set (Fig. 21.4).

21.4.3 Hormonal Control of Fruit Set and Its Relation with Nutrition and Growth

Pollinization has been shown to increase fruit set in *Citrus*. In past decades, growers used to place compatible related genotypes in their orchards to enhance fruit set, fruit size and harvest. However, seediness in *Citrus* has become a non-desirable marketable trait since customers pay premium prices for seedless fruits. Fortunately, many species of *Citrus* genus are autoincompatible and have parthenocarpic ability so that seedless fruit can be produced under non-cross-pollinization condition. This trait has been shown to depend on the endogenous GAs content in the ovary tissues. As an example, 'Nules' Clementine mandarin, growing in solid block, behaves as shy bearing cultivar, but when GA_3 sprays are applied during blossom, fruit set increases dramatically and then productivity reaches profitable levels (Agustí and Almela 1991; Zacarías et al. 1995). Furthermore, even when carbohydrates availability is forced to increase through girdling (Rivas et al. 2006a, b), an improvement of fruit set is hardly achieved, revealing that organic nutrition is not the main limiting factor during the flower to fruitlet transition stage. It has been shown that pollinization increases GAs content (*13*-hydroxy-GAs, GA_{19}, GA_{20}, GA_{29}, GA_8 and the biologically active GA_1) in developing ovaries (Ben-Cheikh et al. 1997), thereby promoting flower to fruit transition, thus encouraging final fruit set. However, it is widely accepted that Satsuma mandarin produces high fruit number without

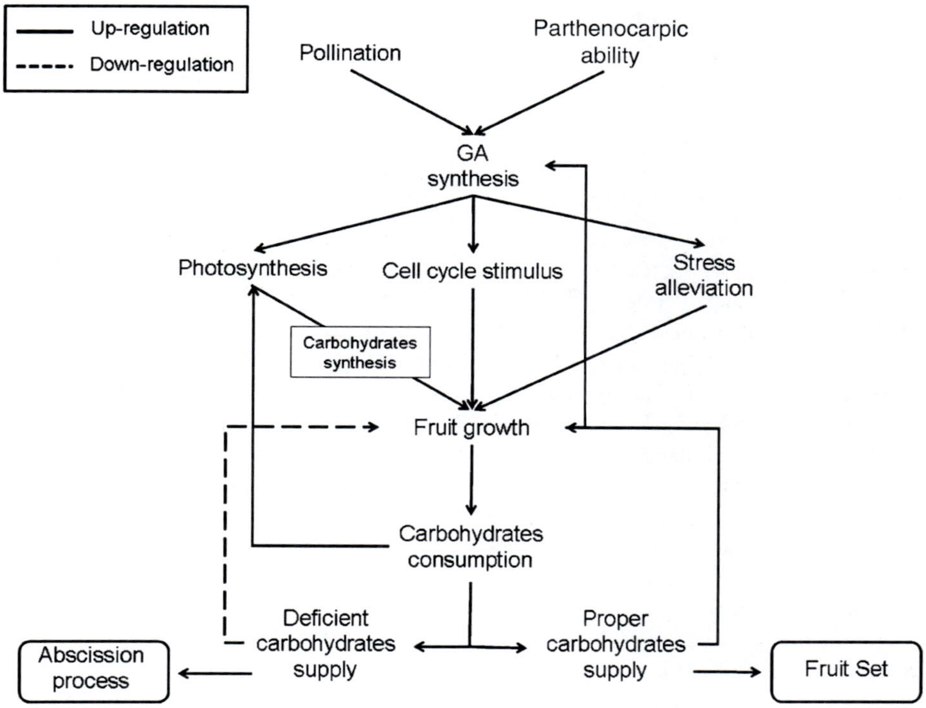

Fig. 21.4 General scheme relating fruit growth and signalling with carbohydrate metabolism during fruit set in *Citrus*

pollination stimulus. When analysed by GC-MS, it has been shown that ovary GAs content of Satsuma is higher than those of non-fertilized Clementine (Talón et al. 1990). Even more, in *Citrus*, when GAs synthesis is disrupted by applying PBZ, it has been shown that fruit growth ceases and fruit set fails (Ben-Cheikh et al. 1997; Rivas et al. 2010). These evidences put forward the leading role of hormone signalling (mainly GAs) for setting fruits in *Citrus*.

Gibberellins are involved in the cell cycle modulation, shortening the interphase and inducing cell at G1 stage to synthesize DNA (Srivastava 2002). Indeed, up-regulation of five cell cycle regulation, DNA processing proteins and cell wall biosynthesis genes (including an endo-1,4-*β*-D-glucanase, a cellulose synthase-like gene and *β*-1,3-glucanases) were found in transgenic sense S23 line of Carrizo citrange (*C. sinensis* [L.] Osb. × *P. trifoliata* [L.] Raf.), overexpressing *CcGA20ox1* and showing higher active GA_1 concentrations (Huerta et al. 2008). In this way, GAs can stimulate cell division rate and consequently promote continuous and faster fruit growth and carbohydrate consumption (sink strength), leading to higher fruit survival chance (Rivas et al. 2007).

The involvement of a hormonal control of photosynthesis by the fruit is feasible since GA_3 sprays have been shown to increase carbon fixation capability by promoting Rubisco activity (Yuan and Xu 2001). In fact, in young leaves of transgenic Carrizo citrange overexpressing *CcGA20oxl*, both stomatal conductance (gs) and transpiration rate (E) increase and consequently so does net photosynthetic CO_2 uptake (Huerta et al. 2008). Increased GA_1 concentrations were also found in developing fruitlets of girdled trees, which presented better fruit set (Talón et al. 2000; Mehouachi et al. 2009). Consequently, it is highly suggestive that GAs-mediated signals from growing fruitlets could elicit the up-regulation of photosynthetic capacity, promoting carbohydrate synthesis and so mediating the source-sink communication in *Citrus* (Rivas et al. 2007).

Abscission comprises the process of natural separation of organs from the parent plant. In *Citrus* and other tree species, it is well accepted that this process is triggered as a way to adjust carbohydrates demand with supply under carbohydrate shortage, lack sink strength or as a response against some environmental stress (i.e. extreme temperatures, water deficit or excess, pest and pathogen attack). Under a biochemical concept, abscission is the result of changes in gene expression, resulting in the synthesis of enzymes which catalyse the loose of adjacent cell walls within the zone and subsequent cell separation. The physiological and biochemical process involved in the abscission stage has attracted significant interest for grower and researchers since premature abscission can lead to significant crop losses in *Citrus*.

Abscission of reproductive structures may occur from early stages of flower development up to a later step of the ovary development, even when a significant growth is achieved. Flower abscission has been proposed to occur by other factor than carbohydrate shortage, whereas fruitlet abscission has been assumed to occur as a consequence of carbohydrate depletion. Although some works report higher carbohydrate reserves in abscising as compared to non-abscising fruit (Guardiola et al. 1984; Ruiz and Guardiola 1994), the same authors claimed the lack of sink demand as the possible explanation for this effect.

Abscission in *Citrus* has also been suggested to be mediated by hormonal signalling. As stated above, the first visual symptoms of abscission are evidenced by fruit growth disruption. Under natural conditions, carbon shortage promotes the cease of growth (Rivas et al. 2006b) which has been associated with increased levels of ABA in developing fruitlets, accompanied by decreased levels of GAs (Zacarías et al. 1995; Rivas et al. 2010). Therefore, ABA may act sensing the intensity of the nutrient shortage and modulating the levels of 1-aminocyclopropane-1-carboxylic acid (ACC) and ethylene, which finally is pointed out as the triggering hormone of the photo-assimilate competition-induced fruit abscission in *Citrus* (Gómez-Cadenas et al. 2000). Therefore, carbohydrates content, fruit growth, hormones content (ABA, GAs, ethylene and auxins) and their crosstalk are finely orchestrated in order to sense accurately the current plant status, hence mediating abscission in *Citrus*. As mentioned, GAs can play a direct role in the abscission process, not only encouraging cell division and growth but also for its metabolic role during stress. It has been recently reported that GA_3 sprays up-regulate the gene expression encoding proteins (dehydrins and cysteine proteinase, inositol-3-phosphate synthase, omega-3 fatty acid desaturase, osmotin-like proteins) involved in the physiological and biochemical defence devoted to overcome abiotic and biotic stress (Huerta et al. 2008). Under this perspective, GAs may also act alleviating stress episodes during fruit set and consequently avoiding in some extent the abscission process.

21.4.4 Mineral Nutrition and Fruit Set

Whilst organic nutrition and hormone regulation has paid most attention, modest interest has been focus on mineral nutrition and its interaction with hormones during the fruit set process. Supply of macro- and micro-nutrients is an essential component to sustain accurate physiological and biochemical processes regarding with CO_2 assimilation, plant stress metabolism, cell division and growth. Some nutrients act as a direct component of cell macromolecules (i.e. proteins, enzymes, DNA, hormones and carbohydrates), and others play a role in mediating catalytic enzymatic activity. For example, boron (B) deficiency leads to starch, glucose, and fructose (but not sucrose) accumulation in *Citrus* leaves, causing feedback regulation of photosynthesis and even poor protection from oxidative damage (Shuang et al. 2008, 2009). Zn plays several critical functions in the cell such as protein metabolism, gene expression, structural and functional integrity of biomembranes, photosynthetic carbon metabolism and plant tolerance to heat stress. It is involved in the IAA pathway which has been related to cell growth in *Citrus*. Ferrous (Fe) and magnesium (Mg), as in other species, play a role as a part of the photochemical apparatus and the electron transport chain of the light reaction phase of photosynthesis. *Citrus* is also a high-demanding K genus. Indeed, fruits remove K more than any other mineral nutrient. Therefore, the improvement of fruit set, size, quality (i.e. ascorbic acid content, total soluble solid content) and storage properties has been achieved by applying K under deficient K soil conditions (Ashraf et al. 2010). These are some examples of the role of nutrient for fruit set and development.

21.4.5 Nutritional and Hormonal Interaction During Fruit Growth

During fruit growth, the nutritional, anatomical, physiological and biochemical mechanisms interact in order to ensure continuous growth of the developing fruits until maturity. Potential fruit growth is established even at early stages of flower development; thus, flower and fruit developmental stages are intimately related and can be handled by growers in order to obtain quality crops.

Fruit growth of *Citrus* follows a sigmoidal curve; at the first stage, growth is the result of continuous ovary cell division. Although organic, mineral and water supply is essential for cell division, this stage is reported to be regulated by hormonal stimulus, mainly cytokinins (CKs) and GAs. Thereafter, fruit starts its linear phase of growth (stage II) based mainly on the enlargement of juice sacs located in the locules (Guardiola et al. 1993). At this secondary stage, most of the final fruit size is achieved, being regulated basically by auxins.

Final fruit size in *Citrus* has been claimed to reside in two core components: (a) initial ovary size and (b) fruit growth rate. Initial ovary size has been reported to be inversely correlated with flowering intensity. Thus, reducing the number of flowers per tree, an increase of the ovary growth rate has been observed, being this effect more intense in leafy inflorescences (Guardiola et al. 1984). It can be assumed that fruit size depends on cell number rather than cell size, so those factors affecting cell division cycle activity assume importance (Fig. 21.5). This observation brings about several horticultural implications to manage fruit size in *Citrus*, since applying GA_3 reduces flower intensity, and redistribution of shoot types is achieved, increasing the percentage of leafy ones, with better ovary size.

Fig. 21.5 Physiological factors determining fruit size in *Citrus*

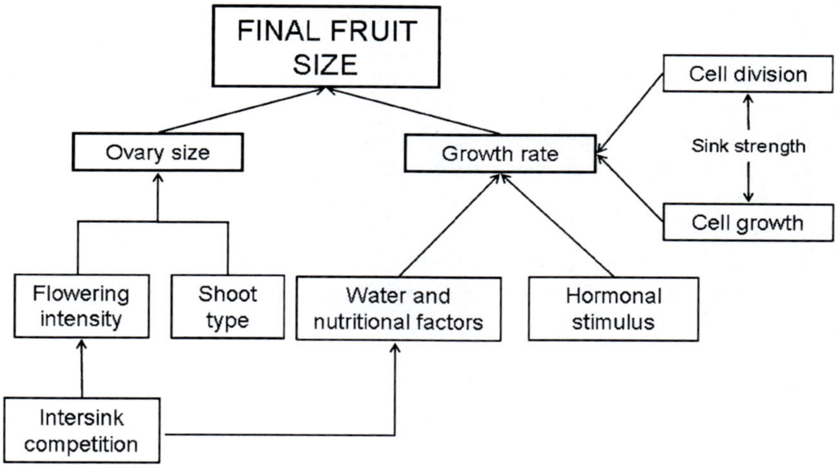

Several reports have linked carbohydrate availability and fruit sink strength, not only as a key requirement for fruit survival but also for fruit growth (Agustí et al. 1994; Goldschmidt and Koch 1996; Mehouachi et al. 1996; El-Otmani et al. 2000). Although carbohydrate demand increases the photosynthetic rate, this increase is limited, and a minimum leaf/fruit ratio is needed to achieve a maximum fruit size (Cohen 1984; Guardiola et al. 1993). Fruit thinning and their positive effect on fruit growth also reveals that intersink competition plays a relevant role determining final fruit size.

It has been widely suggested that sink demand depends on hormonal stimulus and carbohydrate utilization. Seed is one of the main hormonal releasing organs, being GAs a major one. As previously quoted, GAs stimulates fruit set since it promotes cell division and carbohydrates synthesis during the early stages of fruit development. In this way, seeds can increase cell number and carbohydrates availability, thereby improving the potential fruit growth. Many works have positively related fruit seed content with fruit size.

Nutrient and water supply to the growing fruits is considered to be limited by the pedicel vascular area. In this way, in 'Shamouti' sweet orange (*C. sinensis* [L.] Osb.), peduncle diameter increases as a result of leaf/flower ratio increment, which often occurs when flowering is reduced (Erner and Shomer 1996). Therefore, peduncle area has been directly related to final fruit growth (Bustan et al. 1995). However, it has been reported that synthetic auxin sprays promote growth directly by increasing fruit cell size, either than through better vascular connection (El-Otmani et al. 1993; Guardiola et al. 1993; Aznar et al. 1995). Indeed, the application of 2,4-dichlorophenoxy acetic acid (2,4-D) locally to the peduncle, without wetting the fruit, significantly increases peduncle vascular area, without increasing final fruit size (Mesejo et al. 2003). Therefore, during stage II of fruit development, it is assumed that auxins play a central role in fruit enlargement of *Citrus*. It has been suggested that IAA exerts its effect inducing cell wall degradation as a prerequisite to promote cell expansion. Auxins have been also proposed to induce acid invertase activity and so sucrose hydrolysis to hexoses. This osmotic imbalance promotes the passive influx of water to cell, which finally has been pointed as the physical power to carry out cell expansion (Evans 1985).

21.5 Fruit Maturation

Maturation is the developing stage in which fruit acquires the appropriate organoleptic characteristics for consumption. It is a complex process that involves physiological and biochemical changes subjected to environmental and endogenous control.

Citrus fruits behave as non-climacteric since respiration declines slowly throughout the latest stages of fruit development, and ethylene production is constant and extremely low (Aharoni 1968; Eaks 1970; Goldschmidt et al. 1993). Structural and physiological changes occur in the peel and pulp during stage III of fruit development. Both parts develop mostly independently, being the lack of vascular connection between each other one of the reasons to explain this behaviour (Monselise 1977). Main evidence to external maturation is colour development, which results from green pigments reduction, and orange-reddish pigments increase. Internal maturation is characterized by the increment of juice soluble sugars and acidity diminution. Noteworthy is that mature *Citrus* fruits do not contain starch; thus, they reach organoleptic maturity on the tree.

In this section, hormonal and nutritional factors associated to peel colour development will be presented.

21.5.1 Flavedo Pigment Evolution

Quantitative and qualitative changes in pigment composition occur during fruit growth and development. During stages I and II of fruit growth, chlorophylls are the major pigments. At stage III, chlorophylls degradation takes place, evidencing carotenoids presence, which also increase in the peel during this stage (Eilati et al. 1969b; Gross 1981; Alós et al. 2006). Fruit colour not only depends on carotenoid content but also on its physicochemical arrangement in the plastids and on the plastid distribution in the tissue (Gross 1977).

Colour development results from the decrease of *a* and *b* chlorophyll, *β*,*ε*-carotenoids, *β*-carotene, neoxanthin and all *E*-violaxanthin, and the increase of *β*,*β*-xanthophylls (Gross 1981; Rodrigo et al. 2003, 2006; Alós et al. 2006; Gambetta 2009). Carotenoid composition and their concentration in the flavedo at maturity vary among species and cultivars and depend on environmental conditions. However, some evolution patterns have been described, as the phytoene and phytofluene colourless carotenoids decrease from green to orange fruit, remaining constant and in low levels along maturity. Lutein and *α*- and *β*-carotene diminish during colour development, whilst violaxanthin level increases spectacularly (Gross 1981; Rodrigo et al. 2003; Alós et al. 2006; Gambetta 2009). This pigment has been considered not only a basic chloroplast pigment but also a chromoplast one, representing the most abundant pigment in the peel of almost all *Citrus* species (Gross 1977). With the onset of colour break, new pigments could be identified, especially the orange *β*-cryptoxanthin and the red *β*-citraurin, which increase during this stage. Both pigments, which normally present in tiny amounts, impart the typical mature colour of oranges and mandarins (Gross 1981; Rodrigo et al. 2003; Alós et al. 2006). Moreover, when fruits regreen in the tree, their concentrations diminish paralleling flavedo colour, whilst *cis*-violaxanthin level remains constant (Gambetta 2009).

21.5.2 Hormones Associated to Fruit Colour Promotion

Ethylene is the main hormone associated to fruit maturation. Although *Citrus* fruit produces tiny amounts of it, exogenous ethylene stimulates fruit colour change (Pons et al. 1992; Goldschmidt et al. 1993). It promotes chlorophyll degradation by increasing the chlorophyllase de novo synthesis and activity (Purvis and Barmore 1981; Trebitsh et al. 1993; Fujii et al. 2007). Moreover, the application of ethylene action inhibitors (silver nitrate, norbornadiene or *1*-methylcyclopropene) inhibits chlorophyll loss from green harvested orange fruit (Goldschmidt et al. 1993; Porat et al. 1999). Exogenous ethylene also stimulates carotenoid biosynthetic pathway and increases ABA concentration in the flavedo (Rodrigo et al. 2006; Fujii et al. 2007; Rodrigo and Zacarías 2007).

Chlorophyllase, the first enzyme in chlorophyll breakdown, expresses constitutively during *Citrus* fruit development, and its expression does not appear to markedly increase towards maturation. However, exogenous ethylene increases the enzyme level, enhancing chlorophyll breakdown (Jacob-Wilk et al. 1999). Ethylene represses as well the transcription of most genes involved in photosynthesis and chloroplast biogenesis, such as chlorophyll *a*/*b*-binding protein, the photosystem I subunit and ribulose-1,5-bisphosphate carboxylase. Additionally, ethylene down-regulates chlorophyll biosynthesis, suppressing the transcription of magnesium chelatase, which mediates the insertion of Mg^{2+} into protoporphyrin IX, the unique enzyme of the chlorophyll biosynthetic pathway (Fujii et al. 2007). Thus, at the transcriptional level, ethylene is found to play binary roles related to chlorophyll, suppressing its biosynthesis and enhancing its degradation.

Exogenous ethylene could increase fruit total carotenoid concentration or not modify it, but it indeed affects carotenoid composition in the tissue. It up-regulates transcription of most carotenoid biosynthetic pathway genes, increasing 9-*cis*-violaxanthin, *β*-cryptoxanthin and *β*-citraurin and decreasing the concentration of chloroplastic carotenoids (Fujii et al. 2007; Rodrigo and Zacarías 2007). These data indicate that exogenous ethylene reproduces and accelerates the physiological and molecular changes related to carotenoid biosynthesis that naturally occur during *Citrus* fruit maturation.

To sum up, ethylene promotes fruit colour development, repressing transcription of most genes involved in photosynthesis, chloroplast biogenesis and chlorophyll synthesis, and promoting transcription of several genes related to chlorophyll degradation and carotenoid biosynthesis.

Abscisic acid is other hormone related to colour development, as it accumulates in the flavedo of *Citrus* fruits during natural maturation (Goldschmidt 1976; Lafuente et al. 1997; Rodrigo et al. 2003). Additionally, different transcripts of the CsNCED enzyme, which controls a critical step of ABA biosynthesis, increase from green to orange fruit (Rodrigo et al. 2006; Agustí et al. 2007). According to this, fruits of 'Pinalate', a yellow coloured mutant of 'Navelate' orange (*C. sinensis* [L.] Osb.), contain less xanthophylls and ABA concentration in the flavedo than parental fruits (Rodrigo et al. 2003). In addition, when 'Valencia' fruits remain on the tree until regreening, ABA concentration in the flavedo decreases, according to fruit colouration (Gambetta 2009). Nevertheless, Richardson and Cowan (1995) state that full colour development of *Citrus* fruit only occurs when flavedo ABA concentration declines.

Exogenous ABA has been found to increase (Rasmussen 1974) or not modify (Gambetta 2009) flavedo ABA concentration, but in both cases, it does not affect ethylene production or colour development. However, ABA injection through the peduncle increases ethylene production (Rasmussen 1974).

Considering all these results, it seems that the flavedo ABA increment, that attends colour development, is not enough to target the colour break process.

21.5.3 Hormones Associated to Fruit Colour Delay

Gibberellins antisenescence role has been studied during more than four decades; however, little information about endogenous bioactive GAs concentrations in this stage of fruit development has been obtained. Higher GA-like activity has been detected in fruits until 14 weeks after petal fall or the beginning of chlorophyll loss (García-Luis et al. 1985; Ali-Dinar et al. 1988; Murti 1989). The lowest GA-like activity has been cited at maturity (Kuraoka et al. 1977). Recently, it has been demonstrated that GA_1 and GA_4, main bioactive GAs in *Citrus*, rise in the flavedo approximately 15 days before colour change and then fall as rind colouration takes place. Additionally, fruit peduncle girdling, which delays fruit colouration, results in higher flavedo GA_1 and GA_4 concentration than control fruit. Therefore, it is proposed that the presence of GAs in the flavedo prevents the onset of fruit colour change (Gambetta et al. 2011).

Exogenous GA_3 applied previous to colour break delays colour development, reducing chlorophyll degradation (Agustí et al. 1988; Alós et al. 2006) and carotenoid biosynthesis (Lewis and Coggins 1964; García-Luis et al. 1986a). In addition, KGA application promotes flavedo *a* and *b* chlorophyll biosynthesis (Lewis et al. 1964). GA_3 also modifies carotenoid composition, maintaining the typical carotenoids of green fruit (lutein, α-carotene, β-carotene, all *E*-violaxanthins and neoxanthin), and reducing the *cis*-violaxanthin, β-citaurin and β-cryptoxanthin accumulation (Alós et al. 2006; Gambetta 2009). GA_3 counteracts ethylene-induced chlorophyll loss by reducing chlorophyllase mRNA and its activity (Trebitsh et al. 1993; Jacob-Wilk et al. 1999), as well as the carotenoid biosynthetic pathway (Rodrigo and Zacarías 2007; Fujii et al. 2008).

Foliar sprays of prohexadione-calcium, a GAs biosynthesis inhibitor, applied 2 weeks before harvest, enhance rind colour of 'Navelina' sweet orange (*C. sinensis* [L.] Osb.), promoting carotenoid biosynthesis (Barry and van Wyk 2004).

In summer, high levels of active GAs in the flavedo or GA_3 applications prevent colour change, delaying chlorophyll degradation and carotenoid biosynthesis.

Cytokinins effect on *Citrus* fruit colouration is scarce and non-consistent. Benzyladenine has been cited to delay colour break in oranges, associated to endogenous maintenance of GAs levels, prevention of ABA and ethylene increment (Van Staden et al. 1988) or diminishing slightly the ethylene-induced chlorophyllase activity (Trebitsh et al. 1993). On the contrary, no delay on colour change could be found after benzyladenine or kinetin application in Satsuma mandarin (García-Luis et al. 1986a).

Information about the effect of auxins on fruit maturation is scarce when compared with other developmental stages. Exogenous applications delay fruit senescence (Frenkel and Dick 1973) and fruit pre-harvest abscission (Cooper et al. 1968; Tzur and Goren 1977; Agustí et al. 2006). Therefore, 2,4-D or 2,4-DP (2-ethylhexyl ester of 2,4-dichlorophenoxy propionic acid) applications diminish pre-harvest drop in 'Washington navel' and 'Navelate' (Agustí et al. 2006). It has been demonstrated that 2,4-D inhibits cellulase activity (Goren 1993); the same way of action has been suggested for naphthalene acetic acid (NAA), 3,5,6-TPA and 2,4-DP (Anthony and Coggins 2001; Agustí et al. 2006). No information associating auxins with colour development is available.

Polyamines are reported to be effective antisenescence agents in plants, as they retard chlorophyll loss and membrane deterioration and increase RNAse and protease activities (Evans and Malmberg 1989). In lemons, putrescine (Put) application at colour break and at fully coloured fruit enhances fruit firmness and diminishes weight loss during postharvest. In mature fruits, this treatment delays colour development, diminishes endogenous ABA levels and increases Put and spermidine endogenous levels (Valero et al. 1998). Up to now, the role of polyamines is controversial since they are suggested to be a N source either than to have a hormonal role (Arias et al. 2005).

21.5.4 Organic and Mineral Nutrition Associated to Fruit Colour Development

Soluble sugar concentration increases in the flavedo during colour development, mainly as a result of reducer sugars increment (Huff 1984; Fidelibus et al. 2008; Gambetta et al. 2011). Chlorophyll content and sugars in the pericarp of 'Valencia' orange are negatively correlated (Huff 1984). Sucrose supplementation in vitro (Huff 1983) or in vivo (Iglesias et al. 2001) promotes flavedo degreening, increases endogenous sucrose and reduces nitrogen amounts. In addition, application of ethylene, which promotes colour break, down-regulates transcription of most genes involved in photosynthesis, chloroplast biogenesis and sugar metabolism, except for a sucrose transporter and acidic invertase genes which are up-regulated (Fujii et al. 2007). These effects are counteracted by GA_3 application (Fujii et al. 2008), which delays colour development. In maturing fruit of 'Fortune' mandarin (*C. clementina* Hort. ex Tan. × *C. tangerina* Hort. ex Tan), sucrose translocation rather than sucrose synthesis was reported to play a major role in maintaining flavedo sucrose levels (Holland et al. 1999). Exogenous GA_3, which delays colour break, diminishes sugar concentration in the flavedo, without modifying sucrose/hexose ratio, suggesting that colour delay in *Citrus* is partially mediated by fructose and glucose levels in the flavedo (Fidelibus et al. 2008; Gambetta 2009).

On the contrary, N metabolism seems to be involved on fruit colour delay. Late applications or high N fertilizations retard colour break (Jones and Embleton 1959; Sala et al. 1992; Quiñones et al. 2004) and increase the percentage of green fruit at harvest (Koo and Reese 1977). In addition, excessive N fertilizations promote on-tree fruit regreening (Jones and Embleton 1959). Nitrate applications previous to colour break delay chlorophyll loss and modify flavedo carotenoid composition and concentration, comparable to the effect of GA_3 sprays. Nitrogen applications delay the reduction of lutein, neoxanthin and all *E*-violaxanthin and the accumulation of *9-cis*-violaxanthin and *β*-crypoxanthin (Alós et al. 2006). Moreover, in vitro, N is the unique nutrient necessary to regreen *Citrus* epicarp discs (Huff 1983).

However, the association between endogenous N concentration and fruit colour development is not so clear. No changes in total proteins, amino acids or total N concentration during fruit maturation have been cited (Lewis et al. 1967; Huff 1984; Win et al. 2006). Nevertheless, in other experiments, lower total N concentration in mature fruits has been found. Proteinaceous fraction represents 98–99% of the total N in the flavedo, the other fraction is ammonium, as no nitrates or nitrites are detectable in the peel (Sala et al. 1992; Iglesias et al. 2001; Gambetta et al. 2011). In peduncle-girdled fruits, which remain green, N decreases until 30 days after treatment and then remains unchanged in accordance with colour development. Two months after girdling, peduncle-girdled fruits have higher flavedo N and active GA concentration, compared to control fruits, which reached orange colour (Gambetta et al. 2011). In addition, GA_3 applied previous to colour break, that delays fruit colour development, increases flavedo N concentration. These results suggest that N may not be a triggering factor controlling *Citrus* fruit external maturation (Gambetta 2009) and coincide with the proposal of that the abundance of N is important in regulating plastid metamorphosis in *Citrus* fruit, but accumulation of sugars in the epicarp is a major factor affecting seasonal changes in plastid form (Huff 1984).

Information about the association between other mineral nutrients and colour development is scarce and non-consistent. High K levels delay colour change and increase the proportion of green fruit at harvest (Embleton et al. 1967; Koo and Reese 1977). In 'Valencia' orange trees, K increases the proportion of regreened fruit (Reuther and Smith 1952). However, colour index of 'Nules' Clementine mandarin has been improved with three foliar applications of KNO_3 in spring, summer and autumn, with respect to untreated trees (Bañuls et al. 2004).

In general, colour development has been negatively associated to P applications. More intense colouration has been reached in P-deficient trees (Chapman and Reyner 1951), and when foliar concentration of P increase, a slight fruit regreening is observed (Embleton et al. 1971). However, different P-treated trees have no differences in orange (Koo and Reese 1976) or lemon colouration (Embleton et al. 1967).

Calcium is essential in the cell wall maintenance, especially reducing cuticle permeability, and it is efficient in controlling senescence-related physiological disorders (Zaragoza et al. 1996). Calcium chloride treatment, applied previous to colour break, delayed lemon fruit colouration and maintained fruit firmness during postharvest (Valero et al. 1998).

Considering that no mineral nutrient except for N provokes in vitro regreening of *Citrus* epicarp disc (Huff 1983), and the inconsistence of other mineral nutrient effects on fruit colouration, it is suggested that no mineral nutrients have a target role in *Citrus* fruit colour development.

21.5.5 Environmental Factors Associated to Fruit Colour Development

Different dates of *Citrus* fruit maturation and colour intensity obtained among years for the same cultivar in a given location, suggest environmental involvement on fruit colour regulation. Fruit maturation of most *Citrus* cultivars takes place during autumn–winter when tree reduces N absorption and transport, GAs synthesis and carbohydrate competence. In addition, fruits growing under tropical conditions do not reach complete colouration.

Studies carried out on sweet orange indicate that colour break takes place when air temperature diminishes below 12.8°C (Stearns and Young 1942), and that regreening of 'Valencia' orange is promoted by temperatures above 30°C/15°C (day/night) (Coggins et al. 1981). The greatest chlorophyll reduction and xanthophylls increment has been reported in trees exposed to a combination of cool day-air temperatures with cold night-air and soil temperatures (Young and Erickson 1961). Under controlled conditions, optimum soil temperature for the synthesis of violaxanthin, *β*-citraurin and *β*-cryptoxanthin seems to be between 12°C and 14°C (Sonnen et al. 1979). Soil temperature below 15°C or 22°C markedly diminishes root growth and respiration of different *Citrus* rootstocks (Bevington and Castle 1985; Poerwanto et al. 1989; Bryla et al. 2001). A reduction in the translocation of GA-like substances (Eilati et al. 1969a) and N (Wallace 1953) has been reported in *Citrus* plants in response to low temperatures, which could explain the promoting effect on colour development. In addition, in experiments carried out in young plants of citrange 'Carrizo', endogenous

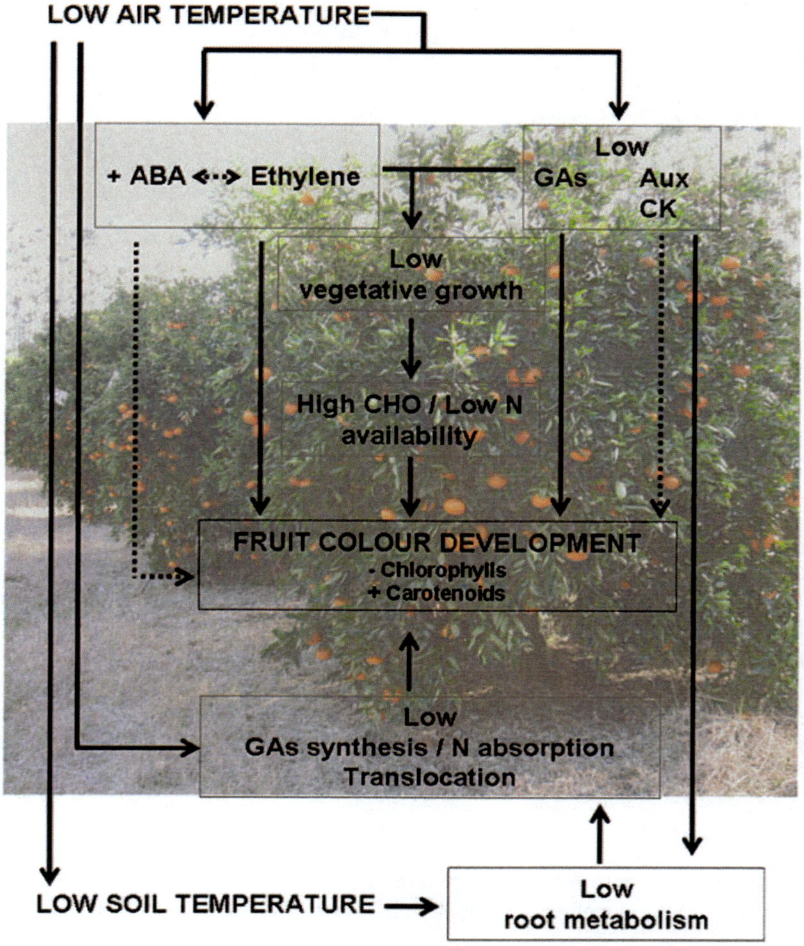

Fig. 21.6 Schematic representation of hormonal and nutritional regulation of the *Citrus* fruit colour development associated to the environmental control. *Solid lines* indicate more accepted regulation and *dotted lines* indicate suggested regulation

GA_1 diminishes in the canopy at 17°C/ 12°C day/night temperature with respect to 32°C/27°C as a consequence of the diminution of *CsGA20ox1* gene expression (Vidal et al. 2003). As air temperature decreases, the tree vegetative activity is reduced, and consequently fruit turns the main sink of carbohydrates; in addition, invertase activity and reducer sugars increase in fruit flavedo (Purvis and Grierson 1982). All these factors are positive associated to fruit colour development. Indeed, reducing soil temperature by covering soil with a white reflective mulch during the last 2 months before harvest, advances rind colour break and consequently harvest date of 'Clemenpons', a precocious Clementine mandarin (Mesejo et al. 2011).

To sum up, it seems that endogenous active GAs should decline in the flavedo, after which colour change may be stimulated by the basal level of endogenous ethylene. This hormonal change stimulates chlorophyll degradation and carotenoids biosynthesis, in parallel with sugar and ABA accumulation and N reduction in the flavedo. Therefore, GAs and ethylene are assumed to play a major role in the endogenous regulation of *Citrus* fruit colour development. Reduction of air and soil temperature is the main environmental signal triggering this physiological process (Fig. 21.6).

21.6 Concluding Remarks

The reviewed information provided in this chapter evidences the extensive bulk of research devoted to gain insight into the intricate hormonal and nutritional interactions, governing the key processes of *Citrus* flowering, fruit growth and development (Fig. 21.7). However, some issues remain rather obscure to achieve solid conclusions about these processes. Elucidating the role of other hormones (i.e. jasmonates, salicylic acid) related to stress and their interaction with further hormones and organic and mineral nutrition will clarify flowering, fruit set and development processes. Exciting ongoing research involving molecular signalling mediating

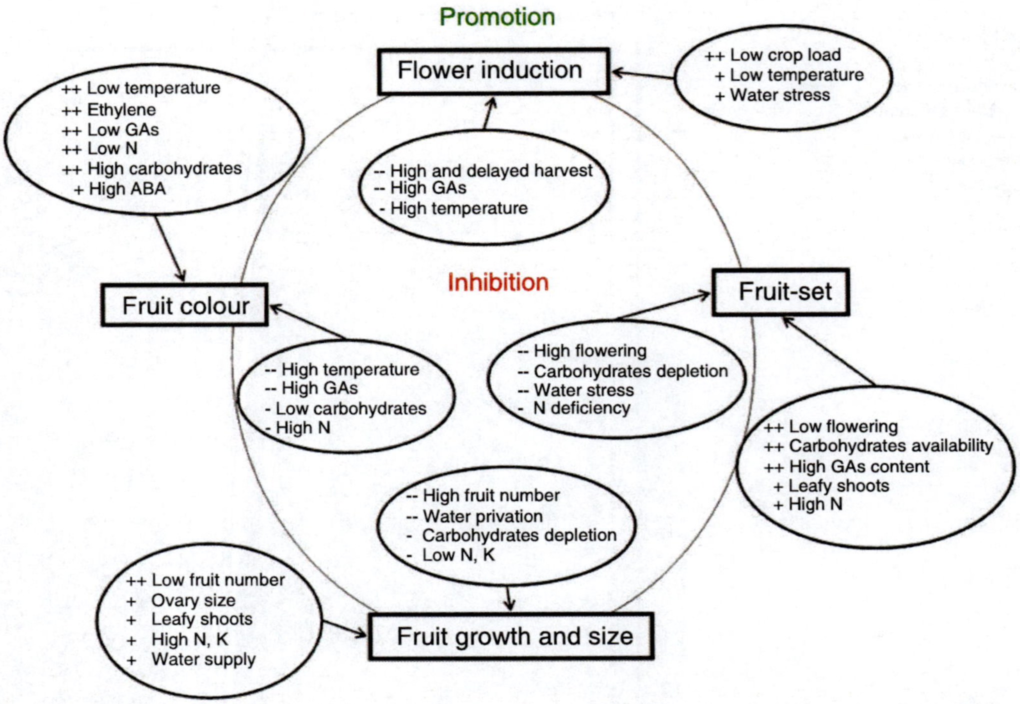

Fig. 21.7 Schematic proposal of hormonal and nutritional regulation of the reproductive process in *Citrus*. ++ Strong positive effect, + light positive effect, –– strong negative effect, –light negative effect

environmental, nutritional and physiological interactions will bring about novel data to expand the knowledge on the control of reproductive cycle in *Citrus*.

References

Abbott Ch (1935) Blossom-bud differentiation in citrus trees. Am J Bot 22:476–485

Agustí M (1980) Biología y control de la floración en el género Citrus. Tesis Doctoral, Universidad Politécnica de Valencia

Agustí M, Almela V (1991) Aplicación de fitorreguladores en citricultura. Ed. Aedos, Barcelona-España

Agustí M, García-Marí F, Guardiola JL (1982) The influence of flowering intensity on the shedding of reproductive structures in sweet orange. Sci Hortic 17:343–352

Agustí M, Almela V, Guardiola JL (1988) Aplicaciones de ácido giberélico para el control de alteraciones de la corteza de las mandarinas asociadas a la maduración. Invest Agr Prot Veg 3(2): 125–137

Agustí M, Almela V, Aznar M et al (1994) The use of 2,4-DP to improve fruit size in citrus. Proc Int Soc Citricult VII Int Citrus Cong Acireale Italia 1:423–427

Agustí M, Juan M, Martínez-Fuentes A et al (2006) Application of 2,4-dichlorophenoxypropionic acid and acid 2-ethylhexyl ester reduces mature fruit abscission in *Citrus* navel cultivars. J Hortic Sci Biotechnol 81(3):532–536

Agustí J, Zapater M, Iglesias DJ et al (2007) Differential expression of putative 9-cis-epoxycarotene dioxygenases and abscisic acid accumulation in water stressed vegetative and reproductive tissues of citrus. Plant Sci 172:85–94

Aharoni Y (1968) Respiration of oranges and grapefruit harvested at different stages of development. Plant Physiol 43:99–102

Ali-Dinar HM, Krezdorn AH, Wheaton TA (1988) The sexual-hormonal relation in citrus during fruit set. Acta Hortic 218:159–174

Alós E, Cercós M, Rodrigo MJ et al (2006) Regulation of color break in Citrus fruits. Changes in pigment profiling and gene expression induced by gibberellins and nitrate, two ripening retardants. J Agr Food Chem 54(13):4888–4895

Anthony FE, Coggins CW Jr (2001) NAA and 3,5,6-TPA control mature fruit drop in California citrus. Hortic Sci 36:1296–1299

Arias M, Carbonell J, Agustí M (2005) Endogenous free polyamines and their role in fruit set of low and high parthenocarpic ability citrus cultivars. J Plant Physiol 162:845–853

Ashraf MY, Gul A, Ashraf M et al (2010) Improvement in yield and quality of Kinnow (*Citrus deliciosa* x *Citrus nobilis*) by potassium fertilization. J Plant Nutri 33:1625–1637

Ayalon S, Monselise SP (1960) Flower bud induction and differentiation in the Shamouti orange. Am Soc Hortic Sci 75:216–221

Aznar M, Almela V, Juan M et al (1995) Effect of the synthetic auxin phenothiol on fruit development of 'Fortune' mandarin. J Hortic Sci 70:617–621

Bañuls J, Quiñones A, Primo-Millo E et al (2004) Complementary effect of foliar spray of KNO_3 on mineral status, yield and fruit quality in citrus orchard. Proc Int Soc Citricult II:686–687

Barry GH, van Wyk AA (2004) Novel approaches to rind colour enhancement of citrus. Proc Int Soc Citricult III:1076–1079

Becerra S, Guardiola JL (1984) Interrelationship between flowering and fruiting in sweet orange, cv. 'Navelina'. Proc Int Soc Citricult 1:190–194

Ben-Cheikh W, Perez-Botella J, Tadeo FR et al (1997) Pollination increases gibberellin levels in developing ovaries of seeded varieties of Citrus. Plant Physiol 114:557–564

Benhamou L, El-Otmani M, Goumari M et al (2004) The potential use of GA_3 and urea to manipulate flowering and reduce alternate-bearing pattern of the 'Nour' Clementine mandarin. Proc Int Soc Citricult I:479–483

Bevington KB, Castle WS (1985) Annual root growth pattern of young citrus trees in relation to shoot growth, soil temperature, and soil water content. J Am Soc Hortic Sci 110:840–845

Borroto CG, López M, Hidalgo O (1981) Efecto del estrés hídrico y la presencia de frutos de la cosecha anterior sobre el rendimiento de los naranjos Valencia. Centro Agrícola 2:43–55

Bryla DR, Bouma TJ, Hartmond U et al (2001) Influence of temperature and soil drying on respiration of individual roots in citrus: integrating greenhouse observations into a predictive model for the field. Plant Cell Environ 24:781–790

Bustan A, Goldschmidt EE (1998) Estimating the cost of flowering in a grapefruit tree. Plant Cell Environ 21:217–224

Bustan A, Erner Y, Goldschmidt EE (1995) Interactions between developing citrus fruits and their supportive vascular system. Ann Bot 76:657–666

Chapman HD, Parker ER (1942) Weekly absorption of nitrate by young bearing orange trees growing out of doors in solution cultures. Plant Physiol 17(8):366–376

Chapman HD, Reyner DS (1951) Effects of various levels of phosphate on the growth, yield, composition and quality of Washington navel oranges. Hilgardia 20:325–358

Coggins CW, Hall AE, Jones WW (1981) The influence of temperature on regreening and carotenoid content of the 'Valencia' orange rind. J Am Soc Hortic Sci 106:251–254

Cohen A (1984) Citrus fruit enlargement by means of summer girdling. J Hortic Sci 59(1):119–125

Cooper WC, Rasmussen GK, Rogers BJ et al (1968) Control of abscission in agricultural crops and its physiological basis. Plant Physiol 43:1560–1576

da Cunha Barros M, Gravina A (2006) Influencia del tipo de brote en el cuajado y crecimiento del fruto del tangor Ortanique. Agrociencia X 1:37–46

Davenport TL (1990) Citrus flowering. Hortic Rev 12:349–408

Eaks IL (1970) Respiratory response, ethylene production, and response to ethylene on citrus fruit during ontogeny. Plant Physiol 45: 334–338

Eilati SK, Goldschmidt EE, Monselise SP (1969a) Hormonal control of colour change in orange peel. Experientia 25:209–210

Eilati SK, Monselise SP, Budowski P (1969b) Seasonal development of external color and carotenoid content in the peel of ripening 'Shamouti' oranges. J Am Soc Hortic Sci 94(4):346–348

El-Otmani M, Agustí M, Aznar M et al (1993) Improving the size of fortune mandarin fruits by 2,4-DP. Sci Hortic 55:283–290

El-Otmani M, Coggins CH, Agustí M et al (2000) Plant growth regulators in citriculture: world current uses. Crit Rev Plant Sci 19(5): 395–447

El-Otmani M, ZharaTaibi F, Lmoufid B et al (2004) Improved use of foliar urea on Clementine mandarin to manipulate cropping in a sustainable production system. Acta Hortic 632:167–175

Embleton TW, Jones WW, Page AL (1967) Potassium and phosphorous effects on deficient Eureka lemon trees and some salinity problems. Proc Am Soc Hortic Sci 91:120–127

Embleton TW, Jones WW, Labanaukas CK (1971) Leaf analysis and phosphorous fertilization of oranges. Calif Citrogr 56:101–104

Erner Y, Shomer I (1996) Morphology and anatomy of stems and pedicels of spring flush shoots associated with citrus fruit set. Ann Bot 77:537–545

Erner Y, Artzi B, Tagari E et al (2000) Carbohydrate and vascular bundle effects on citrus fruit set. Proc Int Soc Citricult 1:693–698

Evans ML (1985) The action of auxin on plant cell elongation. Crit Rev Plant Sci 2:213–265

Evans TE, Malmberg RL (1989) Do polyamines have roles in plant developments?. Ann Rev Plant Physiol/ Plant Mol Biol 40:235–269

Fidelibus MW, Koch KE, Davies FS (2008) Gibberellic acid alters sucrose, hexoses, and their gradients in peel tissues during color break delay in 'Hamlin' orange. J Am Soc Hortic Sci 133(6): 706–767

Foyer CH, Galtier N (1996) Source–sink interaction and communication in leaves. In: Zaminski E, Schaffer AA (eds) Photoassimilate distribution in plants and crops: source–sink relations. Marcel Dekker, New York

Frenkel C, Dick R (1973) Auxin inhibition of ripening in Bartlett pears. Plant Physiol 51:6–9

Fujii H, Shimada T, Sugiyama A et al (2007) Profiling ethylene-responsive genes in mature mandarin fruit using a citrus 22K oligoarray. Plant Sci 173:340–348

Fujii H, Shimada T, Sugiyama A et al (2008) Profiling gibberellin (GA_3)-responsive genes in mature mandarin fruit using a citrus 22K oligoarray. Sci Hortic 116(3):291–298

Gambetta G (2009) Control endógeno y exógeno de la maduración externa de los frutos cítricos. Tesis Doctoral, Universidad Politécnica de Valencia España

Gambetta G, Espino M, Pardo E et al (2008) 'Montenegrina' Mandarin: characterization of the agronomic behaviour and fruit size improvement. Proc Int Soc Citricult I:561–566

Gambetta G, Martínez-Fuentes A, Bentancur O et al (2011) Hormonal and nutritional changes in the flavedo regulating rind colour development in sweet orange [*Citrus sinensis* (L.) Osb.)]. J Plant Growth Regul. doi:10.1007/s00344-011-9237-5

García-Luis A, Agustí M, Almela V et al (1985) Effect of gibberellic acid on ripening and peel puffing in Satsuma mandarin. Sci Hortic 27:75–86

García-Luis A, Fornes F, Guardiola JL (1986a) Effects of gibberellin A3 and citoquinins on natural and post-harvest, ethylene-induced pigmentation of Satsuma mandarin peel. Physiol Plant 68:271–274

García-Luis A, Almela V, Monerri C et al (1986b) Inhibition of flowering 'in vivo' by existing fruits and applied growth regulators in *Citrus unshiu*. Physiol Plant 66:515–520

García-Luis A, Kanduser M, Santamarina P et al (1992) Low temperature influence on flowering in *Citrus*. The separation of inductive and bud dormancy releasing effects. Physiol Plant 86:648–652

García-Luis A, Fornes F, Guardiola JL (1995) Leaf carbohydrates and flower formation in citrus. J Am Soc Hortic Sci 120(2):222–227

Goldschmidt EE (1976) Endogenous growth substances of Citrus tissues. Hortic Sci 11:95–99

Goldschmidt EE (1999) Carbohydrate supply as a critical factor for citrus fruit development and productivity. Hortic Sci 34(6):1020–1024

Goldschmidt EE, Huber SC (1992) Regulation of photosynthesis by end-product accumulation in leaves of plants storing starch, sucrose, and hexose sugars. Plant Physiol 99:1443–1448

Goldschmidt EE, Koch KE (1996) Citrus. In: Zaminski E, Schaffer AA (eds) Photoassimilate distribution in plants and crops: source–sink relations. Marcel Dekker, New York

Goldschmidt E, Monselise SP (1972) Hormonal control of flowering in Citrus and some other woody perennials. In: Carr DJ (ed) Plant growth substances. Springer, New York

Goldschmidt EE, Aschkenazi N, Herzano Y et al (1985) A role for carbohydrate levels in the control of flowering in Citrus. Sci Hortic 26:159–166

Goldschmidt EE, Huberman M, Goren R (1993) Probing the role of endogenous ethylene in the degreening of citrus fruit with ethylene antagonist. Plant Growth Regul 12:325–329

Gómez-Cadenas A, Mehouachi J, Tadeo FR et al (2000) Hormonal regulation of fruitlet abscission induced by carbohydrate shortage in citrus. Planta 210:636–643

Goren R (1993) Anatomical, physiological and hormonal aspects on abscission in citrus. Hortic Rev 15:145–182

Gravina A (2007) Aplicación del ácido giberélico en Citrus: revisión de resultados experimentales en Uruguay. Agrociencia XI 1:57–66

Gravina A, Arbiza H, Juan M et al (1996) Flowering-fruiting interrelationships in the 'Ellendale' tangor under the growing conditions of Spain and Uruguay. Proc Int Soc Citricult II:1081–1085

Gravina A, Arbiza H, Telias A et al (2004) Harvest date effect on fruit quality and return bloom in three citrus cultivars. Proc Int Soc Citricult 1:284–290

Greenberg J, Goldschmidt EE, Goren R (1993) Potential and limitations of the use of paclobutrazol in Citrus orchards in Israel. Acta Hortic 329:58–61

Gross J (1977) Carotenoid pigments in citrus. In: Nagy S (ed) Citrus science and technology, vol I. AVI, Westport, pp 302–354

Gross J (1981) Pigment changes in the flavedo of Dancy tangerine (*Citrus reticulata*) during ripening. Z Pflanzenphysiol 103: 451–457

Guardiola JL, Monerri C, Agustí M (1982) The inhibitory effect of gibberellic acid of flowering in Citrus. Physiol Plant 55:136–142

Guardiola JL, García-Marí F, Agustí M (1984) Competition and fruit set in the Washington navel orange. Physiol Plant 2:297–302

Guardiola JL, Barres MT, Albert C et al (1993) Effects of exogenous growth regulators on fruit development in *Citrus unshiu*. Ann Bot 71:169–176

Harty AR, Van Staden J (1988) Paclobutrazol and temperature effects on lemon. In: Proceedings of the sixth international citrus congress, Aviv, pp 343–353

Holland N, Sala JM, Menezes HC et al (1999) Carbohydrate content and metabolism as related to maturity and chilling sensitivity of Cv. Fortune mandarins. J Agr Food Chem 47:2513–2518

Huerta L, Forment J, Gadea J et al (2008) Gene expression analysis in citrus reveals the role of gibberellins on photosynthesis and stress. Plant Cell Environ 31:1620–1633

Huff A (1983) Nutritional control of regreening and degreening in *Citrus* peel segments. Plant Physiol 73:243–249

Huff A (1984) Sugar regulation of plastid interconversions in the epicarp of *Citrus* fruit. Plant Physiol 76:307–312

Iglesias DJ, Tadeo FR, Legaz F et al (2001) In vivo sucrose stimulation of colour change in citrus fruit epicarps: interactions between nutritional and hormonal signals. Physiol Plant 112:244–250

Iglesias DJ, Lliso I, Tadeo FR et al (2002) Regulation of photosynthesis through source:sink imbalance in citrus is mediated by carbohydrate content in leaves. Physiol Plant 116:563–572

Iwahori S, Oohata J (1981) Control of flowering of 'Satsuma' mandarins (*Citrus unshiu* Marc) with gibberellin. Proc Int Soc Citricult I:247–249

Iwahori S, García-Luis A, Santamarina P et al (1990) The influence of ringing on bud development and flowering in Satsuma mandarin. J Exp Bot 41:1341–1346

Jacob-Wilk D, Holland D, Goldschmidt EE et al (1999) Chlorophyll breakdown by chlorophyllase: isolation and functional expression of the *Chlase1* gene from ethylene-treated *Citrus* fruit and its regulation during development. Plant J 20(6):653–661

Jones WW, Embleton TW (1959) The visual effect of nitrogen nutrition on fruit quality of 'Valencia' oranges. Proc Am Soc Hortic Sci 73:234–236

Jones WW, Embleton TW, Coggins CW (1975) Starch content of rootstock of 'Kinnow' mandarin trees bearing fruit in alternate years. Hortic Sci 10(5):514

Khairi MM, Hall AE (1976) Effect of air and soil temperatures on vegetative growth of Citrus. J Am Soc Hortic Sci 101(4): 337–341

Koo RCJ, Reese RL (1976) Influence of fertility and irrigation treatments on fruit quality of Temple orange. Proc Fla State Hortic Sci Soc 89:49–51

Koo RCJ, Reese RL (1977) Influence of nitrogen, potassium, and irrigation on citrus fruit quality. Proc Int Soc Citricult I:34–38

Koshita Y, Takahara T, Ogata T et al (1999) Involvement of endogenous plant hormones (IAA, ABA, GAs) in leaves and flower bud formation of satsuma mandarin. Sci Hortic 79:185–194

Kozlowski TT, Pallardy SG (1997) Photosynthesis. In: Physiology of woody plants, 2nd edn. Academic, San Diego

Krajewsky A, Rabe E (1995) Citrus flowering: a critical evaluation. J Hortic Sci 70(3):357–374

Kriedemann PE (1969) ^{14}C-translocation in orange plants. Aust J Agr Res 20:291–300

Krueger RR, Arpaia ML (2008) Seasonal uptake of nutrients by mature field-grown 'Valencia' (*Citrus sinensis* O.) trees in California. Proc Int Soc Citricult 1:588–594

Kuraoka T, Iwasaki K, Ishii T (1977) Effects of GA_3 on puffing and levels of GA3-like substances and ABA in the peel of Satsuma mandarin (*Citrus unshiu* Marc.). J Am Soc Hortic Sci 102(5): 651–654

Lafuente MT, Martínez TM, Zacarías L (1997) Abscisic acid in the response of Fortune mandarins to chilling. Effect of maturity and high-temperature conditioning. J Sci Food Agr 73:494–502

Lavon R, Goldschmidt EE, Salomon R et al (1995) Effect of potassium, magnesium and calcium deficiencies on carbohydrate pools and metabolism in *Citrus* leaves. J Am Soc Hortic Sci 120(1):54–58

Layne DR, Flore JA (1995) End-product inhibition of photosynthesis in *Prunus cerasus* L. in response to whole-plant source–sink manipulation. J Am Soc Hortic Sci 120:583–599

Leegood RC (1996) Primary photosynthate production: physiology and metabolism. In: Zaminski E, Schaffer AA (eds) Photoassimilate distribution in plants and crops: source–sink relations. Marcel Dekker, New York

Lewis LN, Coggins CW (1964) The inhibition of carotenoid accumulation in navel oranges by gibberellin A_3, as measured by thin layer chromatography. Plant Cell Physiol 5:457–463

Lewis LN, Coggins CW, Garberg MJ (1964) Chlorophyll concentration in the navel orange rind as related to potassium gibberellate, light intensity and time. J Am Soc Hortic Sci 84:177–180

Lewis LN, Coggins CW Jr, Labanauskas CK et al (1967) Biochemical changes associated with natural and gibberellin A_3 delayed senescence in the Navel orange rind. Plant Cell Physiol 8:151–160

Lord EM, Eckard KJ (1987) Shoot development in *Citrus sinensis* L. (Washington navel orange). II. Alterations of developmental fate of flowering shoots after GA treatment. Bot Gaz 148:17–22

Lovatt C, Zheng Y, Hake K (1988) A new look at the Kraus-Kraybill hypothesis and flowering in citrus. Proc Int Soc Citricult I:475–483

Manzi M (2011) Respuesta metabólica y reproductiva de dos variedades de cítricos bajo estrés hídrico. Tesis de Maestría, Facultad de Agronomía Uruguay

Martínez-Fuentes A, Mesejo C, Juan M et al (2002) Restrictions on the exogenous control of flowering in citrus. Acta Hortic 632:91–98

Martínez-Fuentes A, Mesejo C, Reig C et al (2010) Timing of the inhibitory effect of fruit on return bloom of 'Valencia' sweet orange (*Citrus sinensis* (L.) Osbeck). J Sci Food Agr 90(11): 1936–1943

Mehouachi J, Tadeo FR, Zaragoza S et al (1996) Effects of gibberellic acid and paclobutrazol on growth and carbohydrate accumulation in shoots and roots of citrus rootstock seedlings. J Hortic Sci 71(5):747–754

Mehouachi J, Iglesias D, Agustí M et al (2009) Delay of early fruitlet abscission by branch girdling in citrus coincides with previous increases in carbohydrate and gibberellin concentrations. Plant Growth Reg 58:15–23

Mesejo C, Martínez-Fuentes A, Juan M et al (2003) Vascular tissues development of citrus fruit peduncle is promoted by synthetic auxins. Plant Growth Regul 39:131–135

Mesejo C, Gambetta G, Gravina A et al (2011) Relationship between soil temperature and fruit colour development of 'Clemenpons' Clementine mandarin (*Citrus clementina* Hort ex. Tan). J Sci Food Agr 91. doi:10.1002/jsfa.4600

Monerri C, Fortunato-Almeida A, Molina RV et al (2011) Relation of carbohydrate reserves with the forthcoming crop, flower formation and photosynthetic rate, in the alternate bearing 'Salustiana' sweet orange (*Citrus sinensis* L.). Sci Hortic 129:71–78

Monselise SP (1977) Citrus fruit development, endogenous systems, and external regulation. Proc Int Soc Citricult 2:664–668

Moss GI (1976) Temperature effects on flower initiation in sweet orange (*Citrus sinensis*). Aust J Agr Res 27:399–407

Moss GI (1977) Major factors influencing flower formation and subsequent fruit-set of sweet orange. Proc Int Soc Citricult II: 215–222

Murti GSR (1989) Studies on gibberellin-like substances in acid lime (*Citrus aurantifolia* Swingle). Indian J Plant Physiol 32:57–64

Nishikawa F, Endo T, Shimada T et al (2007) Increased *CiFT* abundance in the stem correlates with floral induction by low temperature in Satsuma mandarin (*Citrus unshiu* Marc.). J Exp Bot 58(14):3915–3927

Okuda H (2000) A comparison of IAA and ABA levels in leaves and roots of two citrus cultivars with different degrees of alternate bearing. J Hortic Sci Biotechnol 75(3):355–359

Okuda H, Kihara T, Iwagaki I (1996) Effects of paclobutrazol application to soil at the beginning of maturation on sprouting, shoot growth, flowering and carbohydrate contents in roots and leaves of satsuma mandarin. J Hortic Sci 71(5):785–789

Paul MJ, Pellny TK (2003) Carbon metabolite feedback regulation of leaf photosynthesis and development. J Exp Bot 54:539–547

Plummer JA, Mullins MG, Vine JH et al (1989) The role of endogenous hormones in shoot emergence and abscission in alternate bearing 'Valencia' orange trees. Acta Hortic 239:341–344

Poerwanto R, Inoue H (1990) Effect of air and soil temperatures on flower development and morphology of Satsuma mandarin. J Hortic Sci 65(6):739–745

Poerwanto R, Inoue H, Kataoka I (1989) Effect of temperature on the morphology and physiology of the roots of trifoliate orange budded with satsuma mandarin. J Jpn Soc Hortic Sci 58(2):267–274

Pons J, Almela V, Juan M et al (1992) Use of ethephon to promote colour development in early ripening clementine cultivars. Proc Int Soc Citricult I:459–462

Porat R, Weiss B, Cohen L et al (1999) Effects of ethylene and 1-methylcyclopropene on the postharvest qualities of 'Shamouti' oranges. Postharvest Biol Technol 15:155–163

Purvis AC, Barmore Ch (1981) Involvement of ethylene in chlorophyll degradation in peel of *Citrus* fruits. Plant Physiol 68:854–856

Purvis AC, Grierson W (1982) Accumulation of reducing sugar and resistance of grapefruit peel to chilling injury as related to winter temperatures. J Am Soc Hortic Sci 107:139–142

Quiñones A, González MC, Montaña C et al (2004) Fate and uptake efficiency of 15N applied with different seasonal distributions in Citrus trees. Proc Int Soc Citricult II:587–592

Rabe E (1994) Yields benefits associated with pre-blossom low-biuret urea sprays on *Citrus spp*. J Hortic Sci 69(3):495–500

Rabe E, van Rensburg P (1996) Gibberellic acid sprays, girdling, flower thinning and potassium applications affect fruit splitting and yield in 'Ellendale' tangor. J Hortic Sci 71(2):195–203

Rasmussen GK (1974) Cellulase activity in separation zones of citrus fruit treated with abscisic acid under normal and hypobaric atmospheres. J Am Soc Hortic Sci 99:229–231

Reuther W, Smith PF (1952) Relation of nitrogen, potassium and magnesium fertilization to some fruit quality of Valencia orange. Proc Am Soc Hortic Sci 59:1–12

Richardson GR, Cowan AK (1995) Abscisic acid content of Citrus flavedo in relation to colour development. J Hortic Sci 70:769–773

Rivas F, Arbiza H, Borges A et al (2006a) Caracterización Del comportamiento reproductivo de la mandarina Nova en el sur de Uruguay. TodoCitrus 32:22–31

Rivas F, Erner Y, Alós E et al (2006b) Girdling increases carbohydrate availability and fruit-set in citrus cultivars irrespective of parthenocarpic ability. J Hortic Sci Biotechnol 81:289–295

Rivas F, Gravina A, Agustí M (2007) Girdling effects on fruit set and quantum yield efficiency of PSII in two Citrus cultivar. Tree Physiol 27:527–535

Rivas F, Martínez-Fuentes A, Mesejo C et al (2010) Efecto hormonal y nutricional del anillado en frutos de diferentes tipos de brotes de cítricos. Agrociencia XIV(I):8–14

Rodrigo MJ, Zacarías L (2007) Effect of postharvest ethylene treatment on carotenoid accumulation and the expression of carotenoid biosynthetic genes in the flavedo of orange (*Citrus sinensis* L. Osbeck) fruit. Postharvest Biol Technol 43(1):14–22

Rodrigo MJ, Marcos JF, Alférez F et al (2003) Characterization of Pinalate, a novel *Citrus sinensis* mutant with a fruit-specific alteration that results in yellow pigmentation and decreased ABA content. J Exp Bot 54(383):727–738

Rodrigo MJ, Alquezar B, Zacarías L (2006) Cloning and characterization of two 9-cis-epoxycarotene dioxygenase genes, differentially regulated during fruit maturation and under stress conditions, from orange (*Citrus sinensis* L. Osbeck). J Exp Bot 57(3):633–643

Ruan YL (1993) Fruit set, young fruit and leaf growth of *Citrus unshiu* in relation to assimilate supply. Sci Hortic 53:99–107

Ruiz R, Guardiola JL (1994) Carbohydrate and mineral nutrition of orange fruitlets in relation to growth and abscission. Physiol Plant 90:27–36

Sala JM, Cuñat P, Collado M et al (1992) Effect on nitrogenous fertilization (quantity and nitrogen form) in precocity of colour change of 'Navelina' oranges. Proc Int Soc Citricult II:598–602

Sanz A, Monerri C, González-Ferrer J et al (1987) Changes in carbohydrates and mineral elements in Citrus leaves during flowering and fruit set. Physiol Plant 69:93–98

Schaffer AA, Goldschmidt EE, Goren R et al (1985) Fruit set and carbohydrate status in alternate and non-alternate bearing Citrus cultivars. J Am Soc Hortic Sci 110:574–578

Shuang H, Li-Song Ch, Huan-Xin J et al (2008) Boron deficiency decreases growth and photosynthesis, and increases starch and hexoses in leaves of citrus seedlings. J Plant Physiol 165:1331–1341

Shuang H, Tang N, Jiang HX et al (2009) CO_2 assimilation, photosystem II photochemistry, carbohydrate metabolism and antioxidant system of citrus leaves in response to boron stress. Plant Sci 176:143–153

Sonnen HD, Lenz F, Gross J (1979) Influence of root temperature on carotenoid development in the peel of *Citrus unshiu* and *Citrus madurensis*. Gartenbauwissenschaft 44:49–52

Southwick SM, Davenport T (1987) Modification of the water stress induced floral response in 'Tahiti' lime. J Am Soc Hortic Sci 112(2):231–236

Spiegel-Roy P, Goldschmidt EE (1996) Biology of Citrus. Cambridge University Press, Cambridge

Srivastava LM (2002) Plant growth and development. Hormones and environment. Academic, London

Srivastava AK, Singh S, Huchche AD (2000) An analysis of citrus flowering – a review. Agr Rev 21(1):1–15

Stearns CR, Young GT (1942) The relation of climatic conditions to color development in citrus fruit. Proc Fla State Hortic Soc 56:39–61

Syvertsen JP, Goñi C, Otero A (2003) Fruit load and canopy shading affect leaf characteristics and net gas exchange of 'Spring' navel orange trees. Tree Physiol 23:899–906

Talón M, Zacarías L, Primo-Millo E (1990) Hormonal changes associated with fruit set and development in mandarins differing in their parthenocarpic ability. Physiol Plant 79:400–406

Talón M, Mehouachi J, Iglesias DJ et al (2000) Citrus fruitlet abscission: physiological bases supporting the "competition hypothesis". Proc Int Soc Citricult I:602–604

Tamim M, Altman A, Goren R et al (1996) Modification of the time and intensity of flowers in Citrus cultivars by water stress, light, low temperatures and growth regulators. Proc Int Soc Citricult II:945–948

Trebitsh T, Goldschmidt EE, Riov J (1993) Ethylene induces *de novo* synthesis of chlorophyllase, a chlorophyll degrading enzyme, in *Citrus* fruit peel. Proc Nat Acad Sci USA 90:9441–9445

Tzur A, Goren R (1977) Reducing preharvest drop in 'Temple' orange fruits by 2,4-D. Role of cellulase in the calyx abscission zone. Sci Hortic 7:237–248

Urban L, Léchaudel M, Lu P (2004) Effect of fruit load and girdling on leaf photosynthesis in *Mangifera indica* L. J Exp Bot 55:2075–2085

Valero D, Martínez-Romero D, Serrano M et al (1998) Influence of postharvest treatment with putrescine and calcium on endogenous polyamines, firmness and abscissic acid in lemon (*Citrus limon* L. Burm Cv. Verna). J Agr Food Chem 46:2102–2109

Valiente J, Albrigo G (2004) Flower bud induction of sweet orange trees *Citrus sinensis* (L.) Osbeck: effect of low temperatures, crop load, and bud age. J Am Soc Hortic Sci 129(2):158–164

Van Staden J, Cook E, Nooden LD (1988) Citokinins and senescense. In: Senescence and aging in plants. Academic, San Diego

Verreynne JS, Lovatt CJ (2009) The effect of crop load on budbreak influences return bloom in alternate bearing 'Pixie' mandarin. J Am Soc Hortic Sci 134:299–307

Vidal AM, Ben-Cheikh W, Talon M et al (2003) Regulation of gibberellin 20-oxidase gene expression and gibberellin content in citrus by temperature and citrus exocortis viroid. Planta 217:442–448

Wallace A (1953) Nitrogen absorption and translocation by citrus cuttings at different root temperatures. Am Soc Hortic Sci 61: 89–94

Win TO, Srilaong V, Kyu KL et al (2006) Biochemical and physiological changes during chlorophyll degradation in lime (*Citrus aurantifolia* Swingle cv. 'Paan'). J Hortic Sci Biotechnol 81(3): 471–477

Wünsche JN, Palmer JW, Greer DH (2000) Effect of crop load on fruiting and gas-exchange characteristics of 'Braeburn'/M.26 apple trees at full canopy. J Am Soc Hortic Sci 125:93–99

Young LB, Erickson LC (1961) Influence of temperature on colour change in Valencia oranges. Proc Am Soc Hortic Sci 78:197–200

Yuan L, Xu DQ (2001) Stimulation effect of gibberellic acid short term treatment on leaf photosynthesis related to the increase in Rubisco content in broad bean and soybean. Photosynth Res 68:39–47

Zacarías L, Talón M, Ben-Cheikh W et al (1995) Abscisic acid increases in non-growing and paclobutrazol-treated fruits of seedless mandarins. Physiol Plant 95:613–619

Zaragoza S, Almela V, Tadeo R et al (1996) Effectiveness of calcium nitrate and GA_3 on the control of peel-pitting of 'Fortune' mandarin. J Hortic Sci 71(2):321–326

Zucconi F, Monselise SP, Goren R (1978) Growth abscission relationships in developing orange fruit. Sci Hortic 9:137–146

Lime-Induced Iron Chlorosis in Citrus: Diagnosis Through Physiological and Metabolic Evidences

22

A. Jiménez, A. Torrecillas, F. Sevilla, M.F. Ortuño, W. Conejero, F. Ferreres, S. Medina, A. Galindo, and A. Gil-Izquierdo

Abstract

Iron is essential for plants nutrition, being required in small but in critical concentrations for plant growth and crop productivity. Despite total iron content in soils is much higher than that required by plants, soil iron bioavailability is low, particularly in calcareous soils. For this, many agricultural crops worldwide growing in semiarid climates and calcareous soils suffer from iron deficiency (iron chlorosis or lime-induced chlorosis). In this sense, chlorotic leaves may contain as much or more total iron than healthy leaves of the same age (*chlorosis paradox*). In this chapter, we discuss the strengths and weaknesses of the leaf visual rating of chlorosis, floral mineral analysis and leaf active iron for diagnosing citrus iron nutritional status. The effect of iron chlorosis on yield and physical and chemical characteristics of citrus fruits was also reviewed. We analyse the metabolic response of citrus to iron deficiency, paying special attention to the inducible mechanisms developed under low-iron stress and to the metabolic changes able to be used to evaluate the iron deficiency. Finally, we reviewed the bioactive compounds in the citrus fruits, paying special attention to the fact that iron deficiency enhances phenolics and, as a consequence, health-promoting effects of the fruits for the human body.

Keywords

Citrus • Iron • Chlorosis • Fruit quality • Plant metabolism • Phenolics • Human nutrition • Health effects • Omics • Genomics • Transcriptomics • Proteomics • Metabolomics

22.1 Introduction

Iron is present in the earth's crust mainly in the form of ferromagnesium silicates and is the fourth most abundant element in the lithosphere. Also, iron is an essential nutrient for plants, classified inside the *trace elements* or *micronutrients* because it is required in small but in critical concentrations for plant growth and crop productivity. Although, in most soils, total iron content is much higher than that required by plants, soil iron bioavailability is low, particularly in calcareous soils (Imsande 1998). Calcium carbonate, present in great amounts in calcareous soils, and the resulting high level of bicarbonate ions, are the main causes of low soil iron bioavailability. For this, many agricultural crops worldwide growing in semiarid climates and calcareous soils, which cover over 25–30% of the land surface of the earth (Boxma 1972), suffer from iron deficiency (commonly known as iron chlorosis or lime-induced chlorosis). In this sense, it is estimated that in the Mediterranean region, iron chlorosis is one of the major abiotic stresses affecting fruit trees (Forner-Giner et al. 2010) since as high as 20–50% of fruit trees

A. Jiménez • A. Torrecillas • F. Sevilla • M.F. Ortuño • W. Conejero • F. Ferreres • S. Medina • A. Galindo • A. Gil-Izquierdo (✉)
Departments of Biología del Estrés y Patología Vegetal, Riego and Ciencia y Tecnología de Alimentos, Centro de Edafología y Biología Aplicada del Segura (CSIC), P.O. Box 164, E-30100 Espinardo (Murcia), Spain
e-mail: ajimenez@cebas.csic.es; atorreci@cebas.csic.es; fsevilla@cebas.csic.es; mfortuno@cebas.csic.es; wenceslao@cebas.csic.es; federico@cebas.csic.es; smescudero@cebas.csic.es; agalindo@cebas.csic.es; angelgil@cebas.csic.es

A.K. Srivastava (ed.), *Advances in Citrus Nutrition*,
DOI 10.1007/978-94-007-4171-3_22, © Springer Science+Business Media B.V. 2012

grown in the Mediterranean basin are reported to suffer on account of iron deficiency (Jaegger et al. 2000).

Economically, fertilizers alone constitute about 20–30% of total cost of citrus production (Srivastava et al. 2008). Therefore, the diagnosis of nutrient constraints and their efficient remediation are two pillars of an effective citrus nutrition programme. Iron chlorosis affects several metabolic processes in roots and leaves, leading mainly to the yellow colour characteristic of chlorotic leaves, a delay in fruit ripening and a decrease in yield and quality of fruits. In this background information, our emphasis is to highlight the major physiological and metabolic aspects of lime-induced chlorosis in citrus, reviewing current methods to diagnose iron deficiency and its effects on metabolism, yield and fruit quality. Besides the final goal of citrus fruit for human consumption, it is also intended to review the evolution of bioactive compounds caused by this type of stress in citrus fruit and its practical application in human health.

Fig. 22.1 Iron deficiency symptoms in lemon tree (*above*) and the same tree after iron chelate application to the soil (*below*)

22.2 Plant Indicators of Iron Nutritional Status

22.2.1 Leaf Visual Rating of Chlorosis

Visible symptoms of iron deficiency in citrus leaves are similar to those found in many other crop plants (Figs. 22.1 and 22.2), namely, green veins sharply distinguished from a less green or yellow mesophyll (Maldonado-Torres et al. 2006). This decrease in chlorophyll content in leaves in response to iron deficiency is probably the oldest known indicator of a nutritional disorder (Gris 1844). When these leaf symptoms are distinguishable, the iron chlorosis can be visually rated using a scale from 1 (no yellowing) to 5 (severe yellowing with some necrosis) according to large number of studies (Rodriguez de Cianzio et al. 1979; Romera et al. 1991; Sanz and Montañés 1997). When the rating is made by trained personnel, it is possible evaluate the chlorophyll content with a good accuracy (Rodriguez de Cianzio et al. 1979). However, visual ratings are subjective measurements, and therefore, accuracy may be reduced if evaluation is performed in different growing environments and by different individuals (Berrang and Steiner 1980; Campbell et al. 1990). Nevertheless, this method has been considered as the simplest and most rapid method (Rodriguez de Cianzio et al. 1979; Mansur 1992) for screening genotypes for their ability to tolerate iron deficiency conditions (Table 22.1).

22.2.2 Leaf and Floral Mineral Analysis

It is recognized that the tree itself is the best indicator of its nutritional status since it integrates all the factors that influence soil nutrient availability and plant uptake. Since the plant nutritional status controls many physiological processes and crop productivity, leaf mineral analysis can be highly useful to monitor the nutritional status of crops and to select the right type and amount of fertilizer for a rational fertilization programme, avoiding negative environmental impacts arising due to excessive fertilizer applications.

In comparison with annual plants, fruit trees are more susceptible to iron chlorosis mainly due to differences in life cycle, plant size, characteristics of the root system, scion-rootstock compatibility, etc. Moreover, iron deficiency in trees affects the nutritional balance in the following year since new growth depends on iron stored in the plant (Pestana et al. 2004b).

Leaf iron analysis presents limitations when applied to diagnosis of lime-induced chlorosis since in many field-grown plants, there is no correlation between leaf total iron concentration and the degree of chlorosis (Table 22.1). Thus, chlorotic leaves may contain as much or more total iron than healthy leaves of the same age (Wallihan 1955; Bar-Akiva 1961; DeKock et al. 1979; Morales et al. 1998). This behaviour has been identified as *chlorosis paradox* by Römheld (2000) and can be related to the localization and binding state of iron in leaves. In this context, a proportion of iron might be insoluble in the apoplasm of leaves and might not

Fig. 22.2 Different levels of iron chlorosis in lemon leaves

be physiologically functional (Mengel and Geurtzen 1988; Römheld 2000).

To avoid this paradox, Hellín et al. (1984b) and Wallace (1990) proposed the use of nutrient ratios to interpret iron foliar contents in citrus. The ratios P:Fe, K:Ca, Fe:Mn and Zn:Fe state nutritional imbalances that appear prominently when iron immobilization takes place (Chouliaras et al. 2004; Pestana et al. 2004b). However, according to Chaney (1984), no absolute values could be established for any of the ratios to enable the precise diagnosis of iron chlorosis.

Sanz et al. (1997a, b) stated that floral analysis could be used for the *subclinical* iron chlorosis diagnosis since the mineral composition of flowers at full bloom is often related to the nutrient content in leaves taken later in the season. Pestana et al. (2004a) showed that the model that suitably predicted iron chlorosis in orange trees was based on the Mg:Zn ratio and proposed critical values for this ratio. A ratio below 100 indicates that trees will develop iron chlorosis, while with flowers having a ratio above 200, the leaves will remain fully green (Table 22.1). These authors also concluded that the use of floral rather than foliar analysis can bring forward the diagnosis of iron chlorosis from July to April, and that an early prognosis of iron chlorosis can benefit growers, since it allows them to detect and correct any deficiencies before fruit set, thus giving sufficient time for nutrient applications to improve yield and fruit quality.

22.2.3 Leaf Active Iron

In spite of the fact that most of plant nutritional disorders induce leaf chlorosis, when it is known that iron deficiency is the cause of a nutritional disorder, the most accurate and useful diagnostic method to quantify the deficiency degree is the spectrometric measurement of leaf chlorophyll (Bar-Akiva and Lavon 1967; Rodriguez de Cianzio et al. 1979; Chen and Barak 1982; Torrecillas et al. 1984). However, this technique possesses many disadvantages because it is labour-intensive, time-consuming and requires destructive sampling of plant material. In contrast, using a portable leaf chlorophyll meter (SPAD, soil plant analysis development meter), it is possible for more rapid, convenient and non-destructive measurement of leaf chlorophyll content (Table 22.1) as per number of studies (Porro et al. 2001; Pestana et al. 2004a, b). This device measures the absorbances of the leaf in the red and near-infrared region, and using these two transmittances, it calculates a numerical SPAD value which is proportional to the chlorophyll present in the leaf.

Taking into account that total concentration of iron in plant leaves is not a valid index for iron nutritional status of crops (Römheld 2000; Sönmez and Kaplan 2004; Fuentes et al. 2012); some authors (Table 22.1) proposed to extract *active iron*, or its derivative, through chemical fractionation procedures using weak acids and some chelating agents, mainly 1 N HCl, diethylene triamine pentaacetic acid (DTPA) or 1.5% *o*-phenanthroline at pH 3 as extractants (Oserkowsky 1933; Llorente et al. 1976; Katyal and Sharma 1980; Basar 2003; Sönmez and Kaplan 2004). Llorente et al. (1976) showed that chlorophyll content and peroxidase activity of lemon leaves are proportional to leaf *active iron* evaluated by extraction with 1 N HCl. Sudahono et al. (1994) in a sand culture study reported that *active iron* was well correlated with chlorophyll content and with visual ratings of chlorosis in citrus rootstock seedlings. Mohammad et al. (1998) indicated

Table 22.1 Main advantages and disadvantages of the most used methods for iron nutritional status diagnosis in citrus trees

Method	Parameter evaluated			References
Leaf visual rating	Green and yellow colour	Advantages	Simple and rapid Accurate when made by trained personnel	Rodriguez de Cianzio et al. (1979) and Mansur (1992)
		Disadvantages	Subjective measurements Reduced accuracy under different individuals and environments	Berrang and Steiner (1980) and Campbell et al. (1990)
Leaf analysis	Leaf total iron	Disadvantages	Chlorosis paradox (no correlation between total iron and degree of chlorosis)	Römheld (2000)
Floral analysis	Leaf Mg:Zn ratio	Advantages	Early prognosis of iron chlorosis (leaves fully green when Mg:Zn > 200)	Pestana et al. (2004a)
Leaf green colour intensity	Leaf chlorophyll content or SPAD units	Advantages	Accurate and useful method to quantify the deficiency degree Able to be precise and rapidly estimated using a non-destructive SPAD meter	Bar-Akiva and Lavon (1967), Chen and Barak (1982), Torrecillas et al. (1984), Porro et al. (2001), and Pestana et al. (2004a, b)
		Disadvantages	It is necessary to know that iron deficiency is the cause of the nutritional disorder Traditional analysis procedures are labour-intensive and time-consuming	Bar-Akiva and Lavon (1967) and Torrecillas et al. (1984)
Leaf iron fractions	Leaf active iron	Advantages	Fe-soluble in 1 N HCl is proportional to chlorophyll content and peroxidase activity Fe-soluble in o-phenanthroline is closely related to the degree of iron chlorosis	Oserkowsky (1933), Llorente et al. (1976), and Mohammad et al. (1998)
		Disadvantages	The extractants remove some non-active iron from leaves because they are not specific for active iron	
Metabolic aspects	Specific enzyme activity, metabolite or immunoassay (peroxidase, chlorophyllase, catalase, cytochromes, cytochrome c oxidase, Fe-SOD, ferredoxin, aldolase and aconitase)	Advantages	Specific enzyme activity depends on the iron available in the plant tissues	Bar-Akiva (1961), Hellín et al. (1984a, 1995), Alcaraz et al. (1985), García et al. (1990), del Río et al. (1991), Garcia and Galindo (1991), Laurie and Manthey (1994), Chouliaras et al. (2004), and Ma (2005)
		Disadvantages	Most of these assays require laboratory facilities that render field assays impossible Most of the analytical procedures are expensive, labour-intensive, time-consuming and sophisticated instrumentation and experienced personnel are required	Torrecillas et al. (1985), García et al. (1990), and Srivastava et al. (2008)

that *active iron* extracted by *o*-phenanthroline and/or the ratio of *active* to total iron was closely related to the degrees of chlorosis in lemon trees. In addition, Köseoglü and Açikgöz (1995) and Sönmez and Kaplan (2004) concluded that hydrochloric acid extraction is the most suitable method to be used for diagnosing iron deficiency in peach and apple tree leaves, respectively, mainly when oven-dry leaf samples are used. In contrast with these results, recently, Neaman and Aguirre (2007) and Neaman and Espinoza (2010) showed that *o*-phenantroline extraction applied on avocado fresh leaves was superior over other methods for diagnosis of iron deficiency. These differences could be due to the fact that these extractants are not specific for *active iron*, and as a consequence, they could remove some *non-active iron* from leaves.

The inability of soil analysis for providing information on the nutrient uptake capacity of the plant and the inability of total iron analysis to indicate the *active* and *inactive* parts of the elements suggested the necessity of incorporating physiological aspects into the diagnostic method by using metabolic processes as enzymatic activities or metabolite concentrations (Bar-Akiva 1972; del Río 1983; Lavon and Goldschimidt 1999). Iron is an important cofactor of many enzymes, including those involved in the biosynthetic pathway of chlorophylls, and thus, iron deficiency affects the biochemistry, morphology and physiology of the whole plant (Sevilla et al. 1984; Almansa et al. 1991; del Río et al. 1991; Hellín et al. 1995; Larbi et al. 2006; González-Mas et al. 2009). Several authors suggested to evaluate the *active iron* in plant leaves using peroxidase tests because its activity depends on the iron available in the plant tissue, and its decrease in activity was found to be specific for iron deficiency (Bar-Akiva 1961; Bar-Akiva et al. 1967; O'Sullivan et al. 1969; Carpena et al. 1976; Llorente et al. 1976; Hellín et al. 1984a; Chouliaras et al. 2004). Also, García and Galindo (1991) proposed the use of chlorophyllase activity as a biochemical indicator of iron deficiency in citrus, while Hellín et al. (1995) and Chouliaras et al. (2004) proposed peroxidase, catalase and some superoxide dismutase activities, that are part of the intrinsic enzymatic defence system required for the detoxification of superoxide radicals, as indicators of iron deficiency in lemon trees. Moreover, other biomacromolecules and enzyme systems can be used for diagnosing iron deficiency in citrus plants such as cytochromes, cytochrome c oxidase, Fe-SOD and ferredoxin (Hellín et al. 1984a; Alcaraz et al. 1985; del Río et al. 1991; García and Galindo 1991; Laurie and Manthey 1994).

Standard peroxidase assays (Bar-Akiva et al. 1967; O'Sullivan et al. 1969) require laboratory facilities that rendered field assays impossible in citrus including other plants. For these reasons, Torrecillas et al. (1985) and García et al. (1990) developed a rapid peroxidase test using leaf discs able to be performed under field conditions, which is adequately reproducible and sensitive for the diagnosis of iron deficiency in citrus, based on the chronometric measurement of the oxidation of ascorbic acid.

For the use of an enzymatic system as a valid indicator of activity of a certain nutrient element in plant tissue, it is essential that the enzyme is specific for the element (Hellín et al. 1995). Thus, metalloenzymes whose enzymatic activities are directly influenced by the metabolic activity of the nutrients are thought to be valid systems (Table 22.1). The enzymatic assays do not give us the exact concentration of the respective mineral element, but the magnitude of the enzymatic activity provides an indication of the deficiency (del Río 1983; Leidi et al. 1987; Lavon and Goldschimidt 1999). As mentioned above, in many cases, iron deficiency provokes an inhibition of the enzymatic activity. An example is the peroxidase enzyme, which is used in the diagnosis of iron for citrus cultivars grown on differentially fertile soils (Hellín et al. 1984a, b; Chouliaras et al. 2004).

Among the metalloenzymes, superoxide dismutases (SODs) are a family of proteins with different metals as part of the enzyme (Cu-Zn, Mn and Fe). Fe-containing metalloenzymes as peroxidase and catalase have been proposed as an indicator of environmental stress, including iron deficiency (Nenova and Stoyanov 2000; Almansa et al. 2002; Chouliaras et al. 2004). In lemon leaves, four Cu, Zn-SOD, two Mn-SOD and three Fe-SOD were described to be present, and the effect of manganese and iron deficiencies on total and cyanide-resistant SOD activities from leaves was studied by Sevilla et al. (1984, 1989). Some of the Fe isoenzymes were described to disappear when iron was deficient and reappeared upon infiltration of the leaves with iron (Sevilla et al. 1984; Almansa et al. 1989). Lime-induced iron deficiency in *Citrus limon* trees led to decreases in leaf chlorophyll levels, subcellular iron fractions and SOD activity, especially Fe-SOD isoenzymes, and it was described that iron resupply to chlorotic leaves restored the iron levels, SOD activity and chloroplast structure (Hellín et al. 1995). Other enzymes as aldolase and aconitase activities have been reported to show reduced activity under iron deficiency conditions while increased under Mn deficiency, so these proteins could also be used to distinguish between both deficiencies (García et al. 1990).

When unknown chlorotic symptoms appear during the growing season, the level of specific enzyme activity has been shown to be a quick and efficient tool to identify a certain mineral deficiency in plants (Lavon and Goldschmidt 1999; Srivastava and Singh 2006), but this method of diagnosis did not find much favour among citrus researchers because of the change in enzymatic activities over time, higher cost compared to conventional leaf analysis due to the sophisticated instrumentation and experienced personnel required (Srivastava et al. 2008). Alternatively, the use of antibodies to measure specific enzymes is allowing the application of immunoabsorbent assays as ELISA as another efficient tool for diagnosis of mineral element deficiencies (Lavon et al. 1999; Chu et al. 2006).

22.3 Metabolic Changes Under Iron Deficiency

Some of the metabolic changes described under iron deficiency in plants include changes in organic acid levels related to changes in the iron homeostasis (Ma 2005). The synthesis of citrate in roots and enhanced citrate extrusion are known to be induced by both Fe and inorganic P deficiency (Neumann et al. 1999). Internal citrate concentrations increased in lateral and cluster roots under iron deficiency in non-graminaceous plants and Fe-deficient transfer cells exhibited an increased synthesis of malic and citric acids, which are used to chelate iron from the soil (Landsberg 1986; McCluskey et al. 2004). It has been described that increased synthesis of citrate correlated with a high number of mitochondria in these cells and with the role that citrate plays in the long distance transport of iron in plants through the xylem (Landsberg 1986). Similarly to roots, leaves of Fe-deficient plants showed increased organic acid levels, mainly citrate and malate to a lesser extent (Wallace 1971). Aconitase catalyses the conversion of citrate into isocitrate, and this activity requires Fe. Under iron deficiency, roots and leaves of lemon fruits were shown to induce organic acids, especially citrate accumulated in the juice vesicle cells (Shlizerman et al. 2007). However, mRNA level of aconitase exhibited no changes under reduced iron concentrations while the analysis of aconitase isozymes demonstrated that out of two aconitase isozymes detected in citrus fruit, only the cytosolic form displayed a reduced activity under low iron concentrations. Based on these studies, it has been suggested then that a situation with limited Fe availability induces reduction in cytosolic aconitase, resulting in a slower rate of citrate breakdown and a concomitant increase in citrate levels.

Citrus trees on many commercial rootstocks do not perform well in high-carbonate soils where micronutrients, including iron, are largely unavailable (Castle et al. 2004). Because of the so-called iron *chlorosis paradox* (Morales et al. 1998; Römheld 2000), the total leaf iron content appears to be not potentially useful in iron nutritional studies, so other metabolic changes are used to evaluate the iron deficiency. Plants can respond to low-iron stress through several inducible mechanisms, including among others, electron release at the root surface (Briat and Lobréaux 1997). Enhanced Fe(III) reduction under iron deficiency by the root Fe(III) chelate reductase (FCR) is the most typical feature of majority of plants which presents the so-called strategy I. In contrast, graminaceous species adopt strategy II consisting in secreting Fe(III)-chelating substances as phytosiderophores (Römheld and Marschner 1986). Thus, determination of FCR activity has been widely used for selecting iron chlorosis-tolerant genotypes. In fact, a protocol has been developed to induce this ferric chelate reductase and can be used in screening assays to select rootstock genotypes tolerant to iron chlorosis (Gogorcena et al. 2000).

An interesting combined strategy has been described by Cesco et al. (2006) which reported that chlorosis-susceptible citrus trees growing on calcareous soil recovered in the presence of grass species due to the use of the iron solubilized from a sparingly soluble source by the phytosiderophores released from graminaceous species. This effect was particularly evident for the susceptible rootstock, and the beneficial effect was evident from the leaf re-greening observed in the Fe-deficient citrumelo 'Swingle' plants, showing that graminaceous cover species can improve the Fe nutrition of fruit trees grown on calcareous soils by enhancing Fe availability. A similar strategy has been described to be effective using the arbuscular mycorrhizal (AM) fungi *Glomus versiforme* (Wang et al. 2007, 2008). The presence of this fungi affected the citrus plant iron uptake, resulting in an increased plant growth, contents of chlorophyll and iron, and root ferric chelate reductase activity and decreased in the ratios of P/Fe and 50 (10P + K)/Fe. For this, inoculation with AM fungi may, therefore, have considerable potential to remedy plant chlorosis due to iron deficiency.

Other strategy used in an attempt to increase iron acquirement in iron starvation conditions is the use of gene transformation. Genes involved in strategies I and II have been cloned and transformed into different plant species in some cases with success as in soybean (Vasconcelos et al. 2006). In this case, transgenic plants expressing the ferric chelate reductase gene *FRO2* increased the FCR activity, which correlated with an increase in iron concentration in the plants when grown under iron-limited conditions. Once iron is absorbed by roots, it is transported to the aerial tissues and is stored in chloroplast for the synthesis of chlorophyll. Because of this, other approaches to improve iron acquisition have been developed with success modifying ferritin levels by gene transformation (van Wuytswinkel et al. 1998).

22.4 Changes in Gene Expression

Differential gene expression analysis has recently been reported in citrus rootstock as a new tool to understand the mechanisms underlying the Fe-deficiency physiopathy and to develop new screening strategies to identify chlorosis-tolerant citrus rootstock genotypes (Forner-Giner et al. 2010). With this technique, some genes induced by iron deficiency in *Poncirus trifoliata* L. have been identified, putatively involved in determining photosynthesis rate and chlorophyll content, cell wall modifications and reducing oxidative stress.

Also, the involvement of these processes in the response to iron deficiency has been confirmed by analysis of the cell walls of plants being thinner than control plants, content of chlorophyll decreasing under iron starvation and measurement of enzymatic activities as peroxidase and catalase showing a decrease in iron-deficient plants. All these studies together with future research on the mechanisms underlying the response of citrus plants to iron deficiency will help towards the identifying molecular markers linked to iron susceptibility to be eventually applied in breeding programmes.

22.5 Effects of Iron Chlorosis on Yield and Fruit Quality

It has been commonly accepted that iron chlorosis is a major yield-limiting concern for citrus trees grown on calcareous soils. In fact, severe reductions of citrus yield are reported (Pestana et al. 2001; Bañuls et al. 2003). Not only citrus yield is adversely affected by iron deficiency but also other tree crops such as olive (Pastor et al. 2002), peach (Tagliavini et al. 2000; Yoshikawa 1988), pear (Elkins et al. 2002) and plum (Yoshikawa et al. 1982). These reductions were associated with decreases in fruit tree load that can be induced by iron deficiency during flower development and fruit set (Álvarez-Fernández et al. 2011). Other factor that could contribute in the reduction of the yield is the fruit size. Fruits from iron-deficient lemon trees had a smaller size, resulting iron in a large decrease in the percentage of commercially acceptable fruits. Concretely, fruits of Balady lime (*Citrus aurantifolia* L.) under the effect of iron chlorosis resulted in smaller than others in good iron conditions (El-Kassas 1984). This reduction in fruit size can be explained by the interplay of two factors: the decrease in photosynthesis caused by leaf chlorosis (Larbi et al. 2006) and the reduction in sink size. It is important to note that the effect of iron chlorosis in yield depends on the severity of iron chlorosis (Tagliavini et al. 2000; Rombolá and Tagliavini 2006) and of the period in which leaf chlorosis develops, often more severe during blooming and fruit set (Rombolá and Tagliavini 2006).

However, iron chlorosis not only had negative effects on yield but also on the fruit quality. These yield losses coupled with declining quality of the fruit can potentially cause significant economic damage. Moreover, the fact that citrus fruits are non-climacteric fruits, the iron deficiency may adversely affect both the harvest date as well as the storage of fruits (Pestana et al. 2001). A field experiment showed that tangerines and orange fruits of Fe-deficient trees, besides being smaller, had lower fresh weight and total juice content (Pestana et al. 1999, 2002).

Fruit concentrations of chemical compounds such as organic acids, vitamins and phenolic compounds are influenced by Fe deficiency. Decreases of total soluble solids and increases in citric acid concentration resulting in delayed ripening usually occur in oranges (Pestana et al. 2001), as well as in other fruits such as tomato (Lyon et al. 1943) and peach (Sanz et al. 1997b). Iron deficiency also leads to higher concentrations of organic anions and phenolic compounds and to lower total sugar/total organic acid ratios in the fruits (Spiegel-Roy and Goldschmidt 1996; Álvarez-Fernández et al. 2011; Mellisho et al. 2011).

The colour and firmness of the fruit are influenced by iron chlorosis. In citrus, there is almost no literature on the influence of iron deficiency on the colour and firmness of fruits. Scarce studies have shown that the colour of the fruits from chlorotic trees was less intense than iron rich trees. In the case of peach fruits, the change of the colour (less red) of the fruit could be induced by the decrease in total photosynthate production (Álvarez-Fernández et al. 2011).

22.6 Effect of Iron Deficiency on Citrus Bioactive Compounds

The final goal of the cultivation of the different species of citrus has mainly led to the consumption of their fruits (Gil-Izquierdo et al. 2001, 2004). In this way, bioactive compounds with biological activity for the human health get increasing interest to know their behaviour and qualitative and quantitative occurrence in citrus fruits (Gil-Izquierdo et al. 2001). In this context, phenolics, vitamin C, carotenoids, alkaloids and terpenoids (like essential oils) are the most important among them. However, only agronomical studies on phenolics have afforded significant results under iron deficiency of citrus tree. Phenolic compounds emerge as a strategy of the citrus plant to iron uptake involving physiological changes caused by the scarcity of Fe. These changes implying excretion into the rhizosphere of protons and organic compounds which enhance Fe(III)-reducing capacity mainly dependent on NAD(P) H-ferric chelate reductase (Mellisho et al. 2011; Zocchi et al. 2007).

It is known that, together with other bioactive compounds such as terpenoids and alkaloids, there is an accumulation of flavonoids like flavonols, flavanones, flavones and other phenolics in root and shoot tissues in response to iron deficiency (Fig. 22.3). This stress caused by low iron available in the citrus tree provokes an increase of flavanones, flavones and flavonols in citrus fruits (around 33% higher compared to citrus trees with normal iron nutritional status) (Fig. 22.3). However, there are no current reports describing the relationship between iron and phenolics in the lemon tree and their possible translocation from the root to the fruit (Mellisho et al. 2011).

Therefore, this significant variation of this group of plant secondary metabolites can be improved to use them as markers

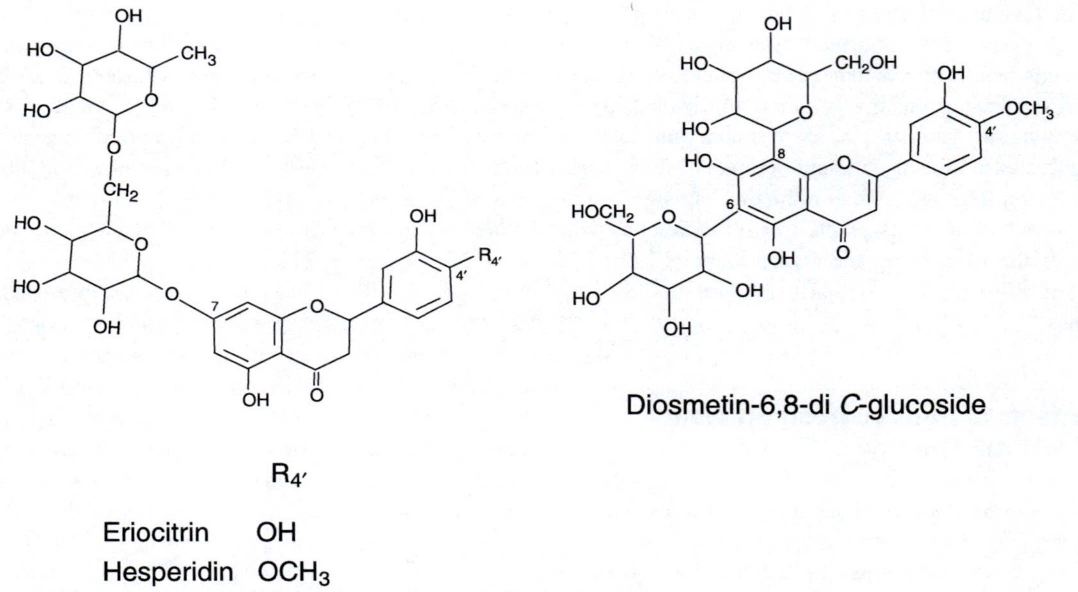

Fig. 22.3 Chemical structures of some of the most relevant compounds in citrus fruits affected by iron deficiency

of iron deficiency in citrus fruits. Particularly, diosmetin-6,8-di-C-glucoside from lemon fruit was selected for the detection of the insufficient concentration of iron in lemon trees (Mellisho et al. 2011).

22.7 Utility of Iron Deficiency in Citrus for Human Nutrition and Health

From a human nutritional point of view, the only edible part of a citrus, the fruit, provides a duality concerning to the iron and bioactive compounds (particularly phenolic compounds).

Polyphenols or phenolic compounds in citrus fruits represent an important and regular contribution of health-promoting compounds to the human body (Gil-Izquierdo et al. 2011). Citrus juices contain high levels of flavonoids, particularly flavanone and flavone glycosides (800–1,500 mg L^{-1}) (Gil-Izquierdo et al. 2003, 2004; Moriguchi et al. 2003; Spreen 2001; Tomás-Barberán and Clifford 2000). Other minor phenolics are hydroxycinnamic acids and flavonols (Gil-Izquierdo et al. 2004; Mellisho et al. 2011; Clifford 1999). All these compounds are regularly present in the body, thanks to the orange juice consumption, staple food product internationally accepted by its organoleptic properties and also its inclusion in the international breakfast at the hotels. The importance of these compounds and, in general, the citrus intake lies in their interesting activities against osteoporosis and osteopaenia, certain types of cancer or the lowering effects on the LDL cholesterol (Habauzit et al. 2011; Cesar et al. 2010; Li et al. 2010). These types of compounds are absorbed at the distal part of the large intestine as aglycones previous hydrolysis of the sugar moiety (normally, rhamnoglucosides or glycosides) by the enzymes of the gut microflora (Manach et al. 2003). The absorption rates and the bioavailability of the phenolics are linked to several factors, but higher concentrations of them may provide higher protection against gastrointestinal disorders and also a possible higher occurrence at physiological level in the human body.

As it has been described in the previous section, iron deficiency enhances phenolics in the citrus fruit. Therefore, the stress condition of the citrus tree caused by the low availability of this chemical element means higher health-promoting phenolics for the human body (Mellisho et al. 2011; Fig. 22.3). On the other hand, the combat of iron deficiencies in human populations by citrus fruits is quite limited, owing to their low concentration themselves (i.e. 8 mg kg^{-1} dry weight) and the occurrence of absorption inhibitors in plant physiological system like phytic acid or the proper phenolic compounds (Frossard et al. 2000). As conclusion, iron deficiency in citrus trees favours healthier fruits and their corresponding juices for human consumption rather than low iron supplementation to the human body caused by the scarcity of iron available for absorption in the fruit.

22.8 Future Research

The future research on iron chlorosis of citrus is mandatory linked to the use of classical diagnostic techniques with high-throughput techniques (–omics). The integration at the same time of genomics, transcriptomics, proteomics and metabolomics can help to know novel early diagnostic tools of iron chlorosis and the pathways affected by this pathology including novel biomarkers to prevent this type of stress in citrus. The analysis of changes in the mRNA expression (transcriptomics) of citrus plant samples with different disorders is the first step in the study of the flow of molecular information from the genome to the proteome and metabolome. Furthermore, proteomics may facilitate the discovery of key proteins whose function is to regulate metabolic pathways, their synthesis and degradation and their modification of specific physiological or physiopathological conditions. So, both technologies provide valuable information about the possible pathways that could be affected. However, the conclusive data and end points will be provided by metabolomics. It is well known that the metabolome is changing constantly, and that it represents the end points of physiological, regulatory processes. Therefore, changes in metabolite concentrations will efficiently contribute for the description of the biochemical state of a biological system of citrus and consequently may be a better measure of gene function, which is obtained from the transcriptome and/or proteome. The conjunction of these techniques will place the sensitivity of citrus *sp.* and cultivars to iron deficiency, provide new tools for the early prevention of chlorosis avoiding the development of visual effects and unravel new physiological pathways and key biomarkers affected by this type of plant pathology. Besides, the same -omics techniques may be assayed in human population (healthy or with different pathologies or disorders) (Gil-Izquierdo et al. 2011) to value the physiological effects on the health by the intake of citrus fruits with higher content of phenolics caused by an agronomical and environmental factors like lime-induced iron chlorosis.

References

Alcaraz CF, Hellín E, Sevilla F et al (1985) Influence of the leaf iron content on the ferredoxin levels in citrus plants. J Plant Nutr 8:603–611

Almansa MS, del Río LA, Alcaraz CF et al (1989) Isoenzyme pattern of superoxide-dismutase in different varieties of citrus plants. Physiol Plant 76(4):563–568

Almansa MS, Palma JM, Yáñez J et al (1991) Purification of an iron-containing superoxide dismutase from a citrus plant, *Citrus limonum* R. Free Radic Res Commun 12–13:319–328

Almansa MS, Hernández JA, Jiménez A et al (2002) Effects of salt stress on superoxide dismutase activity in leaves of *Citrus limonum* in different rootstock-scion combinations. Biol Plant 45:545–549

Álvarez-Fernández A, Melgar C, Abadía J et al (2011) Effects of moderate and severe iron deficiency chlorosis on fruit yield, appearance and composition in pear (*Pyrus communis* L.) and peach (*Prunus persica* L. Batsch). Environ Exp Bot 71:280–286

Bañuls J, Quiñones A, Martín B et al (2003) Effects of the frequency of iron chelate supply by fertigation on iron chlorosis in Citrus. J Plant Nutr 26:1985–1996

Bar-Akiva A (1961) Biochemical indications as a means of distinguishing between iron and manganese deficiencies in Citrus plants. Nature (Lond) 190:647–648

Bar-Akiva A (1972) Leaf analysis: possibilities and limitations. In: Proceedings of 18th international horticultural congress, Tel-Aviv, vol 4, pp 333–345

Bar-Akiva A, Lavon R (1967) Visible symptoms and some metabolic patterns in micronutrient deficient Eureka lemon leaves. Isr J Agric Res 17:7–16

Bar-Akiva A, Kaplan M, Lavon R (1967) The use of a biochemical indicator for diagnosing micronutrient deficiencies of grapefruit trees under field conditions. Agrochimica 11:283–288

Basar H (2003) Analytical methods for evaluating iron chlorosis in peach trees. Commun Soil Sci Plant Anal 34:327–341

Berrang P, Steiner KC (1980) Resistance of pin oak progenies to iron chlorosis. J Am Soc Hort Sci 105:519–522

Boxma R (1972) Bicarbonate as the most important soil factor in lime-induced chlorosis in The Netherlands. Plant Soil 37:233–243

Briat JF, Lobréaux S (1997) Iron transport and storage in plants. Trends Plant Sci 2:187–193

Campbell RJ, Mobley KN, Marini RP et al (1990) Growing conditions alter the relationship between SPAD-502 values and apple leaf chlorophyll. Hort Sci 25:330–331

Carpena O, Llorente S, León A et al (1976) Parámetros de actividad funcional en el diagnóstico de la deficiencia de hierro y manganeso. An Edafol Agrobiol 35:319–332

Castle WS, Grosser JW, Gmitter FG et al (2004) Evaluation of new citrus rootstocks for 'Tahiti' lime production in southern Florida. Proc Fla State Hort Soc 117:174–181

Cesar TB, Aptekmann NP, Araujo MP et al (2010) Orange juice decreases low-density lipoprotein cholesterol in hypercholesterolemic subjects and improves lipid transfer to high-density lipoprotein in normal and hypercholesterolemic subjects. Nutr Res 30:689–694

Cesco S, Rombola AD, Tagliavini M et al (2006) Phytosiderophores released by graminaceous species promote ^{59}Fe-uptake in citrus. Plant Soil 287:223–233

Chaney RL (1984) Diagnostic practices to identify iron deficiency in higher plants. J Plant Nutr 7:47–67

Chen Y, Barak P (1982) Iron nutrition of plants in calcareous soils. Adv Agron 35:217–240

Chouliaras V, Terrizo I, Molassiotis A et al (2004) Iron chlorosis in grafted sweet orange (*Citrus sinensis* L.) plants: physiological and biochemical responses. Biol Plant 48:141–144

Chu MM, Guo ZQ, Muto N et al (2006) Development of ELISA for metallothionein-II allows determination of heavy metal pollution of fresh water. Front Biosci 11:2113–2122

Clifford MN (1999) Chlorogenic acids and other cinnamates-nature, occurrence, and dietary burden. J Sci Food Agric 79:362–372

DeKock PC, Hall A, Inkson RHE (1979) Active iron in plant leaves. Ann Bot 43:737–740

del Río LA (1983) Metalloenzymes as biological markers for the appraisal of micronutrient imbalances in higher plants. Life Chem Rep 2:1–34

del Río LA, Sevilla F, Sandalio LM et al (1991) Nutritional effect and expression of SODs: induction and gene expression; diagnostics; prospective protection against oxygen toxicity. Free Radic Res Commun 12–13:819–828

El-Kassas SE (1984) Effect of iron nutrition on the growth, yield, fruit quality, and leaf composition of ceded balady lime trees grown on sandy calcareous soils. J Plant Nutr 7:301–311

Elkins RB, Meyer RD, Adams CD (2002) Use of urea-sulfuric acid and iron chelate to mitigate iron chlorosis in pear. Acta Hortic 596: 671–676

Forner-Giner MA, Llosa MJ, Carrasco JL et al (2010) Differential gene expression analysis provides new insights into the molecular basis of iron deficiency stress response in the citrus rootstock *Poncirus trifoliata* (L.) Raf. J Exp Bot 61:483–490

Frossard E, Bucher M, Mächler F et al (2000) Potential for increasing the content and bioavailability of Fe, Zn and Ca in plant for human nutrition. J Sci Food Agric 80:861–879

Fuentes M, Ortuño MF, Pérez-Sarmiento F et al (2012) Efficiency of a new strategy involving a new class of natural hetero-ligand iron (III) chelates (Fe(III)-NHL) to improve fruit trees growth in alkaline-calcareous soils. J Sci Food Agric (under revision)

Garcia AL, Galindo L (1991) Chlorophyllase activity as biochemical indicator of Mn and Fe deficiencies in citrus. Photosynthetica 25:351–357

García AL, Galindo L, Sánchez-Blanco MJ et al (1990) Peroxidase assay using 3,3′, 5,5′ tetramethyl benzidine as H-donor for rapid diagnosis of iron deficiency in *Citrus*. Sci Hortic 42:251–255

Gil-Izquierdo A, Gil MI, Ferreres F et al (2001) In vitro availability of flavonoids and other phenolics in orange juice. J Agric Food Chem 49:1035–1041

Gil-Izquierdo A, Gil MI, Tomás-Barberán FA et al (2003) Influence of industrial processing on orange juice flavanone solubility and transformation to chalcones under gastrointestinal conditions. J Agric Food Chem 51:3024–3028

Gil-Izquierdo A, Riquelme MT, Porras I et al (2004) Effect of the rootstock and interstock grafted in lemon tree (Citrus limon (L.) Burm.) on the flavonoid content of lemon juice. J Agric Food Chem 52:324–331

Gil-Izquierdo A, Peñalvo LJ, Gil JI, et al (2011) Soy isoflavones and cardiovascular disease epidemiological, clinical and -omics perspectives. Curr Pharm Biotechnol 13(5):624–631

Gogorcena Y, Abadía J, Abadía A (2000) Induction of *in vivo* root ferric chelate reductase activity in fruit tree rootstock. J Plant Nutr 23(1):9–21

González-Mas MC, Llosa MJ, Quijano A et al (2009) Rootstock effects on leaf photosynthesis in Navelina trees grown in calcareous soil. Hort Sci 44:280–283

Gris E (1844) Nouvelles expériences sur l'action des composés ferrugineux solubles, appliqués a la végétation, et spécialement au traitement de la chlorose et de la débilité des plantes. C R Acad Sci (Paris) 19:1118–1119

Habauzit V, Sacco SM, Gil-Izquierdo A et al (2011) Differential effects of two citrus flavanones on bone quality in senescent male rats in relation to their bioavailability and metabolism. Bone 49:1108–1116

Hellín E, Llorente S, Piquer V et al (1984a) Actividad peroxidasa inducida como indicador de la efectividad de compuestos de hierro en la corrección de la deficiencia de Fe en limonero. Agrochimica 28: 432–441

Hellín E, Ureña R, Sevilla F et al (1984b) Efectividad de complejos orgánicos de hierro en la corrección de la clorosis férrica del limonero. An Edafol Agrobiol 43:1195–1203

Hellín E, Hernández-Cortes JA, Piqueras A et al (1995) The influence of the iron content on the superoxide dismutase activity and chloroplast ultrastructure of *Citrus limon*. In: Abadia J (ed) Iron nutrition in soils and plant. Kluwer Academic Publishers, Dordrecht, pp 247–254

Imsande J (1998) Iron, sulfur, and chlorophyll deficiencies: a need for an integrative approach in plant physiology. Physiol Plant 103:139–144

Jaegger B, Goldbach H, Sommer K (2000) Release from lime induced iron chlorosis by CULTAN in fruit trees and its characterization by analysis. Acta Hort 531:107–113

Katyal JC, Sharma BD (1980) A new technique of plant analysis to resolve iron chlorosis. Plant Soil 55:105–119

Köseoglü AT, Açikgöz V (1995) Determination of iron chlorosis with extractable iron analysis in peach leaves. J Plant Nutr 18:153–165

Landsberg E-C (1986) Function of rhizodermal transfer cells in the Fe stress response mechanism of *Capsicum annuum* L. Plant Physiol 82:511–517

Larbi A, Abadia A, Abadia J et al (2006) Down co-regulation of light absorption, photochemistry, and carboxylation in Fe-deficient plants growing in different environments. Photosynth Res 89:113–126

Laurie SH, Manthey JA (1994) The chemistry and role of metal ion chelation in plant uptake processes. In: Manthey JA, Crowley DE, Luster DE (eds) Biochemistry of metal micronutrients in the rhizosphere. Lewis Publishers, Boca Raton, pp 165–182

Lavon R, Goldschmidt EE (1999) Enzymatic methods for detection of mineral element deficiencies in citrus leaves: a mini-review. J Plant Nutr 22:139–150

Lavon R, Salomon R, Goldschmidt EE (1999) Effect of potassium, magnesium and calcium deficiencies on nitrogen constituents and chloroplast components in Citrus leaves. J Am Soc Hort Sci 124:158–162

Leidi EO, Gómez M, de la Guardia MD (1987) Soybean genetic differences in response to Fe and Mn: activity of metalloenzymes. Plant Soil 99:139–146

Li WQ, Kuriyama S, Li Q et al (2010) Citrus consumption and cancer incidence: the Ohsaki cohort study. Int J Cancer 127:1913–1922

Llorente S, León A, Torrecillas A et al (1976) Leaf iron fractions and their relation with iron chlorosis in citrus. Agrochimica 20:204–212

Lyon CB, Beeson KC, Ellis GH (1943) Effects of micro-nutrient deficiencies on growth and vitamin content of the tomato. Bot Gaz 104:495–514

Ma J (2005) Plant root responses to three abundant soil minerals: silicon, aluminum and Fe. Crit Rev Plant Sci 24:267–281

Maldonado-Torres R, Etchevers-Barra JD, Alcantar-Gonzalez G et al (2006) Morphological changes in leaves of Mexican lime affected by iron chlorosis. J Plant Nutr 29:615–628

Manach C, Morand C, Gil-Izquierdo A et al (2003) Bioavailability in humans of the flavanones hesperidin and narirutin after the ingestion of two doses of orange juice. Eur J Clin Nutr 57:235–242

Mansur LM (1992) Resistance to iron-deficiency chlorosis in PI 437654 soybean. Crop Sci 32:1137–1138

McCluskey J, Herdman L, Skene KR (2004) Iron deficiency induces changes in metabolism of citrate in lateral and cluster roots of *Lupinus albus*. Physiol Plant 121:586–594

Mellisho CD, González-Barrio R, Ferreres F et al (2011) Iron deficiency enhances bioactive phenolics in lemon juice. J Sci Food Agric 91:2132–2139

Mengel K, Geurtzen G (1988) Relationship between iron chlorosis and alkalinity in Zea mays. Physiol Plant 72:460–465

Mohammad MJ, Najim H, Khresat S (1998) Nitric acid- and *o*-phenanthroline-extractable iron for diagnosis of iron chlorosis in Citrus lemon trees. Commun Soil Sci Plant Anal 29:1035–1043

Morales F, Grasa R, Abadia A et al (1998) Iron chlorosis paradox in fruit trees. J Plant Nutr 21:815–825

Moriguchi T, Kita M, Hasegawa S et al (2003) Molecular approach to citrus flavonoid and limonoid biosynthesis. Food Agric Environ 1:22–25

Neaman A, Aguirre L (2007) Comparison of different methods for diagnosis of iron deficiency in avocado. J Plant Nutr 30:1097–1108

Neaman A, Espinoza G (2010) Advances in diagnosis of iron deficiency in avocado. J Plant Nutr 33:38–45

Nenova VR, Stoyanov IG (2000) Effects of some growth regulators on young iron deficient maize plants. Biol Plant 43:35–39

Neumann G, Massonneau A, Martinoia E et al (1999) Physiological adaptations to phosphorus deficiency during proteoid root development in white lupin. Planta 208:373–382

O'Sullivan M, Flynn MI, Cold FJ (1969) A biochemical method for diagnosing micronutrient deficiencies in plants. Ir J Agric Res 8:111–119

Oserkowsky J (1933) Quantitative relation between chlorophyll and iron in green and chlorotic pear leaves. Plant Physiol 8:449–468

Pastor M, Castro J, Hidalgo J (2002) La correzione della clorosi ferrica dell'olivo. Olivae 90:42–45

Pestana M, Gonçalves DA, de Varennes A et al (1999) The recovery of citrus from iron chlorosis using different foliar applications. Effects on fruit quality. In: Anaç D, Martin-Prével P (eds) Improved crop quality by nutrient management. Kluwer Academic Publishers, Dordrecht, pp 95–98

Pestana M, Correia PJ, Varennes A et al (2001) Effectiveness of different foliar iron applications to control iron chlorosis in orange trees grown on a calcareous soil. J Plant Nutr 24:613–622

Pestana M, Correia PJ, Miguel MG et al (2002) Foliar treatments as a strategy o control iron chlorosis in orange trees. Acta Hort 594:223–228

Pestana M, de Varennes A, Goss MJ et al (2004a) Floral analysis as a tool to diagnose iron chlorosis in orange trees. Plant Soil 259:287–295

Pestana M, Faria EA, de Varennes A (2004b) Lime-induced iron chlorosis in fruit trees. In: Dris R, Jain SM (eds) Production practices and quality assessment of food crops, vol 2, Plant mineral nutrition and pesticide management. Kluwer Academic Publishers, Dordrecht

Porro D, Dorigatti C, Stefanini M et al (2001) Use of SPAD meter in diagnosis of nutritional status in apple and grapevine. Acta Hort 564:243–252

Rodriguez de Cianzio S, Fehr WR, Anderson IC (1979) Genotypic evaluation for iron deficiency chlorosis in soybeans by visual scores and chlorophyll concentration. Crop Sci 19:644–646

Rombolá AD, Tagliavini M (2006) Iron nutrition of fruit tree crops. In: Barton LL, Abadía J (eds) Iron nutrition in plants and rhizospheric microorganisms. Springer publisher, Dordrecht, The Netherlands, pp 61–83

Romera FJ, Alcántara E, de la Guardia MD (1991) Characterization of the tolerance to iron chlorosis in different peach rootstocks grown in nutrient solution. II. Iron stress response mechanisms. In: Chen Y, Hadar Y (eds) Iron nutrition and interactions in plants. Kluwer Academic Publishers, Dordrecht

Römheld V (2000) The chlorosis paradox: Fe inactivation as a secondary event in chlorotic leaves of grapevine. J Plant Nutr 23:1629–1643

Römheld V, Marschner H (1986) Evidence for a specific uptake system for iron phytosiderophores in roots of grasses. Plant Physiol 80:175–180

Sanz M, Montañés L (1997) Diagnóstico visual de la chlorosis férrica. Info Téc Eco Agric 93:7–22

Sanz M, Belkhodja R, Toselli M et al (1997a) Floral analysis as a possible tool for prognosis of iron deficiency in peach. Acta Hort 448:241–245

Sanz M, Pascual J, Machín J (1997b) Prognosis and correction of iron chlorosis in peach trees: influence on fruit quality. J Plant Nutr 20:1567–1572

Sevilla F, del Río LA, Hellín E (1984) Superoxide dismutases from a Citrus plant: presence of two iron-containing isoenzymes in leaves of lemon tress. J Plant Physiol 116:381–387

Sevilla F, Almansa MS, Hellín E et al (1989) Influence of the citrus species and varieties on the isoenzyme profile of iron-superoxide dismutases. In: Goren R, Mendal K, Goren N (eds) Citriculture. Balaban, Tel Aviv, pp 561–570

Shlizerman L, Marsh K, Blumwald E et al (2007) Iron-shortage-induced increase in citric acid content and reduction of cytosolic aconitase activity in Citrus fruit vesicles and calli. Physiol Plant 131:72–79

Sönmez S, Kaplan M (2004) Comparison of various analysis methods for determination of iron chlorosis in apple trees. J Plant Nutr 27:2007–2018

Spiegel-Roy P, Goldschmidt EE (1996) Biology of citrus-Biology of horticultural crops. Cambridge University Press, Cambridge, U.K. 230p

Spreen T (2001) FAO citrus symposium [online]. http://www.fao.org/docrep/003/x6732e/x6732e02.htm

Srivastava AK, Singh S (2006) Diagnosis of nutrient constraints in citrus orchards of humid tropical India. J Plant Nutr 29:1061–1076

Srivastava AK, Singh S, Albrigo LG (2008) Diagnosis and remediation of nutrient constraints in citrus. In: Janick J (ed) Horticultural reviews. Wiley, Hoboken, pp 277–364

Sudahono D, Byrne H, Rouse RE (1994) Greenhouse screening of citrus rootstocks for tolerance to bicarbonate induced iron chlorosis. HortSci 29:113–116

Tagliavini M, Abadía J, Rombola AD et al (2000) Agronomic means for the control of iron deficiency chlorosis in deciduous fruit trees. J Plant Nutr 23:2007–2022

Tomás-Barberán FA, Clifford MN (2000) Flavanones, chalcones and dihydrochalcones – nature, occurrence, and dietary burden. J Sci Food Agric 80:1073–1080

Torrecillas A, León A, Del Amor F et al (1984) Determinación rápida de clorofila en discos foliares de limonero. Fruits 39:617–622

Torrecillas A, León A, Del Amor F et al (1985) A rapid chronometric assay of peroxidase activity in citrus leaf discs. Sci Hort 26:273–277

van Wuytswinkel O, Vansuyt G, Grignon N et al (1998) Iron homeostasis alteration in transgenic tobacco overexpressing ferritin. Plant J 17:93–97

Vasconcelos M, Eckert H, Arahana V et al (2006) Molecular and phenotypic characterization of transgenic soybean expressing the Arabidopsis ferric chelate reductase gene, *FRO2*. Planta 224: 1116–1128

Wallace A (1971) Regulation of the micronutrient status of plants by chelating agents and other factors. Edwards Bros Inc, Los Angeles

Wallace A (1990) Nitrogen, phosphorus, potassium interactions on Valencia orange yields. J Plant Nutr 13:357–365

Wallihan EF (1955) Relation of chlorosis to concentration of iron in citrus leaves. Am J Bot 42:101–104

Wang M-Y, Xia R-X, Hu L-M et al (2007) Arbuscular mycorrhizal fungi alleviate iron deficient chlorosis in *Poncirus trifoliata* L. Raf under calcium bicarbonate stress. J Hort Sci Biotech 82(5): 776–780

Wang M, Christie P, Xiao Z et al (2008) Arbuscular mycorrhizal enhancement of iron concentration by *Poncirus trifoliata* L. Raf and Citrus reticulata Blanco grown on sand medium under different pH. Biol Fertil Soil 45:65–72

Yoshikawa FT (1988) Correcting iron deficiency of peach trees. J Plant Nutr 11:1387–1396

Yoshikawa FT, Reil WO, Stromberg LK (1982) Trunk injection corrects iron deficiency in plum trees. Calif Agric 36:13

Zocchi G, De Nisi P, Dell'Orto M et al (2007) Iron deficiency differently affects of metabolic responses in soybean roots. J Exp Bot 58:993–1000

Mycorrhiza in Citrus: Growth and Nutrition

23

Ibrahim Ortas

Abstract

In this review, effect of mycorrhizal inoculation on citrus growth, nutrient and water uptake, and mycorrhizal dependency was searched. Arbuscular mycorrhiza (AM) is symbiotic associations between 90% of higher plants and fungi. Since citrus plants have very few and short root hair, in order to get sufficient nutrient and water, they need mycorrhizal colonization. It has been shown that the host plant was a factor affecting the interaction between mycorrhizal fungi. It has been shown that greater fungal activities in AM hyphae have a significant effect on citrus growth and nutrient uptake.

Mycorrhizal fungi differ in their ability to improve citrus cultivar growth. AM species have different responses to different citrus cultivar's nutrient uptake, particularly less mobile phosphorus (P), zinc (Zn), and copper (Cu). Under arid and semiarid soil conditions, mycorrhizae enhanced acquisition nutrient to the host plants. AMF also improves water relations under drought-stressed conditions.

Effect of different mycorrhizal species under different soil conditions on citrus rootstocks was also evaluated. AMF-colonized citrus seedlings quality is high.

Mycorrhizal inoculation also increases plant resistance against stress conditions such as salt, drought, and temperature as well.

Also mycorrhizal hyphae contribute on soil aggregation; by this way, soil fertility can increase. AM fungus colonization had been shown that enhanced plant growth under drought stress indirectly through affecting the soil moisture retention via glomalin's effect on soil water-stable aggregates development. Studies have shown that mycorrhizal fungal hyphae participate in uptake and transport of water to host plants.

Also, the mechanism behind nutrient uptake and water uptake was evaluated in this review. Arbuscular mycorrhizal fungi also increase the activity of soil enzymes, including dehydrogenase, phosphatase, and urease. Soil phosphatase activity was increased with the increase of AM colonization. And these enzymes help plant to be stronger against stress conditions.

Keywords

Mycorrhizal inoculation • Mycorrhizal dependency • Plant nutrients • Nutrient uptake • Water and salt stress

I. Ortas (✉)
Department of Soil Science and Plant Nutrition,
Faculty of Agriculture, University of Cukurova, Adana, Turkey
e-mail: iortas@cu.edu.tr

A.K. Srivastava (ed.), *Advances in Citrus Nutrition*,
DOI 10.1007/978-94-007-4171-3_23,

23.1 Introduction

Tree plants have an adaptive and/or regulatory processes that control the interaction between nutrition and environmental factors which require integration of nutritional requirements at the whole-plant level, and/or regulatory mechanisms with aboveground processes such as photosynthesis (Lukac et al. 2010; Rennenberg and Schmidt 2010). One of the important features of tree plants is to make use of new resources and to improve the efficiency of nutrient acquisition such as changing rhizosphere environment and symbiosis with mycorrhizal fungi. The most important symbiotic microorganism for the roots of terrestrial plants are mycorrhizal fungi. Mycorrhizal infection is a very common association between plant roots and microorganisms which are responsible for nutrient uptake especially P, Zn, Cu, K, and partly ammonium-N. Arbuscular mycorrhizae (AM) fungi penetrate the living cells of plants without harming them and form the typical organs, such as arbuscules and vesicular in the root. According to a fossil record, the earliest evidence for the existence of *Glomeromycetes* comes from spores and hyphae observed in Ordovician fossils, dating back some 460 million years ago (Redecker et al. 2000). There are several types of mycorrhizae fungi, mainly endo- and ectomycorrhizae. Arbuscular mycorrhizae (endomycorrhizae) are the most widespread plant–root symbiosis, nearly 90% of terrestrial plant species including citrus plant (Gadkar et al. 2001). Endomycorrhizae, specifically AM, have a broad host range, and approximately over 150 species of AM fungi colonize 225,000 species of plant host (Wu et al. 2009). Although *Sclerocystis, Gigaspora, and Acaulospora* species were found in the citrus plant rhizosphere (de Souza et al. 2002; Nemec et al. 1981; Vinayak and Bagyaraj 1990), the main mycorrhizal spore in citrus rhizosphere is *Glomus* species (Davies and Albrigo 1994b). Fidelibus et al. (2000) found *Glomus occultum* as >80% of the total number of AM fungal spores at the Yuma orchard. Also, work (Bhattacharya 1999) on the Nagpur mandarin growing soils of central India's mycorrhizal spore population showed *Glomus* as the predominant species.

23.2 Effect of Mycorrhizal Infection on Citrus Growth

Varied responses of mycorrhiza on growth of citrus have been reported; Verbruggen and Kiers (2010) reported that the root systems of most agronomic crops are colonized by diverse assemblages of arbuscular mycorrhizal fungi, varying in the functional benefits (e.g., nutrient transfer, pathogen protection, water uptake) provided to hosts. One of the most important effects of arbuscular mycorrhiza fungi on the physiology of plants is to increase the absorption of nutrient and water. The main emphasis is when phosphorus is limiting the uptake of phosphorus, and growth is better. The potential of mycorrhiza as a biological entity increases the nutrient uptake efficiency of responsive plants from the native sources in nutrient-depleted soils (Marschner 1995). Once plant is infected with mycorrhizae for a long time period, plants might be carrying out the mycorrhizal infection.

AM fungi's main functions on plants are (1) promoting the absorption of minerals, especially P, Zn, Cu, and NH_4; (2) increase the water uptake; (3) growth is stimulated; (4) high quality of fruits; (5) enhancing resistance to environmental stresses; and (6) enhancing resistance to soil disease.

Citrus are soil and environment selective plants, which mainly grow in light and well-dried soil conditions. Under such conditions, plants get more benefit from mycorrhizae. Read and Fremont (1935) were the first to realize the importance of mycorrhizae for citrus plants. Srivastava et al. (2002) indicated that mycorrhizae are highly effective in low fertility, coarse-textured soils.

Later on Kleinchmidt and Gerdemann (1972) reported that nutrient deficiency in methyl bromide–applied citrus orchard may be caused by elimination of mycorrhizae. In the United States, especially in extensive agricultural area, methyl bromide was used for eliminating soilborne organisms. At the same time, viable indigenous mycorrhizae were killed, and consequently plants were stunted. After using soil fumigation for production of citrus in the North America, nutrient-decency symptoms and stunting were observed (Menge et al. 1978). The stunting and nutrient deficiency were related to the toxic effect of fumigation. Kleinchmidt and Gerdemann (1972) reported that such symptoms were due to the elimination of AM fungi. Citrus plant is so sensitive about indigenous soil organisms that after soil sterilization (Fig. 23.1) plants are not grown. Timmer and Leyden (1978) showed that soil fumigation caused the stunting of citrus seedlings. Mycorrhiza inoculated and non-inoculated citrus seedling growth on sterile soil after 18 mounts showed that non-inoculated plant is not grown and stunt and the inoculated plant grown healthy (Fig. 23.1).

23.2.1 Indigenous Mycorrhizae on Citrus Growth

AM infection can also maintain citrus yield and quality at low inputs of nutrients. Nemec et al. (1981) reported that citrus orchard soils contain communities of AM fungi rather than a single species, and several or all of these species might colonize citrus roots at the same time. It seems that orchard indigenous mycorrhizae have a rich mycorrhizal diversity which may have great contribution on plant growth and nutrient uptake. Michelini et al. (1993) compared site effect on indigenous mycorrhiza, and they found that the significant correlations between measures of AM fungal infection and

Fig. 23.1 Effect of soil sterilization on citrus seedling growth

site characteristics varied between the Caribbean Islands. Different species in different geographic isolates of the same species of AM fungi might vary with respect to their ability to colonize roots and improve plant growth (Camprubi and Calvet 1996; Graham et al. 1996, 1982). According to de Souza et al. (2002), in order to quantify AMF spores present in citrus nurseries and orchards in Rio Grande do Sul, Brazil, soil and root samples were collected at 10 nurseries and 12 citrus orchards and found that AMF species, in decreasing order of occurrence, were *Glomus macrocarpum* > *Scutellospora heterogama* > *Acaulospora scrobiculata* = *Acaulospora birreticulata* > *Glomus invermaium* = *Glomus occultum* = *Entrophospora colombiana* > *Glomus claroideum* = *Glomus constrictum* > *Scutellospora persica*. It seems that there is a great number of mycorrhizal species at the same root area of citrus plant. Thirteen indigenous AMF species were isolated from rhizosphere soil of citrus orchards in Thailand and recultivated and inoculated for citrus seedling, and it has been reported that *G. etunicatum* had significant effect on citrus plant growth (Watanarojanaporn et al. 2011)

Many research works done have documented the effect of AM fungi on citrus growth and physiology, largely been based on differences between plants inoculated with a single common isolate of AM fungi and non-AM plants. Ortas et al. (unpublished data) showed that indigenous mycorrhizae inoculation has high response to citrus plant than single spore inoculation (Fig. 23.2).

Graham et al. (1996) indicated that the relevance of species diversity to the function of AM fungi in the field is poorly understood because data comparing different communities of AM fungi on plant growth and physiology are lacking (Graham 1986). Sharma et al. (2009) isolated, identified, and reinoculated 12 AMF species in citrus orchards grown in nine different plantation areas of northwestern Himalayan region (NWHR) of India, and they found that significant increase in growth parameters such as height, diameter, root length, and leaf area was more evident for the seedlings inoculated with *G. fasciculatum* and *G. mosseae*.

23.2.2 Mycorrhizae and Seedling Growth

Mycorrhizae produce many effects on plants that are horticulturally valued. The fungi can increase seedling survival rate, plant growth rate, and number of flowers produced. Also mycorrhizae increased seedling quality (Fig. 23.3) and better growth after transplanting from greenhouse to field conditions. Wang and Xia (2009) showed that the colonization of *G. versiforme* significantly increased the plant height, stem diameter, leaf numbers, and dry mass. *G. mosseae*–inoculated citrus grafting seedling trifoliate orange/cara cara significantly increased the plant height, stem diameter, leaf area, and shoot length of test seedling (Wu and Xia 2004, 2005). The results of Wu and Xia (2004) showed that arbuscular mycorrhizal fungi inoculation could increase plant growth, such as plant height, stem diameter, leaf area, shoot dry weight, root dry weight, and plant dry weight, when the water content of soil was 20%, 16%, and 12%. Similarly, Tong et al. (2006) searched the effect of several mycorrhizal inoculum on the seedling growth, and they found that AM fungi could be infected effectively, and their shoot and root growth, especially fibrous root growth, was significantly improved, compared with the control. However, Jifon et al. (2002) reported that inoculation with an AM fungus (*G. intraradices*) depressed growth of *Citrus aurantium* seedlings in soil with a high P supply.

Fig. 23.2 Effect of indigenous and selected mycorrhizae on citrus seedling growth

Fig. 23.3 Mycorrhiza-inoculated citrus seedling production

Mycorrhizae were observed to be highly effective in low fertility and coarse-textured soil conditions. Mycorrhizal-treated citrus trees had better plant growth and uptake of nutrients like P, Ca, Zn, Cu, and Fe compared to non-mycorrhizal trees (Srivastava et al. 2002). Wang et al. (2009) reported that arbuscular mycorrhizal fungi could affect the activation on mineral elements and improve the available iron contents through the changes on iron species in soil. Wang and Xia (2009) searched the effects of arbuscular mycorrhizal fungus (*G. versiforme*) on growth and iron uptake of trifoliate orange [*Poncirus trifoliata* (L.) Raf.] at different pH levels of nutrient solution. They found that the colonization of *G. versiforme* significantly increased the plant height, stem diameter, leaf numbers, and dry mass. Also, it is reported that mycorrhizal fungus significantly enhanced the accumulation of chlorophyll, active iron, total iron, and root Fe (III) chelate reductase activity.

23.3 Mycorrhizae Species and Nutrient Uptake

Citrus responses to the introductions of AMF have often been unpredictable, especially in productive agricultural systems and mycorrhizal colonization. The highly significant positive correlation between soil AMF inoculum density and performance of sour orange growth was observed. Soil inoculation with selected AMF species to increase the citrus growth and nutrient uptake has been beneficial. Several mycorrhizae species were used for screening the best inoculum for sour orange under Mediterranean soil conditions. The *G. clarium*–inoculated plants shoot parts weighed 24.15 g plant^{-1}, and the control plants produced 2.53 g shoot dry matter (Table 23.1). Also the *G. clarium*–inoculated plant

Table 23.1 The effect of different arbuscular mycorrhizae inoculation on shoot and root dry weight of sour orange seedling

Mycorrhizal species	Shoot dry weight (g plant^{-1})		Root dry weight (g plant^{-1})		Shoot/root dry weight ratio
Control	2.53b	±0.75	3.51b	±0.99	0.72b
G. mosseae (1)	2.03b	±0.81	1.84b	±1.11	1.10b
G. mosseae (2)	2.70b	±0.69	3.26b	±0.56	0.82b
G. clarium	24.15a	±2.65	19.49a	±6.37	1.24a
G. caledonium	4.93b	±0.71	3.65b	±0.42	1.37a
G. etunicatum	3.17b	±0.89	3.62b	±0.24	0.87b
P	***		***		***

Adapted from Ortas et al. (2002a)

G. mosseae (1) collected from UK; *G. mosseae* (2) collected from Germany, mean (three replicates) bracket is SE (standard error)

* ** and *** Significant at $P<0.05$, 0.01, and 0.001, respectively in table 23.1

Fig. 23.4 Effect of mycorrhizal inoculation on rootstocks' development

produced 19.49 g root dry weight, and non-inoculated plant produced 3.51 g root dry weight. *G. clarium* gave the best improvements in growth and nutrition, resulting in greater leaf area, plant height, stem diameter, and plant biomass, with higher shoot P, Zn, and Cu contents (Ortas et al. 2002a, b).

23.3.1 Effect of Mycorrhizal Inoculation on Rootstocks

Rootstocks differ in their nutrient acquisition and mycorrhizal dependency. If rootstocks root systems are not capable to take up nutrients efficiently, or if there is a less mycorrhizal infection (heavy fungicide and phosphate fertilizer application kill mycorrhiza), plant are showing nutrient deficiency. Nemec (1979) searched the response of six citrus rootstocks with three Glomus species and found that there are significant differences in rootstocks dependency to mycorrhizae. Dutra et al. (1996) found that the *G. intraradices*–inoculated citrus rootstocks sour orange (*Citrus aurantium* L.) and "Carrizo" citrange (*Citrus sinensis* (L.) Osb. × *Poncirus trifoliata* (L.) Raf.) increased root and shoot growth. Rootstocks have their own root growth patterns which can influence nutrient uptake differently. Since the root system of rootstocks is also highly differing from each others', mycorrhizal infection level is supposed to be different as well (Fig. 23.4, unpublished data). As nutrient uptake is highly dependent on root surface area, it is important to select the rootstocks which produce a suitable root system for individual soil. Because of this difference, maybe, every soil and rootstock and variety of citrus should be tested individually (Marchal 1987).

Labanaus and Bitters (1974) concluded that rootstocks differ in their concentration for nutrient elements. Rootstocks also differ in terms of K/Mg ratio which is critical for a balanced nutrition. For example, Cleopatra as rootstock was investigated in North Africa, and it has been found that leaf composition depressed K content and increased Mg very considerably. Based on the selection of citrus rootstocks with high yield potential in nutrient-deficient soils is an agricultural strategy especially for developing countries which are paying high amount of their capital per years for fertilizer inputs. It is well known that citrus rootstocks are quite differing in their ability to take up and utilize mineral nutrient from soils. Citrus variety (inter rootstocks) grown over rootstocks also varies in terms of utilization of mineral nutrients. One of the major points in terms of selection of rootstocks for screening program is adaptation to soil conditions, especially to nutrient status of soils. According to results of Wutscher and Dube (1977), and Wutscher and Shull (1976), Severinia rootstock leaves have high amount of K, Zn, and Cu concentration than other rootstocks.

Vinayak and Bagyaraj (1990) screened 18 AM fungi used for Troyer citrange citrus rootstock for their symbiotic response in unsterilized soil. They found that *Glomus macrocarpum, G. caledonium, G. velum, G. monosporum, and Gigaspora margarita* gave the best improvement in growth and nutrition, resulting in greater leaf area, plant height, stem diameter, and plant biomass, with higher root and shoot P, Zn, and Cu contents. Srivastava et al. (2002) indicated that phosphorus nutrition of mycorrhizal-treated citrus trees was best improved by using rock phosphate as a source of P as opposed to other sources. Camprubi and Calvet (1996) worked on the selection of the most effective arbuscular mycorrhizal (AM) fungi for growth enhancement of citrus cultivars used as rootstocks in citrus nurseries in Spain. They found that the most common AM fungi found in citrus soils in eastern Spain were *G. mosseae and G. intraradices.* And also they indicated that the most effective fungus for growth enhancement of citrus rootstocks was *G. intraradices*. Also, many citrus rootstocks cultivars which provide different tolerance or resistance to diseases, soil factors or environmental factors especially soilborne diseases, salinity, and drought.

23.3.2 Nutrient Uptake

Carbon (C), hydrogen (H), oxygen (O), nitrogen (N), phosphorus (P), potassium (K), calcium (Ca), magnesium (Mg), sulfur (S), iron (Fe), zinc (Zn), manganese (Mn), boron (B), copper (Cu), molybdenum (Mo), chlorine (Cl), and nickel (Ni) are considered to be necessary and essential plant nutrients for the growth of plants: C, H, and O are obtained from carbon dioxide and water. The remaining elements, called the "mineral nutrients (macronutrients and micronutrients)," are obtained from the soil and some are obtained from water. Citrus growth or nutrient deficiency is negatively affected not only by low nutrient availability of soil but also because of ability of tree roots to acquire the nutrients from soil. Citrus are able to grow on diverse soil conditions including high soil pH, low availability of P, K, Zn, Fe, and Mn, and salinity. Under these conditions, nutrition of citrus is one of the major considerations for farmers. It is usually difficult to assess nutritional status of plant level because of different ecological conditions.

It is important to know plant nutrient content before nutrient management. Most of the time, farmers are using heavy fertilizer; consequently, there is an unbalanced nutrient content. Also, heavy fertilizers reduce the mycorrhizal development, especially in citrus plant (Ortas et al. 2002a). Accordingly, it is important to know optimum nutrient requirement and balance nutrient uptake under mycorrhizal treatments. The contribution of mycorrhizae is predicted to be a function of an increase in P uptake due to mycorrhizal infection. As the world fertilizer sources, especially P, are very limited and costly, it is reasonable to use renewable natural sources such as selection of high resistant variety or genotype of crops and mycorrhizal symbiosis as a natural source to agriculture. Phosphorus uptake is defined as the actual rate of phosphorus absorption under certain conditions. And phosphorus demand is defined as the rate of P absorption by which the performance of a plant is measured by growth rate and fitness.

At the moment, one-third of the world's population is getting less Zn and Fe intake through food chain. The availability of phosphorus (P), zinc (Zn), and iron (Fe) in many citrus grown soils is often low, which is limiting an adequate plant growth. Therefore, chemical fertilizer application becomes a common practice. It is reasonable to draw a chart showing that nutrient deficiencies of plant are reflecting nutrient deficiency of soil and affecting human nutrition. Plants are the principle entry points of nutrients into the food chain leading to man. Since citrus as fruit is largely consumed by humans all over the world, accordingly, it is very important to improve citrus nutrition situation in order to increase human intake of mineral nutrients which is addressing the human health.

23.3.2.1 Contribution of Mycorrhizal Infection to Nutrient Uptake and Plant Yield

Citrus roots under orchard condition are normally mycorrhizal (Nemec et al. 1981). AM fungal enhancement of citrus growth has been usually attributed to enhanced phosphorus (P) nutrition of AM plants, but only when soil P supply is limited (Graham 1986). Data by several workers (Auge et al. 1986; Krikun and Levy 1980; Levy and Krikun 1980; Manjunath et al. 1983) showed that the mycorrhiza-inoculated citrus rootstocks had higher P and Zn content than the non-inoculated plants. Timmer and Leyden (1980) reported that mycorrhiza-inoculated plants had higher Cu contents than the non-mycorrhizal plants. To maintain a good nutritional status, it is worth to inoculate citrus seedling roots with mycorrhizae. Under controlled greenhouse conditions, the extraradical mycelium of AM fungi may supply 80–3.133% of P, 25–250% of N, 10–1.011% of K, 70% of Ca, 25–67% of Zn, and 60% of Cu (Marschner and Dell 1994). Tong et al. (2006), under the pot experiment conditions, inoculated axenic pomelo (Citrus grandis cv. Changshou Shatian You) seedlings with arbuscular mycorrhizal (AM) fungi *Gigaspora margarita, G. mosseae, and G. versiforme*, respectively, and their vegetative growth and mineral contents were studied. Also, three AM fungi significantly elevated the N, P, K, Ca, Mg, Zn, Cu, and Mn contents in seedling's leaves. *G. mosseae*–inoculated seedling had the best vegetative growth, the highest mycorrhizal dependence, and N, P, K, Ca, Zn, and Cu contents. Similarly, Wu and Zou (2009b,c) showed that the sole AMF inoculation significantly increased total dry weight, leaf P, K, Ca, Mg, Fe, Cu, and Mn contents, and root P, K, Ca, Fe, Cu, and Zn contents of the seedlings, compared to the non-AMF control.

da Rocha et al. (1994) reported that a mixed population of *Acaulospora morrowae*, G. *clarum*, and G. *etunicatum* up to inoculation promoted better Cleopatra mandarin cv. growth and increased the boron contents in the shoots. Wang et al. (2008b) searched the effect of the arbuscular mycorrhizal (AM) fungus (*G. versiforme*) on iron contents of two citrus rootstocks trifoliate orange (*Poncirus trifoliata* L. Raf.) and red tangerine (*Citrus reticulata* Blanco) was studied in sand culture under different pH conditions, and they found that colonization by *G. versiforme* enhanced plant growth. Also they reported that iron uptake and translocation of citrus plant were enhanced, and AM fungi may be considered as a potential tool for bioremediation of citrus iron deficiency. Srivastava et al. (2002) reported that mycorrhizal-treated citrus trees had better plant growth and uptake of nutrients like P, Ca, Zn, Cu, and Fe compared to non-mycorrhizal trees. The results of Antunes and Cardoso (1991) showed that mycorrhizal inoculation significantly increased dry matter yields and also increased P and K content of citrus. AM-infected citrus plants grow better than non-mycorrhizal one (Hattingh and Gerdemann 1975; Krikun and Levy 1980; Levy et al. 1983; Menge et al. 1982). The results of Menge et al. (1978) showed that the concentration of Zn, Cu, and Mn appears to be influenced in the presence of mycorrhizal fungi. Youpensuka et al. (2008) indicated that AM fungi increased concentrations of P and Mg in leaves of citrus tangerine. When plants are mycorrhizal, the uptake of Mn is generally reduced (Kothari et al. 1991; Marschner 1995), indicating that this effect has been attributed to a lower Mn^{+4} reducing potential in the rhizosphere of mycorrhizal plants, probably because of lower population of manganese reducers organisms.

Ortas et al. (2002a) tested effect of P and Zn levels and mycorrhizal inoculation on citrus growth. Dry matter production in mycorrhizae-inoculated plants was significantly stimulated by mycorrhizal infection. The effects of various levels of P and Zn application on shoot and root dry weight production is shown on Table 23.2. Non-inoculated and non-fertilized P_0Zn_0 treatment resulted in 0.82 g plant^{-1} shoot dry matter; however, inoculated P_0Zn_0 treatment resulted in 25.31 g plant^{-1} shoot dry matter. Highest shoot dry weight was also obtained with Zn supply in inoculated plants. Also root growth was effected by mycorrhizal inoculation. Inoculated plants had a shoot:root ratio less than 1 compared to non-inoculated plants which have shoot:root ratio more than 1 (Table 23.2). Increasing P and Zn supply gradually increased shoot and root dry weight ratio of plants inoculated with mycorrhizae.

Our previous work results showed that mycorrhiza-inoculated citrus plant tissue has greater total P and Zn content. Total P uptake (shoot and root P content) in mycorrhiza-inoculated P_0Zn_0 treatment was 39 mg P plant^{-1}, but non-inoculated P_0Zn_0 treatment was 0.58 mg P plant^{-1} (Figs. 23.5 and 23.6). The effect of various levels of P and Zn supply on Zn content was also calculated, and mycorrhiza-inoculated plant has 11, 24 mg Zn compared to 0.16 mg Zn plant (Fig. 23.7). In non-inoculated plants, increasing P and Zn addition increased P and Zn content; however, contribution of mycorrhiza is much higher than fertilizer application.

Table 23.2 Effect of mycorrhizal inoculation and P and Zn application on leaf, stem and shoot, root dry weight, and shoot and root dry weight ratio

P and Zn supply (mg kg^{-1} soil)	Shoot dry weight (g plant^{-1})		Root dry weight (g plant^{-1})		Shoot: root dry weight ratio
−Mycorrhiza					
P_0Zn_0	0.82	±0.12	1.10	±0.19	0.75
P_0Zn_1	1.41	±0.45	2.71	±2.51	0.82
P_0Zn_2	1.26	±0.27	1.52	±0.32	0.83
P_1Zn_0	1.17	±0.21	1.63	±0.62	0.72
P_1Zn_1	2.24	±0.90	2.74	±1.40	0.82
P_1Zn_2	2.38	±0.31	2.83	±0.05	0.84
P_2Zn_0	4.13	±0.59	4.96	±0.50	0.83
P_2Zn_1	3.52	±0.68	3.62	±1.88	0.97
P_2Zn_2	3.50	±0.80	2.99	±0.56	1.17
+Mycorrhiza					
P_0Zn_0	25.31	±0.55	14.55	±0.03	1.74
P_0Zn_1	24.24	±6.88	13.89	±1.82	1.75
P_0Zn_2	26.52	±3.82	14.94	±1.51	1.77
P_1Zn_0	20.40	±6.79	13.56	±2.72	1.50
P_1Zn_1	23.74	±3.13	16.30	±1.49	1.46
P_1Zn_2	23.28	±1.20	14.33	±0.43	1.62
P_2Zn_0	22.00	±3.13	13.51	±2.21	1.63
P_2Zn_1	21.68	±1.49	14.45	±1.42	1.50
P_2Zn_2	20.57	±1.93	13.71	±1.46	1.50

According to Marschner and Dell (1994), it has been reported that mycorrhizae can increase other nutrients such as B and Fe. Treeby (1992) used two citrus rootstocks differing in mycorrhizal dependency and lime tolerance, rough lemon (*Citrus jambhiri*), and trifoliate orange (*Poncirus trifoliata* Raf.) grown at two soils with different pH, and it was concluded that mycorrhizal fungi may increase the supply of Fe to the host plant in an acid soil, but not in an alkaline soil. The results of Wu and Xia (2006) showed that under water stress conditions, the levels of K^+ and Ca^{2+} in leaves and roots were significantly higher in AM seedlings than those in non-AM seedlings.

23.3.2.2 Mycorrhizae and Rock Phosphate Increase Citrus Plant P Concentration

Phosphate, which is one of the essential three main mineral nutrients, is applied in agriculture for plant growth. The world's rock phosphate sources are limited and nonrenewable. The world's rock phosphate reserves, most of the phosphate mines,

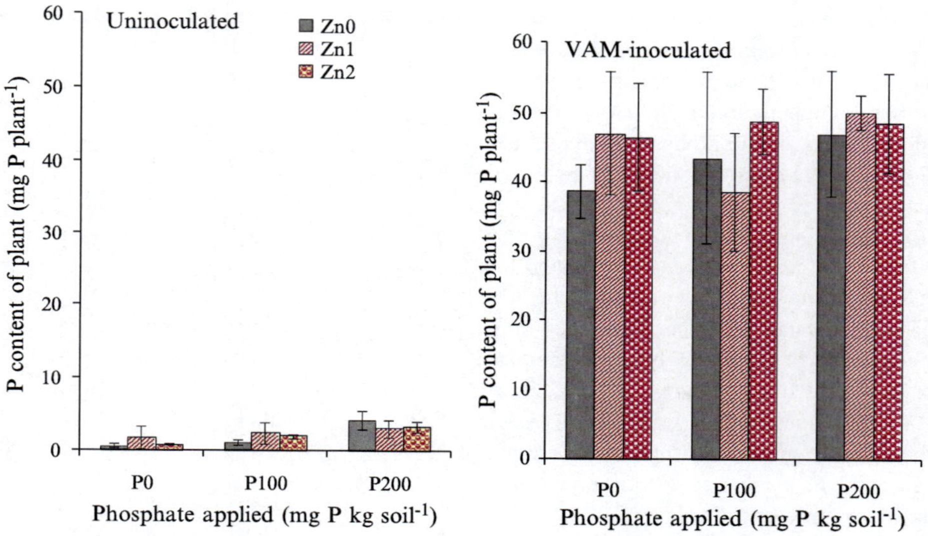

Fig. 23.5 The effect of P and Zn and mycorrhizal inoculation on total P uptake by citrus plants: Zn0 (0 mg Zn kg^{-1} soil), Zn1 (2.5 mg Zn kg^{-1} soil), and Zn2 (5 mg Zn kg^{-1} soil)

Fig. 23.6 Effect of phosphorus and zinc application on citrus growth with and without mycorrhizal inoculation

Fig. 23.7 The effect of P and Zn supply and mycorrhizal inoculation on Zn content in dry weight of citrus plants

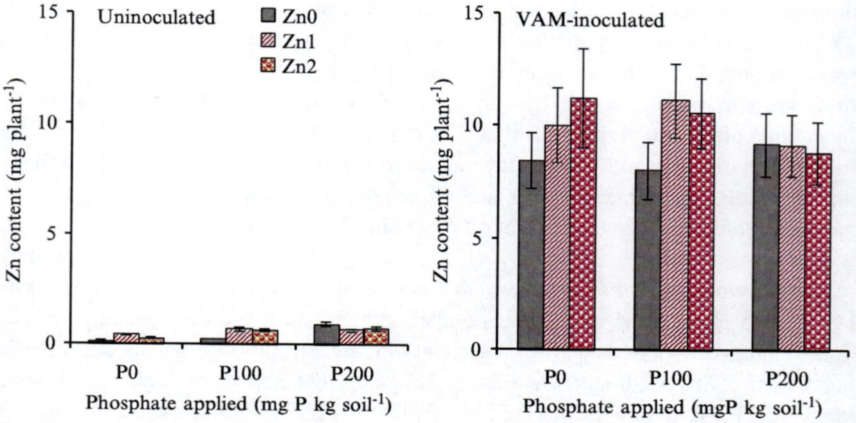

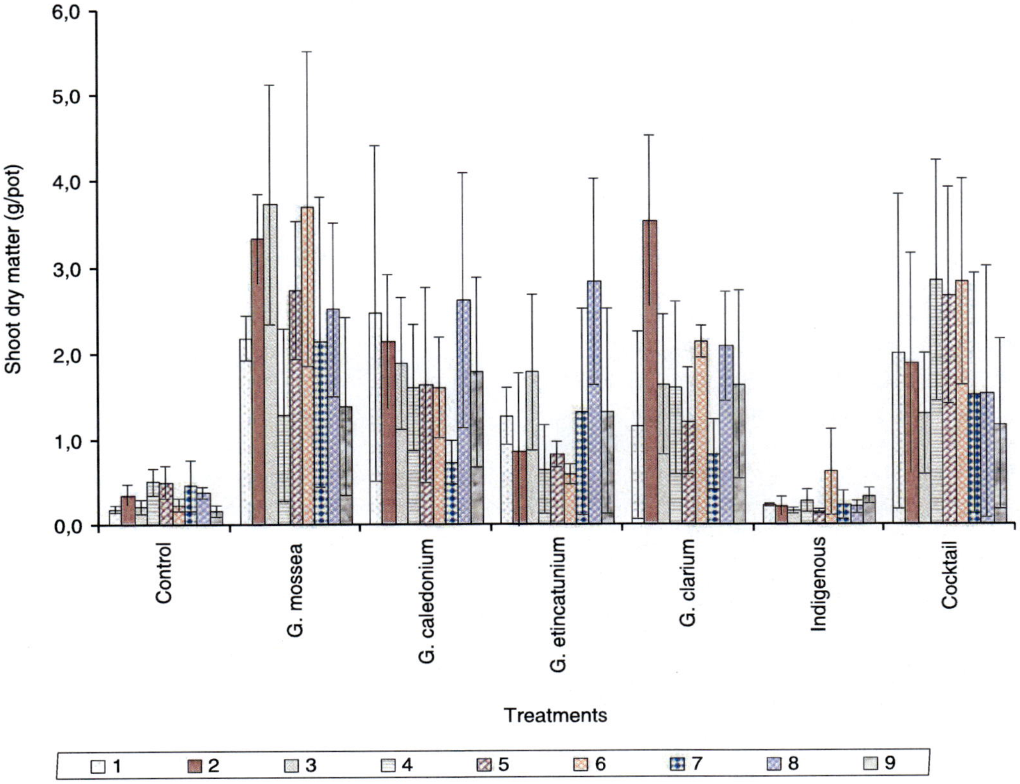

Fig. 23.8 Effect of rock phosphate, compost application and several mycorrhizal inoculation on citrus growth

will be depleted in about 100 years (Herring and Fantel 1993). Since the world phosphate sources are limited, mycorrhiza and other phosphate solubilizing organisms should be important organic sources for sustainable agriculture. It has been estimated that under field conditions, a reduction of 80% of the recommended phosphate fertilizer could be supplemented by inoculation with AM fungi (Jakobsen 1995). Previously, it has been reported that mycorrhizal infection and plant yield were increased with increasing P application either as a soluble P level or rock phosphate addition. Srivastava et al. (2002) reported that phosphorus nutrition of mycorrhiza-treated citrus trees was best improved by using rock phosphate as a source of P as opposed to other sources. The influence of mycorrhizal infection on nutrient uptake is well known; there is however still a need for quantification of the data on the contribution of AM to nutrient uptake of citrus growing under the eastern part of the Mediterranean soils. Mycorrhizae are also considered to be involved in decomposition of mineral and organic materials through the production of organic acids and other substances (MacFall 1994). Especially under poor soil nutrient condition or slow release fertilizer application, mycorrhizal colonization could help plant to dissolve the nutrient in soil. Graham and Syvertsen (1985) reported that under low-nutrient soils condition, mycorrhizae tend to thrive and provide benefit to their host plants only under conditions of low-to-moderate nutrient availability. For this reason, rock phosphate (naturally slow release) has been recommended as a suitable phosphate source to encourage mycorrhizal development in citrus. In this way, rock phosphate can be the main mineral source of phosphorus available to organic farming with mycorrhizae use.

The application of insoluble phosphate, such as rock phosphate (Graham and Timmer 1985), calcium phosphate, and bone powder, is effective to maintain AM fungi. An additional advantage of insoluble phosphate is long-term availability of P, compared to soluble phosphate. Ortas et al. (unpublished data) reported that the combination treatments of AM fungi and rock phosphate have the potential to increase plant growth where phosphorus was limiting plant production. Mycorrhizal inoculation significantly increased citrus shoot dry weight with rock phosphate application (Fig. 23.8). Rock phosphate and compost also have influence on plant growth; however, contribution of mycorrhizae is much higher. In this experiment, indigenous mycorrhizae have less effect. In our previous experiment, we have found significant effect of indigenous mycorrhizae on plant growth.

23.4 Mycorrhizal Dependency

Under the natural habitat, plant species are dependent on the mycorrhizal condition in order to produce its maximum growth or yield at a given level of soil fertility (Gerdemann 1975). Early establishment and growth of mycorrhizae-dependent (MD) plants require mycorrhizal formation. Plant varies with soil fertility especially P, soil type, and species of mycorrhizal fungi (Graham et al. 1982). Especially tree plants such as citrus are obligatory mycorrhizal dependent (Graham 1986; Graham and Eissenstat 1994; Graham and Syvertsen 1985; Krikun and Levy 1980; Menge et al. 1982; Ortas et al. 2002b). In general, rootstocks with long and abundant root hairs are less mycorrhizal dependent than those with short or few root hairs (Wu et al. 2009).

It has been experimentally shown that different rootstocks have large differences of MD. Within citrus, different rootstocks show different degrees of dependency on mycorrhizae (Spiegel-Roy and Goldschmidt 1996). Bevington (2002) reported that commonly used rootstocks in Australia, Troyer citrange, Carrizo citrange, Rough lemon (citronelle), and Sweet orange are known to be mycorrhizal dependent. Poerwanto et al. (1989) showed that trifoliate orange (*Poncirus trifoliata* (L.) Raf.) has relatively short root hairs in field and is strongly mycorrhizae dependent (Wu and Xia 2006). The results of Graham and Syvertsen (1985) showed that rootstocks are strongly deferent in term of mycorrhizal dependency. Also Graham and Eissenstat (1994) postulated that among closely related Citrus genotypes, there is a greater tendency for less dependent species to limit the rate but not the extent of colonization, even in high-P soils. Camprubi and Calvet (1996) reported that sour orange and Cleopatra mandarin were more dependent than Troyer citrange and Swingle citrumelo.

Rootstocks with lower mycorrhizal dependency also generally had greater hydraulic conductivity of roots, greater transpiration, and CO_2 assimilation rates. Selection of AM-dependent citrus rootstocks in terms of water utilization also is an important agricultural strategy for Middle East where water is a limited source. Graham and Eissenstat (1994) concluded that carbon expenditure on the fungus at high P may result in mycorrhizal-induced growth depression. It has been shown that drought stress strongly affects citrus MD. Under greenhouse condition, Wu et al. (2007b) tested the efficacy of five *Glomus species, Glomus mosseae, G. geosporum, G. versiforme, G. etunicatum, and G. diaphanum* to measure their ability to improve water relations of Citrus tangerine under well-watered and drought stress conditions. They found that the ranking of five Glomus species for mycorrhizal dependency of C. tangerine was as follows: *G. mosseae approximate to G. geosporum*>*G. versiforme*> *G. etunicatum*>*G. diaphanum*. And also they reported that both G. mosseae and G. geosporum colonization showed greater transpiration rates and stomatal conductance, and they were more efficient fungi in improving water relations of *C. tangerine,* and *G. etunicatum* was less efficient fungi. On the other hand, Watanarojanaporn et al. (2011) indicated that *G. etunicatum* has an influence on citrus growth and P uptake, suggesting it to be the highly effective strain for mycorrhizal efficiency index. The dependency is mostly related to the enhanced phosphorus acquisition from low phosphorus soils. Menge et al. (1978) and Ortas et al. (2002b) have demonstrated the mycorrhizal dependency of several citrus cultivars in low P soil. Graham and Eissenstat (1994) indicated that the dependency of mycorrhizae is regulated by the host genotype. Also, there are significant differences in mycorrhizal dependency among rootstocks which were confirmed by Ortas et al. (2002a). They tested five different mycorrhizae species effect on mycorrhiza dependency with citrus seedling, and they found the order of effectiveness of species as follows: *G. clarium*>*G. mosseae* (1)>*G. mosseae* (2)>*G. caledonium*>*G. etunicatum*>control. And in the second experiment, the order of effectiveness of species was as follows: *G. clarium*>*G. caledonium*>*G. etunicatum*>*G. mosseae* (2) *G. mosseae* (1)>control. Ortas et al. (2002b) tested mycorrhizal dependency of citrus plant to P and Zn nutrition, and they found that MD was less affected by Zn supply than the P supply (Fig. 23.9). In P_0Zn_0, P_1Zn_0, and P_2Zn_0 treatments, MD was 97%, 94%, and 81%, respectively. Also, with increasing Zn supply, dependency slightly decreased, but decrease in P supply affected much higher. Since the non-inoculated plants did not respond to P and Zn supply, it can be said that citrus plants strongly depended on mycorrhizal infection. As mycorrhizae-inoculated plants have high Zn content compared to the non-inoculated plants, it is not clear whether mycorrhizal inoculation helps plants to have more Zn uptake or it is just a result of high plant growth (dilution effects). Since mycorrhizae inoculation increases the Zn uptake, it is wonder whether mycorrhizal dependency of Zn nutrition is significant rather than P nutrition (Ortas et al. 2002b).

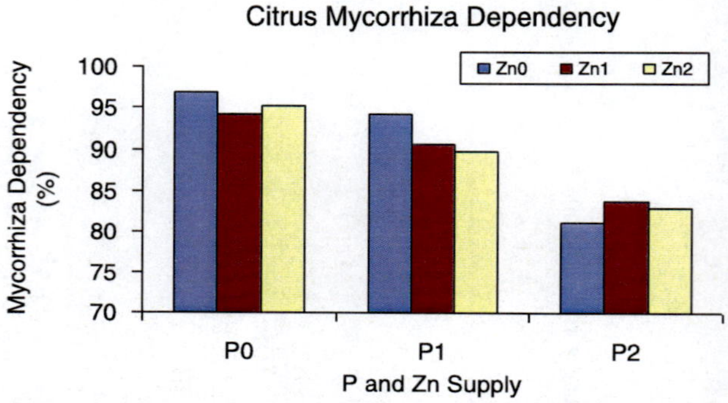

Fig. 23.9 Effect of P and Zn application on mycorrhizal dependency of citrus plant

Camprubi and Calvet (1996) and Tong et al. (2006) reported that *Gigaspora margarita-, G. mosseae-, and G. versiforme–* inoculated citrus seedlings are mycorrhizal dependent; especially the seedlings inoculated with *G. mosseae* had the best vegetative growth, the highest mycorrhizal dependence, and N, P, K, Ca, Zn, and Cu contents. Since the citrus genotypes are varied in terms of mycorrhizal dependency due to soil fertility, it is important to select the mycorrhizal- and non-mycorrhizal-dependent rootstocks in terms of P and Zn nutrition status. It seems that mycorrhizal dependence is an inherent characteristic for which plant nutrient requirement and uptake efficiency are important parameters, especially for P requirement. Considering the importance of mycorrhizal dependence for plant survival, it is of great interest to categorize species according to this characteristic. Wu et al. (2006a) showed that the mycorrhizal infection rate was significantly positively correlated with spore density ($P<0.01$) and negatively correlated with soil available phosphorus content ($P<0.01$), also indicating that higher spore density and lower soil available phosphorus content could accelerate the mycorrhizal infection on citrus roots.

23.5 Role of Mycorrhizae in Citrus Root Growth

The presence of mycorrhizae can have significant effects on the morphology of a plant's root system. Actually many fungi are capable of producing plant growth hormones which change the branching pattern of the root system. Differences among plant species in nutrient acquisition are related to root growth and root colonization with AM fungi (Marschner 1995). AM fungal inoculum did affect citrus morphology that might enhance plant root length or root:shoot ratios. Root growth was usually affected by inoculation (Fig. 23.10). It is difficult to identify potential AM fungal effects on plant function that are not directly related to mycorrhizal enhancement of plant mineral nutrition. However, several researchers (Eissenstat et al. 1993; Graham and Syvertsen 1984; Peng et al. 1993) reported that there is evidence that AM fungi stimulate citrus root growth independent of P nutrition. Sharma et al. (2009) reported that the significant increase in growth parameters, namely, height, diameter, root length, and leaf area was more evident for the seedlings inoculated with *G. fasciculatum* and *G. mosseae*. Similarly, Cho et al.'s (2009) results show that plant growth characteristics, such as plant height, root length, leaf area, number of lateral roots, fresh weight of shoots and roots, and chlorophyll content, were significantly enhanced by AMF-inoculated seedlings compared to non-inoculated seedlings. *G. mosseae*–inoculated citrus seedling's root length, lateral root number, fibrous root number, and length of the seedlings were increased by 22.7%, 35.7%, 160.8%, and 103.2%, respectively (Tong et al. 2006).

Since citrus seedlings with small root systems are vulnerable to desiccation (Davies and Albrigo 1994a), inoculation of citrus roots with AM fungal populations that enhance growth might improve transplant survival of seedling nursery stock into field sites. In our previous studies, Ortas et al. (2002a) have shown that dramatic differences in citrus growth as a function of different mycorrhizal isolates were found. Since soil was partially sterilized, all the growth condition was the same for all inoculums. The causes of differential growth have been suggested to be related to phosphorus limitation (Camprubi and Calvet 1996; Graham et al. 1982) and high carbon costs (Graham et al. 1997; Graham and Eissenstat 1998; Johnson et al. 1997). The results of Tong et al. (2006) showed that the seedlings inoculated with *G. mosseae*, the root length, lateral root number and fibrous root number, and length of the seedlings were increased by 22.7%, 35.7%, 160.8%, and 103.2%, respectively. Espeleta et al. (1999) reported that mycorrhizal roots exhibited lower SRL, lower root/soil respiration, and about 10% lower fine

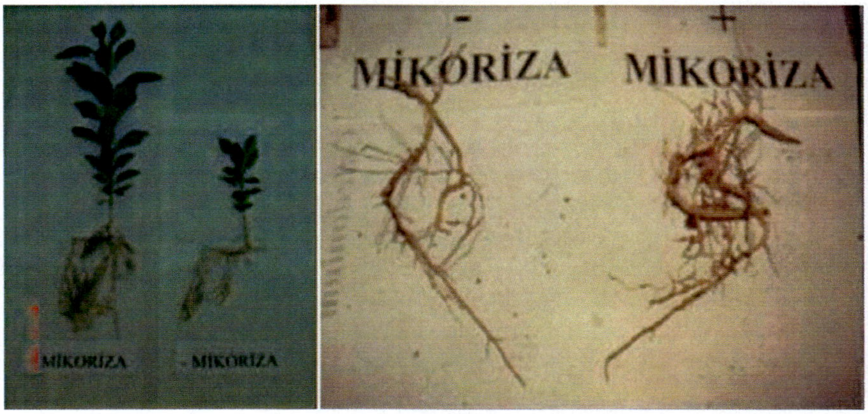

Fig. 23.10 Effect of mycorrhizal inoculation on root growth

root mortality of citrus than non-mycorrhizal roots of citrus. Since root growth is less sensitive than shoot growth to dry soil condition (Hsiao and Jing 1987), most of the time, root growth cannot be great. Root/shoot ratios indicated that rootstock dependency decreased as capacity for root production increased (Nemec 1979).

23.6 Mycorrhizal Hyphae and Citrus Nutrition

Arbuscular mycorrhizae fungi have external hyphae (>10 μm) which enter the intracellular space and often disintegrate to enrich soil fertility. The network of external hyphae is the fungal structure responsible for uptake and transport of nutrient ions. Through this way, plant gets extra benefit from mycorrhizae for making a large mycorrhizosphere to take more nutrient and increase resistance against stress factors. The external hyphae have several functions. Once established, the fungus can begin to spread to all available roots of the plant. The highly branched hyphae have enormous surface area which can absorb water and nutrients. The fungi hyphae link plant and soil, transporting mineral nutrients, especially P, Zn, Cu, and NH_4 to plant and carbon compounds to the soil. Nutrients are taken up by mycorrhizal hyphae with several mechanisms. Nutrient acquisition process involved in nutrient transportation from the soil solution through the membrane of the fungi hyphae is one of the important mechanisms of mycorrhizae. The association between root and mycorrhizal hyphae depends on several soil and environmental factors.

Root colonization is mostly inversely affected by the level of available P content in soils. This level may be differing among different plant communities. Quantitative studies of population of AM fungi have been based on identification of the resting spores under the microscope. It is quite right to know the type of mycorrhizae under the different ecological condition related with plant adaptation mechanisms. Contribution of AM on nutrient, especially P uptake, has been widely studied, and also the dependency on AM varies with soil P level; host plant species and type of AM fungus have been less studied. Hyphae of AM fungi are primarily responsible for helping plant acquire nutrient especially less mobile nutrients. The capacity of mycorrhizal hyphae varies among mycorrhizal fungus; however, the capacity of hyphae on nutrient uptake under the different Zn application and farming systems is not known. The reason is not known why nutrient uptake by the mycorrhizae under the greenhouse conditions is much better than field conditions. May be the question can be asked what is the diverse effect of soil biota under the field conditions on nutrient uptake. Also, mycorrhizae have significant impact on soil physical and biological quality as well. It has been shown that AM hyphae bind soil particles into water-stable aggregates (Tisdall and Oades 1979). Later, Rillig and Mummey (2006) reviewed the relationship between mycorrhiza and aggregate formation and reported that that different species or communities of fungi can promote soil aggregation to different degrees. Wu et al. (2008) have shown that mycorrhizae enhanced>2 mm, 1–2 mm, and>0.25 mm water-stable aggregate fractions, but reduced 0.25–0.5 mm water-stable aggregates. Wright and Upadhyaya (1996) have also shown that this binding due to mycorrhizal hyphae produced glomalin an immunoreactive glycoprotein. Wright and Upadhyaya (1998) indicated that glomalin appears to be as ubiquitous as the AM fungi themselves.

23.7 Mycorrhizae and Plant Protection

Mycorrhizae not only help plants to take up nutrients and water but also increase the resistance of plant against disease and pathogenic organisms. Graham (1986) indicated that mycorrhizae-inoculated citrus plant has more resistance against pathogens. Mycorrhizae also can help control pests such as pathogenic fungi and nematodes by releasing antibiotics into the soil which reduce the risk of infection. It is well known that the presence of mycorrhizal fungi in the plant root system improves plant health and, as a result, can reduce pest damage. When infection has well-established plants are increased resistance to fungal pathogens and nematodes parasites (Sikora 1992). The results of Calvet et al. (1995) showed that in the presence of the nematode, mycorrhizal plants achieved higher values in all growth parameters measured. Since mycorrhizal-inoculated plants are well-fed, plants show resistance against pathogens. When mycorrhizal infection is already established, pathogenic infection is dramatically reduced. Nematode (*Meloidogyne incognita*) inoculation reduced plant yield by 45%, but in case of AM infection, this was reduced up to only 25%. Because of better nutrition, some phytopathological viruses replicate better in mycorrhizal plants than in non-AM plants. Michelini et al. (1993) searched the mycorrhizal presence in the four islands in the Eastern Caribbean, and they found that plant health and degree of VAM fungal infection were related; the healthiest plants were the most highly colonized. Protection of the host plants from soilborne pathogens can be achieved by mycorrhizal inoculation. Using AM fungi increased the resistance of plants against decease and soil and root pathogens.

23.8 Soil and Crop Management

Soil and plant management is directly related with mycorrhizal management under citrus orchard. Use of low solubility fertilizers such as rock phosphate and low concentration of available nutrients can encourage indigenous mycorrhizal fungi colonization. Cover cropping may encourage building up diversity of mycorrhizal population. Also application of mature compost and organic manure avoids excessive nutrient use. This indirectly affects mycorrhizal diversity. On the other hand, excess mechanical cultivation destroys mycorrhizal hyphal networks; excess fertilizer and heavy irrigation also reduce mycorrhizal colonization. Bare ground by using fungicides and mechanical cultivation can cause lack of host plants which can result in a decline in mycorrhizal population. Non-mycorrhizal cover crops may cause decline of MF population also on account of lack of host plants. Use of copper-based fungicides and other fungicides can be directly toxic to mycorrhizae. Incorporation of uncomposted or unmaturated organic matter also can reduce mycorrhizal diversity. Anaerobic soil conditions also can produce toxin compounds which may reduce mycorrhizal spore and subsequent colonization (Burrows and Pfleger 2002; Chen et al. 2004; Douds et al. 1997; Gosling et al. 2006; Ishii and Kadoya 1996; Rutto et al. 2002; Wang et al. 2008a).

In some areas, soils are very low in P nutrition, and on the other hand, in some area as a result of long-term P fertilization with different types of fertilizers, there is a residual P accumulation. Especially under the high soil pH and clay content conditions, availability of P is low and is restricted by formation of calcium phosphate. In Ca-P rich soils, mycorrhizal infection, root growth, and rhizosphere acidification caused by NH_4 nutrition are supposed to be the main factors controlling the P availability in rhizosphere (Ortas and Rowell 2004). There is a considerable fixed P accumulation in the soil, since the farmers are using more P fertilizers without regarding soil and plant analysis results. Heavy application of P fertilizer can inhibit infection formation and spore production. Large imputes of soluble P, associated, for example, with application of superphosphates, can decrease or eliminate mycorrhizal advantages by inhibiting the growth and activity of the vegetative mycelium (Abbott and Robson 1984). Plants growing under the high P fertility in commercial nurseries and greenhouses frequently do not get infected when inoculated with the mycorrhizal fungi. Reduced impute of fertilizer or pesticides will help to contribute to the healthy environment conditions for mycorrhizal development. Use of arbuscular mycorrhiza (AM) as a biological tool for production management of responsive plants, especially the high value crops, is now a reality. Crops and cropping systems which help to enhance the AM potential of soils, in quantity and quality, without compromising the farmers' traditional preference, may help in exploiting the AM potential of soils, provided the native efficient AMF species have not been suppressed by unfavorable agricultural practices.

In a survey work, in citrus orchards grown on northwestern Himalayan region of India, Sharma et al. (2009) indicated that isolated spores were used alone and with combination of N+P fertilize for citrus growth. And they found that *G. fasciculatum* and *G. mosseae* inoculation significantly increased the citrus height, diameter, root length, and leaf area. Plants species and even genotypes within a species mostly differ in their P efficiency. Differences between plant species in P acquisition are related to root growth, rhizosphere acidification, and root colonization with AM fungi. Mycorrhizal infection plays an important role in the utilization of P by strongly exploited soil volume, or infection may modify the mycorrhizosphere soil for higher P availability. Thus, both rhizosphere acidification and mycorrhizal infection may increase P uptake by sparingly solubilizing Ca-P and accumulated P fractions. It has been claimed that in some areas, there is no detectable effect of P application on crop yield because of high amount of P fertilization due to negative effect of phosphorus on mycorrhizal formation. Sometimes, accumulated P becomes fixed by the soil particles which are not easily available to plants. In order to manage the natural mycorrhizal infection, there is a need to know the relationship between the soil P level and the percentage of root colonization for citrus plants. Several agricultural management systems can influence the mycorrhizal infection ratio such as pesticide application, crop rotation, and soil tillage systems (Abbott and Robson 1991).

Pre-crops have significant influence on AMF spore number and infective inoculum density on citrus plant (Panja and Chaudhuri 2004). Barea et al. (1993) reported that productivity of the crops exhibiting considerable mycotrophy was likely to be increased if functionally compatible AM fungi were available to colonize the developing plant root system. Among the many agronomic ways to enhance AMF-crop association reviewed by Thompson (1994), the use of AMF-building hosts in the cropping and fallow cover periods (Douds 1994) has been considered especially worthy of further investigation (Hoffman and Carroll 1995). The presence and diversity of arbuscular mycorrhizal fungi in intensively managed agricultural soil in the Sichuan Province of southwest China with maize, wheat, and sweet orange cultivation was searched by Wang et al. (2008b), and they indicated that the highest mycorrhizal potentials were found in purple soil cropped with maize and citrus, 324 and 278 propagule in 100 g soil, respectively. It is important to manage the abundance and diversity of mycorrhizal fungi strongly influenced by the diversity of vegetation, including ground cover and "weeds" in the orchard, not simply the density of ground cover. Burrows and Pfleger (2002) and Chen et al. (2004) indicated that a more diverse ground cover means more opportunities for root associations with a wider range of mycorrhizal species and tends to reduce soil nitrate levels and produce more biomass, all of which favor mycorrhizae. In a 20-year-old citrus orchard, if cover plant around mandarin plant were constantly cleaned, spore count and diversity are reduced. It may be possible to use mycorrhiza-dependent maize, millet, and onion as pre-crops may help to enhance the subsequent mycorrhizal infection of citrus plants. Panja and Chaudhuri (2004) reported that pre-crop variables individually and cumulatively contributed to the highly significant positive correlation between the AMF potential of the pre-cropped soils and growth of mandarin orange plants through their effect on mycorrhizal root mass development (i.e., extent of mycorrhization) of the mandarin orange plants.

AMF plays an important role in the formation and stability of soil aggregates and contributes to soil physical fertility and quality (Wright and Upadhyaya 1998) by producing glomalin, quantified in soil as glomalin-related soil protein. In general, under no-tillage conditions, mycorrhizal activity in soil is stimulated, which favors nutrient uptake by the plants and contributes to the soil quality. It has been demonstrated that tillage reduced the total AM colonization, spore density, hyphal length density, and phosphatase activity in the no-tillage citrus orchards than tilled orchards in South China (Wang et al. 2011).

23.9 Mycorrhizae-Water Relationship

Citrus plants are so sensitive to drought stress. It has been revealed that drought stress strongly inhibits citrus growth and deleterious fruit quality (Levy et al. 1978, 1979). Wu et al. (2009) indicated that mycorrhizal inoculation has been shown to alleviate symptoms of drought stress. The pioneer effect of mycorrhizae on drought stress of citrus has been started 30 years ago by Levy and Krikun (1980). Results of many experiments were summarized by (Wu et al. 2009) revealing that drought stress also reduced the root colonization. It has been reported that mycorrhizal infection is not only beneficial for mineral nutrition of citrus but also has positive effects on water uptake (Wu and Xia 2004) and utilization (Levy et al. 1983). Srivastava et al. (Srivastava et al. 2002) reported that inoculation of soil with mycorrhizae also helped in regulating the water relations and carbohydrate metabolism of citrus trees. All experiments were done on mycorrhizae-water relation with citrus and confirmed that mycorrhizae-inoculated plants have higher stomatal conductancy, translation, and photosynthesis than non-mycorrhizal one. Relationship between AM symbiosis and water uptake of citrus plant has been extensively studied by Wu et al. (2006b). Also, the effect of mycorrhizae on citrus growth relation with antioxidants was studied. The study of Wu et al. (2006b) also showed that *G. versiforme*–inoculated trifoliate orange (*Poncirus trifoliata*) seedlings potted under well-watered (WW) and water-stressed (WS) conditions had higher shoot dry weight, plant height, and stem diameter in AM than in non-AM seedlings, regardless of the water status.

The potential of arbuscular mycorrhizal (AM) fungi to enhance citrus growth (Menge 1983; Ortas et al. 2002a, b) whole-plant transpiration (Levy et al. 1983), resistance to dry and wet soil condition (Fidelibus et al. 2000), root hydraulic conductivity (Graham and Syvertsen 1984), photosynthesis (Levy and Krikun 1980), and osmotic adjustment (Wu and Xia 2006) has been reported. It was also reported that the enhanced drought resistance of AM plants was independent of plant nutrient uptake, especially P (Auge et al. 1986; Bethlenfalvay et al. 1988). It was concluded that AM fungus colonization enhanced plant growth under drought stress indirectly through affecting the soil moisture retention via glomalin's effect on soil water-stable aggregates, although direct mineral nutritional effects could not be excluded. Auge (2001) reported that mycorrhizal inoculation can enhance drought resistance of plants possibly by maintaining higher growth rates and nutrient uptake including P, Cu, and Zn during drought conditions. Srivastava et al. (Srivastava et al. 2002) indicated that inoculation of soil with mycorrhizae also helped in regulating the water relations and carbohydrate metabolism of citrus trees. Wu and Zou (2009c) reported that the improved nutrient uptake in colonized seedlings demonstrates the potential of AM symbiosis to enhance drought resistance in citrus. Wu et al.'s (2007a) work under greenhouse conditions showed that under water stress, *G. versiforme* inoculation decreased the leaf superoxide anion radical (O_2^-) content by 31%, compared with that under normal water supply. Also, they suggested that the drought resistance of Citrus tangerine leaves was enhanced after *G. versiforme* inoculation. They also indicated that under normal water supply and water stress, *G. versiforme* inoculation increased the leaf P content by 45% and 27% and decreased the leaf malondialdehyde (MDA) and hydrogen peroxide (H_2O_2) contents by 25% and 21% and 16% and 16%, compared with the control, respectively.

Mycorrhizal colonization is strongly affected by the water stress conditions (Wu et al. 2006a). Wu et al. (2007b) showed that development of mycorrhizae was higher under watered than under drought-stressed conditions. They concluded that AM colonization enhances osmotic solute accumulation of trifoliate orange seedlings, thus providing better osmotic adjustment in AM seedlings, under the water stress and other stress factors mycorrhiza may have influence on plant secondary metabolites. Wu and Xia (2006) showed that AM colonization increased the distributed proportions of soluble sugar and NSC to roots. The results of Wu et al. (2006a) suggested that the increased concentrations of antioxidant enzymes and nonenzymatic antioxidants found in AM plants may serve to protect the organism against oxidative damage, enhancing drought tolerance. Also Wu et al. (2007b) in other experiment showed that mycorrhiza-inoculated citrus seedlings accumulated more soluble sugars, soluble starch, and total nonstructural carbohydrates in leaves and roots than corresponding non-AM seedlings regardless of soil-water status. Wu and Xia (2005) indicated the mechanism that AM fungi could enhance the drought tolerance of host plant might be related to the protective system of host plant. Many experiments examining AM effects on drought resistance have shown that when the mycorrhizal symbiosis improves host drought resistance, it does so by aiding drought avoidance (Auge 2001). The results of Wu et al. (2011) showed that higher mycorrhizal hyphal enzymes' activities, soil acid and total phosphatase activities, and plant P contents in AM-colonized seedlings, particularly in *G. mosseae*–colonized seedlings and/or under drought stress, would result in a better growth of the host plants, which might be the basis for enhancing drought tolerance in plants.

23.10 Mycorrhizae and Salt Stress

Under saline environment conditions, mycorrhizae can occur (Hildebrandt et al. 2001; Johnson-Green et al. 1995; Pond et al. 1984). Hildebrandt et al. (2001) reported mycorrhizal colonization as well as a high spore density in Central European salt marshes. It has been reported by several researchers that mycorrhizal root colonization is reduced in the presence of sodium chloride (NaCl), probably due to a direct effect of NaCl on the mycorrhizal fungi (Dixon et al. 1993; Juniper and Abbott 2006; McMillen et al. 1998). Juniper and Abbott (2006) tested the effect of NaCl on spore germination and found that for some fungi, the reduction in the amount of hyphae in the presence of increasing concentrations of NaCl was primarily associated with a delay in spore germination, while for other fungi, the specific rate of hyphal production was also substantially reduced in the presence of NaCl. The application of microorganisms such as arbuscular mycorrhizal fungi to enhance salt resistance is quite well known (Wu et al. 2009). Syvertsen and Levy (2005) reported that mycorrhizae can affect the salt tolerance of citrus roots and may increase chloride (Cl^-) uptake. On the other hand, Na concentration was not affected by mycorrhizae (Syvertsen and Levy 2005). Hartmond et al. (1987) showed that mycorrhiza-inoculated citrus plant leaves have high Cl^- than non-mycorrhizal one. AM fungi can positively influence plant establishment and growth by improving nutrient uptake, increasing tolerance to drought and salt stress, and increasing resistance against soilborne pathogens (Azcon-Aguilar and Barea 1997).

Wu and Zou (2009a) used the non-colonized and colonized seedlings by *G. mosseae* or *Paraglomus occultum* which were exposed to salt stress by irrigation with 100 mM NaCl solutions. And they found that salt stress significantly depressed *G. mosseae* colonization but not P. occultum colonization. Also, they indicated that mycorrhizal association markedly increased both plant performance (leaf number, leaf area, shoot, and root dry weights) and leaf relative water content of citrus seedlings exposed to salt stress. Under controlled conditions Juniper and Abbott (2006) showed that germination of AM spores was delayed in the presence of NaCl treatments. AM symbiosis can increase host resistance to salinity and drought stress in dry area. Since mycorrhizae increase the host resistance, it is important strategy to use in agricultural area. In many studies, AM has been shown to increase plant yield in saline soil (Hirrel and Gerdemann 1980; Ojala et al. 1983) along with some controversial results, mainly work concentrated on reducing the intention of Na, Cl element on root cell, and mycorrhizal hyphae development. The results of Wu and Zou (2009a) showed that mycorrhizae inoculation promotes root concentrations of K^+, Ca^{2+}, and Mg^{2+} at all salinity levels. Also, they figure out that it seems that mycorrhizal inoculation possesses the potential to enhance salt tolerance of citrus. Finally, all the results indicated that there are mycorrhizae in salt soils, and heavy salt can dally root colonization, and also mycorrhizae alleviate the salt effect which is much more, depending on host plant species than on environmental stresses.

23.11 Mycorrhizae and Other Stress Factors

Drought, salinity, heavy metals, or heat are serious problems in many parts of the world, particularly in arid and semiarid areas. Arbuscular mycorrhizal fungi (AMF) have repeatedly been demonstrated to alleviate soil and environmental stress such as heavy metal, salt, drought, salt, and temperature stress on plants. The results of Wu and Zou (2010) also showed that mycorrhizal formation had the beneficial effects on growth, photosynthesis, root morphology, and part nutrient uptake of citrus seedlings grown at moderate temperature, but the beneficial roles of arbuscular mycorrhizae were almost lost at low temperature. Wu and Zou (2010) also showed that mycorrhizae-inoculated citrus seedlings grown at 25°C maintained better stem diameter, plant height, leaf area, root and total dry weights, higher photosynthetic rate, transpiration rate and stomatal conductance, higher root volume, and more uptake of P, Ca, and Mg relative to corresponding non-mycorrhizal control.

23.12 Future Research

- Because of complexity of the microbe-soil-plant system, the effect of several selected and indigenous mycorrhizal species on citrus growth under different soil and climate conditions is needed to be investigated.
- The effect of soil and climate on diversity of AMF colonizing needs to be searched.
- Very recently, mainly under citrus growing regions including the Mediterranean region, the climate is changing. The climate change will have further effect on citrus

growth. There is a need to study the effect of temperature rise and CO_2 emission on citrus growth under arid and semiarid conditions. Further research needs to apply AM fungi to improve plant growth and for biological control of target diseases in agro systems, especially under climate change conditions.

- Effect of mycorrhizal inoculation on carbon storage and sequestration is needed to be searched.
- Since response of mycorrhizal hyphae to drought stress was less studied, there is a need to work on the effect of mycorrhizae and P nutrition on water uptake. Still, the effects of AMF on soil phosphatase activity and P contents in citrus rhizosphere are poorly known, particularly under drought and salty stress conditions.
- Under arid and semiarid conditions, water and fertilizer-use-efficiency are among the major production-related constraints. Fertigation (application of nutrients through the irrigation) should be studied with mycorrhizal inoculation. Effect of fertigation on tree growth, fruit yield, quality, the reserve pool of soil nutrients, and consequently the plant nutritional status should be search. Drip irrigation and deep-drip irrigation need to be searched in mycorrhizal work for root growth and nutrient uptake. Site-specific nutrient management is a dynamic concept and need to be work.
- For better citrus nutrition under mycorrhizal and non-mycorrhizal conditions, an appropriate diagnostic and recommendation tools based on leaf and soil analysis need to be developed. Still, there is need for proper organic and inorganic fertilizer recommendation.
- The role of fertilizer and mycorrhizal inoculation flowering, fruit set, and fruit quality is important.
- Under low input agriculture systems, the native mycorrhizal population through choice of crops and cropping systems can be taken advantage of in ecological management for better inoculation.
- Citrus roots across different soil-management practices need to be deeply search for better management of indigenous mycorrhizal effect on plant growth.
- Apart from soil and crop management, modern soil biotechnology tools can be applied to improve citrus quality and production.

References

Abbott LK, Robson AD (1984) Colonization of the root-system of subterranean clover by 3 species of vesicular arbuscular mycorrhizal fungi. New Phytol 96(2):275–281

Abbott LK, Robson AD (1991) Field management of mycorrhizal fungi. In: Keisterand DL, Cregan PB (eds) The rhizosphere and plant growth. Kluwer Academic, Dordrecht, pp 355–362

Antunes V, Cardoso E (1991) Growth and nutrient status of citrus plants as influenced by mycorrhiza and phosphorus application. Plant Soil 131(1):11–19

Auge RM (2001) Water relations, drought and vesicular-arbuscular mycorrhizal symbiosis. Mycorrhiza 11(1):3–42

Auge RM, Schekel KA, Wample RL (1986) Osmotic adjustment in leaves of va mycorrhizal and nonmycorrhizal rose plants in response to drought stress. Plant Physiol 82(3):765–770

Azcon-Aguilar C, Barea JM (1997) Applying mycorrhiza biotechnology to horticulture: significance and potentials. Sci Hortic 68(1–4):1–24

Barea JM, Azcon R, Azcon-Aguilar C (eds) (1993) Mycorrhiza and crops, vol 9, Advances in plant pathology. Academic, London, pp 167–189

Bethlenfalvay GJ, Brown MS, Ames RN et al (1988) Effects of drought on host and endophyte development in mycorrhizal soybeans in relation to water-use and phosphate-uptake. Physiol Plant 72(3):565–571

Bevington K (2002) Know your rootstocks. In National Citrus Nursery Workshop, Mildura (Australian Citrus Propagation Association Inc.). http://www.auscitrus.com.au/files/Auscitrus_Nursery_Workshop_Proceedings.pdf

Bhattacharya P (1999) Use of biofertilizers in citrus. In: Singh S (ed) Citriculture. National Research Centre for Citrus, Nagpur, pp 194–204

Burrows RL, Pfleger FL (2002) Arbuscular mycorrhizal fungi respond to increasing plant diversity. Can J Bot Revue Canadienne De Botanique 80:120–130

Calvet C, Pinochet J, Camprubi A et al (1995) Increased tolerance to the root-lesion nematode Pratylenchus-vulnus in mycorrhizal micropropagated ba-29 quince rootstock. Mycorrhiza 5(4):253–258

Camprubi A, Calvet C (1996) Isolation and screening of mycorrhizal fungi from citrus nurseries and orchards and inoculation studies. HortScience 31(3):366–369

Chen X, Tang JJ, Fang ZG et al (2004) Effects of weed communities with various species numbers on soil features in a subtropical orchard ecosystem. Agric Ecosyst Environ 102(3):377–388

Cho EJ et al (2009) Effects of AMF inoculation on growth of Panax ginseng CA Meyer seedlings and on soil structures in mycorrhizosphere. Sci Hortic 122(4):633–637

da Rocha MR, de Oliveira E, Correa GC et al (1994) Effect of phosphorus doses and mycorrhizal fungi on mineral nutrition and growth of 'Cleopatra' mandarin (*Citrus reshni* Hort ex Tan). Pesquisa Agropecuaria Brasileira 29(5):725–731

Davies FS, Albrigo LG (1994a) Citrus. In: Atherton J, Rees A (eds) Crop production science in horticulture. CAB International, Wallingford

Davies FS, Albrigo LG (eds) (1994b) Citrus. Crop production science in horticulture, vol 2. CAB International, Wallingford

de Souza PVD, Schmitz JAK, de Freitas RS et al (2002) Identification and quantification of native arbuscular mycorrhizae fungi of citrus in the State of Rio Grande do Sul, Brazil. Pesquisa Agropecuaria Brasileira 37(4):553–558

Dixon RK, Garg VK, Rao MV (1993) Inoculation of Leucaena and Prosopis seedlings with Glomus and rhizobium species in saline soil - rhizosphere relations and seedling growth. Arid Soil Res Rehabil 7(2):133–144

Douds DD (1994) Effects of agricultural practices upon mycorrhizal fungi. Mycorrhiza News 6(2):1–5

Douds DD, Galvez L, Franke-Snyder M et al (1997) Effect of compost addition and crop rotation point upon VAM fungi. Agric Ecosyst Environ 65(3):257–266

Dutra PV, Abad M, Almela V et al (1996) Auxin interaction with the vesicular-arbuscular mycorrhizal fungus *Glomus intraradices* Schenck & Smith improves vegetative growth of two citrus rootstocks. Sci Hortic 66(1–2):77–83

Eissenstat DM, Graham JH, Syvertsen JP et al (1993) Carbon economy of sour orange in relation to mycorrhizal colonization and phosphorus status. Ann Bot 71(1):1–10

Espeleta JF, Eissenstat DM, Graham JH (1999) Citrus root responses to localized drying soil: a new approach to studying mycorrhizal effects on the roots of mature trees. Plant Soil 206(1):1–10

Fidelibus MW, Martin CA, Wright GC et al (2000) Effect of arbuscular mycorrhizal (AM) fungal communities on growth of 'Volkamer' lemon in continually moist or periodically dry soil. Sci Hortic 84(1–2):127–140

Gadkar V, David-Schwartz R, Kunik T et al (2001) Arbuscular mycorrhizal fungal colonization. Factors involved in host recognition. Plant Physiol 127(4):1493–1499

Gerdemann JW (1975) VA mycorrhizae. In: Torrey JG, Clarkson DT (eds) Development and function of roots. Academic, London, pp 575–591

Gosling P, Hodge A, Goodlass G et al (2006) Arbuscular mycorrhizal fungi and organic farming. Agric Ecosyst Environ 113(1–4):17–35

Graham JH (1986) Citrus mycorrhizae – potential benefits and interactions with pathogens. HortScience 21(6):1302–1306

Graham JH, Eissenstat DM (1994) Host genotype and the formation and function of va mycorrhizae. Plant Soil 159(1):179–185

Graham JH, Eissenstat DM (1998) Field evidence for the carbon cost of citrus mycorrhizas. New Phytol 140(1):103–110

Graham JH, Syvertsen JP (1984) Influence of vesicular arbuscular mycorrhiza on the hydraulic conductivity of roots of 2 citrus rootstocks. New Phytol 97(2):277–284

Graham JH, Syvertsen JP (1985) Host determinants of mycorrhizal dependency of citrus rootstock seedlings. New Phytol 101(4): 667–676

Graham JH, Timmer LW (1985) Rock phosphate as a source of phosphorus for vesicular-arbuscular mycorrhizal development and growth of citrus in a soilless medium. J Am Soc Hortic Sci 110(4):489–492

Graham JH, Leonard RT, Menge JA (1982) Interaction of light-intensity and soil-temperature with phosphorus inhibition of vesicular arbuscular mycorrhiza formation. New Phytol 91(4):683–690

Graham JH, Drouillard DL, Hodge NC (1996) Carbon economy of sour orange in response to different *Glomus spp.* Tree Physiol 16(11–12):1023–1029

Graham JH, Duncan LW, Eissenstat DM (1997) Carbohydrate allocation patterns in citrus genotypes as affected by phosphorus nutrition, mycorrhizal colonization and mycorrhizal dependency. New Phytol 135(2):335–343

Hartmond U, Schaesberg NV, Graham JH et al (1987) Salinity and flooding stress effects on mycorrhizal and nonmycorrhizal citrus rootstock seedlings. Plant Soil 104(1):37–43

Hattingh MJ, Gerdemann JW (1975) Inoculation of Brazilian sour orange seed with an endomycorrhizal fungus. Phytopathology 65(9):1013–1016

Herring JR, Fantel RJ (1993) Phosphate rock demand into the next century: impact on world food supply. Nat Resour Search 2:226–246

Hildebrandt U et al (2001) Arbuscular mycorrhizal colonization of halophytes in Central European salt marshes. Mycorrhiza 10(4):175–183

Hirrel MC, Gerdemann JW (1980) Improved growth of onion and bell pepper in saline soils by 2 vesicular-arbuscular mycorrhizal fungi. Soil Sci Soc Am J 44(3):654–655

Hoffman CA, Carroll CR (1995) Xan we sustain the biological basis of agriculture. Ann Rev Ecol Syst 26:69–92

Hsiao TC, Jing J (1987) Leaf and root expansive growth in response to water deficits. In: Cosgrove DJ, Knievel DP (eds) Physiology of cell expansion during plant growth. American Society of Plant Physiology, Rockville, pp 180–192

Ishii T, Kadoya K (1996) Utilisation of vesicular-arbuscular mycorrhizal fungi in citrus orchards. In: Manicom B, Robinson J, Du Plessis SF et al. (eds) Proceedings of the international society of citriculture, South City South, Africa, 12–17 May 1996, pp 777–780

Jakobsen I (1995) Transport of phosphorus and carbon in VA mycorrhizas. In: Varma A, Hock B (eds) Mycorrhiza. Springer, Berlin, pp 297–324

Jifon JL, Graham JH, Drouillard DL et al (2002) Growth depression of mycorrhizal Citrus seedlings grown at high phosphorus supply is mitigated by elevated CO_2. New Phytol 153(1):133–142

Johnson NC, Graham JH, Smith FA (1997) Functioning of mycorrhizal associations along the mutualism-parasitism continuum. New Phytol 135(4):575–586

Johnson-Green P, Kenkel N, Booth T (1995) The distribution and phenology of arbuscular mycorrhizae along an inland salinity gradient. Can J Bot 73:1318–1327

Juniper S, Abbott LK (2006) Soil salinity delays germination and limits growth of hyphae from propagules of arbuscular mycorrhizal fungi. Mycorrhiza 16(5):371–379

Kleinchmidt GD, Gerdemann JW (1972) Stunting of citrus seedlings in fumigated nursery soils related to the absence of endomycorrhizae. Phytopathology 62:1447–1453

Kothari SK, Marschner H, Romheld V (1991) Contribution of the va mycorrhizal hyphae in acquisition of phosphorus and zinc by maize grown in a calcareous soil. Plant Soil 131(2):177–185

Krikun J, Levy Y (1980) Effect of vesicular arbuscular mycorrhiza on citrus growth and mineral-composition. Phytoparasitica 8(3):195–200

Labanaus CK, Bitters WP (1974) Influence of rootstocks and interstocks on macro-nutrients and micro-nutrients in Valencia orange leaves. Calif Agric 28(9):12–13

Levy Y, Krikun J (1980) Effect of vesicular-arbuscular mycorrhiza on citrus-jambhiri water relations. New Phytol 85(1):25–31

Levy Y, Bielorai H, Shalhevet J (1978) Long-term effects of different irrigation regimes on grapefruit tree development and yield. J Am Soc Hortic Sci 103(5):680–683

Levy Y, Shalhevet J, Bielorai H (1979) Effect of irrigation regime and water salinity on grapefruit quality. J Am Soc Hortic Sci 104(3): 356–359

Levy Y, Syvertsen JP, Nemec S (1983) Effect of drought stress and vesicular arbuscular mycorrhiza on citrus transpiration and hydraulic conductivity of roots. New Phytol 93(1):61–66

Lukac M, Calfapietra C, Lagomarsino A et al (2010) Global climate change and tree nutrition: effects of elevated CO2 and temperature. Tree Physiol 30(9):1209–1220

MacFall JS (1994) Effects of ectomycorrhizae on biogeochemistry and soil structure. In: Pfleger FL, Linderman RG (eds) Mycorrhizae and plant health. APS Press, St. Paul, pp 213–237

Manjunath A, Mohan R, Bagyaraj DJ (1983) Response of citrus to vesicular arbuscular mycorrhizal inoculation in unsterile soils. Can J Bot Revue Canadienne De Botanique 61(10):2729–2732

Marchal J (1987) Citrus. In: Prevel PM, Gagnard J, Gautier P (eds) Plant Analysis. Lavoiser Publishing, Newyork, USA

Marschner H (1995) Mineral nutrition of high plants. Academic, London

Marschner H, Dell B (1994) Nutrient-uptake in mycorrhizal symbiosis. Plant Soil 159(1):89–102

McMillen BG, Juniper S, Abbott LK (1998) Inhibition of hyphal growth of a vesicular-arbuscular mycorrhizal containing sodium chloride of infection from fungus in soil limits the spread spores. Soil Bio Biochem 30(13):1639–1646

Menge JA (1983) Utilization of vesicular-arbuscular mycorrhizal fungi in agriculture. Can J Bot Revue Canadienne De Botanique 61(3):1015–1024

Menge JA, Johnson ELV, Platt RG (1978) Mycorrhizal dependency of several citrus cultivars under three nutrient regimes. New Phytol 81(3):553–559

Menge JA et al (1982) Predicting mycorrhizal dependency of troyer citrange on glomus-fasciculatus in California citrus soils and nursery mixes. Soil Sci Soc Am J 46(4):762–768

Michelini S, Nemec S, Chinnery LE (1993) Relationships between environmental-factors and levels of mycorrhizal infection of citrus on 4 islands in the eastern caribbean. Trop Agric 70(2):135–140

Nemec S (1979) Response of 6 citrus rootstocks to 3 species of glomus, a mycorrhizal fungus. Citrus Indus 60(5):5–14

Nemec S, Menge JA, Platt RG et al (1981) Vesicular-arbuscular mycorrhizal fungi associated with citrus in Florida and California and notes on their distribution and ecology. Mycologia 73(1):112–127

Ojala JC, Jarrell WM, Menge JA et al (1983) Influence of mycorrhizal fungi on the mineral-nutrition and yield of onion in saline soil. Agron J 75(2):255–259

Ortas I, Rowell DL (2004) Effect of ammonium and nitrate on indigenous mycorrhizal infection, rhizosphere pH change, and phosphorus uptake by sorghum. Commun Soil Sci Plant Anal 35(13–14): 1923–1944

Ortas I, Ortakci D, Kaya Z (2002a) Various mycorrhizal fungi propagated on different hosts have different effect on citrus growth and nutrient uptake. Commun Soil Sci Plant Anal 33(1–2):259–272

Ortas I, Ortakci D, Kaya Z et al (2002b) Mycorrhizal dependency of sour orange in relation to phosphorus and zinc nutrition. J Plant Nutr 25(6):1263–1279

Panja BN, Chaudhuri S (2004) Exploitation of soil arbuscular mycorrhizal potential for AM-dependent mandarin orange plants by pre-cropping with mycotrophic crops. Appl Soil Ecol 26(3):249–255

Peng SB, Eissenstat DM, Graham JH et al (1993) Growth depression in mycorrhizal citrus at high-phosphorus supply – analysis of carbon costs. Plant Physiol 101(3):1063–1071

Poerwanto R, Inoue H, Kataoka I (1989) Effects of temperature on the morphology and physiology of the roots of trifoliate orange budded with satsuma mandarin. J Jpn Soc Hortic Sci 58(2):267–274

Pond E, Menge J, Jarrell W (1984) Improved growth of tomato in salinized soil by vesicular-arbuscular mycorrhizal fungi collected from saline soils. Mycologia 76:74–84

Read HS, Fremont T (1935) Factors that influence the formation and development of mycorrhizal associations in citrus roots. Phytopathology 25:645–647

Redecker D, Kodner R, Graham LE (2000) Glomalean fungi from the Ordovician. Science 289:1920–1921

Rennenberg H, Schmidt S (2010) Perennial lifestyle-an adaptation to nutrient limitation? Tree Physiol 30(9):1047–1049

Rillig MC, Mummey DL (2006) Mycorrhizas and soil structure. New Phytol 171(1):41–53

Rutto KL, Mizutani F, Moon G et al (2002) The relationship between cultural practices and arbuscular mycorrhizal (AM) activity in orchards under different management systems. J Jpn Soc Hortic Sci 71(5):601–609

Sharma SD, Kumar P, Singh SK et al (2009) Indigenous AM fungi and Azotobacter isolates, and their screening from citrus seedlings at different levels of inorganic fertilizers application. Indian J Hortic 66(2):183–189

Sikora RA (1992) Management of the antagonistic potential in agricultural ecosystems for the biological-control of plant parasitic nematodes. Ann Rev Phytopathol 30:245–270

Spiegel-Roy P, Goldschmidt EE (1996) Biology of citrus. Cambridge University Press, Cambridge

Srivastava AK, Singh S, Marathe RA (2002) Organic citrus: soil fertility and plant nutrition. J Sustain Agric 19(3):5–29

Syvertsen J, Levy Y (2005) Salinity interactions with other abiotic and biotic stresses in citrus. Horttechnology 15(1):100–103

Thompson JP (1994) What is the potential for management of mycorrhizas in agriculture? In: Robson AD, Abbott LK, Malajcuk N (eds) Management of mycorrhiza in agriculture, horticulture and forestry. Kluwer Academic, Dordrecht, pp 191–200

Timmer LW, Leyden RF (1978) Stunting of citrus seedlings in fumigated soils in Texas and its correction by phosphorus fertilization and inoculation with mycorrhizal fungi. J Am Soc Hortic Sci 103(4):533–537

Timmer LW, Leyden RF (1980) The relationship of mycorrhizal infection to phosphorus-induced copper deficiency in sour orange seedlings. New Phytol 85(1):15–23

Tisdall JM, Oades JM (1979) Stabilization of soil aggregates by the root systems of ryegrass. Aust J Soil Res 17(3):429–441

Tong R, Yang X, Li D (2006) Effects of interspecies difference of arbuscular mycorrhizal fungi on Citrus grandis cv. Changshou Shatian You seedlings vegetative growth and mineral contents. Yingyong Shengtai Xuebao 17(7):1229–1233

Treeby MT (1992) The role of mycorrhizal fungi and nonmycorrhizal microorganisms in iron nutrition of citrus. Soil Bio Biochem 24(9):857–864

Verbruggen E, Kiers ET (2010) Evolutionary ecology of mycorrhizal functional diversity in agricultural systems. Evol Appl 3(5–6): 547–560

Vinayak K, Bagyaraj DJ (1990) Vesicular-arbuscular mycorrhizae screened for troyer citrange. Bio Fertil Soils 9(4):311–314

Wang M, Xia R (2009) Effects of arbuscular mycorrhizal fungi on growth and iron uptake of Poncirus trifoliata under different pH. Weishengwu Xuebao 49(10):1374–1379

Wang MY et al (2008a) Arbuscular mycorrhizal enhancement of iron concentration by Poncirus trifoliata L. Raf and Citrus reticulata Blanco grown on sand medium under different pH. Bio Fertil Soils 45(1):65–72

Wang YY et al (2008b) Diversity and infectivity of arbuscular mycorrhizal fungi in agricultural soils of the Sichuan Province of mainland China. Mycorrhiza 18(2):59–68

Wang M, Xia R, Wang P (2009) Effects of arbuscular mycorrhizal fungi on different iron species in Poncirus trifoliata rhizospheric soil. Weishengwu Xuebao 49(10):1347–1352

Wang P et al (2011) Arbuscular mycorrhizal development, glomalin-related soil protein (GRSP) content, and rhizospheric phosphatase activity in citrus orchards under different types of soil management. J Plant Nutr Soil Sci 174(1):65–72

Watanarojanaporn N, Boonkerd N, Wongkaew S et al (2011) Selection of arbuscular mycorrhizal fungi for citrus growth promotion and *Phytophthora* suppression. Sci Hortic 128(4):423–433

Wright SF, Upadhyaya A (1996) Extraction of an abundant and unusual protein from soil and comparison with hyphal protein of arbuscular mycorrhizal fungi. Soil Sci 161(9):575–586

Wright SF, Upadhyaya A (1998) A survey of soils for aggregate stability and glomalin, a glycoprotein produced by hyphae of arbuscular mycorrhizal fungi. Plant Soil 198(1):97–107

Wu Q-s, Xia R-x (2004) Effects of arbuscular mycorrhizal fungi on plant growth and osmotic adjustment matter content of trifoliate orange seedling under water stress. Zhiwu Shengli yu Fenzi Shengwuxue Xuebao 30(5):583–588

Wu Q-s, Xia R (2005) Effects of AM fungi on drought tolerance of citrus grafting seedling trifoliate orange/cara cara. Yingyong Shengtai Xuebao 16(5):865–869

Wu QS, Xia RX (2006) Arbuscular mycorrhizal fungi influence growth, osmotic adjustment and photosynthesis of citrus under well-watered and water stress conditions. J Plant Physiol 163(4):417–425

Wu QS, Zou YN (2009a) Arbuscular mycorrhizal symbiosis improves growth and root nutrient status of citrus subjected to salt stress. Scienceasia 35(4):388–391

Wu QS, Zou YN (2009b) The effect of dual application of arbuscular mycorrhizal fungi and polyamines upon growth and nutrient uptake on trifoliate orange (*Poncirus trifoliata)* seedlings. Notulae Botanicae Hort Agrobotanici Cluj-Napoca 37(2):95–98

Wu QS, Zou YN (2009c) Mycorrhizal influence on nutrient uptake of citrus exposed to drought stress. Philipp Agric Sci 92(1):33–38

Wu QS, Zou YN (2010) Beneficial roles of arbuscular mycorrhizas in citrus seedlings at temperature stress. Sci Hortic 125(3): 289–293

Wu Q, Xia R, Zou Y (2006a) Arbuscular mycorrhizal fungal growth on citrus roots and its correlations with soil available phosphorus content and phosphatase activity. Yingyong Shengtai Xuebao 17(4): 685–689

Wu QS, Xia RX, Zou YN (2006b) Reactive oxygen metabolism in mycorrhizal and non-mycorrhizal citrus (*Poncirus trifoliata*) seedlings subjected to water stress. J Plant Physiol 163(11): 1101–1110

Wu Q-s, Zou Y-m, Xia R-x (2007a) Effects of arbuscular mycorrhizal fungi on reactive oxygen metabolism of Citrus tangerine leaves under water stress. Yingyong Shengtai Xuebao 18(4):825–830

Wu QS, Zou YN, Xia RX et al (2007b) Five Glomus species affect water relations of citrus tangerine during drought stress. Bot Stud 48(2):147–154

Wu QS, Xia RX, Zou YN (2008) Improved soil structure and citrus growth after inoculation with three arbuscular mycorrhizal fungi under drought stress. Eur J Soil Bio 44(1):122–128

Wu QS, Levy Y, Zou YN (eds) (2009) Arbuscular mycorrhizae and water relations in citrus. In: Tennant P, Benkeblia N (eds) Citrus II: Tree and forestry science and biotechnology. University of the West Indies. Jamaica 3:105–112

Wu QS, Zou YN, He XH (2011) Differences of hyphal and soil phosphatase activities in drought-stressed mycorrhizal trifoliate orange (*Poncirus trifoliata*) seedlings. Sci Hortic 129(2): 294–298

Wutscher HK, Dube D (1977) Performance of young nucellar grapefruit on 20 rootstocks. J Am Soc Hortic Sci 102(3): 267–270

Wutscher HK, Shull AV (1976) Performance of Orlando tangelo on 16 rootstocks. J Am Soc Hortic Sci 101(1):88–91

Youpensuka S, Lordkaewb S, Rerkasemc B (2008) Arbuscular mycorrhizal fungi associated with tangerine (*Citrus reticulata*) in Chiang Mai province, northern Thailand, and their effects on the host plant. Scienceasia 34:259–264

Organic Citrus: Soil Fertility and Plant Nutrition Management

24

Stefano Canali, Giancarlo Roccuzzo, Fabio Tittarelli, Corrado Ciaccia, Stagno Fiorella, and Francesco Intrigliolo

Abstract

During the last decade, the organic food and farming (OFF) sector has grown considerably worldwide. Citrus play an important role in organic farming systems, being one of the most highly demanded products on the market for organic produce. In this chapter, the criteria for citrus orchards fertility management and plant nutrition in the organically managed agro-ecosystems are discussed in the light of the most relevant scientific literature. Moreover, two case studies carried out in Southern Italy and aimed at comparing conventional and organic orange management in terms of yield, yield quality and long-term impact on soil fertility are reported. The body of knowledge available and the results presented demonstrate that organic citrus management is a technically feasible option for citrus growers. In addition, the shift to organic farming could contribute to enhance the environmental sustainability of citrus productions in the long term.

Keywords

Organic farming • Organic citrus • Organic fertilisers • Compost • Long-term field experiment • Field survey • Soil organic matter • Soil quality • Soil fertility • N isotope techniques

24.1 Importance of the Organic Citrus Sector in the World

Citrus products, labelled as 'organic', are those certified as having been produced through clearly defined organic production methods (i.e. EC Regulation 834/2007; US NOP Final Rule 2000; JAS 2001). The compliance of the grower with these methods is verified by an independent organisation (generally called certification body) accredited by an authority (e.g. national authority in the producing or importing country).

Organic citrus is still a niche and makes about 1–2% of the global citrus production (Kilcher 2005) and 0.6% of the world citrus cultivated area (FAOSTAT 2009). As no crop details were available for some of the world's leading citrus producers such as China, Brazil and India, it can be assumed that the world figures for area under organic citrus are higher (IFOAM FIBL 2011). However, it is increasing year by year in parallel with the increasing demand for organic products.

Latin and North America, Europe and Near East represent the main areas for organic citrus cultivation, and, more in particular, many tropical countries in the Americas are extending their production (Kilcher 2005). The main markets are the European Community and the United States, which are also the world's largest two producers.

In EU-27, the organic citrus area has grown in all southern European countries and is at 28,000 ha (Willer and Kilcher 2009). Supply has increased significantly in the previous years.

S. Canali (✉) • F. Tittarelli • C. Ciaccia
Soil Plant System Research Centre, Agricultural Research Council, Via della Navicella 2, 00184 Rome, Italy
e-mail: stefano.canali@entecra.it; fabio.tittarelli@entecra.it; corrado.ciaccia@entecra.it

G. Roccuzzo • S. Fiorella • F. Intrigliolo
Citrus and Mediterranean Crops Research Centre, Agricultural Research Council, Corso Savoia 190, 95024 Acireale (CT), Italy
e-mail: giancarlo.roccuzzo@entecra.it; fiorella.stagno@entecra.it; francesco.intrigliolo@entecra.it

A.K. Srivastava (ed.), *Advances in Citrus Nutrition*,
DOI 10.1007/978-94-007-4171-3_24, © Springer Science+Business Media B.V. 2012

The EU-27 market for certified organic fruit and vegetables was estimated at US$ 1.7 billion in 2002 (*FoodNews* 2003), accounting for between 15% and 20% of total retail sales of organic products. Among the different organic produce categories, citrus fruits are the most widely consumed. Orange (*Citrus sinensis* (L.) Osbeck) comes before banana (*Musa spp*. L.) as the most consumed organic fruit in the EU. No data are available for the sales value of organic fresh citrus fruits. However, it has been estimated that they represent between 5% and 7% of fresh organic produce sales, i.e. between US$ 70 and 100 million. In terms of volumes, it was estimated that the EU consumed over 130,000 t of certified organic citrus in 2000. This figure does not take into account organic citrus fruit that was not certified, which may represent one-third of EU organic citrus output. According to Hamm et al. (2002), the EU consumed over 350,000 t of fresh certified organic fruit in 2000, 37% of which represented by citrus fruits (Liu 2003). The EU market for fresh organic oranges, tangerines (*Citrus tangerina L.*) and lemons (*Citrus limon* (L.) Burm.f.) is dominated by Italy and Spain. More in particular, Italy remains the major producer country – in the traditional citrus growing regions of Sicily and Calabria – with a total area of 22,062 ha (Willer and Kilcher 2009), representing the 12.8% of total citrus Italian area (FAOSTAT 2009).

The United States is the country with the largest market for organic foods and beverages in the world, and retail sales of these products were estimated at close to US$ 12 billion in 2003 (International Trade 2002a, b). Fresh organic citrus consumption accounted for 64,000 t by year (Liu 2003).

About the citrus juice origins and cultivation areas, the most important orange juice producers are Brazil (1.4 million of tons, Mt) and the United States (0.6 Mt; USDA Citrus update, 2011). There is some production of organic grapefruit (*Citrus paradisi L.*) juice (e.g. Cuba, Israel and the United States) and lemon juice (Argentina, Spain), but volumes have been very low so far.

The market for organic citrus juices is presently extremely small, accounting for some 0.3% of total citrus juice consumption (Liu 2003). The bulk of organic citrus juice consists of orange juice. Citrus juices other than orange are almost negligible. Similar to the conventional sector, organic orange juice is marketed in two main forms: frozen concentrated (FCOJ) and not from concentrate (NFCJ). While FCOJ has long dominated the market, consumption of NFCJ has increased rapidly in recent years, notably in North America.

24.2 Principles of Soil Management and Plant Nutrition in Organic Farming

Soil fertility is fundamental in determining the productivity of all farming systems, and it is most commonly defined in terms of the ability of a soil to supply nutrients to crops. However, Palm and Swift (2002) suggested that it is more helpful to view soil fertility as an ecosystem concept, integrating the diverse soil functions, including nutrient supply, which promote plant production.

Agriculture, over time, has made a simplification of ecosystems, replacing the rich biological communities with a few plant species grown. From an ecological point of view, agriculture is hence a modification of the natural ecosystem, and the interactions between all the agricultural system's components should be considered when designing and managing the system: this ecology-based approach is the 'agroecology'. At the heart of agroecology, there is the idea that a crop field is an ecosystem in which ecological processes, such as nutrient cycling, predator/prey interactions, competition, commensalism and successional changes, also occur. As a result, a number of researchers (Spedding 1975; Conway 1981a, b, 1985; Ellen 1982; Altieri 1983; Gliessman 1983; Chambers 1983; Lowrance et al. 1984;) have begun to view the agricultural field as a particular kind of ecosystem called agroecosystem. Understanding these processes and relations may lead to management of agroecosystems in a more sustainable way, with fewer negative environmental or social impacts (Altieri 1995; Altieri and Nicholls 2005). Mineral cycles, energy transformations, biological processes and socioeconomic relationships are viewed and analysed as integral parts of a whole system rather than as individual components. Organic farming is an agricultural regime based on agroecology, and it is the only sustainable farming system that is legally defined (Watson et al. 2002a). Indeed, it is a production system that sustains agricultural production by avoiding or largely excluding synthetic fertilisers and pesticides. Whenever possible, external resources are replaced by resources found on or near the farm. These internal resources include solar or wind energy, biological pest controls, biologically fixed nitrogen and other nutrients released from organic matter or from soil reserves. A key issue of organic agroecosystems is hence the increasing or maintaining of the soil organic matter over time. Thus, soil organic matter turnover (i.e. the rate of organic matter accumulation and decomposition processes) is a fundamental life-promoting process.

In more depth, according to the International Federation of Organic Agricultural Movements (IFOAM), basic principles for soil health and quality management in organic farming rely on the returns of microbial, plant and animal organic material to the soil. Moreover, the cultivation techniques are aimed at increasing soil biological activity and nutrients while fertility inputs must be applied in a way that does not harm soil, water and biodiversity (IFOAM 2011). Organic farming systems, hence, emphasise reliance on ecological interactions and biological processes over direct intervention, recognising the complex relationships that exist between different system components (Watson and Stockdale 2000).

24.3 Criteria for Fertility Management and Plant Nutrition in Organically Managed Citrus Orchards

24.3.1 Introduction to Nutrient Cycling in Organically Managed Agroecosystems

The lack in sustainability of most modern conventional agriculture should be identified in the attention to the short-term productivity despite the long-term ecosystem health. The aim is the maximisation of productivity and profitability, without taking into account their effects on natural resources and ecosystem health. On the contrary, in sustainable systems, the main aim is to enhance soil fertility, and its management is strictly connected to this objective. Furthermore, the essential basis for the identification of the additional off-farm desirable rates is to understand the movement of nutrients in the soil-plant system and to identify the real plant requirements. Nutrient uptake is regulated in response to shoot demand (Engels and Marschner 1992; Marschner 1995): correspondingly, supply of plant roots with mineral nutrients has a strong influence on vegetative and reproductive development of the shoots (Jeschke and Hartung 2000). However, it should be highlighted that the degree to which kinetics of nutrient uptake or other potential adjustments are expressed would ultimately depend on soil nutrient availability and soil factors that determine nutrient transport to the root surface (Bassirirad 2000). Management practices strive to optimise diverse biological processes in the soil to create a complex environment that ensures adequate nutrition to the crop. Organic soil fertility programmes are hence designed to maintain adequate levels of nutrients or to increase the nutrients pool as needed. This goal should be partially achieved by the recycling of crop residues (i.e. pruning residues) and the only output nutrients should be considered those contained in harvested yield (Watson et al. 2002b). The quantity and quality of crop residues will clearly influence the build-up of soil organic matter (Jenkinson and Ladd 1981) and the subsequent availability and timing of release of nutrients to following crops (Jarvis et al. 1996). In the case of addition of admitted off-farm input, it should be considered that organic fertilisers often include both a relatively small proportion of soluble nutrients and another nutrient fraction that is either unavailable to the plant or available only gradually over time (Whitmore 2007). The composition (i.e. the C/N ratio) and particle size of the material as the environmental characteristics should be determining factors in the rate of microbial decomposition and nutrient availability (Six et al. 2002; Lavelle et al. 2005; Diaz and Savage 2007). Many soil amendments and fertilisers commonly approved for organic production systems have, indeed, appreciable amounts of nutrients, but only a portion of these nutrients are available to the current crop (Agehara and Warncke 2005). These materials should be applied earlier in anticipation of plant nutritional requirements (Bath 2000).

24.3.2 Soil Fertility Management in Organic Citrus

Organic standards restrict the inputs and management options available to certified organic orchards. It is important to manage field in ways that make them inherently more robust and resilient to diseases, pests and weeds, and that could minimise the orchard's dependence on material inputs and management interventions. Because of its focus on reducing or eliminating chemical inputs and improving soil health, organic management alters many soil parameters. As an example, when conventional and organic soil sampled in citrus orchards in Spain and Italy were analysed, the organically managed soils were found to be higher in such factors as organic matter, humic substances, carbohydrate, aggregate stability, microbial gums, microbial biomass and enzymatic activity (Albiach et al. 1999; Canali 2003). These differences, representing advantages for organic citrus, should be aimed in orchards management. For example, mulch – used by many organic citrus producers to suppress weed growth – conserve soil moisture and protect the soil from temperature extremes. A benefit of using organic material as mulch is that it adds valuable organic matter to the soil as it decomposes (Abouziena et al. 2008).

As far as the nutrient management is concerned, the effective plant requirements should be known and, on this basis, an input/output balance should be adopted (Watson et al. 2002b). Balanced nutrition of plants should be a high-priority management objective for citrus growers. Plants require a balanced nutrition programme formulated to fulfil specific needs for maintenance and for expected production performance. To make it possible, an important aspect is monitoring the nutritional level of plants and the soil content (Embleton et al. 1973; Reuter and Robinson 1986; Intrigliolo et al. 1999). These trends give an indication of the long-term sustainability of the organic orchard. As far as the citrus uptake is concerned, it may reflect the amount of nutrients lost in harvested fruit, abscised fruitlets and flowers, senescent leaves, pruning wood and root turnover, while in young trees, there is an important demand for tree growth (Mattos et al. 2006; Menino et al. 2007). The spring flush depends on an intense remobilization of stored nutrients, since uptake of external nutrients is small during that period (Menino et al. 2007). Similarly, a significant proportion of the nitrogen and other nutritive elements in older leaves are redistributed within citrus trees before leaf drop occurs. A part of these nutrients become locked up in the permanent structure of the tree (trunk, major branches and roots) and are unavailable for the life of the tree. It has been estimated, for example, that in the order of 20 kg ha^{-1} $year^{-1}$ of N is deposited in the permanent structure of mature citrus trees (Dasberg 1987; Boaretto et al. 2006). As reported by Yaseen and Ahmad (2010), citrus plant needs, however, application of nutrients at critical growth stages when plant really has a demand of nutrition. This happens in particular for N during the initial stages of orange fruit setting and development, i.e. in the northern hemisphere, during June through August or September depending on cultivars (Alva et al. 2001). In order to satisfy these requirements, the first step in nutrient management of organic citrus is the fostering of the role of nutrient recycling by the use of green manure and cover crops, incorporation of livestock manure – where appropriate and available – and especially the return of as much orchard residues as possible to the soil. For example, the nutrients incorporated into leaves and fine roots are indeed continually recycled as those organs die and decompose and should be considered in nutritional planning (Dou et al. 1997). The same option could be chosen for pruning and other plant residues (i.e. by composting processes). In other terms, if citrus residues are recycled, the only uptake considered as nutrient output in surface balance should be that concerning the harvested yield.

According to what was reported above, the assessment of nutrient mineralization from soil organic matter (SOM) and organic fertilisers added to soil is an important challenge of organic citrus cultivation. Then, one of the main aims of organic fertilisation programmes consists of synchronising nutrients release with plant uptake by carefully choosing the fertiliser typology. In addition, it has been proved that adequate fertilisation programme may achieve earlier yield and heavier fruit set (Ibrahim et al. 2004; Abd-Allah 2006; Alva et al. 2006).

24.4 Case Studies in Southern Italy

24.4.1 Eastern Sicily Citrus Field Survey

A field survey was carried out in the Catania province, Eastern Sicily (Italy), with aim to compare (1) yield, (2) fruit quality, (3) orchard nutritional status and (4) soil quality.

The 4-year investigation was carried out between 2001 and 2005 on 'Tarocco' and 'Navelina' orange. The climate in the studied area is classified as typical Mediterranean, with average rainfall and temperature of 350 mm and 18.7°C, respectively. A comprehensive description of the orchards and the agricultural practices adopted during the survey are reported in Table 24.1. The cover crops were communities of natural weeds, retained all the year around and disturbed and partially ploughed only when fertilisers were incorporated into the soil. Soil amendments (composts) and organic fertilisers (poultry manure, dairy manure, plant residues, etc.), allowed by Italian and European legislation, were applied to soil in the organic farms. Mineral fertilisers were applied in conventional farms. Data shown are based on the information gathered from the citrus growers and on estimation of the organic C and nutrients content for input materials applied as fertilisers. In both organic and conventional farms, the irrigation water was distributed by micro-sprinklers, at annual rates ranging between 300 and 650 mm.

Table 24.1 Eastern Sicily citrus field survey. Description of the main characteristics and agricultural practices adopted in organic and conventional citrus orchards referred to in the survey

		Organic		Conventional	
Item		Min	Max	Min	Max
Orchard area (ha)		0.40	6.00	0.17	27.00
Orchard age (years)		5	49	8	70
Number of trees (ha^{-1})		400	625	334	625
C input from off-farm fertilisers (kg × ha^{-1})		423	1,235	–	–
Nutrient input from off-farm fertilisers (kg × ha^{-1})	N	47	190	29	300
	P_2O_5	16	118	20	219
	K_2O	13	156	26	250
Tillage (no. of operations)		1	8	0	4
Pruning material recycling (no. of orchards)		11		8	
Cover cropping (no. of orchards)		13		9	
Weed control		Mowing		Herbicides	

Modified from Canali et al. (2009)

24.4.1.1 Yield and Fruit Quality

Total production and fruit quality were determined yearly in each farm. On each fruits sample, physical parameters (firmness, fruit weight, width of the central axis, peel thickness) were measured using standard methods (Wardowski et al. 1979). Each sample of fruits was squeezed, and juice content, total acidity (TA) and total soluble solids (TSS) were determined. Vitamin C was analysed by high-performance liquid chromatography (HPLC) (Rapisarda and Intelisano 1996). In addition, anthocyanins were determined for *Tarocco* orange (Rapisarda et al. 2001). ANOVA was performed and mean values separated with Tukey HSD test (SPSS package ver. 18).

No statistically significant differences in yield were observed between organic and conventional orchards (Table 24.2). Moreover, only negligible differences were noticed in the other fruit quality parameters investigated. More in depth, vitamin C values were significantly higher in organically produced *Tarocco* and *Navelina* oranges with respect to the conventionally produced ones. Conversely, anthocyanin values (present only in blood oranges, i.e. *Tarocco* cultivar) were higher in conventionally managed trees. These data confirm previous observations and seem to be related to the higher level of available mineral N in soil due to the application of soluble fertilisers in conventional farms (Rapisarda et al. 2005).

Table 24.2 Eastern Sicily citrus field survey. Yield and fruit quality (4-year average values)

		Yield (t/ha)	Fruit weight (g)	Juice content (%)	Rind thickness (mm)	Central axis (mm)	Total soluble solids (%)	Total acidity (%)	TSS/TA	Vitamin C (mg/100 ml)	Antocyanins (mg/L)
'Tarocco'	Organic	25.3	193	40	5.4	9.5	11.03	1.32	8.35	69	44
	Conventional	27.1	203	39	5.6	9.8	11.01	1.35	8.15	64	61
										*	*
'Navelina'	Organic	19.9	197	43	5.0	9	11.81	1.23	9.65	65	–
	Conventional	19.0	200	41	5.2	7	11.48	1.13	10.56	59	–
										*	

Note: *$P \leq 0.05$

Table 24.3 Eastern Sicily citrus field survey. Leaf nutritional levels (4-year average values)

		Macronutrients (%)						Macronutrients (mg kg^{-1})		
		N	P	K	Ca	Mg	S	Fe	Zn	Mn
'Tarocco'	Organic	2.49	0.141	0.94	4.32	0.23	0.26	120	22	24
	Conventional	2.52	0.133	0.88	4.24	0.20	0.29	101	24	25
			*					**		
'Navelina'	Organic	2.52	0.170	1.09	4.80	0.21	0.32	132	21	25
	Conventional	2.60	0.153	0.95	4.60	0.19	0.29	118	24	24
			*	**				*		

Note: *$P \leq 0.05$; **$P \leq 0.01$

24.4.1.2 Tree Nutritional Status

Plant nutritional status was determined by foliar analysis performed on 80–100 leaves picked up in October from non-fruit-bearing terminal shoots of the year's spring flush in 40–60 trees in each plot (Embleton et al. 1973; Intrigliolo et al. 1999). The leaves were (1) washed in tap water by rubbing both sides using cheesecloth, (2) rinsed in deionised water, (3) oven dried at 65°C for 72 h, (4) ground and (5) dried at 105°C for 4 h. The concentration of N was determined on 1 g of ground leaf tissue using the micro-Kjeldahl method (Büchi Distillation Unit K370). Another 1 g of ground leaf tissue was ashed in a muffle furnace at 550°C for 12 h. After incineration and extraction with nitric acid (1% v/v), nutrients like P, K, Ca, Mg, Fe, Zn and Mn were determined using inductively coupled plasma optical emission spectrometry (ICP-OES, OPTIMA 2000DV, Perkin-Elmer Italia). ANOVA was performed and mean values separated with Tukey HSD test (SPSS package ver. 18).

Statistically significant differences in nutritional status were observed between organically and conventionally managed oranges (Table 24.3). In particular, P and Fe concentrations were significantly higher in organically managed oranges; K values were higher in organic *Navelina*, but this increase was not statistically significant in *Tarocco* orange. Higher K level in leaf analyses of organic orchards cannot be justified by lower yields (Koo 1985) that were similar – in our case – in the 4-year mean values. No relevant differences were detected for the other investigated nutritive elements in the organically and conventionally managed orchards throughout the trial.

24.4.1.3 Soil Quality

Most of the few data available on soil quality evaluation on organically managed orchards are not referred to semiarid regions (i.e. Reganold et al. 2001). In Brazil, Franca et al. (2007) demonstrated that the soil of organically managed citrus orchards showed higher microbial activity and arbuscular mycorrhizal fungal (AMF) richness and diversity than that under conventional management. Initial results of the introduction of organic farming on soil quality of organically managed citrus orchards in the Mediterranean region were reported by Intrigliolo and coauthors (2000). They reported that organic management induced only slight differences in the main physical and chemical characteristics of conventionally managed soil. Furthermore, Albiach et al. (1999) and Canali et al. (2004), in their study based on the comparison between organic and conventional citrus orchards, found significant differences in soil organic matter content and soil aggregates stability.

According to Karlen et al. (2001), 13 pairs of conventionally managed and organically managed citrus orchards, converted at least 6 years earlier, were selected on the basis of the respect of the homogeneity of inherent soil characteristics. The latter being those determined by parent material, climate and vegetation and which are meaningful in determining the capacity of a soil for a specific land use. According to what is reported above, a preliminary statistical analysis on soil-inherent characteristics of the 13 pairs of orchards was carried out in order to verify the absence of significant differences.

The orchard soils located in the area surrounding the Etna volcano are *Typic Xerorthents, Andic Xerochrepts and/or Ultic Haploxeralfs* (Dazzi 2005). Soil samples were taken 11–12 months after the last fertiliser application to minimise the effects of fresh nutrients on the soil properties to be measured. Chemical and biochemical parameters were used as indicators of soil quality (Park and Seaton 1996; Schloter et al. 2005). Total organic carbon (C_{org}) was measured according to Springer and Klee (1954). The A_s% index, representing the relative sum of the areas of peaks focused at pH > 4.7 in an isoelectric focusing (IEF) profile of extracted humic substances and corresponding to more humified organic matter, was then determined following the procedure reported in Canali et al. (2009). Soil C mineralisation was studied by measuring CO_2–C production in closed jars (Isermeyer 1952). The CO_2 released was determined at the first (C_1), second, fourth, seventh, tenth, 14th, 17th and 21st (C_{21}) day of the incubation period [mg(CO_2–C) /kg_{soil} day^{-1}]. The value of CO_2–C released [mg(CO_2–C) /kg_{soil} day^{-1}] on the 21st day of incubation was assumed as soil basal respiration (C_{basal}). Cumulative CO_2–C released after 21 days (C_{21cum}) was calculated for each soil. The carbon of the soil microbial biomass (C_{mic}) was measured [mg C /kg_{soil}] by the chloroform fumigation-extraction method of Vance et al. (1987). On the basis of the measured soil chemical and biochemical parameters, soil metabolic indexes were calculated. In more depth:

- Metabolic quotient $q(CO_2)$, defined as specific soil respiration of the microbial biomass, was calculated from basal respiration values (after the 21st day) by $q(CO_2) = [(mg\ CO_2—C/mg\ C_{mic})/h]$ (Anderson and Domsch 1985).
- Mineralisation quotient C_{MIN}, defined as mineralised soil C at steady-state conditions (soil microflora basal respiration) with respect to the total amount of soil organic C, was calculated from soil basal C (C_{basal}) by $C_{MIN} = [(mg\ CO_2–C/kg_{soil})/mg\ C_{org}\ day^{-1}]$ (Dommergues 1960).
- The ratio [C_{mic}/C_{org}] % was used as an index of microbial biomass contribution to soil organic C (Anderson and Domsch 1989).

Statistical analysis was carried out comparing two vectors composed of 13 values each, and boxplots were utilised to show the distribution of measured soil parameters. The Wilcoxon signed range test, a nonparametric test for paired data, was used to compare differences between organic and conventional soils (Soliani 2004).

In Fig. 24.1, the boxplots describing the frequency distribution of the main soil physical and chemical properties of the selected pairs of orchards are shown. No significant differences were found for clay, sand, pH and CEC, demonstrating the homogeneity of the inherent properties of each pair of soils.

The C_{org} values of soils under organic management were significantly larger than those of the conventionally managed ones (Fig. 24.2a). These results suggested that the adoption of this farming system caused the increase of soil organic matter under these climatic conditions. Other authors found a trend, though not a significant difference, in the increase of organic C content in organically managed soils with respect to the conventional ones (Mäder et al. 2002; van Diepeningen et al. 2006). According to Clark et al. (1998), our results can be explained with the higher input of C in citrus organic systems which, however, cannot be attributed uniquely to the different patterns of fertiliser use (organic vs inorganic fertilisers) but, presumably, to a range of different agricultural practices (i.e. weed management, green manuring, depth of tillage, animal manures, fertiliser use) and their interactions.

The A_s% values of organically managed soils were significantly higher than those of conventionally managed ones (Fig. 24.2d). Several authors (De Nobili et al. 1990; Ciavatta et al. 1990; Ciavatta and Govi 1993; Alianiello and Fiorelli 1998; Dell'Abate et al. 2002; Canali et al. 2004; Trinchera et al. 2007) demonstrated that humic and humic-like substances purified from soils, peats, composts and organic wastes show a group of characteristic electrophoretic peaks focusing at pH higher than 4.7, used to calculate the A_s%. The significant difference between the A_s% value for organic and conventional soils evidenced a qualitative improvement of humic substances, characterised by lower acidity and increased molecular complexity, in organically managed soils (Dell'Abate et al. 2002).

Boxplots for soil microbial biomass (C_{mic}) values are reported in Fig. 24.3a. The frequency distribution of the 13 pairs of soils showed significantly higher values in the organically managed soils. Soil microbial biomass represents the living organic matter fraction responsible for energy and nutrients cycling and for regulating organic matter transformation (Gregorich et al. 1994). It is a potential source of plant nutrients, and a linear relationship has been found between soil C microflora and total organic C in agricultural soils (Anderson and Domsch 1980). Results obtained in our work confirmed what has been reported above because in organic management, both soil microflora C (C_{mic}) and total organic carbon (C_{org}) were higher than in conventional treatments. In addition, because microbial biomass content is generally considered as one of the indicators of soil fertility, the higher value observed in organically managed citrus soils could be interpreted as an indication of greater soil quality (Franca et al. 2007).

Both microbial biomass and soil respiration are positively correlated with soil organic matter content and, often, with microflora metabolic activity (Alef 1995). In our soils the values of mineralised C after the first day of incubation were higher in organic than in conventional soils, even though the frequency distributions were not statistically different (data not shown). However, the basal respiration C (C_{basal}) was significantly higher under organic than conventional management (Fig. 24.3b). It has been demonstrated that

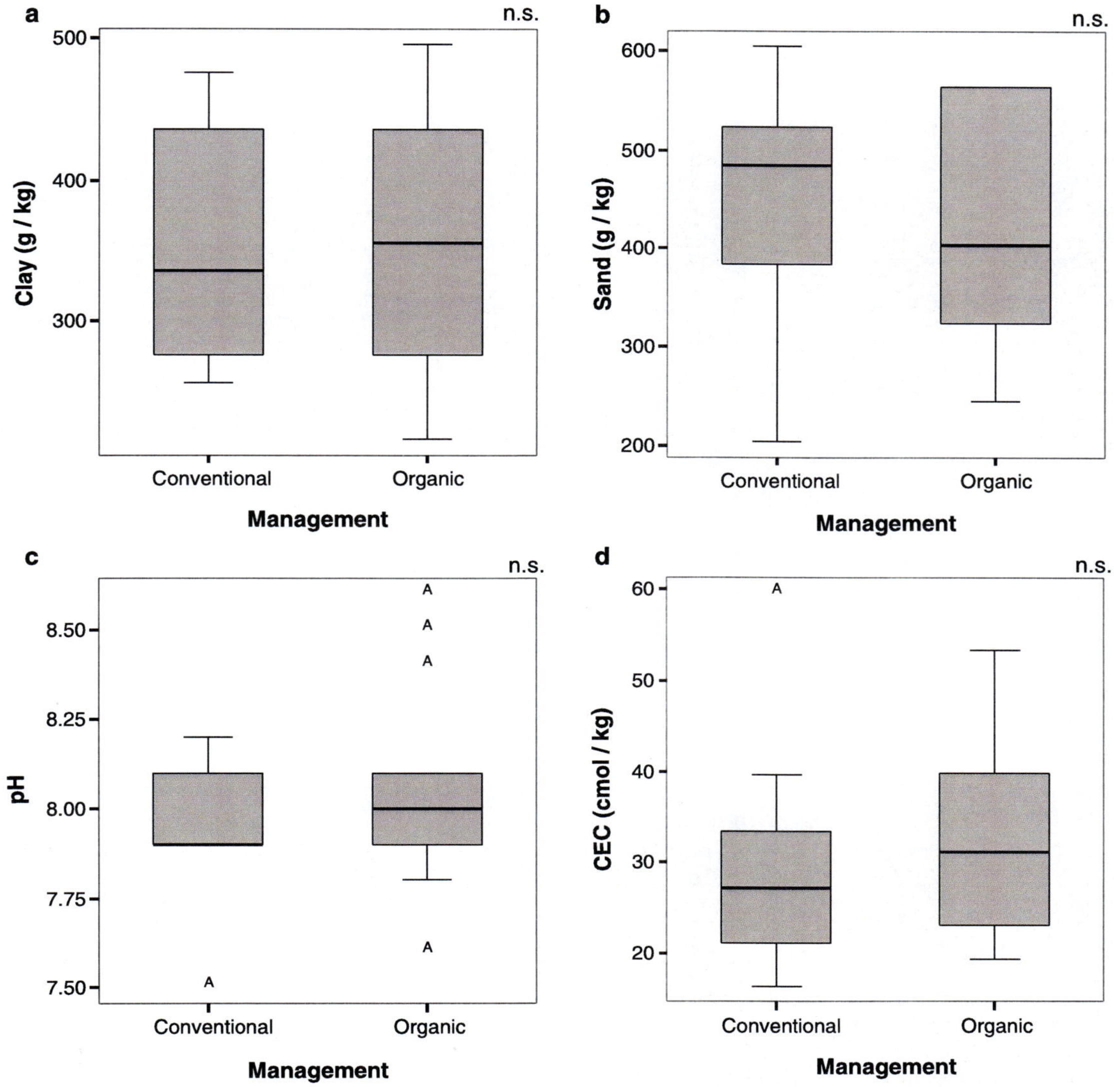

Fig. 24.1 Frequency distribution of clay (**a**), sand (**b**), pH (**c**) and CEC (**d**) of the 13 pairs of orchard soils (organic vs conventional). *n.s.* distributions not different according to the Wilcoxon signed range test. *Note*: Boxplots were utilised in order to graphically represent frequency distribution. *Bars* (or *whiskers*) represent the dispersion of the values below or above the *lower quartile* and the *upper quartile*, respectively; in (**b**) and (**c**), *whiskers* are not absent, but coincident with upper/lower quartile (Adapted from Canali et al. 2009)

measurements of soil respiration can be used to discriminate between different soil management practices (Pankhurst et al. 1995). This is also confirmed by our results which showed that basal respiration is able to discriminate between organic and conventional managements. Higher values for basal respiration in soils under organic management are in accordance with the increase of soil organic matter and soil microbial C biomass. Because basal respiration represents the soil microbial energy requirement at a steady-state condition, the higher values found in organically managed soils indicate that its microbial community increased the energy needs.

The metabolic quotient, qCO_2, links respiratory activity (basal respiration) to the quantity of soil microflora (Anderson and Domsch 1985). In our soils, the distribution of qCO_2 values was significantly higher in conventional treatments (Fig. 24.3c). However, the C_{mic}/C_{org} ratio (Anderson and Domsch 1989) and C_{MIN} values, which link soil basal respiration (C_{basal}) to total soil organic C (C_{org}), were not statistically different in soil from conventional and organic orchards (data not shown). The metabolic quotient (qCO_2) has been widely used as an indicator of soil quality and soil C utilisation efficiency at steady state. In particular, Anderson and Weigel (2003) demonstrated that this parameter is able to discriminate

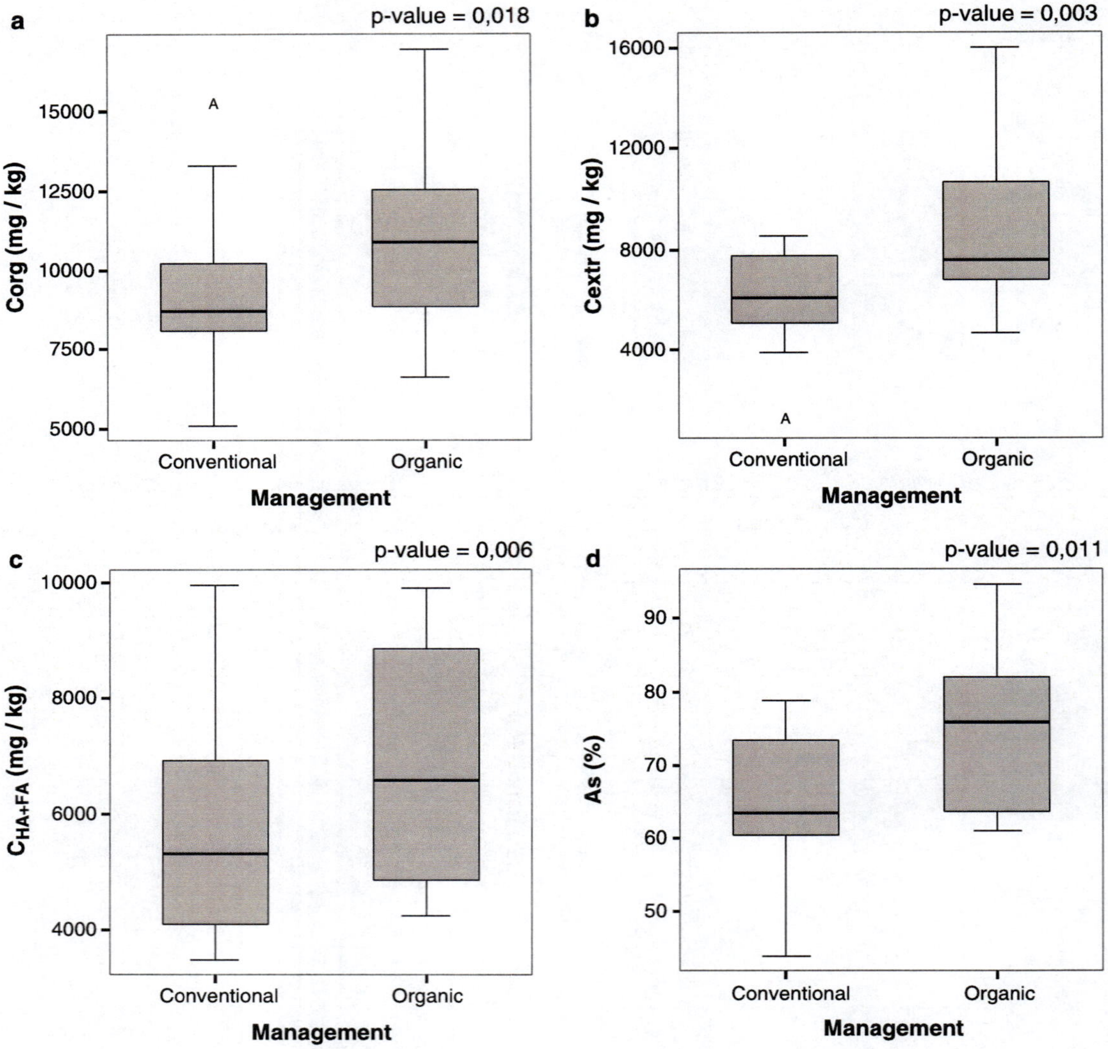

Fig. 24.2 Frequency distribution of total organic carbon soil (C_{org}) (**a**), extractable carbon (C_{extr}) (**b**), humified carbon (C_{HA+FA}) (**c**) and A_s (relative area of IEF peaks focused at pH > 4.5) (**d**) of the 13 pairs of orchard soils (organic vs conventional). *p*-value = significance value of the difference between the distributions according to the Wilcoxon signed range test. *Note*: Boxplots were utilised in order to graphically represent frequency distribution. *Bars* (or *whiskers*) represent the dispersion of the values below or above to the *lower quartile* and to the *upper quartile*, respectively (Adapted from Canali et al. 2009)

between soil management systems. In their study of soil microbial activity in organically and conventionally managed citrus orchards, Franca et al. (2007) found that qCO_2 of organically managed soils was higher than that of the conventionally managed ones. This difference was explained by the greater availability of organic residues with a low C/N ratio in the organically managed soil, causing increased respiration without an equivalent increment in microbial biomass. Conversely, Anderson and Domsch (1990) showed that crop rotations favoured a lower qCO_2 than monocultures, while Fließbach and Mäder (2000) found a lower qCO_2 in organic rather than in conventional soils. Anderson (2003) showed that microbial communities of long-term crop rotation systems are energetically more efficient, having a lower qCO_2 value and higher corresponding C_{mic}/C_{org} ratio (for increased microflora C content) compared to monoculture soil systems. Mäder et al. (2006), in their long-term DOK experiment, found a lower qCO_2 in biodynamic and organic systems compared to the conventional systems. On the basis of this evidence, these authors concluded that the lower values observed demonstrated that soil C utilisation in conventional management is metabolically less efficient. Also, Probst et al. (2008) found a lower qCO_2 in soil of organically managed vineyards compared to that in the conventional

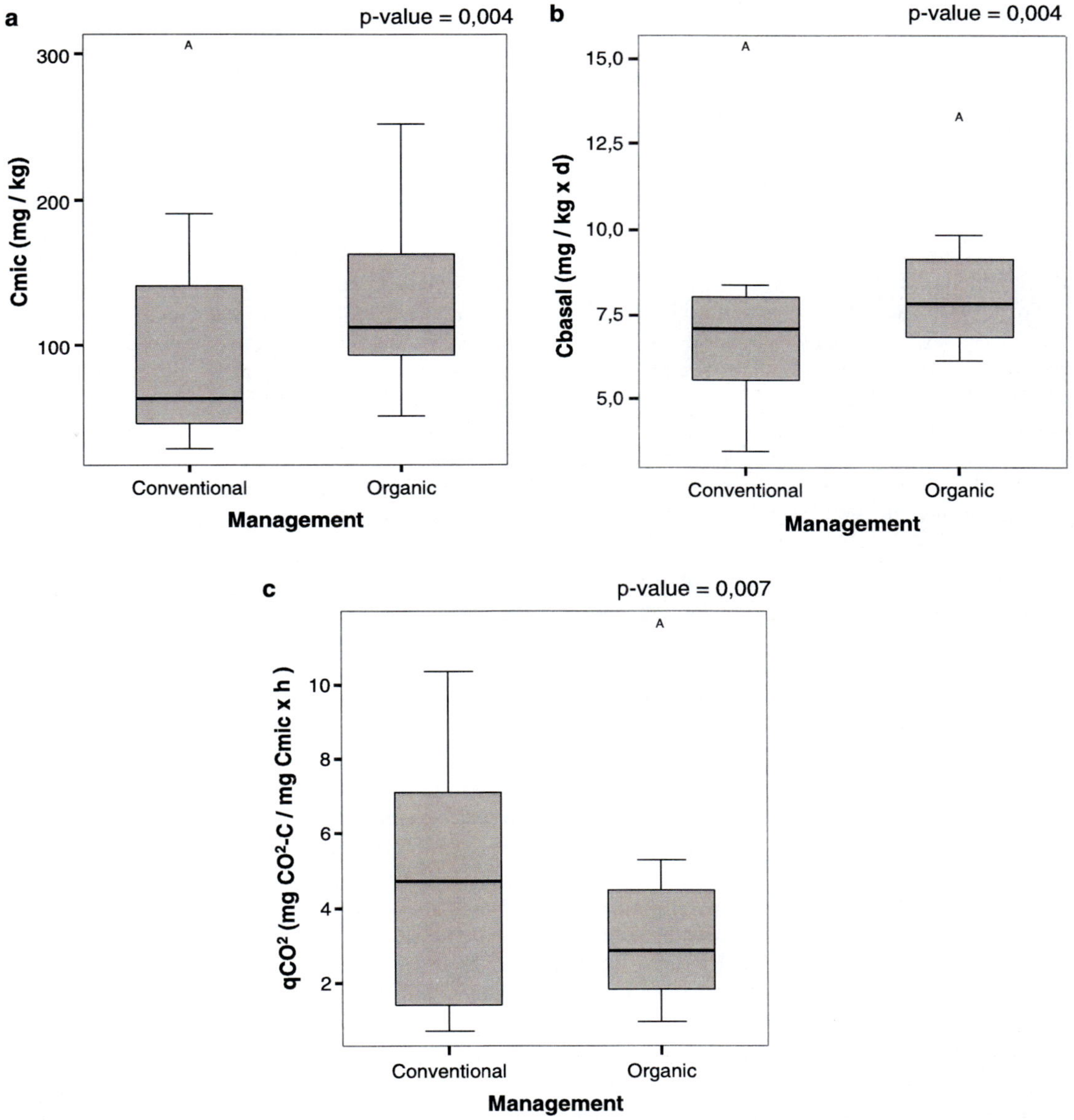

Fig. 24.3 Frequency distribution of soil microbial biomass (C_{mic}) (**a**), soil basal respiration (C_{basal}) (**b**) and soil metabolic quotient (qCO_2) (**c**) of the 13 pairs of orchard soils (organic vs conventional). *p*-value = significance value of the difference between the distributions according to the Wilcoxon signed range test. *Note*: Boxplots were utilised in order to graphically represent frequency distribution. *Bars* (or *whiskers*) represent the dispersion of the values below or above the *lower quartile* and the *upper quartile*, respectively (Adapted from Canali et al. 2009)

vineyards, confirming that the increase in microbial substrate use efficiency cannot be probably attributed to specific agricultural practices but to the overall effect of the organic management of soil.

The same trend was confirmed by our data because the organic citrus orchards were characterised by a significantly smaller qCO_2 and an increase of C_{MIN} and C_{mic}/C_{org} (even if the last parameters were not significantly different) compared to the conventional systems (Fig. 24.3b). According to the results obtained, qCO_2 showed a higher sensitivity, than C_{MIN} and C_{mic}/C_{org} ratio, to soil management.

24.4.2 The PALAP9 Long-Term Field Experiment

A field trial was set up in 1995 in the 'Palazzelli' research farm of the Citrus and Mediterranean Crops Research Centre of the Agricultural Research Council located in

Table 24.4 Organic C and nutritive element inputs (g tree^{-1}) yearly distributed in the PALAP9 long-term experiment

Treatment	C	N	P_2O_5	K_2O	Fe	Zn	Mn
CB	6,643	455	142	244	51	3	4
LW	4,271	455	584	577	17	7	9
PM	3,729	455	547	466	27	1	3
MF	–	455	301	392	–	–	–

Palazzelli (Lentini, SR; Eastern Sicily – 37°17′56″76 N, 14°50′29″76 E) with the aim to study the effects of the organic farming method introduction in an already bearing sweet orange orchard cv. 'Valencia late' grafted on sour orange (*C. aurantium* L.).

In more depth, the experiment was aimed at evaluating the effects of long-term repeated organic fertilisers application on (1) soil fertility; (2) citrus-bearing trees nutritional status by means of leaf analysis; (3) yield and fruit quality, determining parameters currently utilised to evaluate sweet orange performance; and (4) identification and selection of plant and fruit parameters useful for organic citrus monitoring and traceability.

In a randomised block experimental design with three replicates and plots of 60 plants (6×4 m between the rows and within the row, respectively), four different fertiliser treatments were compared, namely, citrus by-products compost (CB), poultry manure (PM), livestock waste compost (LW) and mineral fertiliser (MF) as control. The amount of organic C and nutritive elements distributed yearly are reported in Table 24.4.

In winter 2007, soil was sampled and analysed. Total soil organic carbon (C_{org}) was determined by mineralisation with 2 N $K_2Cr_2O_7$ and 96% H_2SO_4 solutions at 160°C for 10 min according to Springer and Klee (1954). Soil humic and fulvic acids were extracted and purified according to Ciavatta et al. (1990) and Ciavatta and Govi (1993).

Potentially mineralisable C (PMC) was determined by measuring CO_2–C production in closed jars (Isermeyer 1952). Each of the soil samples (25 g, oven dry-weight equivalent) was rewetted to its −33 kPa water tension and incubated at 30°C. The CO_2 evolution was determined at the first, second, fourth, seventh, tenth, 14th, 17th and 21st day of the incubation period [$mg(CO_2–C) \times kg_{soil}^{-1} \times d^{-1}$]. The cumulative CO_2–C vs time was fitted according to the first-order exponential equations $Ct = PMC\ (1-e^{-kt})$. This elaboration allowed the calculation of the PMC pool ($mg\ C\,kg^{-1}{}_{soil}$) for each soil. Total soil nitrogen (N_{tot}, $mg \times kg^{-1}$) was determined by Kjeldahl's procedure (Bremner and Mulvaney 1982).

Potentially mineralisable N (NPM) was estimated by calculating the NH_4^+–N ($mg \times kg^{-1}$) accumulated after 7 days of anaerobic incubation at 40°C, according to Sahrawat and Ponnamperuma (1978), slightly modified by Canali et al. (2004).

Each year, plant nutritional status was determined by foliar analysis performed on spring-cycle leaves collected in October from terminal, non-fruiting shoots (Embleton et al. 1973). The leaves were (1) washed in tap water by rubbing both sides using cheesecloth, (2) rinsed in deionised water, (3) oven-dried at 65°C for 72 h, (4) ground and (5) dried at 105°C for 4 h. The concentration of N was determined on 1 g of ground leaf tissue using the micro-Kjeldahl method (Büchi Distillation Unit K370). Another 1 g of ground leaf tissue was ashed in a muffle furnace at 550°C for 12 h. After incineration and extraction with nitric acid (1% v/v), P, K, Ca, Mg, Fe, Zn and Mn were determined using inductive coupled plasma optical emission spectrometry (ICP-OES; OPTIMA 2000DV, Perkin-Elmer Italia).

Total yield was recorded at harvest, and, on a sub-sample of 40 fruits collected from the outer part of the canopy, the fruit mean weight and fruit physical and chemical parameters were determined. In more detail, firmness, fruit weight, width of the central axis and peel thickness were measured according to Wardowski et al. (1979). Furthermore, for each sub-sample juice content, total acidity (TA) and total soluble solids (TSS) were determined. Vitamin C was analysed by high-performance liquid chromatography (HPLC) (Rapisarda and Intelisano 1996). Synephrine content was determined by the HPLC method described by Rapisarda et al. (2005).

Measurement of the $^{15}N/^{14}N$ ratio of leaves, pulp and amino acids of juice were realised following the methods described by Bricout and Koziet (1987) with slight modification. For the measurement, an isotope ratio mass spectrometer (Delta plus XP ThermoFinnigan, Bremen, Germany) equipped with an elemental analyser (EA Flash 1112 ThermoFinnigan) was used. The values were expressed in δ‰ against international standards (air for $\delta^{15}N$). The isotopic values were calculated against working in-house standards (mainly casein), calibrated against L-glutamic acid USGS 40. The uncertainty (2σ) of measurements was ±0.3‰.

ANOVA was performed and mean values separated with Tukey HSD test (SPSS package ver. 18). Moreover, data were processed by means of linear canonic discriminant analysis (CDA) to evaluate all parameters at the same time and detect those that mostly affect group differentiation.

Soil C_{org} of the different treatments is reported in Table 24.5. The CB treatment showed higher significant values than the control MF treatment of about 30% (2.52 vs 1.92 g C_{org} 100 g^{-1} soil, Table 24.5). This evidence could be explained considering that CB had a lower C/N ratio than the two other organic fertilisers applied (LW and PM), and consequently, since the dosage was established in order to satisfy the N requirements of bearing orange trees, a higher amount of organic C was applied to the soil of the CB-treated plots. Also, the C_{org} content of the PM and LW treatments was higher than the control MF treatments (13% and 20%, respectively). However, these differences were not statistically significant

Table 24.5 Chemical and biochemical soil fertility parameters measured in the PALAP9 long term experiment soil

Treatment	C_{org} (g 100 g^{-1})	$C_{(HA+FA)}$ (g 100 g^{-1})	PMC (mg kg^{-1})	N_{tot} (mg kg^{-1})	PMN (mg kg^{-1})
CB	2.52 a	1.19 a	547 a	2339 a	50 a
PM	2.17 ab	1.03 ab	328 b	1982 b	35 b
LW	2.33 ab	1.04 ab	389 b	2052 b	45 ab
MF	1.92 b	0.90 b	331 b	1723 b	38 ab

Note: Mean separation at 5% level with Tukey HSD test
C_{org} total organic C, $C_{(HA+FA)}$ humic and fulvic C, N_{tot} total N, *PMC* potentially mineralisable C, *PMN* potentially mineralisable N, *CB* citrus by-products compost, *PM* poultry manure, *LW* livestock waste compost, *MF* mineral/synthetic fertilizers

Table 24.6 Average values of leaf analysis (D.M. basis) measured in the PALAP9 long-term experiment

			Macronutrients (g 100 g^{-1})				Micronutrients (mg kg^{-1})		
Treatment	N (g 100 g^{-1})	δ^{15}N leaf (g kg^{-1})	P	K	Ca	Mg	Fe	Mn	Zn
CB	2.71	5.12 b	0.134	0.77	5.14 a	0.42 b	117 a	16 a	15 a
PM	2.54	6.55 a	0.133	0.82	4.80 ab	0.43 b	92 b	12 b	14 ab
LW	2.65	6.90 a	0.128	0.83	4.65 b	0.37 c	91 b	13 ab	15 ab
MF	2.65	2.49 c	0.132	0.75	4.83 ab	0.47 a	110 a	13 ab	13 b

Note: Mean separation at 5% level with Tukey HSD test

either from the control MF treatment or from the soil fertilised with citrus compost (CB).

Similar trend was showed by the humic and fulvic C being the values of the CB treatment significantly higher than the control. PM and LW treatments had intermediate values without statistical significance. It is noticeable that all treatments showed a very close ratio between $C_{(HA+FA)}/C_{org}$, which ranged from a minimum value of 0.44 (LW) to 0.47 (PM).

The CB treatment showed significant higher values of PMC than either of the control MF and the other two fertiliser treatments (PM and LW), while no significant differences were observed among control MF, PM and LW, even if the latter showed higher values of PMC (18%) from both the MF and the PM treatments, which presented very similar values.

As far as N_{tot} is concerned, values reported in Table 24.5 show that the CB treatment had the highest value of all the fertiliser treatment studied and the difference was statistically significant. On the other hand, even if PM and LW presented higher values than the control MF treatment (15% and 19% for PM and FYM, respectively), this difference did not reach the significance.

Results about PMN assumed a different trend with respect to the above-considered soil fertility parameters. In fact, even if the CB treatments had the highest values, the differences among this treatment and the control MF and LW ones were not significant. The lowest values were unexpectedly showed by the PM treatment which was, however, significantly different only from the CB. This result should depend on the typology of PM nitrogen compounds, characterised by a high mineralization rate after application (Eghball et al. 2002), which make N patterns more similar to MF ones (90% of N content in poultry available in the first year after application against 20% for composted manure, Eghball et al. 2002). To confirm this hypothesis, non-statistically significant differences for PMN between PM and MF were recorded (Table 24.5).

As far as the nutritional status is concerned, no significant difference between treatments was noticed for leaf N, K and P content, whereas significant differences for Ca, Mg and micronutrients were observed (Table 24.6). Ca content was higher in CB only with respect to LW; the latter showed Mg values lower than other treatments. All values for macronutrients were in the optimal range according to the international standard for diagnosing nutritional status (Embleton et al. 1973). CB leaves constantly showed higher micronutrient content, in the case of iron compared to PM and LW, for manganese compared to PM and for zinc compared to MF. Even though leaf levels of Mn and Zn were deficient, no deficiency symptoms were observed in field. The comparison between P and K leaf content and related inputs showed clearly the increased nutrient use efficiency in CB treatment in comparison with the other ones. In fact, in the CB treatment, higher values of P and K in the leaves corresponded to lower input of these elements (Table 24.4). This finding is attributable to the increase of organic matter content and the consequent higher biochemical fertility of CB which could have a favourable role in cycling and availability of the two macronutrients (Srivastava et al. 2002; Dick and Gregorich 2004; Toselli 2010). Similarly, also the higher iron leaf levels in CB could only partially be justified from higher input of this element, whereas the increase of its availability could explain the obtained results.

Table 24.7 Average values of yield and main produce quality parameters measured in the PALAP9 long-term experiment

	Yield (kg tree^{-1})	TSS (g 100 g^{-1})	TA (g 100 ml^{-1})	TSS/TA	δ^{15}N juice (g kg^{-1})	δ^{15}N pulp (g kg^{-1})
CB	138.8	9.91 b	1.10 b	9.05	6.27 b	6.51 c
PM	137.4	10.33 a	1.15 a	8.97	7.88 a	8.45 b
LW	139.7	10.22 ab	1.16 a	8.80	8.48 a	8.95 a
MF	138.4	10.35 a	1.16 a	8.96	4.22 c	4.63 d

Note: Mean separation at 5% level with Tukey HSD test
TSS total soluble solids, *TA* total acidity

No significant difference was noticed for yield among the treatments, thus demonstrating that fertiliser treatments did not affect citrus productivity (Table 24.7). Regarding fruit quality parameters, CB treatment showed values of total soluble solids (TSS) and total acidity (TA) lower than other treatments even if this result had no relevance on the maturity index (TSS/TA) because of the compensation between the two parameters. No difference among treatments was recorded for fruit weight, firmness, peel thickness, central axis, juice content, vitamin C and synephrine (data not shown). Despite the fertilisation regimens being sufficient to produce differences between treatments in levels of N-containing compounds (Rapisarda et al. 2005), in our experiment, no differences were observed in synephrine content between organic and conventional fruits probably because the different treatments of the experiment received the same amount of N (Table 24.4) and had the same N nutritional status, as demonstrated by foliar analysis (Table 24.6).

The use of citrus compost (CB) improved main fruit quality characteristics, particularly with respect to mineral fertilisers, confirming previous results (Rapisarda et al. 2005). These findings could be explained considering that the improved soil fertility condition and the consequent ameliorated nutritional status of the long-term quality compost-treated plants (especially for those nutrients which are normally available in suboptimal amounts as P, Fe, Zn, Mn) positively influenced the measured fruit quality parameters (Zook and Lehmann 1968; Mostafa 2006; Abouziena et al. 2008).

The δ^{15}N detected in leaves (Table 24.6) and amino acids of juice (Table 24.7) showed the lower level in MF, an intermediate value in CB and the higher level in animal-derived fertilisers (PM and LW). In the case of δ^{15}N in proteins of pulp, the complete separation among treatments was noticed. In other words, ^{15}N tracing in plots fertilised with animal by-products (PM and LW) showed some differences with the plant-derived fertiliser (CB). However, the most relevant differences were noticed comparing all the organically managed treatments with synthetic mineral fertiliser. These findings are in accordance with Choi et al. (2006) and are a further confirmation of the potential of the δ^{15}N technique to be used as an indicator of agricultural regime, being suitable to discriminate between citrus produced according to the organic or conventional management (Georgi et al. 2005; Bateman et al. 2005, 2007).

Table 24.8 Standardised canonical discriminant function coefficients (PALAP9 long-term experiment)

	Function		
Variables	1	2	3
N	0.677	−1.013	−0.395
P	0.299	−0.064	1.038
K	−0.643	0.743	−0.129
Ca	−0.415	0.600	0.735
Mg	−0.377	0.899	−0.635
Fe	0.231	−0.538	0.293
Zn	0.371	−0.400	0.369
Mn	0.277	−0.193	−0.605
Synephrine	0.173	−0.327	0.597
N juice	−0.143	−0.894	−1.012
δ^{15}N pulp	−1.368	0.018	0.017
Yield	0.092	−0.251	−0.025
Fruit weight	0.435	0.011	0.205
Juice content	0.034	0.263	0.448
TSS	0.260	1.210	0.105
TA	−0.744	−0.255	−0.411
Peel thickness	0.345	0.419	−0.010
Central axis	0.251	0.401	0.465
Firmness	0.205	−0.490	0.001
Vitamin C	0.010	−0.168	0.217
C* skin	0.001	0.704	0.322
C* pulp	0.256	−0.283	−0.151
% Explained variance	96.9	1.8	1.3
% Cumulate variance	96.9	98.7	100.0
Wilks' lambda	0.005	0.290	0.579
Significance	0.000	0.011	0.088

Also, discriminant analysis of overall leaf and fruit analytical data set successfully separated treatments. First, discriminant canonical function explains 96.9% of the variability, with highly significant Wilks' lambda (Table 24.8). Obtained values indicated the weight of each variable on separation between groups for each discriminant function. Pulp δ^{15}N, leaf N and K levels and juice acidity showed the higher relative weights.

Standardised discriminant canonical scores of functions 1 and 2 are plotted in Fig. 24.4. Distribution of points allows to visualise clearly the separation of groups and the predominant effect of function 1. PM and LW treatments were not clearly differentiated. As a matter of fact, cross validation classified correctly all MF and CB samples (100% of cases), whereas PM and LW in a few cases were mixed up.

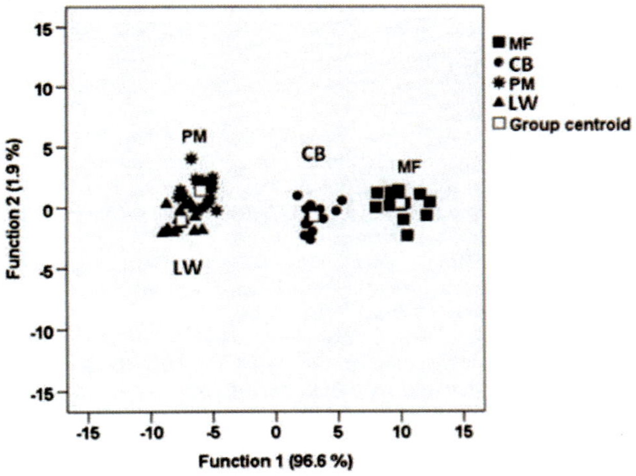

Fig. 24.4 Canonical discriminant functions 1 vs 2 (PALAP9 long-term experiment)

Multivariate approach by means of discriminant analysis succeeded to highlight the effects of fertiliser treatments. $\delta^{15}N$ was confirmed to be a good indicator for management discrimination (Rapisarda et al. 2005, 2010), but other leaf (N, K) and fruit parameters (acidity) affected the separation between data sets, as well.

24.5 Conclusions and Future Scope

The comprehensive evaluation of the above-reported results demonstrated that the organic management determined an increase of soil organic C stock and, at the same time, a higher C utilisation efficiency which contributes to soil organic C conservation in the long term. Lower $q(CO_2)$ values in organic systems supported the hypothesis of an improved efficiency by soil microflora in utilising energy and organic resources, which means an increased tendency of organically managed soils to establish ecological processes able to provide the proper amount of soil nutrients (i.e. phosphorus and micronutrients) and, thus, sustain orchard productivity in the long term. This hypothesis is also supported by the results of leaf analysis of orange trees which showed concentration of nutrients within the optimal ranges either in the organic or in the conventional system and by the lack of significant differences in yield between the organically and conventionally managed orange orchard. Furthermore, the increase of the organic matter level in organic soils could have determined the slight, significant, differences in some foliar nutritional levels (P, K, Fe) and, indirectly, in the fruit quality parameters (vitamin C and anthocyanins contents).

The body of knowledge available and the results presented demonstrate that organic citrus management is a technically feasible option for citrus growers. In addition, the shift to organic farming could contribute to the enhancement of the environmental sustainability of citrus productions in the long term.

Future research should aim to design organic citrus agro-ecosystems based on agro-ecofunctional intensification principles, thus able to give stable and quality yield, reduce the use of off-farm inputs and being resilient and environment friendly (Schmid et al. 2009). This is particularly problematic in highly specialised systems, which do not include livestock (or include livestock to a limited extent) and that, consequently, do not benefit from the favourable effects obtained from the interconnection of plant and animal productions and full exploitation of resources which are internal to the agroecosystems.

The studies carried out in organically managed groves in recent years have enhanced the recycling of organic residues available in citrus agroecosystems. Studies about ground cover, cover crops introduction and proper management could further contribute to shape citrus agroecosystems based on the optimal use of internal resources. In semiarid environments, reduction of water consumption and increase of water utilisation efficiency are a priority, and research work is needed to verify the possible role of dead mulches to increase nutrient recycling and water reserves in soil during the irrigation season. Furthermore, given the soil carbon pool increase as the main goal, the interaction of water and nutrients in the rhizosphere and the rooting system behaviour are open fields of research.

References

Abd-Allah ASE (2006) Effect of spraying some macro and micro nutrients on fruit set, yield and fruit quality of Washington Navel orange trees. J Appl Sci Res 2(11):1059–1063

Abouziena HF, Hafez OM, El-Metwally IM et al (2008) Comparison of weed suppression and mandarin fruit yield and quality obtained with organic mulches, synthetic mulches, cultivation and glyphosate. Hortic Sci 43(3):795–799

Agehara S, Warncke DD (2005) Soil moisture and temperature effects on nitrogen release from organic nitrogen sources. Soil Sci Soc Am J 69:1844–1855. doi:10.2136/sssaj2004.0361

Albiach R, Canet R, Pomares F et al (1999) Structure, organic components and biological activity in citrus soils under organic and conventional management. Agrochimica 43(5/6):235–242

Alef K (1995) Soil respiration. In: Alef K, Nannipieri P (eds) Methods in applied soil microbiology and biochemistry. Academic, London

Alianiello F, Fiorelli F (1998) Iso-electric focusing in soil science: a tool to develop for the knowledge of humic substances. Fresenius Environ Bull 7:523–530

Altieri MA (1983) The question of small development: who teaches whom? Agric Ecosyst Environ 9:40–405

Altieri MA (1995) Agroecology: the science of sustainable agriculture. Intermediate Technology Publications, Westview press, London

Altieri MA, Nicholls CI (2005) Agroecology and the search for a truly sustainable agriculture. United Nations Environment Programme, Mexico. http://www.agroeco.org/doc/agroecology-engl-PNUMA.pdf. Accessed 24 Oct 2011

Alva AK, Paramasivam S, Hostler KH et al (2001) Effects of nitrogen rates on dry matter and nitrogen accumulation in citrus fruits and fruit yield. J Plant Nutr 24(3):561–572. doi:10.1081/PLN-100104980

Alva AK, Mattos DJ, Paramasivam S et al (2006) Potassium management for optimizing citrus production and quality. Int J Fruit Sci 6:3–43. doi:10.1300/J492v06n01_02

Anderson TH (2003) Microbial eco-physiological indicators to asses soil quality. Agric Ecosyst Environ 98:285–293

Anderson TH, Domsch KH (1980) Quantities of plant nutrients in the microbial biomass of selected soils. Soil Sci 130:211–216

Anderson TH, Domsch KH (1985) Maintenance requirements of actively metabolizing microbial populations under in situ conditions. Soil Biol Biochem 17:197–203

Anderson TH, Domsch KH (1989) Ratios of microbial biomass carbon to total organic-C in arable soils. Soil Biol Biochem 21: 471–479

Anderson TH, Domsch KH (1990) Application of eco-physiological quotients (qCO_2 and qD) on microbial biomasses from soils of different cropping histories. Soil Biol Biochem 22:251–255

Anderson TH, Weigel HJ (2003) On the current debate about soil biodiversity. Landbauforschung Völkenrode 4(53):223–233

Bassirirad H (2000) Kinetics of nutrient uptake by roots: responses to global change. New Phytol 147:155–169

Bateman SA, Kelly SD, Woolfe M (2005) Nitrogen isotope relationship between crops and fertilizer: implication for using nitrogen isotope analysis as an indicator of agriculture regime. J Agric Food Chem 53:2664–2670. doi:10.1021/jf050374h

Bateman SA, Kelly SD, Woolfe M (2007) Nitrogen isotope composition of organically and conventionally grown crops. J Agric Food Chem 55:2664–2670. doi:10.1021/jf0627726

Bath B (2000) Matching the availability of N mineralized from crops with the N-demand of field vegetables. Doctoral thesis, Swedish University of Agricultural Science, Upssala, Sweden

Boaretto AE, Ueta FZ, Trivelin PCO et al (2006) Efficiency of nitrogen fertilization on citrus orchards. Acta Hortic 721:331–336

Bremner JM, Mulvaney CS (1982) Nitrogen total. In: Page AL (ed) Methods of soil analysis Part 2, vol 9, 2nd edn, Agron monograph. ASA/SSSA, Madison

Bricout J, Koziet J (1987) Control of the authenticity of orange juice by isotopic analysis. J Agric Food Chem 35(5):758–760. doi:10.1021/jf00077a027

Canali S (2003) Soil quality of organically managed citrus orchards in the Mediterranean area. Organic agriculture: sustainability, markets and policies. OECD/CABI Publishing, Paris/Wallingford

Canali S, Trinchera A, Intrigliolo F et al (2004) Effect of long term addition of composts and poultry manure on soil quality of citrus orchards in Southern Italy. Biol Fertil Soils 40:206–210. doi:10.1007/s00374-004-0759-x

Canali S, Di Bartolomeo E, Trinchera A et al (2009) Effect of different management strategies on soil quality of citrus orchards in Southern Italy. Soil Use Manag 25:34–42. doi:10.1111/j.1475-2743.2008.00191.x

Chambers R (1983) Rural development: putting the last fist. Longman, London

Choi WJ, Arshad SX, Chang SX et al (2006) Grain ^{15}N of crops applied with organic and chemical fertilizers in a four-year rotation. Plant Soil 284:165–174. doi:10.1007/s11104-006-0038-8

Ciavatta C, Govi M (1993) Use of insoluble polyvinylpyrrolidone and isoelectric focusing in the study of humic substances in soils and organic wastes. J Chromatogr 643:261–270. doi:10.1016/0021-9673(93)80560-U

Ciavatta C, Govi M, Vittori Antisari L et al (1990) Characterization of humified compounds by extraction and fractionation on solid polyvynilpyrrolidone. J Chromatogr 509:141–146

Clark MS, Horwath WR, Shermann C et al (1998) Changes in soil chemical properties resulting from organic and low-input farming practices. Agron J 90:662–671

Conway GR (1981a) What is an agroecosystem and why is it worthy of study? Paper presented at the workshop on human/agroecosystem interactions. PESAM/EAPI. Los Banos College, Laguna

Conway GR (1981b) Man versus pests. In: May R (ed) Theoretical ecology. Blackwell Science, Boston

Conway GR (1985) Agroecosystem analysis. Agric Admin 20:1–30

Dasberg S (1987) Nitrogen fertilization in citrus orchards. Plant Soil 100:1–9. doi:10.1007/BF02370928

Dazzi C (2005) I suoli della Sicilia. Notes from lessons. University of Palermo, Palermo (In Italian)

De Nobili M, Bragato G, Alcaniz JM et al (1990) Characterisation of electrophoretic fractions of humic substances with different electrofocusing behaviour. Soil Sci 150:763–770

Dell'Abate MT, Benedetti A, Trinchera A et al (2002) Humic substances along the profile of two Typic Haploxerert. Geoderma 107:281–296. doi:10.1016/S0016-7061(01), 00153-7

Diaz LF, Savage GM (2007) Factors that affect the process. In: Diaz LF (ed) Compost science and technology. Elsevier, Amsterdam

Dick WA, Gregorich EG (2004) Developing and maintaining soil organic matter levels. In: Schjönning P, Elmholt S, Christensen BT (eds) Managing soil quality. Challenges in modern agriculture. CABI Publishing, Wallingford

Dommergues Y (1960) La notion de coefficient de minéralisation du carbone dans les sols. Agron Trop XV(1):54–60 (In French)

Dou H, Alva AK, Khakural BR (1997) Nitrogen mineralization from citrus tree residues under different production conditions. Soil Sci Am J 61:1226–1232

EC Regulation 834/2007 of the Council of 28th June 2007. On organic production and labelling of organic products and repealing Regulation (EEC) No 2092/91. Official J Eur Union L 189, pp 1–23

Eghball B, Wienholdy BJ, Gilleyz JE, Eigenberg RA (2002) Mineralization of manure nutrients. Soil Water Conserv 57(6): 470–473

Ellen R (1982) Environment, substance and systems. Cambridge University Press, New York

Embleton TW, Jones WW, Labanauskas CK et al (1973) Leaf analysis as a diagnostic tool and guide for fertilization. In: Reuther W (ed) The citrus industry, vol III., pp 183–210

Engels C, Marschner H (1992) Adaptation of potassium translocation into the shoot of maize (*Zea mays*) to shoot demand: evidence for xylem loading as a regulating step. Physiol Plant 86:263–268. doi:10.1034/j.1399-3054.1992.860211.x

FAOSTAT (2009) http://faostat.fao.org/site/567/DesktopDefault.aspx?PageID=567#ancor. Accessed 11 July 2011

Fließbach A, Mäder P (2000) Microbial biomass and size-density fractions between soils of organic and conventional agricultural systems. Soil Biol Biochem 32:757–768

FoodNews (2003) Natural selection. Nat Ingredients/Organics, March 2003, Tunbridge Wells, UK

Franca SC, Gomes-da-Costa SM, Silveira APD (2007) Microbial activity and arbuscular mycorrhizal fungal diversity in conventional and organic citrus orchards. Biol Agric Hort 25:91–102

Georgi M, Voerkelius S, Rossmann A et al (2005) Multielement isotope ratios of vegetables from integrated and organic production. Plant Soil 275:93–100. doi:10.1007/s11104-005-0258-3

Gliessman SR (1983) Allelopathic interactions in crop/weed mixture: applications for weed management. J Chem Ecol 9(8):991–999. doi:10.1007/BF00982206

Gregorich EG, Carter MR, Angers DA et al (1994) Towards a minimum data set to assess soil organic-matter quality in agricultural soils. Can J Soil Sci 74:367–385

Hamm U, Gronefeld F, Halpin D (2002) Analysis of the European market for organic food. Dissertation, University of Wales Aberystwyth, Aberystwyth

Ibrahim M, Ahmad N, Anwar SA et al (2004) Micronutrient effects on citrus-fruit yield growing on calcareous soils. In: Ibrahim M, Anwar

SA, Ahmad N et al (eds) Proceedings of the symposium on plant-nutrition management for horticultural crops under water-stress conditions, organised by Soil Science Society of Pakistan at Agricultural Research Institute Sariab, Quetta, Pakistan, 5–6 Oct 2004

IFOAM – International Federation of Organic Agriculture Movements (2011) http://www.ifoam. Accessed 20 Aug 2011

International Trade Centre (2002a) Overview of world markets for organic food and beverages. ITC, Geneva. http://www.intracen.org/. Accessed 18 July 2011

International Trade Centre (2002b) The United States market for organic food and beverages. ITC, Geneva. http://www.intracen.org/. Accessed 18 July 2011

Intrigliolo F, Roccuzzo G, Lacertosa G et al (1999) Agrumi: modalità di campionamento per terreno, foglie, acque d'irrigazione e frutti. In: Intrigliolo F (ed) Valori analitici di riferimento. C.U.E.C.M, Catania, p 86 (In Italian)

Intrigliolo F, Montemurro N, Roccuzzo G et al (2000) Field survey on soil fertility and plant nutritional status in organic and conventional citrus orchards. In: International Association for the Optimisation of Plant Nutrition – IAOPN, National Research Centre – NRC (eds) Book of abstracts, X international colloquium for the optimization of plant nutrition, Cairo

Isermeyer H (1952) Estimation of soil respiration in closed jars. In: Alef K, Nannipieri P (eds) Method in applied soil microbiology and biochemistry. Academic, London

Jarvis SC, Stockdale EA, Shepherd MA et al (1996) Nitrogen mineralization in temperate agricultural soils: processes and measurement. Adv Agron 57:187–235. doi:10.1016/S0065-2113(08), 60925-6

JAS (2001) Japanese Agriculture Standard. http://www.maff.go.jp/e/jas/specific/organic.html. Accessed 24 Oct 2011

Jenkinson DS, Ladd JN (1981) Microbial biomass in soil: measurement and turnover. In: Paul EA, McLaren AD (eds) Soil biochemistry, vol 5. Marcel Dekker, New York

Jeschke WD, Hartung W (2000) Root-shoot interactions in mineral nutrition. Plant Soil 226:57–69. doi:10.1023/A:1026431408238

Karlen DL, Andrews SS, Doran JW (2001) Soil quality: current concepts and applications. Adv Agron 74:1–39

Kilcher L (2005) Organic citrus: challenges in production and trade. In: Cuaderno de Resumenes I Conferencia Internacional de Citricultura Ecologica BIOCITRICS, pp 22–27

Koo RCJ (1985) Potassium nutrition of citrus. In: Munson RS (ed) Potassium in agriculture. American Society of Agronomy, Madison

Lavelle P, Dugdale R, Scholes R et al (2005) Nutrient cycling. In: Hassan R et al (eds) Ecosystems and human well-being: current state and trends: findings of the condition and trends working group of the millennium ecosystem assessment, vol 1. Island Press, Washington, DC

Liu P (2003) World markets for organic citrus and citrus juices – current market situation and medium-term prospects. FAO, Rome. http://www.fao.org/DOCREP/006/J1850E/j1850e00.htm#. Last Accessed 20 sept 2011

Lowrance R, Stinner BR, House GS (1984) Agricultural ecosystems. Macmillan, New York

Mäder P, Fließbach A, Dubois D et al (2002) Soil fertility and biodiversity in organic farming. Science 296:1694–1697. doi:10.1126/science.1071148

Mäder P, Fließbach A, Dubois D et al (2006) The DOK experiment (Switzerland). In: Raupp J, Pekrun C, Oltmanns M et al (eds) Long-term field experiments in organic farming. ISOFAR – Verlag Dr Köster, Berlin

Marschner H (1995) Mineral nutrition of higher plants, 2nd edn. Academic, London

Mattos D, Quaggio JA, Cantarella H et al (2006) Response of young citrus trees on selected rootstocks to nitrogen, phosphorus and potassium fertilization. J Plant Nutr 29:1371–1385. doi:10.1080/01904160600830159

Menino MR, Carranca C, de Varennes A (2007) Distribution and remobilization of nitrogen in young non-bearing orange trees grown under Mediterranean conditions. J Plant Nutr 30:1083–1096. doi:10.1080/01904160701394543

Mostafa FMA (2006) Improving fruit quality of organically produced fruits of Egyptian lime by applying some biofertilizers and gibberellic acid. Assiut J Agric Sci 37(2):83–100

NOP Final Rule (2000) National Organic Program regulation. http://www.ams.usda.gov/AMSv1.0/nop. Accessed 24 Oct 2011

Palm CA, Swift MJ (2002) Soil fertility as an ecosystem concept. In: Accomplishments and changing paradigm towards the 21st century. Proceedings of the 17th world congress of soil science, Bangkok, Thailand, 14–21 Aug 2002, Paper No. 1988

Pankhurst CE, Hawke BG, McDonald HJ et al (1995) Evaluation of soil biological properties as potential bioindicators of soil health. Aust J Exp Agric 35:1015–1028. doi:10.1071/EA9951015

Park J, Seaton RAF (1996) Integrative research and sustainable agriculture. Agric Syst 50:81–100

Probst B, Schuler C, Joergensen RJ (2008) Vineyard soils under organic and conventional management – microbial biomass and activity indexes and their relation to soil chemical properties. Biol Fertil Soils 44:443–450. doi:10.1007/s00374-007-0225-7

Rapisarda P, Intelisano S (1996) Sample preparation for vitamin C analysis of pigmented orange juices. Ital J Food Sci 3:251–256

Rapisarda P, Bellomo SE, Intelisano S (2001) Storage temperature effects on blood orange fruit quality. J Agric Food Chem 49:3230–3235. doi:10.1021/jf0100321

Rapisarda P, Calabretta ML, Romano G et al (2005) Nitrogen metabolism components as a tool to discriminate between organic and conventional citrus fruits. J Agric Food Chem 53:2664–2669. doi:10.1021/jf048733g

Rapisarda P, Camin F, Faedi W et al (2010) New markers for the traceability of organic fruit. Acta Hortic 873:173–183

Reganold JP, Glover JD, Andrews PK et al (2001) Sustainability of three apple production systems. Nature 410:926–930. doi:10.1038/35073574

Reuter DJ, Robinson JB (1986) Plant analysis: an interpretation manual. Inkata Press, Melbourne

Sahrawat KL, Ponnamperuma FN (1978) Measurement of exchangeable NH_4^+ in tropical land soils. Soil Sci Soc Am J 42:282–283

Schloter M, Munch JC, Tittarelli F (2005) Managing soil quality. In: Bloem J, Hopkins DW, Benedetti A (eds) Microbiological methods for assessing soil quality (editorial board Burns RG, Dilly O, Fliessbach A, Lemanceau P, Lynch JM, Nannipieri P, Tittarelli F, Van Elsas JD). CABI Publishing, Cambridge

Schmid O, Padel S, Halberg N et al (2009) Strategic research agenda for organic food and farming. Technological Platform Organics, IFOAM EU Group, Brussels, p 116

Six J, Feller C, Denef K et al (2002) Soil organic matter, biota and aggregation in temperate and tropical soils – Effect of no-tillage. Agronomie (Paris) 22:755–775. doi:10.1051/agro:2002043

Soliani L (2004) Manuale di statistica per la ricerca e la professione, Charter 8, pp 1–34 (In Italian). Available at: www.dsa.unipr.it/soliani/soliani.html. Accessed 12 Oct 2011

Spedding CRW (1975) The biology of agricultural systems. Academic, London

Springer U, Klee J, 1 (1954) Prüfung der Leistungsfähigkeit von einigen wichtigeren Verfahren zur Bestimmung des Kohlemstoffs mittels Chromschwefelsäure sowie Vorschlag einer neuen Schnellmethode. Z Pflanzenernähr Dang Bodenk 64 (In German)

Srivastava AK, Singh S, Marathe RH (2002) Organic citrus: soil fertility and plant nutrition. J Sustain Agric 19:5–29. doi:10.1300/J064v19n03_03

Toselli M (2010) Nutritional implications of organic management in fruit tree production. Acta Hortic 868:41–48

Trinchera A, Tittarelli F, Intrigliolo F (2007) Study of organic matter evolution in citrus compost by isoelectrofocusing technique. Compost Sci Util 15(2):101–110

USDA Citrus update (2011) Citrus: World markets and trade. http://www.fas.usda.gov/htp/2011_July_Citrus.pdf. Last Accessed XXXXX.

van Diepeningen AD, de Vos OJ, Korhtals GW et al (2006) Effects of organic versus conventional management on chemical and biochemical parameters in agricultural soils. Appl Soil Ecol 31:120–135. doi:10.1016/j.apsoil.2005.03.003

Vance ED, Brookes PC, Jenkinson DS (1987) An extraction method for measuring microbial biomass C. Soil Biol Biochem 19:703–707

Wardowski W, Soule J, Grierson W et al (1979) Minimum quality (maturity) standards. In: Florida citrus quality tests. Florida Cooperative Extension Service, IFAS, University of Florida, Gainesville

Watson CA, Stockdale EA (2000) Nutrient budgets on organic farms: a review of published data. http://orgprints.org/8380. Last Accessed 20 sept 2011

Watson CA, Atkinson D, Gosling P et al (2002a) Managing soil fertility in organic farming systems. Soil Use Manag 18:239–247

Watson CA, Bengtsson H, Ebbesvik M et al (2002b) A review of farm-scale nutrient budgets for organic farms as a tool for management of soil fertility. Soil Use Manag 18:264–273

Whitmore AP (2007) Determination of the mineralization of nitrogen from composted chicken manure as affected by temperature. Nutr Cycl Agroecosyst 77:225–232. doi:10.1007/s10705-006-9059-1

Willer H, Kilcher L (eds) (2009) The world of organic agriculture: statistics and emerging trends 2009. IFOAM/FiBL, Bonn/Frick

Yaseen M, Ahmad M (2010) Nutrition management in citrus: effect of multinutrients foliar feeding on the yield of Kinnow at different locations. Pak J Bot 42(3):1863–1870

Zook EG, Lehmann J (1968) Mineral composition of fruits II: nitrogen, calcium, magnesium, phosphorus, potassium, aluminium, boron, copper, iron, manganese and sodium. J Am Diet Assoc 52:225–231

Integrated Nutrient Management in Citrus

25

Anoop Kumar Srivastava

Abstract

Citrus by its avid nutrient-absorbing capacity, is considered a highly nutrient-responsive perennial fruit crop. Differential efficacy of two conventional methods of fertilization (soil versus foliar application), although helped in improving the quality citrus, but of late, continuous fertilization, has failed to sustain the same yield expectancy on a long-term basis due to depletion of soil carbon stock and consequently emerged multiple nutrient deficiencies, irrespective of soil type. The menace of multiple nutrient deficiencies would be further triggered through increase in air temperature via changes in microbial communities and activities within the rhizosphere in the light of climate change. Such changes will dictate adversely on the orchard's productive life in the long run.

Gradual shift from purely inorganic to organic fertilizers started gaining wide-scale use for enhanced biogeochemical nutrient cycling. Long-term data accrued on response of organic manuring versus inorganic fertilizers demonstrated that important soil quality indices like soil microbial diversity, soil microbial biomass nutrient (C_{mic}, P_{mic}, and N_{mic}), and organic carbon partitioning displayed significant changes but without much difference in quantum of fruit yield. The other approaches involving multiple microbial inoculation along with enrichment of organic manures through inorganic fertilizers known as substrate have further been highlighted as a part of INM module to provide an understanding on mechanism involved in C stabilization in soils for regulating soil C sequestration and associated nutrient dynamics under INM-based production system in citrus orchards. Integration of such microbial consortium with organic manures and chemical fertilizers (basis for INM-based nutrient management) in addition to concepts like sensor-based programmable fertigation and precision-oriented site-specific nutrient management exploiting spatial variation in soil fertility and leaf nutrient composition could further provide the much desired niche in the production sustainability through engineering nutrient dynamics within the rhizosphere under changing climatic scenario.

Keywords

Inorganic fertilization • Manuring • Microbial consortium • Soil microbial biomass • Soil microbial biomass nutrients • Fruit yield • Fruit quality

25.1 Introduction

Citrus is globally considered as one of the premier fruit crops, both in terms of area and production. Sub-optimum production due to the prevalence of nutrient constraints is well established

A.K. Srivastava (✉)
Soil Science Laboratory, National Research Centre for Citrus, Post Box. No. 464, Amravati Road, Shankarnagar Post Office, Nagpur 440 010, Maharashtra, India
e-mail: aksrivas2007@gmail.com

A.K. Srivastava (ed.), *Advances in Citrus Nutrition*,
DOI 10.1007/978-94-007-4171-3_25,

Table 25.1 Global distribution of nutrient deficiencies in citrus orchards

Citrus regions	Nutrient deficiencies	References
Argentina (Tucuman)	N, Cu, Fe, Mg, Zn	Aso and Dantur (1970)
Australia (New South Wales, Riverland, Sunraysia)	N, P, Cu, Mn, Zn, B	Halase (1963) and Duncan (1969)
Brazil (Sao Paulo, Parana)	Ca, Mg, P, K, Zn, B	Caetano et al. (1984) and Fidalski and Auler (1997)
Chile (Azapa, Elqui, Limari, Cachapoal)	N, Zn, Mn, P, S	Veregara et al. (1973)
China (Fujian, Sichuan)	Ca, P, Fe, Mn, Zn, Mo	Li et al. (1998) and Yin et al. (1998)
Costa Rica (Atlantic zone)	N, P, K, Ca, Mg, Mn, Zn	Bornemisza et al. (1985), Alvardo et al. (1994), and Araya et al. (1994)
Egypt (Aswan, Beheira, Tahrir)	N, P, Fe, Mn, Zn,	El-Fouly et al. (1984) and Salem et al. (1995)
India (northwest, northeast, south, central region)	N, P, Ca, Mg, Fe, Mn, Zn	Awasthi et al. (1984), Dhatt (1989), Srivastava and Singh (2004a, b), Srivastava and Singh (2006, 2008a), and Srivastava et al. (2005)
Iran (Jiroff valley)	Zn, Mn, Cu	Rao (1993)
Israel (Negev, Sinai, Jordan valley)	Ca, Mg, Fe, Zn	Shaked and Ashkenazy (1984) and Horesh et al. (1986)
Italy (Sicily, Calabria, Basilicata)	N, K, Mg, Cu	Pennisi (1975)
Japan (Shizuoka, Ehime, Kanagawa)	N, P, K, Mg, Zn	Takatsuji and Ishihara (1980), Kozaki (1981) and Wada et al. (1981)
Kenya (Rift valley)	N, P, B, Fe, Zn, Cu, Mn, Mo	Kimani (1984)
Korea (Jeju Island)	N, P, K, Ca, Mg, S, Cu, Zn	Kim et al. (1969) and Moon et al. (1980)
Morocco (Sous valley)	Fe, Mn, Zn	Penkov et al. (1979)
Nepal (Dhankuta, Lamjung, Gorkha)	B, Mg, Cu, Ca, Zn	Gupta et al. (1989) and Tripathi and Harding (2001)
Pakistan (Punjab)	K, Zn, B	Haq et al. (1995)
Sierra Lone (Sierra)	N, P, K, Ca, Mg, Zn	Haque and Godfrey (1976)
Spain (Valencia, Seville, Murcia, Catania)	N, P, K, Ca, Mg, Fe, Mn, Zn	Majorana (1960) and Hellin et al. (1988)
Thailand (Korat Plateau)	Ca, Mg, P Zn	McCall (1965)
Trinidad (Caribbean area)	Mg, Zn, Mn	Weir (1969)
Turkey (Izmir, Aegean region)	Ca, Mg, Fe, Zn	Ercivan (1974) and Saatci and Mur (2000)
USA (Florida, California, Texas)	N, P, K, Fe, Mg, Zn, Mn, Cu, B, Mo	Koo (1982), Zhou and Alva (1993), Tucker et al. (1995), and Zhang et al. (1997)
Venezuela (Carabobo)	N, P, Ca, Mg, Zn	Pinto and Leal (1974)

in citrus, as in any other crop (Srivastava et al. 2008). Malnutrition of citrus orchards in Asian countries like India, Pakistan, Sri Lanka, Thailand, China, Philippines, Nepal, Iran, etc. is more or less a commonality (Ghosh and Singh 1993) with some exceptions. However, the situation by contrast is extremely different in frontline citrus-growing countries like USA, Brazil, Israel, Spain, etc. in terms of rootstock options as per soil conditions, micro-irrigation/fertigation technology, better refined diagnostic techniques, and a larger proportion of orchards being regularly fertilized based on nutrient demand and supply analysis. Single or multiple nutrient deficiency–linked decline in citrus orchard productivity is reported the world over (Table 25.1).

Considering the economics of citrus production, fertilizers alone on average constitute about 20–30% of the total cost of citrus production which is a significant recurring expenditure and a grower needs to invest every year (Srivastava and Singh 2003, 2005, 2008c). Like any other fruit tree, citrus requires 16 essential elements for normal growth, production, and quality irrespective of the source (Zekri 1995).

The present citrus production trends are characterized by either frequent crop failure or recurrence of alternate on-and-off years, setting unaccountable monetary loss to the industry (Jones et al. 1975; Smith 1976; Dass et al. 1998; Rojas 1998; Aiyelagbe 2001). In recent years, nutrient additions have been exclusively in favour of mineral fertilizers due to demographic pressure, demands related to lifestyles, and trade involvement (Reddy et al. 2003). While the quick and substantial response to fruit yield due to mineral fertilizers eclipsed the use of organic manures, the inadequate supply of the latter sources exacerbated this change (Ghosh 2000). Integrated nutrient management (INM) with emphasis on the use of bio-organics is a comparatively recent concept which needs to be vigorously pursued to achieve sustainability in citrus production trend spaced over years. Additionally, crop nutrition, therefore, must respect the prescriptions of INM. The merits of INM-based practices also take into account that the mobilization of unavailable nutrients could also be effected by speeding up the rate of mineralization of various organic substrates.

INM as a dynamic concept of nutrient management that considers the economic yield in terms of fruit yield coupled with quality on one hand and soil physico-chemical and microbiological health on other hand as a marker of resistance against the nutrient mining that arises because of failure to strike a balance between annual nutrient demand versus the quantity of nutrients applied. Soils under citrus differ from other cultivated soils, which remain fallow for 3–6 months every year forcing depletion of SOM (Bhargava 2002). In contrast, biological oxidation of existing C continues in soil covered under citrus (Srivastava et al. 2002). Multiple nutrient deficiencies are considered to have a triggering effect on potential source of atmospheric CO_2. Soil carbon stock is, hence, considered as an important criterion to determine the impact of INM in the longer version of impact assessment (He et al. 1997; Singh et al. 1999; Joa et al. 2006). The amount of accumulated C within the rhizosphere soil does not continue to increase with time with increasing C outputs. An upper limit of C saturation level occurs, which governs the ultimate limit of soil C sink and rate of C sequestration in mineral soils, independent of C input rate. An understanding of the mechanism involved in C stabilization in soils is needed for controlling and enhancing soil C sequestration (Goh 2004).

Recognition of the importance of soil microorganisms has led to increased interest in measuring the quantum of nutrients held in their biomass. An increase in the microbial biomass often goes along with increased nutrient immobilization. Over the years, the concepts of INM and integrated soil management (ISM) have been gaining acceptance, moving away from a more sectoral- and inputs-driven approach. INM advocates the careful management of nutrient stocks and flows in a way that leads to profitable and sustained production. ISM emphasizes the management of nutrient flows but also highlights other important aspects of soil complex such as maintaining organic matter content, soil structure, moisture, and microbial biodiversity. Still, more attention is needed towards integrated soil biological management as a crucial aspect of soil fertility management since providing protection to citrus rhizosphere against the nutrient depletion is of utmost importance for sustained orchard production in which the objectivity of INM could have far-reaching consequences. Exploring microbial diversity perspectives in a citrus crop is, therefore, important and equally useful to arrive at measures that can act as indicators of soil quality and sustainable orchard productivity using biological soil management to be ultimately integrated with INM. Diagnosis of nutrient constraints and their efficient management have, therefore, now shifted in favour of INM through collective use of organic manures, inorganic fertilizers, and beneficial microorganisms and become all the more difficult.

25.2 Soil Fertilization

This is still the most accepted and widely used method of fertilization in commercially productive citrus orchards. The main organ for absorbing water and nutrients by a plant is its roots. Average concentration of micronutrients (mg kg^{-1}) in the fibrous roots of 'Valencia' orange grown in sand culture reported by Smith et al. (1953) were B 25, Cu 157, Fe 1,783, Mn 257, and Zn 462 mg kg^{-1}, whereas the average concentration (Kg ton^{-1} fruit) of different nutrients in fruits was observed as N 1.20, P 0.18, K 1.54, Ca 0.57, Mg 0.12, B 1.63×10^{-3}, Cu 0.39×10^{-3}, Fe 2.1×10^{-3}, Mn 0.38×10^{-3}, and Zn 0.40×10^{-3} (Mattos-Junior et al. 2003). Climate- and soil-related factors such as low temperature, excessive moisture (water infiltration rate), drought, etc., however, disturb nutrient and water uptake during plant growth; the effects of which may vary from a temporary restriction of growth to reduced fruit yield and quality at harvest (Papadakis et al. 2004). The uptake and translocation of iron and zinc of 'Valencia' orange on trifoliate orange (susceptible to iron and zinc deficiencies) and rough lemon rootstock (resistant to iron and zinc deficiencies) were studied by Khadr and Wallace (1964). Under low iron and zinc, rough lemon absorbed and translocated both nutrients more to the top, while under high supply, the difference between rootstocks disappeared for iron. These observations suggested that iron and zinc translocation from roots to leaves may be a more important problem than absorption *per se* (Srivastava and Singh 2008b).

Roots impose differential nutrient demand depending upon the sink strength, in the form of fruits and newly emerging vegetative growth, which eventually dictate the nutrient requirement (Srivastava et al. 2007). Changes in the nutrient content of oranges from young and mature 'Bellamy' navel orange trees throughout fruit development showed that during early growth (fruit dry weight <10 g), the contents of K and B (phloem mobile) and of Ca and Cu (phloem immobile) increased linearly in relation to fruit dry weight. In contrast to K and B, the Ca and Cu content plateaued at a fruit dry weight of 15 g. There was a comparatively greater influx of Ca into the albedo than the pulp during stage I of fruit development. During stage I of fruit development, normalized Ca fluxes into whole fruit and albedo tissue were higher in fruits from young trees than in fruits from mature trees (Storey and Treeby 2002).

The fertilizers are usually given to citrus orchards following three different application techniques. Some of the common techniques are: circle banding (cutting furrow 20 cm wide and 30 cm deep around the tree in circle beneath the outer canopy), strip band application (cutting parallel furrows 20 cm wide and 30 cm deep, between the tree rows), and hole placement (digging 4–5 holes, each of 15–20 cm diameter and 30 cm deep, beneath the outer canopy of each tree).

25.2.1 Macronutrient Requirement

Response of N fertilization in improving the growth, yield, and quality of different citrus cultivars is well recognized under different agroclimatic regions of the countries like

Table 25.2 Optimum macronutrients for different commercial citrus cultivars through soil application

Country	Dose	Crop/citrus spp.	Reference
Algeria	600 g N, 150 g P_2O_5, 600 g K_2O/tree	Clementine	Dris (1997)
Algeria	240 g N, 40 g P_2O_5, 400 g K_2O/tree	Dancy mandarin	Pedrera et al. (1988)
Algeria	450 kg N, 0–180 kg P, 0–30 kg K ha^{-1}	Valencia orange	Sarooshi et al. (1991)
Algeria	120 kg N, 150 kg P, 75 kg S, 6 kg Cu, 0.8 kg Mo, 5.0 kg Zn ha^{-1}	Neck orange	Lim et al. (1993)
Australia	450 kg N, 30 kg P, 180 kg K ha^{-1}	Valencia orange	Bevington (1984)
Australia	22.5–25 kg N, 5–12.5 kg P_2O_5, 10–12.5 kg K_2O mu^{-1}	Satsuma mandarin	Wang (1985)
Brazil	200 kg N, 140 kg P, 210 kg K ha^{-1}	Pera sweet orange	Canteralla et al. (1992)
Brazil	180 kg N, 90 kg P, 180 kg K ha^{-1}	Pera sweet orange	Rodriguez (1980)
Egypt	600 g N, 135 g P, 285 g K/tree	Navel orange and Balady mandarin	El-Hagah et al. (1983)
Egypt	750 g N, 200 g P_2O_5, 500 g K_2O/tree	Egyptian Balady lime	Ahmed et al. (1988)
Egypt	1,500 g N, 400 g P_2O_5, 750 g K_2O/tree	Egyptian Balady lime	Maatouk et al. (1988)
Egypt	500 kg N, 100 kg P, 100 kg K/ha	Satsuma mandarin	Kacharava (1985)
Egypt	100 kg N, 200 kg P_2O_5, 300 kg K_2O ha^{-1}	Valencia orange	Goepfert et al. (1987)
Egypt	475 g N, 320 g P_2O_5, 355 K_2O/tree	Satsuma mandarin	Koseoglu et al. (1995b)
France	180 g N, 90 g P_2O_5, 45 g K_2O, 800 g CaO/tree	Clementine	Aubert and Vullin (1998)
Greece	0.5 kg N, 0.5 kg P_2O_5, 1.0 kg K_2O ha	Grapefruit	Androulakis et al. (1992)
Greece	1.02 kg N, 0.58 kg P_2O_5, 0.55 kg K_2O/tree	Satsuma mandarin	Liu et al. (1994)
Greece	420 g N, 323 g P_2O_5, 355 g K_2O/tree	Satsuma mandarin	Koseoglu et al. (1995a, b)
India	500 g N, 100 g P_2O_5, 400 g K_2O/tree	Mosambi sweet orange	Ghosh (1990)
India	125 g N, 175 g P, 100 g K/tree	Satsuma mandarin	Hong and Chung (1979)
India	160 g N, 320 g P, 480 g K/tree	Valencia late orange	Hernandez (1981)
India	2.72 kg N, 1.81 kg P, 0.60 kg K/tree	Sathgudi sweet orange	Reddy and Swamy (1986)
India	800 g N, 170 g P_2O_5, 391 g K_2O/tree	Mosambi Sweet orange	Desai et al. (1986)
India	250 g N, 250 g P_2O_5, 500 g K_2O/tree	Coorg mandarins	Kannan et al. (1989)
USA	625 kg N, 525 kg K_2O ha	Valencia orange	Tucker et al. (1990)

Brazil, Australia, South Africa, India, etc. (Ghosh et al. 1989; Tachibana and Yahata 1996). Recovery of applied N ranged from 33% to 61% and from 1% to 33% through soil and mature leaves, respectively, with maximum N absorption at the rate of 27 mg/plant/day during the summer. On annual basis, 25% of the total N in the sweet orange tree came from the reserve N of transplanted plant, 16% from soil, and remaining 59% from the urea applied to the soil (Boaretto et al. 1999). The effect of N fertilizers at 168 kg/ha produced the best response on yield of citrus cultivars, viz. 'Valencia', 'Parson Brown', 'Hamlin', and 'Sunburst' sweet orange grown in Hardee county of Florida, USA (Alva et al. 2001).

The mathematical relation between N fertilizer rate and yield using variance analysis showed that application of 1.18 kg N/plant on the medium fertility soil produced a yield of 40.5 kg/plant and 35.9 kg/plant with 1.03 kg N/plant on poor fertility soil (Liu et al. 1994). Contrary to foliar fertilization, soil application of macronutrients proved more efficacious. The optimum requirement of macronutrients for different commercial citrus cultivars (Table 25.2) suggested a large variation in recommendations due to cultivar specificity to nutrient acquisition (by roots, movement across roots to xylem, distribution and remobilization) and final utilization in growth and metabolism in addition to difference in soil and climate.

The reports about the shortage of S in citrus orchards are extremely limited. The response of Ca and Mg application is common as an amendment in citrus orchards established on soils of varying acidities. Lime (up to 12 ton ha^{-1}) and phosphogypsum (up to 4 ton ha^{-1}) incorporation to the surface soil has proved effective in alleviating subsoil acidity, increase Ca and Mg content, and base saturation near 60% down to 60 cm depth in the soil profile. These changes improved the yield of 'Valencia' orange (Quaggio et al. 1992, 1998). Other amendments like gypsum (Anderson 1968), sulphur (Rasmussen and Smith 1959), magnesium sulphate (Mdinaradze and Datuadze 1987), magnesium carbonate (Koo 1966), basic slag (Koo 1964), and phosphogypsum (O'Brien and Sumner 1988) have also shown promising results in citrus orchard. Koo (1971), in two long-term trials testing sources and rates of Mg on 'Marsh' grapefruit and 'Valencia' orange, reported that application of 1.5 ton MgO ha^{-1} increased the yield by 12.6% and total soluble solids by 14.7% over low rate of 0.60 ton MgO ha^{-1}. Dolomite application at much lower rates, 400 kg/ha/year, produced an additional fruit yield of 75 ton ha^{-1} in 'Satsuma' mandarin in 6 years of experiment compared to application at 200 kg/ha/year (Shimorgori et al. 1980).

Table 25.3 Optimum micronutrients for different commercial citrus cultivars through soil application

Country	Dose	Crop/citrus spp.	Reference
Cuba	$MnSO_4$ (483 kg/tree), $ZnSO_4$ (303.8 g/tree)	Valencia orange	Garcia Alvarez et al. (1983)
Georgia, USSR	Zn aldehyde (4–12 kg/ha)	Satsuma mandarin	Mdwaradze (1981)
India	$ZnSO_4$ (500 g/tree)	Sweet orange cv. Blood Red	Khera et al. (1985)
India	$ZnSO_4$, K_2SO_4 (0.5% foliar spray), K_2O as K_2SO_4 (210 g/tree soil application)	Kagzi lime	Singh et al. (1989)
India	$ZnSO_4$ (100 g/tree soil application), (0.5% foliar spray)	Sathgudi orange	Devi et al. (1996)
India	300 g $ZnSO_4$/tree/year	Nagpur mandarin	Srivastava and Singh (2008b)
USA	Fe, Mn, Zn-EDTA (292 g, 292 g, 315 g ha^{-1})	Valencia orange	Alva and Tucker (1992)
USA	$ZnSO_4$ (810 g/tree soil application), $MnSO_4$ (630 g/100 gal foliar spray)	Lemon	Embleton et al. (1966)
USA	Zn-EDTA 30 g/tree	Grapefruit cv. Rio Blood	Swietlik (1996)
USA	Zn-EDTA 2.1 g m^{-2}	Valencia orange	Anderson (1984)
USA	Zn (3 g), B (3 g), Mo (1.5 g/tree)	Valencia orange	Egorashvili et al. (1991)

Anderson (1987) later, comparing the results of a 17-year-old study on the response of 'Valencia' orange to lime application, reported that increase in soil pH from initial value of 5.2–7.0 increased the yield by 50% in the first 7-year period which further improved the yield by 200% in the next 10 years with no significant yield difference between limestone and dolomite. Specialized with slow release nutrients extensively tested all over the citrus-growing countries especially as a method of reducing nitrate leaching (Khalaf and Koo 1983; Ferguson et al. 1988; Wang and Alva 1996; Paramasivam and Alva 1997; Obereza et al. 1999; Schumann et al. 2003) have also attracted citrus researchers. Most of these studies conducted on young trees were short-term experiments focussed mainly on the effect of several controlled-release fertilizers on tree growth with very few on fruit yield (Koo 1986; Zekri and Koo 1992). Controlled-release fertilizers compared to soluble fertilizers have proved to be very effective in increasing growth due to continuous rather than fluctuating nutrient supply (Khalaf and Koo 1983; Koo 1988), besides reducing the rates and number of applications during the growing season (Zekri and Koo 1991).

A large number of commercially exploited coating materials, viz. sulphur, osmocote, isobutylidene diurea, crotonylidene, triazimes, gypsum, phosphogypsum, ureaform, magnesium ammonium phosphates, etc. (Maynard and Lorenz 1979), have been tested in citrus. The research studies with various controlled-release fertilizers (CRF) products showed that nitrate leaching potential could be reduced compared to similar rates of conventional soluble fertilizers. Obreza et al. (1999) reviewed the performance of five CRFs on young 'Valencia' oranges and the economics of using these fertilizers instead of conventional soluble granular products. They found that the CRFs produced similar or better yields but that the cost of fertilizing trees with CRFs alone at the full N rate was four times the conventional fertilization cost, and the return was only 15% greater. They concluded that the high cost of CRFs currently makes them uneconomical for exclusive use in citrus production. For this reason, the current ridge citrus, N BMPs do not account for the use of any CRFs. Most recently, tests on mature 'Hamlin' oranges with CRFs in Florida flatwoods soils have been more encouraging (Obreza and Rouse 2004). These CRFs performed better when applied once a year at 220 kg N ha^{-1} than water-soluble fertilizer applied three times a year at 180 kg N ha^{-1}. These observations suggest that the favourable economic CRFs would undoubtedly be a valuable addition to the current citrus best management practices for not only N-use efficiency but could be an effective supplement to other important nutrients like P, K, Ca, Mg, and S as well (Maynard and Lorenz 1979).

25.2.2 Micronutrient Requirement

The chelates like Fe-EDTA on acid soils and Fe-EDDHA on alkaline soils are most widely used in citrus (Leonard and Stewart 1952). The optimum dose of chelates depends on the tree size, degree of chlorosis, soil type, and management practices (Ghosh and Besra 2000). The studies carried out world over have, therefore, shown some diversity in optimum doses of micronutrients (Table 25.3) due to difference in nutrient-supplying capacity of soil conditioned by soil properties, (e.g. texture, pH, salinity, calcareousness, cation–anion ratio, etc.) nutrient requirement by specific rootstock–scion combination, planting density, irrigation source, region-specific cultural practices, the agroclimate, etc. The combination of soil application and foliar spray has also produced equally good results, e.g. $ZnSO_4$, K_2SO_4 (0.5% foliar spray), K_2O as K_2SO_4 (210 g/tree soil application) for 'Kinnow' mandarin (Singh et al. 1989) and $ZnSO_4$, $FeSO_4$, $MnSO_4$ (50 g/tree each soil application), (0.50% foliar application) for 'Sathgudi' sweet orange (Devi et al. 1996).

25.3 Integration of Organic Manures and Fertilizers

Improvement in fruit eating quality of Satsuma mandarin treated with rapeseed cakes and Qixing organic compound poultry manure (Huang et al. 1995; Shi et al. 2000; Intrigliolo and Stagno 2001), promising role of chicken manure in citrus fertilization program in Florida USA (Fischer 1992), and Urea-N + farmyard manure for better pre-bearing performance of Kinnow mandarin (Dudi et al. 2003) supported integrated basis of nutrient management. As early as 1969, Ozordzadze observed that the average yield of 'Satsuma' mandarin doubled with peat + fertilizer treatment and farmyard manure + P, K nutrients compared to unfertilized trees. A number of studies indicated improved efficacy of fertilizer nutrients with combined use of manures and fertilizers (Prasad and Singhania 1989; Rokba et al. 1995; Kohli et al. 1998) or combined use of inorganic fertilizers (Singh et al. 1993; Medhi et al. 2006) in addition to enzymatic activities (acid phosphatase, dehydrogenase, and urease) of soil (Tiwari 1996).

The response of different rootstock types viz. sour orange, Swingle citrumelo, Troyer citrange, and Cleopatra mandarin seedlings grown on a mixture of peat, vermiculite, and sand showed an increase in root to shoot dry weight ratio by 15% and 21%, with application of 0.5% and 1% root solution, respectively, as bio-stimulant containing humic acids, marine algae extracts, plant metabolite, and vitamin B (Swietlik 1991). According to Hsiung and Iwahori (1984), organic salts of calcium were more effective than inorganic salts in retarding abscission of the explants when added to ethephon solution. In particular, calcium propionate and calcium salicylate were highly effective, followed by calcium acetate. Reddy et al. (1980) showed that animal manure increased soluble phosphorus in soil. Many organic materials like manures, muck, cake fertilizers, and peat mixed with $FeSO_4$ alleviated the Fe deficiency–induced chlorosis on calcareous soils (Loeppert and Clarke 1984; Loeppert 1986; Chen et al. 1990).

The pattern of change in soil organic carbon over the last 8 years was interpreted (Fig. 25.1). It was observed that with all the treatments, the organic carbon content continued to increase, e.g. from an initial level of 0.50–0.68% with FYM (T_1) treatment, 0.51–0.89% with vermicompost treatment (T_2), 0.54–0.72% with poultry manure (T_3), 0.52–0.60% with neem cake (T_4), 0.58–0.86% with sunhemp green manuring (T_5), and from 0.54% to 0.61% with inorganic fertilizers (T_6) during 2008–2009. But, from thereon, the inorganic carbon continued to decline to 0.60% with FYM (T_1), 0.74% with vermicompost (T_2), 0.60% with poultry manure, 0.50% with neem cake and then green manure of fodder cowpea, 0.70% with green manuring of sunhemp, and to 0.54% with inorganic fertilizers during 2010–2011. These observations warranted that increase in organic carbon content in soil can be realized up to certain period, and thereafter it declined, possibly it operates in a cyclic manner. Such cycles differ from one production system to another production system (Table 25.4). On the other hand, changes in soil fertility status in response to different organic manuring treatments showed (Table 25.4) a consistent increase in almost all the nutrients (available nutrients in soil or leaf nutrients concentration) with all the treatments. Green manuring treatment produced the maximum increase in available nutrients followed by poultry manure, vermicompost, neem cake, farmyard manure, and inorganic fertilizers. These responses were well supported by the changes in leaf nutrient concentration. The efficiency of different treatments in relation to leaf nutrients concentration was rated as green manure > vermicompost > poultry manure > neem cake > farmyard manure > inorganic fertilizers. Considering these observations, green manuring and vermicompost treatments have so far produced the best response, although such studies deserve to be interpreted on long-term basis. Inclusion of AM (arbuscular mycorrhiza 500 g/plant) + PSB (phosphate solubilizing bacteria 100 g/plant) + Az (*Azospirillum* 100 g/plant) + Th (*Trichoderma harzianum* 100 g/plant) to 75% RDF improved the fruit yield of 'Kinnow' mandarin on rough lemon at Ludhiana, Punjab (location 1) and in 'Khasi' mandarin at Tinsukia, Assam (location 2) through different INM-based treatments (Table 25.5).

Application of farmyard manure (FYM) in combination with nutrients like N, P, and K improved the leaf area (Pennisi 1971; Motskobilli 1984; Plemanac 1985; Beridze 1990), winter hardiness (Motskobilli 1986) in 'Satsuma' mandarin, canopy volume by substituting up to 50% N with FYM in 'Coorg' mandarin (Mustaffa et al. 1997), and fruit quality (juice, total soluble solids, rag content, etc.) in 'Nagpur' mandarin (Huchche et al. 1998), 'Blood Red Malta' variety of sweet orange treated with combination of $(NH_4)_2$ SO_4 + farmyard manure (Dhillon et al. 1961), 'Matau Peiya' cultivar of pummelo treated with dolomite + organic fertilizer (Chen et al. 1997), and 'Satsuma' mandarin treated with NPK plus farmyard manure + CaO (Pirchalajsvili 1970) in comparison to inorganic chemical fertilizers carrying NPK alone.

Highest fruit yield with improved quality was obtained with 25 kg FYM with 400 g N, 150 g P, and 300 g K/plant in 'Khasi' mandarin on acid red soils (Ghosh and Besra 1997); 150 kg FYM, 1,500 g N, 440 g P_2O_5, and 600 g K_2O/tree in navel orange (El–Koumey et al. 2000); 52 kg FYM, 1.82 kg $(NH_4)_2SO_4$–N/tree in 'Balady' mandarin (Gamal and Ragab 2003); and 15 kg neem cake, 800 g N, 300 g P_2O_5, 600 g K_2O/tree/year in sweet orange (Tiwari et al. 1997). Arsenidze and Chanukuvadze (1988) observed maximum yield of 'Satsuma' mandarin for the trees receiving PK Ca + FYM with N was applied as ammonium nitrate at 100 g/tree to 4- to 5-year-old tree in Western Georgia. The best fruit yield

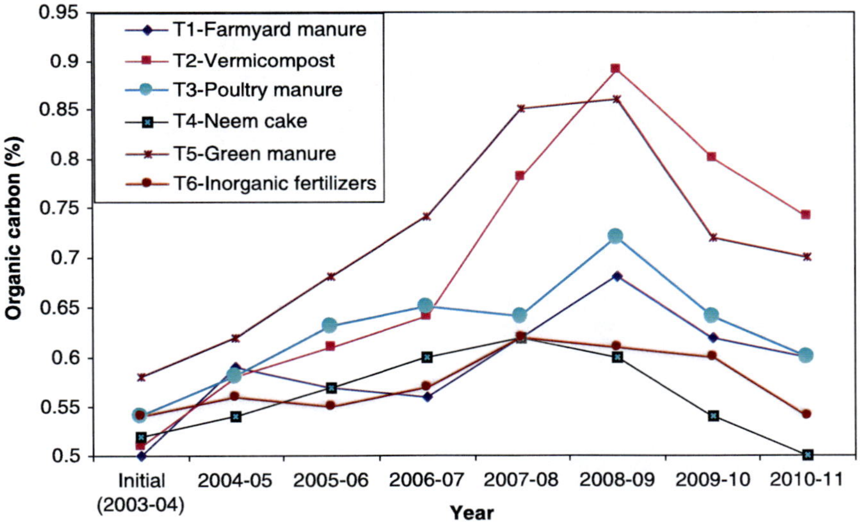

Fig. 25.1 Changes in organic carbon during the last 8 years due to different organic manuring versus inorganic fertilizer treatments. T_1–T_4 were computed on N-equivalent basis T_1 (12–60 kg/tree), T_2 (8–40 kg/tree), T_3 (6–30 kg/tree), T_4 (4–20 kg/tree), T_5 (seeds of *Crotalaria juncea* sown beneath tree canopy), and T_6 (100–600 g N, 50–200 g P_2O_5, 25–100 g K_2O/tree), all computed on N-equivalent basis (Adapted from Srivastava et al. 2011a)

Table 25.4 Change in available nutrients (mg kg^{-1}) in soil as influenced by different organic manuring treatments to 'Nagpur' mandarin in central India

		Macronutrients			Macronutrients			
Treatments		N	P	K	Fe	Mn	Cu	Zn
T_1	Farmyard manure	132.8	14.3	186.2	14.6	12.3	1.12	0.79
		(2.10)	(0.08)	(1.10)	(68.9)	(52.2)	(5.4)	(20.8)
T_2	Vermicompost	152.6	16.1	201.6	17.2	14.4	1.64	1.70
		(2.38)	(0.12)	(1.48)	(92.4)	(61.3)	(6.2)	(25.8)
T_3	Poultry manure	140.7	18.2	192.1	18.4	13.2	1.32	0.98
		(2.24)	(0.10)	(1.24)	(67.8)	(55.4)	(6.2)	(21.6)
T_4	Neem cake	124.4	12.7	174.2	13.8	10.8	1.10	0.80
		(2.01)	(0.09)	(0.90)	(58.3)	(49.4)	(6.1)	(20.8)
T_5	Green manure	126.3	12.4	171.3	13.1	11.0	1.40	0.82
	(sunhemp)	(2.18)	(0.09)	(0.98)	(58.4)	(51.4)	(7.1)	(20.4)
T_6	Inorganic	131.3	16.1	182.7	14.0	11.2	1.82	0.84
	fertilizers (NPK)	(2.30)	(0.10)	(1.20)	(71.3)	(53.2)	(6.2)	(22.3)
	CD (p=0.05)	3.6	1.1	9.1	1.1	2.1	0.69	0.10
		(0.08)	(0.009)	(0.12)	(2.8)	(1.1)	(NS)	(1.2)

Adapted from Srivastava et al. (2011a)
T_1–T_4 were computed on N-equivalent basis T_1 (12–60 kg/tree), T_2 (8–40 kg/tree), T_3 (6–30 kg/tree), T_4 (4–20 kg/tree), T_5 (seeds of *Crotalaria juncea* sown beneath tree canopy), and T_6 (100–600 g N, 50–200 g P_2O_5, 25–100 g K_2O/tree), all computed on N-equivalent basis
Figures in parenthesis indicate leaf nutrient composition (macronutrients %, micronutrients ppm)

and quality of 'Balady' mandarin was obtained with the use of FM (a by-product of sugar industry) at a rate of 120 kg + and inorganic 6.0 kg N/tree (Ebrahiem and Mohamed 2000). The amount of nitrogen fertilizer in a high-density planting of 'Satsuma' mandarin can be reduced to 100 kg/ha/year, if 20 tons rice straw/ha /year, or 200 kg/ha/year when no organic matter is applied, without any yield reduction (Tachibana and Yahata 1996). Apart from higher fruit yield (163.3 kg/tree), reduced fruit drop (19.7%) was also observed with integrated use of pig manure (110 kg/tree) + urea-N (750 g/tree) + muriate of potash (650 g/tree) compared to fruit yield (107 kg/tree) and fruit drop (24.6%) with pig manure (110 kg/tree) alone in Khasi mandarin (Dubey and Yadav 2003).

Microbiological analysis of citrus soil treated with PK (phosphorus and potassium) or PK + different N types

Table 25.5 Response of different INM-based treatments on fruit yield of Kinnow mandarin (location 1: soil order Entisol) and Khasi mandarin (location 2: soil order Alfisol)

Treatments	Location 1: Ludhiana, Punjab		Location 2: Tinksukia, Assam	
	2005–2006	2006–2007	2005–2006	2006–2007
T_1	72.8	46.4	82.8	84.4
T_2	78.2	49.5	72.9	74.8
T_3	82.2	53.6	79.8	78.0
T_4	72.2	44.4	107.1	110.8
T_5	69.6	40.3	87.3	88.7
CD ($p=0.05$)	4.6	2.9	10.7	14.9

Adapted from Medhi et al. (2006), Murti et al. (2008), and Srivastava (2008)
T_1 – RDF (recommended doses of fertilizers, 600 g N, 300 g P_2O_5, 600 g K_2O, 7.5 kg neem, cake/tree); T_2 – 100% RDF + AM (arbuscular mycorrhiza 500 g/plant), PSB (phosphate solubilising bacteria 100 g/plant), Az (*Azospirillum* 50 g/plant); T_3 – 100% RDF, AM, PSB, Az, Th (*Trichoderma harzianum*) 100 g/plant; T_4 – 75% RDF, AM, PSB, Az, Th; and T_5 – 50% RDF, AM, PSB, Az, Th

(Gochelashvili 1984) indicated that the soil microflora in young orchards was more responsive to mineral N rates than in mature orchards. In trials with young lemon trees, cv. 'Meyer', NOMF (an organo-mineral fertilizer prepared from peat enriched with neutral solution of Ca industrial waste, Mg, Al, and some other elements) + NPK applied in spring with N as a top dressing in summer, produced 29–38% higher fruit yield. While other study by Lomtadze and Muradova (1986) reported 15–17% higher yields in the first 3 cropping years than NPK alone with peat + NPK applied in spring and N in summer. Influence of different combinations of organic fertilizers like groundnut cake, neem cake, and farmyard manure along with urea-N as inorganic fertilizer showed higher increase in canopy volume and girth of 'Coorg' mandarin by substituting 25% of urea-N with groundnut cake (Mustaffa et al. 1997). Highest fruit yield and improved quality (Ghosh and Besra 1997) and concentration of different nutrient in index leaves (Singh et al. 1993) were observed with 25 kg organic matter in combination with 400 g N, 150 g P, and 300 g K in Khasi mandarin and sweet orange grown, respectively, on acid red soil under humid tropical climate of northeast India and alkaline sandy loam soil of northwest India.

Beridze (1990) observed highest yield of 6.6 ton ha^{-1} from the trees receiving basal dressing of 150 kg N, 120 kg P_2O_5, 80 kg K_2O, and 25 kg peat ha^{-1} as a mulch plus 55 ton ha^{-1} farmyard manure in a 5-year-old 'Meyer' lemon tree. In Georgia, on a podzolized brown earth soil with a low humus N (2.6–3.16%), P (6.8–7.5%), and K (17.4–22.1%), the growth and yield of lemon trees were greater with NPK than FYM application alone (Marsanija 1970). Application of manures and/or fertilizers increased organic carbon, NH^{4+}–N, NO_3–N, and Olsen P in the soil. Soil pH increased and organic carbon decreased with the increase in the time of incubation. Phosphorus-enriched manures maintained a higher level of phosphorus in soil solution for a longer period than the fertilizer alone (Prasad and Singhania 1989). Addition of manure delayed the hydrolysis of urea and thus reduced the loss of nitrogen. Nitrogen and phosphorus-enriched manures maintained a higher level of available nitrogen and phosphorus in the soil for a longer period than the fertilizer alone. Hsiung and Iwahori (1984), on the other hand, observed that organic salts of calcium were more effective than inorganic salts in retarding abscission of the explants when added to ethephon solution. In particular, calcium propionate and calcium salicylate were highly effective, followed by calcium acetate.

The effect of N viz. 0, 200, 400, 600, and 800 g/plant/year through urea and farmyard manure application on yield, quality, and shelf life were studied by Huchche et al. (1998). Differential N levels were supplemented with doses of P, K, and farmyard manure. The farmyard manure at 25 kg/plant/year alone was kept at the control. Application of N significantly increased the yield of 'Nagpur' mandarin with increasing N rates as compared to farmyard manure alone (Table 25.6). There was no adverse effect of chemical N on physico-chemical properties of fruits. Storage of fruits for 12 days at 20–28°C, 50–70% relative humidity and for 80 days at 5–6°C, 85–90% relative humidity revealed no adverse effect of chemical inorganic N (up to 600 g/plant/year) supplemented with P, K, and farmyard manure on weight loss and decay (%) in 'Nagpur' mandarin. The shelf life of fruits was not significantly better when fruits were grown in farmyard manure alone. According to Gill et al. (1998), the lowest urease activity was recorded in the plot without organic amendment, while the highest activity was observed in wheat straw–amended combined harvested plot followed by N plus farmyard manure plots.

Quality response on various citrus cultivars to combined application of organic manure along with inorganic N, P, and K fertilizers is reported widely (Rokba et al. 1995; Borah et al. 2001). The beneficial effect of supplying 'Balady'

Table 25.6 Response of farmyard manure alone and in combination with inorganic fertilizers on yield, quality, and postharvest weight loss and decay at ambient and refrigerated conditions in 'Nagpur' mandarin (pooled data of three growing seasons)

					Weight loss and decay (%) at			
		Quality			Ambient condition		Refrigerated condition	
Treatment (N:P:K + FYM/tree)	Yield (fruits/tree)	Firmness (kg force)	Juice (%)	TSS/acid	Wt. loss	Decay	Wt. loss	Decay
0:200:100 g + 25 kg	524	2.73	41.15	14.24	11.97	1.15	11.28	11.90
200:200:100 g + 25 kg	556	2.73	44.79	12.12	10.84	1.51	10.43	7.15
400:200:100 g + 25 kg	575	2.82	42.88	13.89	11.74	2.22	10.41	8.83
600:200:100 g + 25 kg	658	2.79	44.97	13.89	11.13	0.30	10.91	9.85
800:200:100 g + 25 kg	706	2.76	43.16	14.32	11.63	2.38	11.35	14.67
FYM (25 kg)	495	2.81	44.69	13.14	12.94	1.99	12.88	10.75
LSD ($p = 0.05$)	112	NS	NS	NS	NS	NS	NS	NS

Adapted from Huchche et al. (1998) and Srivastava (2008)
NS non-significant, *FYM* farmyard manure

mandarin trees grown in sandy soil (during both off or on year bearing state) with filter mud (FM, a by-product of the sugar industry) or farmyard manure (FYM) on growth, nutrition, yield, and fruit quality was studied during three successive seasons of 1997–1999. Supplying trees with organic fertilizers in combination with the mineral N source improved shoot length; N, P, and K content in leaves; fruit set; yield; number of fruits per tree; fruit weight; juice percentage; percentage of total soluble solids; and ascorbic acid content. Application of FM provided overall better results compared to that of FYM. The best yield and fruit quality were obtained with the use of FM at 120 kg + 6.0 kg inorganic N/tree and the rest to be added via any mineral N source (Ebrahiem and Mohamed 2000). Similarly, response of a 20-year-old Washington navel to combination of chemical fertilizers (200 g N as ammonium sulphate, 100 g P_2O_5 as calcium superphosphate, 120 g K_2O/tree as potassium sulphate) and FYM (0.06 Mg^3/tree) during 2002–2004 on sandy soil of El-Tahreer Egypt showed significantly higher fruit yield; leaf N, P, K, Ca; and juice total solids compared to FYM alone (Maksoud and Haggag 2004).

Another study on the effect of various components of INM as application of individual higher doses of farmyard manure (50 kg/tree) and conjoint use of iron pyrites (200 g) and farmyard manure (25 kg – pressmud 2 kg/ha) on acid lime (*Citrus aurantifolia* Swingle) on changes in soil fertility status (Aariff and Begum 2007) showed much higher status of available N (196.7–219.2 kg/ha), P (27.6–35.1 kg/ha), K (342.4–361.3 kg/ha), and S (8.4–17.4 kg/ha). Application of 5 kg vermicompost and 20 kg FYM in the months of November and May under sub-humid tropical climate of central India significantly ($p < 0.05$) improved the fruit yield and quality over either of the individual effect of two manures individually or when applied in combination with inorganic chemical fertilizers (Table 25.7).

Table 25.7 Response of manure versus integrated treatment of manure plus inorganic fertilizers on yield and quality of 'Nagpur' mandarin

		Fruit quality (%)	
Treatment	Fruit yield (ton ha^{-1})	Total soluble solids	Juice
T_1	28.0	9.62	38.7
T_2	30.7	10.52	41.2
T_3	52.5	10.83	43.3
CD ($p = 0.05$)	2.2	0.33	1.03

Adapted from Ramamurthy (2006)
T_1 20 kg FYM (farmyard manure), T_2 recommended doses of fertilizers as 50 kg FYM, 600 N, 200 P_2O_5, 100 K_2O g/tree, T_3 5 kg vermicompost + 20 kg FYM/tree

Joa et al. (2006) investigated the effect of different methods of fertilizers applied for 10 years such as NF (no fertilization); N, P, K (28, 40, 48 kg 10 $acre^{-1}$); 3 N, 3 P, 3 K (84, 120, 84 kg 10 $acre^{-1}$); compost (2 ton 10 $acre^{-1}$); and compost plus N, P, K on the density and biomass C of microorganisms, enzyme activities, and amount of phospholipid fatty acid (PLFA) in Satsuma mandarin orchard. No significant difference in the density of microorganisms and biomass C was observed between treatments. However, the density of acid-resistant bacteria (613×10^3 cfu g^{-1} dry soil) in NF was about twice higher than that in 3 N, 3 P, 3 K (324.8×10^3 cfu g^{-1} dry soil), and the number of *Bacillus spp.* in compost treatment was highest. There was a seasonal fluctuation in the number of microorganisms; fluorescent *Pseudomonas spp.* was dominant in spring and gram-negative bacteria were in fall. The activity of soil acidic phosphatase was observed as 311.5 (NF), 478.4 (N, P, K), 586.3 (3 N, 3 P, 3 K), 548.7 (compost), and 503.3 (compost N, P, K). The activity of CMCase (carboxymethyl cellulase) was observed in the order of 17.5 (NF), 68 (N, P, K), 20.5 (3 N, 3 P, 3 K), 116.5 (compost), and 106 (compost N, P, K). The amount of PLFA was significantly high in compost–N–P–K and compost. Principal components analysis with PLFA data showed that microorganisms in compost and compost N, P, K formed different community from them in other treatments.

Rhizosphere modification through root exudation is an important attribute that regulates not only the availability of nutrients in soil but also their acquisition by plants (Ferguson 1990). Effect of inoculation of AM (*Glomus deserticola*) and *Azotobacter chroococcum* in different combinations with FYM and inorganic fertilizers was studied for six seasons on the performance of Kinnow mandarin on alkaline loam soil. Combined use of AM and FYM reduced the soil pH from 8.5 to as low as 6.4. While AM in comparison to *Azotobacter chroococcum* modified the rhizosphere favourable to improve soil nutrient availability and consequent uptake by plants towards better growth, yield, and quality (Usha et al. 2004).

25.3.1 Integration of Oilcakes and Fertilizers

Various studies involving combination of oilcakes with inorganic fertilizers have shown promising results. These included: combination of 15 kg neem cake and 800 g N, 300 g P, and 600 g K/tree/year in sweet orange on alkaline loam soil (Tiwari et al. 1997); 15 kg neem cake, 600 g N, 300 g P, and 300 g K along with 15 kg neem cake/plant/year in acid lime on alkaline black clay soils (Ingle et al. 2001); 7.5 kg neem cake, 600 g N, 300 g P, and 600 g K/plant/year in 'Khasi' mandarin on acidic loam soil in Tinsukia belt of Assam (Borah et al. 2001); 7.5 kg castor cake, 400 g N, 150 g P, and 300 g K/plant/year in 'Sathgudi' sweet orange on alkaline sandy loam soil in Tirupati, A.P. (Seshadri and Madhavi 2001); and 7.5 kg neem cake, 800 g N, 200 g P, and 300 g K/tree/year in acid lime on black clay soil of Central India (Hiwarale et al. 2004).

Addition of KCl or $MgSO_4$ (0.15 kg/tree) with peanut oilcake (1.75 kg/tree, containing 6.3% N, 1.2% P, and 0.6% K) on red loam soil (pH 6.2, available N 100.6 mg kg^{-1}, P 10.6 mg kg^{-1}, and K 77.0 mg kg^{-1}) significantly increased the yield of *Citrus sinensis* L. Osbeck cultivar (Liu Cheng) by 16.8% or 10.9%, respectively. Similarly, the average fresh fruit weight increased by 10.5% or 7.6%, respectively. Quality-wise with peanut oil cake, the application of KCl or $MgSO_4$ enables citrus fruits to observe 3.99% or 1.77% increase of soluble sugar and 0.39% or 0.11% decrease of citric acid over those of peanut oil cake alone treatment. The °Brix/acid ratio, hence, increases from 11.8 with peanut oil cake to 31.1 with peanut oilcake plus KCl or 15.9 with peanut oilcake plus $MgSO_4$ (Huang and Yang 1990). Application of neem cake at rate or 7.5 kg + 75% of RDF (recommended doses of fertilizers) produced a better build-up in soil fertility with reference to available N, P, and K (Mukherjee et al. 1991).

25.3.2 INM in 'Nagpur' Mandarin: A Case Study

An experiment was carried on Vertic Ustochrept in Central India during 2007–2011 with a total of 11 treatments based on recommended dose fertilizers (RDF)viz. T_1 – 100% RDF; T_2 – *Azotobacter chroococcum, Pseudomonas fluorescens, Bacillus polymyxa, Bacillus mycoides*, and *Trichoderma harzianum*; T_3 – 75% RDF + 25% farmyard manure (computed on N-equivalent basis); T_4 – 50% RDF + 50% farmyard manure (computed on N-equivalent basis); T_5 – 75% RDF + 25% FYM + bioinoculant; T_6 – 50% RDF + 50% FYM + bioinoculant; T_7 – 75% RDF + 25% vermicompost (computed on N-equivalent basis); T_8 – 50% RDF + 50% vermicompost (computed on N-equivalent basis); T_9 – 75% RDF + 25% vermicompost + bioinoculant; T_{10} – 75% RDF + green manuring (biometric sunhemp) + bioinoculant (added after the cutting at just flower initiation stage) and T_{11} – 50% RDF + green manure (sunhemp) + bioinoculant (added after the cutting at just flower initiation stage). Results are briefly summarized below:

25.3.2.1 Biometric Response

Different INM-based treatments produced significant response on canopy volume (Table 25.8). Interestingly, with regard to average increase in canopy volume, sole application of inorganic fertilizers (1.54 m^3) was very well overtaken by either FYM-based INM treatments (1.49–1.95 m^3) or by vermicompost-based INM treatments (1.88–2.34 m^3) but not by green manuring with either 75% RDF + MC (1.32 m^3) or with 50% RDF + MC (0.93 m^3). Amongst different INM-based combinations, 75% RDF as inorganic fertilizers was not so effective with MC (1.37 m^3) or even with 50% or 75% RDF combined with 50% or 25% FYM (1.84 or 1.49 m^3) when compared with 100% RDF, unless all the three components of INM (inorganic fertilizers, organic manures, and microbial biofertilizers) are tested conjointly. For example, T_5 (1.88 m^3) and T_6 (1.95 m^3) were superior over either T_3 (1.51 m^3) or T_4 (1.49 m^3) while T_9 (2.34 m^3) produced significantly higher magnitude of response over either T_7 (1.88 m^3) or T_8 (1.91 m^3). MC when combined with 75% RDF was not as effective as when treatments were supplemented with organic manures, thereby suggesting further that exclusive use of chemical fertilizers was detrimental to multiplication of microbial population in soil due to absence of mineralizable carbon in soil.

25.3.2.2 Response on Fruit Yield and Quality

Data on response to fruit yield in the first year is just notional, since fruiting was not uniformly distributed. But, nonetheless, fruit yield response (Table 25.8) of different treatments displayed equally interesting results commensurating with

Table 25.8 Response of 'Nagpur' mandarin growth parameters in response to different INM-based treatments (pooled data)

Treatments	Canopy volume (m^3)	Net increase over initial (m^3)	Fruit yield (kg/tree)	Fruit quality parameters (%)		
				Juice	TSS	Acidity
T_1 (100% RDF)	4.26 (2.72)	1.54	15.6	46.4	9.2	0.78
T_2 (75% RDF + MC)	4.49 (3.12)	1.37	8.4	47.2	8.9	0.78
T_3 (75% RDF + 25% FYM)	5.17 (3.69)	1.51	9.2	47.8	9.3	0.70
T_4 (50% RDF + 50% FYM)	4.01 (2.52)	1.49	9.8	46.9	9.4	0.71
T_5 (75% RDF + 25% FYM + MC)	5.15 (3.27)	1.88	14.5	46.8	9.2	0.79
T_6 (50% RDF + 50% FYM + MC)	4.68 (2.74)	1.95	12.8	46.9	9.4	0.80
T_7 (75% RDF + 25% Vm)	6.32 (4.44)	1.88	14.4	47.2	9.3	0.74
T_8 (50% RDF + 50% Vm)	5.26 (3.35)	1.91	14.6	48.4	9.6	0.70
T_9 (75% RDF + 25% Vm + MC)	5.77 (3.43)	2.34	18.8	48.8	9.7	0.70
T_{10} (75% RDF + Gm + MC)	4.12 (3.1)	1.32	11.2	47.2	9.2	0.74
T_{11} (50% RDF + Gm + MC)	3.01 (2.09)	0.93	10.6	47.2	9.0	0.74
CD ($p = 0.05$)	0.18 (NS)	0.12	–	–	–	–

Adapted from Srivastava et al. (2011b)
Figures in parenthesis indicate initial values of canopy volume
MC stands for microbial consortium developed by isolating the native microbes from the experimental soil (mixture of *Azotobacter chroococcum*, *Bacillus mycoides*, *Bacillus polymyxa*, *Pseudomonas fluorescens*, and *Trichoderma harzianum*)
FYM, *Vm*, *Gm*, and *RDF* stand for farmyard manure, vermicompost, green manuring, and recommended doses of fertilizers, respectively
Vm (nutrient composition: 2.28% N, 0.11% P, 1.41% K, 3,012 ppm Fe, 118 ppm Mn, 34 ppm Cu, and 58 ppm Zn)
FYM (nutrient composition: 1.20% N, 0.09% P, 0.72% K, 389 ppm Fe, 51 ppm Mn, 30 ppm Cu, and 24 ppm Zn)
Gm(Nutrient composition: 1.10% N, 0.08% P, 0.58% K, 280 ppm Fe, 68 ppm Mn, 28 ppm Cu and 24 ppm Zn).
RDF (200–600 g N, 50–200 g P_2O_5, 100–300 g K_2O, 50–200 g $ZnSO_4$, 50–200 g $FeSO_4$, 50–200 g $MnSO_4$/tree/year)

the responses on incremental growth of canopy volume. Amongst the INM-based treatments, FYM-based treatments T_3–T_6 registered a much lower fruit yield (9.2–12.8 kg/tree) than vermicompost-based treatments (14.4–18.8 kg/tree), while green manuring–based treatments (T_{10}–T_{11}) recorded fruit yield little lower (10.6–11.2 kg/tree) than FYM-based treatments (9.2–14.5 kg/tree)

Evaluation of these treatments on fruit quality response revealed similar results. Vermicompost-based treatments (T_7–T_9) produced much favourable response on fruit quality (47.2–48.8% juice and 9.3–9.7% TSS) than FYM-based treatments T_3–T_6 (46.8–47.8% juice and 8.9–9.4% TSS) but much better than T_1 (46.4% juice and 9.2% TSS) supplying inorganic fertilizers.

25.3.2.3 Response on Soil Fertility Changes

Changes in available supply of nutrients in soil were evaluated based on both macro- as well as micronutrients.

Available Macronutrients

Evaluation of different INM-based treatments demonstrated significant changes on available supply of nutrients in soil (Table 25.9). On equivalent basis, T_3 (137.63 N, 13.8 P, 155.2 K mg kg^{-1}) recorded significantly higher level of macronutrients over T_4 (134.5 N, 11.8 P, 150.4 K mg kg^{-1}), but statistically on par with T_1 (134.1 N, 12.9 P, 152.7 K mg kg^{-1}). These responses were of much lower magnitude with either exclusive use of inorganic fertilizers as T_1 (134.2 N, 12.9 P, 152.7 K mg kg^{-1}) or even when combined with MC as T_2 (128.8 N, 11.0 P, 144.4 K mg kg^{-1}). These observations lend strong support to the fact that microbial biofertilizers are not compatible with exclusive chemical fertilizers unless cushioned with some C-source as manures. Comparison of FYM (T_5) or vermicompost (T_9)-based INM treatments revealed significantly higher magnitude of available macronutrients with the latter (150.9 N, 16.4 P, 168.3 K mg kg^{-1}) over the former treatment (143.1 N, 15.1 P, 159.9 K mg kg^{-1}). Replacement of organic manure up to 50% of RDF as T_6 was not so effective (140.4 N, 16.4 P, 168.7 K mg kg^{-1}), but significantly better than inorganic fertilizers carrying treatment as T_1 (134.2 N, 12.9.P, 152.7 K mg kg^{-1}). On the other hand, green manuring in combination with either 75% RDF + MC as T_{10} (135.8.1 N, 15.6 P, 151.4 K mg kg^{-1}) or with 50% RDF + MC as T_{11} (135.9 N, 16.0 P, 150.82 K mg kg^{-1}) could not match with any of the treatments including inorganic fertilizers but significantly superior over T_2 (128.8 N, 11.0 P, 144.4 K mg kg^{-1}) where MC was used in combination with 75% RDF.

Available Micronutrients

The response on changes in availability of micronutrients was similar to that of macronutrients (Table 25.9). There was no significant response of different treatments, on the availability of Cu. The treatments T_1 100% RDF (8.93 Fe, 6.77 Mn, 0.77 Zn mg kg^{-1}), T_3 with 75% RDF+25% FYM (9.11 Fe, 6.91 Mn, 0.72 Zn mg kg^{-1}) showed no significant difference

Table 25.9 Response of different INM-based treatments on the available nutrients (mg kg^{-1}) in soil (pooled data)

Treatments	N	P	K	Fe	Mn	Cu	Zn
T_1 (100% RDF)	134.2	12.9	152.7	8.93	6.77	1.45	0.77
	(119.2)	(10.2)	(125.0)	(8.03)	(6.20)	(1.49)	(0.72)
T_2 (75% RDF + MC)	128.8	11.0	144.4	7.84	6.11	1.56	0.72
	(120.1)	(9.8)	(124.6)	(8.50)	(7.34)	(1.10)	(0.76)
T_3 (75% RDF + 25% FYM)	137.6	13.8	155.2	9.11	6.91	1.54	0.80
	(117.4)	(9.2)	(126.2)	(8.86)	(6.34)	(1.51)	(0.70)
T_4 (50% RDF + 50% FYM)	134.5	11.8	150.4	8.68	6.83	1.48	0.80
	(116.2)	(10.1)	(119.2)	(9.37)	(6.01)	(1.60)	(0.72)
T_5 (75% RDF + 25% FYM + MC)	143.1	15.1	159.9	9.87	7.50	1.50	0.91
	(118.7)	(9.7)	(116.5)	(7.77)	(7.60)	(1.25)	(0.74)
T_6 (50% RDF + 50% FYM + MC)	140.5	15.0	161.1	9.58	7.61	1.52	0.88
	(117.9)	(10.0)	(116.7)	(8.56)	(7.21)	(1.49)	(0.72)
T_7 (75% RDF + 25% Vm)	142.7	14.8	158.5	9.51	7.77	1.58	0.88
	(115.6)	(9.9)	(122.9)	(8.38)	(7.85)	(1.51)	(0.75)
T_8 (50% RDF + 50% Vm)	143.1	14.7	159.2	9.48	7.91	1.53	0.87
	(117.5)	(9.6)	(126.1)	(7.22)	(6.87)	(1.30)	(0.78)
T_9 (75% RDF + 25% Vm + MC)	150.9	16.4	168.7	11.42	9.38	1.54	0.98
	(119.3)	(9.8)	(128.7)	(7.44)	(7.12)	(1.26)	(0.78)
T_{10} (75% RDF + Gm + MC)	135.8	15.6	151.4	9.95	7.74	1.57	0.90
	(120.4)	(9.4)	(123.6)	(8.21)	(7.08)	(1.18)	(0.70)
T_{11} (50% RDF + Gm + MC)	135.9	16.0	150.8	9.50	7.70	1.55	0.87
	(121.5)	(9.9)	(122.4)	(8.14)	(7.21)	(1.75)	(0.71)
CD ($p = 0.05$)	2.4	0.70	2.7	0.51	0.41	NS	0.06

Adapted from Srivastava et al. (2011b)
Figures in parenthesis represent the initial values
MC stands for microbial consortium developed by isolating the native microbes from the experimental soil (mixture of *Azotobacter chroococcum*, *Bacillus mycoides*, *Bacillus polymyxa*, *Pseudomonas fluorescens* and *Trichoderma harzianum*)
FYM, *Vm*, *Gm*, and *RDF* stand for farmyard manure, vermicompost, green manuring, and recommended doses of fertilizers, respectively
Vm (nutrient composition: 2.35% N, 0.13% P, 1.51% K, 3,002 ppm Fe, 111.4 ppm Mn, 43.4 ppm Cu, and 59 ppm Zn)
FYM (nutrient composition: 1.20% N, 0.09% P, 0.84% K, 489 ppm Fe, 61.03 ppm Mn, 25.46 ppm Cu, and 29.8 ppm Zn)
Gm (nutrient composition: 0.78% N, 0.08%P, 0.70% K, 281 ppm Fe, 63 ppm Mn, 30 ppm Cu, and 26 ppm Zn)
RDF (200–600 g N, 50–200 g P_2O_5, 100–300 g K_2O, 50–200 g $ZnSO_4$, 50–200 g $FeSO_4$, 50–200 g $MnSO_4$/tree/year)

between them, while inferior to T_2 carrying 75% RDF + MC (7.84 Fe, 6.11 Mn, 0.72 Zn mg kg^{-1}) supporting the earlier observation about the ineffectiveness of MC with inorganic fertilizers. While green manuring supplemented with 75% RDF (9.95 Fe, 7.74 Mn, 0.90 Zn mg kg^{-1}) or 50% RDF (9.50 Fe, 7.70 Mn, 0.87 Zn mg kg) with MC was not as effective compared to either FYM- or vermicompost-based INM treatments. On the other hand, effectiveness of MC was more visible when treatments carried both organic manures as well as inorganic fertilizers, e.g. T_5 (9.87 Fe, 7.50 Mn, 0.91 Zn mg kg) was superior over T_3 (9.11 Fe, 6.91 Mn, 0.80 Zn mg kg) or T_6 (9.58 Fe, 7.61 Mn, 0.88 Zn mg kg) being superior over T_4 (8.68 Fe, 6.83 Mn, 0.80 Zn mg kg). Similarly, amongst vermicompost-based treatments, T_9 (11.42 Fe, 9.38 Mn, 0.98 Zn mg kg) proved better than T_7 (9.51 Fe, 7.77 Mn, 0.88 Zn mg kg). The combination of 50% RDF + 50% organic manure and vermicompost at 50% produces higher supply of available Mn (7.91 mg kg) than FYM (6.83 mg kg) with other micronutrients remaining statistically unchanged.

25.3.2.4 Changes in Soil Microbial Population

The soil microbial count in terms of bacterial and fungal count showed significant changes in response to different INM-based treatments (Table 25.10). No difference in bacterial and fungal counts was observed between T_2 incorporating MC with 75% RDF and T_1 having 100% RDF as inorganic fertilizers supporting the fact that as long as good soil productivity is obtained, soil microbial health could be maintained even with inorganic fertilizers. Of course, the magnitude of such response will be of comparatively lower order when compared with organic manures or in combination with inorganic fertilizers plus microbial biofertilizers. Out of two organic manures (FYM versus vermicompost), vermicompost-based treatments produced much favourable response on soil microbial counts as evident from superiority of T_7 (50 and 25 cfu g^{-1} bacterial and fungal counts, respectively) over T_3 (45 and 19×10^3 cfu g^{-1} bacterial and fungal counts, respectively). Similarly, treatment T_8 (53 and 26×10^3 cfu g^{-1} bacterial and fungal counts, respectively)

Table 25.10 Changes in total bacterial and fungal count of soil samples (0–15 cm) in response to different INM-based treatments (pooled data)

	SMB (cfu × 10^3 g^{-1} soil)		SMBN (mg kg^{-1} soil)		
Treatments	Bacterial count	Fungal count	C_{mic}	N_{mic}	P_{mic}
T_1 (100% RDF)	32	16.3	152.1	19.1	16.1
T_2 (75% RDF + MC)	31	16	146.3	19.3	15.2
T_3 (75% RDF + 25% FYM)	45	19	159.1	23.9	15.9
T_4 (50% RDF + 50% FYM)	44	20	164.1	25.6	17.0
T_5 (75% RDF + 25% FYM + MC)	57	25	169.7	29.6	17.6
T_6 (50% RDF + 50% FYM + MC)	48	26	169.2	30.3	19.3
T_7 (75% RDF + 25% Vm)	50	25	169.6	29.2	18.9
T_8 (50% RDF + 50% Vm)	53	26	176.1	29.8	20.4
T_9 (75% RDF + 25% Vm + MC)	68	41	202.5	49.4	24.5
T_{10} (75% RDF + Gm + MC)	50	25	178.6	31.8	18.6
T_{11}(50% RDF + Gm + MC)	44	25	170.7	28.6	17.2
CD (p = 0.05)	3.7	2.2	2.6	1.9	3.0

Adapted from Srivastava et al. (2011b)
MC stands for microbial consortium developed by isolating the native microbes from the experimental soil (mixture of *Azotobacter chroococcum*, *Bacillus mycoides*, *Bacillus polymyxa*, *Pseudomonas fluorescens*, and *Trichoderma harzianum*)
FYM, *Vm*, *Gm*, and *RDF* stand for farmyard manure, vermicompost, green manuring, and recommended doses of fertilizers, respectively
SMB and *SMBN* stand for soil microbial population and soil microbial biomass nutrients, respectively
Vm (nutrient composition: 2.35% N, 0.13% P, 1.51% K, 3,002 ppm Fe, 111.4 ppm Mn, 43.4 ppm Cu, and 59 ppm Zn)
FYM (nutrient composition: 1.20% N, 0.09% P, 0.84% K, 489 ppm Fe, 61.03 ppm Mn, 25.46 ppm Cu, and 29.8 ppm Zn)
Gm (nutrient composition: 0.78% N, 0.08%P, 0.70% K, 281 ppm Fe, 63 ppm Mn, 30 ppm Cu, and 26 ppm Zn)
RDF (200–600 g N, 50–200 g P_2O_5, 100–300 g K_2O, 50–200 g $ZnSO_4$, 50–200 g $FeSO_4$, 50–200 g $MnSO_4$/tree/year)

was better than T_4 (44 and 20 × 10^3 cfu g^{-1} bacterial and fungal counts, respectively). Likewise with MC also, vermicompost-based INM module as T_9 (68 and 41 × 10^3 cfu g^{-1} bacterial and fungal counts, respectively) supports much better soil microbial counts compared to FYM-based INM as T_5 (57 and 25 × 10^3 cfu g^{-1} bacterial and fungal counts, respectively) and T_6 (48 and 26 × 10^3 cfu g^{-1} bacterial and fungal counts, respectively). On the contrary, inclusion of green manuring with 75% RDF plus MC (50 and 25 × 10^3 cfu g^{-1} bacterial and fungal counts, respectively, with T_{10}) and 50% RDF + MC (44 and 25 × 10^3 cfu g^{-1} soil bacterial and fungal counts, respectively, with T_{11}) although brought some favourable changes on soil microbial counts but were more significantly of lower order when compared to other FYM or vermicompost-based INM treatments (e.g. T_1, T_6, T_9, etc.).

25.3.2.5 Microbial Biomass Nutrients

Different INM-based treatments (Table 25.10) were accompanied with significant changes in soil microbial biomass nutrients in terms of microbial carbon (C_{mic}), microbial N (N_{mic}), and microbial P (P_{mic}). Combination of 75% RDF + MC as T_2 produced no significant difference in soil microbial biomass nutrients except N_{mic} (146.3 mg kg^{-1} C_{mic}, 19.3 mg kg^{-1} N_{mic}, and 15.2 mg kg^{-1} P_{mic}) compared to 100% RDF as T_1 (152.1 mg kg C_{mic}, 19.1 mg kg^{-1} N_{mic}, and 16.1 mg kg^{-1} P_{mic}). But incorporation of organic manure either FYM (159.1 mg kg^{-1} C_{mic}, 23.9 mg kg^{-1} N_{mic}, and 15.9 mg kg^{-1} P_{mic} with T_3 or 164.1 mg kg^{-1} C_{mic}, 25.6 mg kg^{-1} N_{mic}, and 17.0 mg kg^{-1} with T_4) or vermicompost (169.6 mg kg^{-1} C_{mic}, 29.2 mg kg^{-1} N_{mic}, and 18.9 mg kg^{-1} P_{mic} with T_7 or 176.1 mg kg^{-1} C_{mic}, 29.8 mg kg^{-1} N_{mic}, and 20.4 mg kg^{-1} P_{mic} with T_8) helped further in recording higher level of soil microbial biomass nutrients. But these changes were still of lower order in the absence of microbial consortium treatments. The treatments like T_5 (169.7 mg kg^{-1} C_{mic}, 29.6 mg kg^{-1} N_{mic}, and 17.6 mg kg^{-1} P_{mic}), T_6 (169.2 mg kg^{-1} C_{mic}, 30.3 mg kg^{-1} N_{mic}, and 19.3 mg kg^{-1} P_{mic}) and T_9 (202.5 mg kg^{-1} C_{mic}, 49.4 mg kg^{-1} N_{mic}, and 24.5 mg kg^{-1} P_{mic}) produced the best response compared to treatments with MC (e.g. T_3, T_4, T_7, and T_8). These responses supported the fact that all the three components of INM are mandatory in order to harness the best effectiveness of different INM modules.

25.3.2.6 Soil Carbon Loading

The treatments evaluated for changes in soil carbon stock (soil organic carbon, soil inorganic carbon as $CaCO_3$, and total carbon) showed significant changes (Table 25.11)

Table 25.11 Response of different INM-based treatments on soil carbon loading (organic + inorganic C) (pooled data)

	Soil carbon fractions (g kg^{-1})			Soil	Soil C:N
Treatments	Organic carbon	Inorganic carbon	Total carbon	Total N	Ratio
T_1 (100% RDF)	6.33	2.05	8.38	0.597	14.0:1
T_2 (75% RDF + MC)	6.14	2.04	8.18	0.575	14.2:1
T_3 (75% RDF + 25% FYM)	6.61	2.09	8.70	0.604	14.4:1
T_4 (50% RDF + 50% FYM)	6.44	2.09	8.53	0.613	13.9:1
T_5 (75% RDF + 25% FYM + MC)	6.77	2.05	8.82	0.640	13.8:1
T_6 (50% RDF + 50% FYM + MC)	6.86	2.05	8.91	0.650	13.7:1
T_7 (75% RDF + 25% Vm)	6.97	2.02	8.99	0.634	14.1:1
T_8 (50% RDF + 50% Vm)	6.96	2.06	9.02	0.649	13.9:1
T_9 (75% RDF + 25% Vm + MC)	7.48	1.88	9.36	0.735	12.7:1
T_{10} (75% RDF + Gm + MC)	6.61	1.92	8.53	0.676	12.6:1
T_{11} (50% RDF + Gm + MC)	6.44	1.94	8.38	0.660	12.7:1
CD ($p = 0.05$)	0.11	0.4	0.13	0.044	–

Adapted from Srivastava et al. (2011b)
Figures in parenthesis represent the initial values
MC stands for microbial consortium developed by isolating the native microbes from the experimental soil (mixture of *Azotobacter chroococcum*, *Bacillus mycoides*, *Bacillus polymyxa*, *Pseudomonas fluorescens*, and *Trichoderma harzianum*)
FYM, *Vm*, *Gm*, and *RDF* stand for farmyard manure, vermicompost, green manuring, and recommended doses of fertilizers, respectively
Vm (nutrient composition: 2.35% N, 0.13% P, 1.51% K, 3,002 ppm Fe, 111.4 ppm Mn, 43.4 ppm Cu, and 59 ppm Zn)
FYM (nutrient composition: 1.20% N, 0.09% P, 0.84% K, 489 ppm Fe, 61.03 ppm Mn, 25.46 ppm Cu, and 29.8 ppm Zn)
Gm (nutrient composition: 0.78% N, 0.08%P, 0.70% K, 281 ppm Fe, 63 ppm Mn, 30 ppm Cu, and 26 ppm Zn)
RDF (200–600 g N, 50–200 g P_2O_5, 100–300 g K_2O, 50–200 g $ZnSO_4$, 50–200 g $FeSO_4$, 50–200 g $MnSO_4$/tree/year)

which eventually affected soil C to N ratio to varying proportions. The highest level of soil carbon fractions (7.48 g kg^{-1}organic carbon, 1.88 g kg^{-1} inorganic carbon, and 9.36 g kg^{-1} total carbon) was observed with T_9 followed by T_8(6.96 g kg^{-1} organic carbon, 2.06 g kg^{-1} inorganic carbon, and 9.02 g kg^{-1} total carbon). All these treatments were vermicompost-based. While on equivalent basis, different FYM-based treatments like T_5 (6.77 g kg^{-1} organic carbon, 2.05 g kg^{-1} inorganic carbon, and 8.82 g kg^{-1} total carbon), T_6 (6.86 g kg^{-1} organic carbon, 2.05 g kg^{-1} inorganic carbon, and 8.91 g kg^{-1} total carbon) and T_4 (6.44 g kg^{-1} organic carbon, 2.09 g kg^{-1} inorganic carbon, and 8.53 g kg^{-1} total carbon) were not so effective and significantly inferior to vermicompost-based treatments. The treatments T_{10} (6.61 g kg^{-1} organic carbon, 1.92 g kg^{-1} inorganic carbon, and 8.53 g kg^{-1} total carbon) and T_{11} (6.44 g kg^{-1} organic carbon, 1.94 g kg^{-1} inorganic carbon, and 8.38 g kg^{-1} total carbon) comprising green manuring with 75% and 50% RDF + MC recorded significant reduction in soil carbon fractions compared to T_1 where exclusive use of inorganic fertilizers were tested. Decimal change in soil C to N ratio was commonly observed. The soil C to N ratio varied from lowest level of 12.7 with T_9 to highest level of 13.9–14.4 with T_3, T_4, T_5, and T_7, with T_1 (having exclusive level of inorganic fertilizers) registering C to N ratio of 14.0. Green manuring–based INM treatments like T_{10} and T_{11} registered C to N ratio of 12.6–12.6.

25.3.2.7 Response on Leaf Nutrient Composition

The concentration of both macro- (NPK) and micronutrients (Fe, Mn, Zn) except Cu was significantly influenced by various INM-based treatments (Table 25.12). Compared to exclusive use of chemical fertilizers as T_1 carrying 100% RDF (2.13% N, 0.088% P, 1.23% K, 30.3 ppm Fe, 28.4 ppm Mn, 19.2 ppm Zn), the treatment T_2 where 25% RDF was replaced by MC (2.08% N, 0.085% P, 1.16% K, 28.7 ppm Fe, 26.8 ppm Mn, 18.6 ppm Zn) and T_3 replacing 25% RDF with FYM (2.13% N, 0.092% P, 1.22% K, 30.8 ppm Fe, 28.2 ppm Mn 19.6 ppm Zn) were not statistically superior. Even replacement of 50% RDF through T_4 (2.15% N, 0.092% P, 1.24% K, 31.3 ppm Fe, 29.4 ppm Mn, 20.1 ppm Zn) was neither better to 100% RDF as T_1.

However, with incorporation of MC brought some encouraging changes in leaf nutrient composition. For example, treatment T_5 (2.21% N, 0.0116% P, 1.32% K, 33.2 ppm Fe, 30.8 ppm Mn, 20.9 ppm Zn) was significantly superior to T_3 (2.13% N, 0.092% P, 1.22% K, 30.8 ppm Fe, 28.2 ppm Mn, 19.6 ppm Zn). Likewise another FYM-based INM treatment, T_6 (2.23% N, 0.0113% P, 1.33% K, 33.5 ppm Fe, 31.2 ppm Mn, 20.6 ppm Zn) was superior to T_4 without MC treatment (2.15% N, 0.092%, 1.24% K, 31.4 ppm Fe, 29.4 ppm Mn,

Table 25.12 Response of different INM-based treatments on leaf macronutrients concentration

Treatments	N (%)	P	K	Fe (ppm)	Mn	Cu	Zn
T_1 (100% RDF)	2.13	0.088	1.23	30.3	28.4	7.3	19.2
	(1.98)	(0.088)	(1.10)	(24.1)	(23.0)	(6.8)	(16.2)
T_2 (75% RDF + MC)	2.08	0.085	1.16	28.7	26.5	7.4	18.6
	(1.92)	(0.084)	(1.10)	(28.6)	(26.1)	(5.9)	(17.4)
T_3 (75% RDF + 25% FYM)	2.13	0.092	1.22	30.8	28.2	7.3	19.6
	(1.98)	(0.082)	(1.12)	(26.0)	(27.2)	(7.1)	(16.3)
T_4 (50% RDF + 50% FYM)	2.15	0.092	1.24	31.4	29.4	7.0	20.1
	(1.96)	(0.084)	(1.00)	(30.3)	(24.1)	(6.7)	(16.6)
T_5 (75% RDF + 25% FYM + MC)	2.21	0.116	1.32	33.2	30.8	6.9	20.9
	(1.96)	(0.084)	(1.10)	(31.1)	(23.4)	(6.5)	(17.3)
T_6 (50% RDF + 50% FYM + MC)	2.23	0.113	1.33	33.5	31.2	7.0	20.6
	(1.98)	(0.084)	(1.06)	(32.3)	(24.1)	(7.1)	(18.3)
T_7 (75% RDF + 25% Vm)	2.20	0.101	1.31	32.8	30.3	6.9	20.4
	(1.94)	(0.088)	(1.08)	(32.8)	(22.7)	(6.2)	(17.8)
T_8 (50% RDF + 50% Vm)	2.21	0.074	1.29	33.2	30.5	6.8	21.4
	(1.96)	(0.089)	(1.20)	(30.5)	(24.3)	(7.3)	(16.9)
T_9 (75% RDF + 25% Vm + MC)	2.28	0.125	1.42	36.4	32.8	7.3	22.6
	(1.94)	(0.078)	(1.12)	(32.0)	(21.9)	(6.4)	(17.2)
T_{10} (75% RDF + Gm + MC)	2.16	0.095	1.30	32.4	30.0	7.2	20.7
	(1.90)	(0.084)	(1.18)	(25.1)	(25.4)	(7.0)	(17.4)
T_{11} (50% RDF + Gm + MC)	2.13	0.092	1.28	32.9	30.1	6.9	21.2
	(1.94)	(0.086)	(1.18)	(23.4)	(23.6)	(6.1)	(17.6)
CD ($p = 0.05$)	0.06	0.004	0.083	1.17	1.25	NS	0.63
				(NS)			(NS)

Adapted from Srivastava et al. (2011b)
Figures in parenthesis represent the initial values
MC stands for microbial consortium developed by isolating the native microbes from the experimental soil (mixture of *Azotobacter chroococcum*, *Bacillus mycoides*, *Bacillus polymyxa*, *Pseudomonas fluorescens*, and *Trichoderma harzianum*)
FYM, *Vm*, *Gm*, and *RDF* stand for farmyard manure, vermicompost, green manuring, and recommended doses of fertilizers, respectively
Vm (nutrient composition: 2.35% N, 0.13% P, 1.51% K, 3,002 ppm Fe, 111.4 ppm Mn, 43.4 ppm Cu, and 59 ppm Zn)
FYM (nutrient composition: 1.20% N, 0.09% P, 0.84% K, 489 ppm Fe, 61.03 ppm Mn, 25.46 ppm Cu, and 29.8 ppm Zn)
Gm (nutrient composition: 0.78% N, 0.08% P, 0.70% K, 281 ppm Fe, 63 ppm Mn, 30 ppm Cu, and 26 ppm Zn)
RDF (200–600 g N, 50–200 g P_2O_5, 100–300 g K_2O, 50–200 g $ZnSO_4$, 50–200 g $FeSO_4$, 50–200 g $MnSO_4$/tree/year)

20.1 ppm Zn). These observations suggested that unless any treatment possesses all the three components of INM, it cannot be so effective.

All the vermicompost-based INM treatments demonstrated much superior response on leaf nutrient composition over FYM-based treatments (Table 25.12). The treatment T_9 (2.28 N, 0.125% P, 1.42% K, 36.4 ppm Fe, 32.8 ppm Mn, 22.6 ppm Zn) registered much higher concentration of all the nutrients compared to T_7 without MC (2.20% N, 0.101% P, 1.31% K, 32.8 ppm Fe, 30.3 ppm Mn, 20.4 ppm Zn) or even T_8 (2.21% N, 0.074% P, 1.29% K, 33.2 ppm Fe, 30.5 ppm Mn, 21.4 ppm Zn). On the other hand, green manuring–based treatments like T_{10} (2.16% N, 0.095% P, 1.30% K, 32.4 ppm Fe, 30.0 ppm Mn, 20.7 ppm Zn) and T_{11} (2.13% N, 0.092% P, 1.28% K, 32.9 ppm Fe, 30.1 ppm Mn, 21.2 ppm Zn) showed statistically compatible effect on leaf nutrient composition compared to treatments comprising either T_1 or T_2 (Table 25.12).

These summarized results are more of interpretative than suggestive, affirming the practice leading to significant reduction in load on the use of inorganic chemical fertilizers under INM. Various components of INM are further summarized (Srivastava et al. 2008) through a flow diagram (Fig. 25.2).

The ultimate rationale of INM is, hence, the judicious use of all its three principal components viz. exploiting the existing synergism between dual purpose microbe (growth promoting as well as biocontrol agent against soil-borne pathogens) types with limited use of inorganic chemical fertilizers, triggering the multiplication of indigenous soil microbial diversity through a suitable substrate of organic origin, in such a way that the nutrients inflow always exceeds the nutrients flow leaving the system, besides ensuring the market favouring production economics (Srivastava and Ngullie 2009). However, still there are many core areas where an urgent redressal is required in order to tag INM, a globally vibrant nutrient management strategy.

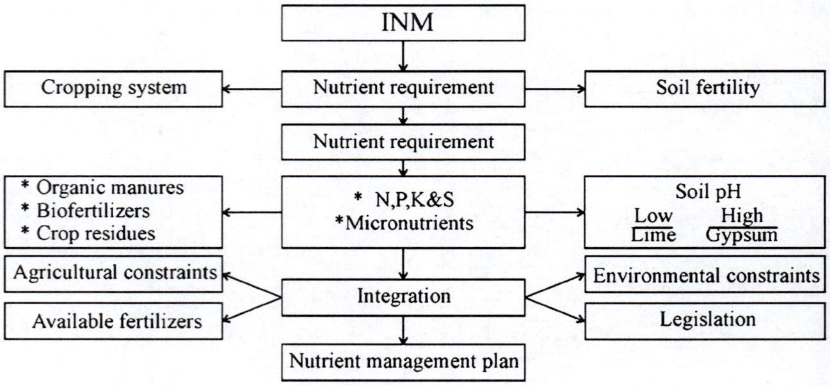

Fig. 25.2 Schematic representation of INM (Adapted from Srivastava 2009)

25.4 Future Research

Despite many cutting-edge technologies addressing a variety of core issues of nutrient management, many more issues are yet to be attempted with respect to INM-based citrus production vis-à-vis rhizosphere dynamics. Studies on biochemical response of citrus in relation to varying nutrient supply systems (through INM modules) especially under agropedological conditions facing multi-nutrient deficiencies, and establishing the cause–effect relationship between the physicochemical and microbiological changes within rhizosphere, to be able to coordinate changes in shoot system (changes in canopy size and fruit yield i.e. yield efficiency), are very much imperative. These are seemingly most sensitive to various combinations of remediative treatments under different INM modules.

Nutrient dynamics is another virgin area where limited attempts have been made using citrus as test crop. Amongst different nutrients, Zn has attracted worldwide investigation from various angles. The changes in rhizosphere bring different simultaneous changes in microbial diversity *vis-à-vis* C_{mic}, N_{mic}, and P_{mic}, and nutrient regime especially for diffusion-limited nutrients like P, Zn, Fe, Mn, etc. has to find serious considerations in any nutrient management program that involves INM-based corrective treatments. Additionally, the conditions under which citrus trees are most likely to respond to corrective Zn treatments are still not fully understood. The role of Zn in flowering, fruit set, fruit quality (external and internal), and juice shelf life; models defining the critical periods of Zn supply to assure sustained response and its uptake for helping the management decision under different citrus-based cropping systems; and devising means for improved Zn uptake efficiency need to be attempted to unravel many of the complexities involved with Zn nutrition under INM-based production management.

Out of different soil properties, the microbial biomass is the one biological property of soil that undergoes immediate change in response to fertilizer-like input. Studies, therefore, need to be undertaken with a view to explore the possibility whether microbial properties could be used as a potential tool for finding out soil fertility constraint instead of available supply of nutrients in soil. Simultaneously, an eye should be kept on long-term changes in total carbon pool of soil to arrive at the logistic conclusion that sequestration of carbon through improved production level could rejuvenate the lost productivity potential of nutritionally depleted soil. However, it remains to be further established that any change in microbial diversity within the rhizosphere brought about by different sources of substrate is whether translocated into improved productivity and if there is any, how the nutrient dynamics is associated with orchard productivity under an effective INM concept.

Impacts due to environmental changes and anthropogenic activity are the potential threats to the conservation of soil quality, while expanding citriculture to marginal soils having a wide range of limitations. With the availability of more technical know-how on efficient use of bulky organic manures, prolonged shelf life of microbial biofertilizers, and better understanding on citrus–mycorrhiza symbiosis with regard to nutrient acquisition and regulating the water relations, a more effective integrated citrus production system could evolve in future. The molecular approach to breeding of mineral deficiency resistance and mineral efficiency would facilitate to produce nutritionally efficient biotypes in order to maximize the quality production on sustained basis. Fertilizer applications are currently managed to protect environmentally sensitive areas by using controlled-release fertilizers (use of organic manures, a befitting option), frequent low-concentration fertigation, multiple applications, and variable rate application technology in order to improve fertilizer use efficiency. However, using newly emerging techniques of nutrient management and geo-informatics linked site-specific management on the principles of INM could be worked out accommodating soil's nature and properties. Simultaneously, concerted efforts would be required to develop INM-based yield monitors and soil quality indicators.

References

Aariff KMA, Begum H (2007) Effect of INM on available nutrient status of young acid lime (*Citrus aurantfolia* Swingle) orchards of A.P., India. Asian J Hortic 2(2):64–67

Ahmed FF, El-Sayed MA, Maatouk MA (1988) Effect of nitrogen, potassium and phosphorus fertilization on yield and quality of Egyptian Balady lime trees (*Citrus aurantifolia*) II. Yield and fruit quality. Ann Agric Sci (Cairo) 33(2):1249–1268

Aiyelagbe IOO (2001) Productivity of intercropped sweet orange orchard in southwestern Nigeria. Biol Hortic Agric 8(4): 317–325

Alva AK, Tucker DPH (1992) Foliar application of various sources of iron, manganese and zinc to citrus. Proc Fla State Hortic Soc 105:70–74

Alva AK, Paramasivam S, Hostler KH et al (2001) Effect of nitrogen rates on dry matter and nitrogen accumulation in citrus fruits and fruit yield. J Plant Nutr 24(3):561–572

Alvardo A, Araya R, Bornemisza E, Hernandez RL (1994) Nutritional studies in eleven citrus cultivars and one local selection in the Atlantic zone of Costa Rica major and minor elements. Agron Coast 18(1):13–19

Anderson CA (1968) Effects of gypsum as a source of calcium and sulfur on tree growth, yields, and quality of citrus. Proc Florida State Hort Soc 81:19–24

Anderson CA (1984) Micronutrient uptake by citrus from soil–applied zinc compounds. Proc Soil Crop Sci Soc Fla Lake Alfred USA 43:36–39

Anderson CA (1987) Fruit yield, tree size and mineral nutrition relationships in Valencia orange trees as affected by liming. J Plant Nutr 10:1907–1916

Androulakis IJ, Loupassaki MH, Beidoun F et al (1992) The effect of NPK fertilizers on the yield and mineral contents of leaves of grapefruit. In: Tribulato E, Gentile A, Refergiato G (eds) Proceedings of International Society of Citriculture, vol 2, Acireale, Italy, 8–13 Mar, pp 624–627

Araya R, Alvardo A, Hernandez RL et al (1994) Nutritional studies in eleven citrus cultivars and one local selection in the Atlantic zone of Costa Rica. II Micronutrients. Agron Coastar 1:21–27

Arsenidze A, Chanukuvadze F (1988) Diagnostic of nitrogen nutrition in young mandarin trees. Subtropicheskie Kul'tury 2:91–96

Aso PJ, Dantur NC (1970) Citrus nutrient deficiencies in Tucuman. Revista Industrial y Agricola de Tucuman 47(2):10–61

Aubert A, Vullin G (1998) Citrus nurseries and planting techniques. CIRAD/GTZ, Montpellier/Eschborn, pp 64–149

Awasthi RP, Singh G, Sharma RC (1984) A survey of the mineral nutrients status of mandarin orchards in Nagpur area of Himachal Pradesh. Punjab Hortic J 24:27–36

Beridze TR (1990) The effect of organic fertilizers on lemon tree productivity. Subtropicheskie Kul'tury 3:83–86

Bevington KB (1984) Effect of nutrition on yield and fruit quality of citrus. Aust Citrus News 60(10):2

Bhargava BS (2002) Leaf analysis for nutrient diagnosis recommendation and management in fruit crops. J Indian Soc Soil Sci 50(4): 352–373

Boaretto AE, Muraoka T, de Oliveira MW (1999) Corrective and preventive foliar fertilization in citrus. Laranja 20(1):233–250

Borah SC, Barbora AC, Bhattacharya D (2001) Effect of organic and inorganic manuring on growth, yield and quality of Khasi mandarin (*Citrus reticulata* Blanco). South Indian Hortic 49(special):115–118

Bornemisza E, Hernandez RL, Chaverri W et al (1985) Annual variation of foliar levels of seven nutrients in Persian lime and San Fernando and Frost nucellar Lisbon lemons in Costa Rica. Agron Costar 9(2):135–141

Caetano AA, Rodriguez O, Hiroce R et al (1984) Nutritional survey forty sweet orange groves in Babedauro, SP, Brazil. In: Montenegro HWS, Moreira CS (eds) Proceedings of International Society of Citriculture, vol 1, Sao Paulo, Brazil, 15–20 July, pp 151–154

Canteralla H, Quaggio JA, Bataglia OC et al (1992) Response of citrus to NPK fertilization in a network of field trials in Sao Paulo State, Brazil. In: Tribulato E, Gentile A, Refergiato G (eds) Proceedings of International Society of Citriculture, vol 2, Acireale, Italy, 8–13 Mar, pp 607–612

Chen M, Zhang KQ, Xiao L (1990) Effects of foliar phosphates spray on quality of satsuma mandarin. In: Huang B, Yang Q (eds) Proceedings of International Society of Citriculture, Guangzhou, China, 5–8 Nov, pp 457–462

Chen H, Huang H, Lu C (1997) Effect of organic fertilizer and dolomite lime on the growth and fruit quality of Matou Peiyu (*Citrus grandis* Osbeck cv Matou Peiyu). Research Bulletin. 38. Taichung District Agricultural Improvement Station, Taichung, pp 69–76

Dass HC, Srivastava AK, Lallan R et al (1998) Flowering behaviour of Nagpur mandarin (*Citrus reticulata* Blanco) as affected by vertical variation in physico–chemical composition. Indian J Agric Sci 68(10):692–694

Desai UT, Choudhari KG, Choudhari SM (1986) Studies on the nutritional requirements of sweet orange. J Maharashtra Agric Univ 11(2):145–147

Devi DD, Srinivasan PS, Balakrishnan K (1996) Carbonic anhydrase activity as an indicator of zinc status of sathgudi orange. Orissa J Hortic 24(1/2):66–68

Dhatt AS (1989) Nutrient management in citrus with special reference to Kinnow. In: Proceedings of citriculture in North–western India. Punjab Agricultural University, Ludhiana, pp 157–167

Dhillon NS, Singh KK, Bakshi JC (1961) The effect of manuring on fruit quality in sweet orange. Punjab Hortic J 1:24–34

Dris R (1997) Effect of NPK fertilization on clementine grown in Algeria. Acta Hortic 448:375–381

Dubey AK, Yadav DS (2003) Response of khasi mandarin (*Citrus reticulata* Blanco) to organic versus inorganic fertilization. Indian J Agric Res 37(3):214–218

Dudi OP, Dal S, Dahiya SS et al (2003) Impact of various levels of N and FYM on growth parameters of Kinnow mandarin. Haryana J Hortic Sci 32(2):29–31

Duncan J (1969) Citrus nutrition survey – M.I.A. Farmer Newsl 105:17–19

Ebrahiem TA, Mohamed GA (2000) Response of Balady mandarin trees growing on sandy soil to application of filter mud and farmyard manure. Assiut J Agric Sci 31(5):55–59

Egorashvili NV, Tavadze AM, Romanadze AD (1991) Determination of optimal rate of micronutrients based on field trials in citrus plantation. Subtropicheskie Kul'tury 5:71–79

El-Fouly MM, Fawzi AFA, Firgany AH et al (1984) Micronutrient status of crops in selected areas of Egypt. Commun Soil Sci Plant Anal 15(10):1175–1189

El-Hagah MH, Higazi AM, El-Niggar SZ et al (1983) Effect of soil and foliar fertilization on growth of Navel orange and Balady mandarin trees. Minufiya J Agric Res 1:261–279

El-Koumey BY, El-Saedy MM, Omran SS et al (2000) The effect of fertilization on the availability of nutrients in soil and their concentrations in leaves. Egypt J Soil Sci 40(1/2):131–140

Embleton TW, Wallihan EF, Goodall GE (1966) Coastal lemon. Zinc and manganese fertilization. Citrog 51, 131, 140, 142–3, bibl.2

Ercivan S (1974) Incidence of zinc deficiency on satsuma mandarin trees in Izmir. J Turkish Phytopathol 3(1/2):51–56

Ferguson JJ, Davies FS, Mathews CH et al (1988) Controlled-release fertilizers and growth of young 'Hamlin' orange trees. Proc Florida State Hort Soc 101:17–20

Fidalski J, Auler PAM (1997) Nutritional survey of orange plantations in northwest Parana. Arquivos de Biologia e Tecnologia 40(2):443–451

Fischer J (1992) Use of chicken manure in citrus fertilization programs. Citrus Indus 73(5):38–40

Gamal AM, Ragab MA (2003) Effect of organic manure source and its rate on growth, nutritional status of the trees. Assiut J Agric Sci 34(6):253–264

Garcia Alvarez N, Haydar E, Ferrer C (1983) Influence of zinc and manganese on the physiological behaviour and yields of Valencia oranges. Centro Agricala 10(2):57–68

Ghosh SN (1990) Nutritional requirement of sweet orange (*Citrus sinensis* Osbeck) cv Mosambi. Haryana J Hortic Sci 19(1–2):39–44

Ghosh SP (2000) Nutrient management in fruit crops. Fertil News 45:71–76

Ghosh SN, Besra KC (1997) Growth, yield and physico–chemical properties of sweet orange cv. mosambi grown in response to organic and inorganic fertilizers under rainfed laterite soils. In: Abstract, National symposium on citriculture, Nagpur, India, 17–19 Nov, p 40

Ghosh SN, Besra KC (2000) Effect of zinc, boron and iron spray on yield and fruit quality of sweet orange cv Mosambi grown under rainfed laterite soil. Indian Agric 44(3):147–151

Ghosh SP, Singh R (1993) Citrus in South Asia. RAPA Publication, FAO, Regional Office, Bangkok, p 21

Ghosh SP, Verma AN, Govind S (1989) Nutritional requirement of bearing Khasi mandarin orange (*Citrus reticulata* Blanco) trees in Meghalaya. Hortic J 2(1):4–11

Gill JS, Khind CS, Singh B, Yadvinder S (1998) Ammonia volatalization under flooded conditions as affected by urease activity of soil amended with crop residue on long term basis. J Indian Soc Soil Sci 46(3):448–450

Gochelashvili ZA (1984) Effect of fertilizers on microflora development in young citrus orchards. Subtropicheskie Kul'tury 5:72–79

Goepfert CF, Saldanha ELSD, Porto ODM (1987) The response of Valencia orange (*Citrus sinensis* Osbeck) to fertilizer levels, average of eight harvests. Agronomia Sulriograndense 23(2):203–215

Goh KM (2004) Carbon sequestration and stabilization in soils: implications for soil productivity and climate change. Soil Sci Plant Nutr 50(4):467–476

Gupta RP, Pandey SP, Tripathi BP (1989) Soil properties and availability of nutrient element in mandarin growing areas of Dhankuta districts. PAC technical paper 113. Pokhbar Agricultural Centre, Dhankuta

Halase NJ (1963) Control of copper, zinc, and manganese deficiencies in fruit trees. J Agric W Aust 4:241–244

Haq IUI, Ghani A, Habibur UR (1995) Nutritional status of citrus orchards in NWFP and the effect of fertilizer application on fruit production. Technical bulletin 1/95. Directorate of Soil and Plant Nutrition, Agriculture Research Institute, Tarnab, Peshawar, NWFP, pp 1–36

Haque A, Godfrey SA (1976) Nutritional survey of citrus orchards in Sierra Leone. Commun Soil Sci Plant Anal 7(19):843–860

He ZL, Yao H, Chen G et al (1997) Relationship of crop yield to microbial biomass in highly weathered soils of China. In: Ando T et al (eds) Plant nutrition for sustainable food production and environment. Kluwer Academic Publishers, Tokyo, pp 745–746

Hellin E, Urena R, Sevilla F et al (1988) Effects of several Fe–chelates on the iron leaf content of Verna lemon trees. In: Goren R, Mendel K (eds) Proceedings of International Citrus Congress, vol 1, Tel Aviv, Israel, 6–11 Mar, pp 555–560

Hernandez J (1981) Effect of N, P and K on yield, fruit quality and nutritional status of Valencia Late orange. In: Matsumoto K, Oogaki C, Kozaki I, Kadoya F, Mizutani F, Yuda E (eds) Proceedings of International Society of Citriculture, vol 2, 9–12 Nov, Tokyo, Japan, pp 564–566

Hiwarale JS, Patil MN, Laharia GS (2004) Effect of organic and inorganic fertilizers on growth, yield and quality of acid lime. Annu Plant Physiol 18(1):28–30

Hong SB, Chung SK (1979) Effects of N P and K fertilizer levels on the growth, yield and fruit quality of satsuma trees. Res Rep, Rural Dev Adm Hortic Agric Eng 21:67–75

Horesh I, Levy Y, Goldschmidt EE (1986) Prevention of lime induced chlorosis in citrus trees by peat and iron treatments to small soil volumes. HortScience 21(6):1363–1364

Hsiung TC, Iwahori S (1984) Prevention of abscission of ponkan, *Citrus reticulata* Blanco leaves by various calcium salts. Mem Fac Agric Kagoshima Univ 20:50–62

Huang Z, Yang S (1990) Effect of organic–inorganic fertilization on citrus grown on the red soils. In: Huang Bangyan, Yang Qian (eds) Proceedings of International Citrus Symposium, Guangzhou, China, International Acac Publishers, China, 5–8 Nov, pp 513–516

Huang FS, Zhang HS, Qian HM et al (1995) Study on the effects of organic manures and leaf spraying P and K on the fruit quality of extra–early satsuma mandarin. China Citrus 24(2):31–32

Huchche AD, Ladaniya MS, Ram L et al (1998) Effect of nitrogenous fertilizer and farmyard manure on yield and shelf–life of Nagpur mandarin. Indian J Hortic 55(2):108–112

Ingle HV, Athawale RB, Ghawde SM et al (2001) Integrated nutrient management in acid lime. South Indian Hortic 49:126–127

Intrigliolo F, Stagno F (2001) Organic fertilizers in citrus crops. Rivista di Frutticoltura e di Ortofloricoltura 63(11):75–79

Joa JH, Han CL, Seung GH et al (2006) Microbial activities in dark brown volcanic ash soils was affected by different fertilizer application methods in citrus orchard. In: Abstract 138–84, 18th World Congress of Soil Science, Pennsylvania

Jones WW, Embleton TW, Coggins CW Jr (1975) Starch content of roots of 'Kinnow' mandarin trees bearing fruit in alternate years. HortScience 10:514

Kacharava ON (1985) Fertilization system in mandarin plantations. Subtropicheskie Kul'tury 6:45–50

Kannan M, Thamburaj S, Seemanthini R (1989) Studies on the manurial requirement of mandarins in Shevroy hills. South Indian Hortic 37(4):203–208

Khadr AA, Wallace A (1964) Uptake and translocation of radioactive iron and zinc by trifoliate orange and rough lemon. Proc Am Soc Hortic Sci 85:189–200

Khalaf HA, Koo RCJ (1983) The use of controlled release nitrogen on container growth citrus seedlings. Citrus Veg Mag 46(9):10–32

Khera AP, Singh HK, Daulta BS (1985) Correcting micronutrients deficiency in citrus cv. Blood red. Haryana J Hortic Res 14:27–29

Kim JC, Chung SM, Kong SJ (1969) Studies on the nutritional status of Korean citrus trees (*Citrus unshiu* Marc.) by means of leaf analysis. Research report. Division of Agriculture Korea 12(2):45–51

Kimani W (1984) The influence of mineral nutrition on the severity of citrus greening disease in Kenya. In: Montenegro HWS, Moreira CS (eds) Proceedings of International Society of Citriculture, vol 1, Sao Paulo, Brazil, 15–20 July, pp 172–173

Kohli RR, Srivastava AK, Shivankar VJ (1998) Organic culture in citrus cultivation. Indian Hortic 43:12–15

Koo RCJ (1964) Some result of fairfield slag. Proc Florida State Hort Soc 77:26–32

Koo RCJ (1966) Potash puts quality into citrus. Better Crops 50(4):8–11

Koo RCJ (1982) Use of leaf, fruit and soil analysis in estimating potassium status of orange trees. Proc Florida State Hort Soc 75:67–72

Koo RCJ (1986) Effect of nitrogen and potassium fertilization on winter injury of citrus trees. Proc Florida State Hort Soc 98:53–56

Koo RCJ (1988) Use of controlled-release nitrogen for citrus in a humid region. In: Goren R, Mendel K (eds) Proceedings of 7th International Citrus Congress, vol 2, Rehevot, Israel, 6–12 Mar, pp 633–641

Koseoglu AT, Colakoglu HC, Kovanci I (1995a) The effect of chemical fertilizers on the fruit yield of young satsumas (*Citrus unshiu* Marc.). Doga Turk Tarim Ve Ormacilik Dergisi 14(1):33–44

Koseoglu AT, Eryuce N, Colakoglu HC (1995b) The effects of N, P, K fertilizers on fruit yield and quality of satsuma mandarin (*Citrus unshiu* Marc.). Acta Hortic 379:89–96

Kozaki I (1981) Introduction to citriculture in Japan. In: Matsumoto K, Oogaki C, Kuzaki I et al (eds) Proceedings of International Society Citriculture, vol 1, Tokyo, Japan, 9–12 Nov, pp XXVII–XXXIII

Leonard CD, Stewart I (1952) Correction of iron chlorosis in citrus with chelated iron. Proc Fla State Hortic Soc 65:20–24

Li J, Shi Q, Zeng WX (1998) Study on rational fertilizer application in citrus orchards in Fujian province by nutritional diagnosis. J Fruit Sci 15(2):145–149

Lim M, Mualsri C, Suthipradit S et al (1993) Nutrient element soil cultivation of neck orange (*Citrus reticulata* Linn.). Kasetsart J Nat Sci 27(4):412–420

Liu YW, Lin MS, Qiu HL (1994) Study of fertilizer Satsumas. China Citrus 23(2):7–10

Loeppert RH (1986) Reaction of iron and carbonates in calcareous soils. J Plant Nutr 9(3–7):195–214

Loeppert RH, Clarke T (1984) Reactions of Fe^{2+} and Fe^{3+} in calcareous soils. Soil Sci Soc Am J 49:597–603

Lomtadze ZK, Muradova MI (1986) Effectiveness of a new organo–mineral fertilizer (NOMF) on lemon plantations in the Kolkhida lowlands. Subtropicheskie Kul'tury 4:43–47

Maatouk MA, Ahmed FF, El-Sayed MA (1988) Effect of nitrogen, potassium and phosphorus fertilization on yield and quality of Egyptian Balady lime trees (*Citrus aurantifolia*) I. Vegetative growth and chemical composition of leaves. Annu Agric Sci (Cairo) 33(2):1233–1247

Majorana G (1960) Symptoms of copper deficiency experimentally confirmed on sweet orange and mandarin trees in Catania province. Teen Agric 12:645–661

Maksoud MA, Haggag LF (2004) Influence of some fertilizing regimes on nutritional status and productivity of Washington navel orange trees grown on sandy soil. J Agric Sci (Cairo) 12(2):669–677

Marsanija II (1970) The effect of the prolonged use of mineral. Fertilizes and FYM on the growth yield of lemon trees. Subtrop Ku'ltury 1:69–75

Mattos-Junior D, Quattio JA, Cantarella H et al (2003) Nutrient content of biomass components of Hamlin sweet orange trees. Sci Agric 60(1):155–160

Maynard DN, Lorenz OA (1979) Controlled release fertilizers for horticultural crops. Hortic Rev 2:79–141

McCall WW (1965) Zinc deficiency of citrus in Thailand. Kasterast J 5:1–9

Mdinaradze TD, Datuadze OV (1987) The effect of magnesium–containing fertilizer on the available zinc content in the soil and uptake by mandarin trees. Subtropicheskie Kul'tury 6:77–80

Mdwaradze TD (1981) Effect of different rates of zinc fertilizers on qualitative indices of mandarin fruit. Subtropicheskie Kul'tury 5:49–51

Medhi BK, Saikai AJ, Bora SC et al (2006) Fertility evaluation of mandarin orange (*Citrus reticulata* Blanco) as influenced by combined application of organic and inorganic fertilizers on a Typic Haplustert soil – a case study of Tinsukia district. Assam J Agric Sci Soc 17:34–38

Moon DY, Kwon HM, Lee WJ et al (1980) Studies on the nutritional diagnosis of Jeju orchards by leaf analysis. Research report. Office Rural Develop, Horticulture and Sericulture, Suwon

Motskobilli NA (1984) Assimilation area of satsuma in relation to mineral nutrition and its effect on productivity. Subtropicheskie Kul'tury 5:83–90

Motskobilli NA (1986) Effect of mineral nutrition on the winter hardiness of Satsumas. Subtropicheskie Kul'tury 2:131–134

Mukherjee D, Mitra S, Das AC (1991) Effect of oil cakes on changes in carbon, nitrogen and microbial population in soil. J Indian Soc Soil Sci 39:457–462

Murti GSR, Patil P, Reena RT (2008) All India coordinated research project and ICAR. In: *Adhoc* schemes on tropical fruits – biennial research report. Indian Institute of Horticultural Research, Bangalore, pp 179–183

Mustaffa MM, Kumar V, Reddy BMC (1997) Studies on the organic and inorganic fertilization on the vigour of Kagzi lime. In: Abstract of National Symposium on Citriculture, Nagpur, Maharashtra, India, 17–19 Nov, p 39

O'Brien LO, Sumner ME (1988) Effects of phosphogypsum on leachate and soil chemical composition. Comm Soil Sci Pl Anal 19:1319–1329

Obreza TA, Rouse RE (2004) Controlled–release fertilizers for Florida citrus production. In: University of Florida IFAS, Soil Water Sci Dept Ext Pub SL, p 214

Obreza TA, Rouse RE, Sherrod JB (1999) Economics of controlled-release fertilizer use on young citrus trees. J Product Agric 12:69–73

Ozordzadze MZ (1969) The effect of peat–mineral–ammonia fertilizers on the growth and productivity of mandarin plantings. Subtropicheskie Kul'tury 3:172–181

Papadakis IE, Protopapadakis E, Dimassi KN et al (2004) Nutritional status, yield and fruit quality of Encore mandarin trees grown in two sites of an orchard with different soil properties. J Plant Nutr 27(9):1505–1515

Paramasivam S, Alva AK (1997) Nitrogen recovery from controlled release fertilizers under intermittent leaching and drying cycles. Soil Sci 162:447–453

Pedrera B, Lambert I, Oviedo D et al (1988) Study of three rates of N application with basal P and K on Dancy mandarin in a red ferrallitic soil II. Effects on growth, fruit quality and annual yield performance. Ciencia Y Tecnica en la Agricultura, Citricos Y Otros Frutales 11(2):17–28

Penkov M, Al-Alami A, Krhistova D (1979) Causes of chlorosis of clementines grown in the Sous river valley in Morocco. Pochoznonic i Agrokhimiya 14(2):81–90

Pennisi C (1971) Citrus manuring. Inf Ortoflorofruttic 19(7):3–5

Pennisi L (1975) Studies on the nutrient status of citrus in experimental fields in Sicily, Calabria and Brasilicata. Annali dell' Instituto Sperimentale per Agrumicoltura 6:323–337

Pinto MR, Leal PF (1974) The nutritional status of some orange orchards in the Valles Altos of Carabobo, Venezuela. Revista de la Facultad de Agronomia Venezuela 8(1):71–80

Pirchalajsvili ES (1970) The effect of some forms of N fertilization on ascorbic acid accumulation in leaves and fruit of Washington Navel oranges. Subtropicheskie Kul'tury 3:71–73

Plemanac M (1985) Effect of nitrogen fertilization on some agrobiological K properties of the orange cultivar Washington Navel. Jugoslovensko Vocarstov 19(73/74/(3/4)):459–467

Prasad RA, Singhania RA (1989) Effect of different types of enriched manures and time of incubation on soil properties. J Indian Soc Soil Sci 37:319–322

Quaggio JA, Cantarella H, Van Raiz B (1998) Phosphorus and potassium soil test and nitrogen leaf analysis as a guide to citrus fertilization. Nutrient Cycling Agroecosys 52(1):67–74

Quaggio JA, Teofilo SJ, Dechen AR (1992) Magnesium influence on fruit yield and quality of Valencia sweet orange on Rangpur lime. In: Tribulato E, Gentile A, Refergiato G (eds) Proceedings of International Society Citriculture, vol 2, Acireale, Italy, Mar 8-12, pp 637–663

Ramamurthy V (2006) Vermicompost application improves the productivity of Nagpur mandarin (*Citrus reticulata* Blanco). Organic News 3(2):1–6

Rasmussen GK, Smith PF (1959) Effects of H-ion concentration on growth of pineapple orange seedlings in alternate solution and water cultures. Proc Am Soc Hort Sci 73:242–247

Rao ACS (1993) Diagnosis of nutrient deficiencies of citrus orange orchards in Jiroff valley of Iran. Agrochimica 37(1/2):41–54

Reddy PS, Swamy GS (1986) Studies on nutritional requirement of sweet orange (*Citrus sinensis* Linn.) variety sathgudi. South Indian Hortic 34(5):288–292

Reddy KR, Khaleel R, Overcash MR (1980) Nitrogen, phosphorus and carbon transformations in a coastal plain soil treated with animal manure. Agri Wastes 2:225–228

Reddy GS, Naidu LGK, Anyaneyulu K (2003) Influence of soil nutrient mining on growth, yield and nutritional disorders in sweet orange (*Citrus sinensis* Osbeck) orchards. Indian J Citric 2:10–13

Rodriguez O (1980) Nutricao e. adubacao dos citros. In: Roudriguez O, Viega F (eds) Citriculturea Brasileira Fundacao Cargill, vol 2, Sao Paulo, Brazil. pp 385–428

Rojas E (1998) Floral response of Tahiti lime (*Citrus latifolia* Tan.) to foliar sprays of hydrogen cyanamide. Fruits (Paris) 53(1):35–40

Rokba AM, Mahamood LT El, Menshawai El E (1995) Effect of organic and inorganic fertilizer on orange trees. Agri Res Rev 53(3):7–14

Saatci N, Mur BY (2000) Relationships between concentration of iron, macro and micronutrients in satsuma mandarin leaves (*Citrus unshiu.* Macf.). J Plant Nutr 23(11/12):1745–1750

Salem SE, Ibrahim TA, Gudy LF et al (1995) Response of Balady mandarin trees to foliar application of iron, zinc, manganese and urea under sandy soil. Bull Fac Hortic, Univ Cairo 46(2):277–288

Sarooshi R, Weir RG, Coote BG (1991) Effect of nitrogen, phosphorus and potassium fertilizer rates on fruit yield, leaf mineral concentration and growth of young orange trees in Sunraysia district. Aust J Expt Agric 31(2):263–272

Schumann AW, Fares A, Alva AK et al (2003) Response of 'Hamlin' orange to fertilizer source, annual rate and irrigated area. Proc Florida State Hort Soc 116:256–260

Seshadri SV, Madhavi M (2001) Effect of organic and inorganic manuring of twenty year old seedlings of sweet orange (*Citrus sinensis* (L.) Osbeck) cv Sathgudi. South Indian Hortic 49:122–125

Shaked A, Ashkenazy S (1984) Single citrumelo as a new citrus rootstock in Israel. In: Montenegro HWS, Moreira CS (eds) Proceedings of International Society of Citricult, vol 1, Sao Paulo, Brazil, 15–20 July, pp 48–50

Shi XG, Cheng SN, Gong JQ et al (2000) Effects of field applying Qixing organic compound fertilizer on the Satsuma mandarin cultivars. South China Citrus 29(2):9–10

Shimorgori Y, Wada M, Hatano H (1980) Soil management on satsuma mandarin orchard derived from volcanic ash soil on the growth and fruit yield. Bulletin 33, Miyazaki, Agricultural Experiment Station, Japan

Singh AR, Maurya VN, Pande NC et al (1989) Role of potash and zinc on the biochemical parameters of Kagzi lime (*Citrus aurantifolia* Swingle). Haryana J Hortic Sci 18(1–2):46–50

Singh C, Sharma BB, Singh C (1993) Leaf nutrient compositions of sweet orange as affected by combined use of bio and chemical fertilizers. South Indian Hortic 3:131–134

Singh S, Srivastava AK, Shivankar VJ (1999) Citrus based cropping system – a review. Agric Rev 20:123–128

Smith PF (1976) Collapse of 'Murcott' tangerine trees. J Am Soc Hortic Sci 101:23–25

Smith PF, Reuther W, Shudder GK Jr (1953) Effect of differential supplies of nitrogen, potassium and magnesium on growth and fruiting of young Valencia orange trees in sand culture. Proc Fla State Hortic Soc 61:38–48

Srivastava AK (2008) Integrated nutrient management, concept and application in citrus in southeast Asia. In: Xiuxin D, Juan X, Shimin L et al (eds) Proceedings of International Society of Citriculture, vol 1, Wuhan, China, 26–30 Oct, pp 726–733

Srivastava AK (2009) Integrated nutrient management: concept and application in citrus. In: Tennant P, Beakebhia N (eds) Citrus II. Tree For Sci Biotechnol 3 (Special Issue 1):32–58

Srivastava AK, Ngullie E (2009) Integrated nutrient management: theory and practice. Dyn Soil, Dyn Plant 3(1):1–30

Srivastava AK, Singh S (2003) Diagnosis and management of nutrient constraints in citrus. Manual no. 2. National Research Centre for Citrus, Nagpur, pp 1–70

Srivastava AK, Singh S (2004a) Zinc nutrition, a global concern for sustainable citrus production. J Sustain Agric 25(3):5–42

Srivastava AK, Singh S (2004b) Zinc nutrition and citrus decline – a review. Agric Rev 25(3):173–188

Srivastava AK, Singh S (2005) Diagnosis of nutrient constraints in citrus orchards of humid tropical India. J Plant Nutr 29(6):1061–1076

Srivastava AK, Singh S (2006) Nutrient diagnostics and management in citrus. In: Technical bulletin no. 8. National Research Centre for Citrus, Nagpur, pp 1–130

Srivastava AK, Singh S (2008a) Citrus nutrition research in India: problems and prospects. Indian J Agric Sci 78:3–16

Srivastava AK, Singh S (2008b) Zinc nutrition in Nagpur mandarin on Haplustert. J Plant Nutr 32(7):1065–1081

Srivastava AK, Singh S (2008c) Citrus decline: soil fertility and plant nutrition. J Plant Nutr 32(2):197–245

Srivastava AK, Singh S, Marathe RA (2002) Organic citrus: soil fertility and plant nutrition. J Sustain Agric 19(3):5–29

Srivastava AK, Huchche AD, Lallan R et al (2005) Performance or irrigated monocropped versus intercropped Nagpur mandarin orchards on basalt derived tropical soils. Trop Agric (Trinidad) 82(2):1–8

Srivastava AK, Huchche AD, Lallan R et al (2007) Yield prediction in intercropped versus monocropped citrus orchards. Sci Hortic 114:67–70

Srivastava AK, Singh S, Albrigo LG (2008) Diagnosis and remediation of nutrient constraints in citrus. Hortic Rev 34:277–363

Srivastava AK, Gupta SG, Rao CN et al (2011a) Organic manuring in Citrus. Annual report. National Research Centre for Citrus, Nagpur, pp 36–39

Srivastava AK, Huchche AD, Dinesh Kumar (2011b) Development of INM module for sustained productivity of Citrus. Annual report. National Research Centre for Citrus, Nagpur, pp 40–43

Storey R, Treeby MT (2002) Nutrient uptake into navel oranges during fruit development. J Hortic Sci Biotechnol 77(1):91–99

Swietlik D (1991) Growth of four citrus rootstocks treated with a natural biostimulant. Subtrop Plant Sci 44:11–14

Swietlik D (1996) Responses of citrus trees in Texas to foliar and soil Zn applications. In: Manicom B, Robinson J, Du Plessis SF et al (eds) Proceedings of International Society of Citriculture, Eighth International Citrus Congress, vol 2, Sun City, South Africa, May 12–17, pp 772–776

Tachibana S, Yahata S (1996) Optimizing nitrogen fertilizer application for a high density planting of satsuma mandarin. J Jpn Soc Hortic Sci 65(3):471–477

Takatsuji T, Ishihara M (1980) Studies on the potassium nutrition of satsuma III. Influence of cold hardiness and leaf composition. In: Bulletin no. 7. Fruit Tree Res State (Yatabe), Ibaraki, Japan, pp 45–62

Tiwari SC (1996) Effects of organic manure and N P K fertilization on enzyme activities and microbial population in subtropical Oxisols. J Hill Res 9(2):334–340

Tiwari US, Rai JSB, Jagtap DD et al (1997) Effect of organic and inorganic fertilizers on yield and quality of sweet orange (*Citrus sinensis* Osbeck). In: Abstract, National Symposium of Citriculture, Nagpur, Maharashtra, India, 17–19 Nov, pp 41–42

Tripathi BP, Harding AH (2001) Nutrient status of mandarin trees in some mandarin growing pockets of Lamzung and Gorkha districts of Nepal. J Indian Soc Soil Sci 49(3):503–506

Tucker DPH, Davis RM, Wheaton TA et al (1990) A nutritional survey of south central southeast and coast flatwoods citrus grooves. Proc Fla State Hortic Soc 103:324–327

Tucker DPH, Alva AK, Jackson LK et al (1995) Nutrition of Florida citrus trees. In: SP 169. University of Florida Cooperative Extension Service, Gainesville, p 61

Usha K, Saxena A, Singh B (2004) Rhizosphere dynamics influenced by arbuscular mycorrhizal fungus (*Glomus deserticoloa*) and related changes in leaf nutrient status and yield of kinnow mandarin [King (*Citrus nobilis*) Willow and Willog leaf (*Citrus deliciosa*)]. Aust J Agric Res 55(5):571–576

Veregara I, Schalscha BE, Ruiz I, Wallihan EF (1973) Nutritional status of representative citrus orchards in Chile as evaluated by leaf and soil analysis. HortScience 8(4):325–326

Wada M, Shimogoori Y, Hatano H (1981) Behaviour of surface applied cations in the soil of a satsuma mandarin orchard. In: Matsumoto K, Oogaki C, Kozaki I et al (eds) Proceedings of International Society of Citriculture, vol 2, Tokyo, Japan, Nov 9–11, pp 544–548

Wang TC (1985) Application of fertilizers to satsumas based on leaf analysis. J Soil Sci 16(6):275–277

Wang FL, Alva AK (1996) Leaching of nitrogen from readily soluble and slow release fertilizers in sandy soils. Soil Sci Soc Am J 60: 454–1438

Weir CC (1969) Investigation on the mineral nutrition of citrus in Caribbean area. In: Technical bulletin no. 2. Citrus Research Unit. University of West Indies, West Indies, p 69

Yin KL, Li YG, Wang CQ et al (1998) Specifically prescribed fertilization for citrus trees and the principle component of analysis of foliar Fe and Zn and other nutrient elements. J Southwest Agric Univ 20(3):1989–1992

Zekri M (1995) Nutritional deficiencies in citrus trees: nitrogen, phosphorus and potassium. Citrus Indus 8:58–60

Zekri M, Koo RCJ (1991) Evaluation of controlled-release fertilizers for young citrus trees. J Am Soc Hort Sci 116(6):987–990

Zekri M, Koo RCJ (1992) Use of controlled release fertilizers for young citrus trees. Scientia Hort 49:233–241

Zhang M, Alva AK, Li YC, Calvert DV (1997) Chemical association of Cu, Zn, Mn and Pb in selected sandy soils. Soil Sci 162:189–192

Zhou B, Alva AK (1993) Distribution of trace metals in some sandy soils under citrus production. J Soil Sci Soc Am 57:350–355

Citrus Best Management Practices

26

Brian Boman

Abstract

Agricultural best management practices (BMPs) are practical, cost-effective actions that agricultural producers can take to reduce the amount of pesticides, fertilizers, sediment, and other pollutants entering water resources. BMPs are designed to benefit water quality while maintaining or even enhancing agricultural production. Implementing BMPs benefits both the farmer and the environment and demonstrates agriculture's commitment to water resource protection and is a key component of agriculture's environmental stewardship role.

Most of the BMPs can be implemented through management actions without great expense. Typical practices include:

- *Nutrient management* – to determine nutrient needs and sources, and manage nutrient applications to minimize impacts to water resources. Practices may include soil testing, tissue testing, quantifying noncommercial nutrient sources (e.g., residual nitrogen from legumes), following recommended application rates based on published research, splitting fertilizer applications, using precision application techniques, and adhering to application setbacks from water bodies and sensitive areas.
- *Irrigation management* – to address the method and scheduling of irrigation to reduce water and nutrient losses. Practices may include the use of high-efficiency irrigation systems, diagnostic tools such as soil moisture sensors, tensiometers, water table observation wells, and weather-related data such as evapotranspiration.
- *Treatment and erosion control* – to reduce or prevent the transport of nutrients and sediments from production areas to water bodies. These practices may include vegetated buffers for streams and wetlands, filter strips to treat field runoff, vegetative cover in nonproduction areas to reduce erosion, and retrofitting ditches in high-velocity areas to prevent scouring.

Along with local governments and private industry, the agricultural community is expected to reduce their water quality impacts. It is important that producers embrace and implement BMPs in their operations to underscore agriculture's participation in and commitment to water resource protection.

Keywords

Nutrient management • Pest management • Sediment • Irrigation • Drainage • Fertilizer • IPM • Runoff

B. Boman (✉)
Indian River Research and Education Center, University of Florida,
2199 South Rock Road, Fort Pierce, FL 34945, USA
e-mail: bjbo@ufl.edu

A.K. Srivastava (ed.), *Advances in Citrus Nutrition*,
DOI 10.1007/978-94-007-4171-3_26,

26.1 Introduction

Best management practices (BMPs) are production systems and management strategies scientifically shown to minimize adverse water quality and other environmental impacts of agricultural production. BMPs can be defined as those agricultural operational procedures designed to achieve greatest agronomic efficiency in food and fiber production, while limiting the off-site effects of agricultural operations and maintaining an economically viable farming operation. Many BMPs have little or no cost and can be implemented quickly as they are management strategies related to cultural operations. BMPs that require modification of orchard features or require the purchase of equipment may require several years to implement so there is not an excessive financial burden on the grower. The development and implementation of BMPs for a particular orchard should be an ongoing process to continually identify and develop improved, science-based BMPs that will enhance and protect water resources.

BMPs have the potential to improve water quality, reduce the off-site transport of nutrients and sediments, and provide more suitable timing of surface water discharges into receiving streams. Improvements will come through diligent, incremental program achievements and efforts that reduce the volume of water and loading of constituents that degrade water quality. There are numerous, significant benefits that can occur with adoption of BMPs, but not all BMPs are applicable for every citrus-producing region and every citrus producer. To achieve maximum benefit, it is important for the overall picture of each citrus operation to be considered prior to adoption of any BMP to ensure each BMP achieves its proper objective. In Florida, programs were developed for three separate coastal (flatwoods) areas (Boman et al. 2000, 2004, 2005) and one for the interior well-drained ridge area (FDACS 2002). Parsons and Boman (2006) describe process for BMPs developed for the Florida citrus industry.

BMPs must protect the environment and be economically viable. Recommendations must be based on factual information and science and must be focused on solutions that work. However, in some cases, there is no definitive research that applies to a particular BMP. In other cases, available research cannot predict the effectiveness of a BMP. In situations like these, BMPs are based upon a combination of available research, best professional judgment, and practical experience. Best professional judgment should allow modification of a particular practice to field conditions in a specific citrus production unit to achieve water quality and quantity objectives. BMPs should be consistent, while providing reasonable flexibility to accommodate local needs and site-specific conditions. Most of all, BMPs should not be a barrier to incentive-based programs or to technical innovation.

26.2 Implementation of BMPs

Citrus producers should maintain records and provide documentation regarding the implementation of all BMPs utilized and applied on their farm as well as document why certain BMPs are not applicable to their specific situation. Adequate records are very important for documentation of BMP implementation. These records are an integral part of any BMP program. The priorities for BMP implementation are:

1. To correct any qualified existing water quality/quantity problems
2. To minimize water quality/quantity problems resulting from land use
3. To improve effectiveness of applied BMPs
4. To seek additional improvement of BMPs based on new, quantifiable information

All growers are encouraged to perform an environmental assessment of their crop production operations. This resource allocation assessment process is a tool that will aid in identifying which BMPs should be considered to achieve the greatest economic and environmental benefit. Among the incentives for adoption of BMPs are:

- Improved crop yield
- Improved crop quality
- Improved worker safety
- Efficient allocation of resources
- Reduced environmental impacts

It is fully recognized that additional research is required to establish the effectiveness of some BMPs. The development, advancement, and use of precision agriculture techniques in citrus production are strongly recommended. These technologies have provided cost-effective, agronomically sound management alternatives to traditional management methods while providing an increased level of resource protection. The rapid change of precision agriculture technology has resulted in frequent modifications and enhancements to existing applications. It is recommended that growers consistently remain abreast of new technology trends. The application of these emerging technologies by growers should be considered on a site-specific basis. Education is a key factor to ensure success of the BMP programs. Training programs are needed to ensure that BMPs are applied and implemented properly. Monitoring programs are needed to show that BMPs are effective in protecting water quality/quantity and to provide measurable data for BMP revisions. Incentive programs are needed when a proposed BMP results in adverse economic impact. The following sections enumerate BMPs in four major areas: pest management, nutrient management, water resources, and sediment management.

26.3 Pest Management

Over the last 20 years, great strides have been made in the development of crop protection (CP) products that are more target specific, less harmful to the environment, and safer to those who handle and apply these products. The development and implementation of responsible farm management practices that promote the proper handling of these products also have contributed significantly to reducing the risk of environmental problems and protecting water resources, pesticide handlers, and agricultural workers.

26.3.1 Integrated Pest Management (IPM)

IPM is an integrated system using a combination of mechanical, cultural, biological, and chemical approaches to best meet the goals of the program. This approach provides better and more economical management of most pests. IPM is a philosophy of managing pests that aims to reduce farm expenses, conserve energy, and protect the environment. It can be defined as a broad, interdisciplinary approach using a variety of methods to systematically manage pests which adversely affect people and agriculture. IPM does not, as many believe, mean that no CP products are used. Rather, it means that CP products are only one weapon against pests, and they should be used judiciously and only when necessary.

The goals of an IPM program are (McCoy et al. 2011):

1. Improved control of pests, through a broad spectrum of practices that work together to keep pest populations below economically significant thresholds
2. More efficient CP product management through less frequent and more selective use of CP products
3. More economical crop protection from reduced chemical costs and more efficient protection
4. Reduction of potential hazards to farmers, workers, consumers, and the environment through reduced CP product exposure

IPM accomplishes these goals using resistant plant varieties, cultural practices, parasites and predators, other biological controls such as *Bacillus thuringiensis* (BT), and other methods including chemical CP products as appropriate.

The basic steps for an IPM program are:

1. Identify key pests and beneficial organisms and the factors affecting their populations.
2. Select preventative cultural practices to minimize pests and enhance biological controls. These practices may include soil preparation, resistant rootstocks/scions, modified irrigation methods, cover crops, augmenting beneficials, etc.
3. Use trained "scouts" to monitor pest populations to determine if or when a control tactic might be needed.
4. Predict economic losses and risks so that the cost of various treatments can be compared to the potential losses to be incurred.
5. Decide the best course and carry out corrective actions.
6. Continue to monitor pest populations to evaluate results and the effectiveness of corrective actions. Use this information when making similar decisions in the future.

The pests associated with citrus IPM include insects, mites, nematodes, fungi, bacteria, viruses, and weeds (McCoy et al. 2011). It is important to recognize that not all organisms found in citrus orchards are considered pests. Many species in the diverse ecosystem of the citrus orchard are neutral they do not cause injury or loss to citrus, while other species are actually beneficial to the production of citrus. Thus, only a small proportion of the organisms should be considered as pests, and only under certain circumstances when their level of existence threatens the health or productivity of citrus trees.

In addition to known pests and for which IPM tactics are available, there is a continuing threat of the introduction and establishment of exotic pests of citrus that may come from other parts of the world. Establishment and spread of exotic mites, insects, nematodes, diseases, or weeds exert additional pressure on the citrus production system, often leading to the need for additional chemical intervention in order to prevent injury or tree loss. New introductions of pests are destabilizing features of IPM systems and must be treated carefully to avoid disruption of the established pest management system. Approaches that can be used in IPM vary from those that prevent development of damaging levels of the pest to those that reduce the pest below levels of concern. Among the tactics that are available that prevent development of pest injury are sanitation and exclusion. When these tactics fail, eradication can be attempted to eliminate the pest organism from the system and prevent repeated injury or damage. Finally, suppression tactics are used to reduce populations or infection below the level that would compromise tree health or fruit yield.

Informed decisions are an important aspect of citrus IPM. As pest populations are encountered in the orchard, proper identification of the pest organisms involved is a crucial step in making appropriate responses. Following identification of the pest(s) involved, implementation of the appropriate cultural practices or other tactics should occur. When it is deemed that CP products are necessary, targeted application of pest control products should be made so as to maximize suppression of the target pest populations while minimizing nontarget impacts. Inappropriate choice of tactics can lead to undesirable outcomes. The presence of a long history of biological control of numerous pests in Florida citrus makes the system vulnerable to upset through misuse or overuse of CP products. Elimination of parasites or predators through multiple applications of CP products can lead to disruption of the balance between pests and their natural enemies, resulting in higher pest populations and increased damage.

26.3.2 Pest Monitoring and Economic Assessment

Scouting procedures are a very effective and efficient tool for the grower. Scouting allows the grower or production manager to visually formulate recommendations that can be very cost-effective for the crop being grown. It gives the grower the decision-making ability to properly apply CP products and nutrients. By scouting the crop, past, current, and future problems can be seen, and scouting allows growers to address them in the proper manner. When scouting for insects, for example, growers should also be watchful for other problems that could be present. For instance, when scouting for mites, growers could find an increasing problem with weed control or an increasing insect population. When practiced properly, scouting can be a valuable management tool for all growers.

Pest monitoring results should be subjected to an economic assessment before reactive measures are taken. The mere presence of a pest or disease is not sufficient to warrant treatment expense. Assessing the potential loss if the pest goes unchecked versus the cost and financial return preserved through various treatments should assist the pest manager in making decisions that are biologically and economically sound. While cost of suppression compared to savings in yield, quality, or tree health is of primary importance, the costs associated with disruption of other pests by the treatment also should be considered.

26.3.3 Product Selection

Read and understand the CP product label. Pay special attention to the "Environmental Hazards" section of the label. Select target-specific active ingredients that consider natural systems in epidemiological cycles and modes of action (i.e., insect growth regulators, botanicals, and biologicals). Agricultural use of CP products should be part of an overall pest management strategy, which includes biological controls, cultural controls, pest monitoring, and other applicable practices, referred to altogether as integrated pest management or IPM. When a CP product is needed, its selection should be based on effectiveness, toxicity to nontarget species, cost, and site characteristics, as well as its solubility and persistence (Nesheim 1998). While the focus of the IPM program is for field populations of mites, insects, nematodes disease pathogens, and weeds, CP products also are prescribed for postharvest maintenance of fruit quality. Some of these situations require preharvest applications as part of the overall management strategy. Due consideration needs to be given to these treatments in the overall crop BMPs.

26.3.4 Minimize Spray Drift

Reduce the potential for drift through appropriate selection of nozzles, spray pressure, and application methods or techniques for the formulation applied and equipment used (Stover et al. 2002a). Use nozzles that produce as large of a droplet size as possible while yielding adequate plant coverage and pest control (Cromwell 1993). Leave a buffer zone according to the crop protection label between the treated field and any sensitive areas.

26.3.5 Application Timing

Time CP product applications in relation to current soil moisture, anticipated weather conditions, and irrigation schedule to achieve greatest efficiency.

26.3.6 Precision Application of CP Products

Proper operation of equipment is important to achieve high application efficiency (Stover et al. 2003b). Use precision applications of reduced amounts of material to smaller trees in order to minimize application of CP products to nontarget areas and result in more efficient utilization of applied materials (Stover et al. 2003a). The method of CP product application, such as ground or aerial spraying, wicking, granules, etc., is important since the degree of drift and volatilization can vary considerably. Some "intelligent" spraying systems (Fig. 26.1) are equipped with three-dimensional range sensors that can map the image of a tree up to 30 m away on either side of the sprayer. These sensors feed the size, height, and location of the tree into an onboard computer that then turns on spray nozzles inches before the sprayer reaches the tree and turns them off inches past the tree. The nozzles are controlled by electric solenoid valves which are set up in zones so that only the foliage detected by the scanner is sprayed. It is important that "intelligent" systems be properly maintained and operated and that equipment operators be trained in their use. Equipment without intelligent systems

Fig. 26.1 Airblast sprayer equipment with sensors for turning nozzles on when a tree is detected

should have nozzle arrangement to avoid overspray based on tree height. This is sometimes referred to as "nozzling-down" to conserve spray materials and ensure application to target areas. Other systems have been developed that utilize sonar for detecting foliage. These systems utilize ultrasonic impulses to detect the presence or absence of trees and plants. Sensors are installed on each side of the sprayer that may be aimed in any desired direction to cover optimal zones. The number of sensors can vary depending on the diversity of tree sizes within the orchard. Regardless of application system, proper training of applicators and maintenance of spray systems are essential to good management.

26.3.7 Maintenance and Calibration

Proper calibration and maintenance of CP product application equipment are essential for the proper application of agricultural chemicals. Calibration is the process of measuring and adjusting equipment performance. Application equipment that must be calibrated includes granule-applying devices; hand, backpack, boom, airblast, and other sprayers; soil fumigation devices; and injection equipment used for chemigation work. Calibration is not difficult, but before you start to calibrate any equipment, first make sure that all components are clean and in good working order. CP product application equipment can deliver the correct amount of CP product to the target site only if it is working correctly. Follow the manufacturer's directions carefully; they usually explain how to adjust the equipment. Pay particular attention to the parts (such as nozzles and hopper openings) that regulate how much CP product is released. If these parts are clogged, not enough product will be released. If they are worn, too much product will be released. Correct measurement will keep you in compliance with the label, reduce risks to applicators, farm workers, and the environment and save you money. Calibrate using clean water and do not calibrate equipment near wells, sinkholes, or surface water bodies. Proper application of CP products will help reduce farm costs. Improper application can result in wasted chemicals, marginal pest control, excessive carry-over, or crop damage. As a result, inaccurate application is usually very expensive. When you measure nozzle output, calculate application rate or do other calibration-related calculations, the acceptable variation is plus or minus 10%. The following are points to consider for calibration.

26.3.8 Speed

For most types of application equipment, the speed at which the equipment moves (or is carried) through the target site is one of the main factors that determines the actual application rate. This means that for most types of application equipment, you must measure actual travel speed in order to correctly calibrate the device. For certain types of granule-applying devices, you do not need to consider actual travel speed when calibrating.

26.3.9 Uniform Release

If the application equipment has more than one nozzle or hopper, measure the output from each one. This insures that each one is releasing the desired amount of CP product. Always determine that each nozzle (or hopper) is delivering within ±10% of the amount desired.

26.3.10 Calibration Methods

No matter what calibration method you use, your main goal is always the same: to measure how much CP product the equipment is actually applying to a specific area. Whenever possible, apply something other than the actual CP product (i.e., plain water) when you make test applications during calibration runs. As a general rule, if the CP product is normally diluted with water, you can use plain water for the test runs. If the CP product is a dust, granule, fumigant, or a liquid diluted with liquid other than water, it is generally better to use the actual CP product in your test runs. In such cases, always be sure to follow the product's label directions for proper handling during mixing and loading. Begin calibration runs at lower equipment settings and adjust upward until the correct rate is achieved.

26.3.11 Calculate the Actual Application Rate

The total amount of CP product spray applied divided by the amount of target area actually covered is your equipment's actual application rate. If your equipment's actual application rate is not within ±10% of the label rate, then your equipment needs to be calibrated.

26.3.12 Check Calibration Often

Sprayers should be calibrated when new or when nozzles are replaced. Once you have calibrated your equipment, do not assume that it will continue to deliver the same rate forever. Clogging, corrosion, and normal wear will eventually cause the delivery rate to change; the best settings on the best equipment still gradually get out of adjustment. Recalibrate equipment periodically to compensate for wear in pumps, nozzles, and metering systems. Calibration should be adjusted for small trees and other variations.

26.3.13 Calibrating Sprayers

To calibrate spray equipment, you must know (Cromwell 1992):

1. The appropriate pump pressure
2. The spray volume needed per unit of area
3. The type of diluents to use (usually water)
4. The type of nozzle tips and strainers
5. Nozzle spacing and height

The amount of chemical solution applied per unit land area depends upon the forward speed, system pressure, size of nozzle, and spacing of nozzles on the boom. A change in any one of these will change the rate of application. Consult the manufacture's operator manual for information on a particular sprayer.

26.3.14 Record Keeping

Maintain accurate CP product records to meet legal responsibilities and to document production methods. CP product record keeping requires you to have current knowledge concerning the application of CP product materials within your area of influence. In addition, Florida law requires that you record the following items to comply with the restricted-use pesticide record-keeping requirement:

- Brand or product name
- Registration number
- Total amount applied
- Location of application site
- Size of area treated
- Crop/variety/target site
- Month/day/year of application
- Name of applicator
- Method of application
- Name of person authorizing the application if the licensed applicator does not own or lease the property

26.3.15 Protect Water Sources During Mixing

Protect the water source by keeping the water pipe or hose well above the level of the CP product mixture. This prevents contamination of the hose and keeps CP products from back-siphoning into the water source. If you are pumping water directly from the source into a tank, use a check valve, antisiphoning device, or backflow preventer to prevent back-siphoning if the pump fails.

26.3.16 Spill Management

Potential for movement of spilled CP products in water is reduced if the spill is controlled, contained, and cleaned up quickly. Establish a plan-for-action and clean up spills as soon as possible. The sooner you can contain, absorb, and dispose of a spill, the less chance there is that it will cause harm. Always use the appropriate personal protection equipment (PPE) as indicated on the MSDS and the label. In addition, consider the following four steps (Howard et al 1998):

1. CONTROL actively spilling or leaking materials by setting the container upright, plugging leak(s), or shutting the valve.
2. CONTAIN the spilled material using barriers and absorbent material.
3. COLLECT spilled material, absorbents, and leaking containers and place them in a secure and properly labeled container.
4. Store the CONTAINERS of spilled material until they can be applied as a CP product or appropriately disposed.

Small liquid spills may be cleaned up by using an absorbent such as cat litter, diluting with soil, and then applying the absorbent to the crop as a CP product in accordance with the label instructions. Farmers, farm managers, and landowners must comply with all applicable regulations regarding spill response training for employees, spill reporting requirements, spill containment, and cleanup. Keep spill cleanup equipment readily available when handling CP products or their containers.

26.3.17 Mix/Load Sites

When practical, use permanent mix/load stations to reduce CP product spillage. A well-designed permanent mix/load facility is convenient and provides a place where spill-prone activities can be performed over an impermeable surface that can be easily cleaned (Stover et al. 2002b). To minimize the risk of CP products accumulating in the environment from repetitive spills, you may wish to construct a permanent mix/load facility with an impermeable surface (such as sealed concrete) so that spills can be collected and managed. A permanently located mixing and loading facility, or chemical mixing center (CMC), is designed to provide a place where spill-prone activities can be performed over an impermeable surface that can be easily cleaned and permits the recovery of spilled materials.

Locate CP product loading stations away from groundwater wells and areas where runoff may carry spilled CP products into surface water bodies. If such areas cannot be avoided, protect wells by properly casing and capping them and use berms to keep spills out of surface waters. It is crucial that a CMC facility be properly designed and constructed.

Another option for preventing contamination of mixing and loading sites is to use a portable mixing center. Some are little more than a pad of very durable material, while others are made of interlocking steel sections with a custom-fitted liner and built-in sump. Portable mixing centers usually have no roof but should be protected from rain. Since the pad may contain CP product residues, the accumulated rainwater may need to be applied as a CP product or disposed of as hazardous waste. A heavy rain can cause the pad to overflow, washing CP products into the environment. A sudden thunderstorm can result in a considerable amount of contaminated runoff or even a spill. Clean portable mixing centers thoroughly immediately after a spill because the liner material could be damaged by the CP product formulation. Where practical, portable pads for mixing and loading should be used away from wells or surface water. Never leave a tank unattended while filling.

26.3.18 Utilize Nurse Tanks for Random Field Mixing

CP product loading areas should be conducted at random locations in the field with the aid of nurse tanks. Nurse tanks are tanks of clean water transported to the field to fill the sprayer. Nurse tanks make it possible to move the mixing and loading operation away from permanent sites to random locations in the field. Mixing chemicals at random sites in the field lessens the chance of a buildup of spilled materials in one place. One variation is a self-contained mix/load trailer with a nurse tank at one end and a mix/load area at the other, where the mixture is pumped directly into the sprayer. Another use is portable containment facilities with nurse tanks to set up a temporary mixing/loading site in a remote field or on leased land where no permanent structure is practical.

26.3.19 Excess Mixture

Mix only the amount of CP products needed during an application period. It is not always possible to avoid generating excess spray material. The appropriate practices to be followed depend on the type of CP product waste. If there is excess CP product material, use it in accordance with the label instructions.

26.3.20 Container Management

Develop and implement procedures to appropriately rinse and dispose of or recycle agricultural chemical containers. If permitted by the label and local ordinances, bags, boxes, and group I pesticide containers may be burned in an open field by the owner of the crops. Group I containers are containers of organic or metallo-organic CP products, except for organic mercury, lead, cadmium, or arsenic compounds.

Try to avoid the need to dispose of CP product containers as wastes by:

- Using containers that are designed to be refilled by the CP product dealer or the chemical company
- Arranging to have the empty containers recycled or reconditioned
- Using soluble packaging when available

When disposal is needed, rinse CP product containers as soon as they are empty. Pressure rinse or triple rinse containers and add the rinse water to the sprayer. Shake or tap nonrinseable containers such as bags or boxes so that all dust and material fall into the application equipment. Always wear the proper personal protective equipment (PPE) when conducting these rinse operations. After cleaning, puncture the CP product containers to prevent reuse (except glass and refillable minibulk containers). Keep the rinsed containers in a clean area, out of the weather, for disposal or recycling. Storing the containers in large plastic bags is one option to protect the containers from collecting rainwater. Recycle rinsed containers in counties where an applicable program is available or take them to a landfill for disposal. Check with your local landfill before taking containers for disposal, as not all landfills will accept them.

26.3.21 Equipment Sanitation and Wash Water Handling

Wash water from CP product application equipment must be managed properly since it may contain CP product residues. If permanent wash stations are not used, excess mixture needs to be properly disposed of or reused:

- Wash the outside of equipment at random places in the field to avoid chemical buildup at a site.

- Avoid washing contaminated equipment in the vicinity of wells or surface water bodies. Dispose of rinse water according to label instructions.
- If permanent wash stations are used, wash water should be reused or properly disposed.

26.3.22 Storage

Design and build CP product storage structures to keep CP products secure and isolated from the surrounding environment. Store CP products in a roofed concrete or metal structure with a lockable door (Dean and Bucklin 1997). Locate this building at least 15 m from other structures (to allow fire department access) and 30 m from surface water and from direct links to groundwater. Keep CP products in a separate facility or at least in a locked area separate from areas used to store other materials, especially fertilizers, feed, and seed. Do not store CP products near burning materials, hot work (welding, grinding), or in shop areas. Avoid storage of CP products in spaces occupied by people or animals. Do not allow smoking in CP product storage areas. Store personal protective equipment (PPE) where it is easily accessible in the event of an emergency but not in the CP product storage area to avoid contamination and since that may make PPE unavailable in time of emergency.

Check the label and the material safety data sheets (MSDS) for the safety equipment requirements. Keep a written CP product inventory and the MSDS file for the chemicals used in the operation on site. Do not store this information in the CP product storage room. Do not store large quantities of CP products for long periods of time. Adopt the "first in, first out" principle, using the oldest products first to ensure that the product shelf life does not expire. Store CP products in their original containers. Do not put CP products in containers that might cause children and others to mistake them for food or drink. Keep the containers securely closed and inspect them regularly for splits, tears, breaks, or leaks.

Arrange CP product containers so that labels are clearly visible and legible. All CP product containers should be labeled. Refasten all loose labeling. Use non-water-soluble glue or sturdy transparent packaging tape to refasten loose labels. Do not refasten labels with rubber bands (these quickly rot and break) or nontransparent tapes such as duct tape or masking tape (these may obscure important product caution statements or label directions for product usage). If a label is damaged, immediately request a replacement from the CP product dealer or formulator. As a temporary supplement to disfigured or badly damaged labels, fasten a baggage tag to the container handle. On the tag, write the product name, formulation, concentration of active ingredient(s), and the date of purchase.

Dry bags should be stored on pallets and covered with plastic to ensure they do not get wet. Do not store liquid materials above dry materials. Store flammable CP products separately from nonflammable CP products. Segregate herbicides, insecticides, and fungicides to prevent cross-contamination and minimize the potential for misapplication. Cross-contaminated CP products often cannot be applied in accordance with the labels of each of the products. This may make it necessary to dispose of the cross-contaminated materials as wastes and could require the services of a consultant and hazardous waste contractor. Use shelving made of plastic or reinforced metal. Keep metal shelving painted (unless stainless steel) to avoid corrosion. Never use wood shelving because it may absorb spilled CP product materials. CP product storage structures should be identified such that the nature of the contents is made known to those approaching the building.

26.3.23 Excess Formulation

When possible, return excess formulated materials to the CP supplier. In most cases, the excess material must be in an unopened, original container. Contact local dealers for their requirements. The single best practice to handle excess CP product material is to use it as a CP product in accordance with the label instructions.

26.3.24 Purchase and Transport

Appropriately planned and timed purchase of CP products can avoid risks associated with protracted storage. Adherence to instructions provided by product manufacturers relating to transport of CP products can minimize risks of spillage and contamination in the event of accident or other container failure. Follow directions for transport provided on product label, taking into consideration exposure to temperature, moisture, UV light, and other variables. Ensure packages and containers are properly closed and secured prior to transport and are retained in original containers and with original product label attached. Consider restrictions imposed by manufacturers or transportation agencies on transport within enclosed spaces and or by personal vehicle. Appropriate spill response materials should always be transported along with CP products to ensure that immediate spill response can be accommodated.

26.3.25 Product Use Training

Training of field operators responsible for handling, loading, and operating spray machinery is essential for effective application of agricultural chemicals. It is essential that information learned at continuing education classes be transferred to application personnel. Special efforts should be taken to ensure that field personnel understand proper handling, loading, and operating techniques.

26.4 Nutrient Management

Good nutrient management is an integral part of a system of agricultural practices that help conserve and protect natural resources. In fact, water and nutrients are oftentimes linked, and the citrus industry has made great strides in converting many existing orchards to low-volume irrigation systems. These conversions allow more precise nutrient management via the use of fertigation. As such, implementing appropriate nutrient management practices helps maintain or improve agricultural productivity while minimizing environmental risk.

Management of nitrogen and phosphorus levels, in particular, is essential in maintaining healthy surface water bodies and natural systems wherever citrus is grown. These nutrients originate from a variety of land uses, including: agricultural, urban, suburban, and natural areas. Excess nutrients stimulate algal blooms and growth of noxious plants in receiving water bodies and wetlands. This stimulation of growth may eventually result in reduced dissolved oxygen concentrations due to excessive decomposition of plant material. Moreover, lower dissolved oxygen concentrations may stress desirable game fish and promote less desirable fish species.

Nitrogen and phosphorus are two of the essential elements for plant and animal growth and are necessary to maintain profitable crop and livestock production. They can also increase the biological productivity of surface waters by accelerating eutrophication, the natural aging of lakes or streams brought on by nutrient enrichment. Although eutrophication is a natural process, it can be accelerated by changes in the land use of a watershed that increase the amount of nutrients added to an aquatic system. Nitrogen and phosphorus both affect eutrophication, but phosphorus is the critical element in most freshwater systems. Complicating the problem is the fact that eutrophication sometimes occurs many miles from where high-nutrient runoff originally enters the surface water system. By the time the water quality effects are noticeable (sometimes years to decades after the runoff occurs), remedial strategies can be difficult and expensive to implement. This is why source control of nutrients used in fertilization programs is so important.

26.4.1 Education

Proper training of the field operators responsible for handling, loading, and operating fertilizer spreading equipment and for correct maintenance of field equipment can help achieve desired placement of fertilizers, avoid waste, and prevent contamination of open waters. Reinforce training with checklists of critical operating points before application of materials. Confirm that each assigned employee is adequately informed about machine operation, rates of discharge, and intended zone of nutrient placement that focuses on "feeding the tree."

26.4.2 Nutrient Management

Develop a nutrient management plan based upon soil, water, plant, and organic material sample analyses and expected crop yields. Nutrient management is: management of the amount, source, placement, form, and timing of the application of nutrients and soil amendments to ensure adequate soil fertility for plant production and to minimize the potential for environmental degradation, particularly water quality impairment.

Nutrient management plans should include the following components, as applicable:

- Aerial site photographs or maps and a soil map.
- Current and/or planned production sequence.
- Soil-test results and recommended nutrient application rates.
- Plant tissue test results, when used for nutrient management.
- A complete nutrient budget for nitrogen, phosphorus, and potassium for the production system.
- Realistic yield goals and a description of how they were determined.
- Quantification of all important nutrient sources (this could include but not be limited to commercial fertilizer, animal manure, and other organic by-products, irrigation water, etc.).
- Planned rates, methods, and timing (month and year) of nutrient application.
- Location of designated sensitive areas or resources (if present on the conservation management unit).
- Guidance for implementation, operation, maintenance, and record keeping.
- Maximum single application rates of nutrients will be determined based on optimum level of production, producer's goals, soil limitations, site factors, and off-site transport potential. Additional conservation practices that keep nutrients in the soil and root zone area should be planned in environmentally sensitive areas.

Realistic yield goals should be set based on soil type, crop variety, tree age and condition, tree density, historical yield data, climatic conditions, and fertilizer costs versus returns. The form of fertilizer and its timing, placement, and method of application can be planned to conform to seasonal variations in nutrient uptake throughout crop development:

- Consider effects of the seasonal water budget on nutrient balance and on the potential loss by surface runoff or leaching into groundwater.

- Evaluate water quality standards and designated use limitations that exist locally or regionwide.
- Avoid excessive or luxury levels of N, P, and K in the soil to reduce the potential for induced deficiencies of micronutrients.
- Maintain proper soil pH to provide optimum availability of applied nutrients.
- Use appropriate application methods and fertilizer formulations that minimize nutrient losses.
- In high water table soils, water table management will affect the availability and movement of nutrients.
- Proper calibration and use of equipment will improve nutrient material application efficiency and will reduce undesirable overapplications.
- Avoid same-place loading/transfer sites to preclude excess contamination of soils in working areas.

26.4.3 Nutrient Management and Utilization of Waste Resources

Use of animal waste and other waste products on land in an environmentally acceptable manner can be helpful in maintaining or improving soil, air, plant, and water resources. Wastes include those from farm, feedlot, and dairy operations, municipal waste treatment plants, and agricultural processing plants. General criteria for using waste products include:

1. Compliance with all local laws is required for all utilization of wastes including liquid, slurry, and solid waste.
2. Waste application should be accomplished in a manner (timing and rate) such that runoff from the application area will not occur due to the application method used.
3. Waste should be applied based on the most limiting nutrient or metal.
4. Specific conditions for land application of domestic wastewater residuals may be required depending on the locality.
5. Crop nutrient removal rates should be based on realistic yields.
6. Waste application setbacks shall be increased from surface water bodies, wells, sink holes, or fractures.
7. Content of waste should be analyzed for nutrient and metal content.

Supplemental fertilizer may be needed to meet the needs of the crop at various stages of plant growth. Apply wastewater with an electrical conductivity (EC) less than 1.5 dS/m to avoid leaf burn. Note that use of wastewater with high EC conductance can accumulate salts and nutrients into drainage systems and possibly affect downstream receiving water bodies.

26.4.4 Utilize Tissue and Soil Analyses

Fertilizer applications based on leaf tissue and soil tests will help avoid overfertilization and subsequent losses of nutrients in runoff water. Application of mobile elements such as N (nitrogen) and K (potassium) should be made on the basis of leaf tissue analysis and production levels (Obreza and Morgan 2008). Elements such as Ca (calcium), Mg (magnesium), and P (phosphorus) should be based on soil testing and leaf analysis, instead of regular applications of specific amounts. The comparison of both types of testing will give production standards for applications which are based on plant need and response, rather than routine applications of standard amounts. Proper fertilization results in high yields and minimal environmental effects.

Because citrus is a perennial crop, it is its own best indicator of appropriate fertilization. The composition of plant tissue reflects tree health and productive potential supported by prior fertilization and production practices. For this reason, tissue sampling is a useful management tool for fertilizer decisions. Considerable research involving tissue testing of citrus has established the reliability of this method. Leaf analysis, in all its stages from sampling to interpretation of results and developing the nutritional program, is a relatively complicated and costly procedure. Sampling guidelines should be followed precisely to insure that analytical results are meaningful. Interpretation of analytical results and development of nutrient recommendations should be made by those trained in horticulture or in consultation with them.

The best indicators of appropriate fertilizer management practices for citrus are leaf nutritional levels. Leaf analysis standards are based on long-term, well-designed field fertilizer experiments and are accepted throughout the world (Obreza and Morgan 2008). The ranges have been modified very little over the years in spite of many fertilizer studies conducted in different countries with different citrus varieties, rootstocks, and management practices. Leaf analysis provides the best available guideline for managing citrus nutritional programs when leaves have been properly sampled, handled, processed, and analyzed.

In addition to assisting with diagnostic problems within the orchard, tissue sampling can serve as an excellent growth and yield monitoring tool. Tissue collected on an annual basis will provide information about nutritional trends with time. If annual test results are found to be in the deficient or low categories, an increase in annual fertilization for that nutrient is recommended. Alternately, if test results fall within the high or excessive ranges, the annual fertilization for that nutrient should be adjusted downward. If a change in fertilization is indicated, the adjustment should be reasonable. The intent is to find the correct nutrient management level that maintains leaf tissue in an optimum range but does not lead to overfertilization and possible adverse environmental and economic results.

Another management tool which links crop nutrient requirements to nutrient recommendations is a calibrated soil test. A soil test is said to be calibrated when a relationship has been demonstrated between extractable soil-test values and a positive crop response. Soil sampling should be based upon the same management unit concept described for tissue sampling. This process helps to minimize errors by grouping similar tree and soil types. Additionally, annual sampling may be completed at the same time as tissue collection. Sampling at this time will reduce the sampling effort and provide information for fall fertilization decisions. As with tissue sampling, 15–20 trees should be randomly selected from within the management unit or indicator block. One 15-cm deep core, about 2–3 cm in diameter, should be taken within the irrigated zone close to the dripline of each sampled tree. Each core should be placed into a plastic bucket and thoroughly mixed with other cores from that unit or block. A small sample taken from this composite sample should be air-dried prior to shipping to a soil testing laboratory.

In some areas, the pH or minerals in the irrigation water may cause changes in the wetted by the irrigation system. In these cases, an additional (separate) soil sample from the 0–15 cm depth may be taken every 2 years from outside the wetted zone for determining soil pH and available Ca content. Soil pH values between the wetted and unwetted zones can differ by as much as 1–2 units.

The main value of soil testing is the identification of changes with time rather than basing management decisions on individual test results. Soil testing labs may use different extraction methods (e.g., Mehlich-1, Mehlich-3, ammonium acetate (pH 4.8 and 7.0), Bray P1, etc.) to determine nutrient availability in soils. The Bray P1 extracts considerably more P than does the Mehlich-1. Because the different extractants yield data that are not identical, growers are advised to utilize the same analytical procedure for comparison of results. As with tissue analysis, annual soil-test measurements are helpful for making adjustments in fertilization, especially for P. Changes in extractable K, Ca, and Mg should be interpreted with caution. If extractable values are in the high and very high ranges, then it is wise to adjust fertilization downward. However, large changes in fertilization rates should be avoided to preclude making costly errors.

26.4.5 Precision Agriculture

Precision fertilizer applicators (Fig. 26.2) can apply variable rates of materials based upon specific site conditions within a given location. To enable this precise application, systems may be controlled remotely, either via global position units (GPS) utilizing geographical information systems (GIS) or by electronic sensors which sense orchard conditions based upon tree size or other conditions which can be measured electronically from the application equipment as it passes through the orchard.

Fig. 26.2 Smart spreader uses infrared or ultrasonic sensors to sense tree and adjust the speed of the application chains that deliver the fertilizer to the fans. In areas where trees are small or weaker, less fertilizer is used, saving money and wasted fertilizers

26.4.6 Use Appropriate Application Equipment

Operate machinery as designed so as to achieve precise and desired placement of nutrient materials at specified rates consistent with the form and source of nutrient materials. Efficient application practices are critical for insuring fertilizer delivery only to target areas and for reducing losses to leaching and runoff. The following is a list of application techniques for different formulations of fertilizers. Growers may adopt a combination of placement methods exploiting their respective advantages in efficiency and cost. The ultimate goal is to focus on "feeding the tree" by placing nutrients within the root zone of individual trees or dripline bands along hedgerows of trees. Avoid placement in areas prone to off-site transport of nutrients.

When materials are applied using traditional methods (without this new technology), the whole block receives the same amount of products. Small trees, declining trees, and blank areas where trees have been removed all receive the same amount of fertilizer. If application rates are calculated on that required for healthy mature trees, weaker trees and reset trees receive more products than are necessary to achieve maximum economical production.

In contrast, precision application technology can result in significant environmental benefits since specific amounts of crop protection products or fertilizers can be applied based upon soil type, tree age, variety, rootstock, etc. As the non-uniformity of trees in a block increases, the benefits of precision application methods increase compared to traditional methods. These benefits include lower material costs and decreased leaching and runoff of applied materials.

26.4.6.1 Dry Material Spreaders

Dry material spreaders should be adjusted to place fertilizers over or within the root zone. Tree skirts can be pruned to facilitate placement of fertilizers and other agrichemicals such as herbicides. Young tree spreaders with manual or electronic discharge devices can deliver rates of fertilizer quite accurately within the root zones of smaller trees so that nutrients are not lost to runoff or to leaching beyond the root zone. Equipment to apply and incorporate granular controlled-release products should be evaluated to reduce surface movement of nutrients by rainfall and wind. When animal manure is applied, spreading equipment should be designed to place the manure over the root zone.

26.4.6.2 Fertigation

Application of liquid fertilizer through irrigation systems (fertigation) enables multiple applications with little added cost, since fertilizing is scheduled with irrigations. An additional advantage of fertigation is that it applies nutrients to the wetted areas of the root zone where the most active roots are located. Boom application of liquid or suspension fertilizers also offers this advantage. Effective nutrient management using fertigation requires:

1. Systems with high irrigation efficiency
2. Routine monitoring of all system components including pumps, injection devices, lines, filters, tanks, and emitters
3. A routine monitoring program of the fertigation process with particular emphasis on start-up and shut-down periods
4. Calibration and rechecking of injection rates and times frequently to ensure proper operation of the systems
5. Prevention of leaks, runoff, excess applications, and application to areas near surface water
6. Flushing of all system components with clean water following each use

During fertigation, it is recommended that the fertilizer solution not increase the EC of the irrigation water by more than 1.5 dS/m. Fertilizers can cause leaf burn, even if relatively low salinity irrigation water is used. Therefore, it is critical to know the specific conductance of the irrigation water prior to the introduction of fertilizers so the combined concentrations do not cause leaf burn. In addition, the use of fertilizer solutions could accumulate salts and nutrients into drainage systems and possibly affect downstream receiving water bodies. It is preferable to inject small dosages frequently than to make fewer applications at higher rates. Proper and legal backflow prevention devices must be used in the irrigation system to prevent fertilizers from being back-siphoned into the water supply.

26.4.6.3 Boom Applications

Boom-applied fertilizers can be placed directly over the root zone and offer the opportunity to apply other agrichemicals such as herbicides, insecticides, and soil fungicides at the same time. Liquid fertilizer itself can provide some burn down of weed foliage, thus complementing residual herbicides. However, care should be taken to avoid incompatibility of materials. Boom application equipment should be modified to withstand corrosive fertilizer materials.

26.4.6.4 Aerial Application

Aerial application of micronutrients and other sprays is an accepted practice but is not recommended for dry fertilizers. Precautions should be taken to prevent drift into surface water bodies and nontarget areas.

26.4.7 Equipment Calibration and Maintenance

Proper calibration and maintenance of fertilizer application equipment is essential to avoid misapplication of nutrients.

26.4.8 Apply Materials to Target Sites

Place nutrients within the root zone of individual trees or dripline bands along hedgerows of trees. Avoid placement in areas prone to off-site transport of nutrients, especially water furrows.

26.4.9 Avoid High-Risk Applications

Do not apply materials under "high-risk" situations, such as before forecasted rainfall. Avoid applications of nutrients during intense rainfall, on bare soils with extreme erosion potential, or when water tables are near the soil surface (Bottcher and Rhue 1984).

26.4.10 Fertilizer Storage

Use caution when storing fertilizer to prevent contamination of nearby ground and surface water. Fertilizer should be stored in an area that is protected from rainfall (Midwest Plan Service Midwest Plan 1991). Always store fertilizers separately from pesticides, solvents, gasoline, diesel, motor oil, or other petroleum products. Many fertilizers are oxidants and can accelerate a fire. Storage of dry bulk materials on a concrete or asphalt pad may be acceptable if the pad is adequately protected from rainfall and from water flowing across the pad. Secondary containment of stationary liquid fertilizer tanks should be used to prevent loss of fertilizer to the environment in case of a leak or spill.

26.4.11 Spilled Fertilizers

Immediately remove any fertilizer materials spilled on ground surfaces and apply at recommended rates to crops. When possible, place a tarp over ground surfaces where fertilizer transfer operations are conducted. Spilled materials should be transferred to the spreader for application to target sites. Spillage can contaminate open waters and thereby cause proliferation of aquatic weeds. Operators of fertilizer spreaders should be trained how to recover spilled materials for spreader application. Removal of some soil with the spilled materials is usually necessary and adequate for proper maintenance of this BMP. By its design, the spreader equipment will apply the fertilizer and soil to the target site. At fixed loading sites, the area can be cleaned by sweeping or vacuuming (or with a shovel or loader, if a large spill) or by washing down the loading area to a containment basin that is specifically designed to permit recovery and reuse of the wash water. Wash water generated should be collected and applied to the target site.

26.4.12 Use Caution When Loading Near Ditches, Canals, and Wells

Minimize the potential for spilled materials to pollute surface waters. When possible, locate mixing and loading activities away (according to local setback requirements) from groundwater wells, ditches, canals, and other areas where runoff may carry spilled fertilizer into surface water bodies. If such areas cannot be avoided, protect wells by properly casing and capping them and use berms to keep spills out of surface waters. Recover and apply spilled materials to intended zone of application. A concrete or asphalt pad with rainfall protection permits easy recovery of spilled material. If this is not feasible, loading at random locations in the field can prevent a buildup of nutrients in one location. In this case, place a tarp on the ground underneath the fertilizer hopper while loading. Do not load fertilizers on a pesticide chemical mixing center (CMC) because of the potential for cross-contamination. Fertilizers contaminated with pesticides may cause crop damage or generate hazardous wastes.

26.4.13 Alternate Loading Operation Sites

Use multiple fertilizer loading and transfer sites to prevent concentration of nutrients in a single area. If this is not feasible, loading at random locations in the field can prevent a buildup of nutrients in one location.

26.4.14 Use Backflow Prevention Devices

Use backflow prevention devices on irrigation and spray tank filling systems to preclude entry of nutrients into surface waters. Never leave a tank that is being filled unattended. Special precautions should be taken when filling tanks using a hose. Maintain an air gap between the filling hose and the liquid tank mixture. Never leave a tank unattended when it is being filled. An antisiphon device is a safety device used to prevent backflow of a mixture of water and chemicals into the water source or vice versa (Boman 2002). In the case of fertigation, the chemicals are fertilizers.

The possible dangers in fertigation include backflow of fertilizers to the water source causing contamination and water backflow into the fertilizer storage tank. Backflow prevention is an extremely important practice in the prevention of both ground and surface water contamination. Backflow to the storage tank can rupture the tank or cause overflow, contaminating the area around the tank and perhaps indirectly contaminating the water source. Safety equipment is available which, when properly used, will protect both the water supply and the purity of the fertilizer in the storage tank.

Any irrigation system designed or used for the application of chemicals should be equipped with the following components (Boman 2002):

1. A functional check valve located in the irrigation supply line between the irrigation pump and the point of injection of chemicals is required. The check valve will prevent water from flowing from a higher elevation or pressure in the irrigation system back into the well or surface water supply. Thus, water contaminated with chemicals cannot flow back into the water supply.
2. A low-pressure drain located on the bottom of the horizontal pipe between the check valve and the water source. It must be located so that the water flow does not drain back to the water source.
3. A vacuum breaker should be installed on the top of the horizontal pipe between the check valve and the irrigation pump and opposite to the low-pressure drain. The vacuum breaker will allow air to enter the pipe when pumping stops so that water flowing back to the pump will not create a suction, pulling additional water and chemicals from the irrigation system with it.

26.4.15 Split Applications Throughout Season

Dividing the annual fertilizer requirement into two or more applications can minimize leaching and help maintain the supply of nutrients over the long growing season. Frequent fertigations can be an efficient method of application for N and K while minimizing the potential for leaching of nutrients during excessive rainfall events (Boman and Obreza 2002). The trade-off between costs vs. fertilizer use efficiency and resource protection must be considered.

26.4.16 Erosion Control

Erosion control practices should be considered to minimize soil loss and runoff that can carry dissolved and attached nutrients on soil particles to surface waters. Vegetative filter strips are effective in reducing the levels of suspended solids and nutrients.

26.4.17 Irrigation Management

Irrigation should be limited to wetting only the root zone where possible. Excessive irrigation can transport nutrients below the root zone through leaching (Bottcher and Rhue 1984). Proper scheduling and uniform water distribution are necessary to assure control.

26.4.18 Use of Organic Materials

Increase the surface application (mulching) of organic materials like horticultural waste and urban plant debris (yard trimmings) when possible to help increase soil organic matter, retain nutrients and moisture, improve biological ecosystems, and supply slowly released nutrition. The surface application of slowly degraded organic waste materials like horticultural waste and urban plant debris can increase soil moisture retention and nutrient-holding capacity. The nutrient additive properties of organic matter support:

1. Economical ways to safely use nonhazardous wastes
2. Maintenance or increases in soil organic matter content
3. Protection of water quality
4. Protection of air quality
5. Reduction of energy used in manufacturing chemical fertilizer

Both microbial mineralization and immobilization can occur during decomposition of high-carbon, low-nitrogen organic materials like horticultural waste. Mineralization occurs when organic forms of a nutrient are converted to inorganic forms. Immobilization is the reverse of this process where microorganisms convert inorganic forms of nutrients to organic forms. The organic forms of the nutrients are not available to plants as they are bound in some part of the soil organic matter. Plants take up nutrients in inorganic forms. Thus, immobilization reduces nutrient (particularly nitrogen) availability, while mineralization increases nutrient availability. Nitrogen-poor organic materials like straw, fresh sawdust, and most fresh horticultural waste cause microorganisms to remove large amounts of inorganic nitrogen from the soil during decomposition, since that nitrogen is required to build new microbial cells. This process decreases nitrogen availability to citrus trees. However, the nitrogen consumed by the microorganisms will be slowly released when microbial cells decompose.

26.4.19 Well Protection

Prevent groundwater contamination by back plugging improperly constructed and/or deteriorated irrigation wells. This practice involves the protection of existing wells and prevention of problems in wells that are being planned. For existing wells, management activities should be aimed at reducing the potential for contamination. This includes evaluating and, if necessary, moving or modifying potential sources of pollution. Such sources of pollution may include fueling areas and/or areas where pesticides and fertilizer are handled or mixed. Antisiphon devices should be attached to all system discharge points so that backflow siphoning does not contaminate the aquifer. When no longer in use, proper decommissioning or plugging of a well prevents the reentry of surface water and transport of contaminants to the groundwater. Wells should be capped or fitted with valves that close tightly when not in use to reduce the potential for contamination (NRCS 2003). Artesian wells should be fitted with control valves so that water flow can be regulated or stopped when water is not needed.

26.4.20 Use Appropriate Sources and Formulations

Reduce the potential for nutrient leaching and off-site movement by choosing appropriate sources and formulations of fertilizer based on nutritional needs, season (rainy vs. dry), and anticipated weather conditions to achieve greatest efficiency and reduce potential for off-site transport. Utilize controlled-release and slow-release formulations when feasible. Nitrogen source materials are grouped into three categories: inorganic, synthetic organic, or natural organic. The inorganics and synthetic organics are usually high-analysis materials that are most economical to use in citrus orchards. These nutrient source materials are readily available to plants unless they have been formulated in a controlled-release form. Natural organic materials are less readily available and are usually lower in nutrient analysis (Obreza and Morgan 2008).

Following is a general description of the types of fertilizer source materials. You should keep in mind that opportunities for nutrient losses by runoff or leaching are greater for the more available (soluble) materials. Thus, dividing total applications limits the amount of readily available materials in the field at any one time. This reduction in the amount per application will limit the losses in runoff or by leaching. Proper selection of fertilizer source materials and formulations should take into account the season and expected rainfall and weather that may occur after application. Choose materials that will be less available for leaching and runoff.

26.4.20.1 Dry

These are usually bulk-blended into multianalysis (N, P, K plus micronutrients) blends for applications in citrus orchards. Uniform particle size is required to prevent segregation of mixtures. Most dry nutrient forms are water soluble and readily available for tree uptake. Dry fertilizers are applied with conventional application equipment.

26.4.20.2 Solutions

Solution fertilizers are usually free of solids and are made by dissolving readily soluble sources of plant nutrients in water. Nutrient sources used to manufacture solutions include ammonium nitrate, urea, potassium nitrate, potassium chloride, ammonium polyphosphate, and phosphoric acid. Solution fertilizers are suited for application through microsprinkler or drip irrigation systems. Caution should be taken when applying solutions containing phosphorus (P) through such systems. If the pH of the fertilizer-water mixture is not kept acidic, solution P can combine with dissolved calcium (Ca) in the irrigation water to form an insoluble precipitate, which can clog irrigation lines.

26.4.20.3 Suspensions

Suspension fertilizers are fluids in which the solids are held in suspension (prevented from settling) by use of a suspension agent, usually a swelling-type clay such as attapulgite or bentonite. Suspensions provide an excellent way to apply fertilizer-herbicide mixtures and small amounts of micronutrients uniformly. Mechanical agitation may be necessary to maintain a uniform suspension. Suspensions are usually applied with a specially designed herbicide-type boom in a band underneath the tree canopy.

26.4.20.4 Foliar N Sources

In areas where groundwater nitrate contamination exists or is seen as a potential problem, urea sprays can be used to provide a portion of the tree N requirements, especially during months when rainfall and potential leaching potential is greatest. However, tank mixing N sources with pesticides, oil, and other products should be approached with caution, as urea is known to be phytotoxic at higher rates, particularly in combination with oil.

26.4.20.5 Slow Release

These materials have limited water solubility which, during decomposition, release plant-available N (e.g., IBDU and ureaform). These fertilizers are grouped according to their mode of nutrient release in soil. For example, release of N from the ureaforms requires both dissolution of the fertilizer and microbial decomposition (nitrification), while release of N from IBDU involves slow dissolution only (in acidic soils). Processed waste products release N through microbial degradation.

26.4.20.6 Controlled Release

Controlled-release fertilizers contain one or more plant nutrients in a coated form that delays their availability for plant uptake after application or that extends their availability to the plant significantly longer than a readily available fertilizer, such as ammonium nitrate, urea, or potassium chloride. Controlled-release fertilizers are characterized by the rate at which nutrients are released into the soil solution, which may be related to soil temperature or water content. The rate at which nutrients are released through the coating is controlled by varying its thickness or chemical/physical characteristics during manufacture. Controlled-release fertilizers, while substantially more expensive than standard materials, are less exposed to leaching and lower salt index. Thus, their use may be appropriate for selected situations such as fertilization of newly planted trees or in environmentally sensitive areas. Many controlled-release fertilizer products have been recently developed, and many contain a complete N-P-K combination with small amounts of micronutrients. Growers are advised to consult the label of each product to determine its characteristics.

26.4.20.7 Biosolids

Biosolids are nutrient-rich, predominately organic materials. They are generated when solids accumulated during domestic sewage processing are treated further to meet regulatory requirements for land application. Biosolids can be a beneficial agricultural resource because they contain nutrients and organic matter that can improve soils and stimulate plant growth with negligible human health or environmental impacts. Soil accumulation of aluminum and other metals present in biosolids of low quality or from sources obtained from industrialized areas must be monitored if used in high quantities. Although biosolids contain many essential plant nutrients, nitrogen and phosphorus are in greatest concentration. Crops use nutrients from biosolids efficiently because they are released slowly throughout the growing season as biological decomposition occurs. Biosolids release N slowly, therefore is unlikely the N will leach. In the absence of local data, availability from biosolids can be assumed to be 50% of the total rate applied during the first 12 months after application. The amount of phosphorus (P) can be relatively high compared to the amount of nitrogen (N) when biosolids or nonpoultry animal waste is used as a fertilizer source. Physical characteristics of the site and soil, such as filter strips, soil organic matter, and clay or coated sand particles, all contribute to reducing the risk of P moving to nearby ground and/or surface water.

26.5 Conservation Buffers and Setbacks

Strategically incorporating vegetative buffers (either naturally occurring ones or planted forbs and grasses) into the citrus orchard design can help to protect water quality by providing biological filtration, increasing residence time and/or residual nutrient uptake. Managed properly, these vegetative areas or conservation buffers may provide pretreatment, formal treatment, and other treatment train opportunities. A treatment train effect is simply a combination of nonstructural and structural BMPs, which are generally effective for reducing or preventing nonpoint source pollution. Generally speaking, there are certain noncropped areas that could qualify as conservation buffers within a typical agroecosystem. Vegetated field borders, tree row middles, water furrows, ditch and ditch banks, wetlands/setback areas, and associated reservoir systems are examples. Depending on the orchard's surface water management system design, buffer areas can contribute significantly and help to manage off-site nutrient impacts. This whole farm management approach ultimately reduces a grower's risk of incurring negative environmental consequences. The BMPs discussed below are intended to give the reader information for the practical application of conservation buffers.

26.5.1 Pretreatment Options

Manage tree row middles by keeping them well grassed and by maintaining a minimum blade height of two inches. Growers should not rotary mow when standing water is present. Growers may also want to investigate the feasibility of incorporating leguminous plants within the middles, as these plants may be used as an additional source of nitrogen. Drainage ditches should also be managed to encourage grass cover in order to help reduce flow velocities, thus providing an opportunity for particulate matter to settle out.

26.5.2 Formal Treatment Options

A riparian buffer is an area of trees and/or shrubs located adjacent to and up-gradient from associated watercourses. Water sheet flow across this type of buffer will be treated before discharging to the watercourse. Air drainage is an important aspect of crop and tree damage during cold periods. Prior to implementing a riparian buffer, consideration should be given to its effects on air drainage.

26.5.3 Dedicated Conservation Buffers

Grassed waterways and/or filter strips are both excellent conservation buffer choices and can be used to convey and treat smaller volumes of discharge water with a moderate degree of success. In general, these passive treatment areas are more effective in removing phosphorus that is attached to soil particles rather than dissolved nitrogen. Orchards that have some topographic relief should consider using grassed waterways or filter strips to treat and discharge surface water runoff.

26.5.4 Treatment Train Effects

Consider using a combination of structural and nonstructural controls to mitigate the potential for off-site nutrient impacts, especially when discharging to sensitive downstream water bodies.

26.6 Water Resource Management

In many areas, citrus orchards need drainage facilities to protect trees from excess and/or high intensity. As a result, discharge volumes and the velocity of water discharged from structures within the watershed are increased compared with the natural condition. Under natural conditions, water from these areas would be cleaned by traveling downstream via tributaries before reaching coastal water bodies.

All types of land uses within the watershed contribute surface water runoff and pollutant loads to the receiving water bodies. Activities in orchards can affect the water resources of the watershed. Wherever feasible, citrus growers should consider implementing surface water management strategies that can provide additional storage and reduce the impacts associated with excessive stormwater discharges. These surface water management strategies can range from improved ditch maintenance and water table management to additional on-site canal storage or the construction of detention reservoirs for holding excess rainfall and tailwater recovery systems. It is important to conduct site-specific evaluations to determine if additional storage can be provided on-site and to plan long-term water management strategies that will minimize off-site discharges during periods of intense rainfall.

26.6.1 Water Table Management

In areas subject to shallow water tables, water table can be managed more efficiently by having sufficient hydraulic capacity in the ditch/canal system, using water control structures on culverts, laser land leveling where appropriate, constructing and maintaining a properly designed drainage system, and actively monitoring the water table. Effective water management requires monitoring the water table depth with enough precision to minimize pumping for irrigation and drainage. Knowledge of the water table depth is essential to ensure that adequate drainage can be provided (Obreza and Admire 1985). Since a significant portion of the tree water requirements can come from upward flux from the water table, water table monitoring should be an essential tool in irrigation management. Water table manipulation, and associated supplemental irrigation reductions, can also assist in salinity management by reducing the use of low-quality groundwater. A water table observation well allows the water level to rise and fall inside it with changes in the groundwater level. The changing water level moves an inserted float with an attached measuring rod. The float assembly provides a quick visual indication of water table depth. With more precise inspections of the observation well, the information required for more detailed water management can be obtained.

26.6.2 Scheduling Irrigation and Drainage

Drainage and irrigation schedules should focus on optimal crop production and promotion of deep rooting by maintaining a constant water table that minimizes water quantity and quality impacts. During intense rainfall periods when drainage rates are insufficient to prevent upward fluctuations of the water table, root damage can occur. Therefore, irrigation and drainage practices should be focused on maintaining a well-defined root zone that can be managed during both drought and wet periods. High-salinity irrigation water can cause an adjustment to the above-mentioned scheduling.

26.6.3 Irrigation

Citrus evapotranspiration (ET) is largely determined by climatic factors. Water requirements vary with soil type, climate, ground cover, cultivation practices, weed control, tree size and age, scion, rootstock, and tree health. The optimum interval between irrigation events depends on the design of the irrigation system, ET, time of year, and soil characteristics as well as tree size and condition. Evapotranspiration rates in citrus orchards are highly dependent on climatic conditions. Higher evaporation and transpiration rates occur on clear, sunny days with low humidity and high winds compared with cool, damp, overcast winter days. Therefore, irrigation frequency must be adjusted through the year. The use of ET-based irrigation scheduling and soil moisture monitoring combined with properly designed irrigations that utilize automated controls can result in very efficient irrigation management. Automated controls range from simple timers to sophisticated systems that make irrigation decisions based on soil moisture and calculated ET for specific conditions in individual zones. These systems can normally be readily equipped with rain sensors that curtail irrigations when rainfall begins. In general, automated systems can reduce the manpower required to irrigate large acreages, but they need to be periodically updated and calibrated to ensure correct irrigation schedules.

26.6.4 Drainage

Effective water management, which includes both irrigation and drainage, is essential for profitable citrus production on poorly drained soils. Drainage systems in orchards may contain some or all of the following components: canals, retention/detention areas, open ditches, subsurface drains, beds, water furrows, swales, and pumps required to move surface water (Boman et al. 2002). These systems require continuous maintenance to minimize the chances of root damage from prolonged exposure to waterlogged soils caused by poor drainage, high-intensity rains, and high water tables.

Drainage systems should be designed to allow water table drawdown of 10–20 cm/day, a rate that should be sufficient to prevent root damage (Ford et al. 1985). During the cooler months, citrus trees can tolerate flooded conditions for much longer periods than in the summer. Water table observation wells (WTOW) are good tools for observing soil-water dynamics. They are a reliable method for evaluating water-saturated zones in sites subject to chronic flooding injury. These wells can also be used to measure the rate of water table drawdown, which is the key to discovering how long roots can tolerate flooding. Observation wells constructed with measuring rods allow water tables to be visually observed while driving by the well site.

Short-term estimates of flooding stress can be obtained by digging into the soil and smelling soil and root samples. Sour odors indicate an oxygen-deficient environment. The presence of hydrogen sulfide (a rotten egg odor) is an indication that feeder roots are dying. Anaerobic bacteria (bacteria that live in the absence of oxygen) will develop rapidly in flooded soils and contribute to the destruction of citrus roots. In a field survey of poorly drained orchards, anaerobic sulfate-reducing bacteria in more than half of the locations formed toxic sulfides. Nitrites, formed by nitrate-reducing bacteria, and organic acids that are toxic to roots were also found in these flooded soils. Good drainage allows air to move into the soil and prevents oxygen-deprived conditions.

Citrus tree stress caused by flooding is usually less when soil water is moving than when it is stagnant. A higher subsoil pH may help to delay, for a few days at least, the death of citrus roots under flooded conditions. With experience, flooding injury can be diagnosed during periods when groundwater levels are high. Even before there are visible tree symptoms, digging into the root zone may give an estimate of future tree response. Indications of problems include high water tables with saturated soils in the root zones, sloughing roots, and sour odors in the soil.

When the water table recedes, visible damage to the trees may become more obvious. New feeder roots appear and grow rapidly on trees that have survived and received adequate irrigation. Symptoms of damage may not occur for several weeks, depending on the severity of root damage and the frequency of rainfall following the damage. Symptoms usually start to appear after the water table drops and the soil dries out. Symptoms of root damage include leaf yellowing, chlorosis, wilting, fruit drop, leaf drop, and twig dieback (Boman et al. 1995).

Often root damage is so severe that trees may go into a wilt even though water furrows are still wet. Because the root system was severely damaged by the flooding, the full extent of damage may not be known for several months or until drought conditions occur. Young trees are often more sensitive to flooding and may develop symptoms resembling winter chlorosis. More subtle symptoms include reduced growth and sparse foliage that can occur at locations only a few inches lower in elevation than the surrounding area.

Harvesting operations in an orchard after recent flooding may also further damage surface roots that have been injured by the flooding. Hot, dry conditions following flooding will hasten the onset of stress and symptom expression. The reduced root system resulting from flooding is incapable of supporting the existing tree canopy. When this occurs, irrigation management becomes critical. Irrigation must provide moisture to a depleted (shallow) root system. Excessive water could compound existing problems. If root system damage is extensive and tree canopy condition continues to deteriorate with permanent wilt and foliage dieback, some degree of canopy pruning may be necessary to reestablish a satisfactory shoot/root balance. Light, frequent irrigations will be required until the root zone has become reestablished. Subsurface moisture should be maintained to promote root growth into the lower root zone. If root damage is severe, frequent irrigation may be required throughout the winter months, especially if dry winds persist.

26.6.5 Irrigation System Evaluation

An evaluation of an irrigation system should incorporate site-specific data about the soil, crop, and irrigation system to identify problems with system design and operation. System improvements to increase uniformity and efficient scheduling can help growers conserve significant amounts of irrigation water while still providing the water required to meet crop needs. Increased irrigation efficiency not only saves water but also reduces the potential for leaching of nutrients and agricultural chemicals. Such leaching may lead to groundwater and/or surface water contamination.

A typical microirrigation system evaluation begins with collecting information on the system type, design and specifications, tree spacing, canopy size, root depth, wetted area, and soils (Nakayama et al. 2006). The irrigation system is then started and operated normally. Flow rates and pressures at the pump station and at the ends of the manifolds are measured. A representative portion of the operating zone fed by one manifold is chosen for intensive study. Flow rates of emitters are measured at 16 locations in a grid pattern throughout the chosen area, and pressures are measured at the inlet and closed ends of the laterals. Observations on leaks, clogged emitters, and malfunctioning field valves are also noted. Calculations of emission uniformity, variation in manifold pressures, average emitter discharge rate, water application per irrigation and per tree, average wetted area, and the percent of cropped area are made.

Water samples should be taken to determine pH, total dissolved solids, chlorides, sulfides, iron, calcium, magnesium, total hardness, and total alkalinity. Based on the collected data and the calculated values, the recommendations and an irrigation schedule that will help operate the system effectively can be developed.

26.6.6 Irrigation System Maintenance and Evaluation

Irrigation maintenance and evaluation is a management plan designed to maintain irrigation system components in good condition, so that the entire system can perform according to manufacturer's specifications. The uniformity of water application and efficiency in water delivery of an irrigation system tends to decrease over time because of aging, weathering, and component breakdown unless proper maintenance is done on the system. The goal of irrigation system maintenance is to maintain and maximize system performance.

Maintenance programs vary according to the type of irrigation system: For example, maintenance of flood or seepage systems may be limited to a preseason operational check of pump stations and ditches. Likewise, maintenance programs for pressurized pipe irrigation systems generally involve filtration, chlorination/acidification, flushing, repair or replacement of clogged emitters, and observation. Irrigation systems that are managed efficiently help ensure citrus tree uniformity, conserve water, and reduce operation and maintenance costs.

26.6.7 General Irrigation Maintenance

1. Determine and record the design operating values. Know the level of irrigation system efficiency defined as the ratio of the volume of water delivered by an irrigation

system and available to the crop, compared to the volume of water that this system uses from a source (e.g., well, pond, etc.).

2. Periodically check the uniformity, defined as the degree to which an irrigation system can apply equal amounts of water throughout different locations in a field. Improper maintenance can decrease system uniformity.
3. Establish a documented maintenance schedule that should include inspection of the mechanical components, irrigation lines, as well as monitoring of the pump and power unit. This can be done by keeping written records of performance and maintenance actions.
4. Use flow meters and pressure gages to determine existing operating parameters and to properly manage the irrigation system.
5. Test water quality at least once a year, since changes in water quality parameters affect maintenance requirements and frequency.
6. Clean and maintain filtration equipment, so that it operates at optimum pressure ranges.
7. Record the flow rate, pressure delivered by the pump, and energy consumption of the power unit frequently.
8. Regularly check control valves, pressure regulators, leaks on seals and fittings, and pressure relief valves for proper operation. Lubricate as necessary.

Microirrigation systems are an efficient tool for grower utilization to timely deliver water efficiently to the crop. When properly designed, operated and maintained precise amounts of water, nutrients, and other materials are delivered to the target site. If the microirrigation systems are improperly maintained, problems can occur due to clogged filters and microsprinklers, malfunctioning pressure regulators, and damaged lateral lines or connectors. Uniformity of water application is affected by system maintenance and operational pressure. When systems are properly maintained by flushing lines and filters, unclogging emitters and in some cases treating the water to minimize clogging, irrigation efficiency and uniformity is improved, thereby conserving water and improving efficiency which enhance yields. Consider the following microirrigation procedures:

1. Use proper filters with at least 200 mesh (74 micron) or equivalent.
2. Check system prior to injection of fertilizers or chemical to assure even distribution of materials within the orchard.
3. Flush lines regularly to minimize sediment buildup.
4. Flush all fertilizers from the lateral lines prior to shutting the irrigation system down when fertigating to avoid clogging problems.
5. If required, chemically treat water to prevent emitter plugging due to microbial growth and/or mineral precipitation, inject chemicals prior to filtration.
6. Adjust pH of irrigation water to reduce potential for chemical precipitation.

Fig. 26.3 Multisensor capacitance probe installed in access tube in citrus orchard

7. If water is from surface sources, treat algae blooms in the pond to minimize clogging.
8. Flush all fertilizer (when fertigating) prior to shutting the irrigation system down. Flushing can be accomplished manually by opening the end of each lateral or by using automatic flushing endcaps on the laterals.

26.6.8 Monitor Soil Moisture

Use tensiometers, capacitance sensors, or other soil water monitoring devices along with water table observation wells for irrigation and drainage management to avoid excess soil moisture depletion and minimize water volume requirements during irrigation cycles. Good irrigation management requires that the status of soil water be accurately evaluated. There are direct and indirect methods to measure soil water content and several alternative ways to express it quantitatively. There is no universally recognized standard method of measurement and no uniform way to compute and present the results. The most commonly used sensors include tensiometers, TDR probes, and capacitance sensors (Fig. 26.3). Whatever sensor used, the sensor needs to be located in a place that represents the soils, rooting patterns, and irrigation coverage that exists in the orchard (Boman et al. 1999).

Since many orchards have several different soil types, more sensors per block mean a better characterization of the orchard soil water status. Sensors should be placed in the root zone where water uptake is most rapid. Since citrus roots are more abundant near the soil surface, and because upper soil layers usually dry out sooner than the deeper soil layers, placement in the top half of the root zone is ideal.

Capacitance sensors provide a near instantaneous measure of volumetric soil water content. Two types of devices are available: One is a portable sensor that is lowered into the soil through an access tube, and the other is more permanently installed with numerous sensors connected to a data logger. An advantage of capacitance-type sensors is their high accuracy, the availability of excellent software for interpreting data, and the speed data can be formatted and delivered. The information can be available in a matter of minutes on the Internet transmitted via radio waves from a probe, which expedites a grower's decision-making process on operating and irrigation system.

26.6.9 Drainage System Maintenance

Maintain a consistent bottom slope on water furrows between beds to achieve uniform drainage. Avoid rutting and sloughing of swales and water furrow areas. Laser- or RTK-GPS-guided systems on maintenance equipment can be very effective in producing uniform slopes in water furrows. Where possible, maintain vegetation management programs that minimize soil movement in the event of heavy rains by keeping a grass or vegetation cover on the soil surface in between tree rows.

26.6.10 Discharge Structures

Structures and/or pumps that regulate off-site water discharge should be adequately designed, constructed, and maintained so that target water table levels within the orchard can be achieved. If safety or operational concerns prevent structures from being adjusted to regulate discharges during storm drainage events, they should be rehabilitated or replaced (e.g., modifying riser-board structures to allow easier water level control).

26.6.11 Detention, Tailwater Recovery, and Surface Water Use

Where possible, on-site detention should be considered to reduce both the rate and volume of off-site discharges following heavy rains. Detention areas allow all or a portion of the drainage water to be temporarily stored on-site. The excess water can be stored for tailwater recovery or released later at low flow rates. Tailwater recovery is the practice of collecting, storing, and reusing water for irrigation. A tailwater recovery system typically consists of collection and storage components (ditches, ponds) and also delivery components (pump stations, pipes). Tailwater recovery and/or surface water reservoir systems may require additional filtering and purification infrastructure, depending on the quality of the water and the type of irrigation system being used. Off-site seepage from the tailwater recovery system should be controlled and managed properly, especially if the system is expected to receive chemical-laden waters. Control may be in the form of dike compaction, natural-soil liners, soil additives, commercial liners, drain tile, or other approved methods.

26.7 Erosion Control and Sediment Management

Sediments or suspended solids are recognized forms of water pollution and often result in the loss of ditch or canal capacity. Unlike many chemical pollutants, sediment is a natural component of water bodies and the resources they support. Excessive amounts of suspended solids or sediments are often a product of erosion from unstabilized or disturbed land areas. These solids originate from four primary sources:

- Soil particles eroded into ditches
- Soil particles eroded from ditches
- Plant material washed into the ditches
- Plant and biological material growing within ditches and canals

Excessive sediments deposited on stream bottoms and suspended in the water column can harm fish spawning and impair fish food sources, reduce habitat complexity, potentially harm public water supply sources, and reduce water clarity. Reduction in water clarity can harm natural resources, such as sea grasses and oysters, in the receiving estuary. Removal of natural vegetation and topsoil increases the potential for soil erosion, which can change runoff characteristics and result in loss of soil and increased turbidity and sedimentation in water bodies. Other site characteristics such as clay-type soils and/or sloped terrain can significantly increase the risk of erosion and off-site sediment transport. The end result is that runoff containing sediments with *sorbed* nutrients and pesticides can negatively affect surface waters or groundwater.

Erosion control begins with limiting the loss of soil from pastures. The primary means of doing this is to minimize the amount of land that is cleared of vegetation. When clearing vegetation to develop new pastures, revegetation with forage should occur as quickly as possible. Additionally, all land-clearing activities should be planned and conducted during the dry season, whenever possible. Examples of other erosion control BMPs are critical area planting, pasture management, filter strips, and the use of silt screens, where appropriate. Controlling sediment transport involves the use of BMPs to limit the movement of sediments downstream. Examples of these types of BMPs include sediment traps and diversions/terraces. Collectively, these practices will reduce the mass load of sediment reaching a water body, which will improve water quality.

Common erosion control devices include filter strips, silt screens, and sediment traps. These BMPs should be employed progressively by beginning with the more passive erosion control devices first and subsequently employing more aggressive measures as the need arises (e.g., sediment basins to capture sediment-laden water and allow enough time for larger particles to settle out). In addition to potential downstream water quality impacts, the buildup of silts and sediments in the ditches on the farm reduces ditch cross section. This reduction in cross-sectional area results in higher water velocities, as compared to an unfilled ditch or canal. This higher water velocity (compared to unfilled ditches/canals) may induce greater amounts of erosion of fine and coarse particles from ditch and canal banks.

The presence of shoals and sandbars is a good indicator of soil losses. Field erosion also results in site degradation resulting in increased costs for ditch cleaning and reshaping of beds and furrows. In order to minimize effects of sediment transport in surface water, efforts should focus on keeping soils in the orchard and along canal and ditch banks. The following sections describe BMPs that are applicable for water conveyances within citrus orchards. The selection and implementation of particular BMPs must be based upon site-specific circumstances and management styles.

26.7.1 Riser-Board Water Control Structures

Water discharge structures are used to control water table levels and surface water levels in drainage ditches within citrus orchards. The type of structure selected can significantly influence the quality of water discharges (Wilson et al. 2002).

With riser-board control structures (Fig. 26.4), water is forced to flow over the top of the boards. This flow path creates a low current area toward the bottom of the structure, which facilitates the deposition of sediments and their accompanying nutrients or pesticides, essentially removing them from the discharges. Conversely, screw-gates structures do not create this dead-current zone. Since they open form the bottom, sediments and their accompanying load are swept out along with the discharge water.

26.7.2 Sediment Settling Basins in Ditches

Create and maintain localized settling basins (sumps) to trap sediments at field ditch connections to lateral canals, at lateral and collector ditch connections, and prior to water discharge points from the farm. Successful sediment traps require site-specific designs, with the following points to consider:

Fig. 26.4 Accumulation of aquatic weeds at a riser-board control structure. Sediments generated by the decay of these plants will settle to the bottom of the ditch and need to be periodically removed

1. Determine runoff volume and intensity.
2. Determine transport and settling rates for sediments of concern.
3. Size traps to allow adequate residence time for natural settling to occur include considerations for allowable storage (fill-up) of trapped sediments.
4. Make provisions for materials removed from the ditches.
5. Maintenance access to settling basin area should be provided.
6. When sediments are removed, materials need to be placed in a manner that prevents material from sloughing back into the waterway.
7. Sediment excavation and removal should be conducted during low stage conditions or during the dry season. This will reduce the likelihood of increasing turbidity and suspended solid loads.

Settling basins or settling ponds are a quick and simple way to remove sediments out of runoff water. Settling basins simply slow down the water, allowing sediments to settle out of the water before the water returns to the receiving water body.

26.7.3 Ditch Construction

Ditches and canals should be constructed with side slopes consistent with soil types. Table 26.1 gives recommended side slopes for various soil types (Schwab et al. 1966). Stabilize bare soils and canal or ditch banks by encouraging coverage by noninvasive vegetation. Vegetation types selected should be adapted to orchard conditions and should provide

Table 26.1 Side slopes for open channels

Soil	Side slopes Shallow channels (up to 1.2 m)	Deep channels (over 1.2 m)
Peat and muck	Vertical	0.25:1
Heavy clay	0.5:1	1:1
Clay or silt loam	1:1	1.5:1
Sandy loam	1.5:1	2:1
Loose sandy	2:1	3:1

maximum stabilization by roots and foliage. Vegetative buffer strips can also serve to reduce the erosion of soil particles. Whenever practical, plant or encourage establishment of native species.

26.7.4 Ditch Bank Vegetation Maintenance

Broadleaf weed control using herbicides or maintenance mowing of slopes and ditch banks increases grass cover and decreases the proliferation of shade-producing shrubs and weeds, thus reducing erosion from wind and rainfall. Points to consider include:

- Mechanical mowing does not uproot vegetation and expose soil.
- The use of herbicides should be conducted with caution and precision to avoid excessively large areas of bare soil.
- Selective herbicides should be used in order to maintain desired vegetation (e.g., remove broadleaf vegetation while maintaining grasses).

26.7.5 Protect Ditch Banks

Protect canal and ditch banks from erosion in areas subject to high water velocities. In areas where water is constricted (usually at discharge points) or at ditch intersections where velocities are high, rip-rap, concrete, headwalls, or other materials that buffer turbulence should be used to protect ditch banks and reduce sediment transport.

26.7.6 Aquatic Plant Management

When removing vegetation from ditch bottoms, avoid disrupting side slopes. If a backhoe without a vented bucket is used to remove aquatic plants from ditches, special precautions should be taken to prevent washouts. Once a bucketful of vegetation is picked up, the bucket should be raised to allow most of the water to drain out over the deeper part of the ditch. The boom should be swung far enough over the ditch bank so that when the vegetation is dumped, remaining water will flow away from the ditch.

26.7.7 Ditch Maintenance Cleaning and Dredging

Develop and implement a systematic management plan for removing sediments from canals and farm ditches on a regular basis. Maintenance dredging of existing ditches, canals, and intake and discharge structures should include the following:

- Spoil material should be removed and deposited on a self-contained, upland spoil site and not placed in a delineated floodplain. This will prevent the movement of the water and excavated spoil material into wetlands or other surface waters.
- Do not remove any more material than is necessary to restore the original design specifications or configurations.
- No significant impacts should occur to previously undisturbed natural areas.
- Erosion and sedimentation control devices (e.g., turbidity screens) should be used to prevent bank erosion, scouring, and to prevent turbidity from discharging into adjacent waters during maintenance dredging.

Removal of excess sediment to the originally designed and constructed cross-sectional area generally increases the canal cross-sectional area and reduces water velocities (compared to same water volume in filled-in systems), thus reducing the potential for bank scouring. Caution should be considered as ditch maintenance, cleaning, and dredging beyond the originally designed and constructed cross-sectional area may result in upstream and/or downstream adverse water resource impacts.

26.7.8 Construction and Temporary Erosion Control Measures

Special measures and/or temporary erosion control measures should be taken during construction and renovation of orchards and facilities, when culverts and control structures are replaced or repaired and when there is a major disruption of established vegetation such as during irrigation system installation or when buried water lines are repaired. Erosion control measures are used to minimize sediment transport and protect the quality of water bodies that receive runoff from disturbed areas. The most common temporary erosion control tools include straw or hay bale barriers, silt screens, and silt fences; however, more permanent control can be obtained through the use of specialized blankets and mats, gabions, and other systems used for soil stabilization. The cost of erosion control options are highly variable, and agricultural producers are encouraged to consider economics and site-specific conditions when selecting the most appropriate erosion control system for a particular action.

26.8 Future Research

1. Controlled-release fertilizers that are economic and have correct release pattern for crop, soil, rainfall patterns, and temperatures
2. Diagnostic tools to assess plant and soil nutrient status quickly and accurately
3. Economic production systems that can capture nutrients that are leached or runoff during storm events
4. Soil amendments to help retain water and nutrients yet allow for adequate drainage to preserve plant health
5. Economical equipment/techniques to apply pesticides and fertilizers at optimum timing, rate, and location
6. Advanced irrigation management tools
7. Cost share programs to assist growers in implementing new technologies
8. Advanced orchard design that achieves economical returns while minimizing energy and agrichemical inputs and reduced off-site environmental impacts
9. Education programs for all agricultural sectors (regulators, growers, manufacturers, sales, service, dealers, etc.) on irrigation, drainage, nutrition programs, and farming system effects on water quality

References

Boman, BJ (ed) (2002) Water and Florida citrus: Use, regulation, irrigation, systems, and management. University of Florida Pub SP-281, Gainesville, p 603

Boman BJ, Obreza TA (2002) Fertigation nutrient sources and application considerations for citrus. University of Florida, Institute of Food and Agricultural Sciences, Cooperative Extension Service Circular 1410, p 15. UF/IFAS EDIS Website: http://edis.ifas.ufl.edu/CH185

Boman BJ, Parsons LR, Tucker DPH et al (1995) Assessing damage from flooding in citrus groves. Citrus Ind 76(12):28–30

Boman BJ, Levy Y, Parsons L (1999) Water management. In: Timmer LW, Duncan LW (eds) Citrus health management. APS Press, Inc, St. Paul, pp 72–81

Boman BJ, Wilson PC, Hebb JW (eds) (2000) Water quality/quantity BMPs for Indian River area citrus groves. Florida Department of Environmental Protection, Tallahassee, p 167

Boman BJ, Wilson PC, Jennings M et al (2002) Detention/retention for citrus stormwater management. University of Florida, Institute of Food and Agricultural Sciences, Cooperative Extension Service Circular 1405, p 19. UF/IFAS EDIS Web site: http://edis.ifas.ufl.edu/AE216

Boman BJ, Gunter DE, Futch SH (eds) (2004) Best management practices for citrus groves in the Peace river and Manasota basins. Florida Department of Agriculture and Consumer Services, Tallahassee, Pub No. 5 M-5.005.10.04. p 128

Boman BJ, Rouse RE, Shukla S et al (eds) (2005) Best management practices for Gulf citrus. Florida Department of Agriculture and Consumer Services, Tallahassee, Pub No. 5 M-7.005.11.05, p 116

Bottcher D, Rhue D (1984) Fertilizer management – key to a sound water quality program. University of Florida, Institute of Food and Agricultural Sciences, Cooperative Extension Service, Pub No. SP-28

Cromwell RP (1992) The 1/128th of an acre calibration method. University of Florida, Institute of Food and Agricultural Sciences, Cooperative Extension Fact Sheet AE-5

Cromwell RP (1993) Agricultural chemical drift and its control. University of Florida, Institute of Food and Agricultural Sciences, Cooperative Extension Circular 1105. UF/IFAS EDIS Web site: http://edis.ifas.ufl.edu/AE043

Dean TW, Bucklin RA (1997) Building plans and management practices for a permanently-sited pesticide storage facility in Florida. University of Florida, Institute of Food and Agricultural Sciences, Cooperative Extension, Pub No. SM-57

FDACS (2002) Nitrogen best management practices for ridge citrus. Florida Department of Agriculture and Consumer Services, Tallahassee, p13. http://www.floridaagwaterpolicy.com/BestManagementPractices.html

Ford HW, Beville CB, Carlisle VW (1985) A guide for plastic tile drainage in Florida citrus groves. University of Florida, Institute of Food and Agricultural Sciences, Cooperative Extension Service Circular 661, p 8

Howard D, Shahane A, Thomas M (1998) Best management practices for agrichemical handling and farm equipment maintenance. Florida Department of Agriculture and Consumer Services/Florida Department of Environmental Protection

McCoy CW, Nigg HN, Timmer LW et al (2011) Florida citrus integrated pest management guide: the use of pesticides in citrus IPM. University of Florida, Institute of Food and Agricultural Sciences, Cooperative Extension Service Fact Sheet CPMG-03. UF/IFAS EDIS Web site: http://edis.ifas.ufl.edu/CG035

Midwest Plan Service (1991) Designing Facilities for pesticide and fertilizer containment. MWPS-37. MidWest Plan Service, 122 Davidson Hall, Iowa State University, Ames, IA 50011. http://www.mwpshq.org/catalog.html

Nakayama FS, Boman BJ, Pitts DJ (2006) Maintenance. In: Lamm FR, Ayars JE, Nakayama FS (eds) Microirrigation for crop production: design, operation, and management. Elsevier, Amsterdam, pp 389–430

Nesheim ON (1998) Interpreting PPE statements on pesticide labels. University of Florida, Institute of Food and Agricultural Sciences, Cooperative Extension Service Pub PL-16

NRCS (2003) Well decommissioning. USDA-NRCS Conservation Practice Code 351. p 3. http://efotg.sc.egov.usda.gov//references/public/IN/well_decommissioning.pdf

Obreza TA, Admire KE (1985) Shallow water fluctuations in response to rainfall, irrigations, and evapotranspiration in flatwoods citrus. Proc Fla State Hortic Soc 98:32–37

Obreza TA, Morgan KT (eds) (2008) Nutrition of Florida citrus trees, 2nd edn. University of Florida, Institute of Food and Agricultural Sciences Pub SL253, Gainesville, p 96

Parsons L, Boman B (2006) Best management practices for Florida citrus. HortTechnol 16(3):389–393

Schwab GO, Frevert RK, Edminster TW et al (1966) Soil and water conservation engineering. Wiley, New York, p 683

Stover, E., J. Salvatore, and P. McClure (2002a) Influence of spray volume, ground speed, and sensor actuated spray manifolds on pest control in fresh grapefruit. Proc Fla State Hortic Soc 115:1–5

Stover E, Hebb J, Motis T (2002b) Pesticide mix-load and sprayer wash-down systems for Florida citrus. University of Florida, Institute of Food and Agricultural Sciences, Cooperative Extension Service Fact Sheet HS-139. UF/IFAS EDIS Web site: http://edis.ifas.ufl.edu/HS139

Stover E, Salvatore J, Wirth F (2003a) Pesticide spray reduction from use of a sensor-actuated spray system in Indian River grapefruit. HortTechnol 13:178–181

Stover E, Scotto D, Wilson C et al (2003b) Pesticide spraying in Indian River grapefruit: II. Overview of factors influencing spray efficacy and off-target deposition. HortTechnol 13:166–177

Wilson PC, Scotto L, Boman BJ et al (2002) Flatwoods citrus best management practice: riser-board structures. University of Florida, Institute of Food and Agricultural Sciences, Cooperative Extension Service Pub SS-409, p 5. UF/IFAS EDIS Web site: http://edis.ifas.ufl.edu/SS409

Economic Analysis of Conventional and Alternative Nutrient Management Approaches

27

Stephen E. Gareau

Abstract

During times of economy uncertainty, such as the current period, all costs of agricultural production become important and worthy of close scrutiny if the threat of farm foreclosures is to be minimized. This concern particularly applies to the cost of plant nutrients, which, under conventional approaches, can typically represent 24–30% (or more) of the total variable cost of production. The purpose of this chapter is to examine the physical effects, economics, profitability, and sustainability of various plant nutrient management strategies—both conventional and alternative one—in an attempt to identify those strategies that can lead to resource optimization, maximization of profits for farm enterprises, and long-term farm sustainability and survival. The results of this analysis include the following: conventional nutrient management systems, using commercial synthetic fertilizers, can show higher profit for most grain crops (with the exception of corn and sorghum) than organic nutrient management alternative nutrient management strategies for increasing farm profitability. A cover crop system can produce higher yield, higher gross margin, and lower crop yield variation, when compared to no-tillage conventional, manure-based, and crownvetch systems. Manure-based systems that do not require purchase or transport of the manure (as in combined animal and crop production systems) can be considerably more profitable than conventional systems. Both manure-based and cover crop systems that do not include the use of commercial fertilizers (i.e. organic systems) hold particular promise due to the output price premiums typically garnered by the organic crops grown under such conditions. Under dry soil conditions, manure-based systems can provide higher levels of soil organic matter and stimulate the growth of soil organisms that are beneficial to plant nutrient uptake and crop yield.

Keywords

Alternative farming • Animal manure • Commercial fertilizer • Chemical fertilizer • Conventional farming • Cover crop • Economic productivity • Effectiveness • Farm enterprise • Green manure • Nutrient management • Organic farming • Profitability • Profit maximization • Sustainability • Viability

27.1 Introduction

Economics is the study of the optimal use of resources to maximize the welfare of people (Zering 2000). Farmers' economic decision-making can, for the most part, be characterized as profit maximization. It can generally be assumed that the quantities of commodities that farmers produce, and

S.E. Gareau (✉)
Educational Technology Program, Department of Computer Information Systems, State University of New York College at Buffalo, Chase Hall 231, 1300 Elmwood Avenue, Buffalo, NY 14222, USA
e-mail: sgareauca@yahoo.com

A.K. Srivastava (ed.), *Advances in Citrus Nutrition*,
DOI 10.1007/978-94-007-4171-3_27,

the quantities of inputs that they use in production, are selected based on expected profit. Internationally, the past decade has been harsh economically. For instance, according to the National Bureau of Economic Research, the US economy entered a recession in March of 2001 (Walden 2002). The situation was repeated in 2009, as the US economy and other economies around the world entered another recession. US President Barack Obama summed up the economic situation in the USA in 2010 by saying, "Our economy is in a deep recession that threatens to be deeper and longer than any since the Great Depression" (US Office of Management and Budget 2010, p. 1).

There have been periodic signs of improvement, but they have been slow, sporadic, and uncertain. For instance, Hamilton (2010) described the US economic conditions in May of 2010 as being "in the recovery phase, but it remains disappointingly sluggish." In times of economic uncertainty and decline, such as the current period, farm profitability becomes even more critical if the threat of farm foreclosures is to be minimized and the historical trends of both a decreasing number of farms and a decreasing number of farm owners are to be halted or reversed (Havlin et al. 1999). During such times, all costs of agricultural production become important and worthy of close scrutiny but particularly the cost of plant nutrients which, under conventional approaches, typically represents 24–30% (or more) of the total variable costs of production (Lu et al. 2000; Bullen and Brown 2001). The purpose of this chapter is to examine the economics and profitability of various plant nutrient management strategies—both conventional and alternative ones—in an attempt to identify those strategies that can lead to financial resource optimization and, ultimately, maximum profits for farm enterprises.

27.2 Rationale

There are a number of important reasons why a better understanding of sustainable crop fertilization practices is important. For instance, according to Miao et al. (2011), in many parts of the world (including China), there is both an overreliance on commercial, synthetic, and chemical fertilizers and an overapplication of nitrogen (N), which leads to a host of problems, including increased surface and groundwater pollution, increased greenhouse emissions, inefficient N-uptake efficiency, and increased soil acidification. Miao et al. (2011) concluded that "current nutrient management practices are not sustainable and more efficient management systems need to be developed." Singh et al. (1998) supported these conclusions, pointing out that the use of commercial synthetic fertilizers can, over time, lead to declining crop yields and deteriorating soil conditions. Takkar (2006) pointed out the use of synthetic fertilizer alone on degraded soil can lead to a deficiency of and/or toxicity by Ca, Mg, S, and/or micronutrients and can lower soil pH.

Brown (2011) pointed out that farmers everywhere are now facing challenges that we have seldom experienced in the past—all of which emphasize a need for more efficient crop nutrient management strategies. These include (1) rising temperatures (caused by global climate change), which are affecting crop yields; (2) spreading water shortages on irrigated lands in many agriculture-oriented countries of the world, largely caused by overpumping of nonrenewable fossil aquifers; and (3) the maximization of crop yields (i.e., yield per acre) in some of the more advanced countries (such as France, Germany, Britain), with no new technologies on the horizon that might be able to further increase crop yields. In regard to farm sustainability, small- and medium-sized farms currently face a range of serious challenges—the number one challenge being economic survival. For instance, as the US Department of Agriculture (1998) pointed out some years ago, "today [in the U.S.], we have 300,000 fewer farmers than in 1979, and farmers are receiving 13% less for every consumer dollar." According to the USDA (2007), between 2005 and 2006, the USA lost 8,900 farms, which equates to a little more than one farm per hour.

A good part of the problem is the increasingly oligopolistic nature of farming nowadays. As Hoppe et al. (2001) pointed out, "although most U.S. farms are classified as small, agricultural production [and profits] are highly concentrated in large and very large farms. These two groups together made up 8% of all farms in 1998, but accounted for 53% of the total production of agricultural products." Proper nutrient management can be an important contributing factor to the profitability of any farm operation. Havlin et al. (1999) pointed out, "higher crop yields represent the greatest opportunity for reducing per-unit production costs" and, by extension, farm profitability.

Nutrient management also affects farm profitability in terms of the cost of inputs to production. For instance, the cost of fertilizer can typically represent 24–30% (or more) of the total variable costs of production (Lu et al. 2000; Bullen and Brown 2001).

27.3 Methodology

A detailed search of the literature was carried out, and a meta-analysis (analysis of analyses) of the various nutrient management studies identified was conducted. The results of the various studies examined have been described and, in some cases, presented in tabular form. Overall findings have been summarized in the closing section of the chapter.

27.4 Measures of Economic Productivity

There are various ways that the economic productivity (i.e., profitability) of an agricultural enterprise can be measured and compared. These include the following measures: crop yield, gross margin, net return per acre, return over variable costs, income above variable costs, and cost-benefit ratio. "Crop yield" refers to the amount of crop produced during one crop cycle, and "crop yield analysis" refers to a comparison of the levels of crop production under various production scenarios. Related to crop yield is the concept of "maximum economic yield," which is somewhat lower than maximum crop yield and refers to the point at which the last increment of an input pays for itself (Havlin et al. 1999). "Gross margin" (also commonly referred to as "profit") represents net revenue after all costs. The "total expected profit" that a farmer can expect to receive from producing a crop is a function of expected crop yield and expected output price (Uri 1999), as well as the total cost of production. Output price is typically beyond the influence of the individual farmer, yet the higher the output price, the more influential yield differences become—this is commonly referred to as "price sensitivity."

The level of a farmer's profit is directly related to output price and inversely related to the cost of production (Zentner et al. 2001). Total cost of production typically includes both fixed and variable costs. Variable costs include the costs of production inputs, such as seed, fertilizer, hauling, herbicides, insecticides, machine labor, irrigation labor, fuel, repair and maintenance, and interest (Parsch et al. 2001). Since 1976, the prices that farmers have paid for many of these production inputs have increased much more than the output prices received (USDA, as cited by Havlin et al. (1999)). Farmers often operate as price-takers in highly competitive markets and have little control over either (1) the prices that they receive for their products or (2) the prices that they pay for their production inputs (Zering 2000). Therefore, as Havlin et al. (1999) pointed out, higher crop yield represents the greatest opportunity for reducing per-unit production costs, and it is imperative that growers achieve optimum productivity through efficient and cost-effective use of those inputs that will ensure adequate returns on their investment.

Practices that increase yield per unit of land lower the cost of producing a unit of crop, since it costs just as much to prepare, plant, and cultivate a low-yielding field as it does a high-yielding field. Another measure of profitability is "net return per acre." Whereas profit measures return per production cost dollar spent, return per acre is a reflection of how optimal the levels of production inputs are. The farmer must select the input levels that will earn the greatest net return per acre. For instance, an insufficient level of plant nutrient application (conventional or alternative), or another needed input, may result in a high unit cost of production (Havlin et al. 1999). Related to net return per acre is another measure of profitability called "return over variable costs" (ROVC) (Marra and Kaval 2000). ROVC is calculated as the product of crop yield per acre (Y) and output price per unit of crop (P) minus total variable costs per acre (TVC). This relationship can be represented as follows: ROVC = Y * P − TVC. ROVC represents total revenues per acre minus total variable costs per acre and excludes fixed costs, such as land, whose value tends to be location specific. The usefulness of ROVC is in being able to compare cropping systems at widely dispersed locations.

Table 27.1 Summary of farm measures of economic productivity

Crop economic indicator	Formula/description
Crop yield	Amount of crop produced during one crop cycle
Yield variability	How much crop yield varies from one season to the next; measured by standard deviation or coefficient of variation
Crop yield analysis	Comparison of levels of crop production under various crop production scenarios (e.g., using different fertilizers, tillage systems)
Maximum economic yield	Point at which the last increment of an input (e.g., fertilizer) pays for itself
Gross revenue (gross return)	Crop yield × price per unit
Gross margin (actual profit, net return)	= Net revenue = gross revenue (GR) − total cost of production (TC) = GR − TC
Total expected profit	(Expected crop yield (EY) × expected output price per unit of crop (EP)) − expected total cost of production
Net return per acre	Actual crop yield per acre (Y)
Return over variable costs (ROVC)	(Crop yield per acre (Y) × output price per unit of crop (P)) − total variable costs per acre (TVC) = Y × P − TVC
Income above variable costs (IAVC)	Gross revenue (GR) − total variable costs (VC) = GR − VC
Net addition to profit (NATP)	Difference between ROVC of two systems being compared = $ROVC_1 - ROVC_2$
Cost-benefit ratio	Gross revenue/total variable costs = GR/VC

Another related profitability measure is "income above variable costs" (IAVC) (Bukenya et al. 2000), which is similar to ROVC, except that IAVC is total income after variable costs for an entire farm enterprise. Another related profitability measure is "cost-benefit ratio," which is simply the ratio of the gross revenue of a particular crop to the total variable costs of production. These are all valid and valuable measures of farm profitability. Yet, as Uri (1999) pointed out, the level of productivity, and associated profitability, that any particular farm can achieve, is typically a function of many site-specific factors, including soil characteristics, local climatic conditions, cropping patterns, and other attributes of the overall farming operation (such as labor, fertilizers, pesticides, seeds, and machinery). Thus, general inferences about production costs are difficult to make, and decisions about inputs (such as which nutrient management strategy to adopt) must be made at the individual farm level and be based on many site-specific factors.

While discussing economic profitability, it is important to note that there are different levels of costs (Table 27.1). For instance, the farm, as an economic unit, is accountable for certain types of costs, notably production costs. But when one moves beyond the farm to a larger economic unit, such as the county, state, or nation, there are other types of costs that come into view, notably environmental costs. In terms of nutrient management, these would include costs associated with the leaching of nitrogen into the groundwater or the volatilization of ammonia into the atmosphere. In both cases, there can be an increase in the level of pollution in the environment and an increase in both the visible costs (i.e., easily observable and measurable) and the invisible (hidden) costs associated with that condition. This consideration of costs, beyond the farm as an economic unit, relates to the concept of "sustainability," for which numerous definitions exist.

27.5 Nutrient Management and Sustainability

The word "sustain," from the Latin word "sustinere," means to keep in existence or maintain and implies long-term support or permanence. Bopp (1995) defined "sustainability" as the capacity of systems to continue working. In regard to agriculture, "sustainable" describes farming systems that are "capable of maintaining their productivity and usefulness to society indefinitely. Such systems must be resource-conserving, socially supportive, commercially competitive, and environmentally sound" (Ikerd, as cited in Duesterhaus 1990). Most would agree that sustainability requires a satisfactory level of profitability. For instance, the net income per hour of a family farmer should be at least equal to the stipulated wages of farm workers, and the earnings on invested capital should be sufficient to cover loan interest (Kumm 2001). Also, the efforts to achieve high production and good profitability should not result in production technology that pollutes the environment or leads to a misuse of natural resources (Kumm 2001). Marra and Kaval (2000) provided a similar definition of sustainability, describing it as "fewer negative environmental impacts from agricultural production, coupled with sufficient net returns for growers to continue farming." In terms of nutrient management sustainability, the nutrient management strategy applied to a farm should not have a negative impact on runoff or leaching, such that surface water or groundwater sources become polluted for human, animal, or plant uses. Similarly, the nutrient management strategy should not lead to excessive amounts of ammonia being released into the atmosphere, such that air quality is negatively affected.

27.5.1 Common Nutrient Management Strategies

A variety of plant nutrient management strategies currently exist. These include (1) the conventional, widely used, commercial synthetic fertilizer approach; (2) a cover crop (green manure) approach; (3) an animal manure-based approach; and (4) an organic approach. For the purposes of this chapter, each of these approaches can be defined and described as follows:

27.5.1.1 Conventional Approach

A conventional nutrient management system is one that uses synthetic chemical fertilizers as the major source of nitrogen (and other essential elements such as phosphorus and potassium), and the fertilizers are typically applied at extension- or supplier-recommended rates (Marra and Kaval 2000). Typical nitrogen-based commercial fertilizers include ammonium nitrate (34% N), ammonium nitrate limestone (20% N), ammonium nitrate sulfate (26% N), ammonium sulfate (21% N), anhydrous ammonia (82% N), and aqua ammonia (16–25% N) (Zublena et al. 2000).

27.5.1.2 Cover Crop Approach

A cover crop (green manure) system is one that uses certain nutrient-rich crops as the major source of nitrogen and are typically planted in the fall and plowed into the soil in the spring before planting. Decomposition of cover crops is usually rapid, and the residual effects are well recognized (Havlin et al. 1999). Cover crops can be useful for adding organic biomass to the soil and are also particularly useful for protecting against soil erosion over the winter months. If fertilized by commercial, synthetic fertilizer, cover crops typically require much less commercial fertilizer than conventional production crops. Cover crops can be a legume, a nonlegume, or a combination grown together (for which there are certain synergistic advantages). Commonly used cover crops (with their associated N content) include hairy vetch (4.62% N; 113 lb/ac), crimson clover (3.67% N; 102 lb/ac), winter pea (4.56% N; 61 lb/ac), and rye (89 lb/ac) (Havlin et al. 1999).

27.5.1.3 Animal Manure-Based Approach

A manure-based system is one that derives plant nutrients from animal manure, such as from hogs, chickens, dairy cows, or beef cattle. Typical nitrogen concentrations of fresh manures are as follows: hog manure (0.6%), chicken manure (1.3–3%), dairy cow manure (0.5%), and beef cattle manure (0.6%) (Bukenya et al. 2000; Zublena et al. 2000). Fresh animal manure is often treated prior to land application, via some type of treatment facility, such as an aerobic or anaerobic lagoon.

27.5.1.4 Organic Approach

An organic nutrient management system is one that is free of commercial, synthetic/petrochemical fertilizers, as well as pesticides (Marra and Kaval 2000). As such, both manure-based systems and cover crop systems (where no accompanying commercial fertilizer is applied) can be considered to be types of organic systems. Another type of organic system is "vermicompost" (VC), which is a type of organic manure used in integrated nutrient management (INM) practices. What distinguishes VC from regular composting is that vermicompost uses (1) decomposing vegetable or food waste (but not meat, dairy, fats, or oils) and (2) various species of worms (such as red wigglers, white worms, and earthworms). The result is a heterogeneous mixture of organic matter, including decomposing food materials, bedding materials, worm castings (vermicast), worm humus, and/or worm manure. Containing a range of water-soluble nutrients, vermicompost is considered to be a very nutrient-rich organic fertilizer and soil conditioner. VC typically contains high quantities of microorganisms which can fix atmospheric nitrogen and secrete phytohormones which, in turn, can increase soil organic matter (SOM) and also help to increase the availability of other nutrients (Pareek et al. 1996).

27.6 Crop Yield Effects and Sustainability of Nutrient Management Strategies

Miao et al. (2011) conducted a meta-analysis of a variety of international long-term studies that examined nutrient management effectiveness and sustainability. They identified problems with synthetic chemical fertilizers (particularly N fertilizers) and organic fertilizers (such as manure). For instance, according to Miao et al. (2011), the number one long-term problem with synthetic fertilizers (such as ammonium- or urea-based nitrogen fertilizers) is overapplication which "can result in soil acidification and changes in soil chemical properties, which can reduce biodiversity, retard nutrient cycling, and release potential toxic metals into water and plants, leading to soil degradation, environmental pollution and reduced crop yield." Overapplication of synthetic nitrogen fertilizers can also lead to nitrate leaching into groundwater and ammonia volatilization into the air. In contrast, the benefits from long-term application of organic fertilizers, such as manure, have long been recognized and include "higher organic matter content [a key indicator of soil fertility], hydraulic conductivity, porosity and aggregate stability, lower bulk density, increased soil biological activity, and a more complete supply of nutrients" (Miao et al. 2011). Yet, the long-term application of organic manure can have its negative side effects as well. These include nitrate leaching (due to increased porosity in manured soils) and an accumulation of phosphorus (P) in the soil. Based on their meta-analysis, Miao et al. (2011) concluded that a combination of organic fertilizers and commercial, synthetic fertilizers is a better practice than applying either type of fertilizer alone and can help to "achieve higher crop yields, improve soil fertility, alleviate soil acidification problems, and increase nutrient-use efficiency." The results of a study by Raun, Barreto, and Westerman (1993) support this conclusion. Their study found that, when growing conditions were not good, plots treated with synthetic NPK fertilizer produced a better yield than with a cattle manure treatment, whereas when growing conditions were optimal, cattle manure treatment produced a better yield. Thus, a combination of both types of fertilizers can lead to more stable results.

Apart from fertilizer type, another potentially effective nutrient management strategy is crop rotation. Norton et al. (2010), in a study of wheat-based cropping systems, found that a crop rotation strategy (using a wheat/barley/peas system) produced 2.5 times the energy equivalent of a continuous wheat system. They theorize that a crop rotation strategy can provide disease breaks between each phase, provide opportunities for alternative weed control strategies, and/or improve soil conditions to support the crop in the following phase. In regard to technology-enhanced nutrient management strategies, Miao et al. (2011) discovered that the use of precision, site-specific, nitrogen management, using crop-sensor technology, can diagnose crop nutrient status and determine crop nutrient demand, which can lead to substantially increased N-use efficiency without loss of crop yield. The practice can also help to prevent the typical negative environmental effects normally associated with the application of commercial, synthetic fertilizers. Miao et al. (2011) pointed out that such crop-sensor technologies can be somewhat complex in nature and need to be farmer-friendly, reliable, easy to use, and cheap, if they are to be suitable for small-scale farming. To go along with such technologies, Miao et al. (2011) advocated the creation of innovative and effective agricultural extension and service systems to assist farmers in the effective and efficient use of such technologies. Gollner et al. (2011) compared the effectiveness of synthetic fertilizer use, animal manure use, and unfertilized conditions, under dry soil conditions, such as those found in temperate regions, such as in parts of Austria. Under such conditions, the mobility of nutrients in soil solution and the availability of nutrients to crops are reduced and can be insufficient. Gollner et al. (2011) pointed out that the symbioses between crop plants and soil organisms (such as arbuscular mycorrhizal fungi) are critical for soil fertility, crop nutrition, and crop yield, particularly in low-input farming systems. According to Jeffries et al. (2003), soil organisms, such as arbuscular mycorrhizal fungi, play a key role in plant nutrition, water relations, and resistance to plant pathogens and diseases. Gollner et al. (2011) analyzed data on fertilizer applications and winter rye crop yields from a 40-year period—from 1960 to 2000—under dry soil conditions. They concluded that the use of animal manure (when compared to the use of synthetic NPK fertilizer) results in higher levels of soil organic matter (SOM); higher levels of potassium (N), phosphorus (P), and magnesium (Mg); and higher levels of arbuscular mycorrhizal colonization and arbuscule frequency—all of which support enhanced crop nutrient uptake which, in turn, promotes crop yield stability and sustainability. These effects are particularly important for low-input agricultural systems.

27.7 Economic Effectiveness of Nutrient Management Strategies

Marra and Kaval (2000) conducted a meta-analysis that examined the relative profitability of various sustainable grain cropping systems, including organic systems (i.e., using green and/or animal manure) and conventional systems (i.e., using chemical synthetic fertilizers applied at extension- or supplier-recommended rates). The meta-analysis included an examination of 120 different plant nutrient management studies. The grain crops included in the meta-analysis were barley, corn, oats, oilseed rape, rye, grain sorghum, summer grains, soybeans, wheat, and winter grains. For purposes of analysis, these crops were grouped into three categories: (1) small grains, (2) oilseeds, and (3) "other crops" (which included corn and sorghum). The measure of profitability that was used to analyze each system was "return over variable costs" (ROVC). The measure that was used to compare the profitability of each system was "net addition to profit" (NATP), which is the difference between the ROVC of each system being compared. The statistical technique of linear regression was also used to compare the two systems. Using the NATP measure, the meta-analysis showed that producing grain using a conventional nutrient management system resulted in higher profit, relative to an organic system, for both small grain and oilseed crops (i.e., NATPs of $9.38 and $6.55, respectively), but showed lower profit for the "other crop" category (i.e., NATP of −$16.09). Using linear regression, these differences were confirmed. A summary of the results of the profitability metaanalysis is given in Table 27.2. The data also showed that organic grain production was significantly more profitable in the central region of the USA than in the eastern and western regions of the country.

There are a number of individual studies that examine the profitability of various types of nutrient management strategies. For instance, Lu et al. (2000) conducted a 4-year study from 1994 to 1997, at Beltsville, Maryland, examining the relative profitability of four different cropping systems in corn, wheat, and soybean production. The four cropping systems examined are as follows:

1. No-tillage, conventional system, using a nutrient management system of commercial synthetic fertilizers applied at normally recommended rates.
2. No-tillage, cover crop system, using a winter annual cover crop of hairy vetch before corn and wheat before soybeans.
3. Manure-based system, using cow manure for corn and wheat, and crimson clover overseeded into soybeans to provide additional green manure for corn.

Table 27.2 Comparison of conventional and organic nutrient management systems

Crop type	Net addition to profit (NATP) conventional synthetic fertilizer systems relative to organic systems
Small grain crops	+$9.38
Oilseed crops	+$6.55
Other crops	−$16.09

Adapted from Marra and Kaval (2000)

4. Crown vetch living mulch system, with crown vetch planted after wheat harvest in the first 2 years and allowed to establish itself during the remainder of those years, rather than growing double-crop soybeans. The crown vetch system included the use of commercial synthetic fertilizers as a source of nitrogen.

The experimental design used in this study was a randomized complete block design, with four blocks. Each block contained four cropping systems assigned permanently to two plots, for a total of 32 plots. Each cropping system followed a 2-year rotation with corn in the first year and a wheat/soybean double crop in the second year.

Lu et al. (2000) examined crop yields and all of the costs of production, using actual costs incurred. Current market costs were used where actual costs were not available and the costs needed to be estimated. The cost of the fertilizer for the no-tillage conventional, crown vetch, and cover crop systems was $174.97, $165.64, and $117.08/$^{-1}$ha, respectively. For the manure-based system, it was assumed that the farmer would have both crop and livestock production; thus, there was no cost for manure. The cost of the hairy vetch seed required for the cover crop system was $16.62 ha^{-1}. Crop yield for the four cropping systems showed differing results (Table 27.3). When averaged over the 4 years of the study, the cover crop system produced the highest average corn yield (7.86 MT ha^{-1}), followed by the no-tillage conventional (7.82 MT ha^{-1}), crown vetch (6.44 MT ha^{-1}), and manure-based (5.66 MT ha^{-1}) systems. The crown vetch system produced the highest average wheat yield (4.07 MT ha^{-1}), followed by the no-tillage conventional (3.15 MT ha^{-1}) and manure-based (3.11 MT ha^{-1}) systems. The cover crop system produced the highest average soybean yield (2.70 MT ha^{-1}), followed by crown vetch (1.58 MT ha^{-1}), no-tillage conventional (1.52 MT ha^{-1}), and manure-based (1.44 MT ha^{-1}) systems.

"Yield variability" (also referred to as "yield variance") is another important indicator of cropping system effectiveness and can be measured using either the standard deviation or the coefficient of variation (the latter being preferred when the arithmetic means differ considerably). Yield variability can be defined as the difference between the actual and planned (budgeted) yield of a crop resulting from a given production process. As Chen et al. (2011) pointed

Table 27.3 Comparison of conventional and organic nutrient management systems

Cropping system	Cost/ha	Corn yield (MT/ha)	Wheat yield (MT/ha)	Soybean yield (MT/ha)	Yield variability (coefficient of variation)	Average gross margin ($/ha)
No-tillage conventional	$174.97	7.82	3.15	1.52	0.49	$233.27
No-tillage cover crop	$117.08	7.86	N/A[a]	2.70	0.46	$238.28
Manure-based	$0	5.66	3.11	1.44	0.60	$217.35
Crown vetch living mulch	$165.64	6.44	4.07	1.58	0.76	$53.34

Adapted from Lu et al. (2000)
[a]Note: The addition of winter annual cover crops into the rotation required the elimination of winter wheat as a rotational crop

out, agricultural yield variability is well known to depend on the weather. For instance, extreme weather events, such as hurricanes and droughts, can have obvious impacts. Other seasonal phenomena, such as the El Niño-Southern Oscillation (ENSO) or freak freeze or heat wave events, can also affect yield variability. According to the European Environment Agency (EEA) (2008), climate and its variability are the main source of variations in crop yield in Europe. "Since the beginning of the twenty-first century, the variability of crop yields has increased as a consequence of extreme climatic events (e.g., the summer heat of 2003 and the spring drought of 2007)" (EEA 2008). As a consequence of current climatic change conditions, such extreme weather events are expected to increase in frequency and magnitude, and, as a result, crop yields can be expected to become more variable.

Yet, factors other than weather can also influence yield variability. For instance, Anderson and Hazell (1987) argued that the adoption of common high-yielding crop varieties, uniform planting practices, and common timing of field operations has caused yields of many crops to become more sensitive to the weather, especially in developing countries. However, as a countermeasure to such effects, the EEA (2008) pointed out that changes in farming practices and land management can act as risk-mitigating measures and can lead to lower yield variability. Such practices might include a diversity of crops and varied planting practices, such as a combination of nutrient management strategies. For instance, in the study conducted by Lu et al. (2000), the cover crop system—an example of a diverse cropping approach—had the highest average corn yield and also had the smallest coefficient of variation (0.46), followed by no-tillage conventional (0.49), manure-based (0.60), and crown vetch (0.76) systems. Across the 4 years of Lu et al.'s (2000) study, the cover crop system produced the greatest average gross margin ($238.28 ha^{-1}), partly because of having the highest average corn yields. The average gross margin of the cover crop system was followed closely by the average gross margins of no-till conventional ($233.27 ha^{-1}) and manure-based ($217.35 ha^{-1}) systems. The crown vetch system had the lowest average gross margin ($53.34 ha^{-1}).

It is important to note that the manure-based system had higher gross margins in 1994 and 1995 than all of the other systems but had the lowest gross margins during 1996 and 1997. Poor crop yields (and associated gross margins) in the last 2 years resulted from increased weed competition. As Lu et al. (2000) pointed out, the manure-based system could potentially become the most profitable of the four systems, if the associated weeds can be controlled. This is because, of the four systems analyzed in the study, the manure-based system is the only one where the crops can be certified as organic and sold at premium prices. In assessing the findings of Lu et al. (2000), and as previously noted, weather can have a significant impact on farm profitability, regardless of the nutrient management strategy utilized. This would be in keeping with Leibig's law of the minimum (Havlin et al. 1999), which places lack of moisture ahead of lack of fertility as a limiting factor in the determination of crop yield. For instance, in the study conducted by Lu et al. (2000), weather was extremely variable and was a major extraneous factor affecting crop yield over time. During 1994 and 1996, when uniform rainfall fell throughout the growing season, crop yields for all cropping systems were higher than those in the dry years of 1995 and 1997. During 1997, there was a drought that was the worst in 50 years. Naturally, such an event could reduce crop yield considerably. In this study, average crop yield for all four cropping systems was 9.58 MT ha^{-1} in 1996 and 2.73 MT ha^{-1} in 1997—a decrease of 71.4% in 1997, the year of the drought. As noted previously in this chapter, farm profitability is sensitive to prices—both output prices received for crop yield and input prices paid out, such as the cost of applied nutrients. For instance, with a conventional nutrient management system, the cost of commercial, synthetic fertilizer is a significant portion of total variable costs (i.e., 24–30%), and any variation in prices can significantly affect profitability. With a cover crop system, cover crop seed price is a relatively low portion of total variable costs (e.g., 2.7%) (Lu et al. 2000), so any variation in seed price is not likely to have a major impact on profitability.

With an animal manure-based system, the price of manure can have a significant effect on profitability. For instance, as Lu et al. (2000) pointed out, if livestock producers have excess manure to dispose of, and have to pay crop growers to haul away their manure, the nutrient cost to the crop grower becomes a negative cost. Under such conditions, an animal manure-based nutrient management strategy can quickly become the most profitable of the nutrient management strategies discussed here.

Table 27.4 Comparison of manure-based and conventional nutrient management systems

Plant nutrient scenario	IAVC of manure-based nutrient system	IAVC of conventional, synthetic fertilizer nutrient system
Free poultry manure + Free transportation	$19,846	$14,792
Free poultry manure + Transportation costs	$16,493	$14,792
Purchased poultry manure + Transportation costs	$12,796	$14,792

Adapted from Bukenya et al. (2000)

Due to this type of benefit, as well as the formidable price premiums typically paid for organic products, the potentially higher profitability of manure-based systems—even when yields are lower and/or labor and capital costs are higher—has been documented. Smolik et al. (1995), Hanson et al. (1997), and Vitterso (1997) (all cited in Kumm 2001) recognized these benefits and suggested that development of sustainable organic farming might favor family farming, slow the trends toward very large farms, and hasten the growth of smaller farms. Other studies have also been conducted examining the relative profitability of various cover crop systems. For instance, Ott and Hargrove (1989, cited in Lu et al. 2000) evaluated the profitability of four different cover crop systems in no-tillage corn production in Georgia. They tested the effectiveness of crimson clover, hairy vetch, winter wheat, and winter fallow and found that no-tillage corn following hairy vetch produced the largest average profit. Hanson et al. (1993, cited in Lu et al. 2000) evaluated the profitability of cover crop systems used in corn production in the Maryland Coastal Plain and Piedmont regions. They found that (1) hairy vetch, with an additional 135 kg ha^{-1} of N applied, was the most profitable type of system in the Coastal Plain region and (2) corn following winter fallow, with an additional 45 kg ha^{-1} of N applied, was the most profitable type of system in the Piedmont region.

Bukenya et al. (2000) conducted a study to compare the relative profitability of a manure-based nutrient management system (i.e., fresh poultry manure), with a conventional synthetic chemical fertilizer system (i.e., commercial ammonium nitrate fertilizer) in the production of cotton. A main purpose of their study was to assist poultry producers in making decisions about how to best manage and utilize the animal waste produced from their operations. As Bukenya et al. (2000) pointed out, in the case of animal manure, there is typically more than one way to economically dispose of the material. For instance, it can (1) be used as a plant fertilizer; (2) be converted to biogas or compost, which may increase its value; or (3) be converted to heat via combustion. The measure of profitability that was used to compare the two systems was "income after variable costs" (IAVC), which is similar to the ROVC measure used in the meta-analysis of Marra and Kaval (2000)—the only difference being that IAVC is total income after variable costs for the entire enterprise.

In their study, Bukenya et al. (2000) examined three different plant nutrient scenarios: (1) Scenario 1, which assumed that the poultry manure was free and required no transportation cost; (2) Scenario 2, which assumed that the poultry manure was free but needed to be transported to the farm; and (3) Scenario 3, which assumed that the poultry manure needed to be purchased, as well as transported to the farm. Each of these scenarios (and their corresponding sets of assumptions) represents different types of production systems in existence, with Scenario 1 being particularly representative of those agricultural enterprises that combine both animal and crop production.

The results of the study are as follows (Table 27.4). Note that for all of the results described, the amount of nitrogen provided by the fresh poultry manure and the ammonium nitrate was 80 lb of nitrogen per acre. Also, the ammonium nitrate price used was $27.20 per acre, and the poultry manure price used was $20.03 per acre. Under Scenario 1, the manure-based system was more profitable in providing nitrogen to cotton than the conventional ammonium nitrate fertilizer system. The IAVC of the manure-based system ($19,846) was 25% higher than the conventional system ($14,792). Under Scenario 2, the manure-based system remained more profitable in providing nitrogen to cotton than the conventional system but less so than as under Scenario 1. The IAVC of the manure-based system ($16,493) was 10% higher than the conventional system ($14,792). Under Scenario 3, where poultry manure is purchased and transported, the conventional system was more profitable in providing nitrogen to cotton than the manure-based system. The IAVC of the conventional system ($14,792) was 13% higher than the manure-based system ($12,796). In analyzing the results of this study, it is worth noting that the results would be highly sensitive to changes in (1) commercial ammonium nitrate fertilizer prices, (2) poultry manure prices, and (3) manure transportation costs.

Dass et al. (2008) conducted a 3-year study examining the effects of various nutrient management systems on the growth of the winter vegetables bell pepper and cabbage. They compared the following seven different nutrient management strategies:

T_1: Normal producer's practices (i.e., use of cow manure (CM) only (2.5 MT ha^{-1}))

T_2: Use of commercial, synthetic, chemical fertilizer at normally recommended rates (RRF)

T_3: Use of 50% RRF + vermicompost (VC) (5 MT ha^{-1})

T_4: Use of 50% RRF + bacteria (phosphorus solubilizing bacteria (PSB) at 5 kg/ha + *Azotobacter* at 5 kg ha^{-1})

T_5: Use of 50% RRF + cow manure (CM) (10 MT ha^{-1})

Table 27.5 Comparison of manure-based, conventional, vermicompost-enhanced, and bacteria-enhanced nutrient management systems

Fertilizer	Average yield (MT ha^{-1})		Average gross revenue (GR) ($US ha^{-1})		Average gross margin (GM) ($US ha^{-1})		Average cost-benefit ratio		Soil organic carbon (%)		Soil nitrogen (kg ha^{-1})	
treatment	Pepper	Cabbage	Pepper	Cabbage	Pepper	Cabbage	Pepper	Cabbage	Pepper	Cabbage	Pepper	Cabbage
T_1	1.67	16.30	$510	$807	$103	$139	1.27	1.20	0.58	0.57	287.3	279.7
T_2	6.35	50.34	$1,954	$2,492	$1,242	$1,710	2.73	3.20	0.62	0.61	367.8	358.4
T_3	8.35	53.51	$2,572	$2,649	$1,612	$1,667	2.67	2.70	0.69	0.68	426.5	415.9
T_4	5.26	38.7	$1,619	$1,916	$927	$1,172	2.30	2.60	0.60	0.58	337.8	328.8
T_5	6.90	51.73	$2,126	$2,561	$1,266	$1,634	2.47	2.77	0.70	0.69	442.5	432.4
T_6	7.45	52.74	$2,294	$2,611	$1,388	$1,658	2.53	2.73	0.69	0.68	436.7	430.3
T_7	6.01	45.06	$1,851	$2,230	$1,094	$1,399	2.43	2.70	0.67	0.66	396.5	416.0

Adapted from Dass et al. (2008)

T_6: Use of 50% RRF + VC (2.5 MT ha^{-1}) + CM (5 MT ha^{-1})

T_7: 50% RRF + PSB (5 kg ha^{-1}) + *Azotobacter* (5 kg ha^{-1}) + CM (5 MT ha^{-1})

The results of the study showed the nutrient management strategy that included both commercial, synthetic fertilizer and vermicompost (i.e., T_3) produced, for both bell pepper and cabbage, the highest average crop yields, the highest average gross revenues, and the highest average gross margins over the 3 years (Table 27.5). Treatment 2 (i.e., T_2—application of synthetic fertilizer at recommended rates (RRF)) produced the highest average cost-benefit ratio for both crops. In regard to soil conditioning, treatment 5 (i.e., T_5—50% RRF + cow manure at 10 Mt/ha) showed a marginal advantage over treatments T_3, T_4, T_6, and T_7, in terms of soil organic carbon (SOC) left in the soil. In terms of nitrogen left in the soil, treatment 5 showed a marginal advantage over treatments T_3 and T_6.

27.8 Conclusions and Future Research

This review of plant nutrient management strategies—aimed at helping farm enterprises to maximize their profitability—indicate the following conclusions.

Conventional nutrient management systems, using commercial, synthetic, chemical fertilizers, have been shown to be more profitable for most grain crops (with the exception of corn and sorghum) than organic nutrient management systems.

Using a combination of synthetic fertilizers and organic fertilizers seems to be a better practice than applying either type of fertilizer alone. Such a strategy can help to improve soil fertility, lessen soil acidification problems, increase crop nutrient—use efficiency—and achieve higher crop yields and associated profitability. For instance, a nutrient management strategy that includes both synthetic fertilizer and vermicompost (organic manure) has been shown to promote relatively high crop yields and associated relatively high gross revenues and gross margins. This nutrient management strategy can also promote healthy soil conditions.

The use of animal manure (when compared to the use of synthetic NPK fertilizer) can result in higher levels of soil organic matter (SOM); higher levels of potassium (N), phosphorus (P), and magnesium (Mg); and higher levels of arbuscular mycorrhizal colonization and arbuscule frequency—all of which can support enhanced crop nutrient uptake which, in turn, can promote crop yield stability and sustainability. These effects are particularly important for the profitability of low-input agricultural systems.

Both cover crop and animal manure-based systems show considerable promise as alternative nutrient management strategies for increasing farm profitability. In the case of cover crops, when compared to a no-tillage conventional (i.e., synthetic fertilizer) system, a manure-based system, and a crown vetch system, the cover crop system produced the highest average corn yield and gross margin per hectare, as well as the smallest coefficient of variation.

In the case of manure-based nutrient management systems, those that do not require the purchase or transport of the manure (as in combined animal and crop production systems) can be considerably more profitable than conventional synthetic, chemical fertilizer systems.

Both manure-based and cover crop systems that do not include the use of commercial synthetic fertilizers—i.e., systems that can be classified as "organic systems"—hold particular promise due to the output price premiums typically garnered by the organic crops grown under such conditions. But, as Kumm (2001) pointed out, while the favorable economic results can provide incentives for organic production to increase, increased production could also conceivably lead to lower price premiums due to the effects of supply and demand. As such, more research needs to be conducted into the long-term sustainability of these systems under conditions where price premiums (or, in some cases, subsidies) have decreased or ceased and also when technological progress has improved the efficiency of organic farming. Also, more research needs to be conducted into identifying alternative, lower cost pest (particularly weed) management techniques (Marra and Kaval 2000). Current practices, which tend to be expensive, can have a great impact on profitability.

While individual farm profitability (and viability) is an important issue, sustainability (which includes both economic and environmental viability) is an equally important issue. The studies described here have identified certain plant nutrient management strategies that can be potentially effective for increasing farm profitability. But clearly, more research needs to be conducted to examine the sustainability of these practices over the long term.

There are various technology- and education-oriented nutrient management support systems that can help farmers to increase their crop yields and associated profitability. For instance, there are precision, site-specific, nutrient management crop-sensor technologies that can diagnose crop nutrient status and determine crop nutrient demand—potentially leading to substantially increased nutrient-use efficiency without loss of crop yield. But, to go along with such technologies, innovative and effective agricultural extension and service systems need to be developed and implemented to educate and assist farmers in the effective and efficient use of such technologies.

References

Anderson JR, Hazell PBR (1987) Variability in grain yields: implications for agricultural research and policy in developing countries. International Food Policy Research Institute/The Johns Hopkins University Press, Baltimore

Bopp M (1995) Aboriginal leadership and management training needs for sovereignty and self-reliance: a summary for Nunavut consideration. The Four Worlds Centre for Development Learning, Calgary

Brown L (2011) The earth policy institute holds a conference call on china's growing food demand. Regional business news. Quarterly Earnings Report, Thomson Financial. Retrieved on 20 May, 2011 from http://web.ebscohost.com/ehost/delivery?sid=1d8a8366–9cb7–44ff-8c77-c883b3e01702

Bullen G, Brown B (2001) Economic evaluation of UNR cotton. Department of Agriculture and Resource Economics, NC State University, Raleigh, NC. Retrieved from http://www2.ncsu.edu:8010/unity/lockers/users/a/abbrown/cotton/presentations.htm

Bukenya JO, Befecadu J, Jones HS et al (2000) Economic feasibility of substituting fresh poultry litter for ammonium nitrate in cotton production. J Sustain Agric 16(1):81–89

Chen CC, McCarl BA, Schimmelpfennig DE (2011) Yield variability as influenced by climate: a statistical investigation. National Assessment of Climate Change, Agricultural Focus Group, U.S. Global Climate Change Office, U.S. Department of Agriculture, Washington, DC, pp 1–25

Dass A, Lenka NK, Patnaik US et al (2008) integrated nutrient management for production, economics, and soil improvement in winter vegetables. Int J Veg Sci 14(2):104–129

Duesterhaus R (1990) Sustainability's promise. J Soil Water Conserv 45:4

European Environment Agency (EEA) (2008) Crop-yield variability (CLIM 032)—Assessment published Sept 2008. Retrieved on 2 June 2011 from http://www.eea.europa.eu/data-and-maps/indicators/crop-yield-variability/crop-yield-variability-assessment-published

Gollner MJ, Wagentristl H, Liebhard P et al (2011) Yield and arbuscular mycorrhiza of winter rye in a 40-year fertilisation trial. Agron Sustain Dev 31(2):373–378

Hamilton J (2010) Current economic conditions. Econbrowser: analysis of current economic conditions and policy. Retrieved on 18 Oct 2010 from http://www.econbrowser.com/archives/2010/06/current_economi_4.html

Havlin JL, Beaton JD, Tisdale SL, Nelson WL (1999) Soil fertility and fertilizers: an Introduction to nutrient management, 6th edn. Prentice-Hall, Upper Saddle River

Hoppe RA, Johnson J, Perry JE et al (2001) Structural and financial characteristics of U.S. farms: 2001 family farm report, ERS agriculture information bulletin no. 768. Economic Research Service, U.S. Department of Agriculture, Washington, DC

Jeffries P, Gianinazzi S, Perotto S et al (2003) The contribution of arbuscular mycorrhizal fungi in sustainable maintenance of plant health and soil fertility. Biol Fertil Soils 37:1–16

Kumm KI (2001) Towards sustainable Swedish agriculture. J Sustain Agric 18(4):27–37

Lu YC, Watkins KB, Teasdale JR et al (2000) Cover crops in sustainable food production. Food Rev Int 16(2):121–157

Marra MC, Kaval P (2000) The relative profitability of sustainable grain cropping systems: a meta-analytic comparison. J Sustain Agric 16(4):19–32

Miao Y, Stewart BA, Zhang F (2011) Long-term experiments for sustainable nutrient management in China. Agron Sustain Dev 31(2):397–414

Norton R, Perris R, Armstrong R (2010) Learning from long-term experiments—What do they teach us? Better Crop 94:20–22

Pareek SK, Srivastava VK, Maheshwari ML et al (1996) Effect of *Azotobacter* culture in relation to nitrogen application on growth, yield and alkaloid composition on opium poppy. Indian J Agron 41(2):321–328

Parsch L, Keisling T, Sauer P et al (2001) Economic analysis of conservation and conventional tillage cropping systems on clayey soil in eastern Arkansas. Agron J 93:1296–1304

Raun WR, Barreto HJ, Westerman RL (1993) Use of stability analysis for long-term soil fertility experiments. Agron J 85:159–167

Singh SR, Prakash S, Singh N et al (1998) Organic farming technology for sustainable vegetable production in hills. In: International symposium on sustainable agriculture in hill areas, HPKV, Palampur, India, pp 29–31

Takkar PN (2006) Handbook of agriculture, 5th edn. ICAR, New Delhi

Uri N (1999) The economic benefits and costs of conservation tillage. J Sustain Agric 15(1):5–27

US Department of Agriculture (USDA) (1998) A time to act: a report of the USDA national commission on small farms. Miscellaneous Publication 1545 (MP-1545). [Online]. Available: www.reeusda.gov/smallfarm/report.htm

US Department of Agriculture (USDA) (2007) Farms, land in farms, and livestock operations, national agriculture statistics service. [Online]. Available: http://usda.mannlib.cornell.edu/usda/current/FarmLandIn/FarmLandIn-02–02–2007.pdf

US Office of Management and Budget (2010) A new era of responsibility: renewing America's promise. Retrieved on 18 Oct 2010 from www.gpoaccess.gov/usbudget/fy10/pdf/fy10-newera.pdf

Walden ML (2002) Economic outlook—general trends national and north carolina. NC State Economist (January/February). Department of Agriculture and Resource Economics, NC State University, Raleigh, North Carolina. Retrieved from http://www.ag- econ.ncsu.edu/extension/economist.htm

Zentner R, Campbell C, Biederbeck V et al (2001) In search of a sustainable cropping system for the semiarid Canadian prairies. J Sustain Agric 18(2/3):117–135

Zering K (2000) How to select an alternative manure treatment system. Paper presented at the North Carolina Pork conference, 12 Jan 2000, Fayetteville, North Carolina

Zublena JP, Baird JV, Lilly JP (2000) Soil facts: nutrient content of fertilize and organic ma-terials (Fact Sheet AG-439–18). North Carolina Cooperative Extension Service, NC State University Station, Raleigh, North Carolina

Assessing Salinity Tolerance in Citrus: Latest Developments

28

Eran Raveh

Abstract

Citrus trees are well known as a salt-sensitive crop. Exposing the trees to salt stress usually causes reduction in growth, followed by leaf defoliation and fruit load reduction. Their salt sensitivity reflects their sensitivity to both the osmotic component and toxicity component of salt stress, and it is strongly affected by rootstock/scion interaction. In the following chapter, we will first characterize trees' salt tolerance as well as the mechanisms behind the osmotic and toxicity components of salt stress. We will describe how the osmotic and toxicity thresholds are characterized, and how they are affected by soil characteristics, irrigation design, and the agrotechniques used for orchard maintenance. We will also explain the dynamic effect of salt stress and how the stress builds up with time. The chapter will end by the description of new potential scientific tools which can help with future understanding of citrus salt tolerance and ideas for future research.

Keywords

Chloride • Osmotic stress • Rootstock • Sodium • Toxicity stress • Yield

28.1 Introduction

For the last 50 years, the world has been drifting toward a global water crisis. Today, about 20% of the world's irrigated areas are salinized. As a result, determining a crop's tolerance to salinity is a key factor in evaluating its economic potential. Citrus is an important crop worldwide that is known for its sensitivity to salinity (Fig. 28.1). Its salt tolerance has been studied for more than a century. Initially, citrus salinity research focused on the osmotic effects of salinity and determining the effects of soil water electrical conductivity (EC) on yield reduction (determining the threshold and slope for yield reduction due to increase in soil EC, Maas 1993, and reference within). Then the focus shifted to the toxicity aspects of salinity (mainly chloride toxicity). Researchers worked on the mechanism underlying chloride uptake as well as on the effects of rootstock on chloride uptake (including ranking rootstocks for their tolerance to chloride, Levy and Syvertsen 2004; Storey and Walker 1999). Many reviews have been written on citrus and salinity (Maas and Hoffmann 1977; Maas 1993; Storey and Walker 1999; Levy and Syvertsen 2004; Grattan and Ferguson 2005). Nevertheless, even after all these years, the data are relatively inconsistent and we still have difficulty estimating the effects of salinity on citrus yield. For example, the rootstock Sour Orange appears in some papers as salt tolerant (Levy and Shalhevet 1990), while in others it is characterized as intermediate tolerant (Oppenheimer 1937; Cooper and Gorton 1952; Cooper et al. 1958; Peynado and Young 1962; Nieves et al. 1990; Boman 1994) or salt sensitive (El Hammady et al. 1995; Garcia Legaz et al. 1992; Levy et al. 1999). Over the years, the agrotechniques and the varieties used by the citrus growers have been updated (e.g., moving from flood irrigation to drip irrigation or the abandonment of the "Shamuti" variety by the growers, a variety that was used for evaluating citrus response to salinity; Maas 1993; Bielorai et al. 1988, and

E. Raveh (✉)
Department of Fruit Tree Sciences, Agricultural Research Organization, Gilat Experimental Station, Mobile Post, Negev 85-280, Israel
e-mail: eran@volcani.agri.gov.il

A.K. Srivastava (ed.), *Advances in Citrus Nutrition*,
DOI 10.1007/978-94-007-4171-3_28,

Fig. 28.1 The appearance of severe salt stress in a commercial Valencia orange (*Citrus sinensis* [L.] Osbeck) orchard is shown. The 7-year-old tree was grafted on salt-sensitive rootstock Troyer citrange [*Poncirus trifoliata* (L.) Raf. × *C.sinensis* (L.) Osbeck.], and it leaf chloride content (before defoliation) was about 1.83% (dry weight) (Picture by E. Raveh)

Levy and Syvertsen 2004). As a result, parts of the knowledge become less relevant, while others are missing. The effort has been made in the chapter to summarize the osmotic and toxicity characteristics of salt stress, and the effects of experimental design, measurement method, and scion characteristics on citrus tolerance to salinity will be discussed.

28.2 Characterization of Salt Tolerance

Reliable data on salt tolerance of citrus rootstock/scion combinations can only be obtained from carefully controlled and well-replicated field experiments conducted across a range of salinity treatments (Shannon 1997). Tests should include mature yielding trees over a long time span (on the order of years) (Levy and Syvertsen 2004) in order to evaluate possible cumulative effects of salinity on yield. Such experiments are expensive and, thus, rare (Hoffman et al. 1989; Prior et al. 2007). Tolerance characteristics are usually associated with the trees' ability to maintain yield (mass per area) while being exposed to a given level of salinity (Maas and Hoffmann 1977, and references therein). However, from the grower's point of view, salt tolerance usually means the ability to maintain income under exposure to a given level of salinity. Since income depends on several variables [e.g., time of harvest, fruit quality, average fruit size, total soluble solids content (TSS), exchange rate], salt tolerant for one grower might be characterized as salt sensitive for another. For example, the income of citrus growers who sell their fruits to the fresh market will depend on total fruit mass production, internal and external fruit quality, fruit size, and time of harvest, while that of growers who sell their fruits to the industry depends mainly on total fruit mass production and, in some cases, also on the Brix TSS per fruit weight (Boman 2005). As a result, growers who sell their product to the fresh market might have a lower income under saline conditions due to the known effect of salinity on fruit size. On the other hand, growers who sell their fruit to the industry and expose their trees to saline conditions might see less of a reduction in income due to the increase in fruit TSS (Boman 2005; Grieve et al. 2007). To date, most of the analyses of the effect of salinity on citrus have focused on the effect on total biomass production, and growers should perform their own adapted estimations with regard to how this will affect their income.

28.3 Osmotic and Toxicity Components of Salt Stress

Salt stress can affect plants through two different mechanisms: osmotic stress and toxicity stress. Osmotic stress reflects the effect of salinity on soil water potential and has a direct effect on plant water uptake. Increasing soil salinity causes an immediate reduction in soil water potential which can limit water uptake by the trees. Trees can adjust their water potential; however, this response is limited and varies between different rootstock/scion combinations (Maas 1993; Bañuls and Primo-Millo 1995). Toxicity stress reflects an excess accumulation of salt ions (usually chloride and sodium) in the plant tissue which has a negative effect on plant growth and development. Also, roots are the first organ to be exposed to the salt ions (Storey and Walker 1987); most of the papers dealing with the toxicity effects of salinity are focused on the ion concentration in the leaves (Maas 1993; Levy and Syvertsen 2004, and reference within). Many papers have demonstrated how each of these components (the osmotic and the toxicity components) affects citrus trees (Syvertsen and Smith 1983; Syvertsen and Yelenosky 1988; Nieves et al. 1991; Zekri 1991). In general, there is an immediate reduction in growth, followed by toxicity symptoms such as "scorching," "firing," or defoliation of leaves and death of the trees (Raveh 2005). It was also demonstrated how the salinity components interact with other biotic (such as presence of nematodes, mycorrhizae, and other pathogens) and abiotic (temperature, light intensity, CO_2 concentration) stresses (Willers and Holmden 1980; Graham and Syvertsen 1989; Ball and Munns 1992; Russo et al. 1993; Combrink et al.

1996; Wu et al. 2010). It is very important to understand how the osmotic and toxicity thresholds for citrus are scientifically characterized.

28.3.1 Characterizing the Osmotic Threshold for Citrus Salt Tolerance

Increased soil salinity is a result of an increase in soluble ions in the soil solution. As the ion concentration in the soil solution increases, its osmotic potential becomes more negative and water uptake by the plant becomes limited. The osmotic characteristic of the soil solution is usually determined by measuring the EC of a soil saturated extract (EC_e), a parameter that correlates to the concentration of the ions dissolved in the water. The EC_e measurement is independent of the salt type present (Bernstein 1961). The reviews report an EC_e threshold of ~1.4 dS/m (Maas and Hoffmann 1977; Maas 1993; Grattan and Ferguson 2005). Above this threshold, each unit of increase in soil EC_e is followed by an average yield decrease of ~15%. Yet, when we try to examine the source of these values, we can see that the data are relatively limited, old, and variable. In fact, these values present the average of only five experiments: Bielorai et al. (1978), who worked on grapefruit; Cerda et al. (1990), who worked on lemon trees; and Harding et al. (1958), Bingham et al. (1974), and Bielorai et al. (1988), who worked on oranges. The EC_e threshold in the different experiments ranged from 0.9 to 2.1 dS/m depending on the variety and the rootstock, and the yield-reduction rate ranged from 10% to 17% (per 1 dS/m above the threshold). It is not clear whether yield response to salinity should be analyzed using the threshold-slope model, as was done by Maas and Hoffmann (1977), or using an S-shaped function (Skaggs et al. 2006) which eliminates the need for subjective judgment (Fig. 28.2). From a scientific point of view, the threshold-slope model, which involves the subjective inclusion and exclusion of data (Maas and Hoffmann 1977), is inferior to the S-shaped function.

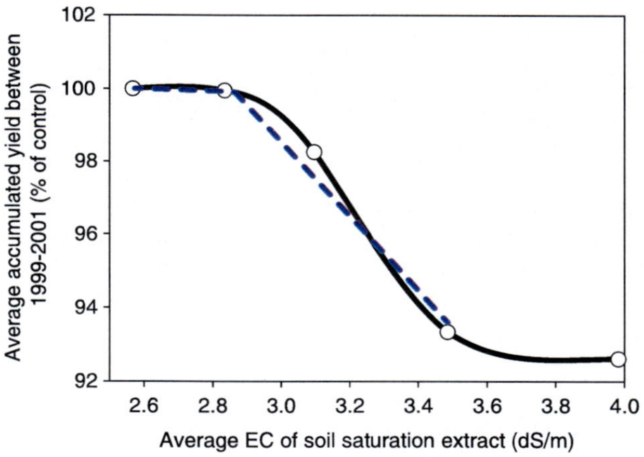

Fig. 28.2 The relationship between soil saturated extract EC and yield of white grapefruit. Data were analyzed using the threshold-slope model (*blue dashed line*) or an S-shaped function (*solid black line*). For the threshold-slope model, we eliminated the last data point. Each yield data point is an average of the accumulated yield between 1999 and 2001 of 90 trees that were grafted on five different rootstocks and exposed to two KNO_3 fertilization regimes. Average leaf chloride concentration of the examined trees was always below 0.4% of dry weight. The average soil EC (dS/m) was measured at the end of the irrigation period (end of October) in all examined years. The upper 90-cm layer was sampled 20 cm away from the dripper and analyzed for its EC using a soil saturated extract (Adapted from Raveh and Levy 2011)

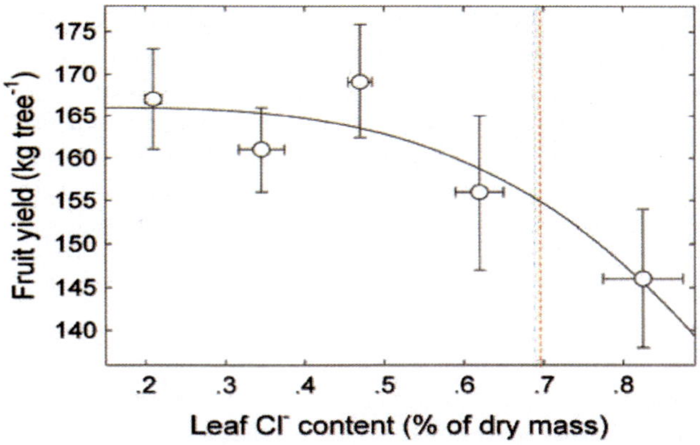

Fig. 28.3 The relationship between leaf chloride (Cl^-) content and fruit yield in a 7-year-old grapefruit tree grafted on Troyer citrange. Tree history, growth conditions, irrigation, and fertilization were as described by Levy et al. (2000) and Raveh and Levy (2011). Salinity irrigation treatments ranged from 230 to 800 mg/L (ppm) chloride. During the first 5 years of the experiment, the correlation between yield and leaf chloride was insignificant (Raveh and Levy 2011). Values are means ± SE (Adapted from Raveh 2005)

28.3.2 The Toxicity Threshold for Citrus Salt Tolerance

To characterize the toxicity threshold for citrus, the relation between leaf chloride (as well as sodium and other salt elements) and yield of bearing trees has to be analyzed. Based on the literature, the toxicity thresholds (excess levels) for leaf chloride and sodium concentrations are 0.7% and 0.23% of dry weight, respectively (Embleton et al. 1973). Unfortunately, these values were characterized based on leaf symptomatology in sand and solution culture experiments as well as on field observation and were not supported by direct measurements of the relation between yield and leaf ion concentration (Embleton et al. 1973). In general, it is very difficult to obtain a significant negative correlation between leaf chloride and yield (Raveh and Levy 2011). However, after several years of exposing trees to saline conditions, when leaf chloride level reaches toxic values, this negative correlation becomes significant (Fig. 28.3; Levy et al. 1992; Raveh 2005). There is no similar correlation between leaf sodium concentration and yield in the citrus literature.

28.4 Experimental Design

As already mentioned, reliable data on salt tolerance of citrus rootstock/scion combinations can be obtained only from carefully controlled and well-replicated field experiments conducted across a range of salinity treatments on mature yielding trees over a long time span (on the order of years). However, experimental design can strongly affect the results, from the soil characteristics through the irrigation design to the agrotechniques use for orchard maintenance.

28.4.1 Soil Characteristics

Plants get their water and nutrients (including salt) from the soil water. The effects of soil type on water and salt have been previously reviewed (Pasternak and De Malach 1995; Mmolawa and Or 2000). In general, different soils are characterized by different water-holding capacities, ion-exchange capacities [usually characterized as sodium adsorption ratio (SAR)], and hydraulic characteristics (Mmolawa and Or 2000). Soil saturated extract electric conductivity (EC_e) of soils that are characterized by the same percent of salt but differ in their water holding capacity will be different (the higher the water holding capacity, the lower the EC_e; Hide 1954). As a result, conducting salinity experiments with different soil types can strongly affect results. Additionally, even if our EC_e values remained constant between irrigations, our actual soil solution EC value (EC, the value that our trees cope with while trying to absorb water from the soil) will vary between irrigation (pending on soil water content; at the lowest just after an irrigation event, followed by a gradual increase until the next pulse of irrigation). Yet, also in smectite-rich soil, EC_e is considered more reliable than EC since it is more reproducible (Hide 1954); from the tree's point of view, the EC value is more important. Indeed, most of the data on salt tolerance have been collected from trees growing in sandy, relatively homogeneous soils. In many commercial groves, this is not the case: the soil is not necessarily sandy and it can be heterogeneous. As a result, even if the soil EC_e in the grove is in the optimal range, the actual soil solution EC, as well as the responses, may differ.

28.4.2 Irrigation Design

Irrigation can be performed using several techniques: *gravity irrigation* (flood irrigation and furrow irrigation), *sprinkler irrigation* (overhead and undercanopy sprinklers), and *drip irrigation* (surface and subsurface drip irrigation). The technique, the irrigation frequency (from biweekly to daily and up to several pulses per day), and the amount and rate of water supply in each irrigation event can affect water and salt distribution in the soil. As a result, irrigation design can affect water and salt uptake by the plants. Originally, salt experiments were conducted while maintaining a homogeneous distribution of water and salt in the soil. But as we have seen, this is not the case in commercial orchards. As a result, it is very difficult to estimate average soil water and salt contents in commercial orchards. Waldo and Schumann (2009) found that to evaluate soil water status in one tree, a few tens of probes are needed. It must be clear that even if we can create a 3-D map of water and salt distribution around the root system, we still would not have any idea of how much salt and/or water the tree will take up, since we have no information on the relative contributions of the different roots in the different zones.

28.5 Agrotechniques for Orchard Maintenance

As mentioned, reliable data on salt tolerance of citrus rootstock/scion combinations can only be obtained from field experiments conducted across a range of salinity treatments on mature yielding trees over a long time span (Levy and Syvertsen 2004). Over the years of such an experiment, the canopies of trees from the salt irrigation and control treatments will be of different sizes and volumes. Under these circumstances, trees with big canopies will be treated with relatively less water and fertilization than the trees subjected to the salt treatment, with their smaller canopies. The leaching fraction (the amount of water needed to leach excess salt around the root system) in control trees will be smaller, and their canopy will be pruned more aggressively than trees from the salt treatment. The ideal situation would be that each tree from each treatment receive its water and fertilization based on its canopy size, and that the trees not be pruned. In reality, individual monitoring and treatment of trees, based on their water and fertilization use (information that can only be achieved by growing the trees in lysimeters), are relatively expensive (Tripler et al. 2007) and are usually not performed (Levy and Syvertsen 2004; Grieve et al. 2007; Raveh and Levy 2011). Moreover, with respect to pruning, a canopy architecture that is different from that found in commercial orchards will result in trees that yield differently from those in commercial orchards.

28.5.1 The Dynamic Effect of Salt Stress

When we try to understand how citrus yield responds to salinity, the response is described as a static process, with a specific and constant response for each soil EC_e (Fig. 28.2).

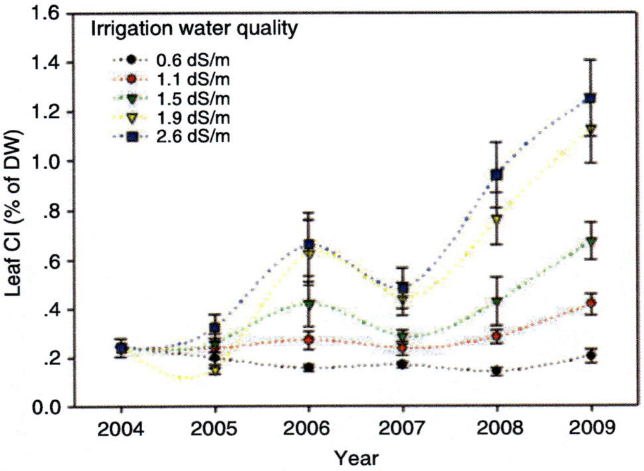

Fig. 28.4 The effect of time and salinity on leaf chloride (Cl^-) concentration of Valencia orange trees grafted on Sour Orange rootstocks. Irrigation with saline water was begun in 2005, when the trees were 6 years old (data for 2004 reflect leaf Cl^- concentration prior to irrigation with saline water). Irrigation water Cl^- concentrations were 110, 292, 438, 630, and 800 ppm, from the lower to higher salinity treatment, respectively. Trees were treated as described by Raveh and Levy (2011). Values are mean ± SE ($n=9$) (Adapted from Raveh and Hillman unpublished data)

In reality, this is not the case, and both the effects of the stress and the response of the trees build up gradually, as in any dynamic process. The effects of the salt stress usually begin with its osmotic effect; at the same time, the toxic ions are slowly accumulating, until the toxicity effect is added. The toxic ions first accumulate in the root system (starting in the distal roots, followed by the proximal ones; Storey and Walker 1987), then they begin to accumulate in the main trunk and only then in the leaves and fruits (Raveh 2005). The process of ion accumulation can take from several days to several years, depending on salinity level, scion/rootstock characteristics, tree age (size), and the existence of other abiotic and biotic stresses (Levy and Syvertsen 2004). Indeed, Bingham et al. (1974) reported that it takes up to 4 years before any salinity effect can be measured on yield. Based on salinity experiments that were conducted over the course of 5 years (and the results of which are currently being analyzed), keeping irrigation water quality at a relatively constant salinity level does not necessarily lead to constant leaf chloride concentration or tree response (Fig. 28.4). The dynamic effect of salinity on plant responses is also a matter of the dynamic acclimation process undergone by trees. This acclimation includes accumulation of nonprotein amino acids, such as proline (Yelenosky 1979; Levy 1980; Syvertsen 1984; Syvertsen and Smith 1984; Walker et al. 1993), which are considered to be osmoprotectants, but do not protect against specific mineral toxicities (Levy and Syvertsen 2004, and references therein).

28.6 Estimating the Effect of Salinity on Commercial Yield

Once we understand that salt stress is characterized by a dynamic effect, and that it mainly affects the trees via osmotic and toxicity stress, it is quite clear that salt-tolerance assessments should focus on parameters that reflect these two components, as well as their dynamic effects on yield.

28.6.1 Estimating the Osmotic Effect of Salinity

Estimations of the osmotic effect of salinity should be based on direct measurement of the effect of salinity on tree water potential (as stem or leaf water potential) or any other physiological parameter that is correlated to tree water potential and reflects the direct effect of salinity on soil water potential. In addition, since we are dealing with a dynamic response, the measurements should be performed often, making the direct measurement of plant water potential using a pressure chamber (Scholander et al. 1965) unfeasible, as this type of measurement is time consuming and cannot be automated. Other physiological parameters which can be related to tree water status are sap flow and stomatal conductance. Sap-flow measurements have been shown to be correlated to tree water potential (Čermák et al. 1980), but this correlation is strongly dependent on canopy area (Bequet et al. 2010). Stomatal conductance cannot be measured automatically, which also makes it irrelevant for our purposes.

Recently, two types of automated measurements have been shown to be correlated to the plant's water potential status: stem water content, measured by time-domain reflectometry (TDR), and leaf turgidity, measured by leaf patch clamp pressure (LPCP) probe. The measurement of stem water content is performed with TDR probes that are inserted into the main stem of the tree (Nadler et al. 2003). The variation in stem water content values is correlated to the variation in stem water potential (Fig. 28.5), and the measurement can be automated and performed as often as needed. In addition, it responds directly to changes in water availability in the soil (such as water withholding or salinity stress), with a response time of less than 4 h (Nadler et al. 2006, 2008). To reduce noise during the measurement, the cable connecting the TDR apparatus to the probes is of limited length; in addition, environmental noise (as emitted radiation from electrical power lines or nearby transmitting antennas) must be taken into account. The TDR measurement can also be replaced by an automated measurement of stem EC. In this case, the output is linearly correlated to the TDR measurements and therefore also reflects the variations in stem water content (Nadler et al. 2008). The stem EC measurement has

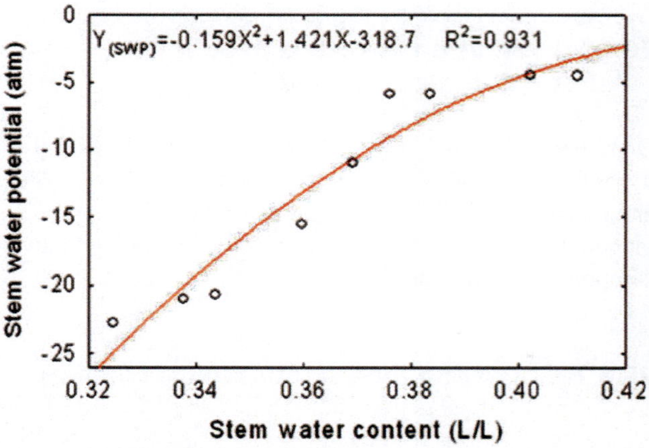

Fig. 28.5 The relationship between stem water potential values (measured with pressure chamber) and stem water content (measured with TDR probes) in lemon trees. Irrigation of the trees was withheld for a period of about 5 weeks, and stem water content and stem water potential were measured in parallel (Adapted from Nadler et al. 2003; Raveh and Nadler 2007)

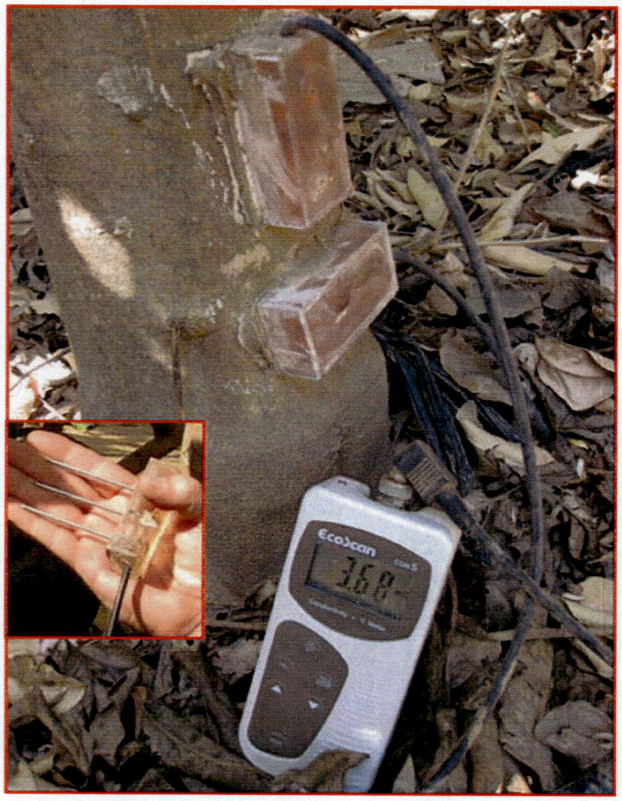

Fig. 28.6 The probes used for stem water content measurement in a pomelo trees (*Citrus maxima* × *Citrus grandis*). The probe can be connected to a TDR or to an EC meter (as seen in the picture). At the *bottom*, we can see the probe with the rods before it was installed to the main stem of the tree. The steel rods are 7 cm in length and can cross the stem from side to side (Picture by E. Raveh and A. Nadler)

Fig. 28.7 The leaf patch clamp pressure (*LPCP*) probe used by Rüger et al. (2010) for measurement changes in tree water status in pomelo trees (*Citrus maxima* × *Citrus grandis*). At the *bottom*, we can see the transmission unit that connected directly (by the *orange cable*) to the probe and used for sending data to the main system (Picture by E. Raveh and Z. Hillman-Bronshtain)

the same accuracy as the TDR measurement, but it is far less complicated and inexpensive (Fig. 28.6).

The LPCP technique makes use of a magnetic probe that measures the pressure-transfer function of an intact leaf, i.e., the attenuation of an externally applied clamp pressure by the leaf tissue (Fig. 28.7). The clamp pressure is generated by magnets. Its magnitude depends on leaf-specific structural features (Zimmermann et al. 2008, 2009; Westhoff et al. 2009). The pressure transfer through a leaf patch is dictated predominantly by turgor pressure. High turgor pressure prevents pressure transfer through the leaf, and, in turn, the output pressure measured by the probe is small. At very low turgor pressure, the transfer function assumes values close to unity, i.e., most of the applied pressure is transferred to the pressure sensor and assumes a maximum value (Zimmermann et al. 2008). The LPCP probe has already been tested on citrus trees, and its output was correlated to stem water potential values (Rüger et al. 2010). Nevertheless, both stem water content measurements with TDR or EC probes and pressure-transfer function of an intact leaf measured by LPCP probe should be examined against yield.

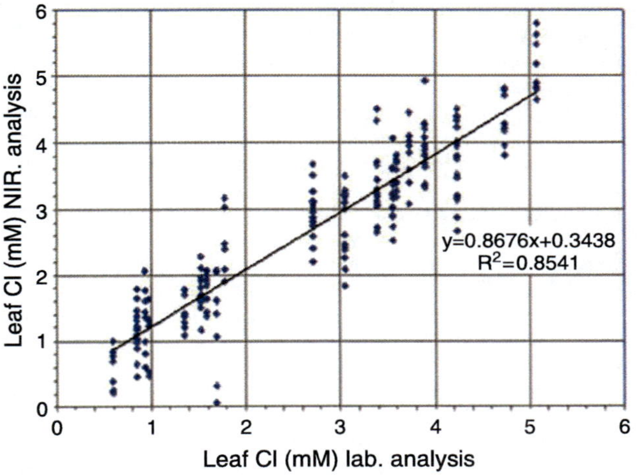

Fig. 28.8 Predicted near-infrared (*NIR*) analysis vs measured chemical (lab.) analysis of leaf chloride (Cl^-) concentration. In the future, data from hyperspectral images will enable prediction of leaf Cl^- concentrations for the early detection of salt stress (Adapted from Raveh and Hillman unpublished data)

28.6.2 Estimating the Toxicity Effect of Salinity

Estimates of toxicity effects of salinity should be based on direct measurement of tree ion concentration. In general, most of the measurements are focused on leaves (Raveh 2005 and references therein). As a transpiring tissue, leaves tend to accumulate toxic ions to a higher concentration than non-transpiring tissues (e.g., roots and stem xylem; Storey and Walker 1987; Storey 1995; Fernández-Ballester et al. 2003; Raveh and Levy 2005). Their leaf ion concentration is usually dependent on leaf age and the transpiration history of the leaf (Storey 1995; Moya et al. 2003). In addition, since we are dealing with a dynamic response, the measurements should be performed at a higher frequency than normally recommended (i.e., more than once a year). Such a requirement makes the direct measurement procedure (a destructive technique which involves chemical analysis of sampled leaves) unfeasible. It has recently been shown that leaf chloride concentration can also be determined by remote-sensing technology (Fig. 28.8; Raveh 2008). This might enable more efficient detection of leaf ion concentration.

28.7 Conclusions

The effect of salt stress on trees is a dynamic process. The stress is a combination of osmotic and toxicity components, each of which affects the trees at different intensities, dependent upon soil-solution salinity, scion/rootstock characteristics, the presence of other biotic and abiotic stresses, and the physiological condition of the tree. As a result, nondynamic models (as presented in Figs. 28.1 and 28.2) cannot be used to predict citrus salt tolerance. A model which combines the osmotic and toxicity effects of salinity in a dynamic way needs to be developed. The data for such a model will not come from potted-tree or short-term experiments, but only from long-term, well-controlled experiments conducted on mature yielding trees. Fortunately, today, new techniques are available for automated and continuous monitoring of the osmotic status of the tree; moreover, a promising technique for automated and continuous monitoring of the toxic ion concentration in leaves is being investigated. Together with the yield data, such information can be used to create a dynamic model which predicts the dynamic salt tolerance of citrus trees.

28.8 Future Research

Improvement of citrus salt tolerance can be achieved by developing new rootstocks or agrotechniques that reduce or help the trees to cope with salt stress. Out of the two options, developing new rootstock research is the preferred solution by the growers since, usually, it is easier and cheaper than adopting new agrotechnique. Yet, developing new rootstock research (at the moment done by conventional classic breeding) is a prolonged procedure (at the scale of 7–15 years; time needed for breeding, screening, and field evaluation). Using modern techniques (for instance, DNA markers for traits associated with salt tolerance) might shorten the time needed for initial screening. Yet, emerging biotechnological approaches should be continuously evaluated for their potential for expediting such breeding efforts. Research for new agrotechniques (included the developing of new and efficient desalination processes) should be cost-effective, affordable by the farmers. Nevertheless, developing both new rootstock research and agrotechnique research must include filed experiments on mature yielded trees.

References

Ball MC, Munns R (1992) Plant responses to salinity under elevated atmospheric concentrations of CO_2. Aust J Bot 40:515–525

Bañuls J, Primo-Millo E (1995) Effects of salinity on some citrus scion-rootstock combinations. Ann Bot 76:97–102

Bequet R, Čermák J, Nadezhdina N et al (2010) Tree water dynamics non-destructively assessed through sap flow measurements and potential evapotranspiration. Biol Plant 54:366–368

Bernstein L (1961) Osmotic adjustment of plant to saline media. I. Steady state. Am J Bot 48:909–918

Bielorai H, Shalhevet J, Levy Y (1978) Grapefruit response to variable salinity in irrigation water and soil. Irrig Sci 1:61–70

Bielorai H, Dasberg S, Erner Y et al (1988) The effect of saline irrigation water on Shamouti orange production. Proc Int Citrus Congr 6:707–715

Bingham ET, Mahler RJ, Parra J et al (1974) Long-term effects of irrigation-salinity management on a Valencia orange orchard. Soil Sci 117:369–377

Boman B (1994) First-year response of 'Ruby Red' grapefruit on four rootstocks to fertilization and salinity. Proc Fla State Hortic Soc 106:12–18

Boman B (2005) Salinity effects on Florida grapefruit in the Indian River region. Hort Technol 15:89–95

Cerda A, Nieves M, Guillen MG (1990) Salt tolerance of lemon trees as affected by rootstock. Irrig Sci 11:245–249

ermák J, Huzulák J, Penka M (1980) Water potential and sap flow rate in adult trees with moist and dry soil as used for the assessment of root system depth. Biol Plant 22:34–41

Combrink NJJ, Labuschagne N, Barnard RO (1996) Predisposing effect of chloride on root disease caused by *Fusarium solani*, *Phytophthora nicotianae* and *Tylenchulus semipenetrans* on four citrus rootstocks. Appl Plant Sci 10:5–9

Cooper WC, Gorton BS (1952) Toxicity and accumulation of chloride salts in citrus on various rootstocks. Proc Am Soc Hortic Sci 59:143–146

Cooper WC, Peynado A, Olson AV (1958) Response of grapefruit on two rootstocks to calcium additions to high-sodium, boron contaminated and high salinity water. Soil Sci 86:180–189

El Hammady AM, Abou Rawash M, Abou Aziz A et al (1995) Impact of irrigation with salinized water on growth and mineral content of some citrus rootstock seedlings. Ann Agric Sci Cairo 40:327–341

Embleton TW, Jones WW, Labanauskas C, Reuther W (1973) Leaf analysis as a diagnostic tool and guide to fertilization. In: Reuther W (ed) The citrus industry. University of California Press, Berkeley, vol 3, pp 183–210

Fernández-Ballester G, García-Sánchez F, Cerdá A et al (2003) Tolerance of citrus rootstock seedlings to saline stress based on their ability to regulate ion uptake and transport. Tree Physiol 23:265–271

Garcia Legaz MF, Garcia Lidon A, Porras Castillo I et al (1992) Behavior of different scion/stock combinations of lemons [*C. limon* (L.) Burm. f.] against Cland Na+ ions. Proc Int Soc Citric 1:379–404

Graham JH, Syvertsen JP (1989) Vesicular arbuscular mycorrhiza increase chloride concentration in citrus seedlings. New Phytol 113:29–36

Grattan SR, Ferguson L (2005) How salinity damages citrus: osmotic effects and specific ion toxicities. Hort Technol 15:95–99

Grieve AM, Prior LD, Bevington KB (2007) Long-term effects of saline irrigation water on growth, yield, and fruit quality of Valencia orange trees. Aust J Agric Res 58:342–348

Harding RB, Pratt PF, Jones WW (1958) Changes in salinity, nitrogen, and soil reaction in a differentially fertilized irrigated soil. Soil Sci 85:177–184

Hide JC (1954) In: U.S. Salinity Laboratory Staff, Richards LA (eds) Diagnosis and improvement of saline and alkali soils. U.S. Department of Agriculture, Washington, DC, p 160

Hoffman GJ, Catlin PB, Mead RM et al (1989) Yield and foliar injury response of mature plum trees. Irrig Sci 10:215–229

Levy Y (1980) Field determination of free proline accumulation and water stress in lemon trees. HortScience 15:302–303

Levy Y, Shalhevet J (1990) Ranking the salt tolerance of citrus rootstocks by juice analysis. Sci Hortic 45:89–98

Levy Y, Syvertsen JP (2004) Irrigation water quality and salinity effects in citrus trees. Hortic Rev 30:37–82

Levy Y, Shalhevet J, Lifshitz J (1992) The effect of salinity on citrus rootstocks and scions. In: Proceedings of the international society of citriculture, VII international citrus congress, Acireale, vol 1, pp 391–396

Levy Y, Lifshitz J, De Malach Y et al (1999) The response of several *Citrus* genotypes to high-salinity irrigation water. HortScience 34:878–881

Levy Y, Raveh E, Lifshitz J (2000) The effect of rootstock and nutrition on the response of grapefruit trees to salinity. In: Proceedings of the international society for citriculture, IX congress, pp 334–337

Maas EV (1993) Salinity and citriculture. Tree Physiol 12:195–216

Maas EV, Hoffmann GJ (1977) Crop salt tolerance – current assessment. ASCE J Irrig Drain Div 103(IR2):115–134

Mmolawa K, Or D (2000) Root zone dynamics under drip irrigation: a review. Plant Soil 222:163–190

Moya JL, Gómez-Cadenas A, Primo-Millo E et al (2003) Chloride absorption in salt-sensitive Carrizo citrange and salt-tolerant Cleopatra mandarin citrus rootstocks is linked to water use. J Exp Bot 54:825–833

Nadler A, Raveh E, Yermiahu U et al (2003) Evaluation of TDR use to monitor water content in stem of lemon trees and soil and their response to water stress. Soil Sci Soc Am J 67:437–448

Nadler A, Raveh E, Yermiahu U et al (2006) Stress induced water content variations in mango stem by time domain reflectometry. Soil Sci Soc Am J 70:510–520

Nadler A, Raveh E, Yermiahu U et al (2008) Detecting water stress in trees using stem electrical conductivity measurements. Soil Sci Soc Am J 72:1014–1027

Nieves M, Martinez V, Cerda A et al (1990) Yield and mineral composition of 'Verna' lemon trees as affected by salinity and rootstock combination. J Hortic Sci 65:359–366

Nieves M, Cerda A, Botella M (1991) Salt tolerance of 2 lemon scions measured by leaf chloride and sodium accumulation. J Plant Nutr 14:623–636

Oppenheimer HR (1937) Injurious salts and the ash composition of fruit trees. Hadar 10:3–16

Pasternak D, De Malach Y (1995) Irrigation with brackish water under desert conditions X. Irrigation management of tomatoes (*Lycopersicon esculentum* Mills) on desert sand dunes. Agric Water Manage 28:121–132

Peynado A, Young R (1962) Performance of nucellar Red Blush grapefruit trees on 13 kinds of rootstocks irrigated with saline and boron-contaminated well water over a 3-year period. J Rio Grande Valley Hortic Soc 16:52–58

Prior LD, Grieve AM, Bevington KB et al (2007) Long-term effects of saline irrigation water on Valencia orange trees: relationships between growth and yield, and salt levels in soil and leaves. Aust J Agric Res 58:349–358

Raveh E (2005) Methods to assess potential chloride stress in citrus: analysis of leaves, fruit, stem-xylem sap, and roots. Hort Technol 15:17–21

Raveh E (2008) Combating salt stress in citrus orchards under semi-arid condition in the Negev desert of Israel. Abstract 228, XIth international citrus congress, Wuhan, China, 26–30 Oct 2008, p 109

Raveh E, Levy Y (2005) Analysis of xylem water as an indicator of current chloride uptake status in citrus trees. Sci Hortic 103:317–327

Raveh E, Levy Y (2011) Effect of KNO_3 fertilization and rootstock on grapefruit response to reclaimed salinized water. Isr J Plant Sci (in press)

Raveh E, Nadler A (2007) Using electrical conductivity (EC) for monitoring stem water content changes in horticultural orchards. In: ASHS annual conference, Scottsdale, AZ. HortSci Conf Issue 42(4):964

Rüger S, Ehrenberger W, Arend M et al (2010) Monitoring of temporal and spatial changes in tree water status using the non-invasive leaf patch clamp pressure probe and the pressure chamber technique. Agric Water Manage 98:283–290

Russo LW, Wutscher HK, Bistline FW (1993) Effects of three water sources on the incidence of Rio Grande gummosis in grapefruit groves. Proc Fla State Hortic Soc 106:35–37

Scholander PF, Hammel HT, Bradstreet ED et al (1965) Sap pressure in vascular plants. Science 148:339–346

Shannon MC (1997) Adaptation of plants to salinity. Adv Agron 60:75–120

Skaggs TH, van Genuchten MTh, Shouse PJ et al (2006) Macroscopic approaches to root water uptake as a function of water and salinity stress. Agric Water Manage 86:140–149

Storey R (1995) Salt tolerance, ion relations and the effect of root medium on the response of citrus to salinity. Aust J Plant Physiol 22:101–114

Storey R, Walker RR (1987) Some effects of root anatomy on K, Na, Cl, loading of citrus roots and leaves. J Exp Bot 38:1769–1780

Storey R, Walker RR (1999) Citrus and salinity. Sci Hortic 78:39–81

Syvertsen JP (1984) Light acclimation in citrus leaves. II. CO2 assimilation and light, water and nitrogen use efficiency. J Am Soc Hortic Sci 109:812–817

Syvertsen JP, Smith ML Jr (1983) Environmental stress and seasonal changes in proline concentration of citrus tree tissues and juice. J Am Soc Hortic Sci 108:861–866

Syvertsen JP, Smith ML Jr (1984) Light acclimation in citrus leaves. I. Changes in the physical characteristics, chlorophyll, and nitrogen content. J Am Soc Hortic Sci 109:807–812

Syvertsen JP, Yelenosky G (1988) Salinity can enhance freeze tolerance of citrus rootstock seedlings by modifying growth, water relations, and mineral nutrition. J Am Soc Hortic Sci 113:889–893

Tripler E, Ben-Gal A, Shani U (2007) Consequence of salinity and excess boron on growth, evapotranspiration and ion uptake in date palm (*Phoenix dactylifera* L., cv. Medjool). Plant Soil 297:147–155

Waldo LJ, Schumann AW (2009) Spatial variability of soil water under citrus tree canopies in central Florida. Acta Hort 824:147–154

Walker RR, Blackmore DH, Qing S (1993) Carbon dioxide assimilation and foliar ion concentrations in leaves of lemon (*Citrus limon* L.) trees irrigated with NaCl or Na_2SO_4. Aust J Plant Physiol 20:173–185

Westhoff M, Reuss R, Zimmermann D et al (2009) A non-invasive probe for online-monitoring of turgor pressure changes under field conditions. Plant Biol 11:701–712

Willers P, Holmden E (1980) The influence of citrus nematode, *Tylenchulus semipenetrans*, on the performance of trees growing under saline conditions. Subtropica 1:13–16

Wu QS, Zou YN, He XH (2010) Contributions of arbuscular mycorrhizal fungi to growth, photosynthesis, root morphology and ionic balance of citrus seedlings under salt stress. ACTA Physiol Plant 32:297–304

Yelenosky G (1979) Accumulation of free proline in citrus leaves during cold hardiness of young trees in controlled temperature regimes. Plant Physiol 64:425–427

Zekri M (1991) Effects of NaCl on growth and physiology of sour orange and Cleopatra mandarin seedlings. Sci Hortic 47:305–315

Zimmermann D, Reuss R, Westhoff M et al (2008) A novel, non-invasive, online-monitoring, versatile and easy plant-based probe for measuring leaf water status. J Exp Bot 59:3157–3167

Zimmermann U, Rüger S, Shapira O et al (2009) Effects of environmental parameters and irrigation on the turgor pressure of banana plants measured using the non-invasive, online monitoring leaf patch clamp pressure probe. Plant Biol 12:424–436

Physiological Responses and Tolerance of Citrus to Aluminum Toxicity

29

Li-Song Chen

Abstract

Aluminum (Al) toxicity is the major factor limiting plant growth in acidic soils, which comprise up to 50% of the world's potentially arable lands. Citrus belong to evergreen subtropical fruit trees and are cultivated in humid and subhumid of tropical, subtropical, and temperate regions of the world mainly on acidic soils. Soil acidification is a major problem in citrus plantations. There has been significant progress in our understanding of the physiological responses and the tolerance of citrus to Al toxicity during the past decade. In this chapter, growth, physiological responses (carbohydrates, photosynthesis, water relation, light energy utilization, photoprotective system, mineral nutrients, and organic acids), and tolerance (genotypic differences, Al uptake, and distribution, and Al-induced secretion of organic acid anions) of citrus to Al toxicity are reviewed. The amelioration of phosphorus (P) and boron (B) on citrus Al toxicity as well as some aspects needed to be further studied are also discussed.

Keywords

Aluminum • Boron • Citrus • Phosphorus • Photosynthesis • Secretion of organic acid anions

29.1 Introduction

Acidic soils comprise up to 50% of the world's potentially arable lands (Kochian et al. 2005). Furthermore, the acidity of the soils is gradually increasing as a result of the environmental problems including some farming practices and acid rain. Soil pH decreased significantly from the 1980s to the 2000s in the major Chinese crop-production areas (Guo et al. 2010). In acidic soils, aluminum (Al) occurs in soluble forms that are phytotoxic, whereas in mildly acidic or neutral soil pH values, it is largely insoluble and biologically inactive. In many acidic soils through the tropics and subtropics, Al toxicity is a major factor limiting crop productivity (Kochian et al. 2004). Citrus belong to evergreen subtropical fruit trees and are cultivated in humid and subhumid of tropical, subtropical, and temperate regions of the world mainly on acidic soils. Soil acidification is a major problem in citrus plantations. Low pH and high Al are the major factors contributing to poor citrus growth and shortened life span of trees (Lin and Myhre 1990). In this chapter, recent progress that has been made in our understanding of A1 toxicity and the mechanisms of A1 tolerance in citrus has been highlighted.

29.2 Effects of Aluminum Toxicity on Growth

Al treatment decreased leaf, stem, and root fresh and dry weight of "Cleopatra" tangerine (*Citrus reshni* Hort. ex Tan.) seedlings, but increased leaf dry weight per area. Leaf and stem fresh and dry weight decreased to a greater degree than root fresh and dry weight in response to Al and resulted in a

L.-S. Chen (✉)
College of Resources and Environmental Sciences, Fujian Agriculture and Forestry University, Fuzhou 350002, PR China
e-mail: lisongchen2002@hotmail.com

A.K. Srivastava (ed.), *Advances in Citrus Nutrition*,
DOI 10.1007/978-94-007-4171-3_29, © Springer Science+Business Media B.V. 2012

Table 29.1 Leaf, stem, and root dry (fresh) weight and leaf dry weight per area of "Cleopatra" tangerine seedlings treated for 8 weeks with 0 (control) or 2 mM Al

	Dry weight (g plant^{-1})			Fresh weight (g plant^{-1})			Leaf dry weight
Treatments	Leaves	Stems	Roots	Leaves	Stems	Roots	per area (g DW m^{-2})
Al	2.00a[1]	1.05a	0.57a	5.07a	4.08a	3.66a	88.0a
Control	3.46b	1.92b	0.81b	9.54b	6.54b	4.49b	72.5b

Adapted from Chen et al. (2005b)
Data are mean of 8 replicates. Within a column, values followed by different letters are significantly different at $P<0.05$

greater ratio of root fresh (dry) weight to shoot fresh (dry) weight (Table 29.1). Increased ratio of root dry weight to shoot dry weight due to Al has been obtained in "sour pomelo" [*Citrus grandis* (L.) Osbeck] plants (Jiang et al. 2008, 2009a, b). dos Santos et al. (2000) reported that Al did not affect the ratio of shoot dry weight to root dry weight of "Swingle" citrumelo (*Citrus paradisi* Mcf. × *Poncirus trifoliata* Raf.) seedlings, but increased leaf dry weight per area. Lin and Myhre (1991a) investigated the effects of Al treatments for 60 days on growth of five citrus rootstocks. At high Al concentration, seedlings of five citrus rootstocks had fewer new and fibrous roots, thickened tips, and root caps covered with gelatinous material. Seedlings of some rootstocks treated with high-level Al had yellow, mottled, and withered new leaves near the end of experiment. New-growth fresh weight of whole plants appeared to be a better indicator for evaluation of Al tolerance than new-growth root or shoot length.

29.3 Physiological Responses of Citrus to Aluminum Toxicity

29.3.1 Leaf Carbohydrates

When expressed on a leaf area basis, no differences were observed in the concentrations of glucose, fructose, sucrose, starch, and total nonstructural carbohydrates (TNC) at either predawn or dusk between Al-treated and control leaves. The concentrations of sucrose, starch, and TNC in control leaves were higher at dusk than at predawn, whereas glucose and fructose concentrations did not change with measurement time. There were no differences in the concentrations of glucose, fructose, starch, and TNC in Al-treated leaves between measurement times, whereas sucrose concentration was greater at dusk than at predawn. When expressed on a leaf dry weight basis, the concentrations of fructose, sucrose, starch, and TNC at dusk and the concentrations of glucose and sucrose at predawn were slightly higher in control than in Al-treated leaves, whereas the other weight-based results were similar to the area-based results (Fig. 29.1; Chen et al. 2005b). It was likely that Al decreased both the utilization of carbohydrates for growth and the translocation of carbohydrates from the leaves, which would explain why Al had little effect on carbohydrate concentration despite causing severe impairment of CO_2 assimilation. Although the activity of ADP-glucose pyrophosphorylase (AGPase, EC 2.7.7.27) was higher in Al-treated leaves than in controls (Fig. 29.2f, m), starch concentration was not increased by Al (Fig. 29.1d, i). It is unclear whether AGPase was in excess or does not catalyze a rate-limiting step in Al-treated leaves.

29.3.2 Photosynthesis and Water Relations

Several studies with potted or solution-cultured citrus plants showed that CO_2 assimilation decreased under Al toxicity. The decrease in CO_2 assimilation in Al-treated citrus leaves was not caused by stomatal limitation because the intercellular CO_2 concentration in Al-treated leaves was higher than or similar to that of control leaves (Chen et al. 2005b; Jiang et al. 2008, 2009a, b; Pereira et al. 2000). The decrease in chlorophyll (Chl) concentration in response to Al was probably not the primary factor limiting CO_2 assimilation because there was a greater decrease in CO_2 assimilation than in Chl concentration (Chen et al. 2005b; Jiang et al. 2008, 2009a, b). Al had no effect on "Cleopatra" tangerine leaf malondialdehyde (MDA, a marker of peroxidative damage) concentration (Chen et al. 2005a, b), meaning that the Al had no effect on the membrane lipid in chloroplast. Therefore, the observed lower CO_2 assimilation in Al-treated leaves cannot be attributed to photooxidative damage. Pereira et al. (2000) investigated the effects of 0, 50, 100, 200, and 400 μM Al on several photosynthesis-related characteristics in four citrus rootstocks, "Cravo" lemon (*Citrus limonia* L. Osbeck), "Volkamer" lemon (*Citrus volkameriana* Hort. ex Tan.), "Cleopatra" tangerine, and "Sunki" tangerine (*Citrus sunki* Hort. ex Tang) seedlings. The CO_2 assimilation was decreased by Al in all the rootstocks, and the "Cravo" lemon seedling was the most affected, with a decrease of 85% at 400 μM Al. Stomatal conductance was not significantly affected by Al in

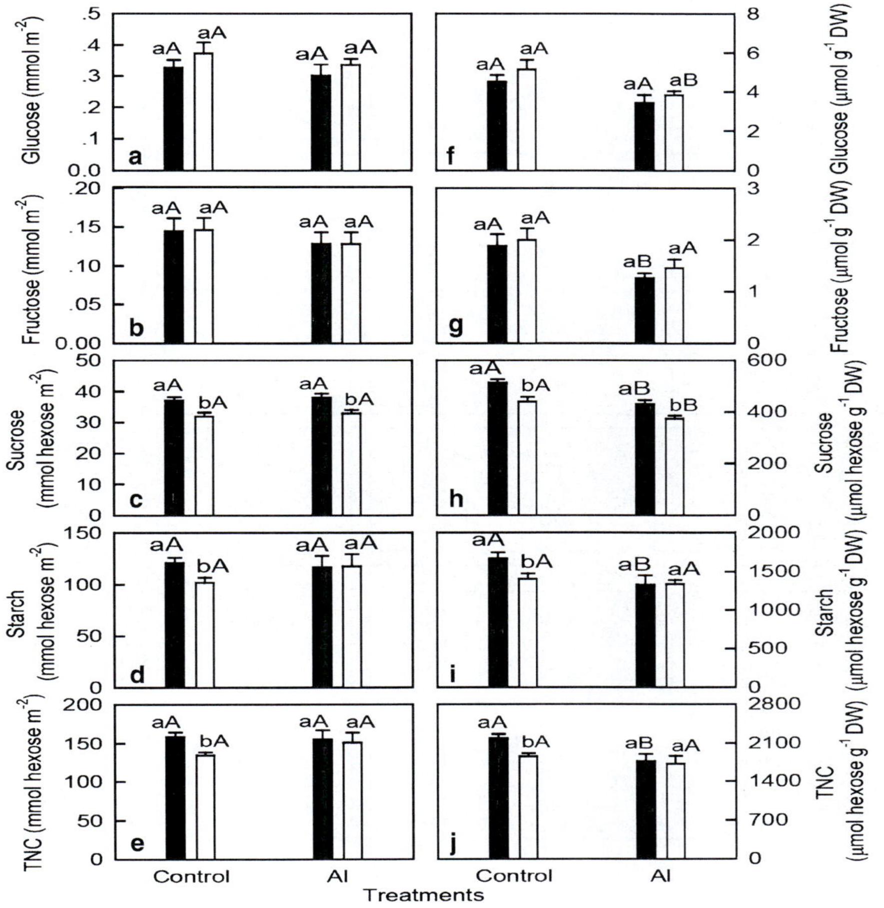

Fig. 29.1 Glucose (**a** and **f**), fructose (**b** and **g**), sucrose (**c** and **h**), starch (**d** and **i**), and total nonstructural carbohydrate (TNC, **e** and **j**) concentrations expressed on a leaf area or a leaf dry weight basis at predawn and at dusk in leaves from "Cleopatra" tangerine seedlings treated for 8 weeks with 0 (control) or 2 mM aluminum (Al). *Black* and *white bars* represent dusk and predawn, respectively. *Bars* represent mean ±SE ($n=5$). Significant difference was tested between predawn/dusk values for the same treatment and between Al treatments within each schedule. Different *small letters* above the *bars* indicate significant difference between predawn/dusk values for the same treatment at $P<0.05$. Different *capital letters* above the *bars* indicate significant difference between Al treatments within each schedule at $P<0.05$ (Adapted from Chen et al. 2005b)

the lemon seedlings but was increased in the tangerine ones. The Al-induced decrease in CO_2 assimilation might result from the structural damage to the thylakoids, as shown by a decrease in the ratio of variable fluorescence to initial fluorescence (F_v/F_o).

Our work showed that Al decreased CO_2 assimilation in "Cleopatra" tangerine (Table 29.2), but Al either increased or had no effect on the activities of enzymes involved in the Calvin cycle (Fig. 29.2). Also, Al did not cause accumulation of leaf carbohydrates (Fig. 29.1). It was suggested that the Al-induced decrease in CO_2 assimilation capacity was not associated with feedback inhibition by carbohydrate accumulation, but was probably caused by a combination of factors such as a reduced rate of electron transport through photosystem II (PSII), increased closure of PSII reaction centers, and increased photorespiration (Chen et al. 2005b). A study with potted "sour pomelo" seedlings showed that shoot growth was more sensitive to Al toxicity than root growth, gas exchange, Chl concentration, Chl a fluorescence (OJIP) transient, and related parameters (Jiang et al. 2008). Leaves from Al-stressed plants showed decreased CO_2 assimilation and Chl concentration, yet intercellular CO_2 concentration increased, and ribulose-1,5-bisphosphate carboxylase/oxygenase (Rubisco EC 4.1.1.39) activity did not change. Both control and Al-treated leaves showed a typical polyphasic rise in Chl a fluorescence (cf. Strasser et al. 1995). Al toxicity resulted in an increase in the heterogeneity of samples. OJIP transients from Al-treated leaves

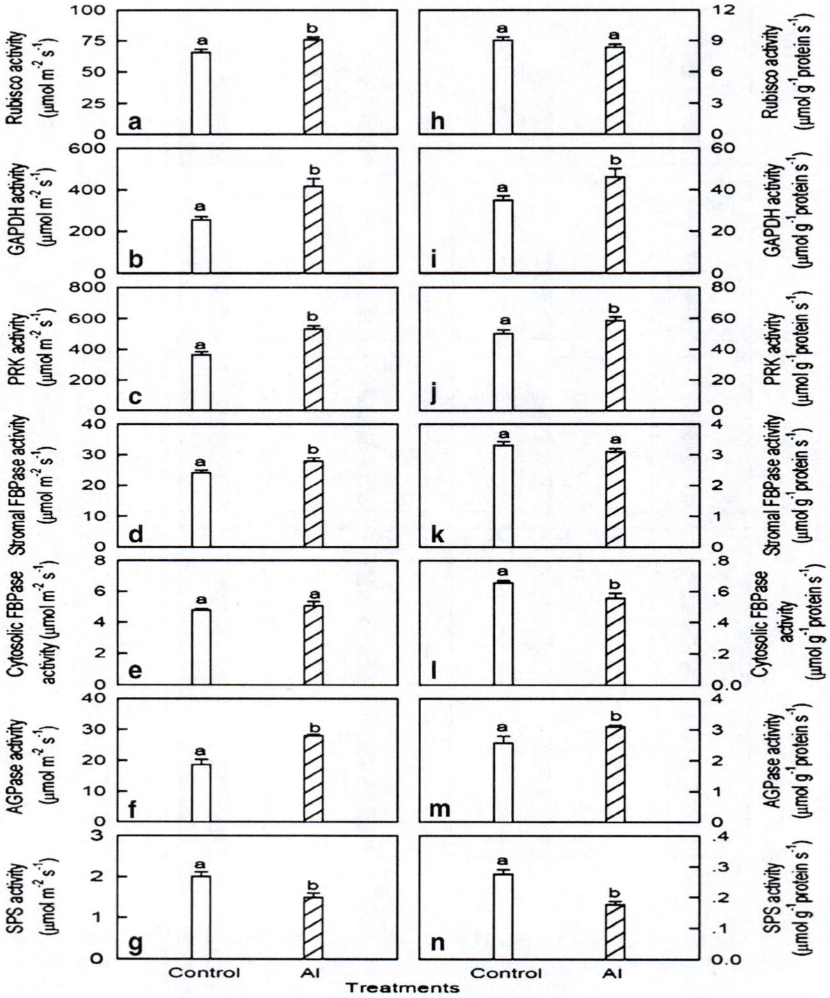

Fig. 29.2 Activities of ribulose-1,5-bisphosphate carboxylase/oxygenase (Rubisco, **a** and **h**), NADP-glyceraldehyde-3-phosphate dehydrogenase (GAPDH, **b** and **i**), phosphoribulokinase (PRK, **c** and **j**), stromal fructose-1,6-bisphosphatase (FBPase, **d** and **k**), cytosolic FBPase (**e** and **l**), ADP-glucose pyrophosphorylase (AGPase, **f** and **m**), and sucrose phosphate synthase (SPS, **g** and **n**) expressed on a leaf area or a leaf-soluble protein basis in leaves from "Cleopatra" tangerine seedlings treated for 8 weeks with 0 (control) or 2 mM aluminum (Al). *Bars* represent mean ±SE ($n=6$). Different letters *above the bars* indicate significant difference at $P<0.05$ (Adapted from Chen et al. (2005b))

Table 29.2 CO_2 assimilation, stomatal conductance, and intercellular CO_2 concentration in leaves from "Cleopatra" tangerine seedlings treated for 8 weeks with 0 (control) or 2 mM Al

Treatments	CO_2 assimilation (μmol m^{-2} s^{-1})	Stomatal conductance (mmol m^{-2} s^{-1})	Intercellular CO_2 concentration (μmol mol^{-1})
Al	3.95a	70.2a	242a
Control	6.55b	101.5b	229b

Adapted from Chen et al. (2005b)
Data are means of 6 replicates. Within a column, values followed by different letters are significantly different at $P<0.05$

showed a large rise in the O-step and a large depression at the P-step (Fig. 29.3). Al-treated leaves displayed positive ΔL-, ΔK-, ΔJ-, and ΔI-bands around 150 μs, 300 μs, 2–4 ms, and 30–100 ms, respectively, and decreased maximum amplitude of IP phase compared with controls (Fig. 29.4). Normalized total complementary area above the OJIP (S_m), performance index (PI_{abs}), total performance index ($PI_{tot,abs}$), efficiency with which a trapped exciton can move an electron into the electron transport chain from Q_A- to the PSI end electron acceptors (ρ_{Ro}), efficiency with which an electron can move

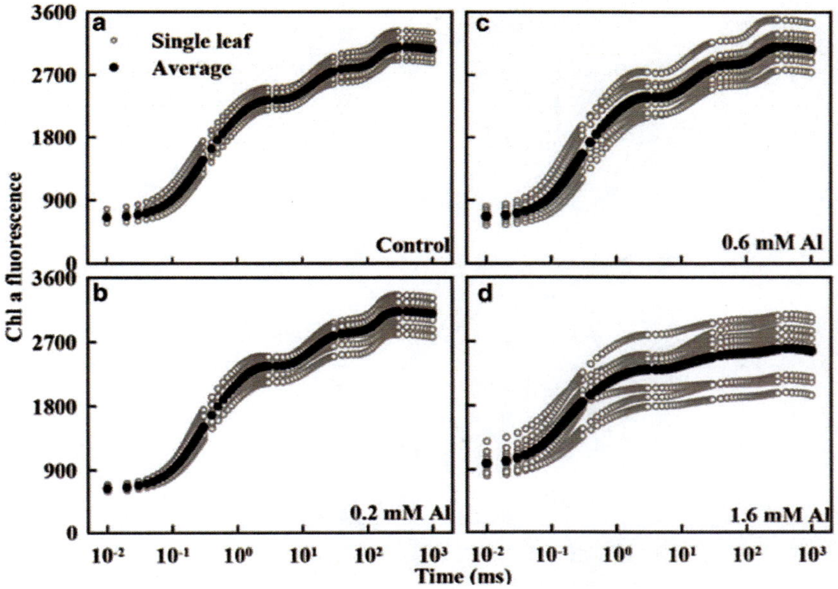

Fig. 29.3 Effects of aluminum (Al) treatments on chlorophyll (Chl) a fluorescence (OJIP) transient of dark-adapted citrus leaves from "sour pomelo" seedlings treated for 5 months with 0 (control), 0.2, 0.6, or 1.6 mM Al plotted on a logarithmic time scale (0.01–1 s). *Gray lines* are the single measurements of different samples and *black line* is the average transient of all measured samples. The heterogeneity of the samples is increasing with increasing Al concentration (Adapted from Jiang et al. 2008)

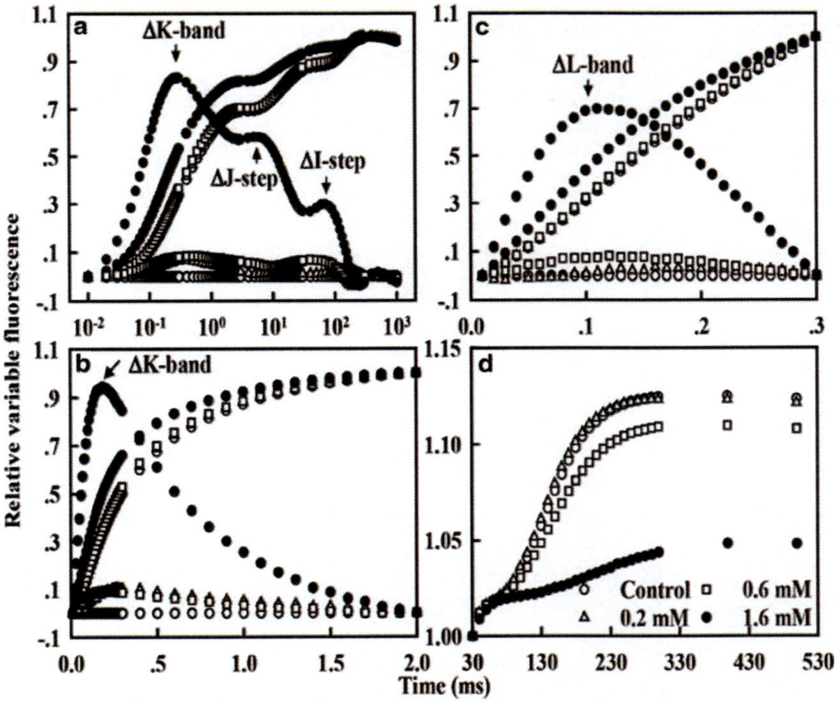

Fig. 29.4 Mean chlorophyll (Chl) a fluorescence kinetics (F_t) as the kinetics of relative variable fluorescence: (**a**) between F_o and F_m: $V_t=(F_t-F_o)/(F_m-F_o)$, (**b**) between F_o and F_J: $W_t=(F_t-F_o)/(F_J-F_o)$, (**c**) between F_o and $F_{300\mu s}$: $(F_t-F_o)/(F_{300\mu s}-F_o)$, and (**d**) between F_o and F_I: $(F_t-F_o)/(F_I-F_o)$. In each of the plots (**a**), (**b**), and (**c**), the differences of the four samples to the reference sample of nonstressed control "sour pomelo" plants treated for 5 months with 0 (control), 0.2, 0.6, or 1.6 mM aluminum (Al) are plotted with an amplication of 5, 6.5, and 6, respectively. The ΔK and ΔL bands are clearly revealed in the plots (**b**) and (**c**), respectively. The IP phase normalized on the F_o to F_I phase $=(F_t-F_o)/(F_I-F_o)$ (Adapted from Jiang et al. 2008)

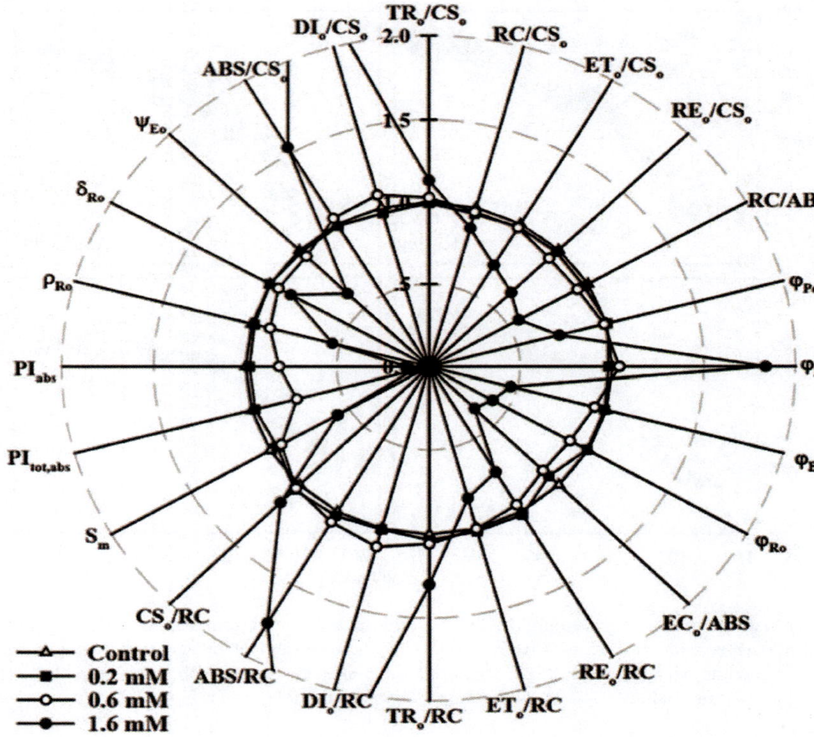

Fig. 29.5 Effects of aluminum (Al) treatments on "sour pomelo" plants, evaluated by the behavior pattern of 24 structural and functional parameters of photosystem, derived by the JIP-test from the chlorophyll (Chl) a fluorescence (OJIP) transient exhibited by dark-adapted leaves of plants treated for 5 months with 0 (control), 0.2, 0.6, or 1.6 mM Al. In each case, the parameters are derived from the corresponding average transient. All the values were expressed relative to the control set as 1. ABS/CS_o absorption flux (ABS) per excited cross section (CS) at $t=0$, DI_o/CS_o dissipated energy flux per CS at $t=0$, TR_o/CS_o trapped energy flux per CS at $t=0$, RC/CS_o amount of active photosystem II (PSII) reaction centers (RCs) per CS at $t=0$, ET_o/CS_o electron transport flux per CS at $t=0$, RE_o/CS_o reduction of end acceptors at PSI electron acceptor side per CS at $t=0$, RC/ABS amount of active PSII RCs per ABS, φ_{Po} maximum quantum yield of primary photochemistry at $t=0$, φ_{Do} quantum yield at $t=0$ for energy dissipation, φ_{Eo} quantum yield for electron transport at $t=0$, φ_{Ro} quantum yield for the reduction of end acceptors of PSI per photon absorbed, EC_o/ABS electron carriers per ABS at $t=0$, RE_o/RC reduction of end acceptors at PSI electron acceptor side per RC at $t=0$, ET_o/RC electron transport flux per RC at $t=0$, TR_o/RC trapped energy flux per RC at $t=0$, DI_o/RC dissipated energy flux per RC at $t=0$, ABS/RC absorption flux per RC, CS_o/RC excited cross section per RC at $t=0$, S_m normalized total complementary area above the OJIP (reflecting multiple-turnover Q_A reduction events) or total electron carriers per RC, $PI_{tot,abs}$ total performance index, measuring the performance up to the PSI end electron acceptors, PI_{abs} performance index (PI) on absorption basis, ρ_{Ro} efficiency with which a trapped exciton can move an electron into the electron transport chain from Q_A^- to the PSI end electron acceptors, δ_{Ro} efficiency with which an electron can move from the reduced intersystem electron acceptors to the PSI end electron acceptors, ψ_{Eo} probability (at time 0) that a trapped exciton moves an electron into the electron transport chain beyond Q_A^- (Adapted from Jiang et al. 2008)

from the reduced intersystem electron acceptors to the PSI end electron acceptors (δ_{Ro}), probability (at time 0) that a trapped exciton moves an electron into the electron transport chain beyond Q_A^- (ψ_{Eo}), amount of active PSII reaction centers (RCs) per excited cross section (CS) at $t=0$ (RC/CS_o), electron transport flux per CS at $t=0$ (ET_o/CS_o), reduction of end acceptors at PSI electron acceptor side per CS at $t=0$ (RE_o/CS_o), amount of active PSII RCs per absorption (ABS) (RC/ABS), maximum quantum yield of primary photochemistry at $t=0$ (φ_{Po}, F_v/F_m), quantum yield for electron transport at $t=0$ (φ_{Eo}), quantum yield for the reduction of end acceptors of PSI per photon absorbed (φ_{Ro}), electron carriers per ABS at $t=0$ (EC_o/ABS), reduction of end acceptors at PSI electron acceptor side per RC at $t=0$ (RE_o/RC), and electron transport flux per RC at $t=0$ (ET_o/RC) were decreased in Al-stressed leaves, whereas absorption flux per CS at $t=0$ (ABS/CS_o), dissipated energy flux per CS at $t=0$ (DI_o/CS_o), quantum yield at $t=0$ for energy dissipation (φ_{Do}) and dissipated energy flux per RC at $t=0$ (DI_o/RC), trapped energy flux per CS at $t=0$ (TR_o/CS_o), trapped energy flux per RC at $t=0$ (TR_o/RC), absorption flux per RC (ABS/RC), and excited cross section per RC at $t=0$ (CS_o/RC) was increased (Fig. 29.5). Regressive analysis showed that relative leaf Al concentration increased linearly with increasing Al supply,

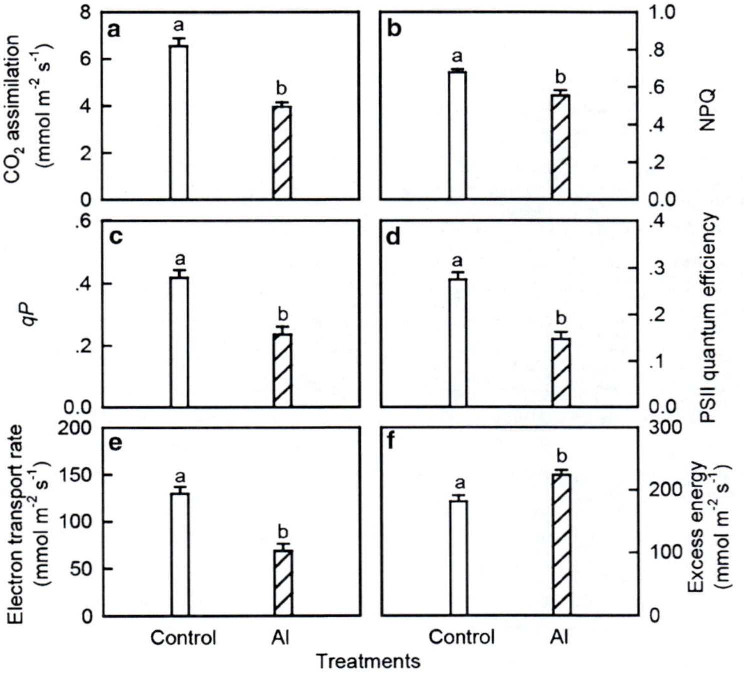

Fig. 29.6 CO_2 assimilation (**a**), non-photochemical quenching (NPQ, **b**), photochemical-quenching coefficient (*qP*, **c**), PSII quantum efficiency (**d**), electron transport rate (**e**), and excess energy (**f**) in leaves from "Cleopatra" tangerine seedlings treated for 8 weeks with 0 (control) or 2 mM aluminum (Al). *Bars* represent mean ±SE ($n = 8$). Different letters *above the bars* indicate significant difference at $P < 0.05$ (Adapted from Chen et al. 2005a)

whereas shoot DW, leaf IP rise from F_{80ms} to F_m, and CO_2 assimilation decreased linearly. Shoot DW decreased linearly with increasing leaf Al concentration, and leaf CO_2 assimilation increased linearly with increasing leaf IP rise from F_{80ms} to F_m (Jiang et al. 2008). We concluded that the impaired electron transport capacity accompanied by the lack of reducing equivalents were the main factor contributing to the decreased CO_2 assimilation in Al-stressed leaves. The Al-induced photoinhibition occurring at both the donor (i.e., the oxygen-evolving complex, OEC) and the acceptor sides of PSII could be associated with growth inhibition. Besides decreased absorption due to reduced Chl concentration, enhanced energy dissipation protected Al-stressed leaves from photooxidative damage in high light. Pereira et al. (2000) reported that Al, at 100, 200, or 400 μM, decreased water use efficiency (WUE) in 3 of the evaluated 4 *Citrus* spp. rootstocks due to the increase in transpiration and decrease in CO_2 assimilation, while Al, at 50 μM, increased WUE because Al increased CO_2 assimilation more than transpiration. However, Chen et al. (2010) observed that both leaf transpiration and WUE in "Cleopatra" tangerine decreased in response to Al, whereas relative water content (RWC) showed nonsignificant change.

29.3.3 Light Energy Utilization and Photoprotective Systems

Chen et al. (2005a) showed that Al-treated "Cleopatra" tangerine leaves only used a smaller fraction of the absorbed light in electron transport since CO_2 assimilation decreased to a greater degree than leaf Chl concentration or leaf light absorption in response to Al. As a result, more excess excitation energy existed in Al-treated leaves compared with controls under high photon flux at midday (Fig. 29.6). It has been suggested that excess absorbed light could be harmlessly dissipated as heat through xanthophyll cycle-dependent thermal energy dissipation in the antenna pigment complexes of PSII (Demmig-Adams and Adams 1996; Niyogi et al. 1998). However, the xanthophyll cycle-dependent thermal energy dissipation could not have a central role in dissipating excess excitation energy in Al-treated "Cleopatra" tangerine leaves since xanthophyll cycle-dependent thermal energy dissipation, measured as non-photochemical quenching (NPQ), was slightly decreased by Al (Fig. 29.6). Interestingly, dissipated energy (DI_o/CS_o, φ_{Do} and DI_o/RC) was higher in Al-treated "sour pomelo" leaves than in controls (Jiang et al. 2008, 2009a, b).

An alternative route for energy dissipation and consumption of photosynthetic electrons is directly in the water-water (Asade) cycle or indirectly in the photorespiration (Asada 1999). As expected, the activities of enzymes [superoxide dismutase (SOD, EC 1.15.1.1), ascorbate peroxidase (APX, EC 1.11.1.11), monodehydroascorbate reductase (MDAR, EC 1.6.5.4), dehydroascorbate reductase (DHAR, EC 1.8.5.1), and glutathione reductase (GR, EC 1.6.4.2)], the concentrations of antioxidant metabolites [ascorbate (AsA), dehydroascorbate (DAsA), reduced glutathione (GSH), and oxidized glutathione (GSSG)] involved in water-water cycle, the activity of catalase (CAT, EC 1.11.1.16, an enzyme involved in scavenging the bulk H_2O_2 generated by photorespiration) (Chen et al. 2005a), and the activity of Rubisco in "Cleopatra" tangerine were significantly increased by Al (Chen et al. 2005b). Closure of PSII RCs will result in formation of toxic-activated oxygen species. The upregulation of enzymatic and nonenzymatic antioxidants also agreed with the increased requirement for scavenging reactive oxygen species in Al-treated leaves due to increased closure of PSII reaction centers, as indicated by decreased qP (Fig. 29.6c). Although superoxide generation and H_2O_2 production increased in Al-treated "Cleopatra" tangerine leaves, no significant difference was found in MDA concentration between the Al treatments (Chen et al. 2005a). These results indicate that upregulation of the antioxidant system provided considerable protection to Al-treated leaves against photooxidative damage. In contrast to NPQ, antheraxanthin (A)+zeaxanthin (Z) expressed on a xanthophyll cycle pool basis or a leaf Chl basis at both predawn and midday was higher in Al-treated than in control "Cleopatra" tangerine leaves. At midday, A+Z expressed on a xanthophyll cycle pool basis or a Chl basis increased, especially in Al-treated "Cleopatra" tangerine leaves, but A+Z only accounted for less 40% of the total xanthophyll cycle pool even in Al-treated leaves, with the balance in violaxanthin (V). The increase in the conversion of V to A and Z in Al-treated "Cleopatra" tangerine might help to quench $^1O_2$ as its production might increase in Al-treated leaves under high light due to increased closure of PSII (Chen et al. 2005a). Z could directly protect the thylakoid membrane against photooxidation (Müller et al. 2001). In addition, Z could play a role in stabilization of the thylakoid membrane (Tardy and Havaux 1997). NPQ was believed to depend on the accumulation of de-epoxidation products of A+Z of the xanthophyll cycle pool, but under Al stress, the changes in NPQ in "Cleopatra" tangerine leaves did not correspond to the changes in A+Z expressed on a xanthophyll cycle pool basis (Chen et al. 2005a). The mechanisms underlying these phenomena need to be addressed in the further research.

29.3.4 Mineral Nutrients

We investigated the effects of 2 mM Al treatment for 8 weeks on mineral nutrients in "Cleopatra" tangerine leaves (Chen 2006; Chen et al. 2010). The concentrations of P, K, Mg, Cu, Zn, and Mo were lower in Al-treated leaves compared with controls, while the concentrations of N, Ca, Fe, Mn, and B did not significantly differ between the Al treatments. A field experiment with 30-year-old trees of *C. sinensis* (L.) Osbeck cv. Hamlin on sour orange (*C. aurantium* L.) rootstock showed that the concentrations of Zn, Fe, and Mn in fibrous roots decreased with increased Al supply to soil, whereas the concentrations of Ca, Mg, K, P, and B were unaffected (Lin and Myhre 1990). In a solution-culture experiment, Lin and Myhre (1991b) determined the effects of seven levels of Al ranging from 4 to 1,655 μM for 60 days on the mineral concentrations in five citrus rootstocks: "Cleopatra" tangerine, rough lemon, "Swingle" citrumelo, sour orange, and "Carrizo" citrange. The concentrations of K, Mg, and P in roots and the concentrations of P and K in shoots increased as Al concentration in nutrient solution increased from 4 to 178 μM, while the concentrations of Ca, Zn, Cu, Mn, and Fe in roots and the concentrations of Ca, Mg, Cu, and Fe in shoots decreased. Yokomizo and Ishihara (1973) observed that the concentrations of Mn and Zn in Natsudaidai (*C. natsudaidai* Hayata) seedlings decreased with increased Al to the nutrient solution.

29.3.5 Organic Acids

The concentrations of malate and citrate in +Al leaves decreased with increasing phosphorus (P) supply, but their concentrations in −Al leaves did not significantly change in response to P supply. The concentrations of malate under 50 μM P and of citrate under 50 and 100 μM P were higher in +Al leaves than in −Al ones, but malate concentration was lower in +Al leaves than in −Al ones under 500 μM P (Fig. 29.7a, b). There was no significant difference in root malate and citrate concentrations among different P and Al combinations except for an increase in malate and citrate under 50 μM P+0 mM Al, and a slight decrease in malate under 50 μM P+1.2 mM Al (Fig. 29.7c, d). The activities of acid-metabolizing enzymes [citrate synthase (CS, EC 2.3.3.1), aconitase (ACO, EC 4.2.1.3), phospho*enol*pyruvate carboxylase (EC 4.1.1.31), NADP-isocitrate dehydrogenase (NADP-IDH, EC 1.1.1.42), phospho*enol*pyruvate phosphatase (PEPP, EC 3.1.3.60), NAD-malate dehydrogenase (NAD-MDH, EC 1.1.1.37), NADP-malic enzyme (NADP-ME, EC 1.1.1.40), and pyruvate kinase (PK, EC 2.7.1.40)] in most cases were less affected by P and Al interactions in roots compared with

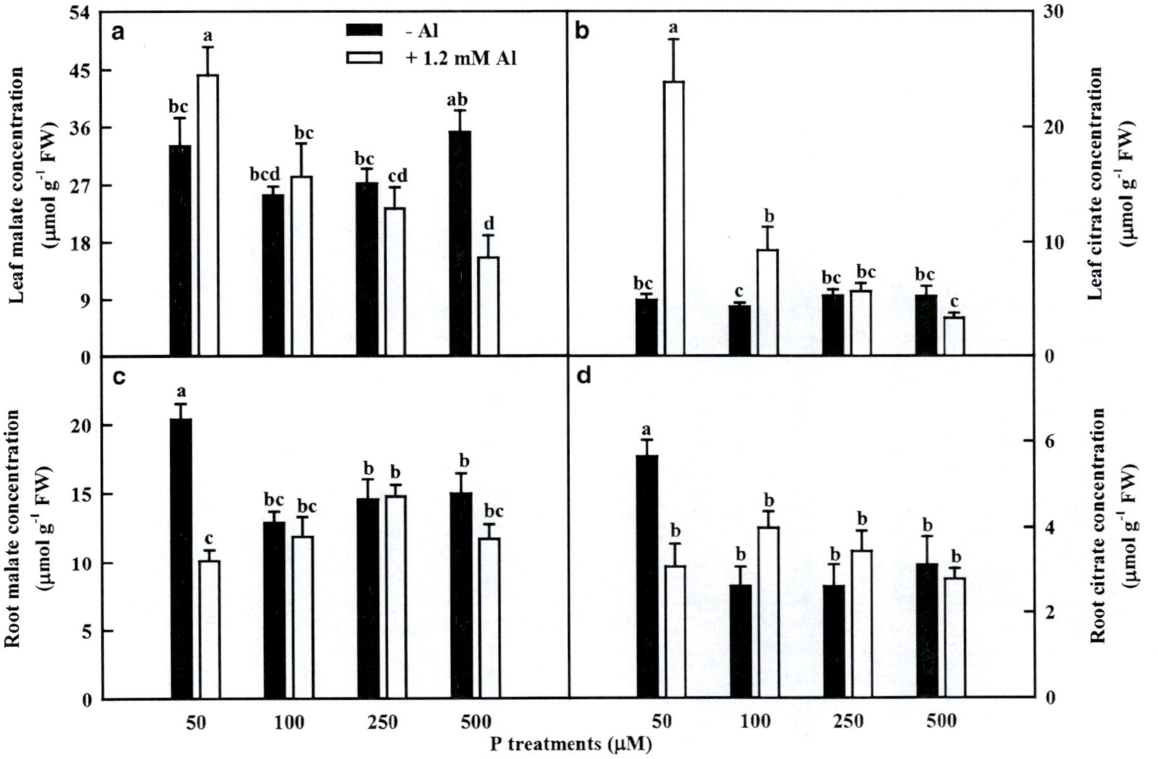

Fig. 29.7 Malate and citrate concentrations in roots and leaves from "sour pomelo" seedlings treated with different phosphorus (P) and aluminum (Al) levels for 18 weeks. *Bars* represent mean ±SE (n=5 or 6). Each variable was analyzed by 2 (Al)×4 (P) ANOVA. P values for Al, P, and the interaction between the two were 0.2158, 0.0015, and 0.0021 (**a**), <0.0001, <0.0001, and <0.0001 (**b**), <0.0001, 0.0194, and 0.0005 (**c**), and 0.8673, 0.0549, and 0.0012 (**d**), respectively. Different letters *above the bars* indicate a significant difference at $P<0.05$ (Adapted from Chen et al. 2009)

leaves (Figs. 29.8 and 29.9). This may be associated with the smaller changes in root Al *versus* leaf Al (Chen et al. 2009; Jiang et al. 2009a, Fig. 29.11). These results demonstrate that changes in organic acid (OA) metabolism differ between roots and leaves of "sour pomelo" in response to P and Al interactions (Figs. 29.7, 29.8, and 29.9).

29.4 Aluminum Tolerance

29.4.1 Genotypic Differences in Aluminum Tolerance in Citrus

Wide genotypic differences in Al tolerance exist in citrus. de Magalhães (1987) compared the tolerance of citrus rootstocks to Al supplied to the soil and observed that Al significantly affected the growth of the roots and shoots, with the "Rugoso da Florida" lemon as the most tolerant to Al, followed by "Cleopatra" tangerine and by "Rangpur" lime. Lin and Myhre (1991a) evaluated the tolerance of five citrus rootstocks to Al. The sequential order of Al tolerance for new-growth fresh weight of whole plants was as follows: "Cleopatra" tangerine > rough lemon > sour orange > "Swingle" citrumelo > "Carrizo" citrange, and the neutral (dividing) Al concentration between beneficial and toxic effects were 371, 193, 189, 178, and <100 μM, respectively, for the above rootstocks. Pereira et al. (2003) investigated the effects of 0, 50, 100, 200, and 400 μM Al treatments for 70 days on the growth of "Rangpur" lime (*Citrus limonia* Osbeck), "Volkamer" lemon (*Citrus volkameriana* Hort. ex Tan.), "Cleopatra" tangerine, and "Sunki" tangerine (*Citrus sunki Hort.* ex Tan.) in hydroponic culture. "Rangpur" lime was the most sensitive rootstock to Al, and "Cleopatra" tangerine tree presented the greatest tolerance to Al. Jiang et al. (2009c) investigated the effects of 0, 0.2, 0.6, 1.0, or 1.6 mM Al treatments for 3 months on seedling growth of 12 citrus species and cultivars. The results showed that *C. sinensis* (L.) Osbeck cv. Xuegan, *C. reticulata* Blanco cv.

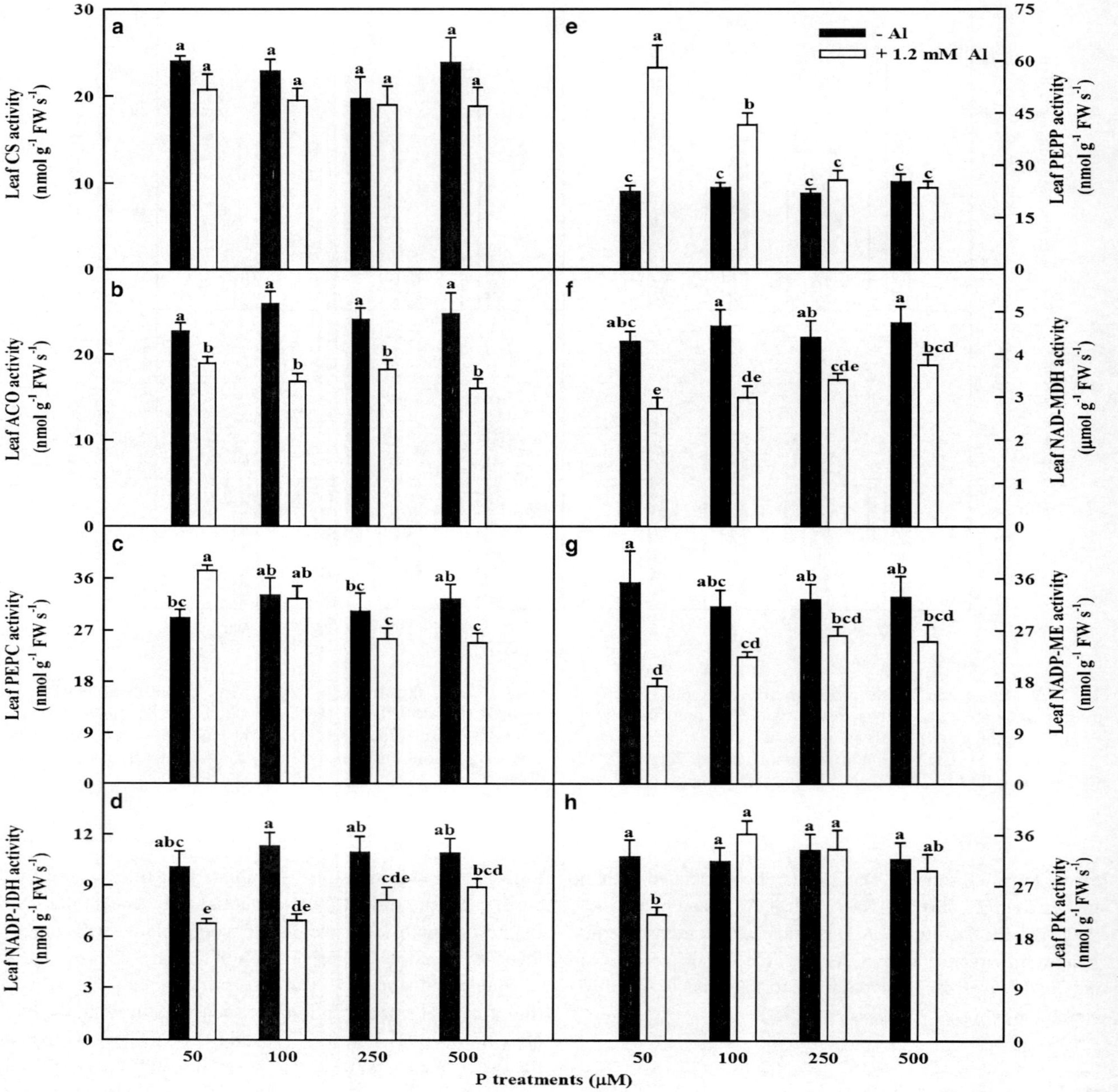

Fig. 29.8 Citrate synthase (**a**, CS), aconitase (ACO, **b**), phospho*enol*pyruvate carboxylase (PEPC, **c**), NADP-isocitrate dehydrogenase (NADP-IDH, **d**), phospho*enol*pyruvate phosphatase (PEPP, **e**), NAD-malate dehydrogenase (NAD-MDH, **f**), NADP-malic enzyme (NADP-ME, **g**), and pyruvate kinase (PK, **h**) in leaves from "sour pomelo" seedlings treated with different phosphorus (P) and aluminum (Al) levels for 18 weeks. *Bars* represent the mean ± SE (n = 5 or 6). Each variable was analyzed by 2 (Al) × 4 (P) ANOVA. P values for Al, P, and the interaction between the two were 0.0357, 0.4953, and 0.7530 (**a**), <0.0001, 0.8958, and 0.1920 (**b**), 0.6585, 0.0574, and 0.007 (**c**), <0.0000, 0.2165, and 0.4609 (**d**), <0.0001, <0.0001, and 0.0002 (**e**), <0.0001, 0.1615, and 0.5681 (**f**), <0.0001, 0.7132, and 0.1851 (**g**), and 0.4345, 0.0914, and 0.0446 (**h**), respectively. Different letters *above the bars* indicate a significant difference at $P < 0.05$ (Adapted from Chen et al. 2009)

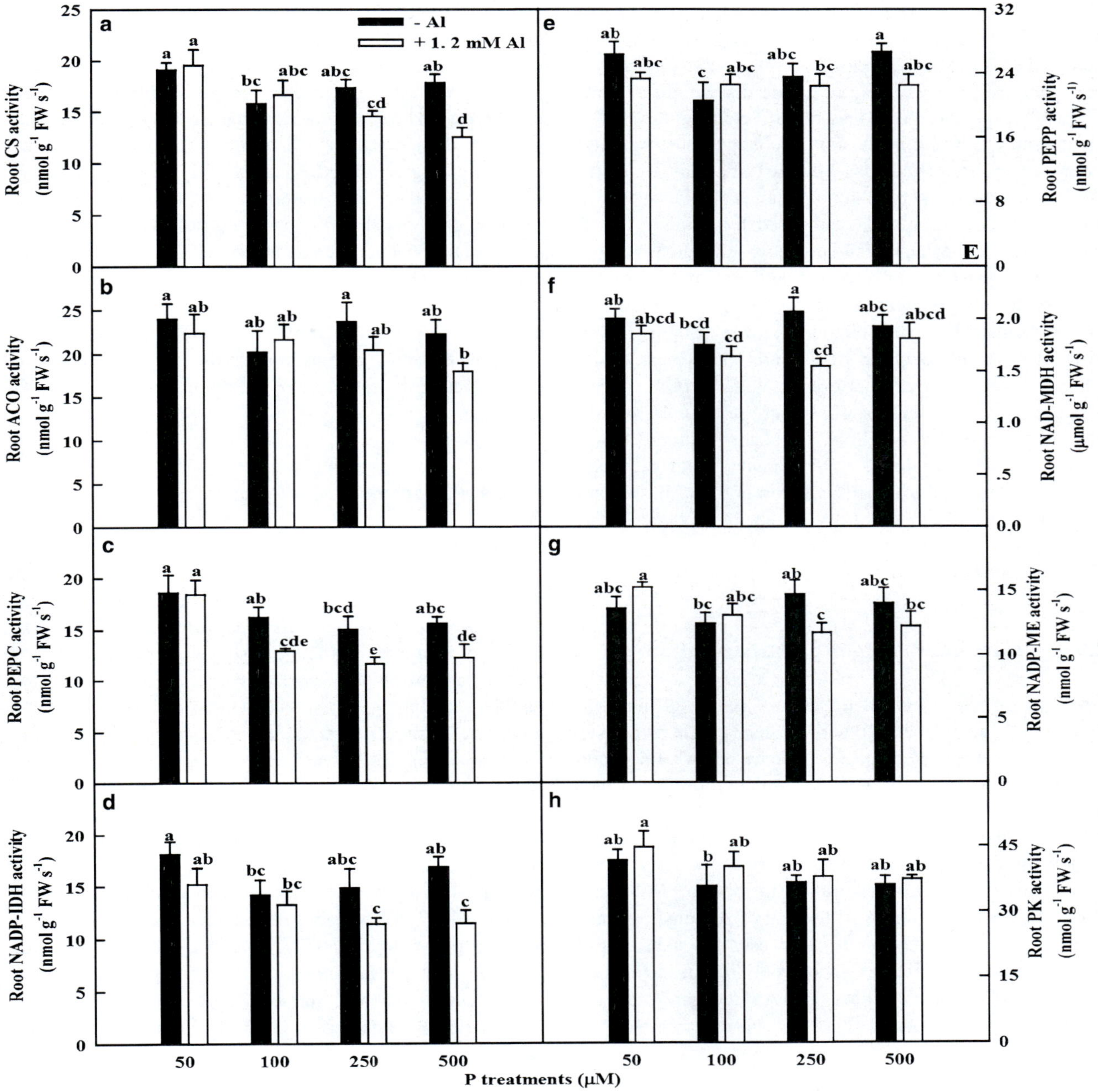

Fig. 29.9 Citrate synthase (**a**, CS), aconitase (ACO, **b**), phospho*enol*pyruvate carboxylase (PEPC, **c**), NADP-isocitrate dehydrogenase (NADP-IDH, **d**), phospho*enol*pyruvate phosphatase (PEPP, **e**), NAD-malate dehydrogenase (NAD-MDH, **f**), NADP-malic enzyme (NADP-ME, **g**), and pyruvate kinase (PK, **h**) in roots from "sour pomelo" seedlings treated with different phosphorus (P) and aluminum (Al) levels for 18 weeks. *Bars* represent the mean ±SE ($n=5$ or 6). Each variable was analyzed by 2 (Al) × 4 (P) ANOVA. *P* values for Al, P, and the interaction between the two were 0.0344, 0.0027, and 0.0222 (**a**), 0.1014, 0.3794, and 0.5137 (**b**), 0.0057, 0.0004, and 0.1349 (**c**), 0.0017, 0.0534, and 0.4168 (**d**), 0.1319, 0.1110, and 0.1916 (**e**), 0.0072, 0.1887, and 0.1903 (**f**), 0.3646, 0.3315, and 0.0711 (**g**), and 0.2496, 0.1468, and 0.9451 (**h**), respectively. Different letters *above the bars* indicate a significant difference at $P<0.05$ (Adapted from Chen et al. 2009)

Chachiensis, *C. reticulata* Blanco cv. Ponkan, *C. reticulata* Blanco cv. Fuju, *C. reticulata* Blanco cv. Sichunongju, *C. limon* (L.) Burm. cv. Eureka, and *Poncirus trifoliata* (L.) Raf. were relative Al tolerant, whose dry weight of roots, shoots, and whole plants was not affected by 1.6 mM Al treatment. *C. aurantium* cv. Goutoucheng (from Zhejiang), sour orange (from Hubei), *C. limonia* Osbeck cv. Red limonia, and citrumelo were relative Al sensitive, whose growth of roots and/or shoots was inhibited by 1.6 mM Al. *C. grandis* (L.) Osbeck cv. Wendanyou was the most Al sensitive of all these citrus species and cultivars, and 0.2 mM of Al significantly decreased the dry weight of roots and whole plants. Al stress resulted in an increase in the ratio of root dry weight to shoot dry weight of "Chachiensis," "Fuju," "Sichunongju," sour orange, and *P. trifoliata* seedlings, but did not significantly affect the ratio of root dry weight to shoot dry weight of "Ponkan," "Guotoucheng," "Eureka," "Red limonia," "Wendanyou," and citrumelo seedlings. No significant difference was found in the ratio of root dry weight to shoot dry weight of "Xuegan" seedlings among different treatments except for a significant decrease under 0.2 mM Al.

29.4.2 Aluminum Uptake and Distribution

Tolerance of plants to Al toxicity is associated not only with low Al uptake but also with relatively little Al translocation from roots to shoots. Evidence shows that some Al-tolerant plant species can accumulate more Al in their roots and less Al in their shoots (Roy et al. 1988). Lin and Myhre (1991a) observed that Al concentration in roots and shoots increased with increasing Al concentration in nutrient solution, and Al concentration in roots of Al-tolerant citrus rootstocks was higher than that of Al-sensitive ones. Our work showed that under Al stress, Al concentration was higher in Al-tolerant "Xuegan" roots than in Al-sensitive "sour pomelo" ones, but lower in +Al "Xuegan" stem and leaves (Yang et al. 2011).

29.4.3 Aluminum-Induced Secretion of Organic Acid Anions

A major mechanism of Al tolerance in plants is the Al-induced secretion of OA anions from roots (Kochian et al. 2004; Ma 2000, 2007). A study from our laboratory showed that Al-tolerant "Xuegan" roots secreted more malate and citrate than Al-sensitive "sour pomelo" ones in response to Al (Yang et al. 2011). Deng et al. (2009) tested the Al tolerance of Yuzu (*C. junos* Sieb. ex Tanaka) based on root elongation, suggesting that Yuzu was an Al-tolerant plant compared with other plant species. Al activated the secretion of citrate from the Yuzu roots. These results indicated that the Al-induced secretion of citrate from roots was involved in the Al tolerance of Yuzu. Al also increased the concentration of citrate, the expression level of mitochondrial citrate synthase (CjCS) gene, and the activity of CjCS. *Nicotiana benthamiana* plants overexpressing a Yuzu *CjCS* displayed enhanced Al tolerance, increased citrate concentration, and secretion of citrate in roots. Zhang et al. (2008) reported that overexpression of a Yuzu malate dehydrogenase in tobacco (*Nicotiana tobacum* cv. Xanthi) conferred Al tolerance. Deng et al. (2008) isolated a mitochondrial dicarboxylate/tricarboxylate carrier gene (*CjDTC*) from Yuzu and observed that its expression was induced by Al, suggesting that this gene protein might be involved in the excretion of OA anions and rhizotoxic Al tolerance.

29.4.4 Phosphorus

P could alleviate Al-induced inhibition on growth of "sour pomelo" seedlings (Fig. 29.10), as found for rice (*Oryza sativa* L., Nakagawa et al. 2003), *Lespedeza bicolor* (Sun et al. 2008), and sorghum [*Sorghum bicolor* (L.) Moench, Tan and Keltjens 1990a, b]. P-induced amelioration of growth inhibition may be related to the following two factors, including (*a*) the formation of Al-P at the root surface and/or in the root tissues and less Al accumulation in stems and leaves (Fig. 29.11a–c) and (*b*) increased P level in the roots, stems, and leaves (Fig. 29.11d–f). Our finding that root Al concentration increased from 50 to 250 μM, then remained unchanged at the highest P supply under Al stress (Fig. 29.11a), indicates that increased insoluble Al-P, which are nontoxic to plants, at the root surface and/or in the root tissues is likely responsible for the lower stem and leaf Al level at high P supplies (Fig. 29.11b–c). This agrees with previous view that insoluble nontoxic Al-P precipitates may accumulate at the root surface and/or in the root tissues (Taylor 1991). Al decreased leaf CO_2 assimilation, Rubisco activity, and Chl concentration, whereas it increased or did not affect intercellular CO_2 concentration. Al affected CO_2 assimilation more than Rubisco and Chl under 250 and 500 μM P. Al decreased leaf maximum quantum yield of primary photochemistry (F_v/F_m), maximum amplitude of IP phase, and total performance index ($PI_{tot,abs}$), but increased leaf minimum fluorescence (F_o), relative variable fluorescence at K- and I-steps. P could alleviate Al-induced increase or decrease for all these parameters (Jiang et al. 2009a, Fig. 29.12). Regressive analysis showed that leaf CO_2 assimilation increased with increasing leaf initial Rubisco activity, maximum amplitude of IP phase, and $PI_{tot,abs}$, respectively (Jiang et al. 2009a). We concluded that P alleviated

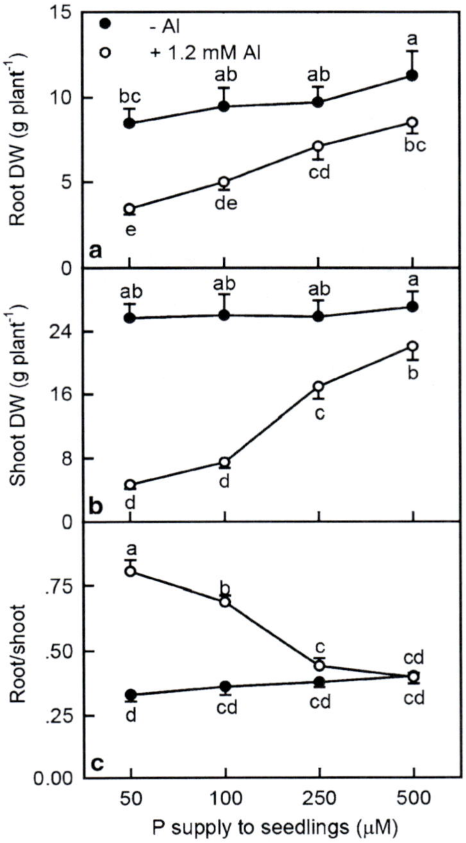

Fig. 29.10 Effects of phosphorus (P) and aluminum (Al) interactions on root dry weight (DW, **a**), shoot DW (**b**), and root/shoot ratio (**c**) of "sour pomelo" seedlings treated with different P and Al levels for 18 weeks. Each *point* represents the mean ±SE ($n=12$). Each variable was analyzed by 2 (Al)×4 (P) ANOVA. (**a**) *P* values for Al, P, and the interactions between the two were <0.0001, <0.0001, and 0.2092, respectively, (**b**) and (**c**) *P* values for Al, P, and the interactions between the two were all <0.0001. Different *letters* indicate significant differences at $P<0.05$ (Adapted from Jiang et al. 2009a)

Al-induced inhibition of growth and impairment of the whole photosynthetic electron transport chain from photosystem II (PSII) donor side up to the reduction of end acceptors of PSI, thus preventing photosynthesis inhibition through increasing Al immobilization in roots and P level in roots and shoots.

29.4.5 Boron

Recently, our work (Jiang et al. 2009b) showed that B could alleviate the inhibitory effects of Al toxicity on growth (Fig. 29.13) and leaf Chl, Chl a, and Chl b concentrations. The ameliorative effects of B was not brought about by an increase in the B concentration of roots, stems, and leaves because B concentration was not lower in +Al than in −Al roots, stems, and leaves (Fig. 29.14). The sequence of the ameliorative effect of B on growth inhibition and Chl decrease in +Al seedlings was 25 μM B > 10 μM B ≥ 50 μM B > 2.5 μM B (Fig. 29.13, Jiang et al. 2009b), indicating that the Al-induced growth inhibition is not due to Al-induced B deficiency. Corrales et al. (2008) reported that Al increased the concentration of GSH in roots of maize (*Zea may* L.) plants growing with adequate B supply but not in those growing in excess B, which, in turn, caused extensive cell damage in the root tips of maize plants even in the absence of Al. The lower root dry weight in 50 μM B + 0 mM Al-treated seedlings (Fig. 29.13a) implies that these plants received excess B. This would explain why the ameliorative effect of 50 μM B was lower than that of 25 μM B because +Al roots, stems, and leaves displayed higher or similar B concentration (Fig. 29.14d–f). No difference for root Al concentration among B treatments (Fig. 29.14a) indicates that the B-induced amelioration of root inhibition was probably caused by B-induced changes in Al speciation and/or subcellular compartmentation rather than by less Al accumulation in root tips. Our finding that Al concentration was the highest in 2.5 μM B-treated stems and leaves under Al stress, followed by 10, 50, and 25 μM B-treated ones (Fig. 29.14b–c) indicates that the B-induced amelioration of shoot inhibition could be due to less Al accumulation in shoots. Al-treated leaves had decreased CO_2 assimilation but increased or similar intercellular CO_2 concentration. Both initial and total Rubisco activity in Al-treated leaves decreased to a lesser extent than CO_2 assimilation. Al decreased F_v/F_m, maximum amplitude of IP phase, and $PI_{tot,abs}$, but increased F_o, K-band, and relative variable fluorescence at J- and I-steps. B could alleviate the Al-induced increase or decrease for all these parameters (Fig. 29.15; Jiang et al. 2009b). We concluded that the Al-induced photosynthesis inhibition was mainly caused by impaired photosynthetic electron transport chain. The B-induced amelioration of photosynthesis inhibition and photoinhibitory damage occurring at both donor and acceptor sides of photosystem II could be due to less Al accumulation in leaves.

29.4.6 Other Mechanism

Lin and Myhre (1991a) observed that some citrus rootstock root tips were covered with a root cap formed by black gelatinous material when the Al concentration in nutrient solution was 308 μM Al or higher. The number of black root caps increased with increasing Al concentration. The more Al-tolerant rootstocks had more of this kind of black root caps. These results indicated that the gelatinous material might be involved in the Al tolerance of citrus rootstocks.

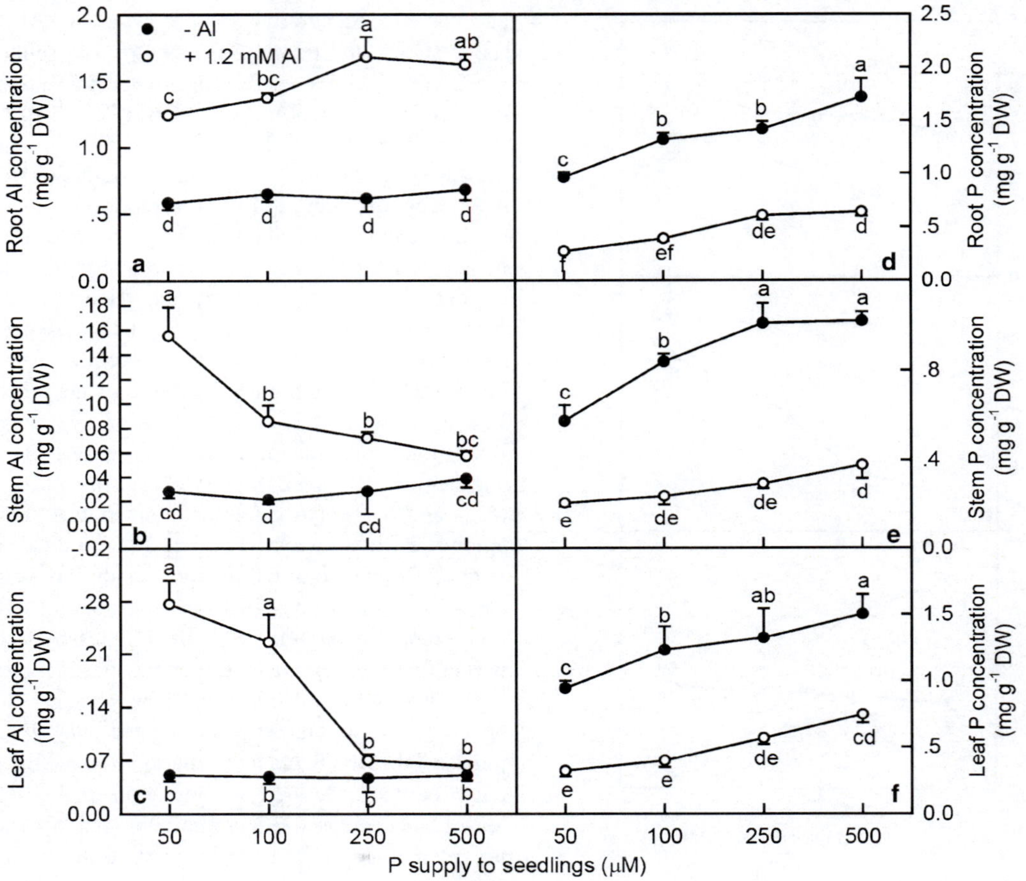

Fig. 29.11 Effects of phosphorus (P) and aluminum (Al) interactions on the concentrations of P and Al in roots, stems, and leaves from "sour pomelo" seedlings treated with different P and Al levels for 18 weeks. Each *point* represents the mean ±SE (n=4–6). Each variable was analyzed by 2 (Al)×4 (P) ANOVA. P values for Al, P, and the interactions between the two were <0.0001, 0.0049, and 0.4744 (**a**), <0.0001, 0.0033, and 0.0001 (**b**), <0.0001, <0.0001, and <0.0001 (**c**), <0.0001, <0.0001, and 0.3022 (**d**), and <0.0001, <0.0001, and 0.0065 (**e**), <0.0001, <0.0001, and 0.3385 (**f**), respectively. Different *letters* indicate significant differences at $P<0.05$ (Adapted from Jiang et al. 2009a)

29.5 Concluding Remarks

There has been significant progress in our understanding of the physiological responses and the tolerance of citrus to Al toxicity during the past decade. However, many questions remain to be answered. For example, how do citrus roots perceive the Al signal and transmit it? Al is an effective inhibitor of photosynthesis, but the mechanisms involved have not been fully understood yet. The Al-induced secretion of OA anions was involved in the Al tolerance of citrus, but further studies are required to understand these mechanisms at the physiological and molecular levels. Further studies are needed to answer the question whether the B-induced amelioration of citrus root inhibition was caused by its well-established positive effect on cell wall and plasma membrane stability, but also by B-induced changes in Al speciation and/or compartmentation. It is not known whether the enhancement of citrus Al tolerance by B is associated with an increased secretion of OA anions from roots or not. Use of Al-tolerant citrus rootstocks is a cheaper and permanent solution to citrus Al toxicity. Breeding and selection for Al-tolerant citrus genotypes will be a potentially rewarding area of research. There are few studies on the Al-induced changes in gene expression, protein synthesis, and metabolites in citrus. Fortunately, new techniques such as genomics, transcriptomics, proteomics, and metabolomics will allow greater progress in our understanding of the physiological responses and tolerance of citrus to Al toxicity.

Acknowledgments This work was financially supported by grants from the National Natural Science Foundation of China (Nos. 30270930; 30771487), the Agricultural Commonweal Industrial Special Fund Program of Department of Agriculture, China (No. nyhyzx07-023), the Natural Science Foundation of Fujian Province of China (No. B0710011), and the earmarked fund for China Agriculture Research System.

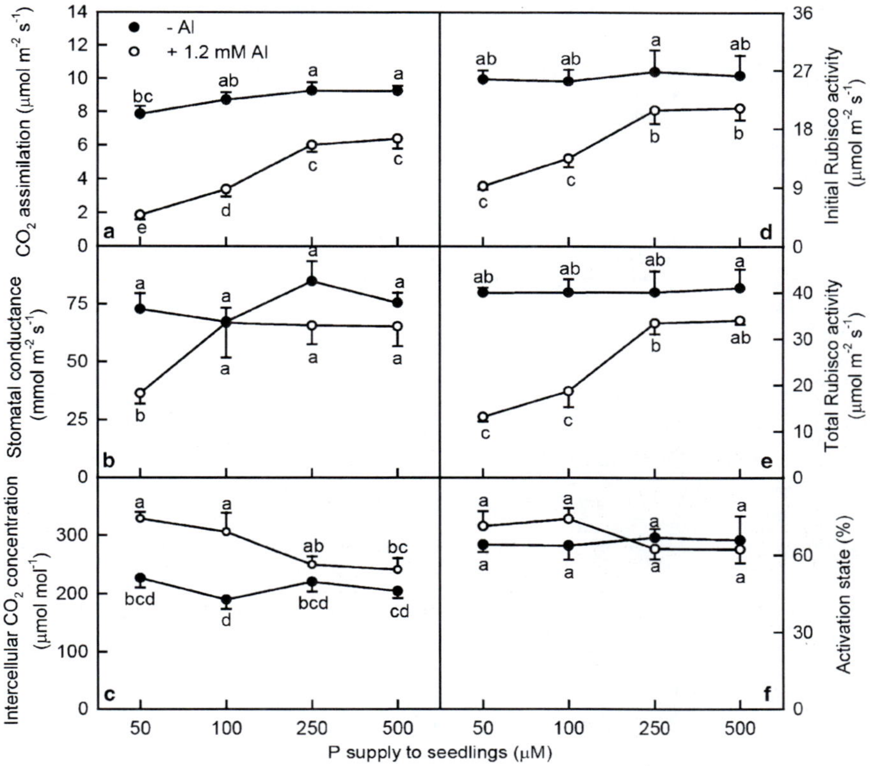

Fig. 29.12 Effects of phosphorus (P) and aluminum (Al) interactions on CO_2 assimilation (**a**), stomatal conductance (**b**), intercellular CO_2 concentration (**c**), initial ribulose-1,5-bisphosphate carboxylase/oxygenase (Rubisco) activity (**d**), total Rubisco activity (**e**), and activation state (**f**) in leaves from "sour pomelo" seedlings treated with different P and Al levels for 18 weeks. Each *point* represents the mean ±SE ($n=5$–7). Each variable was analyzed by 2 (Al)×4 (P) ANOVA. *P* values for Al, P, and the interactions between the two were <0.0001, <0.0001, and 0.0001 (**a**); 0.0014, 0.0629, and 0.0770 (**b**); <0.0001, 0.0113, and 0.0222 (**c**); <0.0001, 0.0025, and 0.0158 (**d**); <0.0001, 0.0051, and 0.0020 (**e**); 0.5646, 0.2018, and 0.9599 (**f**), respectively. Different *letters* indicate significant differences at $P<0.05$ (Adapted from Jiang et al. 2009a)

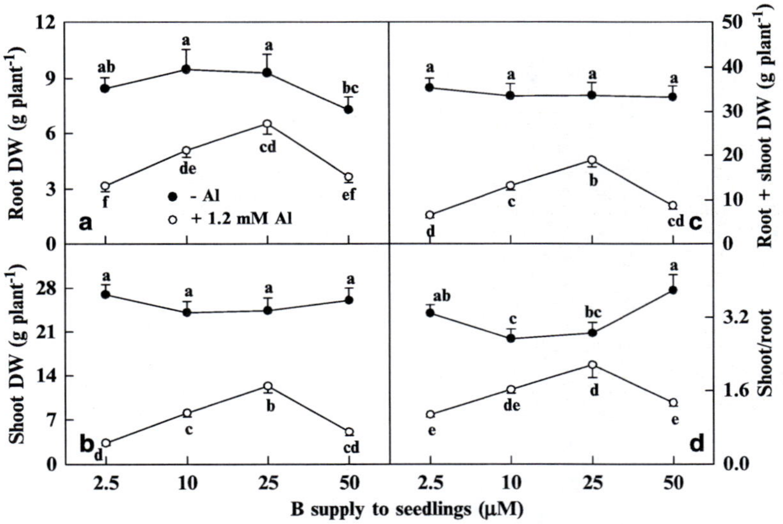

Fig. 29.13 Effects of aluminum (Al) and boron (B) interactions on root (**a**), shoot (**b**), and root + shoot (**c**) dry weight (DW), and shoot/root ratio (**d**) of "sour pomelo" seedlings treated with different B and Al levels for 18 weeks. Each *point* represents the mean ±SE ($n=8$–15). Difference among eight treatments was analyzed by 2 (Al)×4 (B) ANOVA. (**a**) *P* values for Al, B, and the interaction between the two were <0.0001, <0.0001, and 0.3905, respectively; (**b**), (**c**), and (**d**) *P* values for Al, B, and the interaction between the two were all <0.0001. Different *letters* indicate significant differences among eight treatments at $P<0.05$ (Adapted from Jiang et al. 2009b)

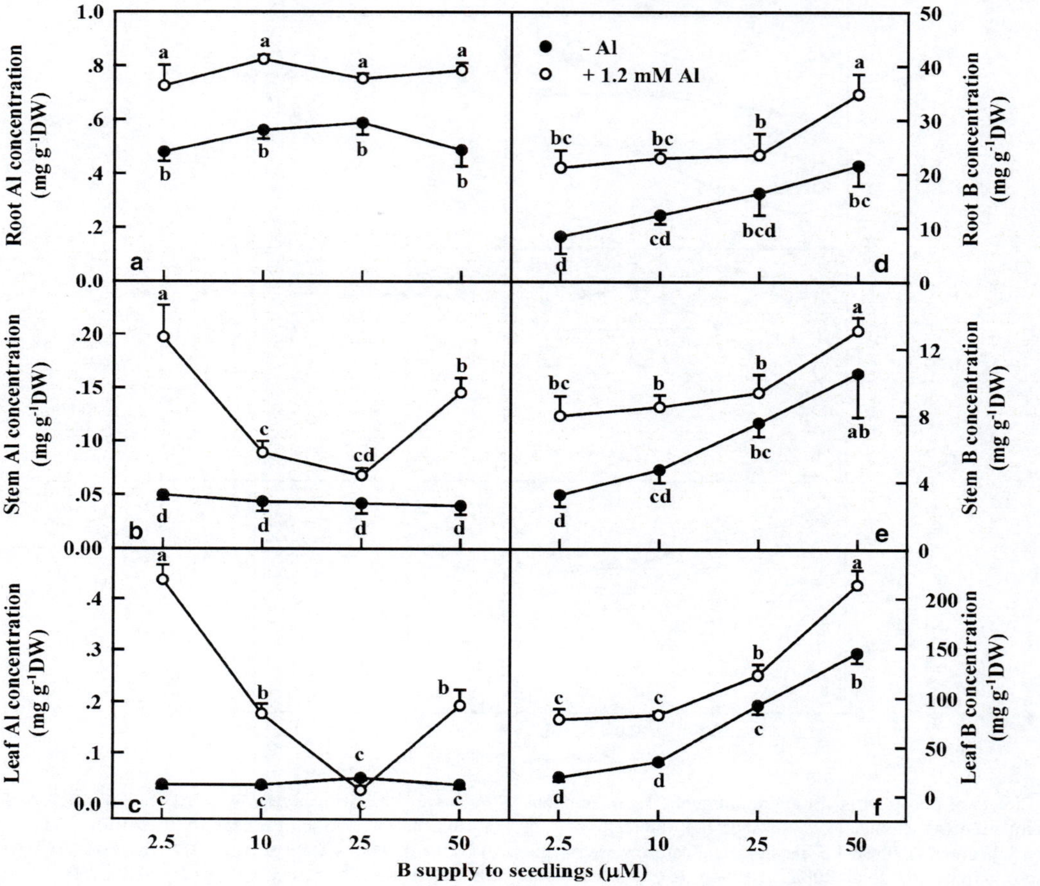

Fig. 29.14 Effects of aluminum (Al) and boron (B) interactions on the concentrations of Al and B in roots, stems, and leaves from "sour pomelo" seedlings treated with different B and Al levels for 18 weeks. Each *point* represents the mean ± SE ($n = 4$–5). Difference among eight treatments was analyzed by 2 (Al) × 4 (B) ANOVA. *P* values for Al, B, and the interaction between the two were 0.0001, 0.3206, and 0.3148 (**a**); <0.0001, <0.0001, and 0.0001 (**b**); <0.0001, <0.0001, and <0.0001 (**c**); 0.0002, 0.0066, and 0.8568 (**d**); and 0.0006, 0.0001, and 0.7807 (**e**); 0.0001, 0.0001, and 0.2002 (**f**), respectively. Different *letters* indicate significant differences among eight treatments at $P < 0.05$ (Adapted from Jiang et al. 2009b)

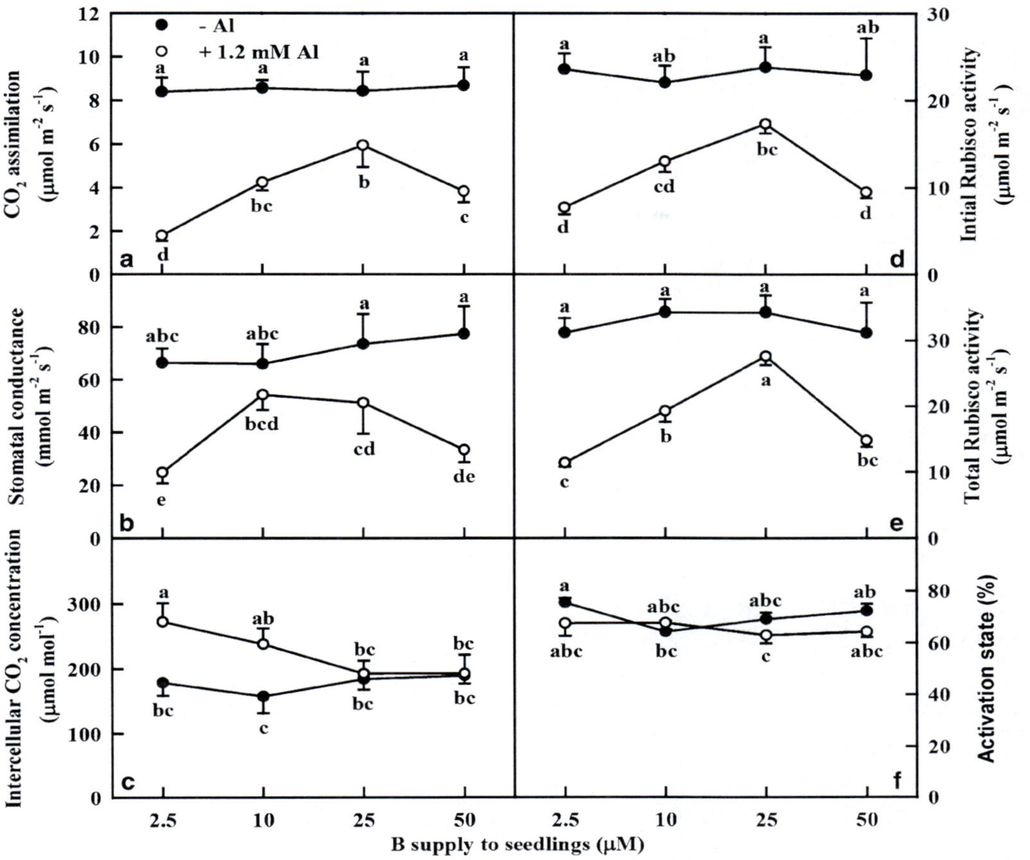

Fig. 29.15 Effects of aluminum (Al) and boron (B) interactions on CO_2 assimilation (**a**), stomatal conductance (**b**), intercellular CO_2 concentration (**c**), initial ribulose-1,5-bisphosphate carboxylase/oxygenase (Rubisco) activity (**d**), total Rubisco activity (**e**), and activation state (**f**) in leaves from "sour pomelo" seedlings treated with different B and Al levels for 18 weeks. Each *point* is the mean ±SE ($n = 4-5$). Difference among eight treatments was analyzed by 2 (Al)×4 (B) ANOVA. *P* values for Al, B, and the interaction between the two were <0.0001, 0.0121, and 0.0492 (**a**); 0.0308, 0.5072, and 0.1252 (**b**); 0.0041, 0.3548, and 0.1155 (**c**); <0.0001, 0.1192, and 0.2326 (**d**); <0.0001, 0.0016, and 0.1244 (**e**); 0.0575, 0.2836, and 0.2166 (**f**), respectively. Different *letters* indicate significant differences among eight treatments at $P < 0.05$ (Adapted from Jiang et al. 2009b)

References

Asada K (1999) The water-water cycle in chloroplasts: scavenging of active oxygen and dissipation of excess photons. Annu Rev Plant Physiol Plant Mol Biol 50:601–639

Chen L-S (2006) Physiological responses and tolerance of plant shoot to aluminum toxicity. J Plant Physiol Mol Biol 32:143–155

Chen L-S, Qi Y-P, Liu X-H (2005a) Effects of aluminum on light energy utilization and photoprotective systems in citrus leaves. Ann Bot 96:35–41

Chen L-S, Qi Y-P et al (2005b) Aluminum-induced decrease in CO_2 assimilation in citrus seedlings is unaccompanied by decreased activities of key enzymes involved in CO_2 assimilation. Tree Physiol 25:317–324

Chen L-S, Tang N et al (2009) Changes in organic acid metabolism differ between roots and leaves of *Citrus grandis* in response to phosphorus and aluminum interactions. J Plant Physiol 166:2023–2034

Chen L-S, Tang N et al (2010) Effects of aluminum on leaf water relation and nutritional elements of citrus seedlings. In: Deng XX, Xu J et al (eds) Proceedings of the international society of citriculture, vol 1. China Agriculture Press, Beijing, pp 650–653

Corrales I, Poschenrieder C, Barceló J (2008) Boron-induced amelioration of aluminum toxicity in a monocot and a dicot species. J Plant Physiol 165:504–513

de Magalhães AFJ (1987) Tolerância de porta-enxertos de citrus ao alumínio. Rev Bras Frutic 9:51–55

Demmig-Adams B, Adams WW III (1996) The role of xanthophyll cycle carotenoids in the protection of photosynthesis. Trends Plant Sci 1:21–26

Deng W, Luo K et al (2008) Molecular cloning and characterization of a mitochondrial dicarboxylate/tricarboxylate transporter gene in *Citrus junos* response to aluminum stress. Mitochondrial DNA 19:376–384

Deng W, Luo K et al (2009) Overexpression of *Citrus junos* mitochondrial citrate synthase gene in *Nicotiana benthamiana* confers aluminum tolerance. Planta 230:355–365

dos Santos CH, Filho HG et al (2000) Influence of different levels of aluminum on the development of citrus rootstock 'Swingle' citrumelo (*Citrus paradisi* Mcf. × *Poncirus trifoliata* Raf.) in nutrient solution. Braz Arch Biol Technol 43:27–33

Guo JH, Liu XJ et al (2010) Significant acidification in major Chinese croplands. Science 327:1008–1010
Jiang H-X, Chen L-S et al (2008) Aluminum-induced effects on photosystem II photochemistry in citrus leaves assessed by the chlorophyll a fluorescence transient. Tree Physiol 28:1863–1871
Jiang H-X, Tang N et al (2009a) Phosphorus alleviates aluminum-induced inhibition of growth and photosynthesis in *Citrus grandis* seedlings. Physiol Plant 137:298–311
Jiang H-X, Tang N et al (2009b) Antagonistic actions of boron against inhibitory effects of aluminum toxicity on growth, CO_2 assimilation, ribulose-1,5-bisphosphate carboxylase/oxygenase, and photosynthetic electron transport probed by the JIP-test, of *Citrus grandis* seedlings. BMC Plant Biol 9:102
Jiang H-X, Chen L-S et al (2009c) Effects of aluminum on the growth of young citrus seedlings. Chin Agric Sci Bull 25(4):167–170
Kochian LV, Hoekenga OA et al (2004) How do crop plant tolerance acid soils? Mechanisms of aluminum tolerance and phosphorous efficiency. Annu Rev Plant Biol 55:459–493
Kochian LV, Piñeros MA et al (2005) The physiology, genetics and molecular biology of plant aluminum resistance and toxicity. Plant Soil 274:175–195
Lin Z, Myhre DL (1990) Citrus root growth as affected by soil aluminum level under field conditions. Soil Sci Soc Am J 54:1340–1344
Lin Z, Myhre DL (1991a) Differential response of citrus rootstocks to aluminum levels in nutrient solutions: I. Plant growth. J Plant Nutr 14:1223–1238
Lin Z, Myhre DL (1991b) Differential response of citrus rootstocks to aluminum levels in nutrient solutions: II. Plant mineral concentrations. J Plant Nutr 14:1239–1254
Ma JF (2000) Role of organic acids in detoxification of Al in higher plants. Plant Cell Physiol 41:383–390
Ma JF (2007) Syndrome of aluminum toxicity and diversity of aluminum resistance in higher plants. Int Rev Cytol 264:225–252
Müller P, Li X-P, Niyogi KK (2001) Non-photochemical quenching, a response to excess light energy. Plant Physiol 125:1558–1566
Nakagawa TO, Mori S, Yoshimura E (2003) Amelioration of aluminum toxicity by pretreatment with phosphate in aluminum-tolerant rice cultivar. J Plant Nutr 26:619–628
Niyogi KK, Grossmn AR et al (1998) *Arabidopsis* mutants define a central role of the xanthophyll cycle in the regulation of photosynthetic energy conversion. Plant Cell 10:1121–1134
Pereira WE, de Siqueira DL et al (2000) Gas exchange and chlorophyll fluorescence in four citrus rootstocks under aluminum stress. J Plant Physiol 157:513–520
Pereira WE, de Siqueira DL et al (2003) Growth of citrus rootstocks under aluminium stress in hydroponics. Sci Agric 60:31–41
Roy AK, Sharma A, Talukder G (1988) Some aspects of aluminum toxicity in plants. Bot Rev 54:145–178
Strasser RJ, Srivastava A et al (1995) Polyphasic chlorophyll a fluorescence transient in plants and cyanobacteria. Photochem Photobiol 61:32–42
Sun QB, Shen RF et al (2008) Phosphorus enhances Al resistance in Al-resistant *Lespedeza bicolor* but not in Al-sensitive *L. cuneata* under relatively high Al stress. Ann Bot 102:795–804
Tan K, Keltjens WG (1990a) Interaction between aluminum and phosphorus in sorghum plants. I. Studies with the aluminum sensitive sorghum genotype TAM428. Plant Soil 124:15–23
Tan K, Keltjens WG (1990b) Interaction between aluminum and phosphorus in sorghum plants. II. Studies with the aluminum tolerant sorghum genotype SC0283. Plant Soil 124:25–32
Tardy F, Havaux M (1997) Thylakoid membrane fluidity and thermostability during the operation of the xanthophyll cycle in higher-plant chloroplasts. Biochim Biophys Acta 1330:179–193
Taylor GJ (1991) Current views of the aluminum stress response: the physiological basis of tolerance. Curr Top Plant Biochem Physiol 10:57–93
Yang L-T, Jiang H-X et al (2011) Mechanisms of aluminum-tolerance in two species of citrus: secretion of organic acid anions and immobilization of aluminum by phosphorus in roots. Plant Sci 180:521–530
Yokomizo H, Ishihara M (1973) Studies on the mineral nutrition of fruit trees by sand and water culture. I. Effects of composition of nutrient solution on the growth of Satsuma mandarin trees. Kaju Shikenjo Hokoku A 12:29–77
Zhang M, Luo X-Y et al (2008) Characterization of malate dehydrogenase gene from *Citrus junos* and its transgenic tobacco's tolerance to aluminium toxicity. Acta Hortic Sin 35:1751–1758

Fruit Yield Estimation Through Multispectral Imaging

30

Xujun Ye and Kenshi Sakai

Abstract

Recent advances in spectral imaging technology have enabled the development of models that estimate various crop parameters from spectral imagery data. The present research investigated the alternate bearing dynamics as well as the estimation of fruit yield in citrus crops using multispectral imaging technology. Canopy features of individual trees were extracted from the multispectral images and were then used to relate to the fruit yield of citrus trees through various modelling techniques. Results showed that the alternate bearing behaves more significantly in terms of the fruit density rather than the total fruit yield on individual trees. The normalized difference vegetation index (NDVI) demonstrated greater relevance than other multiple wavebands in predicting the fruit yield on individual citrus trees. Analysis results confirmed the interrelationships between canopy features and the fruit yield of citrus crops and implied the unmatched energy allocation mechanisms between different leaf types within the canopy of citrus crops. Effective models were developed for fruit yield estimation of citrus from multispectral images acquired before the fruit-growing season.

Keywords

Citrus • Multispectral imaging • Modelling • Yield estimation

30.1 Introduction

The management of alternate bearing has become an important issue in fruit production as the profitability of the orchard depends on the production of reasonable crop every year. During the past couple of decades, fruit thinning (removing young fruits in the early season) and other control methods, such as applying plant growth regulators, etc., have been used as effective ways to prevent alternate bearing and improve fruit quality in citrus (Iwagaki 1997). The yield information of individual trees is a prerequisite for the successful application of these methods. On the other hand, until recently, most fruit orchards were uniformly managed without regard to spatio-temporal variability within the orchard. The uniform application of fertilizers in an orchard may not match the requirements of individual trees and may result in over- or under-application. Therefore, it is important to obtain and map the potential yield information of individual fruit trees for site- and tree-specific implementation of alternate bearing control measures and other management practices.

Remote sensing has been widely used to obtain and map the temporal and spatial variability of crops in fields. Among many remote sensing applications, hyper-spectral imaging collects spectral data in hundreds of spectrally continuous channels, while multispectral imaging is expected to be more effective

X. Ye (✉)
College of Biosystems Engineering and Food Science, Zhejiang University, Hangzhou 310058, Zhejang Province, PR China
e-mail: yezising@zju.edu.cn

K. Sakai
International Center, Environmental and Agricultural Engineering Department, Faculty of Agriculture, Tokyo University of Agriculture and Technology, 3-5-8 Saiwa-Cho, Fuchu, Tokyo 183-8509, Japan
e-mail: kenshi779@gmail.com

A.K. Srivastava (ed.), *Advances in Citrus Nutrition*,
DOI 10.1007/978-94-007-4171-3_30, © Springer Science+Business Media B.V. 2012

for improving the spatial resolution to observe the targets, which allows the study of targets structural information in more detail. In addition, as multispectral imagery contains only a few spectrally discontinuous wavebands, the cost of multispectral images is usually much lower than that of hyper-spectral images. Therefore, multispectral imaging has been more widely applied to various research fields due to its easier availability.

Different multispectral remote sensing applications have proved to be a potential source of reflectance data for estimation of several canopy variables related to biophysical, physiological or biochemical characteristics (Hahlbrock and Grisebach 1979; Ahlrichs and Bauer 1983; Clevers 1991; Boegh et al. 2002; Hall et al. 2003). Various vegetation indices derived from multispectral wavebands were also used to assess and study agricultural crop growth conditions using different tools for model development (Robberecht and Caldwell 1986; Wang and Thai 2003; Zhang et al. 2006). Furthermore, multispectral imaging has demonstrated the ability in the detection of weeds (Goel et al. 2002), diseases (Qin and Zhang 2005) and the discrimination of conventional and conservation agricultural management practices (Hache et al. 2005) within a field or an orchard.

Citrus is a major fruit crop that suffers from alternate bearing around the world. This study attempts to study the alternate bearing of citrus crops using airborne multispectral imaging technology. The specific objectives of this research were (1) to investigate the alternate bearing dynamics in citrus with airborne multispectral imagery and (2) to explore the potential of using the airborne multispectral imagery acquired at an early date to estimate the fruit yield attainable in the coming season.

Fig. 30.1 Location of Nebukawa Agricultural Research Station, Kanagawa Prefecture, Japan

30.2 Methodology

30.2.1 Study Area

The present study was conducted in a citrus orchard located at the Nebukawa Agricultural Research Station, Kanagawa Prefecture, Japan (139°07′44.29″ east and 35°12′13.01″ north) (Fig. 30.1). The area is characterized by a temperate climate. The annual mean temperatures were 15.4°C, 16.5°C and 15.4°C, and the annual precipitations were 2816.5, 2413.5 and 1762.5 mm in 2003, 2004 and 2005, respectively. We chose to study Satsuma mandarin (*Citrus unshiu Marc.*), a native citrus variety in southeast Asia, for this study. It is abundantly grown in Japan, accounting for more than 75% of all citrus fruit produced in Japan in 1994 (Iwagaki 1997). Citrus trees are grown on terraces. The area of the studied orchard was approximately 2,700 m^2 (Fig. 30.2a). Uniform fertilizing and management schemes are implemented for the trees within the orchard, including the timing and amount of fertilizing, weeding, pest and disease control, medium tillage, harvesting, etc. During the study periods (from 2002 through 2006), no severe plant pest and disease infections and climatic disasters such as drought, typhoons and floods occurred in the study area.

30.2.2 Alternate Bearing in Experimental Citrus Orchard

Yield data (number of fruits) on individual citrus trees were collected during the local harvest season (from the last week of November to the early part of December) for four consecutive years from 2002 through 2005. This preliminary analysis was conducted on a plot consisting of thirty-one 19-year-old trees (transplanted in April 1988) and seventeen 8-year-old trees (transplanted in March 1999), both of which are currently in a growing stage with a high fruit-bearing capacity (Fig. 30.2b). Yield data on these 48 trees were illustrated in Fig. 30.3. The majority of the trees show a clear alternate bearing pattern.

30.2.3 Acquisition of Airborne Multispectral Images

Two multispectral sensor systems, consisting of the airborne digital sensor 40 system (ADS40) and the UltraCam digital

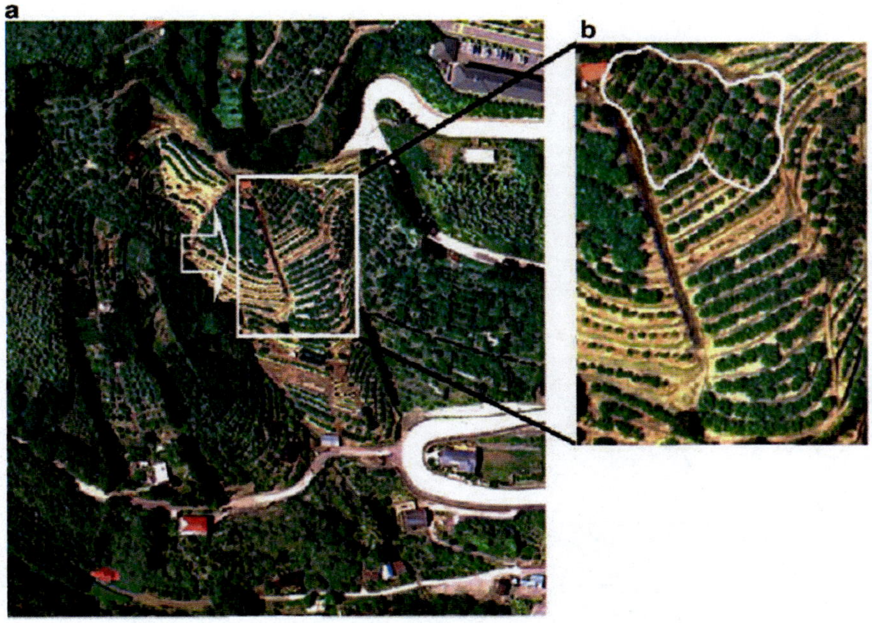

Fig. 30.2 (**a**) The experimental citrus orchard at Nebukawa Agricultural Research Station, Kanagawa Prefecture in Japan. (**b**) Location of 48 trees of same variety category used in the study (*white circle* on image)

camera system (UCD) (Pasco Co., Ltd., Tokyo, Japan), were used to obtain multispectral images over the experimental citrus orchard. Technical specifications of these two sensors are shown in Table 30.1. Images were acquired 11 times: once after the harvest season in 2002 (on December 2) and ten times during the early growing seasons from 2003 through 2006 (Table 30.1). Images were taken at an altitude of approximately 1,000 m above ground during cloud-free periods in the daytime. The images acquired by the ADS40 consist of four composite bands in the blue (430~490 nm), green (535~585 nm), red (610~660 nm) and near-infrared (835~885 nm) bands, with a high spatial resolution of 0.2 m×0.2 m, and the UCD features multispectral images with four composite bands in the blue (410~530 nm), green (490~630 nm), red (585~680 nm) and near-infrared (685~910 nm) bands and with a high spatial resolution of 0.25 m×0.25 m (Table 30.1).

Image preprocessing was performed during and after flight by the provider, Pasco Co., Japan. The image data were captured with a 16-bit per pixel dynamic range and without any grain noise. Images were geometrically corrected using on-board global positioning system/inertial measurement unit (GPS/IMU) data for initial geometric correction and on-ground control points (GCP) with a polynomial method for second-level geometric correction. Root-mean-square error (RMSE) was kept below half the pixel dimension value. The images were geo-rectified to Universal Transverse Mercator (UTM) geographic coordinates. The final output images are in .img format, compatible with major photogrammetric software suppliers.

30.2.4 Image Processing and Data Extraction

30.2.4.1 NDVI Image Generation

The multispectral images were firstly imported into the ERDAS IMAGINE 8.6 software (ERDAS IMAGINE 8.6, Leica Geosystems GIS & Mapping, LLC, Atlanta, Georgia, USA). With the ERDAS IMAGINE 8.6 software, composite images with any three different bands can be generated. In addition, images based on several vegetation indices, such as normalized difference vegetation index (NDVI), transformed normalized difference vegetation index (TNDVI), simple ratio (SR), etc., can be generated. In this study, the images based on NDVI, the most commonly used vegetation index, were generated, using the formula below:

$$\mathrm{NDVI} = \frac{\mathrm{NIR} - \mathrm{R}}{\mathrm{NIR} + \mathrm{R}} \tag{30.1}$$

30.2.4.2 Data Extraction

Two methods were used to extract canopy spectral data for each band of the multispectral (R, G, B and NIR) images as well as the generated NDVI images.

Average Spectral Reflectance (ASR)

The first one utilizes the average spectral reflectance (ASR) of individual canopies. It was done with the aid of ArcMap 8.1 software (ArcMap 8.1, ESRI, Redlands, CA, USA). The software allows counting of pixel numbers and extraction of several statistical properties such as mean, minimum, maximum

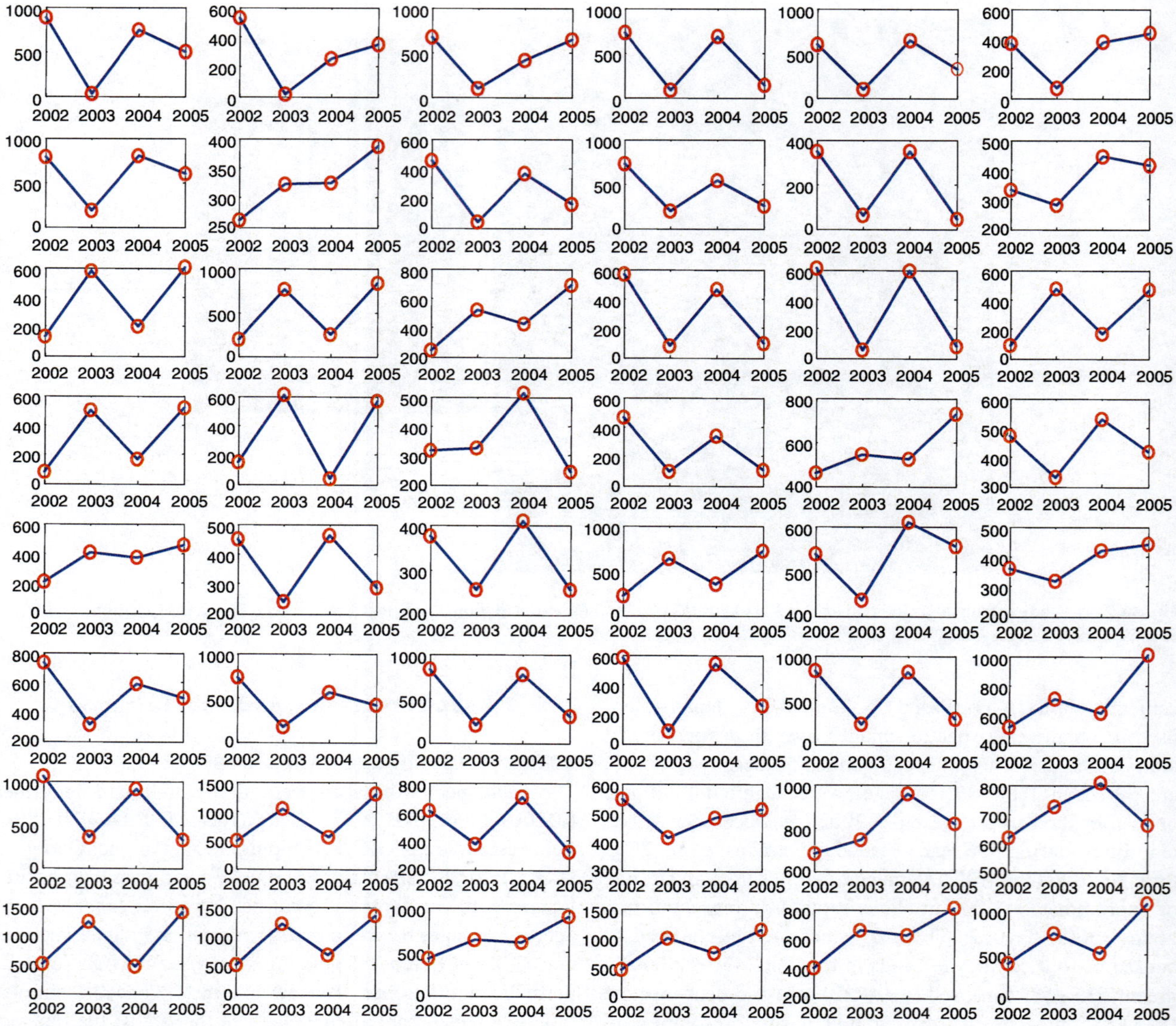

Fig. 30.3 Trend of yield variation on individual trees from 2002 through 2005. Each graph indicates the yield variation for one tree

and standard deviation of the reflectance values for each canopy. However, in this study, only the average property for each canopy was extracted.

The extraction process entails identifying the individual canopies manually, drawing polygons over the canopies to create shape files, and finally running the embedded program to extract statistical average property of the pixels within the drawn polygons. To compare images taken at different time periods during the growing season, a white roof from the images, located close to the experiment field, was used as a reference location. Spectral data extracted from the same reference location in different images were then used to normalize the extracted spectral data for individual canopies on the corresponding images. Such normalization enables the investigation of canopy spectral changes over the growing season.

Thresholded Pixel Count (TPC)

The second method tries to extract the number of pixels that satisfies specified reflectance levels otherwise called thresholds for individual canopies. A program was developed for this purpose in MATLAB R12 environment.

For running the program, the images were first exported from the ERDAS IMAGINE 8.6 software and saved as the format of .bmp. After being loaded in MATLAB R12 environment, the pixel values of the images were normalized into a greyscale from 0 to 1. Subsequently, the data extraction process was carried out on the normalized images, following three

Table 30.1 Specifications of multispectral sensors and image acquisition dates

Specifications	ADS40	UCD	
Multispectral bands (wavelength range)	Blue	430–490 nm	410–530 nm
	Green	535–585 nm	490–630 nm
	Red	610–660 nm	585–680 nm
	NIR	835–885 nm	685–910 nm
Data type		Unsigned 16 bit	Unsigned 16 bit
Data format		img (ERDAS IMAGINE)	img (ERDAS IMAGINE)
Pixel size		0.2 m×0.2 m	0.25 m×0.25 m
Data processing		Ortho-rectified (original data)	Geo-rectified using camera model (level 1)
Map projection	Projection type	UTM	UTM
	Spheroid name	WGS84	WGS84
	Datum name	WGS84	WGS84
	Zone	54 N	54 N
Image acquisition dates		December 2, 2002	June 6, 2005
		April 10, 2003	July 21, 2005
		May 21, 2003	April 22, 2006
		June 5, 2003	
		May 26, 2004	
		July 2, 2004	
		August 14, 2004	
		September 1, 2005	

Adapted from Pasco Co., Japan

sequential steps: (1) recognizing individual canopies on the images, (2) manually drawing polygonal shapes on each canopy to create shape files and finally (3) running the thresholding program to extract the thresholded pixel count (TPC), that is, the number of pixels that satisfy specified thresholds for each canopy. The thresholding process first divides the pixel value range from 0 to 1 into 10 sub-ranges at an increment of 0.1 and then extracts the TPC for various threshold criteria, such as less than a certain threshold level, more than a threshold level or within a threshold range for the drawn polygons.

The above data extraction procedures were applied to individual canopies on all multispectral images. The TPCs with specific reflectance criteria for individual canopies with the time and band information, that is, each acquisition date for each band, were extracted. Figure 30.4 shows the images of thresholded canopy areas generated with different threshold criteria based on the image obtained on April 11, 2003.

30.2.5 Preliminary Data Analysis

30.2.5.1 Investigation of Alternate Bearing in Terms of Fruit Density

Our previous studies have analysed the alternate bearing pattern observed in the experimental citrus orchard. The analysis was conducted by comparing the fruit yield on individual trees over several consecutive years (Ye et al. 2006, 2007, 2008a, b). In this study, the alternate bearing was investigated in terms of fruit density (no. of fruits/canopy size) in individual tree canopies. This analysis would help understand the influence of tree size on the alternate bearing of fruit yields.

30.2.5.2 Correlation Between TPC and Fruit Yield

The TPCs with specific reflectance criteria extracted in the preceding section were used to relate to the fruit yield of citrus trees. Correlation analysis was performed between each of the TPCs from the acquired images and the fruit yield data of three consecutive years, consisting of the present growing season in which the analysed images were acquired and the growing seasons preceding and following the season of image acquisition. The purpose of relating the canopy information extracted from the images to the fruit yield in the present growing season was to examine the correlations between canopy structural features and fruit yield within the same growing season. This serves as the basis for exploring the potential of remotely sensed canopy information in early seasons to estimate fruit yield of citrus. On the other hand, relating those images to the fruit yield in the preceding and the following growing seasons was to investigate the influence of fruit yield on canopy development and conversely the effect of canopy structural features on fruit yield formation in the next growing season. The understanding of interrelationships between canopy structural features and fruit yield within and/or over growing seasons would contribute to revealing the dynamics of alternate bearing phenomenon in fruit tree crops.

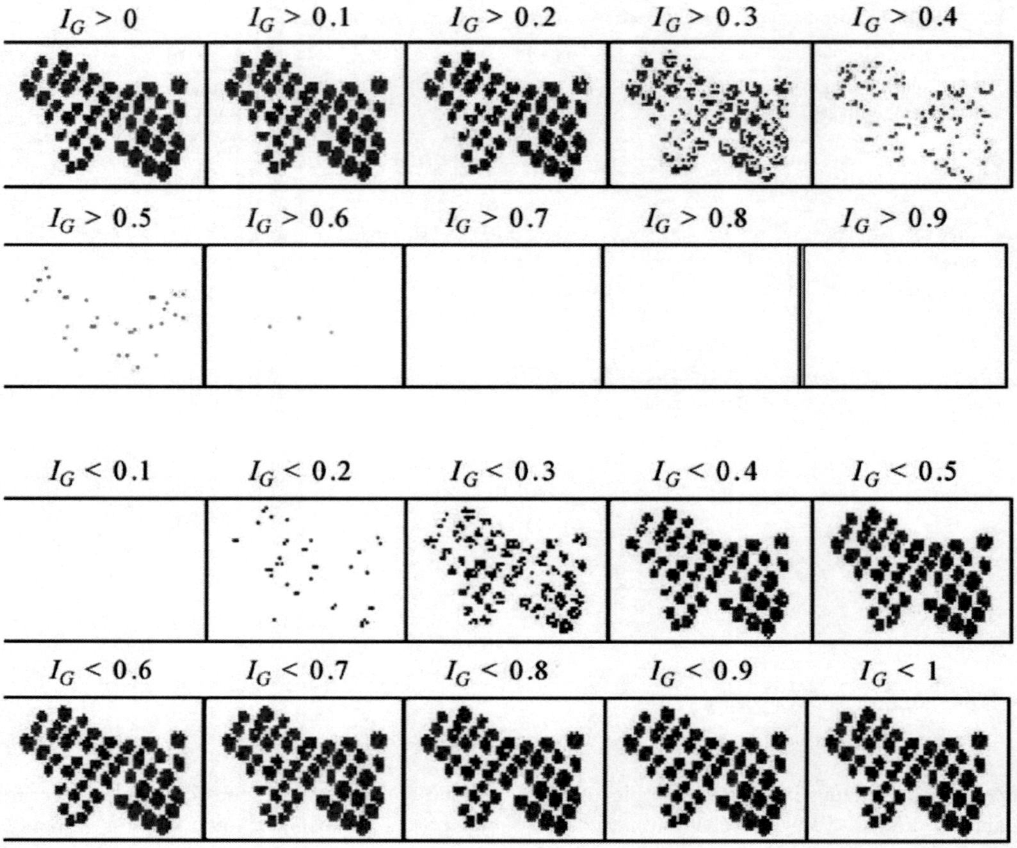

Fig. 30.4 The images of thresholded canopy areas generated with different threshold criteria based on the image obtained on April 11, 2003. I_G indicates the normalized pixel value at G band

30.2.6 Model Development and Validation

Four different approaches were attempted to develop prediction models in this study. In order to evaluate the performance of the developed models, cross-validation procedures were employed. The entire data set was first arranged in ascending order of citrus yield and was then separated into two subsets, one as the training data set (with odd numbers) and the other as the validation data set (with even numbers).

Upon completing the development of models, three indices, that is, the coefficient of determination (R^2), the root mean square of error (RMSE) and the relative root mean square of error (RRMSE), were employed to evaluate and compare the performance of different models:

$$R^2 = \left(1 - \frac{\sum_{i=1}^{N}[y(i) - \hat{Y}(i)]^2}{\sum_{i=1}^{N}[y(i) - \overline{y}]^2}\right) \tag{30.2}$$

$$RMSE = \sqrt{\frac{1}{N}\sum_{i=1}^{N}[y(i) - \hat{Y}(i)]^2} \tag{30.3}$$

$$RRMSE = \frac{\sqrt{\frac{1}{N}\sum_{i=1}^{N}[y(i) - \hat{Y}(i)]^2}}{\sqrt{\frac{1}{N}\sum_{i=1}^{N}[y(i) - \overline{y})]^2}} \tag{30.4}$$

where N is the number of data samples, $y(i)$ is the actual ith value, $\overline{y}$ is the average of the actual values and $\hat{y}$ is the predicted ith value.

30.2.6.1 Models Based on Tree Canopy Size and Previous Year's Yield Data

Previous studies by Noguchi et al. (2003) and Sakai et al. (2003) have demonstrated the possibility of predicting the yield of individual citrus trees for the coming season, using the previous 2 years' yield data of the corresponding trees as the model input variables (R^2=0.78). In this study, the feasibility of predicting citrus yield based on the previous year's yield data and the tree canopy size extracted from the images was examined.

30.2.6.2 Models Based on ASR

This approach uses the ASRs extracted from the images to develop prediction models. The ASRs for each of the multispectral wavebands, R, G, B and NIR, and the generated NDVI were

Table 30.2 Methods for development of prediction models based on ASRs extracted from 2002 and 2003 images

Model classes	Predictor selection methods
I	ASR for a single waveband in a single month
II	Multiple ASRs for all wavebands in the same month
III	Multiple ASRs for the same waveband in all months
IV	Five ASRs randomly selected among all wavebands in all months

used as a single predictor variable to develop prediction models for citrus yield. Various combinations of multiple wavebands were also attempted for the development of prediction models, using a multiple linear regress (MLR) algorithm (Table 30.2). This analysis was performed by predicting the yield of 2003 based on the images acquired in 2002 and 2003.

30.2.6.3 Models Based on ASR and Tree Canopy Size

As stated above, in addition to the alternate bearing mechanism, which results in the difference in canopy features on individual trees, tree canopy size may also have influence on the yield level of individual trees in different years. Therefore, besides the ASRs of single or multiple wavebands for individual canopies, the modelling in this section attempts to include the tree canopy size of individual trees in the model development. In this procedure, the ASRs were selected by the same methods in the preceding section (Table 30.2) and the tree canopy sizes were extracted from the image obtained on May 21, 2003.

30.2.6.4 Models Based on TPC with Specific Reflectance Criteria

One of the main results from the preceding correlation analysis between TPCs and citrus yield is the identification of the specific reflectance criteria, with which the obtained TPCs achieved the highest correlation with citrus yield. The TPCs obtained with the identified reflectance criteria were then used to develop prediction models. If these TPCs achieve a consistent performance both for the training and the validation data sets, the leaf areas corresponding to these TPCs can be potentially used to estimate the fruit yield of citrus.

30.3 Results and Discussion

30.3.1 Investigation of Alternate Bearing in Terms of Fruit Density

Figures 30.5 and 30.6 illustrate scatter plots of the yield of 2004 against the yield of 2003 in terms of the total fruit yield and the fruit density (no. of fruit/m^2) for individual trees, respectively. Correlation analysis was performed between the 2 years using each of the two yield indices. Both of these two yield indices show a negative correlation between the 2 years' data, which agrees with the negative autocorrelations between seed production of acorn, as were done by Norton and Kelly (1988), Koenig et al. (1994) and Sork et al. (1993). Furthermore, it was revealed that the fruit density (r=−0.6562) had a higher correlation than the total yield (r=−0.1341) for individual trees between these 2 years, showing a more significant pattern of alternate bearing in terms of the fruit density index. This result revealed the mechanism of alternate bearing on individual trees and indicated the influence of tree canopy size on fruit yield. As a result of this mechanism, even in an off year, individuals with a larger canopy size may achieve a higher yield than those small-sized on-year trees and vice versa. This may be one of the reasons that the total fruit yield in an orchard sometimes may not show a clear alternate bearing pattern in several consecutive years, though most of the trees individually exhibit a clear alternate bearing pattern, as was observed in 2005 in this study.

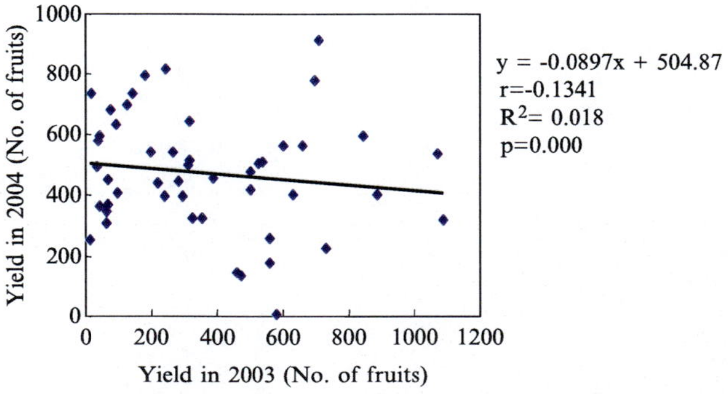

Fig. 30.5 Comparison of fruit yield for individual trees between 2003 and 2004

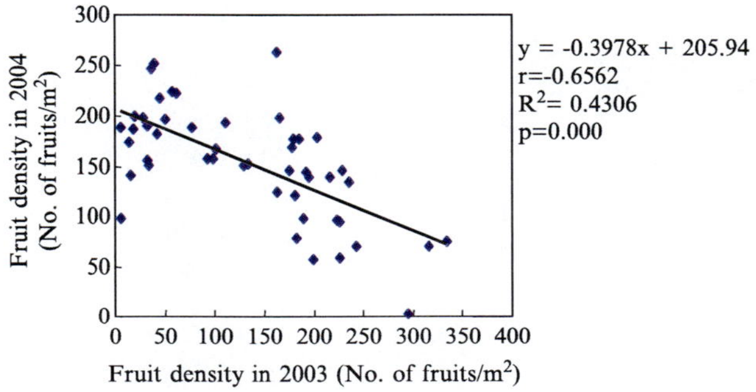

Fig. 30.6 Comparison of fruit density for individual trees between 2003 and 2004

30.3.2 Correlation Between TPC and Fruit Yield

30.3.2.1 2002 and 2003 Images

Pearson's correlation analysis was conducted between each of the TPCs extracted from the four images obtained in 2002 and 2003 and the fruit yield collected in 2002, 2003 and 2004. Table 30.3 shows the correlation coefficients between

Table 30.3 Pearson's correlation coefficients between TPCs from 2002 and 2003 images and fruit yield of citrus in 2003

Image (acquisition dates)	Normalized pixel value (I)	R band	G band	B band	NIR band	NDVI
December 2, 2002	<0.1	0.4376**	0.1179	–	0.1841	0.0731
	<0.2	0.6299***	0.4574**	0.4320**	0.4607***	0.0232
	<0.3	__0.6318__***	0.6481***	0.5669***	0.5102***	–0.0360
	<0.4	0.5969***	__0.6924__***	__0.5772__***	0.5435***	–0.1125
	<0.5	0.5743***	0.6064***	0.5546***	__0.5763__***	–0.2006
	:	:	:	:	:	:
	<0.9	0.5718***	0.5729***	0.5728***	0.5727***	0.5711***
	<1.0	0.5729***	|	0.5729***	0.5729***	__0.5729__***
April 10, 2003	<0.1	0.5750***	0.1884	0.1771	–	–0.1370
	<0.2	__0.7249__***	0.4353**	__0.6263__***	–0.1493	0.0605
	<0.3	0.5500***	__0.7479__***	0.5221***	–0.0992	–0.0453
	<0.4	0.5152***	0.6013***	0.4995***	0.1829	–0.1460
	<0.5	0.5003***	0.5225***	0.5010***	0.3299*	–0.0143
	:	:	:	:	:	:
	<0.9	|	|	|	0.4975***	0.3903**
	<1.0	|	|	|	__0.5010__***	__0.5010__***
May 21, 2003	<0.1	–	–	–	–	0.1001
	:	:	:	:	:	:
	<0.5	0.5636***	0.3351*	__0.5700__***	–0.5188***	0.0969
	<0.6	__0.5701__***	0.5691***	0.5699***	0.0046	0.2224
	<0.7	0.5701***	__0.5699__***	|	0.5462***	0.2854**
	<0.8	0.5699***	|	|	0.5698***	0.2863**
	<0.9	|	|	|	__0.5699__***	0.5598***
	<1.0	|	|	|	|	__0.5699__***
June 5, 2003	<0.1	0.2489	–0.0400	–0.0299	–0.1288	0.1097
	<0.2	0.5266***	0.1505	0.3581*	–0.0825	0.0105
	<0.3	__0.6085__***	0.3952**	0.5084***	0.0681	–0.0338
	<0.4	0.5435***	0.4326**	0.5325***	0.0792	–0.1327
	<0.5	0.5359***	0.5063***	0.5351***	0.1753	–0.0656
	<0.6	0.5366***	__0.5491__***	__0.5356__***	0.2231	0.0217
	:	:	:	:	:	:
	<1.0	0.5356***	|	|	__0.5356__***	__0.5356__***

Notes: *, ** and *** indicate levels of statistical significance at $P<0.05$, $P<0.01$ and $P<0.001$, respectively. '–' indicates that there are no pixels below the specified threshold. ':' indicates the omitted insignificant results. '|' indicates that increasing the threshold value did not add any more pixels to the TPC and hence the correlation coefficient would be identical to that of the previous, lower threshold. Thus, the cells in the tables with filled correlation coefficients indicate the distribution ranges of normalized pixel values in each of the images

each of TPCs obtained with an increasing threshold and the fruit yield in 2003. All the TPCs with a threshold of $I<1.0$, which corresponds to the entire canopy area, are very significantly ($p<0.01$) or extremely significantly ($p<0.001$) correlated with the fruit yield in 2003. This is in agreement with previous studies showing the correlations between fruit yield and tree canopy size (Hill et al. 1987; Zaman et al. 2006). It is interesting that some TPCs with specific thresholds show a much stronger correlation with fruit yield than the entire canopy TPCs. Particularly, those with the thresholds $I_R<0.3$, $I_G<0.4$ and $I_B<0.4$ for the image of December 2, 2002 and $I_R<0.2$, $I_G<0.3$ and $I_B<0.2$ for the image of April 10, 2003 have a high correlation coefficient of 0.6318, 0.6924, 0.5772, 0.7249, 0.7479 and 0.6263, respectively.

Figure 30.7 shows the canopy areas generated with different thresholds from two images acquired on December 2, 2002 and April 10, 2003. The shown canopy areas have a less spectral reflectance but a higher spectral absorbance at the respective visible wavelengths. They are much darker in colour, indicating the higher photosynthesized compound concentrations in these mature leaves. In contrast, the remaining canopy areas, which are not shown in the thresholded images, represent the younger leaves with a light-green colour. They are characterized by a higher spectral reflectance and less spectral absorbance in the visible wavelength range. This spectral absorbance pattern in citrus, that mature leaves absorb more visible light than younger leaves, agrees with the previous studies for other species (Gates 1965; Gates

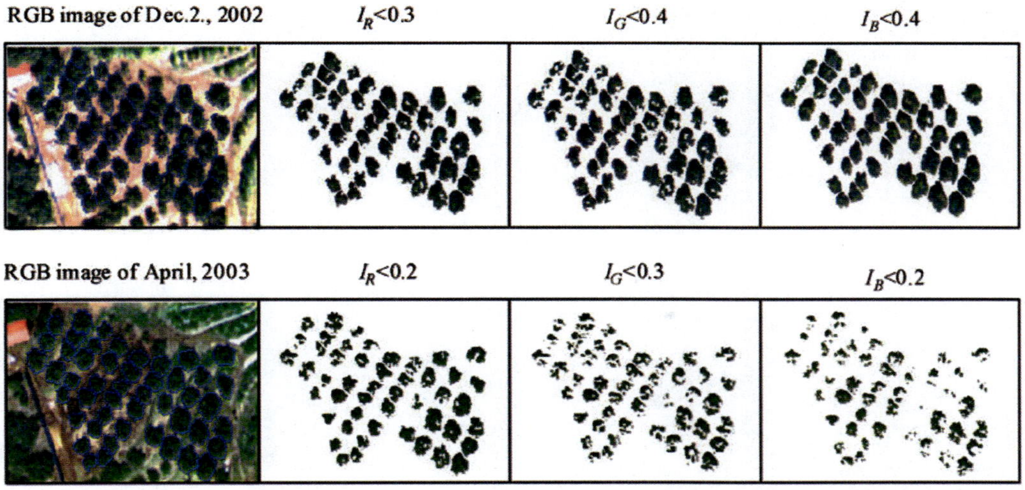

Fig. 30.7 The RGB images acquired on December 2, 2002 and April 10, 2003 and the thresholded images of canopy areas identified as being significantly correlated with the citrus yield of 2003. I_α indicates normalized pixel value at α band

et al. 1965; Gausman et al. 1976; Gupta and Woolley 1971; Qi et al. 2003; Vogelmann and Moss 1993). This may be due to the accumulation of chlorophylls, photosynthesized compounds and the various metabolites in mature leaves that strongly absorb and/or transmit visible or photosynthetically active radiation (Hahlbrock and Grisebach 1979; Jordan 1996; Krauss et al. 1997; Robberecht and Caldwell 1986).

Tables 30.4 and 30.5 summarize the correlation coefficients obtained from the analyses between each of TPCs with a decreasing threshold and the fruit yield in 2002 and 2004, respectively. Similarly, all the TPCs with a threshold of $I>0.0$, which corresponds to the entire canopy area, are significantly ($p<0.05$) or very significantly ($p<0.01$) correlated with the fruit yield in 2002 and extremely significantly ($p<0.001$) correlated with the fruit yield in 2004. This indicates that in addition to the effect on fruit yield in the same growing season, the tree canopy size also affects the fruit yield in the next growing season, and the canopy itself is also influenced by the fruit yield level that was harvested in the previous year.

Similarly, some TPCs with specific thresholds show a much stronger correlation with fruit yields in 2002 and 2004 than the entire canopy TPCs. Specifically, the TPCs obtained with the thresholds $I_R>0.2$, $I_G>0.4$ and $I_B>0.3$ for the image of December 2, 2002 and $I_R>0.2$, $I_G>0.3$ and $I_B>0.2$ for the image of April 10, 2003 have the highest correlations with the fruit yield in 2002. The TPCs with the thresholds $I_R>0.1$, $I_G>0.3$ and $I_B>0.2$ for the image of December 2, 2002 and $I_R>0.1$, $I_G>0.3$ and $I_B>0.2$ for the image of April 10, 2003 were found to be best correlated with the fruit yield in 2004 (Fig. 30.8). It is interesting to note that these thresholds are very close or equal to those that were identified by the correlation analysis between the TPCs and the fruit yield in 2003. The only difference lies in that the leaf areas that are highly correlated with the fruit yield in 2003 correspond to the mature leaves that have a pixel value less than the identified thresholds, while the leaf areas that were found to be best correlated with the fruit yield in 2002 and 2004 correspond to the younger leaves that have a pixel value more than the identified thresholds. In particular, in the green band, the threshold values for the two images, with which the obtained TPCs were found to be best correlated with the fruit yields in 2002, 2003 and 2004 are exactly the same, with the exception of the threshold for the December 2, 2002 image in its correlation with the fruit yield in 2004, which shows a slight shift to the lower threshold value from 0.4 to 0.3. This indicates that with a single green waveband, the thresholds 0.4 for the December 2, 2002 image and 0.3 for the April 10, 2003 image could divide the entire canopy into two leaf types, one representing the mature leaves (with pixel values less than the thresholds) and the other younger leaves (with pixel values more than the thresholds). The mature leaves are highly correlated with the fruit yield in the growing season, whereas the younger leaves are correlated with the fruit yield in the previous and the next growing seasons. These findings suggest that the energy reserves in mature leaves in the early season as well as the new energy in these leaves accumulated by photosynthesis along the later growing season may directly contribute to the fruiting in the same growing season, while the younger leaves in the early season may be influenced by the fruiting in the previous growing season and the energy in these leaves may contribute more to the fruiting in the next growing season.

Some TPCs for the NIR band and NDVI were also found to have higher correlations with citrus yield, but their correlation coefficients were comparatively lower than those

Table 30.4 Pearson's correlation coefficients between TPCs from 2002 and 2003 images and fruit yield of citrus in 2002

Image (acquisition dates)	Normalized pixel value (I)	R band	G band	B band	NIR band	NDVI
December 2, 2002	>0.9	−0.2722	0.0479	−0.2185	0.1268	0.3158*
	:	:	:	:	:	:
	>0.5	−0.0313	0.3576*	−0.1059	0.3631*	**0.3392***
	>0.4	0.0804	**0.6288*****	0.0995	0.3730**	0.3369*
	>0.3	0.2623	0.5752***	**0.4246****	0.3959**	0.3180*
	>0.2	**0.3940****	0.4065**	0.3174*	**0.4068****	0.3091*
	>0.1	0.3301*	0.2975*	0.2970*	0.3212*	0.3023*
	>0.0	0.2970*	0.2970*	0.2970*	0.2970*	0.2970*
April 10, 2003	>0.9	–	–	–	0.1492	−0.0751
	:	:	:	:	:	:
	>0.5	−0.1567	0.3497*	−0.0812	**0.5023*****	0.4157**
	>0.4	0.1356	0.5972***	−0.0799	0.4458**	0.4215**
	>0.3	0.2615	**0.7299*****	0.2982*	0.4103**	**0.4232****
	>0.2	**0.7213*****	0.4654***	**0.6599*****	0.4187**	0.4210**
	>0.1	0.4529**	0.4204**	0.4225**	0.4203**	0.4202**
	>0.0	0.4203**	0.4203**	0.4203**	|	0.4203**
May 21, 2003	:	:	:	:	:	:
	>0.6	−0.0696	0.0592	–	−0.3136*	**0.3871****
	>0.5	−0.1335	0.0009	0.0949	0.1804	0.3681*
	>0.4	−0.0387	0.2354	0.0431	**0.3529***	0.3585*
	>0.3	0.0602	0.3376*	0.0422	0.3426*	0.3484*
	>0.2	0.3405*	**0.3421***	**0.3421***	0.3420*	0.3436*
	>0.1	**0.3421***	|	|	0.3421*	0.3425*
	>0.0	|	|	|	|	0.3421*
June 5, 2003	>0.9	0.1233	–	–	−0.0851	0.0895
	:	:	:	:	:	:
	>0.5	0.1539	**0.4692*****	−0.0999	0.2617	0.3154*
	>0.4	0.2058	0.4400**	−0.0378	0.2761*	0.3287*
	>0.3	**0.4707*****	0.4084**	0.1142	0.3132*	**0.3482***
	>0.2	0.4300**	0.3456*	**0.3978****	0.3199*	0.3461*
	>0.1	0.3491*	0.3287*	0.3097*	0.3306*	0.3464*
	>0.0	0.3461*	0.3461*	0.3461*	**0.3461***	0.3461*

Notes: *, ** and *** indicate levels of statistical significance at $P<0.05$, $P<0.01$ and $P<0.001$, respectively. '–' indicates that there are no pixels above the specified threshold. ':' indicates the omitted insignificant results. '|' indicates that decreasing the threshold value did not add any more pixels to the TPC and hence the correlation coefficient would be identical to that of the previous, higher threshold. Thus, the cells in the tables with filled correlation coefficients indicate the distribution ranges of normalized pixel values in each of the images

obtained from TPCs of the visible bands. This may result from the spectral pattern that chlorophylls, photosynthesized compounds and the various metabolites that are highly related to citrus yield are less sensitive to near-infrared wavelengths than visible or photosynthetically active radiation. In addition, the TPCs extracted from the images acquired in May and June 2003 were not as successful in relating to the fruit yields as those acquired in December 2002 and April 2003. This may be due to the growing of new leaves on the branch tips in the citrus canopy, which reduces the differences in canopy structural features between individual trees that can be detected by the multispectral sensor. This information indicates that the pigments in the citrus canopy after the harvest (December) and before the fast vegetative growing season (May) may provide the best information for citrus yield.

Similar results were obtained from analyses of the multispectral images and the yield data for 2004–2006. These results reveal the interrelationships between canopy structural features and fruit yield and suggest the contribution of mature leaves (dark green in colour) to the fruiting in the same growing season and a delayed contribution of the younger leaves (light green in colour) to the fruiting in the next growing season. In previous studies, the so-called resource budget model proposed by Isagi et al. (1997) has been widely used to understand the mechanism of alternate bearing phenomenon in fruit trees. The model describes a deterministic process of energy storage due to photosynthesis and energy depletion

Table 30.5 Pearson's correlation coefficients between TPCs from 2002 and 2003 images and fruit yield of citrus in 2004

Image (acquisition dates)	Normalized pixel value (*I*)	R band	G band	B band	NIR band	NDVI
December 2, 2002	>0.9	–0.205	–0.0156	–0.1612	0.3437**	0.3411**
	:	:	:	:	:	:
	>0.5	0.0746	0.3238*	0.0416	0.4747***	0.5333***
	>0.4	0.1225	0.5810***	0.1764	0.5553***	**0.5405*****
	>0.3	0.2572	**0.6559*****	0.5061***	**0.6288*****	0.5266***
	>0.2	0.4907***	0.6162***	**0.5417*****	0.6177***	0.5224***
	>0.1	**0.5470*****	0.5184***	0.5179***	0.5434***	0.5183***
	>0.0	0.5179***	0.5179***	|	0.5179***	0.5179***
April 10, 2003	>0.9	–	–	–	0.3685**	0.2077
	:	:	:	:	:	:
	>0.5	–0.059	0.3562*	–0.0258	**0.6233*****	**0.6153*****
	>0.4	0.0649	0.5294***	0.0296	0.6131***	0.6070***
	>0.3	0.1959	**0.7217*****	0.4138**	0.5953***	0.6083***
	>0.2	0.5921***	0.6255***	**0.7433*****	0.6052***	0.6066***
	>0.1	**0.6271*****	0.6065***	0.6086***	0.6064***	0.6064***
	>0.0	0.6064***	0.6064***	0.6064***	|	|
May 21, 2003	>0.9	–	–	–	–	–0.207
	:	:	:	:	:	:
	>0.5	0.0478	0.1322	0.1249	0.4311**	**0.5390*****
	>0.4	0.1471	0.4282**	0.1933	**0.5378*****	0.5347***
	>0.3	0.1059	0.5195***	0.255	0.5240***	0.5279***
	>0.2	**0.5233*****	**0.5219*****	**0.5219*****	0.5220***	0.5230***
	>0.1	0.5219***	|	|	0.5219***	0.5222***
	>0.0	|	|	|	0.5219***	0.5219***
June 5, 2003	>0.9	0.173	–	–	0.1086	–0.0043
	:	:	:	:	:	:
	>0.5	0.2713	0.6237***	0.0019	0.6004***	0.5699***
	>0.4	0.3795**	0.6521***	0.0868	0.6072***	0.5879***
	>0.3	0.5587***	**0.6579*****	0.3209*	**0.6098*****	0.6067***
	>0.2	**0.6289*****	0.6096***	0.6063***	0.6002***	0.6062***
	>0.1	0.6107***	0.6009***	0.5916***	0.6017***	0.6069***
	>0.0	0.6070***	0.6070***	**0.6070*****	0.6070***	**0.6070*****

Notes: *, ** and *** indicate levels of statistical significance at $P<0.05$, $P<0.01$ and $P<0.001$, respectively. '–' indicates that there are no pixels above the specified threshold. ':' indicates the omitted insignificant results. '|' indicates that decreasing the threshold value did not add any more pixels to the TPC and hence the correlation coefficient would be identical to that of the previous, higher threshold. Thus, the cells in the tables with filled correlation coefficients indicate the distribution ranges of normalized pixel values in each of the images

due to flower and fruit production. It was assumed that each tree accumulates photosynthate every year, producing flowers when the energy reserves exceed a threshold level, and sets fruits at a rate limited by the ratio between the cost of flowering and fruiting in individual plants. After a major reproductive event, energy reserves will be depleted and it may take 1 year or several years before levels regain the reproductive threshold. The model was proposed based on the assumption that energy reserves in the plant body contribute to the fruiting of citrus trees as a whole, without taking account of the different energy accumulation and allocation behaviours of different leaf types in the canopy. Although the dynamics of plant energy contribution to the fruiting of citrus is still unknown, the results obtained in this study imply the unmatched energy allocation within the citrus canopy and thus may serve as a basis for investigating the dynamics of the alternate bearing in citrus trees.

30.3.3 Development of Prediction Models

30.3.3.1 Models Based on Tree Canopy Size and Previous Year's Yield Data

Preliminary analysis on the fruit density between the 2003 and 2004 yield data has revealed that the fruit density shows a more significant alternate bearing pattern than the total yield for individual trees (Figs. 30.5 and 30.6). This indicates the importance of tree canopy size on the yield

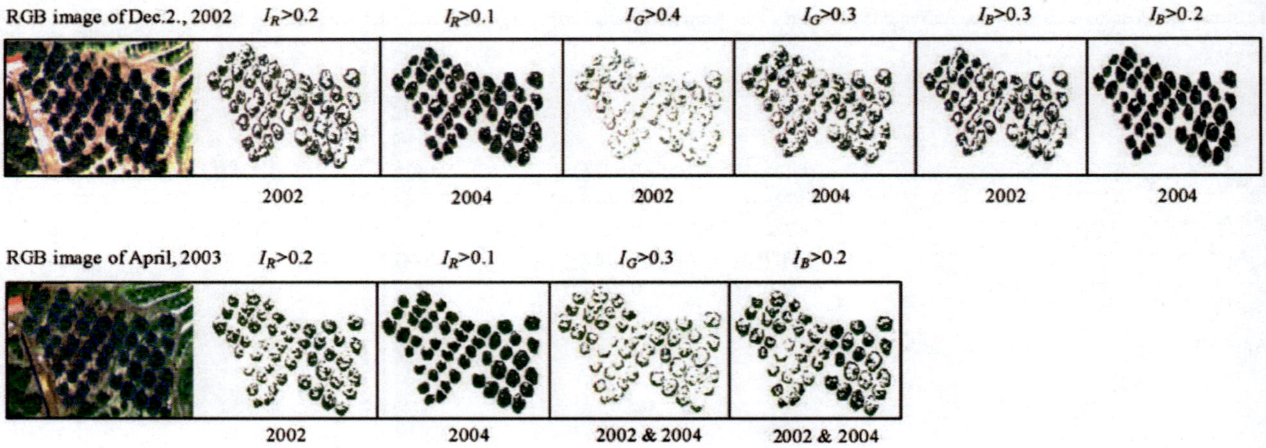

Fig. 30.8 The RGB images acquired on December 2, 2002 and April 10, 2003 and the thresholded images of canopy areas identified as being significantly correlated with the citrus yield of 2002 and 2004. I_α indicates normalized pixel value at α band. The years under the thresholded images indicated that the thresholded canopy areas are significantly correlated with the fruit yield of the indicated year

levels for individual trees, showing the greater relevance of the fruit density than the total yield in studying the alternate bearing of individual citrus trees. Therefore, in this section, in addition to the total fruit yield, the fruit density and the tree canopy size for individual trees were also included as predictor variables in the development of citrus prediction models.

Figure 30.9 shows the results of the models developed with the entire data set for predicting the yield of 2004, using three combinations of predictor variables that were obtained in 2003: (1) the canopy size and the total yield for individual trees in 2003, (2) the canopy size and the fruit density for individual trees in 2003 and (3) the canopy size, the total fruit yield and the fruit density for individual trees in 2003. Result indicated that all the models achieved a reasonable prediction with the R^2 more than 0.5 and the RRMSE less than 0.6. Comparatively, the model based on all the three predictor variables performed the best among the three models, with the R^2 and the RRMSE being 0.665 and 0.5788, respectively.

The models were also examined by cross-validation procedures, as shown in Fig. 30.10. All the models demonstrated a consistent performance in predicting the citrus yield. In comparison, the models based on the canopy size and the total yield achieved the lowest RRMSE (0.8249) for the validation data set among the three models. This shows that a simple combination of the canopy size and the total yield for individual trees in 1 year can provide sufficient information for the yield attainability in the following year. The fruit density plays a role similar to the canopy size in the model development, thus resulting in the slight decrease in the prediction accuracy for the validation accuracy due to the variable redundancy, although the inclusion of it shows a better prediction for the entire data set.

30.3.3.2 Models Based on ASR

A total of 20 ASRs for each of the R, G, B, NIR wavebands and the NDVI obtained or generated from the four multispectral images in 2002 and 2003 were extracted in the preceding data extraction process. Four classes of linear regression models were developed using different methods for selecting predictors from the 20 ASRs (Table 30.2).

Table 30.6 shows the results of linear regression models developed with above four predictor selection methods. A total of 20 models in model class I and 15,182 models in model class IV were developed; however, only five best models of these two classes are shown in Table 30.6. The first three model predictor selection methods either used a single waveband or several wavebands acquired at a certain time period or a single waveband over several time periods to develop prediction models. These predictors are comparatively easy and cheap to obtain in practice. However, the models in these three model classes obtained a low R^2 value for the validation data set, indicating their poor performance in predicting citrus yields. In contrast, the five best models in model class IV performed better, with the R^2 values for both the training and validation data sets improved. However, the predictors involved in these models include several wavebands acquired at several time periods. The application of these models might be limited by the high cost incurred while acquiring the required remotely sensed data with more channels and at more time periods. This indicates the difficulty in predicting the fruit yield of citrus by simply relying on the ASRs for individual tree canopies.

30.3.3.3 Models Based on ASR and Tree Canopy Size

The results of the prediction models developed with ASR and tree canopy size are summarized in Table 30.7. All the

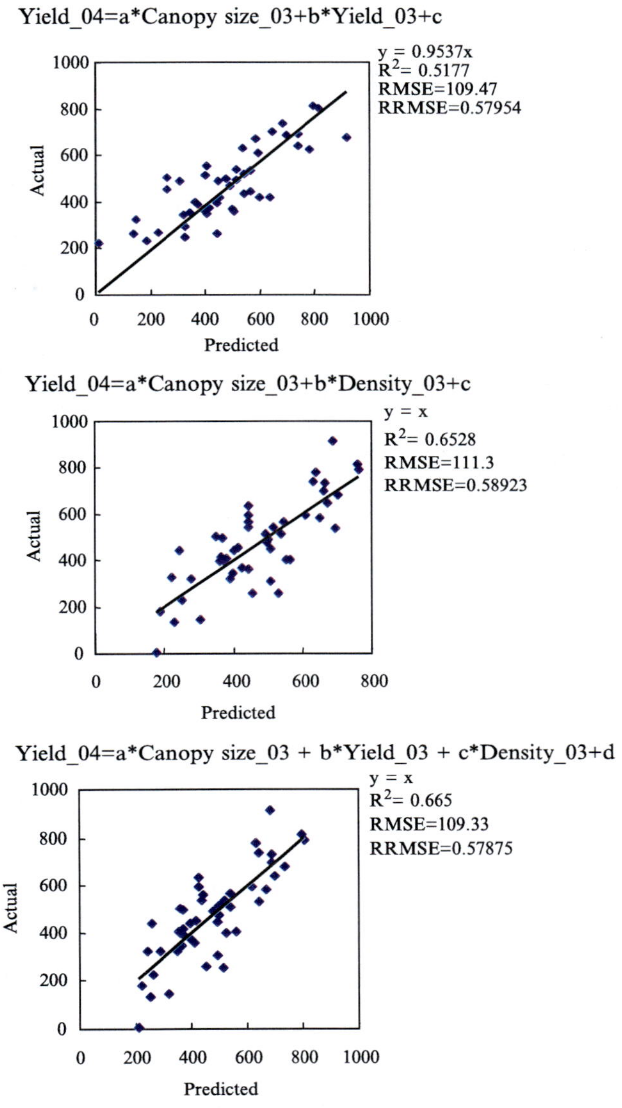

Fig. 30.9 Performance of models developed with previous year's yield, canopy size and fruit density (entire data)

models achieved a better prediction performance than those developed only with ASR in the same model classes. This indicates the importance of tree canopy size in developing models for predicting fruit yield on individual trees. Another important finding from this analysis is that among the multiple wavebands, NDVI in the early season was found to be most significantly correlated with fruit yield on individual trees. Analysis indicated that among the 20 ASRs, NDVIs in April, May and June of 2003 were three of the five ASR predictors that were identified as being best correlated with citrus yield. NIR in May and R in April of 2003 formed the other two ASRs, which are two important wavebands for calculating the NDVIs. The performance of these five best models was illustrated with a scatter plot in Fig. 30.11. This result shows the more relevance of NDVI as well as its component wavebands (NIR and R) in exploring the potential of airborne multispectral imagery to predict citrus yield.

Numerous previous studies have demonstrated the feasibility of estimating a large number of vegetation properties from the NDVI index. Typical examples include the leaf area index, biomass, chlorophyll concentration in leaves, plant productivity, fractional vegetation cover, accumulated rainfall, etc. As these properties are in some way related to the final yield property in crop production, there is a potential of estimating the final yield of crops from the NDVI. The results achieved in this study demonstrated the usefulness of NDVI in estimating fruit yield of citrus on individual trees, which confirmed the similar findings in several other crops expressed in the literature.

30.3.3.4 Models Based on TPC

The TPCs that have been identified as being best significantly correlated with citrus yield for each of the visible wavebands obtained from the images of December 2, 2002, April 10, 2003, June 6, 2005 and April 22, 2006 were further analysed with a modelling and cross-validation procedure. The purpose of this analysis was to explore the potential of using the leaf areas corresponding to the identified thresholds to estimate the fruit yield of citrus. The fruit yield prediction models were first developed with the training data set and were subsequently cross-validated using the validation data set. The performance of the developed models was then evaluated by conducting linear regression analysis between the predicted values and the actual fruit yields for both of the two data sets.

The results for the modelling and cross validation of the yield in 2003, 2002 and 2004 based on the TPCs obtained from the images of December 2, 2002 and April 10, 2003 are shown in Figs. 30.12, 30.13 and 30.14, respectively.

Figure 30.12 shows the model equations and the R^2 values for predicting the fruit yield of 2003. Among each of the three visible band thresholds for the two images, the thresholds $I_G<0.4$ for the December 2, 2002 image, $I_G<0.3$ and $I_R<0.2$ for the April 10, 2003 image obtained the best TPCs that had the highest potential to predict the fruit yield of 2003. Their R^2 values were 0.5283, 0.6586 and 0.6671, respectively. The cross-validation results also revealed the better performance of these three models in estimating fruit yield. Particularly, the models based on the green-band thresholds obtained the best prediction accuracy for each of the two images. This result may originate from the fact that the green band is more sensitive to the chlorophylls and other compounds that are highly correlated to the fruiting in the same growing season.

Figures 30.13 and 30.14 illustrate the modelling and cross-validation results for the models developed to estimate

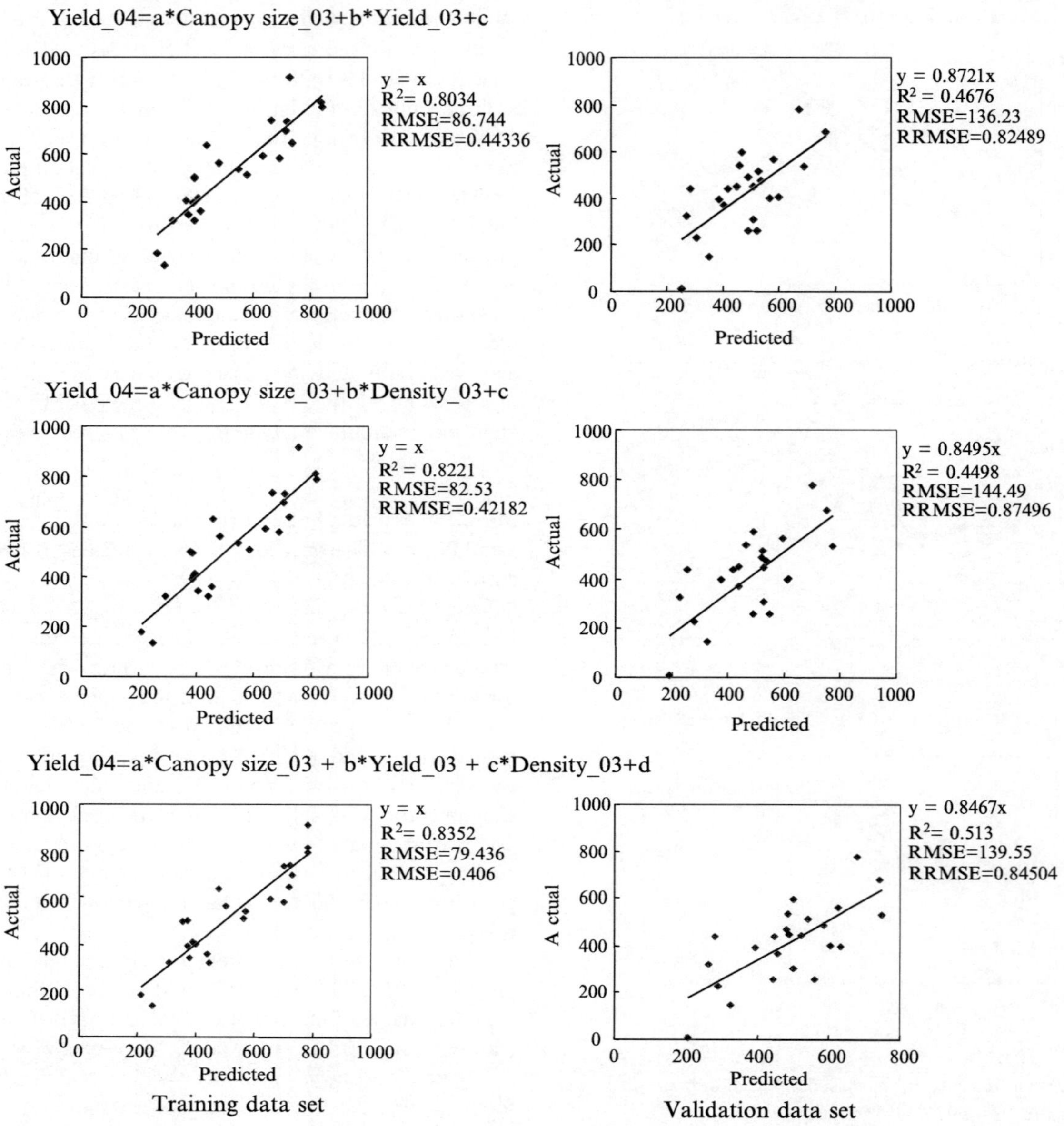

Fig. 30.10 Performance of models developed with previous year's yield, canopy size and fruit density (for both the training data set and validation data set)

the fruit yields of 2002 and 2004, respectively. It was found that the thresholds for the visible wavebands identified in the modelling analysis between their corresponding TPCs and the fruit yield of 2003 also perform well in estimating the fruit yields of 2002 and 2004, except those of the red and blue bands for the December 2, 2002 image, which not only shows a shift of the thresholds but also displays a decline of prediction accuracy. The green band exhibited a good prediction accuracy for the 2002 and 2004 yield estimation, but the red band and the blue band for the April 10, 2003 image were found to have the best prediction accuracy for the 2002 and the 2004 yield estimation (R^2=0.4579 and 0.5686), respectively. This originates from the high correlations between the TPCs extracted with these thresholds (more than 0.9).

Similarly, the modelling and cross validation of the yield in 2005 and 2004 was performed based on the thresholds identified from the image of June 6, 2005, as shown in Figs. 30.15 and 30.16. The TPCs based on both of the thresholds of I_R<0.3 and I_G<0.5 show a significant linear correlation with the citrus yield in 2005, with the R^2 values for the training data set being 0.5082 and 0.6824, respectively. The consistency of the developed models was further confirmed

Table 30.6 Results of linear regression models developed with ASRs for different predictor selection methods

		Training data set			Validation data set		
Model classes	Waveband(s) selected as predictor(s)	R^2	RMSE	RRMSE	R^2	RMSE	RRMSE
I	NIR_{2003}^{May}	0.3310	227.75	0.8179	0.2924	244.02	0.8764
	$NDVI_{2003}^{June}$	0.4422	207.95	0.7468	0.1456	261.63	0.9396
	R_{2003}^{April}	0.3792	219.38	0.7879	0.1178	276.25	0.9921
	$NDVI_{2003}^{May}$	0.2556	240.25	0.8628	0.0978	273.37	0.9818
	$G_{2002}^{Dec.}$	0.2203	245.87	0.8830	0.0944	274.56	0.9860
II	$R_{2002}^{Dec.}, G_{2002}^{Dec.}, B_{2002}^{Dec.}, NIR_{2002}^{Dec.}$ & $NDVI_{2002}^{Dec.}$	0.3097	231.34	0.8308	0.0279	298.14	1.0707
	$R_{2003}^{April}, G_{2003}^{April}, B_{2003}^{April}, NIR_{2003}^{April}$ & $NDVI_{2003}^{April}$	0.4821	200.38	0.7197	0.0915	285.70	1.0261
	$R_{2003}^{May}, G_{2003}^{May}, B_{2003}^{May}, NIR_{2003}^{May}$ & $NDVI_{2003}^{May}$	0.4967	197.55	0.7095	0.2278	275.25	0.9885
	$R_{2003}^{June}, G_{2003}^{June}, B_{2003}^{June}, NIR_{2003}^{June}$ & $NDVI_{2003}^{June}$	0.4895	198.95	0.7145	0.1419	262.51	0.9428
III	$R_{2002}^{Dec.}, R_{2003}^{April}, R_{2003}^{May}$ & $R_{2003}^{Jun.}$	0.5408	188.69	0.6777	0.2158	264.46	0.9498
	$G_{2002}^{Dec.}, G_{2003}^{April}, G_{2003}^{May}$ & $G_{2003}^{Jun.}$	0.2610	239.37	0.8597	0.1168	270.19	0.9704
	$B_{2002}^{Dec.}, B_{2003}^{April}, B_{2003}^{May}$ & B_{2003}^{June}	0.3046	232.20	0.8339	0.0420	288.14	1.0348
	$NIR_{2002}^{Dec.}, NIR_{2003}^{April}, NIR_{2003}^{May}$ & NIR_{2003}^{June}	0.3924	217.04	0.7795	0.3321	234.74	0.8431
	$NDVI_{2002}^{Dec.}, NDVI_{2003}^{April}, NDVI_{2003}^{May}$ & $NDVI_{2003}^{June}$	0.5461	187.60	0.6737	0.1396	274.86	0.9871
IV	$R_{2002}^{Dec.}, R_{2003}^{May}, R_{2003}^{June}, NDVI_{2003}^{June}$ & B_{2003}^{April}	0.6450	165.91	0.5959	0.4909	222.55	0.7993
	$R_{2002}^{Dec.}, R_{2003}^{May}, G_{2003}^{June}, NDVI_{2003}^{June}$ & B_{2003}^{April}	0.6404	166.97	0.5996	0.4893	219.83	0.7895
	$R_{2002}^{Dec.}, R_{2003}^{May}, G_{2003}^{April}, G_{2003}^{June}$ & $NDVI_{2003}^{June}$	0.6376	167.62	0.6020	0.4911	220.43	0.7916
	$R_{2002}^{Dec.}, R_{2003}^{May}, R_{2003}^{June}, NDVI_{2003}^{June}$ & B_{2003}^{June}	0.6976	153.12	0.5499	0.4292	240.50	0.8637
	$R_{2002}^{Dec.}, R_{2003}^{May}, R_{2003}^{June}, G_{2003}^{June}$ & $NDVI_{2003}^{June}$	0.6980	153.03	0.5496	0.4262	239.99	0.8619

Notes: λ_{year}^{month} represents the waveband (λ) obtained in the indicated month of the indicated year

Table 30.7 Results of linear regression models developed with ASRs and canopy size for different predictor selection methods

		Training data set			Validation data set		
Model classes	Waveband(s) selected as predictor(s)	R^2	RMSE	RRMSE	R^2	RMSE	RRMSE
I	$NDVI_{2003}^{June}$	0.6223	175.69	0.6143	0.5814	196.14	0.6858
	$NDVI_{2003}^{May}$	0.5216	197.78	0.6916	0.5047	203.64	0.7121
	NIR_{2003}^{May}	0.5716	187.16	0.6544	0.4959	205.07	0.7171
	R_{2003}^{April}	0.4958	203.05	0.7100	0.4849	206.97	0.7237
	$NDVI_{2003}^{April}$	0.5032	201.55	0.7048	0.4576	212.83	0.7442
II	$R_{2002}^{Dec.}, G_{2002}^{Dec.}, B_{2002}^{Dec.}, NIR_{2002}^{Dec.}$ & $NDVI_{2002}^{Dec.}$	0.7283	145.14	0.5212	0.3266	292.85	1.0517
	$R_{2003}^{April}, G_{2003}^{April}, B_{2003}^{April}, NIR_{2003}^{April}$ & $NDVI_{2003}^{April}$	0.7175	147.99	0.5314	0.4122	220.41	0.7915
	$R_{2003}^{May}, G_{2003}^{May}, B_{2003}^{May}, NIR_{2003}^{May}$ & $NDVI_{2003}^{May}$	0.6215	171.29	0.6151	0.4234	231.24	0.8304
	$R_{2003}^{June}, G_{2003}^{June}, B_{2003}^{June}, NIR_{2003}^{June}$ & $NDVI_{2003}^{June}$	0.6382	167.48	0.6014	0.5469	188.11	0.6755
III	$R_{2002}^{Dec.}, R_{2003}^{April}, R_{2003}^{May}$ & R_{2003}^{June}	0.7050	151.24	0.5432	0.3936	239.92	0.8616
	$G_{2002}^{Dec.}, G_{2003}^{April}, G_{2003}^{May}$ & G_{2003}^{June}	0.6754	158.65	0.5698	0.5186	227.57	0.8173
	$B_{2002}^{Dec.}, B_{2003}^{April}, B_{2003}^{May}$ & B_{2003}^{June}	0.6009	175.90	0.6317	0.4704	226.15	0.8122
	$NIR_{2002}^{Dec.}, NIR_{2003}^{April}, NIR_{2003}^{May}$ & NIR_{2003}^{June}	0.6334	168.59	0.6055	0.4847	227.10	0.8156
	$NDVI_{2002}^{Dec.}, NDVI_{2003}^{April}, NDVI_{2003}^{May}$ & $NDVI_{2003}^{June}$	0.6898	155.09	0.5570	0.4533	215.35	0.7734
IV	$R_{2003}^{June}, G_{2002}^{Dec.}, G_{2003}^{May}, B_{2003}^{May}$ & NIR_{2003}^{May}	0.7443	140.81	0.5057	0.6173	178.78	0.6421
	$R_{2003}^{April}, R_{2003}^{June}, G_{2002}^{Dec.}, G_{2003}^{May}$ & NIR_{2003}^{May}	0.7376	142.64	0.5123	0.6205	177.50	0.6375
	$R_{2003}^{June}, G_{2002}^{Dec.}, G_{2003}^{May}, B_{2003}^{June}$ & NIR_{2003}^{May}	0.7834	129.58	0.4654	0.5703	192.64	0.6918
	$R_{2003}^{April}, G_{2002}^{Dec.}, G_{2003}^{May}, G_{2003}^{June}$ & NIR_{2003}^{May}	0.7374	142.70	0.5125	0.6088	180.73	0.6491
	$R_{2002}^{Dec.}, G_{2002}^{Dec.}, G_{2003}^{April}, G_{2003}^{May}$ & NIR_{2003}^{May}	0.7395	142.12	0.5104	0.6203	181.66	0.6524

Notes: λ_{year}^{month} represents the waveband (λ) obtained in the indicated month of the indicated year

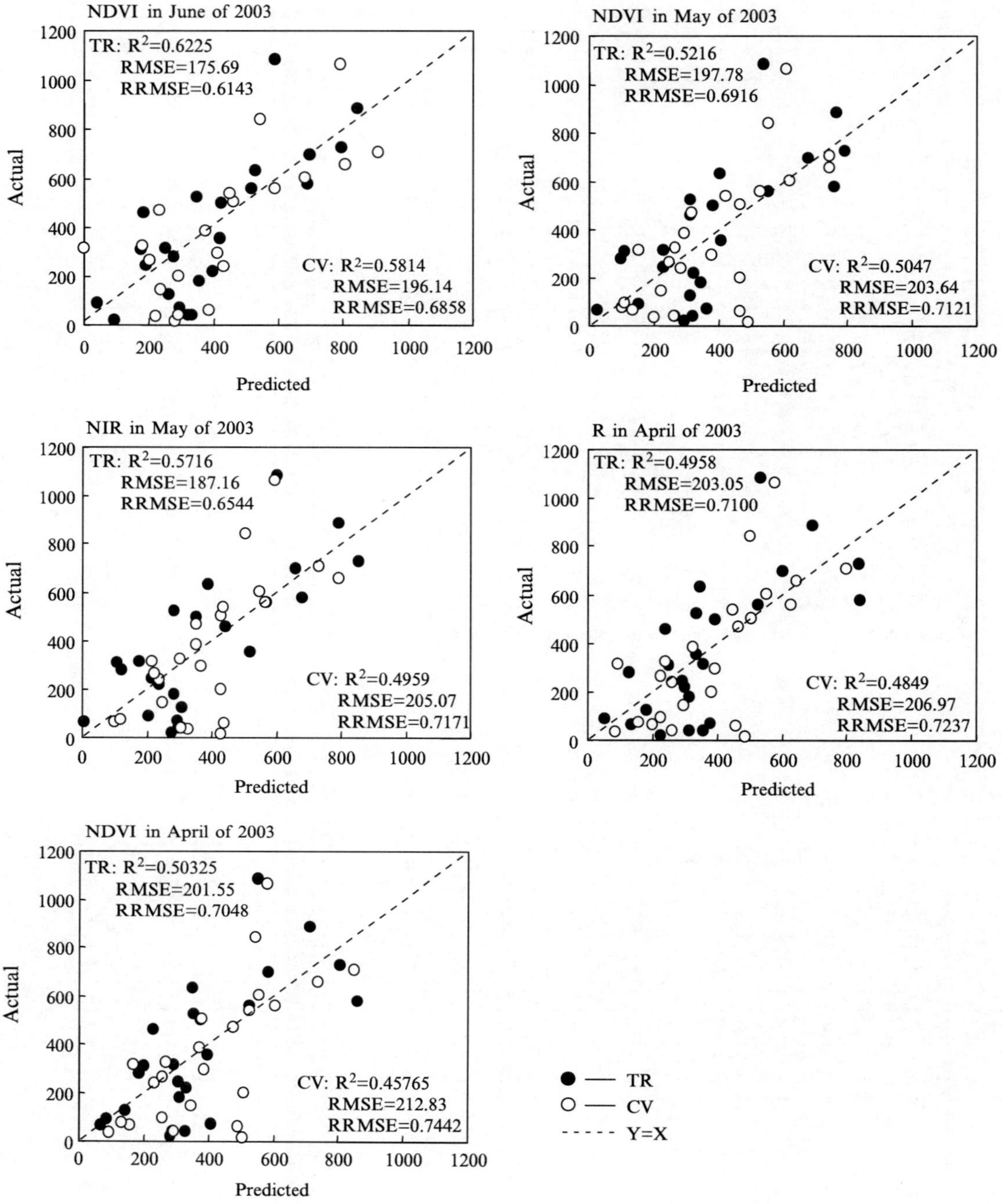

Fig. 30.11 Performance of five best models developed with ASR of a single waveband and tree canopy size (for both the training data set (*TR*) and cross-validation data set (*CV*))

in the cross validation ($R^2=0.6316$ and 0.6049, respectively). A significant correlation was also found between the TPCs obtained with the thresholds of $I_R>0.3$ and $I_G>0.4$ and the fruit yield in 2004.

These results demonstrate that the interrelationships between canopy structural features and fruit yields in citrus exist not only within the same growing season but also within the adjacent growing seasons. It also suggests the potential of using the images acquired at an early date in one growing season to forecast the fruit yield attainable in the same growing season as well as the next growing season.

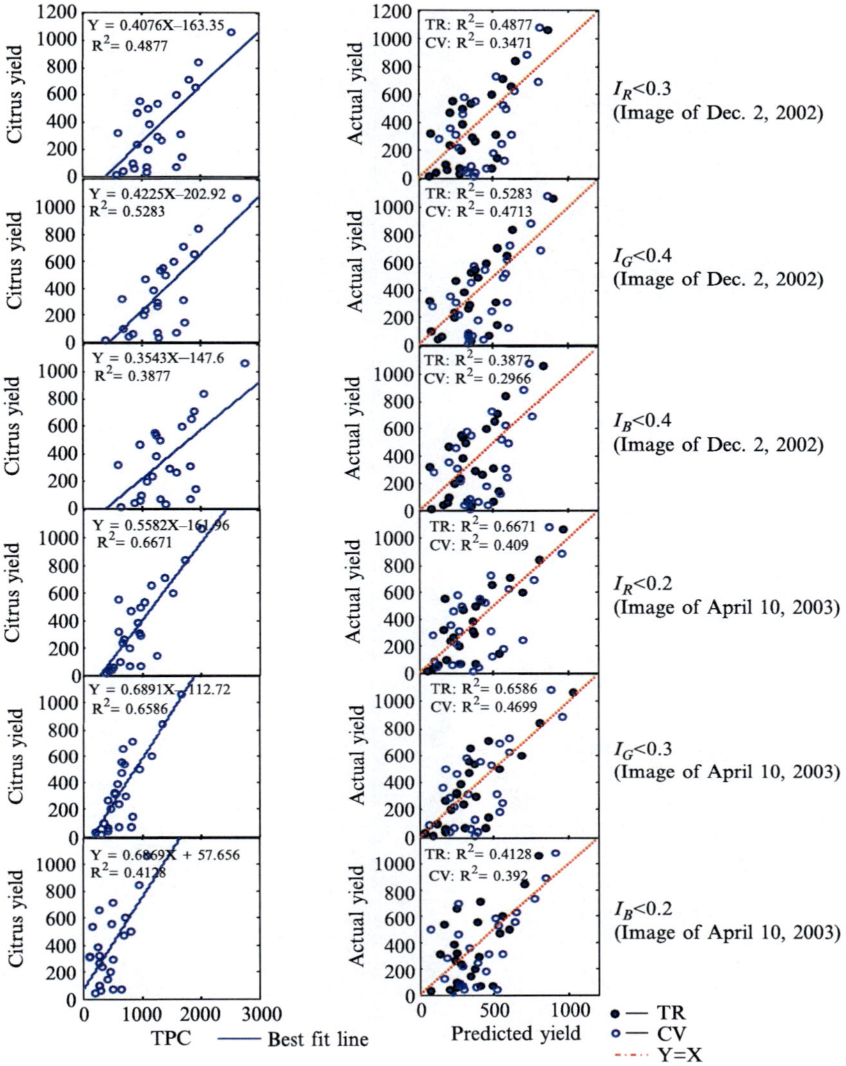

Fig. 30.12 The modelling and validation of the correlation between the thresholded pixel counts (*TPCs*) from 2002 and 2003 images and the fruit yield of citrus in 2003. I_α indicates the normalized pixel value at α band. The *left column* shows the developed models with the training data sets. The *right column* indicates the model performance for both the training data set (*TR*) and the cross-validation data set (*CV*)

30.4 Conclusions

The following conclusions can be drawn from the results obtained:

- The fruit density was found with a higher correlation than the total yield for individual trees in the study of alternate bearing present in consecutive years. This result reveals the mechanism of alternate bearing on individual trees, showing that the alternate bearing behaves more significantly in terms of the fruit density rather than the total fruit yield on individual trees.
- Modelling based on ASRs shows the difficulty in predicting the fruit yield of citrus by simply relying on the ASRs for a single or several wavebands on individual tree canopies. By incorporating tree canopy size into the model, the prediction results were greatly improved. In addition, this study demonstrates the greater relevance of NDVI than other multiple wavebands in predicting the fruit yield on individual citrus trees.
- Significant correlations were found between canopy structural features and fruit yield in citrus trees. Specifically, the mature leaves in canopies (dark green in colour) are significantly correlated with the fruit yield of the same growing season, suggesting an immediate contribution to fruit formation. Conversely, the younger leaves were found to be more significantly correlated with the fruit yield of the following growing season, suggesting a delayed contribution to yield formation. This information implies the unmatched energy allocation mechanisms within the canopy of citrus crops.

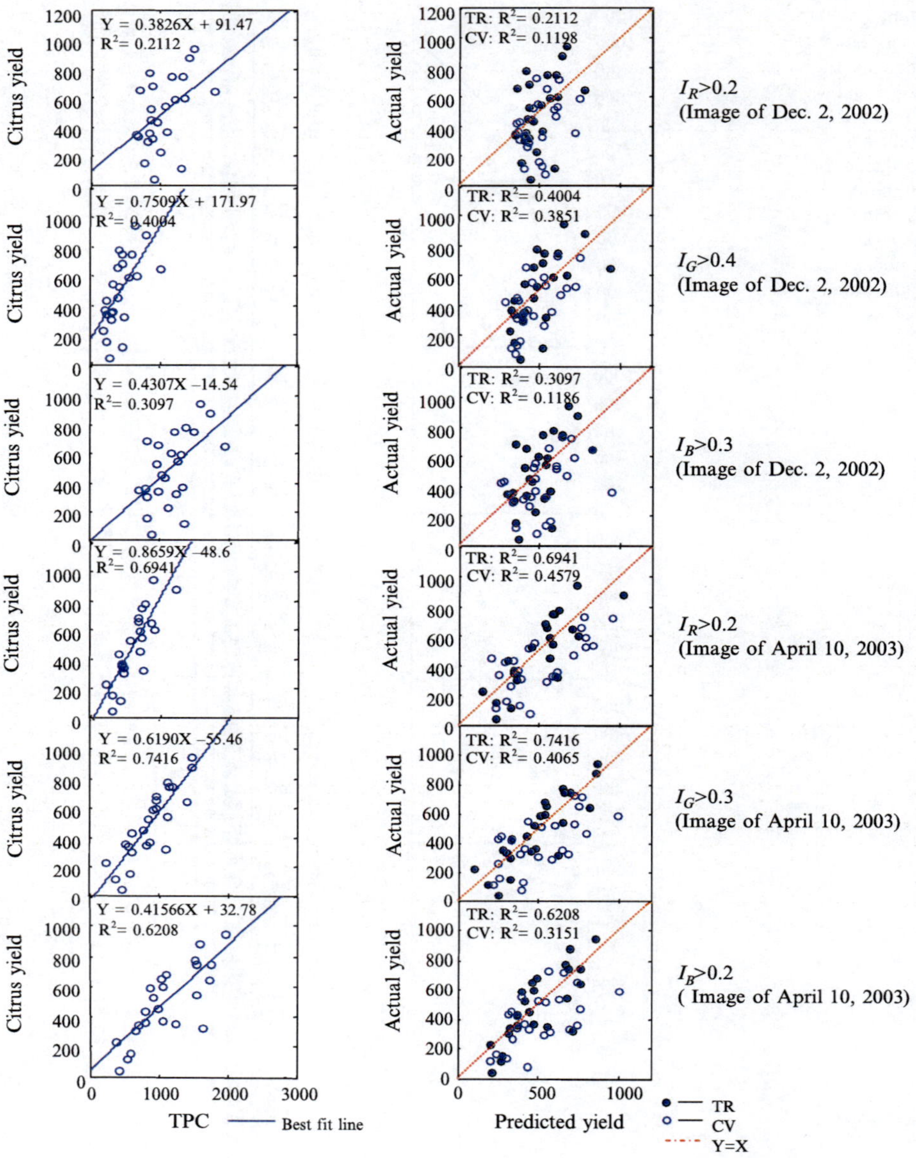

Fig. 30.13 The modelling and validation of the correlation between the thresholded pixel counts (*TPCs*) from 2002 and 2003 images and the fruit yield of citrus in 2002. I_α indicates the normalized pixel value at α band. The *left column* shows the developed models with the training data sets. The *right column* indicates the model performance for both the training data set (*TR*) and the cross-validation data set (*CV*)

- Finally, this study also demonstrated the potential of airborne multispectral imagery acquired at an early date to forecast the fruit yield attainable in coming growing seasons.

30.5 Future Research

Some possible areas for future research based on the results of this study include the following:

- This study revealed the interrelationships between canopy structural features and citrus yield, thus providing new insights for the study of alternate bearing dynamics in tree crops. Based on these findings, it would be possible to develop new theoretical models for alternate bearing dynamics in tree crops.
- In addition to multispectral imaging, hyper-spectral imaging provides a higher spectral resolution spectrum for each pixel in an image. Our previous studies proved the superiority of hyper-spectral images in providing more subtle spectral data, which allows the application of more modelling techniques in estimating fruit yield (Ye et al. 2006, 2007, 2008a, b, 2009; Zaman et al. 2006; Zhang et al. 2006). However, due to the engineering trade-off between spatial and spectral resolutions in optical sensor

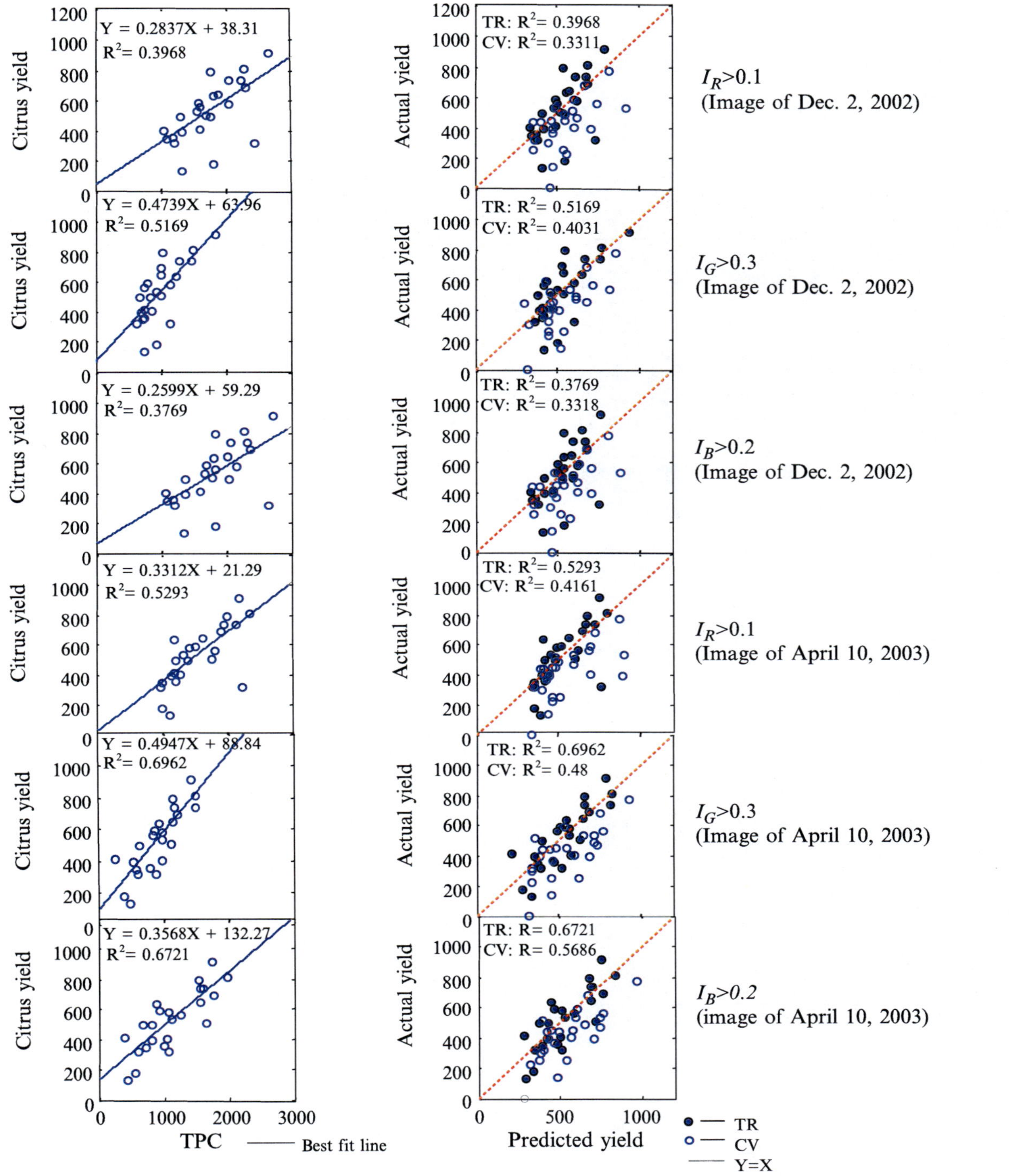

Fig. 30.14 The modelling and validation of the correlation between the thresholded pixel counts (*TPCs*) from 2002 and 2003 images and the fruit yield of citrus in 2004. I_α indicates the normalized pixel value at α band. The *left column* shows the developed models with the training data sets. The *right column* indicates the model performance for both the training data set (*TR*) and the cross-validation data set (*CV*)

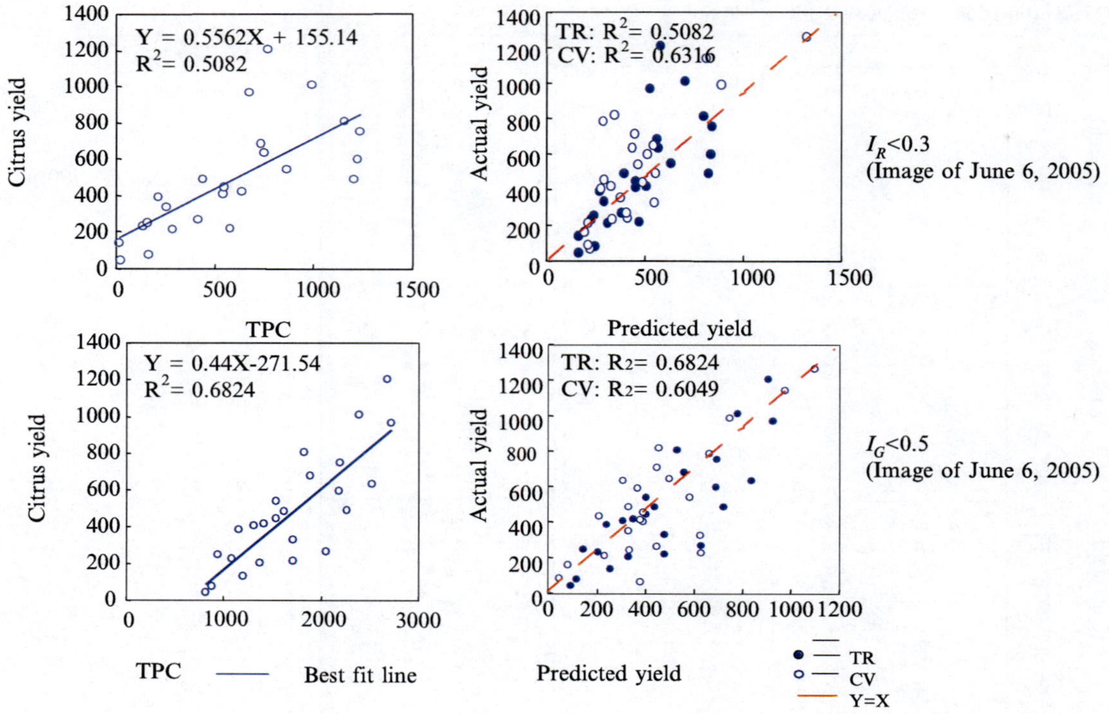

Fig. 30.15 The modelling and validation of the correlation between the thresholded pixel counts (*TPCs*) from 2005 image and the fruit yield of citrus in 2005. I_α indicates the normalized pixel value at α band. The *left column* shows the developed models with the training data sets. The *right column* indicates the model performance for both the training data set (*TR*) and the cross-validation data set (*CV*)

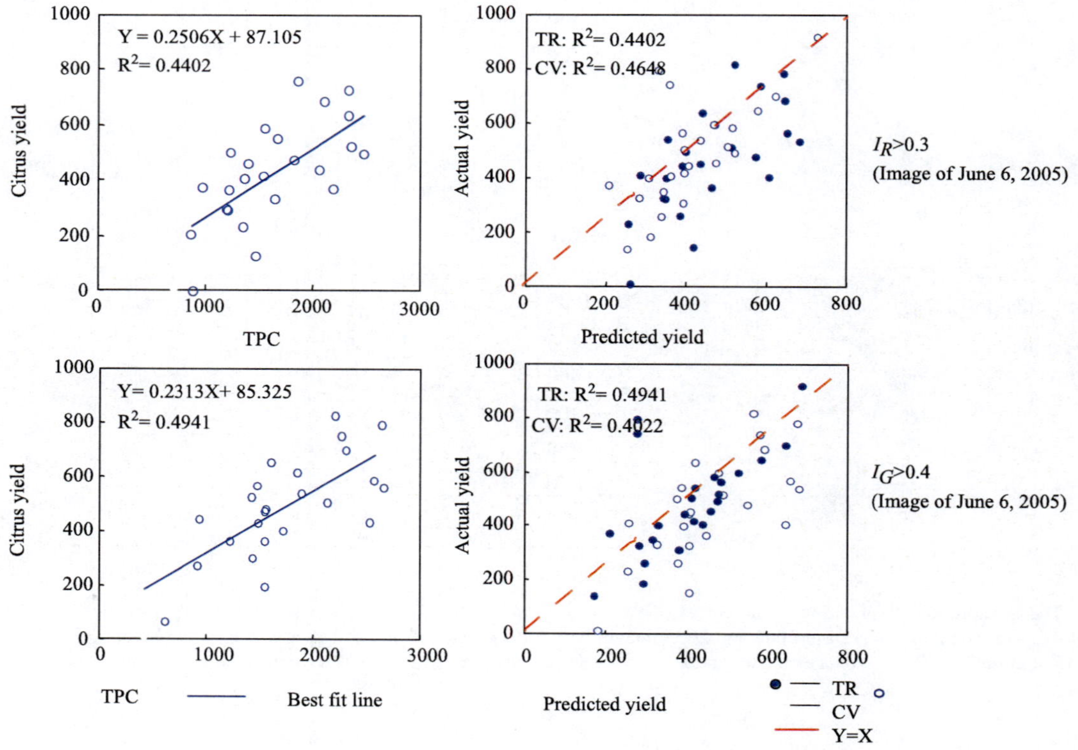

Fig. 30.16 The modelling and validation of the correlation between the thresholded pixel counts (*TPCs*) from 2005 image and the fruit yield of citrus in 2004. I_α indicates the normalized pixel value at α band. The *left column* shows the developed models with the training data sets. The *right column* indicates the model performance for both the training data set (*TR*) and the cross-validation data set (*CV*)

design, hyper-spectral imaging sensors usually exhibit a low spatial resolution. This resulted in the difficulty in the accurate identification of citrus canopies in hyper-spectral images. Recently, the so-called image fusion technique has been developed to produce images with enhanced spatial and spectral resolutions, based on hyper-spectral and multispectral or panchromatic images. Adoption of various image fusion techniques could allow the accurate identification and classification of the observed materials in the image at a fine resolution and thus contribute to the accuracy of the developed models for yield prediction.

- Finally, more research and investigations, including field image acquisition, model selection, yield mapping, etc., are necessary before application of the proposed approach on a field scale can be generalized.

Acknowledgements This research was jointly supported by the JSPS Grants-in-Aid for Scientific Research (nos. 2110, 14360148 and 15658074), the NSFC grant (61071220), the Fundamental Research Funds for the Central Universities, the Educational Commission of Zhejiang Province (Y20084455) and the Doctoral Fund of Ministry of Education of China (20090101120080). We gratefully acknowledge Dr. Bhuweneshwar P. Sah and Mr. Tomoyuki Suhama of Pasco Corporation for the acquisition of multispectral images and Shinichi Asada for the collection of citrus yield data.

References

Ahlrichs JS, Bauer ME (1983) Relation of agronomic and multi-spectral reflectance characteristics of spring wheat canopies. Agron J 75:987–993

Boegh E, Soegaard H, Broge N et al (2002) Airborne multi-spectral data for quantifying leaf area index, nitrogen concentration, and photosynthetic efficiency in agriculture. Remote Sens Environ 81:179–193

Clevers JGPW (1991) Application of the WDVI in estimating LAI at the generative stage of barley. ISPRS J Photogram Remote Sens 46(1):37–47

Gates DM (1965) Energy, plants, and ecology. Ecology 46:1–14

Gates DM, Keegan HJ, Schleter JC et al (1965) Spectral properties of plants. Appl Opt 4(1):11–20

Gausman HW, Rodriguez RR, Richardson AJ (1976) Infinite reflectance of dead compared with live vegetation. Agron J 68:295–296

Goel PK, Prasher SO, Patel RM et al (2002) Use of airborne multi-spectral imagery for weed detection in field crops. Trans ASAE 45(2):443–449

Gupta RH, Woolley JT (1971) Spectral properties of soybean leaves. Agron J 63:123–126

Hache C, Shibusawa S, Sasao A (2005) Discriminating conventional and conservation agricultural management practices with airborne multi-spectral imagery. Agric Ecosyst Environ 111:354–366

Hahlbrock K, Grisebach H (1979) Enzymic controls in the biosynthesis of lignin and flavonoids. Annu Rev Plant Physiol 30:105–130

Hall A, Louis J, Lamb D (2003) Characterising and mapping vineyard canopy using high-spatial-resolution aerial multi-spectral images. Comput Geosci 29:813–822

Hill SJ, Stephenson DW, Taylor BK (1987) Almond yield in relation to tree size. Sci Hortic 33(1–2):97–111

Isagi Y, Sugimura K, Sumida A et al (1997) How does masting happen and synchronize? J Theor Biol 187:231–239

Iwagaki I (1997) Citrus production in Japan: new trends in technology. Food and Fertilizer Technology Centre, Taipei, Taiwan. Available from http://www.agnet.org/library/article/eb440.html. Accessed 20 Nov 2010

Jordan BR (1996) The effects of ultraviolet-B radiation on plants: a molecular perspective. In: Callow JA (ed) Advances in botanical research. Academic, New York, pp 97–162

Koenig WD, Mumme RL, Carmen WJ et al (1994) Acorn production by oaks in central coastal California: variation within and among years. Ecology 75(1):99–109

Krauss P, Markstädter C, Riederer M (1997) Attenuation of UV radiation by plant cuticles from woody species. Plant Cell Environ 20:1079–1085

Noguchi Y, Sakai K, Asada S et al (2003) Modeling of alternate bearing in Satsuma mandarin—linear dynamic model based on ensemble set for harvested fruits number. Jpn J Agric Eng 65(6):55–61

Norton DA, Kelly D (1988) Masting seeding over 33 years by *Dacrydium cupressinum* Lamb. (rimu) (*Podocarpaceae*) in New Zealand: the importance of economics of scale. Funct Ecol 2:399–408

Qi Y, Bai S, Heisler GM (2003) Changes in ultraviolet-B and visible optical properties and absorbing pigment concentrations in pecan leaves during a growing season. Agric For Meteorol 120:229–240

Qin Z, Zhang M (2005) Detection of rice sheath blight for in-season disease management using multi-spectral remote sensing. Int J Appl Earth Obs Geoinf 7:115–128

Robberecht R, Caldwell MM (1986) Leaf UV optical properties of *Rumex patientia* L. and *Rumex obtusifolius* L. in regard to a protective mechanism against solar UV-B radiation injury. In: Worrest RC, Caldwell MM (eds) Stratospheric ozone reduction, solar ultraviolet radiation and plant life, NATO ASI series. Springer, Berlin, pp 251–259

Sakai K, Noguchi Y, Asada S (2003) Nonlinear dynamics in arable land: alternate bearing of tree crops. Technical report IEICE NLP 2002–132

Sork VL, Bramble J, Sexton O (1993) Ecology of mast fruiting in 3 species of North-America deciduous oaks. Ecology 74:528–541

Vogelmann JE, Moss DM (1993) Spectral reflectance measurements in the genus sphagnum. Remote Sens Environ 45:273–279

Wang T, Thai CN (2003) A genetic algorithm based tool for analysis and modeling of multi-spectral data. ASAE annual meeting, paper no. 033126. ASAE, St. Joseph, MI

Ye X, Sakai K, Garciano LO et al (2006) Estimation of citrus yield from airborne hyperspectral images using a neural network model. Ecol Model 198(3–4):426–432

Ye X, Sakai K, Manago M et al (2007) Prediction of citrus yield from airborne hyperspectral imagery. Precis Agric 8(3):111–125

Ye X, Sakai K, Asada S et al (2008a) Application of narrow-band TBVI in estimating fruit yield in citrus. Biosyst Eng 99(2):179–189

Ye X, Sakai K, Sasao A et al (2008b) Potential of airborne hyperspectral imagery to estimate fruit yield in citrus. Chemometrics Intell Lab Syst 90(2):132–144

Ye X, Sakai K, Asada S et al (2009) Interrelationships between canopy features and fruit yield in citrus as detected by airborne multispectral imagery. Trans ASABE 51(2):739–751

Zaman QU, Schumann AW, Hostler HK (2006) Estimation of citrus fruit yield using ultrasonically-sensed tree size. Trans ASAE 22(1):39–44

Zhang J, Wang K, Bailey JS et al (2006) Predicting nitrogen status of rice using multi-spectral data at canopy scale. Pedosphere 16(1):108–117

Index

A.K. Srivastava (ed.), *Advances in Citrus Nutrition*,
DOI 10.1007/978-94-007-4171-3, © Springer Science+Business Media B.V. 2012